## ...ECTED VITAMINS AND MINERALS

| Manganese (mg) | Fluoride (mg) | Chromium ($\mu$g) | Molybdenum ($\mu$g) |
|---|---|---|---|
| 0.3–0.6 | 0.1–0.5 | 10–40 | 15–30 |
| 0.6–1.0 | 0.2–1.0 | 20–60 | 20–40 |
| 1.0–1.5 | 0.5–1.5 | 20–80 | 25–50 |
| 1.5–2.0 | 1.0–2.5 | 30–120 | 30–75 |
| 2.0–3.0 | 1.5–2.5 | 50–200 | 50–150 |
| 2.0–5.0 | 1.5–2.5 | 50–200 | 75–250 |
| 2.0–5.0 | 1.5–4.0 | 50–200 | 75–250 |

National Academy Press, Washington, DC.

# NUTRITIONAL
# BIOCHEMISTRY

# NUTRITIONAL BIOCHEMISTRY

## Second Edition

## TOM BRODY

University of California at Berkeley
Berkeley, California

Academic Press

San Diego  New York  Boston
London  Sydney  Tokyo  Toronto

Academic Press
*a division of Harcourt Brace & Company*
525 B Street, Suite 1900, San Diego, California 92101-4495
http://www.academicpress.com

Academic Press Limited
24-28 Oval Road, London NW1 7DX, UK
http://www.hbuk.co.uk/ap/


Library of Congress Cataloging-in-Publication Data

Brody, Tom.
    Nutritional biochemistry / Tom Brody. -- 2nd. ed.
        p.    cm.
    First published: San Diego  :  Academic Press, 1994.
    Includes bibliographical references and index.
    ISBN 0-12-134836-9 (alk. paper)
    1. Nutrition.    2. Metabolism.    I. Title.
QP141.B853  1998
612.3'9--dc21                                                98-40384
                                                                CIP


PRINTED IN THE UNITED STATES OF AMERICA
98  99  00  01  02  03  QW  9  8  7  6  5  4  3  2  1

*To Shideh and Dawnia*

# CONTENTS

# PREFACE

Nutrition involves the relationship of food and nutrients to health. Biochemistry is the science of the chemistry of living organisms. As implied by the title, this book emphasizes the overlap between problems of nutrition and the techniques of biochemistry.

The nutritional sciences also include many aspects of related disciplines such as physiology, food chemistry, toxicology, pediatrics, and public health. Thus, any given problem in the nutritional sciences may also be a problem in one of these disciplines. Nevertheless, nutrition is a unique discipline because of its specific goal, that is, improving human health by understanding the role of diet and supplying that knowledge in everyday living.

Nutritional sciences employ various experimental techniques. The methods used to assess a deficiency can also be used to determine the requirement for a given nutrient. Dietary deficiency, a technique applied to animals and microorganisms, was used in the discovery of vitamins and in proving the essential nature of certain amino acids and lipids. This book features a strong emphasis on the techniques used to assess both requirements and deficiencies. Two of the most important techniques, those involving nitrogen balance and the respiratory quotient, are covered in some detail.

The book focuses on the details of two or three aspects of problems related to each selected topic. Clinical and research data are used to illustrate these problems, and case studies are frequently presented. Emphasis on primary data is intended to encourage readers to use their own trained judgment when examining data from the literature as well as data from their own research experience.

The ability to organize facts into a hierarchy of importance is useful in understanding the biological sciences. This book encourages the researcher to employ this method of organization. For example, the order of use of energy fuels is described in the chapter on regulation of energy metabolism. The order of appearance of signs of folate deficiency is detailed in the chapter on vitamins. The book also encourages the researcher to accept the potential value of data that are ambiguous or apparently contradictory. For example, the chapter on digestion shows that a barely detectable increase in plasma secretin levels can be physiologically relevant. The section on starvation reveals that the body may suffer from signs of vitamin A deficiency even though substantial amounts of the vitamin are stored in the liver. The section on fiber explains how an undigestible nutrient supplies vital energy to cells of the human body.

Some of the dreaded nutritional diseases of the past — such as scurvy, pellagra, and pernicious anemia — are discussed in this book. Such contemporary problems as infectious diarrhea, xerophthalmia, protein/energy malnutrition, and folate deficiency are discussed, as are diabetes and cardiovascular disease, two of the most significant nutrition-related diseases. The last two conditions can be controlled in part by dietary intervention.

This book stresses the importance of nutritional interactions. Some nutrients are closely related and usually discussed together. Some are antagonistic to each other, whereas others act synergistically. Examples of uniquely related nutrients are bean and rice protein, saturated and monounsaturated fatty acids, folate and vitamin $B_{12}$, vitamin E and polyunsaturated fatty acids, and calcium and vitamin D. Some closely related biological molecules are discussed, including insulin and glucagon, cholecystokinin and secretin, and low- and high-density lipoproteins. Interactions involving multiple organ systems and multiple cell types are stressed. More emphasis is placed on interorgan relationships than in typical biochemistry textbooks.

Drugs that influence nutrient metabolism are discussed in various sections. These drugs include lovastatin, pravastatin, omeprazole, dilantin, methotrexate, allopurinol, warfarin, furosemide, thiouracil, and diphosphonate. Alcohol is also discussed in this context because, depending on its intake, it functions as a food, drug, or toxin.

The **recommended dietary allowances** (RDAs) for various nutrients are discussed. RDAs are the quantities in the diet of all nutrients required to maintain human health. RDAs are established by the Food and Nutrition Board of the National Academy of Sciences, and are published by the National Academy Press. The RDA values are revised periodically on the basis of new scientific evidence.

RDAs are used to define a relationship between various human populations and the nutrients required by the human body at various stages of life. They are intended to serve as a basis for evaluating the adequacy of diets of groups of people rather than of individuals. A comparison between the RDA for a specific nutrient and individual intake of that nutrient can indicate the probability or risk of a deficiency in that nutrient. The actual nutritional status with respect to the nutrient can be assessed only by appropriate tests. These tests are usually of a biochemical nature, but also may be hematological or histological. Nutrient RDAs have been determined for men, women, and children of different ages. In most cases, the RDA differs with body weight and, in some cases, with gender. For convenience, RDA values are sometimes expressed in terms of an ideal or reference subject such as "the 70-kg man" or "the 55-kg woman." The current RDAs for all nutrients are listed on the inside back cover. RDAs have not been set for a number of required or useful nutrients. The estimated safe and adequate intakes of these nutrients established by the Food and Nutrition Board, are listed on the inside front cover.

# ACKNOWLEDGMENTS

## FIRST EDITION

My father was the earliest influence on this work. He introduced me to all of the sciences. This book arose from my teaching notes, and I thank Kristine Wallerius, Lori Furutomo, and my other students for their interest. I thank Professor Mary Ann Williams of the University of California at Berkeley for her comments on writing style and for her friendliness. I thank a number of research professors for answering lengthy lists of questions over the telephone. I thank Clarence Suelter of Michigan State University for comments on Cl and K, and James Fee of the Los Alamos National Laboratory for aid with oxygen chemistry. I thank Sharon Fleming (fiber), Nancy Amy (Mn), and Judy Turnlund (Zn) of the University of California at Berkeley for help with the listed nutrients. I am grateful to Andrew Somlyo of the University of Virginia and Roger Tsien of the University of California at San Diego for help in muscle and nerve biochemistry. I am indebted to Gerhard Giebisch of Yale University for answering difficult questions on renal cell biology. I thank Herta Spencer of the Veterans Administration Hospital in Hines, Illinois, for a lengthy and revealing discussion on calcium nutrition. I thank Steven Zeisel (choline) of the University of North Carolina, Wayne Becker (Krebs cycle) of the University of Wisconsin at Madison, Daniel Atkinson (urea cycle) of the University of California at Los Angeles, and Peter Dallman (Fe) of the University of California at San Francisco for comments on the listed subjects. I am deeply appreciative of Quinton Rogers of the University of California at Davis for his insightful written comments on amino acid metabolism. I would like to thank Michelle Walker of Academic Press for her immaculate work and skillful supervision of the production phase of this book. Finally, I would like to take this opportunity to thank Professor E. L. R. Stokstad for accepting me as a graduate student, for the friendly and lively research environment in his laboratory, and for his encouragement for over a decade.

# ACKNOWLEDGMENTS

## SECOND EDITION

I am grateful to the following researchers on the University of California at Berkeley campus. I thank Gladys Block for patiently answering numerous questions regarding methodology in epidemiology. Ronald M. Krauss answered several questions and provided inspiration for adding further details on atherosclerosis. Maret Traber answered a number of questions on oxidative damage to LDLs, and inspired a change in my focus on this topic. I thank Ernst Henle for several enlightening discussions on DNA damage and repair. I am grateful to Hitomi Asahara for guidance in biotechnology. I thank H. S. Sul and Nancy Hudson for help in fat cell biochemistry and for providing orientation in the field of human obesity.

I acknowledge Penny Kris-Etherton of University of Pennsylvania for helping me with questions regarding dietary lipids. I thank Judy Turnlund of the Western Human Nutrition Center in San Francisco for answering a list of questions about copper and zinc. I thank Paul Polakis of Onyx Pharmaceuticals (Richmond, CA) for his insights on new developments on the APC protein and catenin protein. I am grateful to Pascal Goldschmidt-Clermont of Ohio State University for answering a few questions regarding the MAP kinase signaling pathway and hydrogen peroxide. I thank Ralph Green of the University of California at Davis for sharing his knowledge on gastric atrophy. I am grateful to Paul Fox of the Cleveland Clinic Foundation for advice regarding iron transport, as well as to Anthony Norman of the University of California at Riverside for his insights on vitamin D.

I appreciate the perspective given to me by Jeanne Rader of the Food and Drug Administration in Washington, DC, regarding folate supplements. I thank Dale Schoeller of the University of Wisconsin–Madison for his comments on the energy requirement.

I thank Tim Oliver for his professionalism in editing and typesetting. Finally, I thank Kerry Willis and Jim Mowery for overseeing this project and for their contributions in the final phases of the work.

# ABBREVIATIONS

| | | | |
|---|---|---|---|
| ANP | Atrial natriuretic peptide | NAD | Nicotinamide adenine dinucleotide |
| ATP | Adenosine triphosphate | | |
| A–V difference | Concentration in arterial blood minus that in venous blood | NADP | NAD phosphate |
| | | N balance | Nitrogen balance |
| BCAA | Branched chain amino acid | NTD | Neural tube defect |
| BCKA | Branched chain keto acid | P | Phosphate group |
| BMI | Body mass index | PC | Phosphatidylcholine |
| BMR | Basal metabolic rate | PE | Phosphatidylethanolamine |
| BV | Biological value | PEPCK | Phosphoenolpyruvate carboxylase |
| cAMP | Cyclic AMP | | |
| CCK | Cholecystokinin | PER | Protein efficiency ratio |
| CE | Cholesteryl ester | PLP | Pyridoxal phosphate |
| cDNA | Complementary DNA | PPAR | Peroxisome proliferator activated receptor |
| CoA | Coenzyme A | | |
| CoA-SH | Coenzyme A | PTH | Parathyroid hormone |
| C peptide | Connecting peptide of insulin | PUFA | Polyunsaturated fatty acid |
| CTP | Cytosine triphosphate | RAR | Retinoic acid receptor |
| ECF | Extracellular fluid | RBP | Retinol binding protein |
| EFA | Essential fatty acid | RDA | Recommended dietary allowance |
| ER | Endoplasmic reticulum | | |
| F-1,6-diP | Fructose-1,6-bisphosphate | RQ | Respiratory quotient |
| FAD | Flavin adenine dinucleotide | SAH | $S$-adenosyl-homocysteine |
| FFA | Free fatty acid | SAM | $S$-adenosyl-methionine |
| FIGLU | Formiminoglutamic acid | SREBP | Sterol response element binding protein |
| GLUT | Glucose transporter gene or protein | | |
| | | TG | Triglyceride |
| GTP | Guanosine triphosphate | TPP | Thiamin pyrophosphate |
| Hb | Hemoglobin | TTP | Thymidine triphosphate |
| HDL | High-density lipoprotein | UV light | Ultraviolet light |
| IP3 | Inositol-1,4,5-trisphosphate | VDR | Vitamin D receptor |
| IRS | Insulin-responsive substrate; insulin receptor substrate | VLDL | Very-low-density lipoprotein |
| | | $mM$ | Millimolar ($10^{-3}$ $M$) |
| LCAT | Lecithin cholesterol acyl-transferase | $\mu M$ | Micromolar ($10^{-6}$ $M$) |
| | | $nM$ | Nanomolar ($10^{-9}$ $M$) |
| LDL | Low-density lipoprotein | $pM$ | Picomolar ($10^{-12}$ $M$) |
| mRNA | Messenger RNA | $fM$ | Femtomolar ($10^{-15}$ $M$) |

# 1

## CLASSIFICATION OF BIOLOGICAL STRUCTURES

### OVERVIEW

A review of chemical bonds, acid/base chemistry, and the concept of water solubility is provided first, to assure that readers with various backgrounds begin with the same grounding in beginning chemistry. Then the discussion progresses to molecular structures of increasing complexity, including carbohydrates, nucleic acids, and amino acids. The concept of water solubility is then expanded, and an account of micelles, lipid bilayers, and detergents is presented. A review of the genome and the synthesis of messenger RNA is given. The reader will return to the topics of DNA and RNA in later chapters, in accounts of the actions of vitamin A, vitamin D, thyroid hormone, and zinc, as well as in commentaries on the origins of cancer. The chapter closes with descriptions of protein synthesis, maturation, and secretion and of the properties of several classes of proteins.

### BASIC CHEMISTRY

This section reviews some elementary chemistry to establish a basis for understanding the later material on hydrophilic interactions and on water-soluble and water-insoluble nutrients.

## Structure and Bonding of Atoms

*Atomic Structure*

An atom consists of an inner nucleus surrounded by electrons. The nucleus consists of protons and neutrons. Each proton has a single positive charge. The number of protons in a particular atom, its **atomic number**, determines the chemical nature of the atom. Neutrons have no charge, but the electrons that surround the nucleus each have a single negative charge. Generally, the number of electrons in a particular atom is identical to the number of protons, so the atom has no overall charge. The electrons reside in distinct regions, called **orbitals**, that surround the nucleus. The actual appearance of the electron as it moves about in its orbital might be thought of as resembling a cloud. Addition of one or more additional electrons to a particular atom produces a net negative charge, whereas removal of one or more electrons results in a net positive charge. Atoms with a positive or negative charge are called **ions**. Conversion of a neutral atom (or molecule) to one with a charge is called **ionization**.

The various orbitals available to the electrons represent different energy levels and are filled in an orderly manner. If one were creating an atom, starting with the nucleus, the first electron added would occupy the orbital of lowest energy, the 1s orbital. Since each orbital is capable of holding two electrons, the second electron added also would occupy the 1s orbital. The next available orbital, which has an energy slightly higher than that of the 1s orbital, is the 2s orbital. A completely filled 2s orbital also contains two electrons. After the 1s and 2s orbitals are filled, subsequent electrons fill the $2p_x$, $2p_y$, and $2p_z$ orbitals. These three orbitals (the 2p orbitals) have identical energy levels. The orbitals that are next highest in energy are 3s, $3p_x$, $3p_y$, and $3p_z$. Of still greater energy are the 4s and 3d orbitals, as indicated in Table 1.1. The 4s and 3d orbitals contain electrons at similar energy levels, whereas the 4p orbitals contain electrons of even higher energy. The terms "higher" and "lower" energy can be put into perspective by understanding that lower-energy electrons have a more stable association with the nucleus. They are dislodged from the atom less easily than higher-energy electrons.

The electrons in the filled orbitals of highest energy, are called **valence electrons**. These electrons, rather than those at lower energy levels, take part in most chemical reactions. Table 1.1 outlines the way that electrons fill orbitals in isolated atoms. However, inside molecules, electrons are shared by atoms bonded to each other. These electrons occupy **molecular orbitals**. The orderly manner in which electrons fill molecular orbitals resembles the filling of atomic orbitals, but a description of molecular orbitals is beyond the scope of this chapter.

The number of electrons that fill the orbitals of an atom is generally equal to the number of protons in its nucleus. However, atoms tend to gain or lose electrons to the extent that a particular series of valence orbitals is either full or empty. This condition results in an overall decrease in energy of the other electrons in valence orbitals. In the inert elements (i.e., helium, neon, and argon), the series of valence orbitals is filled completely. For example, the 10 electrons of neon, a stable and chemically unreactive atom, fill all the 1s, 2s, and 2p orbitals (see Table 1.1). On the other hand, sodium, which contains 11 electrons, loses one electron under certain

**TABLE 1.1**  Electronic Structure of Various Atoms

| Atom | Atomic number | Number of electrons filling the atomic orbital | | | | | | | | | |
|------|---------------|------|------|--------|--------|--------|------|--------|--------|--------|------|
|      |               | $1s$ | $2s$ | $2p_x$ | $2p_y$ | $2p_z$ | $3s$ | $3p_x$ | $3p_y$ | $3p_z$ | $4s$ |
| H  | 1  | 1 | | | | | | | | | |
| He | 2  | 2 | | | | | | | | | |
| Li | 3  | 2 | 1 | | | | | | | | |
| Be | 4  | 2 | 2 | | | | | | | | |
| B  | 5  | 2 | 2 | 1 | | | | | | | |
| C  | 6  | 2 | 2 | 2 | | | | | | | |
| N  | 7  | 2 | 2 | 1 | 1 | 1 | | | | | |
| O  | 8  | 2 | 2 | 2 | 1 | 1 | | | | | |
| F  | 9  | 2 | 2 | 2 | 2 | 1 | | | | | |
| Ne | 10 | 2 | 2 | 2 | 2 | 2 | | | | | |
| Na | 11 | 2 | 2 | 2 | 2 | 2 | 1 | | | | |
| Mg | 12 | 2 | 2 | 2 | 2 | 2 | 2 | | | | |
| Al | 13 | 2 | 2 | 2 | 2 | 2 | 2 | 1 | | | |
| Si | 14 | 2 | 2 | 2 | 2 | 2 | 2 | 1 | 1 | | |
| P  | 15 | 2 | 2 | 2 | 2 | 2 | 2 | 1 | 1 | 1 | |
| S  | 16 | 2 | 2 | 2 | 2 | 2 | 2 | 2 | 1 | 1 | |
| Cl | 17 | 2 | 2 | 2 | 2 | 2 | 2 | 2 | 2 | 1 | |
| Ar | 18 | 2 | 2 | 2 | 2 | 2 | 2 | 2 | 2 | 2 | |
| K  | 19 | 2 | 2 | 2 | 2 | 2 | 2 | 2 | 2 | 2 | 1 |
| Ca | 20 | 2 | 2 | 2 | 2 | 2 | 2 | 2 | 2 | 2 | 2 |

conditions. In this state, the sodium atom has a single positive charge and is considered an ion. The stable nature of the $Na^+$ ion arises from its electronic structure, which is the same as that of neon.

## Covalent and Ionic Bonds

Stable interaction between two or more atoms results in the formation of a molecule. Typically, the atoms in a molecule are connected to each other by covalent bonds. In an ordinary covalent bond, each atom involved contributes one electron to form a pair. The two atoms share this pair of electrons. An electron of one atom can be shared with a second atom when the second atom has valence orbitals that are either vacant or half filled. The hydrogen atom, with an atomic number of 1, contains a half-filled $1s$ orbital. In the hydrogen molecule ($H_2$), the sharing of electrons results in formation of a bonding orbital. A single bonding orbital occurring between two atoms is equivalent to a single covalent bond.

The nitrogen atom, with an atomic number of 7, contains filled $1s$ and $2s$ orbitals and half-filled $2p_x$, $2p_y$, and $2p_z$ orbitals. Because of the presence of these three half-filled orbitals, nitrogen atoms tend to form three covalent bonds. In nitrogen gas ($N_2$), the two nitrogen atoms share the electrons in their $2p$ orbitals, resulting in the formation of three covalent bonds. Since these bonds occur between the same two atoms, they constitute a **triple bond**. In ammonia ($NH_3$), the nitrogen atom and three hydrogen atoms share electrons, resulting in the formation of a

**single bond** between the nitrogen atom and each of the hydrogen atoms. Note that, in these compounds, the nitrogen atom also contains a pair of electrons in its own filled $2s$ orbital. Two electrons in a filled nonbonding valence orbital are called a **lone pair**. This lone pair is not directly involved in the covalent bonds just described but contributes to the chemical properties of ammonia.

The oxygen atom, with an atomic number of 8, contains filled $1s$, $2s$, and $2p_x$ orbitals and half-filled $2p_y$ and $2p_z$ orbitals. Because of the two half-filled valence orbitals, oxygen tends to form two covalent bonds. In oxygen gas ($O_2$), the two oxygen atoms share electrons from their $2p_y$ and $2p_z$ orbitals to form two covalent bonds between the same two atoms. This interaction is called a **double bond**. In water ($H_2O$), the oxygen atom forms a single bond with each of the two hydrogen atoms. The oxygen atom contains two lone pairs (in the $2s$ and $2p_x$ orbitals) that contribute to the properties of water.

The electrons of the carbon atom, with an atomic number of 6, fill the $1s$ and $2s$ orbitals and half-fill the $2p_x$ and $2p_y$ orbitals. Since this is the most stable state of the carbon atom, one might expect that, in molecules, the carbon atom would form two covalent bonds. However, carbon generally forms four covalent bonds. This behavior results in promotion of one electron from the $2s$ orbital to give a half-filled $2p_z$ orbital. In this slightly higher energy state, carbon has four half-filled valence orbitals. Formation of four covalent bonds results in a lower energy state for the molecule as a whole. The carbon atoms in such molecules do not contain lone pairs.

The single bonds described in these examples are formed from two shared electrons, one furnished by each of the two bonded atoms. Bonds in which both of the shared electrons are furnished by one of the atoms can form also. Generally, such bonds involve a lone pair from the **donor atom** and an unfilled orbital in the **acceptor atom**, usually a positively charged ion. These bonds are called **electron donor-acceptor bonds**.

When two identical atoms are bonded to each other, the distribution of electrons between them is symmetrical and favors neither atom. However, in bonds involving two different atoms, the electrons may shift toward one end of the bond. In such a case, the bond is said to have **ionic character** and to be an **ionic bond**. The difference between an ionic and a covalent bond is not absolute, because bond types occur with varying degrees of ionic character. An extreme example of an ionic bond is found in sodium chloride (NaCl). In solid crystals of NaCl or in gaseous NaCl, the sodium atom occurs as $Na^+$, whereas the chlorine atom occurs as $Cl^-$. Individual NaCl molecules do not exist; each positive $Na^+$ ion is surrounded by negative $Cl^-$ ions. The attraction between the ions is very strong, but the bonding electrons are shifted almost completely to the $Cl^-$ ions, that is, the bonding is highly ionic in character. A molecule that contains one or more bonds with measurable ionic character is called a polar molecule.

*Hydrogen Bonds*

Bonds involving hydrogen may be fully covalent, as in $H_2$, partially covalent and partially ionic, as in $H_2O$, or nearly completely ionic, as in HCl. In the more ionic bonds, the electrons are distributed unevenly, skewed away from hydrogen toward its partner atom. This partial removal of electrons from the hydrogen atom results in partially vacant valence orbitals of hydrogen. The partial vacancy can be filled by electrons from an atom in a second molecule, resulting in the phenomenon of hydrogen bonding. The hydrogen atoms of water, alcohol, organic acids, and amines can participate in hydrogen bonding. The other atom involved in the hydrogen bond can be the oxygen atom of molecules such as water, ethers, ketones, or carboxylic acids or the nitrogen atom of ammonia or other amines. For example, hydrogen bonds can form between two water molecules:

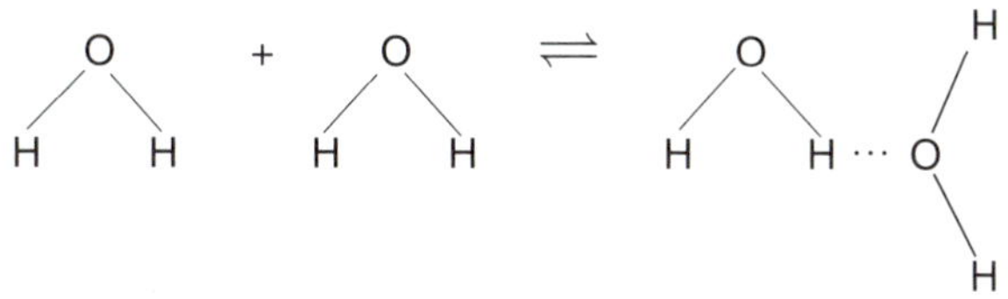

or between water and an ester:

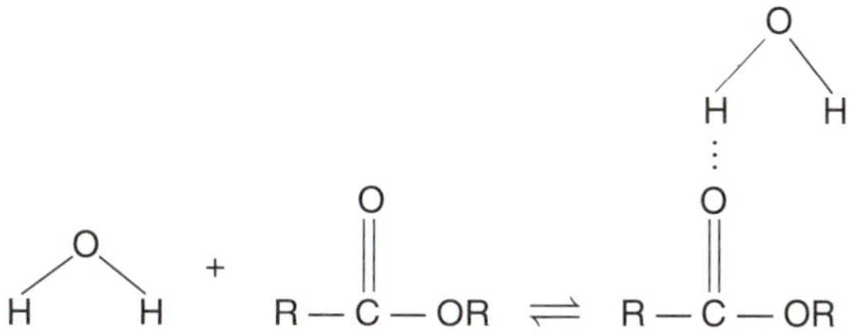

Hydrogen bonds are much weaker than covalent bonds. In aqueous solution, they are broken and re-formed continuously, rapidly, and spontaneously. Note that a water molecule can form hydrogen bonds with up to four other water molecules. In liquid water, hydrogen bonds link together most of the molecules.

## Hydration

The digestion and absorption of organic and inorganic nutrients, as well as all other biochemical processes in living organisms, are influenced by the unique properties of water. Water is an interactive liquid or solvent. Its chemical interactions with solutes are called **hydration**. Hydration involves weak associations of water molecules with other molecules or ions, such as $Na^+$, $Cl^-$, starch, or protein. Because hydration bonding is weak and transitory, the number of water molecules associated with an ion or molecule at any particular moment is approximate and difficult to measure. However, typical indicated **hydration numbers** are: $Na^+$, 1–2; $K^+$, 2; $Mg^{2+}$, 4–10; $Ca^{2+}$, 4–8; $Zn^{2+}$, 4–10; $Fe^{2+}$, 10; $Cl^-$, 1; and $F^-$, 4 (Conway, 1981). Hydration is a consequence of two types of bonding: (1) electron donor–acceptor bonding, and (2) hydrogen bonding. The primary type involved depends on the ion.

Hydration allows water-soluble chemicals to dissolve in water. For example, a crystal of table salt (NaCl) is held together by strong ionic interactions. However, when NaCl is dissolved in water, the $Na^+$ and $Cl^-$ ions become independent hydrated entities. The energy produced by hydration of the $Na^+$ and $Cl^-$ ions more than balances the energy required to remove them from the NaCl crystal lattice. In the $Na^+$ ion, a lone pair of electrons from a water oxygen atom fills an empty valence orbital of $Na^+$ to form an electron donor–acceptor bond. The $Cl^-$ ion interacts electrostatically with water hydrogen atoms, as described in the section on hydrogen bonding.

Not all ionically bonded molecules dissolve in water. For example, silver chloride is virtually insoluble. The energy of hydration of the $Ag^+$ and $Cl^-$ ions is not sufficient to overcome their energy of interaction in the crystal lattice.

## Acids and Bases

In biochemical reactions, an acid is a proton donor, whereas a base is a proton acceptor. In an acid, the bond between the proton ($H^+$) and the parent compound is an ionic bond. In a strong acid the bond has a markedly ionic character. In a weak acid the bond has a more covalent character. When a strong acid, such as HCl, is dissolved in water it dissociates almost completely. Weaker acids, such as acetic acid or propionic acid, dissociate only partially in water. After the parent compound loses its proton it acts as a base, because it can now readily accept a proton.

Conventionally, some chemicals are called acids, whereas others are called bases. This convention is based on the form the chemical takes in its uncharged state or when it is not in contact with water. For example, although the acetate ion that is formed when acetic acid dissociates is a base, acetate ion generally is not called "acetic base."

When an acid (HA) dissociates in water, the dissociated protons do not accumulate as free protons. Instead, each immediately binds to a molecule of water to form a hydronium ion ($H_3O^+$). The proton binds to one of the available lone pairs of the oxygen atom. In this reaction, water serves as a base:

$$HA + H_2O \rightleftharpoons A^- + H_3O^+$$

The equilibrium depicted is extremely rapid. The lifetime of any given molecule of $H_3O^+$ is only $10^{-13}$ seconds (Eigen, 1964). Water is an acid as well as a base. Pure water partially dissociates to form a hydroxide ion and a proton, which binds to another water molecule:

$$H_2O \rightleftharpoons HO^- + H^+$$

The strength of a particular acid is described by its **dissociation constant** (or equilibrium constant; $K$). For water, $K$ is defined by $K = [H^+][HO^-]/[H_2O]$. The symbols in brackets refer to molar concentrations ($M$) of the indicated chemicals. The concentration of pure liquid water is 55.6 $M$. In the human body, the concentrations of $H^+$, $HO^-$, and most other chemicals are far lower than 55 $M$, and are in the range of $10^{-3}$ to $10^{-8}$ $M$.

Because the concentration of water is so high in most aqueous solutions, and because its concentration fluctuates very little in most living organisms, the $[H_2O]$ term conventionally is omitted from the formula for $K$. To omit $[H_2O]$, set the value at 1 to yield a simpler version of the formula: $K = [H^+][HO^-]$. For pure water at 25°C, $[H^+] = 10^{-7}$ $M$ and $[HO^-] = 10^{-7}$ $M$. These two concentrations must be identical since the dissociation of one proton from water results in the production of one hydroxide ion.

## pH is a Shorthand for Expressing the Proton Concentration

The concentration of $H^+$ (which actually occurs as the hydronium ion) in solutions is expressed as the **pH**, defined by the formula: $pH = -\log [H^+]$. To use this formula to describe pure, uncontaminated water, enter the known concentration of $H^+$. This concentration is $10^{-7} M$. Solving the equation gives $pH = 7.0$. As the formula shows, solutions with high $H^+$ levels have a low pH; those with low $H^+$ levels have a high pH. A solution that has a pH of 7.0 is said to be neutral. Solutions with a lower pH are said to be acidic. When considering acidic solutions, biochemists often are concerned with the reactive properties of $H^+$. Solutions with a pH greater than 7.0 are said to be alkaline or basic. Biochemists may be concerned with reactions involving the hydroxide ion ($HO^-$) in such solutions.

## Strong Acids are Highly Dissociated in Water; Weak Acids are Slightly Dissociated in Water

The degree of dissociation of any given acid (HA) in water is expressed in terms of the distinct value of its dissociation constant, $K$, defined by the formula $K = [H^+][A^-]/[HA]$. When comparing weak and strong acids, the strength of the acid conventionally is expressed by its $pK$, defined as $pK = -\log K$.

Strong acids have low $pK$ values; weak acids have high $pK$ values. For example, formic acid is moderately strong: $pK = 3.75$. The $pK$ values for other acids and proton-donating compounds are: phosphoric acid ($H_3PO_4$), 2.14; acetic acid ($CH_3COOH$), 4.76; carbonic acid ($H_2CO_3$), 3.8; ammonium ion ($NH_4^+$), 9.25; and bicarbonate ion ($HCO_3^-$), 10.2. The values for $K$ and $pK$ refer to reactions that are reversible in aqueous solution and have attained a condition of equilibrium.

Consider an imaginary acid, HA, with $K = 0.01$ ($pK = 2.0$). When any quantity of HA is mixed with water, the acid will dissociate to the extent that satisfies the formula $[A^-][H^+]/[HA] = 0.01$. The term "any quantity" refers to a broad range of concentrations far below $55.6 M$. Once a degree of dissociation occurs that results in levels of HA and $A^-$ that satisfy the formula, the net trend toward dissociation stops. Although dissociation continues, reassociation occurs at an equal rate. Thus, an equilibrium situation is reached.

The lone pair electrons of water (O atom), ammonia (N atom), and amino groups (N atom) influences the behavior and concentrations of hydrogen ions ($H^+$) in water. Hydrogen ions, produced either by dissociation of water or by dissociation of acids, do not occur as free entities in aqueous solutions. They associate with the lone pair electrons of other water molecules to form hydronium ions, $H_3O^+$. This association involves the formation of an electron donor–acceptor bond.

Electron donor–acceptor bonds involving nitrogen are stronger than those involving oxygen, so some nitrogen-containing molecules dissolved in water will bind any $H^+$ ions that are present with a greater strength than any single surround-

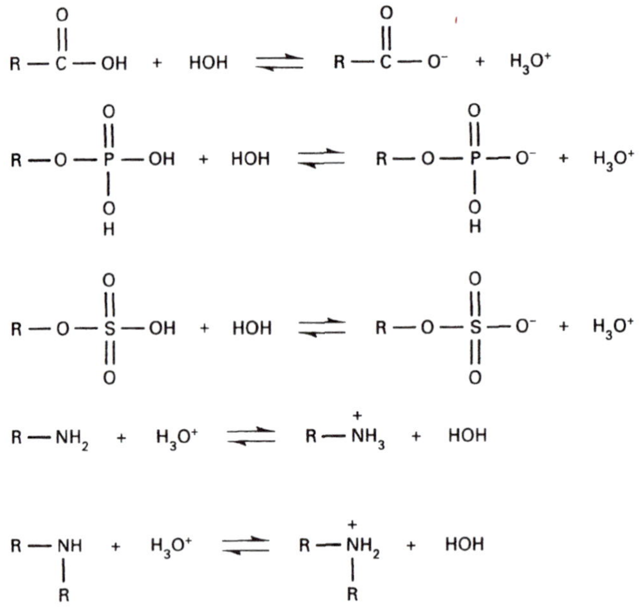

$$R-\overset{\overset{O}{\|}}{C}-OH \ + \ HOH \ \rightleftharpoons \ R-\overset{\overset{O}{\|}}{C}-O^- \ + \ H_3O^+$$

$$R-O-\overset{\overset{O}{\|}}{\underset{\underset{H}{O}}{P}}-OH \ + \ HOH \ \rightleftharpoons \ R-O-\overset{\overset{O}{\|}}{\underset{\underset{H}{O}}{P}}-O^- \ + \ H_3O^+$$

$$R-O-\overset{\overset{O}{\|}}{\underset{\underset{O}{\|}}{S}}-OH \ + \ HOH \ \rightleftharpoons \ R-O-\overset{\overset{O}{\|}}{\underset{\underset{O}{\|}}{S}}-O^- \ + \ H_3O^+$$

$$R-NH_2 \ + \ H_3O^+ \ \rightleftharpoons \ R-\overset{+}{NH_3} \ + \ HOH$$

$$R-\underset{\underset{R}{|}}{NH} \ + \ H_3O^+ \ \rightleftharpoons \ R-\underset{\underset{R}{|}}{\overset{+}{NH_2}} \ + \ HOH$$

**FIGURE 1.1** Ionization of acids and bases. An acid is defined as a chemical that dissociates and donates a proton to water. A base is defined as a chemical that can accept a proton. The double arrows indicate that the ionization process occurs in the forward and backward directions. The term *equilibrium* means that the rate of the forward reaction is equal to the rate of the backward reaction, and that no net accumulation of products or reactants occurs over time.

ing water molecule. For example, if dimethylamine ($CH_3$–NH–$CH_3$) is added to water, it tends to remove $H^+$ from molecules of $H_3O^+$ that may be present:

$$H_3O^+ \ + \ CH_3-\overset{\overset{H}{|}}{N}-CH_3 \ \rightleftharpoons \ H_2O \ + \ CH_3-\underset{\underset{H}{|}}{\overset{\overset{+H}{|}}{N}}-CH_3$$

This transfer results in a decrease in the concentration of $H_3O^+$ in the solution. Therefore, molecules of this type act as bases.

## Chemical Groups

Table 1.2 presents structural formulas of the chemical groups used to classify compounds of biological interest. The common abbreviation for the group, the name of the group, and the name of the class of compounds containing the group are also given. Note that, when a compound contains more than one group, it is named from the group considered most significant. "R" represents the rest of the molecule on the other side of a single covalent bond. In molecules containing more than one R group, "R" represents the same configuration of atoms unless the groups are distinguished as $R_1$, $R_2$, and so forth.

**TABLE 1.2**  Chemical Groups Used to Classify Compounds of Biological Interest

| Structure | Abbreviation | Group name | Compound name |
|---|---|---|---|
| R—O—H | R–OH | Hydroxyl | Alcohol |
| R—C(=O)—H | R—COH | Aldehyde | Aldehyde |
| R—C(=O)—R | RCOR | Keto | Ketone |
| R—C(=O)—OH | RCOOH | Carboxyl | Acid |
| R—C(=O)—O—R | RCOOR | Ester | Ester |
| R—O—R | ROR | Ether | Ether |
| R—N(H)—H | $RNH_2$ | Amino | Primary amine |
| R—N(H)—R | RNHR | Amino | Secondary amine |
| R—C(=O)—N(H)—H | $RCONH_2$ | Amido | Amide |
| R—O—P(=O)(OH)—OH | (P) | Phospho | Phosphate |
| R—O—S(=O)(=O)—OH | $RSO_4H$ | Sulfo | Sulfate |
| R—S—R | RSR | Sulfide | Sulfide |
| R—S—S—R | RS—SR | Disulfide | Disulfide |
| R—N=C(H)—R | RN=CHR | Imido | Imine (Schiff base) |

## Ionic Groups

Compounds that contain carboxyl groups are called acids (or carboxylic acids). As illustrated in Figure 1.1, a carboxyl group in aqueous solution is partially ionized to the carboxylate anion. The degree of ionization depends on the dissociation constant of the acid and the initial pH of the solution. The strength of these acids varies somewhat depending on the attached R group. Esters are formed by reaction of a carboxylic acid with an alcohol. Amides are formed by reaction of a carboxylic acid with an amine.

Inorganic phosphate and organic phosphates are ionized when dissolved in water. Similarly, inorganic sulfate and organic sulfate are ionized when dissolved in water. In inorganic phosphate and sulfate, the R group is a hydrogen atom.

As also illustrated in Figure 1.1, a primary or secondary amine group can function as a weak base. The degree to which the group is protonated to the positive ion depends on the dissociation constant of the molecule to which it is attached and on the initial pH of the solution.

### Counterions

Ionized forms of molecules are nearly always accompanied by counterions of the opposite charge. When the counterion is a proton, the ion and proton complex is called an acid. When the counterion is a different cation, such as a sodium, potassium, or ammonium ion, the complex is called a salt.

## MACROMOLECULES

In biological systems, atoms tend to form very large molecules called **macromolecules**, which can be segregated into four groups: carbohydrates, nucleic acids, proteins, and lipids.

## Carbohydrates

The term carbohydrate refers to a class of polyhydroxy aldehydes and polyhydroxy ketones with the general formula $(CH_2O)_n$. The name derives from the composition of the formula unit, that is, carbon plus water. All carbohydrates are composed of basic units called **monosaccharides**. Polymers containing two to six monosaccharides are called **oligosaccharides**; those with more are called **polysaccharides**. Starch, cellulose, and glycogen are examples of polysaccharides. Monosaccharides and oligosaccharides are also called sugars.

### Monosaccharides

The open-chain structure of a monosaccharide is a straight-chain saturated aldehyde or 2-ketone with three to seven carbon atoms. The carbon atoms are numbered as shown in Figure 1.2. Every carbon, except those of the aldehyde or ketone group, has one hydroxyl group. In biological materials, monosaccharides with five and six carbon atoms are most common.

In many sugars, such as glucose, the carbon chain can cyclize in two different ways, producing the α and β isomers. These rings are formed by reaction of the

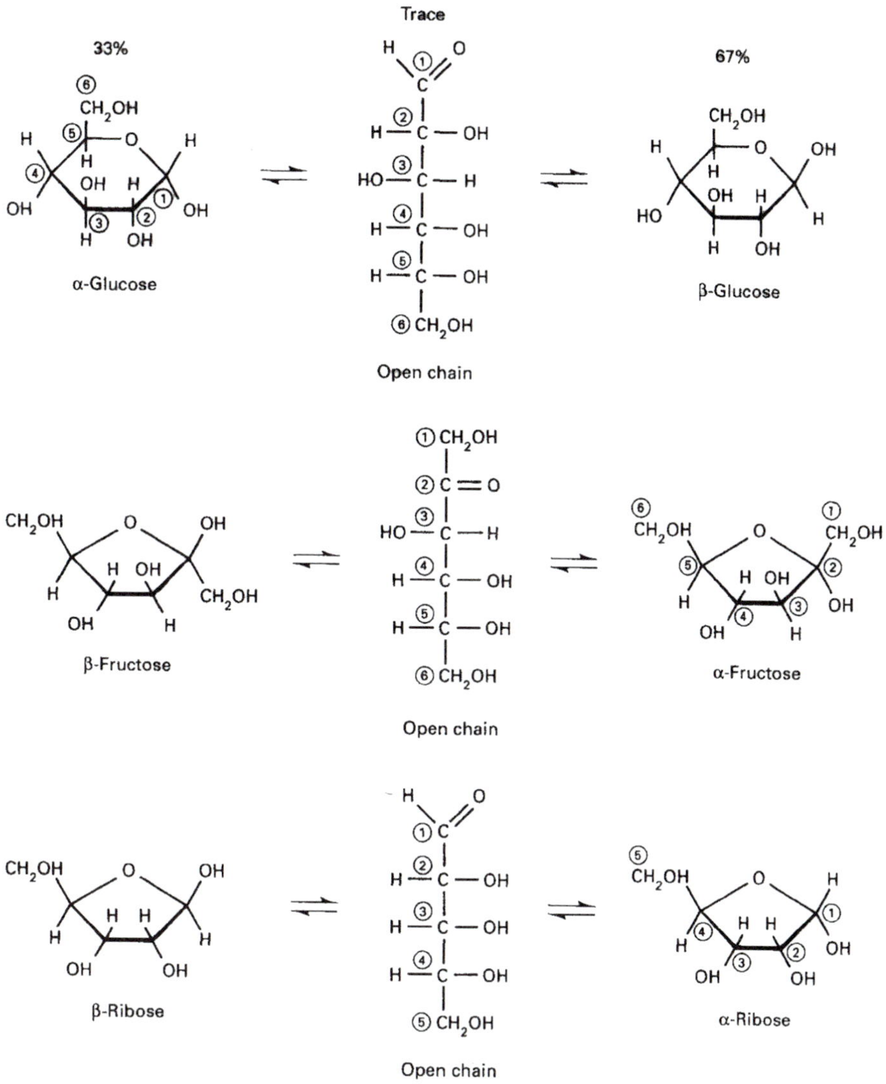

**FIGURE 1.2** Straight chain and cyclic monosaccharides. Glucose and fructose contain six carbons, while ribose contains five carbons. Cyclization requires the participation of an aldehyde group or a keto group.

hydroxyl group of the next to last carbon with the aldehyde or keto group, forming a hemiacetal or hemiketal group, respectively. The carbon atoms retain the numbers assigned to them in the straight-chain form. The two ring forms are in equilibrium when the free monosaccharide molecules are in solution.

Figure 1.2 also shows the ring and open-chain structures of fructose and ribose. The β form of ribose occurs in the ribonucleic acid (RNA). The β form of 2-deoxyribose, a modified form of ribose, occurs in deoxyribonucleic acid (DNA).

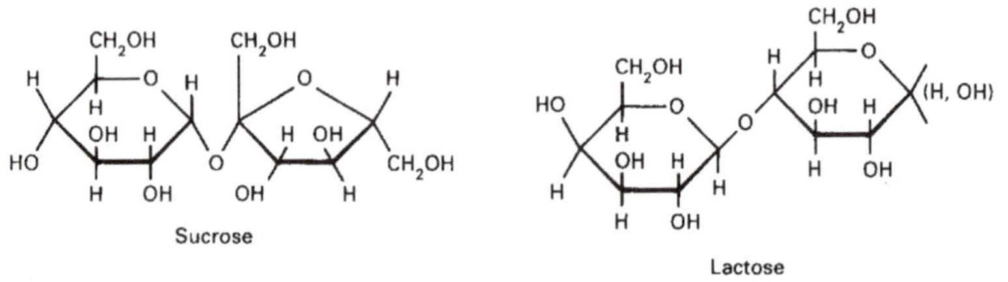

**FIGURE 1.3** Disaccharides: sucrose and lactose. Sucrose occurs in high levels in beets and sugar cane, while lactose occurs only in milk. The two monosaccharides in sucrose, for example, are joined via an oxygen atom.

## Oligosaccharides

The most common oligosaccharides in nature are **disaccharides,** two monosaccharide units joined by a **glycosidic linkage.** Sucrose, for example, contains one unit of fructose and one of glucose. Lactose contains one unit of galactose and one of glucose (see Figure 1.3). The glycosidic linkages are named according to the carbon from each sugar that participates in the bond. For example, the linkage in lactose is $\beta(1 \rightarrow 4)$.

## Polysaccharides

Several polysaccharides are important in biological systems. Starch is a polymer of glucose monomers connected by a glycosidic linkages. **Amylose** is a straight-chain starch containing only $\alpha(1 \rightarrow 4)$ linkages. In **amylopectin,** the chain is branched at approximately 25-monomer intervals by $\alpha(1 \rightarrow 6)$ glycosidic linkages. Glycogen is similar to amylopectin, but branches occur more frequently. Cellulose is a linear polymer of glucose monomers containing $\beta(1 \rightarrow 4)$ glycosidic linkages.

## Nucleic Acids

The two nucleic acids are **deoxyribonucleic acid** (DNA) and **ribonucleic acid** (RNA). As suggested by their names, these compounds occur most commonly in the nucleus of the cell. DNA is the genetic material that contains all the information needed to create a living organism, that is, all the information needed to provide for the structure of an animal; the abilities to reproduce, think, and learn; and some forms of behavior and language. DNA generally consists of two linear polymers (or strands) of **nucleotides,** tightly associated with each other by a series of hydrogen bonds between the two strands. The two DNA strands are twisted around each other, and the overall structure is called a **double helix.** The length of each strand of DNA in each human cell is about 2 m and contains approximately 11 billion nucleotides. DNA actually is only the name of a chemical, while "genome" is the term used to refer to all of the DNA in any particular cell of a specific organism.

When double-stranded DNA (dsDNA) becomes unraveled, the result is two strands of single-stranded DNA (ssDNA). During the normal course of metabolism, short stretches of dsDNA are unraveled in the cell to give regions of the chromosome that consist of ssDNA.

Figure 1.4 shows the nucleosides of DNA: **deoxyadenosine** (dA), **deoxythymidine** (dT), **deoxyguanosine** (dG), and **deoxycytidine** (dC). A nucleoside containing one to three phosphate groups bound to the 5'-carbon of the deoxyribose group is called a **nucleotide**. Also shown are the nucleosides of RNA: **adenosine** (A), **uridine** (U), **guanosine** (G), and **cytidine** (C). RNA consists of single polymer strands, not double strands as found in DNA.

Figure 1.5 shows the manner in which nucleotide units are joined in the polymer strands of DNA. Each phosphate group is bonded to the 3'-carbon of one sugar unit and to the 5'-carbon of the next sugar unit, forming a **phosphodiester linkage**. Adjacent nucleotides of RNA are also joined in this manner.

## Complementation of Bases Maintains the Double-Stranded Structure of DNA, and Allows DNA to Code for a Corresponding Polymer during RNA Synthesis

The aromatic rings connected to the ribose moieties of RNA and the deoxyribose moieties of DNA are called nucleic acid bases. They are bases because they contain nitrogen atoms that bind protons. **Complementation** is the pattern of hydrogen bond formation that occurs between specific pairs of nucleic acid bases. An example is shown in Figure 1.6. Complementation of the bases in nucleic acids is responsible for the maintenance of the double helix structure of DNA. Complementation also guides the transfer of genetic information from the DNA in a parent chromosome to a daughter chromosome, during the process of DNA synthesis, which occurs shortly before the cell divides. Finally, complementation allows DNA to serve as a template during RNA synthesis.

In maintaining the structure of DNA, interactions occur between dA and dT, and between dG and dC. Adenosine and thymine are complementary bases, while guanine and cytosine are complementary bases.

Complementation guides the transfer of information from specific regions of DNA (called genes) during formation of RNA. During the synthesis of RNA, the order of occurrence of the bases in DNA guides the order of polymerization of the ribonucleotides to create RNA. RNA synthesis does not involve a permanent association of DNA with RNA. In RNA synthesis, the association between incoming ribonucleotides is fleeting and temporary. The exposure of successive DNA bases allows the successive process of matching of free ribonucleotides with the deoxyribonucleotides occurring in the strand of DNA. As soon as an incoming free ribonucleotide finds a match (on the DNA template), the free ribonucleotide is covalently attached to the growing strand of RNA.

The following bases of DNA and RNA hydrogen bond to each other and therefore are complementary to each other: dA of DNA to U of RNA; dT of DNA to A of RNA; dG of DNA to C of RNA; and dC of DNA to G of RNA.

A description of the event of DNA synthesis allows us to make use of two of the concepts introduced herein. During DNA synthesis, incoming free deoxyribonucleotides find their match (on the DNA template), but as soon as a match is

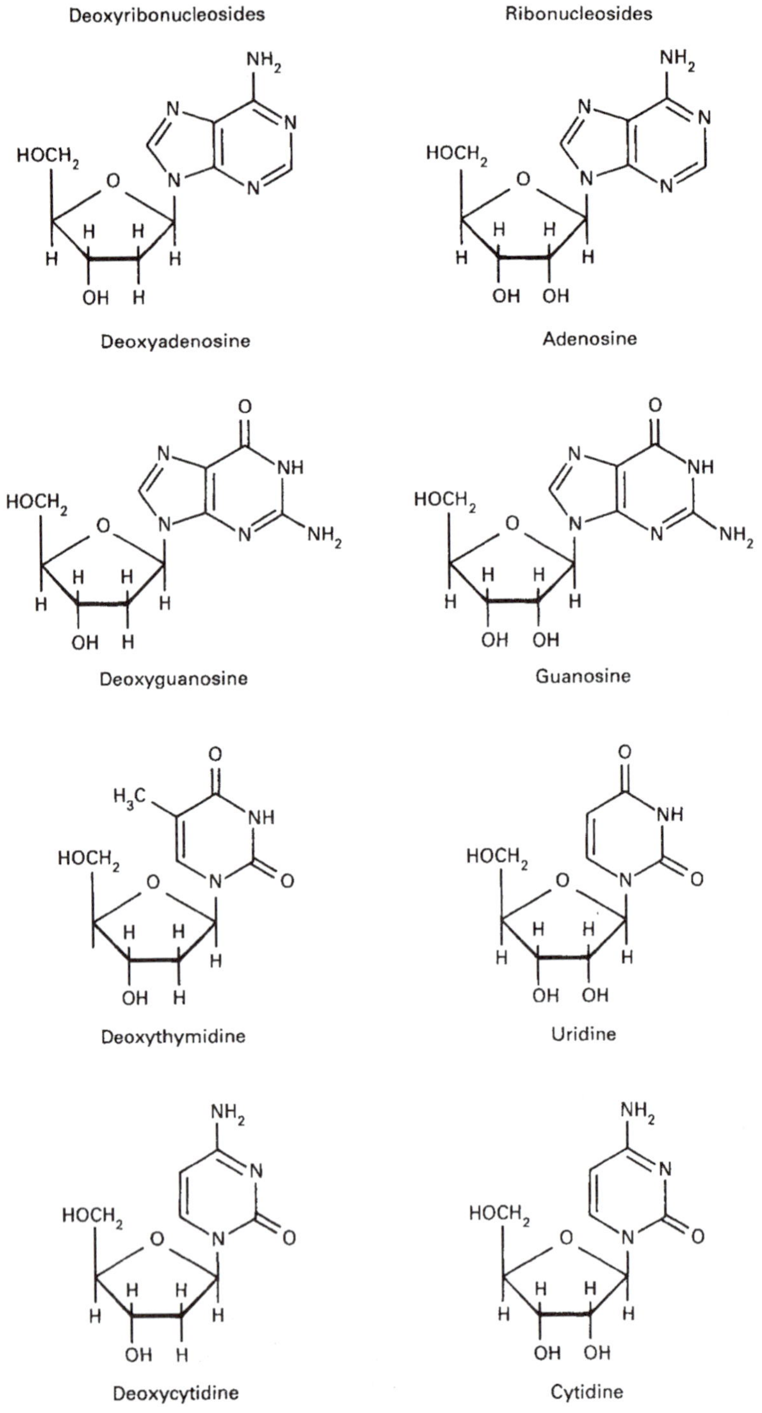

**FIGURE 1.4** Nucleosides of DNA and RNA. The nucleosides are each composed of a base and a 5-membered sugar. The bond from the sugar to the base involves a carbon-to-nitrogen bond. The base is, in fact, a base because of the nitrogen atoms, which can be protonated.

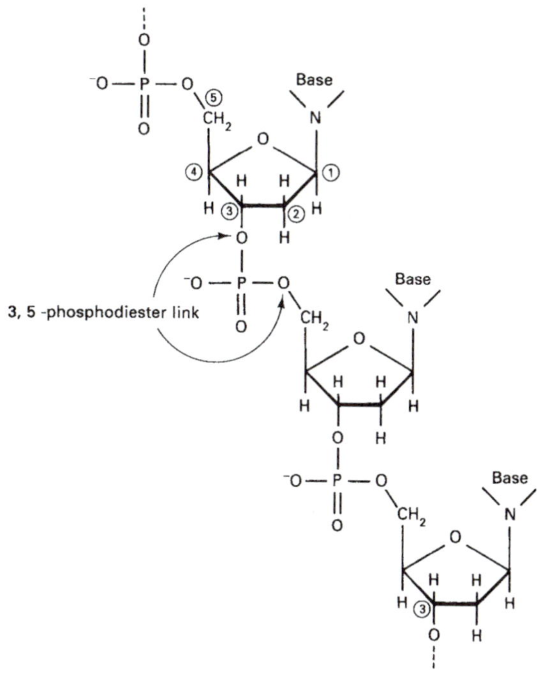

**FIGURE 1.5** Trinucleotide fragment of a DNA chain. A nucleotide is a nucleoside containing a phosphate group. The phosphodiester linkage received its name because it involves an acid (phosphoric acid) linked via an oxygen atom to an R group (the 5-membered sugar). Since two such bonds occur, the entire structure is called a diester.

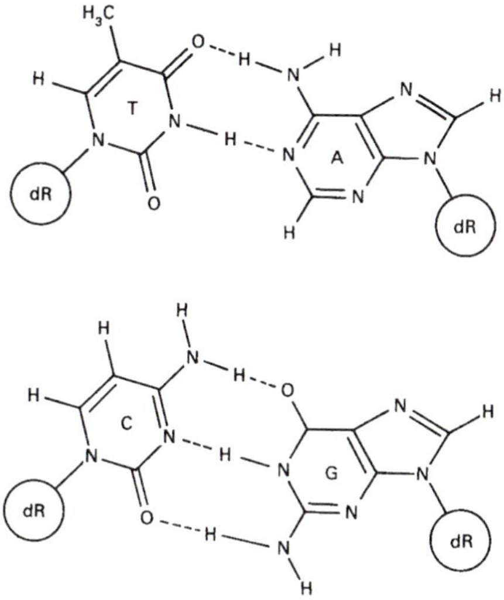

**FIGURE 1.6** Bonding of complementary bases in the DNA double helix. The bases are identified by A, T, G, and C, whereas the deoxyribose groups are identified by dR. Two hydrogen bonds form between dA and dT; three form between dC and dG. This bonding occurs between bases on opposite strands of the double helix. During RNA synthesis (transcription), the DNA strands are separated. This separation is only momentary. The separation allows the RNA bases to hydrogen bond temporarily with the DNA bases, thus governing the order of polymerization of the ribonucleotides. Some genes are transcribed using one strand of the DNA, whereas others are transcribed from the opposite strand.

made the free deoxyribonucleotide is covalently connected to a growing strand of DNA. Thus, descriptions of RNA synthesis and DNA synthesis both make use of the concept of a **template**. The other concept used to describe DNA synthesis is that interactions occur between dA and dT, and between dG and dC.

DNA synthesis occurs during the process of **replication**. Replication is the process by which DNA synthesis causes all the DNA in the nucleus to make a duplicate of itself. In eukaryotic cells, the process of replication causes the cell to change from a diploid cell to one that is temporarily tetraploid. Replication in eukaryotic cells is usually immediately followed by cell division, which results in two cells that are once again diploid.

## Amino Acids and Proteins

Figure 1.7 depicts the general formula for a 2-amino acid, also called an α-**amino acid**. An oligomer consisting of two amino acids is called a **dipeptide**. The amino acids are bound to each other by a **peptide bond**, which involves a keto group and an amino group. Amino acid polymers of moderate length are called **oligopeptides**, whereas longer polymers are called **polypeptides** or **proteins**. A typical protein contains about 300 amino acids and has a molecular weight of about 50,000. The polypeptide chains that constitute proteins are linear and contain no branching.

Generally, specific proteins can bind to each other in the body to form dimers (duplex structures), trimers, tetramers, or even larger multiples. These subunit proteins may be of identical or different structure. The different proteins in these multimeric structures are bound to each other by hydrogen bonds and other weak interactions. These multimers often perform physiological functions that cannot be carried out by the individual separated proteins.

Examination of proteins with an electron microscope reveals that some are somewhat spherical, others asymmetric, and others long and fibrous. The overall shape, function, and chemical properties of a specific oligopeptide or protein are determined by the identities and order of polymerization of its constituent amino acids.

*Classical Amino Acids*

The classical amino acids are those that are incorporated into proteins during polymerization in the cell. Table 1.3 lists the 20 classical amino acids in order of one

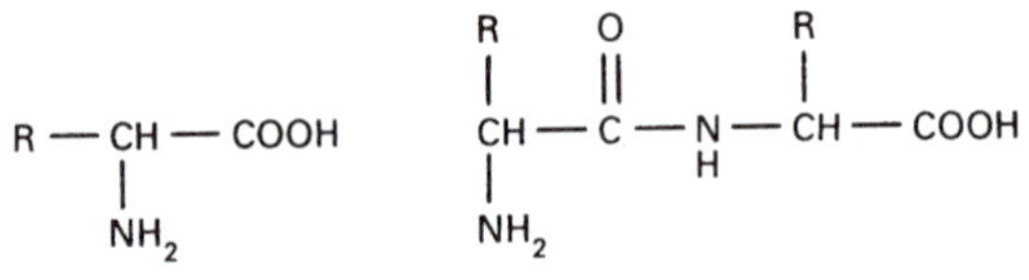

**2-Amino Acid**                    **Dipeptide**

**FIGURE 1.7** Generic structure of an amino acid (left) and generic structure of two amino acids linked by a peptide bond (right).

**TABLE 1.3** The Twenty Classical Amino Acids Listed in Order of Their Approximate Relative Increasing Hydrophilicity[a]

| Amino acid | Abbreviation |
| --- | --- |
| Isoleucine | Ile |
| Valine | Val |
| Leucine | Leu |
| Phenylalanine | Phe |
| Cysteine | Cys |
| Methionine | Met |
| Alanine | Ala |
| Glycine | Gly |
| Threonine | Thr |
| Tryptophan | Trp |
| Serine | Ser |
| Tyrosine | Tyr |
| Proline | Pro |
| Histidine | His |
| Glutamic acid | Glu |
| Glutamine | Gln |
| Aspartic acid | Asp |
| Asparagine | Asn |
| Lysine | Lys |
| Arginine | Arg |

*Source*: Kyte and Doolittle (1982).

[a]Isoleucine has the most lipophilic side chain and arginine has the most hydrophilic side chain.

property of the R group (side chain), namely, increasing hydrophilicity (decreasing lipophilicity). This order does not reflect an absolute property to which a numerical value might be ascribed, but reflects the observed tendency of the amino acid to occur in the lipophilic core or on the hydrophilic surface of a protein. These observations were made by examining the three-dimensional structures of many proteins with somewhat spherical or globular structures. Amino acids with ionizable or hydrogen-bonding R groups are more hydrophilic, whereas those with alkane or aromatic R groups are more lipophilic. Isoleucine, valine, and leucine are the most lipophilic amino acids; arginine, lysine, and asparagine are the most hydrophilic. Lysine, for example, contains a protonated amino group at the end of its side chain.

Of all the classical amino acids, with the exception of glycine, only the 2-carbon atom is bonded to four different groups, that is, a carboxyl, hydrogen, amino, and R group. Bonding of four different groups to any carbon atom can occur in two different isomeric arrangements. In amino acids, these are called the L-amino acids and D-amino acids. Essentially, all the amino acids in the diet and in the body occur as the L-isomer. In this textbook, all references to amino acids are to the L-isomer unless otherwise specified. D-Amino acids occur in small quantities in certain molecules synthesized by invertebrates and bacteria. Sometimes isomeric mixtures containing equal proportions of a certain amino acid in the L and D forms are given to animals as supplements to the diet. Such mixtures may be given because they are less expensive than a supplement containing the pure L-amino acid. Baker (1984) discussed that some D-amino acids can be converted in the body (isomerized) to the L isomer, whereas others tend to be broken down rather than isomerized.

The simplest amino acids are **glycine** and **alanine**. The R group of glycine is a hydrogen atom; the R group of alanine is a methyl group.

$$HCH - COOH \qquad CH_3CH - COOH$$
$$| \qquad\qquad\qquad |$$
$$NH_2 \qquad\qquad\qquad NH_2$$
$$\text{Glycine} \qquad\qquad \text{Alanine}$$

**Serine** and **cysteine** are also relatively simple.

$$HO - CH_2CH - COOH \qquad HS - CH_2CH - COOH$$
$$| \qquad\qquad\qquad\qquad |$$
$$NH_2 \qquad\qquad\qquad\qquad NH_2$$
$$\text{Serine} \qquad\qquad\qquad \text{Cysteine}$$

Serine, glycine, and cysteine are dispensable (or nonessential) amino acids because they can be biosynthesized from precursors that are readily available in the body. Serine can be made from or converted back to glucose, and also is used in the synthesis of cysteine. The pathways for these conversions are detailed in Chapter 8.

**Threonine** and **serine** contain hydroxyl groups that sometimes serve as a point of attachment for a string of sugar molecules. These oligosaccharide strings usually include mannose and glucose as well as other sugars. The hydroxyl group of serine also may serve as a point of attachment for a phosphate group. The milk protein casein may contain up to 10 phosphoserine residues. These negatively charged phosphate ions serve as binding sites for positive calcium ions.

$$CH_3 \qquad\qquad\qquad\qquad OH$$
$$| \qquad\qquad\qquad\qquad\qquad |$$
$$HO - CHCH - COOH \qquad {}^-O - P - O - CH_2CH - COOH$$
$$| \qquad\qquad\qquad\qquad\qquad \| \qquad\qquad |$$
$$NH_2 \qquad\qquad\qquad\qquad O \qquad\qquad NH_2$$
$$\text{Threonine} \qquad\qquad\qquad \text{Phosphoserine}$$

**Isoleucine, valine,** and **leucine** are the **branched-chain amino acids** (BCAAs). They are indispensable (essential), but the risk of developing a dietary deficiency is low because they are plentiful in most diets. The branched-chain amino acids, in addition to phenylalanine, are the most lipophilic of the amino acids:

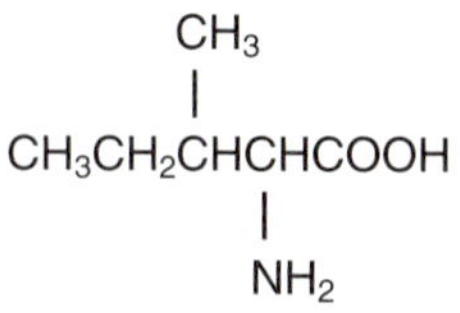

$$CH_3$$
$$|$$
$$CH_3CH_2CHCHCOOH$$
$$|$$
$$NH_2$$

Isoleucine

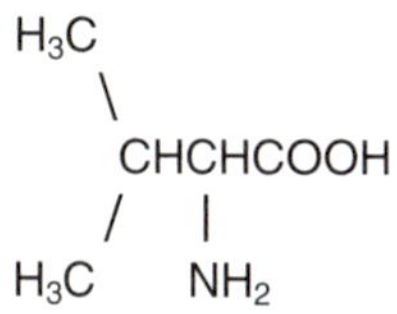

$$H_3C$$
$$\backslash$$
$$CHCHCOOH$$
$$/ \quad |$$
$$H_3C \quad NH_2$$

Valine

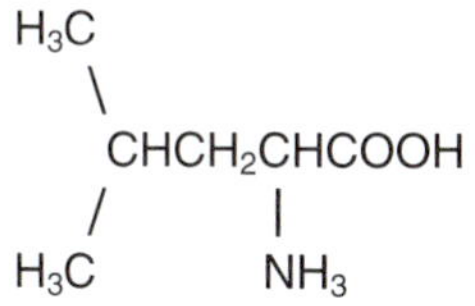

$$H_3C$$
$$\backslash$$
$$CHCH_2CHCOOH$$
$$/ \qquad |$$
$$H_3C \qquad NH_3$$

Leucine

$$CH=CH$$
$$/ \qquad \backslash$$
$$CH \qquad C-CH_2CHCOOH$$
$$\backslash\backslash \quad // \qquad |$$
$$CH-CH \qquad NH_2$$

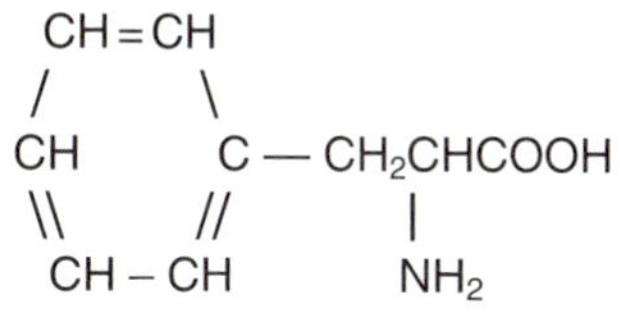

Phenylalanine

**Methionine** and **cysteine**, the sulfur-containing amino acids, are related meta-bolically. Methionine can be converted, in the body, to cysteine. In other words, methionine is the source of the sulfur atom in the synthesis of cysteine in the body. Methionine nutrition is of occasional concern, because legume proteins have a relatively low methionine content. Therefore, legume-based diets, including those based on beans and peas, may not result in maximum growth rates for infants or animals.

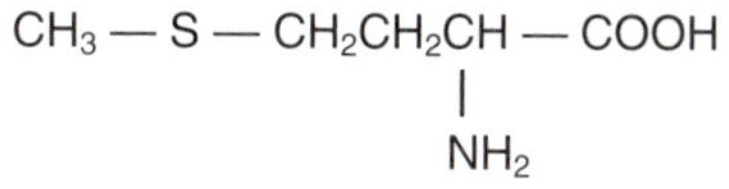

$$CH_3-S-CH_2CH_2CH-COOH$$
$$|$$
$$NH_2$$

Methionine

**Tyrosine** and **tryptophan**, as well as **phenylalanine**, are the aromatic amino acids. The body can convert phenylalanine to tyrosine. Thus, tyrosine is a dispensable (nonessential) amino acid.

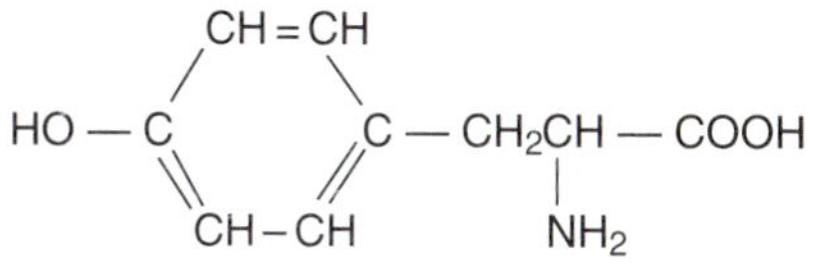

$$CH=CH$$
$$/ \qquad \backslash$$
$$HO-C \qquad C-CH_2CH-COOH$$
$$\backslash\backslash \qquad // \qquad |$$
$$CH-CH \qquad NH_2$$

Tyrosine

$$CH=CH$$
$$/ \qquad \backslash$$
$$CH \qquad C-C-CH_2-CH-COOH$$
$$\backslash\backslash \qquad // \backslash \qquad |$$
$$CH-C \qquad CH \qquad NH_2$$
$$\backslash \quad /$$
$$N$$
$$H$$

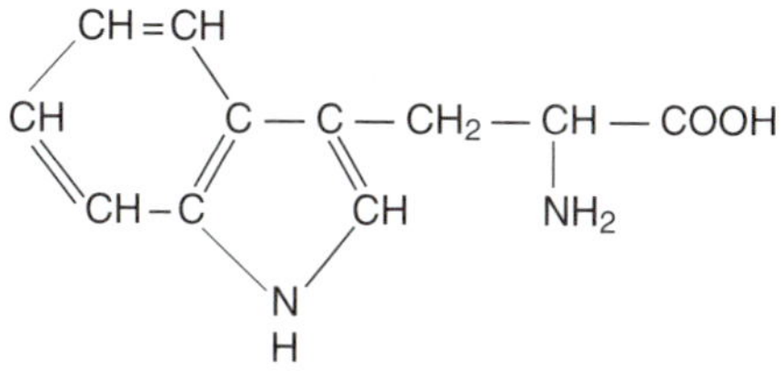

Tryptophan

The acidic amino acids are **glutamic acid** and **aspartic acid**. These amino acids also occur in amide forms as **glutamine** and **asparagine**. Glutamate is the amino acid present in greatest abundance in a variety of dietary proteins.

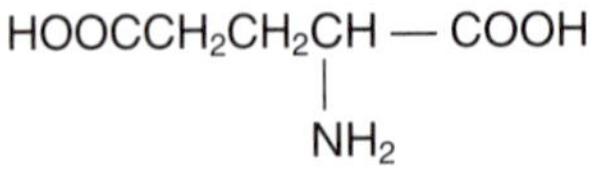

$$HOOCCH_2CH_2CH-COOH$$
$$|$$
$$NH_2$$

Glutamic Acid

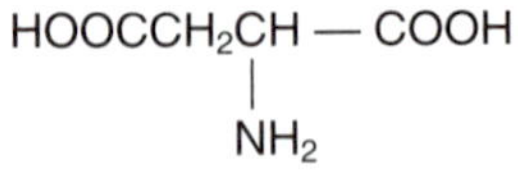

$$HOOCCH_2CH-COOH$$
$$|$$
$$NH_2$$

Aspartic Acid

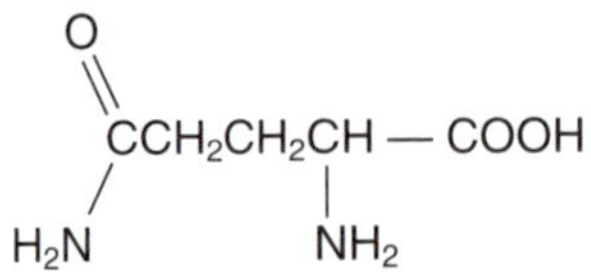

$$\overset{O}{\overset{\|}{C}}CH_2CH_2CH-COOH$$
$$H_2N \qquad\qquad | $$
$$NH_2$$

Glutamine

$$\overset{O}{\overset{\|}{C}}CH_2CH-COOH$$
$$H_2N \qquad\quad | $$
$$NH_2$$

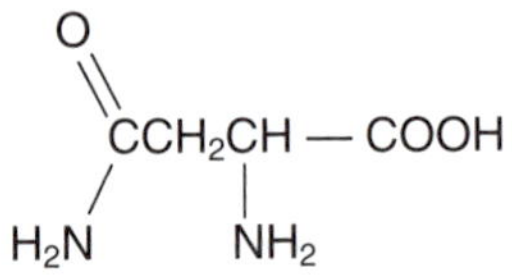

Asparagine

**Lysine, arginine**, and **histidine**, with glutamine and asparagine, are the basic amino acids. Lysine is the amino acid in lowest supply in grains; hence, lysine shortage is of concern when raising animals on wheat, corn, or rice-based diets.

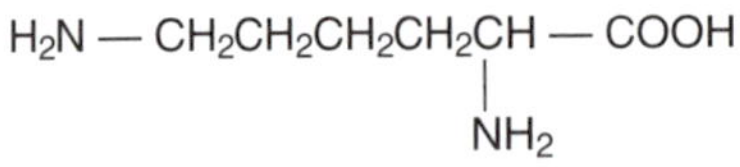

$$H_2N-CH_2CH_2CH_2CH_2CH-COOH$$
$$|$$
$$NH_2$$

Lysine

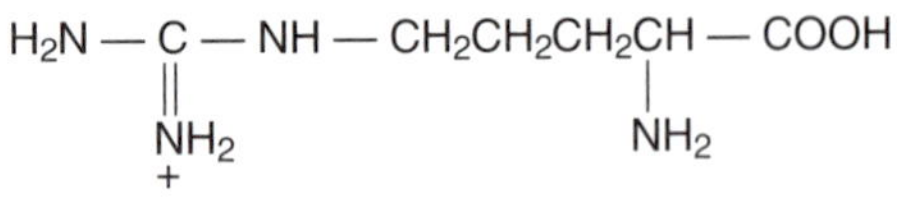

$$H_2N-C-NH-CH_2CH_2CH_2CH-COOH$$
$$\overset{\|}{NH_2} \qquad\qquad\qquad\quad |$$
$$+ \qquad\qquad\qquad\qquad\quad NH_2$$

Arginine

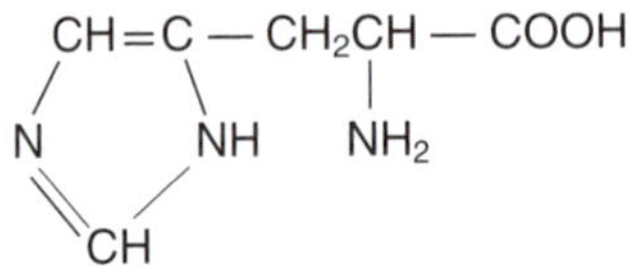

$$CH=C-CH_2CH-COOH$$
$$N \quad\;\; NH \quad NH_2$$
$$CH$$

Histidine

$$CH=C-CH_2CH-COOH$$
$$N \quad\;\; NH \quad NH-C-CH_2CH_2NH_2$$
$$CH \qquad\qquad\qquad \overset{\|}{O}$$

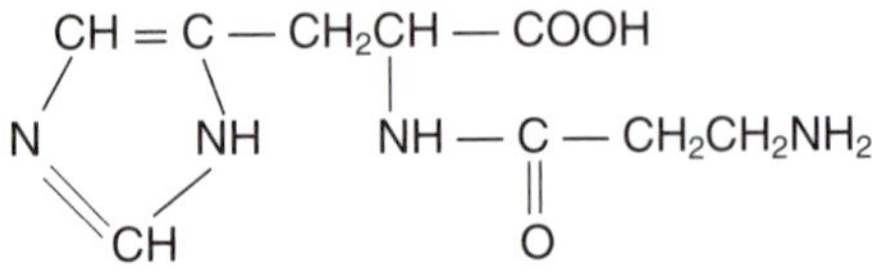

Carnosine

Arginine takes part in the cyclic pathway of reactions known as the **urea cycle**. The urea cycle is a series of reactions that is used to package ammonium ions. The ammonium ion is generated during the normal, hour-by-hour breakdown of proteins in the body. This ion must be efficiently packaged and excreted, since it is toxic to nerves. A very small proportion of arginine in the body is broken down in a pathway that results in the formation of nitric oxide (NO). NO is a hormone used for the regulation of blood flow through certain vessels and used to regulate blood pressure. In short, NO is synthesized in the endothelial cells that line blood vessels. It then diffuses out of the cells and provokes nearby muscle cells to relax, resulting in dilation of the vessel. Vessel dilation also is provoked by other hormones, such as acetylcholine and serotonin.

Histidine is an indispensable amino acid. Its requirement is easily demonstrated in young, growing animals but is difficult to show in adults. Apparently, the signs of deficiency fail to materialize when adult animals are fed a histidine-free diet because of histidine stored in muscle in the form of a related compound, **carnosine**. The availability of carnosine as a source of histidine varies among species. Human and rat muscle contain carnosine, but mouse muscle does not. Fish muscle contains a methylated form of carnosine, called anserine, that does not seem to be available to the fish as a source of histidine.

**Proline** is an unusual amino acid because the amine group is part of a cyclic structure:

Proline                    Hydroxyproline

Although the amino group appears to be "tied up," it still can participate in the formation of a polypeptide chain. **Hydroxyproline**, a modified form of proline, is not one of the 20 classical amino acids but is found in structural proteins such as those of connective tissues.

## Modified Amino Acids

As stated earlier, only classical amino acids are built into polypeptides during amino acid polymerization. However, other amino acids, classical amino acids that have been modified after incorporation into the chain, are found in proteins. Some of the modified amino acids found in proteins are listed in Table 1.4. Vitamins are required for the synthesis of some of the modified amino acids. For example, vitamin C is required for conversion of proline to hydroxyproline. This and other **vitamin cofactors** are listed in Table 1.4.

An unusual amino acid behaves like the classical amino acids. The amino acid **selenocysteine** is incorporated into the polypeptide chain during amino acid polymerization. The story of selenocysteine is revealed in the Selenium section in Chapter 10.

Modified amino acids that are not part of polypeptides may be formed by modification of one of the classical amino acids. Among the many examples of this type of modified amino acid are creatine (a modified form of glycine), ornithine

**TABLE 1.4**  Some Modified Amino Acids

| Parent amino acid | Modified form | Vitamin cofactor |
|---|---|---|
| Serine | Phosphoserine | None |
| Tyrosine | Phosphotyrosine | None |
|  | Sulfotyrosine | None |
| Glutamic acid | γ-Carboxyglutamic acid (GLA) | Vitamin K |
| Lysine | Aminoadipic semialdehyde | Ascorbic acid |
|  | Biotinyllysine | None |
|  | Trimethyllysine | None |
| Proline | Hydroxyproline | Ascorbic acid |
| Glycine | Amidated amino acid | Ascorbic acid |

**TABLE 1.5** Classification of Amino Acids According to Dietary Need

| Indispensable | Dispensable |
|---|---|
| Leucine | Aspartic acid |
| Isoleucine | Asparagine |
| Valine | Glutamic acid |
| Phenylalanine | Glutamine |
| Tryptophan | Glycine |
| Histidine | Alanine |
| Threonine | Serine |
| Methionine | Cysteine |
| Lysine | Proline |
| | Arginine |
| | Tyrosine |

(modified arginine), and homocysteine (modified methionine). Creatine, ornithine, and homocysteine have well-established functions in the body. Kynurenine
and formiminoglutamic acid are modified forms of tryptophan and histidine,
respectively, and are broken down in the body to simpler molecules and result in
waste products.

## Indispensable and Dispensable Amino Acids

In Table 1.5 the classical amino acids are segregated according to their necessity in
the diet. Those that are required to maintain life are called indispensable (*essential*)
amino acids. Those that may be present in the diet but can be omitted without
threatening life are called dispensable (*nonessential*) amino acids. The proteins of
the most value, from a nutritional point of view, are the ones that contain all the
indispensable amino acids as well as a variety of dispensable amino acids.

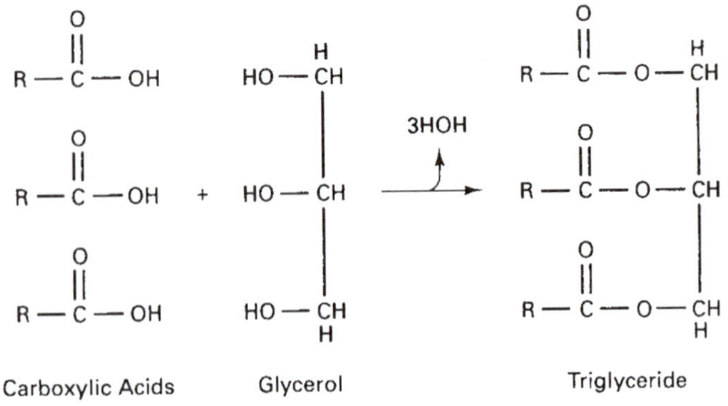

**FIGURE 1.8** Formation of a triglyceride. A triglyceride consists of a backbone of glycerol
that is linked, via ester bonds, to three carboxylic acids.

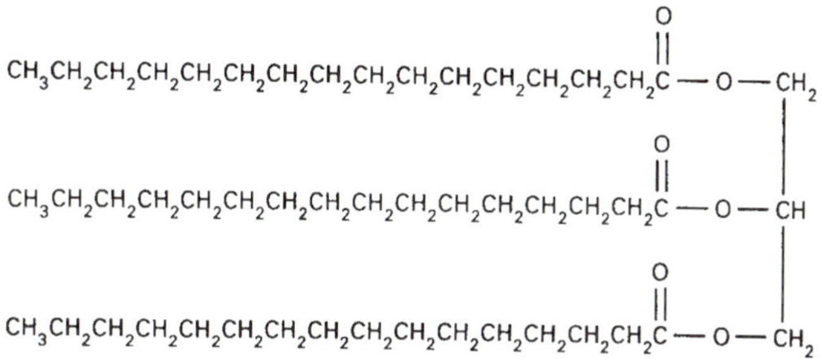

$$CH_3CH_2CH_2CH_2CH_2CH_2CH_2CH_2CH_2CH_2CH_2CH_2CH_2CH_2CH_2\overset{\displaystyle O}{\overset{\|}{C}} - O - CH_2$$

$$CH_3CH_2CH_2CH_2CH_2CH_2CH_2CH_2CH_2CH_2CH_2CH_2CH_2CH_2CH_2\overset{\displaystyle O}{\overset{\|}{C}} - O - CH$$

$$CH_3CH_2CH_2CH_2CH_2CH_2CH_2CH_2CH_2CH_2CH_2CH_2CH_2CH_2CH_2\overset{\displaystyle O}{\overset{\|}{C}} - O - CH_2$$

**FIGURE 1.9** Tripalmitate. Tripalmitate is a triglyceride consisting of a glycerol backbone that is linked, via ester bonds, to three molecules of palmitic acid, a 16-carbon carboxylic acid.

## Glycogenic and Ketogenic Amino Acids

Amino acids can be classified as **glycogenic** or **ketogenic**. This classification refers to the products of **catabolism** (breakdown) of the amino acid in the body. Glycogenic amino acids can be converted to glucose, whereas ketogenic amino acids form ketone bodies. This classification is discussed in the Protein chapter.

# Lipids

## Fats and Oils

The structure of glycerol (1,2,3-trihydroxypropane) is given in Figure 1.8. When this molecule forms a triester with three carboxylic acids (molecules ending in a carboxyl group), the product is called a triglyceride. Figure 1.9 shows the structure of tripalmitate, the triglyceride of palmitic acid. Carboxylic acids in which the R group is of the saturated or unsaturated long-chain aliphatic type shown in Figure 1.10 are called fatty acids. Triglycerides of fatty acids are called **fats** if they are solid at room temperature or **oils** if they are liquid.

## Phospholipids

The most common phospholipid, **phosphatidylcholine**, contains two molecules of fatty acid and one molecule of choline phosphate attached to a glycerol backbone (see Figure 1.11). Like all other phospholipids, it is amphipathic. The water-soluble end features a phosphate group, an amino group, and two keto groups.

## Micelles, Bilayer Sheets, and Vesicles

Phospholipids can form organized structures, as shown in Figure 1.12, when suspended in water solutions. The small circles represent the ionic water-soluble ends of the phospholipid molecules, containing phosphate and amino groups. The

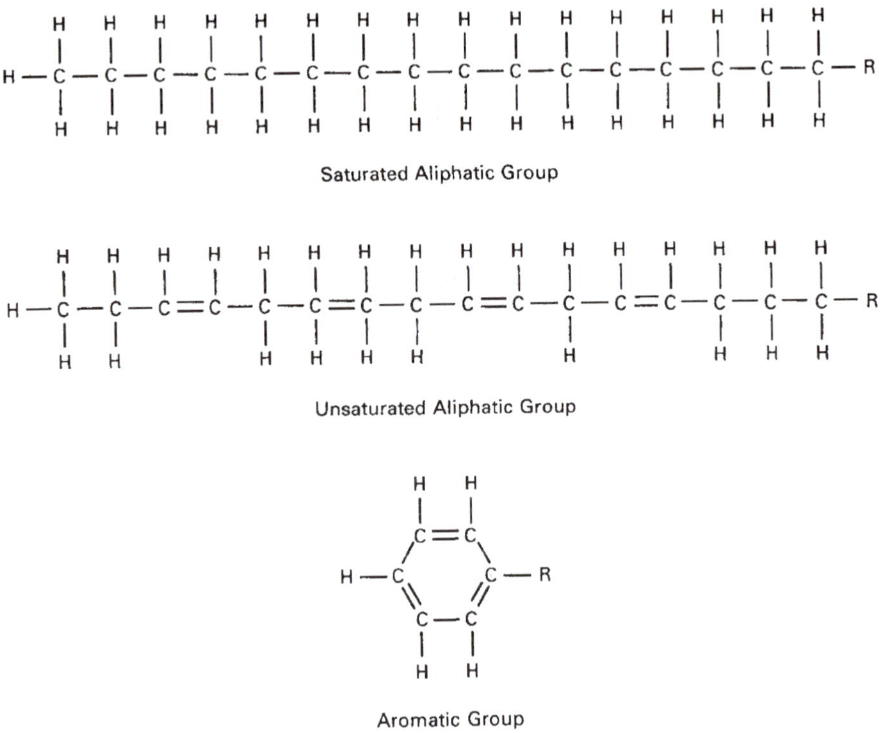

Saturated Aliphatic Group

Unsaturated Aliphatic Group

Aromatic Group

**FIGURE 1.10** Aliphatic and aromatic groups. A 16-carbon aliphatic group (top), 16-carbon aliphatic group bearing several desaturations (center), and an aromatic group (bottom) are shown.

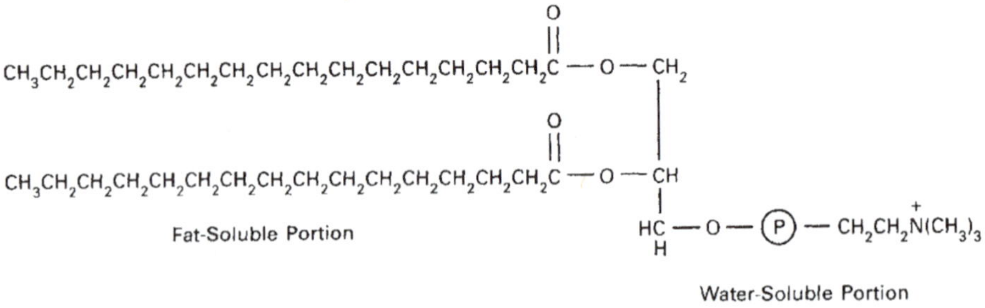

$CH_3CH_2CH_2CH_2CH_2CH_2CH_2CH_2CH_2CH_2CH_2CH_2CH_2CH_2CH_2C$ —— O —— $CH_2$

$CH_3CH_2CH_2CH_2CH_2CH_2CH_2CH_2CH_2CH_2CH_2CH_2CH_2CH_2CH_2C$ —— O —— CH

Fat-Soluble Portion

HC —— O —— Ⓟ —— $CH_2CH_2\overset{+}{N}(CH_3)_3$

Water-Soluble Portion

**FIGURE 1.11** Phosphatidylcholine: an amphipathic molecule. Phosphatidylcholine is a diglyceride. It contains a glycerol backbone, two molecules of fatty acid, and a molecule of choline phosphate. The choline phosphate group is not linked to the central carbon of glycerol.

lines represent the lipophilic "tails." In a **micelle**, the tails face inward and the water-soluble ends face outward, making the surface hydrophilic and the interior lipophilic.

The **bilayer sheet**, also shown in Figure 1.12, is like a sandwich. The "bread" is composed of water-soluble groups and the "filling" of alkane tails. The alkane groups associate with one another but have little contact with the water above and below.

The third structure in Figure 1.12 is a **vesicle**. Vesicles are larger than micelles and contain water in the interior. The membrane, or lipid-containing portion, is a bilayer sheet that is curved to form a spheroid.

## SOLUBILITY

Solubility refers to an interaction between a solute (which may be a solid, liquid, or gas) and a solvent (which is a liquid). A material that is soluble can be dispersed

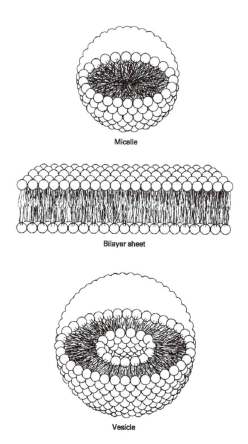

Micelle

Bilayer sheet

Vesicle

**FIGURE 1.12** Phospholipid structures. Most biological membranes take the form of a bilayer sheet. In addition to containing phosopholipids, biological membranes contain proteins and cholesterol. Vesicles are used in biology for transporting and delivering biochemicals from the interior of the cell to the membrane, or to the extracellular fluids. (Reprinted by permission from Darnell *et al.*, 1990.)

on a molecular level within the solvent. A liquid, such as water, that can act as both an electron donor and an electron acceptor is called a **polar solvent**. A liquid that does not engage in such interactions, like gasoline or vegetable oil, is called a **nonpolar solvent**. Compounds that are polar or have a charged or ionic character are soluble in polar solvents but not in nonpolar solvents. Nonpolar compounds are soluble in nonpolar solvents but not in polar solvents.

The relative solubility of materials is relevant to most of the subjects covered in this book. The biochemical "machinery" used for digestion, absorption, and transport of nutrients throughout the body depends on whether the nutrients are water soluble or fat soluble. Relative solubilities are also crucial to the function, composition, and architecture of cells and their surrounding membranes.

A nutrient is classified as water soluble if it can be dissolved in water. However, this property is relative, not absolute, since all materials dissolve to some degree in all solvents. Therefore, one might arbitrarily choose a concentration limit of 1 millimolar (1.0 m$M$) to define water solubility, that is, any compound whose saturated solution in water contains more than 1.0 mmol per liter (1.0 mol per 1000 liters) is considered water soluble. Cholesterol, for example, is definitely not water soluble. A saturated solution is only 0.001 m$M$. A nutrient is classified as fat soluble if it can be dissolved in fats or oils. Again, limits of solubility must be established arbitrarily.

## Amphipathic Molecules

Parts of large molecules can exhibit the properties of the atoms of those parts of the molecule. For example, the end of a long molecule containing charges or ionic bonds can exhibit **hydrophilic** properties, that is, behave as though the molecule were water soluble. If the rest of the molecule is an alkane chain with no charged groups, this other end of the molecule can exhibit **lipophilic** properties, that is, behave as though the molecule were fat soluble. Large molecules are said to be amphipathic if they have one hydrophilic end and one lipophilic end.

## Water-Soluble and Fat-Soluble Nutrients

Water-soluble nutrients usually contain one or more of the following polar or ionizable groups: carboxyl, amino, keto, hydroxyl, or phosphate (see Table 1.2). Molecules that contain several hydroxyl groups, such as sugars, may be very soluble. Amino acids are water soluble, although some are more soluble than others. Glutamic acid, with two carboxyl groups, is very soluble. Compounds containing ionic groups, either positive or negative, interact well with water and tend to be water soluble.

A molecule that does not contain polar or ionizable groups is not likely to be soluble in water. The structures of biological molecules that are not water soluble generally contain only aromatic or aliphatic (alkane-like) components. These groups do not interact well with water and are said to be lipophilic. If an alkane (e.g., octane) is added to a container of water, it will not associate with the water but will form a separate layer on top of the water. (If the molecule were more dense than water, it would sink to the bottom to form a separate layer.)

If a piece of fat is added to the container, it will float on the water layer and absorb some of the alkane. Some of the alkane will "dissolve" into the fat. If an aromatic liquid is added (e.g., benzene), it also will associate with the piece of fat. These materials are fat soluble or lipophilic. Fat-soluble nutrients associate with fat, not because they are forced away from water by some repulsive force but because their molecules are attracted by those in the fat. Water does not "repel" fat. Figure 1.13 shows the structures of two molecules that are not water soluble. The structure at the left is octane, an alkane. The structure in the center is benzene, an aromatic molecule, which usually is simplified as shown on the right.

An aromatic compound can be water soluble if several polar or ionic (water-soluble) groups are bonded to its aromatic ring. The structure of vitamin $B_6$ is based on an aromatic ring, but the ring contains aldehyde, phosphate, hydroxyl, and amine groups. Consequently, vitamin $B_6$ is water soluble.

Some lipophilic nutrients are amphipathic because their molecules contain water-soluble groups at one end. A fatty acid has a long alkane "tail" with a carboxyl group at one end. A bile salt has a large aromatic structure with a carboxyl group at one end.

## Effective Water Solubility of Fat-Soluble Molecules

Bile salts act as "detergents" in nature to maintain insoluble (fat-soluble) compounds in water solution. The bile salts form **mixed micelles** that consist of amphipathic bile salt molecules surrounding the lipophilic (fat-soluble) molecules. The hydrophilic ends of the amphipathic molecules face outward, forming a lipophilic environment in the interior of the micelle. Bile salt molecules contain acid groups, such as carboxyl and sulfonyl groups, that usually are ionized under physiological conditions.

In water solutions, bile salt molecules associate to form micelles only when present in sufficient concentration. The critical concentrations for micelle formation are in the low millimolar range. The structures of these micelles are not particularly stable. They continually change in size and shape by fusing and blending with nearby micelles.

Figure 1.14 presents a cross-section of a bile salt micelle and the way in which the micelle solubilizes a molecule of fatty acid. The bile salt shown is taurocholate and the fatty acid is palmitic acid. Within the micelle, the fatty acid remains in solution and can diffuse over some distance. The micelle that results from addition of the fatty acid molecule is called a mixed micelle because it contains more than one type of molecule. In the absence of the bile salt, the fatty acid might adhere to

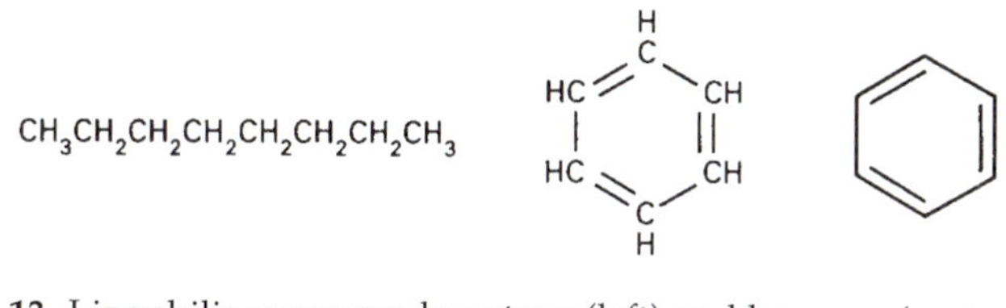

FIGURE 1.13  Lipophilic compounds: octane (left) and benzene (center and right).

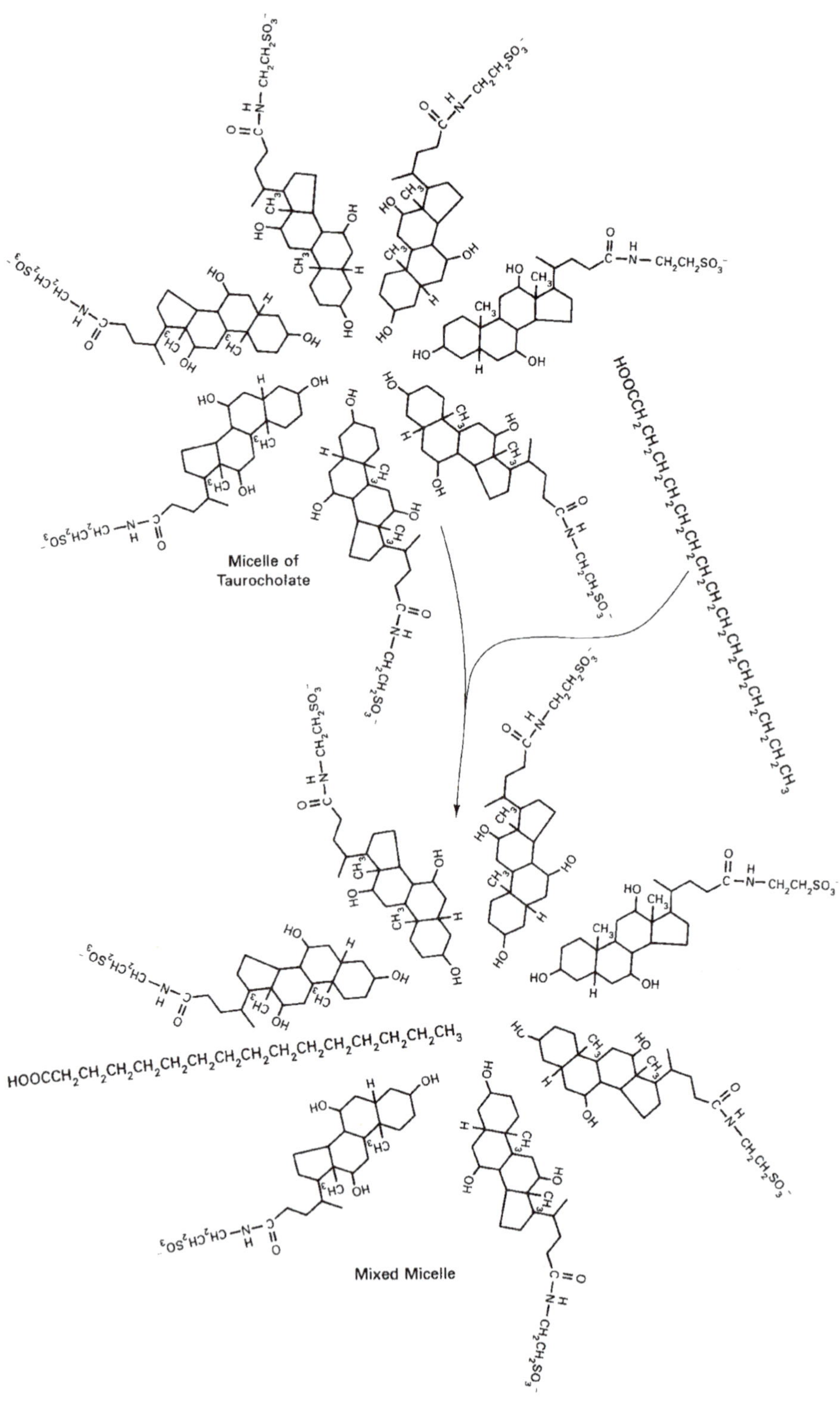

**FIGURE 1.14** Formation of a mixed micelle. Cross-sections of the micelle and mixed micelle are shown. The micelle is shown acquiring a molecule of a 16-carbon carboxylic acid.

a particle of food fat and not efficiently reach an intestinal cell. Detergent micelles are used by the body to solubilize dietary fats, oils, fatty acids, cholesterol esters, and fat-soluble vitamins.

## Ionization and Water Solubility

Just as bile salt micelles can influence the environment in which fat-soluble compounds are found, ionization can change the environment preferred by a water-soluble material. Many water-soluble groups can be ionized to carry positive or negative charges. A nonionized carboxyl group, R—COOH, is fairly soluble in water because it is polar. However, an ionized carboxyl group, R—COO⁻, interacts much more avidly with water. Amino groups, R—NH$_2$, do not ionize easily by giving up a proton to form R—NH⁻ but readily add a proton to form R—NH$_3^+$. A nonionized amino group interacts with water via the lone pair of electrons on the nitrogen atom, but a protonated group has an even stronger affinity for water.

The stronger or weaker affinity of a molecule for water has two consequences. First, the molecule with stronger affinity can be dissolved to a higher concentration. Second, the molecule with stronger affinity is more likely to remain in the water phase in an environment that contains both aqueous and lipid phases. For example, a lipid phase might be introduced to a water solution by submerging a piece of fat or floating a layer of oil on the surface. A molecule with little or no affinity for water is more likely to transfer to the lipid phase.

Before considering the effect of charged groups on large structures in a cell, consider the effect of ionization on the behavior of a small molecule. Figure 1.15 shows acetic acid in its protonated (acetic acid) and unprotonated (acetate ion) forms. In water solution, these two forms are in equilibrium. At neutral pH (7.0), the rate of dissociation of the proton is greater than the rate of reassociation. Hence, the concentration of acetate ion is greater than that of protonated acetic acid. The larger arrowhead is used to indicate this condition.

Figure 1.16 depicts the equilibrium between ionized and nonionized acetic acid in an aqueous environment. The figure also shows the tendency of the nonionized form to leave the aqueous solution. The ionized form interacts strongly with water and has little tendency to leave the aqueous phase for the atmosphere. No equilibrium is established, because the atmosphere has an infinite volume compared to that of the beaker of water solution. In a closed system in which the beaker is sealed at the top, a true equilibrium will form between the aqueous phase and the gaseous phase containing vaporized acetic acid.

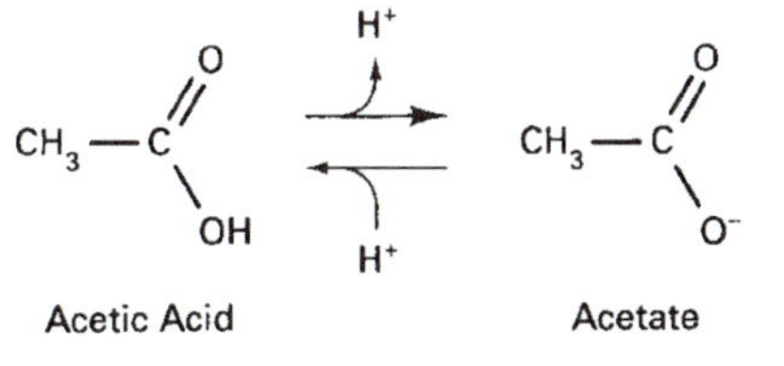

**FIGURE 1.15** Deprotonation of acetic acid and protonation of acetate.

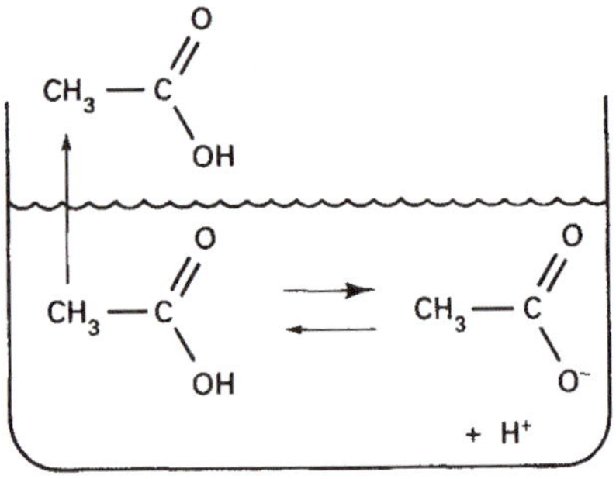

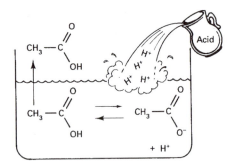

**FIGURE 1.16** Dilute solution of acetic acid in water. Acetic acid and acetate are in rapid equilibrium with each other. In this equilibrium, each molecule of acetic acid (or acetate) becomes deprotonated (or protonated) millions of times per second. To be exact, the rates of deprotonation and protonation are each about one trillion times per second (Eberson, 1969). Acetate tends not to leave the aqueous phase, since the ionized acid group interacts strongly with water. Acetic acid has a greater tendency to leave the water phase, since it does not bear this charge. The rate of loss of acetic acid into the atmosphere is negligible compared to the rates of protonation and deprotonation. Thus, it is reasonable to describe the events occurring in the aqueous solution an equilibrium situation.

The equilibrium depicted in Figure 1.16 can be shifted by adding a strong acid, such as HCl, to the beaker as shown in Figure 1.17. This increase in the concentration of H+ ($H_3O^+$) ions shifts the equilibrium to increase the concentration of acetic acid and decrease the concentration of acetate ion. The increase in the concentration of molecules that leave the aqueous phase increases the rate of volatilization from the beaker to the air. Addition of strong acid does not increase the tendency of any particular molecule of acetic acid to enter the atmosphere. The strong acid merely increases the overall concentration of **protonated** acetic acid (at the expense of the **ionized** acetic acid). Protonated acetic acid has a much greater tendency than ionized acetic acid to leave the water phase.

**FIGURE 1.17** Disturbance of an equilibrium situation by addition of a strong acid. The addition of hydrochloric acid forces the protonation of the acetate groups, resulting in an increase in acetic acid. Acetic acid binds to water to a lesser extent than acetate, and the end-result of adding HCl is an increased rate of loss of the total acetic acid (acetic acid + acetate) into the atmosphere.

$$CH_3 \!-\! NH \qquad\qquad CH_3 \overset{H^+}{\underset{|}{-\!-\!NH}}$$

$$\underset{CH_3}{|} \qquad\qquad\qquad \underset{CH_3}{|}$$

Nonprotonated         Protonated

**FIGURE 1.18** Dimethylamine. Dimethylamine is a foul-smelling chemical made by bacteria residing on fish. Adding acid to the cooked fish prior to eating causes the fishy smell to remain associated with the fluids on the food and to vaporize at a lesser rate.

This simple example is relevant to many topics discussed in subsequent chapters, for example, the behavior of bicarbonate and carbon dioxide. Also, recognition of the difference between reversible (equilibrium) and irreversible (nonequilibrium) reactions is important to an understanding of the role of enzymes in catalyzing various reactions.

Food science provides another simple but practical example of the effect of protonation. Dimethylamine, produced during the spoiling of saltwater fish, can exist in the protonated and nonprotonated forms shown in Figure 1.18. The uncharged form has only a moderate affinity for water, tends to volatilize, and contributes to the unpleasant odor of spoiled fish. The protonated (charged) form has a higher affinity for the fluids of the fish and does not contribute to the odor. Applying acid increases the proportion of molecules present in the protonated form, which is one reason people flavor fish with lemon juice or malt vinegar.

## CELL STRUCTURE

The cell is the unit of life. The cells of all organisms are distinguished by the following properties: the ability to compartmentalize themselves from the environment, the ability to utilize nutrients from the environment for the production of energy, the ability to maintain a relatively constant internal environment, the ability to store genetic information in a form of DNA, and the ability to reproduce.

The cells of mammals and other animals contain the following structures. A plasma membrane (PM), which is the outer border of the cell, has a structure similar to the bilayer sheet shown in Figure 1.12. The PM contains phospholipids and many membrane-bound (embedded) proteins used to facilitate the transport of nutrients and minerals into and out of the cell. The outside of the PM of some cells is coated with polysaccharides for protection. The outside of the PM of other cells bears proteins that control which cells are chosen neighbors. Generally, the material bound to the outside of the PM is synthesized by the cell itself rather than derived from other cells. The cytoplasm is the fluid contained and bounded by the plasma membrane. This fluid has a gel-like consistency because it contains a high concentration of proteins. Most of the biochemical reactions that occur within the cell take place in the cytoplasm. The remainder take place within various organelles.

Organelles of the cell include the nucleus, the endoplasmic reticulum, the sarcoplasmic reticulum, secretory vesicles, lysosomes, and mitochondria. The **mitochondria** are used to generate energy. More specifically, this organelle catalyzes the breakdown of organic nutrients and, in a manner that is coupled with oxygen utilization, the synthesis of **adenosine triphosphate** (ATP). ATP is a small molecule that is used as a source of energy by thousands of different reactions in the cell. Each molecule of ATP may be thought of as a portable battery that can be used anywhere in the cell to drive a specific chemical reaction. Utilization of the ATP results in its conversion to **adenosine diphosphate** (ADP) plus inorganic phosphate ($P_i$). The "energy battery" is then recharged as follows. The ADP travels into a mitochondrion, where it is converted to ATP again. The main purpose of mitochondria in the cell is to recharge molecules of ATP.

All proteins of the cell are synthesized on structures called **ribosomes**. The proteins of the cytoplasm are synthesized on ribosomes that float free in the cytoplasm, whereas those of the plasma membrane are synthesized on ribosomes that bind to the outside of the **endoplasmic reticulum** (ER), a network of interconnected tubules in the cytoplasm resembling a nest of hollow noodles. During polymerization of the amino acids, a nascent protein is driven into the ER. From there it is shunted into secretory vesicles, some of which insert proteins into the PM, while others deliver different proteins to the outside of the cell (into the extracellular fluid).

The **sarcoplasmic reticulum** is an organelle of muscle cells that resembles the ER in structure. It is used for rapid release and uptake of calcium ions. The resultant changes in cytosolic $Ca^{2+}$ concentrations control muscle contraction.

**Lysosomes** are large vesicles that reside in the cytoplasm. They receive material from the outside of the cell, digest (hydrolyze) it into small molecules, and release end-products into the cytoplasm. Material is transferred from the extracellular fluid to the lysosomes via endocytic vesicles.

The **nucleus** contains the genetic material of the cell. During the course of the day, and on a minute-by-minute basis, the genetic material is used for synthesis of new molecules of messenger RNA (mRNA), a process called **transcription**. Each molecule of mRNA resides momentarily in the nucleus, where it undergoes biochemical grooming. This grooming process involves the removal of stretches of excess mRNA, called **introns**, in a process called **splicing** (Mount, 1996). The grooming process also involves the attachment of a small compound called a **cap** at one end of the mRNA, and the polymerization of a moderately sized compound, called a **poly A tail**, at the other end of the mRNA. Eventually, the mRNA enters the cytoplasm, where it is used for coding the synthesis of polypeptides.

## GENETIC MATERIAL

The genetic material controls the properties and functions of the cell and determines, for example, whether the cell is a brain cell, a liver cell, a blood cell, or even a cancer cell. The genetic material controls the location of the eyes in the body, how fast we grow and how much we weigh, and, in some cases, personality traits.

The following paragraphs describe the activities in the cell, which can be summed up by the phrase *DNA makes RNA makes protein*. Although DNA is only one of many molecules used to make RNA, this terse phrase is accurate, since DNA is the central molecule in the process, and since DNA contains the information needed to formulate the sequence of ribonucleotides in the RNA. The study of all the steps in the process where "DNA makes RNA makes protein," and the study of the process of DNA replication, is called **molecular biology**.

The DNA that makes up our genetic material is divided into 46 pieces within the cell. These pieces are called **chromosomes**. People have 22 **numbered chromosomes**, each of which occurs in duplicate, and 2 **sex chromosomes**. The sex chromosomes of females consist of two X chromosomes, while the sex chromosomes of males consist of one X chromosome and one Y chromosome. Where a cell contains duplicate copies of most, or all, of its chromosomes, the cell is called a **diploid cell**. Human genetic material contains an estimated 80,000 genes (Collins *et al.*, 1997). For comparison, the genetic material of the yeast contains 6400 genes. Not counting duplication of all DNA in the cell, which exists because of the occurrence of diploid chromosomes, each human cell contains about 3 billion nucleotides in its DNA (Schuler *et al.*, 1996).

Our sex chromosomes determine whether we are male or female. It is interesting to point out that, in human females, one of the two X chromosomes is inactivated, and is not much used to code for mRNA (Migeon, 1994). In human males, the Y chromosome is largely inactive, and is not much used to code for mRNA, while the X chromosome is active (Rice, 1996). An end-result of these inactivation events is that, to some extent, all the cells of human males and females behave as though they contained only one X chromosome (and not two sex chromosomes).

The inactivation of the genes of the sex chromosomes, and of some of the genes of numbered chromosomes, results, in part, from the attachment of methyl groups ($—CH_3$) to the DNA. Methyl group attachment is thought to be the only type of intentional alteration to the chemistry of human DNA. The methyl groups attached to our DNA reside on residues of cytosine.

The **genetic code** refers to specific sequences of RNA bases (or DNA bases) that encode specific amino acids. Each of these sequences is composed of a triplet of bases (three bases in a row). Hence, any mRNA molecule can be considered a continuous polymer of successive triplets. For example, the triplet UUU codes for phenylalanine, CAU encodes histidine, GAG encodes glutamate, AAA encodes lysine, and AUG codes for methionine. DNA is used for information storage, while mRNA is used for information transfer.

## Directionality of Nucleic Acids

During transcription, the mRNA molecule is created, starting from the 5′-end, and finishing at the 3′-end. The numbers 5-prime and 3-prime refer to the ends of the nucleic acid polymer, where the indicated carbon atom of the ribose is free, and not involved in a phosphodiester bond (see Figure 1.2). During translation, the 5′-end of the coding region is first read. The 3′-end of the coding region of the mRNA is read last. All enzymes that are involved in nucleic acid synthesis are able to sense the 5′-direction and 3′-direction. For example, when RNA polymerase catalyzes the synthesis of RNA, the enzyme moves only in one direction along the surface of the DNA molecule. RNA polymerase never reads codons in the backwards direction when it polymerizes nucleotides to create mRNA. RNA polymerase is a protein. Since RNA polymerase catalyzes a reaction, it is also an enzyme.

## Transcription

Messenger RNA (mRNA) is the primary type of ribonucleic acid in the cell. Other types include **transfer RNA** (tRNA) and **ribosomal RNA** (rRNA). Messenger RNA occurs as a long, linear strand and contains information needed for the synthesis of a particular protein. tRNA also consists of a long, linear strand of RNA, but it spontaneously folds and twists to form an oblong structure. A ribosome consists of a large complex of proteins and rRNA molecules.

A simplified diagram of transcription is shown here. Double-stranded DNA (dsDNA) is shown as a long rectangle. RNA polymerase binds to a region of DNA called a **core promoter**. The core promoter consists of all the DNA between the TATA sequence and the transcription start site. The core promoter is a short stretch of DNA that serves to bind and orient RNA polymerase, and the basal transcription factors. The TATA sequence, which is composed of only four nucleotides (T, A, T, and A), usually occurs about 25 base pairs upstream of the transcription start site. The TATA-binding protein is a special protein that binds to the TATA sequence prior to initiating transcription.

The promoter has a sequence that is variable and indistinct. Not all promoters contain the TATA sequence. Some promoters contain the GC sequence, or other simple sequences. To highlight the important function of these simple sequences, they are usually called "boxes," i.e., the TATA box or GC box. The GC box, which is only two nucleotides long, usually occurs 40–70 base pairs upstream of the transcription start site. The GC box is recognized by a special transcription factor called SP1.

Some promoters contain a TATA box, while others contain both a TATA box and GC box, or the GC box only, or other types of boxes.

Even though the transcription start site represents part of a biochemical process that is central to all biology, the structure of this site is variable and indistinct. In general, the transcription start site consists of adenine, followed by a few pyrimidines (A–Py–Py–Py–Py). The bases of the transcription start site occur in the DNA molecule, as well as in the first few ribonucleotides of the RNA molecule. The reader must not confuse the transcription start site with the translation start site. The translation start site, which occurs in messenger RNA, occurs downstream of the transcription start site, i.e., from 20 base pairs to several hundred base pairs downstream of the transcription start site (Roeder, 1996; Zawel and Reinberg, 1995; Kornberg, 1996).

The **promoter** consists of the *core promoter plus various types of regulatory elements*. Among these regulatory elements are the **hormone response elements**, which are detailed in the Vitamin chapter. The following diagram illustrates the relative position of special regions that occur in most genes. These regions are: (1) the core promoter, (2) the response element, (3) the RNA polymerase binding site, and (4) the transcribed region:

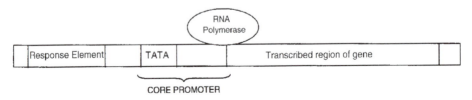

The second part of the diagram shows RNA polymerase at a point where it has catalyzed the synthesis of about half of the mRNA molecule. The term **gene** can be defined as follows. For any polypeptide existing in nature, the corresponding gene consists of the transcribed region plus special regions of DNA that are used for regulating the rate of initiating transcription:

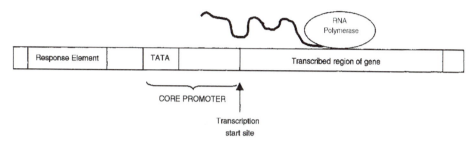

The preceding discussion applies to the RNA polymerase of eukarya that is called **RNA polymerase II**. RNA polymerase II catalyzes the synthesis of the vast number of mRNAs that occur in the cell. The human body is expected to contain over 50,000 different types of mRNA. Variations in the preceding theme apply to the actions of RNA polymerase I and RNA polymerase III. **RNA polymerase I** is used to catalyze the synthesis of ribosomal RNA (rRNA). This type of RNA occurs, complexed with ribosomal proteins, to form ribosomes. **RNA polymerase III** catalyzes the synthesis of transfer RNA (tRNA) and 5S RNA. 5S RNA occurs as part of the ribosome.

## Illustration of the Use of Response Elements Using the Example of Hexokinase

Hexokinase is an enzyme that is mentioned at an early point in most biochemistry courses, since this enzyme is required for the first step in the body's utilization of glucose. The promoter for the hexokinase gene is shown here. This promoter consists of about 160 base pairs, and contains a TATA box (AATAA; underlined), a CCAAT box (also underlined), and a short sequence of DNA that is regulated by a hormone. The hormone-regulated sequence is CCACGTCA (underlined). This short sequence of DNA is one of many types of "response elements" that occur in the genome. Specifically, CCACGTCA is called the **cyclic AMP response element**, for reasons that are explained later. This response element occurs in the hexokinase promoter, just a few nucleotides beyond the CCAAT box (Osawa *et al.*, 1996):

```
5'-GGGCTCTGGGCGCTGATTGGCTGTGGACTGCGGGCGGGCAGCCGGAGAGCGC-
   ACACACCCTCTTCCCGCAGCCAATGAGCGCGCCCACGTCACTGTCTTGGGCGGCCC-
     AAAGAGCCGGCAGCCCCTCAATAAGCCACATTGTTGCACCAACTCCAGTGCTAGAGT-3'
```

Please note that only one strand of the DNA is shown, not both strands. Note also that the TATA box has an atypical sequence: it occurs as AATAA rather than as TATA. The transcription start site occurs immediately after the promoter sequence, and the first few bases that are transcribed are shown in boldface type. CCACGTCA binds a special regulatory protein, as mentioned in what follows, while the CCAAT box binds a special regulatory protein called NF-Y (Printz *et al.*, 1997).

## Hormone Response Elements in the Genome

How does CCACGTCA work? Briefly, during prolonged exercise, the concentration of the hormone glucagon increases in the bloodstream. At elevated levels, this hormone binds to muscle cells, thus activating the enzyme adenylyl cyclase. Adenylyl cyclase catalyzes the synthesis of a small molecule called cyclic AMP. Cyclic AMP binds to an enzyme called protein kinase A, and activates it. Activated protein kinase A catalyzes the attachment of a phosphate group to a special protein that can bind to CCACGTCA, resulting in an increase in this event of binding. This increased binding to CCACGTCA provokes an increase in the rate of transcription of the nearby gene (the hexokinase gene). This scenario results in an increase in the amount of hexokinase in the muscle cell, with the consequent increase in energy production from glucose. The goal of this sketch is to pinpoint the role of *promoters*, by way of example, in the physiology of the organism. Next time you begin training for a marathon race, you might take a moment to realize that your hexokinase promoter is being stimulated. We shall return to matters such as cyclic AMP and protein kinase A at later points in the text.

## Transcription Termination

What makes RNA polymerase terminate its transcription of a particular gene? Clearly, RNA polymerase does not remain bound to the genome after catalyzing the synthesis of a molecule of RNA, nor does it catalyze the synthesis of an infinitely long piece of RNA. There might be expected to exist certain signals in the genome that provoke RNA polymerase to conclude its task, though these signals have eluded detection. Evidence suggests that, in a few genes, certain unusual sequences in the DNA, such as GTGTGTGTGTGTGTGTGTGTGTGT, can provoke RNA polymerase to conclude transcription at a specific point (Hong *et al.*, 1997).

## Translation

Translation is the act of polymerization of amino acids into polypeptides using mRNA as a template. In translation, the mRNA binds to a ribosome. The tRNA functions as biochemical "tongs," holding each amino acid in the form of an amino acyl-tRNA. For each classical amino acid, one or more types of tRNA molecule exist that are intended specifically for this use. The mRNA guides the order of selection of amino acyl-tRNAs, and thus guides the order of polymerization of amino acids. The ribosome serves as a biochemical "anvil," aligning the mRNA and different amino acyl-tRNAs as the polypeptide chain is created. The mRNA molecule is not a permanent component of the cell. It is used over and over for synthesis of the same protein but is eventually degraded, perhaps within a few hours of being made.

Because of the nature of the peptide linkage, a polypeptide chain of any length has a free amino group at one end and a free carboxyl group at the other. The first peptide bond is formed between the carboxyl group of the first amino acid and the amino group of the second amino acid, the second peptide bond is formed between the carboxyl group of the second amino acid and the amino group of the third, and so on.

All mRNA molecules contain a central region, called the coding region, which codes for a protein. The coding region is bordered by untranslated regions (UTRs), which are called 5'-UTR and 3'-UTR, as indicated in the following diagram. Nearly all polypeptides begin with methionine, and hence the first triplet occurring in the coding region of nearly all mRNAs is AUG (Kozak, 1996). Polypeptide synthesis continues until the ribosome encounters a stop codon (UAA, UAG, or UGA) (Tate and Brown, 1992). The poly A tail is also shown:

5'-UTR          CODING REGION          3'-UTR          poly A

Students of elementary biology learn that the first amino acid of polypeptide chains contains a free amino group, i.e., $H_2N–R$, and that the first amino acid is methionine. However, in eukaryotic cells, the N-terminal amino acid may be modified, shortly after translation, by the attachment of an acetyl group (Scaloni et al., 1992; Dormann et al., 1993). In some cases, the N-terminal methionine may be cleaved, yielding a protein that is one amino acid shorter than that expected by the coding region of the mRNA (Flinta et al., 1986; Arfin and Bradshaw, 1988).

Just for purposes of orientation, one might note that a typical protein consists of about 300 amino acids. The first amino acid, bearing the free amino group, is called the **amino terminus** (N terminus) of the polypeptide chain. The opposite end, is called the **carboxyl terminus** (C terminus).

## Genetic Code

The genetic code is shown in Table 1.6. A total of 64 different combinations of the four DNA bases can occur, and 61 of these possible combinations are actually used to specify amino acids. Many of the amino acids are designated by more than one type of codon. This redundant situation is called **degeneracy**. The genetic code is thus degenerate. ATG codes for methionine. Methionine occurs at various positions in most proteins, and occurs as the first amino acid in essentially all proteins. For this reason, the codon ATG occurs at the beginning of coding regions of nearly all genes. ATG is called the **start codon**. At the very end of all coding regions, there occurs one stop codon. There exists three different stop codons, and these are TAA, TAG, and TGA. In mRNA, where the start and stop codons actually perform their function, the corresponding codons are AUG (start codon), UAA, UAG, and UGA (stop codons). With rare exceptions, stop codons never code for an amino acid. The sequence of codons that begins with ATG and ends with a stop codon is often called an **open reading frame** (ORF). The genetic code is the same for eukarya and bacteria, but differs somewhat for archae.

**TABLE 1.6**  The Genetic Code

| First base in codon | Second base in codon | | | | | | | |
|---|---|---|---|---|---|---|---|---|
| | T | | C | | A | | G | |
| T | TTT | Phe | TCT | Ser | TAT | Tyr | TGT | Cys |
| | TTC | Phe | TCC | Ser | TAC | Tyr | TGC | Cys |
| | TTA | Leu | TCA | Ser | TAA | Stop | TGA | Stop |
| | TTG | Leu | TCG | Ser | TAG | Stop | TGG | Trp |
| C | CTT | Leu | CCT | Pro | CAT | His | CGT | Arg |
| | CTC | Leu | CCC | Pro | CAC | His | CGC | Arg |
| | CTA | Leu | CCA | Pro | CAA | Gln | CGA | Arg |
| | CTG | Leu | CCG | Pro | CAG | Gln | CGG | Arg |
| A | ATT | Ile | ACT | Thr | AAT | Asn | AGT | Ser |
| | ATC | Ile | ACC | Thr | AAC | Asn | AGC | Ser |
| | ATA | Ile | ACA | Thr | AAA | Lys | AGA | Arg |
| | ATG | Met | ACG | Thr | AAG | Lys | AGG | Arg |
| G | GTT | Val | GCT | Ala | GAT | Asp | GGT | Gly |
| | GTC | Val | GCC | Ala | GAC | Asp | GGC | Gly |
| | GTA | Val | GCA | Ala | GAA | Glu | GGA | Gly |
| | GTG | Val | GCG | Ala | GAG | Glu | GGG | Gly |

The genetic code indicates the amino acids that are coded for by the codons appearing in DNA and mRNA. To acquire a genetic code for the codons in mRNA, change each thymine (T) to uracil (U). In actual practice, scientists usually refer to a table that contains the DNA codons, in analyzing genetic information, and rarely use a table that contains the codons appearing in mRNA.

The leap from information in mRNA to the sequence of amino acids in a polypeptide chain is bridged by transfer RNA. Transfer RNA molecules are relatively small, when compared to mRNA and proteins, and consist of only about 40 ribonucleotides. There exist about 40 distinct types of tRNA, and these share the task of aligning the 20 amino acids according to the sequence of ribonucleotide bases occurring in any molecule of mRNA. Since there exist more types of mRNA molecules (about 40) than amino acids (20), one can see that the collection of tRNA molecules is also redundant or degenerate.

## Events Occurring after Translation

*Translation Occurs on Free Ribosomes and on Ribosomes Bound to the ER*

Most proteins are synthesized on ribosomes that float freely in the cytoplasm. After synthesis of the polypeptide chain, the completed protein dissociates from the ribosome and begins to function. Some of the proteins of the cytosol contain special sequences (or clusters) of amino acids that guide the entire protein into the mitochondrion or nucleus to perform functions specific to that organelle. Many other proteins are synthesized on ribosomes that are bound to the endoplasmic reticulum. Polypeptides that are manufactured on the ER contain, as part of their mRNA molecule, information that promotes its binding to the outside surface of the ER. The polypeptides synthesized on the surface of the ER are generally those destined for secretion from the cell or for insertion into the plasma membrane.

## How Proteins Are Incorporated into the Plasma Membrane or Secreted from the Cell

The synthesis and transportation of polypeptides made at the endoplasmic reticulum (ER) are depicted in Figures 1.19–1.21. Three fates are depicted, and these are:

1. Insertion of the polypeptide into the membrane (Fig. 1.19).

2. Insertion into the membrane, with temporary residence in the lumen of the ER, and packaging in a secretory vesicle (Fig. 1.20).

3. Insertion into the membrane, with temporary residence in the lumen of the ER and packaging as a component of the membrane of a secretory vesicle (Fig. 1.21).

Figure 1.19 depicts only early events in the history of a secreted protein or membrane-bound protein, namely, the polymerization of amino acids and inser-

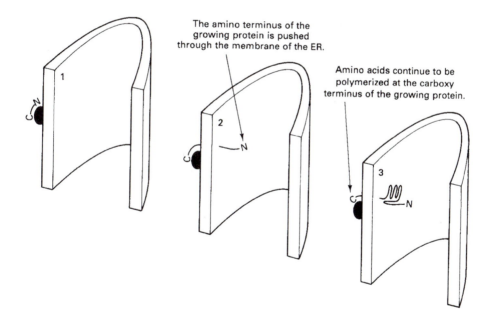

**FIGURE 1.19** Synthesis of a polypeptide on the endoplasmic reticulum (ER). The black oval represents a ribosome; the line represents a newly made polypeptide chain. (1) Ribosome arrives at ER. (2) Amino terminus of the growing protein is pushed through the membrane of the ER. (3) Amino acids continue to be polymerized at the carboxy terminus of the growing protein.

tion of the polypeptide through the membrane of the ER. The polypeptide travels through a special pore, made of a protein complex called the **translocon** (Powers and Walter, 1997).

Figure 1.20 depicts an available next step of the sequence, where the step leads to secretion into the **extracellular fluid**. Polypeptides of this type include albumin, polypeptide hormones, blood clotting proteins, lipoproteins, and antibodies.

Figure 1.21 depicts an alternative pathway. Following the events of Figure 1.19, the polypeptide is inserted into the membrane, packaged into a secretory vesicle, and inserted into the plasma membrane via fusion of the vesicle with the PM. Membrane-bound proteins include nutrient transport proteins, hormone receptors, ion pumps, and proteins that transmit impulses along the length of a nerve or muscle fiber.

As synthesis of the protein begins, the ribosome associates with the ER, as shown in Step 1 of Figure 1.19. With continued polymerization of the amino acids, the nascent (growing) chain is pushed into the interior, or **lumen**, of the ER (Step 2). As the polypeptide chain lengthens, it continues to enter the ER and fold into a three-dimensional structure unique to that particular protein (Step 3). After the entire polypeptide has been synthesized, it dissociates from the membrane and floats, in soluble form, within the lumen of the ER (Step 4A). If the protein is destined to be secreted, it is packaged into a secretory vesicle. Packaging begins with the budding of part of the membrane of the ER (Step 5A). The completed secretory vesicle containing the secretory protein is shown in Step 6A. The secretory protein is in soluble form. The event of secretion involves fusion of the vesicle with the plasma membrane, followed by expulsion of the secretory protein from the cell. This process resembles a reversal of Steps 4A through 6A.

The plasma membrane of the cell is a lipid bilayer sheet in which membrane-bound proteins are embedded. Steps 4B–6B of Figure 1.21 illustrate some events in the production of a membrane-bound protein. After synthesis of the protein, the ribosome on which it was formed dissociates from the membrane but the protein remains bound to the membrane (Step 4B). This binding is mediated by a short stretch of lipophilic amino acids that may occur near the C terminus, as shown in Figure 1.21, or near the N terminus in the case of other proteins. Subsequently, part of the ER membrane forms a bud that breaks off (Step 5B) to form a secretory vesicle (Step 6B). The continued association of the entire membrane-bound protein during the budding process and during subsequent events is maintained by the special lipophilic sequence. Eventually, the secretory vesicle fuses with the plasma membrane in a process that resembles a reversal of Steps 4B–6B. After completion of the insertion of the membrane-bound protein into the plasma membrane, its N terminus is in contact with the extracellular fluid and its C terminus is in contact with the cytoplasm, at least for the protein depicted in Figure 1.21.

## Maturation of Proteins

The specific order of the various amino acids in a polypeptide is determined by the mRNA that encoded its synthesis. A newly made polypeptide may or may not be modified by changing one or more of its constituent amino acids, as discussed.

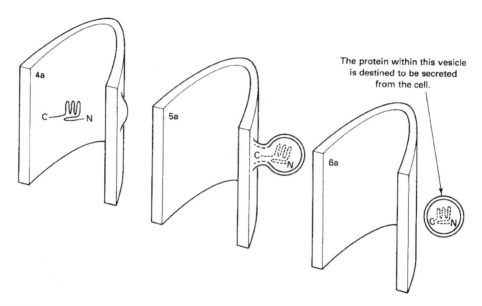

The protein within this vesicle is destined to be secreted from the cell.

**FIGURE 1.20** Packaging of the newly made polypeptide into a secretory vesicle. The protein in the vesicle is destined to be secreted from the cell.

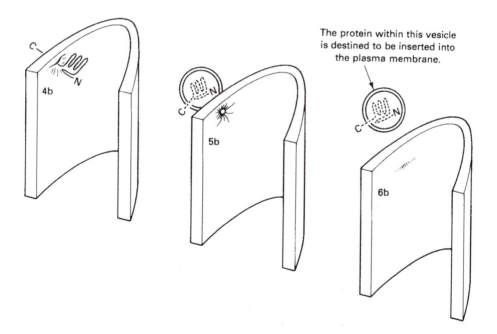

The protein within this vesicle is destined to be inserted into the plasma membrane.

**FIGURE 1.21** Packaging of a newly made membrane-bound protein into a secretory vesicle. The protein in the vesicle is destined to be inserted into the plasma membrane.

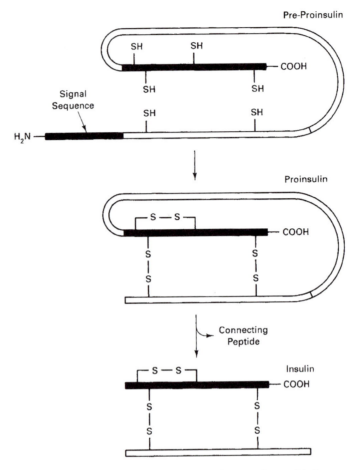

**FIGURE 1.22** Maturation pathway of insulin, a polypeptide hormone.

Another form of modification is cleavage of one or more of the peptide bonds to produce two or more smaller polypeptides. Polypeptides that are destined for secretion from the cell are generally first cleaved at very specific positions in the chain.

The hormone **insulin** is an example of such a polypeptide. Like other polypeptides destined for secretion from the cell, insulin contains a stretch of amino acids called a **signal sequence**. The location of the signal sequence in a newly formed preproinsulin molecule is shown in Figure 1.22. This sequence guides the molecule to the interior of the ER. As the preproinsulin molecule enters the organelle, the signal sequence is cleaved, resulting in conversion of the polypeptide to proinsulin (Figure 1.22). The sole function of the signal sequence is to guide the newly made polypeptide into the interior of the organelle. Once inside the ER, further maturation occurs. The proinsulin molecule is cleaved at two places, removing a center stretch called the **connecting peptide**. The mature insulin molecule and the connecting peptide are packaged into secretory vesicles. Such transfers of secretory proteins (proteins destined for secretion) are one function of the endoplasmic reticulum. The concepts of protein maturation by cleavage and of secretory vesicles are featured in discussions of stomach physiology in Chapter 2 (Digestion and Absorption).

## Enzymes

Enzymes are proteins that catalyze biochemical reactions. A catalyst is a substance that greatly accelerates the rate of a particular reaction without being used up or permanently altered. In the real world, most catalysts eventually deteriorate and no longer function as a catalyst. In the cell, all enzymes are eventually degraded and converted back to their constituent amino acids plus, in some cases, byproducts of oxidation or other types of damage. Proteins do not have some unique magical property that allows them to function as enzymes. For certain activities nucleic acids also participate in the chemistry of catalysis. For example, mRNA can catalyze certain types of RNA splicing.

The specific order of the amino acids in an enzyme governs the reaction it catalyzes. The polypeptide chain folds, largely spontaneously, to form a definite but somewhat flexible three-dimensional structure. The region of the enzyme directly involved in catalysis is called the **active site**. This region contains amino acid R groups that are arranged to bind specific **substrates** or **reactants**. In some enzymes, the active site functions by promoting bending of the substrate in such a way that the rate of a specific reaction is enhanced greatly. In other enzymes, the active site contains amino acid R groups that attack or chemically react with the substrate and thus enhance the rate of a specific reaction. Following the reaction, the products are released from the surface of the enzyme. Each cell of the body contains several thousand different types of enzymes. Enzymes occur in the cytosol at concentrations between 0.01 and 10 μM.

### Enzymes Have a Three-Dimensional Structure

Figure 1.23 shows the overall structures of two enzymes, **chymotrypsin** and **dihydrofolate reductase**. Chymotrypsin catalyzes the digestion of dietary proteins, while dihydrofolate reductase catalyzes a step in vitamin metabolism. In these diagrams, a ribbon is a form of shorthand that represents a linear polymer of amino acids. The enzymes in Figure 1.23 are roughly spherical, not long and filamentous. Such spherical proteins have an inside core and an outer surface. The amino acids in parts of the peptide chain near the inner core tend to have lipophilic R groups. Those with hydrophilic R groups tend to occur on the outer surface.

As a polypeptide is synthesized in a cell, its amino acid sequence might appear random (i.e., a few hydrophilic acids, followed by several lipophilic acids, then a number of hydrophilic ones, and so on). However, as the polypeptide folds and coils into a three-dimensional shape, it becomes apparent that this structure is controlled by the sequence of R groups. The polypeptide chain folds so that the

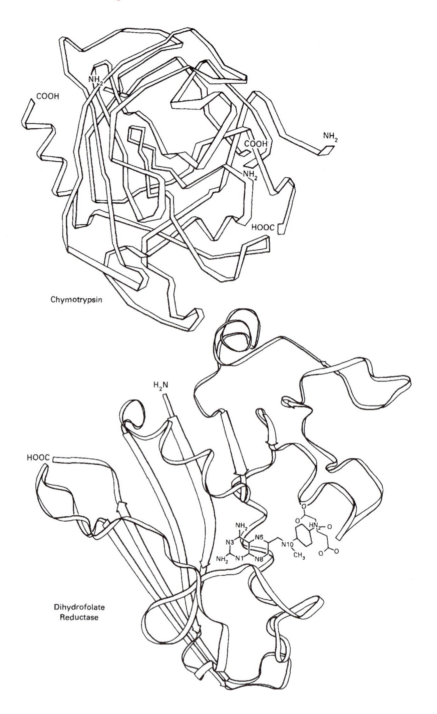

**FIGURE 1.23** Three-dimensional structures of two proteins, chymotrypsin (top) and di-hydrofolate reductase (bottom). Dihydrofolate reductase is shown not in its natural state, but with a drug molecule bound to its surface. [Reprinted by permission from *Nature* **214**, 652–656 (copyright © 1967 Macmillan Magazines Limited) and *Science* **197**, 452 (copyright © 1977 by the AAAS (London)).]

lipophilic amino acid R groups associate with one another in the core and the hydrophilic groups remain on the outer surface, where they associate with the surrounding water. Compare this configuration with those of vesicles and micelles.

## Conventions for Depicting Substrates and Products

Some conventions in representing the catalytic activity of enzymes are as follows. The **substrates** of the enzyme are shown to the left of an arrow, while the **products** are shown to the right. The example of hexokinase is shown here. The substrates are glucose and ATP, while the products are glucose-6-phosphate and ADP. It is common for the student or scientist to write the name of the enzyme above or below the arrow, or to write the name of necessary cofactors above or below the arrow:

Hexokinase

Glucose + ATP ⟶ Glucose-6-P + ADP

The main goal of hexokinase is to convert glucose to glucose-6-phosphate, and not to convert ATP to ADP. For this reason, it is sometimes conventional to write the preceding reaction as

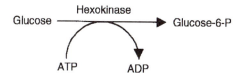

How is ATP handled, when an enzyme utilizes ATP to drive a particular reaction? In general, any given enzyme uses only one molecule of ATP at a time, not several, when it catalyzes one occurrence of a reaction. Enzymes do not gobble up several ATP molecules and then proceed to catalyze a reaction.

What about a situation where ATP contains much more energy than is needed to drive a particular reaction? Whenever ATP is used in a reaction, its energy becomes completely spent. There probably does not exist any situation in biology where part of ATP's energy is used to drive one reaction, with the remaining portion of the ATP's energy utilized to drive a separate reaction. The way biology works is that, in catalyzing an ATP-requiring reaction, an enzyme uses the energy it needs, and the remaining energy of the ATP (if any remains) is discharged as heat. In speaking about the energy stored in a molecule of ATP, it is more accurate to refer to the energy stored in its phosphodiester bonds. Further details of ATP occur in the Energy Requirement chapter and in the Calcium and Phosphate section.

*Reversibility of Enzyme-Catalyzed Reactions*

Most or all enzyme-catalyzed reactions are reversible. The reversibility of any particular reaction is indicated by a bidirectional arrow:

Hexokinase

Glucose + ATP ⟷ Glucose-6-P + ADP

Another way of expressing the concept of bidirectionality is as follows. If a researcher placed, in a test tube, a solution containing hexokinase, glucose-6-P, and ADP, one would expect to find the production of measurable amounts of glucose and ATP. The amount of glucose formed would account for a very small proportion of total glucose (glucose plus glucose-6-P). For all enzyme-catalyzed reactions, the direction of the reaction represents an approach to chemical equilibrium. As equilibrium is approached and then reached, for the reaction of hexokinase, one would find only a tiny fraction of the glucose occurring as glucose, with most of it occurring as glucose-6-P. The proportion of all molecules present, once equilibrium has been reached, is exactly the same where the enzyme is placed in a test tube containing the substrates (glucose + ATP) or placed in a test tube containing the products (glucose-6-P + ADP).

Although it is conventional to represent the reaction of hexokinase, and of many other enzymes, using a unidirectional arrow, this convention is not scientifically accurate. The unidirectional arrow is used only to make it convenient to remember that the equilibrium lies far in the direction of product formation (glucose-6-P formation). In general, most ATP-utilizing reactions are driven in the direction of product formation, and are therefore said not to be freely reversible.

*What Makes Metabolism Go Forward?*

A simplified diagram of "metabolism" is shown in Figure 1.24. Glucose is a major energy source for most animals. The digestion of starches of sugars in the small intestines releases glucose. The glucose is transported through the wall of the gut,

**FIGURE 1.24** Simplified diagram of metabolism. Dietary glucose passes through cell membranes through special pores, called transporters. Once inside the cell, glucose is oxidized to acetate groups. The acetate groups are further oxidized to carbon dioxide, in a process that is coupled with the generation of ATP. The carbon dioxide does not build up in the body; it is exhaled via the lungs to the atmosphere.

through the bloodstream, and into various cells. One factor that makes glucose go into cells is that its concentration may be higher on the outside of the cell than inside. Another factor is that glucose, as well as its immediate breakdown products, are modified (by attachment of a phosphate group) to give a form that cannot pass through the cell membrane and cannot exit the cell. The eventual metabolism of glucose to carbon dioxide is the ultimate force that drives the continued uptake of glucose by the cell.

The breakdown of glucose to acetate groups, and then to $CO_2$, is driven by two factors: (1) the release of heat, which occurs at many steps of the multi-step breakdown pathway, and (2) the generation of many small breakdown products from one molecule of glucose. As glucose contains six carbon atoms, one can see that its complete breakdown may produce six molecules of carbon dioxide. To be brief, metabolism is driven by increases in enthalpy (heat production) and by increases in entropy (degrees of free movement). The types of movement available to six molecules of carbon dioxide is greater than that available to the same six atoms when tethered together in a molecule of glucose. The goal of metabolism is not simply to break down the nutrients of food. The goal is not simply to create heat. The goal is to capture some of the released energy as ATP, in order that the body can do other useful things with the nutrients of food. Please note that the curved arrow indicates ATP synthesis (Figure 1.24). By including a point of ATP synthesis in the preceding scheme (Figure 1.24), some of the energy that ordinarily that would be lost as heat is captured as chemical energy. The ATP, in turn, is used as the immediate energy source that drives muscle contraction, activities of the nervous system, the continual turnover, and replacement of most parts of the body, growth, and reproduction.

One of the properties of life is the constancy of the environment within, that is, the internal climate. In animals, the various metabolites in the bloodstream, and within all cells, are held at a constant level. In the case of some metabolites, the concentrations are strictly regulated, while in other cases fluctuations can easily be tolerated. This quality of life was originally proposed by Claude Bernard (1865).

## Membrane-Bound Proteins

The amino acids of a protein control its location in the cell. Some proteins are water soluble, whereas others are bound to the cell membrane (plasma membrane), the mitochondrial membrane, and the membranes of the endoplasmic reticulum and nucleus. The association of a protein with a membrane is maintained by a stretch of lipophilic amino acids. Insertion of this stretch into the membrane occurs as the protein is synthesized. Water-soluble proteins are formed on ribosomes that "float" free in the cytoplasm. Membrane-bound proteins are formed on ribosomes that associate with the endoplasmic reticulum (ER). As the amino acids are polymerized in the vicinity of the ER, a stretch of lipophilic acids becomes inserted into the membrane of the ER. This anchoring of the protein is maintained when it is shuttled from its location in the ER to its desired location in the plasma membrane.

Figure 1.25 presents part of the amino acid sequence of a membrane-bound protein. The amino acids embedded within the membrane are those with lipophilic R groups. The amino acids residing in the fluids surrounding the membrane are mostly those with ionic R groups.

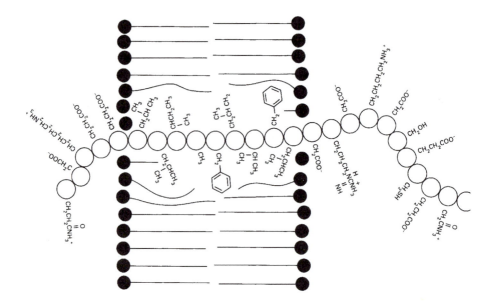

**FIGURE 1.25** Polypeptide chain passing through a biological membrane. The line of open circles represents the backbone of the polypeptide chain. The filled circles represent the hydrophilic groups of the phospholipids in the membrane, and the long lines represent their alkane chains. Amino acids with lipophilic R groups tend to locate within the membrane, whereas those with hydrophilic R groups tend to locate in the aqueous regions on either side of the membrane.

A variety of membrane-bound proteins are of vital interest to the medical and nutritional scientist, because defects or changes in these proteins can cause such problems as lactose intolerance, cardiovascular disease, cystic fibrosis, and diabetes. Sucrase-isomaltase, an enzyme of the small intestine, is a membrane-bound protein, bound to the plasma membrane of the enterocyte (gut cell). Part of the production of this enzyme is depicted in Figure 1.26. In Step 1, the polypeptide chain is polymerized on the ribosome (shown in black). In Step 2, part of the amino acid chain near the N terminus crosses the membrane of the ER into the lumen but some of the amino acids at the N terminus remain outside. Step 3 shows the protein assuming a three-dimensional shape within the lumen; both the C and N

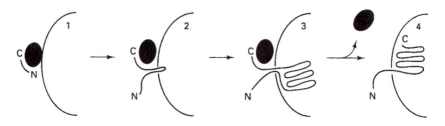

**FIGURE 1.26** Sequence of events during insertion of sucrase-isomaltase into the endoplasmic reticulum.

H₂N–Met–Ala–Lys–Arg–Lys–Phe–Ser–Gly–Leu–Glu–Ile–Thr–<u>Leu–Ile–Val–Leu–Phe–</u>

<u>Val–Ile–Val–Phe–Ile–Ile–Ala–Ile–Ala–Leu–Ile–Ala–Val–Leu–Ala</u>–Thr–Lys–Thr–

**FIGURE 1.27** First 35 amino acids of sucrase-isomaltase.

termini are still outside. In Step 4, the ribosome has dissociated and the C terminus has slipped through the membrane into the lumen, but the association of the protein with the membrane is maintained by a stretch of lipophilic amino acids near the N terminus.

The first 35 amino acids in the sequence of sucrase-isomaltase are shown in Figure 1.27. The stretch of lipophilic acids near the N terminus that anchors the enzyme to the membrane is underlined. The first few are leucine, isoleucine, valine, and isoleucine.

EXERCISE

Determine the following:

1. Which branched chain amino acids occur in the lipophilic stretch?

2. Which acidic and basic amino acids occur at either end of the hydrophilic stretch?

3. Which sulfur-containing amino acids are present in the entire sequence given in Figure 1.27?

## Glycoproteins

Glycoproteins consist of a polypeptide with one or more oligosaccharides connected to it. The point of attachment is almost always at residues of asparagine, serine, or threonine. Asparagine-linked oligosaccharides are also called **N-linked**, while those connected to serine or threonine are called **O-linked**. Glycoproteins tend to be secreted proteins, membrane-bound proteins, and lysosomal proteins. These particular locations are in keeping with the fact that the oligosaccharide moiety is attached to newly made polypeptides that are created on the endoplasmic reticulum (ER). Lysosomal proteins, for example, are made on the ER, after which they migrate through the lumen of the ER, and then to the Golgi, and finally to tiny vesicles which deliver them to the lysosomes.

To focus on the structure of N-linked glycoproteins, the growing oligosaccharide chain is manufactured while covalently attached to a lipid called **dolichol-phosphate**. The structure of this chain is shown in Figure 1.28. Man indicates a residue of mannose, Glc is glucose, and GlcNAc is *N*-acetyl-glucosamine. The R indicates the attachment point to dolichol-phosphate. The R also indicates the attachment point (after transfer to the protein) to an asparagine residue of a polypeptide.

The glycoproteins of the bloodstream include thrombin and fibrinogen (proteins used for blood clotting) and the antibodies. The glycoproteins of the plasma

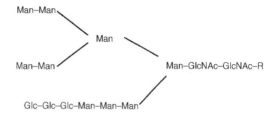

**FIGURE 1.28.** Structure of N-linked oligosaccharide.

membrane include the glucose transporter, the thrombin receptor (used for blood clotting), the LDL receptor (used to take up lipids), and the transferrin receptor (used to take up iron). Proteins that hug the outside surfaces of cells, also called extracellular matrix proteins, are also glycosylated. These include collagen, heparan, chondroitin, and mucus.

A small number of soluble, intracellular proteins are glycoproteins, and these include some of the transcription factors (proteins used for regulating the rate of transcription).

The most obvious function of oligosaccharides is their barrier function, that is, to protect their associated protein or an associated cell surface. The oligosaccharide component of mucus makes it slippery and viscous. The mucus of the stomach and intestines protects the cells from the harsh environment (Rhodes, 1997).

The oligosaccharides attached to lysosomal enzymes are responsible for their particular location in the cell, i.e., for directing the newly made enzyme to the lysosomal compartment.

In some glycoproteins, the oligosaccharide takes an active part in the biological activity of the protein, while in many other cases, there appears to be absolutely no function of the oligosaccharide group (Varki, 1993).

## Antibodies

A knowledge of antibodies, white blood cells, and other aspects of the immune system is required for an understanding of how the body combats infection. A familiarity with the immune system is also essential for an understanding of the mechanisms of diabetes and atherosclerosis. Hence, this chapter will conclude with a sketch of the immune system.

Antibodies are proteins that are synthesized by cells of the immune system. The polypeptide chains of antibodies are folded into a conformation that is relatively spherical or globular, when compared to proteins such as collagen, and thus they are named **immunoglobulins**. There exist five types of immunoglobulins, i.e., immunoglobulin G (IgG), IgA, IgM, IgD, and IgE. Most types of antibodies consist of two different types of polypeptide chains, and these are called the **light chain** and **heavy chain**. The light chain has a molecular weight of about 25 kDa and consists of about 200 amino acids, while the heavy chain is about 50 kDa and consists of about 450 amino acids. The antibody itself consists of two light chains and two heavy chains. Thus, the overall molecular weight is 150 kDa.

An example of an IgG molecule appears in Figure 1.28. All four polypeptide chains are connected to each other via disulfide bonds. The disulfide bonds (R—S—S—R) are represented by heavy black bars. In viewing the diagram, one can see that the C-terminal end of the light chain has a residue of cysteine, where cysteine's sulfhydryl group is bound to that of a cysteine residue occurring near the N-terminal end of a heavy chain. One can also see that the heavy chain of this particular antibody contains two cysteine residues situated near the center of the polypeptide chain, and that these cysteine residues are connected to each other via disulfide bonds (Brody, 1997).

The region of the antibody where the two heavy chains are connected to each other via a disulfide bond is called the **hinge** (Figure 1.29). In looking at the N-terminal regions of the light and heavy chains, one can see that each light chain is nearby a region of a heavy chain. These regions occur tightly associated or complexed with each other, and form the antigen-recognizing site. An **antigen** is any molecule that is recognized and bound by an antibody. Antibodies actually function as two identical antibodies, which are held together by the hinge region. When binding to an invading bacterium, for example, the two half-antibodies are able to move about on the flexible hinge, facilitating their grasp on two nearby target sites on the bacterium's surface.

Antibodies are synthesized by the body for the purpose of recognizing foreign molecules, and for inactivating the infectious organisms that are the source of the foreign molecules. The goal of recognizing this foreignness is to combat infecting viruses, bacteria, and protozoans. How does the body know to create a specific type of antibody? The immune system consists of several types of cells. These are generally called white blood cells. The white blood cells called lymphocytes are used in the antibody manufacturing process. Certain types of lymphocytes function to take up foreign proteins. The foreign proteins are then processed by these cells, and then presented to other types of lymphocytes. The presentation process stimulates the recipient cell to divide and multiply, and to secrete one type of antibody. The secreted antibody corresponds to the processed foreign protein that was originally presented. What does the term "correspond" mean in this commentary? It means that the secreted antibody specifically recognizes and tightly binds to the original foreign protein. As mentioned earlier, the region of the antibody that binds to foreign proteins occurs at the far N-terminal part of the heavy chain and light chain. The relevant parts of the heavy and light chain are called the **variable region**. The remainder of the antibody is called the **constant region**, as this region tends to have the same sequence of amino acids, where the antibody recognizes a variety of foreign macromolecules.

## Antibodies as a Tool for Studying Enzymology and Physiology

The biochemist can take advantage of the immune response by stimulating the immune system to synthesize antibodies that recognize biochemicals of interest. In brief, a goat or rabbit is injected with small quantities of the polypeptide or protein of interest. A month or so later, the blood is removed from the animal. The antibody is then purified from the blood, using the technique of column chromatography, and then stored in a vial for future use. The physiologist interested in

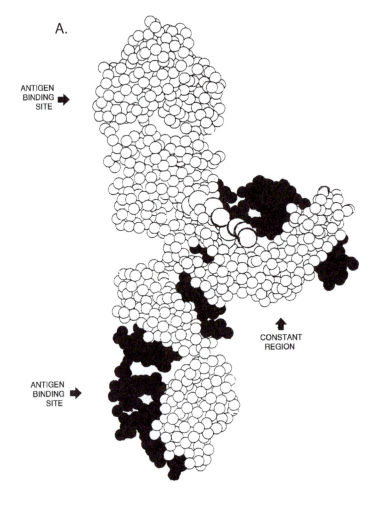

A.

ANTIGEN
BINDING
SITE

CONSTANT
REGION

ANTIGEN
BINDING
SITE

B.

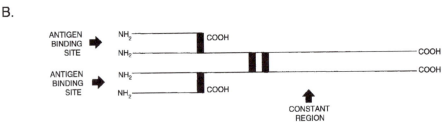

ANTIGEN
BINDING
SITE

ANTIGEN
BINDING
SITE

CONSTANT
REGION

**FIGURE 1.29.** Antibody structure. **A.** Three-dimensional structure and schematic structure of an antibody. Each sphere represents an amino acid. The large spheres near the center of the three-dimensional structure represent the sugar residues of the N-linked oligosaccharide chain. One oligosaccharide is covalently attached to each heavy chain, and the binding site is in the region of the hinge. One of the two heavy chains is filled in black. Within the constant region, the two heavy chains are intertwined with each other, but within the variable region (the region containing the antigen binding site) the two heavy chains are separate from each other. **B.** The schematic diagram shows the two light chains and two heavy chains. The region of the antibody where the two heavy chains are connected, via the disulfide bonds, is called the *hinge.* The hinge allows the two combining sites to flop about, thus facilitating their ability to reach and bind the two totally independent target antigens. (Redrawn with permission from Silverton *et al.,* 1977.)

testing the function of a specific hormone in the bloodstream may inject antibodies into the test animal. The result is that the antibody recognizes and tightly binds to the hormone, creating a stable antibody/hormone complex that continues to circulate in the bloodstream. Any alteration in animal physiology occurring in the minutes (or possibly hours) following the injection is expected to result from the lowering of the effective concentration of active hormone in the bloodstream.

The biochemist interested in testing the function of a specific protein in a complex mixture, such as a tissue homogenate, may use an antibody in the following way. Reactions may be conducted in two different test tubes. A typical experiment may involve the study of a reaction that requires the participation of a dozen or so separate proteins. The researcher may place a small quantity of an antibody, known to combine with one of the proteins, in one (not both) of the test tubes before starting the reaction. To both test tubes the researcher may then add identical quantities of tissue extract and specific salts or reagents that are needed to support the reaction. The researcher then allows 10 minutes to pass in order to allow the accumulation of products. Finally, reactions in both test tubes are terminated, and the amount of product in each test tube is measured. Any difference in the amount of product is justifiably attributed to inactivation of the target protein, in the complex mixture.

For some analytical purposes, a half-antibody may be more useful than an intact antibody. Half-antibodies, which contain only one antigen-binding site, can easily be prepared by standard methods.

## SUMMARY

The first part of Chapter 1 contains a brief review of the chemistry and biochemistry background required to understand many of the topics covered in this book. For example, the biochemical apparatus used for the digestion and transport of a specific nutrient depends on whether the nutrient is fat soluble or water soluble. Although the actual solubility of a nutrient in water is controlled by its chemical groups, the effective solubility of lipophilic substances can be increased by incorporation in micelles. Similarly, the ionization of water-soluble nutrients in the body can change depending on the surrounding environment.

Genetic material was defined, as were the events of transcription and translation. Genetic material contains the information that specifies the sequence of amino acids in all proteins, as well as information that regulates the rate of transcription. In studying the quantity of any protein in the cell, one must realize that the level is controlled by at least four conceptually different processes: (1) the rate of RNA synthesis (transcription), (2) the rate of mRNA degradation, (3) the rate of protein synthesis (translation), and (4) the rate of protein degradation (protein turnover).

A number of genetic diseases are described in the text. Since most genes occur in duplicate (one on each of the diploid chromosomes), it is expected that the genetic disease will be less severe where a mutation affects only one of the two copies, while the genetic disease will be more severe where the mutations affects both copies. Any genetic disease that results in sickness and that is discovered in people represents an "in-between land." Many genetic changes are not detected because they do not result in sickness, and thus are not studied. Other genetic changes result in a defect that is so severe that it kills the embryo before it develops. Hence, all detected genetic diseases might be considered to represent a "survivable insult" to the organism.

The "knock-out" technique is a standard genetic technique used to delete one or both copies of a specific gene. This technique, usually applied to mice, allows the researcher to increase the repertoire of defective genes, over that which is available from the mutations that have naturally been detected in the human and rodent populations.

Extra attention was devoted to proteins and enzymes, as they have been the traditional "bread and butter" of the nutritional biochemist. The amino acids that form proteins can be classified in several ways: (1) relatively hydrophilic or lipophilic, (2) dispensable or indispensable (nonessential or essential), (3) glycogenic or ketogenic, and, finally, (4) classical or modified. Only a minority of the amino acids in proteins are modified after incorporation into the polypeptide chain.

Proteins also can be classified as those that are soluble and remain in the cell, those that are membrane-bound in the cell, and those that are soluble and secreted from the cell. The specific sequence and nature of the amino acids in the polypeptide chain determine the eventual location of the polypeptide in the cell. Insulin, a secreted polypeptide, is of great interest to nutritional scientists and the medical profession. Sucrase-isomaltase is a membrane-bound protein of occasional interest in nutrition.

The term *polypeptide* can refer either to a large protein or to a polypeptide that is too small to be considered a protein. Small polypeptides include insulin and cholecystokinin. The distinction between large and small polypeptides is somewhat arbitrary. Proteins include enzymes, structural proteins, and signaling proteins. Enzymes catalyze biochemical reactions, and include hexokinase, pyruvate dehydrogenase, and DNA polymerase. Structural proteins maintain the shape of the cell, and its adherence to specific neighboring cells, and include collagen, proteoglycans, integrin, cadherin, and laminin. Signaling proteins include receptors in the plasma membrane as well as water-soluble proteins which reside in the cytoplasm and nucleus. The signaling proteins include the insulin receptor, the light receptors in the eye, G proteins, protein kinases, protein phosphatases, cyclic AMP-binding protein, estrogen receptor, vitamin A receptor, and vitamin D receptor. The signaling proteins that are able to catalyze reactions are also classed as enzymes.

Protein kinases are ATP-using enzymes that catalyze the attachment of phosphate groups to other molecules, i.e., to proteins. Phosphatases are enzymes that catalyze the hydrolysis of phosphate groups from other molecules.

## REFERENCES

Arfin, S. M., and Bradshaw, R. A. (1988). Cotranslational processing and protein turnover in eukaryotic cells. *Biochemistry* **27**, 7984–7990.

Baker, D. H. (1984). Equalized versus ad libitum feeding. *Nutr. Rev.* **42**, 269–273.

Bernard, C. (1865). "Introduction a l'Étude de la Medicine Experimentale." J.B. Baillière et Fils, Paris.

Brody, T. (1997). Multistep denaturation and hierarchy of disulfide bond cleavage of a monoclonal antibody. *Anal. Biochem.* **247**, 247–256.

Collins, F. S., Guyer, M. S., Chakravarti, A. (1997). Variations on a theme: Cataloging human DNA sequence variation. *Science* **278**, 1580–1581.

Conway, B. E. (1981). "Ionic Hydration in Chemistry and Biophysics." Elsevier, New York.

Dormann, P., Borchers, T., Korf, U., Hojrup, P., Roepstorff, P., and Spencer, F. (1993). Amino acid exchange and covalent modification by cysteine and glutathione explain isoforms of fatty acid-binding protein occurring in bovine liver. *J. Biol. Chem.* **268**, 16286–16292.

Eberson, L. (1969). Acidity and hydrogen bonding of carboxyl groups. *In* "The Chemistry of Carboxylic Acids and Esters" (S. Patai, ed.), pp. 211–293. Interscience, New York.

Eigen, M. (1964). Proton transfer, acid-base catalysis, and enzymatic hydrolysis. *Angewandte Chemie International Edition in English* **3**, 1–19.

Flinta, C., Persson, B., Jornvall, H., and Heijne, G. (1986). Sequence determinants of cytosolic N-terminal protein processing. *Eur. J. Biochem.* **154**, 193–196.

Hong, S.-B., Kim, S. J., Noh, M. N., Lee, Y. M., Kim, Y., and Yoo, O. J. (1997). Identification of the transcription termination site of the mouse *nkx*-1.2 gene: Involvement of sequence-specific factors. *Gene* **198**, 373–378.

Kornberg, R. D. (1996). RNA polymerase II transcription control. *Trends Biochem. Sci.* **21**, 325–326.

Kozak, M. (1996). Interpreting cDNA sequences: Some insights from studies on translation. *Mammalian Genome* **7**, 563–574.

Kyte, J., and Doolittle, R. F. (1982). A simple method for displaying the hydropathic character of a protein. *J. Mol. Biol.* **157**, 105–132.

Mathews, D. A., Alden, R. A., Bolin, J. T., Freer, S. T., Hamlin, R., Xuong, N., Kraut, J., Poe, M., Williams, M., and Hoogsteen, K. (1977). Dihydrofolate reductase: X-ray structure of the binary complex with methotrexate. *Science* **197**, 452–455.

Migeon, B. R. (1994). X-chromosome inactivation: Molecular mechanisms and genetic consequences. *Trends Genet.* **10**, 230–235.

Mount, S. M. (1996). AT–AC introns: An attack on the dogma. *Science* **271**, 1690–1718.

Osawa, H., Robey, R., Printz, R., and Granner, D. (1996). Identification and characterization of basal and cyclic AMP response elements in the promoter of the rat hexokinase II gene. *J. Biol. Chem.* **271**, 17296–17303.

Powers, T., and Walter, P. (1997). A ribosome at the end of the tunnel. *Science* **278**, 2072–2126.

Printz, R. L., Osawa, H., Ardehali, H., Koch, S., and Granner, D. K. (1997). Hexokinase II gene: Structure, regulation and promoter organization. *Biochem. Soc. Trans.* **25**, 107–112.

Rhodes, J. M. (1997). Colonic mucus and ulcerative colitis. *Gut* **40**, 807–808.

Rice, W. R. (1996). Evolution of the Y sex chromosome in animals. *Bioscience* **46**, 331–338.

Roeder, R. G. (1996). The role of general initiation factors in transcription by RNA polymerase II. *Trends Biol. Sci.* **21**, 327–335.

Scaloni, A., Jones, W. M., Barra, D., Pospischil, M., Sassa, S., Popowicz, A., Manning, L. R., Schneewind, O., and Manning, J. M. (1992). Acylpeptide hydrolase: Inhibitors and some active site residues of the human enzyme. *J. Biol. Chem.* **267**, 3811–3818.

Schuler, G. D., Boguski, M. S., Stewart, E. A., Stein, L. D., and Gyapay, G. (1996). A gene map of the human genome. *Science* **274**, 540–544.

Silverton, E. W., Navia, M. A., and Davies, D. R. (1977). Three-dimensional structure of an intact human immunoglobulin. *Proc. Natl. Acad. Sci. U.S.A.* **74**, 5140–5144.

Tate, W. P., and Brown, C.M. (1992). Translation termination: "Stop" for protein synthesis or "pause" for regulation of gene expression. *Biochemistry* **31**, 2443–2450.

Varki, A. (1993). Biological roles of oligosaccharides: All of the theories are correct. *Glycobiology* **3**, 97–130.

Zawel, L., and Reinberg, D. (1995). Common themes in assembly and function of eukaryotic transcription complexes. *Annu. Rev. Biochem.* **64**, 533–561.

BIBLIOGRAPHY

Balch, W. E. (1989). Biochemistry of interorganelle transport. *J. Biol. Chem.* **264**, 16965–16968.

Cavener, D. R., and Ray, S. C. (1991). Eukaryotic start and stop translation sites. *Nucl. Acids Res.* **19**, 3185–3192.

Ellis, J. R., and Vies, S. M. (1991). Molecular chaperones. *Annu. Rev. Biochem.* **60**, 321–347.

Fisher, J. M., and Scheller, R. H. (1988). Prohormone processing and the secretory pathway. *J. Biol. Chem.* **263**, 16515–16518.

Hildebrand, J. H. (1979). Is there a "hydrophophobic effect"? *Proc. Natl. Acad. Sci. U.S.A.* **76**, 194.

Kelly, R. B. (1985). Pathways of protein secretion in eukaryotes. *Science* **230**, 25–31.

Lee, S. P., Park, H. Z., Madani, H., and Kaler, E. W. (1987). Partial characterization of a nonmicellar system of cholesterol solubilization in bile. *Am. J. Physiol.* **252**, G374–G383.

Lievremont, J., Rizzuto, R., Hendershot, L., and Meldolesi, J. (1997). BiP, a major chaperone protein of the endoplasmic reticulum lumen, plays a direct and important role in the storage of the rapidly exchanging pool of $Ca^{2+}$. *J. Biol. Chem.* **272**, 30873–30879.

Lodish, H. F. (1988). Transport of secretory and membrane glycoproteins from the rough endoplasmic reticulum to the Golgi. *J. Biol. Chem.* **263**, 2107–2110.

Matthews, B. W., Sigler, P. B., Henderson, R., and Blow, D. M. (1967). Three-dimensional structure of tosyl-α-chymotrypsin. *Nature (London)* **214**, 652–656.

Nigg, E. A., Baeuerle, P. A., and Luhrmann, R. (1991). Nuclear import–export: In search of signals and mechanisms. *Cell* **66**, 15–22.

Radzicka, A., Pedersen, L., and Wolfenden, R. (1988). Influences of solvent water on protein folding. *Biochemistry* **27**, 4538–4541.

Rademacher, T. W., Parekh, R. B., and Dwek, R. A. (1988). Glycobiology. *Annu. Rev. Biochem.* **57**, 785–838.

Roda, A., Hofmann, A. F., and Mysels, K. J. (1983). The influence of bile salt structure on self-association in aqueous solutions. *J. Biol. Chem.* **258**, 6362–6370.

Roise, D., and Schatz, G. (1988). Mitochondrial presequences. *J. Biol. Chem.* **263**, 4509–4511.

Schatz, G., and Dobberstein, B. (1996). Common principles of protein translocation across membranes. *Science* **271**, 1519–1539.

Simon, S., and Blobel, G. (1991). A protein-conducting channel in the endoplasmic reticulum. *Cell* **65**, 371–380.

Wahl, M., and Sundaralingham, M. (1997). C-H...O hydrogen bonding in biology. *Trends Biol. Sci.* **22**, 97–102.

Wolfenden, R., Andersson, L., Cullis, P. M., and Southgate, C. C. B. (1981). Affinities of amino acid side chains for solvent water. *Biochemistry* **20**, 849–855.

# 2

# DIGESTION AND ABSORPTION

## OVERVIEW

Nutritionists, physicians, and related health care professionals often must advise heart, kidney, and diabetic patients on food choices to prolong life, must educate consumers on prudent choices for their diets, and must suggest healthful and appetizing ways to prepare food for the table. To discuss with patients the reasons for choosing some foods rather than others, the professional must know how food is digested and absorbed. The focus of this chapter is on the biochemical effects of hormones and enzymes on the digestion and absorption of food in the body.

## DIGESTIVE TRACT

### Components

The path through the digestive tract begins at the mouth, proceeds through the esophagus to the stomach, and through the pyloric sphincter into the small intestine. The small intestine consists of three sections: the **duodenum**, the **jejunum**, and the **ileum**, which empties into the large intestine or colon. The colon also contains three sections: the ascending colon, the transverse colon, and the descending colon, which empties into the rectum.

The upper portion of the stomach is called the **fundus**; the lower portion is called the **antrum**. The inside of the small and large intestines is called the gut. The **mucosa** are the surfaces of an organ that are wet and slippery because of secretion of mucus. A continuous layer of cells that lines the surfaces of the body and its organs is called an **epithelial membrane**. The most common cell of the epithelial membrane of the gut is the **enterocyte**. The epithelial membrane of the gut also contains mucus-secreting glands called **goblet cells**. The epithelial surface of the stomach is smooth, whereas that of the small intestine has a rough appearance because of finger-like protuberances called **villi**, whereas the epithelial surface of the large intestine is flat and lacks villi. Each villus is a finger-like projection about 0.025 mm high. Enterocytes coat each villus, and the surface of these enterocytes, which faces the food (the lumenal surface), contains hundreds of small villi called **microvilli**. The villi and microvilli greatly enhance the absorptive surface area of the small intestine. Because of its appearance, the mucosal surface of the small intestine is also called the **brush border**.

Other organs of the digestive system are the salivary glands, pancreas, liver, and gall bladder. The **salivary glands** produce saliva that enters the mouth and is mixed with food during chewing. The fluid produced by the **pancreas** enters the pancreatic duct. The pancreatic duct joins the bile duct near the duodenum, forming a short common channel called the **ampulla of Vater**. The pancreas also secretes chemicals directly into the bloodstream. The fluid produced by the **liver**, called **bile**, flows to the **gall bladder**. The gall bladder stores the bile and releases it to the duodenum when needed at this location.

Digestion is controlled by the autonomic branch of the nervous system. The autonomic nervous system of the gut consists mainly of the parasympathetic nervous system (which includes the vagus nerve) and the enteric nervous system, and to a small extent the sympathetic nervous system. The **vagus nerve**, also called the vagal nerve, transmits signals from the brain to the stomach, pancreas, and other digestive organs. This nerve is one of the twelve cranial nerves, which branch out from the brain stem rather than from the spinal cord. The vagus nerve is so named because it seems to wander around in the body (i.e., it is "vague" in where it goes). Figure 2.1 indicates how one of the trunks of the vagus nerve contacts the stomach and branches out to various parts of the antrum and fundus. The vagus nerve also transmits signals from various organs back to the brain.

The gut contains a collection of nerves called the enteric nervous system. Various activities of the gut, such as peristaltic contractions, and certain activities of the pancreas and gall bladder, are controlled in a manner that is relatively independent of the central nervous system. The **central nervous system** consists of the brain and spinal cord. Five types of activity are controlled by the enteric nervous system: (1) contraction of smooth muscles that create the peristaltic waves used to mix and propel food through the intestines; (2) release of juices by secretory cells; (3) release of hormones from endocrine cells of the gut; (4) patterns of blood flow through the arteries of the gut [variations in blood flow occur because of the opening of blood vessels (**vasodilation**) or the closing of blood vessels (**vasoconstriction**)]; and (5) activities of immune cells of the gut (Goyal and Hirano, 1996).

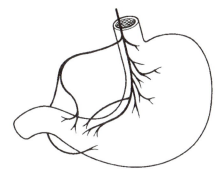

**FIGURE 2.1** Innervation of the stomach.

## Enzymes and Zymogens

The chemical reactions associated with the digestion of nutrients are catalyzed by many different enzymes. Some enzymes are secreted in their catalytically active form, whereas others are secreted as the corresponding zymogens (given in parentheses). Zymogens require chemical modification to be converted to their catalytically active forms.

The salivary glands secrete a very small amount of α-amylase. The tongue of the rat secretes lingual lipase. The stomach secretes pepsin (pepsinogen); the human stomach also secretes gastric lipase. Enzymes secreted by the pancreas include α-amylase, trypsin (trypsinogen), chymotrypsin (chymotrypsinogen), carboxypeptidase A (procarboxypeptidase A), carboxypeptidase B (procarboxypeptidase B), elastase (proelastase), phospholipase (prophospholipase), ribonuclease, deoxyribonuclease, pancreatic lipase, and colipase. The digestive enzymes of the gut that are made by the enterocytes include enterokinase, sucrase–isomaltase, and alkaline phosphatase. To summarize, the significant sources of digestive enzymes are the stomach, pancreas, and enterocytes.

## Hormones

A subsequent section of this chapter details the actions of three hormones of the gastrointestinal tract: **gastrin**, **secretin**, and **cholecystokinin** (CCK). (A variety of other gut hormones exist, including vasoactive intestinal polypeptide, substance P, motilin, neurotensin, gastrin-releasing peptide, somatostatin, and tyrosine-tyrosine.) Gastrin, which is secreted from the **G cells** located in the antrum of the stomach, provokes the **parietal cells** to secrete gastric acid. When gastric acid leaves the stomach and enters the duodenum, it provokes the release of secretin, which in turn stimulates the pancreas to release fluid and bicarbonate into the gut. Dietary fats or proteins entering the small intestine provoke the release of CCK into the bloodstream, which stimulates the pancreas to release enzymes and zymogens into the gut. CCK also stimulates the gall bladder to release bile salts into the gut. The locations of the specialized cells of the gastrointestinal tract are specified in subsequent sections of this chapter.

## Bile Acids

Bile acids are acidic compounds containing steroid rings. Common bile acids of the body include cholic acid, chenodeoxycholic acid, deoxycholic acid, and litho-cholic acid. The general pathway for bile acid synthesis, which occurs in the liver, is shown in Figure 2.2. The conversion of cholesterol to 7α-hydroxycholesterol is catalyzed by **cholesterol 7α**-hydroxylase. As indicated in Figure 2.2, bile acids may be modified by the removal of 7α-hydroxyl groups by the gut microflora in the lumen of the small intestine. Bile acids play an important role in the transport and absorption of the products of lipid digestion, as discussed in the section on digestion and absorption of lipids.

Bile acids are easily dissolved in water, and after dissolving in water they can be neutralized by adding an alkali, such as sodium bicarbonate or sodium hydrox-ide. After neutralization, these compounds no longer occur as an acid. Instead, they occur as a salt. Often, discussions regarding the metabolism and function of these compounds use the terms *bile acid* and *bile salt* interchangeably.

## SECRETION OF DIGESTIVE MATERIALS

### Salivary Glands

Saliva is secreted by the salivary glands in response to various stimuli. This fluid contains α-amylase, to which lingual lipase is added by the Von Ebner glands of the tongue in humans and certain animals (Hof *et al.*, 1997).

### Fundus of the Stomach

The gastric glands are located in the wall of the fundus of the stomach. As illustrated in Figure 2.3, inside each gland are parietal cells, chief cells, and goblet cells. Hormone-secreting cells are also present. The **chief cells** secrete **pepsinogen**; the goblet cells secrete large amounts of mucus. **Parietal cells** (also called **oxyntic cells**) have a characteristic triangular shape, with the base of the triangle facing the bloodstream and the apex facing the lumen of the stomach. These cells secrete a solution containing 0.1 $N$ hydrochloric acid and **intrinsic factor**, a protein required for the absorption of vitamin $B_{12}$. Intrinsic factor is not much damaged by the harsh conditions in the stomach. However, the hydrochloric acid causes denaturation of most dietary proteins, facilitating attack by various proteases. Proteases catalyze the hydrolysis of peptide bonds. The acid also provokes the autoactivation of pepsinogen to create **pepsin**. In other words, upon contact with gastric acid, pepsinogen is converted to **pepsin**.

### Antrum of the Stomach

Gastrin, a polypeptide hormone, is secreted into the bloodstream from the G cells located in the antrum of the stomach. The gastrin travels through the bloodstream

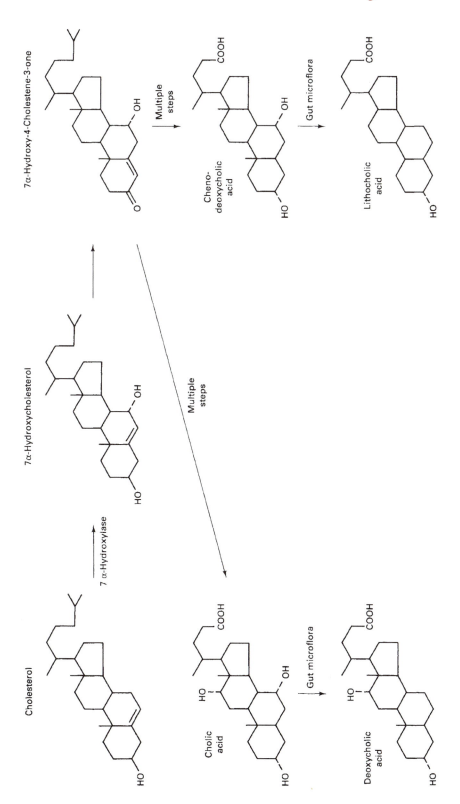

**FIGURE 2.2** Conversion of cholesterol to bile salts and further metabolism of bile salts by bacteria in the intestines.

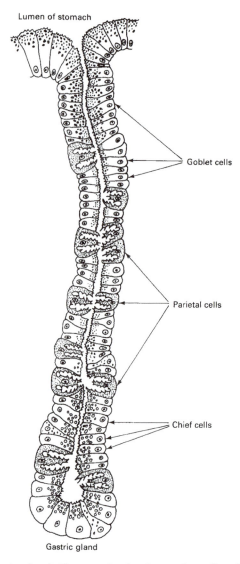

Lumen of stomach

Goblet cells

Parietal cells

Chief cells

Gastric gland

FIGURE 2.3 The gastric gland. The gastric gland contains cells which serve a variety of functions. [Redrawn with permission from Ito and Winchester (1963) and from the *Journal of Cell Biology*, 1963, pp. 543–557. By copyright permission of the Rockefeller University Press.]

to reach the parietal cells of the fundus, and it then stimulates the parietal cells to secrete gastric acid.

## Pancreas and Small Intestines

The pancreas consists of about 82% acinar cells by volume. These cells secrete various zymogens and enzymes into the pancreatic duct in a relatively small volume of fluid. The pancreas contains about 3.9% **duct cells**, which secrete fluid and bicarbonate into the pancreatic duct. The acinar and duct cells are **exocrine cells**. The pancreas contains only 1.8% endocrine cells by volume. The **endocrine cells** secrete glucagon and insulin into the bloodstream. Glucagon and insulin are small proteins that are hormones. Blood vessels account for about 3.7% of the volume of the pancreas. The ionic composition of pancreatic juice is approximately 120 m$M$ bicarbonate, 140 m$M$ sodium, 70 m$M$ chloride, 5 m$M$ potassium, and 2 m$M$ calcium with a pH of 7.2–7.4. Generally, the exocrine cells secrete about 1000 ml of fluid into the duodenum per day.

Release of the zymogens into the lumen of the small intestine results in their exposure to a new environment. This environment is not acidic (unlike that of the stomach), and it contains **enterokinase** (also called **enteropeptidase**), a protease of the small intestine. Enterokinase is constitutively present (always present) in the small intestine. Studies with pig intestines revealed that enterokinase is present only in the duodenum, not the jejunum or ileum, and is bound to the outside of the enterocyte. Enterokinase catalyzes the cleavage of one specific peptide bond in trypsinogen, resulting in its conversion to **trypsin**. The point of action is between one residue of lysine and one of isoleucine, as shown in Figure 2.4.

Trypsin catalyzes the activation of a variety of other pancreatic enzymes. As does enterokinase, trypsin catalyzes hydrolysis of pancreatic zymogens only at specific sites. This enzyme cleaves the zymogens at sites immediately following residues of *arginine* and *lysine*, that is, on the carboxyl side of these amino acids. Trypsin catalyzes the activation of trypsinogen, thus amplifying the effect of enterokinase. The action of trypsin on trypsinogen is called an **autocatalytic reaction**. Trypsin catalyzes the activation of the following pancreatic zymogens:

> Trypsinogen
> Chymotrypsinogen
> Procarboxypeptidase A
> Procarboxypeptidase B
> Proelastase
> Prophospholipase

When trypsin activates chymotrypsin, it cleaves chymotrypsinogen at the peptide bonds between amino acids Arg 15 and Ile 18. The numbers indicate the amino acid residue sequence numbers, counting from the N-terminal end of the protein.

As stated earlier, enterokinase catalyzes the activation of trypsinogen, which begins a cascade of events resulting in the activation of all the pancreatic enzymes

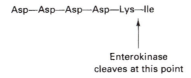

Asp—Asp—Asp—Asp—Lys—Ile

Enterokinase
cleaves at this point

**FIGURE 2.4** Cleavage of trypsinogen by enterokinase. The cleavage is via a hydrolytic reaction.

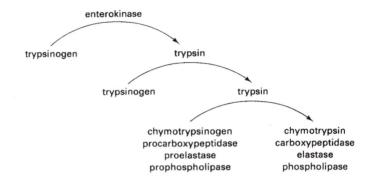

**FIGURE 2.5** Activation cascade involving digestive enzymes of the small intestine. The position of the word "enterokinase" on the first reaction arrow indicates that this enzyme activates trypsinogen.

(Figure 2.5). Such activation schemes, called **activation cascades**, are encountered periodically in biological systems. (Other examples discussed in this book are the action of glycogen phosphorylase, the blood clotting cascade, and the MAP kinase signaling cascade.) One advantage of an activation cascade is that a small signal is amplified into a large response that involves many proteins. Another advantage is that the amplification is rapid, i.e., perhaps reaching a maximal rate of activity within 20–30 seconds.

The pancreas also secretes a number of enzymes in *active form*, including α-amylase, lipase, deoxyribonuclease (DNase), and ribonuclease (RNase). DNase catalyzes the hydrolysis of DNA, resulting in the liberation of deoxyadenosine monophosphate (dAMP), deoxythymidine monophosphate (dTMP), deoxyguanosine monophosphate (dGMP), and deoxycytosine monophosphate (dCMP).

## Liver and Gall Bladder

Bile salts are synthesized in the **hepatocyte** of the liver, and are released into channels called **bile canaliculi**. These channels lead from their termini in the liver to a large vessel called the bile duct that directs the bile salts to the gall bladder for storage, and from the gall bladder to the duodenum.

Before excretion from the liver, most of the bile salts are modified by the attachment of glycine or taurine to form conjugated bile salts, as shown in Figure 2.6. The bonds of attachment resemble peptide bonds because they involve the carboxyl group of cholic acid and the amino group of glycine or taurine. The conjugated bile salts are formed via the pathway outlined in Figure 2.7. In the first step, cholate forms a thiol ester bond with the sulfhydryl group of coenzyme A. In this form, the cholate molecule is activated and can react readily with glycine, as shown, or with taurine. In the second step, the amino group of glycine attacks the thiol ester bond, displacing the molecule of coenzyme A, to form cholylglycine. The same enzyme catalyzes the conjugation of either glycine or taurine with the bile salts.

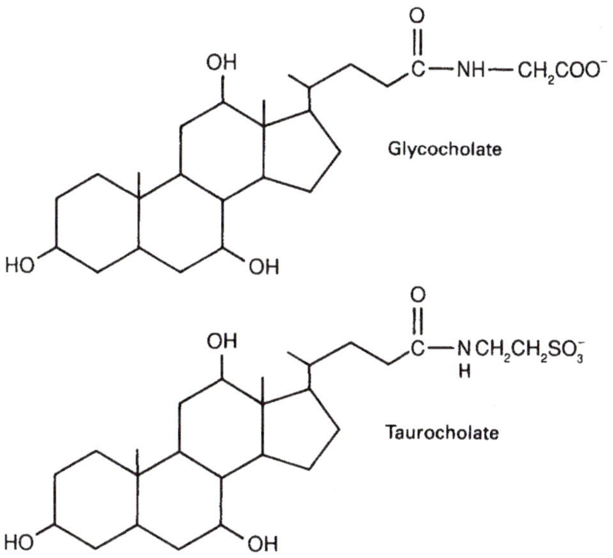

**FIGURE 2.6** Glycocholate and taurocholate. Glycocholate contains a molecule of glycine, while taurocholate contains a molecule of taurine.

Let us review some of the conventions used, when talking about enzyme-catalyzed reactions. The first reaction depicted in Figure 2.7 involves three **substrates** (cholate, HS–CoA, and ATP) and three **products** (cholyl–S–CoA, AMP, and PP$_i$). The substrates are shown on the left-hand side of the arrows, while the products are shown on the right-hand side. ATP is used to drive the reaction, i.e., to make it occur in the direction of the arrow. Since consumption of ATP is not the goal of this particular reaction, ATP may be called a **co-substrate**. The abbreviation HS–CoA means coenzyme A. Cholate is the ionized form of cholic acid.

## Small Intestine

Secretin and CCK are secreted by cells of the duodenum, the segment of the small intestine directly connected to the stomach. The endocrine cells of the duodenum include the **I cells** and **S cells**. The I cells secrete CCK, a hormone that enters the bloodstream, through which it reaches a variety of organs. Some of these organs, such as the stomach, gall bladder, and pancreas, contain CCK receptors on their plasma membranes. CCK binds to these receptors and generates a number of

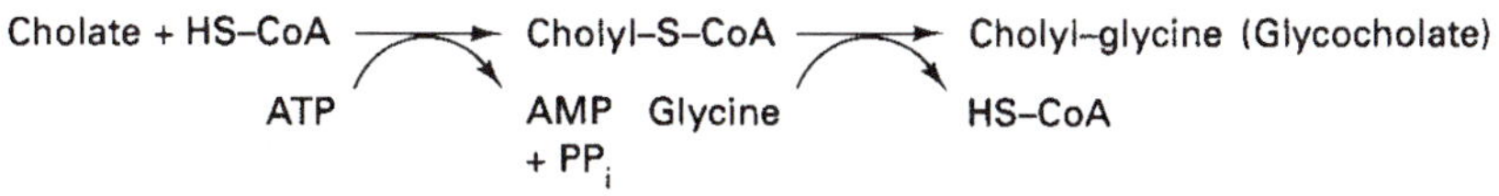

**FIGURE 2.7** Conjugation of glycine with cholate to form glycocholate.

effects, the nature of which depends on the target cell or organ. The S cells release secretin into the bloodstream. The release of secretin from the S cells of the duodenum is stimulated by acid entering the duodenum. This acid can take the form of a mixture of gastric juice and partially digested food. Secretin enters the bloodstream and travels to the pancreas, where it binds to secretin receptors located in the plasma membrane of the cells of the pancreatic duct (duct cells) and provokes these cells to release bicarbonate and large amounts of fluid into the pancreatic duct and, thus, into the duodenum. The bicarbonate neutralizes gastric acid. Many of the digestive enzymes of the small intestine require a near-neutral pH to be catalytically active.

Another important hydrolytic enzyme of the gut is **acid phosphatase**. Like enterokinase, it is bound to the enterocyte facing the lumen and is present in the duodenum, jejunum, and ileum. **Alkaline phosphatase**, a zinc metalloenzyme, also occurs in the gut. Acid phosphatase and alkaline phosphatase catalyze the removal of phosphate groups from a wide variety of compounds in foods, for example, sugar phosphates, triose phosphates, nucleotides such as AMP, ADP, and ATP, pyrophosphate, and phosphorylated amino acids. A number of sugar and triose phosphates are described in the section on glycolysis in Chapter 4.

Figure 2.8 illustrates the general distribution of some of the hormone-secreting cells of the gastrointestinal tract. These cells do not occur in clusters to form visible glands but are dispersed among other cells, mainly enterocytes. This situation has made it difficult for researchers to elucidate the behavior and functions of these cells. This problem is being approached by cloning the hormone-producing cells, thus obtaining a population containing only the endocrine cell of interest. Cloning may be accomplished by isolating a single cell of interest under a microscope, transferring it to a culture medium containing the required nutrients, and allowing the cell to replicate, thereby producing a large quantity of identical cells.

## STIMULATION OF THE DIGESTIVE SYSTEM

### Cephalic Phase

Biochemical and physiological events in the body that are induced by thinking of, smelling, tasting, or chewing food belong to the **cephalic phase** of digestion. The term *cephalic* is derived from the Greek word meaning *head*, implying involvement of the nervous system. The major nerve involved in controlling digestion is the vagus, which stimulates the parietal cells to secrete gastric acid.

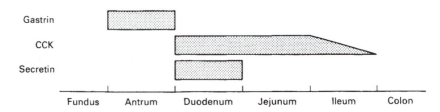

**FIGURE 2.8** Distribution of hormone production over the gastrointestinal tract.

In a legendary experiment, Pavlov demonstrated that a dog could be stimulated to secrete gastric acid solely by so-called "mental" acts. He demonstrated that the sound of a bell could provoke stomach secretions in dogs that had been trained to expect food shortly after that sound. Pavlov used the **esophageal fistula** technique to divert food eaten by the dog from its stomach. A fistula created by surgery diverts fluid from an organ to the outside of the body. The dog was permitted to eat food, but the food contacted neither the stomach nor the intestines. This is called a **sham meal**. Pavlov demonstrated that a sham meal provoked the dog's stomach to secrete gastric acid, and also caused secretion of fluids from the pancreas into the small intestines. The pancreatic secretions were found to commence prior to those of the stomach, demonstrating that the pancreatic secretions were not caused by those of the stomach.

## Effect of Sham Meals

The cephalic phase of digestion stimulates only a fraction of the maximum possible levels of gastric and pancreatic secretions. This phase does not seem to produce a rise in the levels of gastrin and CCK. Cephalic stimulation of the pancreas, as mediated by the vagus nerve, provokes release of pancreatic enzymes into the small intestine. Cephalic stimulation of the parietal cells, as mediated by the vagus nerve, provokes release of gastric acid into the lumen of the stomach. In humans, the cephalic phase does not seem to result in release of bicarbonate into the lumen of the small intestine.

The most apparent component of the cephalic phase may be the sensation of the release of saliva upon smelling or seeing an appealing meal, for example, a large steaming turkey moist with aromatic juices. People who like jalepeno peppers may experience sudden salivary secretions when passing cans of them in a grocery store. However, cutting the vagus nerve fibers that lead to the stomach or pancreas prevents responses of these organs to sham feeding. This surgical procedure once was used to minimize the production of gastric acid in patients suffering from stomach ulcers.

## Gastric Acid Experiment

The following experiment describes the release of gastric acid during the cephalic phase. The study involved human subjects who volunteered to have a tube inserted into a nostril and pushed down into the stomach, a standard procedure for collecting stomach juices. The subjects were instructed to be silent, to discuss sports, or to discuss food. Samples were removed at 15-minute intervals during the course of the experiment. The concentration of hydrogen ions ($H^+$) in the samples was used to calculate the rate of acid release.

Figure 2.9 illustrates the rate of acid secretion during an initial hour of silence, followed by discussions about sports or food. The rate of secretion during the silent period is used as the basal rate, that is, the rate occurring with no stimulation. The black bar in Figure 2.9 indicates the 30-minute period during which the discussions occurred. The basal rate of secretion as well as the rates occurring during the discussions are recorded in Table 2.1.

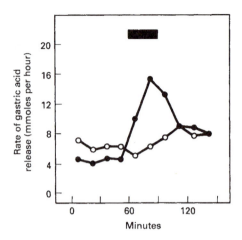

**FIGURE 2.9** Gastric acid release during an hour of silence followed by half-hour discussions of sports (○) or food (●). The rate of secretion during the silent period is used as the basal rate, that is, the rate occurring with no stimulation. The black bar indicates the 30-min period during which the discussions occurred. (Redrawn with permission from Feldman and Richardson, 1986.)

Other subjects volunteered for sham feeding (chewing food then spitting it out without swallowing). Another group of subjects received an injection of gastrin, a polypeptide hormone produced by the G cells of the stomach. This hormone causes the parietal cells of the stomach to secrete acid. In all cases, the rate of gastric acid production was monitored via samples withdrawn through a stomach tube. The results appear in Table 2.1.

Note that gastrin injection produced the maximal rate of acid output, followed by sham feeding and by talking about food. If the latter responses are to be calculated as percentages of this maximum, the basal rate must be subtracted from all three values first. This manipulation is a simple example of the preliminary corrections that must be applied to experimental data before more complicated questions can be answered correctly.

**TABLE 2.1** Rate of Gastric Acid Release under Various Conditions

| Stimulus | Rate of gastric acid release (mmol/hr) |
| --- | --- |
| Basal acid output | 4 |
| Talking about sports | 4 |
| Talking about food | 15 |
| Sham feeding | 20 |
| Injection of gastrin | 38 |

*Source*: Feldman and Richardson (1986).

## Chemical Phase

Digestion also involves a chemical phase. In this phase, components of food directly provoke responses from cells of the gastrointestinal tract. For example, dietary protein stimulates the G cells of the stomach to release gastrin into the circulatory system. Gastrin returns to the stomach, where it prompts the parietal cells to produce gastric acid. Thus, the parietal cells are activated by both nervous and hormonal stimulation. The role of gastrin in digestion is revealed in studies in which its level is increased indirectly (by eating food) or directly (by injection of gastrin).

The chemical phase of digestion also involves the hormone cholecystokinin (CCK). Dietary fats and proteins elicit the release of CCK from cells of the intestines. The versatility of this hormone in digestion is revealed, later in this chapter, via studies involving dogs, rats, and humans. These studies address the influence of CCK on the release of pancreatic enzymes, bile salts, and pancreatic bicarbonate into the lumen of the small intestines.

# MOLECULES IMPORTANT TO DIGESTION AND ABSORPTION

## Gastrin

*Structure*

Human gastrin is a polypeptide hormone containing 34 amino acids (see Figure 2.10). The N-terminal residue is a cyclized form of glutamic acid called **pyroglutamate**, which is represented by the abbreviation GLP (see Figure 2.11). The C-terminal residue is an amidated form of phenylalanine. The origin of the amide group is a residue of glycine. The formation of amidated polypeptides is described

GLP—LEU—GLY—PRO—GLN—GLY—PRO—PRO—HIS—LEU—VAL—ALA—ASP—PRO—SER—LYS—

LYS—GLN—GLY—PRO—TRP—LEU—GLU—GLU—GLU—GLU—GLU—ALA—TYR—GLY—TRP—

MET—ASP—PHE— $NH_2$

Human gastrin

TYR—ILE—GLN—GLN—ALA—ARG—LYS—ALA—PRO—SER—GLY—ARG—VAL—SER—MET—ILE—

LYS—ASN—LEU—GLN—SER—LEU—ASP—PRO—SER—HIS—ARG—ILE—SER—ASP—ARG—ASP—

TYR*—MET—GLY—TRP—MET—ASP—PHE—$NH_2$

Porcine (pig) CCK<br>*(The tyrosine residue contains a sulfate group<br>covalently bound to the ring oxygen)

HIS—SER—ASP—GLY—THR—PHE—THR—SER—GLU—LEU—SER—ARG—LEU—ARG—ASP—SER—

ALA—ARG—LEU—GLN—ARG—LEU—LEU—GLN—GLY—LEU—VAL—$NH_2$

Porcine secretin

**FIGURE 2.10** Structures of gut hormones. Gastrin, CCK, and secretin are all polypeptide hormones. Each of these hormones contains an amino group at the C terminus.

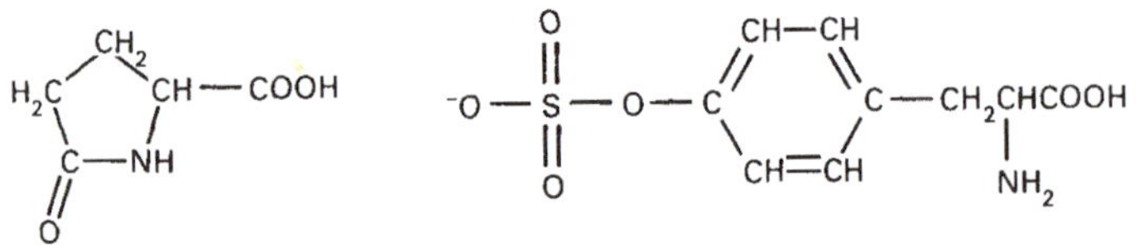

FIGURE 2.11 Structures of pyroglutamate and sulfotyrosine. The occurrence of pyroglutamate (GLP) and sulfotyrosine, on any particular hormone, would be expected to vary from species to species.

in the section on vitamin C. Several different polypeptide hormones contain pyroglutamate at the N terminus and amide groups at the C terminus.

The sequence of amino acids at the C terminus of gastrin is absolutely required for its hormonal action, that is, for transmitting its signal to the target cell. Shorter versions of gastrin have the same effect as the 34-amino acid version in provoking the release of gastric acid. The C-terminal portion of gastrin controls the destination of the hormone by binding to the hormone receptor on the plasma membrane of the target cell. The amino acids at the N terminus do not appear to be vital for hormonal activity.

Gastrin occurs naturally in different sizes. In addition to "big gastrin," which has 34 amino acids, smaller versions with 17 and 14 amino acids are also present in the circulatory system. These versions appear to produce similar biological effects.

EXERCISE

In Figure 2.10, examine the order of amino acids at the C termini of gastrin and cholecystokin. How many amino acids are identical in the two hormones? Are any other identical sequences shared by these hormones? These questions are relevant to the concept that a number of hormones with different functions are actually part of the same family of hormones. A number of proteins, such as various enzymes, hormone receptor proteins, and DNA-binding proteins, also share similar primary sequences and thus may be considered members of specific families of proteins.

## Gastrin Physiology

The experiment described in the following section illustrates some of the broad properties of gastrin. The study shows that the levels of serum gastrin and stomach acid secretion both increase after a meal in a human subject. In isolation, this information does not provide evidence that the rise in serum gastrin generates the release of gastric acid. Additional evidence is needed to make a firm connection between hormonal levels and the behavior of the parietal cells. Therefore, the study also includes an experiment of an invasive nature, that is, infusion of gastrin. **Infusion** means a slow injection, perhaps over the course of an hour.

The experiment also includes a control study in which the subject is infused with salt water only. The tests include infusions of low, moderate, and high

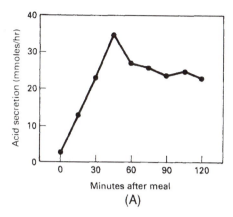
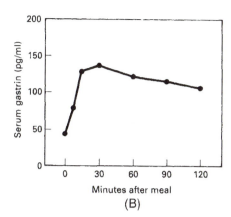

**FIGURE 2.12** Acid secretion and serum gastrin level after a meal. After a human subject was given a meal consisting of a hamburger and water, a stomach tube was used to remove samples at the times indicated (A). To facilitate removal of samples, the hamburger was homogenized in a blender and transferred into the stomach with a tube. Blood samples, taken at the times indicated (B), were used to measure gastrin levels. The data suggest that the basal level of acid secretion is low compared with the maximal rate (A). However, the basal level of serum gastrin is significant compared with the maximal level reached (B). A plateau in serum gastrin concentration was reached about 10 min after the hamburger was consumed (B). A plateau in the rate of acid release was not reached until 30 min after consumption (A). (Reprinted with permission from Blair *et al.*, 1987.)

concentrations of the hormone. In all cases, consequent changes in serum gastrin level and rates of gastric acid release are measured.

The human subject was given a meal consisting of a hamburger and water. A stomach tube was used to remove samples at the times indicated in Figure 2.12A. To facilitate removal of samples, the hamburger was homogenized in a blender and transferred into the stomach with a tube. Blood samples, taken at the times indicated in Figure 2.12B, were used to measure gastrin levels.

In this experiment, the basal level of acid secretion was low compared with the maximal rate (Figure 2.12A). However, the basal level of serum gastrin was significant compared with the maximal level reached (Figure 2.12B). A plateau in serum gastrin concentration was reached about 10 min after the hamburger was consumed (Figure 2.12B). A plateau in the rate of acid release was not reached until 30 min after consumption (Figure 2.12A).

The data presented so far seem to indicate no close relationship between serum gastrin concentration and gastric acid release because of the following discrepancies. First, the basal level of gastrin is about one-third the maximal observed level. Nevertheless, the basal level of gastrin does not seem to be associated with a significant level of basal acid secretion. In other words, the basal level of gastrin is not extremely far below the maximal level, but the basal level of acid secretion is very far below the maximal level. Second, the rate of acid release continues to increase when the concentration of serum gastrin does not increase (30–45 minutes). Therefore, it does not seem possible to draw firm conclusions from these data.

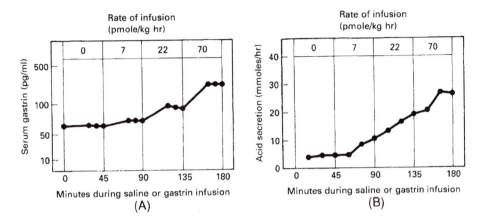

**FIGURE 2.13** Serum gastrin level and acid secretion during infusions. The researchers measured serum gastrin during the gastrin infusion to confirm that gastrin levels did in fact increase (A). An increased rate of stomach acid production occurred in response to the elevated levels of plasma gastrin (B). (Redrawn with permission from Blair *et al.*, 1987.)

However, the data presented in Figure 2.13 and Table 2.1 directly address the relationship between gastrin and acid secretion, and indicate that the hormone does provoke gastric acid production. A human subject was infused with saline over a period of 45 minutes, and subsequently was infused with low, moderate, and high concentrations of gastrin, each for a 45-minute period (Figure 2.13A). During saline infusion, the concentration of serum gastrin remained at the basal level. However, progressively higher concentrations were detected as higher concentrations of gastrin were infused (Figure 2.13). These data demonstrate that the researchers did not simply assume that gastrin infused at higher concentrations would lead to higher concentrations in the bloodstream, but actually measured the resultant levels. Figure 2.13B shows that the rate of acid release increased in a progressive manner with increasing levels of serum gastrin.

EXERCISE

During the infusion of gastrin (Figure 2.13A), did the level of serum gastrin rise above the maximal level obtained after feeding (Figure 2.12B)? In evaluating any study in which hormones or metabolites are introduced into an animal or cell culture medium, the concentrations used should be compared with those that occur in nature. One reason for this precaution is that secondary effects may take place when the introduced chemical is at supranormal levels. An effect generated by such supranormal levels is called a **pharmacologic effect**, even if the chemical occurs naturally in the body and is not a drug.

Reaching firm conclusions about the effects of hormones requires several related experiments, some with living animals and others with simpler systems. These systems, in order of decreasing complexity, include isolated organs, cells in tissue culture, isolated cell membranes or organelles, and purified enzymes.

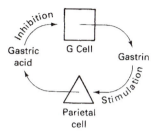

**FIGURE 2.14**  Feedback inhibition.

Gastrin affects the motility, or movement, of the stomach and appears to cause the chief cells to secrete pepsin. However, the chief cells respond to a number of different stimuli, and no single stimulant has been found to be of primary importance.

## Feedback Control of Release

The feedback control loop, depicted in Figure 2.14, normally prevents the release of excessive amounts of gastric acid into the lumen. The sequence involves three steps. First, gastrin is released from the G cells in response to stimulation by dietary peptides and amino acids. The vagus nerve also may stimulate the G cells to release gastrin; however, studies have shown that serum gastrin levels may or may not increase during sham feeding. Distention of the stomach by food also provokes the release of gastrin. Second, increased levels of gastrin in the bloodstream induce the parietal cells to release gastric acid (0.1 $N$ HCl) into the lumen of the stomach. The pH of 0.1 $N$ HCl is 1.0, so release of the acid into the lumen may bring the pH of the contents of the stomach to as low as 2.0. Third, the presence of strong acid in the lumen of the stomach inhibits the release of gastrin by the G cells. This inhibition represents the completion of the feedback loop. Gastrin secretion begins to be inhibited when the pH in the lumen falls to 3.0–3.5. Secretion is inhibited completely when the pH falls to 1.5.

## Chronic Presence and Chronic Absence of Gastrin

In addition to its hormonal effect in provoking acid secretion, gastrin is a growth hormone. Gastrin can provoke the growth of parietal cells, as well as other cells that bear the gastrin receptor in their plasma membrane. Chronic elevation of gastrin may occur, in humans, in a disease called pernicious anemia. This disease results in the destruction of the parietal cells. How could the body's destruction of its own parietal cells be connected somehow to elevated gastrin? The consequent lack of stomach acid interrupts the normal feedback loop for gastrin secretion. With the feedback loop no longer operating, the result is continual secretion of gastrin by the G cells of the stomach, a condition which very occasionally promotes certain types of stomach cancers (Koh *et al.*, 1997).

The role of gastrin for maintaining the health of the parietal cells was revealed in patients who had the antrum of their stomachs surgically removed. The result, in some cases, is atrophy of the parietal cells. A way of reducing gastrin levels, other than cutting out tissue bearing the G cells, is by creating an animal lacking the gene coding for gastrin. Using standard genetic techniques, "knockout mice" lacking this gene were prepared. Surprisingly, these mice lived and thrived. However, in the context of this chapter, the important point is that the population of parietal cells dropped by about 35%, and the mice's stomachs were less acidic (Koh *et al.*, 1997).

### Questions to Ask of All Hormones

An excess or absence of gastrin is not one of the major health issues facing our population. However, the preceding examples begin to illustrate concepts that can be applied to all hormones: Does the hormone have more than one type of influence on target cells? Does it have secondary effects? Can we learn of new targets of the hormone by identifying cells that bear the hormone's receptor? Is complete absence of the hormone compatible with life? Which regulatory systems take over and compensate for complete absence of the hormone?

## Pepsin

### Structure

The chief cells secrete pepsinogen, a moderately sized zymogen protein with a molecular weight of 40,400. Pepsin, an enzyme with a molecular weight of 32,700, is formed in the acidic environment of the stomach when pepsinogen loses its activation peptides. These activation peptides range in length from three to six, or even more, amino acids. (The term peptide is sometimes used when referring to small polypeptides.) As is the case with most secretory proteins, pepsinogen initially contains a signal sequence or signal peptide that is removed during conversion to pepsinogen, as shown in Figure 2.15. Further conversion to pepsin requires removal of additional peptides, which explains the lower molecular weight of the active enzyme.

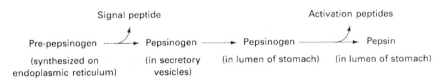

**FIGURE 2.15** Maturation and activation of pepsin.

## Action of Pepsin

As is true of all proteases, pepsin catalyzes the hydrolysis of peptide bonds. Pepsin tends to recognize a specific family of peptide bonds, namely those occurring between lipophilic amino acids. These amino acids, which frequently occupy the interior or core of proteins, are exposed under the denaturing conditions of the stomach. The products of the pepsin-catalyzed reactions generally are partially digested proteins and polypeptides rather than free amino acids.

The activation of pepsinogen is thought to occur by two different mechanisms: intramolecular, in which the zymogen acts on itself to generate pepsin, and intermolecular, in which one molecule of pepsin acts on a molecule of pepsinogen to convert it to pepsin.

The experiment depicted in Figure 2.16 used purified preparations of enzymes to illustrate the course of generation of pepsin over a period of 1 hour. At the time indicated by "0 seconds," the solution was neutral (pH 7.0). Then acid was added to yield a mildly acidic solution (pH 4.0). In the experiment illustrated by the upper curve, acid was added at time 0 to a solution of pepsinogen. Following a delay of about 800 seconds, increasing amounts of pepsin were generated, illustrating an intramolecular reaction facilitated by the addition of acid. In the experiment depicted by the lower curve, acid was added at time 0 to a solution of pepsinogen that contained a small amount of pepsin. The presence of pepsin generated an immediate production of more pepsin, without the 800-second delay. The lower curve illustrates the effect of the intermolecular mechanism of pepsinogen activation.

The gastric phase of protein digestion is not absolutely required for health. Patients lacking gastric function because of a gastrectomy (surgical removal of the

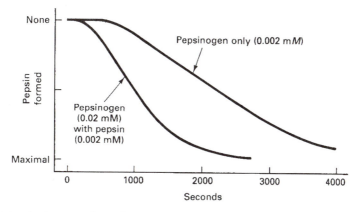

**FIGURE 2.16** Activation of purified pepsinogen after acidification. In the upper curve, acid was added at time zero to a solution of pepsinogen. Following a delay of about 800 sec, increasing amounts of pepsin were generated, illustrating an intramolecular reaction provoked by the addition of acid. In the experiment depicted by the lower curve, acid was added at time zero to a solution of pepsinogen containing a small amount of pepsin. The presence of pepsin provoked an immediate production of more pepsin, without the 800-sec delay. The lower curve illustrates the effect of the intermolecular means of pepsinogen activation. (Redrawn with permission from McPhie, 1972.)

stomach) or because of pernicious anemia often still can digest and utilize dietary proteins adequately.

## Cholecystokinin

### *Structure*

CCK is a gut hormone synthesized by the I cells of the small intestine. This hormone occurs in a variety of sizes: versions of 58, 39, 33, and 8 amino acids are found in the bloodstream. Whether the slight differences in effects of larger or smaller versions of CCK are physiologically important has not been clearly established. Because of its chemical simplicity, the 8-amino acid version has been used in many research studies.

### *Effect of Cholecystokinin on the Gall Bladder*

CCK stimulates the gall bladder to contract, thereby discharging the bile salts stored inside through the bile duct to enter the duodenum. Bile salts are required for digestion and efficient absorption of dietary fats. Bile salts also are required for the efficient absorption of the fat-soluble vitamins.

The effect of CCK on the gall bladder was demonstrated by the study on dogs outlined in Figures 2.17, 2.18, and 2.19. The effects of food on plasma CCK levels and on the volume of the gall bladder were measured. A decrease in gall bladder volume means that the contents of the organ are being released into the duodenum. During the control period, the dog was fed saline (Figure 2.17). During one

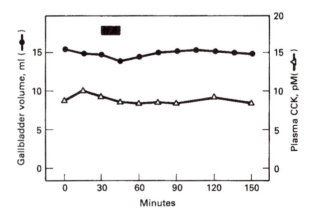

**FIGURE 2.17** Gall bladder volume and plasma CCK after saline feeding. The control animal, a dog fed saline, was fed by a tube leading to the duodenum. The volume of the gall bladder was measured by surgically implanting another tube into that organ. This tube was used to suck out the gall bladder contents to measure its volume and to replace the contents. The results show that neither the level of plasma CCK (Δ) nor the gall bladder volume (●) changed over the course of the study. The black bar indicates the feeding interval. (Redrawn with permission from Shiratori *et al.*, 1986.)

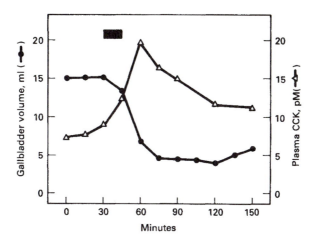

FIGURE 2.18 Gall bladder volume and plasma CCK after corn oil feeding. The gall bladder volume (●) decreased promptly after corn oil feeding. Plasma CCK levels (Δ) increased from a basal level of 7 pM to about 20 pM with feeding. This result demonstrates an association with contraction of the gall bladder, but does not imply any causal effect. (Redrawn with permission from Shiratori *et al.*, 1986.)

experimental period, the dog was fed corn oil (Figure 2.18). In other experiments, the dog was fed corn oil but was also infused with a solution with or without antibodies to CCK (Figure 2.19). The antibody to CCK recognizes the hormone in

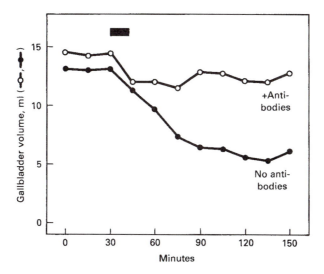

FIGURE 2.19 Gall bladder volume after corn oil feeding. The decrease in gall bladder volume is shown after the dog was fed corn oil (●) or fed corn oil and infused with the anti-CCK antibody (○). The use of antibodies prevented the contraction of the gall bladder. This result strongly implicates CCK as a hormone that provokes the release of bile salts during feeding. (Redrawn with permission from Shiratori *et al.*, 1986.)

the bloodstream and binds tightly to it, thereby preventing it from exerting any biological effect.

Figure 2.17 shows results from the control animal (the dog fed saline). Feeding was by a tube leading to the duodenum. The volume of the gall bladder was measured by surgically implanting another tube into that organ. This tube was used to suck out the gall bladder contents for measuring its volume, and to replace the contents. The results show that neither the level of plasma CCK nor the gall bladder volume changed over the course of the study. The black bar indicates the feeding interval.

Figure 2.18 shows the results from corn oil feeding. In this experiment, the gall bladder volume decreased promptly after the meal. Plasma CCK levels increased from a basal level of 7 pM to about 20 pM with feeding. This result demonstrates an association with contraction of the gall bladder but does not prove any causal effect. These picomolar levels of hormone might be compared with those of glucagon and insulin, about 50 pM, and that of plasma glucose, 100 million times greater (5 mM).

Figure 2.19 shows the decrease in gall bladder volume when the dog was fed corn oil or fed corn oil and infused with the anti-CCK antibody. The dog was fed corn oil to provoke an increase in plasma CCK, but the antibody prevented the effect of the hormone, that is, prevented contraction of the gall bladder. This result strongly implicates CCK as a hormone that provokes the release of bile salts during feeding. An antibody used in this way is a powerful tool for proving that one specific hormone has a specific effect in the body.

A species difference should be pointed out. CCK does not provoke gall bladder contraction in rats, since rats do not have gall bladders. The rat liver, however, does release bile salt into the duodenum. The rat probably lacks a gall bladder because it tends to nibble its food slowly over the course of a night. The dog, on the other hand, "wolfs" its food and needs a big supply of bile salts to contend with the sudden entry of fat into the gut.

## Effect of Cholecystokinin on the Pancreas

CCK stimulates the release of pancreatic enzymes into the duodenum, including those used for hydrolysis of fats (lipase and colipase), phospholipids (phospholipase), starches (amylase), proteins (various zymogens), and nucleic acids. These proteins are made and stored in the acinar cells of the pancreas. The acinar cells are arranged in a clump (resembling a bunch of grapes) surrounding one end of a duct. The other end of the duct leaves the pancreas and joins many similar ducts to form a large collecting duct, which in turn leads, via the **ampulla of Vater**, to the duodenum.

The effect of CCK in inducing the exocrine pancreas to release enzymes is revealed by the following experiment with rats. Infusion of CCK into the rat is shown to elicit release of a zymogen from the pancreatic duct. Figure 2.20 depicts the rate of release of trypsinogen before and after an infusion of the octapeptide version of CCK (CCK-8). The results demonstrate a prompt increase in the rate of zymogen release with CCK-8 infusion. Other data, not shown, demonstrated that amylase was also released, simultaneously with trypsinogen. The hormone infu-

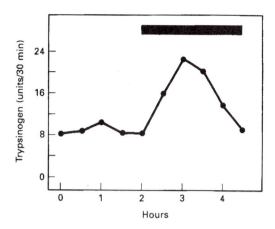

**FIGURE 2.20** Rate of trypsinogen release from the pancreas before and after infusion of CCK-8. The interval of hormone infusion is indicated by the black bar. The results demonstrate a prompt increase in the rate of zymogen release with CCK-8 infusion. (Redrawn with permission from Folsch *et al.*, 1987.)

sion was sufficient to raise the plasma CCK concentration from a basal level of about 5 pM to about 30 pM.

CCK also acts as a growth hormone. It can stimulate the growth of the pancreas. Studies with rats revealed that feeding high-protein diets provoked an increase in size of the pancreas, and an increase in output of digestive enzymes. These changes are thought to be due mainly to increases in plasma CCK (Logsdon, 1995).

EXERCISE

The data in Figure 2.20 indicate that, with prolonged infusion of CCK, trypsinogen levels decline to the basal level. This drop may cause some concern, since the text states that CCK provokes trypsinogen release. Propose one or two possible reasons for the decline in zymogen output.

## Effect of Cholecystokinin on Bicarbonate Release by the Pancreas

CCK also stimulates the release of sodium bicarbonate and large amounts of fluid from cells of the pancreatic duct. These epithelial cells line the lumen of the duct. The bicarbonate neutralizes gastric acid entering the duodenum through the pylorus. The effect of CCK on the duct cells is described later in this chapter.

## Effect of Cholecystokinin on the Stomach

CCK influences gastric motility, movements of the gastrointestinal tract that influence the passage of food. CCK stimulates contraction of the **pyloric sphincter** or pylorus. A sphincter is a ring of muscles that controls the rate of passage of material through an opening or channel in the body. The pylorus is located at the junction between the stomach and the duodenum. The effect of CCK on the pylorus is part of a feedback regulatory system involving four steps: (1) dietary protein and fat enter the duodenum; (2) this food stimulates the release of CCK into the bloodstream; (3) CCK levels are elevated in the bloodstream; and (4) CCK causes the pylorus to close, thus controlling or limiting the rate of passage of food from the stomach into the small intestine.

### Effect of Cholecystokinin on the Brain

CCK acts on the brain to control the appetite. Most studies involving CCK and eating behavior show that increases in the plasma levels of CCK are closely associated with satiety. CCK is concentrated in specific regions in the brain, where it acts locally. The term *locally* in this instance means that the hormone is produced by the brain and acts only in the brain. Further details on appetite physiology appear in the Obesity chapter.

## Secretin

### Structure

Secretin, a polypeptide hormone of 27 amino acids, occurs in only one form, not in large and small versions.

### Effect of Secretin on the Duct Cells

Secretin enters the bloodstream and travels to the pancreas, where it stimulates the duct cells to release bicarbonate and large volumes of fluid into the pancreatic duct. The duct cells are located in the walls of the pancreatic duct. This duct leads from the acinar cells of the pancreas to the ampulla of Vater. The bicarbonate neutralizes the gastric acid entering the duodenum, providing an environment suitable for the functioning of enzymes and absorptive processes that require a neutral pH.

Elucidation of the function of secretin has been particularly difficult, because clear and consistent increases in the concentration of plasma secretin generally do not occur after feeding. Several studies have shown that plasma secretin levels increase only slightly, and sometimes only sporadically in spikes, after a meal.

The following study illustrates the slight changes in plasma secretin after a meal. Human subjects consumed a liquid meal consisting of 20 g protein, 20 g fat, and 80 g carbohydrate in a volume of 400 ml. The meal was consumed at the time indicated by the arrow in Figure 2.21 (0 min). As shown, subsequent changes in the concentration of plasma secretin were only slight compared with the increases of plasma CCK. Figure 2.21 also shows the rate of appearance of trypsin in the duodenum. Recall that the release of trypsinogen from the pancreas is prompted by CCK, not by secretin.

The function of secretin has been difficult to establish for a second reason. Research has indicated that the effects of secretin on the pancreas require the

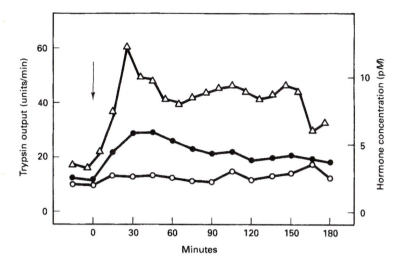

**FIGURE 2.21** Trypsin output (Δ), secretin (O), and CCK (●) plasma levels after meal. Human subjects consumed a liquid meal consisting of 20 g protein, 20 g fat, and 80 g carbohydrate in a volume of 400 ml. The meal was consumed at the time indicated by the arrow (0 min). As shown, subsequent changes in the concentration of plasma secretin were only slight compared with the increase of plasma CCK. Also shown is the rate of appearance of trypsin in the duodenum. [Redrawn with permission from Beglinger *et al.* (1985) and the *Journal of Clinical Investigation*, 1985, pp. 1471–1476. By copyright permission of the American Society for Clinical Investigation.]

presence of CCK. The two hormones appear to act synergistically, that is, the effect of secretin is potentiated by CCK, as shown by the data in Figure 2.22. This study involved human subjects infused with saline or with hormones.

Secretin is thought to influence the motility of the gastrointestinal tract. Experiments with humans have demonstrated that infusions can inhibit emptying of the stomach. This effect may represent a feedback loop in which gastric acid gives rise to the release of secretin and secretin prevents stomach emptying. Prevention of stomach emptying would decrease the rate of entry of gastric acid into the small intestine. A lower rate of entry of gastric acid into the duodenum would result in less stimulation of the S cells.

## Intrinsic Factor

The function of intrinsic factor is to allow the absorption of vitamin $B_{12}$. The vitamin $B_{12}$ liberated from dietary proteins during cooking, or in the harsh conditions of the stomach, binds tightly to the intrinsic factor that is secreted by the stomach. Although intrinsic factor is a protein, its complex with vitamin $B_{12}$ remains intact and is not broken down by the potentially denaturing effects of gastric acid or by the presence of pepsin. The complex enters the small intestine and travels through the duodenum and jejunum to the ileum, as illustrated in Figure 2.23. In the ileum, the vitamin is released from the intrinsic factor. Here, vitamin $B_{12}$ is absorbed by the gut cell and enters the bloodstream, where it binds

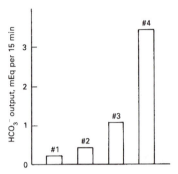

**FIGURE 2.22** Rate of bicarbonate production with different hormonal stimulations. The numbers designate control subjects infused with saline (1) and experimental subjects infused with CCK only (2), secretin only (3), and both CCK and secretin (4). The results reveal that bicarbonate output was maximal with infusion of both hormones (4). The rate of bicarbonate release with both hormones (4) was greater than the sum of the rates released with each of the two hormones alone (2,3). This effect is called synergism, the cooperative effect of two or more stimuli. (Redrawn with permission from You *et al.*, 1983.)

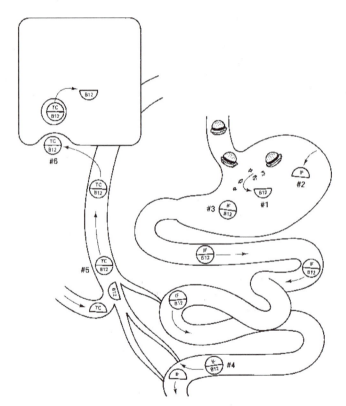

FIGURE 2.23 The absorption of vitamin $B_{12}$, an event dependent on intrinsic factor. The absorption of vitamin $B_{12}$ from the gut and its delivery to various cells of the body involves the following steps. (1) Vitamin $B_{12}$ is liberated from meat. (2) Intrinsic factor is secreted from parietal cells in the stomach. (3) A tight complex between intrinsic factor and the vitamin is formed. (4) The complex travels to the ileum, where the vitamin dissociates from the intrinsic factor at the membrane of the enterocyte. (5) Transcobalamin, a protein circulating in the blood, forms a complex with the vitamin. (6) The complex travels to various organs, diffuses from the capillary to the plasma membrane of a cell, and is bound by receptor proteins. The complex is then taken into the cell by endocytosis.

to another protein, **transcobalamin II**. The complex of vitamin $B_{12}$ and transcobalamin circulates in the bloodstream to be absorbed by various organs of the body.

Absorption of this complex involves binding to a specific receptor protein located in the plasma membrane of a cell. A large complex is formed, and this complex is composed of vitamin $B_{12}$, transcobalamin, and the receptor. The large complex is engulfed (enclosed in a formed pocket) by an invagination of the plasma membrane of the cell, resulting in generation of a vesicle within the cell. This process, which results in transport of a molecule into a cell, is called **endocytosis**. The vesicle containing the vitamin $B_{12}$ complex fuses with a lysosome in the cell. This fusion results in delivery of the contents of the vesicle to the interior of the lysosome. Lysosomes, small organelles that contain a number of digestive enzymes in an acidic environment, resemble tiny stomachs. The enzymes liberate the vitamin from the complex, permitting movement of the vitamin from the interior of the lysosome to the cytoplasm. Vitamin $B_{12}$ appears to be unique among most of the nutrients of the diet, because it is taken up by the process of endocytosis.

Defects in the parietal cells can result in a lack of production of intrinsic factor and thus in failure of vitamin $B_{12}$ to be absorbed by the gut. Such defects occur in the disease pernicious anemia, in which the parietal cell becomes a target of antibodies produced by the body. These antibodies may recognize and bind to various proteins of the parietal cell, including intrinsic factor, resulting in destruction of the cell. The major purpose of the immune system is to produce antibodies that recognize and bind to infecting agents, such as bacteria, protozoans, and viruses. The binding of the antibody to the infecting agent generally is followed by phagocytosis by a white blood cell and digestion of the complex. However, in an autoimmune disease such as pernicious anemia antibodies bind to normal components of the body. Pernicious anemia also results in a failure to produce gastric acid.

A species difference should be noted. Intrinsic factor is produced by the parietal cells in humans, rabbits, cats, monkeys, and guinea pigs, but by the chief cells and parietal cells in rats, and by mucous cells and parietal cells in hogs.

## Gastric Acid

### Production by the Parietal Cell

Figure 2.24 shows a parietal cell in the resting state and in the secreting state. Note the somewhat triangular shape of the cell. In the nonsecreting state, the cell contains many vesicles, each of which contains **proton pumps**. These proton pumps, located in the membranes of the vesicles, are the "molecular engines" used to deliver acid into the lumen of the stomach. However, they are not active while they reside in vesicles within the cell.

When the parietal cells are stimulated by gastrin, the round and oval **tubulovesicles** fuse with that part of the plasma membrane of the cell that faces the stomach lumen (the apical side). The vesicles do not fuse with the opposite side of the cell (the basolateral side). The direction of migration and fusion of the vesicles is controlled by structural proteins in the cell that make up the cytoskeleton. Fusion of the vesicles inserts the proton pumps into the plasma membrane. In the secreting state (Figure 2.24, *right*), the cell contains few vesicles. Most have been inserted into the plasma membrane, increasing its surface area by about tenfold and activating the proton pumps.

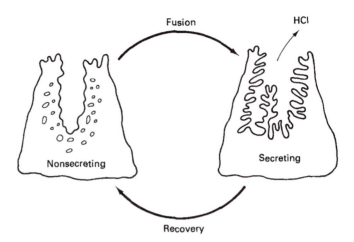

**FIGURE 2.24** The parietal cell in its nonsecreting and secreting states. Note the somewhat triangular shape of the cell. In the nonsecreting state, the cell contains many vesicles, each of which contains proton pumps. These proton pumps, located in the membranes of the vesicles, are the "molecular engines" used to deliver acid into the lumen of the stomach.

After the stomach has reached a sufficiently low pH to denature food proteins and to allow the full activity of pepsin, the proton pumps are retrieved from the plasma membrane. This retrieval occurs in the same manner as the process of endocytosis, except that membrane-bound proteins rather than the soluble contents in the interior of the vesicles are being retrieved. In short, parts of the plasma membrane bud off, reforming the tubulovesicles.

### Ion Exchange and Transport Systems of the Tubulovesicles and Plasma Membrane

Figure 2.25 illustrates some of the changes in the nature of the plasma membrane that result from fusion with tubulovesicles. These vesicles contain a protein that mediates exchange of protons with potassium ions. The overall goal, following the fusion event, is secretion of protons into the lumen of the stomach. Without fusion, exchange within the isolated vesicle can accomplish nothing of value. The plasma membrane contains a potassium chloride transport system that permits potassium cations and chloride anions to leave the cell and enter the lumen of the stomach. Before the vesicles fuse with the plasma membrane, the KCl transport system has no noticeable effect. However, fusion with the plasma membrane inserts the proton/potassium exchanging mechanism into the plasma membrane and completes the assembly of the gastric acid pump.

The four steps that occur during gastric acid production are outlined in Figure 2.26. The goal is the secretion of HCl, a strong acid. Step 1 shows the membrane-bound protein that facilitates the release of protons from the cell. Transport of protons from the cell is potentially difficult, because their concentration *in the cell* is very low (about $10^{-7}$ M). The secreted fluid has a proton concentration that is about one million times higher, $10^{-1}$ M (i.e., pH 1.0).

Step 2 shows an exchange of protons for potassium ions. This exchange maintains the electroneutrality of the cell. For every positively charged ion leaving the

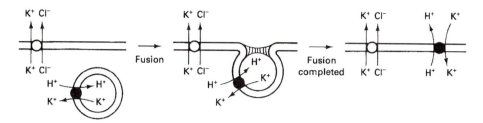

**FIGURE 2.25** Mechanism of gastric acid synthesis. (*Left*) A proton pump exchanging protons for potassium ions in the membrane of a vesicle. Also shown is a mechanism for transporting potassium chloride. (*Center*) The vesicle begins to fuse with the plasma membrane. (*Right*) The completed mechanism for producing gastric acid. The net effect is the secretion of HCl while potassium ions are recycled in and out of the cell.

cell, a positively charged ion enters the cell. Living cells cannot achieve a net secretion of only positively charged ions to any great extent; however, they can easily achieve the exchange of one cation for another. The exchange of $H^+$ for $K^+$ is catalyzed by a membrane-bound protein called **H,K-ATPase** that catalyzes the transfer of two potassium ions and two protons for each molecule of ATP hydrolyzed to ADP and inorganic phosphate.

The mechanism of Step 2 indicates that extracellular potassium ions are required for acid secretion. Food is a reliable source of potassium ions; however, a basal level of acid secretion may occur in the absence of food and the cephalic phase can result in acid secretion in the absence of food. Therefore, the cell must supply its own potassium ions, as shown in Step 3.

To maintain electroneutrality during potassium secretion, the ion transport mechanism involves co-transport of $K^+$ and $Cl^-$, as shown in Step 4. Details of the

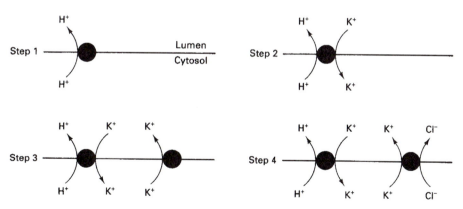

**FIGURE 2.26** Mechanism of acid secretion explained in four steps of increasing complexity. These events take place at the plasma membrane of an acid-producing cell. Separation of these events into four steps is intended to facilitate understanding of acid secretion and is not related to any physiological sequence of events. (1) The H,K-ATPase is shown secreting $H^+$. (2) The H,K-ATPase exchanges $H^+$ for $K^+$. (3) The KCl transporter secretes $K^+$. (4) The KCl transporter secretes $K^+$ and $Cl^-$, thus allowing the secretion of HCl and the recycling of potassium ions.

mechanism depicted for production of gastric acid remain to be proven. Acid can be produced by a number of other cells, including those used for the acidification of urine. Acid is produced within the lysosome. The mechanisms for production of acid in other cells may be different from the one shown here for the production of gastric acid.

## Source of Protons and Chloride Ions

The mechanism presented for gastric acid production did not reveal the source of protons and chloride ions. Continued production of gastric acid by this mechanism would result in depletion of the chloride ions in the cell and alkalinization of the cell because of the loss of protons. Figure 2.27 shows that the source of chloride ions is the bloodstream and the source of protons is carbonic acid, $H_2CO_3$. As detailed in a later section, carbonic acid is produced by the action of **carbonic anhydrase**, the enzyme that catalyzes the reversible condensation of a molecule of water with a molecule of carbon dioxide. Carbonic acid ionizes to produce a

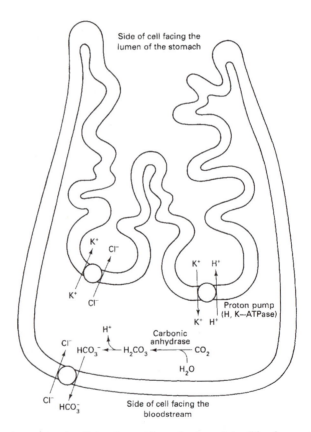

**FIGURE 2.27** The parietal cell in the acid-producing state. The large circles represent membrane-bound proteins. The apical membrane contains the H,K-ATPase and the KCl transporter shown in Figure 2.26. Also shown is cytoplasmic carbonic anhydrase and the chloride–bicarbonate exchanger in the basal membrane of the cell. The terms "apical" and "basal" refer to opposite sides of the plasma membrane of epithelial cells.

proton and a bicarbonate anion, $HCO_3^-$. Figure 2.27 shows that the exchange of chloride anions for bicarbonate anions serves two purposes: (1) it brings chloride ions into the cell from the bloodstream for use in the secretion of HCl, and (2) it eliminates bicarbonate ions, a byproduct of acid production, from the cell. The overall process of acid production is driven in part by the proton pump. The activity of the proton pump, also called H,K-ATPase, is driven by hydrolysis of ATP to ADP and inorganic phosphate.

### Proton Pump and Treatment of Gastric Ulcers

Gastric ulcer formation may be precipitated by overproduction of gastric acid. It was the practice in the past to treat gastric ulcers by altering the diet; however, the pharmacological approach is far more effective, i.e., use of omeprazole, cimetidine, and other drugs (Garner *et al.*, 1996). **Omeprazole** is a heterocyclic compound that binds tightly to H,K-ATPase, inhibiting its activity and preventing the cell from secreting gastric acid. The use of this drug results in healing of the ulcers and disappearance of pain in human patients. The experiment depicted in Figure 2.28 brings into sharper focus the role of H,K-ATPase in the secretion of gastric acid.

Any account of gastric ulcers should mention the fact that colonization of the stomach by the bacterium *Helicobacter pylori* is a contributing factor to many ulcers. This organism lives below the layer of mucus that covers the gastric mucosa. Hence, gastric ulcers are often treated and healed by simultaneous therapy with omeprazole and an antibiotic (Harris and Misiewicz, 1996).

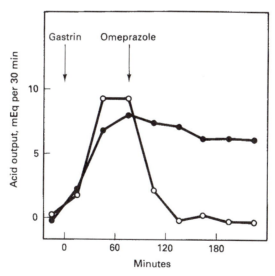

**FIGURE 2.28** Gastric acid production after gastrin injection only (●) and gastrin, with subsequent omeprazole injection(○). Gastrin was injected into two groups of dogs at the time indicated by 0 min; the rate of output of gastric acid was measured at subsequent intervals. Gastric acid production appeared to be maximal within 1 hr of the injection. The first group served as the control and did not receive omeprazole. The second group received omeprazole at 90 min. Gastric acid production continued on a plateau in the control group, but promptly ceased in the drug-treated animals. (Redrawn with permission from De Graf and Woussen-Colle, 1986.)

## DIGESTION AND ABSORPTION OF PROTEINS

### General

The initial steps in the assimilation of dietary proteins include homogenization of food by chewing, denaturation of proteins under the acidic conditions of the stomach, and partial digestion of proteins by pepsin. The appearance of proteins and peptides in the duodenum stimulates release of CCK into the bloodstream, which in turn provokes the pancreas to release a variety of zymogens into the small intestine. The secretions of the stomach and pancreas are also under the control of the vagus nerve. The proteases are activated by a cascade initiated by enterokinase. These proteases catalyze further hydrolysis of the dietary proteins, resulting in a mixture consisting of about 50% free amino acids and 50% oligopeptides from two to eight amino acids in length.

Both pepsin and gastric acid are required for the digestion of dietary proteins in the stomach. Gastric acid is required for activation of pepsin. It also denatures proteins, facilitating their hydrolysis by pepsin. (Cooking also denatures dietary proteins and aids their subsequent enzymatic hydrolysis in the body.) *Protein digestion* and *proteolysis* are terms that mean the hydrolysis (cleaving) of the peptide bonds of a protein. Partial hydrolysis yields various polypeptides; complete hydrolysis yields the component amino acids. Denaturation and hydrolysis also release vitamins and minerals that may be bound tightly to dietary proteins.

Stomach enzymes produce partial digestion of dietary proteins, significant digestion of dietary fats, and little digestion of carbohydrates. Pepsin, the protease of the stomach, is not specific in its site of action. The acidic conditions in the stomach facilitate the hydrolytic activity of pepsin. **Pepsin** catalyzes the partial digestion of proteins by cleaving them at a variety of amino acid residues. **Trypsin**, on the other hand, is quite specific. This enzyme cleaves the zymogens secreted by the pancreas at residues of arginine and lysine, and cleaves dietary proteins at arginine and lysine as well. **Chymotrypsin** cleaves proteins at aromatic acid residues.

**Elastase** is not very specific. It cleaves peptide bonds between a variety of amino acids, although it may prefer to act on amino acids with a small uncharged R group, such as glycine, alanine, valine, and phenylalanine. **Elastin** is a protein of the extracellular matrix that gives tissues its elastic quality. This extracellular protein is distinguished in that it contains cross-links joining different polypeptide chains, and that the protein is extremely lipophilic. The cross-linking and lipophilic amino acids causes elastin to resist the action of most proteases (Mecham *et al.*, 1997).

EXERCISE

An infant was diagnosed as having a deficiency in enterokinase; the gut was not making the enzyme. Please guess the outward symptoms of this disease and devise a treatment. The patient and treatment are described in Haworth *et al.* (1975).

### Peptide and Amino Acid Absorption

A small fraction of the polypeptides contacting the brush border of the small intestine is hydrolyzed by membrane-bound enzymes attached to the outside of the enterocyte. Although a variety of peptidases are bound to the brush border, the most abundant is **aminopeptidase N**, which catalyzes the hydrolysis of amino acids from the N terminus of short polypeptides. This hydrolysis is followed by absorption of the amino acids. A larger fraction of the polypeptides is transported intact into the enterocyte, where complete hydrolysis to free amino acids occurs. Dietary proteins enter the circulation as free amino acids, but also as short peptides.

Different transport systems are used for the uptake of free amino acids and peptides. In the case of the four amino acid transport systems, transport of the amino acid is coupled with transport of an atom of sodium, that is, entry of the amino acid into a cell requires the co-transport of sodium. The four sodium-dependent amino acid transport systems are those specific for:

1.  Acidic amino acids, glutamate, and aspartate.

2.  Basic amino acids, such as lysine and histidine.

3.  Neutral amino acids, such as alanine, glycine, and serine.

4.  Glycine, proline, and hydroxyproline.

The transport of glucose by the small intestine is also sodium dependent, but glucose transport systems also exist that are not sodium dependent. The concept that the passage of any nutrient through a cell membrane depends on sodium is discussed in more detail in Chapter 10.

The digestion of dietary protein results in production of small peptides, including, di-, tri-, and tetrapeptides, as well as of free amino acids, as the end-products of hydrolysis. The dipeptides and tripeptides are absorbed by a special transporter called PEPT1 (Leibach and Ganapathy, 1996). This transporter, and a related transporter called PEPT2, are unique among most of the nutrient transporters in that peptide transport is driven by and coupled with the influx of $H^+$ ions (protons). In contrast, co-transport with $Na^+$ ions drives many other nutrient transporters. Transport of peptides by PEPT1 and PEPT2 is forced to occur when the pH within the extracellular fluid is lower than the pH within the enterocyte. PEPT1 is a membrane-bound protein that is 708 amino acids long. PEPT2 is a related protein that occurs in the renal tubule and is used for reabsorbing the di- and tripeptides in the developing urine, and for reclaiming them for the body.

## Classical and Recent Experiments on Peptide Feeding

The following experiment, involving the lactic acid bacterium *Lactobacillus casei*, illustrates the concept that peptides, such as dipeptides, can have greater nutritional value than free amino acids. This work also illustrates the point that many concepts of nutritional biochemistry were studied in detail first using microorganisms. A more recent experiment involving amino acid and peptide transport by rabbit intestines is also described. The results of both studies are consistent with the facts that short polypeptides are taken up preferentially by the enterocyte during digestion of dietary proteins and that complete hydrolysis is not necessary.

Figure 2.29 gives results from an experiment involving bacteria growing on a defined culture medium containing sugar, salts, vitamins, and a variety of amino acids. The bacterium *L. casei* requires serine. Cell growth was initiated by adding a small drop of bacteria to test tubes containing the indicated growth medium. The tested range of serine supplement concentrations extended from 10 to 1000 µM. The test tubes were incubated for 18 hours to permit growth of the bacteria. After the incubation period, growth was assessed by measuring the weight of bacterial

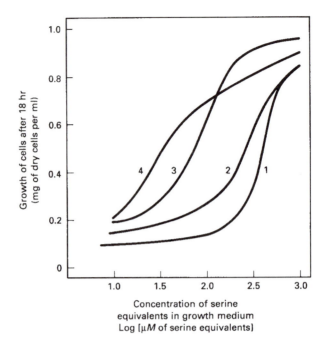

**FIGURE 2.29** Weight of bacterial cells produced with growth in various concentrations of serine supplement. Bacteria were grown under a variety of conditions, with serine supplied as free serine (1) or as the dipeptides Gly–Ser (2), Ser–Ser (3), and Ser–Ala (4). (Redrawn with permission from Kihara and Snell, 1960.)

cells in each tube. The results demonstrate that maximal growth could be stimulated by concentrations of dipeptide far lower than those of the free amino acid, that is, Ser–Ser and Ser–Ala were considered better foods by the bacteria than free serine.

The classical work involving *L. casei* addressed the overall nutritional effects of dipeptides compared with free amino acids, but did not directly address the source of the differences. Figure 2.30 shows the uptake of glycine and of the glycine peptide Gly–Pro by the intestines of rabbits. Glycine or Gly–Pro was added to the lumen of the intestines, and the rate of uptake was measured. The movement of the amino acid or dipeptide from the lumen into the enterocyte was measured using radioactively labeled compounds. In brief, the presence of tritium-labeled glycine, [³H]Gly, in the tissue of the intestines was measured by quickly rinsing the intestines, homogenizing the tissue in water, and measuring the amount of radioactivity in the water extract using a liquid scintillation counter. The experiment was repeated using fetal, neonatal, and mature rabbits.

The results shown in Figure 2.30 illustrate that the uptake of the dipeptide was considerably greater than that of free glycine, especially in the immature animal. The high transport activity in infant tissue seems to be especially valuable, in light of the fact that the digestive apparatus of the stomach is poorly developed in the infant. The stomachs of human infants have low levels of pepsin and a pH that is relatively neutral. The unusually high levels of peptide transport found in the young rabbits may compensate, in part, for the moderate proteolytic activity in the stomach of a newborn.

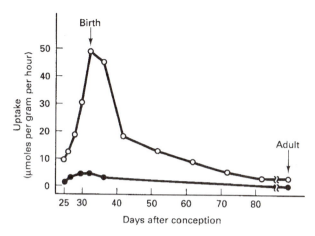

**FIGURE 2.30** Rate of transport by intestines taken from animals of various ages. The uptake of glycine (●) and of the glycine peptide Gly–Pro (○) by the intestines of rabbits is shown. Glycine or Gly–Pro was added to the lumen of the intestines and the rate of uptake was measured. The movement of the amino acid or dipeptide from the lumen into the enterocyte was measured using radioactively labeled compounds. The experiment was repeated using fetal, neonatal, and mature rabbits. [Redrawn with permission from Guandalini and Rubino (1982), © by Williams & Wilkins.]

## DIGESTION AND ABSORPTION OF LIPIDS

### General

The lipids of the diet include triglycerides, phospholipids, cholesterol, cholesteryl esters, and fat-soluble vitamins. Lipids are soluble in water, but only at extremely low concentrations many thousands of times lower than those attainable by water-soluble nutrients. Lipids are called fat-soluble because they tend to adhere to, migrate into, or be easily dispersed into fats and oils. (The term oil-soluble also may be used to describe lipids.) A lipid that is solid at room temperature generally is called a fat; one that is liquid at room temperature is called an oil.

### Chemical Structures

Triglycerides consist of a backbone of glycerol esterified — that is, covalently connected via an ester bond — with three molecules of fatty acid. The fatty acids may be alkane-like with no double bonds (e.g., palmitic or stearic acid) or alkene-like with one or more double bonds (e.g., palmitoleic, oleic, or linoleic acid).

Phospholipids may consist of a backbone of glycerol esterified with two fatty acids and one molecule of choline phosphate in the 3 position. In other phospholipids, the 3 position may be occupied by a molecule of ethanolamine phosphate, serine phosphate, or inositol phosphate. A phospholipid containing choline phosphate is called **phosphatidylcholine**; one containing ethanolamine phosphate is called **phosphatidylethanolamine**. Generally, the fatty acids esterified as triglycerides contain up to one double bond, whereas those esterified as phos-

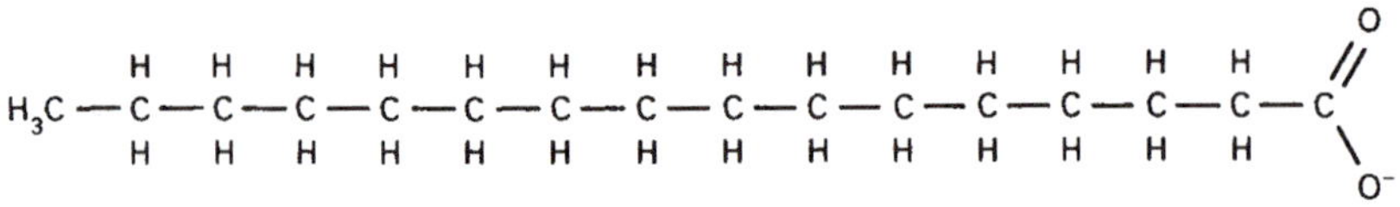

FIGURE 2.31 Structure of the 16-carbon fatty acid palmitic acid.

pholipids contain up to four double bonds. Fatty acids containing a large number of double bonds impart a more liquid character to the resultant fat or phospholipid membrane.

Figure 2.31 shows the structure of the 16-carbon fatty acid palmitic acid. In the triglyceride and phospholipid structures shown in Figure 2.32, $R_1$, $R_2$, and $R_3$ represent the alkane "tails" of fatty acids. The fatty acid compositions of the phospholipids and triglycerides of various foods are listed in Table 2.2.

EXERCISE

Fatty acids with odd numbers of carbon atoms occur as minor components of the lipids in some biological materials. Which of the fatty acids listed in Table 2.2 contain an odd number of carbons? (This is a trick question.)

EXERCISE

To devise an artificial milk for human infants from a vegetable oil, should coconut oil or corn oil be used? Assume that the goal is to formulate milk that most closely resembles human milk.

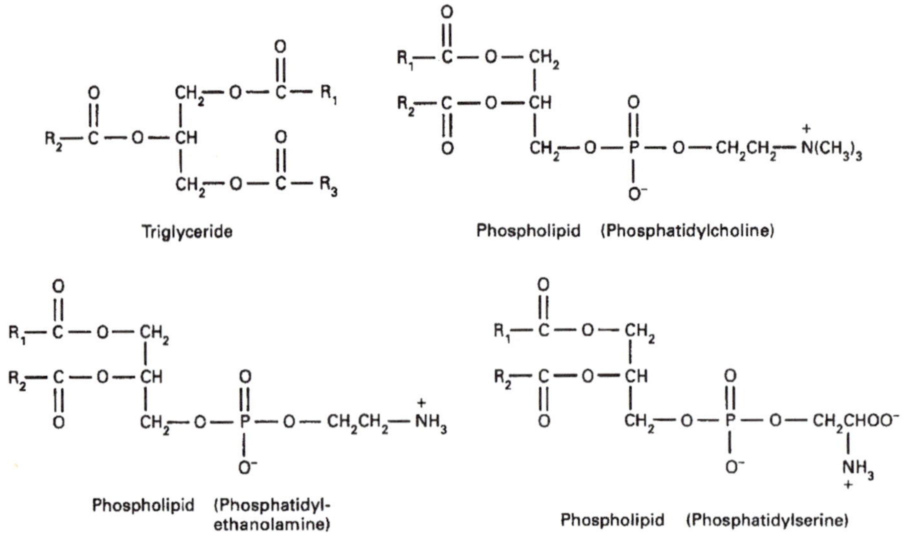

FIGURE 2.32 Structures of triglycerides and phospholipids.

**TABLE 2.2**  Fatty Acids that Occur in Food Triglycerides

| Fatty acid | No. of carbons | Percentage of total fatty acids | | | | |
|---|---|---|---|---|---|---|
| | | Coconut oil | Corn oil | Soy oil | Human milk | Beef fat |
| Caproic | 6 | 0.5 | – | – | – | – |
| Caprylic | 8 | 8.0 | – | – | – | – |
| Capric | 10 | 6.4 | – | – | – | – |
| Lauric | 12 | 48.5 | – | – | 4.6 | – |
| Myristic | 14 | 17.6 | – | 0.1 | 6.4 | 3.0 |
| Palmitic | 16 | 8.4 | 12.2 | 11.1 | 23.4 | 26.0 |
| Palmitoleic | 16 | – | 0.1 | 0.1 | 3.7 | 3.5 |
| Stearic | 18 | 2.5 | 2.2 | 4.0 | 8.6 | 19.5 |
| Oleic | 18 | 6.5 | 27.5 | 23.4 | 33.3 | 40.0 |
| Linoleic | 18 | 1.5 | 57.0 | 53.2 | 12.0 | 4.5 |
| Linolenic | 18 | – | 0.9 | 7.8 | 1.8 | – |

*Source*: Babayan (1987).

## Food Sources

The triglycerides of foods are located mainly in the energy storage sites of the animal or plant: specialized cells called *adipocytes* in meat (muscle), microscopic particles called *lipoproteins* in milk, and *seeds* in plants.

The phospholipids of foods are located mainly in the membranes that surround the cell (the plasma membrane), mitochondria, nucleus, lysosomes, and endoplasmic reticulum. Phospholipids also occur in the thin coatings around lipoprotein particles.

Free fatty acids (FFAs) are fatty acids that are not esterified to glycerol or to some other molecule. FFAs are a concern to the food scientist. These fatty acids are not a major source of energy in the diet but are considered undesirable contaminants in vegetable oils. The term *virgin olive oil* means that the oil is of high quality and does not contain free fatty acids. The free fatty acids present in the circulatory system are bound to albumin, but not covalently.

## Hydrolysis and Absorption

The fats and oils of the diet are digested prior to absorption by the gut. Complete hydrolysis of all three fatty acid groups from the glycerol backbone, as shown in Figure 2.33, occurs in the gut, but only to a very limited extent. Instead, the major route of breakdown involves hydrolysis of only the fatty acids esterified to the 1-carbon and 3-carbon of the triglyceride, as shown in Figure 2.34. 2-Monoglycerides and free fatty acids then are absorbed from the lumen of the proximal small intestine. Diglycerides may accumulate as intermediates during triglyceride hydrolysis.

$$\text{Triglyceride} \xrightarrow[\;3\;H_2O\;]{} \text{Glycerol} + 3\text{ Free fatty acids}$$

**FIGURE 2.33**  Hydrolysis of a triglyceride.

Digestion (hydrolysis) of fats and oils starts in the stomach, catalyzed by the **gastric lipase**. This lipase is secreted from glands in the fundus of the stomach. Gastric lipase resists the proteolytic activity of pepsin as well as the denaturing effects of gastric acid. This enzyme is thought to be responsible for hydrolysis of 10–30% of dietary triglycerides (TGs). The lipase of the pancreatic juice (**pancreatic lipase**) is probably not sufficient to produce maximal digestion and absorption of dietary fats. Gastric lipase may be especially important for infant nutrition, because levels of pancreatic lipase are low at birth, whereas gastric lipase occurs at the same concentrations found in adults. Gastric lipase is active in the stomach of the infant and continues to be active in the duodenum. According to Bernback *et al.* (1990), the TGs found in fat droplets of milk tend to resist digestion by both pancreatic lipase and the lipase found in mother's milk. Partial hydrolysis of the fat droplets by gastric lipase facilitates subsequent complete hydrolysis by the other lipase. Gastric lipase seems to have the unique ability of initiating breakdown of these droplets, which consist of a core of TGs covered by a thin layer of phospholipid and protein. TGs account for about 98% of the fat droplet or globule. The milk of various species contains lipase, although the importance of this enzyme compared with that of pancreatic lipase is not clear. A species difference has been noted. The lipase present in the stomach of humans, pigs, dogs, and rabbits is produced by glands in the stomach, whereas that present in the stomach of rats and mice is released by glands in the tongue (von Ebner glands) (Abrams *et al.*, 1988; Borel *et al.*, 1991).

Pancreatic lipase is secreted in the pancreatic juice, in conjunction with other pancreatic zymogens and enzymes. The enzyme catalyzes the hydrolysis of TGs to 2-monoglycerides (2-acylglycerols). Activity of the enzyme is stimulated by a small protein called **colipase**, secreted in the pancreatic juice at a concentration similar to that of the enzyme (about 10 $\mu M$ in humans). Bile salt detergents bind to dietary fats, forming a layer of bile salts around the fat, but do not effectively solubilize the TGs. This bile salt layer limits the accessibility of the fat to pancreatic lipase, but colipase overcomes this problem by anchoring lipase to the fat globule in the vicinity of the triglyceride. The products of hydrolysis, 2-acylglycerol and free fatty acid, are maintained in solution by the bile salts. Without the bile salts, the products of fat digestion would not be able to migrate or diffuse to the absorptive surface of the gut mucosa.

$$\text{Triglyceride} \xrightarrow[\;2\;H_2O\;]{} \text{2—Monoglyceride} + 2\text{ Free fatty acids}$$

**FIGURE 2.34**  Partial hydrolysis of a triglyceride.

*Events that Occur after the Absorption of Polar Lipids*

Once inside the enterocyte, polar lipids are converted back to triglycerides, as shown in Figure 2.35. The free fatty acids are not esterified "as is" to the 2-monoglycerides. Instead, they form esters with molecules of coenzyme A. These esters

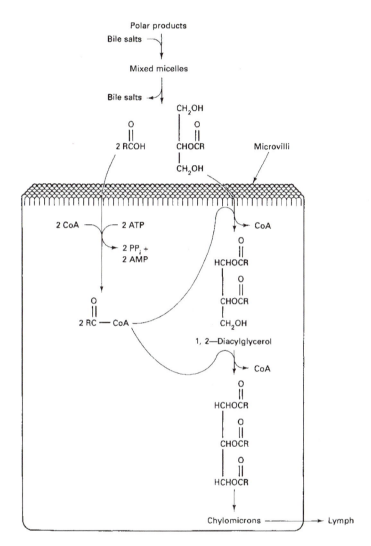

**FIGURE 2.35** Entry and metabolism of polar products in the enterocyte. The polar products of triglyceride digestion include fatty acids and 2-monoacylglycerol. These products are released from mixed micelles at the surface of the enterocyte, pass through the plasma membrane, and enter the cell. The fatty acids are activated by attachment of CoA, with reattachment to 2-monoacylglycerol, yielding a triglyceride. These triglycerides, in addition to other lipids, are packaged into particles called chylomicrons that enter the lymph. Lymph, an extracellular fluid that is ordinarily clear and straw colored, becomes opaque and white when it contains chylomicrons. The triglycerides in the chylomicrons largely retain the fatty acid that originally occurred in the 2-position (in the food triglyceride) (Zock *et al.*, 1995).

are not typical esters involving oxygen atoms, but **thiol esters** involving sulfur atoms (see Table 1.2). A high-energy bond of ATP is broken for the formation of each molecule of thiol ester. Some of this energy is stored in the thiol ester bond and then used to drive the final reaction of the fatty acid with the glycerol backbone. The endoplasmic reticulum (ER) of the enterocyte becomes laden with fat droplets with the re-formation of triglycerides in the cell. These droplets are transferred to secretory vesicles, in preparation for transfer to the extracellular fluid.

Special proteins, called **apolipoproteins**, are required for handling and transport of lipid droplets. These proteins are synthesized on the ER and enter the lumen of the ER, where they are assembled into large macromolecular structures. The relevant proteins include apolipoprotein A (apo A) and apolipoprotein B (apo B). Apo A and apo B combine with lipid droplets to form structures called **chylomicrons**, microscopic particles with large cores of lipid coated with a thin shell of protein. The chylomicrons are transferred to secretory vesicles, which migrate through the cytoplasm to the basal membrane of the cell. Here the vesicles fuse with the membrane, resulting in the expulsion of chylomicrons from the cell. (If the vesicles fused with the apical membrane of the enterocyte, the effect would be a futile transfer of the dietary lipids back to the lumen of the small intestine.)

The secreted chylomicrons diffuse through the spaces between surrounding cells, through pores and channels, and enter a structure called a **lacteal**, shown in Figure 2.36. A lacteal is a capillary-like structure located at the center of the intestinal villus. One end is at the tip of the villus; the other end joins a network of branching tubules that eventually leads to the large **thoracic duct**. The chylomicrons are forced slowly through the lymphatic system, up the thoracic duct to the top (anterior) end, located near the left shoulder (see Figure 2.37). At this point, where the thoracic duct is connected to the left subclavian vein, the chylomicrons enter the bloodstream.

## Bile Salts

Bile salts are required for supporting the activity of pancreatic lipase as well as for maintaining the polar products of fat hydrolysis in solution. While in the lumen of the small intestine, a fraction of the bile salts is modified by the bacteria present. If the taurine or glycine moieties are removed by microbial enzymes, they are replaced in the liver after reabsorption of the bile salts in the distal ileum.

Cholate precipitates under mild acidic conditions, destroying its detergent effect. However, the protonated amino groups of glyco- and taurocholate maintain the solubility of these bile salts in acidic environments. The highly charged sulfonic acid group of taurocholate further enhances its hydrophilicity and maintains its solubility even at pH 1.0. (The term *highly charged* describes a group that maintains its positive or negative charge over a wide pH range, or that it does not resonate between charged and uncharged structures.) A biological difference has been noted in the behavior of glyco- and taurocholate. Some glycocholate appears to be absorbed by diffusion at any point in the small intestine, but taurocholate is highly dependent on the bile salt transport system in the distal ileum. The peptide link of conjugated bile salts resists hydrolysis by human proteases and peptidases, but is cleaved readily by enzymes produced by the gut microflora.

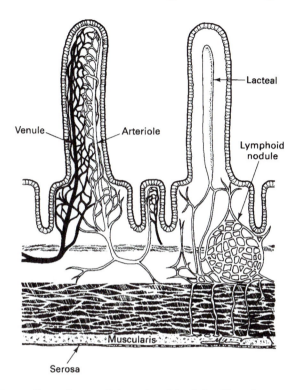

**FIGURE 2.36** Two villi in the small intestine. The left villus shows components of the bloodstream. Oxygenated blood enters via the arteriole, travels through the capillaries, and exits via the venule. The blood leaving by the venule contains water-soluble nutrients absorbed from the diet. These nutrients travel to the liver in the portal vein, and eventually are distributed throughout the body. The right villus illustrates components of the lymphatic system. The central lacteal, which ends near the tip of the villus, receives lipid-soluble nutrients, absorbed from the diet, that are packaged in chylomicrons. These chylomicrons travel slowly through the lymphatic system to a main vessel, the thoracic duct, which directs the lymph to the bloodstream. The lacteals are so named because, after a fat-containing meal, they contain a milky fluid. Lymphoid nodules occur in clusters called Peyer's patches and as isolated nodules. All nodules are part of the immune system discussed in Chapter 3. (Redrawn with permission from Weiss, 1983.)

### Enterohepatic Circulation

Bile salts, after being used to support the migration of polar lipids in mixed micelles to the enterocyte, may diffuse back to food particles for reuse. Eventually, however, they travel down the small intestine to the far end of the ileum, called the distal ileum. Here the bile salts are absorbed into the bloodstream and travel through the portal vein to the liver. Absorption in the ileum is via the **Na$^+$/bile acid co-transporter** (Oelkers *et al.*, 1997). Sodium ions occur at higher concentrations in the gut lumen than in the enterocyte, and it is this gradient that drives the bile acid through the apical membrane of the entrocyte, and into the enterocyte. The bile salts in the bloodstream are absorbed by the liver, and the liver directs them back to the gall bladder for storage and eventual release back into the

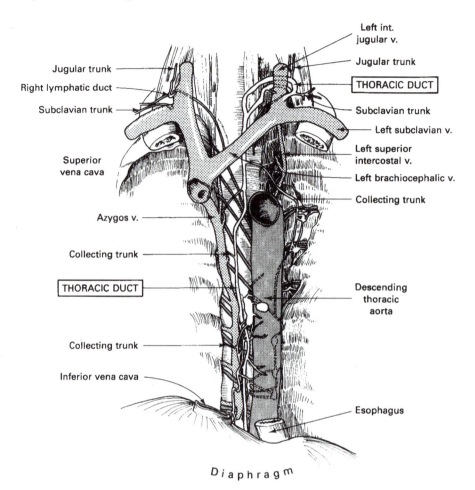

**FIGURE 2.37** Thoracic duct and other vessels of the thorax. Lymphatic capillaries are most numerous just beneath body surfaces, such as the skin and the mucus membranes of the gastrointestinal and respiratory tracts. The mucus membrane of the gastrointestinal tract is called the gut mucosa. The general function of these capillaries is to absorb interstitial fluid that has leaked from the circulatory system and to return it to the bloodstream. The function of the lymphatic capillaries that end in the lacteals of the small intestine is to transport absorbed dietary lipids. These capillaries coalesce and eventually deliver their contents to the thoracic duct. The lymph collected from other parts of the body, as indicated by the "collecting trunk," also is transferred to the thoracic duct. [Redrawn with permission, from "Grant's Atlas of Anatomy," Williams & Wilkins Co., Baltimore, 1978.]

duodenum. This recycling of bile salts is called **enterohepatic circulation**. The total quantity of bile salts in the human body is 2–4 g, but the equivalent of 20–30 g enters and leaves the small intestine each day via enterohepatic circulation, that is, each bile salt molecule enters and leaves the lumen of the gut about 10 times each day. The enterohepatic circulation of bile salts is illustrated in Figure 2.38.

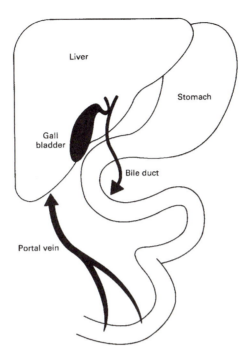

**FIGURE 2.38** Enterohepatic circulation of bile salts. Bile salts from the liver enter the duodenum at the ampulla of Vater, the same opening through which material from the exocrine pancreas is released. Most of the gall bladder is hidden within the lobes of the liver. After varying degrees of utilization and reutilization for lipid absorption, the bile salts reach the distal ileum, where they are reabsorbed and returned to the liver via the portal vein. The portal vein is very important because it also carries water-soluble nutrients absorbed from the diet.

### Enterohepatic Cycle and Cardiovascular Disease

A molecule of bile salt may circulate many times per day through the enterohepatic system. This process can be interrupted by consumption of a special powdered resin that binds to the bile salts. Neither the resin itself nor the complex of resin and bile salt is absorbed by the intestines. Consequently, the complex is excreted in the feces. This resin, **cholestyramine**, is a synthetic compound used as a drug to lower serum cholesterol levels in the treatment of cardiovascular disease.

The human body turns over about 800 mg of cholesterol per day. Most of this turnover (synthesis, degradation, or loss from the body) involves bile salts. More specifically, about 400 mg cholesterol is used to manufacture new bile salts to replace those that have been lost in the feces. About 80 mg cholesterol is lost through the skin; about 50 mg is used for synthesis of steroid hormones. Cholestyramine can stimulate the loss of much more than the equivalent of 400 mg, and can produce clinically significant decreases in serum cholesterol. Cholestyramine alone does not drastically lower serum cholesterol, because the liver senses any depletion and responds by increasing its rate of cholesterol biosynthesis. However, use of the drug in combination with other drugs that inhibit this biosynthetic pathway has proven useful with many patients. To place this drug in perspective, cardiovascular disease is treated by: (1) diet low in saturated fat; (2) statin (a drug); (3) nicotinic acid (a drug); and (4) cholestyramine, where the treatments are listed in order of importance.

### Experiments Demonstrating Lipid Absorption and Bile Salt Recirculation

The experiments discussed in this section demonstrate a role for bile salts in promoting the absorption of cholesterol, as well as the fact that bile salts are reabsorbed in the distal part of the small intestine. The experiments focus on ducts leaving or entering the small intestine. The first study involves measurement of material passing through the thoracic duct. The second study involves measurement of material passing through the bile duct.

Table 2.3 presents the results of an experiment involving **cannulation** of the thoracic ducts of rats. Cannulation is the insertion of a tube into the duct to collect the fluid flowing out. Various test meals were fed to the rats. Each rat received only one type of meal. Lymph was collected for a 24-hour period; the amount of cholesterol in the collected material was measured. Following a test meal of albumin only, the collected lymph contained 10 mg cholesterol (Expt. 1). With the albumin plus cholesterol meal, 8 mg cholesterol was collected (Expt. 2). The amount of cholesterol collected clearly was elevated when the test meal contained taurocholate, albumin, and cholesterol (Expt. 6). This experiment demonstrated that including bile salt in the test meal could enhance the absorption of dietary cholesterol.

EXERCISE

Was the cholesterol entering the thoracic duct in Experiments 1, 2, 4, and 7 absorbed by the gut with or without the aid of bile salts? (This may seem to be a trick question, but the answer should be obvious.)

**TABLE 2.3**  Amount of Cholesterol Collected from the Thoracic Duct

| Experiment | Test meal | Cholesterol collected (mg/24 hr) |
|---|---|---|
| 1 | Albumin in 3 ml salt water | 10 |
| 2 | Albumin + cholesterol (50 mg) | 8 |
| 3 | Albumin + taurocholate | 11 |
| 4 | Albumin + oleic acid | 9 |
| 5 | Albumin + taurocholate + oleic acid | 14 |
| 6 | Albumin + cholesterol + taurocholate | 17 |
| 7 | Albumin + oleic acid + cholesterol | 15 |
| 8 | Albumin + taurocholate + oleic acid + cholesterol | 27 |

*Source*: Vahouny *et al.* (1957).

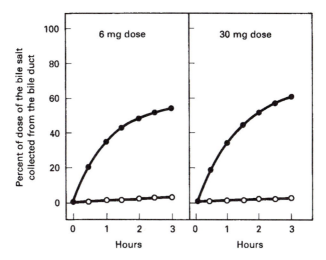

**FIGURE 2.39** Release of bile salts through the bile duct after placing bile salts in different regions of the small intestine. Pieces of thread were tied around different regions of the small intestine to isolate the jejunum from the ileum. The thread prevented the mixing of the contents of the jejunum with those of the ileum, but permitted the flow of nutrients from the gut up the portal vein to the liver. A solution of taurocholate was injected into the lumen of the jejunum (○) or ileum (●). Recovery of the bile salt in the bile duct was measured over the course of a few hours. The experiment was repeated with small and large doses of bile salt. (Redrawn with permission from Weiner and Lack, 1962.)

EXERCISE

Cholesterol is thought to be absorbed more efficiently from a mixed micelle containing bile salts and fatty acids than from a micelle of bile salts only. Give evidence for the stimulatory effect of fatty acids.

Figure 2.39 describes an experiment involving cannulation of the bile ducts of guinea pigs to permit collection of bile salts flowing from the liver. The data demonstrate two points. First, bile salt introduced into the gut can be transported to the liver and then secreted through the bile duct. Second, the bile salt is absorbed not by the proximal small intestine, but by the distal small intestine.

## Taurine

Taurine is a rather specialized topic in the nutritional biochemistry of digestion. Taurine, 2-aminoethane sulfonic acid, is synthesized from cysteine, as shown in Figure 2.40. The taurine molecule contains an amino group as well as an acidic group. Taurine has not been found in proteins, although some evidence suggests that it occurs in certain polypeptides. Taurine occurs as a component of bile salts and plays an important role in the transport and absorption of lipids.

Taurine is found in animals and thus in animal foods and milk. It is not found in plants. Most animals, including humans, can synthesize sufficient taurine for their metabolic needs. The major use of taurine is attachment to (conjugation of)

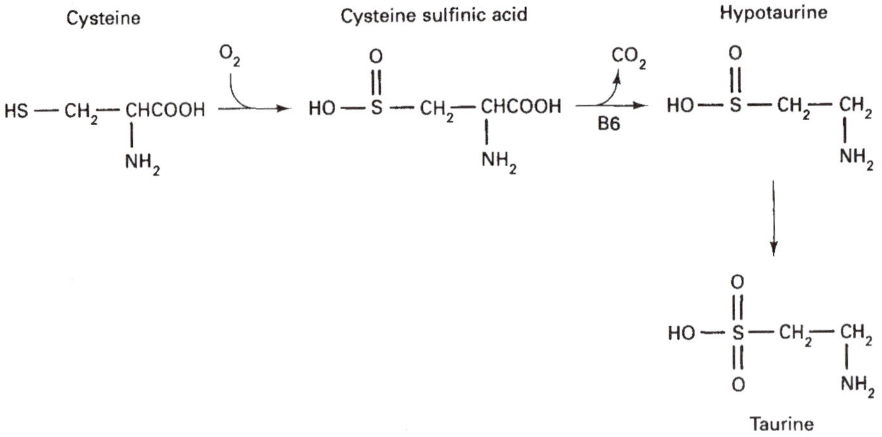

Cysteine            Cysteine sulfinic acid            Hypotaurine

$$HS - CH_2 - CHCOOH \xrightarrow{O_2} HO - \overset{O}{\underset{\|}{S}} - CH_2 - CHCOOH \xrightarrow[B6]{CO_2} HO - \overset{O}{\underset{\|}{S}} - CH_2 - CH_2$$

Taurine

FIGURE 2.40 The major pathway for taurine biosynthesis in the liver. First, cysteine is converted to cysteine sulfinic acid in an oxygen-requiring reaction catalyzed by an iron metalloenzyme. The second step, catalyzed by a vitamin $B_6$-requiring enzyme, is a decarboxylation reaction. The final step appears to be catalyzed by a copper metalloenzyme and to require oxygen. Apparently, about one-fourth of the cysteine in the liver eventually is converted to taurine.

bile salts. Glycine is used for bile salt conjugation also. In the adult human, the glycocholate/taurocholate ratio in the bile varies from 1.0 to 4.0. Unconjugated bile salts, such as cholic acid (cholate), account for a very small proportion of the total.

The human neonate has a limited ability to synthesize taurine. Despite this limitation, neonates tend to conjugate bile salts exclusively with taurine. Mother's milk is a rich source of taurine. Feeding a taurine-free diet to human infants causes a decrease in the proportion of bile salts that occur as taurocholate, with a corresponding increase in glycocholate. Whether the shift from tauro- to glycocholate results in changes in bile salt function, that is, in their ability to support the digestion and absorption of lipids, is not clear. Controlled studies involving two groups of human infants, fed taurine-free and taurine-containing diets, revealed little or no effect on absorption of dietary fats (Galeano et al., 1987).

Cats, unlike humans, dogs, and rats, have little ability to biosynthesize taurine because of low levels of the enzyme that catalyzes the first step of the pathway. However, cats conjugate their bile salts almost exclusively with taurine. Cat bile salts contain about 80% taurocholate, 15% taurodihydroxycholate, 1% glycocholate, and 1% cholate. Feeding cats a taurine-free diet was shown to result in a 50% drop in the concentration of bile salts in the gall bladder, but only a small increase in the concentration of glycocholate. For comparison, the bile salts present in the human infant consist roughly of 23% taurocholate, 19% taurochenodeoxycholate, 8% taurolithocholate, 34% glycocholate, and 17% glycochenodeoxycholate. These proportions differ widely among different infants and among children of various ages (Challacombe et al., 1975).

Since humans manufacture taurine, the data in cats might be of little interest. However, although cats remain in general good health with the consumption of a taurine-free diet, there are a few exceptions:

1. They become blind. A taurine deficiency results in degeneration of the retinas; the tapetum lucidum, a layer of cells behind the retina, degenerates.

2. The reproductive capacity of female cats is compromised. A taurine deficiency provokes frequent fetal resorptions and abortions.

3. The survival of the newborn kittens is poor, and they have retinal and neurological abnormalities.

Changes in gait and reflexes indicates malfunctions of the **cerebellum** (Sturman *et al.*, 1985). (The cerebellum is the part of the brain used to control the coordination of movements of the extremities.) Apparently, taurine serves some unique functions in the nervous system. These findings have prompted food companies to include taurine in their infant formulas and in solutions intended for parenteral feeding. However, whether humans can benefit from taurine in the diet is not clear.

The concentration of free taurine in certain tissues of the body may be higher than that of the classical amino acids. Typical concentrations are muscle, 50–80 m$M$; the retina, 42 m$M$; liver, 10 m$M$; and plasma, 0.02–0.1 m$M$. Although taurine in the liver might be expected to be used mainly for bile salt conjugation, its use in muscle is not clear. Evidence suggests that high intracellular levels of taurine help regulate osmotic pressure within the cell. (Osmotic pressure is a function of the rates of entry and exit of water from the cell.) Taurine appears to play a key role in controlling the pressure of the fluids within the eye (Vinnakota *et al.*, 1997).

# DIGESTION AND ABSORPTION OF CARBOHYDRATES

## General

Carbohydrates generally supply about 45% of our energy requirement. Dietary carbohydrates include monosaccharides (such as glucose and fructose), disaccharides (such as sucrose and lactose), and the longer-chain oligo- and polysaccharides. (Polysaccharides with fewer than 15 monosaccharide units are classed as oligosaccharides.) Dietary polysaccharides include starches, such as amylose and amylopectin, and some of the dietary fibers.

Sucrose is a disaccharide composed of a molecule of glucose and a molecule of fructose. Lactose, or milk sugar, is composed of glucose and galactose, and maltose of two molecules of glucose. Amylose is a straight-chain polymer of glucose residues. Amylopectin is also a polymer of glucose units, but contains many branch points, giving rise to additional chains of glucose. (Amylopectin should not be confused with pectin, a polymer consisting of galacturonic acid residues.) The structures of some of these carbohydrates appear in Figures 2.41 and 2.42. The three commonly used modes of depicting monosaccharide structures are shown for glucose at the top of Figure 2.41.

As shown in Tables 2.4, 2.5, and 2.6, fruits and vegetables are excellent sources of carbohydrates. Sucrose accounts for about half the solid material of a beet or peach. **Sucrase**, a membrane-bound enzyme of the small intestine, catalyzes the hydrolysis of sucrose, yielding one molecule each of glucose and fructose. Meat is a poor source of carbohydrate. However, milk, an animal product, has a high

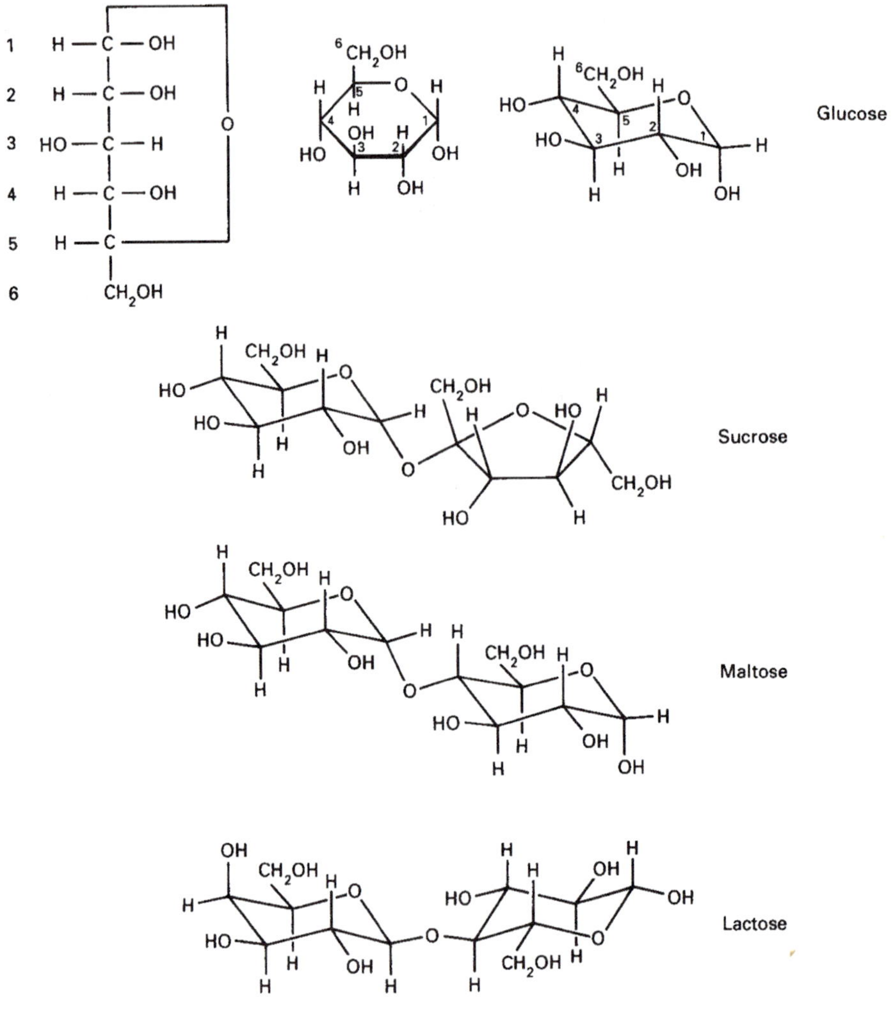

FIGURE 2.41  Sugars.

content of lactose. The lactose content of human milk is 50–70 g per liter; longer-chain tri- and tetrasaccharides occur at lower concentrations (about 10 g per liter). **Lactase**, a membrane-bound enzyme of the small intestine, catalyzes the hydrolysis of lactose to glucose and galactose. (Lactose should not be confused with the 3-carbon carbohydrate lactic acid.) The monosaccharide galactose generally is not found in the diet. However, this sugar may be liberated from lactose by the bacterial enzyme β-galactosidase, present in yogurt because it is produced during culturing of the product. Hydrolysis of lactose to glucose and galactose probably occurs not in the yogurt itself, but in the human gut after yogurt consumption. This hydrolysis may account for the lower incidence of lactose intolerance with the consumption of yogurt (derived from milk) than with milk itself. Lactose is not present in significant amounts in plants.

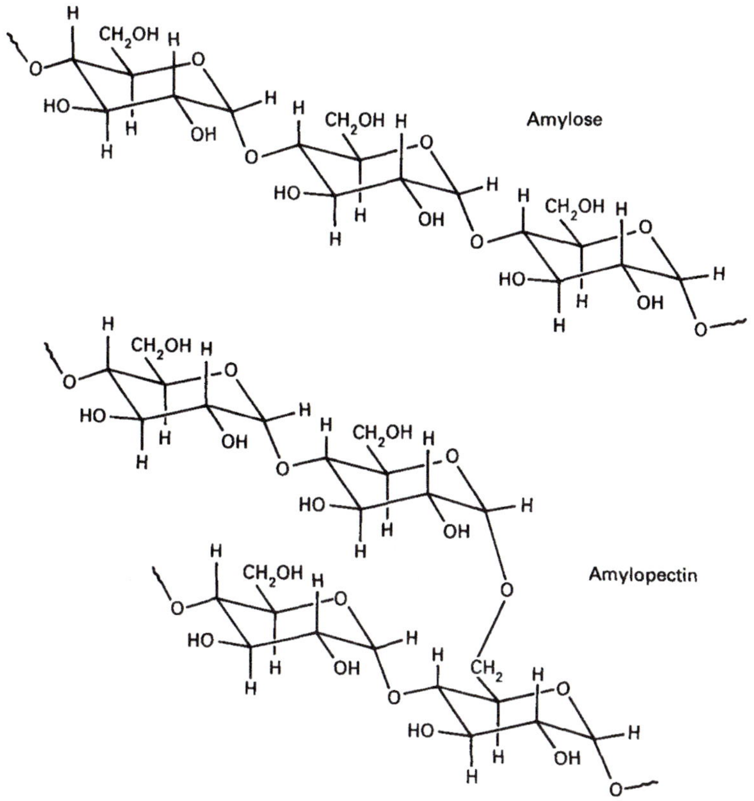

FIGURE 2.42  Starches.

**TABLE 2.4  Sugar Content of Plant Foods**

| Food | Total solids (%) | Glucose | Fructose | Sucrose | Maltose | Raffinose | Stachyose |
|------|------------------|---------|----------|---------|---------|-----------|-----------|
| | | | | Sugar content (% of solids) | | | |
| Apple | 16 | 7.3 | 37.8 | 23.7 | – | – | – |
| Apricot | 14 | 12.0 | 8.9 | 40.4 | – | – | – |
| Peach | 13 | 7.1 | 9.2 | 54.1 | 0.9 | – | – |
| Pear | 14 | 7.0 | 49.9 | 11.9 | 2.3 | – | – |
| Plum | 18 | 19.4 | 8.5 | 27.5 | 0.8 | – | – |
| Beet | 11 | 1.6 | 1.4 | 54.6 | – | – | – |
| Carrot | 12 | 7.1 | 7.1 | 35.3 | – | – | – |
| Sweet corn | 23 | 1.5 | 1.4 | 13.4 | – | – | – |
| Squash | 13 | 7.3 | 8.9 | 12.3 | – | – | – |
| Sweet potato | 23 | 1.5 | 1.3 | 15.0 | – | – | – |
| Lima bean | 27 | 0.1 | 0.3 | 9.7 | – | 2.5 | 2.2 |
| Snap bean | 8 | 13.9 | 15.4 | 3.2 | – | 1.4 | 2.4 |
| Pea | 26 | – | 0.3 | 11.7 | – | 0.2 | 0.2 |

*Source*: Anderson (1982).

**TABLE 2.5** Composition of Various Foods

| | Percentage by weight | | | | |
|---|---|---|---|---|---|
| Food | Starch | Protein | Fat | Moisture | Other |
| Corn | 60 | 9 | 4.0 | 16 | 11 |
| Potato | 18 | 2 | 0.1 | 78 | 2 |
| Wheat | 64 | 13 | 2.0 | 14 | 7 |
| Cassava root | 26 | 1 | 0.3 | 66 | 7 |
| Rice | 78 | 8 | 0.5 | 12 | 2 |

*Source*: Vam Beynum and Roels (1985).

Some beans contain significant quantities of the sugars raffinose and stachyose (Table 2.4). These sugars are not digestible by enzymes produced by humans, although they can be used by the gut microflora, producing gaseous waste products. The increase in microbial metabolism provoked when undigested sugars reach the large intestine results in an increase in the rate of production of various gases in the gut lumen. Carbon dioxide is the gas produced in the greatest amount.

Starches are the major carbohydrates of such foods as corn, wheat, and rice (Table 2.5). Potatoes and legumes are also rich in starches. Most starches are a mixture of amylose and amylopectin, as shown in Table 2.6. In general, amylose constitutes about one-third of the starch in various foods; the remainder is amylopectin. Of the dry matter of the plants listed in Table 2.4, 70–90% consists of starch. These plants are called starchy foods. Take special note of the general constituents of the cassava root. (Tapioca is made from cassava.) This food has the lowest protein/carbohydrate ratio of any of the foods listed in Table 2.5, about 1/25 for cassava compared with about 1/5 for wheat.

The starch in plants occurs in small microscopic granules that have a core of starch surrounded by a network of protein. In the granules, amylopectin has an orderly crystalline structure, whereas amylose is somewhat amorphous. The orderly structure of the starch molecules may impair digestion by pancreatic amylase. The occurrence of starch in granules, whether in wheat, potatoes, bananas, or other foods, may prevent the digestion of up to 50% of the starch in these foods

**TABLE 2.6** Proportion of Food Starch that Occurs as Amylose and Amylopectin

| Food | Amylose (%) | Amylopectin (%) |
|---|---|---|
| Corn | 28 | 72 |
| Potato | 21 | 79 |
| Wheat | 28 | 72 |
| Cassava root | 17 | 83 |
| Rice | 17 | 83 |

*Source*: Van Beynum and Roels (1985).

when they are consumed in the raw state. Cooking results in hydration of the starch molecules and in the swelling and gelatinization of the starch granules, increasing the susceptibility of the starch to enzymatic hydrolysis, with a consequent increase in digestibility.

## Starches and Other Polysaccharides Used by the Food Industry

Starches and modified starches are used by the food industry for cake mixes, pie fillings, frozen and canned foods, and jellies (Chinachoti, 1995). When starch granules are heated at 56–68°C (or higher) in water, all of the hydrogen bonds between its chains lose contact with each other and, instead associate with water. The result of heating is a viscous suspension, or gelatination. To prevent the re-formation of a solid gel, with heating, the food industry adds phosphate groups to the hydroxyl residues of starch, to add a repelling force between adjacent chains of sugars, thus maintaining the starch in gel form.

Polymers from seaweed, such as agar, carrageenan, and alginates, are also used to form gels in foods, by the food industry. Agar and carrageenan are polymers of **galactose**, while alginates are polymers of two different sugars, **mannuronic acid** and **glucuronic acid**. Alginate gels are distinguished in that they may consist of nearly 100% water, i.e., 99% water and 1.0% polymer. Alginates have found use in making jelly that spreads easily, pie fillings that do not crack, ice cream that stays smooth with storage, and pudding that rapidly gels. Gums are polymers acquired from shrubs and trees from Africa and Asia. Gums are complex polysaccharides made of units of **galactose** and **mannose**. Gums are used by the food industry to prevent ice crystal formation in ice cream and candy, to prevent separation of oils and water in various beverages, and to prevent the leakage of water from processed cheese.

### Modified Starch

The starches purified from some plants, such as cassava (tapioca starch) or potato, are too viscous for use by the food industry. They create a gummy texture in foods and reduce their palatability. These root starches develop a rubbery texture when cooked. Some starches, such as corn starch and wheat starch, form the desired texture when cooked, but with cooling these starches form a gel that eventually loses its water-binding properties. This process, which results in water separation, is called **syneresis**. Syneresis occurs with freezing or storage at low temperatures. The foregoing problems (gumminess and syneresis) can be avoided by modifying food starch prior to adding it to the food. Modification can be effected by acid treatment, hypochlorite treatment (oxidation), and by drying the starch and heating it (Wurzburg, 1995).

Acid treatment results in partial hydrolysis of the sugar-to-sugar bonds in the starch, i.e., an event similar or identical to that catalyzed, during digestion, by amylase. Hypochlorite treatment results in the conversion of hydroxyl groups to carboxyl groups, aldehyde groups, or ketones.

In making candy (gumdrops), corn starch is added to the hot, liquified product. Pouring this liquid into the candy mold is difficult, unless the starch is first processed by acid treatment. Oxidized starches are mainly used in breaded foods. This additive improves adhesion of the batter to fish or meat. Dry heat converts starch to a product called **dextrin**. Dextrins are used for coating candies, for imparting a gloss to baked goods, and for encapsulating oils and oily flavors to form a dry powder (Wurzburg, 1995).

## Carrageenan and Agar

**Carrageenan** is a generic term for various polysaccharides isolated from red seaweed. These seaweeds include *Hyphae* and *Eucheuma*, which come from the shores of Indonesia and the Philippines, as well as *Gigartina* and *Chondrus* (Irish Moss), which come from the shores of the northern Atlantic Ocean. Several types of carrageenan, which differ by the amount and distribution of sulfate groups, are used by the food industry. The sulfate groups are attached to the hydroxyl residues of the galactose groups.

λ-Carrageenan (lambda-carrageenan) is unique in that it works as a thickening agent *without causing gel formation*, even under extreme conditions (high sugar or salt content in the food), and is used in salad dressings, sauces, and fruit drinks. κ-Carrageenan (kappa-carrageenan), in contrast, *forms strong gels* and is used to make milk gels (custards), yogurt, chocolate milk shakes, ice cream, and meat preserves. Carrageenans are added to foods to give concentrations of 0.03 to 3.0%.

Carrageenan and agar are both made of repeating disaccharides, consisting of β-galactose groups linked with α-galactose groups. However, the former polymer occurs as units of repeated β-D-galactose–α-D-galactose, while the latter polymer is made of units of repeated β-D-galactose–α-L-galactose.

**Agar**, which is also extracted from various red seaweeds, is used in cake frosting, donut glazes, candy, meat and dairy products, and beverages. When used in cake frosting (icing) and glazes, agar prevents the frosting from becoming sticky. Here the polymer is used at 0.2–1.0%. Agar has the property of heat resistance (it does not dissolve except at high heat), and hence is used (rather than carrageenan) at a level of 2% in the broth of canned meat and fish, since it resists breakdown during heat sterilization (Stanley, 1995).

## Digestibility

Alginate may be partially, though not completely, digested, as indicated by the finding that the consumption of 15 g alginate has a laxative effect in humans (McNeely and Kovacs, 1975). Carrageenan is thought not to be digested or absorbed by humans or animals. The typical daily intake of carrageenan is only 25–35 mg/day. Hence, there is little reason to believe that our intake of this seaweed polysaccharide contributes to or detracts from normal metabolism (Stancioff and Renn, 1975).

## Enzymes Used to Digest Carbohydrates

Saliva and pancreatic juice both contain α-amylase. The activity of this enzyme in saliva is not significant compared with that in the pancreatic juice released into the gut. Release of amylase from the pancreas is controlled by a mechanism similar or identical to the one that stimulates the release of trypsinogen, namely by the influence of CCK on the exocrine pancreas. Amylase catalyzes the hydrolysis of starch at interior positions rather than at the ends of the polymer. This activity yields products such as maltose and longer-chain-length oligomers of glucose. Also, amylase does not catalyze the hydrolysis of starches at branching points. Therefore, small branched-chain structures called **dextrins** are formed that are not hydrolyzed by the enzyme.

Dextrins are hydrolyzed by a membrane-bound enzyme, isomaltase, which occurs in the same polypeptide chain as sucrase, the enzyme that hydrolyzes sucrose. Two active sites (catalytic sites) reside on one polypeptide chain. The entire protein is called **sucrase–isomaltase**. Enzymes containing more than one active site on one polypeptide chain are called **multifunctional**. The orientation of sucrase–isomaltase in the gut cell, or enterocyte, is shown in Figure 2.43. Both active sites are situated in the lumen of the gut; the N-terminal region is anchored in the membrane. Each of the active sites of sucrase–isomaltase is capable of hydrolyzing maltose. Perhaps a better, although cumbersome, name for the enzyme would be sucrase/maltase–isomaltase/maltase. The isomaltase catalytic site is closest to the membrane, whereas the sucrase site is the C-terminal portion of the enzyme.

Lactase, which catalyzes the hydrolysis of lactose, is also a bifunctional enzyme. Lactase resides on the same polypeptide chain as **phlorizin hydrolase**. The entire protein is called lactase–phlorizin hydrolase. The N-terminal end of the polypeptide resides in the lumen, and bears the lactase activity. The central portion of the

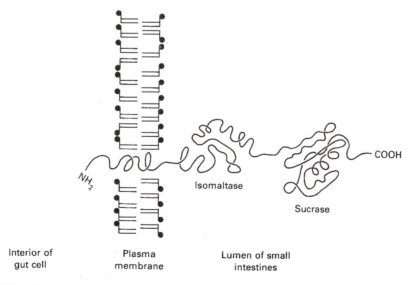

Interior of          Plasma          Lumen of small
gut cell            membrane          intestines

**FIGURE 2.43** Orientation of sucrase–isomaltase, a membrane-bound enzyme.

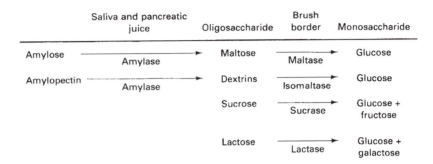

FIGURE 2.44 Carbohydrate-hydrolyzing enzymes. Starch-hydrolyzing enzymes act prior to the oligosaccharide- and disaccharide-hydrolyzing enzymes.

polypeptide bears phlorizin hydrolase activity, while the C-terminal end crosses one time through the plasma membrane (Keller *et al.*, 1995). Phlorizin hydrolase catalyzes the hydrolysis of sugars bound to lipids. Sugar-containing lipids, called **ceramides**, occur in milk and other foods. They are discussed in the section on sphingosine-based lipids. The activities of some of the enzymes discussed in this section are summarized in Figure 2.44.

The regulation of the activities of sucrase and lactase has been the subject of several studies. A study of rats, depicted in Figure 2.45, addressed differences in

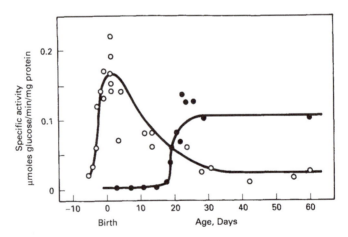

FIGURE 2.45 Activity of sucrase and lactase in samples of intestines taken from animals of different ages. The intestines of rats were removed at various stages of life and used for the sucrase- (●) and lactase- (○) sensitive assays plotted in this figure. Lactase activity is greatest at birth, which is consistent with the milk-drinking behavior of rats prior to weaning. In contrast, sucrase activity is low at birth but increases dramatically at 3 weeks of age, the time of weaning in rats. Separate experiments revealed that, if milk feeding was continued past the normal weaning time, sucrase activity still increased on schedule and lactase activity still declined. Lactase activity also declines in children as they grow older. As in rats, lactase activity in humans cannot be maintained at high levels or restored in adults by feeding milk or lactose-containing diets. (Redrawn with permission from Henning and Kretchmer, 1973.)

the activities of the two enzymes at various stages of development. The results are relevant to the loss of lactase activity known to occur with development in humans. The intestines of rats were removed at various stages of life and used for the sucrase- and lactase-sensitive assays depicted in Figure 2.45. Lactase activity is greatest at birth, which is consistent with the milk-drinking behavior of rats prior to weaning. In contrast, sucrase activity is low at birth but increases dramatically at 3 weeks of age, the time of weaning in rats. Separate experiments revealed that if milk feeding was continued past the normal weaning time, sucrase activity still increased and lactase activity still declined. Lactase activity also declines in children as they grow older. As in rats, lactase activity in humans cannot be maintained at high levels or restored in adults by feeding milk or lactose-containing diets. Sucrase activity may change somewhat in response to the diet, as indicated by a study involving adult rats. Enzyme activity was assessed in two different regions of the small intestine: the jejunum and the ileum (see Figure 2.46).

EXERCISE

Would ceramide-hydrolyzing activity change in response to the level of sucrose in the diet? Why or why not?

## Absorption of Carbohydrates

Glucose, a monosaccharide, does not require hydrolysis prior to absorption by the gut. Its absorption is followed by increases in the level of sugar in the plasma,

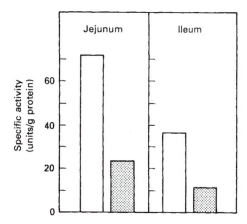

FIGURE 2.46  Activity of sucrase. Rats were fed diets that were high in sucrose (open bars) or were carbohydrate free (filled bars). The sucrose diet contained 650 g sucrose per kilogram of food. The carbohydrate-free diet contained a mixture of nonnutritive fiber and corn oil rather than sucrose. The animals were fed the diets for 4 days; then the intestines were removed for assay of enzyme activity. The results demonstrate that enzyme activity was about threefold greater with the sucrose diet than with the carbohydrate-free diet. Separate experiments revealed that maximal adaption of sucrase activity to the diet required 0.5–1.0 days. (Redrawn with permission from Riby and Kretchmer, 1984.)

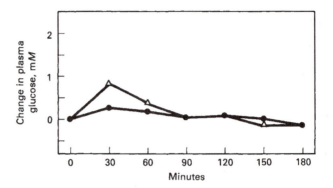

**FIGURE 2.47** Change of plasma glucose concentration after meals of wheat (Δ) and cassava (●) starch. Samples of blood, withdrawn at 30-min intervals, were analyzed for plasma glucose. The data indicate slight increases in the plasma glucose levels; wheat starch produced a slightly greater increase than cassava starch. (Redrawn with permission from Bornet *et al.*, 1989.)

particularly when large quantities are consumed. The rise in plasma glucose is illustrated in Chapter 4. For example, consumption of a dose of 75 g glucose by a human can produce an increase in plasma glucose from a normal fasting level of about 4.0 m*M* to a level of about 6.0 m*M*.

Figures 2.47 and 2.48 show the effects of starch digestion and absorption on human subjects. In Figure 2.47, a meal consisting of 35 g raw wheat starch or cassava starch was consumed in a suspension of water. Samples of blood, withdrawn at 30-minute intervals, were analyzed for plasma glucose. The data indicate slight increases in the plasma glucose levels; wheat starch perhaps produced a

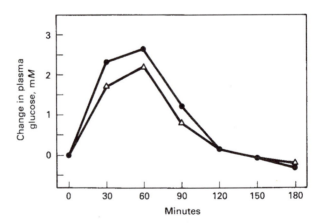

**FIGURE 2.48** Change of plasma glucose concentration after meals of boiled wheat (Δ) and cassava (●) starch. These data demonstrate a dramatic increase in plasma glucose levels 1 hr after the meal. These results are consistent with the fact that uncooked starch resists digestion in the small intestine. (Redrawn with permission from Bornet *et al.*, 1989.)

greater increase than cassava starch. Figure 2.48 shows a similar study using cooked starch, consumed after it was boiled in water. These data demonstrate a dramatic increase in plasma glucose level 1 hour after the meal. These results are consistent with the fact that uncooked starch resists digestion in the small intestine.

Interest in the rise in plasma glucose levels following consumption of different starchy foods arose because of health concerns for diabetes. Certain diabetics require a specific schedule of energy intake throughout the day. These patients require a constant supply of carbohydrate but must avoid drastic increases or fluctuations in the concentration of plasma glucose. Consequently, their nutritional treatment includes instructions to avoid rapidly absorbed sugars (mono- and disaccharides) and to consume the more slowly absorbed starches.

EXERCISE

Which diet would result in a more rapid absorption of carbohydrate from the gut: a dry baked potato or a baked potato flavored with a generous amount of butter or oil? Why?

## Special Topic: Sugar Transporters

Sugar transport proteins are membrane-bound proteins that are required to allow the passage of glucose, and other sugars, across phospholipid membranes. During passage from the lumen of the gut and into the bloodstream, dietary glucose must cross two membranes. Glucose crosses the apical membrane of the enterocyte via one type of transporter. Once inside the enterocyte, the glucose travels through the cytosol and exits across the basolateral membrane via a different transporter, followed by passage through the portal vein to the liver (Figure 2.49). These two transporters, respectively, are called the **$Na^+$-dependent glucose transporter** and **GLUT-2**.

The $Na^+$-dependent glucose transporter consists of 665 amino acids (Lee *et al.*, 1994). When $Na^+$ occurs at greater concentrations outside the cell than inside, the concentration gradient drives the passage of glucose across the membrane and into the enterocyte. In the absence of sodium, the driving force for glucose transport is absent.

The $Na^+$-dependent glucose transporter also occurs in the kidney, i.e., in the renal tubule. The renal tubule is a microscopic structure that resembles a tiny intestinal tract. To introduce renal (kidney) physiology, the kidney creates urine by using a filter that retains the blood cells and proteins. The **immature urine** is rich in amino acids, salts, and sugars. During passage through the renal tubule, various transport proteins serve to reclaim (reabsorb) all of these nutrients through the renal tubule cell and back to the bloodstream. The resulting urine, which is poor in nutrients, is called **mature urine**. GLUT-2 occurs in the basolateral membrane of the renal tubule cell. In the enterocyte and renal tubule cell, one can see that two different transport proteins work hand-in-hand to shuttle glucose across an epithelial cell membrane.

**GLUT-5** is the fructose transporter. It was named GLUT-5 before its true function was known. It occurs in the apical membrane of the enterocyte, allowing dietary fructose to be absorbed. GLUT-5 also occurs in skeletal muscle, adipocytes,

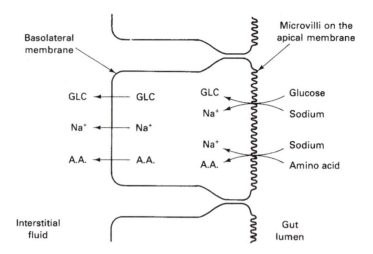

**FIGURE 2.49**  An absorptive cell of the villus. The part of the plasma membrane facing the lumen is the apical membrane, whereas that facing the blood supply is the basal and lateral (basolateral) membrane. The membrane-bound proteins used to mediate the uptake of a variety of nutrients requires the simultaneous co-transport of sodium ions. The diagram reveals that the transport of glucose and amino acids is dependent on sodium ions. Sodium-independent transport systems also exist for many nutrients. The sodium depicted in the figure is supplied by intestinal secretions and need not be supplied by any particular diet.

and sperm cells. Once absorbed into the enterocyte, fructose is thought to cross over into the bloodstream via one of the other GLUT transporters.

## Sugar Transport into Brain Cells

The goal of sugar is not simply to cross layers of epithelial cells, but to enter cells and be oxidized to produce energy. **GLUT-1** and **GLUT-3** work hand-in-hand for this purpose in the brain. GLUT-1 occurs in the blood capillaries in the brain and allows glucose to exit the capillaries and enter the extracellular space, while GLUT-3 occurs in the membranes of nerve cells, allowing them to acquire their preferred energy food (glucose) (Zeller *et al.*, 1995; Seatter *et al.*, 1997; Taha *et al.*, 1995).

## Sugar Transport into Muscle during Rest and Exercise

**GLUT-1** is responsible for the transport of glucose into muscle under basal conditions, while another transporter, **GLUT-4**, allows the increased transport that occurs when muscle is stimulated by hormones or by a signal from nerves (Hansen *et al.*, 1995). GLUT-1 occurs in a wide variety of tissues, including skeletal muscle, the blood vessels of the brain, and red blood cells. GLUT-4 occurs only in tissues where glucose transport is stimulated by insulin. These tissues are skeletal muscle, heart muscle, and adipocytes (fat cells). GLUT-4 is distinguished in that it occurs in the membranes of tiny vesicles that reside in the cytoplasm. With stimulation by insulin or by a nervous impulse, the vesicles fuse with the plasma membrane, thereby increasing the number of glucose transporters in this location. Eventually, the vesicles are re-formed, removing the GLUT-4 from the plasma membrane, thus returning the rate of glucose transport to the basal level.

## *Sugar Transport in and out of the Liver*

The liver serves a "motherly function" in the body. One of its motherly functions is that it helps makes sure that the other organs acquire suitable meals, and at the proper time. To do this, the liver takes up glucose from the bloodstream via **GLUT-2**, and converts the glucose to animal starch (glycogen). During fasting, the liver breaks down the glycogen back to glucose units, and transports the glucose back into the bloodstream via GLUT-2. During starvation, the liver manufactures glucose from other molecules, and transports the glucose out via GLUT-2. Glucose is a nutrient required by the brain and central nervous system. Other organs, such as muscle, do not have this sort of reliance on a steady supply of glucose.

GLUT-2 is a 524-amino acid protein. In a rare genetic disease, a mutation in the GLUT-2 gene results in a nonfunctional protein of about half the normal length (Santer *et al.*, 1997; Efrat, 1997). What are some of the metabolic consequences of this disease? *With feeding*, the patients have abnormally high levels of plasma glucose. This situation occurs because of impaired glucose uptake by the liver. *With fasting*, the patients have an abnormally low plasma glucose level since the liver cannot adequately release glucose. These results highlight the view that GLUT-2 mediates the transport of glucose in, as well as out, of the cell.

**GLUT-7** occurs in the liver and allows the passage of glucose through an intracellular membrane, i.e., into the endoplasmic reticulum (Klip *et al.*, 1994).

The proteins GLUT-1, GLUT-2, GLUT-3, GLUT-4, GLUT-5, and GLUT-7 form a family of membrane-bound proteins having a similar structure. GLUT-6 exists only as a gene, and there is no corresponding protein. This type of gene is called a **pseudogene**. Pseudogenes that correspond to a large number of functional genes occur within the human genome. There is no reason to believe that they have any immediate value to the organism.

## Issues in Carbohydrate Nutrition

Carbohydrate nutrition is different in nature from the nutrition of amino acids, fats, vitamins, and minerals. Humans have no dietary requirement for any specific type of carbohydrate. All carbohydrates of the body can be synthesized from dietary glucose or fructose, the major sugars in the diet. The body stores sugar in the form of glycogen, a polysaccharide composed of glucose units.

Ribose is a vital sugar used for molecules of energy transfer, as it is a component of the ATP molecule. Ribose is also an integral part of a molecule used for information transfer (RNA), while deoxyribose occurs as a part of DNA, the molecule used for information storage.

**Glycoproteins** occur as two types. The first type might be considered to be a "garden variety" glycoprotein that consists of typical proteins bearing a small chain of sugars attached to residues of arginine, serine, or threonine. Nearly all of the proteins of the bloodstream, and of the plasma membrane, are glycoproteins. The second type of glycoprotein, which is secreted from cells and is extracellular, consists of a linear protein backbone bearing vast networks of sugar chains. This second type of glycoprotein includes proteoglycans (used for controlling cell-to-cell contact) and mucus (used for protecting cells from the external environment, and for lubricating surfaces).

Ribose and the sugars of glycoproteins and proteoglycans can be synthesized from a number of dietary sugars, such as glucose, fructose, and galactose. The conversion of glucose to ribose is shown in the Thiamin section in Chapter 9.

Most aspects of carbohydrate nutrition are simpler than those of other nutrients. (For example, fat nutrition is complicated by the fact that the metabolism of fats requires bile salts to maintain solubility during digestion and lipoproteins and albumin during distribution in the body.) On the other hand, the nutrition of the carbohydrates that take the form of dietary fibers is very complicated. This complexity is due to the fact that they are metabolized by enzymes of the gut microflora.

Reliance on carbohydrates as the only source of energy can lead to a number of problems. Omitting fats from the diet for a prolonged period can cause a deficiency of essential fatty acids. Omitting protein for prolonged periods of time while continuing to consume carbohydrates can lead to an imbalance in the breakdown of the body's emergency source of protein, namely muscle tissue. Normally, the breakdown of muscle, during fasting and starvation, can replace the gradual net oxidation and the loss of amino acids that occurs. However, the imbalance provoked by a diet that contains carbohydrate but is protein-free results in a condition called **kwashiorkor**.

Carbohydrates are more plentiful and constant in food supplies throughout the world when compared to other nutrients, such as proteins, vitamin A, folic acid, and iodine. A naturally occurring deficiency specifically in carbohydrates is unknown. However, deliberate omission of carbohydrates from the diet with continued consumption of fat as an energy source can lead to specific problems. Glucose is required as an energy source by the central nervous system. When there is a deficiency of glucose, the body adjusts its metabolism to provide ketone bodies, nutrients derived from fat, which can be utilized by the brain and other parts of the central nervous system. However, excessive production of the ketone bodies can result in acidosis, a lowering of the pH of the blood, which is potentially toxic.

Carbohydrates are of major interest in food science. The monosaccharides and starches present in natural and processed foods have a marked effect on their color, texture, consistency, and palatability. Lactose, the major carbohydrate of milk, can limit its acceptability as a food for those with lactose intolerance. Slowly digestible carbohydrates are used in the diets of certain diabetics, who must eliminate or restrict their intake of foods containing rapidly absorbed carbohydrates such as candies, honey, syrup, and jam.

# ABSORPTION PHYSIOLOGY

## Crypt and Villus: Structures of the Mucosa of the Small Intestine

In the small intestine, absorption occurs through the lumenal face of the mucosa, which is covered with finger-like projections called **villi**. Each villus is bordered by several pouches or invaginations called **crypts**. On the average, each villus is surrounded by eight crypts. The cells of the villus are primarily absorptive and are used for absorbing dietary nutrients, material originating from pancreatic secretions, and damaged cells sloughed off from the mucosa. The cells covering the villus, mainly enterocytes, contain enzymes used for triglyceride and chylomicron synthesis, as well as membrane-bound and intracellular digestive enzymes and membrane-bound transport proteins. These transport proteins include those used for the co-transport of sodium and glucose and of sodium and amino acids (see Figure 2.49).

The transport of sodium ions is coupled to both glucose and amino acids. The passage of sodium ions from the gut lumen, through the apical membrane, through the cytoplasm, and out of the basolateral membrane drives the co-transport of the coupled nutrient. The several different sodium/amino acid co-transport proteins include those that are specific for basic, acidic, and neutral amino acids. Sodium/nutrient co-transporters also exist for bile salts and for glucose. Further details of sodium metabolism are given in the section on sodium, potassium, and water in Chapter 10. For example, the driving force for transit of sodium ions through the cell membrane is the activity of Na,K-ATPase (the sodium pump).

The cells of the crypt secrete water and bicarbonate. Water aids in the dispersion and solubilization of the solid material in food. Bicarbonate neutralizes gastric acid. The cells of the crypt rapidly divide and proliferate and are enriched with enzymes used for the synthesis of DNA and other cell components. These cells move up out of the crypt and up the surface of an adjacent villus. At the same time, they differentiate, develop microvilli, and increase their content of hydrolytic enzymes. Once situated on a villus, the cells continue to migrate toward the top of the villus where they are sloughed off eventually. The time from cell division through travel up the villus to sloughing off is 2–3 days. This process takes place constantly throughout life. The migrating cells covering the villus undergo little cell division.

The crypt contains a group of about 150 rapidly dividing cells. Each cycle of cell division requires 9–13 hours. Studies with mouse guts revealed that each crypt contains 25–30 cells, and that 25–30% of these cells are in the process of dividing at any given point in time (if they are in the small intestines), and that only 8–9% of the cells are in the process of dividing (if they are in the large intestines) (Fleming *et al.*, 1994).

Cell division and the subsequent differentiation give rise to several different types of cells, including the enterocyte and goblet cells. Other types that account for only about 1.0% of the epithelial cells of the gut include endocrine cells, which produce hormones, and **Paneth cells**, which produces lysozyme, an antibacterial enzyme. The reason for the rapid turnover of the epithelial surface of the gut is the necessity of maintaining its function in the harsh environment of pancreatic enzymes.

The composition of the plasma membrane of the epithelial cells of the crypt and villus has been a subject of interest. The rapidly changing function of these cells during their differentiation provides an opportunity to examine the possible roles of plasma membrane (PM) composition in changing or controlling its activities. The PM of the crypt cell is more flexible or "fluid" than that of the villus cell. All biological membranes, including the PM, mitochondrial membrane, and nuclear membrane, contain phospholipids. Generally, only the PM contains a significant amount of cholesterol, which increases the rigidity of the membrane. The PM of the villus cell contains a larger cholesterol/phospholipid ratio (0.85) than the PM of the crypt cell (0.60) and is more rigid (Meddings *et al.*, 1990).

The normal functioning of the cells of the crypts and villi is disrupted in certain malabsorptive diseases such as **celiac disease**, which involves a flattening of the intestinal mucosa. This flattening results from disappearance of the villi, as shown in Figure 2.50. The depth of the crypt may or may not change. (Figure 2.50 indicates a slight increase in crypt depth.) Extensive disappearance of intestinal villi leads to malabsorption and serious malnutrition.

Another nutritional issue concerns the crypts. Certain bacterial and viral infections can provoke vast increases in the secretory activity of the crypts, resulting in excessive losses of salts and water, as well as diarrhea. Secretory diarrheas that continue for a week or longer may be life threatening, as discussed in Chapter 10 in the section on Sodium, Potassium, and Water.

## Sodium and Chloride Absorption by the Gut

Adequate dietary intakes of sodium and chloride for the adult are estimated to be 1.1–1.3 and 1.7–5.1 g per day, respectively. These dietary salts are needed to replace obligatory losses in the urine and small losses in the sweat. Most of the sodium and chloride ions in the diet are absorbed by the jejunum and ileum; only about 5% is lost in the feces.

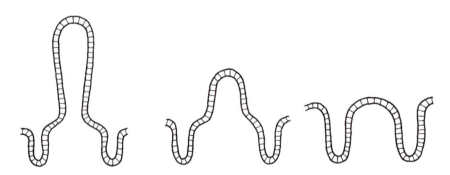

**FIGURE 2.50** Crypts and villus. (*Left*) A normal crypt and villus. Two crypts are shown on either side of the villus. Malabsorptive diseases may involve degeneration of the villi. (*Center*) A short villus. (*Right*) Complete disappearance of a villus. The disease may involve an enlargement of the crypts, which is shown. The figure does not show the lacteal and other structures of the villus. (Redrawn with permission from Riecken, 1988.)

The mechanisms of absorption of Na and Cl differ in different organs and in different regions of the small and large intestines. The transport proteins responsible for mediating transfer of the ions across cell membranes are only beginning to be understood. The mechanism thought to operate in the ileum shares the following features with the acid pump of the parietal cell: (1) electroneutral exchange of protons for potassium ions; (2) use of carbonic anhydrase to generate protons and bicarbonate ions; and (3) electroneutral exchange of chloride anions for bicarbonate anions.

The mechanisms for Na and Cl absorption is shown in Figure 2.51. The top of the enterocyte (A), the apical side of the cell, faces the lumen of the gut. The bottom of the cell, which faces the interstitial fluid and the bloodstream, is the basal side. To the right and left are other enterocytes. Figure 2.51A depicts a simple process: exchange of $Na^+$ for a proton. This process is electroneutral because an ion with a positive charge is exchanged for another ion with the same charge. In a more complicated diagram (Figure 2.51B), carbonic anhydrase supplies the cell with protons by catalyzing the reversible reaction of water with $CO_2$. The resulting carbonic acid dissociates to produce a proton and a bicarbonate anion. The proton can be used by the Na/H exchange mechanism shown in the apical membrane. Figure 2.51C reveals that the function of the bicarbonate anion is permitting electroneutral exchange for a chloride anion. This event explains the transport of sodium and chloride ions from the lumen into the enterocyte, but the mechanism is not yet complete. The complete mechanism is shown in the most complicated drawing (Figure 2.51D). This figure depicts the passage of Na and Cl through channels in the basal membrane of the cell, and also shows $CO_2$ diffusion through the cell membrane. Passage of these chemicals through the basal membrane maintains their concentrations in the cell at fairly constant levels.

EXERCISE

Acetazolamide can be used to inhibit the activity of carbonic anhydrase. Would treatment with this drug impair sodium transport only, chloride transport only, or both? (See Turnberg *et al.*, 1970.)

EXERCISE

In a study of salt absorption, a solution of NaCl was placed in the lumen of the gut with a tube, and the absorption of ions was monitored by withdrawing samples of fluid at various times. Devise a mechanism by which the enterocyte can absorb sodium ions from a solution of sodium sulfate in the lumen. Sulfate ions are absorbed poorly by the gut, so the mechanism should depict sulfate ions staying in the lumen. Would the contents of the lumen become more acidic or more basic as more and more sodium is absorbed? (See Turnberg *et al.*, 1970).

## Passage of Water Through Membranes

Water molecules are able to diffuse through phospholipid membranes, i.e., bilayers composed of phospholipids and cholesterol. The term **permeability** is used to express the degree to which water can cross a particular membrane. Laboratory

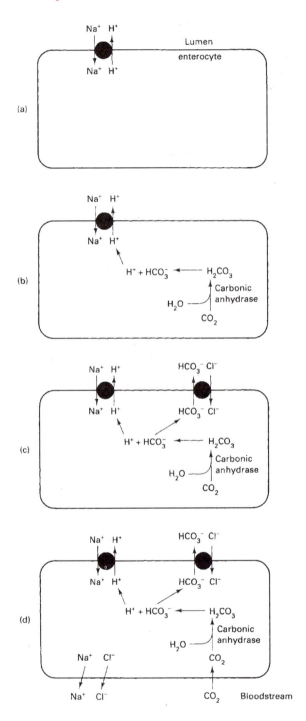

**FIGURE 2.51** A mechanism for sodium chloride uptake, as explained in four steps of increasing complexity: (a) The sodium/proton exchanger. (b) The sodium/proton exchanger with catalysis of acid production by carbonic anhydrase. (c) The sodium/proton exchanger with catalysis of acid production and utilization of bicarbonate by the bicarbonate/chloride exchanger. (d) The finished model for transport of sodium and chloride ions.

studies with various biological membranes, such as layers of epithelial cells, revealed that the permeability to water could be up to 100-fold greater than that of an artificial phospholipid membrane. This increased permeability is due to the presence of specific membrane-bound proteins called **water channels** or **aquaporins**.

Aquaporins are on the small side, consist of about 260 amino acids, and have a molecular weight of about 30 kDa. They weave six times in or out of the plasma membrane (Preston *et al.*, 1994; Raina *et al.*, 1995; Deen *et al.*, 1994). The aquaporins are related to each other, and constitute a family of proteins. This means that if one were to line up all the polypeptides, one on top of the other, one would find that about half of all amino acids occur at identical positions, when viewing their order from the N terminus to the C terminus. The technical term for measuring relatedness in any protein is **sequence homology**.

**Aquaporin-1** (AQP-1) and **AQP-3** are used for basal water transport in many tissues. The term **constitutive** is used to refer to any basal activity. The term **regulated** is used to refer to any activity that is increased or decreased by hormones or other components of the environment. **AQP-5** is required to produce the water in tears and saliva. **AQP-4** occurs in the brain, where it is used in the mechanism that senses if one needs to drink more or less water.

In epithelial cells, **AQP-2** resides in the apical membrane, while **AQP-3** occurs in the basolateral membrane. AQP-2 is distinguished by the fact that it is regulated by the hormone vasopressin. During times of dietary water deficiency, vasopressin provokes an increase in AQP-2 activity in the renal tubule, resulting in an increased rate of water reabsorption. If AQP-2 did not exist, our daily output of water (in urine) would be many times greater than normal. To focus on the mechanism of regulation, AQP-2 resides in tiny vesicles in the cytoplasm of the renal tubule cell. Vasopressin provokes the fusion of these vesicles with the apical membrane, thus causing the insertion of more AQP-2 in the membrane facing the immature urine, with the consequence that more water is reabsorbed (Ishibashi *et al.*, 1997; Kanno, 1995; Marples *et al.*, 1996).

Further details on renal physiology and water regulation appear in the Sodium, Potassium, Chloride, and Water section.

# BIOCHEMICAL MECHANISMS

The process of digestion and the activity of carbonic anhydrase are major everyday concerns in the medical profession. A clearer understanding of the concepts of protein digestion and carbonic anhydrase activity is provided by simplified descriptions of the relevant biochemical mechanisms. Protein digestion involves the addition of water to a peptide bond. Carbonic anhydrase activity involves the addition of water to $CO_2$. A related process, the phosphorolysis of glycogen, is also described.

## Hydrolysis and Phosphorolysis

**Hydrolysis** means the reaction of a molecule of water with a chemical bond that results in the breaking of that bond. The process of hydrolysis results in the lysis of the target bond, association of the hydroxyl group of water with the structure on one side of that bond, and association of the proton of water with the structure on the other side.

Proteins, nucleic acids, and starches are among the biological polymers hydrolyzed in the body. Proteases catalyze the hydrolysis of proteins, DNases and RNases catalyze the hydrolysis of nucleic acids, and amylase catalyzes the hydrolysis of starches. Sucrase, maltase, and isomaltase catalyze the hydrolysis of specific sugars.

**Phosphorolysis** of glycogen is one notable exception to the generalization that biological polymers are broken down by hydrolysis. Glycogen, a polymer consisting of glucose units, occurs in the liver and muscle. This polymer is one of the energy stores of the body. When the body needs to mobilize or liberate the units of glucose, the mechanism of hydrolysis is not used. Instead, the enzyme **glycogen phosphorylase** catalyzes the reaction of a molecule of phosphate with each of the bonds connecting the glucose units. The mechanism of phosphorolysis is similar to that of hydrolysis, as shown in Figure 2.52.

Simplified mechanisms of peptide hydrolysis and glycogen phosphorolysis are outlined as follows:

Step 1:   Water molecules (HOH) are partially dissociated into hydroxyl groups (HO$^-$) and protons (H$^+$). The first step in enzyme-catalyzed hydrolysis may involve the attack of a proton on the target bond. Molecules of phosphoric acid ($H_3PO_4$) dissociate to give phosphate anions ($PO_4^{3-}$) and protons. The first step in enzyme-catalyzed phosphorolysis may involve the attack of a proton.

Step 2:   The attack of the proton on the target bond results in the development of a positive charge, which attracts an electron from an adjacent atom of the polymer. In the case of the peptide bond, the electron is drawn away from the carbonyl carbon. In the case of the carbohydrate (sugar) bond, the electron is drawn away from the carbon adjacent to the oxygen. The result of this electron withdrawal is breaking of the bond, resulting in formation of a carbonium ion (shown at the left) and a stable byproduct (shown at the right).

Step 3:   Carbonium ions are unstable reactive molecules. As the carbonium ion is formed, the hydroxyl group (HO$^-$) or phosphate anion quickly condenses with it.

In the case of the phosphorolysis of glycogen, condensation of the phosphate group with the carbonium ion results in production of a molecule of phosphorylated glucose. The glucose phosphate is ready to participate in the energy-generating pathway of glycolysis. If hydrolysis were used instead of phosphorolysis, a separate step would be required to phosphorylate the glucose molecule. This step would require ATP and thus would be energy expensive. Another reason for phosphorolysis is that glucose (and other molecules) in phosphorylated form does not tend to cross phospholipid membranes and is thus retained in the cell more readily.

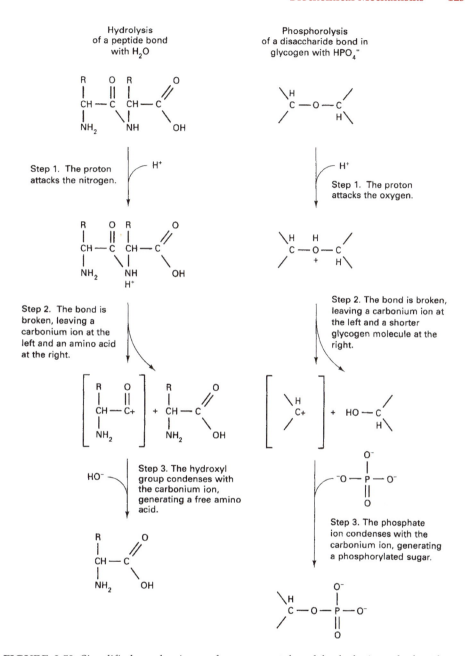

**FIGURE 2.52** Simplified mechanisms of enzyme-catalyzed hydrolysis and phosphoro-lysis. The end-result of both mechanisms is the splitting of a covalent bond.

## Action of Proteases

This discussion of chymotrypsin illustrates several biological concepts: (1) En-zymes greatly increase the rate at which a particular reaction occurs over the rate of a spontaneous reaction in the absence of a catalyst. (2) The amino acids that constitute a particular enzyme occur in a specific arrangement to form a catalytic system. The catalytic system may consist of several sites on the enzyme that may be hydrogen bonded to each other and to the substrate. The various sites act together (in concert) during the event of catalysis. (3) Enzymes that catalyze similar, or identical, reactions may be members of the same family of enzymes. If two different enzymes catalyze a similar reaction, the amino acids that form the catalytic systems may be identical. However, the amino acids that govern which substrates bind to the catalytic site may differ. In short, the events of substrate binding and of catalysis may be governed by different regions of an enzyme.

Chymotrypsin contains serine at residue 195 (Ser 195) from the N terminus. The hydroxyl group of Ser 195 is reactive and participates in the chemistry of catalysis. Asp 102 (asparagine) and His 57 (histidine) act together to draw off the proton from the hydroxyl group of Ser 195, as shown in Figure 2.53. The hydroxyl group of serine normally is not acidic; in free serine, it has a very high $pK$ ($pK = 14$). However, in chymotrypsin, the proton of Ser 195 is drawn off by His 57. In turn, a proton from His 57 is drawn off by Asp 102. These combined actions, called a charge-relay system, are activated by the binding of a substrate (a dietary protein). The binding of substrate to enzyme, and operation of the charge-relay system, is shown in Figure 2.53A–D.

Chymotrypsin contains glycine at position 193 (Gly 193). The amino group of Gly 193 attracts electrons from a carbonyl oxygen of the substrate. This attraction, which involves hydrogen bonding, promotes the development of a positive charge on the carbonyl carbon of the substrate (Figure 2.53C). In short, the carbonyl group becomes more highly polarized than usual. This condition enables Ser 195 to attack the substrate. This attack results in the splitting in two of the dietary protein (Figure 2.53E,F). One part is discharged from the enzyme, while the other part remains (temporarily) covalently bound.

Finally, a molecule of water participates in the chemistry of catalysis. The HO⁻ group attacks the covalently bound part of the substrate (dietary protein), resulting in its discharge from the chymotrypsin molecule (Figure 2.53F). The H⁺ of water then binds to Ser 195, regenerating the serine group to its initial non-ionized state.

Proteins are hydrolyzed very slowly with storage in water at neutral pH. However, addition of proteases can increase the rate of hydrolysis about 10 billion times over the spontaneous rate. The chymotrypsin mechanism depicted in Figure 2.53 is shared by trypsin and elastase. These three proteases are members of a family called the serine proteases (named after Ser 195). Carboxypeptidase A and pepsin catalyze peptide hydrolysis by different mechanisms and are not part of this family.

EXERCISE

Carboxypeptidase A contains a zinc ion at its active site. This $Zn^{2+}$ functions in a manner similar to Gly 193 in chymotrypsin, but its action is more powerful. Explain more fully the function of zinc in this enzyme.

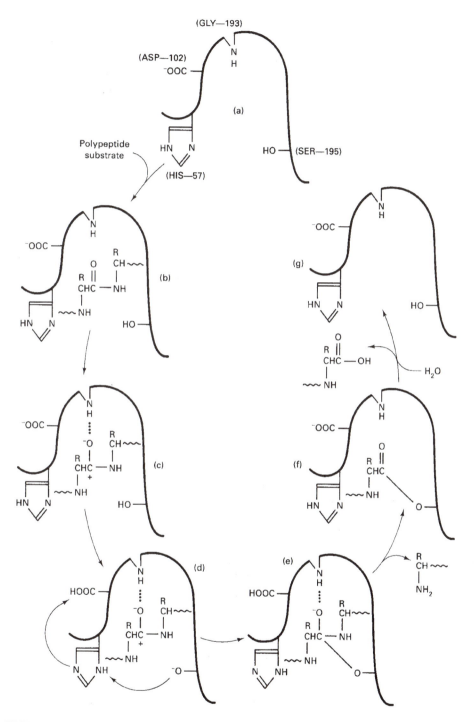

**FIGURE 2.53** Catalytic mechanism for hydrolysis of a peptide bond by chymotrypsin. (a) Four amino acids in the polypeptide chain of chymotrypsin have been shown to participate in the chemistry of catalysis: His 57, Asp 102, Gly 193, and Ser 195. (b) Binding of the substrate, a dietary polypeptide, to the active site of chymotrypsin. (c) The carbonyl group of the target peptide bond of the substrate becomes more polarized. (d) A charge relay

## Addition of Water to Carbon Dioxide

Carbonic anhydrase is an enzyme of central importance to the production of gastric acid. This enzyme, which contains zinc, accelerates the naturally occurring reversible reaction of $CO_2$ with water. Before considering the mechanism of action of the enzyme, note the resonating structure of $CO_2$. The carbon-to-oxygen bonds are polar; the molecular structure can be represented by the resonance hybrids shown in Figure 2.54. Note that in two of the three forms shown the central carbon atom has a positive charge.

Figure 2.55 shows a simplified mechanism for the action of carbonic anhydrase. This diagram depicts the events that occur at the active site of the enzyme. The active site contains a zinc ion and a residue of histidine, part of the polypeptide chain that constitutes the enzyme. The positively charged zinc ion attracts the oxygen atom of a water molecule. The oxygen atom of water carries a partial negative charge. Histidine contains a nitrogen in its side chain group. One property of nitrogen atoms is that they can be protonated (accept $H^+$). As shown in Figure 2.55, the enzyme mechanism involves three steps:

Step 1:    A water molecule is attracted to the zinc ion at the active site of carbonic anhydrase. The positively charged zinc ion displaces a proton from the water molecule. The displaced proton finds a new place of residence — the histidine residue. This histidine residue probably aids in the removal of the proton from the water molecule, in concert with the action of the zinc ion. Combination of the zinc ion ($Zn^{2+}$) with the hydroxyl group does not form a complex with the structure $Zn^+OH$. The zinc atom does not change its valence (its number of charges). Instead, the complex has the structure $Zn^{2+}(OH^-)$. [Calcium ions ($Ca^{2+}$) behave similarly to zinc ions. In contrast, iron and copper ions readily change their valences when they participate in biochemical reactions.]

Step 2:    A molecule of carbon dioxide approaches the active site of the enzyme. The carbon atom of $CO_2$ attracts the hydroxyl group associated with the zinc ion. The hydroxyl group leaves the zinc ion for the $CO_2$ molecule.

system involving Asp 102 and His 57 provokes dissociation of a proton from the hydroxyl group of Ser 195. (e) Ser 195 attacks the hyperpolarized carbonyl group of the substrate and forms a covalent bond with the substrate. (f) The peptide bond is broken and a product is discharged. Expressed as "paper chemistry," the negative charge on the oxygen flips down to the carbon atom. Since a carbon atom cannot have five bonds, the bond between the carbon and nitrogen is broken and the nitrogen has a negative charge. At the same time, the charge relay reverses itself and the negatively charged nitrogen immediately receives a proton from His 57. (g) Water enters the active site and is used as a substrate. The OH of water attacks the carbonyl carbon, forming a carboxylic acid, and a second product is discharged from the enzyme. The $H^+$ of water immediately binds to Ser 195. At this point, the active site is identical to that in (a).

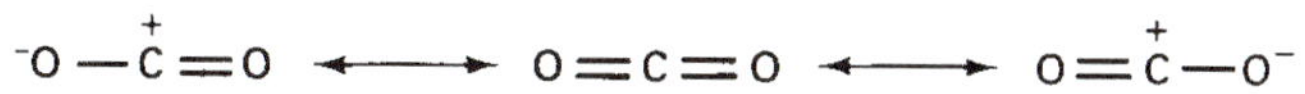

$$^-O-\overset{+}{C}=O \longleftrightarrow O=C=O \longleftrightarrow O=\overset{+}{C}-O^-$$

**FIGURE 2.54** Resonance of carbon dioxide.

Step 3:   Condensation of the hydroxide anion with $CO_2$ yields a bicarbonate anion ($HCO_3^-$). The bicarbonate anion leaves the active site of the enzyme. The proton bound to the histidine residue dissociates, producing a proton and a bicarbonate ion. Combination of these two products produces carbonic acid ($H_2CO_3$).

As discussed earlier, carbonic acid is a source of protons in gastric acid (HCl) production. The bicarbonate anion, which might be considered a byproduct of gastric acid production, is transported out of the parietal cell into the bloodstream.

## Family of Carbonic Anhydrases

Mammalian cells contain not one, but seven, carbonic anhydrases. All of these enzymes catalyze the same reaction, but they have slightly different structures, occur in different tissues, and serve different physiological roles. These roles include the dehydration of bicarbonate in the lungs (before exhalation as $CO_2$), acid production by the stomach, blood pH regulation by the kidney, and reclamation of bicarbonate from the urine (Baird *et al.*, 1997).

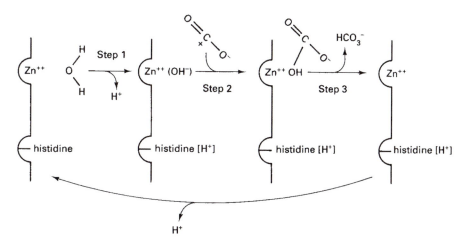

**FIGURE 2.55** The active site of carbonic anhydrase and simplified mechanism of action of the enzyme. The enzyme is shown catalyzing the hydration of $CO_2$, although the dehydration of bicarbonate also is catalyzed freely by the enzyme.

## SUMMARY

The digestive and absorptive system of the human body includes the digestive tract and the organs that provide the chemicals needed for its action and control. These chemicals include enzymes synthesized by the salivary glands, the stomach, the pancreas, and the liver; gastric acid to produce the proper pH for denaturation of proteins in the stomach; and bicarbonate to neutralize the acid to the neutral pH required by the enzymes of the small intestine. Some enzymes are secreted in active form, others as zymogens that are activated in the stomach or small intestine. Some enzymes are present at all times; others are secreted only as required. Secretion is stimulated by hormones that travel in the bloodstream or by signals from the brain that are transmitted through the vagus nerve. Another class of chemicals, the bile salts synthesized by the liver, is required for transport and absorption of insoluble nutrients. Three of the most important hormones in digestion are gastrin, cholecystokinin, and secretin. Ion exchange mechanisms are required for gastric acid production as well as salt absorption.

Specific chemical reactions and catalysts are required for digestion of the three major classes of nutrients — proteins, fats (lipids), and carbohydrates. Others are required for the absorption of digestive products through the wall of the small intestine. Hydrolysis, phosphorolysis, and the condensation of $CO_2$ with water are important biochemical mechanisms in digestion and absorption.

## REFERENCES

Abrams, C. K., Hamosh, M., Lee, T. C., Ansher, A. F., Collen, M. J., Lewis, J. H., Benjamin, S. B., and Hamosh, P. (1988). Gastric lipase. *Gastroenterology* **95**, 1460–1464.

Anderson, T. A. (1982). Recent trends in carbohydrate consumption. *Annu. Rev. Nutr.* **2**, 113–132.

Babayan, V. K. (1987). Medium chain triglycerides and structural lipids. *Lipids* **22**, 417–420.

Baird, T., Waheed, A., Okuyama, T., Sly, W. S., and Fierke, C. A. (1997). Catalysis and inhibition of human carbonic anhydrase IV. *Biochemistry* **36**, 2669–2678.

Beglinger, C., Fried, M., Whitehouse, I., Jansen, J. B., Lamers, C. B., and Gyr, K. (1985). Pancreatic enzyme response to a liquid meal and to hormonal stimulation. *J. Clin. Invest.* **75**, 1471–1476.

Bernback, S., Blackberg, L., and Hernell, O. (1990). The complete digestion of milk triacylglycerol in vitro requires gastric lipase, pancreatic colipase dependent lipase, and bile salt-stimulated lipase. *J. Clin. Invest.* **85**, 1221–1226.

Blair, A. J., Richardson, C. T., Walsh, J. H., and Feldman, M. (1987). Variable contribution of gastrin to gastric acid secretion after a meal in humans. *Gastroenterology* **92**, 944–949.

Borel, P., Armand, M., Senft, M., Andre, M., Lafont, H., and Lairon, D. (1991). Gastric lipase. *Gastroenterology* **100**, 1582–1589.

Bornet, F., Fontvieille, A. M., Rizkalla, S., Colonna, P., Blayo, A., Mercier, C., and Slama, G. (1989). Insulin and glycemic responses in healthy humans to native starches processed in different ways. *Am. J. Clin. Nutr.* **50**, 315–323.

Challacombe, D. N., Edkins, S., and Brown, G. A. (1975). Duodenal bile acids in infancy. *Arch. Dis. Child.* **50**, 837–843.

Chinachoti, P. (1995). Carbohydrates: Functionality in Foods. *Am. J. Clin. Nutr.* **61**, 922S–929S.

Deen, P., Verdijk, M., Knoers, N., Wieringa, B., Monnens, L., Os, C., and Oost, B. A. (1994). Requirement of human renal water channel aquaporin-2 for vasopressin-dependent concentration of urine. *Science* **264**, 92–95.

De Graf, J., and Woussen-Colle, M. (1986). Influence of the stimulation state of the parietal cells on the inhibitory effect of omeprazole on gastric acid secretion in dogs. *Gastroenterology* **91**, 333–337.

Efrat, S. (1997). Making sense of glucose sensing. *Nature Genet.* **17**, 249–250.

Feldman, M., and Richardson, C. T. (1986). Role of thought, sight, smell, and taste of food in the cephalic phase of gastric acid secretion in humans. *Gastroenterology* **90**, 428–433.

Fleming, S. E., Youngman, L. D., and Ames, B. N. (1994). Intestinal cell proliferaion is influenced by intakes of protein and energy, aflatoxin, and whole-body radiation. *Nutr. Cancer* **22**, 11–30.

Folsch, U. R., Cantor, P., Wilms, H. N., Schafmayer, A., Becker, H., and Creutzfeldt, W. (1987). Role of cholecystokinin in the negative feedback control of pancreatic enzyme secretion in conscious rats. *Gastroenterology* **92**, 449–458.

Galeano, N. F., Darling, P., Lepage, G., Leroy, C., Collet, S., Giguere, R., and Roy, C. C. (1987). Taurine supplementation of a premature formula improves fat absorption in preterm infants. *Pediatr. Res.* **22**, 67–71.

Garner, A., Gadlallah, H., and Parsons, M. E. (1996). 1976 and all that! — 20 years of antisecretory therapy. *Gut* **39**, 784–786.

Goyal, R., and Hirano, I. (1996). The enteric nervous system. *New Engl. J. Med.* **334**, 1106–1115.

Guandalini, S., and Rubino, A. (1982). Development of dipeptide transport in the intestinal mucosa of rabbits. *Pediatr. Res.* **16**, 99–103.

Hansen, P., Gulve, E., Marshall, B., Gao, J.,, Pessin, J., Holloszy, J., and Mueckler, M. (1995). Skeletal muscle glucose transport and metabolism are enhanced in transgenic mice overexpressing the glut4 glucose transporter. *J. Biol. Chem.* **270**, 1679–1684.

Harris, A., and Misiewicz, J. J. (1996). Treating *Helicobacter pylori* — the best is yet to come? *Gut* **39**, 781–783.

Haworth, J. C., Hadorn, B., Gourley, B., Prasad, A., and Troesch, V. (1975). Intestinal enterokinase deficiency. *Arch. Dis. Child.* **50**, 277–282.

Henning, S. J., and Kretchmer, N. (1973). Development of intestinal function in mammals. *Enzymes* **15**, 3–23.

Hof, W., Blankenvoorde, M., Veerman, E., and Amerongen, A. (1997). The salivary lipocalcin Von Ebner's gland protein is a cysteine proteinase inhibitor. *J. Biol. Chem.* **272**, 1837–1841.

Ishibashi, K., Sasaki, S., Fushimi, K., Yamamoto, T., Kuwahara, M., and Marumo, F. (1997). Immunolocalization and effect of dehydration on AQP3, a basolateral water channel of kidney collecting ducts. *Am. J. Physiol.* **272**, F235–F241.

Ito, I., and Winchester, R. J. (1963). The fine structure of the gastric mucosa in bat. *J. Cell Biol.* **16**, 543–557.

Kanno, K., Sasaki, S., Hirata, Y., Ishikawa, S., Fushimi, K., Nakanishi, S., Bichet, D. G., and Marumo, F. (1995). Urinary excretion of aquaporin-2 in patients with *diabetes insipidus*. *New Engl. J. Med.* **332**, 140–1545.

Keller, P., Semenza, G., and Shaltiel, S. (1995). Disposition of the carboxy-tail of rabbit lactase–phlorizin hydrolase elucidated with protein kinase A in vitro and in tissue culture. *FEBS Lett.* **368**, 563–567.

Kihara, H., and Snell, E. E. (1960). Peptides and bacterial growth. *J. Biol. Chem.* **235**, 1409–1414.

Klip, A., Tsakiridis, T., Marette, A., and Ortiz, P. (1994). Regulation of expression of glucose transporters by glucose: A review of studies in vivo and in cell cultures. *FASEB J.* **8**, 43–53.

Koh, T., Goldenring, J., Ito, S., Mashimo, H., Kopin, A. S., Varro, A., Dockray, G. J., and Wang, T. C. (1997). Gastrin deficiency results in altered gastric differentiation and decreased colonic proliferation in mice. *Gastroenterology* **113**, 1015–1025.

Lee, W., Kanai, Y., Wells, R. G., and Hediger, M. A. (1994). The high affinity $Na^+$/glucose cotransporter. *J. Biol. Chem.* **269**, 12032–12039.

Leibach, F. H., and Ganapathy, V. (1996). Peptide transporters in the intestine and the kidney. *Annu. Rev. Nutr.* **16**, 99–119.

Logsdon, C. D. (1995). Pancreatic duct cell cultures: There is more to ducts than salty water. *Gastroenterology* **109**, 1005–1009.

Marples, D., Frokiaer, J., Dorup, J., Knepper, M. A., and Nielsen, S. (1996). Hypokalemia-induced downregulation of aquaporin-2 water channel expression in rat kidney medulla and cortex. *J. Clin. Invest.* **97**, 1960–1968.

Mecham, R. P., Broekelmann, T., Fliszar, C., Shapiro, S., Welgus, H., and Senior, R. M. (1997). Elastin degradation by matrix metalloproteinases. *J. Biol. Chem.* **272**, 18071–18076.

McNeely, W. H., and Kovacs, P. (1975). The physiological effects of alginates and xanthan gum. *In* "Physiological Effects of Food Carbohydrates" (A. Jeanes and J. Hodge, eds.), pp. 269–281. Am. Chem. Soc., Washington, DC.

McPhie, P. (1972). A spectrophotometric investigation of the pepsinogen–pepsin conversion. *J. Biol. Chem.* **247**, 4277–4281.

Meddings, J. B., DeSouza, D., Goel, M., and Thiesen, S. (1990). Glucose transport and microvillus membrane physical properties along the crypt–villus axis of the rabbit. *J. Clin. Invest.* **85**, 1099–1107.

Oelkers, P., Kirby, L., Heubi, J., and Dawson, P. (1997). Primary bile acid malabsorption caused by mutations in the ileal sodium-dependent bile acid transporter gene (SLC10A2). *J. Clin. Invest.* **99**, 1880–1887.

Preston, G. M., Smith, B. L., Zeidel, M. L., Moulds, J. J., and Agre, P. (1994). Mutations in aquaporin-1 in phenotypically normal humans without functional CHIP water channels. *Science* **265**, 1585–1587.

Raina, S., Preston, G. M., Guggino, W. B., and Agre, P. (1995). Molecular cloning and characterization of an aquaporin cDNA from salivary, lacrimal, and respiratory tissues. *J. Biol. Chem.* **270**, 1908–1912.

Riby, J. E., and Kretchmer, N. (1984). Effect of dietary sucrose on synthesis and degradation of intestinal sucrase. *Am. J. Physiol.* **246**, G757–G763.

Riecken, E. O. (1988). Derangements of mucosal growth and differentiation in small intestinal diseases. *Scand. J. Gastroenterol., Suppl.* **23**(151), 86–93.

Santer, R., Schneppenheim, R., Dombrowski, A., Gotze, H., Steinmann, B., and Schaub, J. (1997). Mutations in GLUT2, the gene for the liver-type glucose transporter, in patients with Fanconi–Bickel syndrome. *Nature Genet.* **17**, 324–326.

Seatter, M., Kane, S., Porter, L., Arbuckle, M., Melvin, D., and Gould, G. (1997). Structure–function studies of the brain-type glucose transporter, GLUT3. *Biochemistry* **36**, 6401–6407.

Shiratori, K., Watanabe, S., Chey, W. Y., Lee, K. Y., and Chang, T. M. (1986). Endogenous cholecystokinin drives gallbladder emptying in dogs. *Am. J. Physiol.* **251**, G553–G558.

Stancioff, D. J., and Renn, D. W. (1975). Physiological effects of carrageenan. *In* "Physiological Effects of Food Carbohydrates" (A. Jeanes and J. Hodge, eds.), pp. 282–295. Am. Chem. Soc., Washington, DC.

Stanley, N. F. (1995). Agars. *In* "Food Polysaccharides and Their Applications" (A. M. Stephen, ed.), pp. 187–202. Dekker, New York.

Sturman, J. A., Moretz, R. C., French, J. H., and Wisniewski, H. M. (1985). Taurine deficiency in the developing cat. *J. Neurosci. Res.* **13**, 405–416.

Taha, C., Mitsumoto, Y., Liu, Z., Skolnik, E., and Klip, A. (1995). The insulin-dependent biosynthesis of GLUT1 and GLUT3 glucose transporters in L6 muscle cells is mediated by distinct pathways. *J. Biol. Chem.* **270**, 24678–24681.

Turnberg, L. A., Bieberdorf, F. A., Morawski, S. G., and Fordtran, J. S. (1970). Interrelationships of chloride, bicarbonate, sodium, and hydrogen transport in the human ileum. *J. Clin. Invest.* **49**, 557–567.

Vahouny, G. V., Fawal, I., and Treadwell, C. R. (1957). Factors facilitating cholesterol absorption from the intestine via lymphatic pathways. *Am. J. Physiol.* **188**, 342–346.

Van Beynum, G. M. A., and Roels, J. A. (1985). "Starch Conversion Technology." Dekker, New York.

Vinnakota, S., Qian, X., Egal, H., Sarthy, V., and Sarkar, H. K. (1997). Molecular characterization and in situ localization of a mouse retinal taurine transporter. *J. Neurochem.* **69**, 2238–2250.

Weiner, I. M., and Lack, L. (1962). Absorption of bile salts from the small intestines in vivo. *Am. J. Physiol.* **202**, 155–157.

Weiss, L. (1983). "Histology," 5th ed. Elsevier Biomedical, New York.

Wurzburg, O. B. (1995). "Modified Starches in Food Polysaccharides and Their Applications" (A. M. Stephen, ed.), pp. 67–97. Dekker, New York.

You, C. H., Rominger, J. M., and Chey, W. Y. (1983). Potentiation effect of cholecystokinin–octapeptide on pancreatic bicarbonate secretion stimulated by a physiologic dose of secretin in humans. *Gastroenterology* **85**, 40–45.

Zeller, K., Duelli, R., Vogel, J., Schrock, H., and Kuschinsky, W. (1995). Autoradiographic analysis of the regional distribution of glut3 glucose transporters in the rat brain. *Brain Res.* **698**, 175–179.

Zock, P. L., Vries, J. H, Fouw, N. J., and Katan, M. B. (1995). *Am. J. Clin. Nutr.* **61**, 48–55.

## BIBLIOGRAPHY

Aldini, R., Montagnani, M., Roda, A., Hrelia, S., Biagi, P., and Roda, E. (1996). Intestinal absorption of bile acids in the rabbit: Different transport rates in jejunum and ileum. *Gastroenterology* **110**, 459–468.

Aponte, G. W., Park, K., Hess, R. Garcia, R., and Taylor, I. L. (1989). Meal-induced peptide tyrosine inhibition of pancreatic secretion in the rat. *FASEB J.* **3**, 1949–1955.

Borgstrom, B., Erlanson-Albertsson, C., and Wieloch, T. (1979). Pancreatic colipase: Chemistry and physiology. *J. Lipid Res.* **20**, 805–816.

Buller, H. A., Montgomery, R. K., Sasak, W. V., and Grand, R. J. (1987). Biosynthesis, glycosylation, and intracellular transport of intestinal lactase–phlorizin hydrolase in rat. *J. Biol. Chem.* **262**, 17206–17211.

Burman, P., Mardh, S., Norberg, L., and Karlsson, F. A. (1989). Parietal cell antibodies in pernicious anemia inhibit H,K-adenosine triphosphatase, the proton pump of the stomach. *Gastroenterology* **96**, 1434–1438.

Chijiiwa, K., and Linscheer, W. G. (1987). Distribution and monomer activity of cholesterol in micellular bile salt: Effect of cholesterol level. *Am. J. Physiol.* **252**, G309–G314.

Cornish-Bowden, A., and Wharton, C. W. (1988). "Enzyme Kinetics." IRL Press, Washington, DC.

Crapo, P. A. (1985). Simple versus complex carbohydrate use in the diabetic diet. *Annu. Rev. Nutr.* **5**, 95–114.

Dourish, C. T., Rycroft, W., and Iversen, S. D. (1989). Postponement of satiety by blockade of brain cholecystokinin (CCK-B) receptors. *Science* **245**, 1509–1511.

Erikson, R. H., and Kim, Y. S. (1990). Digestion and absorption of dietary protein. *Annu. Rev. Med.* **41**, 133–139.

Freeman, H. J., and Kim, Y. S. (1978). Digestion and absorption of protein. *Annu. Rev. Med.* **29**, 99–116.

Friedman, H. I., and Nylund, B. (1980). Intestinal fat digestion, absorption, and transport — A review. *Am. J. Clin. Nutr.* **33**, 1108–1139.

Hafkenscheid, J. C., and Hectors, M. P. (1975). An enzymatic method for the determination of the glycine/taurine ratio of conjugated bile acids in bile. *Clin. Chim. Acta* **65**, 67–74.

Hardison, W. G., and Grundy, S. M. (1983). Effect of bile acid conjugation pattern on bile acid metabolism in normal humans. *Gastroenterology* **84**, 617–620.

Hartsuck, J. A., and Lipscomb, J. A. (1970). Carboxypeptidase A. *In* "The Enzymes," 3rd ed. (P. D. Boyer, ed.), Vol. 3, pp. 1–56. Academic Press, New York.

Henning, S. J. (1985). Ontogeny of enzymes in the small intestines. *Annu. Rev. Physiol.* **47**, 231–245.

Hunziker, W., Spiess, M., Semenza, G., and Lodish, H. F. (1986). The sucrase–isomaltase complex: Primary structure, membrane orientation, and evolution of a stalked, intrinsic brush border protein. *Cell* 46, 227–234.

Kelley, M., and Vessey, D. A. (1994). Determination of the mechanism of reaction for bile acid: CoA ligase. *Biochem. J.* **304**, 945–949.

Knickelbein, R. G., Aronson, P. S., and Dobbins, J. W. (1988). Membrane distribution of sodium–hydrogen and chloride–bicarbonate exchanges in crypt and villus cell membranes from rabbit ileum. *J. Clin. Invest.* **82**, 2158–2163.

Knopf, K., Sturman, J. A., Armstrong, M., and Hayes, K. C. (1978). Taurine: An essential nutrient for the cat. *J. Nutr.* **108**, 773–778.

Kraut, J. (1977). Serine proteases. *Annu. Rev. Biochem.* **46**, 331–358.

Krejs, G. J. (1986). Physiological role of somatostatin in the digestive tract: Gastric acid secretion, intestinal absorption, and motility. *Scand. J. Gastroenterol., Suppl.* **21**(119), 47–53.

Leblond, C. P., and Stevens, C. E. (1948). The constant renewal of the intestinal epithelium in the albino rat. *Anat. Res.* **100**, 357–371.

Louvard, D., Maroux, S., Baratti, J., and Desnuelle, P. (1973). On the distribution of enterokinase in porcine intestine and on its subcellular localization. *Biochim. Biophys. Acta* **309**, 127–137.

Nicholl, C. G., Polak, J. M., and Bloom, S. R. (1985). The hormonal regulation of food intake, digestion, and absorption. *Annu. Rev. Nutr.* **5**, 213–239.

Shu, R., David, E., and Ferraris, R. P. (1997). Dietary fructose enhances intestinal fructose transport and GLUT5 expression in weaning rats. *Am. J. Physiol.* **272**, G446–G453.

Small, D. M. (1997). Point mutations in the ileal bile salt transporter cause leaks in the enterohepatic circulation leading to severe chronic diarrhea and malabsorption. *J. Clin. Invest.* **99**, 1807–1808.

Vessey, D. A. (1978). The biochemical basis for the conjugation of bile salts with either glycine or taurine. *Biochem. J.* **174**, 621–626.

Weller, A., Smith, G. P., and Gibbs, J. (1990). Endogenous cholecystokinin reduces feeding in young rats. *Science* **247**, 1589–1591.

Wolfe, M. M., and Soll, A. H. (1988). The physiology of gastric acid secretion. *N. Engl. J. Med.* **319**, 1707–1715.

# 3

# NUTRIENTS THAT RESIST OR ESCAPE DIGESTION

## OVERVIEW

Nutrients that may resist or escape digestion include proteins (specifically maternal antibodies), carbohydrates (lactose, starches), dietary fibers, and — in malabsorption syndromes — lipids. This chapter begins with a discussion of the antibodies of mother's milk, a description of that part of the immune system located in the gut, and a recommendation for avoiding food allergies. Problems related to lactose and sucrose intolerance, that is, the inability to hydrolyze dietary disaccharides, are then discussed. Data revealing that hydrogen gas can be produced in the gut introduces the topic of the ecology of the gut microflora. The gut microflora metabolize lactose and dietary fibers. Dietary fiber consists of a diverse group of organic polymers that, with the exception of lignin, contain units of various sugars. Their role in digestion and health is discussed. Finally, malabsorption syndromes involving undigested lipids are addressed.

## RESISTANCE OF PROTEINS TO DIGESTION

Dietary proteins may escape digestion in the mildly acidic stomach of the human newborn. After feeding, the pH of the infant's stomach may remain near neutral or drop to only pH 4.0, in contrast with the pH of 2.0 reached in children and adults after eating. The output of pepsin, the protease of the stomach, also may remain low for the first several months of life. Consequently, proteins may escape acid denaturation and proteolysis in the stomach of the newborn. Therefore, proteolysis by *pancreatic* proteases assumes increased importance.

The lack of digestive powers of the neonatal stomach may have the benefit of allowing increased survival of the antibodies in mother's milk. Survival of these antibodies during passage through the newborn's digestive tract is required to help prevent infections in the lumen of the gastrointestinal tract. Several classes of antibodies, including immunoglobulin A (IgA) and immunoglobulin G (IgG), are present in the milk produced during the first few days of lactation. The antibody-rich milk produced during the first few days of lactation, whether from humans or animals, is called **colostrum**. The antibody present in highest concentration is IgA, which can recognize and bind to bacteria, aiding in their eventual phagocytosis by white blood cells. For example, IgA binds to *Escherichia coli*, a common pathogen of the neonate. IgG is also present, but at a lower concentration than IgA. For comparison, human milk contains 12 g total protein per liter, 1–2 g IgA/liter, and 70 g lactose/liter.

As shown in Figure 3.1, the concentrations of IgA and IgG decline in human breast milk in the weeks following birth. These antibodies tend to survive digestion in the infant's stomach because of its poor digestive powers and because the antibodies have structures that resist proteolysis. Breast-fed infants are more resistant to enteric (intestinal) infections than formula-fed infants. Breast milk is espe-

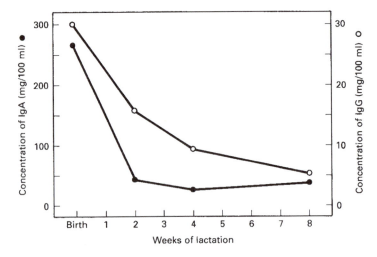

**FIGURE 3.1** Concentration of antibodies in milk. The concentrations of IgA (●) and IgG (○) decline in breast milk in the weeks following birth. (Redrawn with permission from Miranda *et al.*, 1983.)

cially important for this reason to infants in underdeveloped countries, where the water supply may be contaminated with animal and human feces. Enteric infections and the associated diarrhea are thought to be responsible for one-fourth of infant deaths on a global level. These infections can be minimized by public health measures and by breast-feeding.

Antibodies are released by the mammary gland both as free antibody molecules and as white blood cells that synthesize antibodies in the milk. The antibodies consumed by the human infant probably are not absorbed by the infant's intestines and transferred to the bloodstream, but pass through to fight bacterial infections within the lumen of the gut. However, the bloodstream of the newborn does contain antibodies, such as IgG, and these were transferred to the fetus via the placenta prior to birth. Hence, the newborn receives antibodies from the mother both before and after birth. This situation is in contrast with that of calves. The cow does not transfer antibodies to the fetal calf, so the entire transfer of immunoglobulins occurs through the colostrum. The newborn calf is able to absorb these immunoglobulins into the bloodstream, but only during the first day of life. After the first day of life, the ability of the calf to absorb antibodies is curtailed sharply. The mortality of calves due to infections is a chronic farm problem, so it is imperative that a newborn calf receive colostrum during its first day.

## *Immunology of the Intestinal Tract and the Breast*

What controls the specificity of the antibodies released by mother's milk? How can a molecule (an antibody) released by the mammary gland be able to recognize and bind to bacteria of the gut? What factor is responsible for this relationship between two seemingly remote parts of the body, the breast and the gut lumen? The small intestine contains a variety of types of cells: enterocytes, goblet cells, endocrine cells, and white blood cells. The white blood cells occur in clusters called **Peyer's patches**. Approximately 20 such clusters occur in the small intestine. Another term for this cluster is a **follicle**. The white blood cells face the lumen of the gut and are exposed to its contents. They interact with bacteria in the gut and synthesize antibodies against components of the bacterial cells. The white blood cells in the mother's Peyer's patches can leave their location in the gut mucosa and migrate to other parts of the mother's body, for example, to the mammary gland. The mammary gland is able to secrete antibodies that recognize and bind to structures of the gut bacteria because of the migration of the white blood cells from the gut to the breast.

Each Peyer's patch consists of a group of various cells, including **M cells**, lymphocytes, and macrophages. The M cells serve to take up foreign proteins and microorganisms from the lumen of the gut, so that the lymphocytes can respond and create antibodies (Kerneis *et al.*, 1997). Peyer's patches acquire lymphocytes from the bloodstream by a specific homing or targeting mechanism. In brief, the lymphocyte contains a membrane-bound protein called **integrin**. The integrin recognizes and binds to proteins of the extracellular matrix of the Peyer's patch, and other target locations, such as breast tissue. Actually, two different integrins are involved, and these are integrin $\alpha 4\beta 7$ and integrin $\alpha E\beta 7$ (Wagner *et al.*, 1996). The first integrin acts as a lasso, and loosely tethers the lymphocyte to the surface cells (epithelial cells) of the Peyer's patch. Then the second type of integrin develops a strong bond with the extracellular matrix. Eventually, the lymphocyte pushes itself in between the epithelial cells and makes its home inside of the Peyer's patch.

As mentioned earlier, the lymphocytes within the Peyer's patches eventually leave, migrate to other parts of the body via the bloodstream, and bind to sites dispersed throughout the gut mucosa, to the female genital tract, to the lungs, and to the female breast during pregnancy (Kagnoff, 1993). After taking residence in these new locations, the lymphocytes produce antibodies that protect these vulnerable areas from infection. The preceding scenario is sometimes called **enteric immunization**.

## Food Allergies

An unfortunate consequence of the survival of proteins in the gastrointestinal tract of the infant is the development of food allergies. Some of the proteins fed to an infant may come from mother's milk, whereas others may have a nonhuman origin, such as cow's milk. A foreign protein, once hydrolyzed to small peptides and amino acids, is no longer able to provoke an immune response. The small peptides and amino acids are no longer recognized as having a foreign origin and do not stimulate the synthesis of antibodies. However, when foreign proteins are fed to the infant, the proteins can escape digestion and retain enough "foreignness" to generate an immune response. This response may take the form of an allergic reaction.

Cow's milk is a food known to give rise to allergies. Perhaps 0.3% of infants develop allergies to cow's milk because of consumption at a young age. After the age of 6 months, the danger of developing new allergies ceases to be a problem in infant nutrition. In general, continuing breast-feeding infants to the age of 6 months is thought to be a wise practice, especially if food allergies tend to "run in the family."

## CARBOHYDRATES THAT ESCAPE DIGESTION

### Lactose

Lactose, the primary sugar found in milk, is a disaccharide composed of glucose and galactose. With age, many children lose their ability to digest large amounts of lactose, leading to **lactose intolerance**. This permanent condition, which occurs in adolescence and adulthood, is not an important health concern, although it may lead to an aversion to milk, an important food.

The enzyme lactase catalyzes the hydrolysis of lactose in the lumen of the gut. Generally, intestinal lactase activity occurs at a maximal level from birth through early childhood. Then activity declines to a residual level of 5–10% by the age of 5 or later in most populations of the world. In persons of European descent, lactase activity can remain at a high level throughout adulthood. Low levels of lactase result in lactose intolerance, evidenced as diarrhea and flatulence after consumption of milk or after a test dose of lactose. The symptoms of lactose intolerance can be avoided by drinking milk at a slower rate than usual.

Lactose intolerance is not an absolute condition. Some persons display symptoms after drinking one glass of milk, whereas others can tolerate up to a quart before suffering from abdominal cramps or diarrhea. The degree of lactose intolerance can be assessed by measuring the release of hydrogen gas in the breath following consumption of lactose. This hydrogen, a product of bacterial metabolism, is produced in the large intestine. Part of the gas is absorbed in the bloodstream, allowing it to be exhaled in the breath.

High incidences of lactose intolerance are found throughout East Asia and Southeast Asia and among Australian aborigines and Native Americans. Anthropological work has demonstrated that lactose intolerance is most common in regions of the world where adults do not drink milk. One particularly interesting study was conducted in Nigeria. This country contains tribes that traditionally have been either milk drinkers or non-milk drinkers. Members of the Fulani tribe in the north lead a pastoral lifestyle, raising livestock. Apparently, members of this tribe have consumed milk for thousands of years and tend to be lactose tolerant. However, the Yoruba in western Nigeria are an agricultural people. They tend not to raise cattle, are not milk drinkers, and show high incidences of lactose intolerance. Figure 3.2 shows the results of a study of the lactose tolerance of Australian aborigines compared with that of European students living in Australia. A rise in the concentration of hydrogen gas ($H_2$) in breath samples was used as an indication that lactose had not been digested (lactose intolerance). The results reveal striking differences between persons of different ancestral backgrounds; they also show considerable variations among individuals with the same background.

The study shown in Figure 3.3 addresses the biochemical mechanism of lactose intolerance and uses the technique of breath collection and $H_2$ analysis. Breath $H_2$ rose after a lactose-intolerant child consumed a dose of lactose, but did not rise after a dose of equal weights of glucose and galactose. These results are consistent with the fact that lactose intolerance arises from a deficiency in lactase.

An interesting phenomenon is seen in the far reaches of the northern hemisphere, where **sucrose intolerance** occurs among Eskimos. About 10% of the native population of Greenland is sucrose intolerant. The occurrence of sucrose intolerance among these populations is consistent with the lack of availability of sucrose-containing staples. People in Greenland traditionally have consumed meat as the major food source, and not fruits or sugar cane.

EXERCISE

Devise a test for lactose intolerance that involves the measurement of plasma glucose levels.

EXERCISE

Fatty foods may prolong the gastric emptying time. How might the consumption of fat with milk affect lactose intolerance?

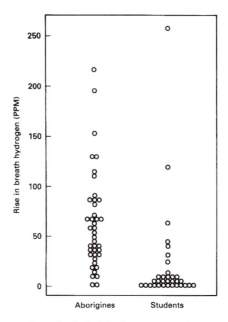

**FIGURE 3.2** Rise in concentration of exhaled hydrogen gas after a test meal. Australian aborigines and European students living in Australia consumed a test dose of 50 g lactose. Breath samples were taken before and 2 hr after the test dose. Gas analysis revealed that, whereas most of the aborigines experienced a rise in breath hydrogen, most of the students exhibited no rise at all. The data reveal that strikingly different responses to milk sugar can occur in persons of different ancestral backgrounds. The data also show that considerable variations in $H_2$ production occur in different individuals. (Redrawn with permission from Brand *et al.*, 1983.)

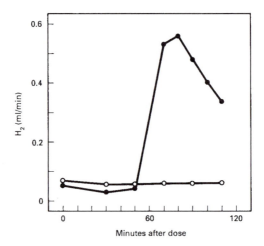

**FIGURE 3.3** Rate of exhalation of hydrogen gas. A lactose-intolerant child consumed a test dose of lactose (●) or a test dose containing equal weights of glucose and galactose (□). The rate of release of hydrogen gas in the breath was measured at intervals of time after each of the test doses. The data demonstrate that breath $H_2$ increased following the dose of lactose but not after consumption of the two monosaccharides. These results are consistent with the fact that lactose intolerance arises from a deficiency in lactase. (Redrawn with permission from Fernandes *et al.*, 1978.)

**TABLE 3.1**  Change of Starch and Sugar Content of Bananas during Ripening

| Carbohydrate | Percentage of dry matter of a banana | | |
| --- | --- | --- | --- |
| | Green (unripe) | Yellow (ripe) | Black (overripe) |
| Starch | 82 | 26 | 3 |
| Sugar | 7 | 63 | 88 |

Reprinted with permission from Englyst and Cummings (1986).

## Resistant Starch

Starch is the major energy source in the diet. Since starch consists of polymers of glucose residues joined by α-glycosidic linkages, one might expect starch to be completely digested and absorbed by the small intestine. However, its digestibility is influenced by the physical form of the starch in the food. About 15–20% of the starch in the diet is amylose; the remainder is amylopectin.

The starch content in plant foods changes with the age of the plant, that is, it decreases during the ripening process, with the concomitant production of free glucose. Starch may account for 3–82% of the dry matter of bananas, depending on ripeness (Table 3.1). Apparently, bananas are good sources of carbohydrate, whether unripe or overripe. However, are unripe bananas really good sources of carbohydrates? The ripening process involves the hydrolysis of the starch in the ripening plant. This activity is a form of autolysis (self-digestion). The starch also can be broken down after consumption via the action of pancreatic α-amylase, as well as other enzymes. The effect of amylase in cleaving banana starch is illustrated in Table 3.2. The data show the effects of attempts to digest banana starch in a test tube. The starch was mixed with hog pancreas amylase and incubated. Then the extent of hydrolysis was measured. As indicated in the table, different treatments were used. About half the starch in a raw banana resisted hydrolysis, but all the starch in a cooked banana was hydrolyzed. On cooling a cooked banana, a small fraction of the starch lost its susceptibility to hydrolysis.

The starch in starchy foods, such as bananas, potatoes, and rice, occurs in structures called grains. The starch molecule occurs in an organized arrangement in such grains. The grains are not soluble in water and are somewhat resistant to attack by digestive enzymes. The starch in the grains is hydrolyzed during the process of ripening (in the case of bananas) or with storage (in the case of potatoes).

**TABLE 3.2**  Percentage of Banana Starch Resisting Hydrolysis by Hog Pancreatic α-Amylase

| | |
| --- | --- |
| Raw banana | 54 |
| Freshly cooked banana | 0 |
| Cooked and cooled | 8 |
| Cooked, cooled, and reheated | 5 |

*Source*: Englyst and Cummings (1986).

Cooking disrupts the structure of the starch molecules in the grains, resulting in a marked increase in their digestibility. With cooling, the starch molecules regain a more orderly structure, resulting in a slight increase in resistance to enzymatic hydrolysis, as illustrated in the study with banana starch. The reversion to an organized structure with cooling is called the "crystallization" of starch. Although this process may not result in a significant effect on the nutritional value of the food, it may detract from its aesthetic value. For example, freshly baked bread may be quite appealing. With storage for a day or so, the bread loses its appeal and is called "stale." The original consistency of the bread can be revived by heating it. The staling effect occurs because of the crystallization of the bread starch. The revival by heating results because of the disruption of its orderly structure.

## DIETARY FIBERS

The average daily intake of fiber in the United States is about 12 g. The Food and Nutrition Board has not established a requirement for fiber but states that fiber intake should be derived from fruits, vegetables, legumes, and cereals, not from fiber concentrates. If a recommended dietary allowance for fiber is established, it will refer to specific types of fiber rather than to fiber in general.

The American Dietetic Association recommends an intake of 25–30 g fiber/day for adults, on the basis of the demonstrated benefit of reducing the risk for diverticular disease, and possible benefit of reducing the risk for colorectal cancer and breast cancer (Marlett and Slavin, 1997).

The National Cancer Institute recommends an intake of 20–30 g fiber/day because of its association with reduced risk for certain types of cancer (Labarthe, 1996).

### Classes, Sources, and Biochemistry

Dietary fiber includes all the polysaccharides and lignin of the diet that are not digested by human enzymes. Plant cell walls are the major source of dietary fibers. The major fibers of foods are cellulose, β-glucans, hemicellulose, and pectins. Gums are minor fibers of foods. Lignin is another minor food fiber. Dietary fibers may be classified as resistant starches, nonstarch polysaccharides, and lignin. About 75% of the fibers in cereals are polysaccharides other than cellulose, 17% are cellulose, and 7% are lignin. In fruits, the fiber consists of roughly 63% carbohydrates other than cellulose, 20% cellulose, and 17% lignin. In vegetables, the fiber consists of roughly 68% polysaccharides other than cellulose, 31% cellulose, and 3% lignin (Slavin, 1987). Note that vegetables are relatively rich in cellulose.

Resistant starch resists hydrolysis not because of its chemical structure but because of its physical structure, as illustrated by the study of bananas. Resistant starch occurs in raw starch granules and can accumulate during freeze-drying of foods or during prolonged cooling of some cooked foods. This form of starch escapes digestion in the small intestine but may be degraded substantially by bacterial enzymes in the large intestine. Various studies with human subjects have shown that about 18% of the starch in beans escapes breakdown in the small intestine. The values for other starch sources are cornbread, 6%; oat bread, 8%; white bread, 10-20%; and potato, 13% (Cummings and Englyst, 1987).

Nonstarch polysaccharides include pectins, gums, hemicelluloses, cellulose, β-glucans, and other polysaccharides. The nonstarch polysaccharides are not hydrolyzed by human enzymes but may be degraded substantially by the gut microflora, especially during passage through the large intestine. Nonstarch polysaccharides constitute about 2.0% by weight of foods such as brown rice, potatoes, apples, cabbage, and carrots. Of the weight of whole wheat flour, oats, and barley, 7–10% consists of nonstarch polysaccharides; the value for wheat bran is about 40%.

Figure 3.4 shows the structures of pectin and cellulose. Also shown are the sugars that make up the polysaccharide backbone and side chains of the heterogeneous compound hemicellulose.

Pectin is a polymer of galacturonic acid, with side chains that contain a number of other sugars such as glucose, galactose, and rhamnose. As shown in Figure 3.4, pectin contains methoxy groups ($-OCH_3$). In plants, pectin occurs as a layer in between cells. The linear chain of pectin contains an occasional residue of the sugar rhamnose (which interrupts the chain), and, where the rhamnose is attached, there may grow a branch of polygalacturonic acid. Pectin is used to control the movement of water in plant fluids, especially during the growth of the plant. The pectin polymers reside in between fibers of cellulose and may even be covalently linked at various points to the cellulose. The softening of some fruits, such as tomatoes,

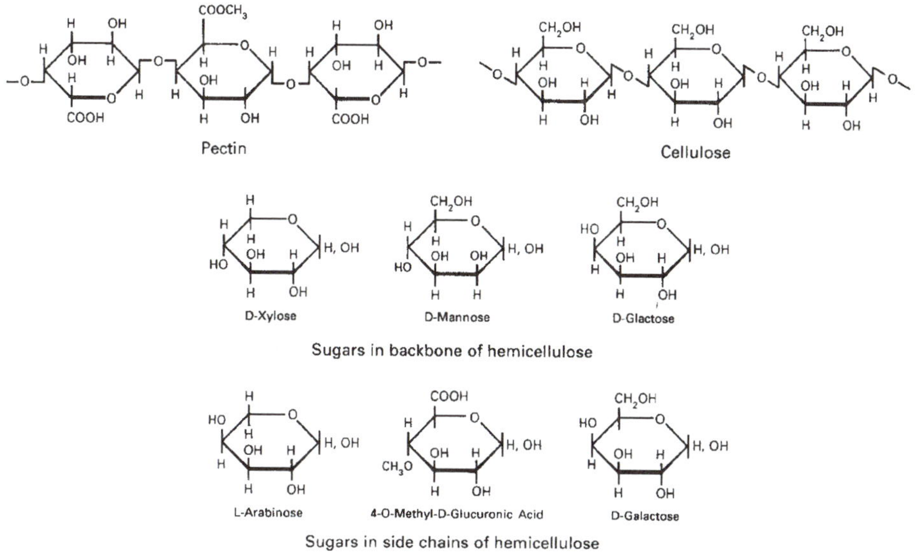

FIGURE 3.4  Structures of the dietary fibers pectin and cellulose. Also shown are the sugars that make up the polysaccharide backbone and side chains of the heterogeneous compound hemicellulose.

is due to the activity of the enzyme polygalacturonase (Thakur *et al.*, 1997). The pectin found in citrus fruits and apples is degraded almost completely by the gut microflora. The methoxy groups of pectin are metabolized by the gut microflora to generate methane gas.

Cellulose, the most abundant organic molecule in nature, is a linear polymer of several thousand glucose units connected by $\beta(1 \rightarrow 4)$ linkages. Cellulose occurs in tightly packed aggregates in plants and tends to resist microbial hydrolysis. The extent of degradation of cellulose may range from 25 to 75%.

Hemicelluloses and gums are heterogeneous polymers that occur with a variety of chemical structures. Hemicellulose, a major fiber of corn and wheat bran, is 50–80% degraded by the gut bacteria. Gums, which are degraded substantially, are not major fibers of the diet. Gums are used as thickening agents in the food industry. For example, the gum isolated from the Indian cluster bean is used as a stabilizer for ice cream.

Lignins, three-dimensional networks built of units of phenylpropane (Figure 3.5), are components of the tough or woodlike portions of plants and seeds. Bran and other cereal products are the main sources of dietary lignin. This fiber is distinguished by the fact that it is not degraded by enzymes produced by the human or by the gut microflora (Adlercreutz, 1984).

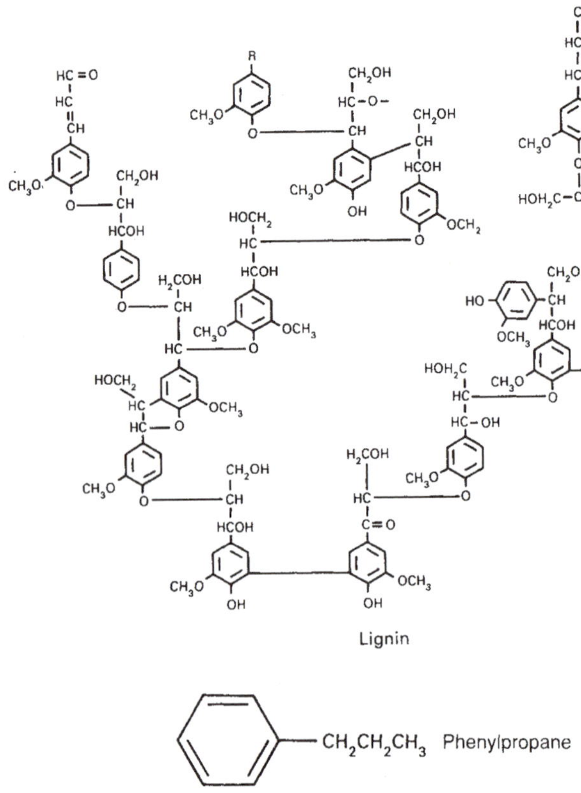

FIGURE 3.5 Lignins, three-dimensional networks built of units of phenylpropane.

## Fuel for the Host

The primary fate of dietary fibers is digestion and catabolism by the gut microflora to short-chain fatty acids and carbon dioxide. The major products of this microbial metabolism — acetic, propionic, and butyric acid — are important sources of energy for ruminants (sheep, cows). Dietary fiber is retained in a chamber of their gastrointestinal tracts, called the **rumen**, where it is converted to short-chain fatty acids by the gut microflora. The fatty acids produced may supply 35–75% of the energy requirement of the ruminant.

Short-chain fatty acid production is minimal in carnivores (dogs), in which the gastrointestinal tract is relatively short and the transit time of material through the gut is short also. Production is moderate in omnivores (humans, pigs), in which these acids are produced throughout the large intestine (Bergman, 1990). Short-chain fatty acids supply 50–75% of the energy requirement of the colonocytes of animals such as rats and humans, but only about 5% of the overall energy requirement of the animal itself.

The fluids of the colon and feces contain short-chain fatty acids at an overall concentration of 100–240 m$M$. Acetic, propionic, and butyric acids are present at molar ratios of about 60/25/15. Studies with rats suggests that butyrate is used preferentially by the colonocyte, whereas acetate and propionate preferentially appear in the bloodstream to be used by organs of the body. Less than 10% of the short-chain fatty acids produced in the gut are excreted in the feces.

The initial step in the metabolism of short-chain fatty acids, whether in cells of the gut lining or in the liver, is conversion to the coenzyme A derivative. For example, acetate is converted to **acetyl coenzyme A** (acetyl CoA). The acetyl CoA formed in the cytoplasm can be used for the synthesis of fatty acids, whereas that formed in the mitochondria can be used for immediate oxidation. **Propionyl CoA** can be metabolized as shown in Figure 8.7 in Chapter 8. Butyric acid can enter the mitochondria for conversion to **butyryl CoA** and oxidation in the pathway of fatty acid oxidation.

## Fiber as a Nutrient and Protective Agent on the Dairy Farm

Fiber is a major energy source for ruminants, as mentioned earlier, but fiber also serves to protect the health of dairy cows. Fiber is needed in the diets of dairy cows and, to a lesser extent, in the diets of beef cows. Reduced levels of fiber can lead to impaired health, i.e., acidosis (lowered blood pH), liver damage, erosion of the lining of the rumen and a lower fat content in the milk (Mertens, 1997). The rumen is an organ that is roughly equivalent to the stomach. The goal of the dairy farmer is to feed as much cereal grain as possible, and thus acquire as much milk as possible. However, cereal grain has the effect of stimulating the metabolism of various rumen bacteria, with the consequence that large quantities of short-chain fatty acids are created. The short-chain fatty acids are absorbed through the lumen, appear in the bloodstream, and are absolutely required by the cow as a source of energy. However, overproduction of this nutrient can cause acidosis and possibly even death. High levels of short-chain fatty acids also impair the metabolism of rumen bacteria that degrade cellulose (Russell and Wilson, 1996; Weimer, 1996). These bacteria are *Ruminococcus albus*, *Ruminococcus flavefaciens*, and *Fibrobacter succinogens.* The foregoing problems can be solved by feeding dairy cows a diet containing about 20% fiber. Alfalfa, dried hay, and beet pulp are used as sources of dietary fiber. Grain cereal is somewhat low in fiber. It contains only 10% cellulose (fiber). With under 20% fiber, the farmer risks harming the cows. With over 20% fiber, the farmer risks lesser milk production.

## Dietary Fiber and Health of the Small and Large Intestines

Dietary fiber can be classed according to its molecular structure, but also according to its solubility or dispersibility. Pectins, gums, and certain hemicelluloses are soluble fibers. Soluble fibers tend to have little effect on fecal bulk. They decrease the rate of passage of material through the upper gastrointestinal tract and tend to delay the rate of absorption of nutrients, probably because of their ability to form viscous solutions. Soluble fibers may decrease the rate of absorption of glucose by the small intestine and may reduce the rate of rise of plasma glucose that follows a meal. Also, they may reduce plasma cholesterol levels. Fruits, oats, barley, and legumes are relatively high in soluble fibers.

Cellulose, lignin, and many hemicelluloses are insoluble fibers. These fibers result in greater fecal bulk and an increase in the rate of passage of material through the large intestine. This effect on passage rate is called a "decrease in the transit time." Vegetables, wheat, and most grain products are relatively high in insoluble fibers.

Dietary fiber is commonly used for the treatment of various maladies of the gastrointestinal tract, such as constipation, diverticular disease, irritable bowel syndrome.

### Dietary Fiber and Constipation

Insoluble fiber is used to treat **constipation**, defined as defecation less frequent than every third day. Constipation may arise from a number of causes, such as excessive bedrest, depression, and confusion.

A major complication of constipation in the elderly is **fecal impaction**, a potentially serious condition that can result from ignoring impulses to defecate. Impaction is the buildup of fecal mass that becomes too large to expel. Increasing dietary fiber is part of the management program for chronic constipation of the elderly. Addition of bran (6–20 g/day) increases fecal weight and increases the frequency of defecation. When wheat bran is not effective, the laxative **lactulose** is recommended. Lactulose is a disaccharide composed of galactose and fructose connected by a $\beta(1 \rightarrow 4)$ bond. This compound is absorbed poorly in the small intestine because of the absence of enzymes that catalyze its hydrolysis.

### Dietary Fiber and Diverticular Disease

Dietary fiber is used to prevent and treat **diverticular disease**. This disease occurs in the human colon and involves outpouchings (diverticuli) where small regions of the gut mucosa slip in between belts of muscle and extend into the body (Figure 3.6). The belts of muscle constitute a layer in the wall of the large intestines.

## Lumen of Colon

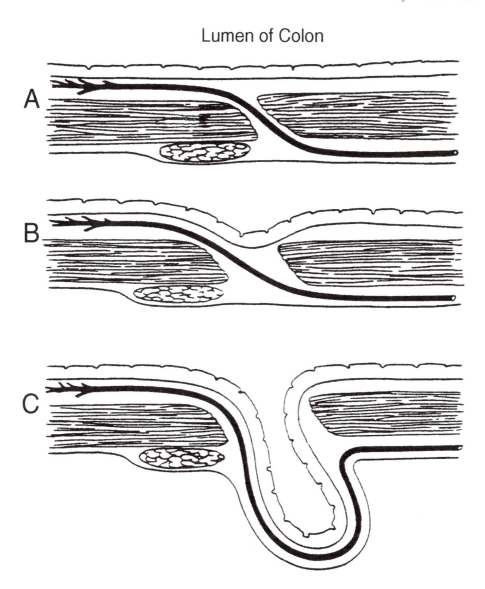

**Figure 3.6.** Formation of a diverticulum. (A) A cross-section of the wall of the colon reveals the mucosal layer (top surface of diagram), part of the belt of muscle that encircles the gut, and an artery passing through a naturally occurring gap in the muscle. (B) As middle age approaches, weak spots may develop in places where there exist gaps between muscles. (C) Diverticuli develop. The artery follows the path of the outpouching, and it is this part of the artery that may rupture and bleed and produce rectal bleeding in persons with diverticular disease. (Redrawn with permission from Meyers *et al.*, 1976.)

Another diagram of this layer of muscle appears in the Diet and Cancer chapter. Diverticuli occur because of weaknesses in the wall of the colon, and because of internal pressure due to contractions occurring both upstream and downstream of the weak area.

Diverticular disease consists of **diverticulosis** (presence of at least one diverticulum) and **diverticulitis** (diverticulosis with inflammation). Diverticulosis is a precursor of diverticulitis (Quirk and Barry, 1997). Diverticular disease occurs in about half the persons over 50 years of age in the United States, with the proportion increasing with further aging (Deckmann and Cheskin, 1993). It is thought that only about 20% of persons with the disease are ever aware of it. The disease presents by severe abdominal pain, sometimes with painless bleeding from affected areas. The disease can be correctly diagnosed by a barium enema along with an X-ray, or by looking into the colon with a tube (endoscope). Fiber is used to prevent and to treat diverticular disease. Fruit and vegetable fiber seem to prevent the disease, while cereal fiber (especially bran) is used to reduce the symptoms of existing diverticular disease (Aldoori *et al.*, 1994). Five slices of whole wheat bread can supply an effective amount of bran (10–25 grams of wheat bran). Diverticuli are not the same things as adenomas, another pathological structure of the colon, which may also be sites of bleeding. Adenomas are growths that extend into the colon, and that begin to be common in persons over 50 years old (see Diet and Cancer chapter).

### Dietary Fiber and Irritable Bowel Syndrome

**Irritable bowel syndrome**, which results in constipation in the young and middle aged, involves severe abdominal pain, nausea and vomiting, and rock-hard feces. Fiber may or may not be effective in improving the course of this disease. Irritable bowel syndrome is not a distinct disease, but a collection of symptoms. Dietary fiber is often used to treat some of these symptoms. In addition to abdominal pain, the syndrome involves a pattern of alternating constipation and diarrhea. With attempted defecation, a sense of incomplete evacuation may occur, and this is accompanied by continuing pain, perhaps for some hours, until feces are actually produced. In irritable bowel syndrome, the feces may be like hard marbles, due to extensive water removal during periods of constipation, or they may occur as long ribbons, due to rectal spasms. Fiber (12–16 grams of bran/day) is used to treat the constipation. Drugs are used to treat the diarrhea and spasms. Various stress-reduction techniques, such as relaxation exercises, are a vital part of the treatment of irritable bowel syndrome (Schuster, 1989).

### Dietary Fiber or Short-Chain Fatty Acids Are Needed for Health of Enterocytes

Whether dietary fiber is required for the health of the colonocytes has not been proven, although evidence suggests such a requirement. Absorption of salts and water is a major function of the large intestine. Short-chain fatty acids stimulate the absorption of sodium, chloride, and water in the colon (Hoverstad, 1986). In the absence of short-chain fatty acids, the mucosa of the colon may become inflamed or atrophied.

**Total parenteral nutrition** (TPN) means supplying all nutrients to a patient through a liquid diet fed intravenously. With TPN, food is not consumed, although one might expect continued sloughing off of enterocytes into the lumen. TPN often is used in the treatment of pancreatic diseases and in the care of very-low-birth-weight infants. (The intestinal tracts of such infants may not be able to accept a full daily diet.) Patients with severe loss of functioning intestines may require TPN to remain alive. Those with less than 0.6 meters small intestine require TPN. The cells of the colon atrophy with TPN, possibly because of the lack of dietary fiber and the resultant lack of short-chain fatty acid production. This atrophy is a medical problem, since it results in diarrhea when oral feeding is resumed. This diarrhea may result in part from the inability of the atrophied colon to perform its major function, water absorption.

## Dietary Fiber and Diabetes

Fiber may help in the management of diabetes. Diets that are high in fiber and digestible starches may be used to minimize the elevation of blood sugar that generally occurs after a meal. In some patients, diets that supply 55–60% of the body's energy from these carbohydrates result in improvement in the clinical picture, that is, they improve the control of plasma glucose levels and reduce the requirement for injected insulin. The soluble fibers used in these diets are supplied by guar gum, wheat bran, and apples. Additional details on diabetes appear in Chapter 4.

## Dietary Fiber and Colon Cancer

Epidemiological studies have indicated that dietary fiber is a contributing factor in lowering the incidence of cancer of the colon. Colon cancer is the leading cause of cancer death in the United States. The increase in the bulk of material in the colon provided by insoluble fibers has been speculated to reduce the concentration of chemicals in the colon that promote the conversion of a normal colonocyte to a cancerous one. An increase in flow rate through the colon, generated by dietary fibers, would reduce the time of exposure of the colonocytes to these chemicals. These effects should reduce the levels of cancer-producing chemicals supplied directly by the diet, as well as those produced by microbial metabolism. Further details are revealed in the Diet and Cancer chapter.

# MICROORGANISMS AND DIGESTION

## Gut Microflora

The gut microflora consist of microorganisms, mainly bacteria, of the gastrointestinal tract. The small intestine of the adult human is about 4.0 m long. The large intestine, or colon, is about one-third this length (1.5 m). The colon receives 1.5–2.0 liters of water per day, most of which is absorbed. Only 150–200 ml of water is lost in the feces. The colon also absorbs sodium and chloride originating from the diet and from secretions of the small intestine. The lumenal surface of the large intestine secretes mucus, as does the small intestine. However, the mucosa of the colon contains crypts but lacks villi.

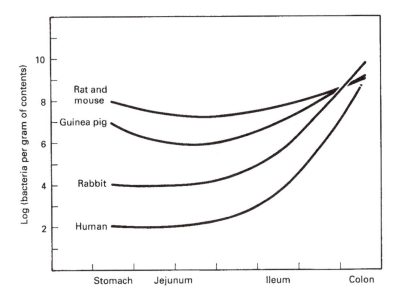

**FIGURE 3.7** Approximate concentrations of bacteria in various regions of the gastrointestinal tract. The density of bacteria in the human stomach and small intestine is relatively low compared with that in the large intestine, unlike other species shown. The human stomach contains about 100 bacteria per g contents, whereas the large intestine contains about one billion per gram. For comparison, a sample of pure bacterial cells contains about one trillion cells per gram. (Redrawn with permission from Rowland, 1988.)

In humans, the density of bacteria in the stomach and small intestine is relatively low compared with that in the large intestine, as shown in Figure 3.7. The human stomach contains about 100 bacteria per gram of contents, whereas the large intestine contains about one billion per gram. For comparison, a sample of pure bacterial cells contains about one trillion cells per gram.

The microflora of the gastrointestinal tract, which develop shortly after birth, are found in the lumen, associated with remnants of food, on the mucosal surface, and in the crypts. These microorganisms, including prokaryotic and eukaryotic species, tend not to use oxygen in their energy metabolism since they exist in an environment that is essentially anaerobic. Over 99.9% of the bacterial cells are anaerobic, meaning they are unable to use oxygen even when it is provided for them. Of these bacteria, 5–20% are unable to tolerate $O_2$ and die in its presence.

The human large intestines contains at least 400 different species of bacteria. Most are anaerobes, such as *Bacteroides*, *Eubacterium*, and *Bifidobacterium*. *Lactobacilli* and *Clostridia* are also present. Anaerobes cannot utilize oxygen gas as an electron acceptor. Protozoa, yeasts, and other fungi occur only in very low numbers in the human gut (Cummings and Macfarlane, 1991). The most highly publicized bacterium of the gut is *Eschericia coli*, which is a species of *Enterobacteria*. However, the *Enterobacteria* occur in numbers that are 100 to 10,000-fold less than the *Bacteroides*. *E.coli* is able to use $O_2$ as an electron acceptor, but it is also able to grow fermentatively (in the absence of $O_2$). The *Enterobacteria* consume the available oxygen and help produce an oxygen-free environment that is required by many of the anaerobes. Methane-producing organisms (methanogens) also occur in the gut. As one might recall, the three forms of life which exist on earth are *Eucarya*, *Bacteria*, and *Archae*. Methanogens are *Archae*.

The human infant's gut microflora begins to establish itself after birth, and consists mainly of *Bifidobacteria* when its food is breast milk. However, with formula feeding the gut microflora consists of a mixture of bacterial species resembling that in the adult gut (Gibson and Roberfroid, 1995).

The relationship between different microbial species in the gut is an intriguing topic. These species include carbohydrate-fermenting bacteria, pectinolytic bacteria, and methanogens (methane-generating bacteria). The gases produced by bacterial metabolism include $CO_2$, $H_2$, and $CH_4$ (methane). Analysis for these gases has proven useful in assessing the extent to which various dietary fibers are used by the gut microflora and in assessing the variability in microfloral ecology in different human subjects.

Different species of gut bacteria can supply nutrients to each other. For example, the pectinolytic bacterium *Bacteroides* produces methanol during the course of pectin degradation, which originates from the methoxy groups of pectin. This methanol is taken up by methanogenic bacteria and reduced to methane gas. In this reaction, methanol serves as an electron acceptor. The methanogens utilize $H_2$ as a reductant during this conversion, which is provided by another group of bacteria, the carbohydrate fermenters. Some highlights in the ecology of the gut microflora appear in Figure 3.8. Some of the hydrogen gas produced by the carbohydrate fermenters may be released unchanged from the host. Methane production is variable in humans. About one-third of adult humans are methane producers.

The major gaseous product of the gut bacteria is $CO_2$, a gas also produced by the metabolism of the human host. Hydrogen gas is produced exclusively by the gut bacteria and not by the host. The studies of lactose metabolism shown in Figures 3.2 and 3.3 revealed that the rate of $H_2$ production can be used to assess lactase status in human subjects. In a study involving two human subjects (shown in Figure 3.9), $H_2$ production was used to follow the time of passage of dietary gum through the small intestine and its eventual attack by bacterial enzyme. The

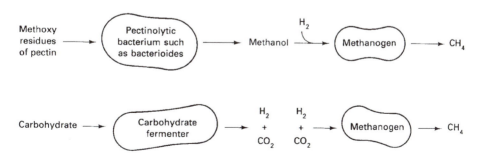

**FIGURE 3.8** Pathways of gas formation due to fiber fermentation in the intestines. Some of the hydrogen gas produced by the carbohydrate fermenters may be released unchanged from the host. Methane production is variable in humans. About one-third of adult humans are methane producers. Some of the $H_2$ may be used by methanogens for $CH_4$ production.

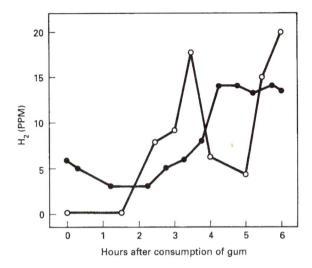

**FIGURE 3.9** Concentration of hydrogen in the breath. In two human subjects, $H_2$ production ($\bigcirc$,$\bullet$) was used to follow the time of passage of dietary gum through the small intestine and its eventual attack by bacterial enzymes. The subjects each consumed 25 g gum arabic. The concentration of $H_2$ in their exhaled gases was measured at various times afterward. The study revealed that the initial rate of $H_2$ production was different in the two subjects, that fermentation of the gum started to take place within 3 hr of the test dose, and that the rate of gas production fluctuated over the course of time. (Redrawn with permission from Ross *et al.*, 1983.)

results indicate that fermentation of the gum started within 3 hours of consumption of the test dose. The initial rate of $H_2$ production was different in the two subjects, and fluctuated significantly during the experiment. In general, studies of the passage, absorption, or excretion of various nutrients yield results that cannot be reproduced closely in different experimental subjects.

### Microbe–Host Interactions

The gut microflora produce a vast array of compounds in addition to gases and short-chain fatty acids, including vitamins (biotin and vitamin K) that are absorbed by the intestines and supply part of the dietary requirement. The functions of biotin and vitamin K are detailed in the Vitamin chapter.

**Queuine** is included in the list of potentially useful products synthesized in the gut. Queuine is a nucleoside base that resembles a modified form of guanine. Queuine is incorporated into certain types of tRNA in a reaction catalyzed by **tRNA-guanine transglycosylase**, resulting in the ribonucleotide **queuosine** (Figure 3.10). The enzyme catalyzes the replacement of an existing residue of guanine by queuine, where the targets of the reaction are tRNAs used for carrying the amino acids Asn, Asp, His, and Tyr (Slany *et al.*, 1993). The existence of this unusual base in transfer RNA should not be regarded as unusual, since a total of 80 types of modified bases have been found in the tRNAs of eucarya and bacteria. These bases include altered versions of adenine, uracil, guanine, and cytosine. tRNA molecules have a greater density of modified bases than other types of nucleic acids. The function of queuosine is not established, but evidence suggests that it is required for the biosynthesis of tyrosine (Marks and Farkas, 1997). If a vital function for queuine is found, it may be classed as a vitamin.

The gut microflora metabolize the bile acids produced by the human host. Their enzymes catalyze the deconjugation of the bile acids, that is, removal of the taurine and glycine groups. These enzymes also catalyze dehydroxylation, converting cholic acid to deoxycholic acid. Gut bacteria convert bile acids to compounds that appear to be carcinogenic and possibly contribute to cancer of the colon.

## MALABSORPTION SYNDROMES

Several diseases, called **malabsorption syndromes**, result in the lack of absorption of a number of nutrients. Their consequent loss in the feces clearly is not desirable. Malabsorption of lipids, which occurs more frequently than malabsorption of any other class of nutrient, can result in diarrhea and steatorrhea. **Steatorrhea**, production of feces with a high fat content, derives its name from "stearic acid," a fatty acid. With severe steatorrhea, over 30 g fat may appear in the feces per day. The problem can be elicited by liver diseases that prevent the production of bile salts and thus impair the absorption of fats.

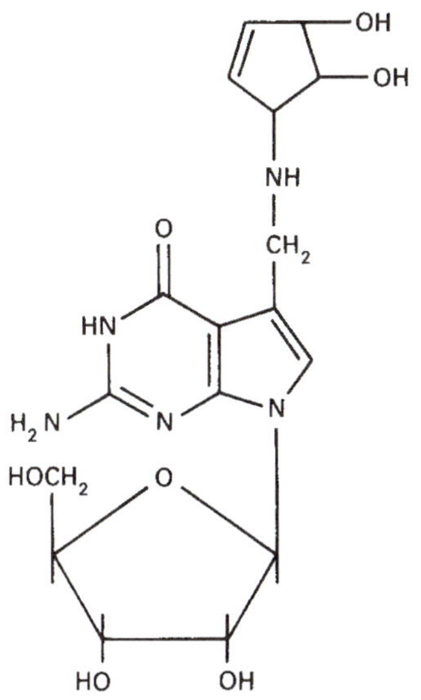

**FIGURE 3.10** Structure of queuosine. Queuosine is incorporated into certain types of transfer RNA. The importance of queunosine to human health is not certain. A great quantity of compounds are synthesized by the gut microflora, including vitamin $B_{12}$, vitamin K, and biotin (a vitamin). In some cases, these sources of vitamin contribute to the nutritional requirements of the host animal. Essentially all of our vitamin $B_{12}$ is acquired from bacteria, but indirectly via meat. Beef cattle acquire their vitamin $B_{12}$ from their gut bacteria, and people, in turn, acquire this vitamin from meat.

Several malabsorption diseases involve defects of the intestinal mucosa. **Celiac disease** results from damage to the intestinal mucosa, provoked by cereal proteins, in susceptible individuals. The symptoms of the disease include watery diarrhea, bleeding of the intestines, and iron deficiency (due to iron malabsorption or to the iron lost in the blood) (Arranz *et al.*, 1993; Fine, 1996). Abnormally low pancreatic secretions can also occur in a third of persons with celiac disease, resulting in abnormally low digestion, lack of absorption of nutrients, and a lowered growth rate (Carroccio *et al.*, 1997). The disease, which involves atrophy of the villi on the enterocytes, is diagnosed by taking a biopsy of the jejunum. Biopsies are taken before and after treatment with a diet that is free of cereal protein. The term for a partially purified preparation (solubilized in alcohol) of the cereal protein in question is **gluten**. Gluten consists of proteins called gliadin, and other proteins. Hence, the diet is called a gluten-free diet. Celiac disease can be treated with the gluten-free diet. With treatment, the weight loss and watery diarrhea experienced by sufferers is prevented in most patients (Gray, 1997).

Celiac disease is due to an abnormal immunological response to the gliadin proteins, especially to α-gliadin (Tighe and Ciclitira, 1995). Cereals that contain gliadin, or related proteins, include wheat, barley, rye, and oats. Nontoxic cereals include rice and corn. The disease can also be diagnosed by measuring the abnormally high secretion of the antibodies IgM and IgA into the lumen of the small intestines. These antibodies recognize and bind to gliadin.

**Crohn's disease** involves inflammation of the gut. Its cause is not known, but evidence suggests that infection with *Mycobacterium paratuberculosis* may be a contributing factor (Hermon-Taylor, 1993). The inflammation results in steatorrhea, diarrhea, abdominal pain, and weight loss. Crohn's disease is usually treated with drugs (corticosteroids), but it may also be treated by the elimination of a normal diet and its replacement with an elemental diet. The elemental diet consists of oligosaccharides, amino acids, and short-chain fatty acids. This diet is not very palatable, but it can result in relief of symptoms and a reduction in inflammation of the intestines. Two weeks of dietary treatment may be sufficient to allow inflammation to subside (Lochs *et al.*, 1991).

**Tropical sprue** is a malabsorptive disease thought to be generated by specific types of infections. The disease involves diarrhea and is associated with the malabsorption of folate. Hence, its treatment may involve supplements of folic acid.

Nutrient malabsorption also occurs in the genetic disease **cystic fibrosis**. This disease arises from a mutation in the chloride ion transporter, and results in pancreatic insufficiency as well as respiratory difficulties. Treatment of the malabsorption symptoms of cystic fibrosis involves supplementation with pancreatic enzymes and synthetic water-soluble versions of the fat-soluble vitamins. The labored breathing and respiratory infections due to the disease cannot be treated by dietary intervention.

## SUMMARY

The antibodies supplied by mother's milk serve the useful purpose of combating intestinal diseases in the infant. This role is important when the infant is raised in an environment contaminated with fecal matter, as often occurs in underdeveloped countries. In infants, intestinal diseases resulting from a contaminated food or water supply can involve severe vomiting and diarrhea, which, if continued for several weeks, can lead to death by dehydration.

Lactose intolerance is not a serious condition. This condition tends to be present in late childhood in persons of non-European descent. The symptoms can be provoked by quickly drinking several cups of milk.

The digestibility of starch depends on source, physical condition, and food preparation. With the exception of lignin, dietary fibers may be partially or fully hydrolyzed by enzymes of the microflora present in the colon. Some products are absorbed into the bloodstream, but most serve as energy sources for the colonocytes. Specific dietary fibers can influence fecal bulk, the rate of passage of material through the intestines, and certain maladies such as constipation.

## REFERENCES

Adlercreutz, H. (1984). Does fiber-rich food containing animal lignin precursors protect against both colon and breast cancer? *Gastroenterology* **86**, 761–766.

Aldoori, W. H., Giovannucci, E. L., Rimm, E. B., Wing, A. L., Trichopoulos, D. V., and Willett, W. C. (1994). A prospective study of diet and the risk of symptomatic diverticular disease in men. *Am. J. Clin. Nutr.* **60**, 757–764.

Arranz, E., and Ferguson, A. (1993). Intestinal antibody pattern of celiac disease: Occurrence in patients with normal jejunal biopsy histology. *Gastroenterology* **104**, 1263–1272.

Bergman, E. N. (1990). Energy contributions of volatile fatty acids from the gastrointestinal tract in various species. *Physiol. Rev.* **70**, 567–590.

Brand, J. C., Gracey, M. S., Spargo, R. M., and Dutton, S. P. (1983). Lactose malabsorption in Australian aborigines. *Am. J. Clin. Nutr.* **37**, 449–452.

Carroccio, A., Iacono, G., Lerro, P., Cavataio, F., Malorgio, E., Soresi, M., Baldassarre, M., Notarbartolo, A., Ansaldi, N., and Montalto, G. (1997). Role of pancreatic impairment in growth recovery during gluten-free diet in childhood celiac disease. *Gastroenterology* **112**, 1839–1844.

Cummings, J. H., and Englyst, H. N. (1987). Fermentation in the human large intestine and the available substrates. *Am. J. Clin. Nutr.* **45**, 1243–1255.

Cummings, J. H., and Macfarlane, G. T. (1991). The control and consequences of bacterial fermentation in the human colon. *J. Appl. Bacteriol.* **70**, 443–459.

Deckmann, R. C., and Cheskin, L. J. (1993). Diverticular disease in the elderly. *J. Am. Geriatr. Soc.* **40**, 986–993.

Englyst, H. N., and Cummings, J. H. (1986). Digestion of the carbohydrates of banana (*Musa paradisiaca sapientum*) in the human small intestines. *Am. J. Clin. Nutr.* **44**, 42–50.

Fernandes, J., Vos, C. E., Douwes, A. C., Slotema, E., and Degenhart, H. J. (1978). Respiratory hydrogen excretion as a parameter for lactose malabsorption in children. *Am. J. Clin. Nutr.* **31**, 597–602.

Fine, K. D. (1996). The prevalence of occult gastrointestinal bleeding in celiac sprue. *New Engl. J. Med.* **334**, 1163–1167.

Gibson, G. R., and Roberfroid, M. B. (1995). Dietary modulation of the human colonic microbiota: Introducing the concept of prebiotics. *J. Nutr.* **125**, 1401–1412.

Gray, G. (1997). Persistence of diarrhea in treated celiac sprue: Refractory disease of another organ's malfunction? *Gastroenterology* **112**, 2146–2147.

Hermon-Taylor, J. (1993). Causation of Crohn's disease: The impact of clusters. *Gastroenterology* **104**, 643–646.

Hoverstad, T. (1986). Studies of short-chain fatty acid adsorption in man. *Scand. J. Gastroenterol.* **21**, 257–260.

Kagnoff, M. F. (1993). Immunology of the intestinal tract. *Gastroenterology* **105**, 1275–1280.

Kerneis, S., Bogdanova, A., Kraehenbuhl, J.-P., and Pringault, E. (1997) Conversion by Peyer's patch lymphocytes of human enterocytes into M cells that transport bacteria. *Science* **277**, 949–952.

Labarthe, D. R. (1996). Dietary fiber. *Circulation* **94**, 2696–2698.

Lochs, H., Steinhardt, H., Klaus-Wentz, B., Zeitz, M., Vogelsang, H., Sommer, H., Fleig, W., Bauer, P., Schirrmeister, J., and Malchow, H. (1991). Comparison of enteral nutrition and drug treatment in active Crohn's disease. *Gastroenterology* **101**, 881–888.

Marks, T., and Farkas, W. (1997). Effect of a diet deficient in tyrosine and queuine on germfree mice. *Biochem. Biophys. Res. Commun.* **230**, 233–237.

Marlett, J. A., and Slavin, J. L. (1997). Position of the American Dietetic Association: Health implication of dietary fiber. *J. Am. Diet. Assoc.* **97**, 1157–1159.

Mertens, D. R. (1997). Creating a system for meeting the fiber requirements of dairy cows. *J. Dairy Sci.* **80**, 1463–1481.

Meyers, M. A., Alonso, D. R., Gray, G. R., and Baer, J. W. (1976). Pathogenesis of bleeding colonic diverticulosis. *Gastroenterology* **71**, 577–583.

Miranda, R., Saravia, N. G., Ackerman, R., Murphy, N., Berman, S., and McMurray, D. N. (1983). Effect of maternal nutritional status on immunological substances in human colostrum and milk. *Am. J. Clin. Nutr.* **37**, 632–640.

Quirk, D. M., and Barry, M. J. (1997). Physician specialty and cost-effectiveness of diverticulitis care: A difficult knot to untangle. *Gastroenterology* **112**, 2147–2150.

Ross, A. H., Eastwood, M. A., Anderson, J. R., and Anderson, D. M. (1983). A study of the effects of dietary gum arabic in humans. *Am. J. Clin. Nutr.* **37**, 368–375.

Rowland, I. R. (1988). "Role of the Gut Flora in Toxicity and Cancer." Academic Press, New York.

Russell, J. B., and Wilson, D. B. (1996). Why are ruminal cellulolytic bacteria unable to digest cellulose at low pH? *J. Dairy Sci.* **79**, 1503–1509.

Schuster, M. M. (1989). Irritable bowel syndrome. *In* "Gastrointestinal Disease," 4th ed. (M. H. Sleisenger and J. S. Fordtran, eds.), pp. 1402–1418. Saunders, Philadelphia.

Slany, R. K., Bosl, M., Crain, P., and Kersten, H. (1993). A new function of *S*-adenosylmethionine: The ribosyl moiety of AdoMet is the precursor of the cyclopentenediol moiety of the tRNA wobble base queuine. *Biochemistry* **32**, 7811–7817.

Slavin, J. L. (1987). Dietary fiber: Classification, chemical analyses, and food sources. *J. Am. Diet. Assoc.* **87**, 1164–1171.

Thakur, B. R., Singh, R. K., and Handa, A. K. (1997). Chemistry and uses of pectin — a review. *CRC Crit. Rev. Food Sci. Nutr.* **37**, 47–73.

Tighe, R., and Ciclitira, P. (1995). Molecular biology of coeliac disease. *Arch. Dis. Child.* **73**, 189–191.

Wagner, N., Lohler, J., Kunkel, E., Ley, K., Leung, E., Krisansen, G., Rajewsky, K., and Muller, W. (1996). Critical role for β7 integrins in formation of the gut-associated lymphoid tissue. *Nature* **382**, 366–370.

Weimer, P. J. (1996). Why don't ruminal bacteria digest cellulose faster? *J. Dairy Sci.* **79**, 1496–1502.

## BIBLIOGRAPHY

Almy, T. P., and Naitove, A. (1983). Diverticular disease of the colon. *In* "Gastrointestinal Disease," 3rd ed. (M. H. Sleisenger and J. S. Fordtran, eds.), pp. 896–912. Saunders, Philadelphia.

Anderson, I. H., Levine, A. S., and Levitt, M. D. (1981). Incomplete absorption of the carbohydrate in all-purpose wheat flour. *N. Engl. J. Med.* **304**, 891–892.

Armentano, L., and Pereira, M. (1997). Measuring the effectiveness of fiber by animal response trials. *J. Dairy Sci.* **80**, 1416–1425.

Atkins, F. M., and Metcalfe, D. D. (1984). The diagnosis and treatment of food allergy. *Annu. Rev. Nutr.* **4**, 233–255.

Bell, R. R., Draper, H. H., and Bergan, J. G. (1973). Sucrose, lactose, and glucose intolerance in northern Alaskan eskimos. *Am. J. Clin. Nutr.* **26**, 1185–1190.

Bingham, S. (1987). Definitions and intakes of dietary fiber. *Am. J. Clin. Nutr.* **45**, 1226–1231.

Calloway, D. H. (1966). Respiratory hydrogen and methane as affected by consumption of gas-forming foods. *Gastroenterology* **51**, 383–389.

Greenwald, P., Lanza, E., and Eddy, G. A. (1987). Dietary fiber in the reduction of colon cancer risk. *J. Am. Diet. Assoc.* **87**, 1178–1188.

Harig, J. M., Soergel, K. H., Komorowski, R. A., and Wood, C. M. (1989). Treatment of diversion colitis with short-chain fatty acid irrigation. *N. Engl. J. Med.* **320**, 23–28.

Jorgensen, N. (1984). From first colostrum to weaning. *Hoard's Dairyman* **129**, 591.

Koruda, M. J., Rolandelli, R. H., Settle, R. G., Zimmaro, D. M., and Rombeau, J. L. (1988). Effect of parenteral nutrition supplemented with short-chain fatty acids on adaptation to massive small bowel resection. *Gastroenterology* **95**, 715–720.

Kretchmer, N. (1971). Memorial lecture: Lactose and lactase — A historical perspective. *Gastroenterology* **61**, 805–813.

Kripke, S. A., Fox, A. D., Berman, J. M., Settle, R. G., and Rombeau, J. L. (1989). Stimulation of intestinal mucosal growth with intracolonic infusion of short-chain fatty acids. *J. Parenteral Enteral Nutr.* **13**, 109–116.

Mariadason, J. M., Barkla, D. H., and Gibson, P. R. (1997). Effect of short-chain fatty acids on paracellular permeability in caco-2 intestinal epithelium model. *Am. J. Physiol.* **272**, G705–G712.

Moe, S. T., Draget, K.I., Skjak-Braek, G., and Smidsrod, O. (1995). "Alginates in Food Polysaccharides" (A. M. Stephen, ed.), pp. 245–286. Dekker, New York.

Nocek, J. E. (1983). Colostrum isn't always of value to calves. *Hoard's Dairyman* **128**, 1340–1341.

O'Dwyer, S. T., Smith, R. J., Hwang, T. L., and Wilmore, D. W. (1989). Maintenance of small bowel mucosa with glutamine-enriched parenteral nutrition. *J. Parenteral Enteral Nutr.* **13**, 579–585.

Piculell, L. (1995). Gelling carrageenans. *In* "Food Polysaccharides" (A. M. Stephen, ed.), pp. 205–244. Dekker, New York.

Pomare, E. W., Branch, W. J., and Cummings, J. H. (1985). Carbohydrate fermentation in the human colon and its relation to acetate concentrations in venous blood. *J. Clin. Invest.* **75**, 1448–1454.

Read, N. W., and Timms, J. M. (1987). Constipation: Is there light at the end of the tunnel? *Scand. J. Gastroenterol., Suppl.* **22**(129), 88–96.

Rowe, W. A., and Bayless, T. M. (1992). Colonic short-chain fatty acids: Fuel from the lumen? *Gastroenterology* **103**, 336–339.

Siragusa, R. J., Cerda, J. J., Baig, M. M., Burgin, C. W., and Robbins, F. L. (1988). Methanol production from the degradation of pectin by human colonic bacteria. *Am. J. Clin. Nutr.* **47**, 848–851.

Stave, U. (1978). "Perinatal Physiology." Plenum Medical, New York.

Theander, O. (1987). Chemistry of dietary fiber components. *Scand. J. Gastroenterol., Suppl.* **22**(129), 21–28.

Trock, B., Lanza, E., and Greenwald, P. (1990). Dietary fiber, vegetables, and colon cancer: Critical review and meta-analyses of the epidemiological evidence. *J. Natl. Cancer Inst.* **82**, 650–661.

# 4

# REGULATION
# OF ENERGY
# METABOLISM

## OVERVIEW

The subject of energy regulation involves a huge array of interrelated facts. These include symbiotic relationships between the brain, liver, muscle, pancreas, and energy stores such as fat in adipose tissue and glycogen in muscle and liver. These relationships change with the body's activities, including feeding, brief exercise, prolonged exercise, fasting, and starvation. Relationships both within and between these organs change during these activities. The complexity of these interactions is depicted in Figure 4.1, which shows some of the main pathways of energy flux.

The overall (net) direction of energy flux is controlled mainly by two hormones: glucagon and insulin. The phrase *direction of energy flux* can mean, for example, towards storage or oxidation. Both hormones are polypeptides, and both are secreted into the bloodstream by the pancreas. Glucose is the main regulator of insulin secretion, while certain amino acids such as arginine have a minor effect in provoking insulin secretion. The nervous system (i.e., cephalic phase) also is a minor stimulus of insulin secretion. Glucagon secretion is inhibited

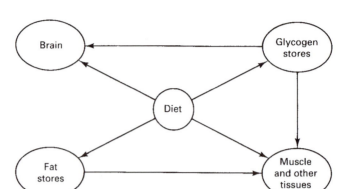

**FIGURE 4.1** Thoroughfares of energy flux. The arrows show some of the main thoroughfares of energy flux. For example, the brain may receive energy directly from the diet as well as from glucose liberated from glycogen stored in the liver. No arrow points from fat stores to the brain, because the brain does not derive an appreciable amount of energy from the triglycerides stored in fat. In contrast, muscle and other tissues receive energy directly from the diet, from glycogen, and from the fat stores, as indicated by the arrows.

by glucose and insulin and stimulated by amino acids (Philippe, 1991; Teff *et al.*, 1991). Other hormones of importance in regulating energy metabolism include norepinephrine, epinephrine, and thyroid hormone.

### The Brain Needs Glucose

A major topic of this chapter is the supply and maintenance of the brain's source of energy. The brain stores very little of its own energy, and relies almost exclusively on a constant supply of glucose from the bloodstream. The concentration of plasma glucose is finely controlled by glucagon and insulin. The role of these two hormones is demonstrated in this chapter by experiments involving human subjects. The goal of maintaining the brain's supply of glucose is used as a reference point in the discussions that follow.

### Various Pathways May Have a Twin, Which Operates in the Reverse Direction

First considered are metabolic pathways that occur with a reverse version of the same pathway. Separate pathways exist for the degradation and synthesis of glycogen. Separate pathways exist for the degradation of glucose-6-phosphate (**glycolysis**) and for the synthesis of glucose-6-phosphate (**gluconeogenesis**). Separate pathways also exist for the synthesis and catabolism of fatty acids. All these pathways are controlled in a coordinated manner by glucagon and insulin. The effects of these hormones are most easily understood by considering not their concentrations in the plasma, but the ratio of their concentrations, (glucagon)/(insulin).

*Energy Metabolism Takes Various Styles*

Different types of energy metabolism are illustrated by the different interactions that occur between the liver, muscle, and adipose tissue during the states of rest, exercise, and carbohydrate consumption. Aerobic metabolism and anaerobic metabolism (fermentation) are two types of metabolism. Two other styles are illustrated by that of the liver, which tends not to consume much glucose for its own purposes but combusts fat, and that of the brain, which usually consumes only glucose and is not capable of burning fat. Skeletal muscle uses different types of energy metabolism under different conditions. The energy fuels used by skeletal muscle may be arranged in a hierarchy, with those depleted in the early stages of exercise occurring first and those used in later stages of exercise last. This list includes creatine phosphate, glycogen, fat, and ketone bodies.

This chapter focuses on a number of issues involving the aforementioned fuels. Topics include regulation of fatty acid synthesis at the enzymatic and genetic levels, the role of carnitine in fatty acid oxidation, long- and medium-chain fatty acids, and serum albumin as a fatty acid carrier. The concept that oxygen consumption is related to the rate of energy use is illustrated by a feeding experiment involving food oils. The precise quantitative nature of the relationships between oxygen consumption and fuel combustion are revealed in the next chapter, on Energy Requirement.

*Glycolysis Produces Pyruvate; The Krebs Cycle Produces CO2*

The glycolysis pathway comprises a series of reactions involving the breakdown and partial oxidation of carbohydrates and the conversion of some of our food energy to ATP and heat. Glycolysis results in the conversion of sugars to pyruvate, a 3-carbon metabolite. During oxidative metabolism, pyruvate is transported into the mitochondrion, where it is converted, via pyruvate dehydrogenase, to a 2-carbon metabolite, the acetate group. The acetate group is bound to a molecule of coenzyme A and thus occurs as acetyl-coenzyme A. The acetate group then enters the Krebs cycle, where it is completely oxidized to $CO_2$. The Krebs cycle comprises the series of reactions required for converting most of our food energy to ATP and heat. The enzymes that catalyze these reactions occur in the mitochondrion. Nutrients enter the Krebs cycle and, in a stepwise manner, are oxidized and broken down to yield carbon dioxide. Nearly all of the carbon from fats and carbohydrates enters the Krebs cycle in the form of a 2-carbon unit, the acetate group. A number of nutrients, such as citrate and malate, can enter the Krebs cycle intact (not as acetate groups) for oxidation and breakdown.

*Water Is an End-Product of the Respiratory Chain*

The electrons released from nutrients, in turn, are transferred to oxygen, resulting in its reduction to water. These electrons, which initially have a high reducing power, are not directly transferred from nutrients to oxygen. Instead, the reducing power of the electrons is captured in a series of discrete steps catalyzed by a series of proteins known as the **respiratory chain**. This process is called **electron trans-**

**port**. The electrons reduce oxygen to water at the terminus of the respiratory chain, where they reach their lowest level of reducing power. Organisms that require oxygen for life do so because of this particular step in metabolism.

### The Final Goal of the Respiratory Chain Is to Produce ATP

The operation of the respiratory chain is coupled with the transport of hydrogen ions (protons) out of the mitochondrion. This transport, which is mediated by specific transport proteins embedded in the membrane, results in a higher concentration of $H^+$ ions outside the organelle. Concomitantly, the protons reenter the organelle by means of a separate membrane-bound protein complex. This complex is called **ATP synthase**. The passage of protons through ATP synthase is driven by the higher concentration of $H^+$ that is maintained outside the mitochondrion. The protons seek to equilibrate their concentrations on both sides of the membrane, resulting in the synthesis of ATP.

Most of the electrons removed from fuels during energy metabolism are transferred via nicotinamide adenine dinucleotide (NAD). NAD collects electrons from many different energy fuels in reactions catalyzed by specific enzymes. These enzymes are dehydrogenases. Reduced NAD, in turn, shuttles the electrons to the respiratory chain. Flavin adenine dinucleotide (FAD) also acts as an electron shuttle. In each reaction involving NAD (or FAD), two electrons are transferred; that is, two electrons are carried or shuttled. NAD and FAD are small molecules with molecular weights of 663 and 785 and are manufactured in the body from the vitamins niacin and riboflavin, respectively. These molecules are called N.A.D. and F.A.D., not "*nad*" or "*fad*."

## THE BRAIN AND ENERGY METABOLISM

The brain and central nervous system are of highest priority relative to energy metabolism. The relationship between the brain and the energy stores of the body is unique. The central nervous system is fastidious in its requirement for glucose. Unlike most other organs, the brain does not use **free fatty acids** (FFAs) for energy. Also unlike other organs, it cannot regulate its nutrient uptake systems or reduce its energy requirements in response to deficiencies in the diet and in the fuel supply. These, and other points, are summarized as follows:

1.  The brain uses a large amount of energy. With the human body at rest, it accounts for about 20% of total energy consumption. The energy consumption of the brain is disproportionately large, on a per-weight basis, compared with that of other organs.

2.  The brain does not have appreciable reserves of energy. It contains only a small amount of glycogen and is dependent on a continuous supply of glucose from the bloodstream.

3.  The brain is dependent on glucose as an energy source. It cannot take up fatty acids, which are abundant in the circulation.

4. The brain, apparently, is not sensitive to glucagon or insulin. Changes in the glucagon/insulin ratio in the bloodstream serve to coordinate responses of various organs to changes in the energy supply, but the brain does not seem to be one of these organs.

5. The brain, however, is not completely inflexible. After a few days of fasting and long after the body's supplies of glycogen have been depleted, the brain readily makes use of a new fuel that appears in the bloodstream: **ketone bodies**. With prolonged fasting, the brain can derive an increasing proportion of its energy requirement from this fuel; the greatest proportion that has been measured in humans is 50%.

Ketone body metabolism is a normal energy pathway for the neonatal infant. The supply of glucose may be somewhat precarious, and the infant may not have large supplies of glycogen (glucose stores) in its tissues. Infants, who may not have food available during their first day of life, more readily form ketone bodies with brief fasting than adults. Use of these ketone bodies by an infant's brain protects its central nervous system during times when the supply of blood glucose is borderline.

## GLUCAGON AND INSULIN

The central nervous system requires a constant supply of glucose, and we have noted the importance of maintaining constant levels of this nutrient in the plasma. But before detailing the styles of energy metabolism of other organs, we shall view the hormones used to control plasma glucose levels, and document the mechanisms used to control entry of glucose, from the plasma, into muscle and other organs.

The pancreas plays a central part in controlling plasma glucose levels. The pancreas detects changes in the concentration of plasma glucose, and responds by altering its rates of secretion of glucagon and insulin into the bloodstream. These hormones are used to maintain plasma glucose at a constant level. Glucagon stimulates its target organs to mobilize and liberate fuel stores and to increase their rate of fuel oxidation. Insulin stimulates its target organs to store and conserve energy fuels and to decrease their rate of fuel oxidation.

### What Glucagon Does

In many cases, glucagon is the primary effector of biochemical changes within a cell or tissue. Glucagon induces an increase in the concentration of cyclic adenosine monophosphate (cAMP) in the cell, with the consequent activation of a number of **protein kinases**. Protein kinases are enzymes that catalyze covalent attachment of a phosphate group to specific target enzymes, resulting in their activation or inactivation. The source of these phosphate groups is the terminal phosphate group of adenosine triphosphate (ATP). Cyclic AMP is a slightly modified version of AMP where the phosphate group on the 5' position of adenosine forms an ester linkage with the hydroxyl group at the 3' position. The preceding relationships are outlined in the following diagram. The diagram is meant as a prelude to a more detailed commentary, and illustrates the influence of a hormone (glucagon), which acts at the cell membrane, on changes in the activities of various enzymes within the cell:

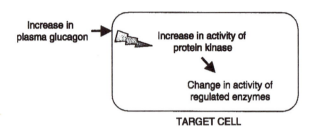

TARGET CELL

## What Insulin Does

Insulin antagonizes the effects of glucagon by inducing a reduction in the level of cAMP and thus preventing activation of the protein kinases. The action of insulin is dependent on the glucagon–cAMP–kinase pathway, where insulin's role is to act as an antagonist; however, insulin also has a variety of effects that are independent of glucagon, that is, actions that do not oppose the effects of glucagon. One such independent effect is activation of glucose transport systems. Insulin stimulates the transport of glucose from the plasma into adipose tissue and muscle, but not into liver, brain, or red blood cells.

Insulin transmits a signal into its target cell. The signal takes the form of a series of events involving the modification of specific proteins and phospholipids. These proteins may be altered by phosphorylation or by de-phosphorylation, depending on the protein. The phospholipid that is specifically altered is phosphatidylinositol (PI). PI is the lipid that has gained fame and renown as the source of inositol-1,4,5-trisphosphate (IP3). IP3 is a modified sugar that has proven to be a versatile and often utilized molecule that participates in signaling events within most types of cells in the body.

The following diagram illustrates some themes that will be developed later on, namely that a hormone (insulin), which acts at the membrane, can influence other events in the membrane (nutrient transport), as well as events in the cytosol (fat synthesis):

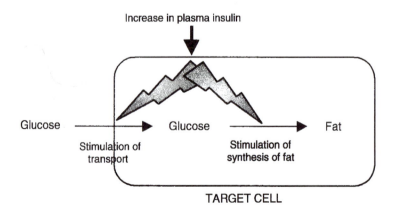

TARGET CELL

## Hormonal Picture with Exercise and Fasting

The hormonal picture, as far as glucagon and insulin are concerned, is similar during exercise and fasting. The concentration of plasma glucagon increases during both states, and this elevation stimulates the synthesis of cAMP in cells and the phosphorylation of various enzymes involved in energy metabolism. The arrays of enzymes that become phosphorylated are largely identical for both conditions, and the resultant patterns of activation or inactivation of enzymes are similar or identical.

## Hormonal Picture with Feeding

Feeding provokes an increase in the concentration of plasma insulin. This elevation counteracts the effects of glucagon mentioned earlier. Insulin also has a variety of effects independent of glucagon action and of cAMP. The most well-known glucagon-independent effect is an increase in glucose transport into muscle for the purpose of glycogen synthesis and into adipose tissue for the purpose of fatty acid synthesis.

During feeding, insulin counteracts the effects of glucagon on the fatty acid biosynthetic pathway by stimulating the activity of the first enzyme of fatty acid biosynthesis, **acetyl-CoA carboxylase**. During feeding, transport of glucose into adipose tissue is stimulated, which is consistent with the enhanced activity of acetyl-CoA carboxylase. This enhanced glucose transport supplies the adipose tissue with more carbons from carbohydrate, and the greater acetyl-CoA carboxylase activity allows the adipose tissue to process glucose, and to convert it to fat at a greater rate. Because of its role in shifting energy metabolism toward the synthesis of energy stores, insulin might be thought of as the "after-dinner hormone."

## Synthesis of Glucagon and Insulin

The actions of glucagon and insulin at the cell membrane have been introduced. Before continuing with this theme, let us view the sources of these two hormones. The pancreas is both an **exocrine** and an **endocrine** gland. The exocrine cells (**acinar cells**) constitute about 80% of the volume of the organ. These cells synthesize vast amounts of digestive enzymes and release copious quantities of fluid. The endocrine cells, which occur in clusters called **islets of Langerhans** ("the islets," for short), account for only about 2.0% of the volume of the pancreas. Each islet (a few cells) contains both α-cells and β-cells. The α-cells are responsible for the synthesis and secretion of glucagon; the β-cells synthesize and secrete insulin. The smaller volume of the endocrine cells is reasonable, as they synthesize substances (glucagon and insulin) present in only trace amounts in the bloodstream.

The arrangement of the acinar cells and of the α-cells and β-cells in an islet is shown in Figure 4.2. Because both α-cells and β-cells are present in each islet and each islet is surrounded by acini, it is difficult for researchers to remove any one type of cell from the pancreas. An alternative approach for selectively manipulating the function of the β-cells has, however, been devised. This method involves chemical, not surgical, removal of these cells. Injection of the drug **streptozotocin**

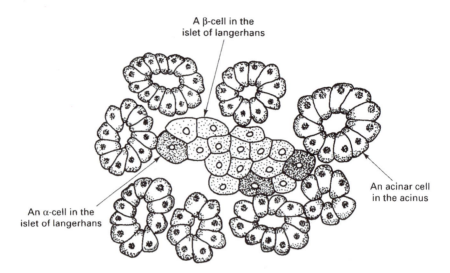

**FIGURE 4.2** Arrangement of the cells of the pancreas. The exocrine cells (acinar cells) of the pancreas occur in groups called *acini* (singular, *acinus*). The endocrine cells (α- and β-cells) occur in groups called *islets of Langerhans*. The islets, which constitute only 2.0% of the organ's volume, are dispersed in the acini. (Redrawn with permission from Guyton, 1971.)

into an experimental animal causes chemical ablation of the β-cells, resulting in cessation of insulin synthesis and secretion. This drug thus chemically induces diabetes, a disease in which the primary defect is an impairment in insulin function. The goal of inducing diabetes in an animal is not to produce illness, of course, but to learn how the body responds to the lack of insulin and how to treat the disease.

Streptozotocin is isolated from a fungus. Its structure, shown in Figure 4.3, resembles that of glucose. Prior to the development of streptozotocin, alloxan was used to induce diabetes in animals; however, streptozotocin is superior because it has fewer side effects.

## Changes in Levels with Exercise and Fasting

The normal basal concentration of glucagon in the bloodstream is about 30 pM; that of insulin is about 100 pM. The normal basal concentration of plasma glucose is about 5 mM, 50 million times greater than the concentrations of the hormones. An appreciation of metabolite quantities is needed to understand data acquired in the research and clinical settings.

The following discussion of the behavior of plasma glucagon and insulin is presented prior to more detailed descriptions of the enzymes and pathways that are regulated by these hormones. Five studies, all involving human subjects, are presented. The first concerns exercise; the second, fasting; the third, feeding in the form of a glucose infusion; the fourth, response to an injection of glucagon; and the fifth, response to an infusion of insulin. An infusion is a "slow injection," taking

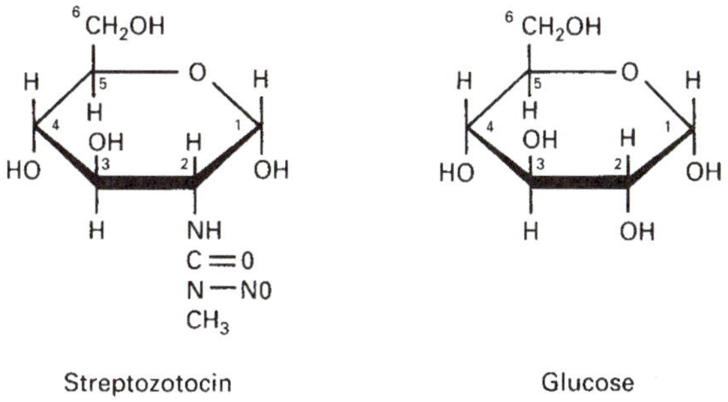

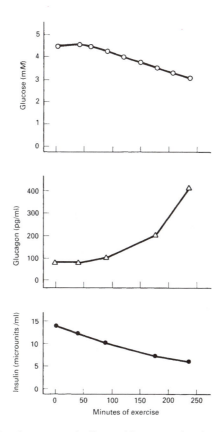

FIGURE 4.3  Structures of streptozotocin and glucose.

place over the course of several minutes or hours. Infused fluids are usually introduced into a vein; injections may be intramuscular or intravenous.

The data plotted in Figure 4.4 show the effect of 250 minutes of exercise on concentrations of plasma glucose, glucagon, and insulin. Plasma glucose remained

FIGURE 4.4  Changes in plasma metabolite and hormone levels with exercise. The human subjects were at rest at time zero. Exercise was continued for 250 min, and concentrations of plasma glucose (O), glucagon (Δ), and insulin (●) were measured at the indicated times. (Redrawn with permission from Felig and Wahren, 1975.)

nearly constant during the first 2 hours and fell slightly thereafter. Plasma glucagon remained fairly stable during the first 2 hours but increased dramatically thereafter. Plasma insulin appeared to decrease at a constant rate from the onset through termination of exercise.

In summary, the data demonstrate that plasma glucose levels are maintained during an activity that greatly increases the use and oxidation of glucose stores within the body. The data also reveal that exercise is associated with an increase in the glucagon/insulin ratio and, a more subtle point, that plasma glucagon levels may fluctuate in a manner independent of insulin levels.

EXERCISE

During the course of exercise, one of the body's major energy stores was nearly completely depleted. Did that store consist of fat or of carbohydrate? Hint: Have you ever heard of anyone depleting all of their body fat during a marathon race?

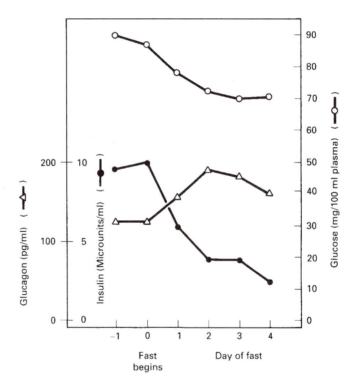

FIGURE 4.5 Changes in plasma metabolite and hormone levels during fasting. The concentrations of plasma glucose (O), glucagon (Δ), and insulin (●) in a human subject were measured before and on each day during a 4-day fast. Plasma glucose levels decreased somewhat early in the fast, but then stabilized at about 80% of the normal value. Plasma glucagon and insulin levels changed in opposite directions during the first 2 days of the fast, resulting in a sharp increase in the glucagon/insulin ratio. This ratio continued to increase with further fasting as a result of a continued drop in insulin levels. (Redrawn with permission from Aguilar-Parada *et al.*, 1969.)

The study depicted in Figure 4.5 involved fasting human subjects over a period of 4 days. Plasma glucose levels decreased somewhat early in the fast, but then stabilized at about 80% of the normal value. Plasma glucagon and insulin levels changed in opposite directions, resulting in a sharp increase in the glucagon/insulin ratio during the first 2 days, followed by a slower increase during the last 2 days.

## Changes in Insulin Levels in Response to Glucose

The study reported in Figure 4.6 involved infusion of glucose into human subjects. This experiment represents a refined version of consumption of a diet containing starch or sugar. A solution of glucose was introduced into a vein at a controlled, defined rate for 200 minutes, and concentrations of plasma insulin and **C-peptide** were measured in blood samples taken at regular intervals. Basal levels of C-peptide are about tenfold greater than those of insulin. Infusion of glucose provoked a dramatic increase in the plasma levels of both molecules, indicating that a decrease in the glucagon/insulin ratio had occurred.

C-peptide (connecting peptide) is originally part of the insulin molecule and is packaged in secretory vesicles, along with insulin, in β-cells. It is secreted into the bloodstream along with insulin, but has no known biological function and may represent a waste product. Recent research has shown that large doses (above the physiological range) of C-peptide seem to have various effects beneficial to health (Steiner and Rubenstein, 1997). C-peptide is described in more detail under Maturation of Proteins in Chapter 1.

In clinical practice, concentrations of C-peptide can give a more accurate picture of pancreatic function than those of insulin itself, because levels of plasma insulin are more subject to changes in the health status of another organ, the liver. Most of the insulin in the circulation is rapidly taken up by the liver and degraded,

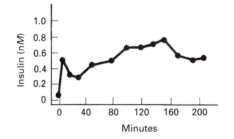 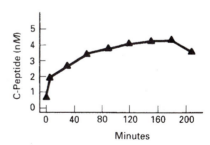

**FIGURE 4.6** Plasma insulin and C-peptide concentrations during infusion of glucose. A glucose solution was introduced into a vein of a human subject at a controlled, defined rate for 200 minutes. Concentrations of plasma insulin and C-peptide were measured in blood samples taken at the indicated times. The basal levels of insulin (●) and C-peptide (▲) were 0.046 nM (6.4 μU/ml) and 0.54 nM, respectively. Basal levels of C-peptide are about tenfold greater than those of insulin. Infusion provoked a dramatic increase in the plasma levels of both molecules, indicating that a decrease in the glucagon/insulin ratio had occurred. Glucagon levels were not available from the study. (Redrawn with permission from Rudenski et al., 1988.)

whereas C-peptide is not rapidly taken up by the liver. Its slower catabolism probably takes place in the kidneys.

EXERCISE

By how many fold did the concentrations of insulin and C-peptide increase with the glucose infusion (Figure 4.6)?

EXERCISE

Patients suffering from alcoholic cirrhosis may have a markedly reduced amount of functional liver tissue. Would you expect the concentrations of insulin to be high or low in these patients?

## Effect of a Glucagon Injection

Figure 4.7 gives results of a relatively manipulative study in which both normal and diabetic human subjects were injected with glucagon. The resulting changes in plasma glucose and C-peptide levels were measured before and at various intervals during the 30 minutes following the injection. The rise in plasma glucose induced by the glucagon injection might conceivably result from one of two events: an increase in the rate of release of glucose into the bloodstream or a decrease in its rate of uptake from the bloodstream. Glucagon stimulates an increase in the rate of release of glucose, and this is illustrated by the continued increase in the plasma glucose level during the 20–30 min after injection. The data also demonstrate a second point: plasma glucose levels were higher in the diabetic

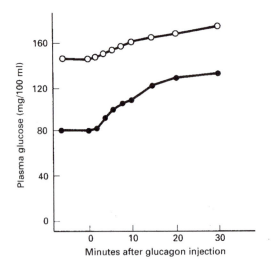

**FIGURE 4.7** Plasma glucose levels after a glucagon injection. Normal (●) and diabetic (○) human subjects were injected with 1.0 mg of glucagon. Plasma glucose and C-peptide levels were measured before and at the indicated times following the injection. (Redrawn with permission from Marchesini *et al.*, 1985.)

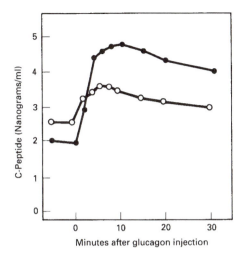

**FIGURE 4.8** Plasma C-peptide levels after a glucagon injection. The increases in plasma C-peptide reflect changes in the rate of release of insulin from the pancreas. The C-peptide level more than doubled in the normal subject (●), and increased more moderately in the diabetic (○). (Redrawn with permission from Marchesini *et al.*, 1985.)

than in the normal subject. These high levels result from the impairment in the uptake of glucose that occurs in this disease.

Figure 4.8 illustrates the changes in plasma C-peptide that occur after an injection of glucagon. These changes reflect changes in the rate of release of insulin from the pancreas. The level of C-peptide more than doubled in the normal subject, whereas a more moderate increase occurred in the diabetic. The increases in insulin secretion did not result directly from the injected glucagon, but from the glucagon-induced rise in plasma glucose.

## Effect of an Insulin Injection

The study depicted in Figure 4.9 continues the discussion of the delicate relationships between plasma levels of glucose, glucagon, and insulin. In this study, normal human subjects were infused with insulin for the 1-hour period indicated by the black bar at the top. Glucose and hormone levels were measured prior to, during, and after the infusion.

Shortly after infusion commenced, the plasma insulin concentration increased well above its basal level. The simultaneous, precipitous drop in plasma glucose may have resulted, in part, from stimulation by insulin of glucose transport into muscle and adipose tissue. There may have another reason for this drop: insulin counteracts glucagon's stimulation of glucose synthesis (or glycogen mobilization) in the liver. The data cannot be used to distinguish between these possibilities. The increase in plasma glucagon occurs in response to **hypoglycemia** (low plasma glucose).

EXERCISE

Compare the rise in plasma insulin that occurred with insulin infusion (Figure 4.9) with the rise that occurred with glucose infusion (Figure 4.6). Was the rise in insulin supplied by

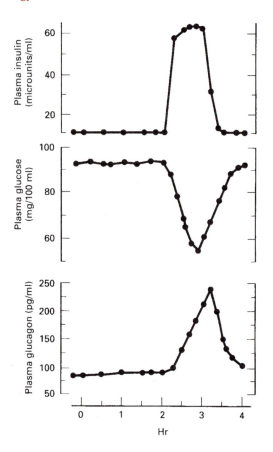

**FIGURE 4.9** Plasma glucose and hormone levels before, during, and after insulin infusion. Normal human subjects were infused with insulin for the 1-hour period indicated by the black bar. Glucose, insulin, and glucagon levels were measured at the indicated times. While plasma insulin increased to six times its basal level, plasma glucose fell precipitously and plasma glucagon rose. (Redrawn with permission from Bolli *et al.*, 1984.)

exogenous hormones (Figure 4.9) similar to the rise that could be provoked by a glucose infusion? In short, did the insulin infusion result in plasma insulin levels within the physiological range?

EXERCISE

Cancer is a disease involving an increased rate of division and multiplication of a particular type of cell. In one type of cancer, an increased number of β-cells in the pancreas may result in overproduction of insulin. Would you expect the brain to have a deficient or overabundant supply of energy? (See Townsend and Thompson (1985).)

EXERCISE

An increased concentration of a metabolite in the bloodstream (hyperemia) often results in an increased concentration in the urine (hyperuria). If a rat is injected with streptozotocin, how might the flavor of the rat's urine be affected?

# Diabetes Mellitus

An overview of the disease **diabetes mellitus** will be used to supply motivation prior to starting our journey through the pathways of energy production. Diabetes mellitus is a major health concern on a global basis. This disease takes two forms, i.e., **insulin-dependent diabetes mellitus (IDDM)** and **non-insulin-dependent diabetes mellitus (NIDDM)**. Nutritional intervention is used in the treatment of diabetes, as revealed later in this chapter and in the Obesity chapter. IDDM is easily treated by injections of insulin, while NIDDM is more difficult to treat. NIDDM is treated with insulin and by weight reduction in obese patients. Use of soluble fiber in the diet may be used in the treatment of both types of diabetes.

## *How Does Diabetes Make You Feel Sick?*

Diabetes mellitus is a major cause of blindness, renal failure, and cause for surgical amputation of legs or feet. The risk of cardiovascular disease is doubled in a person with diabetes mellitus. Thus, diabetes is a serious health hazard. How does diabetes first present itself? The initial symptoms of the disease include excessive thirst, excessive food consumption, excessive urination, weight loss, and blurred vision. Diabetes often results in **neuropathies** leading to impotence (in males) and numb feet. Foot ulcers may occur, and in extreme cases, the foot must be amputated. Diabetes is a leading cause of blindness in adults. Diabetes often accelerates the development of cardiovascular disease, but the mechanism of this acceleration is not clear.

## *Tests for Diabetes*

Diabetes can be detected by elevated levels of glucose in the plasma and in the urine. High urinary glucose was probably first suspected when it was noted that the urine attracted flies. A more sensitive test for diabetes involves a test meal of glucose and measurement of the resultant increase in plasma glucose, with attention paid to the body's ability to control or deal with the rise in plasma glucose. In this **glucose tolerance test**, the patient slowly drinks a solution of 75 g of glucose in water, with blood samples withdrawn before and at intervals afterward. In the healthy subject, plasma glucose levels rise and then promptly fall back to normal, because rising insulin levels stimulate uptake of the consumed glucose into various tissues. In a diabetic, the rise in plasma glucose may be much greater than normal, and its subsequent fall may take much longer than normal.

## *Frequency of IDDM and NIDDM Throughout the World*

The frequency of IDDM in Europe and in the United States is 0.2 to 2.4 occurrences per 100 persons. Diabetes occurs much less frequently in Japan and is very rare in tropical countries. NIDDM is quite common. Its frequency in adults is 3 to 6% in Europe and the United States. Its frequency is much higher among certain Polynesian populations, Native Americans (Pima Indians), Mexican-Americans, and Australian aborigines.

IDDM, also called Type I diabetes, generally presents in persons by the age of 30 years, and accounts for only 5–10% of all persons with diabetes mellitus. The symptoms of IDDM occur somewhat abruptly and require prompt treatment. The disease results in the loss of β-cells of the pancreas and the consequent absence of the insulin that is synthesized by the β-cells. IDDM is an autoimmune disease. This means that the body produces antibodies that recognize and bind to various proteins of the β-cells. This is followed by the infiltration of the β-cells with white blood cells. The white blood cells destroy the β-cells gradually, over the course of several years. IDDM makes the patient feel sick only after most of the β-cells have been destroyed. This disease can be easily controlled with drugs. Insulin is commercially available, and persons with the disease can easily learn to administer the injections at home (Bach, 1994, 1997; Tisch and McDevitt, 1996).

NIDDM, also called Type II diabetes, generally presents in persons over 30 years of age. NIDDM is associated with obesity. Perhaps two-thirds of all persons with NIDDM are also obese. The overall defect in NIDDM is that the muscle and adipose tissue tend not to respond to insulin. NIDDM is a product of multiple biochemical defects that result in impaired glucose metabolism. Apparently, each of these defects accounts for no more than 5–10% of the population of those with NIDDM. NIDDM develops initially because of defects in insulin's ability to transmit a signal, rather than by a lack of insulin in the bloodstream, and for this reason the disease is called non-insulin-dependent diabetes mellitus.

### Mechanism of IDDM, As Revealed by a Case Study of a Child

The study depicted in Figure 4.10 concerns a child examined periodically over the course of a decade. The gradual onset and development of diabetes in this child are demonstrated by progressively more abnormal glucose tolerance curves. The child became less and less able to control the rise in plasma glucose that followed the test dose. The reduction in plasma glucose concentration following a test dose is called the clearance of glucose. In short, the child developed an impaired clearance rate. A normal clearance rate is shown in the test conducted in 1969, but abnormal rates were found in 1979 and 1980. Figure 4.11 shows plasma insulin levels measured from samples taken during the glucose tolerance tests depicted in Figure 4.10. A normal insulin response occurred in 1969, but dampened insulin responses occurred in 1979 and 1980.

The mechanism of this child's diabetes was as follows. Impaired insulin production resulted from destruction, by antibodies, of many of the β-cells of the pancreas. Antibodies are proteins synthesized by cells of the immune system that recognize and tightly bind to macromolecules to form antibody–target complexes. These complexes are then destroyed by white blood cells. A normally acting immune system produces only antibodies that recognize foreign structures; however, this child produced antibodies that recognized components of the child's own β-cells.

A disease in which antibodies that recognize proteins made by one's own body are produced is called an **autoimmune disease**, and the undesirable antibodies

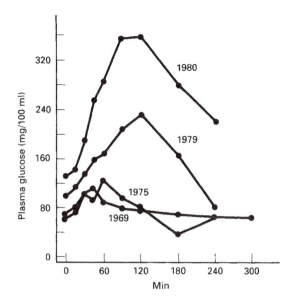

**FIGURE 4.10** Glucose clearance curves. Glucose tolerance tests were performed on the same human child over the course of a decade. The gradual onset and development of diabetes are demonstrated by progressively more abnormal glucose tolerance curves. The reduction in plasma glucose concentration following a test dose of glucose is called *the clearance of glucose*. A normal clearance rate is shown for 1969, but abnormal rates are shown for 1979 and 1980. (Redrawn with permission from Srikanta *et al.*, 1983.)

produced are called **autoantibodies**. Tests can detect autoantibodies in the bloodstream, just as they can detect other types of protein. These tests detected no anti-islet antibodies in the child's plasma in the years between 1967 and 1973, but revealed their presence in 1975 and thereafter.

EXERCISE

Which autoimmune disease results in damage to the parietal cell of the stomach?

## Mechanism of NIDDM

The progress of NIDDM is divided into three steps:

1. First, insulin resistance occurs. Muscle and adipose tissue fail to respond normally to this hormone. The development of insulin resistance tends to be enhanced by obesity and by lack of exercise. This step begins at about the age of 30, i.e., after childhood.

2. With the development of insulin resistance, the pancreas compensates by increasing its secretion of insulin into the bloodstream, resulting in increased plasma insulin and in normal levels of plasma glucose. At this stage of the disease, one may not be aware that one has NIDDM.

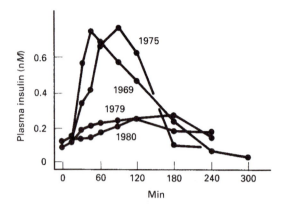

**FIGURE 4.11** Plasma insulin levels during glucose tolerance tests. Plasma insulin levels were measured during the glucose tolerance tests shown in Figure 4.10. A normal insulin response occurred in 1969, but insulin responses were dampened in 1979 and 1980. (Redrawn with permission from Srikanta *et al.*, 1983.)

3. Over the course of many years, the β-cells of the pancreas fail, which leads to insulin deficiency (Patti and Kahn, 1996). The disease presents itself as the β-cells begin to fail.

Non-insulin-dependent diabetes mellitus seems to result from a sizable number of defects. Perhaps no one defect accounts for more than 10% of all persons suffering from NIDDM. In humans with NIDDM, mutations have been detected in the genes coding for several proteins. Just to document the variety of affected proteins, mutations in human genes have been found in those coding for IRS, glucose transporters, glucokinase, transfer RNA (tRNA[LEU]), and for glycerol-P dehydrogenase (Gerbitz *et al.*, 1996). Mutations have also been found in the insulin receptor, but this exact mutation rarely occurs. The notion has developed that NIDDM is not one disease but many, where the major defect differs in different people. A specific single genetic defect responsible for most cases of NIDDM, if any, has not been found.

The term **mutation** is defined. All genes consist of a sequence of nucleotide bases (A, T, G, and C), where successive groups of three bases (codons) code for the sequence of amino acids in the corresponding polypeptide. Any change in the exact sequence of bases can rightfully be called a mutation. Where a change to occur in an amino acid that is designated by the codon, the mutation can result in a change in one or more of the amino acids in the polypeptide. Mutations and changes in an amino acid often result in little or no change in the function of the polypeptide. A mutation that results in a drastic change in the functioning of a polypeptide is very likely to be detected as a genetic disease.

### Diet Therapy in Diabetes

Controlling plasma glucose levels is an issue with both IDDM and NIDDM. Elevations in plasma glucose can be controlled with drugs, as well as by dietary tactics. Eating many small meals during the day (rather than 3 meals) can help reduce plasma glucose levels throughout the day. Consuming soluble fibers (guar gum and oat gum) may have a small influence on controlling plasma glucose. Soluble fibers form a viscous gel in the gut and slow the rate of gastric emptying and intestinal transit. Soluble fibers have a more significant, but still modest, effect on lowering plasma lipid levels (Franz *et al.*, 1994). Diet therapy plays a major part in the treatment of NIDDM, where the patient is obese (see Obesity chapter).

## Insulin at the Cell Membrane

The mechanisms by which insulin antagonizes the effects of glucagon and exerts its glucagon-independent effects are not clear. Much attention has been given to the insulin receptor, a protein embedded in the plasma membranes of insulin-sensitive cells. Binding of plasma insulin to a receptor results in phosphorylation of a tyrosine residue present in the polypeptide chain of the receptor protein. It also results in activation of kinases that catalyze phosphorylation of serine residues of specific proteins. Another effect is a decline in the degree of phosphorylation of a number of other proteins.

Insulin stimulates the rate of transport into various cells, such as those of adipose tissue and muscle. How does a rise in this hormone in the plasma provoke an increase in flow of glucose to the cell's interior? Insulin stimulates the fusion of vesicles with the plasma membrane. These vesicles contain membrane-bound proteins that can transport glucose; thus, the fusion event results in the insertion of glucose transporters into the plasma membrane, with the consequent increase in glucose transport into the cell. This scenario is illustrated in Figure 4.12, in which the signal transmitted into the cell by insulin is represented by a spark.

The effect of insulin in stimulating glucose transport is illustrated in the study reported in Table 4.1. Data from both normal and diabetic subjects are used to compare the functions of hormones in health and disease. Adipocytes were taken from living human subjects and glucose transport was assessed by measuring the rate at which these cells absorbed radioactive glucose from the surrounding cell medium. The data show that addition of insulin to the medium containing cells from normal subjects stimulated a threefold increase in the transport rate, from the basal rate of 43 to 134. Addition of insulin to the diabetic cell medium doubled the transport rate, from 20 to 42. The data indicate that there may have been more than one defect in the adipocytes from the diabetics. First, the basal rate of glucose transport was subnormal; second, the stimulability by insulin was impaired.

The experiment was repeated using another marker for glucose transport, the glucose analogue 3-O-methylglucose. Again, transport was assessed by measuring the rate of accumulation of radioactivity in the adipocytes, after 3-O-methylglucose labeled with radioactive carbon was added to the medium. Results paralleled those obtained with radioactive glucose. The analogue was used to avoid the interference expected with radioactive glucose. A fraction of the glucose would be expected to be oxidized to carbon dioxide. Discharge of radioactive $CO_2$ into the cell medium could interfere with the measurement of glucose uptake. The expected consequence would be an artificially low value for glucose transport. Use of the nonmetabolizable glucose analogue avoids this potential spurious result.

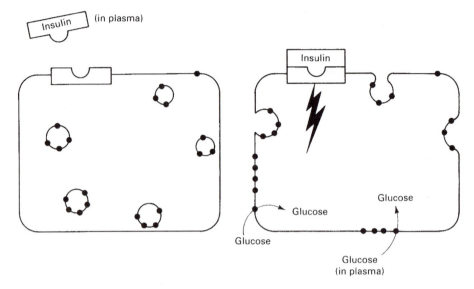

**FIGURE 4.12** Action of insulin at a target cell. The figure shows three separate processes: (1) formation of insulin/receptor complex; (2) stylized version of intracellular signal; and (3) insertion of glucose receptors into the plasma membrane. Insulin and its receptor are represented by rectangles with tabs. The signal transmitted into the cell is represented by an imaginary spark. The transporters are represented by black dots.

EXERCISE

Compare the transport data from studies using glucose and 3-O-methylglucose (Table 4.1). Do the results from these two studies yield similar or different conclusions?

## Fine Details of the Mechanisms of Insulin Release and Insulin Action

Figure 4.13A,B lets us step back and review the big picture, whereby an increase in plasma glucose (due to feeding) provokes a rise in plasma insulin, which in turn provokes an the insertion of glucose transporters into the plasma membrane.

**TABLE 4.1** Transport of Glucose and a Glucose Analogue into Cells

| Subjects | Transport of glucose[a] | | Transport of analogue | |
|---|---|---|---|---|
| | Basal | +Insulin | Basal | +Insulin |
| Normal | 43 | 134 | 57 | 152 |
| Diabetic | 20 | 42 | 25 | 55 |

Source: Foley et al. (1983).

[a]The transport rate is the amount of glucose or analogue contained within one femtoliter per cell per second.

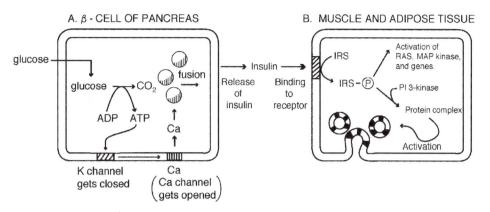

**FIGURE 4.13.** Release of insulin from the pancreas, travel through bloodstream, and action of insulin at a target cell. (A) Mechanism by which elevated plasma glucose provokes release of insulin from the β-cell. Glucose enters the cell, where it is oxidized by the pathways of glycolysis and the Krebs cycle (detailed later in this chapter). The result is an increase in the concentration of ATP in the cell. ATP is a sort of molecular battery that can be used by thousands of different enzymes in the cell, in order to drive their enzymatic reactions. In the β-cell, the increase in ATP levels provokes a change in a membrane-bound protein called the **potassium channel** (K channel). This change provokes a change in another membrane-bound protein, the **calcium channel** (Ca channel). The end-result is a momentary increase in calcium ions inside the cell, which in turn provokes the fusion of insulin-containing vesicles with the plasma membrane, and the release of this hormone into the bloodstream. (B) Mechanism by which insulin binding to its receptor stimulates the transport of glucose into the cell. Insulin binds to its receptor, but does not enter the cell. The event of binding provokes the activation of the receptor. The receptor is actually an enzyme. It is a kinase, meaning that it utilizes ATP as a source of a phosphate group, and catalyzes the attachment of this phosphate group to a target protein. The target, or substrate, of the insulin receptor is a protein called IRS. Once phosphorylated, IRS provokes two activation pathways: (1) the activation of a number of other proteins, resulting in the increased transcription of several genes (see arrow pointing upwards); and (2) the fusion of vesicles (containing glucose transporters) into the plasma membrane (see arrow pointing downwards), resulting in the stimulation of glucose transport into the hungry muscle cell or into the glutinous adipose tissue cell.

Figure 4.13A depicts events that occur only in the pancreas. Figure 4.13B depicts events occurring in all insulin-sensitive tissues, and reveals some of the events that are represented by the spark. The spark appeared in Figure 4.12.

## Insulin Release from the Pancreas

How does an increase in plasma glucose provoke a rise in plasma insulin? Feeding (eating food) elevates the level of plasma glucose, provided that the food contained sugar or starch. This increase results in an increased entry of glucose into many cells of the body, including the β-cells of the pancreas. Consequent metabolism of the glucose results in an increase in ATP levels in the cell. The increased ATP has a direct influence on the regulation of a special pore in the plasma membrane of the β-cell, the potassium channel. Further details appear in the legend to Figure 4.13A. The mechanism involves an increase in calcium ions in the cell, which in turn provokes the fusion of vesicles with the plasma membrane. These vesicles contain insulin. Thus, the fusion event results in the contents of the vesicles (insulin) being dumped into the extracellular fluid (Figure 4.13A). The insulin circulates about the body in the bloodstream.

## Insulin Action at the Plasma Membrane of Muscle and Other Tissues

How does insulin exert its effects on insulin-sensitive tissues? Insulin binds to receptors in the membranes of the skeletal muscle and adipose tissue. The insulin receptor has two α-subunits (each 135 kDa) and two β-subunits (each 95 kDa). Insulin binds to the α-subunit, which is extracellular, and provokes the β-subunit, which contacts the cytoplasm, to catalyze its own phosphorylation. Phosphorylation occurs on residues of tyrosine. This sequence is summarized by the following flowchart:

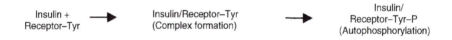

After the insulin receptor catalyzes its own phosphorylation, the receptor is activated in a way that provokes it to phosphorylate other proteins. The main substrate is a cytoplasmic protein called IRS. IRS stands for **insulin responsive substrate**. IRS has a molecular weight of 131,000 (131 kDa). The phosphorylation of IRS is represented by (Mothe and Obberghen, 1996):

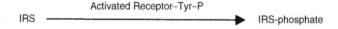

After being phosphorylated, IRS provokes the activation of two separate pathways within the cell. From the IRS branching point come two signaling pathways, in the manner of two branches. These two branches are the **PI 3-kinase signaling pathway** and the **RAS signaling pathway** (Lawrence and Roach, 1997; Taha *et al.*, 1995) (Figure 4.13B). The branching point is represented by

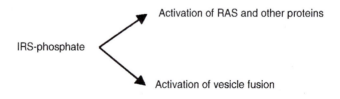

*Details of the RAS Signaling Pathway*

Once IRS occurs as IRS-phosphate, the GRB2/SOS complex also binds to it. This event of binding results in activation of such proteins as RAS, MAP kinase, and other signaling proteins. The end-result of this branch includes the activation of *glycogen synthase*, the enzyme that catalyzes the conversion of glucose to its storage form, glycogen. The GRB2/SOS complex is a complex of two proteins. Our goal, in this chapter, is not to reveal the functions of GRB2/SOS, RAS, and MAP kinase, but only to reveal that a distinct pathway for relaying insulin's signal has been discovered. A cure for diabetes will require a knowledge of these, and other signaling proteins.

*Details of the PI 3-Kinase Signaling Pathway*

Once IRS occurs as IRS-phosphate, PI 3-kinase binds to it. (PI means phosphatidylinositol, a common lipid.) The activity of PI 3-kinase is increased by over tenfold when it binds to IRS-phosphate. Evidence suggests that PI 3-kinase acts on the membranes of intracellular vesicles that contain GLUT-4 (Heller-Harrison *et al*, 1996; Elmendorf *et al.*, 1997). The association of the activated PI 3-kinase with these vesicles catalyzes the addition of phosphate groups to the 3 position of phosphatidylinositol, with the consequence that the vesicles are stimulated to fuse with the plasma membrane. These events are depicted by the following diagram. The diagram depicts events occurring in the lower half of Figure 4.13B:

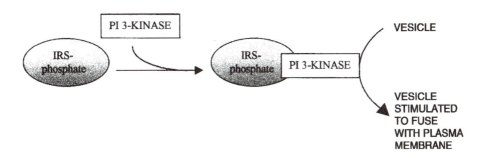

## Glucagon at the Cell Membrane

Increased levels of plasma glucagon result in increased binding to the glucagon receptor. The glucagon receptor is a membrane-bound protein that weaves seven times through the plasma membrane. The glucagon receptor is a polypeptide of 485 amino acids. The N-terminal third of the protein is entirely extracellular, and is used to bind the hormone (Carruthers *et al.*, 1994). The glucagon receptor is coupled to another enzyme, called adenylyl cyclase (also called adenylate cyclase). Adenylyl cyclase is a large protein of about 1100 amino acids, which weaves in or out of the plasma membrane twelve times (Cooper *et al.*, 1995). The glucagon receptor stimulates adenylyl cyclase to catalyze the synthesis of cyclic AMP.

This stimulation occurs via an intermediary protein called **G protein**. In general, all 7-transmembrane domain proteins, including the glucagon receptor, act directly on a protein called **G protein**. With activation, G protein can stimulate adenylyl cyclase, or one of a variety of other enzymes. The orientation of the three membrane-bound proteins, mentioned earlier, is shown in the following diagram:

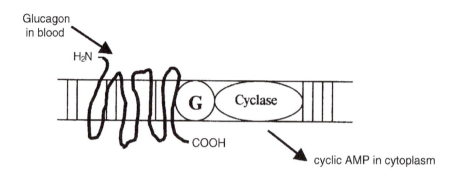

*Summary*

Insulin and glucagon have opposite influences on energy metabolism. Insulin is the "after-dinner hormone," as it promotes the storage of the energy acquired from the diet. Glucagon is used for the mobilization of energy stores, and provokes the release of carbohydrate and fat stored in the body, for use during exercise and fasting. The opposite actions of these two hormones should be kept in mind when viewing their influence on the attachment (and removal) of phosphate groups from various enzymes, as revealed on the succeeding pages. The covalent attachment of phosphate groups (to enzymes) is used throughout all biology in the regulation of various events in the cell. A view of the disease diabetes mellitus provides motivation for exploring the major pathways of energy metabolism in the body. These pathways are: (1) glycolysis; (2) glycogen synthesis and degradation; (3) the Krebs cycle; and (4) fatty acid synthesis and fatty acid oxidation.

## DIFFERENT TYPES OF ENERGY METABOLISM

The digestion of food, its absorption into the bloodstream, and the regulation of glucose uptake has been described. Now it is time to embark on the pathways of oxidation of glucose, fatty acids, and other nutrients.

### Several Types of Energy Metabolism

The major fuel for most organs is fatty acids. Events that control their liberation from triglycerides (TGs) and their oxidation are described in subsequent sections of this chapter. A variety of control points encountered during the complete oxidation of glucose and of fatty acids to $CO_2$ are revealed. A number of well-

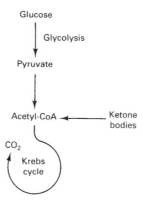

**FIGURE 4.14** Styles of energy metabolism — complete oxidation of glucose to carbon dioxide, and complete oxidation of ketone bodies to carbon dioxide.

known intermediates occur in these oxidation pathways, including phosphoenolpyruvate, pyruvate, lactate, acetyl-CoA, and citrate.

The way in which energy is metabolized differs with the organ and under different conditions. The brain derives energy by the conversion of glucose to carbon dioxide via the pathway of glycolysis followed by the Krebs cycle, as shown in Figure 4.14. During fasting, the brain also uses ketone bodies, which are produced in elevated amounts at this time. The neonatal infant naturally produces and uses ketone bodies, whether fasting or not.

The liver uses fatty acids as its main fuel by the pathway shown in Figure 4.15: conversion to units of acetyl-CoA followed by oxidation to $CO_2$. The liver also derives a small amount of its energy requirement from glucose, as depicted in Figures 4.16 and 4.17; however, a main function of the liver is to synthesize, store, and release glucose for the benefit of other organs, not for itself. As far as glucose consumption is concerned, the liver does not have a "sweet tooth."

In the third type of energy metabolism, partial oxidation of glucose by the pathway of glycolysis (Figure 4.16), glucose is oxidized only as far as pyruvate. The pyruvate is reduced to lactic acid, which is released into the bloodstream for further processing by the liver. This type of metabolism is called **fermentation**, which is derived from a French verb meaning "to close"; fermentation occurs in a closed environment where there is no supply of oxygen. An organ that takes up glucose from the bloodstream and converts it to lactate that it releases into the

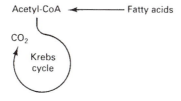

**FIGURE 4.15** Styles of energy metabolism — complete oxidation of fatty acids to carbon dioxide.

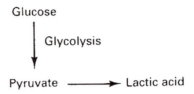

FIGURE 4.16 Styles of energy metabolism — partial oxidation of glucose by the pathway of glycolysis.

bloodstream is utilizing fermentative metabolism. Under certain conditions, muscle utilizes fermentative metabolism, rather than the styles illustrated in Figures 4.15 and 4.17.

The fourth type of metabolism is complete oxidation of glucose and fatty acids (Figure 4.17). Exercising muscle generally uses this style of metabolism, whereas resting muscle tends to derive energy through the complete oxidation of fatty acids (Figure 4.15). Most tissues derive their energy primarily by the latter scheme.

An increase in the glucagon/insulin ratio, as occurs in exercise or fasting, provokes the coordinated actions used to mobilize energy stores. In most cases, the key enzymes involved in the relevant biochemical pathways are regulated by phosphorylation. Specific pathways, their functions, and their responses to an increase in glucagon/insulin ratio are listed in Table 4.2. The arrangement of the pathways in pairs stresses the fact that these pairs have physiologically opposite functions. Simultaneous operation at maximal rates of both pathways of any particular pair leads to simultaneous storage and mobilization of energy, an event known as a "futile cycle."

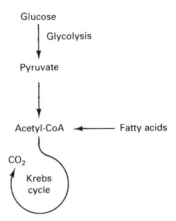

FIGURE 4.17 Styles of energy metabolism — complete oxidation of glucose and fatty acids. The role of oxygen in this pathway, as well as in those pathways shown in Figures 4.14, 4.15, and 4.16, is as follows. The Krebs cycle and, to some extent, glycolysis produce reduced molecules of NAD. These molecules are reoxidized by mitochondrial proteins (cytochromes), resulting in reduced cytochromes. The reduced cytochromes, in turn, catalyze the reduction of oxygen to water.

**TABLE 4.2** Pathways with Roughly Opposite Functions

| Pathway or enzyme | Function | Change in activity with an increase in [glucagon]/[insulin] |
|---|---|---|
| Glycogen phosphorylase | Breaks down glycogen to liberate glucose units | Increase |
| Glycogen synthetase | Synthesizes glycogen from glucose units | Decrease |
| Gluconeogenesis | Synthesizes glucose from lactic acid, the carbon skeletons of amino acids, and glycerol | Increase |
| Glycolysis | Oxidizes glucose units to pyruvate | Decrease |
| Hormone-sensitive lipase | Liberates fatty acids from the TGs stored in vesicles in adipocytes | Increase |
| Lipoprotein lipase | Transfers fatty acids from TGs circulating in lipoproteins to cells | Decrease |
| Carnitine acyltransferase | First and limiting step in the oxidation of free fatty acids | Increase |
| Fatty acid synthetase | Synthesizes fatty acids from units of acetyl-CoA | Decrease |

## Synthesis and Degradation of Glycogen

Carbohydrate is stored in the liver and muscle in the form of **glycogen**. Glycogen is called **animal starch**, though its content in organs is less than that of plant starch in plants. The synthesis of glycogen, catalyzed by **glycogen synthetase**, and release of glucose units from glycogen, catalyzed by **glycogen phosphorylase** (phosphorylase, for short), are shown in Figure 4.18. Glycogen is broken down in the mechanism of **phosphorolysis**, rather than by hydrolysis, as previously discussed.

The immediate substrate of glycogen synthetase is **UDP-glucose**, an activated form of glucose. The immediate source of glucose-1-phosphate, used for the synthesis of muscle glycogen, is glucose derived from the blood plasma. The immediate sources of most of the glucose-1-phosphate, used for the synthesis of liver glycogen, are **gluconeogenic precursors**, converted in the liver via the intermediates of the glucogenic pathway. In other words, the pathway resulting in energy storage involves the following sequence: gluconeogenic precursors and

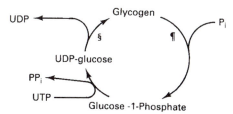

**FIGURE 4.18** Synthesis and breakdown of glycogen. The synthesis of glycogen (§) is catalyzed by glycogen synthetase; the release of glucose units from glycogen (¶) is catalyzed by glycogen phosphorylase.

glucose → glucose-1-phosphate → UDP-glucose → glycogen. The activity of glycogen synthetase and glycogen phosphorylase is controlled in a coordinated manner. An increase in plasma glucagon stimulates a simultaneous decrease in glycogen synthetase activity and an increase in phosphorylase activity. These changes result from the activation of adenylate cyclase (a membrane-bound protein) by the hormone, and the consequent conversion of cytoplasmic ATP to cyclic AMP and inorganic pyrophosphate (PP$_i$). The cyclic AMP, in turn, binds to a protein kinase in the cytoplasm and activates it. The activated protein kinase catalyzes the phosphorylation of a number of proteins, including the enzymes indicated in Figure 4.19.

### Catalytic and Regulatory Subunits Bind or Dissociate

How does cyclic AMP (cAMP) provoke the activation of protein kinase? A few details of the mechanism may prove intriguing. Protein kinase in resting muscle, for example, occurs in the inactive form. It occurs as a complex of catalytic and regulatory subunits. When intracellular cAMP increases, cAMP binds to the regulatory subunits. This event provokes a sharp drop in the binding of the two types

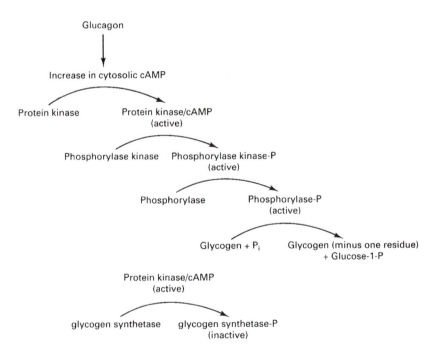

**FIGURE 4.19** Cyclic AMP-dependent signaling pathways of glycogen. The sequence of events can be summarized as follows: (1) Glucagon increases in concentration in the bloodstream. (2) Glucagon binds to glucagon receptors of target cells. (3) Binding results in activation of adenylate cyclase, the enzyme that catalyzes the synthesis of cyclic AMP. (4) Cyclic AMP, in increased concentrations in the cell, binds to protein kinase and activates it. (5) Activated protein kinase catalyzes the phosphorylation of phosphorylase, resulting in its activation and consequent breakdown of the glycogen stored in the cell.

of subunits for each other, and the complex dissociates (Sowadski *et al.*, 1984). The free catalytic subunit is active (it does not bind cAMP) and catalyzes the phosphorylation of various target proteins.

## *Glycogen Phosphorylase Occurs as a Dimer or Tetramer*

To complete the preceding short story, a few words should be added regarding glycogen phosphorylase. In its nonphosphorylated form, the enzyme is not active. Glycogen phosphorylase ("phosphorylase" for short) occurs as a dimer (two subunits), and each subunit consists of 841 amino acids. **Protein kinase** catalyzes the phosphorylation of an amino acid residue near the amino terminus of phosphorylase (serine 14), resulting in the association of two dimers to form a tetramer. The phosphate group provokes shifts in the overall structure of the enzyme, resulting in the exposure of new amino acid residues at the surface of the protein and the materialization of sites that promote the adhesion of the two dimers. In its tetrameric form, glycogen phosphorylase is catalytically active (Barford *et al.*, 1991).

We have been discussing the sequence of events occurring with an increase in plasma glucagon. What happens with a decline in plasma glucagon? A drop in glucagon results in a decrease in intracellular cAMP, the removal of the covalently bound phosphate groups on phosphorylase, and a drop in the rate of glycogen breakdown.

**Epinephrine** and **norepinephrine**, hormones called **catecholamines**, are released from the adrenal medulla during exercise. Epinephrine and, to a lesser extent, norepinephrine stimulate effects in muscles and the liver similar to those produced by glucagon. Epinephrine activates adenylate cyclase, resulting in such events as the breakdown of glycogen and fatty acids.

## Glycolysis and Gluconeogenesis

The term *glycolysis* refers to the pathway by which glucose-6-phosphate is converted to pyruvate. This term is also used to refer to the conversion of glucose to pyruvate. Gluconeogenesis refers to the series of reactions resulting in the conversion of lactate, pyruvate, glycerol, malate, oxaloacetic acid, or related compounds to glucose. Some of the enzymes in the gluconeogenic pathway are identical to those of the glycolytic pathway, but other enzymes are not shared. Keeping in mind that glycolysis and gluconeogenesis do not represent strict reversals of each other, one might summarize the effects of the two pathways by the simple diagram shown in Figure 4.20.

Glycolysis probably takes place in all cells of the body, but gluconeogenesis takes place only in the liver and kidneys. The liver's capacity for gluconeogenesis is about 10 times that of the kidneys. In the liver, glycogen can be converted to glucose-6-phosphate, which can be used in glycolysis or hydrolyzed to glucose, which is transported into the bloodstream. In muscle, glycogen can be converted to glucose-6-phosphate for use in glycolysis, but cannot be hydrolyzed to glucose. Thus, muscle cannot supply glucose to the bloodstream.

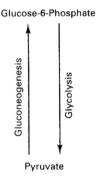

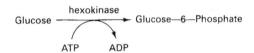

**FIGURE 4.20** Simplified diagram of the gluconeogenic and glycolytic pathways.

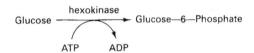

**FIGURE 4.21** Conversion of glucose to glucose-6-phosphate. This process takes place in glucose-using organs.

The glycolytic and gluconeogenic pathways are depicted in Figures 4.23 to 4.25. The glucose-6-phosphate (G-6-P) molecule, shown in Figure 4.24, has several origins. In glucose-using organs such as the brain and muscle, glucose entering the organ is converted to G-6-P by the action of hexokinase (Figure 4.21). Although the liver also contains hexokinase, it is mainly a fat-burning organ, not a glucose-burning organ. The origin of the phosphate group of G-6-P is the terminal phosphate of ATP.

As shown in Figure 4.23, G-6-P can materialize within a tissue from two other sources. The G-1-P, released during breakdown of glycogen, can be converted to G-6-P. It can be produced via the gluconeogenic pathway involving conversion of pyruvate to OAA, to PEP, to glyceraldehyde-3-phosphate, and finally to G-6-P.

*Glucose-6-Phosphate and the Liver's Role in Supplying Free Glucose to Other Tissues*

In the liver, G-6-P is hydrolyzed to glucose by the action of glucose-6-phosphatase (see Figure 4.22). The free glucose produced in this way can enter the bloodstream

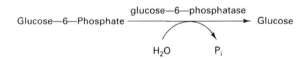

**FIGURE 4.22** Conversion of glucose-6-phosphate to glucose. This process takes place in the liver.

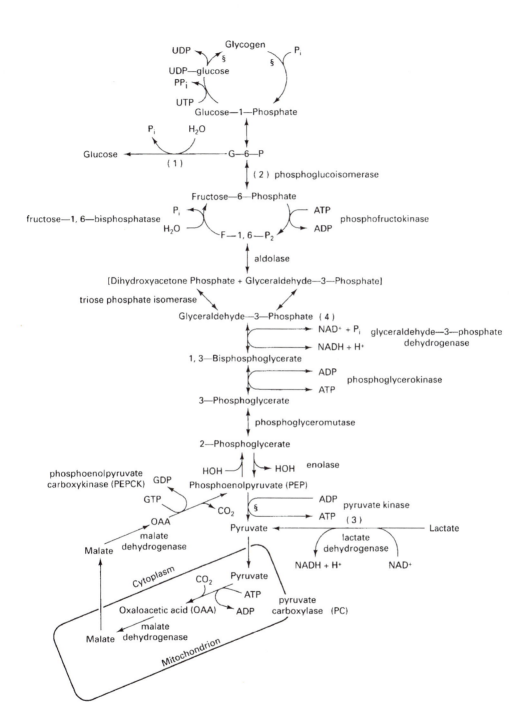

**FIGURE 4.23** Pathways of glycolysis and gluconeogenesis.

for use by other tissues. Glucose-6-phosphatase is distinguished from other enzymes of the gluconeogenic pathway in that it is a membrane-bound enzyme. It is bound to the endoplasmic reticulum with its active site facing the lumen of the organelle. Cytosolic G-6-P enters the endoplasmic reticulum, where its phosphate group is removed by hydrolysis. Glucose-6-phosphatase activity increases with an increase in the glucagon/insulin ratio.

The numbers in Figure 4.23 identify four points: (1) the point at which hepatic glucose leaves the liver; (2) the point at which G-6-P enters glycolysis for conversion to pyruvate; (3) the point where lactic acid from the bloodstream enters the liver for conversion to pyruvate; and (4) the point at which every molecule of glucose entering the pathway yields two molecules of each of the substances listed later.

The pyruvate produced at point (3) can be used to produce energy within the liver or for gluconeogenesis. Pyruvate produces energy when it is oxidized in the mitochondria by the Krebs cycle. It is used for gluconeogenesis when it is converted in the mitochondria to OAA, then to malate, which, after exit from the mitochondria, is converted to PEP and eventually to glucose.

### Muscle Can Supply the Bloodstream with Lactate but not Glucose

Figure 4.23 illustrates the patterns of metabolism in the liver. In muscle, the directions of nutrient flux at points (1) and (3) are opposite those shown; that is, glucose derived from the bloodstream is converted to G-6-P, then to pyruvate. This pyruvate can then be completely oxidized in the mitochondria or can enter the bloodstream in the form of lactate. Muscle does not contain glucose-6-phosphatase and thus cannot export glucose units stored as muscle glycogen.

The enzymes marked with the section symbol (§) are regulated by covalent phosphorylation. In short, changes in the plasma glucagon/insulin ratio produce changes in the phosphate group content of these enzymes. Addition or removal of the phosphate group regulates the enzyme's activity, in accordance with the body's metabolic needs.

### Structures of Intermediates

Figure 4.24 shows the molecular structures of compounds in the glycolytic pathway. The same compounds occur in gluconeogenesis. Generally, G-6-P and pyruvate are considered to be the first and final compounds, respectively, of glycolysis. The triose phosphate intermediate, **dihydroxyacetone phosphate**, is of special interest to nutritional scientists, because both dietary fructose and glycerol enter the pathway at this point.

The 3-carbon intermediate compound 3-phosphoglycerate is also of interest in nutrition, because it is a precursor to serine. In the body, conversion occurs as follows: the 2-hydroxyl group is oxidized to a 2-keto group, which when converted to a 2-amine yields phosphoserine. This conversion is catalyzed by a vitamin $B_6$-dependent enzyme. Finally, hydrolysis of the 3-phosphate by a phosphatase yields serine.

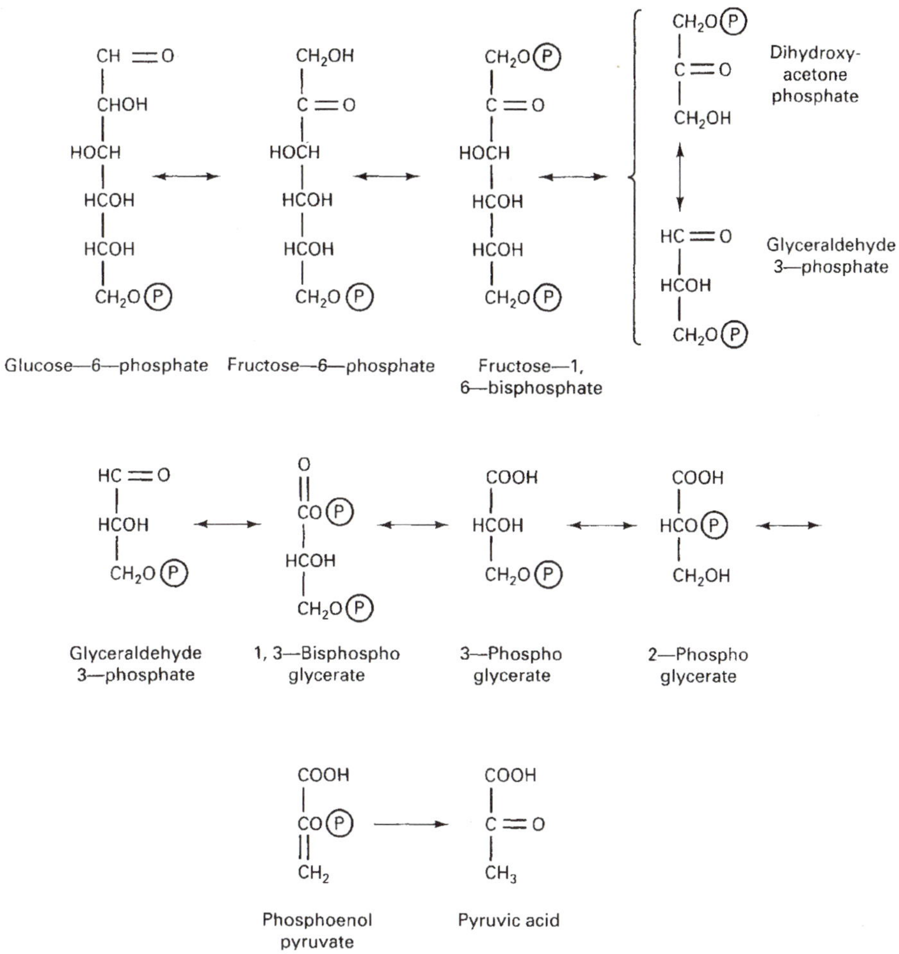

Glucose—6—phosphate    Fructose—6—phosphate    Fructose—1, 6—bisphosphate    Dihydroxyacetone phosphate    Glyceraldehyde 3—phosphate

Glyceraldehyde 3—phosphate    1, 3—Bisphospho glycerate    3—Phospho glycerate    2—Phospho glycerate

Phosphoenol pyruvate    Pyruvic acid

**FIGURE 4.24** Structures of the metabolites of glycolysis and gluconeogenesis.

*An Increase in Plasma Glucagon Does This*

Complete oxidation of glucose involves conversion to pyruvate, followed by entry into the mitochondrion for conversion to carbon dioxide. An increase in the glucagon/insulin ratio stimulates the following changes in the liver (Figure 4.25): glycogen synthetase (1), phosphofructokinase (2), and pyruvate kinase (3) activities decrease. Glycogen phosphorylase (4) and PEPCK (5) activities increase. All of these changes result in a shift of energy metabolism from the utilization of glucose in the liver (for oxidation or for glycogen synthesis) to the mobilization and production of glucose (for use by other organs).

The pathway of glucose breakdown is indicated in Figure 4.25 by heavy dots. Steps that are inactivated by a rise in the glucagon/insulin ratio are indicated by boldface Xs. As explained, PEPCK (more accurately, cytosolic PEPCK) activity is controlled by cellular levels of cAMP in the following manner. Cyclic AMP combines with a special protein (CREBP) to form a complex that binds to DNA. Where does the cAMP/CREBP complex bind? The complex binds to a short stretch of

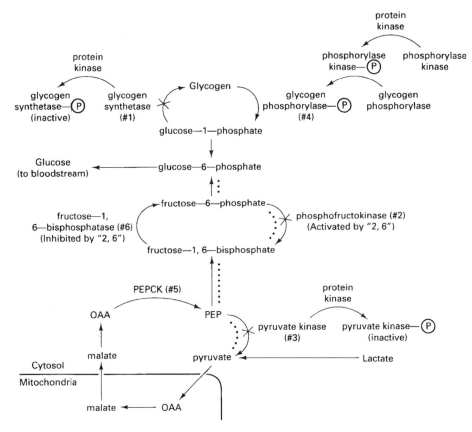

**FIGURE 4.25** Regulation of various steps of glycolysis. The pathway of glucose breakdown is indicated by heavy dots. Steps that are inactivated by a rise in the glucagon/insulin ratio are indicated by boldface Xs. The activity of PEPCK (more accurately, cytosolic PEPCK) appears to be controlled by cAMP as follows: A rise in cellular cAMP results in formation of a complex between cAMP and a special protein. This complex appears in the nucleus, where it binds to regions of DNA next to genes that respond to the cAMP-protein complex. The event of binding provokes an increase in the rate of synthesis of the mRNA of the target gene, resulting eventually in an increase in the enzyme coded by the mRNA. The PEPCK gene (and a number of other genes) responds in accordance to changes in cellular cAMP levels.

DNA called the **cAMP response element**, which resides near the gene coding for PEPCK. The binding provokes an increase in the rate of transcription of the gene, resulting in an increase in the synthesis of PEPCK. The rate of transcription of a number of genes are controlled by cAMP, and one might expect to find a cAMP response element residing in the chromosome nearby the regulated gene.

EXERCISE

A patient with a rare genetic disease lacks glucose-6-phosphatase and is dependent on exogenous sources of glucose. The patient cannot maintain normal levels of plasma glucose by means of gluconeogenesis or glycogen breakdown. Would you recommend that the patient consume fruit juice, cooked barley, or raw barley prior to going to sleep at night? (See Smit *et al.*, 1988.)

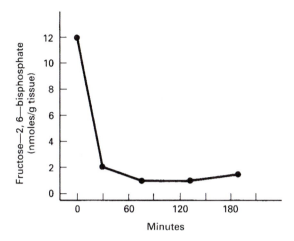

FIGURE 4.27  Liver tissue levels of fructose-2,6-bisphosphate during running. Concentrations of F-2,6-P$_2$ were measured in the livers of running rats after 0 to 3 hours of exercise. A dramatic drop occurs during the first half-hour. (Redrawn with permission from Dohm *et al.*, 1985.)

A species specificity might be pointed out with respect to the location of PEPCK, an enzyme used for gluconeogenesis. OAA is converted to PEP by PEPCK in the liver cytoplasm of rats and mice, because PEPCK is a cytosolic enzyme in these animals. In the livers of chickens and rabbits, however, OAA is converted to PEP in the mitochondria, because PEPCK is mitochondrial in these animals. PEP must then be transported to the cytosol. PEPCK is evenly distributed between the mitochondria and the cytosol in human and cow livers (Pilkis *et al.*, 1988).

## Special Topic: Running Rat Experiment and F-2,6-P$_2$

The coordinated regulation of glycolysis and gluconeogenesis can be more easily understood when related to something concrete, such as an exercising rat. It is well established that changes in cellular cAMP levels result in changes in the state of phosphorylation of various proteins. One of the most interesting fine points is that cAMP-dependent protein phosphorylation controls the enzyme that converts F-6-P to fructose-2,6-bisphosphate (F-2,6-P$_2$). Although F-2,6-P$_2$ is present at low levels, compared with the main intermediates in the glycolytic pathway, it is an important

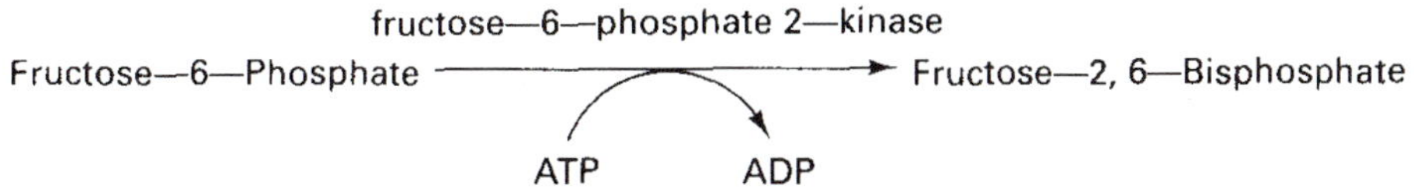

Fructose—6—Phosphate $\xrightarrow{\text{fructose—6—phosphate 2—kinase}}$ Fructose—2, 6—Bisphosphate

ATP        ADP

FIGURE 4.26  Synthesis of fructose-2,6-bisphosphate (F-2,6-P$_2$).

regulator of energy metabolism. For convenience of reading, this regulatory compound will be called "2-6."

The synthesis of "2-6" is shown in Figure 4.26. The following studies, involving living rats, liver cells in culture, and purified enzymes, illustrate the multiple approaches that may, and sometimes must, be taken to yield a convincing biological story. The first study, depicted in Figure 4.27, involves exercising (running) rats. The concentration of "2-6" in the liver after 0 to 3 hours of running shows a dramatic drop after the first half-hour. The fact that the concentration of a metabolite in plasma or an organ changes does not in itself suggest that the metabolite is a regulatory molecule. For example, the concentration of fatty acids can change with exercise, but fatty acids are not regulatory molecules. Their major function is to provide energy.

The second study concerns the enzyme that catalyzes the synthesis of "2-6," fructose-6-phosphate 2-kinase. The liver is composed of several types of cells. The major cell, the **hepatocyte**, constitutes about 80% of the cell population and contains about 99% of the protein. Hepatocytes carry out most of the metabolic functions of the liver. Casteno et al. (1979) added glucagon to hepatocytes in culture and found that fructose-6-phosphate 2-kinase decreased within the first minute of addition of the hormone. Addition of the hormone was intended to simulate, in part, hormonal changes that occur in the body during exercise. The observed decrease in enzyme activity is consistent with the dramatic drop in "2-6" that occurred in the running rat experiment. The work described thus far shows that "2-6" levels in the liver increase or decrease under certain conditions. These facts, though interesting, do not directly indicate a function for "2-6."

The third study involves the purified enzymes fructose-1,6-bisphosphatase and phosphofructokinase. This study brings discussion of "2-6" to a conclusion by demonstrating its effects on these enzymes. Fructose-1,6-bisphosphatase, an enzyme of the gluconeogenic pathway, catalyzes the conversion of $F-1,6-P_2$ to F-6-P. Ekdahl and Ekman (1984) showed that the activity of this enzyme is maximal in the absence of or presence of low concentrations of "2-6." These researchers showed that its activity is inhibited by "2-6" at concentrations of 10 micromolar and higher. The running rat experiment showed that the resting rat liver contained "2-6" at a concentration of about 12 $\mu M$. Their data also indicated that this level of "2-6" is sufficient to inhibit the activity of the enzyme, but that the levels found in the running rat would be expected to permit near-maximal activity.

Phosphofructokinase, an enzyme of the glycolysis, catalyzes a reaction that is, in effect, the reverse of that catalyzed by fructose-1,6-bisphosphatase. Uyeda et al. (1981) examined phosphofructokinase activities at various levels of "2-6" and found activity to be very low in the absence of "2-6" but high at about 4.0 $\mu M$ "2-6." This suggests that phosphofructokinase is inactive in the liver during exercise but active during resting. These data are consistent with the picture that the liver is geared toward gluconeogenesis, not toward glycolysis, during exercise.

Please be aware that the levels of "2-6" that occur in the cell are in the micromolar range, and not in the millimolar range. Generally, hormones occur in the micromolar concentrations, or less, while enzymes and metabolites used for major metabolic pathways occur in the millimolar range.

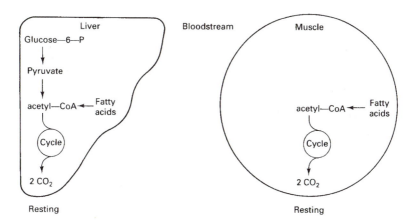

**FIGURE 4.28** Oxidation of carbohydrate and fatty acids for the production of energy. In the resting state, glycogen in the liver is broken down to glucose units; however, both liver (*left*) and muscle (*right*) derive most of their energy from fatty acids released from adipose tissue into the bloodstream and oxidized via the Krebs cycle. When carbohydrate is oxidized in the liver, it tends to be released into the bloodstream as pyruvate and lactate.

## Relationships between the Liver and Other Organs

*What Happens When You Rest*

Figures 4.28, 4.29, and 4.30 can be used as a guide to understanding the changes in energy metabolism occurring in transitions between rest and exercise and between rest and feeding. In Figure 4.28, resting state styles of energy metabolism

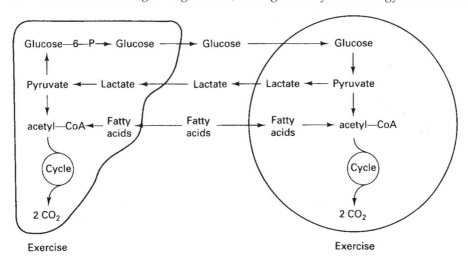

**FIGURE 4.29** Oxidation of carbohydrate and fatty acids and operation of the Cori cycle during exercise. Conversion of glucose to lactate in muscle (*right*) and subsequent conversion of lactate back to glucose in the liver (*left*) constitute the Cori cycle. During exercise, fatty acids are still a major energy fuel, but consumption of glucose by muscle increases. Some is oxidized to carbon dioxide, some only to pyruvate. Most of the glucose, oxidized in exercising muscle, arises from the muscle's own glycogen.

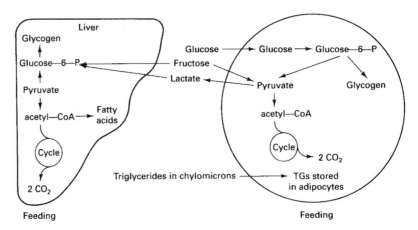

**FIGURE 4.30** Synthesis of glycogen and fatty acids and storage of energy fuels after feeding. Much of the dietary glucose taken up by muscle (*right*) is converted to glycogen; some is metabolized via glycolysis; and some is transferred, as lactate, to the liver (*left*) for conversion to glucose. Relatively little glucose from the diet is used by the liver for glycogen synthesis. The lipoproteins, VLDLs and chylomicrons, are defined in Chapter 6.

in the liver are shown at the left, those in muscle at the right, and those in the bloodstream in between. In the resting state, liver and muscle derive most of their energy from fatty acids released from adipose tissue into the bloodstream. The fatty acids bind to albumin and are carried throughout the circulatory system, eventually to be taken up by the liver, muscle, and other organs. Some of the fatty acids taken up by the liver enter the mitochondria for oxidation; others are converted to TGs, packaged into **very-low-density lipoproteins** (VLDLs), and sent back to the bloodstream. The TGs in the VLDLs are then cleaved as they pass through the capillaries in various tissues, liberating the fatty acids.

In the resting state, glycogen in the liver is broken down to glucose units. It is not clear what proportion of these units enters the bloodstream as glucose and what proportion is burned in the liver itself; however, the former is probably greater than the latter. Here, the term "resting" means the most typical state of our metabolism — the interval between feeding and fasting.

## What Happens When You Exercise

Figures 4.29 and 4.30 depict some of the dramatic shifts in energy metabolism that occur with transitions to exercise or to feeding. Fatty acids continue to be a major energy fuel in exercise, as in rest, but the overall consumption of glucose by muscle increases. Some of this glucose is oxidized to carbon dioxide, some only to pyruvate. Most of the glucose oxidized in exercising muscle arises from the muscle's own glycogen. A portion of the glucose oxidized to pyruvate may be converted to lactic acid. This lactate enters the bloodstream, is carried to the liver, and is used for the synthesis of glucose, which reenters the bloodstream (see Figure 4.29).

The conversion of glucose to lactate in muscle, the subsequent conversion of lactate back to glucose in the liver, and the reentry of glucose into the blood constitute the **Cori cycle**. It should be noted that two molecules of lactate are required for synthesis of one molecule of glucose.

## What Happens When You Eat

Events that occur after feeding are shown in Figure 4.30. Much of the dietary glucose taken up by *muscle* is converted to glycogen, some is metabolized via glycolysis, and some exits the muscle in the form of lactate for transport to the liver and conversion to glucose. Apparently, relatively little of the glucose entering the bloodstream from the diet is used by the *liver* for glycogen synthesis. This is consistent with the liver's tendency to provide glucose to other tissues, rather than use it for its own purposes.

With the breaking of a fast, dietary carbohydrates are used first, mainly for the synthesis of muscle glycogen. Following the replenishment of this energy store, liver glycogen begins to be restored. Following replenishment of liver glycogen, the remaining dietary carbohydrate may be converted to fat. The liver is the main site of fat biosynthesis in some animals; adipose tissue is the main site in others (Bergman, 1990; Louveau *et al.*, 1991). Lipid synthesis occurs mainly in the liver in humans and birds. In ruminants and pigs, it occurs almost exclusively in adipose tissue. In rodents, fat synthesis occurs in both tissues. The proportion of dietary carbohydrate converted to fat is thought to be small compared with that used for glycogen synthesis. Dietary fat is the major source of the fat in our fat stores. Fat is stored in specialized cells called **adipocytes** and, in the form of droplets, in other cells.

## EXERCISE NUTRITION

The different fuels used by exercising muscle are discussed in subsequent sections. These fuels may be arranged in the following "hierarchy," where the order of appearance approximates relative importance during exercise: (1) creatine phosphate, (2) muscle and liver glycogen, (3) gluconeogenesis, and (4) fatty acids.

Fatty acids are the major fuel of muscle at all times. Creatine phosphate is important only during the early stages of exercise (i.e., during sudden bursts of work). Glycogen tends to become depleted during exercise of moderate duration. With this depletion, the importance of gluconeogenesis as a source of glucose increases. During prolonged exercise, FFAs supply most of the energy requirements of muscle. Ketone bodies, which are made by the liver, are also used as fuel by exercising muscle, but their contribution is quite small compared with fatty acids.

Elucidation of muscle energy metabolism is complicated by several factors. Muscle can receive glucose and fatty acids from internal sources as well as from the bloodstream. The relative importance of internal versus external sources can be difficult to determine. Muscle is a heterogeneous tissue. Some muscle types carry out mainly aerobic forms of energy metabolism; other types are fermentative; and still others are composed of mixtures of cells, some aerobic and some fermentative.

One technique for assessing energy metabolism involves measurement of oxygen consumption; another involves measurement of arteriovenous (AV) differences. (A third technique, measurement of the respiratory quotient, is discussed in detail in Chapter 5.)

An experiment monitoring the use of muscle creatine phosphate, muscle glycogen, liver glycogen, and glucose produced by gluconeogenesis in the liver, muscle fat, and adipose tissue fat during a 5-hour marathon would be instructive. No such experiment has been performed, and such data are not available. Parts of this experiment have, however, been performed, and the available results are presented here.

The first study concerns the use of external sources of fatty acids and glucose by exercising muscle. The main external source of fatty acids is adipose tissue. The main external sources of glucose are liver glycogen and hepatic gluconeogenesis. The experiment depicted in Figure 4.31 involves human subjects exercising on a stationary bicycle. Concentrations of oxygen and of energy fuels were measured in blood samples withdrawn at various times during the indicated period. Samples were taken from two sources: blood entering (arterial) and exiting (venous)

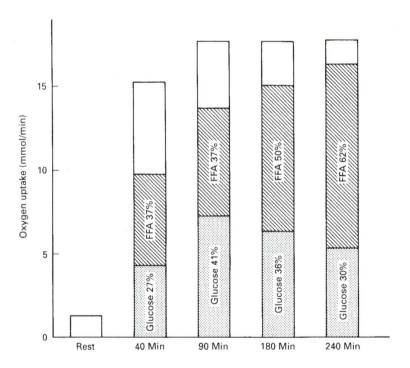

**FIGURE 4.31** Energy fuels oxidized by muscle metabolism during prolonged exercise. Concentrations of oxygen, free fatty acids, and glucose were measured in blood samples withdrawn from human subjects exercising on a stationary bicycle. Samples were taken from two sources, blood entering (arterial) and blood exiting (venous) the muscle. The rate of blood flow through the artery entering the muscle was also measured. Oxygen uptake was calculated from concentrations in arterial blood minus simultaneous concentrations in venous blood, arteriovenous differences, and flow rates. (Redrawn with permission from Ahlborg et al., 1974.)

the muscle. Metabolite analyses revealed the concentrations of FFAs and glucose. The concentration in arterial blood minus the simultaneous concentration in venous blood is the **AV difference**. The rate of blood flow through the artery entering the muscle was also measured.

AV differences and blood flow rates were used to calculate the overall rates of use of oxygen, arterial FFAs, and arterial glucose by exercising leg muscles. Rates were calculated using the formula:

$$\text{rate of use} = \text{rate of blood flow} \times \text{AV difference}$$
$$\text{(mole/min)} \qquad \text{(ml/min)} \qquad \text{(m}M)$$

The AV differences used in this calculation are:

$$[\text{arterial O}_2] - [\text{venous O}_2]$$
$$[\text{arterial FFAs}] - [\text{venous FFAs}]$$
$$[\text{arterial glucose}] - [\text{venous glucose}]$$

Using brackets around the name of a metabolite indicates that the "concentration of the metabolite" is meant.

Experimental results indicate that the overall rate of oxygen uptake increased by about 15-fold with exercise. At rest, $O_2$ uptake was 1.25 mmol/min. With bicycling, $O_2$ uptake increased to about 15 mmol/min, as shown in the histogram. Oxygen uptake by a tissue means that a fuel from an internal store or bloodstream is undergoing oxidation. The histogram indicates what percentage of oxygen uptake is due to oxidation of each of the following fuels: internal stores (open segment); plasma FFAs (hatched segment); plasma glucose (stippled segment).

At rest, the muscle's main fuel is its store of fat. At rest, the brain is the major consumer of glucose in the body. With mild exercise (at about 25% maximal $O_2$ utilization), most of the muscle's energy (80%) is supplied by the adipose tissue, that is, in the form of free fatty acids liberated from this tissue and released into the bloodstream. With moderate exercise (about 65% maximal $O_2$ utilization), the muscle's energy is supplied in roughly equal proportions from muscle glycogen, fat stores within the muscle, and from the free fatty acids liberated from adipose tissue. Where moderate exercise is continued for two hours, or longer, the energy supplied by adipose tissue becomes increasingly important. With short bursts of extreme exercise (85% maximal $O_2$ utilization) muscle glycogen supplies most (60%) of the energy for muscle (Holloszy and Kohrt, 1996).

To summarize, the more intense the exercise, the greater the importance of muscle glycogen; the more mild the exercise, the greater the importance is adipose tissue.

## Measuring Body Fuel Use in Specific Organs

Rates of energy fuel use can be determined in several ways. Internal energy fuel use can be measured by a **tissue biopsy**. A special probe is inserted into the tissue, enabling the researcher to cut out and withdraw a small sample. Biopsies are sometimes taken from liver and muscle by exercise physiologists. Biopsies performed during exercise or recovery from exercise can be used to estimate concentrations of an internal energy store in the tissue.

AV differences can be used to measure rates of use of external energy fuels. This technique involves withdrawal of samples of arterial and venous blood and determination of the concentration of fuel in each. AV difference measurements are sometimes repeated at intervals over time to track consumption of external fuels such as glucose and FFAs.

AV differences also can be used to measure the rate of use of oxygen by a tissue. Oxygen is an external component; however, its rate of use by a tissue or organ is a function of the rates of consumption of both external and internal energy fuels. Oxygen is consumed by the respiratory chain in the mitochondria. An increase in the rate of oxidation of fatty acids or of carbohydrates occurs with increased use of either internal or external energy fuels. All that is required is that the fuel enter the mitochondrion. However, the use of oxygen consumption measurements has a slight drawback, in that it is sensitive only to energy fuel consumption that is oxidative. It is not sensitive to fermentative energy metabolism.

## Arteriovenous Differences

Knowledge of metabolite levels in the bloodstream allows assessment of many changes in metabolism. For example, during fasting or exercise, the concentration of FFAs may increase from a basal level of about 0.5 m$M$ to about 1.0 m$M$. With exercise, the concentration of lactic acid can increase from 1.0 to 2.0 m$M$. These changes alone, however, reveal little definite information about either the organs that produce the metabolites or the organs that use them. An increase in plasma FFAs could result from an increase in lipolysis in adipose tissues and their release into the bloodstream. This increase occurs with exercise, but an increase also can occur with a sudden decrease in their rate of use, as in a sudden cessation of exercise.

The technical term for the rate of uptake of a metabolite by an organ is its **clearance rate**. The clearance rate of a specific metabolite may change in response either to a change in regulation of the relevant biochemical pathway or to a change in blood flow through the organ. For example, the clearance rate of bloodborne energy fuels in muscle may increase with exercise because of an increase in the rate of blood flow through the muscle. Epinephrine, one hormone that controls blood flow through specific tissues, stimulates the constriction of blood vessels in the skin, restricting blood flow to this tissue during exercise; however, it also stimulates relaxation of the arterioles in skeletal muscle, resulting in dilation and an increase in blood flow.

Determination of clearance rates requires measurement of both AV differences and blood flow rates. The AV difference alone can demonstrate only whether there is a net uptake or release of a metabolite from an organ. The blood flow rate is also needed if the data are to be expressed in terms of the quantity of metabolite used per minute.

AV differences have been measured for a variety of organs (e.g., brain, forearm, leg, and liver). To assess brain metabolism, it is customary to take blood samples from the carotid artery and the hepatic vein. Forearm metabolism can be examined using blood samples from the brachial artery and a deep vein. Leg metabolism can

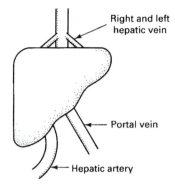

**FIGURE 4.32** Major blood vessels of the liver. Blood enters the liver via the portal vein and hepatic artery. Blood leaves the liver via the right and left hepatic vein, which direct the blood into the vena cava.

be assessed with blood samples taken from the femoral artery and femoral vein. Measurement of AV differences across the liver is, however, quite difficult because the liver is in an inaccessible part of the body and because blood enters from the two majors sources, not just one. As shown in Figure 4.32, blood from the general circulation enters the liver via the hepatic artery and blood from the intestines enters via the portal vein. The portal vein contains water-soluble nutrients absorbed from the diet, and accurate assessment of nutrient uptake by the liver requires measurement of metabolite concentrations in the hepatic artery, portal vein, and hepatic vein. The hepatic artery branches out of the aorta (just below the diaphragm) from the small coeliac trunk. The portal vein drains blood from the gut, as well as from the pancreas and spleen.

## Special Topic: Nitric Oxide and Vasodilation

No discussion of blood flow during exercise can be complete without an account of nitric oxide. Nitric oxide is one of the hormones involved in controlling the rate of flow of blood through arteries that supply nutrients to muscles, brain, lungs, and other organs. In all cases, nitric oxide (NO) provokes the dilation of the arteries. About 1.0% of the arginine in the body is used per day for making nitric oxide (Castillo *et al.*, 1996). NO also plays a role in signaling in the nervous system (Morris, 1996).

Nitric oxide (NO) is made from the amino acid arginine by the catalytic action of NO synthase. Citrulline contains the carbon skeleton and α-amino acid of arginine, and is a byproduct of the reaction, as shown here:

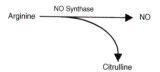

NO synthase is a fascinating enzyme to researchers interested in cofactors. The enzyme utilizes four different cofactors, i.e., FAD, FMN, heme, and biopterin (Saura *et al.*, 1996). Cofactors are accessories that help many enzymes do their job, and are detailed in the Vitamin chapter.

The following study clearly indicates the contribution of NO production to the regulation of blood flow. Hickner *et al.* (1997) studied humans exercising on a stationary bicycle. These researchers provided evidence that NO is utilized as a signal that provokes much of the increase in blood flow through the leg. Blood flow through the leg was measured by inserting a probe into the leg muscle. The possible role of NO in controlling blood flow was studied by injecting an inhibitor of NO synthase into the muscle. The inhibitor was an analogue of arginine. The result of the injection was that the rise in blood flow occurring with cycling was only half the expected value.

To view the numbers, resting blood flow was 14 ml/min. Exercising blood flow was 400 ml/min — an increase of about 25-fold. But exercise with administration of the enzyme inhibitor resulted in a rate of blood flow of 200 ml/min. These numbers refer to the rate of blood flow through 100 grams of muscle.

In addition to provoking the dilation of the muscles that encircle blood vessels, NO plays a part in controlling glucose uptake by the muscle. As mentioned elsewhere in this chapter, glucose uptake by muscle can be stimulated under two entirely different physiological states, where the immediate stimulants in these two conditions are: (1) elevated insulin in the bloodstream; and (2) contraction of muscle. In both cases, stimulation results in a glucose transporter, called GLUT-4, being inserted into the plasma membrane. NO is used as an intermediary only in the case of contraction-stimulated contraction (Roberts *et al.*, 1997).

Nitric oxide is made in the endothelial cells of the blood vessels. However, NO does not provoke a change in these cells. The hormone, occurring at concentrations of 10–1000 n$M$, diffuses through the endothelial cell membrane and into nearby smooth muscle which encircles the blood vessel (Beckman and Koppenol, 1996). Once inside the muscle cell, NO binds to guanylate cyclase, and activates it. **Guanylate cyclase** catalyzes the conversion of GTP to cyclic GMP (cGMP), as shown here:

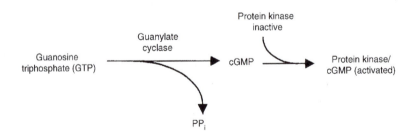

Cyclic GMP, at increased levels in the cell, binds to a protein kinase, and activates it. This protein kinase is called **cGMP-activated protein kinase**. The kinase catalyzes the attachment of a phosphate group to a protein called the **IP3 receptor**. The IP3 receptor, in its modified state, then provokes a decline in the level of calcium in the cytoplasm of the muscle cell. The decline in intracellular calcium ions provokes relaxation of the artery's muscle. (The role of IP3, the IP3 receptor, and calcium in regulating muscle contraction is detailed in the Calcium section.) The biochemistry of cyclic GMP, and its role in intracellular signaling, is quite similar to that of cyclic AMP. However, cAMP is used for signaling in a great variety of cells, whereas cGMP is used for only a few types of specialized cells. For this reason, cGMP and nitric oxide are detailed along with our discussion of smooth muscle dilation and exercise.

## CREATINE PHOSPHATE

**Creatine phosphate** (creatine-P) serves as an "energy buffer" in muscle. (A *buffer* is a chemical that maintains a near-constant pH in a solution or fluid, even when acid or base is added.) Creatine-P helps maintain a constant concentration of ATP in muscle during sudden bursts or exercise that would otherwise deplete ATP concentration in the cell. A sudden burst of exercise or a brief period of exhaustive movements may deplete cellular ATP before hormonal changes can activate glycogen phosphorylase or hormone-sensitive lipase. Plasma glucose is a readily available energy source that might be used during sudden or exhaustive exercise. However, a decrease in plasma glucose is not desirable, because glucose is re-quired by the central nervous system.

Creatine is a small molecule synthesized in the liver from glycine by the biosynthetic pathway depicted in Figure 4.33. Glycine is modified by addition of a guanidino group from arginine and a methyl group donated by **S-adenosyl-methionine** (SAM). Creatine biosynthesis is thought to represent the primary use of methyl groups donated by SAM. (SAM metabolism is discussed further in the section on Folate in Chapter 9.)

The creatine synthesized in the liver is transported through the bloodstream to skeletal and heart muscle. It enters the mitochondria, where it is phosphorylated to creatine-P. Creatine kinase catalyzes this reversible addition of a phosphate group, as shown in Figure 4.34. Creatine-P is unique in that its only known function is as an energy buffer. The creatine-P formed in the mitochondria travels to the contractile proteins in the cytoplasm of the muscle fiber. The polymer, or complex, of contractile proteins is called a **myofibril**. Contraction of a myofibril is coupled to the hydrolysis of ATP to ADP. The immediate replenishment of ATP is catalyzed by a second creatine kinase, residing on the myofibril, that catalyzes the conversion of creatine-P to creatine. This reversal of the reaction takes place in the

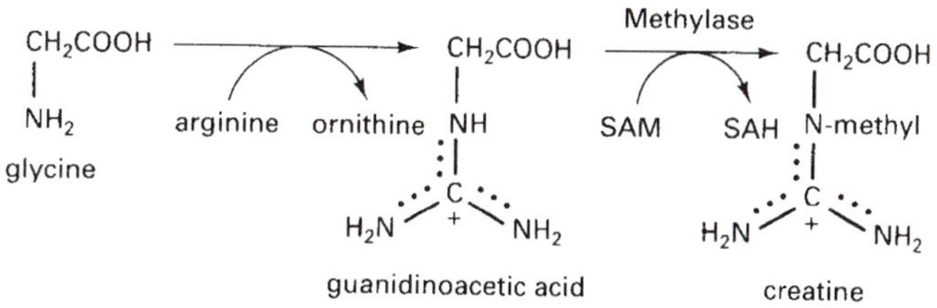

**FIGURE 4.33** Biosynthesis of creatine from glycine. Arginine donates the guanidinium group; SAM donates the methyl group.

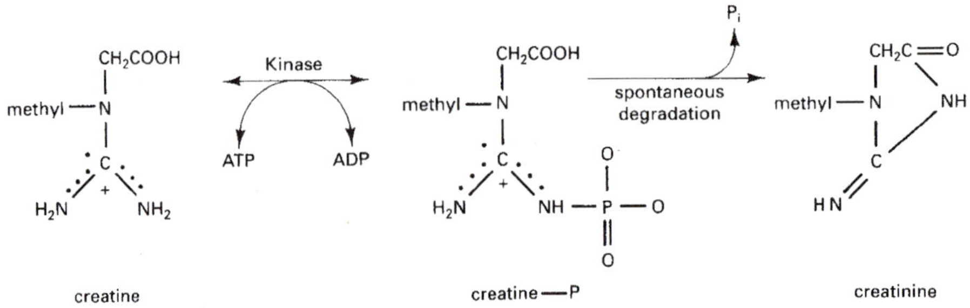

FIGURE 4.34  Use of creatine as an energy buffer and degradation of creatine phosphate.

mitochondrion. This reaction, as well as the product formed by gradual spontaneous degradation of creatine-P, is shown in Figure 4.34.

The scenario depicted in Figure 4.35, conversion of the creatine-P energy buffer,
bound to myofibrils, to creatine and regeneration of creatine-P in the mitochondrion, is called the **creatine phosphate shuttle**.

The time course of use and replenishment of creatine-P is shown in the study
of exercising human subjects reported in Figure 4.36. The subjects exercised on a
stationary bicycle that was adjusted so that exhaustion would be reached after 6
minutes. Muscle biopsies, taken before and after exercise, from the quadriceps
femoris muscle of the leg were used for analysis of creatine-P. The data demonstrate that the reservoir of creatine-P, substantially depleted by the end of the
exercise period, was rapidly repleted during the subsequent resting period.

## CREATININE

Generally, about 1.6% of the body's content of creatine spontaneously breaks down
daily to form creatinine. This creatinine enters the bloodstream and is excreted by

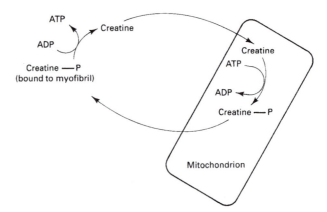

FIGURE 4.35  The creatine phosphate shuttle. Conversion of creatine phosphate bound to
myofibrils to creatine, followed by regeneration of creatine phosphate in the mitochondria,
constitutes the creatine phosphate shuttle.

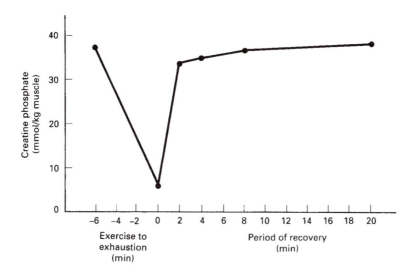

**FIGURE 4.36** Creatine phosphate concentrations before and after exercise. Human subjects exercised on a stationary bicycle, adjusted so that exhaustion was reached after 6 min. Muscle biopsies, taken from the quadriceps femoris muscle of the leg, were used for analyses of creatine phosphate. (Redrawn with permission from Harris *et al.*, 1976.)

the kidneys into the urine. As muscle contains 3 to 5 g of creatinine per kilogram of tissue, the muscle mass can be estimated from the quantity of creatine excreted. That is, the amount of creatine excreted closely reflects the body's skeletal muscle mass; 1 g of urinary creatine per day represents about 18 kg of muscle. These measurements are useful in estimating the body composition of healthy persons as well as those suffering from malnutrition.

A fraction of urinary creatinine arises from the diet. Meat contains muscle and therefore contains creatine. Cooking meat results in the conversion of creatine to creatinine, which, after ingestion, is rapidly excreted into the urine. The human body generates and excretes about 1.7 g of creatinine per day. Dietary creatinine may reach 0.5 g per day. Thus, researchers studying creatinine excretion should control or limit meat intake by their subjects.

Creatinine excretion is intimately related to a relatively constant part of the body, the muscle mass. When studying the excretion of any metabolite, such as urea, calcium, or riboflavin, it might be undesirable to relate the amount of urinary metabolite to bodily weight, because the body contains compartments that are of minimal importance relative to metabolism. The body of an obese person contains a large amount of tissue (adipose tissue) that is metabolically and biochemically irrelevant to the metabolism of compounds such as urea, calcium, and riboflavin. A person whose body contains excess body fluids also has extra mass that may be irrelevant to the metabolism of these as well as other compounds. A meaningful comparison of excretion data from different subjects or from one subject at different times is facilitated by relating the data to urinary creatinine.

The data in Tables 4.3, 4.4, and 4.5 illustrate the broad applicability of urinary creatinine measurements in the nutritional sciences. These studies concern the excretion of urinary nitrogen (Table 4.3), calcium (Table 4.4), and riboflavin (Table

**TABLE 4.3** Excretion of Urinary Nitrogen

| Subject | Urinary N loss (mg/mg creatinine) |
|---|---|
| C.G. | 2.2 |
| M.S. | 2.0 |
| E.D. | 2.3 |
| L.I. | 2.0 |

*Source*: Scrimshaw *et al.* (1976).

**TABLE 4.4** Excretion of Urinary Calcium

| Condition | Calcium excreted (mol/mol creatinine) |
|---|---|
| Urinary calcium (baseline) | 0.10 |
| Increase in urinary calcium after ingestion of calcium phosphate | 0.19 |
| Increase in urinary calcium after ingestion of calcium citrate | 0.49 |

*Source*: Schuette and Knowles (1988).

4.5). In all cases, the amount of metabolite is expressed per milligram of creatinine excreted, rather than on a "per-volume" or "per-day" basis. Another reason for expressing excretion data in terms of creatinine is that, although a researcher might want to determine excretion data on a per-day basis, collection of an entire day's urinary output may be difficult or inconvenient. This difficulty can be overcome by using creatinine measurements.

No discussion of creatinine is complete without reference to the **glomerular filtration rate** (GFR). Up to one-quarter of the heart's blood output is delivered to the kidneys. The rate of blood flow through the kidneys is about 1200 ml/min. Much of this blood passes through tufts, or circles, of special capillaries that reside within a structure called the **glomerulus** (Figure 4.37). A proportion of the water, ions, creatinine, glucose, and amino acids that pass through these capillaries leaks into an area called **Bowman's space**. Blood cells and medium- and large-sized

**TABLE 4.5** Excretion of Urinary Riboflavin

| Age group | Riboflavin excretion (mg/g creatinine) |
|---|---|
| 10–12 | 0.33 |
| 13–15 | 0.34 |
| 16–19 | 0.33 |

*Source*: Ajayi and James (1994).

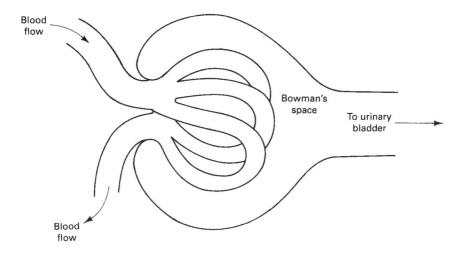

FIGURE 4.37  The glomerulus. Much of the blood that flows through the kidneys passes through tufts, or circles, of special capillaries that reside within a structure called the **glomerulus**. Part of the plasma water, ions, creatinine, glucose, and amino acids leaks into an area called Bowman's space, in the first step in the creation of urine. Tubules leading from all the glomeruli function to reabsorb nearly all of the valuable nutrients that occur in the new urine. These nutrients include glucose, amino acids, minerals, and water. Creatinine is not reabsorbed by the renal tubules. As the urine matures, it becomes more and more depleted of nutrients that the body wishes to retain. The tubules of the kidney coalesce and direct the mature urine to ureters and, thus, to the bladder.

proteins do not leak into Bowman's space. Tubules leading from all the glomeruli coalesce and direct fluid to the **ureters** and, thus, to the **bladder**.

The capillaries in the glomerulus are more permeable than most other capillaries. Their pore size is about 10 nm. For comparison, a red blood cell is about 8000 nm in diameter. The volume of fluid entering Bowman's space and flowing into the renal tubules is about 20% of the total flow of blood plasma through the capillaries, 125 ml/min. This rate is called the GFR.

If all the fluid entering Bowman's space traveled to the bladder, the volume of urine would be well over 100 liters/day. Clearly, producing 100 liters of urine per day would be quite an unusual feat. Most of the water, along with metabolites such as ions, glucose, and amino acids, are reabsorbed by transport mechanisms in the walls of the **renal tubules**. These tubules form a continuous channel (or duct) from Bowman's space to collecting regions in the kidney that direct urine to the ureter. The reabsorptive mechanisms of the renal tubules limit daily urine production to about 2 liters. Glucose, ions, and other metabolites are reabsorbed with varying degrees of efficiency.

Creatinine, a notable exception, is not reabsorbed by the tubules. All of the creatinine entering Bowman's space enters the bladder, without being reabsorbed and reentering the bloodstream. The amount of creatinine found in the urine thus accurately reflects the GFR. In medical practice, plasma and urinary creatinine levels are used on a day-to-day basis for detection of renal diseases. Figure 4.37, a diagram of the glomerulus, illustrates the flow of fluids to the urinary bladder.

## GLYCOGEN

The body's stores of glycogen may be depleted by about an hour of strenuous exercise. Therefore, glycogen is listed after creatinine phosphate in the hierarchy of energy fuels. A "typical 70-kg man" contains about 350 g of muscle glycogen, 80 g of liver glycogen, and 20 g of glucose in extracellular fluids.

The study depicted in Figure 4.38 illustrates the use and replenishment of muscle glycogen stores in human subjects. The subjects exercised for 2 hours (black bar) by swimming, skiing, running, or exhaustive bicycling; then, during the recovery period, they consumed typical lunches and dinners and engaged only in slow walking. Biopsies taken from the thigh muscle at the indicated times demonstrate that the glycogen content decreased to less than 20% of its initial, resting concentration during the 2 hours of exercise. Much of the glycogen was restored over the course of half a day; full restoration required about 2 days.

### Carbohydrate Loading

Exhaustion, during prolonged exercise, occurs coincidentally with the depletion of muscle glycogen. There is some thought that with exhaustion there is some depletion of the intermediates of the Krebs cycle, thus impairing the oxidation of the available free fatty acids. The athlete motivated to perform beyond the expected point of exhaustion might consider it reasonable to eat sugar, during

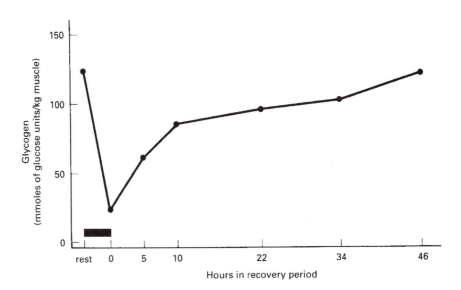

**FIGURE 4.38** Concentration of glycogen before and after exercise. Human subjects exercised for 2 hr (black bar) by swimming, skiing, running, or exhaustive bicycling consumed typical lunches and dinners, then engaged only in slow walking. Glycogen concentrations were determined from biopsies taken from the thigh muscle. Much of the glycogen was restored in the first half-day; full restoration required about 2 days. (Redrawn with permission from Piehl, 1974.)

running or cycling, in order to take the place of the depleted muscle glycogen. However, this tactic does not work, and is not consistent with how the body is regulated. A variation of this tactic does work, that is, temporarily shifting to a mild rate of exercise (or complete rest), to allow the re-synthesis of muscle glycogen, and then resuming vigorous exercise.

The practice of exercising, then resting and eating sugar, has the end-result of building up muscle glycogen stores to twice the usual size. This practice has the benefit of increasing endurance during subsequent prolonged and vigorous exercise. This practice is called **carbohydrate loading** or **glycogen supercompensation**. Glycogen supercompensation occurs only in muscles that have been recently exercised. If you cycle for an hour with one leg only, on an exercycle, and then carbohydrate load, the resting leg will not acquire glycogen stores that are greater than normal (Holloszy and Kohrt, 1996).

Let us view more details of carbohydrate loading procedure. With fasting after exercise, muscle glycogen is restored, but only at a slow rate. A maximal rate of muscle glycogen restoration occurs with glucose or sucrose consumption. Fructose consumption can also restore muscle glycogen, but only at half the rate produced by glucose or sucrose.

The American Dietetic Association (Smith, 1987) recommends the following carbohydrate loading procedure for maximizing muscle glycogen deposits during the week before athletic competition. This procedure is beneficial only to athletes participating in endurance events lasting more than 1.5 hours. About 350 g of carbohydrate is consumed for each of 3 to 4 consecutive days; this amount is increased to about 550 g for the 3 days prior to the event. The athlete should begin a tapered rest period at the start of the week and have complete rest the day before the event. To put things in perspective, one should realize that water nutrition (not sugar) is the major concern during prolonged athletic events.

EXERCISE

During exercise, is glycogen phosphorylase in its phosphorylated or nonphosphorylated form?

EXERCISE

Controlled studies of recovery from exercise demonstrated that the maximal rate of muscle glycogen replenishment is close to 7.0 mmol of glucose units (as glycogen) per hour. How does this rate compare with that of the subjects fed the uncontrolled diets described in Figure 4.38? (See Blom *et al.*, 1987.)

## GLUCONEOGENESIS

Gluconeogenesis occurs primarily in the liver. It can also occur in the kidney, but this contribution is relatively small. The carbon skeletons used for glucose synthesis can be derived from lactate, glycerol, or amino acids. During exercise, the lactate produced and released by fermentative muscle is taken up by the liver and

TABLE 4.6  Metabolism of the Liver before and during Exercise

|  | Rest | Exercise | |
|---|---|---|---|
|  |  | 40 min | 240 min |
| Overall glucose production (mmol/min) | 0.82 | 1.86 | 1.46 |
| Uptake of glucogenic nutrients (glucose equivalents, mmol/min) |  |  |  |
| Lactate | 0.13 | 0.23 | 0.29 |
| Pyruvate | 0.01 | 0.02 | 0.03 |
| Glycerol | 0.02 | 0.10 | 0.21 |
| Amino acids | 0.05 | 0.09 | 0.12 |

*Source*: Ahlborg *et al.* (1974).

used for glucose production via the reactions of the Cori cycle. The glycerol released from TGs by the action of lipase in adipose tissue is also glucogenic. Glycerol is not used by adipose tissue, but enters the bloodstream, where it may accumulate during exercise. Therefore, in many studies of exercise, glycerol measurements have been used to assess the rate of lipolysis in adipose tissue.

Amino acids are also used for gluconeogenesis. The amino acids released from skeletal muscle and sent to the liver are converted to glucose during exercise. For example, the carbon skeleton of alanine enters the gluconeogenic pathway at the point of pyruvate; aspartic acid enters at the point of oxaloacetic acid. (The breakdown of amino acids is covered in some detail in Chapter 8.) It may be surprising that muscle is broken down (to amino acids) during exercise to help maintain plasma glucose concentration. The muscle proteins broken down during exercise appear not to be those of the myofibrils (i.e., not contractile proteins), which resist the action of intracellular proteases.

The role of the liver in gluconeogenesis during exercise is outlined in Table 4.6. The data were taken from a study of human subjects exercising on a stationary bicycle. Catheters inserted into vessels entering and leaving the liver were used to withdraw blood samples for analysis of various metabolites. These analyses provided data for calculating AV differences. During rest, the net production of glucose by the liver was 0.82 mmol/min; after 240 minutes of exercise, it was 1.46 mmol/min. The values for net uptake of lactate, pyruvate, glycerol, and amino acids were used to calculate their contributions to glucose formation. During rest, the estimated contribution of lactate to glucose production was equivalent to 0.13 mmol of glucose/min, about one-sixth the total glucose output. With prolonged exercise, plasma lactate taken up by the liver also accounted for about one-sixth of total glucose output.

These data suggest that, during rest, the glucogenic nutrients taken up by the liver account for only about one-fifth of glucose output; however, with 240 minutes of bicycling, this proportion increases dramatically to about 50% of glucose output. During both rest and exercise, the remaining glucose output results from release of glucose units from hepatic glycogen stores.

## Alanine and Transport of Waste Nitrogen to Liver

Net breakdown of muscle can occur with either exercise or prolonged fasting. The mechanisms that control the breakdown of the various types of protein found in muscle are not well understood. It has, however, been established that while the **branched-chain amino acids** (BCAAs) released tend to be oxidized for energy in the muscle cell, other released amino acids enter the bloodstream for catabolism, and perhaps gluconeogenesis, in the liver. Examination of the amino acids released from skeletal muscle reveals an apparent anomaly; alanine accounts for only about 6% of the amino acids of muscle, but for about 35% of the amino acids released from muscle during exercise.

The initial step in the catabolism of BCAAs, as well as of most amino acids, involves removal of the α-amino group. As depicted in Figure 4.39, **BCAA aminotransferase** catalyzes transfer of the α-amino group of BCAA to α-ketoglutarate, producing glutamate. The α-amino of glutamate, in turn, can be transferred to pyruvate, producing alanine. Hence, the amino group that initially was part of a BCAA becomes part of alanine. As indicated, aminotransferases require vitamin $B_6$ as a cofactor. (The "α" refers to the 2 position on the carbon chain of any linear organic molecule.)

Figure 4.40 depicts the overall directions of alanine carbon movement in muscle and liver during exercise. Fermentative catabolism of glucose in muscle yields pyruvate. Pyruvate acquires an amino group, with catabolism of BCAAs and other amino acids, forming alanine. Alanine travels through the circulation to the liver, where it is converted back to pyruvate. In the hepatocyte, pyruvate then enters the gluconeogenic pathway, in which two molecules are needed to form one molecule of glucose. (The fate of the amino group released from alanine is covered under The Urea Cycle in Chapter 6.) The contribution of the body's protein stores to energy requirements during exercise is relatively small: about 5% of the total energy expanded.

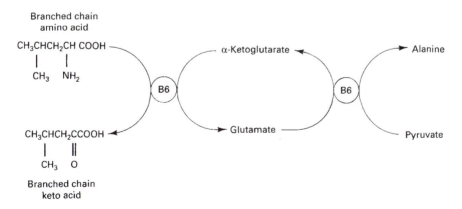

**FIGURE 4.39** First step in catabolism of BCAAs and transfer of α-amino groups to pyruvate.

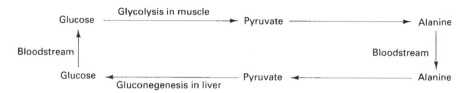

FIGURE 4.40  Use of alanine for carrying waste nitrogen from muscle to the liver.

## Glucose Transport into Muscle during Exercise: A Contradiction?

The rate of consumption of external glucose can increase during exercise. This increase may be supported by dilation of the blood vessels supplying the working muscles with oxygen and energy fuels. The increase may also be supported by an increase in the rate of transport of glucose into these muscle cells. Glucose transport is stimulated by the increase in the number of glucose transporters in the plasma membrane that occurs after eating, when the glucagon/insulin ratio decreases. The purpose of the increased glucose transport is to facilitate the synthesis of glycogen; however, because the glucagon/insulin ratio increases during exercise, the hormonal picture might be expected to impair glucose transport.

The study reported in Table 4.7 involved a skeletal muscle dissected from a rat. The data illustrate the role of insulin and electrical impulses in stimulating glucose transport into muscle. Glucose transport was assessed using a 3-O-methylglucose, a nonmetabolizable analogue of glucose whose rate of transport can be followed without interference by metabolism in the cell. Transport was measured under three conditions: the nonstimulated basal state, the electrically stimulated state, and the insulin-stimulated state. Electrical stimulation involved provocation of muscle contraction once a minute for 10 minutes prior to measurement of glucose transport. The data indicate that electrical stimulation and insulin treatment produce a doubling of the transport rate. The results show that an exercising muscle can increase its use of plasma glucose, even when the glucagon/insulin ratio increases. The increase in transport of glucose into muscle during exercise does not conflict with the hormonal picture, as transport is stimulated by the nervous system.

TABLE 4.7  Glucose Transport into Skeletal Muscle of a Rat

| Condition | Transport rate (μmol/ml per hr) |
|---|---|
| Nonstimulated | 1.2 |
| Electrically stimulated, no insulin | 3.3 |
| Plus insulin | 3.2 |

*Source*: Wallberg-Hendriksson and Holloszy (1985).

## FRUCTOSE

The material presented in the preceding sections concerns mainly carbohydrate metabolism during exercise. Before considering fatty acid metabolism in exercise, a special carbohydrate should be covered — fructose.

Fructose, present in fruits and vegetables, is a major carbohydrate of the diet. Daily intake is estimated at about 7 g from foods and 10 g from high-fructose corn syrup. Comparison of glucose metabolism and fructose metabolism represents a study in contrasts. Fructose does not provoke a release of insulin by the pancreas; glucose does. However, ingestion of fructose may be followed by a rise in plasma insulin, because conversion of fructose to glucose in the liver results in a rise in plasma glucose levels. This conversion is quite rapid. Entry of glucose and fructose into the liver occurs freely and seems not to be hormonally regulated. In contrast, entry of glucose, but not fructose, into muscle and adipose tissue is regulated. Another difference is that glucose contains an aldehyde group, which can condense with amines in the bloodstream to form a Schiff base. This reaction can result in the formation of small amounts of glucosylated hemoglobin and other plasma proteins. Fructose does not contain an aldehyde group and cannot react with amines in this way.

EXERCISE

Explain how ingestion of a dose of fructose can lead to a rise in plasma glucose higher than the rise in plasma fructose. Hint: the explanation might take this form: "The rate of metabolism by pathway X is greater than metabolism by pathway Y."

As outlined in Figure 4.41, fructose is phosphorylated in the cell to produce fructose-1-phosphate (F-1-P). F-1-P, in turn, can then be cleaved to form glyceraldehyde plus 1,3-dihydroxyacetone phosphate. Glyceraldehyde can be converted to glyceraldehyde-3-phosphate. The products mentioned are intermediates in the pathways of glycolysis and gluconeogenesis, and hence might be expected to be converted to either pyruvic acid (glycolysis) or to glucose (gluconeogenesis) in the cell. F-1-P can be metabolized by an alternative pathway, namely by phosphorylation to produce fructose-1,6-bisphosphate. A rare genetic disease resulting in a deficiency in F-1-P aldolase (Figure 4.41) leads to sickness and death in affected infants, especially when fed dietary fructose (Gopher *et al.*, 1990).

The initial metabolism of fructose tends to occur in liver, not in muscle. The opposite situation holds for glucose. The study reported in Table 4.8, involving human subjects, compares the metabolism of glucose with that of fructose. Fasting

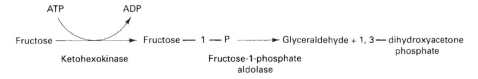

**FIGURE 4.41** Pathway for introduction of fructose into the pathway of glycolysis.

**TABLE 4.8**  Effects of Glucose and Fructose Infusions on Human Subjects

| | Glycogen concentration in tissues (mmol of glucose units/kg tissue) | | | |
| | Liver | | Muscle | |
| | Fasting range | Increase w/ infusion | Fasting range | Increase w/ infusion |
| --- | --- | --- | --- | --- |
| Glucose | 260–280 | 76 | 88–93 | 24 |
| Fructose | 260–280 | 275 | 88–93 | 23 |

*Source*: Nilsson and Hultman (1974).

subjects were infused with solutions of glucose or fructose. They received 23 mmol sugar/kg body weight over the course of 4 hours. Each sugar had several options, upon infusion, including oxidation to $CO_2$, conversion to liver glycogen, or conversion to muscle glycogen. Liver and muscle biopsies were taken before and after infusion, and the contents of tissue glycogen was measured. The most striking result is the value showing that the *fructose* infusion resulted in a doubling of liver glycogen.

A mild fructose malabsorption syndrome may occur in certain individuals. These persons experience abdominal pain and diarrhea following consumption of high-fructose foods. These symptoms are thought to be caused by poor absorption of the sugar. The consequent high osmolality of the gut contents draws fluids into the lumen, which provokes abdominal distension, cramps, and an increase in the motility. Glucose appears to have a protective effect. Although the mechanism is not clear, it is thought that glucose can enhance the rate of absorption of fructose. Therefore, apple juice, which is relatively high in fructose and low in glucose, may provoke abdominal cramps to a greater extent than other foods.

## FATTY ACIDS

### Hormone-Sensitive Lipase

The hydrolysis of TGs to FFAs and glycerol is catalyzed by a lipase in adipocytes called **hormone-sensitive lipase**. This enzyme catalyzes the hydrolysis of fat deposits in adipose tissue or wherever adipocytes are found. The rate of liberation of fatty acids and of the byproduct glycerol from the adipose tissue may be greater than the rate at which they can be oxidized by other tissues. Under these conditions, the concentrations of FFAs and glycerol increase in the plasma. The data in Table 4.9 are from a study of human subjects who exercised on a stationary bicycle. Blood samples withdrawn during rest and at the indicated times during exercise show that FFA concentrations increased from 0.43 mM at rest to nearly 2.0 mM after 3 hours of work.

The glycerol released during lipolysis in adipose tissue can be used for the synthesis of TGs or oxidized within the adipose tissue. Both of these uses would

**TABLE 4.9**  Concentrations of Metabolites and Hormones before and during Exercise

|  | Rest | 40 min | 90 min | 120 min | 180 min |
|---|---|---|---|---|---|
| Free fatty acids (mM) | 0.43 | 0.66 | 0.83 | 1.11 | 1.92 |
| Glycerol (mM) | 0.05 | 0.24 | 0.38 | 0.46 | 0.57 |
| Glucagon (pg/ml) | 77 | 66 | 111 | 158 | 257 |
| Epinephrine (pg/ml) | 75 | 262 | – | 627 | 754 |

*Source*: Ahlborg and Felig (1982).

be expected to be minimized during exercise, when the pathway of gluconeogene-sis becomes increasingly important. With exercise, most of this glycerol travels to the liver for conversion to glucose. The glycerol released by hydrolysis of TGs stored in muscle cannot be used directly by the muscle, because muscle lacks glycerokinase, the enzyme required for introduction of glycerol into the glycolysis pathway. Table 4.9 demonstrates a dramatic tenfold increase in the glycerol con-centration during the course of the bicycling.

Table 4.9 also shows increases in plasma glucagon and epinephrine, hormones that stimulate an increase in the rate of cAMP synthesis in the cell. Cyclic AMP acts as an intracellular signal to provoke the phosphorylation of a number of proteins, resulting in an increased mobilization of energy fuels. Glucagon in-creases with both exercise and fasting, but epinephrine increases only during exercise. Only moderate changes in plasma epinephrine levels have been noted with fasting (Jensen *et al.*, 1987) and after eating (Thorne and Wahren, 1989).

EXERCISE

Muscle glycogen cannot directly supply glucose to the bloodstream. This is due to the lack of what enzyme in muscle?

Figure 4.42 depicts the relationship between hormone-sensitive lipase (phos-phorylated and active), represented by a black dot, and the fatty acids released into the bloodstream. The data in Table 4.10 illustrate the changes in plasma FFAs that

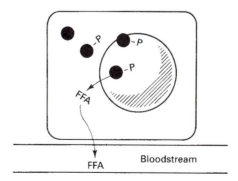

**FIGURE 4.42**  Adipocyte, with lipid droplet inside, and capillary. Hormone-sensitive li-pase, represented by black dots, in its active (phosphorylated) state catalyzes the release of FFAs from the lipid droplet (large circle) inside an adipocyte into the bloodstream.

**TABLE 4.10** Effect of Insulin Infusion on Concentrations of Plasma Insulin, Free Fatty Acids, and Palmitate

|  | Insulin ($\mu U/Ml$) | Total FFAs (m$M$) | Palmitate (m$M$) |
|---|---|---|---|
| Control[a] | 4.3 | 0.45 | 0.11 |
| Insulin infused | 9.3 | 0.12 | 0.03 |

Source: Jensen et al. (1989).
[a]Control subject was infused with saline.

occur with an increase in plasma insulin. Hence, the results from this study might be expected to be opposite those obtained from the bicycling experiment, in which an increase in plasma glucagon occurred. Human control subjects were infused with saline, whereas experimental subjects received insulin. One might consider that insulin infusion mimics the hormonal changes that occur with feeding. Blood samples were withdrawn and used for assays of insulin, FFAs, and palmitate (one of the major fatty acids). The data show a striking decrease in FFA levels, consistent with the general view that insulin counteracts the effects of glucagon.

## Glycerol

Increased mobilization of TGs in adipose tissue with exercise is implied by increases in plasma concentrations of both FFAs and glycerol. Glycerol, but not fatty acids, can be used for gluconeogenesis. As shown in Figure 4.43, glycerol is converted to glycerol phosphate by the action of glycerol kinase; it is then oxidized to dihydroxyacetone phosphate by the action of glycerophosphate dehydrogenase. Dihydroxyacetone phosphate enters the gluconeogenic pathway at the point catalyzed by aldolase or is converted to pyruvate for oxidation in the Krebs cycle.

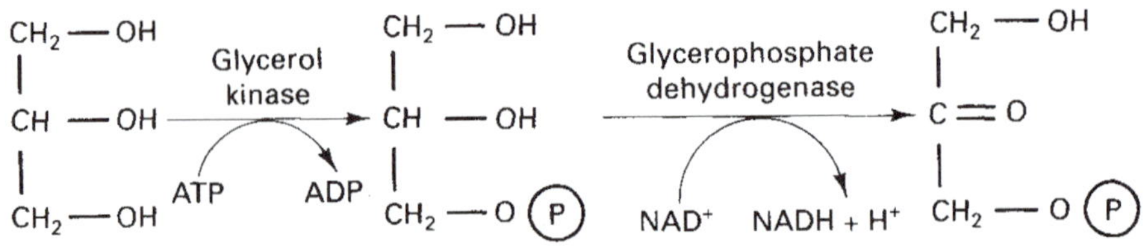

FIGURE 4.43 Pathway for introduction of glycerol into the pathway of glycolysis.

## Difference between Hormone-Sensitive Lipase and Lipoprotein Lipase

Peripheral tissues such as muscle, adipose tissue, and the lactating breast contain both hormone-sensitive lipase and lipoprotein lipase. This second lipase, mentioned in the lipoprotein section of Chapter 6, requires apolipoprotein C-II as a cofactor. The lipase is secreted by cells and migrates through the interstitial fluid to the capillary, where it becomes bound to the membrane with its active site exposed to the bloodstream. This binding to the luminal wall of the capillary occurs by attachment to a polymer called **glycosaminoglycan**.

Figure 4.44 depicts the relationship between lipoprotein lipase and the FFAs liberated from TGs in chylomicrons and VLDLs in the plasma. The enzyme is represented by black squares. The FFAs are packaged into TGs within the cell which are deposited in a lipid droplet.

Lipoprotein lipase may be activated by changes in plasma glucagon and insulin. This activation appears to occur by an increase in migration of the enzyme to the capillary wall. For example, a decrease in the glucagon/insulin ratio, as with feeding, may provoke an increase in the proportion of enzyme located on the luminal wall and a decrease in the proportion that is intracellular. Feeding appears to provoke activation of the enzyme in adipose tissue, but not of the enzyme located in muscle or other tissues.

## Serum Albumin and the Transport of Fatty Acids

Transport and metabolism of lipids are complicated by the fact that lipids are not soluble (or only sparingly soluble) in water. Nature has devised a number of mechanisms for facilitating the handling of lipids, including the use of bile salts and apolipoproteins. The use of "retinol-binding protein" for carrying a fat-soluble vitamin is described under the Starvation section of this chapter and under Vitamin A in Chapter 9. Albumin is used in the transport of FFAs in the bloodstream.

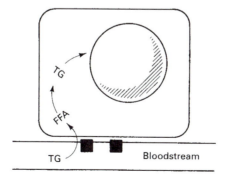

**FIGURE 4.44** Adipocyte, with lipid droplet inside, and capillary. Lipoprotein lipase, represented by black squares, catalyzes the conversion of plasma TGs to FFAs, which enter an adipocyte, are packaged into TGs, and deposited in a liquid lipid droplet.

**TABLE 4.11** Solubility of Fatty Acids and Cholesterol in Water

| Fatty acid (no. of carbons) | Maximal solubility (mM) |
|---|---|
| 4 | 575 |
| 6 | 107 |
| 8 | 19.8 |
| 10 | 3.7 |
| 12 | 0.68 |
| 14 | 0.13 |
| 16 | 0.024 |
| 18 | 0.0044 |
| 20 | 0.00081 |
| 22 | 0.00015 |
| Cholesterol | 0.00004 |

*Source*: Westergaard and Dietschy (1976).

Table 4.11 lists the maximum solubilities of various fatty acids in salt water. Butyric acid, a short-chain fatty acid with 4 carbons, is quite soluble in water and can be dissolved, in the laboratory, to a concentration of about 600 mM. Octanoic acid (8 carbons) is slightly soluble and can be dissolved to about 20 mM. Fatty acids containing 8 to 10 or 12 carbons are **medium-chain fatty acids**. Palmitic (16 carbons), oleic (18 carbons), and longer fatty acids are **long-chain fatty acids**. The highest concentration attainable for long-chain fatty acids range from 0.1 μM to 0.1 mM.

Data presented in previous sections revealed that the concentration of FFAs in plasma may reach 2.0 mM during exercise. How is this possible when the highest attainable concentration in water is only about 0.1 mM? This problem was re-solved by nature by use of albumin as a vehicle for the transport of FFAs within the circulation. Albumin constitutes about 60% of the protein of blood plasma. It is a major carrier of FFAs, other metabolites, hormones, and drugs. Serum albumin has the capacity to carry several fatty acids. Figure 4.45 shows results from an experiment using purified albumin. The number of fatty acid molecules bound per protein molecule is plotted versus the concentration of unbound fatty acids in solution. The study, conducted with lauric acid (12 carbons) and myristic acid (14 carbons), demonstrates that one protein molecule is able to bind at least 8 or 9 molecules of fatty acid. Albumin has a molecular weight of 69 kDa and occurs in human plasma at a concentration of about 0.6 mM (40 mg/ml) (Halliwell, 1988).

## Control of Fatty Acid Synthesis and Oxidation

A number of enzymes have been described that are regulated via phosphorylation. Acetyl-CoA carboxylase is now added to the following list:

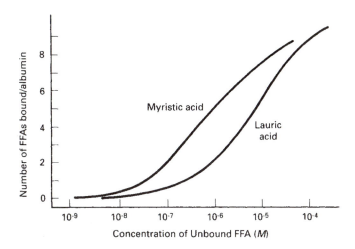

**FIGURE 4.45** Number of bound fatty acid molecules per albumin molecule at various concentrations of fatty acid. Several fatty acid molecules can bind to one serum albumin molecule. The number of bound lauric or myristic acid molecules per molecule of purified albumin is plotted versus the concentration of unbound fatty acid in solution. (Redrawn with permission from Spector, 1975.)

Phosphorylase kinase

Glycogen synthetase

Fructose-6-phosphate 2-kinase

Pyruvate kinase

Hormone-sensitive lipase

Acetyl-CoA carboxylase

Figure 4.46 shows the first step (acetyl-CoA carboxylase catalyzed) in the fatty acid synthesis pathway. The enzyme is biotin-requiring, and the product is malonyl-CoA. Note that the activities of about 100 different enzymes have been found to be controlled by phosphorylation (Shacter *et al.*, 1986). In all cases, the phosphorylation is reversible. The phosphate donor may be ATP or GTP.

The rate of fatty acid synthesis is controlled, at this point, by the activity of acetyl-CoA carboxylase. Activity is regulated at the "enzyme level" by phosphorylation and, at the "genetic level" by control of the rate of synthesis of mRNA coding for the enzyme. Regulatory changes occurring at the enzyme level generally can be effected more promptly than those at the genetic level.

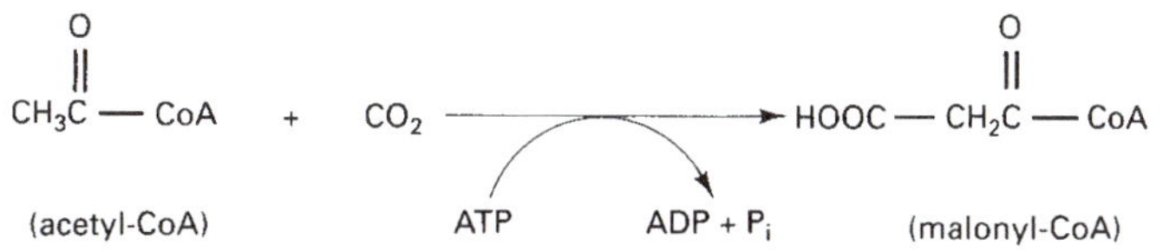

$$CH_3\overset{O}{\overset{\|}{C}}-CoA \;+\; CO_2 \xrightarrow[\text{ATP} \quad \text{ADP} + P_i]{} HOOC-CH_2\overset{O}{\overset{\|}{C}}-CoA$$

(acetyl-CoA)          ATP     ADP + P$_i$        (malonyl-CoA)

**FIGURE 4.46** First step of the fatty acid synthesis pathway.

Glycolysis and gluconeogenesis are coordinately regulated by changes in the concentration of "2-6," as well as by phosphorylation. The pathways of fatty acid synthesis and oxidation are also regulated in a coordinated manner. This means that an increase in the activity of the carboxylase (and in the rate of fatty acid synthesis) coincides with a decrease in the rate of fatty oxidation. What is responsible for this coordination of energy flux? Acetyl-CoA carboxylase catalyzes the synthesis of malonyl-CoA, a compound that, once formed, is committed to fatty acid synthesis. In increased concentrations in the cell, malonyl-CoA slows the oxidation of fatty acids by inhibiting their transport from the cytoplasm into the mitochondrion. These events are summarized in Figure 4.47.

The rat study shown in Figure 4.48 illustrates regulation of acetyl-CoA carboxylase by phosphorylation. After fasting for up to 2 days, livers were removed and the activity and phosphate group content of the enzyme measured. Enzyme activity was highest prior to fasting and progressively decreased during the fast. The number of phosphate groups increased, particularly during the second day of the fast. In this study, phosphorylation of the enzyme occurred *in vivo*. *In vitro* studies using purified enzyme confirmed the fact that phosphorylation controls the activity of the enzyme.

Acetyl-CoA carboxylase levels are controlled at the genetic level. This means that conditions such as feeding and fasting can change the rate of synthesis of mRNA coding for the enzyme. In general, the greater the number of mRNA molecules in the cell, the greater the amount of protein coded by the mRNA synthesized. Therefore, changes in regulation at the genetic level can be measured indirectly by measuring changes in the activity of the enzyme coded by the mRNA. This method cannot, however, distinguish between changes in activity caused by enzyme activation and changes caused by activation of transcription. Regulation at the genetic level can be assessed more directly by measuring changes in the number of mRNAs in the cell.

The mRNA coding for any particular enzyme can be determined by the dot blot technique, described in detail in Appendix B. The data take the form of photographic images on film. A dense image means that the cell or tissue extract, used

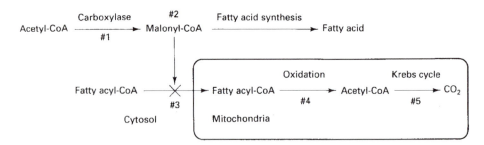

**FIGURE 4.47** Regulation of fatty acid oxidation by malonyl-CoA. Carboxylase catalyzes the first step of fatty acid synthesis (#1). An increase in its activity results in an increase in the levels of malonyl-CoA in the cell (#2). Increased levels of malonyl-CoA inhibit the transport of fatty acids into the mitochondria (#3), which limits or controls their oxidation. The fatty acid oxidation pathway involves the degradation to units of acetyl-CoA (#4), followed by conversion to $CO_2$ in the Krebs cycle (#5).

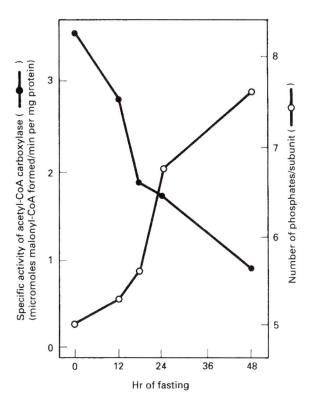

**FIGURE 4.48** Activity of acetyl-CoA carboxylase and number of covalently bound phosphate groups at various times during fasting. During fasting for up to 2 days, rat livers were removed at the indicated times, and the activity (●) and phosphate group content (○) of acetyl-CoA carboxylase measured. (Redrawn with permission from Thampy and Wakil, 1988.)

as the source of the mRNA, contained large amounts of the message. A faint image means that relatively small amounts of message are present.

The following experiment illustrates use of the dot blot technique for measuring changes in the amount of message coding for acetyl-CoA carboxylase. Each rat was exposed to one of three different conditions, all mRNAs were extracted from its liver, and the amount of mRNA coding for acetyl-CoA carboxylase was determined. The first group of rats were fed, the second group were fasted for 2 days, and the third group were fasted for 2 days then fed on the third day. The results (Figure 4.49) demonstrate that the livers of the fed and fasted animals contained about the same levels of message. This indicates that the rate of transcription of the mRNA coding for the enzyme was about the same in both groups. However, the dot blot test image produced by the mRNA extracted from the third group was darker. This means that the rate of transcription of mRNA for the enzyme is greatly increased during feeding after a fast. Generally, an increase in mRNA coding for a particular protein results in increased production of that protein in the cell. These results indicate that packaging of dietary carbohydrate in the form of body fat is "upregulated" with the conclusion of a fast.

## Transport of Fatty Acids into the Mitochondria

The following brief overview describes the modes of transport of FFAs in the body and serves as a preview to the description of the mitochondrial transport system. FFAs are generated by lipolysis of TGs stored in lipocytes of adipose tissue (adipocytes) and in lipocytes of other organs. Fatty acids are also released by the action of lipoprotein lipase in the bloodstream. Special membrane-bound proteins mediate the transfer of FFAs across the membranes of various cells, such as enterocytes and hepatocytes. Within the cell, FFAs are carried on special proteins called fatty acid-binding proteins. These proteins have low molecular weights, about 15,000, and can account for 5% of cytosolic protein (Bernlohr *et al.*, 1997).

A unique mechanism used for mediating the transfer of fatty acids into the mitochondrion involves the conversion of FFAs to fatty acyl-CoA. Here, the fatty acid molecule is connected to a molecule of coenzyme A in the thiolase-catalyzed reaction shown in Figure 4.50. The structure of coenzyme A is shown in Figure 4.51. Fatty acids enter the mitochondrion, not as fatty acyl-CoA, but as fatty acyl-carnitine. A special transferase catalyzes the exchange of the coenzyme A moiety for carnitine (Figure 4.52). The bond between the fatty acid and coenzyme A is a thiol ester linkage, whereas that involving carnitine is an oxygen ester.

Carnitine is used mainly for facilitating the transport of long-chain fatty acids into the mitochondria. As shown in Figure 4.53, this transport system requires the participation of two different carnitine acyltransferases. One is located on the outside of the mitochondrial membrane, the other on the inner side. Once fatty acyl-carnitine is inside the organelle, its carnitine is released. A separate transport system is used to transport this carnitine from the interior of the mitochondrion back to the cytoplasm for reuse.

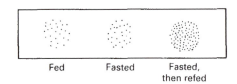

Fed          Fasted          Fasted,
then refed

**FIGURE 4.49**  The dot blot test. The dot blot test is used to measure the amount of mRNA coding for a specific protein. The test requires the use of a unique tool provided by the methods of molecular biology, the DNA probe. The probe consists of a molecule of radioactive DNA having a sequence complementary to the mRNA coding for the protein of interest. The complementary sequence makes possible the hybridization (or binding) of the DNA probe to the mRNA of interest, even when this mRNA is mixed with hundreds of different types of mRNAs in a cell homogenate. The DNA probe does not bind strongly to mRNAs coding for other proteins. The method involves several steps. The total population of mRNAs is extracted from the tissue and connected to a solid matrix. The DNA probe is then mixed with the matrix, where it is able to mingle freely with all of the different types of mRNA, and given enough time to allow maximal binding to the specific mRNA to which it is complementary. This binding is tight and resists attempts to wash it away. Then, after nonbound DNA probe is washed away, high levels of remaining bound DNA probe produce a strong signal in the dot blot test. If no mRNA coded for the protein of interest is present, little or no DNA probe will bind, resulting in a weak signal or no signal. (Redrawn with permission from Bai *et al.*, 1986.)

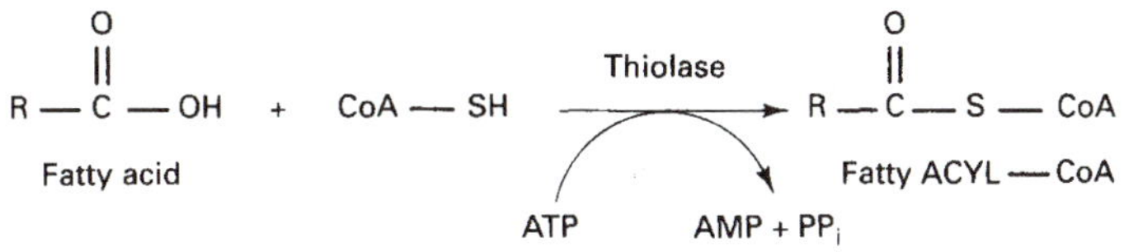

**FIGURE 4.50** Conversion of a free fatty acid to fatty acyl-CoA.

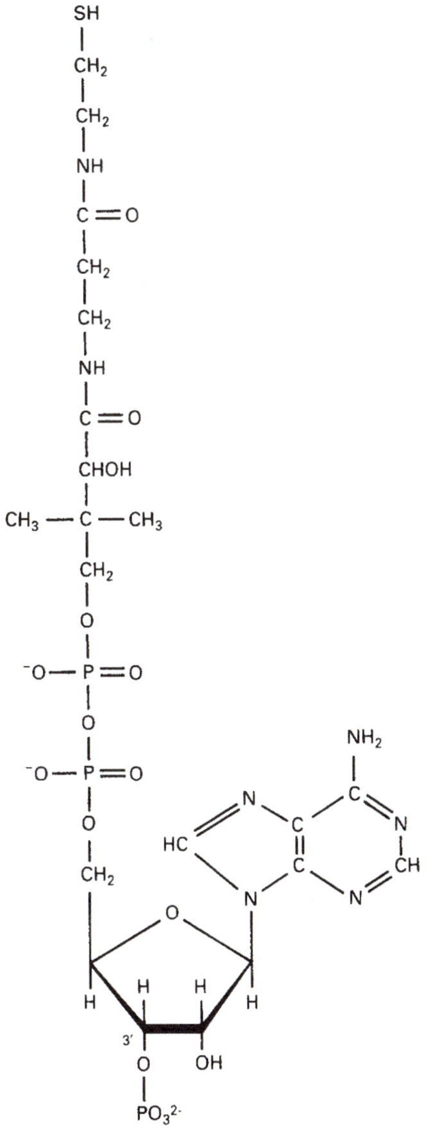

**FIGURE 4.51** Coenzyme A. The vitamin, pantothenic acid, occurs as part of the structure of coenzyme A.

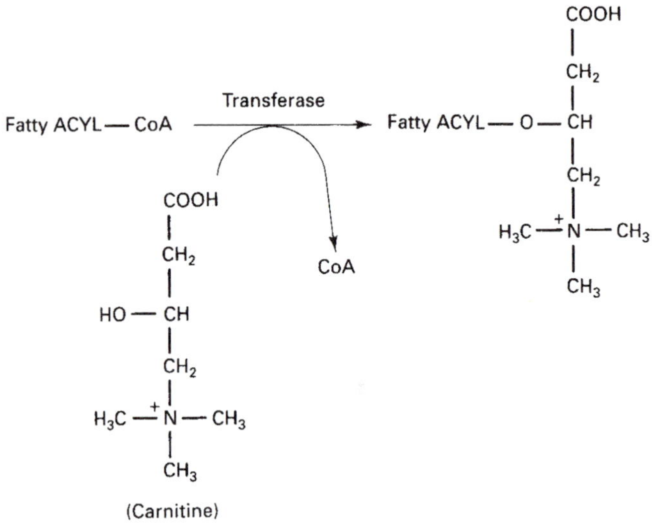

COOH
|
CH$_2$
|
Transferase
Fatty ACYL — CoA  ——————→  Fatty ACYL — O — CH
|
CH$_2$
|
COOH                                    H$_3$C — $\overset{+}{N}$ — CH$_3$
|                                          |
CH$_2$          CoA                       CH$_3$
|
HO — CH
|
CH$_2$
|
H$_3$C — $\overset{+}{N}$ — CH$_3$
|
CH$_3$

(Carnitine)

FIGURE 4.52 Exchange of a coenzyme A moiety for carnitine.

## *Malonyl-CoA Acts as a Regulatory Molecule*

Transport of fatty acids into the mitochondrion is regulated by a mechanism that plays a major role in controlling the overall rate of oxidation of fatty acids. This mechanism is in "communication" with the pathway for fatty acid synthesis. The fatty acid transport system is sensitive to the concentration of one fatty acid synthesis intermediate, malonyl-CoA.

The study described in Figure 4.54 illustrates the role of cytosolic malonyl-CoA in controlling the entry of fatty acids into the mitochondrion. The activity of carnitine acyltransferase was measured after adding various concentrations of fatty acyl-CoA (0–250 µM) to a suspension of mitochondria both with and without malonyl-CoA. Transferase activity increased with increasing concentrations of fatty acyl-CoA until a plateau was reached. At this point, the transferase was saturated and unable to operate at a faster rate when presented with higher concentrations of its substrate. The activity of the transferase was impaired in the presence of malonyl-CoA. This indicates that an increase in the rate of fatty acid synthesis, with the resultant increase in concentration of malonyl-CoA, will impair the oxidation of fatty acids. Therefore, the pathways of fatty acid oxidation and synthesis are coordinated. Figure 4.54 also indicates that malonyl-CoA inhibition can be overcome by higher levels of fatty acyl-CoA. Other studies have revealed

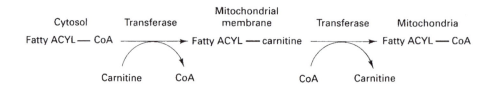

FIGURE 4.53 Transport mechanism for introducing fatty acids into the mitochondrion.

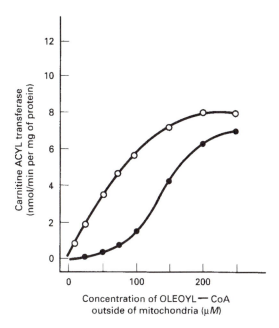

**FIGURE 4.54** Activity of carnitine acyl transferase versus concentration of fatty acyl-CoA, with and without added malonyl-CoA. This study illustrates the role of cytosolic malonyl-CoA in controlling the entry of fatty acids into the mitochondrion. The activity of carnitine acyltransferase was measured after adding various concentrations of fatty acyl-CoA (0–250 μM) to a suspension of mitochondria. The fatty acid used (oleic) was added in the form of oleoyl-CoA. Incubations were carried out in the absence (○) and presence (●) of malonyl-CoA. (Redrawn with permission from Cook, 1984.)

that the malonyl-CoA levels present in living animals are similar to that used in this study. In a fasted rat, the malonyl-CoA concentration in the liver is about 6 mmol/kg. This level should permit a maximal rate of fatty acid oxidation, because, in the fasted state the liver converts carbohydrates to fatty acids at a low rate (i.e., for storage as TGs). The concentration of malonyl-CoA in the liver of a fed rat has been found to reach about 25 mmol/kg. This level could inhibit carnitine acyl-transferase and thus prevent or limit the oxidation of the fatty acids newly synthesized from dietary carbohydrates.

## Experimental Incubations with and without Carnitine

The study reported in Table 4.12 illustrates the role of carnitine in fatty acid oxidation and introduces the topic of medium-chain fatty acids. This study, conducted before the role of carnitine in fatty acid transport was realized, involved addition of radioactive fatty acids to suspensions of liver mitochondria. The fatty acids used included [$^{14}$C]octanoic acid (medium-chain) and [$^{14}$C]palmitic acid (long-chain). The suspensions were incubated for 30 minutes to permit uptake of the fatty acids, their subsequent oxidation, and discharge of radioactive carbon dioxide. The $^{14}CO_2$ produced in the Krebs cycle diffuses out of the mitochondria into the surrounding fluid.

**TABLE 4.12** Effects of Carnitine on Fatty Acid Oxidation

| $^{14}$C-labeled fatty acid | $^{14}CO_2$ produced (nmol) | Carnitine-induced change |
| --- | :---: | :---: |
| [$^{14}$C]Butyric acid | 1478 | −9% |
| [$^{14}$C]Octanoic acid | 698 | −7% |
| [$^{14}$C]Lauric acid | 166 | +16% |
| [$^{14}$C]Palmitic acid | 218 | +78% |
| [$^{14}$C]Stearic acid | 34 | +66% |

*Source*: Fritz (1959).

The incubations were conducted under two conditions: in the presence and in the absence of carnitine. When evaluating the data, one should assume that the mitochondrial suspensions contained small amounts of carnitine in the medium, and within the organelle, derived from the mitochondrial preparation itself. The data indicate that 698 nmol of radioactive $CO_2$ was released with the incubation of [$^{14}$C]octanoic acid. Adding carnitine to the incubation mixture resulted in very little change in the amount of $^{14}CO_2$ released. Incubation with [$^{14}$C]palmitic acid resulted in the release of 218 nmol of $^{14}CO_2$. Here, adding carnitine resulted in the stimulation of fatty acid oxidation. Experiments performed in the 1970s revealed that long-chain fatty acids required the participation of carnitine acyltransferase to enter mitochondria, whereas medium- and short-chain fatty acids were able to pass into these organelles without being transiently modified with carnitine (McGarry and Foster, 1980).

EXERCISE

Why was ATP added along with the $^{14}$C-labeled fatty acids (Table 4.12)? What would the results have been if ATP had been omitted?

## Sources of Carnitine

Carnitine is required for transport of long-chain fatty acids into mitochondria; hence, carnitine plays an essential role in normal oxidative metabolism as well as in the formation of ketone bodies. The concentration of free carnitine in muscle is about 4.0 mmol/kg. The concentration of carnitine bound to long-chain fatty acids (fatty acyl-carnitine) is lower, about 0.2 mmol/kg. Short-chain fatty acids, including acetic, are also esterified to carnitine, but the functions of these complexes are not clear. There is some indication that **keto forms of BCAAs (BCKAs)** can also be esterified to carnitine. These complexes can then be transported into the mitochondria for complete oxidation of the BCKAs. The importance of this mode of BCKA transport is not clear (Takakura *et al.*, 1997).

Carnitine acts catalytically and might be considered to be a cofactor of carnitine fatty acyltransferase. Unlike most cofactors, however, carnitine is not a vitamin and is not derived from a vitamin. Generally, all of the carnitine needed by the body can be synthesized by the body. The biosynthesis of carnitine begins in an unusual manner. The starting material, the lysine residues of a variety of proteins, undergo post-translational methylations, as shown in Figure 4.55. The methyl donor is *S*-adenosylmethionine. Trimethyllysine, liberated from the protein by intracellular hydrolysis, is hydroxylated and then converted to carnitine in three

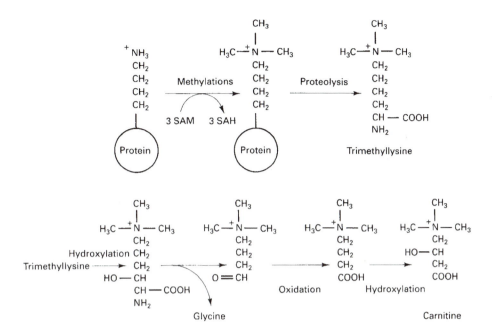

**FIGURE 4.55** Carnitine biosynthetic pathway.

more steps. The muscle content of protein-bound trimethyllysine is about 0.4 mmol/kg; that of free trimethyllysine is about 0.02 mmol/kg. The final step in carnitine synthesis is catalyzed by an iron metalloenzyme. This iron is not part of a heme cofactor. The enzyme uses oxygen as a cosubstrate and requires vitamin C to maintain the iron in its reduced state. Several proteins identified as containing residues of trimethyllysine are myosin, actin, calmodulin, cytochrome $c$, and histones. The function of the trimethyllysine residue in these proteins is not clear.

The typical Western diet supplies about 100 mg of carnitine per day. Meat and milk are good sources; beef contains about 500 mg/kg, and cow milk contains 5 to 40 mg/kg of fluid. Hen eggs and plant food contain little or no carnitine. The trimethyllysine present in food proteins contributes, to a very small extent, to the carnitine made in the body. There is, however, some concern about a transient deficiency of carnitine in newborn infants fed carnitine-free diets. Soy-based infant formula may lack carnitine, and the neonate seems to have a lesser capacity to synthesize the cofactor than adults. Breast milk contains high levels of carnitine; after about a month of lactation, the content declines to about half of its initial value. There is little evidence that the lowered levels of plasma carnitine found in infants fed soy formulas are associated with an impairment of the oxidation of long-chain fatty acids.

Another topic of occasional concern is the carnitine status of people in the Middle East and India who consume cereal-based diets. Although these diets supply only about 15 mg of carnitine per day, there is little reason to believe that the rate of carnitine synthesis in these persons is inadequate to supply the body's needs.

Lysine and methionine are both required for the synthesis of carnitine. This situation has sparked interest among researchers about the possibilities that a deficiency in either of these two indispensable amino adds can cause a carnitine deficiency and that carnitine supplementation can reduce the requirements for these amino acids. Results from studies of these possibilities have been variable. The relative deficiency of lysine in cereal grains may contribute, in part, to the low levels of plasma carnitine found in people consuming diets composed solely of grains.

## Long-Chain versus Medium-Chain Fatty Acids

Triglycerides are a major component of the diet. Most people consume 100 g of fat and oil per day. The fatty acid constituents of these TGs are mainly long-chain (i.e., 16–24 carbons long). Medium-chain TGs are minor components of the diet. Even in foods known to contain medium-chain TGs, such as coconut milk and mother's milk, long-chain TGs predominate.

Differences in bodily absorption and metabolism of medium- and long-chain TGs and FFAs are summarized in Table 4.13. In some cases, differences are relative, not absolute. The properties of fatty acids 12 and 14 carbons in length are between those of the medium- and long-chain fatty acids. The most striking difference between medium- and long-chain TGs is that the former are more water soluble than the latter. There has been some interest in supplying persons suffering from fat-malabsorption syndromes with medium-chain TGs as a source of energy. Medium-chain TGs are normally absorbed by mechanisms independent of those used by long-chain TGs that bypass steps involving bile salts, chylomicrons, and the lymphatic system. Another difference is the greater tendency of medium-chain fatty acids to be metabolized immediately in the mitochondria, rather than stored as fat. Oxidation of any nutrient in the mitochondria entails consumption of

**TABLE 4.13**  Properties of Medium- and Long-Chain Fatty Acids

| Medium-chain FFAs (8–10 carbons long) | Long-chain FFAs (16–24 carbons long) |
|---|---|
| TGs are hydrolyzed rapidly in the gut, at a rate perhaps fivefold greater than that of long-chain TGs. | TGs are hydrolyzed relatively slowly in the intestines. |
| FFAs are absorbed at a faster rate than long-chain FFAs, possibly twofold faster. | FFAs are absorbed at a relatively slow rate. |
| FFAs enter the enterocyte, but they are not reesterified to glycerol. | FFAs enter the enterocyte where they are converted back to TGs. |
| FFAs are transported from the gut via the portal vein to the liver, along with other water-soluble nutrients. | TGs are packaged into chylomicrons and transported through the lymphatic system. |
| Fatty acids tend to bind to albumin on entering the bloodstream. | TGs enter the bloodstream as a component of chylomicrons. |
| Fatty acids are thought not to be esterified to coenzyme A on entering the cell. This lack of formation of the CoA derivative prevents their incorporation into TGs, minimizes their storage as TGs in adipose tissue, and promotes their immediate oxidation in the mitochondria. | TGs entering the body in chylomicrons may be hydrolyzed by lipoprotein lipase and taken up by skeletal muscle, adipose tissue, and other tissues of the body for storage or immediate oxidation. Fatty acids entering the liver may be packaged into VLDLs, as TGs, and shunted back to the bloodstream to be used for oxidation or storage by other tissues. |
| FFAs tend not to be converted to the carnitine derivative. They can pass freely into the mitochondrion, bypassing the carnitine-dependent step. | FFAs require esterification to CoA and transfer to carnitine, prior to entry into the mitochondrion. |

oxygen. The experiment depicted in Figure 4.56 illustrates differences in $O_2$ consumption caused by eating foods containing long- or medium-chain TGs.

In this study, human subjects consumed a meal containing 45 g of a long-chain TG oil (corn oil) or medium-chain TG oil. The basal rate of oxygen consumption as measured prior to the meal was 250 ml/min. During rest after the meal, the rate

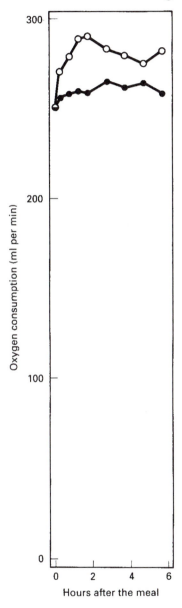

**FIGURE 4.56** Rate of oxygen consumption after a meal. Human subjects consumed a meal containing 45 g of long-chain TG oil (corn oil) (●) and medium-chain TG oil (○), then rested. The basal rate of oxygen consumption, as measured prior to the meal, was 250 ml/min. Rates were determined at the indicated intervals after the meals. About 13% of the medium-chain TGs were completely oxidized; only 4.0% of the long-chain TGs were oxidized. (Redrawn with permission from Seaton *et al.*, 1986.)

of $O_2$ consumption increased slightly in the subjects fed the corn oil; larger increases occurred in those fed the medium-chain TG oil. These results are consistent with the postulate that dietary medium-chain TGs tend to be oxidized immediately, whereas long-chain TGs are first stored as fat. In the 6 hours following the test meals, about 13% of the medium-chain TGs and 4.0% of the long-chain TGs were completely oxidized. This indicates that differences in the fates of long- and medium-chain fatty acids are relative, not absolute.

## THE KREBS CYCLE

The Krebs cycle is a series of reactions catalyzed by seven enzymes in mitochondria. Its function is to catalyze removal of electrons from nutrients and to transfer them to NAD$^+$ and FAD, producing NADH plus H$^+$, and FADH$_2$, respectively. These reduced cofactors exist only momentarily in their reduced (or oxidized) forms as they continually accept and then donate electrons to the respiratory chain. The respiratory chain, composed of a number of cytochromes, uses electrons for reduction of $O_2$ to water. This reduction process is accompanied by or coupled with the regeneration of ATP, that is, conversion of ADP back to ATP. The overall effect may be summarized thus: *The Krebs cycle and respiratory chain are used for oxidizing nutrients to $CO_2$ and for the production of energy.*

The Krebs cycle, which is also called the **citric acid cycle** or the **tricarboxylic acid cycle**, was discovered by Hans Krebs in 1937, at Cambridge University in Great Britain (Holmes, 1993).

The Krebs cycle is presented in three figures. The first (Figure 4.57) depicts the cosubstrates and cofactors used in each reaction. The second (Figure 4.58) gives names for the intermediates. The third (Figure 4.59) lists the names of the enzymes.

Carbohydrates are introduced into the Krebs cycle at the point where pyruvate dehydrogenase catalyzes conversion of pyruvate to acetyl-CoA with the concomitant reduction of NAD. Citrate synthase catalyzes introduction of the 2-carbon unit of acetyl-CoA into the Krebs cycle. Pyruvate can arise from glucose, fructose, lactate, alanine, and glycerol. Acetyl-CoA can arise from pyruvate, as well as from fatty acids. Oxidation of fatty acids results in production of acetyl-CoA, which enters the Krebs cycle at the point catalyzed by citrate synthase. Breakdown of ketogenic amino acids also results in the production of acetyl-CoA, which enters the Krebs cycle at this point. Citrate and malate occur in high concentrations in certain fruits and vegetables. These chemicals directly enter the Krebs cycle at the indicated points.

A number of amino acids can be broken down to produce α-ketoglutarate, succinyl-CoA, or oxaloacetic acid, intermediates of the Krebs cycle. In this process, the carbon skeletons of the amino acids are broken down to form intermediates of the Krebs cycle. The glucogenic amino acids also can be metabolized in this way.

Aconitase catalyzes the isomerization of citrate to isocitrate, isocitrate dehydrogenase catalyzes the oxidative decarboxylation of isocitrate to α-ketoglutarate, and α-ketoglutarate dehydrogenase catalyzes the oxidative decarboxylation of α-ketoglutarate to succinyl-CoA. Succinyl-CoA and the remaining intermediates are the 4-carbon intermediates of the Krebs cycle. Succinyl thiokinase catalyzes the release of coenzyme A from succinyl-CoA and the production of GTP. Succinate dehydro-

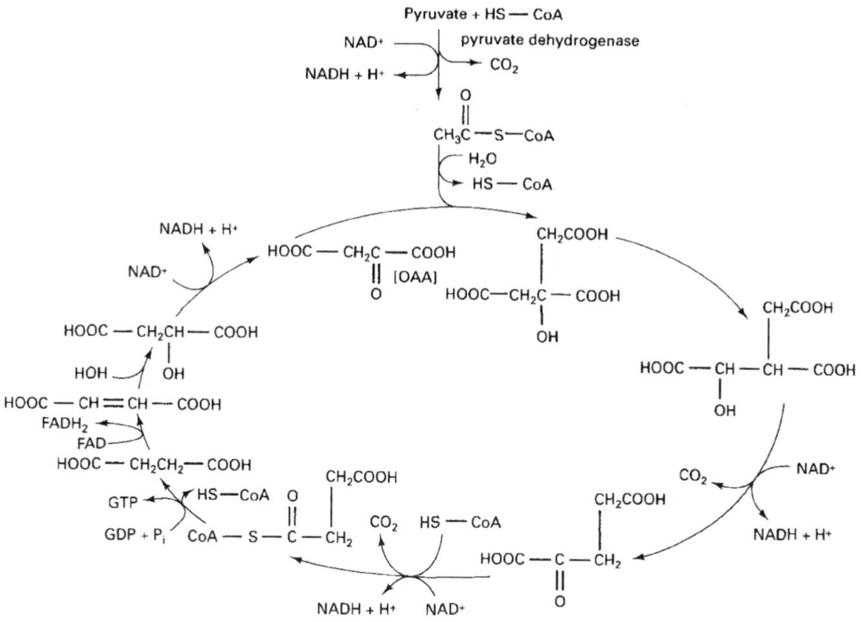

FIGURE 4.57 The Krebs cycle — cofactors and cosubstrates. The energy-producing steps of the cycle function by producing chemical transformations of cofactors and cosubstrates. Much of the energy discharged during the oxidation of substrates entering the Krebs cycle is collected in the form of GTP, reduced NAD, and reduced FAD. Both $FADH_2$ and reduced NAD donate energy to cytochromes for the production of ATP.

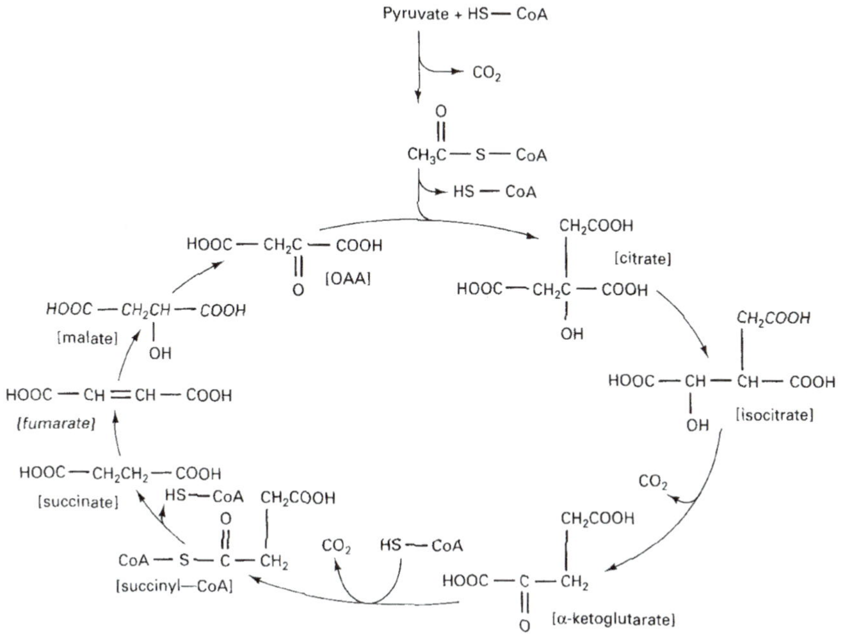

FIGURE 4.58 The Krebs cycle — names of intermediates. The intermediates in the cycle are citrate, isocitrate, α-ketoglutarate, succinyl-CoA, succinate, fumarate, malate, and oxaloacetate. Acetyl-CoA is used to introduce the acetyl group to the Krebs cycle. Carbon dioxide is its final product.

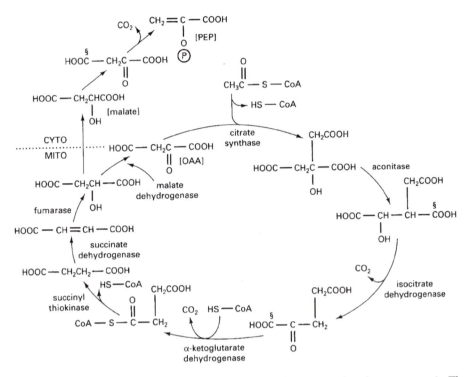

**FIGURE 4.59** Enzymes of the Krebs cycle and removal of malate for gluconeogenesis. The enzymes that catalyze each step of the Krebs cycle are named; all are mitochondrial. The section symbol (§) indicates the two points in the Krebs cycle, and a gluconeogenesis step occurring in the cytoplasm, where carbon is lost as $CO_2$.

genase catalyzes the oxidation of succinate to fumarate, fumarase catalyzes the hydration of fumarate to malate, and malate dehydrogenase catalyzes the oxidation of malate to OAA.

OAA appears at the beginning and end of each turn of the Krebs cycle; however, the carbons of the starting OAA are not the same atoms as those in OAA at the end. This point is obvious from the molecular formulas in the diagrams of the cycle (Figures 4.57–4.59). The 2-carbon unit entering the cycle from acetyl-CoA survives two decarboxylation reactions and constitutes two of the carbons of the OAA molecule at the end of the cycle.

During the complete oxidation of pyruvate, carbon dioxide is discharged at three points: the reactions catalyzed by pyruvate dehydrogenase, isocitrate dehydrogenase, and α-ketoglutarate dehydrogenase. Reducing power, in the forms of NADH + H$^+$ and FADH$_2$, is produced at five points: the reactions catalyzed by pyruvate dehydrogenase, isocitrate dehydrogenase, α-ketoglutarate dehydrogenase, succinate dehydrogenase, and malate dehydrogenase. The section symbol (§) in Figure 4.59 marks the specific carbon in a molecular structure that is discharged in the next enzymatic step of the cycle. Note that the two carbons introduced by acetyl-CoA are not discharged during their first "trip" around the Krebs cycle; however, they are discharged during later turns of the cycle.

## Demographics of Metabolites and Enzymes

The pathways of glycolysis and gluconeogenesis are cytosolic. All of the enzymes are water soluble. An exception is glucose-6-phosphatase, which occurs in the endoplasmic reticulum. The concentrations of the enzymes in the cytosol (of muscle) are generally 50 to 1000 $\mu M$, depending on the enzyme (Srivastava and Bernhard, 1986). The approximate levels of the following metabolites in resting muscle are indicated. The unit of concentration in all cases is mmol/kg of dried whole muscle. This level might be expected to be very roughly similar to the actual intracellular m$M$ concentration: glucose (2.5); glucose-6-phosphate (0.6); glucose-1,6-bisphosphate (0.15); fructose-1,6-bisphosphate (0.03); phosphoenolpyruvate (0.7); pyruvate (1.3); lactic acid (9.0); citric acid (1.5); $\alpha$-ketoglutarate (1.3) (Salmons *et al.*, 1996). One use of the preceding numbers is to compare the concentration of a specific metabolite to that of the enzyme that uses the metabolite as a substrate.

The mitochondrion is bounded by two phospholipid membranes. The outer membrane is freely permeable to molecules, including water, with a molecular weight of up to about 5000. The inner membrane is rich in membrane-bound proteins and consists, in terms of membrane area, of 50% phospholipid and 50% protein (Lenaz, 1988). Pyruvate dehydrogenase, a mitochondrial enzyme, is water soluble. The proteins of the respiratory chain, as well as ATP synthase, are all bound to the inner mitochondrial membrane. The enzymes of the Krebs cycle are water soluble, with the exception of succinate dehydrogenase. This enzyme is bound to the mitochondrial membrane, where it directly funnels electrons, via FAD, to the respiratory chain.

## Krebs Cycle Supplies Electrons to the Respiratory Chain

The proteins of the respiratory chain are NADH dehydrogenase, cytochrome $b$, cytochrome $c_1$, cytochrome $c$, cytochrome $a_1$, and cytochrome $a_3$. Cytochromes $a_1$ and $a_3$ form a complex known as **cytochrome $c$ oxidase**. The proteins are listed in the order in which they are used in the electron transport pathway. The proteins are all membrane-bound proteins, though cytochrome $c$ is only weakly bound to the outer surface of the inner membrane. Its polypeptide chain is not inserted in the membrane. Electrons are delivered in pairs, via NAD, to the respiratory chain. It is thought that three protons are driven out of the mitochondrion for each pair of electrons from NAD passing down the respiratory chain and through cytochrome oxidase to oxygen (Cross, 1981; Hatefi, 1985). Further details are given in Chapter 5 and under Iron in Chapter 10.

ATP synthase catalyzes the synthesis of most of the ATP produced in the body. This enzyme is a membrane-bound protein of the mitochondrion. The passage of protons, through the mitochondrial membrane (via ATP synthase) is required for driving the synthesis of one ATP (by ATP synthase) (Ferguson, 1986; Weber and Senior, 1997). Oxygen reduction and ATP synthesis are not directly coupled. The coupling is indirect, in that operation of electron transport is required for pumping protons out of the mitochondrion, and the passive return of protons back into the organelle is used to drive the synthesis of ATP. The full name of ATP synthase is $F_1F_0$-**ATP synthase**. ATP synthase consists of about 16 different polypeptides. Some of these are grouped to form the $F_1$ unit, while others form the $F_0$ unit. $F_0$ resides within the inner mitochondrial membrane, while $F_1$ resides in the interior of the mitochondrion and contacts the aqueous environment within this organelle. The $F_0$ and $F_1$ subunits are connected with a stalk. Further details are in Chapter 5.

## Modes of Flux of 2-Carbon Units in the Krebs Cycle

The Krebs cycle is an intriguing group of reactions that can be discussed from a number of viewpoints. The major function of the cycle is catalytic oxidation of 2-carbon units produced during oxidative breakdown of dietary and stored fats and of carbohydrates, both dietary and stored as glycogen. Two relatively minor functions are oxidation of the carbon skeletons of amino acids undergoing catabolism and conversion of amino acids to glucose (i.e., gluconeogenesis). Gluconeogenesis from an amino acid involves entry of its carbon skeleton into the cycle in the form of OAA, succinyl-CoA, or α-ketoglutarate. These molecules are converted to malate, which leaves the Krebs cycle and enters the cytoplasm at the point indicated in Figure 4.59. Cytoplasmic malate can be converted to glucose as discussed.

One point of special interest to the nutritionist is that fatty acids cannot be used for the synthesis of glucose (gluconeogenesis). The 2-carbon unit entering the Krebs cycle is depicted in Figure 4.58 as the "top part" of citric acid, to increase its visibility. This 2-carbon unit remains intact during the two steps that result in the discharge of $CO_2$: those of isocitrate dehydrogenase and α-ketoglutarate dehydrogenase. Note that for gluconeogenesis this 2-carbon unit leaves the mitochondrion for the cytoplasm. This would seem to indicate that fatty acids can be used for gluconeogenesis; however, the following material explains why this is not so.

The argument that fatty acids cannot contribute to the net synthesis of glucose can be stated in several ways. The main goal of following these arguments is to gain some understanding of the Krebs cycle:

1. For every 2-carbon unit entering the cycle, two carbons leave as $CO_2$. Thus, the Krebs cycle is not an efficient conduit for the conversion of acetyl groups to malate groups. The carbons entering as acetyl groups are, in effect, only lost as $CO_2$.

2. The introduction of each 2-carbon unit ties up one molecule of OAA, because the acetyl group condenses with OAA to form citric acid, which then travels through the cycle and becomes malate. Malate then leaves for the cytosol, and OAA is not regenerated. Consequently, continued use of acetyl groups for gluconeogenesis would lead to depletion of some of the intermediates of the Krebs cycle. This point introduces the concept that the intermediates in the Krebs cycle act in a catalytic manner in the conversion of energy fuels to $CO_2$.

3. Use of acetyl groups for gluconeogenesis results, in effect, in withdrawal of intermediates from the Krebs cycle. If continued, this withdrawal would result in cessation of the cycle. The total quantities of these intermediates present in the liver are relatively small, compared with the amount of protein in the body available for breakdown and conversion to glucose. Hence, it would make little sense to deplete the intermediates in the liver for use in gluconeogenesis.

Glutamate is a glucogenic amino acid. Its conversion to α-ketoglutarate is detailed in Chapter 8. Introduction of the carbon skeleton of glutamate into the Krebs cycle followed by its withdrawal as malate does not result in depletion of intermediates. The 5-carbon skeleton of this amino acid enters the cycle at α-ketoglutarate and still exits at malate, which has four carbons. Note that two molecules of glutamate are required for the synthesis of one molecule of glucose.

## FERMENTATIVE AND AEROBIC ENERGY METABOLISM

Skeletal muscle is composed of two different types of cells: white and red muscle fibers. Some muscles are fairly homogeneous and contain mainly one type of fiber; others contain mixtures of both types. White muscle fibers contain relatively few mitochondria, are mainly glycolytic, and produce lactic acid as an end-product of fermentation. Red muscle fibers are relatively rich in mitochondria and are well adapted to oxidizing fatty acids and glucose to $CO_2$. Their energy metabolism is dependent on an ample supply of oxygen.

Lactate dehydrogenase is closely associated with the final step in fermentative metabolism: conversion of pyruvate to lactate. Citrate synthase is closely associated with oxidative metabolism, because this enzyme catalyzes the introduction of acetyl groups into the Krebs cycle. Continued operation of the Krebs cycle is dependent on the continued transfer of electrons from reduced NAD and FAD to the respiratory chain, which catalyzes the reduction of $O_2$ to $H_2O$. Table 4.14 lists lactate dehydrogenase/citrate synthase activity ratios for muscles that contain a large proportion of white or red muscle fibers. The data demonstrate that the ratio is relatively high in white cell muscles but low in red cell muscles.

EXERCISE

Would you expect carnitine, which is made in the liver, to be more important to cardiac muscle or to white skeletal muscle?

TABLE 4.14   Enzyme Activity Ratios for Muscles that Contain Large Proportions of White or Red Muscle Fibers

| Muscle | Ratio of activities of lactate dehydrogenase/ citrate synthase |
|---|---|
| White skeletal | |
| Psoas | 1100 |
| Adductor magnus | 1090 |
| Red skeletal | |
| Soleus | 18 |
| Masseter | 16 |
| Cardiac muscle | 6 |

Source: Bass et al. (1969).

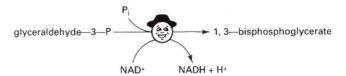

FIGURE 4.60 Oxidation of an aldehyde group to a carboxylic acid group. The smiling face represents the glyceraldehyde-3-phosphate dehydrogenase, happily catalyzing its reaction in the presence of an ample amount of NAD$^+$.

EXERCISE

If you were a long-distance marathon runner, would you expect to have more highly developed white muscle fibers or red muscle fibers? Hint: Consult Figure 4.31. (Also see Gollnick and Hodgson, 1986.)

## Special Function of Lactate Dehydrogenase

One of the more interesting steps in glycolysis is that catalyzed by glyceraldehyde-3-phosphate dehydrogenase. This reaction involves oxidation of an aldehyde group to a carboxylic acid group, with the concomitant reduction of NAD$^+$. The enzyme is depicted in Figure 4.60 as a smiling face. Continued action of this enzyme in the cytoplasm might deplete the supply of NAD$^+$. The problem at hand is then to regenerate NAD$^+$. In Figure 4.61, the enzyme is shown in an unhappy state, because lack of the cosubstrate, NAD$^+$, prevents it from catalyzing the oxidation of glyceraldehyde-3-phosphate.

The concentration of cytoplasmic NAD$^+$ can be maintained by the action of lactate dehydrogenase (LDH) or by the malate–aspartate shuttle. With fermentative metabolism, LDH regenerates NAD$^+$ by catalyzing the reduction of pyruvate. As depicted in Figure 4.62, this permits the continued action of glyceraldehyde-3-phosphate dehydrogenase. This action contributes to the overall flux of energy metabolism, because the electrons removed from glyceraldehyde-3-phosphate during fermentative metabolism in muscle are transported, in the form of lactate, to the liver, where they are used to reduce NAD. The NADH + H$^+$, produced by this circuitous route in the liver can then be used to produce ATP in the mitochondria.

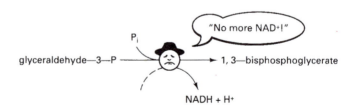

FIGURE 4.61 Oxidation of an aldehyde group to a carboxylic acid group. The unhappy face indicates that the reaction cannot continue without NAD$^+$.

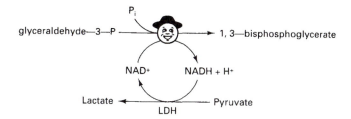

**FIGURE 4.62** Regeneration of NAD⁺. The concentration of cytoplasmic NAD⁺ can be maintained by a reaction, by the concomitant reduction of pyruvate, catalyzed by lactate dehydrogenase (LDH). This permits continued action of glyceraldehyde-3-phosphate dehydrogenase. Once again, the enzyme is happy.

## Special Function of Malate Dehydrogenase

$NAD^+$ is reduced by glyceraldehyde-3-phosphate dehydrogenase during both fermentative and oxidative metabolism. In oxidative metabolism, the electrons of the $NADH + H^+$ reduced in the cytoplasm are transferred to the mitochondrion by the malate–aspartate shuttle. As shown in Figure 4.63, $NADH + H^+$ is not directly transferred into the organelle. Transfer of its reducing power involves a number of components, including cytoplasmic and mitochondrial malate dehydrogenase (MD), glutamate-oxaloacetate aminotransferase ($B_6$), and membrane-bound proteins used for exchanging metabolites.

The overall effect of the malate–aspartate shuttle is to transfer the equivalent of two electrons from the cytoplasm to the mitochondrion. The cycle is thought to be driven by cytoplasmic acid ($H^+$). The concentration of protons in the cytoplasm is greater than that in the mitochondrion, which has an alkaline interior. This concentration gradient is thought to drive the membrane-bound glutamate/aspartate exchanger.

*Tumors*

Tumors are composed of cancer cells, clustered together in a visible mass. The blood supply of certain tumors may be poor, due to a lack of blood vessels within the tumor. Tumor cells may utilize a unique style of energy metabolism. They can employ glycolysis, rather than the Krebs cycle, for making ATP. Metabolic studies conducted with tumor cells, dispersed in a layer in a culture dish, have shown that glycolysis may be preferably used, even when given an ample supply of oxygen. This style of energy metabolism is called **aerobic glycolysis**. In aerobic glycolysis, the two shuttles that normally transport hydrogen to the mitochondrion (the malate–aspartate and glycerol-3-phosphate shuttles) occur at a level that is lower than in normal cells. This forces the cells to use lactate dehydrogenase in order to regenerate NAD. The tendency to rely mainly on glycolysis to supply ATP is thought to ensure survival of migrating tumor cells, as they invade various parts of the body and to survive, prior to acquiring their own complement of arterioles and venules (Mazurek *et al.*, 1997).

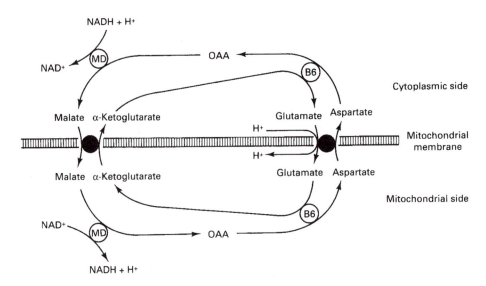

**FIGURE 4.63** Malate–aspartate shuttle. The net effect is the oxidation of cytoplasmic NADH and the reduction of mitochondrial NAD⁺. Amino groups are transferred from amino acids to keto acids by vitamin $B_6$-requiring enzymes (indicated by $B_6$). Membrane-bound transport proteins are indicated by large black dots. The shuttle is driven by a proton gradient, as detailed later. The mitochondrion is bounded by a phospholipid membrane that is not permeable to hydrophilic metabolites. Specialized proteins embedded in the membrane allow certain molecules to pass through the membrane. An important transport protein is the ATP/ADP exchanger. For each molecule of cytoplasmic ADP that enters the mitochondrion a molecule of ATP leaves the organelle for the cytoplasm, thus enabling ATP-requiring reactions in the cytoplasm, in the nucleus, and at the plasma membrane to continue. Two transport systems in the figure are the α-ketoglutarate antiport and the glutamate/aspartate exchanger. The antiport allows the strict exchange of α-ketoglutarate for malate. The exchange is passive and does not require the expenditure of energy, that is, of ATP. The glutamate/aspartate exchanger permits the exit of one molecule of aspartate with the entry of one molecule of glutamate. The exchange is driven by, or coupled to, the entry of a proton. A number of other mitochondrial transport reactions are driven by the entry of protons. This entry occurs spontaneously because of the pH gradient between the interior and exterior of the organelle.

## Ketone Bodies

The molecular structures of the ketone bodies acetoacetate, β-hydroxybutyrate, and acetone are shown in Figure 4.64. Acetone is formed by nonenzymatic break-down of acetoacetate. Between 20 and 75% of the acetone formed can be metabo-lized to usable energy by humans. Experiments with rats determined that acetone is metabolized via acetyl-CoA (Kosug *et al.*, 1986). Unmetabolized acetone is excreted in the urine or released in the breath. The odor of acetone on a person's breath reveals that the person either has uncontrolled diabetes or is fasting. The acetoacetate and β-hydroxybutyrate formed in the body can be an important source of energy; acetone is relatively unimportant.

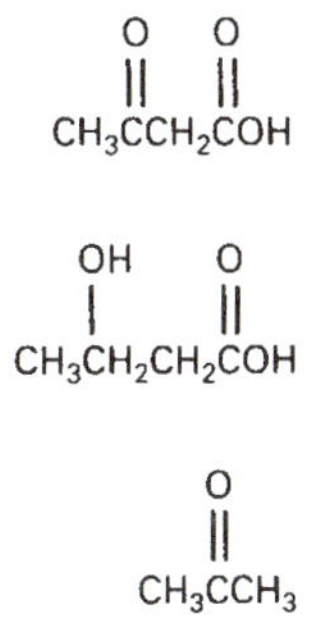

$$\underset{CH_3CCH_2COH}{\overset{O\quad O}{\overset{\|\quad\|}{}}} \qquad \text{Acetoacetate}$$

$$\underset{CH_3CH_2CH_2COH}{\overset{OH\quad\ \ O}{\overset{|\quad\ \ \|}{}}} \qquad \text{β-Hydroxybutyrate}$$

$$\underset{CH_3CCH_3}{\overset{O}{\overset{\|}{}}} \qquad \text{Acetone}$$

**FIGURE 4.64** Structures of ketone bodies. The term "ketone" is accurate, as the structures shown are ketones. However, the term "body" is not relevant to any aspect of the structure or metabolism of these compounds, and is not meant to indicate any unusual property of the compounds.

The most important property of the major ketone bodies, acetoacetate and β-hydroxybutyrate, is their availability as an energy source for the brain. In times of fasting, the supply of glucose stored as glycogen is depleted, and the central nervous system must rely on the glucose formed from glycerol and from amino acids via gluconeogenesis. These latter sources of glucose are limited and must be supplemented by energy from ketone bodies. Fat is the major energy source during fasting. Fatty acids derived from TGs are broken down to acetyl-CoA, which is then used for synthesis of ketone bodies.

The starting material for ketone body synthesis and catabolism, shown in Figure 4.65, is acetyl-CoA. Ketogenesis occurs in the mitochondria of the liver. Hence, ketone body synthesis is, for acetyl-CoA, an alternate fate to immediate oxidation in the Krebs cycle. This pathway results in the formation of acetoacetate and β-hydroxybutyrate. Both appear in the bloodstream (the latter at higher concentrations) and are taken up by various organs, such as the brain and muscle. Here, they are converted back to acetyl-CoA and then oxidized in the Krebs cycle.

The structure of the intermediate, hydroxymethylglutaryl-CoA (HMG-CoA), is shown in Figure 4.66. The mechanism of formation of this compound from acetoacetyl-CoA and acetyl-CoA may not be obvious. Under Thiol Esters in this chapter, the 2-carbon of thiol esters is shown to acquire a negative charge. The negative charge on the 2-carbon of acetyl-CoA attacks a carbonyl carbon of acetoacetyl-CoA, generating HMG-CoA. The HMG-CoA formed here is mitochondrial and should not be confused with that formed in the cytoplasm by the cholesterol biosynthetic pathway.

The rise in plasma ketone bodies is illustrated by studies with both humans (Figure 4.67) and rats (Figure 4.68). Plasma concentrations of FFAs, β-hydroxybutyrate, and acetoacetate are shown at the indicated times before and during fasting. Under nonfasting conditions, ketone body levels are negligible. However, during the course of fasting, the concentration of β-hydroxybutyrate becomes larger than the concentrations of the FFAs. Note that these FFAs are noncovalently bound to albumin in the bloodstream, whereas the ketone bodies are freely soluble.

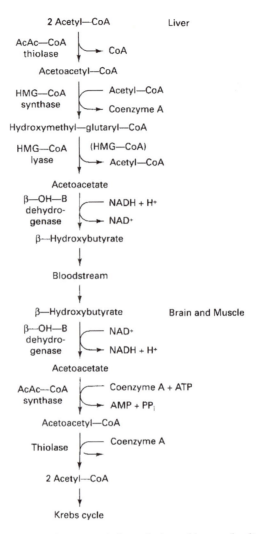

FIGURE 4.65 Synthesis, circulation, and degradation of ketone bodies. Ketone body metabolism involves synthesis in the mitochondria of the liver, distribution via the bloodstream, and oxidation by the Krebs cycle in various organs, such as the brain and muscle.

A comparison of fatty acid and ketone body levels in the bloodstream may overemphasize the importance of the latter compounds as energy sources for the following reasons. First, fatty acid molecules contain 16 to 18 carbons, whereas ketone bodies contain only four carbons. Therefore, the weight of 1 mmol of fatty acid is several times that of 1 mmol of β-hydroxybutyrate. Second, fatty acids are highly reduced, alkane-like structures, whereas ketone bodies are partially oxidized. This means that on a per-weight basis complete oxidation of fatty acids yields more energy than complete oxidation of ketone bodies.

Under normal conditions, glucose is the exclusive fuel of the brain; however, when ketone bodies accumulate in the bloodstream, they are also used by the

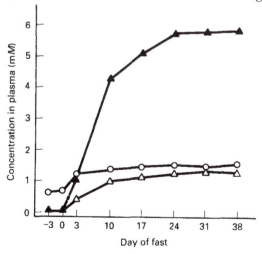

FIGURE 4.66 Formation and reaction of the intermediate, hydroxymethylglutaryl-CoA (HMG-CoA). The arrow pointing from the 2-carbon of acetyl-CoA to the carbon-3 of acetoacetyl-CoA depicts the position of attack of the 2-carbon on the carbonyl carbon during the reaction.

brain. The study reported in Table 4.15 involved obese human subjects fasting for a lengthy period. After 1 month, extraction of fuels by the brain was measured by the AV difference technique. Samples of blood entering and leaving the brain were taken and AV differences of the nutrients determined. The concentration of glucose in venous blood was 0.26 m$M$ less than that in arterial blood, indicating its removal and catabolism by the brain. The concentration of lactate was higher in the venous

FIGURE 4.67 Plasma concentrations of ketone bodies in fasting humans. Plasma concentrations of FFAs (O), β-hydroxybutyrate (▲), and acetoacetate (Δ) are shown before and during fasting. (Redrawn with permission from Owen $et$ $al.$, 1969.)

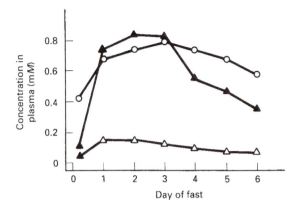

**FIGURE 4.68** Plasma concentrations of ketone bodies in fasting rats. Plasma concentrations of FFAs (○), β-hydroxybutyrate (▲), and acetoacetate (△) are shown before and during fasting. (Redrawn with permission from Parrilla, 1978.)

blood, indicating its production (from glucose) by the brain. These AV differences indicate that in fasting subjects the brain may derive up to about half of its energy from ketone bodies.

A major factor in stimulating ketone body production is increased availability of FFAs. An increased rate of FFA mobilization from the adipose tissue, with the consequent increase in FFA levels in the liver, may be sufficient to enhance ketone body formation. Increased release of FFAs occurs when the glucagon/insulin ratio increases, as in fasting.

The newborn infant is especially prone to ketone body formation because of its limited supply of glycogen. The rate of ketone body turnover in a newborn infant fasted for several hours may be equivalent to that in an adult fasted for several days. The term "turnover" refers to the rates of synthesis and oxidation. Comparison is on a per-kilogram-of-body-weight basis. With either brief or extended fasting in an adult, about 30% of the FFAs released from adipose tissue is converted to ketone bodies in the liver for use by other organs. Another factor leading to ketone body utilization in the infant is the large proportion of the energy in milk

**TABLE 4.15** Brain Arteriovenous Differences

|  | mmol/liter whole blood |
| --- | --- |
| Glucose | 0.26 |
| Lactate | −0.20 |
| Pyruvate | −0.03 |
| β-Hydroxybutyrate | 0.34 |
| Acetoacetate | 0.06 |

*Source*: Reprinted with permission from Owen *et al.* (1967).

supplied in the form of fat. About 55% of the energy in human milk, and 69% in rat milk, is in the form of TGs. It is thought that the relatively large proportion of fat in milk contributes to ketone body formation in newborns.

A few more words might be said concerning the regulation of ketone body synthesis. An increase in FFA levels, in itself, may not be sufficient to provoke the synthesis of ketone bodies. Regulatory changes may occur in the liver, as well as in adipose tissue, to promote ketone body synthesis. Fasting induces an increase in the plasma glucagon/insulin ratio, which stimulates the rate of transport of fatty acids into the mitochondria. This stimulation results from the drop in cytosolic malonyl-CoA. Increased transport of fatty acids into the mitochondria, and the consequent increase in production of acetyl-CoA, may overwhelm the Krebs cycle. Production of acetyl-CoA at a rate greater than it can be utilized by citrate synthase can stimulate its channeling into alternate pathways of metabolism.

An additional factor may result in the use of acetyl-CoA for ketone body synthesis rather than for oxidation by the Krebs cycle. Fasting may result in a decrease in the concentration of OAA in liver mitochondria. Such a drop in OAA, a cosubstrate of citrate synthase, would impair the use of acetyl-CoA by this enzyme. Thus, a drop in OAA may provoke the synthesis of the ketone bodies.

The glucagon/insulin ratio can rise under certain pathological conditions (i.e., insulin-dependent diabetes). A small percentage of diabetics develop **ketoacidosis**, a condition that results from the overproduction and underutilization of ketone bodies. Increased concentrations of β-hydroxybutyrate and acetoacetate, which are acids, can cause a drop in the pH of the blood. This acidification, known as **acidosis**, can impair the ability of the heart to contract and result in a loss of consciousness and coma, which, in rare cases, may be fatal. Diabetic ketoacidosis may manifest as abdominal pain, nausea, and vomiting. A subject may hyperventilate (breathe quickly and deeply) to correct acidosis, as described under Sodium, Potassium, and Water in Chapter 10. It is the responsibility of the clinician, when confronted with a subject whose breath smells of acetone or who is hyperventilating, to facilitate prompt treatment.

Elevated ketone body levels in the bloodstream (**ketosis**) can be caused by fasting, exercise, or consumption of a high-fat diet. A diet that provides over 85% of its energy as fat can result in ketosis. Intentional induction of ketosis with a ketogenic diet is sometimes useful in the treatment of cases of epilepsy resistant to drugs. Epilepsy is not a rare disease. The effect of ketosis was discovered during times of starvation, when it was noticed that ketotic epileptics were relieved of the symptoms of their disease (i.e., seizures). Drug-resistant forms of epilepsy are treated with medium-chain TGs. The medium-chain fatty acids liberated form these TGs are not stored as fat but are immediately transported into the mitochondria. Apparently, the high rate of influx caused by this diet overwhelms the catabolic capacity of the Krebs cycle, resulting in an alternative fate — formation of ketone bodies. The medium-chain TG diet is most often prescribed for epileptic children 2 to 5 years of age. One problem with the diet is its unpalatability.

Evidence suggests that ketone bodies reduce epileptic seizures by changing the metabolism of certain amino acids (glutamate and aspartate) in the brain (Yudkoff *et al.*, 1997). These two amino acids are neurotransmitters, that is, they act as hormones that relay signals from one nerve to another adjacent nerve.

**TABLE 4.16**  Masses of Utilizable Energy Stores of a Standard 70-kg Man

| Energy store or source | Mass (kg) | Energy (kJ) |
|---|---|---|
| Fat | 15 | 590 |
| Protein (muscle) | 6 | 100 |
| Glycogen (muscle) | 0.15 | 2.51 |
| Glycogen (liver) | 0.075 | 1.25 |
| Plasma glucose | 0.020 | 0.33 |
| Plasma triglycerides | 0.003 | 0.125 |
| Plasma free fatty acids | 0.0003 | 0.012 |

*Source*: Cahill (1970).

## Body Fuels Available during Fasting

The preceding discussion of ketone bodies leads to the topic of fasting. Prolonged fasting produces symptoms of deficiencies in many nutrients, most notably water (if water is omitted), energy, protein, and potassium. An evaluation of energy supplies available during fasting requires information on body composition. Tables 4.16 and 4.17 present data on the body composition of the standard 70-kg man. Masses of the different energy stores or usable sources appear in Table 4.16; Table 4.17 lists the weights of the organs. Fat (15 kg) constitutes 80 to 85% of the energy stores in the nonobese man. During starvation, this energy source can sustain life for 50 to 60 days. An obese person can live over 100 days using internal fuels as the sole energy source. At rest, skeletal muscle accounts for 35 to 40% of the body's total consumption of oxygen. Resting muscle depends mainly on fat, not carbohydrates, as an energy fuel.

About 20% of the body's muscle mass (28 kg) consists of protein (6 kg). Muscle protein accounts for about 17% of the body's energy reserves; however, not all muscle protein can be used for energy production. It is thought that the use of more than 50% of the 6 kg available is fatal. Muscle mass decreases during fasting.

**TABLE 4.17**  Organ Masses of a Standard 70-kg Human

| Organ | Mass (kg) |
|---|---|
| Skeletal muscle | 28.0 |
| Adipose tissue | 15.0 |
| Skeleton | 10.0 |
| Skin | 4.9 |
| Liver | 1.8 |
| Brain | 1.4 |
| Heart | 0.33 |
| Kidneys | 0.31 |

*Source*: Forbes (1987).

Noncontractile proteins are catabolized first, because contractile fibers resist proteolysis and breakdown. With prolonged starvation, however, contractile fibers are broken down, and it is not clear what mechanisms control the catabolism of muscle proteins.

Glycogen is a minor energy store. The amount stored in the body is sufficient to supply the energy needs of the brain for only 3 days of fasting. Clearly, the energy value of the nutrients in the bloodstream itself is minimal.

## STARVATION

Severe malnutrition is a complex problem. Factors involved include the overall availability of food; the availability only of starchy, protein-deficient foods; and contaminated water supplies. Infants and young children are special targets because they are rapidly growing, are unusually susceptible to infections, and may not be able to find food for themselves.

There are two general types of severe malnutrition. The first, **marasmus**, is an overall lack of energy and protein and results in overall wasting of the body. The second type occurs with chronic consumption of a diet supplying ample amounts of carbohydrate but little or no protein. This diet results in preservation of fat stores but, unfortunately, also in a tendency to preserve the stores of amino acids, the proteins of skeletal muscle. This failure to mobilize muscle protein results in continued turnover of proteins that are vital to survival, such as serum albumin and immunoglobulins (antibodies). Continued failure to consume protein leads to gradual depletion of these essential proteins, a condition called **kwashiorkor**. Malnutrition can occur as marasmus, kwashiorkor, or a range of intermediate forms. All forms, both mild and severe, are called **protein-energy malnutrition**.

Severe protein-energy malnutrition often occurs after weaning, the transition from nursing to the consumption of foods from other sources (i.e., solid foods). The marasmus common in Latin America is caused by early weaning, followed by use of overdiluted commercial milk formulas. Kwashiorkor occurs in Africa, where babies are fed starchy roots, such as cassava, that are low in protein. It also occurs in the Caribbean, where babies are fed sugar cane. The major symptom of kwashiorkor is edema, mainly of the feet and legs. **Edema** is the condition produced when water normally held in the bloodstream by osmotic pressure leaks into other extracellular spaces. It can result from reduced osmotic pressure in the bloodstream caused by catabolism and depletion of serum albumin.

Kwashiorkor is thought to involve "confused hormones." An adequate supply of carbohydrates, in the form of starchy roots or sugar cane, can result in maintenance of high levels of plasma insulin, which can prevent the net proteolysis that otherwise would supply the liver and immune system with amino acids. The mechanism by which insulin controls muscle turnover is not well understood; however, in marasmus the body's stores of fat and muscle are broken down and mobilized in a more coordinated fashion, resulting in a relatively gradual decline in overall health.

All proteins in the body, whether occurring in the plasma, as membrane-bound proteins, or inside the various compartments of the cell, are degraded or hydrolyzed. The rate of hydrolysis of any specific protein is characteristic of and unique for that protein. Albumin, transferrin, thyroxine-binding protein, and retinol-binding protein (RBP) are blood plasma proteins. Each has a specific half-life in the circulation. The **half-life** of albumin is about 19 days, which means that any given molecule of albumin in the bloodstream has a 50% chance of being degraded within any 19-day period. The relevance of this degradation rate to malnutrition is that a total halt in protein synthesis by the liver would be expected to result in a 50% drop in the concentration of plasma albumin in a 19-day period.

Transferrin, an iron-transport protein in the plasma, has a half-life of about 8 days; thyroxine-binding protein, 2 days; and retinol-binding protein, 0.4 days. Because of the relatively long half-life of albumin, the serum albumin level is not a particularly sensitive indicator of dietary protein status. Therefore, retinol-binding protein and thyroxine-binding protein have been used to assess malnutrition in poorly nourished populations. Note that these two proteins occur together as a complex in the bloodstream.

The study reported in Table 4.18 reveals the decline in plasma proteins, including retinol binding protein (RBP), that occurs with severe malnutrition. The liver secretes a great variety of proteins into the plasma, including RBP. RBP is required for the transport of one form of vitamin A (retinol) from its storage site in the liver to other needy tissues. The study involved children in Egypt; similar studies have been conducted in Senegal, India, Indonesia, and Central America. Table 4.18 lists serum protein levels in healthy children and in those suffering from kwashiorkor. The overall concentrations of serum protein were 7.2 and 4.0 g/100 ml, respectively. Levels of albumin and RBP were reduced to similar extents by malnutrition. The concentration of vitamin A was also reduced.

A consequence of the lower levels of plasma RBP produced by protein malnutrition is impairment of vitamin A transport from the liver to other tissues; however, vitamin A deficiency can also result from heavy reliance on foods that are extremely poor sources of vitamin A, such as rice or cassava. Both factors (lack of RBP or of vitamin A) can lead to deterioration of the epithelial cells of the gastrointestinal tract, respiratory tract, and eye, all signs of vitamin A deficiency. Another sign is nightblindness. Table 4.19 illustrates how feeding protein to malnourished children can result in an improvement in vitamin A status, even when the source of proteins contains little or no vitamin A. Some children with kwashiorkor were fed skim milk. The concentrations of RBP and vitamin A measured in the serum of these children before and after feeding show an increase in plasma RBP. A dramatic rise in plasma vitamin A also occurred, probably as a result of the liver's

**TABLE 4.18** Effect of Severe Malnutrition on Plasma Composition

|  | Healthy | Kwashiorkor |
| --- | --- | --- |
| Number of children | 15 | 21 |
| Average age (yr) | 1.6 | 1.8 |
| Serum protein (g/100 ml) | 7.2 | 4.0 |
| Albumin (g/100 ml) | 4.0 | 1.9 |
| Retinol-binding protein (mg/100 ml) | 2.4 | 1.5 |
| Vitamin A (µg/100 ml) | 22 | 10 |

*Source*: Smith *et al.* (1973).

**TABLE 4.19**  Effect of Feeding Protein to Malnourished Children

|  | Retinol-binding protein (mg/100 ml) | Vitamin A (µg/100 ml) |
|---|---|---|
| Before | 1.3 | 13 |
| After | 3.9 | 39 |

*Source*: Smith *et al.* (1973).

increased ability to synthesize the transport protein used for distributing the vitamin throughout the body.

Children suffering from malnutrition sometimes accumulate deposits of fat in the liver. This condition, called a **fatty liver**, may result from failure to synthesize specific proteins (apolipoproteins) used for the packaging and export of fat from the liver. These proteins are components of the VLDLs, as discussed in Chapter 6.

The issues of starvation and malnutrition are raised again in Chapter 8, where foods rich in energy but poor in protein are discussed. Malnutrition is discussed under Sodium, Potassium, and Water in Chapter 10, where the problem of infant diarrhea is outlined. This problem often occurs with weaning in underdeveloped countries, when pathogenic bacteria and parasites are inadvertently introduced into the infant's diet. The problem is compounded when basic supplies, such as firewood for boiling drinking water, are missing. Malnutrition and infection are also discussed under Vitamin A in Chapter 9.

## ALCOHOL

Alcohol accounts for about 5% of energy intake when averaged over the entire American population. Consumption of 5 to 10% of energy intake as alcohol is common practice, but consumption of 35 to 50% or more, on a chronic basis, is self-destructive. This practice is called **alcoholism**. Most of the ethanol consumed in the United States, about 50%, is in the form of beer; 40% as distilled spirits; and the remainder as wine. Alcohol may be completely used as a useful energy source, with intakes of less than 25% of the total energy intake. At higher levels, alcohol tends to be metabolized by pathways that do not produce ATP. Here, it is partially metabolized by pathways that generate mainly heat. More than 90% is metabolized, mainly in the liver; and 2 to 10% is eliminated unchanged via the lungs or urine.

The energy content of pure ethanol is 29.7 kJ/g (7.1 kcal/g). One can of American beer (12 oz, 354 ml) contains 13.4 g of alcohol, about 400 kJ of energy. The alcohol in three cans can provide 1200 kJ, nearly 10% of the daily energy requirement of the adult male. A typical serving of wine, about 150 ml, contains 16 g of alcohol; one jigger of whisky (45 ml), 18 g. These values are given greater meaning in Chapter 5. (Ethyl alcohol is another name for ethanol, $CH_3CH_2OH$.)

Alcohol is metabolized by the catalytic action of alcohol dehydrogenase, aldehyde dehydrogenase, and the ethanol-oxidizing system of the endoplasmic reticulum (ER). Alcohol dehydrogenase is a zinc metalloenzyme containing two subunits; each subunit has a molecular weight of 40,000 and contains two zinc atoms. One zinc atom, bound to the enzyme via residues of histidine and cysteine, is required for catalytic activity. The function of the second zinc atom, bound to residues of cysteine only, is structural.

Alcohol dehydrogenase, a cytosolic enzyme, catalyzes the oxidation of ethanol to acetaldehyde with the generation of NADH. The energy from this NADH is transferred to mitochondria, primarily by means of the malate–aspartate shuttle. When an alcoholic consumes over 50% of his or her energy in the form of alcohol, this shuttle becomes vastly important. Most of the acetaldehyde is oxidized in the mitochondria by mitochondrial aldehyde dehydrogenase, though a cytosolic version of the enzyme also exists. These reactions are shown in Figure 4.69.

An interesting sidelight, Asians are often deficient in aldehyde dehydrogenase. About 50% of Japanese and Chinese people lack the mitochondrial enzyme. About 40% of native South Americans also lack the enzyme. This results in a number of adverse reactions to drinking alcohol, including facial flushing from dilation of peripheral blood vessels and increased heart rate. Apparently, these reactions are provoked by an unusual buildup of plasma acetaldehyde.

## Family of Alcohol Dehydrogenases

Alcohol dehydrogenase contains two atoms of zinc. One atom participates in the catalytic mechanism, while the other serves a structural role. The enzyme actually is a dimer, and is composed of two subunits. Several types of alcohol dehydrogenase occur in the body. The enzymes have closely related amino acid sequences, and thus consist of a family of related enzymes.

Class I alcohol dehydrogenase catalyzes the metabolism of most dietary ethanol. This enzyme occurs in the liver. The class IV enzyme occurs in the mucosa of the stomach, and thus may begin catabolism of dietary ethanol even before the alcohol is absorbed by the gut. However, in comparing the weight of the human stomach mucosa (50 g), the weight of the human liver (1500 g), and the levels of enzyme activity in these two organs, it appears that the liver has 300 times the capacity for ethanol oxidation than does the stomach (Yin et al., 1996). Class III alcohol dehydrogenase occurs in all tissues and tends not to oxidize ethanol, but instead is used in vitamin A metabolism. Specifically, this enzyme catalyzes the conversion of retinol (an alcohol) to retinaldehyde. The class IV enzyme requires glutathione as a co-substrate, and catalyzes the oxidation of formaldehyde to formic acid. Please note that formaldehyde is not an alcohol, but an aldehyde. This activity serves the useful purpose of eliminating a naturally occurring toxic com-

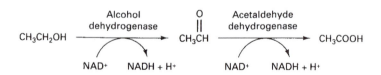

**FIGURE 4.69** Formation and oxidation of acetaldehyde.

pound (formaldehyde) from the body. The enzyme is also called **formaldehyde dehydrogenase** (Uotila and Koivusalo, 1996).

### Alcohol Catabolism in the Endoplasmic Reticulum

A second pathway of alcohol oxidation becomes important with consumption of large amounts of alcohol or with chronic alcoholism. This pathway, called the **ethanol-oxidizing system of the ER**, involves the heme enzyme cytochrome P450 and one other protein. Cytochrome P450 alone cannot catalyze the oxidation. As homogenization of the liver results in disruption of the ER to form tiny vesicles called microsomes, the system has also been called the **microsomal ethanol-oxidizing system**.

At moderate concentrations of alcohol, oxidation is mostly catalyzed by alcohol dehydrogenase. At high alcohol concentrations, this enzyme becomes saturated and cannot function at a greater rate. The ethanol-oxidizing system becomes increasingly important at these high concentrations, because it is not saturated by moderate concentrations of alcohol and is induced by chronic alcohol consumption. Electron microscopy shows that this induction involves an increase in the amount of ER present in the cell. Induction of the ethanol-oxidizing system in the regular drinker can result in more than a 50% increase in the rate of ethanol metabolism, which may account, in part, for the greater tolerance of alcohol by regular drinkers. Of greater importance is the adaption of the nervous system to alcohol, which allows the chronic alcoholic to remain sober despite blood alcohol levels (220 m$M$) that would ordinarily prove fatal.

The conversion of ethanol to acetic acid, by sequential action of the ethanol-oxidizing system of the ER and acetaldehyde dehydrogenase, is shown in Figure 4.70. The conversion of acetic acid to acetyl-CoA, as catalyzed by thiokinase, is shown in Figure 4.71. Thiokinase accepts short-chain fatty acids, as well as acetic acid, as its substrate. Acetyl-CoA is a substrate of the Krebs cycle.

### Human Subject Study with Alcohol Feeding

The experiment reported in Figure 4.72 illustrates production of acetaldehyde and acetate from ethanol. Human subjects received a dose of ethanol via a stomach tube (0.15 g of ethanol/kg of body weight). Blood samples indicated that plasma ethanol reached a peak concentration of 1.5 m$M$ after a half-hour. Acetate reached

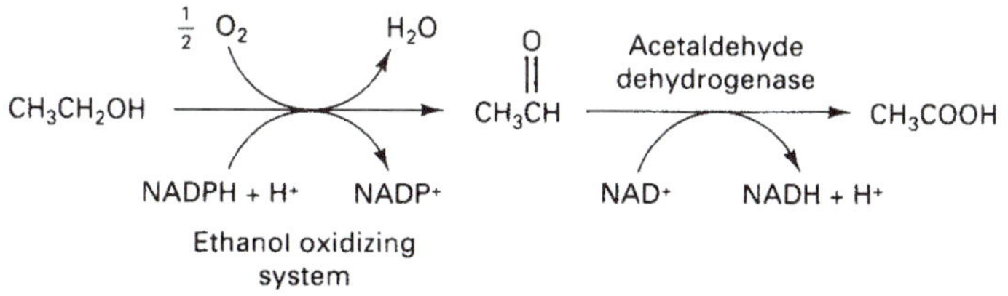

**FIGURE 4.70** Reaction catalyzed by the ethanol-oxidizing system.

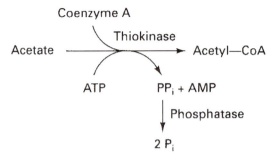

FIGURE 4.71  Short-chain fatty acid reactions catalyzed by thiokinase.

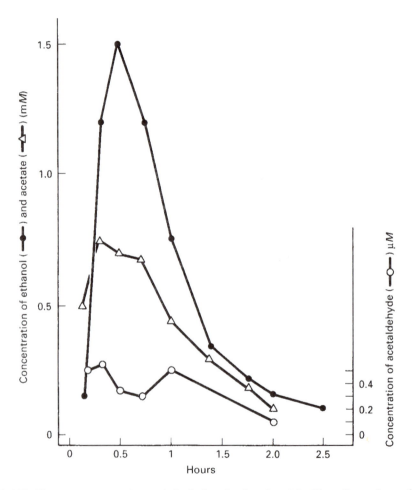

FIGURE 4.72  Plasma concentrations of alcohol and related metabolites after a dose of alcohol. Human subjects received a dose of ethanol via a stomach tube (0.15 g of ethanol/kg body weight). Plasma concentrations of ethanol (●), acetaldehyde (○), and acetate (△) were measured at the indicated times. Acetaldehyde concentrations were one-thousandth those of ethanol, about 1.0 μM. (Redrawn with permission from DiPadova *et al.*, 1987.)

somewhat lower levels. Acetaldehyde concentrations were one-thousandth those of ethanol, about 1.0 $\mu M$.

## Alcohol Toxicity

Evidence suggests that some of the toxic effects of alcohol may result from the generation of acetaldehyde, and that this acetaldehyde can condense with the amino groups of DNA and proteins (Singh and Kim, 1995). Clearly, these types of condensation products would be expected to be harmful to the cell. However, though the low concentrations of acetaldehyde which occur in the body may make it difficult for the researcher to pinpoint the nature of any consequent damage.

As is the case for other aldehydes, acetaldehyde can condense with amino groups to form a Schiff base. Schiff base formation occurs in a number of pathological conditions. For example, breakdown of polyunsaturated fatty acids may generate malondialdehyde, which reacts with proteins to produce damage. To review another example, the elevated plasma glucose levels in diabetes results in an increased condensation of glucose, via a Schiff base, with various proteins in the bloodstream. Aldehyde damage to proteins has been reviewed by Stadtman (1990).

EXERCISE

In the experiment reported in Figure 4.72, was the amount of ethanol consumed large or small? How would you define a large or small amount of dietary ethanol?

EXERCISE

Imagine a situation where the alcohol is first oxidized to acetaldehyde in the cells of the stomach, and then the acetaldehyde moves to the liver for oxidation to acetate. Do the data in Figure 4.72 support or refute this scenario?

A distinction should be made between the actions of alcohol dehydrogenase and the ethanol-oxidizing system in the production of usable energy. The former results in generation of NADH + H$^+$, which can be used to produce ATP in the mitochondria; however, the latter does not produce a product that can be used to generate ATP. This distinction may account for the fact that chronic alcoholics can sometimes consume more energy than required without gaining weight. In a controlled study, human subjects who doubled their normal energy intake by consuming ethanol did not gain weight, but other subjects who doubled their energy intake by consuming a large amount of chocolate gained 3 kg over the course of 2 weeks (Lieber, 1973).

A number of problems are associated with drinking alcohol. A blood alcohol concentration greater than 5.4 mM produces mild intoxication in nonalcoholics; 22 mM may cause neurological dysfunctions. In the United States, intoxication is legally defined as a blood alcohol level greater than 21.7 mM. Concentrations above 109 mM may be fatal, because of respiratory failure. The major problem related to alcohol consumption is death from automobile accidents. About 50,000 fatal auto accidents occur in the United States each year. More than half involve a driver who had recently been drinking alcohol. Disease problems of alcoholism include alcoholic hepatitis, cirrhosis of the liver, **fetal alcohol syndrome** (FAS), and pancreatitis. Alcohol drinking is also associated with an increased risk of cancer of the mouth and esophagus, and breast cancer in women. For example, the rate of breast cancer may be about 30% greater than normal in women who consume one drink per day (Thun *et al.*, 1997).

Alcoholic liver disease has three stages. In the first stage, fat accumulates in the liver, producing a fatty liver. Fatty liver and the associated enlargement of the liver are completely reversible by discontinuing alcohol consumption. Histological examination of an enlarged liver may reveal a ballooning effect, in which the hepatocytes increase in diameter two- to threefold. Cellular disorganization and mitochondrial deformities are other features of the livers of alcoholics. The fatty liver condition is illustrated by the experiment in Table 4.20. For 18 days, human subjects consumed a nonalcoholic diet from which about 25% of the energy was derived from protein, 25% from fat, and 50% from carbohydrate. The protein foods were ground beef, fish, eggs, and turkey. Then, for about 2 weeks the same diet was consumed, except that alcohol replaced the carbohydrates; that is, pure alcohol replaced sugar, bread, cereals, and fruits. Liver biopsies taken after each dietary period were used for measurement of liver TGs. The results demonstrated a striking increase in liver TGs when 50% of the energy had been supplied by ethanol.

The second stage of alcoholic liver disease, **alcoholic hepatitis**, is characterized by the death of a number of liver cells and inflammation of the affected areas. The damage in this stage sometimes proves fatal.

In the third stage, **cirrhosis** (fibrous structures and scar tissue) forms in the liver. In humans, development of cirrhosis requires 5 to 20 years of steady consumption of a high-alcohol diet. It is not clear whether the hepatitis stage is required for formation of cirrhotic tissue. About 75% of the cases of cirrhosis of the liver result from alcoholism. Alcoholism does not always cause cirrhosis. Only about one-quarter of heavy drinkers develop cirrhosis.

TABLE 4.20  Effect of an Alcohol Diet on the Fat Content of the Liver

| Subject | Triglyceride (mg TG/100 mg protein) | |
| --- | --- | --- |
| | Nonalcoholic diet | Alcohol diet |
| 1 | 3.2 | 48.3 |
| 2 | 2.5 | 13.8 |
| 3 | 3.1 | 10.4 |
| 4 | 6.7 | 54.4 |
| 5 | 1.5 | 10.7 |

*Source*: Lieber and Rubin (1968).

The fatty liver condition can be described as follows. With alcohol consumption, the liver's main source of energy (fatty acids) is replaced by alcohol. This results in accumulation of unused fatty acids, in the form of triglycerides, with the consequent deposition of these TGs as fat. This mechanism is different from the fatty liver provoked by kwashiorkor, where a fatty liver results from the failure to synthesize the apolipoproteins needed for exporting TGs from the liver. The fatty liver of kwashiorkor is not associated with cirrhosis, but that of alcoholism is associated with conversion of stellate cells to cells resembling those of connective tissue, i.e., fibroblasts. Stellate cells are fat-storing cells that occur in the interstitial space between the capillaries and hepatocytes, called the **space of Disse**.

**Fetal alcohol syndrome** (FAS) occurs in the fetuses of chronic alcoholic mothers who drink heavily during pregnancy. The frequency of FAS births ranges from 2.5 to 10% of alcoholic mothers. For the pregnant woman, heavy drinking is defined as occasional consumption of 5 drinks, or 45 drinks per month. Consumption of one drink per day has not yet been associated with FAS or with milder forms of the syndrome. FAS is characterized by growth retardation, characteristic facial abnormalities, and mental retardation. It is a major cause of mental retardation, and its effects do not correct themselves as the child grows older.

A minority of alcoholics develop nutrient deficiencies. In Western countries, alcoholics represent the largest population segment that can benefit from dietary intervention. Alcoholics are at risk for deficiencies in folate, thiamin, riboflavin, vitamin $B_6$, vitamin A, and magnesium, particularly when the intake of these substances is low. In some cases, absorption of the nutrient is impaired; in others, catabolism of the nutrient is increased. Thiamin deficiency is a firmly established consequence of alcoholism, as discussed in the Thiamin section.

The case of vitamin A is particularly interesting. Alcoholism induces an increase in the level of cytochrome P450 in the liver, which in turn catalyzes the conversion of vitamin A to inactive products. Drugs such as phenobarbital can also induce increased production of cytochrome P450. Thus, an alcoholic who chronically takes this drug may be at risk of developing a vitamin A deficiency.

Alcohol can stimulate the absorption of iron, probably because it increases gastric acid secretion. Gastric acid enhances the solubility of ferric iron, thereby enhancing its absorbability. Some alcoholics suffer from iron-deficiency anemia; others absorb too much iron, resulting in excess iron stores.

EXERCISE

Why might one expect the activity of the malate–aspartate shuttle to be changed by alcohol consumption? (See Sugano *et al.*, 1990.)

EXERCISE

The action of cytochrome P450 can result in conversion of components of cigarette smoke to cancer-producing chemicals, called **carcinogens**. Explain how alcoholism can increase one's risk of developing cancer. (See Lieber, 1988.)

*Alcohol and the Nervous System*

Drinking alcohol affects the nervous system in five different ways: (1) intoxication (loss of physical coordination); (2) memory loss (blackouts; amnesia); (3) tolerance (loss of ability to become intoxicated); (4) addiction (uncontrolled craving); and (5) withdrawal symptoms (seizures and tremors) (Diamond and Gordon, 1997). These five effects may all be due to the interaction of ethanol with membrane-bound proteins of the brain. Ethanol is unique among most metabolites, in that it is both water soluble and lipid soluble. Hence, it has the ability to pass directly from the bloodstream into various regions of the brain.

Interest has focused on possible interactions between alcohol and various ion channels, various G proteins, and the receptors for various neurotransmitters. An ion channel is like an ion transporter, except that it facilitates the temporary accumulation of an ion, such as $Ca^{2+}$, in the cytoplasm, for the purpose of transmitting a nervous impulse. In contrast, an ion transporter may serve to maintain the flow of an ion across a membrane, continuously over the course of several hours.

Glutamate (the amino acid) is a neurotransmitter. The receptor for this neurotransmitter is a membrane-bound protein that occurs in a complex with a calcium channel. The binding of glutamate to its receptor provokes the momentary opening of the Ca-channel, allowing extracellular calcium to enter the cell and to provoke the activation of various enzymes within the cell. Ethanol inhibits one type of glutamate receptor, called the **NMDA receptor** (Diamond and Gordon, 1997). Chronic and continual drinking results in the continual inhibition of the NMDA receptor in the brain. Ethanol's inhibition of the NMDA receptor is thought to be responsible for the amnesia that occurs in alcoholism.

Tyrosine kinase is an enzyme that catalyzes the attachment of phosphate groups to target proteins. One type of tyrosine kinase, called **Fyn kinase**, is involved in the intoxicating effects of alcohol. Fyn kinase catalyzes the attachment of phosphate to the NMDA receptor. As indicated earlier, alcohol makes certain nerves less active. The purpose of Fyn kinase is to overcome the inhibiting action of alcohol on the NMDA receptor. Studies with mice have revealed that alcohol causes them to become briefly intoxicated. Studies with mice that are genetically lacking in Fyn kinase have shown that these mice remain intoxicated for a longer period of time (Pennisi, 1997). Hence, changes in the NMDA receptor may be responsible for the ability of alcohol to provoke intoxication.

Attention has also focused on the influence of alcohol on other neurotransmitters, such as γ-aminobutyric acid, serotonin, and dopamine, and their receptors. The study of alcohol and the nervous system is a puzzle whose pieces are only beginning to be known.

## METABOLIC ZONING IN THE LIVER

Metabolic zoning has proven to be a crucial concept for researchers studying hepatic metabolism over the course of time, and under different metabolic states. Metabolic zoning involves the sharing of biochemical tasks by different regions within one organ. This form of compartmentation has been documented for glycolysis and gluconeogenesis (Pilkis *et al.*, 1988) and ammonia metabolism (Haussinger, 1986).

Water-soluble metabolites are absorbed by the gut and initially carried, via the portal vein, to the liver. The hepatocytes that surround the portal vein and its arterioles are the first cells of the body (beyond the enterocytes) that are able to metabolize or store the nutrients of the diet. These hepatocytes are called **periportal**. The portal vein accounts for 80% of the blood entering the liver, while the hepatic artery accounts for the remaining 20%. The hepatic artery brings blood from the general circulation into the liver. The blood acquired from the portal vein and hepatic artery continues to pass through the network of arterioles in the liver.

The hepatocytes that are the last to be exposed to the blood traveling through the liver are called **perivenous**. Finally, the perivenous blood leaves the liver via a vessel called the vena cava. Researchers who ask questions about metabolic zonation study the properties of the periportal and perivenous hepatocytes by the techniques of biochemistry, molecular biology, and histology.

Gluconeogenesis tends to occur in the periportal zone. PEPCK and fructose-1,6-bisphosphatase occur mainly in the periportal zone. With eating food, any lactic acid, pyruvate, and alanine acquired from the gut tends to be converted to glucose-6-P, and then stored as glycogen, within the periportal hepatocytes. The glucose acquired from the diet tends to be stored in muscle as muscle glycogen. With fasting, periportal glycogen is hydrolyzed and released into the bloodstream as glucose. The perivenous cells tend to be the glucose-oxidizing cells of the liver, and here glucose catabolism occurs to the point of pyruvate, with the release from the liver in the form of lactate. This lactate enters the general circulation, and may eventually reenter the liver for conversion back to glucose.

To review the curious behavior of the liver, and its handling of glucose: when the liver synthesizes glycogen, it does so from gluconeogenic precursors. But when the liver oxidizes glucose, it tends to do so only to the point of pyruvate.

## Metabolic Zonation of Nitrogen Metabolism

Amino acid catabolism is zoned as follows. Periportal cells convert ammonia to urea. Carbamyl phosphate synthase is periportal. Some of the ammonia escaping this form of packaging continues to pass through the liver, and then acquired by perivenous cells, where it is packaged as glutamine and then released as glutamine into the bloodstream. Glutamine synthase is perivenous (Jungermann and Kietzmann, 1996). The meaning of this material should become more apparent after reading the Protein chapter.

# THIOL ESTERS

The thiol ester linkage has been mentioned at various points in this text, most frequently in the material on energy metabolism. This bond occurs when a carboxylic acid is conjugated with a sulfhydryl group, such as the one on coenzyme A. The thiol ester linkage is an important concern to those studying the mechanisms of action of acetyl-CoA carboxylase, citrate synthase, propionyl-CoA carboxylase, and HMG-CoA synthase.

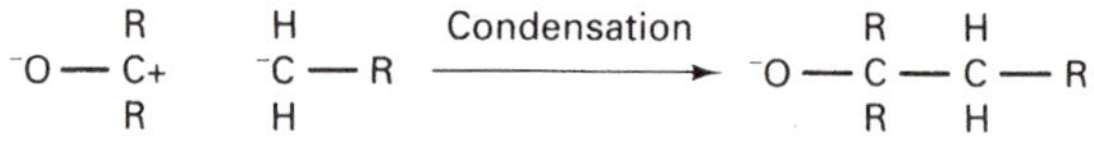

FIGURE 4.73 Condensation of positive and negative carbons to form a covalent bond. The carbanion does not normally occur in solution. It is formed at the active site of enzymes where it exists only momentarily.

Acetyl-CoA carboxylase and propionyl-CoA carboxylase catalyze the condensation of $CO_2$ with the 2-carbon of the thiol ester substrate. Citrate synthase catalyzes the condensation of OAA at the 2-carbon of the thiol ester substrate. These condensation reactions involve attack of an atom with a partial positive charge (the carbon of $CO_2$ or the keto group of OAA) on an atom with a negatively charged character (the 2-carbon of the substrate). Condensation of positively and negatively charged carbons to form a stable, covalent bond is shown in Figure 4.73.

The central role of thiol ester compounds in nutritional biochemistry is illustrated by considering some features of the mechanisms of the reactions catalyzed by citrate synthase and acetyl-CoA carboxylase. A brief background on the properties of the thiol ester bond is first presented. This material has been simplified and is intended only to clarify why the reactions employ thiol esters of carboxylic acids, rather than oxygen esters or the free, unesterified carboxylic acid.

Carboxylic acids resonate in the manner shown in Figure 4.74, allowing the negative charge to be shared by both oxygen atoms. Oxygen esters (Figure 4.75) tend to resonate in a similar manner. The oxygen atom of the ester linkage contains a pair of electrons not involved in covalent bond formation. These electrons can participate in the resonance shown, though the degree of resonance is much less than that of a free carboxylic acid. Thiol esters (Figure 4.76), however, do not tend to resonate, as indicated by the large X. Although the sulfur atom has an electronic structure similar to that of the oxygen atom, its atomic weight is twice that of oxygen. It is thought that sulfur's larger size prevents it from resonating with the keto group. This lack of resonance is useful, because it permits the keto group of the thiol ester to participate in electron shifts involving the R group of the molecule.

The reactions catalyzed by citrate synthase and acetyl-CoA carboxylase are thought to begin with dissociation of a proton from the 2-carbon of the acetyl moiety of acetyl-CoA (Figure 4.77). This produces a carbanion (a carbon atom carrying a negative charge) centered on the 2-carbon. This carbanion is stabilized by the resonance with the keto group also shown in Figure 4.77. Stabilization of the carbanion facilitates removal of the proton. Occurrence of the acetyl group as a free acid or oxygen ester would be expected to prevent resonance and thus deter

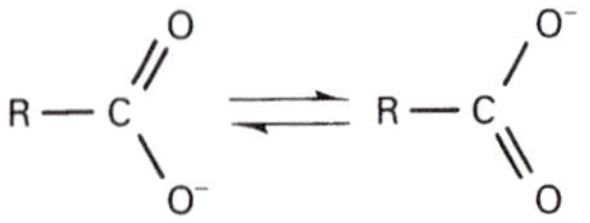

FIGURE 4.74 Resonance of a carboxylic acid.

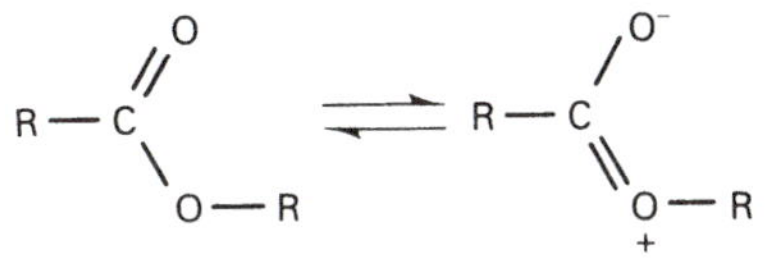

FIGURE 4.75  Resonance of an oxygen ester.

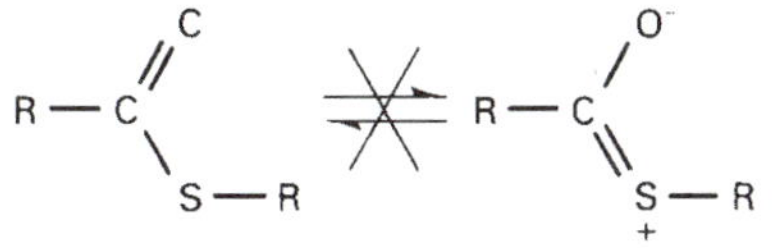

FIGURE 4.76  No resonance in a thiol ester.

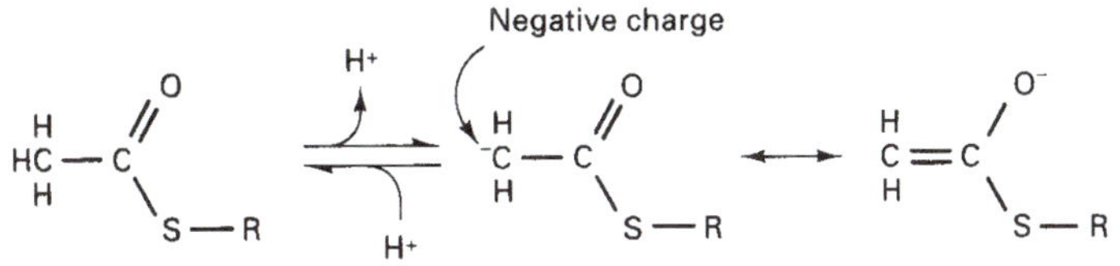

FIGURE 4.77  Development of a negative charge at the α-carbon of a thiol ester.

proton removal. Resonance involving the 2-carbon of a keto group is called **eno-lization**.

Stabilization of the 2-carbanion by enolization is thought to be a critical step in the reactions catalyzed by citrate synthase and acetyl-CoA carboxylase. With the former enzyme, the carbanion attacks an incoming molecule of OAA. As shown in Figure 4.78, condensation of acetyl-CoA with OAA forms citryl-CoA. Hydrolysis of citryl-CoA then produces citric acid and acetyl-CoA. In the case of acetyl-CoA carboxylase, the carbanion attacks an incoming molecule of $CO_2$, resulting in formation of malonyl-CoA (Figure 4.79).

## SUMMARY

The glucagon/insulin ratio was used to introduce the chapter and to provide a theme or reference point for the more complex matters to come. This ratio is of central importance to those interested in energy regulation during working, fasting, feeding, and diabetes. Simple diagrams were used to introduce the major types of energy metabolism, and these were followed by diagrams of biochemical pathways and some of the loci where regulation occurs.

Glucagon and insulin are secreted by endocrine cells of the pancreas into the bloodstream in response to fluctuations in plasma glucose. These hormones, in turn, act at various target organs to influence energy metabolism. Glucose metabolism is regulated at the points of glucose transport and at various points in the pathways of glycolysis and gluconeogenesis, including glycogen synthesis and breakdown. Triglyceride (TG) metabolism is regulated at the points of TG hydroly-

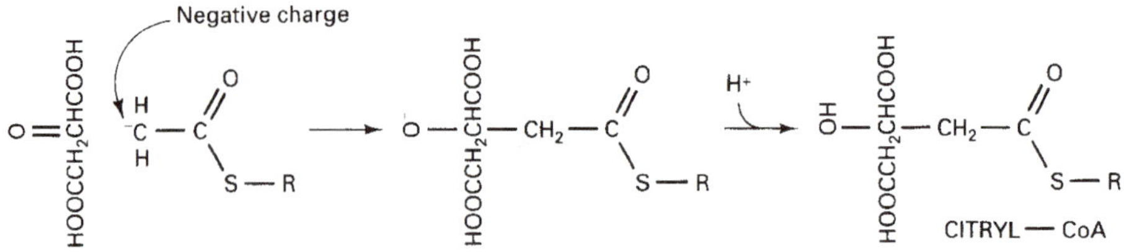

FIGURE 4.78 Condensation of a keto group carbon at the α-carbon of a thiol ester.

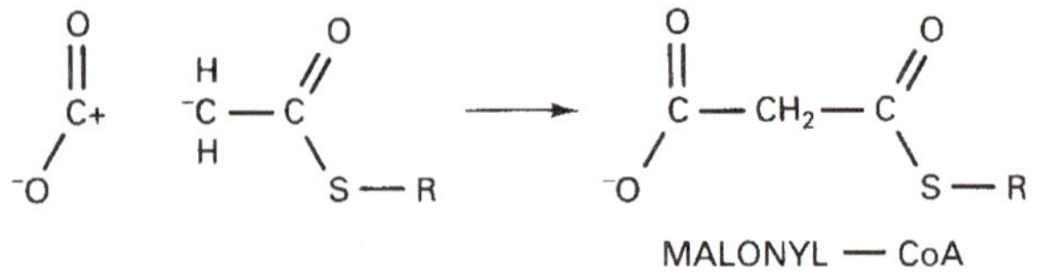

FIGURE 4.79 Condensation of carbon dioxide at the α-carbon of a thiol ester.

sis in the cell, fatty acid synthesis, and fatty acid transport into the mitochondrion, as well as at other points.

The various types of regulation described should be viewed within the following framework. Enzyme regulation can occur by the reversible, noncovalent binding of small molecules such as cyclic AMP. Regulation can also occur by the reversible, covalent binding of small molecules such as phosphate. A more involved form of regulation involves the reversible binding of a regulatory protein, as in the case of the cAMP-dependent mechanism. Some enzymes are simultaneously regulated by both noncovalent and covalent mechanisms, or even by two or more types of noncovalent effectors.

Metabolites that accumulate in the cell to high levels may bind to regulatory sites of enzymes that catalyze their synthesis, resulting in a slowdown of enzyme activity. This type of regulation results in a drop of metabolite levels back to a more normal concentration. This is called **feedback inhibition**. In some cases, the target enzyme is the enzyme that synthesized the metabolite. But if the metabolite is the end-product of a multistep pathway, the target enzyme may be that responsible for the first step of the pathway. In other cases, a metabolite that accumulates to an undesirable or deleterious level in the cell can bind to the regulatory site of an enzyme that catalyzes its removal. Where the binding provokes an increase in enzyme activity, the event is called **activation**.

The region of binding of a regulatory metabolite is, in some cases, within the catalytic site. Here, the bound metabolite can block the entry of the substrate, resulting in inhibition of enzyme activity. In other cases, the binding may occur at a part of the enzyme remote from the active site. Here, the bound regulatory metabolite influences the overall shape of the enzyme in a manner that changes the structure about the active site, that is, the orientation of amino acid residues. This change in shape, which can involve twisting or tilting of different regions of the polypeptide chain, is called a **conformational change**. The form of regulation involving conformational changes is called **allosteric regulation**. Regulation of protein kinase by cAMP, of phosphorylase by phosphorylation, and of a handful of other enzymes has been demonstrated to be allosteric.

The complete oxidation of glucose involves glycolysis, which is cytosolic, and the Krebs cycle, which is mitochondrial. The hepatocyte, for example, contains a few hundred mitochondria. Complete oxidation of TGs involves glycolysis, for the glycerol moiety, and the β-oxidation pathway (Chapter 5) and the Krebs cycle, for the fatty acid groups. In some cases, the oxidation of specific fuels is shared by different organs. In fermentative metabolism, glucose is catabolized to the point of lactate, then released into the bloodstream. Because mammals, as a group, are not fermentative organisms, the lactate is then oxidized in other tissues to carbon dioxide. In ketone body metabolism, ketone bodies are produced from fatty acids in the liver, released into the bloodstream, and then oxidized in muscle and brain to carbon dioxide. In the case of liver glycogen, this energy store can be burned in the liver, but mainly is released into the blood for use by other organs. In the case of muscle glycogen, this energy store cannot be released as free glucose but is used for glycolysis in muscle.

The sharing of the tasks of energy metabolism within the cell involves the mitochondria. The Krebs cycle and the β-oxidation pathway of fatty acids operate in these organelles. These pathways result in the reduction of the cofactors NAD and FAD. In other words, oxidation of energy fuels results in the transfer of their electrons to these cofactors. These cofactors occur in low concentrations in the cell and act in a catalytic manner. They do not store electrons, but transfer them to the respiratory chain. The respiratory chain consists of a series of proteins tightly bound to the membrane of the mitochondrion. The chain accepts electrons from reduced NAD and FAD and shuttles them to oxygen, resulting in the reduction of oxygen to water. The preceding process drives the transport of protons out of the organelle to form a pH gradient, where the interior of the mitochondrion is relatively alkaline. The reentry of protons across the mitochondrial membrane occurs at the point of a membrane-bound protein, ATP synthase, and drives the synthesis of ATP (see Chapter 5).

The thermodynamics of energy metabolism may be summarized. The breakdown of glucose, fatty acids, and other metabolites results in their conversion to products of lower energy. This means that the products either have more stable bonds (lower-energy bonds) or more entropy (more freedom of movement, as when reactants are split into smaller more numerous products). The energy released with metabolite breakdown may be released only as heat. In other cases, the energy released is captured by the concomitant formation of a high-energy bond, such as the phosphate-phosphate bond (phosphodiester bond) formed with the conversion of ADP + $P_i$ to ATP. Glycolysis and the Krebs cycle both contain steps where this type of high-energy bond is formed; however, most of the ATP produced in the body is synthesized by ATP synthase on the mitochondrial membrane and, hence, is coupled to the continued operation of the respiratory chain. This intimate relationship between the respiratory chain and ADP phosphorylation is called **oxidative phosphorylation**, "ox–phos" for short. Most of the ATP produced in mitochondria is transported out to the cytoplasm and elsewhere for the synthesis of proteins, nucleic acids, lipids, and other metabolites, as well as for driving the transport of ions across membranes.

## REFERENCES

Aguilar-Parada, E., Eisentraut, A. M., and Unger, R. H. (1969). Effects of starvation on plasma pancreatic glucagon in normal man. *Diabetes* **18**, 717–723.

Ahlborg, G., and Felig, P. (1982). Lactate and glucose exchange across the forearm, legs, and splanchnic bed during and after prolonged leg exercise. *J. Clin. Invest.* **69**, 45–54.

Ahlborg, G., Felig, P., Hagenfeldt, S., Hendler, R., and Wahren, J. (1974). Substrate turnover during prolonged exercise in man. *J. Clin. Invest.* **53**, 1080–1090.

Ajayi, O. A., and James, O. A. (1984). Effect of riboflavin supplement on riboflavin nutriture of a secondary school population in Nigeria. *Am. J. Clin. Nutr.* **39**, 787–791.

Bach, J.-F. (1994). Insulin-dependent diabetes mellitus as an autoimmune disease. *Endocrine Rev.* **15**, 516–542.

Bach, J.-F. (1997). Autoimmunity and type I diabetes. *Trends Endocrinol. Metab.* **8**, 71–74.

Bai, D. H., Pape, M. E., Lopez-Casillas, F., Luo, X. C., Dixon, J. E., and Kim, K.-H. (1986). Molecular cloning of cDNA for acetyl-coenzyme A carboxylase. *J. Biol. Chem.* **261**, 12395–12399.

Barford, D., Hu, S., and Johnson, L. N. (1991). Structural mechanism for glycogen phosphorylase control by phosphorylation and AMP. *J. Mol. Biol.* **218**, 233–260.

Bass, A., Brdiczka, D., Eyer, P., Hofer, S., and Pette, D. (1969). Metabolic differentiation of distinct muscle types at the level of enzymatic organization. *Eur. J. Biochem.* **10**, 198–206.

Beckman, J. S., and Koppenol, W. H. (1996). Nitric oxide, superoxide, and peroxynitrite: The good, the bad, and the ugly. *Am. J. Physiol.* **271**, C1424–C1437.

Bernlohr, D., Simpson, M., Hertzel, A., and Banaszak, L. J. (1997). Intracellular lipid-binding proteins and their genes. *Annu. Rev. Nutr.* **17**, 277–303.

Bergman, E. N. (1990). Energy contributions of volatile fatty adds from the gastrointestinal tract in various species. *Physiol. Rev.* **70**, 567–590.

Blom, P. C., Hostmark, A. T., Vaage, O., Kardel, K. R., and Maehlum, S. (1987). Effect of different post-exercise sugar diets on the rate of muscle glycogen synthesis. *Med. Sci. Sports Exercise* **19**, 491–496.

Bolli, G., DeFeo, P., Perriello, G., Cosmo, S. D., Compagnucci, P., Santeusanio, F., Brunetti, P., and Unger, R. H. (1984). Mechanisms of glucagon secretion during insulin-induced hypoglycemia in man. *J. Clin. Invest.* **73**, 917–922.

Cahill, G. F. (1970). Starvation in man. *N. Engl. J. Med.* **282**, 668–675.

Carruthers, C., Unson, C., Kim, H., and Sakmar, T. P. (1994). Synthesis and expression of a gene for the rat glucagon receptor. *J. Biol. Chem.* **269**, 29321–29328.

Casteno, J. G., Nieto, A., and Feliu, J. E. (1979). Inactivation of phosphofructokinase by glucagon. *J. Biol. Chem.* **254**, 5576–5579.

Castillo, L., Beaumier, L., Ajami, A. M., and Young, V. R. (1996). Whole body nitric oxide synthesis in healthy men determined from [$^{15}$N]arginine-to-[$^{15}$N]citrulline labeling. *Proc. Natl. Acad. Sci. U.S.A.* **93**, 11460–11465.

Cook, G. (1984). Differences in the sensitivity of carnitine palmitoyltransferase to inhibition by malonyl-CoA are due to differences in $K_i$ values. *J. Biol. Chem.* **259**, 12030–12033.

Cooper, D. M., Mons, N., and Karpen, J. (1995). Adenylyl cyclases and the interaction between calcium and cAMP signalling. *Nature* **374**, 421–424.

Cross, R. L. (1981). The mechanism and regulation of ATP synthesis by F1-ATPases. *Annu. Rev. Biochem.* **50**, 681–714.

Diamond, I., and Gordon, A. S. (1997). Cellular and molecular neuroscience of alcoholism. *Physiol. Rev.* **77**, 1–20.

DiPadova, C., Worner, T. M., Julkunen, R. J., and Lieber, C. S. (1987). Effects of fasting and chronic alcohol consumption on the first-pass metabolism of ethanol. *Gastroenterology* **92**, 1169–1173.

Dohm, G. L., Kasperek, G. J., and Barakat, H. A. (1985). Time course of changes in gluconeogenic enzyme activities during exercise and recovery. *Am. J. Physiol.* **249**, E6–E11.

Efrat, S., Tal, M., and Lodish, H. F. (1994). The pancreatic β-cell glucose sensor. *Trends Biol. Sci.* **19**, 535–538.

Ekdahl, E. N., and Ekman, P. (1984). The effect of fructose-2,6-bisphosphate and AMP on the activity of phosphorylated and unphosphorylated fructose-1,6-bisphosphatase from rat liver. *FEBS Lett.* **167**, 203–209.

Elmendorf, J. S., Damrau-Abney, A., Smith, T., David, T., and Turinsky, J. (1997). Phosphatidylinositol 3-kinase and dynamics of insulin resistance in denervated slow and fast muscles *in vivo. Am. J. Physiol.* **272**, E661–E670.

Felig, P., and Wahren, J. (1975). Fuel homeostasis in exercise. *N. Engl. J. Med.* **293**, 1078–1084.

Ferguson, S. J. (1986). The ups and downs of P/O ratios. *Trends Biochem. Sci.* **11**, 351–353.

Foley, J. E., Kashiwagi, A., Verso, M. A., Beaven, G., and Andrews, J. (1983). Improvement in *in vitro* insulin action after one month of insulin therapy in obese noninsulin-dependent diabetes. *J. Clin. Invest.* **72**, 1901–1909.

Forbes, G. B. (1987). "Human Body Composition," p. 171. Springer-Verlag, New York.

Fritz, I. B. (1959). Action of carnitine on long chain fatty acid oxidation by liver. *Am. J. Physiol.* **197**, 297–304.

Franz, M. J., Horton, E. S., Bantle, J. P., Beebe, C., Brunzell, J., Coulston, A., Henry, R. R., Hoogwerf, B., and Stacpoole, P. W. (1994). Nutrition principles for the management of diabetes and related complications. *Diabetes Care* **17**, 490–522.

Gerbitz, K., Gempel, K., and Brdiczka, D. (1996). Mitochondria and diabetes. *Diabetes* **45**, 113–126.

Gollnick, P. D., and Hodgson, D. R. (1986). The identification of fiber types in skeletal muscle: A continual dilemma. *Exercise Sports Sci. Rev.* **14**, 81–104.

Gopher, A., Vaisman, N., Mandel, H., and Lapidot, A. (1990). Determination of fructose metabolism pathways in normal and fructose-intolerant children. *Proc. Natl. Acad. Sci. U.S.A.* **87**, 5449–5453.

Guyton, A. C. (1971). "Textbook of Medical Physiology," 4th ed., p. 915. Saunders, Philadelphia.

Halliwell, B. (1988). Albumin — an important extracellular antioxidant? *Biochem. Pharmacol.* **37**, 569–57.

Harris, R. C., Edwards, R. H., Hultman, E., Nordesjo, L., Nylind, B., and Sahlin, K. (1976). The time course of phosphorylcreatine resynthesis during recovery of the quadriceps muscle in man. *Pfluegers Arch.* 367, 137-142.

Hatefi, Y. (1985). The mitochondrial electron transport and oxidative phosphorylation system. *Annu. Rev. Biochem.* **54**, 1015–1069.

Haussinger, D. (1986). Regulation of hepatic ammonia metabolism: The intercellular glutamine cycle. *In* "Advances in Enzyme Regulation" (G. Weber, ed.), Vol. 25, pp. 159–180. Pergamon, Oxford.

Hickner, R. C., Fisher, J. C., Ehsani, A., and Kohrt, W. (1997). Role of nitric oxide in skeletal muscle blood flow at rest and during exercise in humans. *Am. J. Physiol.* **273**, H405–H410.

Holloszy, J., and Kohrt, W. (1996). Regulation of carbohydrate and fat metabolism during and after exercise. *Annu. Rev. Nutr.* **16**, 121–138.

Holmes, F. L. (1993). "Hans Krebs: Architect of Intermediary Metabolism, 1933–1937," Vol. 2. Oxford Univ. Press, New York.

Jensen, M. D., Haymond, M. W., Gerich, J. E., Cryer, P. E., and Miles, J. M. (1987). Lipolysis during fasting. *J. Clin. Invest.* **79**, 207–213.

Jensen, M. D., Haymond, M. W., Rizza, R. A., Cryer, P. E., and Miles, J. M. (1989). Influence of body fat distribution on free fatty acid metabolism in obesity. *J. Clin. Invest.* **83**, 1168–1173.

Jungermann, K., and Kietzmann, T. (1996). Zonation of parenchymal and nonparenchymal metabolism in liver. *Annu. Rev. Nutr.* **16**, 179–203.

Kosug, K., Scofield, B. F., Chandramouli, V., Kumaran, K., Schumann, W. C., and Landau, B. B. (1986). Pathways of acetone's metabolism in the rat. *J. Biol. Chem.* **261**, 3952–3957.

Lawrence, J. C., and Roach, P. J. (1997). New insights into the role and mechanisms of glycogen synthase activation by insulin. *Diabetes* **46**, 541–547.

Lenaz, G. (1988). Role of mobility of redox components in the inner mitochondrial membrane. *J. Membr. Biol.* **104**, 193–209.

Lieber, C. S. (1973). Liver adaptation and injury in alcoholism. *N. Engl. J. Med.* **288**, 356–362.

Lieber, C. S. (1988). Biochemical and molecular basis of alcohol-induced injury to liver and other tissues. *N. Engl. J. Med.* **319**, 1639–1650.

Lieber, C. S., and Rubin, E. (1968). Alcoholic fatty liver in man on a high-protein and low-fat diet. *Am. J. Med.* **44**, 200–206.

Louveau, I., Chaudhuri, S., and Etherton, T. D. (1991). An improved method for isolating RNA from porcine adipose tissue. *Anal. Biochem.* **196**, 308–310.

Marchesini, C., Melli, A., Checchia, G. A., Mattioli, L., Capelli, M., Cassarini, S., Zoli, M., and Pisi, E. (1985). Pancreatic β-cell function in cirrhotic patients with and without overt diabetes. *Metabolism* **34**, 695–701.

Mazurek, S., Michel, A., and Eigenbrodt, E. (1997). Effect of extracellular AMP on cell proliferation and metabolism of breast cancer cell lines with high and low glycolytic rates. *J. Biol. Chem.* **272**, 4941–4952.

McGarry, J. D., and Foster, D. W. (1980). Regulation of hepatic fatty acid oxidation and ketone body production. *Annu. Rev. Biochem.* **49**, 395–420.

Morris, B. J. (1995). Stimulation of immediate early gene expression in striatal neurons by nitric oxide. *J. Biol. Chem.* **270**, 24740–24744.

Mothe, I., and Obberghen, E. V. (1996). Phosphorylation of insulin receptor substrate-1 on multiple serine residues 612, 632, 662, and 731 modulates insulin action. *J. Biol. Chem.* **271**, 11222–11227.

Nilsson, L., and Hultman, E. (1974). Liver and muscle glycogen in man after glucose and fructose infusion. *Scand. J. Clin. Lab. Invest.* **33**, 5–10.

Owen, O. E., Morgan, A. P., Kemp, H. G., Sullivan, J. M., Herrera, M. G., and Cahill, G. F. (1967). Brain metabolism during fasting. *J. Clin. Invest.* **46**, 1589–1595.

Owen, O. E., Felig, P., Morgan, A. P., Wahren, J., and Cahill, G. F. (1969). Liver and kidney metabolism during prolonged starvation. *J. Clin. Invest.* **48**, 574–583.

Parrilla, R. (1978). Flux of metabolic fuels during starvation in the rat. *Pfluegers Arch.* **374**, 3–7.

Patti, M.-E., and Kahn, C. R. (1996). Lessons from transgenic and knockout animals about noninsulin-dependent diabetes mellitus. *Trends Endocrinol. Metab.* **7**, 311–319.

Pennisi, E. (1997). Enzyme linked to alcohol sensitivity in mice. *Science* **278**, 573.

Philippe, J. (1991). Structure and pancreatic expression of the insulin and glucagon genes. *Endocrine Rev.* **12**, 252–271.

Piehl, K. (1974). Time course for refilling of glycogen stores in human muscle fibres following exercise-induced glycogen depletion. *Acta Physiol. Scand.* **90**, 297–302.

Pilkis, S. J., El-Maghrabi, M. R., and Claus, T. H. (1988). Hormonal regulation of hepatic gluconeogenesis and glycolysis. *Annu. Rev. Biochem.* **57**, 755–783.

Roberts, C. K., Barnard, R., Scheck, S., and Balon, T. W. (1997). Exercise-stimulated glucose transport in skeletal muscle is nitric oxide dependent. *Am. J. Physiol.* **273**, E220–E225.

Rudenski, A., Hosker, J. P., Burnett, M. A., Matthews, D. R., and Turner, R. C. (1988). The beta cell glucose stimulus–response curve in normal humans assessed by insulin and C-peptide secretion rates. *Metabolism* **37**, 526–534.

Salmons, S., Jarvis, J., Mayne, C., Chi, M., Manchester, J., McDougal, D., and Lowry, O. (1996). Changes in ATP, phosphocreatine, and 16 metabolites in muscle stimulated for up to 96 hours. *Am. J. Physiol.* **271**, C1167–C1171.

Saura, M., Perez-Sala, D., Canada, F. J., and Lamas, S. (1996). Role of tetrabiopterin availability in the regulation of nitric oxide synthase expression in human mesangial cells. *J. Biol. Chem.* **271**, 14290–14295.

Schuette, S. A., and Knowles, J. B. (1988). Intestinal absorption of $Ca(H_2PO_4)$ and Ca citrate compared by two methods. *Am. Clin. Nutr.* **47**, 884–888.

Scrimshaw, N. S., Perera, W. D., and Young, V. R. (1976). Protein requirements of man: Obligatory urinary and fecal nitrogen losses in elderly woman. *J. Nutr.* **106**, 665–670.

Seaton, T., Welle, S. L., Warenko, M. K., and Campbell, R. G. (1986). Thermic effect of medium chain and long chain triglycerides in man. *Am. J. Clin. Nutr.* **44**, 630–634.

Shacter, E., Chock, P. B., Rhee, S. G., and Stadtman, E. (1986). Cyclic cascades and metabolic regulation. *In* "The Enzymes" (P. D. Boyer and E. G. Krebs, eds.), 3rd ed., Vol. 17, Part A, pp. 23–25. Academic Press, San Diego.

Smit, G. P., Ververs, M. T., Belderok, B., Van Rijn, M., Berger, R., and Fernandes, J. (1988). Complex carbohydrates in the management of patients with glycogenesis caused by glucose-6-phosphatase deficiency. *Am. J. Clin. Nutr.* **48**, 95–97.

Smith, F. R., Goodman, D. S., Zaklama, M. S., Gabr, M. K., Maraghy, S. E., and Patwardhan, V. N. (1973). Serum vitamin A, retinol-binding protein, and prealbumin concentrations in protein-calorie malnutrition. *Am. J. Clin. Nutr.* **26**, 973–981.

Smith, M. (1987). Position of the American Dietetic Association: Nutrition for physical fitness and athletic performance for adults. *J. Am. Diet. Assoc.* **87**, 933–939.

Sowadaski, J. M., Xuong, N., Anderson, D., and Taylor, S. S. (1984). Crystallization studies of cAMP-dependent protein kinase. *J. Mol. Biol.* **182**, 617–620.

Spector, A. A. (1975). Fatty acid binding to plasma albumin. *J. Lipid Res.* **16**, 165–179.

Srikanta, S., Ganda, O., Eisenbarth, G. S., and Soeldner, J. S. (1983). Islet-cell antibodies and beta-cell function in monozygotic triplets and twins initially discordant for type I diabetes mellitus. *N. Engl. J. Med.* **308**, 322–325.

Srivastava, D. K., and Bernhard, S. A. (1986). Metabolite transfer via enzyme–enzyme complexes. *Science* **234**, 1081–1086.

Stadtman, E. R. (1990). Covalent modification reactions are marking steps in protein turnover. *Biochemistry* **29**, 6323–6331.

Steiner, D. F., and Rubenstein, A. H. (1997). Proinsulin C-peptide — biological activity? *Science* **277**, 531–532.

Sugano, T., Handler, J. A., Yoshihara, H., Kizaki, Z., and Thurman, R. G. (1990). Acute and chronic ethanol treatment *in vivo* increases malate–aspartate shuttle capacity in perfused rat liver. *J. Biol. Chem.* **265**, 21549–21553.

Taha, C., Mitsumoto, Y., Liu, Z., Skolnik, E., and Klip, A. (1995). The insulin-dependent biosynthesis of GLUT1 and GLUT3 glucose transporters in L6 muscle cells is mediated by distinct pathways. *J. Biol. Chem.* **270**, 24678–24681.

Takakura, H., Yamamoto, T., and Sherman, F. (1997). Sequence requirement for trimethylation of yeast cytochrome *c*. *Biochemistry* **36**, 2642–2648.

Teff, K. L., Mattes, R. D., and Engelman, K. (1991). Cephalic phase insulin release in normal weight males. *Am. J. Physiol.* **261**, E430–E436.

Thampy, K. G., and Wakil, S. J. (1988). Regulation of acetyl-coenzyme A carboxylase. *J. Biol. Chem.* **263**, 6454–6458.

Thorne, A., and Wahren, J. (1989). Diet-induced thermogenesis in well-trained subjects. *Clin. Physiol.* **9**, 295–305.

Thun, M. J., Peto, R., Lopez, A., Monaco, J., Henley, J., Heath, C., and Doll, R. (1997). Alcohol consumption and mortality among middle-aged and elderly U.S. adults. *N. Engl. J. Med.* **337**, 1705–1714.

Tisch, R., and McDevitt, H. (1996). Insulin-dependent diabetes mellitus. *Cell* **85**, 291–297.

Townsend, C. M., and Thompson, J. C. (1985). Surgical management of tumors that produce gastrointestinal hormones. *Annu. Rev. Med.* **36**, 111–124.

Uyeda, K., Furuya, E., and Luby, L. J. (1981). The effect of natural and synthetic D-fructose-2,6-bisphosphate on the regulatory kinetic properties of liver and muscle phosphofructokinase. *J. Biol. Chem.* **256**, 8394–8399.

Wallberg-Hendriksson, H., and Holloszy, J. O. (1985). Activation of glucose transport in diabetic muscle: Responses to contraction and insulin. *Am. J. Physiol.* **249**, C233–C237.

Weber, J., and Senior, A. E. (1997). Catalytic mechanism of $F_1$-ATPase. *Biochim. Biophys. Acta* **1319**, 19–58.

Westergaard, H., and Dietschy, J. M. (1976). The mechanism whereby bile acid micelles increase the rate of fatty acid and cholesterol uptake into the intestinal mucosal cell. *J. Clin. Invest.* **58**, 97–108.

Yin, S.-J., Han, C.-L., Liao, C.-S., and Wu, C.-W. (1996). Expression, activities, and kinetic mechanisms of human stomach alcohol dehydrogenase. *In* "Enzymology and Molecular Biology of Carbonyl Metabolism," Vol. 6 (H. Weiner, R. Lindahl, D. Crabb, and T. Flynn, eds.), pp. 347–355. Plenum, New York.

Yudkoff, M., Daikhin, Y., Nissim, I., Grunstein, R., and Nissim, I. (1997). Effects of ketone bodies on astrocyte amino acid metabolism. *J. Neurochem.* **69**, 682–692.

## BIBLIOGRAPHY

*Brain Energy*

Bray, G. A., Teague, J., and Lee, C. K. (1987). Brain uptake of ketones in rats with differing susceptibility to dietary obesity. *Metabolism* **36**, 27–30.

Ferre, P., Satabin, P., Decaus, J.-F., Escriva, F., and Girard, J. (1983). Development and regulation of ketogenesis in hepatocytes isolated from newborn rats. *Biochem. J.* **214**, 937–942.

Hawkins, R. A., Mans, A. M., Davis, D. W., Vina, J. R., and Hibbard, L. S. (1985). Cerebral glucose use measured with [$^{14}$C]glucose labeled in the 1, 2, or 6 position. *Am. J. Physiol.* **248**, C170–C176.

Levitsky, L. L., Fisher, D. E., Paton, J. B., and Delannoy, C. W. (1977). Fasting plasma levels of glucose, acetoacetate, D-β-hydroxybutyrate, glycerol, and lactate in the baboon infant. *Pediatr. Res.* **11**, 298–302.

Mehta, S., Kalsi, H. K., Nain, C. K., and Menkes, J. H. (1977). Energy metabolism brain in human protein-calorie malnutrition. *Pediatr. Res.* **11**, 290–293.

Mellerup, E. T., and Rafaelsen, O. J. (1969). Brain glycogen after intracisternal insulin injection. *J. Neurochem.* **16**, 777–781.

Siegel, G. J., Agranoff, B. W., Albers, R. W., and Molinoff, P. B. (1989). "Basic Neurochemistry," 4th ed., p. 570. Raven Press, New York.

*Regulation of Glycolysis and Gluconeogenesis*

Acheson, K. J., Flatt, J. P., and Jequier, E. (1982). Glycogen synthesis versus lipogenesis after a 500 gram carbohydrate meal in man. *Metabolism* **31**, 1234–1240.

Acheson, K. J., Schutz, Y., Bessard, T., Ravussin, E., Jequier, E., and Flatt, J. P. (1984). Nutritional influences on lipogenesis and thermogenesis after a carbohydrate meal. *Am. J. Physiol.* **246**, E62–E70.

Argaud, D., Kirby, T. L., Newgard, C., and Lange, A. (1997). Stimulation of glucose-6-phosphatase gene expression by glucose and fructose-2,6-bisphosphate. *J. Biol. Chem.* **272**, 12854–12861.

Bergman, E. N. (1990). Energy contributions of volatile fatty acids from the gastrointestinal tract in various species. *Physiol. Rev.* **70**, 567–590.

Blom, P. C., Hostmark, A. T., Vaage, O., Kardel, K. R., and Maehlum, S. (1987). Effect of different post-exercise sugar diets on the rate of muscle glycogen synthesis. *Med. Sci. Sports Exercise* **19**, 491–496.

Blom, P. C., Vollestad, N. K., and Costill, D. L. (1986). Factors affecting changes in muscle glycogen concentration during and after prolonged exercise. *Acta Physiol., Suppl.* **128**(556), 67–74.

Casteno, J. G., Nieto, A., and Feliu, J. E. (1979). Inactivation of phosphofructokinase by glucagon. *J. Biol. Chem.* **254**, 5576–5579.

Dohm, G. L., Kasperek, G. J., and Barakat, H. A. (1985). Time course of changes in gluconeogenic enzyme activities during exercise and recovery. *Am. J. Physiol.* **248**, E6–E11.

Ekdahl, E. N., and Ekman, P. (1984). The effect of fructose-2,6-bisphosphate and AMP on the activity of phosphorylated and unphosphorylated fructose-1,6-bisphosphatase from rat liver. *FEBS Lett.* **167**, 203–209.

Felig, P., and Wahren, J. (1975). Fuel homeostasis in exercise. *N. Engl. J. Med.* **293**, 1078–1084.

Ferrannini, E., Bjorkman, O., Reichard, G. A., Pilo, A., Olsson, M., Wahren, J., and De-Fronzo, R. A. (1985). The disposal of an oral glucose load in healthy subjects. *Diabetes* **34**, 580–588.

Hargreaves, M., Costill, D. L., Fink, W. J., King, D. S., and Fielding, R. A. (1987). Effect of pre-exercise carbohydrate feeding on endurance cycling performance. *Med. Sci. Sports Exercise* **19**, 33–36.

Haussinger, D. (1986). Regulation of hepatic ammonia metabolism: The intercellular glutamine cycle. *In* "Advances in Enzyme Regulation" (G. Weber, ed.), Vol. 25, pp. 159–180. Pergamon, Oxford.

Hers, H. G., and Hue, L. (1983). Gluconeogenesis and related aspects of glycolysis. *Annu. Rev. Biochem.* **52**, 617–653.

Hultman, E., and Bergstrom, J. (1967). A study of the glycogen metabolism during exercise in man. *Scand. J. Clin. Lab. Invest.* **19**, 218–228.

Katz, J., and McGarry, J. D. (1984). The glucose paradox. *J. Clin. Invest.* **74**, 1901–1909.

Louveau, I., Chaudhuri, S., and Etherton, T. D. (1991). An improved method for isolating RNA from porcine adipose tissue. *Anal. Biochem.* **196**, 308–310.

Magnusson, I., Chandramouli, V., Schumann, W. C., Kumaran, K., Wahren, J., and Landau, B. R. (1987). Quantitation of the pathways of hepatic glycogen formation on ingesting a glucose load. *J. Clin. Invest.* **80**, 1748–1754.

Mair, T., and Muller, S. C. (1996). Traveling NADH and proton waves during oscillatory glycolysis *in vitro*. *J. Biol. Chem.* **271**, 627–630.

Mitchell, J. B., Costill, D. L., Houmard, J. A., Fink, W. J., Pascoe, D. D., and Pearson, D. R. (1989). Influence of carbohydrate dosage on exercise performance and glycogen metabolism. *J. Appl. Physiol.* **67**, 1843–1849.

Neufer, P. D., Costill, D. L., Flynn, M. G., Kirwan, J. P., Mitchell, J. B., and Houmard, J. (1987). Improvements in exercise performance: Effects of carbohydrate feedings and diet. *J. Appl. Physiol.* **62**, 983–988.

Pilkis, S. J., El-Maghrabi, M. R., and Claus, T. H. (1988). Hormonal regulation of hepatic gluconeogenesis and glycolysis. *Annu. Rev. Biochem.* **57**, 755–783.

Richter, E. A., Ploug, T., and Galbo, H. (1985). Increased muscle glucose uptake after exercise. *Diabetes* **34**, 1041–1048.

Scheele, G., Adler, G., and Kern, H. (1987). Exocytosis occurs at the lateral phase membrane of the pancreatic acinar cell during supramaximal secretatogogue stimulation. *Gastroenterology* **92**, 345–353.

Schacter, E., Chock, P. B., Rhee, S. G., and Stadtman, E. (1986). Cyclic cascades and metabolic regulation. *In* "The Enzymes" (P. D. Boyer and E. G. Krebs, eds.), 3rd ed., Vol. 17, Part A, pp. 23–25. Academic Press, San Diego.

Smith, M. (1987). Position of the American Dietetic Association: Nutrition for physical fitness and athletic performance for adults. *J. Am. Diet. Assoc.* **87**, 933–939.

Thorne, A., and Wahren, J. (1989). Diet-induced thermogenesis in well-trained subjects. *Clin. Physiol.* **9**, 295–305.

Uyeda, K., Furuya, E., and Luby, L. J. (1981). The effect of natural and synthetic D-fructose-2,6-bisphosphate on the regulatory kinetic properties of liver and muscle phosphofructokinase. *J. Biol. Chem.* **256**, 8394–8399.

Winder, W. W., Yang, H. T., and Arogyasami, J. (1988). Liver fructose 2,6-bisphosphate in rats running different treadmill speeds. *Am. J. Physiol.* **255**, R38–R41.

### Diabetes and Insulin Signaling

Accili, D. (1997). Insulin receptor knock-out mice. *Trends Endocrinol. Metab.* **8**, 101–104.

Bach, J.-F. (1994). Insulin-dependent diabetes mellitus as an autoimmune disease. *Endocrine Rev.* **15**, 516–542.

Bach, J.-F. (1997). Autoimmunity and type I diabetes. *Trends Endocrinol. Metab.* **8**, 71–74.

Baly, D. L., and Horuk, R. (1987). Dissociation of insulin-stimulated glucose transport from the translocation of glucose carriers in rat adipose cells. *J. Biol. Chem.* **262**, 21–24.

Clark, C. M., and Lee, D. A. (1995). Prevention and treatment of the complications of diabetes mellitus. *N. Engl. J. Med.* **332**, 1210–1217.

Clausen, J. O., Hansen, T., Bjorbaek, C., Echwald, S. M., Urhammer, S. A., Rasmussen, S., Andersen, C., Hansen, L., Almind, K., Winther, K., Haraldsdottir, J., Borch-Johnsen, K., and Pedersen, O. (1995). Insulin resistance: Interactions between obesity and a common variant of insulin receptor substrate-1. *Lancet* **346**, 397–402.

Consoli, A., Nurjhan, N., Capani, F., and Gerich, J. (1989). Predominant role of gluconeogenesis in increasing hepatic glucose production in NIDDM. *Diabetes* **38**, 550–557.

Daniels, M. C., Claraldi, T. P., Nikoulina, S., Henry, R. R., and McClain, D. A. (1996). Glutamine:fructose-6-phosphate amidotransferase activity in cultured human skeletal muscle cells. *J. Clin. Invest.* **9**, 1235–1241.

DeFronzo, R. A., Sherwin, R. S., and Kraemer, N. (1987). Effect of physical training on insulin action in obesity. *Diabetes* **36**, 1379–1385.

Eck, M. J., Dhe-Paganon, S., Trub, T., Nolte, R. T., and Shoelson, S. E. (1996). Structure of the IRS-1 PTB domain bound to the juxtamembrane region of the insulin receptor. *Cell* **85**, 695–705.

Groop, L. C., Bonadonna, R. C., DelPrato, S., Ratheiser, K., Zyck, K., Ferrannini, E., and DeFronzo, R. A. (1989). Glucose and free fatty acid metabolism in non-insulin-dependent diabetes mellitus. *J. Clin. Invest.* **84**, 205–213.

Hariharan, N., Farrelly, D., Hagan, D., Hillyer, D., Arbeeny, C., Sabrah, T., Treloar, A., Brown, K., Kalinowski, S., and Mookhtiar, K. (1997). Expression of human hepatic glucokinase in transgenic mice liver results in decreased glucose levels and reduced body weight. *Diabetes* **46**, 11–16.

Kahn, C. R. (1994). Insulin action, diabetogenes, and the cause of type II diabetes. *Diabetes* **43**, 1066–1084.

Karnieli, E., Zarnowski, M. J., Hissin, P. J., Simpson, L. A., Salans, L. B., and Cushman, S. W. (1981). Insulin-stimulated translocation of glucose transport systems in isolated rat adipose cell. *J. Biol. Chem.* **256**, 4772–4777.

Kolb-Bachofen, V., Epstein, S., Kiesel, V., and Kolb, H. (1988). Low-dose streptozotocin-induced diabetes in mice. *Diabetes* **37**, 21–27.

Kuglin, B., Gries, F. A., and Kolb, H. (1988). Evidence of IgG autoantibodies against human proinsulin in patients with IDDM before insulin treatment. *Diabetes* **37**, 130–132.

Lernmark, A., Freedman, Z. R., Hofmann, C., Rubenstein, A. H., Steiner, D. F., Jackson, R. L., Winter, R. J., and Traisman, H. S. (1978). Islet-cell-surface antibodies in juvenile diabetes. *N. Engl. J. Med.* **299**, 375–380.

Madsen, O. D. (1987). Proinsulin-specific monoclonal antibodies. *Diabetes* **36**, 1203–1211.

Marshall, S., Heidenreich, K. A., and Horikoshi, H. (1985). Stoichiometric translocation of adipocyte insulin receptor from the cell surface to the cell interior. *J. Biol. Chem.* **260**, 4128–4135.

McEvoy, R. C., Witt, M. E., Ginsberg-Fellner, F., and Rubinstein, P. (1986). Anti-insulin antibodies in children with type I diabetes mellitus. *Diabetes* **35**, 634–641.

Mondon, C. E., Dolkas, C. B., and Reaven, G. M. (1980). Site of enhanced insulin sensitivity in exercised-trained rats at rest. *Am. J. Physiol.* **239**, E169–E177.

Nelson, R. G., Bennett, P. H., Beck, G. J., Tan, M., Knowler, W., Mitch, W., Hirschman, G. H., and Myers, B. D. (1996). Development and progression of renal disease in Pima Indians with non-insulin-dependent diabetes mellitus. *N. Engl. J. Med.* **335**, 1636–1642.

O'Brien, R., and Timmins, K. (1994). The role of oxidation and glycation in the pathogenesis of diabetic atherosclerosis. *Trends Endocrinol. Metab.* **5**, 329–334.

Osawa, H., Sutherland, C., Robey, R., Printz, R., and Granner, D. K. (1996). Analysis of the signaling pathway involved in the regulation of hexokinase II gene transcription by insulin. *J. Biol. Chem.* **271**, 16690–16694.

Osei, K. (1990). Predicting type II diabetes in persons at risk. *Annu. Intern. Med.* **113**, 905–907.

Pawson, T. (1995). Getting down to specifics. *Nature* **373**, 477–478.

Sharp, G. W. (1996). Mechanisms of inhibition of insulin release. *Am. J. Physiol.* **271**, C1781–C1799.

Sun, S. J., Wang, L.-M., Zhang, Y., Yenush, L., Myers, M., Glasheen, E., Lane, W. S., Pierce, J. H., and White, M. F. (1995). Role of IRS-2 in insulin and cytokine signalling. *Nature* **377**, 173–177.

Waters, S., Chen, D., Kao, A., Okada, S., Holt, K., and Pessin, J.E. (1996). Insulin and epidermal growth factor receptors regulate distinct pools of Grb2-SOS in the control of ras activation. *J. Biol. Chem.* **271**, 18224–18230.

Whitesell, R. R., and Abumrad, N. A. (1985). Increased affinity predominates in insulin stimulation of glucose transport in the adipocyte. *J. Biol. Chem.* **260**, 2894–2899.

Young, D. A., Uhl, J. J., Cartee, G. D., and Holloszy, J. O. (1986). Activation of transport in muscle by prolonged exposure to insulin. *J. Biol. Chem.* **261**, 16049–16053.

Zinker, B. A., Allison, R. G., Lacy, D. B., and Wasserman, D. H. (1997). Interaction of exercise, insulin, and hypoglycemia studied using euglycemic and hypoglycemic insulin clamps. *Am. J. Physiol.* **272**, E530–E542.

## Gluconeogenesis from Amino Acids

Butterfield, G. (1987). Whole-body protein utilization in humans. *Med. Sci. Sports Exercise* **19**, S157–S165.

Dohm, G. L. (1986). Protein as fuel for endurance exercise. *Exercise Sports Sci. Rev.* **14**, 143–173.

Felig, P., Marliss, E., Pozetsky, T., and Cahill, G. F. (1970). Amino acid metabolism in the regulation of gluconeogenesis in man. *Am. J. Clin. Nutr.* **23**, 986–992.

Felig, P., Pozefsky, T., Marliss, E., and Cahill, G. F. (1970). Alanine: Key role in gluconeogenesis. *Science* **167**, 1003–1004.

Harper, A. E., Miller, R. H., and Block, K. P. (1984). Branched-chain amino acid metabolism. *Annu. Rev. Nutr.* **4**, 409–454.

Kasperek, G. J., Dohm, G. L., and Snider, R. D. (1985). Activation of branched-chain keto acid dehydrogenase by exercise. *Am. J. Physiol.* **248**, R166–R171.

Souba, W. W. (1987). Interorgan ammonia metabolism in health and disease: A surgeon's view. *J. Parenteral Enteral Nutr.* **11**, 569–579.

Wasserman, D. H., Williams, P. E., Lacy, D. B., Green, D. R., and Cherrington, A. D. (1988). Importance of intrahepatic mechanisms to gluconeogenesis from alanine during exercise and recovery. *Am. J. Physiol.* **254**, E518–E525.

Wolfe, R. R. (1987). Does exercise stimulate protein breakdown in humans? *Med. Sci. Sports Exercise* **19**, S172–S178.

*Arteriovenous Differences*

Ahlborg, G., and Felig, P. (1982). Lactate and glucose exchange across the forearm, legs, and splanchnic bed during and after prolonged leg exercise. *J. Clin. Invest.* **69**, 45–54.

Ahlborg, G., Felig, P., Hagenfeldt, L., Hendler, R., and Wahren, J. (1974). Substrate turnover during prolonged exercise in man. *J. Clin. Invest.* **53**, 1080–1090.

Hagenfeldt, L., and Wahren, J. (1971). Human forearm muscle metabolism during exercise. *Scand. J. Clin. Lab. Invest.* **27**, 299–306.

Nestel, P. J., Ishikawa, T., and Goldrick, R. B. (1978). Diminished plasma free fatty acid clearance in obese subjects. *Metabolism* **27**, 589–596.

Owen, O. E., Morgan, A. P., Kemp, H. G., Sullivan, J. M., Herrara, M. G., and Cahill, G. F. (1967). Brain metabolism during fasting. *J. Clin. Invest.* **46**, 1589–1595.

*Arginine, Nitric Oxide, and Vasodilation*

Beckman, J. S., and Koppenol, W. H. (1996). Nitric oxide, superoxide, and peroxynitrite: The good, the bad, and the ugly. *Am. J. Physiol.* **271**, C1424–C1437.

Clarkson, P., Adams, M. R., Powe, A., Donald, A., McCredle, R., Robinson, J., McCarthy, S., Keech, A., Celermajer, D., and Deanfield, J. (1996). Oral L-arginine improves endothelium-dependent dilation in hypercholesterolemic young adults. *J. Clin. Invest.* **97**, 1989–1994.

Drewett, J. G., and Garbers, D. L. (1994). The family of guanylyl cyclase receptors and their ligands. *Endocrine Rev.* **15**, 135–162.

Hickner, R., Fisher, J., Ehsani, A., and Kohrt, W. (1997). Role of nitric oxide in skeletal muscle blood flow at rest and during dynamic exercise in humans. *Am. J. Physiol.* **273**, H405–H410.

Huang, P. L., Huang, Z., Mashimo, H., Bloch, K., Moskowitz, M., Bevan, J., and Fishman, M. (1995). Hypertension in mice lacking the gene for endothelial nitric oxide synthase. *Nature* **377**, 239–242.

Komalavilas, P., and Lincoln, T. M. (1996). Phosphorylation of the inositol 1,2,5-trisphosphate receptor. *J. Biol. Chem.* **271**, 21933–21938.

Smith, J. A., Francis, S., Walsh, K., Kumar, S., and Corbin, J. (1996). Autophosphorylation of type Ib cGMP-dependent protein kinase increases basal catalytic activity and enhances allosteric activation by cGMP or cAMP. *J. Biol. Chem.* **271**, 20756–20762.

Wickelgren, I. (1997). Biologists catch their first detailed look at NO enzyme. *Science* **278**, 389.

*Creatine Phosphate*

Bessman, S. P., and Geiger, P. J. (1981). Transport of energy in muscle: The phosphorylcreatine shuttle. *Science* **211**, 448–452.

Lente, F. V., and Suit, P. (1989). Assessment of renal function by serum creatinine and creatinine clearance: Glomerular filtration rate estimated by four procedures. *Clin. Chem.* **35**, 2326–2330.

Levey, A. S., Perrone, R. D., and Madias, N. E. (1988). Serum creatinine and renal function. *Annu. Rev. Med.* **39**, 465–490.

Sahlin, K., Harris, R. C., and Hultman, E. (1975). Creatine kinase equilibrium and lactate content compared with muscle pH in tissue samples obtained after isometric exercise. *Biochem. J.* **152**, 173–180.

Walser, M. (1987). Creatinine excretion as a measure of protein nutrition in adults of varying age. *J. Parenteral Enteral Nutr.* **11**, 73S–78S.

*Glucose Transport*

Cartee, G. D., Young, D. A., Sleeper, M. D., Zierath, J., Wallberg-Hendriksson, H., and Holloszy, J. O. (1989). Prolonged increase in insulin-stimulated glucose transport in muscle after exercise. *Am. J. Physiol.* **256**, E494–E499.

Chasiotis, D., Edstrom, L., Sahlin, K., and Sjoholm, H. (1986). Activation of glycogen phosphorylase by electrical stimulation of isolated fast-twitch and slow-twitch muscles from rats. *Acta Physiol. Scand.* **123**, 43–47.

Fushiki, T., Wells, J. A., Tapscott, E. B., and Dohm, G. L. (1989). Changes in glucose transporters in muscle in response to glucose. *Am. J. Physiol.* **256**, E580–E587.

Gardemann, A., Strulik, H., and Jungermann, K. (1987). Nervous control of glycogenolysis and blood flow in arterially and portally perfused liver. *Am. J. Physiol.* **253**, E238–E245.

Heller-Harrison, R., Morin, M., Guilherme, A., and Czech, M. P. (1996). Insulin-mediated targeting of phosphatidylinositol 3-kinase to GLUT4-containing vesicles. *J. Biol. Chem.* **271**, 10200–10204.

Hendriksson, J., Salmons, S., Chi, M., Hintz, C. S., and Lowry, O. H. (1988). Chronic stimulation of mammalian muscle: Changes in metabolite concentration in individual fibers. *Am. J. Physiol.* **255**, C543–C551.

Hughes, V. A., Fiatarone, M., Fielding, R., Kahn, B. B., Ferrara, C., Shepherd, P., Fisher, E. C., Wolfe, R. R., Elahi, D., and Evans, W. J. (1993). Exercise increases muscle GLUT-4 levels and insulin action in subjects with impaired glucose tolerance. *Am. J. Physiol.* **264**, E855–E862.

Wallberg-Hendriksson, H. (1986). Repeated exercise regulates glucose transport capacity in skeletal muscle. *Acta Physiol. Scand.* **127**, 39–43.

*Fructose*

Gelfand, R. A., and Sherwin, R. S. (1986). Nitrogen conservation in starvation revisited: Protein sparing with intravenous fructose. *Metabolism* **35**, 37–44.

Gopher, A., Vaisman, N., Mandel, H., and Lapidot, A. (1990). Determination of fructose metabolism pathways in normal and fructose-intolerant children. *Proc. Natl. Acad. Sci. U.S.A.* **87**, 5449–5453.

Hallfrisch, J. (1990). Metabolic effects of dietary fructose. *FASEB J.* **4**, 2652–2660.

Kneepkens, C. M. (1989). What happens to fructose in the gut? *Scand. J. Gastroenterol., Suppl.* **24**(171), 1–8.

Pozza, G., Galansino, G., Hoffeld, H., and Foa, P. P. (1958). Stimulation of insulin output by monosaccharides and monosaccharide derivatives. *Am. J. Physiol.* **192**, 497–500.

Tappy, L., Randin, J.-P., Felber, J.-P., Chiolero, R., Simonson, D. C., Jequier, E., and DeFronzo, R. A. (1986). Comparison of thermogenic effect of fructose and glucose in normal humans. *Am. J. Physiol.* **250**, E718–E724.

Truswell, A. S., Seach, J. M., Diet, D. N., and Thorburn, A. W. (1988). Incomplete absorption of pure fructose in healthy subjects and the facilitating effects of glucose. *Am. J. Clin. Nutr.* **48**, 1424–1430.

Youn, J. H., Kaslow, H. R., and Bergman, R. N. (1987). Fructose effect to suppress hepatic glycogen degradation. *J. Biol. Chem.* **262**, 11470–11477.

*Medium-Chain Fatty Acids*

Cook, G. A., and Gamble, M. S. (1987). Regulation of carnitine palmitoyltransferase insulin results in decreased activity and decreased apparent $K_i$ values for malonyl-CoA. *J. Biol. Chem.* **262**, 2050–2055.

Delafosse, B., Viale, J., Pachiaudi, C., Normand, S., Goudable, J., Bouffard, Y., Annat, G., and Bertrand, O. (1997). Long- and medium-chain triglycerides during parenteral nutrition in critically ill patients. *Am. J. Physiol.* **272**, E550–E555.

Johnson, R. C., and Cotter, R. (1986). Metabolism of medium-chain triglyceride lipid emulsion. *Nutr. Int.* **2**, 150–158.

Kelly, D. P., Whelan, A. J., Ogden, M. L., Alpers, R., Zhang, Z., Bellus, G., Gregersen, N., Dorland, L., and Strauss, A. W. (1990). Molecular characterization of inherited medium-chain acyl-CoA dehydrogenase deficiency. *Proc. Natl. Acad. Sci. U.S.A.* **87**, 9236–9240.

Lysiak, W., Lilly, K., DiLisa, F., Toth, P. P., and Bieber, L. L. (1988). Quantitation of the effect of L-carnitine on the levels of acid-soluble short-chain acyl-CoA and CoASH in rat heart and liver mitochondria. *J. Biol. Chem.* **263**, 1151–1156.

Odle, J. (1997). New insights into the utilization of medium-chain triglycerides by the neonate: Observatonss from a piglet model. *J. Nutr.* **127**, 1061–1067.

Sann, L., Mathiew, M., Lasne, Y., and Ruitton, A. (1981). Effect of oral administration of lipids with 67% medium chain triglycerides on glucose homeostasis in preterm infants. *Metabolism* **30**, 712–716.

Sarda, P., Lepage, G., Roy, C. C., and Chessex, P. (1987). Storage of medium-chain triglycerides in adipose tissue of orally fed infants. *Am. J. Clin. Nutr.* **45**, 399–405.

Sills, M. A., Forsythe, W. L., Haidukewych, D., MacDonald, A., and Robinson, M. (1986). The medium-chain triglyceride diet and intractable epilepsy. *Arch. Dis. Child.* **61**, 1168–1172.

## Lipases

Ahlborg, G., Felig, P., Hagenfeldt, L., Hendler, R., and Wahren, J. (1974). Substrate turnover during prolonged exercise in man. *J. Clin. Invest.* **53**, 1080–1090.

Chernick, S. S., Spooner, P. M., Garrison, M. M., and Scow, R. O. (1986). Effect of epinephrine and other lipolytic agents on intracellular lipolysis and lipoprotein lipase activity in 3T3-L1 adipocytes. *J. Lipid Res.* **27**, 286–294.

Elia, M., Zed, C., Neale, G., and Livesay, G. (1987). The energy cost of triglyceride fatty acid recycling in nonobese subjects after an overnight fast and four days starvation. *Metabolism* **36**, 251–255.

Jensen, M. D., Haymond, M. W., Gerich, J. E., Cryer, P. E., and Miles, J. M. (1987). Lipolysis during fasting. *J. Clin. Invest.* **79**, 207–213.

Kaminsky, L. A., Knowlton, R. G., Perkins, R. M., and Hetzler, R. K. (1986). Relationship of aerobic capacity and percent body fat with plasma free fatty acid following walking. *Am. J. Clin. Nutr.* **44**, 603–609.

Kirchgessner, T. G., LeBoeuf, R. C., Langner, C. A., Zollman, S., Chang, C. H., Taylor, B. A., Schotz, M. C., Gordon, J. L., and Lusis, A. J. (1989). Genetic and developmental regulation of the lipoprotein lipase gene. *J. Biol. Chem.* **264**, 1473–1482.

Lithell, H., Jacobs, L., Vessby, B., Hellsring, K., and Karlsson, J. (1982). Decrease of lipoprotein lipase activity in skeletal muscle in man during a short-term carbohydrate-rich regime. *Metabolism* **31**, 994–998.

Reardon, M. F., Sakai, H., and Steiner, G. (1982). Roles of lipoprotein lipase and hepatic triglyceride lipase in the catabolism *in vivo* of triglyceride-rich lipoproteins. *Arteriosclerosis* **2**, 396–402.

Vannier, C., Deslex, S., Pradines-Figuere, S., and Ailhaud, G. (1989). Biosynthesis of lipoprotein lipase in cultured mouse adipocytes. *J. Biol. Chem.* **264**, 13199–13205.

## Malate–Aspartate Shuttle

Bissell, M. J., Rambeck, W. A., White, R. C., and Bassham, J. A. (1976). Glycerol phosphate shuttle in virus-transformed cells in culture. *Science* **191**, 856–858.

Fillingame, R. H. (1980). The proton-translocating pumps of oxidative phosphorylation. *Annu. Rev. Biochem.* **49**, 1079–1113.

Greenhouse, W. V., and Lehninger, A. L. (1976). Occurrence of the malate–aspartate shuttle in various tumor-types. *Cancer Res.* **36**, 1392–1396.

Indiveri, C., Kramer, R., and Palmieri, F. (1987). Reconstitution of the malate/aspartate shuttle from mitochondria. *J. Biol. Chem.* **262**, 15979–15983.

Lanoue, K. F., and Williamson, J. R. (1971). Interrelationships between malate–aspartate shuttle and citric acid cycle in rat heart mitochondria. *Metabolism* **20**, 119–140.

## Carnitine

Borum, P. R. (1983). Carnitine. *Annu. Rev. Nutr.* **3**, 233–259.

Bowyer, B. A., Miles, J. M., Haymond, M. W., and Fleming, C. R. (1988). L-Carnitine therapy in home parenteral nutrition patients with abnormal liver tests and low plasma carnitine concentrations. *Gastroenterology* **94**, 434–438.

Carroll, J. E., Carter, A. L., and Perlman, S. (1987). Carnitine deficiency revisited. *J. Nutr.* **117**, 1501–1503.

Mitchell, M. E. (1978). Carnitine metabolism in human subjects. *Am. J. Clin. Nutr.* **31**, 293–306.

Rebouche, C. J., and Engel, A. G. (1980). Tissue distribution of carnitine biosynthetic enzymes in man. *Biochim. Biophys. Acta* **630**, 22–29.

Rebouche, C. J., Lehman, L. J., and Olson, A. L. (1986). ε-N-Trimethyllysine availability regulates the rate of carnitine biosynthesis in the growing rat. *J. Nutr.* **116**, 751–759.

Rovamo, L. M., Salmenpera, L., Arjomaa, P., and Raivio, K. O. (1986). Carnitine during prolonged breast feeding. *Pediatr. Res.* **20**, 806–809.

Schwenk, W. F., Hale, D. E., and Haymond, M. W. (1988). Decreased fasting free fatty acids with L-carnitine in children with carnitine deficiency. *Pediatr. Res.* **23**, 491–494.

## Ketone Bodies

Bougneres, P. F., Lemmeil, C., Ferre, P., and Bier, D. M. (1986). Ketone body transport in the human neonate and infant. *J. Clin. Invest.* **77**, 42–48.

Fery, F., and Balasse, E. O. (1985). Ketone body production and disposal in diabetic ketosis. *Diabetes* **34**, 326–332.

Fery, F., and Balasse, E. O. (1986). Response of ketone body metabolism to exercise during transition from postabsorptive to fasted state. *Am. J. Physiol.* **250**, E495–E501.

Ferre, P., Statbin, P., Decaux, J.-F., Escriva, F., and Girard, J. (1983). Development and regulation of ketogenesis in hepatocytes isolated from newborn rats. *Biochem. J.* **214**, 937–942.

Foster, D. W. (1984). From glycogen to ketones — and back. *Diabetes* **33**, 1188–1199.

Girard, J. (1986). Gluconeogenesis in late fetal and early neonatal life. *Biol. Neonate* **50**, 237–258.

Johnson, B. H., Walton, J. L., Krebs, H. A., and Williamson, D. H. (1969). Post-exercise ketosis. *Lancet II*, December, pp. 1383–1385.

Kosug, K., Scofield, B. F., Chandramouli, V., Kumaran, K., Schumann, W. C., and Landau, B. B. (1986). Pathways of acetone's metabolism in the rat. *J. Biol. Chem.* **261**, 3952–3957.

Krall, L. P. (1988). "World Book of Diabetes in Practice," Vol. 3, pp. 209–213. Elsevier, New York.

Levitsky, L. L., Fisher, D. E., Paton, J. B., and Delannoy, C. W. (1977). Fasting plasma levels of glucose, acetoacetate, D-β-hydroxybutyrate, glycerol, and lactate in the baboon infant. *Pediatr. Res.* **11**, 298–302.

McGarry, J. D., and Foster, D. W. (1980). Regulation of hepatic fatty acid oxidation and ketone body production. *Annu. Rev. Biochem.* **49**, 395–420.

Reed, W. D., Baab, P. J., Hawkins, R. L., and Oxand, P. T. (1984). The effects of insulin and glucagon on ketone-body turnover. *Biochem. J.* **221**, 439–444.

Taylor, R., and Agius, L. (1988). The biochemistry of diabetes. *Biochem. J.* **250**, 625–640.

Walton, J. (1972). The effect of exercise upon acetoacetate metabolism in athletes and non-athletes. *Quart. J. Exp. Physiol.* **57**, 73–79.

### Fasting and Starvation

Badger, T. M., and Tumbleson, M. E. (1974). Protein-calorie malnutrition in young miniature swine: Serum free amino acids. *J. Nutr.* **104**, 1339–1347.

Brasel, J. A. (1980). Endocrine adaptation to malnutrition. *Pediatr. Res.* **14**, 1299–1303.

Cahill, G. F. (1970). Starvation in man. *N. Engl. J. Med.* **282**, 668–675.

Garrow, J. S., Fletcher, K., and Halliday, D. (1965). Body composition in severe infantile malnutrition. *J. Clin. Invest.* **44**, 417–425.

Kelleher, P. C., Phinney, S. D., Sims, E. A., Bogardus, C., Hortan, E. S., Bistrian, B. R., Amatruda, J. M., and Lockwood, D. H. (1983). Effects of carbohydrate-containing and carbohydrate-restricted hypocaloric and eucaloric diets on serum concentrations of retinol-binding protein, thyroxine-binding prealbumin, and transferrin. *Metabolism* **32**, 95–101.

Leiter, L. A., and Marliss, E. B. (1982). Survival during fasting may depend on fat as well as protein stores. *J. Am. Med. Assoc.* **248**, 2306–2307.

Lunn, P. G., and Austin, S. (1983). Dietary manipulation of plasma albumin concentration. *J. Nutr.* **113**, 1791–1802.

### Alcohol

Agarwall, D. P., and Goedde, H. W. (1989). Human aldehyde dehydrogenases: Their role in alcoholism. *Alcohol* **6**, 517–523.

Alderman, J., Takagi, T., and Lieber, C. S. (1987). Ethanol-metabolizing pathways in deermice. *J. Biol. Chem.* **262**, 7497–7502.

Charness, M. E., Simon, R. P., and Greenberg, D. A. (1989). Ethanol and the nervous system. *N. Engl. J. Med.* **321**, 442–454.

Danielsson, O., Atrian, S., Luque, T., Hjelmqvist, L., Gonzalez-Duarte, R., and Jornvall, H. (1994). Fundamental molecular differences between alcohol dehydrogenase classes. *Proc. Natl. Acad. Sci. U.S.A.* **91**, 4980–4984.

Diamond, I., and Gordon, A. S. (1997). Cellular and molecular neuroscience of alcoholism. *Physiol. Rev.* **77**, 1–20.

Dreon, D. M., and Krauss, R. M. (1996). "Alcohol, Lipids, and Lipoproteins in Alcohol and the Cardiovascular System," Research Monograph 31 (S. Zakhari and M. Wassef, eds.), pp. 369–391. U.S. Dept. of Health and Human Services, Washington, DC.

Haber, P., Gentry, R., Mak, K., Mirmiran-Yazdy, S., Greenstein, R., and Lieber, C. (1996). Metabolism of alcohol by human gastric cells: Relation to first-pass metabolism. *Gastroenterology* **111**, 863–870.

Iber, F. L. (1980). Fetal alcohol syndrome. *Nutr. Today*, September–October, pp. 4–11.

Karl, P. I., Gordon, B. H., Lieber, C. S., and Fisher, S. E. (1988). Acetaldehyde production and transfer by the perfused placental cotyledon. *Science* **242**, 273–275.

Lands, W. (1995). Alcohol and energy intake. *Am. J. Clin. Nutr.* **62**, 1101–1106.

Lieber, C. S. (1973). Liver adaptation and injury in alcoholism. *N. Engl. J. Med.* **288**, 356–362.

Mitchell, M. C., and Herlong, H. F. (1986). Alcohol and nutrition. *Annu. Rev. Nutr.* **6**, 457–474.

Pennisi, E. (1997). Enzyme linked to alcohol sensitivity in mice. *Science* **278**, 573.

Potter, J. D. (1997). Hazards and benefits of alcohol. *N. Engl. J. Med.* **337**, 1763–1764.

Rosett, H. L., and Weiner, L. (1985). Alcohol and pregnancy. *Annu. Rev. Med.* **36**, 73–80.

Shaw, S., Jayatilleke, E., and Lieber, C. S. (1988). Lipid peroxidation as a mechanism of alcoholic liver injury: Role of iron mobilization and microsomal induction. *Alcohol* **5**, 135–140.

Singh, N., and Khan, A. (1995). Acetaldehyde: Genotoxicity and cytotoxicity in human lymphocytes. *Mutat. Res.* **337**, 9–17.

Stadtman, E. R. (1990). Covalent modification reactions are marking steps in protein turnover. *Biochemistry* **29**, 6323–6331.

Thun, M. J., Peto, R., Lopez, A., Monaco, J., Henley, J., Heath, C., and Doll, R. (1997). Alcohol consumption and mortality among middle-aged and elderly U.S. adults. *N. Engl. J. Med.* **337**, 1705–1714.

Uotila, L., and Koivusalo, M. (1996). Expression of formaldehyde dehydrogenase and *S*-formylglutathione hydrolase activities in different rat tissues. *In* "Enzymology and Molecular Biology of Carbonyl Metabolism," Vol. 6 (H. Weiner, R. Lindahl, D. Crabb, and T. Flynn, eds.), pp. 365–371. Plenum, New York.

Windham, C. T., Wyse, B. W., and Hansen, R. G. (1983). Alcohol consumption and nutrient density of diets in the nationwide food consumption survey. *J. Diet. Assoc.* **82**, 364–373.

Yamauchi, M., Potter, J. J., and Mezey, E. (1988). Characteristics of alcohol dehydrogenase in fat-storing (Ito) cells of rat liver. *Gastroenterology* **94**, 163–169.

Yin, S.-J., Han, C.-L., Liao, C.-S., and Wu, C.-W. (1996). Expression, activities, and kinetic mechanisms of human stomach alcohol dehydrogenase. *In* "Enzymology and Molecular Biology of Carbonyl Metabolism," Vol. 6 (H. Weiner, R. Lindahl, D. Crabb, and T. Flynn, eds.), pp. 347–355. Plenum, New York.

*Thiol Esters*

Bove, J., Martin, R. O., Ingraham, L. L., and Stumpf, P. K. (1959). Studies of the mechanism of condensing enzyme. *J. Biol. Chem.* **234**, 999–1006.

Gilbert, H. F. (1981). Proton transfer from acetyl-coenzyme A catalyzed by thiolase I from porcine heart. *Biochemistry* **20**, 5643–5649.

Jencks, W. P. (1969). "Catalysis in Chemistry and Enzymology," pp. 517–523. McGraw-Hill, New York.

Kramer, P. R., and Miziorko, H. M. (1983). 3-Hydroxy-3-methylglutaryl-CoA lyase: Catalysis of acetyl coenzyme A enolization. *Biochemistry* **22**, 2353–2357.

Myers, J. A., and Boyer, P. D. (1984). Oxygen and deuterium exchanges show reversal of catalytic steps of citrate synthase: Catalytic cooperativity is not observed. *Biochemistry* **23**, 1264–1269.

# 5

# ENERGY REQUIREMENT

## OVERVIEW

The main goal of this chapter is to learn how to determine the body's overall style of energy metabolism, via respiratory quotient (RQ) measurements, and to derive the daily energy requirement. A view of the stoichiometries of the glycolytic pathway, Krebs cycle, and pathways of fatty acid synthesis and oxidation will allow RQ calculations for each individual pathway. Glycolysis and the Krebs cycle were presented in Chapter 4. Fatty acid synthesis and oxidation are detailed here. The locations, at points along various metabolic pathways, where $CO_2$ is produced (in the Krebs cycle) and $O_2$ is consumed (in the respiratory chain), are points of focus in this chapter.

Presented first are definitions of energy and power, followed by a discussion of the energy content of different classes of foods. A description of the respiratory gases and tabulation of the stoichiometry of these gases in chemical and biochemical formulas constitute a background for understanding data on gas metabolism in humans. Respiratory gas data for persons at rest, during exercise, and with overeating are shown and interpretations presented. A description of the bomb calorimeter reveals the connection between respiratory gas

273

data and the food energy requirement. This machine measures heat released from food burning in pure oxygen. The food energy requirement can also be determined with no knowledge of respiratory gases, by measuring the heat produced by a human subject, who is in a calorimetric chamber, during the course of several hours or days. The doubly labeled water technique, which represents an alternate technique to the calorimetric chamber, is outlined.

## UNITS OF ENERGY

The energy requirement is expressed as the number of kilojoules that must be consumed per day to support growth and maintenance. The **joule** (J) is a unit of energy. Another unit of energy, the **calorie**, is being phased out in the scientific community. One calorie is equal to 4.184 J. The daily energy requirement of an adult 70-kg man is 11,300 kJ. This requirement is expressed in terms of power (the rate of energy consumption). It can be expressed in terms of joules per second or joules per day, for example. The unit of **power** is the **watt** (W). One watt is equal to 1 J/sec. A roaring bonfire might be considered to be powerful, as it rapidly uses wood for the production of heat and light. A steaming locomotive might be considered to be powerful, as it rapidly uses coal for the production of heat and work. The term power cannot be used to describe a cold, resting locomotive, as it is not actively consuming energy.

EXERCISE

How powerful is a man compared with a light bulb? The rate of energy consumption is usually inscribed on the glass bulb.

An animal consuming and absorbing an amount of energy, as food, identical to the amount of energy used has achieved **energy balance**. An animal that consumes more energy than used is in **positive energy balance**. An animal in positive energy balance is one that is growing, is pregnant, or is depositing energy in the form of fat. An animal consuming less energy than used is in **negative energy balance**. In negative energy balance, an animal loses weight.

The use of energy fuels by an animal may be compared with the use of energy fuels by a fire. In both cases, the fuel donates its electrons to oxygen; the fuel is converted to $CO_2$ and the oxygen to water. Heat is produced in both cases, and the amounts of heat produced per gram of fuel oxidized are identical.

The energy content of food carbohydrate and food protein is 17 kJ/g, whereas that of food fat is 38 kJ/g. The energy content of ethanol is 30 kJ/g. Though fairly accurate, these values are only generalizations. For example, consider the energies of different carbohydrates: 17.2 kJ/g of starch, 16.6 kJ/g of sucrose, 15.9 kJ/g of glucose.

EXERCISE

Ten potato chips contain 8 g of fat and 10 g of carbohydrate. How many potato chips can satisfy the daily energy requirement of the adult woman? (See Table 5.5.)

The use of energy fuels by animals and by a flame may be compared by considering the stoichiometry of the chemical reactions. They may also be compared by noting the amount of energy produced with complete oxidation of the fuel in question. The following discussion begins with the stoichiometry of combustion of simple compounds — such as paraffin, methane, and methanol — and then continues onto the combustion of more complex materials — such as glucose and fat. The study of the stoichiometry of various reactions requires the counting of the number of molecules of $CO_2$ produced per molecule of fuel combusted, the number of water molecules produced, and the number of oxygen atoms (or molecules) consumed.

The stoichiometry involved in the combustion of fuels can be used to calculate a very interesting ratio: the number of $CO_2$ molecules discharged from the body per number of oxygen molecules consumed. This ratio ($CO_2/O_2$) is called the **respiratory quotient (RQ)**. The biochemical events of $CO_2$ production and $O_2$ utilization are a direct result of the oxidation of various fuels such as fat and glucose. The amount of $O_2$ used is also influenced by the rate of synthesis of new fat in the body. Respiratory gases can be measured and analyzed quite easily. These measurements can be used to calculate the amount of $CO_2$ produced and $O_2$ used by the body over any given period. Calculating the RQ for a subject is quite easy and can provide remarkable insight into the overall behavior of energy fuels in the body. The RQ can provide only limited information on metabolism, but it is a powerful measurement.

The RQ is different for different fuels. For example, the RQ for the complete combustion of glucose is 1.0, and that for the complete combustion of fat is 0.7. These values are determined in three ways:

1. On paper, by balancing chemical formulas, with no knowledge of metabolic pathways.

2. By consideration of the biochemical pathways of the body, providing that these pathways have been memorized or are available in some format.

3. Experimentally, by use of a furnace. The furnace contains an atmosphere of pure oxygen. A mechanism for igniting the fuel or food in question, as well as devices for measuring the amount of $CO_2$ produced and $O_2$ consumed, is available.

Knowledge of the RQ for different energy fuels can be used to determine the predominant fuel used by a particular organ of the body, such as the muscle or brain. To determine the RQ for a particular organ, the blood entering and exiting that organ must be analyzed. Most of the oxygen entering and exiting the organ is bound to hemoglobin. Most of the $CO_2$ entering and exiting the organ is in the form of bicarbonate.

EXERCISE

Would you expect the RQ of resting muscle to be closer to 1.0 or 0.7?

EXERCISE

Which should be higher, the RQ of a person engaging in an exhaustive 10-minute race or that of a person completing a 5-hour marathon?

## CALCULATING THE RESPIRATORY QUOTIENT FROM CHEMISTRY

A brief consideration of bacteriology introduces a simple calculation of the RQ. A variety of microorganisms can use long-chain alkanes (paraffin wax) or short-chain alkanes as an energy source (Van Eyk and Bartels, 1968). The complete oxidation of alkanes has the same outcome, whether catalyzed by a flame at a candlewick, a spark inside a furnace, or enzymes within a living cell. Complete oxidation of hexane means that all of the carbon atoms are oxidized to carbon dioxide:

$$CH_3CH_2CH_2CH_2CH_2CH_3 \rightarrow 6\ CO_2 \qquad (5.1)$$

For simplicity, the discharged hydrogen atoms and consumed oxygen molecules are not written in Eq. (5.1). The discharged hydrogen atoms are shown in the next reaction. For convenience in working through the calculations, they may be written as $H_2$:

$$CH_3CH_2CH_2CH_2CH_2CH_3 \rightarrow 6\ CO_2 + 7\ H_2 \qquad (5.2)$$

The reducing equivalents released from aerobically consumed fuels are transferred to oxygen, producing water. Including water in the formula for hexane oxidation yields

$$CH_3CH_2CH_2CH_2CH_2CH_3 \rightarrow 6\ CO_2 + 7\ H_2O \qquad (5.3)$$

The final step is to add the correct number of oxygen molecules. The right-hand side of Eq. (5.3) shows 19 oxygen atoms. This means that there must be 19 oxygen atoms on the left-hand side. As the atmosphere contains $O_2$, rather than oxygen atoms, it is more appropriate to write the 19 oxygen atoms as $19/2\ O_2$, or $9.5\ O_2$:

$$9.5\ O_2 + CH_3CH_2CH_2CH_2CH_2CH_3 \rightarrow 6\ CO_2 + 7\ H_2O \qquad (5.4)$$

The RQ of hexane, as calculated from Equation (5.4), is $CO_2/O_2 = 6/9.5 = 0.63$. Hexane is an alkane. Fatty acids are quite similar in structure to alkanes. The RQ for the complete combustion of fatty acids has a value close to 0.6.

Methane and methanol provide more simple examples of fuel oxidation. Methane, $CH_4$, is natural gas. Methanol, $CH_3OH$, is a liquid sometimes used as a cooking fuel. The steps in the calculation of the RQ for methane combustion follow:

1. Balance the carbons. With complete combustion, one molecule of methane produces one molecule of $CO_2$:

$$CH_4 \rightarrow CO_2 \qquad (5.5)$$

2. Balance the hydrogens. Four hydrogens appear on the left-hand side of the equation. Thus, four hydrogens must be written on the right-hand side:

$$CH_4 \rightarrow CO_2 + H_2 + H_2 \qquad (5.6)$$

3. Add oxygen. The end-product of fuel combustion in animal cells is not $H_2$, but water:

$$CH_4 \rightarrow CO_2 + H_2O + H_2O \qquad (5.7)$$

4. Balance the oxygens. This involves calculation of the number of oxygen molecules consumed during the complete oxidation of methane:

$$2\,O_2 + CH_4 \rightarrow CO_2 + H_2O + H_2O \qquad (5.8)$$

The RQ for methane, as determined by these calculations, is 0.5. This value is lower than that for fat.

Methanol is partially oxidized compared with methane. Hence, one might expect that the RQ of methanol differs from that of methane. What does it mean to state that "methanol is more oxidized than methane"? The extent of oxidation of a particular carbon may be described by its oxidation number or state. The oxidation number is calculated according to the types of atoms bonded to the carbon. A hydrogen atom contributes −1 to the oxidation number of the carbon. A carbon atom makes no contribution to the number. Hydroxyl groups contribute +1, and keto groups contribute +2. The oxidation states of the carbons in a variety of organic molecules are listed in Table 5.1. The steps in the calculation of the RQ for methanol combustion follow:

1. Balance the carbons. Methanol contains only one carbon. Thus, only one molecule of $CO_2$ is produced:

$$CH_3OH \rightarrow CO_2 \qquad (5.9)$$

2. Balance the hydrogens. Methanol contains four hydrogens. Thus, one might write two $H_2$s on the right-hand side:

$$CH_3OH \rightarrow CO_2 + H_2 + H_2 \qquad (5.10)$$

3. Add oxygen to the hydrogens. This yields two molecules of water:

$$CH_3OH \rightarrow CO_2 + H_2O + H_2O \qquad (5.11)$$

**TABLE 5.1** Oxidation States of Specific Carbons in Various Organic Molecules

| Molecule | Oxidation state of each carbon | Structure of molecule |
|----------|-------------------------------|-----------------------|
| Methane | –4 | $CH_4$ |
| Ethane | –3 | $CH_3$—$CH_3$ |
| Ethylene | –2 | $CH_2$═$CH_2$ |
| Ethanol | –3 and –1 | $CH_3$—$CH_2$—$OH$ <br> ↑     ↑ <br> –3    –1 |
| Acetic acid | –3 and +3 | $CH_3$—$C$ <br> ↑     ↑ <br> –3   +3  $OH$ |
| Formic acid | +2 | $HC$ <br> $OH$ |
| Carbon dioxide | +4 | $O$═$C$═$O$ |

4. Balance the oxygens. Four atoms of oxygen occur on the right-hand side of the equation. Methanol contains one atom of oxygen. Thus, three more atoms of oxygen must be added to the left-hand side:

$$1.5\ O_2 + CH_3OH \rightarrow CO_2 + H_2O + H_2O \tag{5.12}$$

The RQ for methanol, as determined by these calculations, is 0.666.

The stoichiometries involved in the complete oxidation of glucose and of fat are:

$$C_6H_{12}O_6 + 6\ O_2 \rightarrow 6\ CO_2 + 6\ H_2O \qquad RQ = 1.0 \tag{5.13}$$
(glucose)

$$C_{55}H_{104}O_6 + 78\ O_2 \rightarrow 55\ CO_2 + 52\ H_2O \qquad RQ = 0.7$$
(fat)

These stoichiometries are the same whether oxidation occurs in a clean burning flame or inside a living cell. The fat in formula (5.13) is the triglyceride palmitoyl-stearoyl-oleoyl-glycerol.

## ELECTRON TRANSPORT AND ATP PRODUCTION

Food and internal energy stores serve as electron donors. Oxygen functions as the final electron acceptor in metabolism. The overall scheme of energy metabolism

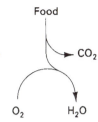

**FIGURE 5.1** Food is an electron donor.

involves the transfer of electrons from food to $O_2$. This transfer occurs in many small steps, some of which result in the capture of energy and its storage in a useful, versatile form, namely ATP. Many of the steps also result in the production of heat. A simplified scheme of energy metabolism is shown in Figure 5.1. This figure illustrates the use of food and oxygen and the production of $CO_2$ and water.

The following material focuses on the reduction of oxygen to water and ATP production, events that take place at the mitochondrial membrane. Energy fuels are electron donors. The coenzymes $NAD^+$ and FAD accept these electrons and mediate their transfer to the respiratory chain. The respiratory chain is a series of iron-containing proteins associated with the mitochondrial membrane. Transfer of four electrons through this series of proteins can result in the reduction of one molecule of oxygen to water. This electron transport chain is shown in Figure 5.2.

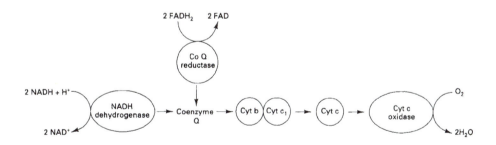

**FIGURE 5.2** Diagram of the respiratory chain. Electrons are donated to the respiratory chain by reduced NAD and reduced FAD. Each reduced cofactor donates four electrons to the respiratory chain, a series of proteins associated with the inner membrane of the mitochondrion. Electrons, donated by either or both cofactors, are transferred from protein to protein with their eventual transfer to water. The proteins do not seem to occur as a single complex, but move about in a more or less random manner in the phospholipid bilayer of the membrane. Collisions between the different proteins allow the electron to pass down the entire chain (Lenaz, 1988). Coenzyme Q is a lipid-soluble coenzyme with a structure similar to that of vitamin K. Coenzyme Q, also called *ubiquinone*, is synthesized in the body and is not a vitamin. NADH dehydrogenase catalyzes the transfer of electrons from reduced $NAD^+$ to coenzyme Q. Coenzyme Q reductase catalyzes the transfer of electrons from the FAD-using dehydrogenases, such as succinate dehydrogenase and that in fatty acid oxidation, to coenzyme Q. The sequential transfer of electrons from one protein to another, as shown in this figure, is called **electron transport**.

The four electrons can all arise from two molecules of reduced $NAD^+$, two molecules of reduced FAD, or a combination of these two cofactors.

The electrons from reduced $NAD^+$ are delivered to NADH dehydrogenase. At this point, the electrons are of relatively high energy. With transfer to ubiquinone, to the complex of cytochromes $b$ and $c_1$, to cytochrome $c$, to cytochrome $c$ oxidase, and finally to water, the energy level of the electrons progressively decreases. The energy released in the step-by-step transfer of electrons between the different components of the respiratory chain is used to do work. This work takes the form of transport of protons out through the inner mitochondrial membrane. The points at which proton transport occurs are shown in Figure 5.3. One consequence of this transport is that the interior of the mitochondrion becomes somewhat more alkaline than the cytoplasm. More important, protons are allowed to flow back through the inner mitochondrial membrane, where this inflow is coupled to the

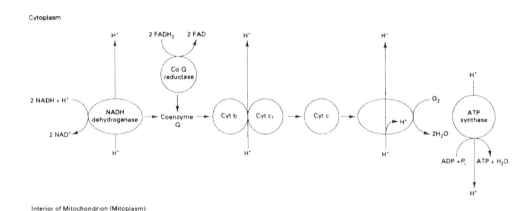

**FIGURE 5.3** Diagram of respiratory chain, proton pumps, and ATP synthase. The diagram represents a more complete version of Figure 5.2. The flow of electrons is used as the source of energy for driving protons out, through the inner mitochondrial membrane, at the indicated proteins or protein complexes. The return of the protons, via ATP synthase, is used as the energy source to drive the synthesis of ATP, which occurs inside the organelle. The mitochondrion is bounded by two membranes, not just one membrane. The proteins of electron transport, as well as ATP synthase, are all bound to the *inner* mitochondrial membrane. The outer membrane allows the passage of many metabolites, including those of the glycerol–phosphate shuttle (dihydroxyacetone-P and glycerol-P). The inner membrane does not allow passage of these molecules. ATP synthase is composed of two major subunits, $F_0$ and $F_1$. $F_0$ acts as a pore and allows protons to flow into the organelle. $F_1$ is the catalytic subunit. $F_1$ couples the flow of protons to the synthesis of ATP. $F_1$ itself is a complex and is composed of five different proteins. The catalytic subunit ($F_1$) is situated at the inside face of the membrane. The two subunits are connected to each other by a stalk, i.e., a connector protein. Protons moving through the $F_0$ subunit provoke a conformational change in this subunit, which is relayed over a distance of about 8 nm through the stalk, to the $F_1$ subunit. This relayed energy is needed to drive ATP synthesis. Surprisingly, the relayed energy is not directly used to connect $P_i$ to ADP. ADP and $P_i$ spontaneously connect when they bind to the $F_1$ subunit, and the ATP produced remains bound to the $F_1$ subunit. The energy coming through the stalk is used to discharge the newly made ATP, for use by the cell (Weber and Senior, 1997).

synthesis of ATP. The net result is that electron transport drives the synthesis of ATP. This roundabout production of ATP was discovered by Peter Mitchell in England (Futai *et al.*, 1989).

## Extrusion of Protons from the Mitochondrion to the Cytoplasm

Protons are extruded from the mitochondrion at the sites of NADH dehydrogenase, at the cytochrome $b/c_1$ complex, and at cytochrome $c$ oxidase. The exact number of protons driven out per electron at each of these steps has been difficult to determine. The ratio can vary under different experimental conditions. Extrusion of a proton at any particular step does not seem to be tightly coupled to the passage of an electron down the respiratory chain; however, it is generally accepted that the passage of two electrons down the entire respiratory chain results in the translocation of 12 protons.

## Passage of Protons from the Cytoplasm Back into the Mitochondrion

The passage of protons through the mitochondrial membrane, via ATP synthase, is also shown in Figure 5.3. Four protons reenter the mitochondrion for the synthesis of each molecule of ATP. Three of these protons are used to drive the continued process of ATP synthesis. Another proton is used to drive the ATP/ADP exchanger. The exchanger allows the entry of one molecule of ADP for the exit of one molecule of ATP. Inorganic phosphate enters the organelle by a separate phosphate carrier. Assuming that 2 electrons result in the exit of 12 protons, one can calculate that 2 electrons can be used to drive the synthesis of 3 ATPs.

## Summary

To summarize, the reduction of oxygen to water involves donation of four electrons and four protons. The electrons arrive via the respiratory chain; the protons are supplied by the catalytic action of cytochrome $c$ oxidase (Figure 5.3). The exact number of protons that pass through ATP synthase per molecule of ATP synthesized is actually not clear. The number could be three or four protons per ATP molecule. Accurate knowledge of this number is important for calculating the amount of energy that can be captured (as ATP, rather than heat) per gram of glucose or fat that is oxidized to $CO_2$ (Weber and Senior, 1997). Another confusing factor is that most of the protons extruded via the respiratory chain return via ATP synthase; however, some protons leak back into the mitochondrion due to the slight permeability of the phospholipid bilayer. There have been claims that about 20% of our food energy is used to drive the extrusion of protons from the mitochondria, which then leak back without the production of ATP (Rolfe and Brand, 1996; Porter *et al.*, 1996).

*Accounting for Water Molecules*

The overall scheme for glucose oxidation was given in Eq. (5.13). Reduction of 6 $O_2$ in the body results in the production of 12 molecules of water. However, Eq. (5.13) shows that only 6 $H_2O$ are produced. Why are 12 $H_2O$ not listed? The catabolism of one molecule of glucose results in the production of two molecules of pyruvate. Complete oxidation of *two pyruvates* results in the use of *six molecules of water*. Water is used by citrate synthase for the hydrolysis of coenzyme A from acetyl-CoA, by succinyl thiokinase for the hydrolysis of coenzyme A from succinyl-CoA, and by fumarase for the hydration of fumarate.

An additional point might be added concerning water. Water is produced at the steps involving the production of ATP and GTP:

$$ADP + P_i \rightarrow ATP + H_2O \quad and \quad GDP + P_i \rightarrow GTP + H_2O$$

Water is produced in this manner at such points as succinyl thiokinase and ATP synthase. The water molecules involved are not included in the calculation of the waters of oxidation.

Complete oxidation of one molecule of glucose results in the production of the equivalent of 38 molecules of ATP. This results in the production of 38 molecules of water, because formation of the phosphodiester bond between ADP and inorganic phosphate involves the splitting off of a molecule of water. These water molecules, however, are not included in the RQ calculations. This is because ATP occurs in catalytic amounts in the cell; it is continuously formed in the glycolytic pathway and mitochondria, and continuously hydrolyzed in various ATP-requiring reactions in the body (Figure 5.4). Glucose does not act catalytically. When it is oxidized it is not recycled back to glucose.

## TALLY OF THE ATPs PRODUCED

The number of ATPs produced by complete oxidation of a molecule of glucose can be calculated at this point. This calculation is not directly relevant to determination of the energy requirement. NADH from the Krebs cycle is responsible for the production of many more ATPs than is glycolysis. In glycolysis, an ATP is expended at the points of hexokinase and phosphofructokinase. For each molecule

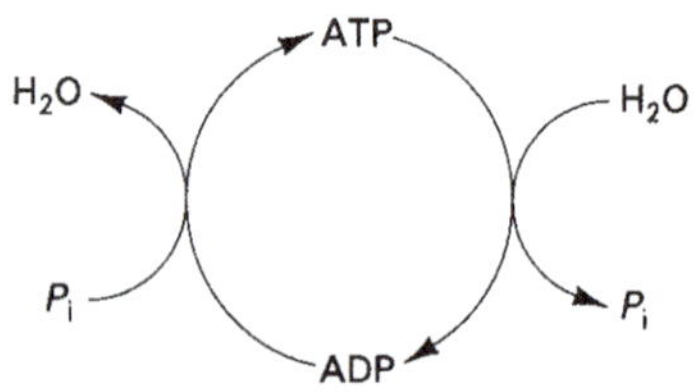

**FIGURE 5.4** One molecule of water is involved in the hydrolysis and synthesis of ATP. Continued synthesis of ATP during the course of the day does not result in net production of water in the body. This is because ATP is continuously hydrolyzed over the course of the day when it participates in ATP-requiring reactions.

of glucose entering glycolysis, four ATPs are generated at the points of phospho-glycerokinase and pyruvate kinase. For each molecule of glucose entering gly-colysis, two NADHs are produced at the point of glyceraldehyde-3-phosphate dehydrogenase. These two NADHs can result in the generation of six ATPs in the mitochondrion. The tally of ATPs produced at this point is eight ATPs.

For every molecule of glucose entering metabolism, the action of pyruvate dehydrogenase results in the reduction of two NADs to two NADHs. Two NADHs are equivalent to six ATPs. The $NAD^+$ and FAD reducing steps of the Krebs cycle produce the equivalent of 22 more molecules of ATP per molecule of glucose. The production of two GTPs, at the point of succinyl thiokinase, yields the equivalent of two molecules of ATP per molecule of glucose. The energy contained in the phosphate bonds of GTP is equivalent to that in ATP. Thus, oxidation of two molecules of pyruvate to $CO_2$ produces 30 ATPs. The sum of the ATPs produced per molecule of glucose via glycolysis (8 ATPs) and those produced in the mito-chondria (30 ATPs) is 38 ATPs.

A slight adjustment of the number 38 is required. The transport of electrons from the cytoplasm to the mitochondria requires a small amount of energy. This energy is used to drive the malate–aspartate shuttle. It is thought that 0.5 mole-cules of ATP are consumed with each turn of the malate–aspartate shuttle. Hence, transport of the electrons from the two NADHs generated by glycolysis into the mitochondria requires the input of 1 ATP. The sum of the ATPs (or ATP equiva-lents) produced by complete oxidation of glucose to $CO_2$ is 37.

The malate–aspartate shuttle is important in the *liver*. The malate–aspartate shuttle (see Figure 4.63 in Chapter 4), allows the delivery of electrons to the very first step in the respiratory chain (this step is NADH dehydrogenase) (Figure 5.3). Another shuttle, the glycerol–phosphate shuttle, is more important in *muscle*. The glycerol–phosphate shuttle does not yield the cell as much energy as the malate–aspartate shuttle, since the glycerol–phosphate shuttle delivers electrons at the point of coenzyme Q reductase, not at the point of NADH dehydrogenase (Figure 5.3). Reduced NAD in the mitochondrion is worth 3 ATPs, whereas reduced FAD is worth only 2 ATPs. Hence, complete oxidation of each glucose in *muscle* results in the generation not of 38 ATPs, but of 36 ATPs.

The glycerol–phosphate shuttle is shown in Figure 5.5. In the electron's journey, the starting point is reduced NAD (in the cytoplasm), while the ending point is at CoQ reductase. With reduction of the FAD cofactor of CoQ reductase, the electrons are, in turn, transferred to the respiratory chain. The FAD cofactor remains bound to the mitochondrial G3P dehydrogenase, when it changes back and forth, from FAD to $FADH_2$ and back, though the cofactor is not bound covalently to the enzyme (Brown et al., 1994, Garrib and McMurray, 1986).

## Complete Oxidation of Glucose to $CO_2$

Complete oxidation of glucose to $CO_2$ yields 6 $CO_2$ and 12 $H_2$, as shown in Table 5.2. The 12 $H_2$ are transferred to 6 $O_2$ by cytochrome *c* oxidase. Therefore, the RQ for the complete oxidation of glucose to $CO_2$ in the cell is 6.0/6.0 = 1.0.

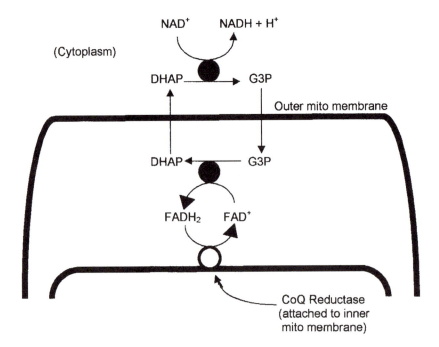

**FIGURE 5.5** Glycerol–phosphate shuttle. The glycerol–phosphate shuttle enables energy produced by glycolysis, in muscle, to be utilized for ATP synthesis. The shuttle serves to transfer electrons, initially occurring in reduced cytoplasmic NAD, to reduced mitochondrial FAD. The black dots represent glycerol phosphate dehydrogenase, which occurs as a soluble form in the cytoplasm, and as a membrane-bound version in the mitochondrion. The open circle represents coenzyme Q reductase, which occurs along with the respiratory chain in the inner membrane. Dihydroxyacetone phosphate (DHAP) and glycerol-3-phosphate (G3P) can pass freely through the outer mitochondrial membrane, but not through the inner membrane.

**TABLE 5.2** Points in Glycolysis and the Krebs Cycle at which Cofactors Are Reduced and Carbon Dioxide Is Released

| Enzyme | NADH + H⁺ or FADH₂ produced per glucose | $CO_2$ produced per glucose |
|---|---|---|
| Glyceraldehyde-3-phosphate dehydrogenase | 2 NADH + H⁺ | 0 |
| Pyruvate dehydrogenase | 2 NADH + H⁺ | 2 |
| Isocitrate dehydrogenase | 2 NADH + H⁺ | 2 |
| α-Ketoglutarate dehydrogenase | 2 NADH + H⁺ | 2 |
| Succinate dehydrogenase | 2 FADH₂ | 0 |
| Malate dehydrogenase | 2 NADH + H⁺ | 0 |
| | Equivalent of 12 $H_2$ | 6 $CO_2$ |

**TABLE 5.3** Points in the Krebs Cycle at which Cofactors Are Reduced and Carbon Dioxide Is Released

| Enzyme | NADH + H⁺ or FADH₂ produced per acetyl-CoA | CO₂ produced per acetyl-CoA |
|---|---|---|
| Isocitrate dehydrogenase | 1 NADH + H⁺ | 1 |
| α-Ketoglutarate dehydrogenase | 1 NADH + H⁺ | 1 |
| Succinate dehydrogenase | 1 FADH₂ | 0 |
| Malate dehydrogenase | 1 NADH + H⁺ | 0 |
| | Equivalent of 4 H₂ | 2 CO₂ |

## Complete Oxidation of the Acetyl Group of Acetyl-CoA to CO₂

The steps in the Krebs cycle generating the equivalent of an $H_2$ (NADH + H⁺ or FADH₂) or producing $CO_2$ are listed in Table 5.3. The RQ of the acetyl group introduced into the Krebs cycle is 1.0. The value of 1.0 was calculated from the knowledge that 2 $CO_2$ are produced, and 2 $O_2$ are required to accept the equivalent of 4 $H_2$. These hydrogens are produced in the steps catalyzed by dehydrogenases.

## Complete Oxidation of Free Fatty Acids to CO₂

Free fatty acids are converted to fatty acyl-CoA by fatty acid thiokinase (Figure 5.6). The fatty acids are broken down, not when in their free form but in the thiol ester form. Breakdown of fatty acids results in step-by-step discharge of two-carbon units as acetyl-CoA. This breakdown also results in the production of FADH₂ and NADH + H⁺. For each two-carbon unit liberated, one molecule each of FAD and NAD⁺ is reduced (Figure 5.7).

The fatty acid oxidation pathway comprises a sequence of steps frequently encountered in biology: (1) oxidation of an alkane to produce an alkene; (2) hydration of the alkene to form a hydroxyl group; and (3) oxidation of the hydroxyl group to form a keto group. This three-step sequence is also found in the Krebs cycle and the isoleucine catabolic pathway.

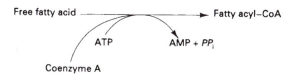

**FIGURE 5.6** Fatty acid thiokinase.

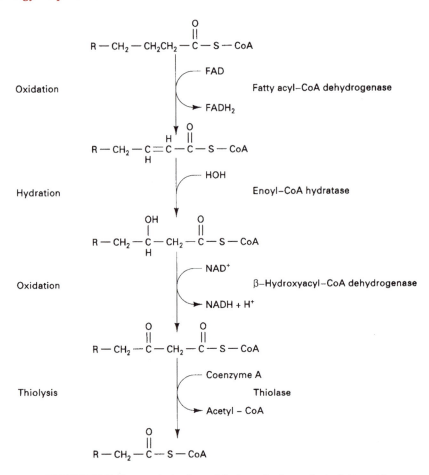

**FIGURE 5.7** One cycle in the oxidation of a long-chain fatty acid.

The final step in fatty acid catabolism involves the attack of a molecule of coenzyme A at the 3-carbon of the fatty acid, resulting in the discharge of a two-carbon unit (acetyl group). The sole breakdown product of even-carbon-numbered fatty acids, such as palmitic acid, is acetyl-CoA. A small fraction of the fatty acids encountered in the diet have an odd number of carbons. The catabolism of these fatty acids yields a number of molecules of acetyl-CoA plus one molecule of propionyl-CoA.

The cycle of reactions resulting in the discharge of a two-carbon unit from a fatty acid as acetyl-CoA produces no $CO_2$ and the equivalent of 2 $H_2$. These 2 $H_2$ can be used for the reduction of *one molecule of oxygen*. Hence, activation of the fatty acid oxidation pathway in a tissue or animal would be expected to result in a decrease in the RQ for the organism. The breakdown of fatty acids followed by the oxidation of acetyl-CoA in the Krebs cycle results in the production of 2 $CO_2$ and the consumption of 3 $O_2$. Thus, the RQ for an organ or animal relying primarily on fatty acid oxidation as an energy source would be expected to be 2.0/3.0, or 0.66. The RQ for fat is about 0.66.

*Summary of Three Methods for Acquiring RQ Values*

The method for calculating the RQ from chemistry was illustrated by simple examples involving paraffin wax and methanol. Although the details for calculating the RQs of glucose and fat were not shown, these RQs are known to be 1.0 and 0.70, respectively. Calculation of the RQs of glucose and a fatty acid on the basis of biochemical pathways yielded values of 1.0 and 0.66, respectively. The RQ can also be determined from analysis of gases produced by the combustion of fuels in a furnace or in a living animal. Next, data from human subjects at rest and during exercise will be presented. One might be surprised to see that the RQ of a human being changes, over the course of time, during prolonged rest and prolonged exercise. These RQ values change with no alteration in the overall rate of energy consumption or in the metabolic rate. These RQ value changes are due to shifts in the types of body fuels being oxidized.

## THE RESPIRATORY QUOTIENT DURING REST AND EXERCISE

The following studies involve human subjects. The first study depicts the results from periodic RQ measurements during rest after an overnight fast. The subjects spent the entire day in bed. Breath samples were taken at hourly intervals. The data illustrate a steady decrease in the RQ over the course of 4 to 5 hours (Figure 5.8). From the chemical and biochemical considerations presented earlier, one might deduce that the RQ data indicate a trend away from the use of carbohydrate fuels and toward the combustion of fat fuels, as the initial RQ was about 0.80 and the final RQ was about 0.75. The RQ never reached 0.70, indicating that fat was never the sole source of energy. In the second study, the RQ was measured prior to exercise, during a 45-minute period of bicycling, and during a subsequent 30-minute recovery period. The results indicate that the major source of energy at rest was fat, whereas that during the initial stages of exercise was glycogen. The

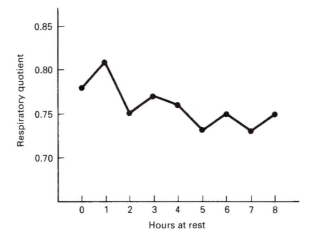

**FIGURE 5.8** Respiratory quotient during rest after an overnight fast. (Redrawn with permission from Leff *et al.*, 1987.)

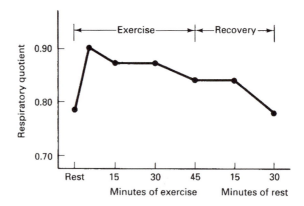

**FIGURE 5.9** Respiratory quotient prior to exercise, during bicycling, and during a subsequent recovery period. The RQ increases with exercise, indicating a shift from fat oxidation to glycogen oxidation. (Reprinted with permission from Murray *et al.*, 1977.)

data in Figure 5.9 illustrate a trend back toward fat oxidation and away from glycogen oxidation with prolonged exercise.

## FATTY ACID BIOSYNTHESIS

The RQ is influenced by the net rate of fatty acid biosynthesis. Fatty acid synthesis involves the conversion of carbohydrate, via the acetyl-CoA intermediate, to long-chain fatty acids. The synthesis of fatty acids requires reduced $NADP^+$ as a cofactor. It involves the consumption of two molecules of $NADPH + H^+$ for each 2-carbon unit incorporated into the fatty acid. The $NADPH + H^+$ is supplied by two separate pathways: the **pentose phosphate pathway** (PPP) and the **malic enzyme/citrate lyase pathway**.
    The overall stoichiometry of the pentose phosphate pathway is:

$$\text{Glucose-6-P} + 12\ NADP^+ + 7\ H_2O \rightarrow 6\ CO_2 + 12\ NADPH + 12\ H^+ + P_i \qquad (5.14)$$

The pathway is cytoplasmic and results in the production of $CO_2$, but no consumption of oxygen. The $NADPH + H^+$ is used in fatty acid synthesis. Hence, an increase in activity of the PPP would be expected to result in an increase in the value of the RQ. The two enzymes of the PPP directly involved in $NADP^+$ reduction are glucose-6-phosphate dehydrogenase and 6-phosphogluconate dehydrogenase, as shown under Thiamin in Chapter 9.
    The malic enzyme/citrate lyase pathway is shown in Figure 5.10. The 2-carbon units (acetyl groups) for fatty acid synthesis are supplied by the activity of citrate lyase, which may be considered an enzyme of fatty acid biosynthesis. The reduced $NADP^+$ is supplied at the point of malic enzyme. Figure 5.10 reveals no net production or utilization of $NAD^+$ in the cytoplasm. The $NADPH + H^+$ generated in the cytoplasm is used for fatty acid synthesis, which regenerates $NADP^+$. One molecule of $CO_2$ is produced in the cytoplasm. The diagram reveals no net production or utilization of $CO_2$ in the mitochondrion. One molecule of $NAD^+$ is

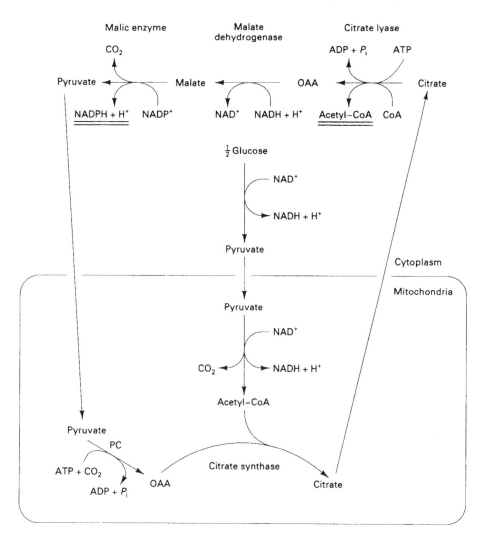

**FIGURE 5.10** Malic enzyme/citrate lyase pathway. Citrate lyase is used for producing acetyl-CoA in the cytoplasm, the site of fatty acid synthase. Acetyl-CoA is used by fatty acid synthase for the synthesis of fatty acids. Malic enzyme catalyzes the reduction of $NADP^+$, which is required as a cofactor by fatty acid synthase.

reduced in the mitochondrion. Each 2-carbon unit used for fatty acid synthesis results in the net production of one molecule of $CO_2$ and one NADH + $H^+$. A shift toward this type of metabolism would be expected to result in a rise in the RQ.

EXERCISE

The synthesis of an 18-carbon fatty acid requires the supply of nine 2-carbon units by citrate lyase. This would be expected to result in the concomitant generation of nine molecules of NADPH + $H^+$. Would you expect the cycle (Figure 5.10) to supply most, or only a small fraction, of the reduced $NADP^+$ required for fatty acid synthesis?

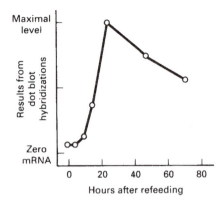

**FIGURE 5.11** Increase in mRNA coding for glucose-6-phosphate dehydrogenase over time. (Redrawn with permission from Prostko *et al.*, 1989.)

## Regulation of NADPH Production

Before continuing with the matter of RQs and the energy requirement, the importance of malic enzyme and the PPP in supplying reduced NADP$^+$ for fatty acid synthesis should be emphasized. Two studies are explored. They involve the fasting of animals for 2 days, followed by refeeding with a fat-free high-carbohydrate diet. This particular manipulation results in maximal increases in the rate of fatty acid biosynthesis and in the activity of acetyl-CoA carboxylase. The response to dietary changes was monitored using the dot blot method. Figure 5.11 provides data from an experiment with rats. The data show the increases in mRNA coding for glucose-6-phosphate dehydrogenase over the course of time (O). The response to feeding was also monitored in chickens, as shown in Figure 5.12; these data show increases in mRNA coding for malic enzyme, but no increase in the amount of mRNA coding for glyceraldehyde-3-phosphate dehydrogenase. The dot blot specific for the latter enzyme was intended as a control. The control indicated that

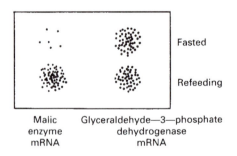

**FIGURE 5.12** Dot blot assay for concentrations of specific types of messenger RNA (consult Appendix 2 for details). The amount of mRNA coding for malic enzyme is low during fasting (since there would be no purpose in synthesizing fat during this period), and increases with refeeding (since this is the time to restore energy stores by synthesizing fat from dietary sugars). (Redrawn with permission from Xiao-Jun *et al.*, 1990.)

refeeding resulted in increases in the mRNA coding for enzymes of fatty acid synthesis, rather than all types of mRNA.

## Fatty Acid Synthase

The reactions catalyzed by fatty acid synthase are shown in Figure 5.13. Synthesis of fatty acids involves the step-by-step addition of two-carbon units to a growing chain. Initially, fatty acid synthase contains an acetyl group (#1). Acetyl-CoA carboxylase catalyzes the ATP-dependent synthesis of malonyl-CoA, which is transferred to fatty acid synthase (#2). At this point, an acetyl group resides at one site of the enzyme, bound to a sulfhydryl group, whereas a malonyl group resides at a second site of the enzyme, also bound to a sulfhydryl group. In moving from Step 2 to Step 3, the acyl group at the first sulfhydryl group attacks the malonyl group. The attack occurs at the central carbon of the malonyl group. This results in the discharge of $CO_2$ and elongation of the growing fatty acid chain by two carbons. At this point, the chain is four carbons long.

The next few steps involve reduction of the four-carbon chain. The process takes the form of a reduction step, a dehydration step, and another reduction step (#3–#6). The reduced chain is transferred to the first sulfhydryl group (#6, #7). Finally, another malonyl group is transferred to the enzyme (#7, #8). The malonyl group is transferred to the second sulfhydryl group. Transfer of the four-carbon unit residing at the first sulfhydryl group to the malonyl group generates a six-carbon unit. The process is continued until a long-chain fatty acid is formed.

The final step in fatty acid synthesis is the discharge of the fatty acid from the sulfhydryl group of fatty acid synthase. This discharge involves the attack of a molecule of coenzyme A, resulting in the release of the fatty acid as fatty acyl-CoA, as shown in Figure 5.14.

In a discussion of fatty acid synthase, the swinging arm of the enzyme should not be overlooked. One of the sulfhydryl groups of the enzyme takes the form of a residue of cysteine. But the other takes the form of **phosphopantetheine**. Phosphopantetheine is covalently bound to the enzyme. It has a structure similar to that of coenzyme A. Both phosphopantetheine and coenzyme A contain sulfhydryl groups. Phosphopantetheine, however, is covalently bound to a residue of serine. The growing fatty acid chain is bound to phosphopantetheine during Steps 3 through 6. The phosphopantetheine acts as a swinging arm, allowing the growing fatty acid to come in contact with the many active sites of fatty acid synthase.

The RQ for the conversion of glucose to fatty acids can be calculated from the biochemical considerations presented so far. This RQ would not be expected to occur in any living tissue or in any animal, because this RQ represents an extreme case in which there is no oxidation of fatty acids or of carbohydrates in the Krebs cycle. Where the rate of fatty acid synthesis is equal to the rate of fatty acid oxidation, the RQ would be expected to be 1.0. Where the rate of fatty acid synthesis is twice as great as the rate of fatty acid oxidation, the RQ would be greater than 1.0. One would not, however, expect to encounter a tissue in which there is only fatty acid synthesis and no fatty acid oxidation.

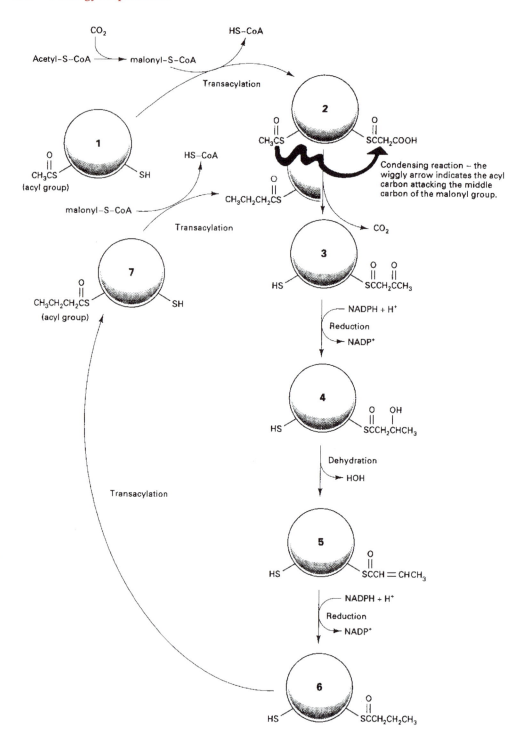

**FIGURE 5.13** Fatty acid synthase catalyzes the addition of 2-carbon units to the acyl end of a growing long-chain fatty acid. The cyclic nature of the fatty acid biosynthetic pathway is indicated by the return of that enzyme, at step 7, back to step 2. (Redrawn with permission from Wakil and Stoops, 1983.)

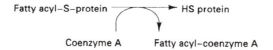

FIGURE 5.14 Removal of the finished fatty acid from the enzyme.

## RESPIRATORY QUOTIENT DATA FROM HUMAN STUDIES

*RQ Values from Humans Consuming Excess Energy*

An RQ greater than 1.0 indicates net synthesis of fatty acids and triglycerides. Under these conditions, triglycerides are mobilized from adipose tissue and fatty acids are oxidized in various tissues. But the rate of fatty acid synthesis from carbohydrates is greater than the rate of fatty acid oxidation. The net synthesis of fatty acids means that there is an accumulation of fat in the body. The accumulation of fat may seem relatively innocuous, but it can have deleterious effects. For instance, when intravenous feeding includes excessive glucose for prolonged periods, fat may accumulate in the liver. This pathological condition is known as "fatty liver." Although reversible, it may trigger various problems.

The following study involves human subjects who were fed intravenously with a low-energy diet (64 kJ/kg body weight) for 4 days, a medium-energy diet (157 kJ/kg) the next 4 days, and then a high-energy diet (245 kJ/kg) for the final 4 days. Breath samples taken on each day were used for analysis of $CO_2$ production and $O_2$ utilization and, hence, calculation of the RQs. The data (Figure 5.15) demonstrate that the RQs rose above 1.0 with the medium- and high-energy diets. This indicates that there was a net synthesis of fat in the body. The increased amount of fat was not synthesized from the body's stores of glycogen, but from the glucose

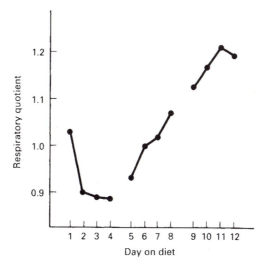

FIGURE 5.15 Increase in respiratory quotients with increase in energy content of diet. (Redrawn with permission from Elwyn *et al.*, 1979.)

in the diet. The data indicate the value of RQ measurements in helping patients to achieve an energy balance, particularly patients who are not able to feed themselves.

### RQ Data from Humans Consuming Equal-Energy Diets that Are High in Carbohydrate or in Fat

A variation of the preceding study involved the transition from a carbohydrate-based diet to a fat-based diet, where both diets supplied equal amounts of energy per day (Schrauwen *et al.*, 1997). As indicated in Figure 5.16, the subjects consumed the carbohydrate-based diet up to and including day 6 of the feeding period, and then consumed the fat-based diet for the next 7 days. Both diets supplied 10.0 MJ/day to each human subject. The body weights of the subjects did not change over the course of the 2 week feeding period. On the other hand, the RQ values for the subjects, which were initially about 0.88, declined to about 0.80 with feeding of the high-fat diet. The RQ of the subjects were found to be similar to the **food quotient**. The food quotient is the RQ of the food, as determined by burning the food in an atmosphere of pure oxygen.

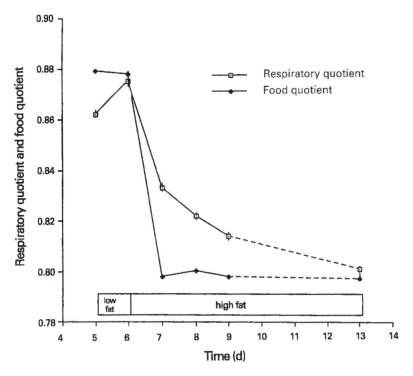

**FIGURE 5.16** Respiratory quotient measurements of human subjects (○); food quotient values (●). Human subjects consumed a low-fat diet for 6 days, followed by a high-fat diet for 7 days. Subjects lived inside separate respiratory chambers for days 4–10, then left the chambers, and returned for days 12–14. Each respiratory chamber was a small room ventilated at a measured rate with air. In-going air was sampled and measured every 15 min, and outgoing air was sampled and measured every 5 min, for the entire period that the subjects lived in the chambers. The RQ values for each subject was calculated for each day, as indicated in the figure. (Redrawn with permission from Schrauwen *et al.*, 1997.)

A question that arises is: "Why did the abrupt switch in diet result in only a gradual change in the RQ?" The researchers thought that the gradual change in RQ was due to the body's temporary increase in glycogen combustion during the transition period.

Horton *et al.* (1995) performed still another variation of the preceding sort of study. All human subjects were fed a baseline or maintenance diet for one week. The baseline diet was designed to result in a maintenance of body weight. Then, the diet was changed to a diet that provided an additional 50% of energy as fat, or an additional 50% of energy as carbohydrate. This study is similar to the first study described earlier (Elwyn *et al.*, 1979), in that overfeeding was used. This study is similar to the second study detailed (Schrauwen *et al.*, 1997), in that two groups were fed diets of equivalent energy. RQ measurements were taken. However, researchers were interested in determining more than simply the style of energy metabolism used by the body. They wanted to determine if the switch to overfeeding resulted in the accumulation of fat via fatty acid biosynthesis or in the accumulation of fat via direct deposition of the fats in the diet. The body's total energy expenditure was calculated from the RQ and from $O_2$ consumption, as detailed later in this chapter, and body composition was determined by the underwater weighing technique (see Obesity chapter).

Horton *et al.* (1995) found that, with carbohydrate–overfeeding, much of the excess energy (75-85%) seemed to be stored as fat. But with fat–overfeeding, nearly all (90-95%) of the excess energy was stored as fat. Surprisingly, the fat storage which occurred with carbohydrate overfeeding seemed to be *due to a decline* in the ongoing basal level of fat oxidation and *not to an increase* in conversion of carbohydrate to fat.

## *What Happens in the Body When Overeating Leads to Obesity?*

The picture that seems to be emerging regarding overeating and obesity is as follows: Eating too much fat can result in obesity because of the direct deposit of the fat from the diet, while eating too much carbohydrate can result in obesity because of a tendency to spare the fat that exists in the diet, and direct deposition of this dietary fat in the adipose tissue. The preceding conclusions extend somewhat beyond the bounds of this chapter, but are intended to illustrate the power of RQ measurements and related techniques.

## PROTEIN OXIDATION AND THE RESPIRATORY QUOTIENT

The influence of protein oxidation on the RQ is generally ignored in clinical and experimental studies involving respiratory measurements. Hence, the RQ determined in these studies is sometimes called the **nonprotein RQ**. The contributions of $CO_2$ produced and $O_2$ consumed by protein oxidation are small compared with those made by fat and carbohydrate oxidation. The hydrolysis of proteins to amino acids does not contribute to the RQ. The RQ associated with the oxidation of mixtures of different amino acids is 0.80. The oxidation of protein (70 g) produces 2.4 moles of $CO_2$ and consumes 3.0 moles of $O_2$ (King *et al.*, 1984). The RQ associated with the oxidation of leucine, for example, can be calculated as follows.

The calculation assumes that all of the nitrogen released is excreted as urea. Urea, as mentioned in Chapter 8, is the excretory form of most of the nitrogen resulting from protein and amino acid catabolism.

1. Oxidation of leucine results in the production of $CO_2$ and urea:

$$2\ CH_3CHCH_2CHCOOH \rightarrow 1\ H_2N\!\!-\!\!\overset{\displaystyle O}{\overset{\displaystyle \|}{C}}\!\!-\!\!NH_2 + 11\ CO_2 \qquad (5.15)$$
$$\underset{\text{(Leucine)}}{\overset{\displaystyle |}{CH_3}\ \ \overset{\displaystyle |}{NH_2}} \qquad\qquad \underset{\text{(Urea)}}{}$$

2. The hydrogens may be balanced by adding water to the products:

$$2\ \text{Leucine} \rightarrow 1\ \text{urea} + 11\ CO_2 + 11\ H_2O \qquad (5.16)$$

3. The oxygens can be balanced by adding $O_2$ to the substrates:

$$15\ O_2 + 2\ \text{leucine} \rightarrow 1\ \text{urea} + 11\ CO_2 + 11\ H_2O \qquad (5.17)$$

4. The RQ for the oxidation of leucine is:

$$CO_2/O_2 = 11/15 = 0.73.$$

The total volumes of $O_2$ consumed and $CO_2$ produced with the oxidation of 1 g of glucose, fat, and protein are listed in Table 5.4. With protein catabolism, 1 g of urinary nitrogen is associated with the consumption of 6.04 liters of $O_2$ and the production of 4.89 liters of $CO_2$.

Respiratory measurements alone can be used to determine the amounts of carbohydrate and fat oxidized in the body over a given period. These amounts can be more accurately determined by including data from urinary nitrogen measurements in the calculations. According to Frayn (1983), the grams of carbohydrate (c) and fat (f) oxidized per minute can be calculated from the formulas

$$c = 4.55\ V_{CO_2} - 3.21\ V_{O_2} - 2.87n$$
$$f = 1.67\ V_{CO_2} - 1.67\ V_{O_2} - 1.92n$$

TABLE 5.4 Total Volumes of $O_2$ Consumed and Produced upon Oxidation of Glucose, Fat, and Protein

|  | Volume of $O_2$ consumed (liters) | Volume of $CO_2$ formed (liters) |
| --- | --- | --- |
| Glucose | 0.746 | 0.746 |
| Fat | 2.03 | 1.43 |
| Protein | 0.966 | 0.782 |

Source: Frayn (1983).

where $n$ is the grams of urinary nitrogen excreted per minute and $V_{O_2}$ and $V_{CO_2}$ are the volumes of $O_2$ consumed and $CO_2$ produced per minute (liters/min), respectively.

EXERCISE

Calculate the RQ for the complete oxidation of glutamic acid where all of the nitrogen is excreted in the form of ammonia. Calculate the RQ where all of the nitrogen is excreted in the form of urea.

## CALCULATING ENERGY EXPENDITURES

The amount of energy expended during an entire day can be determined by measuring the liters, or moles, of oxygen consumed by a subject for the entire day. The subject can maintain an energy balance by consuming an amount of food with an energy content similar to that expended. Energy expenditure can be measured during specific activities, such as sleeping, walking, tennis playing, and coal mining. In these cases the liters or moles of oxygen consumed over the course of a 10-minute period might be determined. The amount of energy can be determined fairly accurately by measuring the oxygen consumed. A more accurate estimate can be made using the rate of oxygen consumption and the RQ during the test activity.

The concentrations of $CO_2$ and $O_2$ in the atmosphere are fairly constant at sea level. However, the concentrations of these molecules in exhaled air differs from that in the inhaled air. The concentration of $CO_2$ is greater, while that of $O_2$ is lesser. The rate of $CO_2$ production can be calculated from the formula:

<table>
<tr><td align="center">Difference in CO<sub>2</sub> concentration in<br>inhaled and exhaled air<br>(moles/liter)</td><td align="center">×</td><td align="center">Liters of air exhaled per minute<br>(liters/min)</td></tr>
</table>

The unit "liter(s)" will cancel when the two quantities are multiplied, with the consequence that the rate of $CO_2$ production is expressed in moles/min. This rate may also be expressed in liters of $CO_2$ per minute; one mole of a gas has a volume of 22.4 liters.

The rate of $O_2$ consumption can be calculated by formulas similar to the preceding. A portable machine called a respirometer can be used to collect samples of exhaled gas during various types of work. The samples are automatically collected at intervals and stored in a bag. The machine also records the total volume of gas exhaled during the time it is worn. An agricultural worker is shown wearing the machine in Figure 5.17.

How is the rate of oxygen consumption and the RQ used to calculate the energy used by the body? The connection between complete combustion of glucose (or fat) to carbon dioxide and energy is determined with a **bomb calorimeter**. The bomb calorimeter is a machine used for the combustion of various foods in an atmosphere of pure oxygen and, consequently, measurement of the heat produced. The amount of heat produced is expressed in joules. Any fuel — such as paraffin, methane, starch, pork, or beans — can be placed in the bomb calorimeter. A spark

FIGURE 5.17 Agricultural worker wearing a portable respirometer.

ignites the fuel. The fuel rapidly and completely burns in the oxygen atmosphere, producing carbon dioxide and water. The water jacket surrounding the combustion chamber absorbs the heat released from the combustion. The rise in temperature of the water and the amount of energy released from the test food are related by the following definition: One kilocalorie (4.184 kJ) is the amount of heat necessary to raise 1 kg of water by 1°C.

The technique of bomb calorimetry has revealed that complete combustion of 1 g of glucose results in the discharge of 15.9 kJ, and complete combustion of 1 g of fat results in the discharge of 38 kJ. A diagram of a bomb calorimeter is shown in Figure 5.18. Use of the chemical considerations outlined at the beginning of this

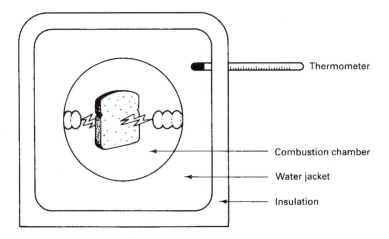

FIGURE 5.18 Bomb calorimeter. The cartoon depicts ignition of a dried food sample in the bomb calorimeter.

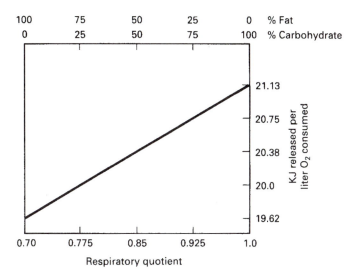

**FIGURE 5.19** Data from oxygen consumption measurements and RQ calculations can be used to determine the internal energy fuels being used.

chapter can reveal the amount (moles or liters) of $O_2$ consumed per gram of fuel combusted.

Combination of all of the preceding material leads to values for the amount of energy discharged per liter of oxygen consumed when glucose is combusted: *21 kJ/liter of $O_2$*. The preceding information can also be used to calculate the amount of energy discharged per liter of oxygen consumed when fat is combusted: *19.5 kJ/liter of $O_2$*. In periods when the body has an RQ of 1.0, it can be inferred that each liter of oxygen consumed represents the burning of 21 kJ in the body. In periods when the body has an RQ of 0.7, it can be inferred that each liter of oxygen consumed represents the burning of 19.5 kJ. In periods of rest or work when the RQ is between 0.7 and 1.0, Figure 5.19 may be used to calculate the energy expenditure from the rate of oxygen consumption.

The amount of energy expended by the body varies from about 19 to 21 kJ/liter of oxygen consumed by the body. For orientation, one might recall that nearly all of the oxygen in the body is consumed in a reaction catalyzed by cytochrome oxidase. Because of the small difference in energy expended per liter of $O_2$ consumed with different types of energy metabolism, a shortcut can be used. This shortcut consists of measurement only of the amount of $O_2$ consumed, while ignoring the amount of $CO_2$ produced and assuming that each liter of $O_2$ consumed is equivalent to about 20 kJ.

The methods described for measuring the rate of $O_2$ consumed by the body, correlating it to the energy released and oxygen consumed in the bomb calorimeter, and calculating the amount of energy consumed by the body constitute **indirect calorimetry**. It is called "indirect calorimetry" because the amount of energy consumed by the body is measured only indirectly; the calculation was dependent on data from the complete combustion of glucose and fat in the bomb calorimeter.

## DIRECT CALORIMETRY

The energy expenditure of an animal or human may also be determined by the method of **direct calorimetry**. Direct calorimetry requires the use of an insulated room, chamber, or suit for the human or animal. The enclosure contains a water jacket. The water passes from one end of the jacket to the other, maintaining the room, chamber, or suit at a constant temperature. The temperature of the water leaving the jacket is used to calculate the energy expended by the subject. The principles behind the use of the chamber are identical to those behind the use of the bomb calorimeter. The major difference is that in bomb calorimetry combustion is catalyzed by a small spark. In addition, in the bomb calorimeter oxygen is present at a high pressure to facilitate combustion. With direct calorimetry, combustion is catalyzed by enzymes. This combustion proceeds more slowly than that catalyzed by a spark, and the temperature of the subject does not increase much over the normal resting body temperature with the various activities.

Figure 5.20 illustrates the general principles behind direct calorimetry. The subject (an animal or human subject) resides in the calorimeter and engages in a number of activities, such as sleeping, resting, and exercising. The rise in temperature in the surrounding water jacket during a given period is used to calculate the energy discharged (fuel oxidized) during that period, as in direct calorimetry.

The methods of indirect and direct calorimetry may not always result in the same values for energy expenditure. Indirect calorimetry is a measure of the heat produced by oxidative processes. Direct calorimetry measures the rate of dissipation of heat from the body. An increase in the rate of heat production, as with exercise, may not always result in an immediate, measurable increase in heat released by the body (from the skin). Instead, the increase in heat production may result in a *rise in body temperature*. That part of the energy requirement used to raise

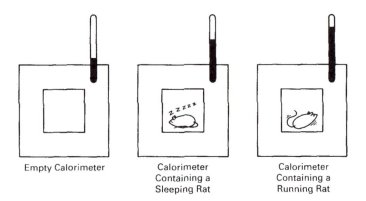

|  Empty Calorimeter | Calorimeter Containing a Sleeping Rat | Calorimeter Containing a Running Rat |

**FIGURE 5.20**  Direct calorimetry. The cartoon depicts an empty calorimeter, one containing a sleeping rat, and one containing an exercising rat. Data from temperature measurements can be used to calculate the energy requirement per unit of time during any sort of physical activity. In actual practice, the temperature of the calorimeter is maintained at a fairly constant temperature by water flowing through the insulating jacket. The rate of rise in temperature of the outflow water is measured and used to calculate the energy discharged per unit of time.

the body temperature over its initial, resting temperature can be detected by indirect calorimetry, but not by direct calorimetry. To obtain an accurate value for energy expenditure with exercise using the method of direct calorimetry, one would have to *continue measuring heat output until the body reached its original, resting temperature* (Jequier, 1987). When the body temperatures immediately prior to and after the interval of interest are identical, the methods of indirect and direct calorimetry should yield nearly identical values.

## METABOLIC RATE

Calorimetric methods were used for determining the **recommended daily allowances** (RDAs) for food energy (Table 5.5). The energy expended during typical daily activities (Table 5.6) was totaled and used to calculate the RDAs for persons of various ages (Table 5.5). RDAs were calculated for light or moderate activities. Those who are immobile should consume less food energy, and those with physically strenuous occupations should consume more food energy, to remain in energy balance. The reference point in addressing the energy needs of an individ-

**TABLE 5.5** Recommended Daily Energy Intake for Persons with Lightly to Moderately Active Lifestyles Having the Median Weight and Height for the U.S. Population

| Category | Age (years) | Weight kg | Weight lb | Height cm | Height in. | Energy needs (kJ/day) |
|---|---|---|---|---|---|---|
| Infants | 0–0.5 | 6 | 13 | 60 | 24 | 2720 |
|  | 0.5–1.0 | 9 | 20 | 71 | 28 | 3556 |
| Children | 1–3 | 13 | 29 | 90 | 35 | 5439 |
|  | 4–6 | 20 | 44 | 112 | 44 | 7531 |
|  | 7–10 | 28 | 62 | 132 | 52 | 8368 |
| Males | 11–14 | 45 | 99 | 157 | 62 | 10460 |
|  | 15–18 | 66 | 145 | 176 | 69 | 12552 |
|  | 19–24 | 72 | 160 | 177 | 70 | 12133 |
|  | 25–50 | 79 | 174 | 176 | 70 | 12133 |
|  | 51+ | 77 | 170 | 173 | 68 | 9623 |
| Females | 11–14 | 46 | 101 | 157 | 62 | 9205 |
|  | 15–18 | 55 | 120 | 163 | 64 | 9205 |
|  | 19–24 | 58 | 128 | 164 | 65 | 9205 |
|  | 25–50 | 63 | 138 | 163 | 64 | 9205 |
|  | 51+ | 65 | 143 | 160 | 63 | 7950 |
| Pregnancy | 1st trimester |  |  |  |  | +0 |
|  | 2nd trimester |  |  |  |  | +1255 |
|  | 3rd trimester |  |  |  |  | +1255 |
| Lactation | 1st 6 months |  |  |  |  | +2090 |
| 2nd 6 months |  |  |  |  |  | +2090 |

*Source*: Food and Nutrition Board (1989).

**TABLE 5.6** Approximate Rates of Energy Expenditure for Various Activities

| Activity | Times greater than the BMR |
|---|---|
| Lying at rest | 1.0 |
| Very light: typing, driving, painting, sewing, cooking, playing cards | 1.5 |
| Light: walking on a level surface, carpentry, house cleaning, golf, ping-pong | 2.5 |
| Moderate: garden work, tennis, carrying heavy boxes, dancing (country dancing or clogging) | 5.0 |
| Heavy: carrying heavy boxes uphill, chopping down trees, digging a deep hole, basketball, football | 7.0 |

*Source*: Food and Nutrition Board (1989).

ual is the **basal metabolic rate** (BMR). The BMR is measured under standard conditions. Another reference point is the **resting metabolic rate** (RMR). The RMR is measured under resting conditions within an hour or a few hours of consuming a meal. The RMR is somewhat higher than the BMR but is more variable. It is more convenient to measure the RMR than the BMR when surveying populations.

The basal metabolic rate (BMR) is the rate of use of the body's energy stores. The BMR is determined while at rest after an overnight fast. Generally, the BMR is determined soon after awakening and 14 hours after the last meal. The BMR is fairly constant over the course of a day, as illustrated in Figure 5.21. These data are from a study of a human subject. Respiratory gas samples were taken at various intervals and used for $O_2$ and $CO_2$ analyses. The results of the analyses were used to calculate the rate of $O_2$ consumption, the RQ, and the metabolic rate at the indicated times. The results demonstrate that the BMR during the first hour of rest was about 190 kJ/hr.

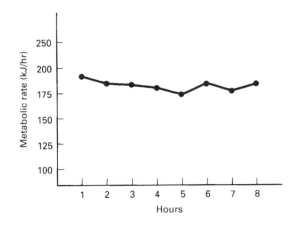

**FIGURE 5.21** Basal metabolic rate. (Redrawn with permission from Leff *et al.*, 1987.)

EXERCISE

The rate of energy consumption of an aerobic tissue is roughly proportional to its rate of oxygen consumption. What is the rate of energy consumption of resting muscle compared with exercising muscle?

EXERCISE

Skeletal muscle represents about 40% of the mass of an adult 70-kg man. Imagine that the man spent 1 to 2 hours per day in strenuous exercise and engaged in sedentary activities for the remainder of the day. Comment on the difference in the expected energy requirements of a sedentary person and a person who exercises 1 to 2 hours per day.

EXERCISE

Express the BMR of the subject in the preceding study (see Figure 5.21) in watts.

The metabolic rate changes under different conditions. For example, in the event that the subject falls asleep during an attempt to measure the BMR, the metabolic rate may fall by 10%, as reflected by a decrease in the rate of $O_2$ consumption. An increase of about 30% may occur when changing from a lying to a standing position. An increase of 5 to 10% may occur shortly after consumption of a meal. This increase is called the **specific dynamic action** of the food. It is due to the mobilization of energy from the body's stores for the secretion of gastric acid and pancreatic juices. The metabolic rate can change quite dramatically with physical activity, as shown in Table 5.6. The metabolic rate changes with aging, as reflected in changes in the energy requirements for persons of different ages (Table 5.5).

# DOUBLY LABELED WATER TECHNIQUE FOR DETERMINING ENERGY EXPENDITURE

The use of indirect calorimetry is not a convenient technique, since it requires a specially constructed room and the confinement of human subjects for a day or longer in the room. The **doubly labeled water technique** offers a convenient alternative, providing that one has a machine to perform **isotope ratio mass spectrometry**. The doubly labeled water technique is used to measure the rate of total $CO_2$ production in the body. This number alone is not sufficient to allow one to calculate the total energy expenditure. But the value for $CO_2$ production (moles $CO_2$/day), along with the RQ, allows one to calculate the oxygen consumption using the following formula:

$$\text{Moles of } O_2 = [\text{moles } CO_2 \text{ produced}] / [\text{RQ}]$$

A knowledge of $O_2$ consumption and RQ can be used to calculate the total amount of megajoules expended per day, as outlined earlier in this chapter.

The RQ can be determined directly, by taking breath samples during the course of the day. However, this method of determining the RQ is confining, and defeats the advantage of the doubly labeled water technique, that is, the fact that subjects can engage in their usual daily activities without being confined. Alternatively, the RQ (of the human subject) can be estimated by determining the food quotient. In this way, analysis of samples of food can be used to estimate the RQ values of human subjects. One might ask, "Why not determine human energy expenditure simply by using a bomb calorimeter and measuring the energy content of the food itself?" This method would work, but it would be imprecise, due to difficulties in tabulating the exact quantities of food consumed, by free-living subjects, over the course of several days.

The doubly labeled water method is used to determine the moles of $CO_2$ produced as follows. The human subject drinks a known quantity of deuterated water ($^2H_2O$) and oxygen-18 labeled water ($H_2^{18}O$). Hydrogen-2 is deuterium. Samples of blood are taken every day for a week. The amount of deuterium and oxygen-18 found in aliquots of blood are measured in each of the samples. The rates of loss of deuterium (hydrogen-2) and oxygen-18 over the course of the one week study are then calculated. The hydrogen-2 can leave the body as water in the urine, or as vapor from the lungs. This fractional rate is called "K2." The oxygen-18 can escape in two totally different guises, i.e., as the water in urine or lung vapor, or as carbon dioxide from the lungs. This fractional rate is called "K18."

The value of K18 must be greater than the value of K2, since K18 represents the oxygen escaping from the body as water and the oxygen escaping from the body as $CO_2$. K2 only represents the oxygen escaping from the body as water.

Subtraction is then performed. This subtraction (K18 – K2) yields the fractional rate of loss of $H_2^{18}O$ that is due to escape in the form of carbon dioxide. The value of (K18 – K2) represents the rate of loss of $H_2^{18}O$ due to total $CO_2$ production and total body $CO_2$ loss from the lungs. Please note that carbon dioxide contains two oxygen atoms, but that water contains only one oxygen atom. Because of this, (K18 – K2) represents the fractional loss of water, but [(0.5) (K18 – K2)] represents the fractional loss of carbon dioxide.

The fractional loss of carbon dioxide in an aliquot of blood is not really the number that we are pursuing. We are really interested in a value for the rate of production (same value as the rate of loss) of total $CO_2$ in the entire body. The rate of production of total $CO_2$ in the body is

$$(N) \ (0.5) \ (K18 - K2)$$

where $N$ is the total body water in moles.

*Total Body Water*

Total body water is calculated by comparing the specific activity of the oxygen-18 in the sample of $H_2^{18}O$ used for dosing, with the specific activity of the oxygen-18 in the sample of bodily fluid (in a blood sample taken 4 hours after drinking the $H_2^{18}O$ (Goran *et al.*, 1994). For example, imagine that 100% of the oxygen in the 100-ml swallowed dose of water is oxygen-18. Also, imagine that oxygen-18 was found to account for only 1/100th of the oxygen in the sample of bodily fluid. One can use these numbers to calculate that the total body water is $100 \times 100$ ml = 10 liters.

*Equilibrium of Total Body Water with CO₂ in Bodily Fluids*

An equilibrium that occurs in the body is shown in the following. The oxygen-18 introduced into the body as $H_2^{18}O$ is shown, in the equilibrium, to become rapidly incorporated into the carbon dioxide that is dissolved in the bodily fluids. The $CO_2$ ($C^{16}O^{16}O$) that is produced in the body via the Krebs cycle is not expected to contain any oxygen-18. However, one would expect an equilibrium to be reached between this $CO_2$ within moments, because of the action of carbonic anhydrase.

Injection      Krebs cycle                    Sample of blood assayed for oxygen-18 in the $CO_2$

$$H_2^{18}O + C^{16}O^{16}O \rightleftharpoons H_2^{16}O + C^{16}O^{18}O$$

Please view an imaginary example to help understand how changes in the body's production of $CO_2$ can result in a change in the amount of $C^{16}O^{18}O$. Imagine that one could step up one's oxidation of fat in the mitochondria. The increased $CO_2$ would act to dilute the oxygen-18 in the $H_2^{18}O$, and thus sweep out the body's $H_2^{18}O$ at a greater rate. In this way, the researcher looking at changes in the amount of water that contains oxygen-18, with repeated samplings of blood, can acquire a value for the total body's production of $CO_2$.

Using the value for $CO_2$ production (moles $CO_2$/day) and the RQ, the rate of oxygen consumption can be calculated using the formula

$$\text{Moles of } O_2 = [\text{moles } CO_2 \text{ produced}]/[RQ]$$

The doubly labeled water technique measures $CO_2$ production, and not $O_2$ production. Because the heat created per mole of expired $CO_2$ differs between the three major fuels (carbohydrate, fat, protein), the calculation of energy expenditure requires a knowledge of the internal fuel being oxidized. Thus, the doubly labeled water technique, coupled with information from RQ measurements, may be used instead of the method of indirect calorimetry (using an insulated chamber to measure heat production) to determine total energy production over the course of several days.

Since the doubly labeled water technique can be used to determine total body water, the method can be used to estimate the fat-free mass, as defined in the Obesity chapter. In brief, with a knowledge of the amount of total body water, and the assumption that normal subjects have a fat-free mass that is about 73% water, one can calculate the amount of fat-free mass. Further details are in the Obesity chapter.

# ISSUES IN ENERGY NUTRITION

The study of energy nutrition is highlighted by a number of interesting issues, such as obesity, lactation, and diseases of the thyroid gland. One aspect of obesity is the question of the energy cost involved in the formation of energy stores. Deposition of various nutrients into energy stores requires the expenditure of energy. Energy is required for the synthesis of glycogen and triglycerides. The energy cost of depositing dietary fats into adipose tissue is relatively low; the energy used for this process is equivalent to about 3.0% of the total amount of energy contained in the triglyceride. The cost of depositing dietary carbohydrates as glycogen is equivalent to about 7.0% of the energy contained in the carbohydrate. The energy cost of converting dietary carbohydrate to triglycerides is, however, relatively high; it is equivalent to about 23% of the energy of the carbohydrate (Sims and Danforth, 1987). These figures indicate that a person desiring to build up his or her stores of fat could more efficiently do so by consuming excessive amounts of fat rather than excessive amounts of glucose. The mechanisms that control the three major fates of carbohydrates — immediate use for oxidation, glycogen synthesis, and fat synthesis — are of obvious interest to the nutritionist.

Of particular interest are the energy requirements for pregnancy and lactation. The energy requirements during these two periods are increased by about 14 and 23%, respectively (Table 5.5). Food consumed is used directly for the synthesis of new maternal and fetal tissues, is oxidized to provide energy for the increase in biosynthetic activity, and is used for the deposition of maternal fat. The total weight gain of a woman during pregnancy may be about 11 kg. The infant represents about one-third (3.3 kg) of this weight gain. The increase in maternal fat stores may also be about 3.3 kg (King, 1985). This amount of fat represents 125,000 kJ (30,000 kcal) of energy.

The increase in the energy requirement during lactation is 2090 kJ/day (500 kcal/day). The energy content of the milk produced per day is about 2400 kJ. The typical volume of milk produced is 800 ml/day. As measured in the bomb calorimeter, the energy content of milk is about 300 kJ per 100 ml, yielding a value of 2400 kJ/day. One might expect, in addition, the need for more energy to support the biosynthetic and secretory events in the lactating breast. How is it possible that the consumption of the recommended energy supplement is 2090 kJ/day can support the production of milk containing 2400 kJ, mainly in the form of the triglycerides in milk globules? One might also wonder about the origin of the extra energy needed to support biosynthesis and secretion of the milk. The answer to this question is simple. Apparently, the extra fat deposited during pregnancy is mobilized during lactation to provide about 800 kJ/day (Prentice and Prentice, 1988).

Diseases of the thyroid gland are also of interest to those studying energy metabolism. The BMR was once commonly used in the diagnosis of diseases involving the thyroid gland. At present, the measurement of plasma thyroid hormone levels is the diagnostic technique of choice. Persons with hypothyroidism have BMRs ranging from 60 to 80% of the normal value. Those with hyperthyroidism usually have BMRs from 120 to 150% the normal value. Hyperthyroidism results in an abnormally high rate of fuel oxidation and heat production. The BMR is sometimes used in evaluating a patient's response to antithyroid drugs or to replacement therapy with thyroid hormones (Becker, 1978). For example, a person with hypothyroidism may be given replacement therapy with thyroid hormone. Analysis of plasma thyroid hormone levels may reveal that the therapy resulted in normal levels of the hormone; however, it would not reveal if the person's metabolism was corrected by the treatment. Measurement of the BMR would provide this information.

EXERCISE

Do you think that the rate of energy oxidation of energy fuels in an infant is increased by 5, 50, or 500% when crying? (See Murlin *et al.*, 1925.

# REFERENCES

Becker, D. V. (1978). Metabolic indices. *In* "The Thyroid" (S. C. Werner and S. H. Ingbar, eds.), 4th ed., pp. 347–363. Harper & Row, New York.

Brown, L., MacDonald, M., Lehn, D., and Moran, S. M. (1994). Sequence of rat mitochondrial glycerol-3-phosphate dehydrogenase cDNA. *J. Biol. Chem.* **269**, 14363–14366.

Elwyn, D. H., Gump, F. E., Munro, H. N., Iles, M., and Kinney, J. M. (1979). Changes in nitrogen balance with increasing infusion to glucose. *Am. J. Clin. Nutr.* **32**, 1597–1611.

Food and Nutrition Board (1989). "Recommended Dietary Allowances," 10th ed. National Academy Press, Washington, DC.

Frayn, K. N. (1983). Calculation of substrate oxidation rates *in vivo* from gaseous exchange. *J. Appl. Physiol.* **55**, 628–634.

Futai, M., Noumi, T., and Maeda, M. (1989). ATP synthase (H-ATPase). *Annu. Rev. Biochem.* **58**, 111–136.

Garrib, A., and McMurray, W. (1986). Purification and characterization of glycerol-3-phosphate dehydrogenase (flavin-linked) from rat liver mitochondria. *J. Biol. Chem.* **261**, 8042–8048.

Goran, M., Poehlman, E., and Danforth, E. (1994). Experimental reliability of the doubly labeled water technique. *Am. J. Physiol.* **29**, E510–E515.

Jequier, E. (1987). Measurement of energy expenditure in clinical nutritional assessment. *J. Parenteral Enteral Nutr.* **11**, 86S–89S.

King, J. C. (1985). Obesity in pregnancy. *In* "Dietary Treatment and Prevention of Obesity" (R. T. Frankle, J. Dwyer, L. Moragne, and A. Owen, eds.), pp. 185–191. John Libbey & Co., London.

King, R. F., Almond, D. J., Oxby, C. B., Holmfield, J. H., and McMahon, M. J. (1984). Calculation of short-term changes in body fat from measurement of respiratory gas exchanges. *Metabolism* **33**, 826–832.

Horton, T. J., Drougas, H., Brachey, A., Reed, G. W., Peters, J. C., and Hill, J. O. (1995). Fat and carbohydrate overfeeding in humans: Different effects on energy storage. *Am. J. Clin. Nutr.* **62**, 19–29.

Leff, M. L., Hill, J. O., Yates, A. A., Cotsonis, G. A., and Heymsfield, S. B. (1987). Resting metabolic rate: Measurement reliability. *J. Parenteral Enteral Nutr.* **11**, 354–359.

Lenaz, G. (1988). Role of mobility of redox components in the inner mitochondrial membrane. *J. Membr. Biol.* **104**, 193–209.

Murlin, J. R., Conklin, M. S., and Marsh, M. E. (1925). Energy metabolism of normal newborn babies. *Am. J. Dis. Child.* **29**, 1–28.

Murray, F. T., Zinman, B., McClean, P. A., Denoga, A., Albisser, A. M., Leibel, B. S., Nakhooda, A. F., Stokes, F. F., and Marliss, E. B. (1977). The metabolic response to moderate exercise in diabetic men receiving intravenous and subcutaneous insulin. *J. Clin. Endocrinol. Metab.* **44**, 708–720.

Porter, R., Hulbert, A., and Brand, M. (1996). Allometry of mitochondrial proton leak: Influence of membrane surface area and fatty acid composition. *Am. J. Physiol.* **271**, R1550–R1560.

Prentice, A. M., and Prentice, A. (1988). Energy costs of lactation. *Annu. Rev. Nutr.* **8**, 63–79.

Prostko, C. R., Fritz, R. S., and Kletzien, R. F. (1989). Nutritional regulation of hepatic glucose-6-phosphate dehydrogenase. *Biochem. J.* **258**, 295–299.

Rolfe, D., and Brand, M. (1996). Contribution of mitochondrial proton leak to skeletal muscle respiration and to standard metabolic rate. *Am. J. Physiol.* **271**, C1380–C1389.

Schrauwen, P., Lichtenbelt, W., Saris, W., and Westerterp, K. R. (1997). Changes in fat oxidation in response to a high-fat diet. *Am. J. Clin. Nutr.* **66**, 276–282.

Sims, E., and Danforth, E. (1987). Expenditure and storage of energy in man. *J. Clin. Invest.* **79**, 1019–1025.

Van Eyk, J., and Bartels, T. J. (1968). Paraffin oxidation in *Pseudomonas aeruginosa. J. Bacteriol.* **96**, 706–712.

Wakil, S. J., and Stoops, J. K. (1983). Structure and mechanism of fatty acid synthetase. *In* "The Enzymes" (P. D. Boyer, ed.), pp. 3–61. Academic Press, San Diego.

Weber, J., and Senior, A. E. (1997). Catalytic mechanism of $F_1$-ATPase. *Biochim. Biophys. Acta* **1319**, 19–58.

Xiao-Jun, M., Salati, L. M., Ash, S. E., Mitchell, D. A., Klautsky, S. A., Fantozzi, D. A., and Goodridge, A. G. (1990). Nutritional regulation and tissue-specific expression of malic enzyme gene in the chicken. *J. Biol. Chem.* **265**, 18435–18441.

Yao, K.-W., and Schulz, H. (1996). Intermediate channeling on the trifunctional β-oxidation complex from pig heart mitochondria. *J. Biol. Chem.* **271**, 17816–17820.

## BIBLIOGRAPHY

Acheson, K. J., Flatt, J. P., and Jequier, E. (1982). Glycogen synthesis versus lipogenesis after a 500-gram carbohydrate meal in man. *Metabolism* **31**, 1234–1240.

Bandini, L. G., Schoeller, D. A., Edwards, J., Young, V. R., Oh, S. H., and Dietz, W. H. (1989). Energy expenditure during carbohydrate overfeeding in obese and nonobese adolescents. *Am. J. Physiol.* **256**, E357–E367.

Christenssen, K., Chwalibog, A., Henckel, S., and Thorbek, G. (1988). Heat production in growing pigs calculated according to the RQ and CN methods. *Comp. Biochem. Physiol. A* **91**, 463–468.

D'Alessio, D. A., Kavle, E. C., Mozzoli, M. A., Smalley, K. J., Polansky, M., Kendrick, Z. V., Owen, L. R., Bushman, M. C., Boden, G., and Owen, O. E. (1988). Thermic effect of food in lean and obese men. *J. Clin. Invest.* **81**, 1781–1789.

Dulloo, A. G., Ismail, M. N., Ryall, M., Melas, G., Geisller, C. A., and Miller, D. S. (1988). A low-budget and easy-to-operate room respirometer for measuring daily energy expenditure in man. *Am. J. Clin. Nutr.* **48**, 1367–1374.

Elia, M., and Livesay, G. (1988). Theory and validity of indirect calorimetry during net lipid synthesis. *Am. J. Clin. Nutr.* **47**, 591–607.

Elia, M., Zed, C., Neale, G., and Livesay, G. (1987). The energy cost of triglyceride fatty acid recycling in nonobese subjects after an overnight fast and four days of starvation. *Metabolism* **36**, 251–255.

Ferrannini, E. (1988). The theoretical bases of indirect calorimetry: A review. *Metabolism* **37**, 287–301.

Jequier, E., Acheson, K., and Schutz, Y. (1987). Assessment of energy expenditure and fuel utilization in man. *Annu. Rev. Nutr.* **7**, 187–208.

King, J. C. (1985). Obesity in pregnancy. *In* "Dietary Treatment and Prevention of Obesity" (R. T. Frankle, J. Dwyer, L. Moragne, and A. Owen, eds.), pp. 185–191. John Libbey & Co., London.

Kinney, J. M. (1987). Indirect calorimetry in malnutrition: Nutritional assessment or therapeutic reference? *J. Parenteral Enteral Nutr.* **11**, 90S–94S.

Pellet, P. L. (1990). Food energy requirements in humans. *Am. J. Clin. Nutr.* **51**, 711–722.

Ravussin, E., Burnand, B., Schutz, Y., and Jaquier, E. (1982). Twenty-four-hour energy expenditure and resting metabolic rate in obese, moderately obese, and control subjects. *Am. J. Clin. Nutr.* **35**, 566–573.

Roberts, S. B., Fuss, P., Heyman, M. B., and Young, V. R. (1995). Influence of age on energy requirements. *Am. J. Clin. Nutr.* **62**, 1053S–1058S.

Rumpler, W. V., Seale, J. L., Conway, J. M., and Moe, P. W. (1990). Repeatability of 24-h energy expenditure measurements in humans by indirect calorimetry. *Am. J. Clin. Nutr.* **51**, 147–152.

Webb, P. (1980). The measurement of energy exchange in man: An analysis. *Am. J. Clin. Nutr.* **33**, 1299–1310.

Weindruch, R., and Sohal, R. S. (1997). Caloric intake and aging. *N. Engl. J. Med.* **337**, 986–994.

*Doubly Labeled Water*

Calazel, C. M., Young, V., Evans, W., and Roberts, S. (1993). Effect of fasting and feeding on measurement of carbon dioxide production using doubly labeled water. *J. Appl. Physiol.* **74**, 1824–1829.

Goran, M., Poehlman, E., and Danforth, E. (1994). Experimental reliability of the doubly labeled water technique. *Am. J. Physiol.* **29**, E510–E515.

Lifson, N., Gordon, G., Visscher, M., and Nier, A. O. (1949). The fate of utilized molecular oxygen and the source of the oxygen of respiratory carbon dioxide, studied with the aid of heavy oxygen. *J. Biol. Chem.* **180**, 803–811.

Roberts, S. B., Coward, W., Schlingenseipen, K., Nohria, V., and Lucas, A. (1986). Comparison of the doubly labeled water ($^2H_2^{18}O$) method with indirect calorimetry and a nutrient balance study for simultaneous determination of energy expenditure, water intake, and metabolizable energy intake in preterm infants. *Am. J. Clin. Nutr.* **44**, 315–322.

Schoeller, D. A. (1983). Energy expenditure from doubly labeled water: Some fundamental considerations in humans. *Am. J. Clin. Nutr.* **38**, 999–1005.

Schoeller, D. A. (1988). Measurement of energy expenditure in free-living humans by using doubly labeled water. *J. Nutr.* **118**, 1278–1289.

Schoeller, D. A., and Hnilicka, J. M. (1996). Reliability of the doubly labeled water method for the measurement of total daily energy expenditure in free-living humans. *J. Nutr.* **126**, 348S–354S.

Schoeller, D. A., and Santen, E. (1982). Measurement of energy expenditure in humans by doubly labeled water method. *J. Appl. Physiol.* **53**, 955–959.

Sheng, H.-P., and Huggins, R. A. (1979). A review of body composition studies with emphasis on total body water and fat. *Am. J. Clin. Nutr.* **32**, 630–647.

# 6

## LIPIDS

## OVERVIEW

Dietary lipids are used as an energy source, as a structural component in the membranes of cells, as structural compo- nents of a small fraction of the proteins in the cell, and, in the case of choles- terol, for the synthesis of detergents that facili- tate digestion and absorption of dietary lipids. Unlike most other nutrients, lipids are distin- guished by their occurrence as energy storage deposits (fat), which are tapped during the course of the day, and which can allow sur- vival for many weeks without eating food.

The simplest topic covered in this chapter is that of **choline**, a compound with a 2-carbon backbone, which occurs in the diet and in the body mainly in the form of the structural lipid **phosphatidylcholine**. Choline is an issue in the metabolism of other lipids, namely, of **phosphatidylserine**, **phosphatidylethanola- mine**, and **sphingomyelin**. **Phosphatidylino- sitol** is a lipid that has received enormous at- tention from cell biologists because of its role in transmitting signals within the cell. **Sphin- gomyelin** is a structural lipid in the plasma membrane, but it has received increased atten- tion because of its role in regulating the growth, differentiation, and death of cells.

**Cholesterol** is a structural lipid that is not required in the diet, since ample amounts are synthesized in the body. This lipid is a required component of most cells of the body. Cholesterol has received a great deal of attention in the nutritional and biochemical communities because of its strange metabolism in certain regions of the body, for example, the arteries that supply oxygen and other nutrients to the heart muscle. This unusual metabolism may result in the pathological accumulation of cholesterol in the artery wall, which gradually creates a sequence of defects that lead to atherosclerosis and cardiovascular disease.

The topic of **lipoproteins** is the most complicated issue presented in this chapter. Lipoproteins are complexes of specific proteins and lipids. Each lipoprotein contains different proportions of various lipids. The constant component of any one type of lipoprotein is the protein; hence, the structure or function is described by first naming the protein. Lipoproteins are synthesized primarily in the intestines and liver and are secreted into the plasma, where their function is to transport various lipids. Lipoproteins are water soluble because of their outside coat of proteins and the hydrophilic phosphate groups of their phospholipids. This water solubility enables lipoproteins to transport the triglycerides and cholesteryl esters that reside within their cores. An understanding of lipoproteins is useful to individuals interested in energy metabolism and essential to those concerned with cardiovascular disease.

Various types of lipoproteins exist in the bloodstream. Two of these have entered the vocabulary of the layperson, where they are known by their initials. These two lipoproteins are the HDLs and LDLs. Both of these lipoproteins contain protein and cholesterol. The HDLs are said to contain "good cholesterol," whereas LDLs are said to contain "bad cholesterol." These terms are, of course, gross simplifications because the cholesterol in both LDLs and HDLs is absolutely essential for life. On the other hand, a high concentration of LDLs in the bloodstream is a risk factor for acquiring cardiovascular disease. The converse is true of HDLs. The LDLs act as biochemical taxicabs for transporting cholesterol from one part of the body to another. However, abnormally high levels of LDLs can sometimes result in the pathological deposit of cholesterol in the arteries of the heart. A high level of HDL-cholesterol is associated with a lower incidence of cardiovascular disease. HDLs act as biochemical vacuum cleaners to remove cholesterol from various organs of the body. Evidence suggests that increased concentrations of HDLs in the bloodstream can counteract the pathological deposit of cholesterol in the walls of arteries in the heart.

When speaking about health, the terms "LDL" and "LDL-cholesterol" mean the same thing, and the terms "HDL" and "HDL-cholesterol" mean the same thing. But to the biochemist, the terms "LDL" and "HDL" mean a specific combination of proteins, cholesteryl esters, triglycerides, and phospholipid — and not just the cholesterol component.

## PHOSPHOLIPIDS

The major components of cell membranes are proteins, cholesterol, lipids with a glycerol backbone, and lipids with a sphingosine backbone. The glycerol-based lipids are phospholipids with fatty acids esterified at the 1 and 2 positions of the glycerol, and a molecule of phosphocholine, phosphoethanolamine,

**TABLE 6.1**  Phospholipid Composition of One Type of Plasma Membrane

| Lipid | Lipid (µg)/protein (mg) |
| --- | --- |
| Cholesterol | 43 |
| Total phospholipid | 180 |
| Phosphatidylcholine | 90 |
| Phosphatidylethanolamine | 45 |
| Phosphatidylserine | 16 |
| Phosphatidylinositol | 10 |
| Sphingomyelin | 5 |
| Other phospholipids | 12 |

*Source*: Thi-Dinh *et al.* (1990).

phosphoserine, or phosphoinositol at the 3 position. The sphingosine-based lipid sphingomyelin is also a phospholipid, since it contains phosphocholine.

**Phosphatidylcholine** (PC) and **phosphatidylethanolamine** (PE) are the major phospholipids of cell membranes. Table 6.1 lists the amount of lipid in cell membranes, expressed as micrograms of lipid per milligram of membrane-bound protein. The membranes were isolated from cells of adipose tissue from rats that had been raised on a diet containing sunflower oil (100 g oil/kg diet) as the source of lipids. The diet was essentially free of cholesterol and phospholipids because plants do not contain cholesterol and the phospholipids of vegetable oils are removed during the refining process.

PC can come from the diet or be synthesized in the body. Dietary PC is cleaved by a pancreatic enzyme, **phospholipase B**, that catalyzes hydrolysis of the fatty acid from carbon-2 (C-2) of the glycerol backbone. This hydrolysis produces a free fatty acid and 1-acyl-3-phosphocholine glycerol. The latter compound is called **lysolecithin**, because its parent compound PC is known as lecithin. Lysolecithin and the free fatty acid are absorbed by the enterocyte, converted back to PC, and then packaged into **chylomicrons** that are released into the lymphatic system. A small fraction of the choline derived from dietary PC enters the bloodstream as free choline.

PC is synthesized in the body by two different pathways, the CDP-choline pathway (Figure 6.1) and the PE methylation pathway. Most of the enzymes of phospholipid biosynthesis are bound to the endoplasmic reticulum (ER), with their active sites facing or exposed to the cytosol. Choline kinase is cytosolic, but cytidylyltransferase is both cytosolic and of the endoplasmic reticulum. Regulation of the activity of the enzyme is thought to involve binding of the cytosolic enzyme to the surface of the ER, which activates the enzyme. Phosphocholine transferase is bound to the ER. **Cytidine triphosphate** (CTP) also is used in the biosynthesis of a variety of other phospholipids. The energy released by hydrolysis of the phosphoanhydride linkages of CTP drives CTP-dependent reactions. Energy is released with the hydrolysis of the linkages between the phosphate groups of CTP, just as it is when ATP is hydrolyzed.

The CDP-choline pathway to PC (Figure 6.1) occurs in a number of tissues, but the PE methylation pathway (Figure 6.2) occurs mainly in the liver and accounts

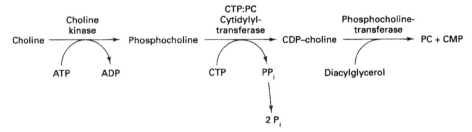

FIGURE 6.1 Phosphatidylcholine (PC) synthesis by the CDP-choline pathway. First, choline is converted to phosphocholine; then the CMP moiety of CTP is transferred to phosphocholine to form CDP-choline. Cleavage of the pyrophosphate ($PP_i$) produced by phosphatases markedly reduces its concentration in the cell and ensures that the reaction is driven in the forward direction. The phosphocholine moiety of CDP-choline then is transferred to diacylglycerol, generating PC. The total number of phosphate–phosphate bonds (phosphodiester bonds) broken in this pathway is three. This number determines the energy cost of PC synthesis. All cells containing a nucleus can make PC via the CDP-choline pathway.

for 20–30% of the PC synthesized there. This reaction involves transfer of three methyl groups to the nitrogen of PE; the function of the methyl donor **S-adenosyl methionine** (SAM) is similar to that of CDP-choline, that is, just as the nucleotide moiety of CDP-choline is required for participation of choline in choline transfer reactions, the nucleotide moiety of SAM is required for involvement of the methyl group of methionine in methyl transfer reactions.

PE can be synthesized by several pathways, including one that is dependent on CTP. The CDP-ethanolamine pathway shown in Figure 6.3 is similar in outline to the CDP-choline pathway of PC synthesis. The ethanolamine can originate from the diet or from catabolism of **phosphatidylserine**. PE also can be synthesized from phosphatidylserine (PS) by two different pathways. In the first, shown in Figure 6.4, an enzyme requiring calcium ions for activity catalyzes a "base exchange" reaction. The second, shown in Figure 6.5, is catalyzed by the vitamin $B_6$-requiring enzyme, **PS decarboxylase**. The contributions to PE synthesis by the

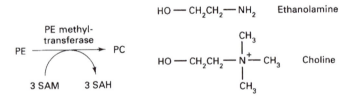

FIGURE 6.2 Phosphatidylcholine (PC) synthesis by the CDP-ethanolamine pathway. Structures of ethanolamine and choline. The nucleotide moiety S-adenosyl methionine (SAM) is required to transfer the methyl group of methionine to phosphatidylethanolamine (PE) to form PC. In the process, SAM is converted to S-adenosyl homocysteine (SAH). The nitrogen atom of PC has four covalent bonds and is called a quaternary amine. It bears a positive charge that is not influenced by changes in the pH of the surrounding fluids. The PE methyltransferase pathway of PC synthesis occurs only in the liver.

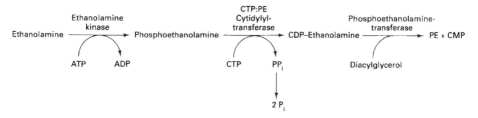

FIGURE 6.3 Phosphatidylethanolamine (PE) synthesis by the CDP-ethanolamine pathway.

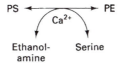

FIGURE 6.4 Phosphatidylethanolamine (PE) synthesis by base exchange. Calcium ion is required for catalytic activity.

base exchange and PS decarboxylation pathways are minor compared to that by the CDP-ethanolamine pathway. The base exchange reaction is freely reversible. However, decarboxylation reactions generally are not reversible. Phosphatidyl-serine, a minor component of cell membranes, can be derived from the diet or it can be synthesized by the base exchange reaction depicted in Figure 6.4. A similar reaction involving PC also occurs. A CDP-serine pathway for PS biosynthesis has not been found, although a pathway for PS synthesis involving CDP-diacyl-glycerol occurs in microorganisms.

The pathway shown in Figure 6.6 could generate unlimited supplies of PC in the body. Phosphatidylserine is not used up in this pathway because it is used only in catalytic amounts in the cycle of reactions. The 2-carbon skeleton of the choline of the PC generated originates from the carbons 2 and 3 of serine. This pathway is thought to be used for making fine adjustments in the phospholipid composition of the body, but does not operate at a sufficient rate to generate the required amounts of PC, the major phospholipid of the body.

CDP-diacylglycerol is also part of the **phosphatidylinositol** (PI) synthesis pathway, shown in Figure 6.7. The PI formed can be phosphorylated on the inositol group to produce **phosphatidylinositol-4,5-diphosphate**. This lipid is a minor

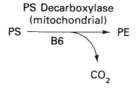

FIGURE 6.5 Phosphatidylethanolamine (PE) synthesis by decarboxylation of phosphatidylserine (PS). Vitamin $B_6$ is required for catalytic activity.

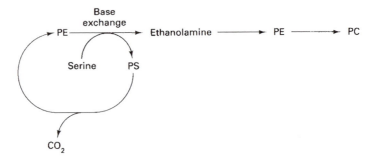

**FIGURE 6.6** Phosphatidylcholine (PC) synthesis by phosphatidylserine (PS) decarboxylation and base exchange. Serine serves as the source of the carbon skeleton of choline.

component of the plasma membrane, but it plays a vital role in the transmission of some hormonal signals.

## CHOLINE AND ETHANOLAMINE

Choline, supplied as dietary PC or as free choline, is required in the diet by rats. Although it has not been established that choline is required by humans, it is probably an essential nutrient and may, in the future, be classified as an essential amine or vitamin. Its possible requirement is a concern to clinicians feeding patients by **total parenteral nutrition** (TPN). In this type of feeding, which may be used for a year or longer, the patient is sustained intravenously with an artificial, chemically defined diet. The choline in foods occurs mainly as PC rather than as free choline. PC is a more desirable dietary component because, when free choline is consumed in large amounts, it is degraded by the gut bacteria to produce trimethylamine, an odoriferous compound (Magil *et al.*, 1981).

The concentrations of choline in various foods are given in Table 6.2. This nutrient occurs as free choline, PC, and sphingomyelin. Concentrations are expressed on a molar basis, allowing direct comparison of choline levels, because 1 mol PC or sphingomyelin contains 1 mol choline.

Ethanolamine occurs in the diet mainly as PE, not as free ethanolamine. PC and PE are present in similar quantities in some foods. PE can be converted to PC in the liver. Therefore, the PE normally present in the diet should reduce the require-

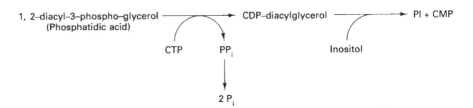

**FIGURE 6.7** Phosphatidylinositol synthesis.

**TABLE 6.2**  Content of Free and Bound Choline in Foods

| Food | Approximate choline content (µmol/kg food) | | |
|---|---|---|---|
| | Choline | Phosphatidylcholine | Sphingomyelin |
| Apple | 27 | 280 | 15 |
| Banana | 240 | 37 | 20 |
| Beef liver | 5831 | 43500 | 1850 |
| Beef steak | 75 | 6030 | 506 |
| Butter | 42 | 1760 | 460 |
| Cauliflower | 1306 | 2770 | 183 |
| Coffee | 1010 | 15 | 23 |
| Corn oil | 3 | 12 | 5 |
| Egg | 42 | 52000 | 2250 |
| Human milk | 200 | 200 | 200 |
| Lettuce | 2930 | 132 | 50 |
| Orange | 200 | 490 | 24 |
| Peanut butter | 3895 | 3937 | 9 |
| Whole wheat bread | 968 | 340 | 11 |

*Source*: Zeisel (1990).

ment for PC, although this possibility does not appear to have been tested by researchers. The significance of the PS decarboxylation/base exchange cycle in supplying the body's need for choline and ethanolamine is not known.


## Choline Deficiency

The effects of choline deficiency can be demonstrated easily using animals. One of the earliest is a **fatty liver**. Feeding rats a choline-free diet for 1 day doubles the fat (triglyceride) content of their livers. Rats that survive the continued consumption of such diets can accumulate over 50 times the normal level of hepatic fat. This condition results from impairment of the normal secretion of fat-containing proteins by the liver. These proteins, called **lipoproteins**, consist of a large core of triglycerides (TGs) coated with a thin shell of phospholipid and protein.

One explanation of these results is that choline deficiency causes production of lipoproteins with abnormal structures, which results in an inability of the liver to export TGs to the bloodstream. Another theory is that choline deficiency produces defects in the structure of the endoplasmic reticulum, the network of tubules used for the synthesis of proteins and protein complexes destined for secretion from the cell. The membrane walls of the ER consist of phospholipids. Because all membranes contain phosphatidylcholine, a choline deficiency might be expected to result in defects in all membranes of the body, including those of the ER.

Electron microscopy studies have revealed that choline deficiency results in disorganization of the structures of the ER and of the Golgi, a small organelle required for the secretion of proteins from the cell. Proteins destined for secretion are synthesized in the ER. Most such proteins are modified by covalent attachment of oligosaccharide chains before transfer to the Golgi, where the structures of the oligosaccharide chains may be altered by the addition or subtraction of sugar residues. Then the proteins are packaged in secretory vesicles and leave the Golgi. These secretory vesicles travel to, then fuse with, the plasma membrane to release their contents into the bloodstream. Lombardi (1971) showed that choline deficiency produced defects in the structure of the ER and Golgi, a 50% decrease in the amount of PC in the ER, and an impairment in the secretion of a number of different proteins.

Lombardi and colleagues (1968) studied the uptake of fatty acids by the liver, packaging as TGs in lipoproteins, and release back into the bloodstream of fatty acids in choline-sufficient and choline-deficient rats. This study involved injection of a radioactive fatty acid (tagged with $^{14}$C) into a blood vessel. After a 30-minute waiting period, the lipoproteins of the plasma were analyzed. The appearance of radioactivity in the lipoproteins *was impaired in the deficient rats*. This experiment is illustrated in Figure 6.8. Additional material on the lipoproteins is presented later in this chapter.

Choline deficiency produces renal as well as hepatic damage. This damage takes the form of degeneration of the renal tubules and bleeding, which may lead to death of the animal within 1 week of feeding the diet (Keith and Tryphonas, 1978). Under certain conditions, the liver and kidneys develop scar tissue. **Cirrhosis** is a term used to describe the conversion of normal tissues to scarlike fibrous tissues.

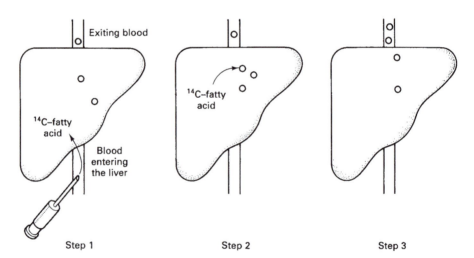

FIGURE 6.8 Packaging of fatty acids by the liver. (1) The radioactive fatty acid is injected into a vein and enters the cells of the liver. Most of the cells of the liver manufacture lipoproteins (circles). (2) Fatty acids in the liver, including the injected radioactive fatty acid, are converted to triglycerides and packaged into the lipoproteins. (3) Lipoproteins are secreted continuously by the liver into the bloodstream. Most of the lipoproteins in the blood have been in the circulatory system for several hours. Blood sample removal after various time periods following injection of the radioactive tracer facilitates generating a minute-by-minute picture of the secretory processes, both in normal animals and in those that are choline deficient.

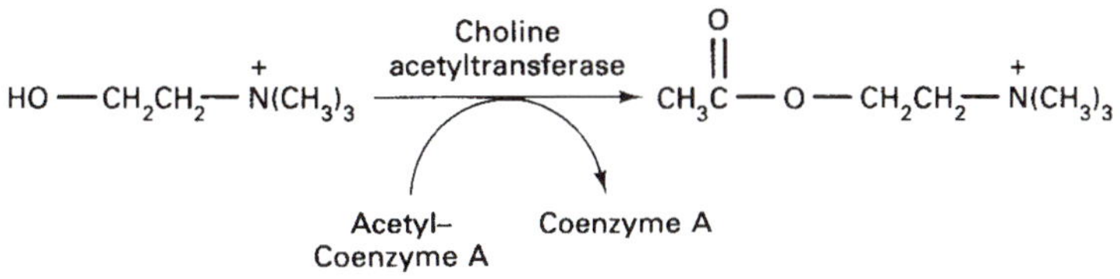

$$\text{HO} - \text{CH}_2\text{CH}_2 \overset{+}{-} \text{N(CH}_3)_3 \xrightarrow[\substack{\text{Acetyl--} \\ \text{Coenzyme A}}]{\substack{\text{Choline} \\ \text{acetyltransferase}}} \underset{}{\overset{\overset{\text{O}}{\parallel}}{\text{CH}_3\text{C}}} - \text{O} - \text{CH}_2\text{CH}_2 \overset{+}{-} \text{N(CH}_3)_3$$

Coenzyme A

**FIGURE 6.9** Synthesis of acetylcholine.

## Acetylcholine

A very small fraction of the body's choline is used to synthesize the neurotransmitter acetylcholine by the pathway outlined in Figure 6.9. Acetyl coenzyme A acts as a donating agent in the transfer of the acetyl group to choline. Acetylcholine is stored in secretory vesicles located in nerve cells in the region known as the nerve ending, where the nerve contacts a muscle fiber. The arrival of an electrical impulse at the nerve ending provokes fusion of the vesicles with the plasma membrane, which results in release of acetylcholine into the extracellular fluid in the synaptic cleft. The acetylcholine diffuses to receptors in the membrane of the muscle cell, causing electrical impulses to travel along the muscle cell. The electrical impulses that travel along the nerve cells are chemically and electrically similar to those that travel along muscle cells. The role of acetylcholine as an intermediary in the stimulation of muscle cells by nerve cells is outlined in Figure 6.10.

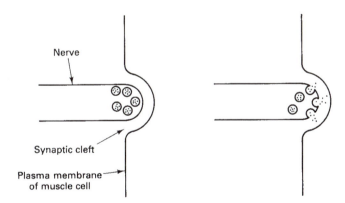

**FIGURE 6.10** The neuromuscular junction. The region of contact between one nerve and another nerve, or between a nerve and a muscle cell, is called a synaptic cleft. The secretory vesicles are represented by circles in which acetylcholine is represented by dots. The nerve impulse provokes the entry of calcium ions (not shown) from the extracellular fluid into the nerve cell. Calcium ions act as a signal that stimulates the fusion of vesicles with the plasma membrane, releasing acetylcholine into the extracellular fluid. Acetylcholine binds to membrane-bound proteins (acetylcholine receptors) on the plasma membrane of the muscle cell, resulting in stimulation of the muscle cell. Acetylcholinesterase of the neuromuscular junction catalyzes the destruction of acetylcholine in the moments after transmission of the nerve impulse. The enzyme is extracellular and is bound to proteoglycan, a molecule of extracellular matrix.

## INFLUENCE OF DIETARY FATS ON PHOSPHOLIPIDS IN THE BODY

Dietary fats and oils are important for several reasons. They are excellent sources of energy and generally supply at least 30–40% of the body's energy requirement. They contribute to the palatability of the diet. A diet consisting solely of protein and carbohydrate would not be accepted readily by most persons. Dietary fats and oils are deposited in **adipocytes** for storage as triglycerides and in membranes as structural phospholipids. Adipocytes, the major cell type of adipose tissue, are found in other organs also. Phospholipids are structural elements of plasma membranes and membranes of the various organelles.

Dietary triglycerides (TGs) are packaged into chylomicrons in the enterocyte and delivered to the circulatory system via the thoracic duct. Normally, the lacteals and thoracic duct contain a clear, yellowish fluid. After a fat-containing meal, however, they are filled with an opaque white liquid that contains chylomicrons laden with TGs. A rise in the plasma concentration of chylomicrons can be detected after a meal of fat or oil by assays for particulate matter or for one constituent of the chylomicrons, the TGs. The rise in chylomicron concentration usually is followed by a fall to very low basal concentrations. However, defects in the mechanisms that facilitate this decline are revealed when the levels of particulate matter or of plasma TGs remain elevated. These defects, which can present a serious health risk, are illustrated by data from patients suffering from **hypertriglyceridemia**. (The prefix "hyper" means higher than normal; "hypo" means lower than normal.)

Fatty acids are classed as short chain, medium chain, or long chain. Short–chain fatty acids are 2-6 carbon atoms long. One such acid, acetic, is designated 2:0 because it is 2 carbons long and contains no double bonds. Medium-chain fatty acids are 8 to 12 carbons long. Long-chain fatty acids are 14–24 carbons long. Palmitic acid, a long-chain fatty acid, is designated 16:0 because it contains 16 carbons and no double bonds. Oleic acid (18:1) contains 18 carbons and 1 double bond. Linoleic acid (18:2) contains 18 carbons and 2 double bonds.

The nature of a fatty acid influences its fate. Short- and medium-chain fatty acids tend to be oxidized immediately to carbon dioxide, rather than deposited as TGs or phospholipids. The presence of double bonds ("unsaturations") in long-chain fatty acids influences the immediate fate of the acid. Some evidence suggests that unsaturated fatty acids, such as 18:2, tend to be oxidized at a slightly faster rate in the hours following a meal than saturated fatty acids, such as 18:0 (Jones *et al.*, 1985; Jones and Schoeller, 1988). More specifically, about 2% of a test meal of 18:0 may be oxidized in the 9 hours following ingestion, whereas about 10% of a test meal of 18:2 may be oxidized in the same period. The mechanisms that influence the fates of unsaturated and saturated fatty acids are only beginning to be understood.

The nature of the fatty acids in the TGs stored in adipose tissue and those in the phospholipids in membranes can be influenced by the diet. The rat study reported in Table 6.3 assessed the plasma membranes of the liver. Diets containing 10% oil or fat by weight were fed to young and rapidly growing rats. Rapidly growing

**TABLE 6.3** Effect of Diet on the Fatty Acid Content of Plasma Membrane Phospholipids

| Fatty acid | Diet content (10% lipid) | | Membrane fatty acid content | |
|---|---|---|---|---|
| | Corn oil | Butterfat | Corn oil diet | Butterfat diet |
| 12:0 | 0 | 3.7 | 0 | 0 |
| 14:0 | 0 | 12.6 | 0 | 0 |
| 14:1 | 0 | 1.5 | 0 | 0 |
| 16:0 | 10.7 | 33.3 | 24.2 | 27.5 |
| 16:1 | 0.2 | 2.2 | 1.7 | 3.5 |
| 18:0 | 1.5 | 12.8 | 16.7 | 19.0 |
| 18:1 | 25.2 | 28.2 | 15.0 | 22.3 |
| 18:2 | 62.4 | 3.4 | 15.6 | 5.0 |
| 20:4 | 0 | 0 | 19.3 | 12.9 |
| 22:5 | 0 | 0 | 1.9 | 0.3 |
| 22:6 | 0 | 0 | 2.1 | 5.4 |

*Source*: Hamm *et al.* (1988).

animals are influenced more by changes or deficiencies in their diets than are adult animals. After 4 weeks on the test diet, the livers of the rats were removed and the plasma membranes isolated for analysis. The fatty acids present in the dietary TGs and in the membrane phospholipids are listed in Table 6.3.

Several observations can be made. One of the major fatty acids of butter is 14:0. Obviously, this fatty acid was not incorporated into the phospholipids of the membrane. Fatty acids with 14 carbons were essentially absent from the membranes. The food lipids did not contain long-chain fatty acids of 20 or 22 carbons; however, the membranes contained substantial amounts of these acids. For example, about 20% of the fatty acid content of the membranes of the corn oil-fed rat was 20:4.

Certain lipids, although present in the diet, may not be incorporated into membranes. However, other lipids that are present from the diet may be present. A closer look at the data reveals a striking difference in the 18:2 content of the membranes. The 18:2 content of corn oil is 18 times greater than that of butterfat. However, the 18:2 content of the membranes of the corn oil-fed rats was not 18 times greater than that of the membranes of butterfat-fed rats, but was only about three times greater. This result indicates that dietary fat can influence the composition of membranes, but an extreme change in the diet may result in only a moderate change in the membrane.

Other changes should be noted. The data in Table 6.3 show that the 20:4 membrane content is greater with corn oil feeding. This increase results from the fact that animals can convert 18:2 to 20:4. The data also demonstrate that, when greater amounts of 18:2 are present in the diet, greater amounts of 20:4 accumulate in the membranes. The explanation for this result is that animals cannot synthesize 18:2. This fatty acid must be supplied in the TGs or phospholipids of the diet. Conversion of 18:2 to 20:4 involves the addition of a 2-carbon group and introduction of two additional double bonds into the molecule. The responses of membrane lipids to the diet as specified in Table 6.3 are similar (although not identical) to changes found in other studies.

## SPHINGOSINE-BASED LIPIDS

Sphingosine-based lipids contain a backbone of **sphingosine** to which a fatty acid and, in some cases, phosphocholine, a monosaccharide, an oligosaccharide, or a sulfate group are attached. Sphingosine-based lipids have names that reflect structures of the nervous system. A sphingosine-based lipid that contains phosphocholine is called **sphingomyelin** (SPM) and is considered a phospholipid. Myelin is the lipid coating that covers nerves. If the lipid contains a monosaccharide, it is called a **cerebroside**. Cerebral means "pertaining to the brain." If the lipid contains an oligosaccharide consisting of a number of different sugars including sialic acid, it is called a **ganglioside**. A ganglion is a special type of nerve cell from which nerve fibers branch.

Sphingosine-based lipids are not present just in the nervous system. They are components of the plasma membranes of most cells and are involved in regulation of cell growth, cell-to-cell adhesion, and transmission of signals from an outside stimulus to the interior of the cell. The study of these lipids has been complicated by the fact that they exist in over 100 forms in mammalian tissues.

### Biosynthetic Pathway for Sphingolipids

The biosynthetic pathway for the sphingolipids lipids uses readily available components such as serine, palmitic acid, and a variety of sugars. The initial three steps (Figure 6.11) are catalyzed by enzymes that utilize cofactors made from three vitamins ($B_6$, niacin, and riboflavin). The required cofactors are pyridoxal phosphate, NADPH, and FAD, respectively. As shown in Figure 6.12, sphingosine can be modified by a long-chain fatty acid, donated by acyl CoA, to yield **ceramide**. Ceramide can be further modified to sphingomyelin by donation of a phosphocholine group from phosphatidylcholine. Ceramide can also be modified to produce cerebrosides, gangliosides, or sulfated gangliosides.

### Sphingolipids in Food and Milk

Sphingosine-based lipids do not appear to be required in the diet. They are catabolized in the body and provide only a minimal amount of the energy requirement. Dietary SPM is a valuable source of choline. The sulfated gangliosides are a source of sulfate, although sulfate is not required in the diet. Beef contains about 13 g PC per kg meat and nearly the same amount (9 g/kg) of SPM (Weihrauch and Son, 1983). The corresponding figures for chicken breast (4 g PC/kg; 0.5 g SPM/kg) and for tuna muscle (1.6 g PC/kg; 2.1 g SPM/kg) are lower than for beef. Hen eggs contain about 27 g PC/kg and 0.8 g SPM/kg. Essentially all the lipid of eggs is found in the yolk rather than in the white.

Human milk contains about 38 g lipid/kg, nearly all in the form of TGs. Approximately 0.25% of the lipids of the fat globule of milk are phospholipids, about one-third each of phosphatidylcholine, phosphatidylethanolamine, and sphingomyelin. Triglycerides make up the bulk of the milk globule, forming its core, whereas phospholipids coat the surface. The sphingomyelin of milk has very-long-chain fatty acids 20, 22, or 24 carbons in length bound to its amino

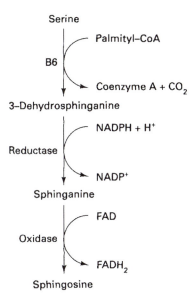

FIGURE 6.11 The sphingosine biosynthetic pathway. As is the case for most biochemical pathways, this pathway utilizes several cofactors that are made from vitamins. The vitamins pantothenic acid, vitamin $B_6$, niacin, and riboflavin are used to make the cofactors indicated in the figure.

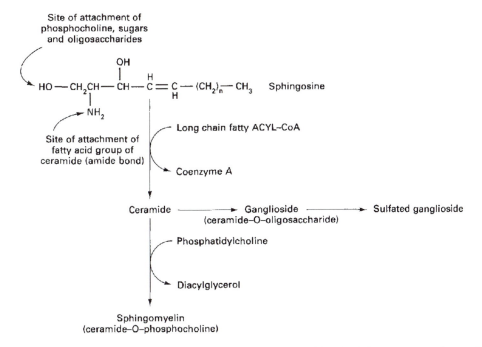

FIGURE 6.12 The biosynthetic pathway for sphingosine-based lipids. Sphingosine can be modified by a long-chain fatty acid, donated by acyl-CoA, to yield ceramide. Ceramide can be modified further to sphingomyelin by donation of a phosphocholine group from phosphatidylcholine. Ceramide also can be modified to produce cerebrosides, gangliosides, or sulfated gangliosides.

group. These long-chain fatty acids have been proposed to be important to the newborn rat, whose brain undergoes very rapid growth and development in the weeks following birth, as mentioned in the section on Essential Fatty Acids in Chapter 9.

### Sphingosine as a Structural Element in Membranes

The plasma membrane of most or all cells contains sphingosine. The example of epithelial cells is discussed. **Epithelial cells**, by definition, are polar, i.e., one side (apical side) faces an outside environment, while the other side (basolateral side; basal and lateral sides) faces an organ or tissue. The apical side of the enterocyte's plasma membrane, which faces food particles, is enriched in sphingosine, while the basolateral side has a sphingosine typical of most other types of cells. To view the numbers, the contents of sphingosine and phospholipid are equal, on a molar basis (Schulthess and Hauser, 1995). The membrane on the opposite side of the enterocyte (basolateral membrane), which faces the capillaries that carry nutrients away from the intestines, has a sphingosine/phospholipid ratio of 1 to 2.6. Sphingosine molecules in membranes form a hydrogen-bonded network with each other, which adds strength to the membrane. In other words, sphingolipids strongly self-assemble, in contrast to the picture with other lipids. The following statement applies to all cells. The plasma membranes of all cells consist of a lipid bilayer. In general, sphingosine resides in the outer leaflet of the membrane, and contacts the extracellular fluid. In contrast, phosphatidylserine is a phospholipid that generally resides in the inner leaflet, and contacts the cytosol.

### Sphingolipids and Cell Signaling

The most well-known components of the various signaling pathways in the cell include G protein, inositol-1,4,5-trisphosphate, cyclic AMP, MAP kinase, and protein kinase C, as discussed at various points in this text. Sphingosine represents a newcomer to the science of cell signaling.

As mentioned above, sphingomyelin consists of a ceramide group (sphingosine backbone plus a long-chain fatty acid) and a phosphocholine group. **Sphingomyelinase** is an enzyme that catalyzes the hydrolysis of the phosphocholine group, resulting in its release into the extracellular fluid. The ceramide byproduct then serves to activate various enzymes that reside within the plasma membrane. These events represent the heart of the sphingomyelin-based cell signaling pathway.

The free ceramide group activates various kinases and phosphatases. These target enzymes have names that are easy to understand, i.e., ceramide-activated protein kinase (CAPK) and ceramide-activated protein phosphatase (Testi, 1996; Reyes et al., 1996). Activation of the above protein kinase can result in the phosphorylation of various proteins, including RAF, MAPK, and JUN. These proteins are discussed in the Diet and Cancer chapter. The eventual result of ceramide-dependent signaling may be a change in the rate of growth of the cell, a change in the cell's state of differentiation, or in the activation of apoptosis (Santana et al., 1996; Hannun, 1996). Apoptosis is a regulated event, which can be activated in most cells, and results in the cell's own death.

A variety of external events can provoke the activation of sphingomyelinase. These include hormones, such as tumor necrosis factor, interleukin, interferon, and vitamin D$_3$ (Zhang *et al.*, 1997; Reyes *et al.*, 1996). An integrated and well-rounded picture of sphingolipids and cell signaling is not yet available.

## COVALENT BINDING OF LIPIDS TO PROTEINS

A number of lipids are used for the covalent modification proteins (Gelb, 1997). The modifications occur shortly after the synthesis of the polypeptide chain and tend to be a permanent part of the protein. The lipids used are palmitic acid (16:0), myristic acid (14:0), the farnesyl group, the geranylgeranyl group, and phosphatidylinositol (PI). PI is connected to the protein by means of a linker composed of a tetrasaccharide and ethanolamine. The ethanolamine is bound to the C-terminal carboxyl group of the protein in an amide bond. In general, the above lipids are thought to be used to maintain adhesion of the protein to the cell membrane, but a role in supporting specific protein-to-protein interactions is also possible (Marshall, 1993). In some cases, a protein is membrane-bound because of transmembrane regions in its polypeptide chain where the protein also contains the covalently bound lipid. In other cases, the protein contains bound lipid, but remains soluble in the cytoplasm. The function of the lipid is not clear in many cases. A small number of proteins modified by lipids have been discovered, and some of these are described below.

The PI anchor maintains adhesion of acetylcholinesterase (of the red blood cell) and of some proteoglycans (sulfated proteins of the extracellular matrix) to the cell membrane. Palmitic acid is bound via thiol-ester bonds to Cys 322 and Cys 323 of rhodopsin (see the section on vitamin A), a 327-amino-acid protein. The polypeptide chain of rhodopsin loops in and out of the membrane several times, leaving the possible function of the lipid as an anchor in question. Myristic acid is bound to the catalytic subunit of the cAMP-dependent protein kinase, though this protein is cytosolic and soluble.

### Myristic Acid Is Bound to the α-Subunit of G Protein

G Protein is a class of proteins, occurring at the plasma membrane, that participates in transmitting signals from the extracellular fluid to various enzymes within the cell. Myristic acid is bound to **G protein**. Specifically, myristic acid is attached to the α subunit of G protein. This lipid is bound via an amide linkage to the N-terminal glycine. A geranylgeranyl group is attached via a thiol-ester bond to a cysteine residue on the γ subunit of G protein (Casey and Seabra, 1996).

### Farnesyl Is Bound to RAS

RAS is a type of G protein that is also used in cell signaling. A farnesyl group is bound to RAS protein. RAS protein, which helps control the rate of cell growth, is important to the study of cancer because specific mutations in its gene result in slight changes in its polypeptide sequence. Some of these mutations in RAS, which naturally occur during human life, result in continuous generation of abnormal signals that cause the cell to become a cancer cell. There is some interest in designing inhibitors of the farnesylation reaction, which in turn would impair the activity of the mutated RAS protein and thus prevent it from causing cancer. In short, the inhibitor is expected to be effective as a drug for curing cancer. Some RAS proteins contain a geranylgeranyl group, rather than a farnesyl group.

*Cholesterol Is Bound to Hedgehog*

Cholesterol is covalently bound to some proteins, as revealed by studies of various animals (Porter *et al.*, 1996). Specifically, cholesterol is attached to **hedgehog**. Hedgehog is a 44-kDa protein that is used to control or guide the formation of the embryo. It is signaling protein that is secreted by various cells. Although the protein might be expected to exist in hedgehogs (a small animal), it received this trivial name in order to make it easy to remember. A nontrivial issue is that in the human population mutations occur naturally in hedgehog. These mutations result in severe birth defects (cyclops eye) and affect 1 in 16,000 newborn infants (Belloni *et al.*, 1996; Roessler *et al.*, 1996).

## CHOLESTEROL

Cholesterol is a lipid that occurs in food and in the body mainly as free cholesterol or as a covalent complex with a fatty acid bound to its hydroxyl group. Such complexes are called **cholesterol esters** or **cholesteryl esters**. Cholesterol is used by many organisms as a structural element in membranes and as the starting material for synthesizing bile salts and steroid hormones, including aldosterone, estrogen, testosterone, and vitamin D. Although most cells seem to synthesize most of the cholesterol that they need, cells that make cholesterol-based hormones have an increased tendency to acquire cholesterol from the bloodstream. Plants do not contain cholesterol, but do synthesize similar compounds such as **sitosterol**, **ergosterol**, and **digitalis**. A small portion of the body's cholesterol is obtained from food, but the molecule is completely dispensable in the diet. Dietary cholesterol is absorbed fairly well (60–80%), but plant sterols are absorbed poorly (<5%).

The average daily intake of total dietary cholesterol is 400–500 mg. Cholesterol also enters the gastrointestinal tract via the bile. Between 800 and 1200 mg of free cholesterol is secreted in the bile per day. By weight, bile consists of 92% water, 6% bile salts, 0.3% cholesterol, and small amounts of bilirubin, fatty acids, phosphatidylcholine, and salts. The cholesteryl esters of the diet are hydrolyzed to free cholesterol and a fatty acid by pancreatic cholesterol esterase. After entry into the enterocyte, the free cholesterol is converted back to cholesteryl esters by acyl CoA:cholesterol acyltransferase. Some evidence suggests that the absorption of dietary cholesterol (from the bile salt micelles) is mediated by a membrane-bound transport protein of the brush border (Thurnhofer and Hauser, 1990).

*Regulation of the Quantity of Cholesterol in the Body*

Plasma cholesterol levels are not changed very much by changes in the quantity of cholesterol in the diet for a number of reasons. First, reduction of dietary intake below typical levels results in a change in available cholesterol that is small compared with the rate of endogenous synthesis: 9–12 mg/kg body weight per day. Hence, a 70-kg man synthesizes about 700 mg per day. Second, a reduction in the dietary intake may result in an increase in the rate of cholesterol biosynthesis in the body. This increase is due to an increase in the activity of one of the enzymes of the cholesterol biosynthetic pathway, **hydroxymethylglutaryl CoA reductase** (HMG-CoA reductase). Third, increasing dietary cholesterol above its typical level may not result in a corresponding increase in absorption. Apparently, the greatest amount that can be absorbed by the gut is about 1.0 g per day. An increase in dietary cholesterol causes a decrease in HMG-CoA reductase activity, resulting in less endogenous synthesis.

## Cholesterol in Cell Membranes

The most vital function of cholesterol in animals is as a structural element in cell membranes. About 90% of the unesterified cholesterol in the cell is located in the plasma membrane. Only a small proportion occurs in internal membranes and in the lipid droplets of the cytoplasm. For example, the cholesterol:phospholipid ratio in the plasma membrane (PM) is about 1.0:1.0 but that in the ER is about 0.1:1.0 (Straka *et al.*, 1990). The cholesterol:phospholipid ratio controls the viscosity of the membrane. Higher ratios result in a more viscous, less fluid membrane. The ER consists of a network of branching tubules. A fluid-like behavior has been observed in the ER of living cells observed under a microscope. Individual branches may move along the tubules and migrate over the surfaces of neighboring branches (Lee and Chen, 1988).

In the laboratory, synthetic membranes can be made from phospholipids alone. They cannot be made from cholesterol alone, but can be made from combinations of phospholipid and cholesterol. Cholesterol stabilizes the structure of a membrane. A minor form of cholesterol, cholesterol sulfate, may prevent the membrane from fusing with other membranes. Cholesterol sulfate is especially prevalent in the PM of sperm (Cheetham *et al.*, 1990).

The major sites of cholesterol synthesis are the liver and the intestines. Generally, about 1/3 of our cholesterol arises from the diet, while 2/3 is made in the body (Jones, 1997). Most of the cholesterol in the body is manufactured by extra-hepatic tissues. This statement applies to most animals, except for rats and mice, where the liver makes most of the body's cholesterol (Dietschy, 1997). Nearly all the cholesterol synthesized in the liver is used for bile salt synthesis. The high contribution of the intestines to the body's synthesis of cholesterol is due to their large surface area and rapid turnover, as discussed in the section on the crypt and villus in Chapter 2.

## Intermediate Biochemicals in the Cholesterol Biosynthetic Pathway

The multistep pathway of cholesterol biosynthesis is shown in Figure 6.13. The acetyl group of the starting material, **acetyl CoA**, is acquired readily by breakdown of dietary sugars or fats. Three molecules of acetyl CoA are used in the synthesis of hydroxymethylglutaryl CoA (HMG-CoA), which is reduced to **mevalonic acid** in a reaction catalyzed by HMG-CoA reductase using the cofactor **NADPH** that is used in the synthesis of fatty acids also. (HMG-CoA is synthesized in the mitochondrion also, but in that organelle the purpose is to make the ketone bodies.) Mevalonic acid is converted to farnesyl pyrophosphate (FPP) at a branching point in the cholesterol synthetic pathway.

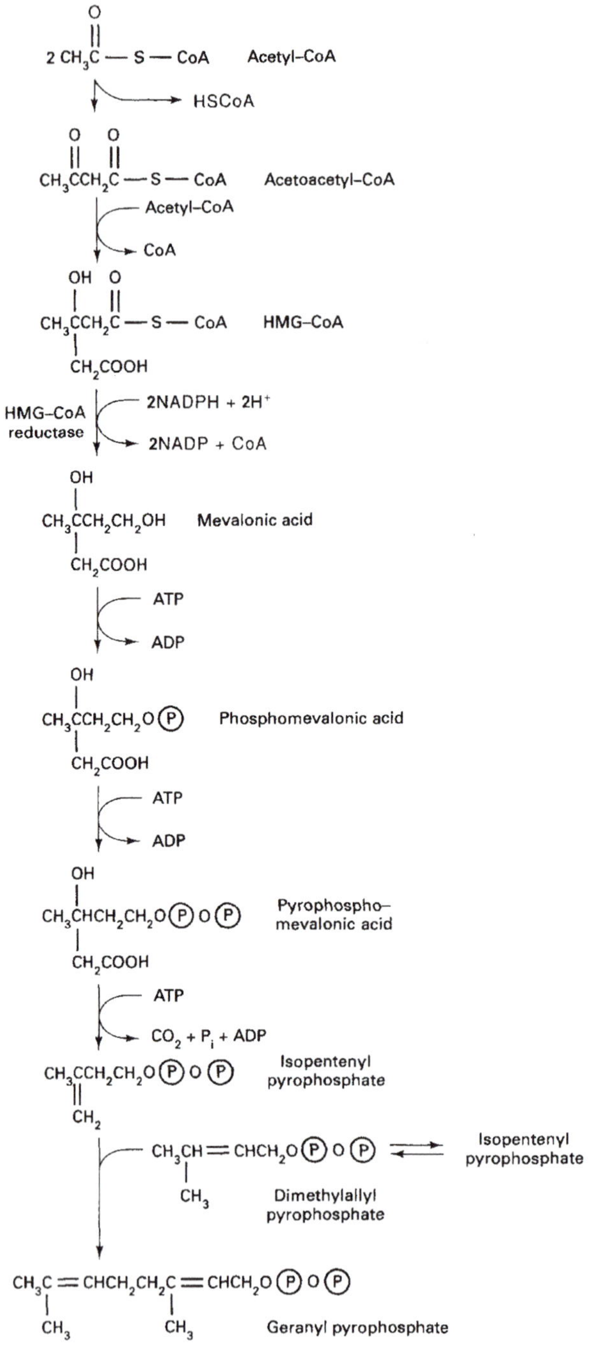

**FIGURE 6.13** The cholesterol biosynthetic pathway. Two molecules of acetyl-CoA, readily acquired by breakdown of dietary sugars or fats, undergo a condensation reaction to produce acetoacetyl-CoA. A third molecule of acetyl-CoA is used in the synthesis of hydroxymethylglutaryl-CoA (HMG-CoA). The keto group of HMG-CoA is reduced to an alcohol group, generating mevalonic acid. This reaction is catalyzed by HMG-CoA reductase, which, like the enzymes that catalyze subsequent reactions, is a membrane-bound enzyme of the ER. However, the pathway

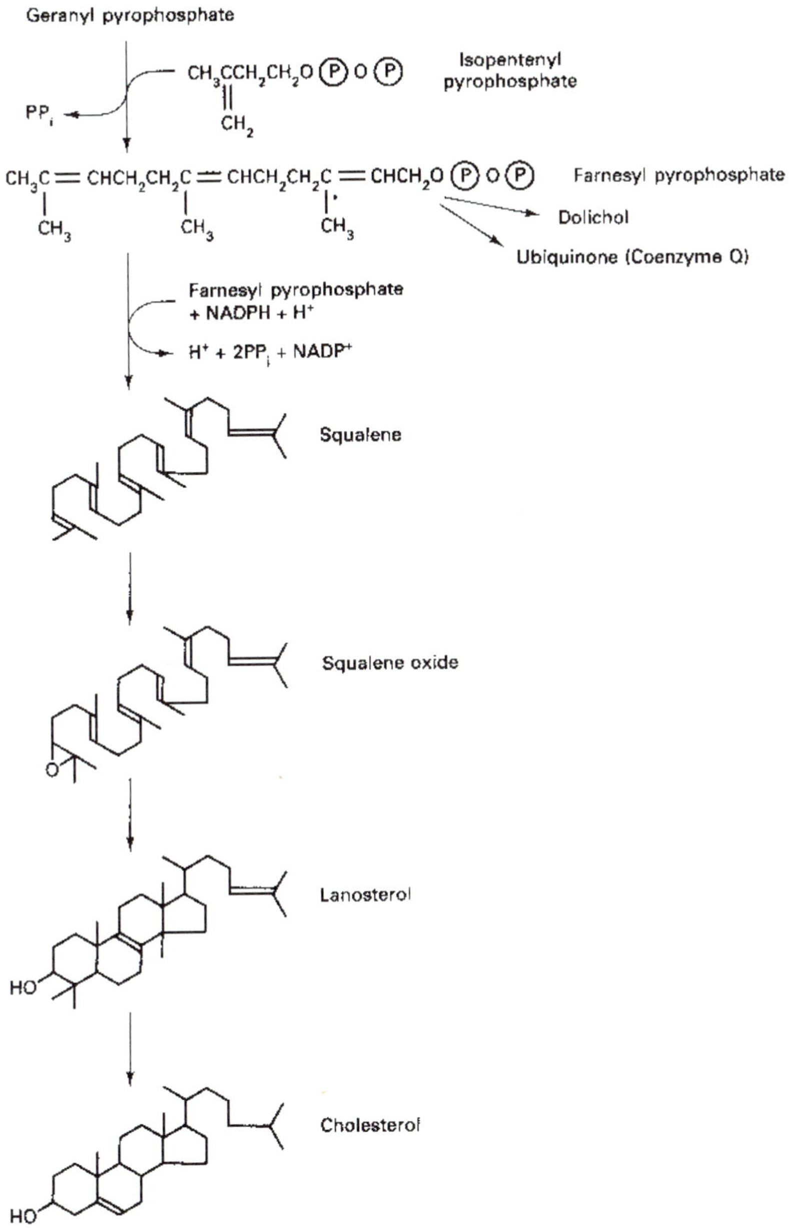

Geranyl pyrophosphate

$CH_3CCH_2CH_2O\ \textcircled{P}\ O\ \textcircled{P}$ — Isopentenyl pyrophosphate

$PP_i$

$CH_3C=CHCH_2CH_2C=CHCH_2CH_2C=CHCH_2O\ \textcircled{P}\ O\ \textcircled{P}$    Farnesyl pyrophosphate

Dolichol

Ubiquinone (Coenzyme Q)

Farnesyl pyrophosphate
+ NADPH + H⁺

H⁺ + 2PP$_i$ + NADP⁺

Squalene

Squalene oxide

Lanosterol

Cholesterol

is cytosolic because the catalytic sites of the enzymes are immersed in the cytosol of the ER, not the lumen. As a source of reducing power, HMG-CoA reductase uses NADPH, a cofactor used in reactions involving the transfer of two electrons. Mevalonic acid is modified further by phosphorylation and decarboxylation to yield isopentenyl pyrophosphate. The size of the intermediate then doubles with condensation of two 5-carbon intermediates to yield geranyl pyrophosphate, a 10-carbon compound. Another 5-carbon compound is added to form a 15-carbon compound, farnesyl pyrophosphate (FPP). FPP is formed at a branching point in the cholesterol synthetic pathway. Two molecules of FPP can condense to form the 30-carbon compound squalene. Alternatively, FPP can be used to synthesize ubiquinone or dolichol. As shown, squalene is converted to lanosterol in two steps. The subsequent conversion of lanosterol to cholesterol involves 19 steps (not shown).

Farnesyl pyrophosphate is used for the synthesis of ubiquinone, dolichol, and squalene. **Ubiquinone** is a cofactor in the respiratory chain of the mitochondrion. **Dolichol phosphate** serves as a biochemical "handle," and is used to hold the core oligosaccharide, and to facilitate its transfer to newly made proteins in the endoplasmic reticulum, to form glycoproteins. **Squalene** is the product of condensation of two FPP molecules.

As shown in Figure 6.13, squalene is converted in two steps to **lanosterol**. Conversion of lanosterol to cholesterol involves 19 steps that are not shown. (Although cholesterol contains several rings, it is not an aromatic compound because its rings do not resonate.)

### Regulation of Hydroxymethylglutaryl-CoA Reductase

Cholesterol synthesis is regulated at the step involving HMG-CoA reductase. The enzyme activity is regulated at the transcriptional level, that is, by changing the rate of synthesis of the mRNA encoding the enzyme. HMG-CoA reductase is regulated by phosphorylation and dephosphorylation also. Phosphorylation, which inactivates the enzyme, appears to be controlled by two different signaling systems, cyclic AMP and a calcium-dependent system. An increase in the level of cyclic AMP in the cell stimulates the activity of a **kinase** that phosphorylates HMG-CoA reductase. (Kinases are enzymes that catalyze the phosphorylation of enzymes using ATP as the source of the phosphate group.) An increase in the level of calcium ions stimulates a number of kinases, including the one that acts on HMG-CoA reductase. The physiological relevance of these changes in HMG-CoA reductase activity is not clear.

HMG-CoA reductase is inhibited by the drug **lovastatin**, a natural product synthesized by a fungus (Figure 6.14). Lovastatin does not inhibit the enzyme directly; it is converted to a compound (similar to the structure of HMG-CoA) that inhibits the enzyme. Lovastatin, as well as several related compounds in a family of chemicals called **statins**, have found use in the treatment of cardiovascular disease throughout the world. The study outlined in Figure 6.15 shows the effectiveness of lovastatin in the treatment of cardiovascular disease. The drug lowers

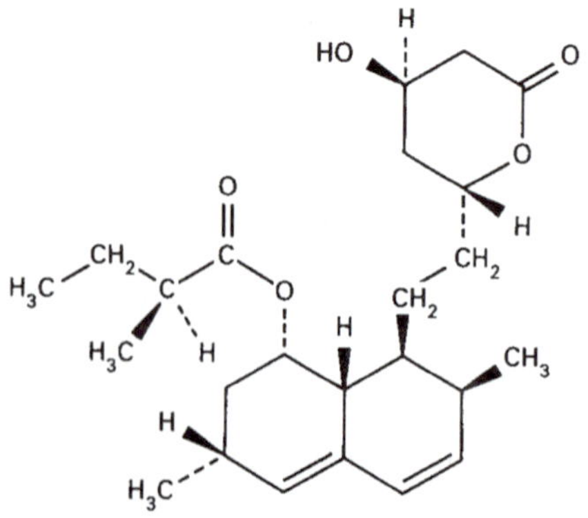

**FIGURE 6.14** Lovastatin. Lovastatin is one of the "statin" drugs.

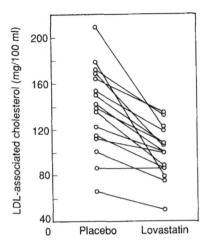

**FIGURE 6.15** Concentration of cholesterol that is associated with low-density lipoproteins. The data show that an inhibitor of HMG-CoA reductase can be used as a drug to lower LDL-cholesterol levels in humans. (Redrawn with permission from Garg and Grundy, 1988).)

the concentration of cholesterol in the bloodstream. Cholesterol, which is not water soluble, circulates as cholesterol and cholesteryl esters in complexes with proteins and other lipids called **lipoproteins**. In this study, blood samples from 16 patients were analyzed for plasma cholesterol associated with **low-density lipoproteins** (LDLs) after treatment with a placebo for 1 month (data on the left). Then, after treatment with lovastatin for 1 month, the quantity of cholesterol associated with LDLs in blood samples was measured (data on the right). The data reveal the ability of the enzyme inhibitor to reduce the concentration of cholesterol (associated with LDL) in the bloodstream.

## Various Fates of Cholesterol in the Liver

In the liver, cholesterol has three major fates: conversion to bile acids, secretion into the bloodstream (packaged in lipoproteins), and insertion into the plasma membrane. Conversion of cholesterol to cholic acid, one of the bile acids, requires about 10 enzymes. The rate of bile synthesis is regulated by the first enzyme of the pathway, **cholesterol 1α-hydroxylase**, one of the cytochrome P450 enzymes (see the section on Iron in Chapter 10). Cholesterol, mainly in the form of cholesteryl esters, is exported to other organs, after packaging in particles called very-low-density lipoproteins. Synthesis of cholesteryl esters is catalyzed by acyl CoA:cholesterol acyltransferase, a membrane-bound enzyme of the ER. Free cholesterol is used in membrane synthesis, where it appears as part of the walls of vesicles in the cytoplasm. These vesicles travel to the plasma membrane, where subsequent fusion results in incorporation of their cholesterol and phospholipids into the plasma membrane.

## LIPOPROTEINS

The lipids of the diet include TGs, phospholipids, cholesteryl esters, cholesterol, and the fat-soluble vitamins. These nutrients require special types of biochemical machinery to facilitate their assimilation and distribution within the body. The biochemical apparatus used includes bile salts, apolipoproteins, serum albumin, and vitamin-binding proteins. Apolipoproteins are the primary subject of this section. The term *apolipoprotein* is used when referring only to the protein, whereas the term *lipoprotein* refers to the complex of apolipoprotein and lipid.

Lipoproteins are quite large, ranging in molecular weight from 150,000 to 30 billion. For comparison, the molecular weight of a typical enzyme subunit is about 50,000. The diameter of the largest lipoprotein is 0.0005 mm. Although smaller than the starch granules in plant food and smaller than blood cells, lipoproteins are large enough to be called particles. One definition of a particle is a piece of matter that, when suspended in a fluid or gas, scatters light (appears cloudy). The molecular weights and general compositions of lipoproteins are listed in Table 6.4.

Lipoproteins are assembled in two organs, the small intestine and the liver. The lipoproteins assembled in the intestine contain the lipids assimilated from the diet. These lipoproteins, called **chylomicrons**, leave the enterocyte and enter the bloodstream via the lymphatic system. The lipoproteins assembled in the liver contain lipids originating from the bloodstream and from *de novo* synthesis in the liver. The term *de novo* simply means "newly made from simple components" as opposed to "acquired from the diet" or "recycled from preexisting complex components." These lipoproteins, called **very-low-density lipoproteins** (VLDLs), are secreted from the liver into the bloodstream. The liver also synthesizes and secretes other lipoproteins called **high-density lipoproteins** (HDLs), which interact with the chylomicrons and VLDLs in the bloodstream and promote their maturation and function. The data in Table 6.4 show that chylomicrons contain a small proportion of protein, whereas HDLs have a relatively high protein content. Of greater interest is the identity and function of the proteins that constitute these particles. These proteins confer specific properties to lipoprotein particles, as detailed later in this chapter.

EXERCISE

Is there any correlation between lipid content and density of lipoproteins?

**TABLE 6.4**  Gross Composition of Lipoprotein Particles

| Lipoprotein | Molecular weight | Weight % | |
|---|---|---|---|
| | | Protein | Lipid |
| Chylomicron | $30 \times 10^9$ | 1 | 99 |
| Very-low-density lipoprotein | $5–10 \times 10^6$ | 10 | 90 |
| Low-density lipoprotein | $2.2–3.5 \times 10^6$ | 25 | 75 |
| High-density lipoprotein | $1.5–4.0 \times 10^5$ | 50 | 50 |

EXERCISE

Would you expect chylomicrons to be more dense or less dense than VLDLs?

Although various lipoproteins have specific names, such as "chylomicrons" and "VLDLs," they are not static unchanging particles. Chylomicrons receive apolipoprotein C-II on entry into the thoracic duct and bloodstream. Apo C-II functions in the bloodstream, and is required for the activity of a special lipase that catalyzes removal of TGs from the chylomicrons. During the progressive loss of TGs from a chylomicron, it becomes smaller until it reaches a stage where it is called a **chylomicron remnant**. A chylomicron remnant can no longer deliver dietary TGs to tissues, because it has been largely depleted of this nutrient. The fate of a chylomicron remnant is that it too is taken up by a tissue, the liver. Chylomicron remnants do not return to the gut cells to be used over again, but are hydrolyzed in the liver to yield free amino acids and lipids. Uptake of a chylomicron remnant by the liver requires two proteins. One, **apolipoprotein E**, resides on the remnant itself. The other resides on the plasma membrane of the hepatocyte and is called the **apo E receptor**. After a chylomicron remnant is bound to an apo E receptor, the resulting complex undergoes endocytosis.

## Properties of the Individual Lipoproteins

The complement of apolipoproteins present in each of the major lipoproteins is listed in Table 6.5. The chylomicrons contain **apolipoprotein B-48 (apo B-48)**, **apo C-II**, and **apo A-I**. (Apo B-48 is also known as the *intestinal form of apo B*.) The VLDLs contain **apo B-100**, **apo C-II**, and **apo E**. The HDLs contain **apo A-I** and

**TABLE 6.5**  Major Apolipoproteins in Lipoprotein Particles

| Lipoprotein | Major constituent apolipoproteins |
|---|---|
| Chylomicron | apo B-48, apo C-II, apo A-I |
| Very-low-density lipoprotein | apo B-100, apo C-II, apo E |
| Low-density lipoprotein | apo B-100 |
| High-density lipoprotein | apo A-I, apo A-II |

Two of the proteins in the table, Apo-B48 and Apo B-100, are discussed. Apo B-48 is synthesized only in the **small intestines** of humans, but in rats, mice, dogs, and horses it is also produced in the liver. Apo B-48 contains 2152 amino acids and is identical to the N-terminal half of apo B-100. Apo B-100 does bind to the LDL receptor, but apo B-48 does not bind to it. Apo B-100 is synthesized only in the liver and is one of the largest proteins known, as it contains 4536 amino acids. The mRNAs coding for apo B-48 and apo B-100 are identical to each other, immediately after transcription. However, subsequent processing of the mRNA can result in a shortening of the coding sequence, with the consequent production of the shorter polypeptide of apo B-48. A mechanism called *mRNA editing* is used to create the shorter polypeptide. A special enzyme catalyzes the conversion of a residue of cytidine to uridine. This results in the change of a codon specifying glutamine (CAA) to a stop codon (UAA). The mRNA editing enzyme requires zinc (Xiong *et al.*, 1992; Innerarity *et al.*, 1996). To compare some sizes, Apo B-48 (which is *smaller* than Apo B-100) is used to make chylomicrons (which are *larger* than VLDLs).

**apo A-II**. The reader might skim quickly through the detailed information in the following paragraphs, read through the section on Life History of Chylomicrons and VLDLs (and subsequent sections), and then return to these paragraphs. Most of the genetic diseases mentioned below are rare.

### Apo B-48 and Apo B-100 are the Defining Proteins of the Chylomicron and VLDL

Apolipoprotein B-48 (apo B-48) and apo B-100 are the main structural components of the lipoproteins secreted by the intestines and liver, respectively. Apo B is not absolutely essential for human life, as a rare genetic disease in humans results in smaller versions, or in the complete absence, of the protein, yet allows survival (Linton *et al.*, 1993; Farese *et al.*, 1995). The smaller versions of the protein have names, such as Apo B-37, Apo B-39, Apo B-74, and Apo B-89. The names refer to the percentage of their length, relative to that of Apo B-100 (100%). The genetic disease results in the malabsorption of dietary fats and in neurological disease. About a third of the fat consumed in the diet is not absorbed, resulting in steatorrhea (greasy feces). The consumption of 100 g of fat, over several hours, results in nausea, vomiting, and diarrhea (fatty food intolerance).

### Apo C-II Is Needed for the Transfer of Energy from Lipoproteins to Muscle and Other Tissues

Apo C-II functions in the transfer of energy from lipoproteins to various tissues of the body. This energy (food energy) consists of TGs. These TGs are hydrolyzed, from the lipoprotein particles, at the walls of capillaries passing through the tissues, thus releasing free fatty acids as energy sources. This TG hydrolysis is catalyzed by the enzyme **lipoprotein lipase**. Lipoprotein lipase alone cannot act on TGs or diglycerides, but requires apo C-II as a cofactor. The letter "C" in apo C-II should serve as a reminder of its function as a *cofactor*.

Apo C-II activates lipoprotein lipase. A diagram of a chylomicron particle and a VLDL particle about to pass through a capillary is shown below. Chylomicrons are relatively large, having a diameter of 500 nm, while VLDLs are smaller, with a diameter of 100 nm. Lipoprotein lipase is shown as small dots adhering to the lumen of the capillary, and the chylomicron is shown contacting one of the lipases (Goldberg, 1996). The released free fatty acids (FFA) are taken up and used by cells that reside near the capillary:

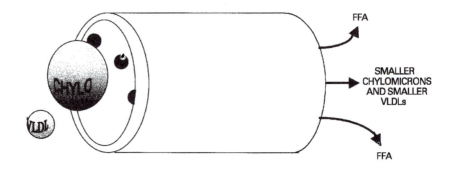

The enzyme sticks to a protein of the extracellular matrix of the endothelial cell, called a **proteoglycan**. Proteoglycan is a glycoprotein, containing chains of sulfated sugars. More specifically, the proteoglycan adhering to the capillaries is heparan sulfate, which contains the sugars glucosamine and glucuronic acid or iduronic acid (Goldberg, 1996). The major organs that produce lipoprotein lipase are muscles, adipose tissue, and mammary gland (Camps *et al.*, 1991).

Apo C-II is a very small protein, as it consists of only 79 amino acids (Streicher *et al.*, 1996). A rare genetic disease involving mutations at single nucleotides results in the conversion of a codon specifying an amino acid to a codon that specifies a translation stop signal. The consequence of these mutations is the lack of detectable Apo C-II in plasma, high levels of chylomicrons (even with fasting), and high plasma TGs. The patient is aware of the disease because of episodes of abdominal pain. Xanthomas and inflammation of the pancreas (pancreatitis) may also occur (Parrott *et al.*, 1992).

Persons with genetic mutations in the gene coding for lipoprotein lipase may or may not be at greater risk for atherosclerosis, depending on the exact nature of the mutation. Severe deficiency in the enzyme results in premature atherosclerosis, abdominal pain, xanthomas, and sometimes pancreatitis. Mild deficiencies in the enzyme appear to occur in about 5% of white people (European descent). Over 60 different naturally occurring mutations have been found in people, where the most common seems to be a genetic mutation leading to the conversion of Asn 291 to Ser 291 (Reymer *et al.*, 1995; Benlian *et al.*, 1996).

## Apo B and Apo E are Used For Uptake of Lipoprotein Particles by Organs

Apo B-100 is the hepatic form of apo B, which facilitates the uptake of its associated lipoprotein by the liver. Apo E also facilitates the uptake of its associated lipoprotein by the liver. The rate of uptake mediated by apo E is greater than that mediated by apo B. Uptake of lipoproteins by the liver is by **endocytosis**, a process that involves binding of a protein or particle to a cell via a receptor located on the cell membrane. This binding is followed by invagination of part of the membrane to form a vesicle that surrounds the complex of protein and receptor. The vesicle then buds off and freely enters the cytoplasm.

Although apo B and apo E may be constituents of certain lipoproteins and may be used for endocytosis, their mere presence on the surface of a lipoprotein particle does not guarantee that the particle will be taken up promptly by the liver or other organs that have apo E or apo B receptors. The uptake of the various lipoproteins tends to occur at specific times during the life of the particle, as it matures in the bloodstream.

Apo E consists of 299 amino acids. A genetic disease, called **type III hyperlipoproteinemia**, results from naturally occurring mutations in the apo E gene. The disease results in high plasma cholesterol, high plasma TGs, premature atherosclerosis, and xanthomas. Xanthomas are lumpy accumulations of cholesterol in the wrists, elbows, knees, and other parts of the body. The most common mutation is one that results in the conversion of Arg 158 to Cys 158 (Lohse *et al.*, 1991). Other naturally occurring mutations result in a truncated polypeptide, or in the complete absence of apo E.

*Apo A-I*

Apo A-I, which consists of 243 amino acids, is the major protein of the HDLs (Schmidt *et al.*, 1995). Apo A-I accounts for about 70% of the protein in the HDL. This protein serves as a cofactor for lecithin:cholesterol acyltransferase (LCAT), and also to facilitate the binding of HDLs to plasma membranes, and to facilitate the removal of cholesterol from tissues. A number of genetic mutations in apo A-I have been discovered. Where the genetic mutation results in the complete absence of apo A-I, the result is atherosclerosis and xanthomas. For example, one patient's mutated apo A-I was truncated and consisted of only the first 84 amino acids, of the normal 243 amino acid polypeptide (Matsunaga *et al.*, 1991).

*Apo A-II*

Apo A-II accounts for about 20% of the protein in HDL. Apo A-II's function is not clear. When rare genetic mutations in the gene coding for apo A-II do occur, they seem not to lead to atherosclerosis. In fact, humans and mice lacking apo A-II have reduced levels of plasma cholesterol (Weng and Breslow, 1996). Apo A-II seems to function to regulate the size of the HDL particles.

## Life History of Chylomicrons and VLDLs

Chylomicrons contain a large core that consists of thousands of TG molecules and cholesteryl esters. The cholesterol is esterified to fatty acids such as palmitic or oleic acid. Long-chain TGs and cholesterol esters are among the most highly lipophilic biological molecules. The core of a chylomicron is sometimes described as greasy or oily, although it is much too small to feel with the fingers. The core of a chylomicron is surrounded by a shell of more hydrophilic compounds — phospholipids, free cholesterol, and apolipoproteins. The hydrophilic parts of the phospholipid molecules include phosphate, choline, and ethanolamine groups. The hydrophilic parts of the cholesterol molecules are the hydroxyl groups. Fat-soluble vitamins of the diet are absorbed via chylomicrons. The general structure of a chylomicron, as well as that of other lipoproteins, is depicted in Figure 6.16. The proteins that reside on the outside surface of the particle constitute a small fraction (1.0%) of the weight of the chylomicron. In contrast, protein constitutes a substantial fraction (50%) of the weight of another type of lipoprotein, the HDL.

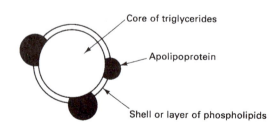

**FIGURE 6.16** Simplified cross-section of a lipoprotein.

Chylomicrons and VLDLs share a common fate once they enter the bloodstream. Both particles acquire additional apolipoproteins and both are attacked by lipoprotein lipase, resulting in partial or substantial loss of the TGs carried within the particle.

## HDLs Donate Apo C-II

As chylomicrons enter the lymphatic system and the bloodstream, they encounter other lipoprotein particles, such as HDLs. HDLs have a number of functions. One HDL protein, apolipoprotein C-II (Apo C-II), is transferred from an HDL to a chylomicron after it leaves the enterocyte. Apo C-II is a cofactor of lipoprotein lipase, an enzyme that resides on the wall of the capillaries in tissues such as muscle and adipose. This enzyme is loosely bound to the lumenal side of the capillary, exposed to the bloodstream. In conjunction with the apo C-II of a chylomicron, it catalyzes the hydrolysis of the TGs of the chylomicron. The free fatty acid products then pass through the wall of the capillary and enter the tissue.

The free fatty acids that enter muscle tissue can undergo immediate oxidation to be used as an energy source. Those that enter adipose tissue undergo immediate conversion back to TGs for storage in the adipocyte. These stored TGs are mobilized eventually for use as an energy source. Some are destined for immediate removal, whereas others may remain stored as TGs for more than one year. The fatty acids of chylomicrons also are released into lactating breast tissue. Here, they are converted to TGs and secreted in the form of milk lipoproteins.

## HDLs Donate Apo E, a Molecular Traffic Signal

Passage of a chylomicron through the bloodstream results in the eventual depletion of most of its complement of TGs and its conversion to a **chylomicron remnant**. Maturation of the chylomicron also involves acquisition of apo E, which may be donated by HDLs during initial passage through the lymph from the gut to the circulatory system. Apo E is required by the chylomicron remnant because it mediates the uptake of this lipoprotein by the hepatocyte. Apo E binds to the chylomicron surface and acts as a traffic signal, directing the nutrient-depleted particle to enter the liver. After entry, the particle is degraded and the cholesteryl esters that may still be present are hydrolyzed to yield cholesterol and used for the synthesis of the bile acids.

Fatty acids occurring as TGs in the VLDLs are derived from three sources: (1) free fatty acids taken up by the liver from the bloodstream; (2) TGs derived from chylomicron remnants, intermediate-density lipoproteins (IDLs), LDLs, and HDLs taken up by the liver via the process of endocytosis; and (3) fatty acids synthesized *de novo* in the liver from carbohydrates. Note that the free fatty acids in plasma are tightly, although not covalently, associated with plasma albumin.

On secretion, the VLDLs acquire apo C-II, which is donated by HDLs. Partial depletion of TGs from a VLDL via action of lipoprotein lipase results in its conversion to a VLDL remnant, also called an IDL. This IDL circulates in the bloodstream until it is taken up by the liver. The apo E that a VLDL receives from an HDL resides on the surface of the lipoprotein and is recognized by receptor proteins bound to the plasma membrane of hepatocytes. The interaction between this receptor and apo E mediates the uptake of IDLs.

The fact that animals do not need to eat continuously throughout the course of the day indicates that chylomicrons are not necessarily actively distributing nutrients during the entire day. Free fatty acids are supplied to tissues via the chylomicrons, but also from the **adipose tissue**, which supplies energy to the body throughout the day. The dry matter of adipose tissue consists of 5% protein and over 90% lipids. About 95% of the lipids are TGs; the rest consists of 1–2% diglycerides, 0.25% phospholipids, and 0.25% cholesterol. Adipose tissue may contain 5–30% water by weight.

The TGs stored in human adipose tissue provide sufficient energy to sustain life in a fasting human for 1–2 months, depending on the initial quantity of fat present. In contrast, the TGs stored by rats are sufficient to maintain survival during fasting for only 1 week. Mobilization of the TGs in adipose tissue does not occur by the release of intact TGs into the plasma. Instead, **hormone-sensitive lipase** catalyzes release of free fatty acids and the glycerol backbone from fat stores. The fatty acids and glycerol are released into the circulation and are taken up by tissues of the muscle, liver, kidney, and other organs. When the body is at rest, the liver takes up about one-third of these free fatty acids. Most are used for the liver's own energy needs, but the excess fatty acids are packaged as TGs in VLDLs and secreted back into the bloodstream. To repeat, a small fraction of the free fatty acids taken up by the liver is converted back to TGs, packaged into VLDLs, and then secreted into the plasma.

### Conversion of VLDLs to VLDL Remnants in the Bloodstream

How do VLDLs remain in the bloodstream long enough to donate their lipids to needy cells before being taken up by the liver? The apo C of the VLDLs appears to prevent their uptake by the liver by covering or obscuring apo E. (The apo C of chylomicrons also prevents their uptake before they are converted to chylomicron remnants.) However, with continued residence in the circulation, the apo C eventually is transferred to HDLs. This loss of apo C exposes both apo E and apo B to the environment. Each VLDL remnant molecule contains several molecules of apo E, which, when exposed, bind avidly to LDL receptors. The VLDL remnant then is taken up by the liver within minutes or hours.

An alternative metabolic pathway is available to a VLDL. In this process, the particle loses its apo E as it matures to an IDL. Without apo E, the particle is not efficiently taken up by the liver but continues to circulate in the bloodstream. Continued removal of TGs from an IDL produces a lipoprotein particle called an LDL that is enriched in cholesteryl esters. After conversion to an LDL, only a single apo B molecule remains on the surface of the particle. This protein binds only relatively weakly to an LDL receptor. Consequently, LDLs have residence times in the circulation of about 3 days. Eventually, LDLs are taken up by various tissues. The receptor that binds an LDL, called an LDL receptor, is similar, or perhaps identical, to the receptors that bind circulating IDLs and mediate their entry into hepatocytes.

The events described are illustrated in Figure 6.17, which shows the acquisition of apo E and apo C-II by chylomicrons and VLDLs. The figure also shows cleavage

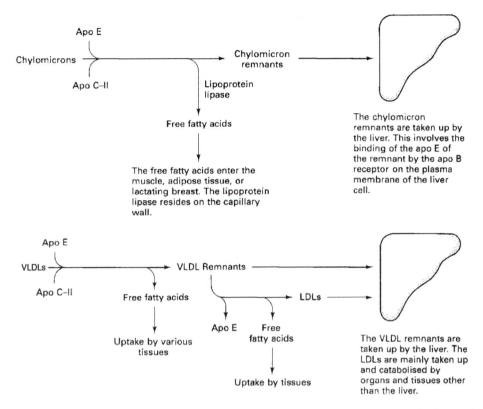

**FIGURE 6.17** Pathways of maturation of the chylomicrons and VLDLs and of release of the fatty acids from these particles.

of the TGs, release of free fatty acids, and uptake by the liver of particles that are relatively poor in TGs but rich in cholesteryl esters.

Several of the concepts described in this section are illustrated on the lipoprotein map (Figures 6.18 and 6.19). The map depicts the synthesis of chylomicrons in the gut and their flow up the thoracic duct to the circulation. The figure also shows the release of fatty acids from these particles in various tissues and the uptake of the remnants by the hepatocytes of the liver. A similar fate is shown for the VLDLs synthesized in the liver. As shown, they mature to LDLs, which deliver cholesterol to various tissues.

## HDLs and the Cycling of Cholesterol

*Apo A-I Is the Defining Protein of the HDL*

HDLs are the smallest of the lipoproteins. The HDL particle is synthesized mainly by the liver, and also by the intestines. When excess cholesterol occurs in extrahepatic tissues, it is picked up by HDLs by a process called **reverse cholesterol transport**. Apo A-I is the vital and defining protein of the HDL. Each HDL particle contains 2–4 molecules of Apo A-I. When secreted by the liver, HDLs are lipid-poor

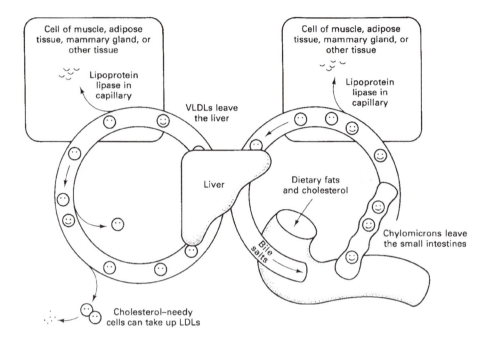

**FIGURE 6.18** The lipoprotein map illustrates the fact that tissues can derive energy from lipoproteins formed by cells of the gut as well as by cells of the liver (hepatocytes). The chylomicrons and VLDLs are mixed together throughout the circulatory system. Not shown is the contribution of apo C-II and apo E to these particles by the HDLs. The map shows that cholesterol is taken up by the peripheral tissues from the LDLs, but not to a large extent from the chylomicrons or the chylomicron remnants. The map also shows that TGs are removed from lipoproteins by lipoprotein lipase, whereas cholesterol is removed after endocytosis of the particle. Also shown is the delivery of bile, which contains bile salts, cholesterol, and phospholipids, to the small intestine. The drawing is stylized and does not closely represent the anatomy.

and contain only a few molecules of phospholipid and sphingomyelin (Fielding and Fielding, 1995) (Figure 6.20). HDLs are involved in transferring cholesterol from cells that contain excessive amounts to the liver or to tissues that are growing and need cholesterol (Huston, 1997). HDLs are involved only in that part of the cholesterol transport and cycling pathway called reverse cholesterol transport.

### HDL Changes from a Disk to a Sphere

The cholesterol in cell membranes occurs as free cholesterol, rather than as cholesteryl esters. As illustrated in Figure 6.21, this cholesterol can be removed and transferred to circulating HDLs in a reaction catalyzed by **lecithin:cholesterol acyltransferase** (LCAT). Continued insertion of cholesteryl esters into an HDL changes its initial **disk shape** to a **sphere shape**. The lipid-poor HDLs (disk-shaped HDLs) account for only a few percent of the HDLs in the bloodstream.

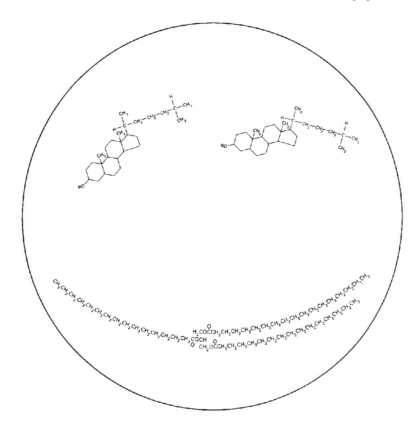

**FIGURE 6.19** The lipoprotein map. The eyes represent cholesteryl esters and the smile represents TGs. The R group on the cholesterol molecule represents the esterified fatty acid.

When the disks become 10 nm wide, the LCAT that is bound by Apo A-I becomes activated by Apo A-I, and the LCAT springs into action. LCAT is an enzyme of 416 amino acids, and thus is a protein of a typical size (Kuivenhoven *et al.*, 1997). LCAT catalyzes the transfer of a fatty acid from the carbon-2 of the glycerol moiety (in lecithin) to the 3-HO group of cholesterol, producing cholesteryl ester. With the packing of cholesteryl esters into the core of the HDL, the HDL acts as a "cholesterol-sponge." LCAT is required to support this sponge function. HDLs pick up most of their free cholesterol from special structures (invaginations) on the plasma membrane called **caveolae** (Oram and Yokoyama, 1996). LCAT is not absolutely needed for human life. A rare genetic disease (fish-eye disease) exists where people do not have active LCAT. The disease results in corneal opacity and blindness, and in reduced levels of HDL-cholesterol.

The cholesteryl esters that reside in HDLs eventually are transferred to VLDLs, which are processed to form IDLs, then LDLs. Cells in need of cholesterol express LDL receptors on their plasma membranes, enabling them to take up an entire LDL, including its cholesteryl esters.

A series of reactions resulting in removal of excessive cholesterol from some cells and delivery of it to cholesterol-hungry cells is shown in Figure 6.22. For example, an HDL may pick up cholesterol from a **macrophage**, a type of white

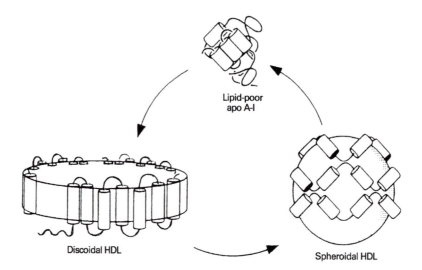

**FIGURE 6.20** Changes in the shape of the HDL particle. The defining protein of the HDL particle is Apo A-I. This means that Apo A-I is always present during the life cycle of the HDL. When the HDL is initially secreted by the liver, it occurs in the lipid-poor form, as shown at the apex of the figure. Apo A-I is the only protein occurring in the lipid-poor form. Lipid accounts for 10–40% of this particle. During passage through the circulatory system, the lipid-poor form acquires phosphatidylcholine and changes to the disk form. With continued residence time in the bloodstream, the disk form is changed to the spherical HDL. Most of the HDL particles in the bloodstream are of the spherical form. Once in its spherical form, several reactions can result in its conversion back to the lipid-poor form. Hepatic lipase, an enzyme loosely bound to the lumen of blood vessels, probably catalyzes most of the depletion of lipid from the spherical form. (Redrawn with permission from Fielding and Fielding, 1995.)

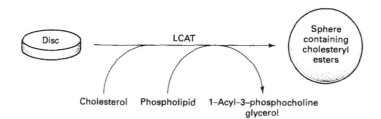

**FIGURE 6.21** Transfer of cholesterol to an HDL. Cholesterol can be removed from cell membranes and transferred to circulating HDLs in a reaction catalyzed by lecithin:cholesterol acyltransferase (LCAT). The substrates of LCAT are cholesterol and phospholipid; the products are the cholesteryl ester and 1-acyl-3-phosphocholine glycerol. In the course of the reaction, the membrane-associated cholesterol is changed to an HDL-associated cholesteryl ester. Continued insertion of cholesteryl esters into an HDL changes its initial disk shape to spherical. The disk is a bilayer of phospholipids and protein; the sphere is a monolayer of phospholipids and protein surrounding a core of cholesteryl esters. HDL contains two major proteins, where 70% is Apo A-I and 20% is Apo A-II. Apo A-I is used to extract or remove lipids from cells. Apo A-I has 243 amino acids. The α-helical regions of this protein are used to bind to the outside surfaces of cells. Each helix is a coil that is 22 amino acids long. These regions are used to bind cells, while other regions of Apo A-I are used to activate LCAT.

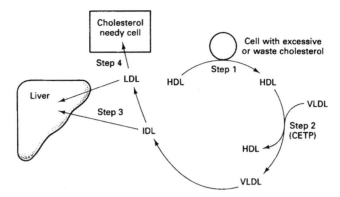

**FIGURE 6.22** Shuttling of cholesterol from one type of cell to another. An HDL may pick up cholesterol from a macrophage, a white blood cell that phagocytizes debris in the bloodstream (e.g., dead red blood cells). A dead red blood cell contains cholesterol, since it contains a plasma membrane. (1) The macrophage can donate the cholesterol (that it has "eaten") to a passing HDL. (2) The cholesteryl ester that is formed is then transferred, in the circulation, to a VLDL. This transfer is catalyzed by an enzyme in the bloodstream called cholesteryl ester transfer protein (CETP). (3) Eventually, the cholesteryl ester can be delivered to the liver and excreted as a bile salt or (4) delivered to a cholesterol-needy cell. This cell may be a premature red blood cell that is engaging in membrane synthesis and mitosis.

blood cell that phagocytizes (eats) debris in the bloodstream. The cholesteryl ester that is formed then might be transferred to a VLDL. The transfer of cholesteryl esters from HDLs to VLDLs is catalyzed by an enzyme in the bloodstream called **cholesteryl ester transfer protein** (CETP). CETP is not absolutely essential for human life. About 1% of persons in Japan are deficient in this enzyme (Tall, 1993).

EXERCISE

Please explain why persons deficient in CETP may be protected from atherosclerosis. (See Tall, 1993.)

## HDLs Have Two Different Fates in the Bloodstream

The following diagram summarizes the intermediate position of the HDL during the event of reverse cholesterol transport. The flow of cholesterol is shown by the arrows of the diagram. The bifurcating arrow indicates that HDLs may be processed by two different means: (1) uptake of HDLs by the liver; and (2) catalytic removal of the cholesteryl esters by CETP.

Most tissues are thought to be able to synthesize all the cholesterol they need, and do not need to rely on the liver or diet as sources of this vital molecule.

**TABLE 6.6** Components of Various Lipoproteins

| | | Weight % | | | |
|---|---|---|---|---|---|
| Lipoprotein | Protein | Phospho-lipid | Trigly-ceride | Free cholesterol | Cholesteryl ester |
| Chylomicron | 1–2 | 3–8 | 80–95 | 1–3 | 2–4 |
| VLDL | 10 | 15–20 | 55–65 | 10 | 5 |
| IDL | 18 | 22 | 31 | 7 | 23 |
| LDL | 25 | 20 | 4 | 9 | 42 |
| HDL | 42–55 | 25–30 | 1–2 | 3–5 | 16–20 |

*Source*: Vance and Vance (1985).

However, growing and regenerating tissues require large amounts of cholesterol. Organs (e.g., the adrenal gland and gonads) that synthesize steroid hormones also require cholesterol because it is a starting material for steroid hormone biosynthesis. These cells express extra LDL receptors on their plasma membranes to take up more LDLs from the bloodstream. A cell adjusts the number of LDL receptors on its membrane according to its need for cholesterol.

## Lipids in Lipoproteins

Table 6.6 gives compositions for several of the lipoprotein types that have been discussed in this chapter. The high TG contents of chylomicrons (80–95%) and VLDLs (55–65%) are consistent with their function in distributing this energy-rich nutrient to various tissues. The low TG content of HDLs implies that these particles are not used to supply energy to cells.

Figure 6.23 depicts the morphology of a lipoprotein. All lipoproteins are spherical except HDLs (prior to acquiring cholesteryl esters), which are discoidal. The

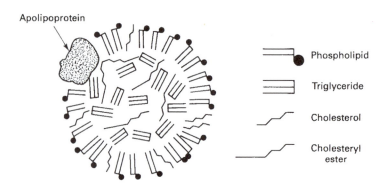

**FIGURE 6.23** Cross-section of a lipoprotein. The proteins, phospholipids, and free cholesterol of a lipoprotein reside on its surface with their hydrophilic components facing the plasma. The core consists of a relatively disorganized mass of TGs and cholesteryl esters.

proteins, phospholipids, and free cholesterol reside on the surface of the particle with their hydrophilic components facing the plasma. The core of the particle consists of a relatively disorganized mass of TGs and cholesteryl esters.

## STUDIES ON THE BEHAVIOR OF LIPOPROTEINS

Concepts concerning the behavior of lipoproteins are somewhat abstract. Therefore, four relevant studies are presented in this section. The first study concerns the acquisition of apo E by newly made chylomicrons that are entering the bloodstream. The high-density lipoproteins donate apo E to these nascent chylomicrons. The second study concerns the detection of differences in the amount of apo E in the plasma of normal persons and in that of those suffering from hyperlipoproteinemia. This study addresses the question of why the disease involves elevated levels of lipoproteins in the bloodstream. The third study concerns a disease in which apo E is present in normal quantities, but has a defective structure. The final study concerns very-low-density lipoproteins isolated from the blood of normal persons and of those suffering from hypertriglyceridemia (high plasma TGs). High plasma TGs is an important risk factor for cardiovascular disease. The study addresses the question of what provokes the associated rise in plasma TGs (TGs contained in VLDLs). This study also illustrates the ability of the HDLs to transfer apo C-II to VLDLs. In one experiment, the deficiency in apo E is shown using a type of chromatography called **electrophoresis**. Although this is a powerful technique, it may fail to reveal subtle defects in the apo E of some patients with hyperlipoproteinemia. In another experiment, the function of apo E is tested using **competitive binding**. The competitive binding test can reveal biological defects that are too subtle to detect by electrophoresis or by common chemical assays. The final experiment, concerning the function of apo C-II, uses enzyme assays to illustrate the role of apo C-II in supporting the activity of lipase. Before considering the above studies, some general properties of plasma lipids will be revealed by feeding experiments.

## Appearance of Chylomicrons in the Bloodstream Following a Meal of Fat or Oil

The experiments depicted in Figure 6.24 involved human subjects. In the first study, subjects consumed a breakfast containing 50 g fat. Blood samples were taken at the indicated times; the concentration of particulate matter present was assessed by measuring the amount of light scattering by the plasma. Light scattering in layperson's terms is cloudiness. The results from four normal subjects are shown by the shaded area of Figure 6.24. The results from four different patients suffering from hyperlipoprotenemia are shown also. The rise in light scattering following the breakfast was the result of the appearance of chylomicrons in the bloodstream.

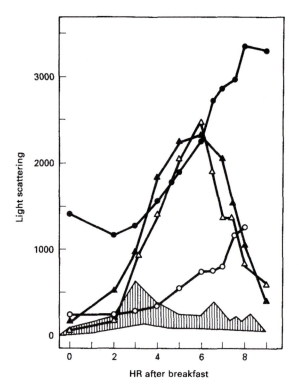

Light scattering

HR after breakfast

**FIGURE 6.24** Cloudiness of serum samples taken at various times. Comparison of four normal subjects (shaded area) and four hyperlipoproteinemic subjects (symbols). (Redrawn with permission from Berger, 1986.)

EXERCISE

At what time after breakfast did the plasma of the normal subjects scatter the most light (Figure 6.24)? State evidence that the disease hyperlipoproteinemia involves defects in the ability of tissues of the body to process chylomicrons.

EXERCISE

A woman suffered a gunshot wound just below the left side of the neck. A ceaseless leakage of fluid issued from the unclosed wound. After eating, the leaking fluid became milky. The patient lost 5 pounds per week over the course of 1 month. What structure had been ruptured? (Consult Crandall *et al.*, 1943.)

In the study depicted in Figure 6.25, normal and hypertriglyceridemic subjects consumed a meal consisting of corn oil. The blood plasma of both subjects contained substantial amounts of triglyceride prior to the test meal. This triglyceride was contained in lipoproteins such as the VLDLs and LDLs. The rise in plasma TG levels following the test meal was due to an increase in the rate of chylomicron synthesis in the enterocyte and their subsequent passage through the lymphatic system to the circulation. Note that corn oil contains long-chain fatty acids in its TGs.

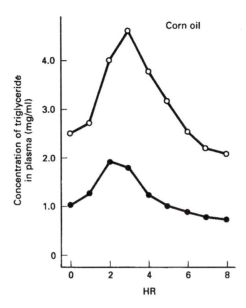

**FIGURE 6.25** Concentration of TGs in blood plasma after a meal of corn oil. Normal (●) and hypertriglyceridemic (○) subjects consumed a meal of corn oil. The TG concentrations in blood samples withdrawn before and at the indicated times after the meal are plotted. (Redrawn with permission from Barr *et al.*, 1985.)

Results of a similar study using medium-chain triglyceride oil are depicted in Figure 6.26. Here, plasma TG levels did not increase after the meal. Medium-chain TG oil consists of TGs as does long-chain TG oil (corn oil). However, the mode by which the body absorbs medium-chain TGs is different from that used for long-chain TG absorption, as described in the section on long-chain and medium-chain triglycerides in Chapter 4. Long-chain TGs are hydrolyzed to polar products in the gut, reformed to make other TGs, and then packaged into chylomicrons. Medium-chain TGs also are hydrolyzed in the gut, but most of the medium-chain free fatty acid products are not resynthesized into TGs. Instead, they directly enter the bloodstream bound to albumin rather than being shuttled up the thoracic duct.

## Transfer of Apo E from One Particle to Another after Eating

In this study, blood samples taken before and after human subjects consumed a breakfast of corn oil (100 g) were analyzed for the total TG and apo E content. A more subtle parameter was measured also, namely the proportion of apo E associated with HDLs and with chylomicrons. This parameter can be measured by **molecular sieve chromatography**. (Details of this method are presented in the next section.) Results from three normal subjects appear in Table 6.7. The data in Table 6.7 demonstrate that the plasma TG levels increased **postprandially** (after eating), as expected. The data also show that the overall concentration of apo E in the bloodstream was fairly constant before and after eating. This constancy should not be interpreted to mean that apo E is not changing in any way. The proportion of

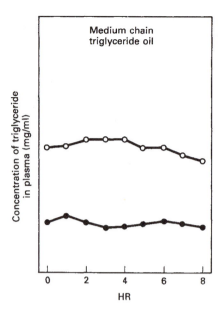

**FIGURE 6.26** Concentration of TGs in blood plasma after a meal of medium-chain TG oil. Normal (●) and hypertriglyceridemic (○) subjects consumed a meal of medium-chain TG oil. The TG concentrations in blood samples withdrawn before and at the indicated times after the meal are plotted. (Redrawn with permission from Barr *et al.*, 1985.)

apo E in the bloodstream that is associated with chylomicrons and with other proteins changes after eating the meal of oil. For example, in the first subject, the fraction of apo E associated with chylomicrons increased from 35 to 60%.

Figure 6.27 presents data from a molecular sieve chromatographic separation of plasma proteins according to their molecular weights. This experiment supplies evidence that the change in location of apo E resulted from transfer of this protein from HDLs to chylomicrons. The data show that, in the fasting subject (●), the amount of apo E associated with HDLs was somewhat greater than that associated with chylomicrons. However, with corn oil feeding (○), the amount of apo E

**TABLE 6.7** Effect of Eating on Plasma Triglycerides and Apolipoprotein E

| Subject | | Total TGs (mg/100 ml plasma) | Apo E (μg/100 ml plasma) | Apo E with chylomicrons (%) |
|---|---|---|---|---|
| 1 | Fasting | 70 | 32 | 35 |
| | After oil | 188 | 32 | 60 |
| 2 | Fasting | 123 | 41 | 51 |
| | After oil | 183 | 38 | 61 |
| 3 | Fasting | 133 | 31 | 42 |
| | After oil | 207 | 35 | 65 |

*Source*: Blum (1982).

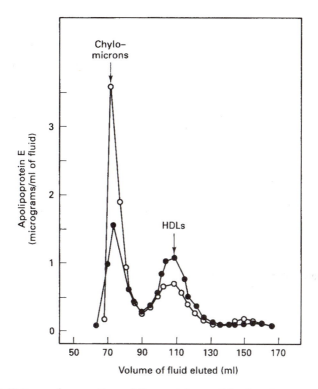

**FIGURE 6.27** Pattern of separation of lipoprotein particles by chromatography using a molecular sieve column. A sample was introduced into one end of a cylinder packed with the molecular sieves. Then a flow of water was maintained through the cylinder. The effluent was collected at the opposite end into a succession of hundreds of test tubes, each receiving an identical volume of effluent. Chylomicrons were separated from HDLs, as indicated by the arrows. (Redrawn with permission from Blum, 1982.)

associated with chylomicrons increased dramatically. Closer inspection of the data indicate that chylomicron particles were eluted from the cylinder when about 75 ml fluid had flowed through; HDLs were eluted when about 110 ml fluid had flowed through. As detailed in the next section, proteins with larger molecular weights migrate more rapidly through the cylinder and are recovered after a smaller volume of fluid has passed through. Proteins with smaller molecular weights are retained by the column and elute only after a greater volume of fluid has passed.

## Molecular Sieve Chromatography

Molecular sieve chromatography, useful for separating proteins of different sizes, operates as follows. The protein sample solution is placed in the top of a cylinder (or column) packed with **molecular sieves,** as shown at the left of Figure 6.28. The sieves consist of tiny balls (or beads) that contain small pores (or holes) of uniform size. The size is small enough to permit the entry (or passage) of a small protein

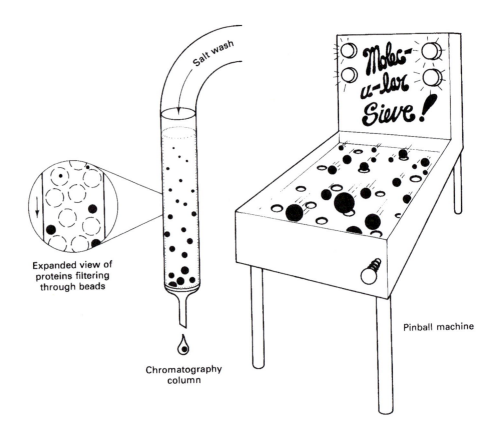

**FIGURE 6.28** Molecular sieve column chromatography. A protein sample solution is placed in the top of a cylinder (or column) packed with molecular sieves, tiny balls (or beads) that contain small pores (or holes) of uniform size. The hole size is small enough to permit entry of a small protein molecule but not of a large protein molecule. Then, as a salt solution flows through the column, the smaller protein molecules are trapped momentarily in the pores or holes, resulting in their slower rate of passage through the column. The larger protein molecules pass right over the holes, avoiding momentary entrapment, and pass out of the bottom of the column prior to the smaller proteins, yielding a separation of proteins according to their molecular weights. The pinball machine (*right*) illustrates the ability of holes to separate spheres according to their size. The machine contains small holes or dimples on its surface. The larger spheres pass right over the dimples and tend to travel faster, whereas the smaller spheres tend to become trapped momentarily, leading to a separation of the spheres according to their size. The enlargement (*left*) shows the proteins (black dots) filtering around and within the beads. The smaller proteins tend to become trapped within the beads momentarily, impairing their rate of passage in the direction indicated by the arrow.

molecule but not of a large protein molecule. The larger protein molecules, excluded from the holes, are then washed through the column by a steady flow of liquid, typically a solution of salts and buffer. The smaller proteins are repeatedly and momentarily trapped in the pores, and take a longer time to pass through the column.

## Genetic Diseases Involving Apo E or Apo C-II

The functions of various apolipoproteins can be brought into focus by case histories on the genetic diseases that alter these proteins. To review, a change in one or more of the nucleotides in a gene coding for any protein can result in a change in the amino acid sequence of the protein. Most genes occur in two copies in the genome, since two copies of each of the numbered chromosomes exist in the cell. Where the change in DNA sequence *occurs in only one of the chromosomes*, the disease is often of a mild variety. Where the change in DNA sequences *occurs on both chromosomes*, the disease may be severe. A change in a single base within the sequence of any gene can result in no change at all in the polypeptide that is coded, in a switch of one amino acid for another, in a shortened version (truncated) of the polypeptide, or in the complete absence of the polypeptide.

One genetic disease involves the production of a defective mutant form of apo E or, in some cases, the complete absence of apo E. The fact that life is possible despite the absence of apo E production suggests that other proteins can perform the function of apo E, but with lesser efficiency.

The following case history concerns apo E. The absence of apo E prevents efficient uptake of **chylomicron remnants** and **IDLs** by the liver; thus, the concentrations of these particles and their constituent TGs and cholesteryl esters can increase dramatically in the plasma. The following comment concerns LDLs, and not the remnants of IDLs. The uptake of LDLs is not affected by the disease, because their uptake is mediated by interactions with the apo B receptor (LDL receptor).

A normal subject and a patient with high levels of plasma lipids (hyperlipoproteinemia) were examined. Blood samples were taken and the lipoproteins in blood samples analyzed by a type of chromatography called **gel electrophoresis**. This method, like molecular sieve chromatography, separates proteins according to their molecular weight. In molecular sieve chromatography, proteins are examined after they have been eluted from the solid matrix. However, in electrophoretic methods such as gel electrophoresis, proteins in a fluid are forced by an electric field to migrate through a solid gel matrix and usually are examined while they still reside in the solid matrix. The term "hyperlipoproteinemia" reveals only a little about the patient; but analysis of plasma proteins by electrophoresis reveals quite a bit more.

The amount of apo E associated with VLDLs, IDLs, and LDLs was analyzed separately, as indicated in regions 1, 2, and 3 of Figure 6.29. The positions of migration of apo B and apo E on the electrophoretic gel are indicated by arrows. The data from the *normal subject* show that the VLDLs and IDLs contained readily detectable amounts of apo E, whereas the LDLs contained very little. The data from the *hyperlipoproteinemic patient* demonstrate that no detectable amount of apo E was associated with any of the three types of particles. This lack of apo E accounts for the hyperlipoproteinemic condition (for the disease).

## Tests of Apo E from Patients with Cardiovascular Disease

Detection of the presence of a specific apolipoprotein in a blood sample is a relatively simple task, because gel electrophoresis is a routine procedure. How-

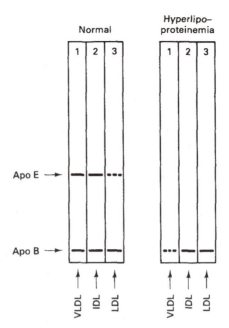

**FIGURE 6.29** Separation of apo E from apo B by electrophoresis. The test samples were separated VLDLs, IDLs, and LDLs taken from a normal subject (*left*) and from a patient with hyperlipoproteinemia (*right*). The migration positions of apo B and apo E on the electrophoretic gel are indicated by arrows. The samples from the normal subject show readily detectable amounts of apo E in the VLDLs and IDLs, but very little in the LDLs. The samples from the hyperlipoproteinemic patient showed no detectable amount of apo E associated with any of the three types of particles. (Redrawn with permission from Ghiselli *et al.*, 1981.)

ever, clinicians and researchers may also be interested in the presence of proteins that are defective in function because of a mutation in the gene. Gel electrophoresis generally cannot distinguish normal from genetically defective proteins, but this distinction can be made with techniques known as functional tests that are based on the biological function of the protein. One such technique used to study the membrane-binding properties of apo E is the **competitive binding assay**.

An understanding of the competitive binding assay can provide a clear picture of the alterations that can pose a risk for cardiovascular disease. In the study depicted in Figure 6.30, apo E preparations isolated from the plasma of one normal subject and two patients were subjected to competitive binding assays. In beginning the study, the researchers used standard methods for connecting atoms of radioactive iodine (iodine-125) to apolipoprotein E.

Iodine-125, phosphorus-32, phosphorus-33, carbon-14, and tritium are isotopes that are commonly used in the research and clinical laboratory. However, it should be noted that iodine-125 and phosphorus-32 are particularly dangerous when handled at close range.

The data demonstrate that adding apo E from the normal subject effectively prevented the [125]I-labeled apo E from binding. In contrast, the apo E from patient

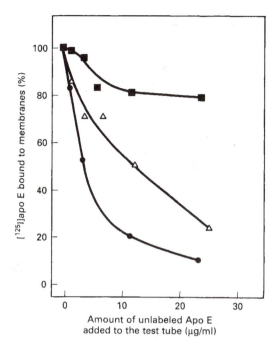

**FIGURE 6.30** Competitive binding assay. Apo E preparations isolated from the plasma of one normal subject (●) and two patients, 1 (■) and 2 (△), were subjected to competitive binding assays. Different concentrations of the apo E samples were mixed in a series of test tubes with solutions containing the same quantity of liver membranes. After a brief waiting period, an equal quantity of normal [$^{125}$I]apo E was added to each test tube. The apo E from the human subjects and the radioactively labeled apo E were allowed to compete for binding to the membrane. (Redrawn with permission from Hui *et al.*, 1984.)

1 competed very poorly with the $^{125}$I-labeled apo E, allowing near-maximal binding to the membranes. The apo E from patient 2 competed, but not as effectively as that from the normal subject. Therefore, the results indicate that the structure of the apo E from patient 2 may have been abnormal, but strongly suggest that the structure of the apo E from patient 1 was abnormal.

## Competitive Binding Assay

The three components of a competitive binding assay are (1) the test protein under scrutiny, (2) the same protein from a normal human subject, chemically tagged (or labeled) with a radioactive atom, and (3) a suspension of cells to which both proteins can bind. The assay, carried out in a test tube filled with buffer and salts to mimic the physiological milieu, requires the two main steps shown in Figure 6.31. The mechanism of the assay is illustrated by its effect on two test proteins: apo E from a normal subject and apo E from a subject with cardiovascular disease. The binding efficiency of a test sample of apo E is determined by measuring the amount of $^{125}$I-labeled apo E bound to the cell membranes. A small amount of bound iodine-125 indicates efficient normal binding of the protein being tested; a

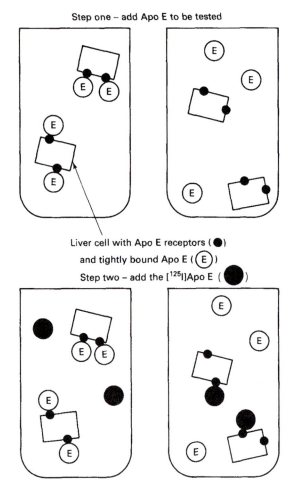

**FIGURE 6.31** Competitive binding assay. The assay mechanism is illustrated by its effect on two test proteins: apo E from a normal subject (*left*) and apo E from a subject with cardiovascular disease (*right*). The rectangles represent liver cells or liver cell membranes; the small dots represent apo E receptors; the big circles represent the apo E proteins to be tested for binding activity. The large dots represent radioactively tagged, normally functioning apo E. (1) The test protein is mixed with a suspension of cells, such as hepatocytes, that contain apo E receptors in their membranes. The test protein from the normal subject (*left*) binds avidly to the receptor, whereas that from the patient (*right*) binds poorly. (2) The radioactively tagged apo E, prepared from normal human subjects, is added to the both suspensions. (The radioactive atom used is iodine-125, so the radioactive protein is designated [$^{125}$I]apo E.) The radioactive protein binds poorly to the cells shown at the left, because it is in competition with the test protein. However, the [$^{125}$I]apo E binds quite well to the cells shown at the right because, in this case, the test protein taken from the patient does not bind well to the membranes and, hence, does not compete. The amount of [$^{125}$I]apo E bound to the cell membranes can be determined easily by subsequent experimental steps.

large amount of bound iodine-125 indicates inefficient abnormal binding. Hence, the quality of apo E from patients suspected of having cardiovascular disease can be assessed.

## Molecular Defects and Binding Properties of Apo E

Apolipoprotein E purified from plasma from a variety of patients with hyperlipo-proteinemia has been studied in detail. Amino acid sequences were determined and reported (position numbers increase from the N terminus to the C terminus). When the disease involved a failure of apo E to bind to its receptor, the genetic mutations involved changes in the amino acids located at positions 142, 145, 146, and 158. Figure 6.32 shows a silhouette of the apo E molecule, with these geneti-cally altered amino acid positions highlighted and numbered. All these positions contain amino acids that are positively charged and occur at the apo E binding site, that is, at that portion of the molecule recognized by an apo E receptor. One form of hyperlipoproteinemia involves production of apo E with a cysteine residue in place of arginine at position 145. Another form of the disease involves replacement of the lysine at position 146 with glutamine. A third involves replacement of the arginine at position 158 with cysteine.

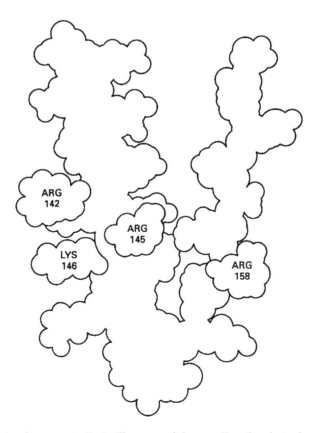

**FIGURE 6.32** Apolipoprotein E. A silhouette of the apo E molecule is shown. The amino acid positions that have been found to be altered genetically in patients with cardiovascular diseases are highlighted and numbered. All the highlighted amino acids are positively charged and occur at the apo E binding site, that is, that portion of the molecule recognized by an apo E receptor. (Redrawn with permission from Innerarity *et al.*, 1984.)

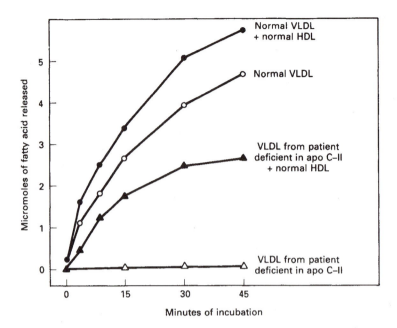

**FIGURE 6.33** Lipase-catalyzed release of fatty acids. This study, conducted in a series of test tubes, compared the action of normal lipase on two different substrates. The substrates were VLDLs from two different people, that is, VLDLs from a normal subject (○,●) and VLDLs from a patient with hypertriglyceridemia (△,▲). Lipase was added to the test tubes containing a solution of VLDLs. The test tubes were incubated for the indicated times (0 to 45 min) to allow catalytic action. Then the amounts of free fatty acids liberated were determined. The results demonstrate a steady increase in free fatty acids from the normal VLDLs but no detectable release from the abnormal VLDLs. The entire protocol was repeated with an identical amount of normal HDLs added to each test tube before incubation. The results demonstrate that the HDL supplement provoked a slight increase in the rate of liberation of fatty acids from the normal VLDLs (●) but a dramatic increase in fatty acid liberation from the abnormal VLDLs (▲). (Redrawn with permission from Breckenridge *et al.*, 1978.)

## Details of the Function of Apolipoprotein C-II

Apolipoprotein C-II serves as a cofactor for lipoprotein lipase. This situation resembles that of colipase, which is required for the activity of pancreatic lipase. When chylomicrons or VLDLs pass through the capillaries of an organ, they encounter lipoprotein lipase. About half the fatty acids liberated by the action of this enzyme are taken up by the tissue, whereas the rest remain in the circulation and return (bound to albumin) to the liver. Apo C-II is part of the structure of chylomicrons and VLDLs.

The study depicted in Figure 6.33 illustrates the attack of lipase on VLDLs isolated from the plasma of normal subjects and from patients suffering from hypertriglyceridemia. The VLDLs from the normal subject would be expected to contain TGs as well as apo C-II. Those from the patient with hypertriglyceridemia would be expected to contain TGs, but little or no apo C-II. The results of the study show a steady liberation of free fatty acids from the normal VLDLs after addition of lipase but no detectable release from the abnormal VLDLs.

When the entire protocol was repeated with addition of identical amounts of normal HDLs to each sample, a slight increase in the rate of liberation of fatty acids from the normal VLDLs was observed but a *dramatic increase was observed from the abnormal VLDLs*. These results illustrate the role of HDLs in transferring apo C-II to VLDLs that have been newly secreted into the bloodstream by the liver.

## CARDIOVASCULAR DISEASE

Cardiovascular diseases account for about half of all deaths in the United States each year. The main risk factors include heredity, male gender, cigarette smoking, high blood pressure, obesity, and a sedentary lifestyle. The main biochemical risk factors are high plasma LDL-cholesterol, low plasma HDL-cholesterol, and high plasma triglycerides. Diabetes is also an important risk factor.

The study of cardiovascular disease can be divided into two parts: (1) mechanisms of atherosclerosis; and (2) mechanisms of blood clotting. Atherosclerosis involves the accumulation of cholesterol in arteries, the narrowing of the arteries, and the formation of abnormal surfaces on the lumenal surface of the arteries. Regions of the artery containing the more severe changes in this scenario are called **atherosclerotic lesions**. A term that is similar, but different, from atherosclerosis is *arteriosclerosis*. This latter term refers to a late step in atherosclerosis, where the lesions acquire deposits of calcium salts and become hard. The narrowing of arteries results in cardiovascular disease, as it can result in **ischemia** (lack of oxygen to heart muscle). Blood clotting is required for a healthy life, but pathological blood clotting can occur, as a consequence of atherosclerosis, to create blood clots within the bloodstream. The blood clots do not close a bleeding wound. Instead they occlude the lumen of the affected artery, prevent the flow of blood, and cause a heart attack. The more or less separate existence of these two parts is demonstrated by the fact that **pravastatin** is used as a drug to prevent the accumulation of cholesterol in arteries, while **warfarin** is used as a drug to prevent blood clots (see Vitamin K section). Warfarin does not reduce the lesions of atherosclerosis, but it does reduce the tendency of an advanced lesion to trigger clot formation.

A few of the genes and proteins involved in lipid metabolism were mentioned above. The genetic diseases resulting from mutations in some of these genes were also revealed; however, the diseases mentioned afflict only a very small portion of the population. In contrast, the early stages of atherosclerosis occur in essentially all adults, and a large proportion of adults eventually die of cardiovascular disease. As far as the general human population is concerned, the specific genes and proteins that govern who will acquire severe atherosclerosis and who will not remain undiscovered. It is expected that many of these genes will be found to regulate the behavior and structure of endothelial cells, the immune system, and the extracellular matrix. Researchers are presently conducting large-scale screening tests in an effort to discover all of the genes that influence the course of atherosclerosis.

*Steps in Atherosclerosis*

The coronary artery is one of the arteries which deliver oxygen and other nutrients to the heart muscle. Although atherosclerotic lesions may occur in a variety of blood vessels, those that occur in the coronary artery are most often responsible for fatal heart attacks. To summarize the steps resulting in atherosclerosis:

1. ***LDLs Are Especially Susceptible to Damage***. VLDLs circulate in the bloodstream, and some of these particles are eventually converted into LDLs. With continued residence of LDLs in the circulatory system, damage may occur to the lipid component, as well as to the protein component. Damage may involve oxidative damage to the lipids, resulting in byproducts that attack and damage the protein component. This oxidative damage involves free radicals. Damage may also occur directly to the protein component, for example, by the condensation of glucose with lysine residues of the protein component, generating a Schiff base, or by reaction with hypochlorite (a chemical made by some white blood cells).

Diabetics are at special risk for cardiovascular disease apparently because diabetes results in chronically elevated levels of plasma glucose. Glucose exists in ring and open-chain forms that are in equilibrium. The open-chain form has an aldehyde group that readily can condense with amino groups such as those of proteins. The condensation product formed by an aldehyde and an amino group is called a **Schiff base**.

The glucose/apo B condensation product in the bloodstream is easy to detect. Studies with diabetics have revealed that 2–5% of the lysine residues of apo B are glucosylated (Curtiss and Witztum, 1985). Glucosylation at this level, and at higher levels, progressively impairs the ability of apo B to be bound by the LDL receptor. Plasma LDLs modified in this way in diabetics are avidly taken up by the Kuppfer cells (macrophages) in the liver, and also by macrophages in the arterial walls. Uptake of LDLs bearing a damaged protein coat, by macrophages in arterial walls, is an ominous early step in the pathway to atherosclerosis.

Further details on oxidative damage to LDLs, and the possible role of hypochlorite and lipid breakdown products, are revealed in the Vitamin E section.

2. ***Damaged LDLs Are Phagocytized via a Receptor Called Scavenger Receptor A***. The LDLs containing a damaged protein component may be removed from the circulation, eventually, by white blood cells called macrophages. Macrophages reside in the liver, where they are called Kuppfer cells, as well as elsewhere in the body. LDLs that bear an oxidative damaged protein coat are essentially undetectable free in the circulation. The LDLs bearing an oxidatively damaged protein coat accumulate within the layers of cells that constitute the artery. Macrophages take up these damaged LDLs by **scavenger receptor A** (Steinberg, 1997a,b). The LDL receptor is not used for this process. **Scavenger receptor A** is not regulated by the amount of cholesterol in the cell, unlike the LDL receptor, which responds to the cell's need for cholesterol. LDLs containing damaged amino acids are rapidly cleared from the bloodstream via scavenger receptor A, i.e., within *a few minutes* (Steinberg, 1997a,b). This is in contrast to the residence time of an undamaged LDL particle, which is *a few days*.

3. ***LDLs Enter the Endothelial Cell Layer, and Become Oxidized***. As LDLs circulate through the bloodstream, they pass in between the cells of the endothelial layer of the arteries, where they are exposed to toxic oxygen released by the endothelial cells and by white blood cells. These white blood cells make their home

Lumen of artery

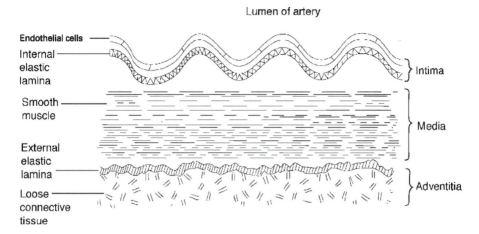

FIGURE 6.34 Cross-section of an artery, and participants in atherosclerosis. The lumen of the artery contains the blood and circulating blood cells. The tissue of the artery contains three major layers: (1) the intima; (2) the media; and (3) the adventitia. The **intima** is the layer that contacts the flowing blood, and consists of endothelial cells and a layer of connective tissue. The endothelial cells are distinguished in that they occur only as a monolayer, i.e., the layer is one cell thick (Ross, 1993). The **internal elastic lamina** separates the intima from the media. The **media** is composed of numerous layers of smooth muscle cells. The **external elastic lamina** separates the media from the adventitia. The **adventitia** is a loose network of connective tissue. During the early steps of atherosclerosis, monocytes cross the epithelial layer and invade the intima, where they adhere and differentiate into macrophages. Eventually, the internal elastic lamina fragments and smooth muscle cells from the media invade the intima. In atherosclerosis, it is the intima that accumulates fat, and eventually becomes calcified (Schwartz *et al.*, 1995; Steinberg, 1997a,b). The above scenario, as described, should be used as a reference point, rather than an established fact, since the events described probably will never be directly observed (in the human coronary artery) in real time. The internal elastic lamina and external elastic lamina do not consist of cells; they consist mainly of extracellular proteins (extracellular matrix proteins).

among the endothelial cells. These white blood cells enter from the bloodstream, squeeze in between the endothelial cells, and reside just under the layer of endothelial cells (Figure 6.34). These white blood cells produce and secrete two types of toxic oxygen, i.e., HOOH and hypochlorite.

4. *Damaged LDLs Provoke the Arrival of More Monocytes and Their Differentiation.* The presence of the damaged LDLs attracts white blood cells. The type of white blood cells attracted by the damaged LDLs are the **monocytes**. The monocytes bind and adhere to the epithelial cell layer by the following special interactions. Monocytes contain integrins in their cell membrane, as do many other types of cells. Integrins are membrane-bound proteins that are used for cell-to-cell recognition. The monocyte membrane contains integrin $\alpha_4\beta_1$ and integrin $\alpha_L\beta_2$ (Kling *et al.*, 1995; Tian *et al.*, 1997). These integrins mediate the binding and adhesion of the monocytes to proteins of the extracellular matrix, occurring on the lumenal surface of the endothelial cells. The damaged LDLs can indirectly provoke the monocytes to differentiate and be converted to macrophages (the macrophages

have increased amounts of scavenger receptor A). Scavenger receptor A does not much take up undamaged LDLs. The big picture that has been proposed is one of a vicious circle: White blood cells provoke oxidative damage to LDLs; the damaged LDLs attract more white blood cells into the lining of the artery, resulting in still greater rates of damage to LDLs.

5. *Foam Cells Consist of a Mixed Population of Endothelial Cells, Macrophages, and Smooth Muscle Cells*. The increased uptake of damaged LDLs results in accumulation of vast quantities of cholesteryl esters, which form lipid droplets within the macrophages. At this point, the macrophages are called **foam cells**. The endothelial cells are lined by the **internal elastic lamina**, and just beyond this layer resides the **media**. The media consists of numerous layers of smooth muscle cells (Figure 6.34). After monocytes invade the endothelial cells, and after the monocytes differentiate into macrophages, the internal elastic lamina becomes fragmented, and smooth muscle cells from the media invade the intima (Steinberg, 1997a,b). The foam cells in the arterial wall form what is called a **fatty streak**. The fatty streak, which often develops in childhood, is itself harmless, but over the years it may become converted to a pathological structure, the **atherosclerotic lesion**. About half of all children in their early teens contain fatty streaks in their coronary arteries (Ross, 1993). Eventually, most or all adults acquire fatty streaks in this location. As the atherosclerotic lesion progresses, the region just under the endothelial cells contains alternating layers of lipid-filled macrophages and smooth muscle cells (Ross, 1993).

6. *The Fibrous Plaque Is Formed*. The fatty streak is a precursor to a structure called the **fibrous plaque**. The cells of the fatty streak produce collagen, an extracellular protein. Collagen is a fibrous protein consisting of polypeptide chains twisted and linked together. An artery containing an atherosclerotic lesion in the fibrous plaque stage contains a narrowed lumen, which obstructs the flow of blood. Where the lumen is 50% obstructed, for example, one finds a corresponding decrease in blood flow to the heart muscle, a reduction in the body's ability to do exercise, and cardiac arrhythmias. Atherosclerotic lesions may also contain crystals of cholesterol and deposits of calcium phosphate that cause the artery to lose its usual elastic structure and become stiff and brittle. (Details of collagen appear in the section on ascorbic acid; details of calcium phosphate deposits appear in the sections on Vitamin D, Calcium, and Magnesium).

7. *Heart Attacks and Brain Attacks*. Constrictions caused by fibrous plaque can result in stagnation of blood flow, which is dangerous because blood clots tend to form in slow-moving blood. Formation of a clot that blocks the flow in arteries supplying nutrients to the heart results in a **heart attack**. Blood clots can arise in areas of stagnant blood but also from exposure to deteriorating tissues. Digestive enzymes (metalloproteases) may damage the surface of the fibrous plaque, resulting in exposure of underlying structures. This event is called the **erosion of the fibrous cap**. The resulting uneven surface can provoke the formation of a blood clot. Once formed, the clot may travel through the bloodstream and block an artery supplying nutrients to the heart or to the brain. A clot that blocks nutrients from part of the brain, resulting in a loss of function of that part of the brain, causes a **stroke** or "brain attack." A blood clot is called a **thrombus**; a traveling clot is called an **embolus**.

# EFFECTS OF DIET AND DRUGS ON ATHEROSCLEROSIS

The events leading to atherosclerotic lesions and cardiovascular disease are manifold and not known with certainty. Therapy has been based on the goals of reducing the concentration of LDL-cholesterol and increasing that of HDL-cholesterol in the plasma. Reduction of LDL-cholesterol is expected to limit the uptake of cholesterol by cells in fatty streaks. An increase in HDL-cholesterol is considered to be of value because its presence indicates an increased rate of removal of cholesterol from the walls of arteries.

Diet therapy and drugs often result in dramatic reductions in LDL-cholesterol levels and a striking and definite reduction in the rate of heart attacks. Studies have shown that lowering cholesterol shrinks the lesions (the plaques) of the coronary artery, but that the effect of shrinking is only slight (Levine *et al.*, 1995). The size of these lesions can be measured in living patients by **angiography**, and monitored during the course of diet and drug therapy. Apparently, a slight reduction of cholesterol content of existing lesions leads to a significant reduction in the tendency of the lesions to weaken, tear, and rupture. A ruptured plaque produces turbulence (uneven blood flow) in the bloodstream, which directly provokes the formation of a blood clot. Drugs or dietary changes that result in a reduction of irregularities in the lumen of blood vessels would be expected to reduce the rate of clot formation.

## Diet Can Influence LDL-Cholesterol Levels

A typical American diet contains about 40% of its total energy content as fat. About half this energy is in the form of **saturated fatty acids**, one-fourth as **monounsaturated fatty acids**, and one-fourth as **polyunsaturated fatty acids** (PUFAs) (Ginsberg *et al.*, 1990). The quantity and nature of dietary fats can influence the level of plasma cholesterol. A reduction in total fat from the typical 40% of energy intake to 30% can result in a decrease in plasma cholesterol.

High levels of saturated fats in the diet result in an increase in plasma LDL-cholesterol. The mechanism of this effect is not clear, but it appears that these fats interfere with the ability of LDL receptors to clear LDLs from the bloodstream. Further details on the mechanism appear at a later point in this chapter.

### *Monounsaturates Can Reduce LDL-Cholesterol*

Maintenance of the 40% contribution of fat to energy intake with substitution of monounsaturated for saturated fats can lead to a reduction in plasma cholesterol also. The level of LDL-cholesterol is reduced, and the level of HDL-cholesterol is unchanged. Olive oil is a well-known source of monounsaturated fatty acids. This oil contains about 12% 16:0 and 2% 18:0 (saturated), 75% 18:1 (monounsaturated), 2% 16:1 (monounsaturated), and 8% 18:2 (PUFA).

## PUFAs Can Reduce LDL-Cholesterol

Dietary PUFAs decrease the plasma LDL-cholesterol level. Vegetable oils contain high levels of PUFAs such as linoleic acid (18:2), which constitutes about 25, 50, 63, and 75% of the fatty acids in peanut, soy, sunflower, and safflower oils, respectively. The quantity of 18:2 in beef and pork fat is only 5–10% and is under 3% in tropical oils. PUFAs produce decreases in LDL-cholesterol, possibly by the same (unknown) mechanism as monounsaturates. The question of whether dietary PUFAs have a greater or similar effect on LDL-cholesterol remains unsettled.

## Comparison of Dietary Saturated Fatty Acids

Palmitic acid (16:0) is the primary saturated fatty acid in most diets. This compound constitutes about 25% of the fatty acids of beef or pork fat, but only 6–10% of the fatty acids of sunflower, safflower, peanut, or soy oils (see Table 6.8). Dietary palmitic acid increases LDL-cholesterol (Grundy and Denke, 1990). Myristic acid (14:0) is present at high levels in butterfat and in the "tropical oils" palm oil and coconut oil. Although myristic acid elevates LDL-cholesterol, it is generally a rather minor component of the diet. Stearic acid (18:0) is also a major component

**TABLE 6.8** Fatty Acids Occurring in the Triglycerides of Various Foods

|  | Chicken breast[a] | Chicken skin[b] | Beef fat[c] | Olive oil[c] | Canola oil[c] | Corn oil[c] | Soy-bean oil[c] | Saf-flower oil[c] |
|---|---|---|---|---|---|---|---|---|
| 14:0 | 0.8 | 0.3 | 3.7 | – | – | – | 0.1 | 0.1 |
| 16:0 | 21.4 | 32.3 | 24.9 | 11.0 | 4.8 | 10.9 | 10.3 | 6.2 |
| 16:1 | 4.9 | 1.9 | 4.2 | 0.8 | 0.5 | – | 0.2 | 0.4 |
| 18:0 | 5.9 | 22.4 | 18.9 | 2.2 | 1.6 | 1.8 | 3.8 | 2.2 |
| 18:1 | 46.3 | 19.9 | 36.0 | 72.5 | 53.8 | 24.2 | 22.8 | 11.7 |
| 18:2 | 17.1 | 0.1 | 3.1 | 7.9 | 22.1 | 58.0 | 51.0 | 74.1 |
| 18:3 | 1.4 | – | 0.6 | 0.6 | 11.1 | 0.7 | 6.8 | 0.4 |
| 20:0 | 0.8 | 2.8 | – | – | – | – | – | – |
| 20:1 | – | – | 0.3 | 0.3 | 0.1 | – | 0.2 | – |
| 21:0 | – | 0.6 | – | 0.3 | 0.1 | – | – | – |
| 22:0 | – | 0.2 | – | – | – | – | – | – |

[a]Ahn *et al.* (1995).
[b]Wertz *et al.* (1986).
[c]McNamara (1992).

Compare the content of saturated fatty acids (12:0, 14:0, and 16:0) in the various foods. The vegetable oils are notable for having the lowest content of these fatty acids, where canola and safflower oil have the lowest levels. Compare the mono- and unsaturated fatty acids. Olive oil is highest in the monounsaturate 18:1, while safflower oil is highest in the polyunsaturate 18:2. In the event that monounsaturates prove to be best for reducing atherosclerosis, then olive oil may be an energy source of choice. In the event that polyunsaturates are found to be the best source of fat, with the goal of reducing atherosclerosis, then safflower oil may be the oil of choice. Presently, it is recommended to replace saturated fats with a mixture of fats containing monounsaturates and PUFAs. One can see, for example, that at least 80% of the fatty acids of corn oil, canola oil, soybean oil, olive oil, and safflower oil occur as the recommended fatty acids (monos and PUFAs).

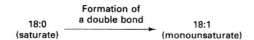

FIGURE 6.35 Conversion of stearic acid to an 18:1 fatty acid. Through formation of a double bond, stearic acid (18:0), a saturate, is converted to acid (18:1), a monounsaturate.

of many fats and oils. It constitutes 15–20% of the fatty acids of beef or pork fat, but only 4–5% of those in vegetable oils. Stearic acid seems not to increase LDL-cholesterol levels. One explanation is that, after absorption, this molecule is converted rapidly to the monounsaturate 18:1 by the mechanism shown in Figure 6.35.

Palmitic acid (16:0) is not much converted in the body to 16:1. However, palmitic acid (16:0) is converted readily to 18:1 by the pathway shown in Figure 6.36. This conversion is much slower than that of dietary 18:0 to 18:1, and thus accounts for the LDL-cholesterol raising effect of palmitic acid.

### HDL-Cholesterol Levels Can Be Altered by the Diet

LDL-cholesterol ("bad cholesterol") is reduced when saturated fats are replaced with PUFAs or with carbohydrates. However, HDL-cholesterol ("good cholesterol") also tends to be reduced when saturated fats are replaced with PUFAs or with carbohydrates. These responses indicate that dietary monounsaturates may be of more value than PUFAs or carbohydrates. However, this statement should be viewed with caution because it has not been established that an elevated HDL-cholesterol level in humans prevents formation of atherosclerotic lesions and cardiovascular disease. High levels of saturated fatty acids in dietary fats are a risk factor for cardiovascular disease. The percentage of energy derived from fat is recommended to be reduced to 30%, but not much below this level.

### Do We Need to Watch Dietary Cholesterol?

The foods that are high in saturated fats tend to be animal products, and thus contain cholesterol. Foods that are high in unsaturated fats tend to be plant products, and thus contain no cholesterol. Apparently, these relationships have led to the notion that reducing the intake of cholesterol will result in lower levels of plasma LDL-cholesterol. As discussed earlier, altering the level of dietary cholesterol generally has little influence on the plasma cholesterol level, although it

FIGURE 6.36 Conversion of palmitic acid to an 18:1 fatty acid. Palmitic acid (16:0), a saturate, is converted in the body to the monounsaturate acid (18:1) by a two-step process involving initial conversion to stearic acid (18:0).

should be noted that some people are more cholesterol sensitive than the general population. In humans, the amount and type of fatty acids consumed is more important than the amount of cholesterol consumed. Dietary cholesterol has a much greater influence on the lipoprotein picture in *animals* than in *humans* (Mustad *et al.*, 1996).

EXERCISE

Anabolic steroid hormones are sometimes self-administered by power lifters and body-builders. This practice results in changes in the plasma levels of both LDL-cholesterol and HDL-cholesterol. Each of these changes creates an increased risk for atherosclerosis. Are these changes increases or decreases? (Consult Webb *et al.*, 1984.)

## The Mediterranean Diet

Ancel Keys noticed, during the early 1950s, that populations living in southern Italy, Greece, and Crete tended to have unusually low rates of cardiovascular disease (Keys, 1995; Nestle, 1995). Upon conducting epidemiological studies, he traced this effect to what is known as the "Mediterranean diet." This diet is largely vegetarian, where olive oil is used for cooking, where bits of meat are eaten only occasionally during the week, and where sprinkles of cheese are used as flavoring. The Mediterranean diet can be contrasted with the American version of Italian cooking, where pasta is smothered with ground beef, where pizzas are thick with cheese and pepperoni, and where bread is coated with butter. The Mediterranean diet also includes 1–2 glasses of wine per day.

Epidemiological data on cardiovascular disease, biochemical data on plasma lipoproteins, and nutritional data on dietary energy sources have generated information on the biochemical and dietary risk factors for cardiovascular disease. Three risk factors for cardiovascular disease can be arranged in a hierarchy:

1. Elevated LDL-cholesterol is a risk factor that is highly associated with cardiovascular disease. Epidemiologists have estimated that for each 10% increase in LDL-cholesterol there is a 20–30% increase in risk for cardiovascular disease (Gaziano *et al.*, 1996).

2. High plasma TGs are moderately associated.

3. Lowered HDL-cholesterol are only somewhat established as a risk factor. The composition of the plasma lipids and lipoproteins vary according to our genetic makeup, but in most people can be altered by diet.

Diets rich in saturated fat provoke an increase in LDL-cholesterol, increase TGs, and lower HDL-cholesterol — these diets worsen all three risk factors. This statement applies to the saturated fatty acids 12:0 (lauric acid), 14:0 (myristic acid), and 16:0 (palmitic acid). In realizing the tendency of these three fatty acids to provoke cardiovascular disease, a question often asked is, "Should we replace the saturated fats in the diet with other saturated fatty acids (such as stearic acid; 18:0), with fats containing monounsaturated fatty acids (such as oleic acid; 18:1), with fats containing polyunsaturated fatty acids (PUFAs such as linoleic acid), or with carbo-hydrates (sugars and starches)?"

When compared to dietary 12:0, 14:0, and 16:0, dietary stearic acid (18:0) lowers LDL-cholesterol. With this substitution, a 25% drop in LDL-cholesterol occurs in many people. Clearly, this picture would be expected to be favorable to long term health. However, the substitution of stearic acid for 12:0, 14:0, and 16:0 can also result in a drop in HDL-cholesterol ("good cholesterol"). Hence one might take caution in being overly enthusiastic about diets high in stearic acid (Aro *et al.*, 1997; Kris-Etherton and Yu, 1997).

When compared to 12:0, 14:0, and 16:0, oleic acid (18:1) or linoleic acid (18:2) provokes a drop in LDL-cholesterol of about 15%. When directly comparing monounsaturated fatty acids with PUFAs, one finds that monounsaturated fatty acids seem to raise plasma TG levels. A slight *increase* in plasma TG may occur with the oleic acid. In addition, PUFAs seem to be more consistent at raising HDL-cholesterol than oleic acid. In these respects PUFAs may be more desirable than monounsaturated fatty acids (Kris-Etherton and Yu, 1997).

## *Trans* Fatty Acids

Triglycerides containing saturated fatty acids easily form a compact structure, when placed together, and form a solid at room temperature. Where the fatty acids contain double bonds in the usual *cis* conformation, the *cis* double bonds create bends in the fatty acid, which impair close packing, and result in a liquid (oil). Where the fatty acids are in the *trans* conformation, the ability to pack closely is in between triglycerides with saturated and *cis* fatty acids. Oleic acid, a *cis* fatty acid, is 18:1, as noted earlier. The corresponding 18:1 *trans* fatty acid is called **elaidic acid**. Elaidic acid is abbreviated as *trans* 18:1.

The fat in beef, mutton, milk, and cheese contains 2–8% *trans* fatty acids. These naturally occurring *trans* fatty acids are formed in the rumen of the stomach (of ruminants) by the action of bacterial enzymes. Most of the *trans* fatty acids of the diet arise from the industrial hydrogenation of food oils. During this process, most of the unsaturated *cis* fatty acids are converted to saturated fatty acids, but a fraction is converted to *trans* fatty acids. In relation to all fatty acids present in the indicated food, salad oils contain 8–17% *trans* fatty acid, shortening contains 14–60%, and margarines 16–70% (Simopoulos, 1996). Overall, about 6% of our dietary fatty acids are *trans* fatty acids, where most of these are elaidic acid (McKeigue, 1995).

As far as the lipoprotein profile is concerned, *trans* fatty acids have an effect in between that of saturated fatty acids and *cis* unsaturated fatty acids. Most scientific studies have shown that *trans* fatty acids raise LDL-cholesterol, while some studies have shown that they lower HDL-cholesterol (Katan *et al.*, 1995b). *Trans* fatty acids constitute a minor component of the diet and, as a health issue, are of lower importance than avoiding smoking, obesity, and saturated fatty acids. The prudent consumer who avoids the regular consumption of butter, lard, shortening, and hard margarine (stick margarine) will also, by this practice, avoid major sources of *trans* fatty acids (shortening, hard margarine) (Willet and Ascherio, 1995).

## Carbohydrates and Atherosclerosis

*The Question of Replacing Saturated Fats with a Non-Fat Energy Source*

Let us turn to the issue of carbohydrates. Replacing 12:0, 14:0, and 16:0 with carbohydrates lowers LDL-cholesterol levels. Hence, it is easy to argue that carbohydrates are a better energy source than 12:0, 14:0, and 16:0 for long-term health. Replacing 12:0, 14:0, and 16:0 with carbohydrates does lower LDL-cholesterol, but it may also provoke a decrease in HDL-cholesterol and an increase in fasting plasma TG levels.

What about the choices between carbohydrates, monounsaturated fatty acids, and PUFAs? Studies on **monounsaturated fatty acids versus carbohydrates** have shown that *monounsaturated fatty acids provoke a decrease in plasma TGs and an increase in HDL-cholesterol*, with not much difference in effect on LDL-cholesterol. Studies on **PUFAs versus carbohydrates** have revealed that *PUFAs result in lower LDL-cholesterol*, with not much difference in effect on plasma TGs or HDL-cholesterol.

The above trends were acquired from studies of humans fed naturally occurring oils and fats. The studies were not conducted with oils containing homogeneous fatty acids because these types of oils do not yet exist in any plant. Although a food oil containing only monounsaturated fatty acids or PUFAs may some day become available, one might wonder if the complete elimination of saturated fatty acids from the diet might lead to undesirable changes in membrane structure or energy metabolism.

There is some genuine interest in promoting monounsaturated fatty acids and PUFAs over carbohydrates, where one has an eye solely on preventing cardiovascular disease (Kris-Etherton and Yu, 1997; Katan *et al.*, 1997). How much carbohydrate should be reduced, and replaced by these oils? Some researchers believe that the quantity involved should be equivalent to 10% of our daily energy requirement (2 tablespoons of oil), while others believe that this amount should be equivalent to 15% of our daily energy requirement (3 tablespoons of oil). As discussed in the section on Vitamin E in Chapter 9, there is some thought that monounsaturated fatty acids may be better for long-term health than polyunsaturated fatty acids.

## Food Choices in Everyday Living

*How Should I Choose Foods for my Menu?*

The recommended effective diet contains 30 percent or less of energy from total fat, with under 7% of this from saturated fat, and under 200 mg cholesterol per day. Since high LDL-cholesterol is more prevalent than high plasma triglycerides, it is generally recommended that saturated fat be reduced by replacement with carbohydrate foods. This can be accomplished, for example, by eliminating frying during cooking, and instead using baking or roasting.

For persons with high LDL-cholesterol and *also high plasma triglycerides*, one might prefer to replace some saturated fat with monounsaturated fatty acids or polyunsaturated fatty acids. A desired level of vegetable oil intake might be 2 tablespoons per day, while maintaining the contribution of fats and oils at about 30% of our energy intake per day. The oil can be used for stir-frying, in making salad oil, in spaghetti sauce, in gravy, and to make muffins. Since it is not yet clear if 18:1 or 18:2 is the most preferred fatty acid, one might alternate between corn oil, olive oil, and safflower oil.

### How Should I Choose Foods for My Child's Menu?

Another question that arises is whether one should limit the fat in the diets of children. One trend in thought suggests that for children dietary fat should not be restricted to below 30% of energy intake. Restriction of fat below a level of 30% may inhibit growth. There is some agreement that the fat intake should account for 50% of energy for infants and toddlers, with a gradual decrease to 30% towards the arrival of adolescence. Hence, a diet containing meat, eggs, dairy products, cereals, fruits, and vegetables is desirable for children (Krauss *et al.*, 1996; Olson, 1995; Lifshitz and Tarim, 1996; Kleinman *et al.*, 1996).

### Alcohol and Cardiovascular Disease

In addition to the effects of fats, the composition of the Mediterranean diet suggests a relationship between alcohol and lower cardiovascular disease. Scientific studies have conclusively demonstrated that moderate alcohol intake (1–2 drinks/day) results in increases in HDL levels. No consistent change, or a possible decline, in LDL levels occurs with moderate alcohol intake. It might be reasonable to recommend that the general population drink 1–2 glasses of wine or beer per day. On the other hand, it is not reasonable to expect many people, upon hearing this recommendation, to limit their intake to only two glasses of alcohol. A proven problem with high alcohol intake is automobile accidents and fetal alcohol syndrome. A potential problem with moderate or high alcohol intake is the increased risk for certain types of cancer. Specifically, cancers of the breast (in women), mouth, esophagus, larynx, and liver increase with daily consumption of one or more drinks of alcohol per day (Thun *et al.*, 1997).

## Mechanism of the Influence of Dietary Fatty Acids on LDL-Cholesterol

The concentration of LDL-cholesterol in the blood is a function of its rate of production and rate of removal from the bloodstream. Diet influences LDL-cholesterol by changing its rate of removal or clearance, according to the following scenario. Let us briefly return to the topics of proteins and DNA. A transcription factor called **sterol response element binding protein** (SREBP) binds to a stretch of DNA called the **sterol response element**. The sterol response element has the structure CACCCCAC. In discussing any sequence of DNA, it is conventional to

show the sequence of only one strand of the DNA double helix. The entire structure of the sterol response element is

CACCCCAC<br>
GTGGGGTG

The sterol response element occurs about 60 base pairs upstream of the transcription start site for the LDL receptor gene. SREBP is a protein of 1150 amino acids
that is attached to the endoplasmic reticulum (ER). Most of SREBP resides within
the cytoplasm, but part of its polypeptide chain loops in and out of the ER. This
loop tethers SREBP to the membrane of the ER. SREBP is activated when cells
sense that they contain reduced levels of free cholesterol, i.e., when they need more
free cholesterol. What does it mean to "activate" the SREBP? In sterol-depleted
cells, a special enzyme catalyzes the cleavage of SREBP at a specific site, releasing
the protein from its tether (Brown and Goldstein, 1997). The freed portion of
SREBP travels to the nucleus, where it finds the sterol response element and binds
to it. This binding provokes an increase in rate of transcription of several genes,
including those for the LDL receptor and 3-hydroxy-3-methylglutaryl-CoA synthase (Sakai *et al.*, 1996). An increase in LDL receptor transcription results in an
increase in the amount of this protein in the plasma membrane, and a consequent
increase in acquisition of LDLs from the bloodstream.

A mechanism has been proposed for the influence of various dietary fatty acids
on plasma LDL-cholesterol levels. The scenario where oleic acid (18:1) can provoke
a decline in LDL-cholesterol is as follows (Woollett and Dietschy, 1994; Rumsey *et
al.*, 1995):

1. An increase in dietary 18:1 leads to an increase in 18:1 delivered to the liver
   cell.

2. An increase in 18:1 in liver cells results in an increased availability of this
   fatty acid to acyl-CoA:cholesterol acyltransferase (ACAT). ACAT is the enzyme that catalyzes the attachment of free fatty acids to cholesterol, creating
   cholesteryl esters.

3. Oleic acid may have a unique action in increasing the activity of ACAT.

4. An increased activity of ACAT provokes a decline in the concentration of
   free cholesterol in the liver cell.

5. Decreased free cholesterol in the cell provokes cleavage of SREBP from its
   tether (cholesteryl esters do not influence the activation of SREBP).

6. An increase in free SREBP stimulates transcription of the gene coding for
   the LDL receptor, resulting in an increase in LDL receptor in the plasma
   membrane, and the consequent decline in plasma LDLs.

## Effect of Drugs

Drugs and diet counseling are routinely used for preventing atherosclerosis in
persons over the age of 40 in whom biochemical tests have revealed the presence
of high LDL-cholesterol. The three main drugs that are used are inhibitors of

HMG-CoA reductase (the "statins"), nicotinic acid, and bile-acid binding resins (cholestyramine).

Clinical studies report various effects of these drugs, and the physician needs to determine which effect is most relevant to long-term health of the patient. A hierarchy of importance is as follows. A drug that lowers LDL-cholesterol is important, but a drug that reduces lesions in the coronary artery is even more important. A drug that reduces lesions is important, but a drug that is shown to reduce the occurrence of heart attacks (or fatal heart attacks) is even more important. In other words, two different drugs may be found to reduce LDL-cholesterol levels by 15%, but only one of these may have the effect of reducing the rate of heart attacks.

Drugs that inhibit HMG-CoA reductase have a potent cholesterol-reducing action, and are relatively free of side effects. These drugs include **pravastatin**, **lovastatin**, and **simvastatin** (the "statins"). Therapy with pravastatin results in an improvement in biochemical risk factors, a reduction in thickness in atherosclerotic lesions, and a reduction in the rate of heart attacks (Levine *et al.*, 1995; Treasure *et al.*, 1995; Sacks *et al.*, 1996). The option of including nicotinic acid or resin with statin therapy is usually decided by the physician. Studies with patients with high cholesterol have shown that, when more than one of these drugs are taken, the cholesterol-reducing effects of the drugs are *additive*.

An interesting species difference might be pointed out. The above HMG-CoA reductase inhibitors inhibit HMG-CoA reductase from a variety of animals, and reduce serum cholesterol in humans, rabbits and dogs, but do not reduce serum cholesterol in rats and mice (Sakono *et al.*, 1996).

Nicotinic acid (the vitamin) is another drug of choice. Therapy with nicotinic acid alone improves the biochemical risk factors for cardiovascular disease (Illingworth *et al.*, 1994; McKenney *et al.*, 1994; Holvoet and Collen, 1995; Schectman and Hiat, 1996). Useful doses of this drug are 1.5–4.0 grams/day. Nicotinic acid is not the drug of choice for patients with non-insulin dependent diabetes (NIDDM). This is because niacin can provoke a worsening of the diabetes. During therapy with nicotinic acid, the patient needs to be monitored carefully, as this drug may provoke liver damage in up to half of treated patients. In addition, nicotinic acid therapy leads to flushing of the skin (vasodilation) and gastrointestinal upset in many patients. Often the flushing effect disappears with continued therapy.

Therapy with nicotinic acid plus bile acid binding resin revealed improvements in the biochemical risk factors, and also proved to reduce the thickness of atherosclerotic lesions (Blankenhorn *et al.*, 1993).

Bile-acid-binding resin results in an interruption of the enterohepatic circulation of bile salts and in their increased excretion in the feces. This drug therapy can produce a 20–25% decrease in plasma LDL-cholesterol. A side effect of cholestyramine is constipation, but this problem can be relieved by a high fiber diet.

One might note the dose differences for the above drugs: statins (20–40 mg/day); nicotinic acid (1.5–4.0 g/day); and resin (30 g/day).

Some gender differences for adults may be of interest. Where a given man and woman have the same values for LDL- and HDL-cholesterol, they will have the same risk for cardiovascular disease. However, women (before menopause) generally have higher HDL-cholesterol and lower LDL-cholesterol values than men. After menopause, women's LDL-cholesterol levels tend to rise to men's levels, or higher. After menopause, women's HDL-cholesterol levels fall slightly.

### Guide to Preventing Cardiovascular Disease

The American Heart Association recommends the following behaviors in order to reduce the risk for cardiovascular disease (Grundy *et al.*, 1997):

1. Do not smoke cigarettes.

2. Prevent obesity.

3. Encourage regular physical exercise, such as walking or moderate-intensity exercise.

4. Reduce dietary intake of cholesterol and saturated fats. Replace saturated fats with unsaturated vegetable oils.

5. If high blood pressure is present, use various methods to reduce blood pressure to the desired level. [Further details on high blood pressure (hypertension) appear in the Sodium and Potassium section.]

6. Use drug therapy where biochemical risk factors suggest a benefit. Statin drugs are the drugs of choice. Where statin therapy alone does not work, it is recommended that statin be used in combination with one other drug. The primary goal is to reduce LDL-cholesterol to levels lower than 160 mg/100 ml. The secondary goal is to reduce plasma triglycerides to levels lower than 200 mg/100 ml, and to raise HDL-cholesterol to levels over 35 mg/100 ml (Grundy *et al.*, 1997).

## Newly Discovered Risk Factors

High plasma LDL is firmly established as a risk factor for cardiovascular disease. High plasma HDL has been somewhat less firmly established as a protective factor. A clearer protective role for HDLs was provided by recent experiments with mice. Edward Rubin and co-workers used standard genetic techniques to create a mouse having increased amounts of Apo A-1, and increased amounts of HDLs (Castro *et al.*, 1997). Remember that Apo A-1 is the defining protein of HDLs. Feeding normal mice diets high in saturated fat and cholesterol produces fatty streaks. Recent studies with the mice containing elevated Apo A-1 and elevated HDLs revealed that the extra Apo A-I protected the mice from acquiring atherosclerotic lesions.

Newly emerging predictors for atherosclerosis concern the sizes of the lipoprotein particles, and the heterogeneity of Apo E. People can differ according to the size (diameter) of their LDL particles and HDL particles. The size of the LDL particles has been related to risk for cardiovascular disease. Persons with larger LDLs (26–27 nm) are at lesser risk, while those with smaller LDLs (24–25 nm) are at greater risk (Dreon *et al.*, 1994; Stampfer *et al.*, 1996; Tribble and Krauss, 1993). Furthermore, people with smaller LDLs tend, with eating a low fat diet, to acquire a more healthy plasma picture (Krauss and Dreon, 1995). The results suggest that diet therapy will be of greater benefit to people who naturally have smaller LDLs. A similar, but weaker, prediction may be made with regard to the type of Apo E contained in any person. Apo E exists as Apo E2, Apo E3, or Apo E4 in different persons. These forms of Apo E differ only in the amino acids at residues 112 and 158 in the polypeptide chain. People with Apo E2 tend to have low levels of LDL-cholesterol, while those with Apo E4 tend to have high LDL-cholesterol levels (Kallio *et al.*, 1997). Apo E4 people tend to respond to diet, while Apo E2 people tend not to respond. To propose a consistent "big picture" for good health, it would be one with low LDL-cholesterol, high HDL-cholesterol, low plasma TGs, large LDLs, and the Apo E2.

## REFERENCES

Ahn, D., Wolfe, F., and Sim, J. (1995). Dietary linolenic acid and mixed tocopherols, and packaging influences on lipid stability in broiler chicken breast and leg muscle. *J. Food Sci.* **60**, 1013–1018.

Aro, A., Jauhianinen, M., Partanen, R., Salminen, I., and Mutanen, M. (1997). Stearic acid, *trans* fatty acids, and dairy fat: Effects on serum and lipoprotein lipids, apolipoproteins, lipoprotein(a), and lipid transfer proteins in healthy subjects. *Am. J. Clin. Nutr.* **65**, 1419–1426.

Barr, S. I., Kottke, B. A., and Mao, S. J. (1985). Postprandial distribution of apolipoproteins C-II and C-III in normal subjects and patients with mild hypertriglyceridemia. *Metabolism* **34**, 983–992.

Belloni, E., Muenke, M., Roessler, E, Traverso, G., Siegel-Bartelt, J., Frumkin, A., Mitchell, H., Donis-Keller, H., Helms, C., Hing, A., Heng, H., Koop, B., Martindale, D., Rommens, J., Tsui, L., and Scherer, S. (1996). Identification of Sonic hedgehog as a candidate gene responsible for holoprosencephaly. *Nature Genet.* **14**, 353–356.

Benlian, P., Gennes, J. L., Foubert, L., Zhang, H., Gagne, S., and Hayden, M. (1996). Premature atherosclerosis in patients with familial chylomicronemia caused by mutations in the lipoprotein lipase gene. *N. Engl. J. Med.* **335**, 848–854.

Berger, G. M. (1986). Clearance defects in primary chylomicronemia: A study of tissue lipoprotein lipase activities. *Metabolism* **35**, 1054–1061.

Blankenhorn, D. H., Selzer, R., Crawford, D., Barth, J. D., Liu, C., Liu, C., Mack, W., and Alaupovic, P. (1993). Beneficial effects of colestipol–niacin therapy on the common carotid artery. *Circulation* **88**, 20–25.

Blum, C. B. (1982). Dynamics of apolipoprotein E metabolism in humans. *J. Lipid Res.* **23**, 1308–1316.

Breckenridge, W. C., Little, J. A., Steiner, G., Chow, A., and Poapst, M. (1978). Hypertriglyceridemia associated with deficiency of apolipoprotein C-II. *N. Engl. J. Med.* **298**, 1261–1273.

Brown, M. S., and Goldstein, J. L. (1997). The SREBP pathway: Regulation of cholesterol metabolism by proteolysis of a membrane-bound transcription factor. *Am. J. Clin. Nutr.* **89**, 331–340.

Camps, L., Reina, M., Llobera, M., Bengtsson-Olivecrona, G., Olivecrona, T., and Vilaro, S. (1991). Lipoprotein lipase in lungs, spleen, and liver: Synthesis and distribution. *J. Lipid Res.* **32**, 1877–1888.

Casey, P. J., and Seabra, M. C. (1996). Protein prenyltransferases. *J. Biol. Chem.* **271**, 5289–5292.

Castro, G., Nihoul, L., Dengremont, C., Geitere, C., Delfly, B., Tailleux, A., Fievet, C., Duverger, N., Denefle, P., Fruchart, J.-C., and Rubin, E. (1997). Cholesterol efflux, lecithin–cholesterol acyltransferase activity, and pre-β particle formation by serum from human apolipoprotein A-I and apolipoprotein A-I/apolipoprotein A-II transgenic mice consistent with the latter being less effective for reverse cholesterol transport. *Biochemistry* **36**, 2243–2249.

Cheetham, J. J., Epand, R. M., Andrews, M., and Flanagan, T. D. (1990). Cholesterol sulfate inhibits the fusion of Sendai virus to biological and model membranes. *J. Biol. Chem.* **265**, 12404–12409.

Crandall, L. A., Barker, S. B., and Graham, D. G. (1943). A study of the lymph flow from a patient with thoracic duct fistula. *Gastroenterology* **1**, 1040–1048.

Curtiss, L. K., and Witztum, J. L. (1985). Plasma apolipoproteins AI, AII, B, CI, and E are glucosylated in hyperglycemic diabetic subjects. *Diabetes* **34**, 452–461.

Dietschy, J. M. (1997). Theoretical considerations of what regulates low-density-lipoprotein and high-density-lipoprotein cholesterol. *Am. J. Clin. Nutr.* **65**, 1581S–1589S.

Dreon, D., Fernstrom, H., Miller, B., and Krauss, R. M. (1994). Low-density lipoprotein subclass pattern and lipoprotein response to a reduced-fat diet in men. *FASEB J.* **8**, 121–126.

Farese, R. V., Ruland, S., Flynn, L., Stokowski, R., and Young, S. (1995). Knockout of the mouse apolipoprotein B gene results in embryonic lethality in homozygotes and protection against diet-induced hypercholesterolemia in heterozygotes. *Proc. Natl. Acad. Sci. U.S.A.* **92**, 1774–1778.

Fielding, C. J., and Fielding, P. E. (1995). Molecular physiology of reverse cholesterol transport. *J. Lipid Res.* **36**, 211–228.

Garg, A., and Grundy, S. M. (1988). Lovastatin for lowering cholesterol levels in non-insulin-dependent diabetes mellitus. *N. Engl. J. Med.* **318**, 84–86.

Gaziano, J., Hebert, P., and Hennekens, C. H. (1996). Cholesterol reduction: Weighing the benefits and risks. *Ann. Intern. Med.* **124**, 914–918.

Gelb, M. H. (1997). Protein prenylation, et cetera: Signal transduction in two dimensions. *Science* **275**, 1750–1751.

Ghiselli, G., Schaefer, E. J., Gascon, P., and Brewer, H. B. (1981). Type III hyperlipoproteinemia associated with apolipoprotein E deficiency. *Science* **214**, 1239–1241.

Ginsberg, H. M., Barr, S. L., Gilbert, A., Karmally, W., Deckelbaum, H., Kaplan, K., Ramakrishnan, R., Holleran, S., and Dell, H. S. (1990). Reduction of plasma cholesterol levels in normal men on an American Heart Association step 1 diet or a step 1 diet with added monounsaturated fat. *N. Engl. J. Med.* **322**, 574–579.

Goldberg, I. J. (1996). Lipoprotein lipase and lipolysis: Central roles in lipoprotein metabolism and atherogenesis. *J. Lipid Res.* **37**, 693–707.

Grundy, S. M., and Denke, H. A. (1990). Dietary influences on serum lipids and lipoproteins. *J. Lipid Res.* **31**, 1149–1172.

Grundy, S., Balady, G., Criqui, M., Fletcher, G., Greenland, P., Hiratzka, L., Houston-Miller, N., Kris-Etherton, P., Krumholz, H., LaRosa, J., Ockene, I., Pearson, T., Reed, J., Washington, R., and Smith, S. C. (1997). Guide to primary prevention of cardiovascular diseases. *Circulation* **95**, 2329–2331.

Hamm, M. W., Sekowski, A., and Ephrat, R. (1988). Dietary fat ratios and liver plasma membrane lipid composition. *Lipids* **23**, 829–833.

Hannun, Y. A. (1996). Functions of ceramide in coordinating cellular responses to stress. *Science* **274**, 1855–1859.

Holvoet, P., and Collen, D. (1995). Lipid lowering and enhancement of fibrinolysis with niacin. *Circulation* **92**, 698–699.

Hui, D. Y., Innerarity, T. L., and Mahley, R. W. (1984). Defective hepatic lipoprotein receptor binding of very-low-density lipoproteins from type III hyperlipoproteinemic patients. *J. Biol. Chem.* **259**, 860–869.

Huston, L. (1997). Receptor offers clues to how "good" cholesterol works. Science **278**, 1228.

Illingworth, D. R., Stein, E., Mitchel, Y., Dujovne, C., Frost, P. H., Knopp, R. H., Tun, P., Zupkis, R., and Greguski, R. A. (1994). Comparative effects of lovastatin and niacin in primary hypercholesterolemia. *Arch. Intern. Med.* **154**, 1586–1595.

Innerarity, T. L., Weisgraber, K. H., Arnold, K. S., Rall, S. C., and Mahley, R. W. (1984). Normalization of receptor binding of apolipoprotein E2. *J. Biol. Chem.* **259**, 7261–7267.

Innerarity, T. L., Boren, J., Yamanaka, S., and Olofsson, S.-O. (1996). Biosynthesis of apolipoprotein B48-containing lipoproteins. *J. Biol. Chem.* **271**, 2353–2356.

Jones, P. J. (1997). Regulation of cholesterol biosynthesis by diet in humans. *Am. J. Clin. Nutr.* **66**, 438–446.

Jones, P. J., and Schoeller, D. A. (1988). Polyunsaturated:saturated ratio of diet fat influences energy substrate utilization in the human. *Metabolism* **37**, 145–151.

Jones, P. J., Pencharz, P. B., and Clandinin, M. T. (1985). Whole body oxidation of dietary fatty acids: Implications for energy utilization. *Am. J. Clin. Nutr.* **42**, 769–777.

Kallio, M., Salmenpera, L., Siimes, M., Perheentupa, J., Gylling, H., and Miettinen, T. A. (1997). Apoprotein E phenotype determines serum cholesterol in infants during both high-cholesterol breast feeding and low-cholesterol formula feeding. *J. Lipid Res.* **38**, 759–764.

Katan, M. B., Zock, P. L., and Mensink, R. P. (1995a). Dietary oils, serum lipoproteins, and coronary heart disease. *Am. J. Clin. Nutr.* **61**, 1368S–1373S.

Katan, M. B., Zock, P. L., and Mensink, R. P. (1995b). *Trans* fatty acids and their effects on lipoproteins in humans. *Annu. Rev. Nutr.* **15**, 473–493.

Katan, M. B. (1997). High-oil compared with low-fat, high-carbohydrate diets in the prevention of ischemic heart disease. *Am. J. Clin. Nutr.* **66**, 974S–979S.

Keith, M. O., and Tryphonas, L. (1978). Choline deficiency and the reversibility of renal lesions in rats. *J. Nutr.* **108**, 434–446.

Keys, A. (1995). Mediterranean diet and public health: Personal reflections. *Am. J. Clin. Nutr.* **61**, 1321S–1323S.

Kleinman, R., Finberg, L., Klish, W., and Lauer, R. (1996). Dietary guidelines for children: U.S. recommendations. *J. Nutr.* **126**, 1028S–1030S.

Kling, D., Fingerle, J., Harlan, J., Lobb, R. R., and Lung, F. (1995). Mononuclear lymphocytes invade rabbit arterial intima during thickening formation via CD18- and VLA-4-dependent mechanisms and stimulate smooth muscle migration. *Circ. Res.* **77**, 1121–1128.

Krauss, R. M., and Dreon, D. M. (1995). Low-density-lipoprotein oxidation and response to a low-fat diet in healthy men. *Am. J. Clin. Nutr.* **62**, 478S–487S.

Krauss, R. M., Deckelbaum, R. J., Ernst, N., Fisher, E., Howard, B., Knopp, R., Kotchen, T., Lichtenstein, A., McGill, H., Pearson, T., Prewitt, E., Stone, N. J., Horn, L., and Weinberg, R. (1996). Dietary guidelines for healthy American adults. *Circulation* **94**, 1795–1800.

Kris-Etherton, P. M., and Yu, S. (1997). Individual fatty acid effects on plasma lipids and lipoproteins: Human studies. *Am. J. Clin. Nutr.* **65**, 1628S–1644S.

Kuivenhoven, J., Pritchard, H., Hill, J., Frohlich, J., Assman, G., and Kastelein, J. (1997). The molecular pathology of lecithin: Cholesterol acyltransferase (LCAT) deficiency syndromes. *J. Lipid Res.* **38**, 191–205.

Lee, C., and Chen, L. B. (1988). Dynamic behavior of endoplasmic reticulum in living cells. *Cell* **54**, 37–46.

Levine, G. N., Keaney, J. F., and Vita, J. A. (1995). Cholesterol reduction in cardiovascular disease. *N. Engl. J. Med.* **332**, 512–521.

Lifshitz, F., and Tarim, O. (1996). Considerations about dietary fat restrictions for children. *J. Nutr.* **126**, 1031S–1041S.

Linton, M., Farese, R., and Young, S. (1993). Familial hypobetalipoproteinemia. *J. Lipid Res.* **34**, 521–541.

Lohse, P., Mann, W. A., Stein, E. A., and Brewer, H. B. (1991). Apolipoprotein E-4. *J. Biol. Chem.* **266**, 10479–10484.

Lombardi, B. (1971). Effects of choline deficiency on rat hepatocyte. *Fed. Proc.* **30**, 139–142.

Lombardi, B., Pani, P., and Schlunk, F. F. (1968). Choline-deficiency fatty liver: Impaired release of hepatic triglycerides. *J. Lipid Res.* **9**, 437–446.

Magil, S. G., Zeisel, S. H., and Wurtman, R. J. (1981). Effects of ingesting soy or egg lecithins on serum choline, brain choline and brain acetylcholine. *J. Nutr.* **111**, 166–170.

Marshall, C. J. (1993). Protein prenylation. *Science* **259**, 1865–1866.

Matsunaga, T., Hiasa, Y., Yanagi, H., Maeda, T., Hattori, N., Yamakawa, K., Yamanouchi, Y., Tanaka, I., Obara, T., and Hamaguchi, H. (1991). Apolipoprotein A-I deficiency due to a codon 84 nonsense mutation of the apolipoprotein A-I gene. *Proc. Natl. Acad. Sci. U.S.A.* **88**, 2793–2797.

McKeigue, P. (1995). *Trans* fatty acids and coronary heart disease: Weighing the evidence against hardened fat. *Lancet* **345**, 269–270.

McKenney, J. M., Proctor, J. D., Harris, S., and Chinchili, V. M. (1994). A comparison of the efficacy and toxic effects of sustained- vs immediate-release niacin in hyper-cholesterolemic patients. *JAMA* **271**, 672–677.

McNamara, D. J. (1992). Dietary fatty acids, lipoproteins, and cardiovascular disease. *Adv. Food Nutr. Res.* **36**, 253–351.

Mustad, V. A., Ellsworth, J. L., Cooper, A., Kris-Etherton, P. M., and Etherton, T. D. (1996). Dietary linoleic acid increases and palmitic acid decreases hepatic LDL receptor protein and mRNA abundance in young pigs. *J. Lipid Res.* **37**, 2310–2323.

Nestle, M. (1995). Mediterranean diets: Historical and research overview. *Am. J. Clin. Nutr.* **61**, 1313S–1320S.

Nicolosi, R. J. (1997). Dietary fat saturation effects on low-density lipoprotein concentrations and metabolism in various animal models. *Am. J. Clin. Nutr.* **65**, 1617S–1627S.

Olson, R. E. (1995). The dietary recommendations of the American Academy of Pediatricians. *Am. J. Clin. Nutr.* **61**, 271–273.

Oram, J. F., and Yokoyama, S. (1996). *Apolipoprotein-mediated removal of cellular cholesterol and phospholipids. J. Lipid Res.* **37**, 2473–2491.

Parrott, C. L., Alsayed, N., Rebourcet, R., and Santamarina-Fojo, S. (1992). ApoC-II: A premature termination mutation in the signal peptide of apoC-II resulting in the familial chylomicronemia syndrome. *J. Lipid Res.* **33**, 361–367.

Porter, J., Young, K., and Beachy, P. (1996). Cholesterol modification of hedgehog signaling proteins in animal development. *Science* **274**, 255–259.

Reyes, J., Robayna, I., Delgado, P., Gonzalez, I., Aguiar, J., Rosas, F. Fanjul, L., and Galarreta, C. (1996). c-Jun is a downstream target for ceramide-activated protein phosphatase in A431 cells. *J. Biol. Chem.* **271**, 21375–21380.

Reymer, P. W., Gagne, E., Groenemeyer, B., Zhang, H., Forsyth, I., Jansen, H., Seidell, J., Kromhout, D., Lie, K., Kastelein, J., and Hayden, M. R. (1995). A lipoprotein lipase mutation (Asn 291 Ser) is associated with reduced HDL cholesterol levels in premature atherosclerosis. *Nature Genet.* **10**, 28–34.

Roessler, E., Belloni, E., Gaudenz, K., Jay, P., Berta, P., Scherer, S., Tsui, L., and Muenke, M. (1996). Mutations in the human Sonic hedgehog gene cause holoprosencephaly. *Nature Genet.* **14**, 357–360.

Ross, R. (1993). The pathogenesis of atherosclerosis: A perspective for the 1990s. *Nature* **362**, 801–809.

Rumsey, S. C., Galeano, N., Lipschitz, B., and Deckelbaum, R. J. (1995). Oleate and other long-chain fatty acids stimulate low-density-lipoprotein receptor activity by enhancing acyl coenzyme A:cholesterol acyltransferase activity and altering intracellular regulatory cholesterol pools in cultured cells. *J. Biol. Chem.* **270**, 10008–100016.

Sacks, F. M., Pfeffer, M., Moye, L., Rouleau, J., Rutherford, J. D., Cole, T. G., Brown, L., Warnica, J., Arnold, J., Wun, C.-C., Davis, B. R., and Braunwald, E. (1996). The effect of pravastatin on coronary events after myocardial infarction in patients with average cholesterol levels. *N. Engl. J. Med.* **335**, 1001–1009.

Sakai, J., Duncan, E. A., Rawson, R., Hua, X., Brown, M. S., and Goldstein, J. L. (1996). Sterol-regulated release of SREBP-2 from cell membranes requires two sequential cleavages, one within a transmembrane segment. *Cell* **85**, 1037–1046.

Sakono, M., Ibi, T., Nagao, K., Ikeda, I., Yamamoto, K., and Imaizumi, K. (1996). Lymphatic transport of cholesterol in normocholesterolemic rats treated with pravastatin, an inhibitor of HMG-CoA reductase. *Atherosclerosis* **124**, 95–102.

Santana, P., Pena, L., Haimovitz-Friedman, A., Martin, S., Green, D., McLoughlin, M., Cordon-Cardo, C., Schuchman, E., Fuks, Z., and Kolesnick, R. (1996). Acid sphingomyelinase-deficient human lymphoblasts and mice are defective in radiation-induced apoptosis. *Cell* **86**, 189–199.

Saucier, S. E., Kandutsch, A. A., Gayen, A. K., Swahn, D. K., and Spencer, T. A. (1989). Oxysterol regulators of 3-hydroxy-3-methylglutaryl-CoA reductase in liver. *J. Biol. Chem.* **264**, 6863–6869.

Schectman, G., and Hiat, J. (1996). Dose–response characteristics of cholesterol-lowering drug therapies. *Ann. Intern. Med.* **125**, 990–1000.

Schmidt, H., Remaley, A., Stonik, J. A., Ronan, R., Wellmann, A., Thomas, F., Zech, L. A., Brewer, H., and Hoeg, J. M. (1995). Carboxyl-domain truncation alters apolipoprotein A-I *in vivo* catabolism. *J. Biol. Chem.* **270**, 5469–5475.

Schulthess, G., and Hauser, H. (1995). A unique feature of lipid dynamics in small intestinal brush border membrane. *Mol. Membr. Biol.* **12**, 105–112.

Schwartz, S. M., de Blois, D., and O'Brien, E. (1995). The intima: Soil for atherosclerosis and re-stenosis. *Circ. Res.* **77**, 445–465.

Simopoulos, A. P. (1996). *Trans* fatty acids. *In* "Handbook of Lipids in Human Nutrition" (G. A. Spiller, ed.), pp. 91–99. CRC Press, Boca Raton, FL.

Stampfer, M. J., Krauss, R. M., Ma, J., Blanche, P., Holl, L. G., Sacks, F., and Hennekens, C. H. (1996). A prospective study of triglyceride level, low-density lipoprotein particle diameter, and risk of myocardial infarction. *JAMA* **276**, 882–888.

Steinberg, D. (1997a). Low-density lipoprotein oxidation and its pathobiological significance. *J. Biol. Chem.* **272**, 20963–20966.

Steinberg, D. (1997b). Oxidative modification of LDL and atherogenesis. *Circulation* **95**, 1062–1071.

Straka, M. S., Junker, L. H., Zacarro, L., Zogg, D. L., Dueland, S., Everson, G. T., and Davis, H. A. (1990). Substrate stimulation of 7α-hydroxylase, an enzyme located in the cholesterol-poor endoplasmic reticulum. *J. Biol. Chem.* **265**, 7145–7149.

Streicher, R., Geisel, J., Weisshaar, C., Avci, H., Oette, K., Muller-Wieland, D., and Krone, W. (1996). A single nucleotide substitution in the promoter region of the apolipoprotein C-II gene identified in individuals with chylomicronemia. *J. Lipid Res.* **37**, 2599–2607.

Tabas, I., Marathe, S., Keesler, G. A., Beatini, N., and Shiratori, Y. (1996). Evidence that the initial up-regulation of phosphatidylcholine biosynthesis in free cholesterol-loaded macrophages is an adaptive response that prevents cholesterol-induced cellular necrosis. *J. Biol. Chem.* **271**, 22773–22781.

Tall, A. H. (1986). Plasma lipid transfer proteins. *J. Lipid Res.* **27**, 361–367.

Tall, A. R. (1993). Plasma cholesteryl ester transfer protein. *J. Lipid Res.* **34**, 1255–1274.

Testi, R. (1996). Sphingomyelin breakdown and cell fate. *Trends Biol. Sci.* **21**, 468–471.

Thi-Dinh, K. L., Demarne, Y., Nicolas, C., and Lhuillery, C. (1990). Effect of dietary fat on phospholipid class distribution and fatty acid composition in rat cell plasma membrane. *Lipids* **25**, 278–283.

Thun, M. J., Peto, R., Lopez, A., Monaco, J., Henley, J., Heath, C. W., and Doll, R. (1997). Alcohol consumption and mortality among middle-aged and elderly U.S. adults. *N. Engl. J. Med.* **337**, 1705–1714.

Thurnhofer, H., and Hauser, H. (1990). Uptake of cholesterol by small intestinal brush border membrane is protein-mediated. *Biochemistry* **29**, 2142–2148.

Tian, L., Yoshihara, Y., Mizuno, T., Mori, K., and Gahmberg, C. G. (1997). The neuronal glycoprotein telencephalin is a cellular ligand for CD11a/CD18 leukocyte integrin. *J. Immunol.* **158**, 928–936.

Treasure, C. B., Klein, J., Weintraub, W., Talley, J., Stillabower, M., Kosinski, A., Zhang, J., Boccuzzi, S. J., Cedarholm, J., and Alexander, R. W. (1995). Beneficial effects of cholesterol-lowering therapy on the coronary endothelium in patients with coronary artery disease. *N. Engl. J. Med.* **332**, 481–487.

Tribble, D. L., and Krauss, R. M. (1993). HDL and coronary artery disease. *Adv. Intern. Med.* **38**, 1–29.

Vance, D. E., and Vance, J. E. (1985). "Biochemistry of Lipids and Membranes." Benjamin/Cummings, Menlo Park, CA.

Weng, W., and Breslow, J. L. (1996). Dramatically increased high-density-lipoprotein cholesterol, increased remnant clearance, and insulin hypersensitivity in apolipoprotein A-II knockout mice suggest a complex role for apolipoprotein A-II in atherosclerosis susceptibility. *Proc. Natl. Acad. Sci. U.S.A.* **93**, 14788–14794.

Webb, O. L., Laskarewski, P. M., and Glueck, C. J. (1984). Severe depression of high-density-lipoprotein cholesterol levels in weight lifters and bodybuilders by self-administered exogenous testosterone and anabolic androgenic steroids. *Metabolism* **33**, 971–975.

Weihrauch, J. L., and Son, Y.-S. (1983). The phospholipid content of foods. *J. Am. Oil Chem. Soc.* **60**, 1971–1978.

Wertz, P. W., Stover, P. M., Abraham, W., and Downing, D. T. (1986). Lipids of chicken epidermis. *J. Lipid Res.* **27**, 427–435.

Willet, W. C., and Ascherio, A. (1995). Response to the International Life Sciences Institute report on *trans* fatty acids. *Am. J. Clin. Nutr.* **62**, 524–526.

Woollett, L., and Dietschy, J. (1994). Effect of long-chain fatty acids on low-density-lipoprotein cholesterol metabolism. *Am. J. Clin. Nutr.* **60**, 991S–996S.

Xiong, W., Zsigmond, E., Gotto, A. M., Reneker, L. W., and Chan, L. (1992). Transgenic mice expressing full-length human apolipoprotein B-100. *J. Biol. Chem.* **267**, 21412–21420.

Zhang, P., Liu, B., Jenkins, G., Hannun, Y., and Obeid, L. (1997). Expression of neutral sphingomyelinase identifies a distinct pool of sphingomyelin involved in apoptosis. *J. Biol. Chem.* **272**, 9609–9612.

Zeisel, S. H. (1990). Choline deficiency. *J. Nutr. Biochem.* **1**, 332–349.

## BIBLIOGRAPHY

### Identification of Fats in Foods and Biological Tissues

Ahn, D. U., Wolfe, F. H., and Sim, J. S. (1995). Dietary α-linoleic acid and mixed tocopherols, and packaging influences on lipid stability in broiler chicken breast and leg muscle. *J. Food Sci.* **60**, 1013–1018.

Aro, A., Jauhiainen, M., Partanen, R., Salminen, I., and Mutanen, M. (1997). Stearic acid, *trans* fatty acids, and dairy fat: Effects on serum and lipoprotein lipids, apolipoproteins, lipoprotein(a), and lipid transfer proteins in healthy subjects. *Am. J. Clin. Nutr.* **65**, 1419–1426.

Baggio, G., Pagnan, A., Muraca, M., Martini, S., Opportuno, A., Bonanome, A., Ambrosio, G., Ferrari, S., Guarini, P., Piccolo, D., Manzato, E., Corrocher, R., and Crepaldi, G. (1988). Olive-oil-enriched diet: Effect on serum lipoprotein levels and biliary cholesterol saturation. *Am. J. Clin. Nutr.* **47**, 960–964.

Bracco, U. (1994). Effect of triglyceride structure on fat absorption. *Am. J. Clin. Nutr.* **60**, 1002–1009.

Camps, L., Reina, M., Llobera, M., Bengtsson-Olivecrona, G., Olivecrona, T., and Vilaro, S. (1991). Lipoprotein lipase in lungs, spleen, and liver: Synthesis and distribution. *J. Lipid Res.* **32**, 1877–1888.

Farese, R. V., Ruland, S., Flynn, L., Stokowski, R., and Young, S. (1995). Knockout of the mouse apolipoprotein B gene results in embryonic lethality in homozygotes and protection against diet-induced hypercholesterolemia in heterozygotes. *Proc. Natl. Acad. Sci. U.S.A.* **92**, 1774–1778.

Goldberg, I. J. (1996). Lipoprotein lipase and lipolysis: Central roles in lipoprotein metabolism and atherogenesis. *J. Lipid Res.* **37**, 693–707.

Gurr, M. I. (1992). "Role of Fats in Food and Nutrition," 2nd ed., pp. 21–35. Elsevier Applied Science, London.

Hamm, M. W., Sekowski, A., and Ephrat, R. (1988). Dietary fat ratios and liver plasma membrane lipid composition. *Lipids* **23**, 829–833.

Innerarity, T. L., Boren, J., Yamanaka, S., and Olofsson, S.-O. (1996). Biosynthesis of apolipoprotein B48-containing lipoproteins. *J. Biol. Chem.* **271**, 2353–2356.

Kris-Etherton, P. M., and Yu, S. (1997). Individual fatty acid effects on plasma lipids and lipoproteins: Human studies. *Am. J. Clin. Nutr.* **65**, 1628S–1644S.

MacDonald, R., Zhang, W., Zhang, J.-P., and Sun, G. Y. (1996). Brain neutral lipids and phospholipids are modified by long-term feeding of beef tallow vs. corn oil diets. *J. Nutr.* **126**, 1554–1562.

McNamara, D. J. (1992). Dietary fatty acids, lipoproteins, and cardiovascular disease. *Adv. Food Nutr. Res.* **36**, 253–351.

Rao, C. V., Zang, E., and Reddy, B. (1993). Effect of high fat corn oil, olive oil, and fish oil on phospholipid fatty acid composition in male F344 rats. *Lipids* **28**, 441–447.

Schulthess, G., and Hauser, H. (1995). A unique feature of lipid dynamics in small intestinal brush border membrane. *Mol. Membr. Biol.* **12**, 105–112.

Thi-Dinh, K. L., Demarne, Y., Nicolas, C., and Lhuillery, C. (1990). Effect of dietary fat on phospholipid class distribution and fatty acid composition in rat cell plasma membrane. *Lipids* **25**, 278–283.

Wertz, P. W., Stover, P. M., Abraham, W., and Downing, D. T. (1986). Lipids of chicken epidermis. *J. Lipid Res.* **27**, 427–435.

## Lipid Metabolism

Bouhours, J. F., and Bouhours, D. (1981). Ceramide structure of sphingomyelin from human milk fat globule membrane. *Lipids* **16**, 721–711.

Datko, A. H., Aksamit, R. R., and Mudd, S. H. (1990). Phosphatidylcholine synthesis in the rat: The substrate for methylation and regulation by choline. *Lipids* **25**, 135–142.

Dietschy, J. M., Turley, S. D., and Spady, D. K. (1993). Role of liver in the maintenance of cholesterol and low-density lipoprotein homeostasis in different animal species, including humans. *J. Lipid Res.* **34**, 1637–1659.

Esko, J. D., and Raetz, C. R. (1983). Synthesis of phospholipids in animal cells. *In* "The Enzymes" (P. D. Boyer, ed.), Vol. 16, pp. 208–253. Academic Press, New York.

Goldberg, I. J. (1996). Lipoprotein lipase and lipolysis: Central roles in lipoprotein metabolism and atherogenesis. *J. Lipid Res.* **37**, 693–707.

Gillotte, K., Davidson, W., Lund-Katz, S., Rothblat, G. H., and Phillips, M. C. (1996). Apolipoprotein A-I structural modification and the functionality of reconstituted high-density lipoprotein particles in cellular cholesterol efflux. *J. Biol. Chem.* **271**, 23792–23798.

Hannun, Y. A., and Bell, R. M. (1989). Functions of sphingolipids and sphingolipid breakdown products in cellular recognition. *Science* **243**, 500–507.

Jenson, R. G., Clark, R. M., and Ferris, A. M. (1980). Composition of the lipids in human milk: A review. *Lipids* **15**, 345–355.

Jones, P. J., Pencharz, P. B., and Clandinin, M. T. (1981). Whole body oxidation of dietary fatty acids: Implications for energy utilization. *Am. J. Clin. Nutr.* **42**, 769–777.

Mixiorko, H. M., Laib, F. E., and Behnke, C. E. (1990). Evidence for substrate channeling in the early steps of cholesterolgenesis. *J. Biol. Chem.* **265**, 9606–9609.

Mookerjea, S., Park, C. E., and Kiksis, A. (1975). Lipid profiles of plasma lipoproteins of fasted and fed normal and choline-deficient rats. *Lipids* **10**, 374–382.

Pelech, S. L., and Vance, D. E. (1984). Regulation of phosphatidylcholine biosynthesis. *Biochim. Biophys. Acta* **779**, 217–251.

Vance, J. E., and Vance, D. E. (1988). Does rat liver Golgi have the capacity to synthesize phospholipids for lipoprotein secretion? *J. Biol. Chem.* **263**, 5898–5909.

Yao, Z., Jamil, H., and Vance, D. E. (1990). Choline deficiency causes translocation of CTP:phosphocholine cytidyltransferase from cytosol to endoplasmic reticulum in rat liver. *J. Biol. Chem.* **265**, 4326–4331.

Zeisel, S. H. (1981). Dietary choline: Biochemistry, physiology, and pharmacology. *Annu. Rev. Nutr.* **1**, 95–121.

Zeisel, S. H., Char, D., and Shepard, N. F. (1986). Choline, phosphatidylcholine and sphingomyelin in human and bovine milk and infant formulas. *J. Nutr.* **116**, 50–58.

## Lipoproteins and Cardiovascular Disease

Brown, M. S., and Goldstein, J. L. (1983). Lipoprotein metabolism in the macrophage. *Annu. Rev. Biochem.* **52**, 223–261.

Chong, K. S., Nicolosi, R. J., Rodger, R. F., Arrigo, D. A., Yuan, R. W., MacKey, J. J., Georas, S., and Herbert, P. N. (1981). Effect of dietary fat saturation on plasma lipoprotein and high-density lipoprotein metabolism of the rhesus monkey. *J. Clin. Invest.* **79**, 675–683.

Connor, W. E., and Connor, S. L. (1997). The case for a low-fat, high-carbohydrate diet. *N. Engl. J. Med.* **337**, 562–563.

Havel, R. J. (1988). Lowering cholesterol. *J. Clin. Invest.* **81**, 1653–1660.

Kane, J. P., and Havel, R. J. (1986). Treatment of hypercholesterolemia. *Ann. Med.* **37**, 427–435.

Katan, M. B. (1997). High-oil compared with low-fat, high-carbohydrate diets in the prevention of ischemic heart disease. *Am. J. Clin. Nutr.* **66**, 974S–979S.

Mahley, R. W. (1988). Apolipoprotein E: Cholesterol transport protein with expanding role in cell biology. *Science* **240**, 622–630.

Ross, R. (1986). The pathogenesis of atherosclerosis: An update. *N. Engl. J. Med.* **314**, 488–500.

Ross, R. (1993). The pathogenesis of atherosclerosis: A perspective for the 1990s. *Nature* **362**, 801–809.

Steinberg, D., Parthasarathy, S., Carew, T. E., Khoo, J. C., and Witztum, J. T. (1989). Beyond cholesterol. *N. Engl. J. Med.* **320**, 915–923.

Stossel, T. P. (1988). A multicenter comparison of lovastatin and cholestyramine therapy for severe primary hypercholesterolemia. *JAMA* **260**, 359–366.

Tall, A. H. (1986). Plasma lipid transfer proteins. *J. Lipid Res.* **27**, 361–367.

# 7

# OBESITY

## OVERVIEW

The body weight of the typical person is stable throughout much of adult life. The yearly change in weight, in adults between 20 and 60 years of age, has been estimated at one half pound per year (for males). Since the amount of energy used to deposit this weight is about 1560 kcal, and since one eats about 900,000 kcal per year, one can calculate that the amount of energy utilized to gain the half pound (per year) represents under 0.2% of the energy eaten over the course of the year (Weigle, 1994). This calculation can be used to convince oneself that there exist mechanisms in the body that maintain the body weight. Once convinced, one might wonder if some of these mechanisms are altered in obesity. Some of these mechanisms will be outlined in the introduction, prior to detailing methods of measuring and diagnosing obesity in humans.

## Classification of Fat

Fat can be classed in several ways: (1) as white fat or brown fat; (2) as upper body fat (android fat) or lower body fat (gynoid fat); (3) as subcutaneous fat or visceral fat; and (4) as hypertrophic fat or hyperplastic fat. It is customary for people to self-diagnose obesity, i.e., whether one is underweight or overweight. A self-diagnosis might be expected to be inaccurate because the measurement of body fat requires special tools and because of psychological influences. At its simplest, the tool consists of a tape measure. At its most complex, the tool is computed tomography.

## Diseases Associated with Obesity

Obesity is a risk factor for non-insulin-dependent diabetes mellitus, athero-sclerosis, and hypertension. This statement is currently being refined to the following form: android type obesity, rather than gynoid type obesity, is a greater risk factor for these diseases. The question of why obesity is a risk factor for these diseases remains largely a mystery.

The question of why obesity occurs in humans is a mystery. Some partial answers are that obesity is due to: (1) overeating; (2) eating too much high-fat food and too few high-carbohydrate foods; and (3) sedentary habits, and (4) genetics.

The definition of obesity has been refined to yield the following form. Obesity can be defined as a process where fat accumulates over a long period of time due to an increased rate of storage of triglycerides in adipose tissue, and the consequent need to overeat slightly to balance the amount of energy diverted to storage. To continue with the definition, a plateau in weight is finally reached where the percentage of body fat is maintained, and where any attempt to lose weight (or to gain still more weight) is resisted by a powerful drive to return to the weight plateau (Flatt, 1995; Leibel *et al.*, 1995).

An unanswered question regarding obesity is, "Why do some fat people make valiant attempts to become thinner by consuming diets of celery sticks and carrots, but fail to lose weight, whereas some thin persons consume high-fat diets of steaks, ice cream, and pies with flaky crusts, but remain slender?" Another unanswered question is, "It is obvious that some obese persons overeat, but what compels them to do so?"

## Early Studies of Obesity in Rodents

Clues to answers of these questions may eventually be found with continued studies of obesity in rodents. Kennedy (1950) suggested that the adipose tissue produces a hormone that regulates body weight. This suggestion gave rise to the "lipostat theory" of body weight control. The lipostat theory states that each body has a special mechanism that maintains the amount of body fat at a certain percent of the body weight. Hervey (1959) produced obesity in a rat by damaging its hypothalamus. The rat overate and became obese. Hervey proposed that the hypothalamus normally responded to a "satiety factor" circulating in the bloodstream, that the rat with the damaged hypothalamus could not respond, and that this rat's adipose tissue produced *greater and greater levels of the satiety factor*. Hervey then used surgery to connect the bloodstreams of a normal rat to that of a brain-damaged rat. How did the normal rat respond to receiving blood from the obese rat? The normal rat responded by *losing weight and starving to death*. Converse experiments also suggested a role of the hypothalamus in regulating weight.

Injection of hormones that are specifically made by this part of the brain (hypo-thalamus) into rats provokes food seeking and obesity (Wang *et al.*, 1997).

Subsequent work revealed that the "satiety factor" is a protein called leptin and that its target is a receptor in the hypothalamus. Attention to the cause of human obesity has focused on changes in the structure of the leptin receptor, defects in the mechanism of transport of serum leptin to the brain (Schwartz *et al.*, 1996; Caro *et al.*, 1996), and on mutations that affect the signaling proteins that respond to the leptin receptor.

## Short-Term and Long-Term Controls of Body Weight

### *Cholecystokinin Seems Involved in Short-term Control of Body Weight*

Various feedback mechanisms exist to control eating and body weight. Short-term feedback mechanisms are activated by eating food. Long-term feedback mecha-nisms are produced by adipose tissue. The various signals are gathered by the brain, processed, and used to increase or decrease food intake. These signals include various polypeptide hormones that are made in the brain and in the intestines. Cholecystokinin (CCK) is a short-term satiety signal. Injecting animals with CCK during feeding results in the animal eating a smaller meal. However, the animal compensates by eating more meals during the day, thus maintaining long-term energy intake (Figlewicz *et al.*, 1996; Woods *et al.*, 1998).

### *Neuropeptide Y and Leptin Seem Involved in Long-Term Body Weight*

Neuropeptide Y (NPY) and leptin are long-term hunger and satiety signals. NPY is a hormone that is made in the hypothalamus, where its synthesis increases when an animal is deprived of food. Repeated injection of NPY into the brain results in eating larger meals, in weight gain, and in obesity (Figlewicz *et al.*, 1996). Leptin is a polypeptide that is made in adipose tissue and secreted into the bloodstream. The chronic, artificial treatment of animals with leptin provokes a decline in feeding and a loss in body weight. Leptin acts at the hypothalamus to reduce the synthesis of NPY.

## Human Psychology of Obesity

Special techniques for the measurement of obesity are used for the diagnosis of individuals, as well as for data-taking in epidemiological studies. Although many epidemiological studies use questionnaires, assessment of obesity by means of a questionnaire would not be reliable, because many people of normal body com-position (especially girls and women) tend to view themselves as obese. From where might this unfortunate view arise? There is an occasional tendency to view thinness in women as a sign of competence, motivation, and sexual attraction, and to view obesity as a sign of laziness and self-indulgence (Wilfley and Rodin, 1995). This tendency has contributed somewhat to the practice of "dieting." "Dieting" usually means the consumption of a low-energy diet with the intention of losing weight. Obese persons may be stigmatized or discriminated against in present-day Western society (Maurer and Sobel, 1995). The stigma attached to obesity is thought to be to blame for the occurrence of such eating disorders as **anorexia nervosa** and **bulimia nervosa**.

A number of studies have revealed that obese persons tend not to have any greater psychological problems (anxiety, depression, rigidity, suspicion) than the general public (Wadden and Stunkard, 1985). However, psychological problems do tend to emerge when an obese person attempts to diet. Prolonged dieting is associated with the acquisition of eating disorders, such as **anorexia nervosa** and **bulimia**. Among young female adults, for example, the prevalence of anorexia nervosa has been found to be about 0.28% of the population in America, while the prevalence of bulimia is about 1.0% (Hoek, 1995). The prevalence of eating disorders in males is only 10% that of females (Anderson, 1995).

Anorexia nervosa involves dieting, induced vomiting, laxative abuse, and obsessive exercising. The desire to become thinner is thought to arise from the victim's perception that her body is obese or fat. People suffering from anorexia nervosa are usually about 85% their normal, desired weight. Bulimia involves binging, i.e., consuming large amounts of food in short periods, followed by "purging" via vomiting and laxative abuse, and finally a period of strict dieting. Persons with bulimia may be of normal weight, obese, or underweight. About 15% of those with these mental illnesses eventually die because of metabolic imbalances caused by their abnormal eating behavior. At best, only 50% of victims of anorexia nervosa or bulimia nervosa can benefit from antidepressant drugs, such as fluoxetine or desipramine (Walsh and Devlin, 1998).

## Summary of Techniques Used for Measuring Adipose Tissue

Several techniques have been devised for measuring fat and its distribution in living humans. Terms used to refer to fat's distribution include subcutaneous fat, visceral fat, android (male type) obesity, and gynoid (female type) obesity. The science of measurement of body size, weight, and proportions is called **anthropometry**. Hence, the following techniques are methods of anthropometry.

1. *Height and weight.* These are measured using a ruler and balance, or scale.

2. *Waist to hip ratio.* This ratio is acquired using a tape measure and by dividing: (circumference at the waist) / (circumference at the hip). This method is commonly used by nutrition researchers to assess if a subject has a male-type obesity or female-type obesity.

3. *Body mass index (BMI) or Quetelet's Index.* The BMI is (weight) / (height)$^2$, and has the units of kg/m$^2$. The BMI is a measure of total body fat. This method is commonly used by nutrition researchers, especially in epidemiological surveys.

4. *Skinfold measurements.* Skinfold thickness is measured by pinching the skin and using calipers to measure the layers of skin and subcutaneous fat that are located to the side of the pinch. The skinfold measurement is a fairly direct measure of the subcutaneous fat at the location of the pinch, but the usual goal of the method is to gain an indication of total body fat.

5. *Densitometry*. Densitometry involves underwater weighing. By weighing the subject in air, and when totally submerged in water, one can calculate the density of the subject's entire body. Work over the course of the decades has produced a formula that relates this density to the total amount of body fat.

6. *Ultrasound*. The technique of ultrasound is familiar to pregnant women who have had their fetus examined, i.e., to determine the growth and position of the fetus. In nutritional studies, ultrasound is used to determine the thickness of subcutaneous fat. The technique gives results that are similar, but not quite identical, to those acquired from skinfold measurements. Skinfold measurements tend to be difficult to acquire in very obese subjects, and here ultrasound may be the preferred technique.

7. *Bioelectrical impedance*. This technique involves connecting electrodes to the hands and feet and passing a mild electric current through the body. Most organs and tissues conduct the electric current, as they contain water and salts. However, adipose tissue contains only about 14% water, by weight, and does not readily conduct electricity. The results give an indication of total body fat.

8. *Computed tomography (CT)*. Computed tomography utilizes X-rays to acquire cross-sectional views of the body. The result is a photograph with areas of dark and light, corresponding to the different tissues, including adipose tissue. The analyst can easily use the pictures for assessing the percentage of fat that is just under the skin (subcutaneous fat) or deeper within the body (visceral fat).

## TYPES OF FAT

### White Fat and Brown Fat

The terms "fat" and "adipose tissue" are used interchangeably in this chapter, though one might recall that, to the biochemist, fat means triglycerides, which form a solid at room temperature. Adipose tissue, or fat, consists of a mass of adipocytes (fat cells) that are loosely held together by collagen fibers. Adipose tissue occurs in two forms — **white fat** and **brown fat**. White adipose tissue is colored white or yellow and has relatively few nerves and blood vessels. Each fat cell contains a single large droplet of triglycerides that is coated with a protein called **perilipin**. The droplet is not surrounded by a bilayer of phospholipids, and thus cannot be called a vesicle. White fat is used as a site for storing energy for physical activity.

Brown fat contains relatively more nerves and blood vessels. Each brown fat cell contains several small droplets of triglyceride, rather than one large droplet. Brown fat is used only for heat production. It occurs in all newborn mammals, including humans, apparently to assure that they keep warm. Brown fat is absent in adult humans, but it occurs in adult hibernating animals and in animals that are active in cold weather, such as rats. Brown fat does not respond much to a low-energy diet or to overeating. Brown fat cells have large numbers of mitochondria. Although these mitochondria utilize oxygen and pump protons (pump $H^+$

outward through the mitochondrial membrane), this process tends not to be linked to ATP production. In typical mitochondria, the flow of protons through the membrane, and then back into the mitochondrion, drives ATP synthesis. But in the mitochondria of brown fat, the flow of protons out and back through the mitochondrial membrane produces mainly heat. The influx of protons through the membrane occurs through channels that contain a special 32-kDa protein called **thermogenin**. The futile cycle of proton pumping and reentry generates heat for the hibernating animal (Trayhurn, 1996).

## Subcutaneous Fat and Visceral Fat

Fat deposits may be classed as subcutaneous and visceral. **Subcutaneous fat** occurs as a more or less continuous layer throughout the body. A measure of subcutaneous fat can be acquired by pinching the skin. Typically, the pinch consists of two layers of skin (each about 1.5 mm thick) and two layers of subcutaneous fat (each about 11 mm thick) (Craig, 1969). In human females, subcutaneous fat is especially thick in the buttocks. In contrast, **visceral fat** (deep fat) is distributed similarly in human males and females. The visceral fat in the abdomen is frequently measured in studies of obesity. Intraabdominal fat occurs in three areas: retroperitoneal, mesenteric, and omental.

Pregnant women have provided a striking illustration of subcutaneous fat (Taggart *et al.*, 1967). The pinch test or, more accurately the skinfold test, was performed on 84 women during the course of their pregnancy. Data were taken at the indicated times, i.e., from week 10 of the pregnancy until 8 weeks after birth (Figure 7.1). Skinfold measurements were acquired at various sites of the body,

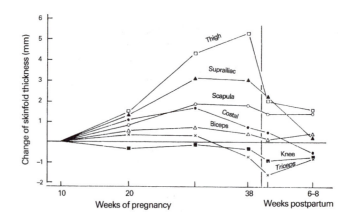

**FIGURE 7.1.** Use of anthropometry to measure changes in subcutaneous fat at various parts of the body during pregnancy. The changes that occur following week 10 of pregnancy are shown. The skinfold measurement test at the thigh revealed that a 5-mm increase had occurred by the 38th week of pregnancy, and that the skinfold had shrunk back to the original 10-week thickness at the timepoint of 2 months after the birth. (Redrawn with permission from Taggart *et al.*, 1967.)

including the thighs, scapula (on the back), costal (over the lowest rib), and triceps (on the arm). Early in the study, at the 10-week timepoint, the triceps skinfold thickness was measured and found to be 15.7 mm (scapula, 11.8 mm; costal; 11.4 mm; thigh, 34.8 mm; knee 11.6 mm). Figure 7.1 reveals the deviations (increases or decreases) from these thickness values over the course of time. The study reveals that subcutaneous fat increases at some sites, but not at others, during the course of pregnancy.

EXERCISE

Is there any evidence that fat, accumulating at any site during pregnancy, can remain accumulated in the mother for at least two months after the birth?

## Android and Gynoid Fat

Obesity means an excess of fat in the body when compared to lean body mass. Lean body mass is defined in the Protein chapter in the section on Growth. The distribution of fat in the body is classed as **upper body fat** (male-type, android, central type) or **lower body fat** (female-type, gynoid, peripheral type). As men become obese, they tend to accumulate fat in the android pattern. Male type fat occurs in the abdomen. As women become obese they may accumulate fat in either the android pattern or gynoid pattern. The gynoid distribution involves accumulation of fat around the hips and thighs. It may also involve accumulation of fat on the arms, to give a bat-wing appearance. To summarize, fat men generally acquire the android fat pattern, while fat women may have acquired the android or gynoid fat pattern. Android and gynoid obesity patterns in adult female subjects are depicted in Figure 7.2.

Measurement of the waist-to-hip ratio has proven to be useful in diagnosing male-type or female-type obesity. This ratio is acquired by measuring the circumference around the waist, and about the hips, and performing division. The units of each measurement is centimeters, while the ratio has no unit. A waist/hip ratio of 1.0 or greater (for men) or 0.8 or greater (for women) is associated with a distinct increase in health risk for non-insulin-dependent diabetes mellitus (NIDDM), hypertension, and atherosclerosis (Bjorntorp, 1985).

A clear illustration of android and gynoid fat distribution patterns is shown in a study of 330 obese subjects and 77 persons of normal adiposity (Krotkiewski *et al.*, 1983). Figure 7.3 depicts the thickness at lower body sites and abdominal sites. To be more specific, the lower body sites were "gluteal" and "femoral." The abdominal sites were "hypogastric" and "epigastric." The lower bar indicates results from the normal weight subjects, while the upper bars are from obese subjects. The data reveal an android fat distribution in men, as the men had the thickest fat layers about the belly (hypogastric and epigastric sites). The data also reveal that the population of women had a gynoid fat distribution, as they contained the thickest fat layer in the buttocks (gluteal). Please look at the figure and

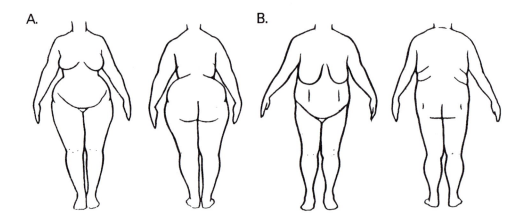

**FIGURE 7.2.** Drawings of obese women with (A) gynoid (lower body) type obesity and (B) android (upper body) type obesity. Gynoid obesity involves excess accumulation of fat on the hips and thighs, while android obesity involves excess abdominal and upper body fat. With android obesity in the female, the waistline is often absent. Abdominal fat tends to accumulate above the umbilicus in males, and below it in females. These trends begin to take place in adolescence. The reference contained only drawings of women. (Redrawn with permission from Craig and Bayer, 1967.)

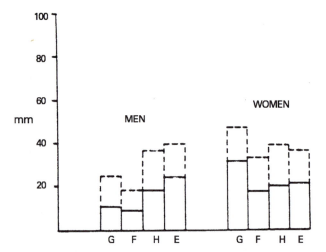

**FIGURE 7.3.** Adipose tissue thickness in people in the general population (solid bars) and in obese persons (dashed bars). Adipose tissue thickness was measured by ultrasound measurements (not by the usual technique of skinfold measurements) at the gluteal site (G), femoral site (F), hypogastric site (H), and epigastric site (E). The epigastric site that was used was one-third the distance down from the tip of the sternum to the umbilicus. This site was somewhat thicker in men than in women. The femoral site was one-third the distance down from the anterior superior iliac spine (wide part of the pelvis) to the patella (knee). This site was much thicker in women than in men. Biopsies were also taken at the indicated sites and examined under the microscope to acquire fat cell size. (Redrawn with permission from Krotkiewski *et al.*, 1983.)

decide if you think that the obese men were concerned about having too much fat on their thighs.

## Hypertrophic and Hyperplastic Obesity

Fat may be classified according to the time course of development of obesity. In this regard, there appears to be two types of obesity, i.e., where the increase in fat is due mainly to increases in fat cell number, or to fat cell size. There is some indication that, with the development of obesity in humans, the *size* of individual cells increases until the body contains about 30 kg of total body fat. With further increases in body fat, there is little further increase in adipocyte size. Instead, the *number* of fat cells increases (Sjostrom, 1980). The type of obesity involving mainly an increase in cell size is called **hypertrophic**, while that involving an increase in cell number is called **hyperplastic**.

### Studies with Rats and Humans on Hypertrophic Fat and Hyperplastic Fat

Studies with rats have revealed that, with overfeeding, existing adipocytes increased in mass until they reached a critical, maximal size (0.8 micrograms). When fat cells reached this mass, adipocyte precursors differentiated to form new, mature adipocytes, and these too accumulated triglycerides up to the critical size. Studies with humans have revealed a similar pattern, where the critical mass was 0.6–0.7 µg. One study involved humans of a range of adiposities (Krotkiewski *et al.*, 1983). The study involved 930 obese persons and 77 persons of normal weight. The study was not an overfeeding study, but only acquired subjects as is. The percent body fat was calculated using a whole body counter to measure potassium-40. This method, which is detailed in the Protein chapter, provides a more or less direct number regarding lean body mass. The body fat was then calculated by subtracting lean body mass (kg) from body weight (kg). The thickness of subcutaneous fat was determined by ultrasound, and the size and mass of the adipocytes were measured on biopsies.

The results of the study, though simple to comprehend, provide an exciting view of how adipocytes behave *in vivo*. The thickness of gluteal (buttocks) (Figure 7.4A) and hypogastric (belly) (Figure 7.4B) fat are shown. Data are shown for male (open circles) and female (closed circles) subjects. The x-axis shows the mass (kg) of total body fat, while the y-axis indicates the thickness of subcutaneous fat. For example, the results show that the subjects with the greatest fat at the buttocks were women (7 cm of fat), while those with the greatest amount of belly fat were men (10 cm of fat). Data on fat cell size reveal, as one views data from subjects of greater and greater fat mass, that the size of fat cells increases from about 0.5 µg to about 0.7 µg, after which a plateau in fat cell mass occurs (Figure 7.5). The results indicate a trend where persons who are slightly obese store their fat in existing cells, creating large adipocytes, while persons who are greatly obese contain greater numbers of adipocytes to accommodate the extra triglyceride.

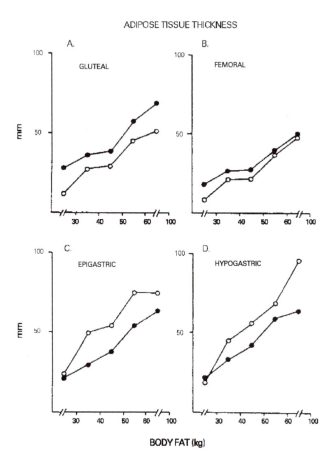

**FIGURE 7.4.** Adipose tissue thickness versus body fat. Body fat was measured by subtracting lean body mass (acquired from $^{40}K$ measurements) from total body weight. Data from men (circles) and women (dots) are shown separately. Data from subjects of differing total body fat were presented separately, as indicated by the X-axis. The results show that fat thickness increased with total body fat. Women tended to have thicker gluteal fat than men, when subjects were matched for total body fat. Men tended to have thicker epigastric fat than women, when subjects were matched for total body fat. (Redrawn with permission from Krotkiewski *et al.*, 1983.)

## TECHNIQUES FOR MEASURING FAT

### Measurement of Body Fat by Body Mass Index

Various ways of classifying fat, as well as the techniques of the skinfold test and waist-to-hip ratio, were discussed earlier. The following material explains in detail a number of techniques for fat measurement, and illustrates their use via several case studies. A direct measurement of total body fat can only be made by analyzing cadavers (dead people). The amount of fat was correlated with the cadaver's height, width, underwater weight, and skin surface area. These results have resulted in the creation of several techniques for indirectly assessing body fat in

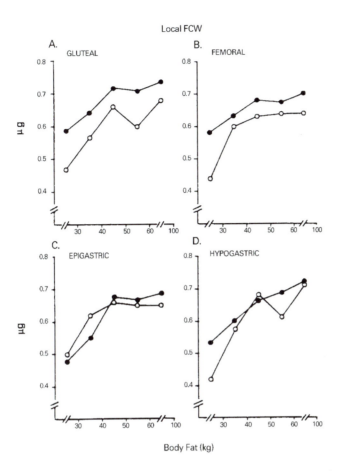

**FIGURE 7.5.** Fat cell weight at various sites for men and women. The conditions were as in Figure 7.4. Data from men (circles) and women (dots) are shown separately. The data reveal that fat cell weight (or size) tends to increase to a specific plateau in size, despite a continued increase in adipose tissue thickness. This finding supports the theory that an increase in fat cell size to a certain point provokes the generation of new adipocytes. (Redrawn with permission from Krotkiewski *et al.*, 1983.)

living people. Some of the tests are easy to perform in primitive countries or in epidemiological studies. Other available tests, which are relatively accurate, can only be performed in laboratories and on small numbers of subjects. One simple test involves a ratio of weight to height. This test is called Quetelet's Index, and results in a value that is called the Body Mass Index (BMI). The test involves the use of the following formula:

$$BMI = (weight) / (height)^2$$

EXERCISE

Please calculate the BMI of a 60-kg female who is 1.60 meters tall. Please calculate your own BMI.

The BMI can be calculated using a scale, ruler, and pocket calculator. BMI values of 20–25 kg/m$^2$ are associated with the least health risk. A BMI of 30 kg/m$^2$ is associated with serious risk for premature death due to obesity. BMI values under 16 kg/m$^2$ suggest an eating disorder such as anorexia nervosa or bulimia.

### Example of Use of Body Mass Index in a Study of Diabetics

The use of the BMI is illustrated by the following example. In a broader sense, the example emphasizes the important connection between obesity and diabetes. The importance of weight loss in reducing the severity of non-insulin-dependent diabetes (NIDDM) has long been recognized. However, the actual physiological relationship between obesity (or weight loss) and the severity of diabetes have been difficult to pinpoint in the laboratory. A careful study of obese and lean subjects revealed one real difference. In this study, lean subjects (with NIDDM) were found more able to take up circulating glucose than were obese subjects (with NIDDM) (Campbell and Carlson, 1993).

Persons of varying obesity were examined, as indicated by the BMI values on the axis of Figure 7.6. Insulin was infused into all patients at a rate of 1.0 microunits per kilogram of body weight per minute. At the same time as the insulin infusion, glucose was also infused. The researchers adjusted the rate of glucose infusion to maintain constant levels of blood glucose. In this way, the researchers were able to assess the rate of glucose uptake by all of the tissues of the body. The results (Figure 7.6) indicate a trend where lean subjects were able to utilize a higher rate of glucose infusion but where the obese subjects were able to utilize glucose infused only at lesser rates. The phrase "able to utilize" means "able to accept a higher rate of glucose infusion but without a rise in plasma glucose concentration." The findings support the notion that obesity can sometimes contribute to insulin resistance in persons with NIDDM.

### Diabetes and Exercise with Weight Loss

Let us dwell on the topic of obesity and weight loss for a bit longer. The American Diabetes Association has recommended that adults with NIDDM exercise for at least 20 min/day at 50% of the maximal O$_2$ consumption rate (Ruderman and Schneider, 1992). This recommendation was designed to promote weight loss, since about 85% of people with NIDDM are also obese. However, evidence also suggests that exercise alone, even without weight loss, can improve the body's sensitivity to insulin.

### Diabetes and Exercise without Weight Loss

This concerns the influence of exercise alone, with no weight loss. Braun *et al.* (1995) studied a group of obese female diabetics. The subjects walked on a treadmill to provoke 50% of their maximal rate of oxygen utilization. Six episodes of walking were performed over the course of 2 days. A total of 3138 kJ (750 kcal) of energy was expended by the 2-day exercise program. The body's response to glucose was measured on the third day.

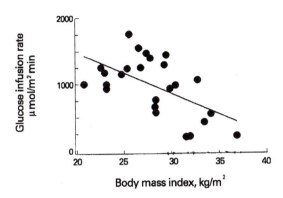

**FIGURE 7.6.** Rate of infusion of glucose into human bloodstream required to maintain constant levels of plasma glucose. All subjects were infused at a constant rate with insulin. All subjects were diabetics. Insulin acts by provoking the uptake of plasma glucose, and its storage as glycogen. By careful infusion of glucose, and constant monitoring, the re-searchers were able to maintain plasma glucose at a constant level. The study illustrates the use of the BMI, and reveals that persons with a greater BMI require lesser rates of glucose infusion to maintain constant plasma glucose levels. The conclusion is that persons with a greater BMI are more resistant to insulin, and thus have one of the characteristics of diabetes. Control studies of healthy thin and obese subjects seemed not to reveal the trend depicted. In clinical practice, obesity is widely believed to worsen the insulin resis-tance of NIDDM. (Redrawn with permission from Campbell and Carlson, 1993.)

In detail, glucose was infused into each woman at a rate of 4 mg glucose/kg fat-free mass per min. Blood samples were taken at intervals of 15 min. Analysis of the blood samples revealed that plasma glucose levels increased at the start of the infusion, as one might expect (Figure 7.7). In taking a close look at the data, one can see that glucose rose and then declined during the infusion.

The decline occurred with the rise in insulin levels, and with the response by insulin-sensitive tissues, such as muscle. At 0 minutes, the glucose levels in the diabetic women were about 12 mM. Towards the end of the 3-hour glucose infusion, the glucose levels in the nonexercised group (open circles) declined to about 11 mM, while the glucose levels in the exercised group (black dots) de-creased further — to about 9 mM. Separate work revealed that this effect of exercise wears off after 4–5 days. One might note that the same women were used for the study of exercised and nonexercised subjects. The results, therefore, dem-onstrate an effect of short-term exercise on glucose control in diabetics. Similar convincing results were provided by Hughes *et al.* (1993).

## Example of Use of Body Mass Index in a Study of Cardiovascular Disease

Let us return to the issue of techniques used to assess obesity and body composi-tion. A second example also reveals how the BMI is used. The example is an

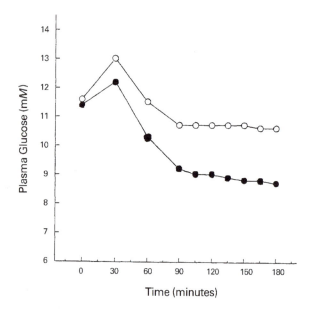

**FIGURE 7.7.** Plasma glucose concentrations during the course of an infusion of glucose. The study addressed the question, "Can exercise without weight loss improve health?" Eight obese diabetic women were studied under two conditions, i.e., with and without a 2-day exercise program. The exercise was not enough to result in the usual benefits of a long-term rigorous exercise program, i.e., a reduction in body fat, increases in muscle capillary density, increases in glycogen stored in muscle. On the third day, the women were infused with glucose for 3 hours. Blood samples were removed every 15 min. Insulin was also infused during the 3-hour period at a rate of 40 mU/$m^2$ per min. The unit "$m^2$" means square meter of skin area. Exogenous insulin was infused because the women had an impaired ability to produce insulin, and because the goal of the researchers was to study the body's sensitivity to insulin. Without the insulin infusion, the researchers would have been forced to study the body's sensitivity to very low levels of insulin. (Open circles, no exercise; black dots, with exercise). (Redrawn with permission from Braun *et al.*, 1995.)

epidemiological survey called the Nurses' Health Study. The study was directed to the question of whether only severe obesity poses a health risk, or if moderate obesity also presents a health risk. The study involved 115,195 nurses. The nurses were examined via questionnaires over a period of 16 years. During this period, 881 subjects died from cardiovascular disease, 2586 from cancer, and 1259 from other causes.

The data were analyzed and presented in several ways (Manson *et al.*, 1995). Figure 7.8A shows the relative risk from dying of cardiovascular disease versus BMI. Figure 7.8B shows the relative risk from dying of cancer versus BMI. The results show that the rates of death from cardiovascular disease were fourfold greater for women with high BMI values (BMI ≥ 29.0) than for the leanest group of women. Rates of death from cancer for the obese women were twice those for the leanest women. The *y*-axis has the units of "relative risk." Relative risk was calculated by dividing as follows:

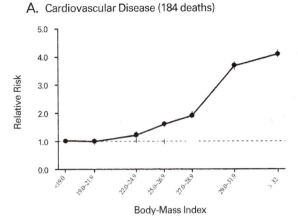

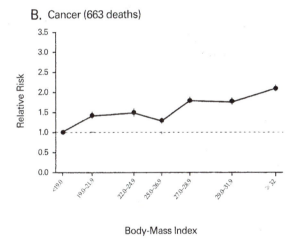

**FIGURE 7.8.** Relative risk for (A) cardiovascular disease and (B) cancer versus body mass index. The subjects were 115,195 nurses who participated in a study over 16 years. During this period, a total of 4726 nurses died — 881 from cardiovascular disease, 2586 from cancer, and 1259 from other causes. Only data from nurses who had never smoked are presented in the figure. The relative risk of those with a BMI under 19.0 was set at unity. The results show that death due to cardiovascular disease was four times higher in the obese women, when compared to the rate for the leanest women. There seemed to be some association of obesity with cancer, though this association was clearly less dramatic than for cardiovascular disease. (Redrawn with permission from Manson *et al.*, 1995.)

$$\frac{\text{(death rate for group nurses with a specific BMI value)}}{\text{(death rate for the leanest group of nurses)}}$$

The authors of the study concluded that a weight gain of 10 kg or more (occurring after the age of 18) results in an increased rate of death by middle adulthood (Manson *et al.*, 1995).

## Measurement of Body Fat by Skinfold Measurements

The skinfold measurement test was introduced earlier in the chapter during the illustration of subcutaneous fat in the course of pregnancy. A more comprehensive picture of this test follows. Skinfold measurements are made by pinching the skin and pulling slightly to form a pinch with roughly parallel sides. The pinch contains two layers of skin and two layers of subcutaneous adipose tissue. A caliper is then applied to the skinfold. Care should be taken to take the reading at four seconds after applying the caliper to the skinfold. Care should be taken to place the caliper tips a distance of 1 cm distal to the finger and thumb. The dietician should also be sure to perform the pinch at the correct regions of the body.

### Selection of Sites on the Body for Skinfold Measurements

A number of different sites have been used to apply the pinch. The most commonly used site is over the triceps muscle of the arm. Hence, the site is called the "triceps skinfold" site. The measurement is taken halfway between two bones, the scapula and ulna. More specifically, the measurement is taken halfway between the *acromion process* of the scapula and the *olecranon process* of the ulna (Figure 7.9). A picture of a dietician taking a triceps skinfold measurement is shown in Figure 7.9.

One problem with the triceps skinfold test, and other single-site measurements, is that the results cannot be used to estimate the amount or percentage of body fat. Data from single-site measurements can only be used where data from one person (your patient) are compared with data taken from a large number of subjects. The result here is knowledge that your patient is fatter than 60%, for example, of the measured population. No useful equation has yet been devised for estimating total body fat from a *single* skinfold test.

Multiple skinfold tests can be used to estimate total body fat (Lee and Nieman, 1996, pp. 223–288). Two-site skinfold measurement involving the triceps and subscapular sites has proven to be useful. The test simply involves taking the triceps measurement and a measurement just below the lower border of the scapula, and adding the measurements. The result is a number having the unit of "millimeters." The sum is then input into a special table. Separate tables have been devised for use with boys (Figure 7.10A) and girls (Figure 7.10B). Use of the table gives an estimate of the percent of the child's body that is fat. A child with the indicated percentage of body fat (over 25% for boys; over 32% for girls) is considered obese.

## Measurement of Body Fat by Densitometry

Skinfold measurements reflect the thickness of layers of fat at specific points of the body. In contrast, the technique of **densitometry** is sensitive to the proportion of fat in the entire body. Density, whether measured by chemists, physicists, or dieticians always means mass/volume (grams/ml). The density of a human being is often measured by using underwater weighing. This technique requires a tub of water, a scale for measuring the subject's weight, and a special chair that is attached to the scale. The subject needs to have defecated and urinated shortly

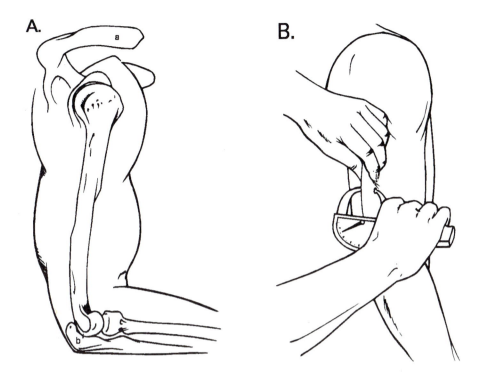

**FIGURE 7.9.** (A) Anthropometry by skinfold measurements. The triceps skinfold measurement is taken halfway between the lateral projection of the acromion process of the scapula (point a) and the inferior margin of the olecranon process of the ulna (point b). The ulna is one of the two bones of the forearm. (B) When taking the measurement, the subject's arm should hang loosely at the side. The dietician may mark the halfway point using a pen, then pinch the skinfold using the left hand, while holding the caliper with the right hand. The tips of the caliper should be placed about 1 cm down from the pinched area to prevent the pressure from the fingers from influencing the measured value. In very obese subjects, it may be difficult to raise the skin by pinching. In this case, the dietician should grasp the obese person's skin with both hands while a partner positions the caliper and takes a measurement. (Redrawn with permission from Lee and Nieman, 1996.)

before the measurement is taken. The subject then sits on the chair. The chair is lowered into the water, and the subject is completely submerged. The subject must exhale as much as possible and remain still for 5–10 seconds while the dietician records the weight of the underwater subject. The weight of the subject in air (not in water) is measured shortly before or after the underwater weight is taken. The results are then placed in the following formula (Brodie, 1988):

$$\text{Density} = \frac{[\text{weight in air}]}{[(\text{weight in air} - \text{weight in water}) / (\text{density of water})] - \text{RV } 100 \text{ ml}}$$

The term "RV" in the density formula means "residual volume." RV is the amount of air remaining in the lungs after maximal exhalation. The RV can be measured at the same time as the underwater weighing. A technique for measuring the air

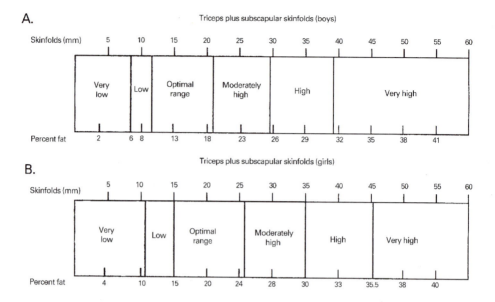

**FIGURE 7.10.** Standards for total body fat for (A) boys and (B) girls (ages 6–17). The graphs relate two-site skinfold measurements with percent of total body fat. This and similar tables were acquired by dissecting cadavers and measuring either the total amount of visible adipose tissue, or by extracting the body with ether and measuring the dissolved triglycerides. The cutoff points for very low body fat, high obesity, and very high obesity were developed by the life insurance industry (Metropolitan Life Insurance Company) by using height and weight data, and health data, of policyholders. (Redrawn by permission from Lohman, 1987.)

space in the lungs (with maximal exhalation) has been described (Wilmore *et al.,* 1980). In brief, the subject sits in the chair, exhales as much as possible, and is completely submerged in water and weighed. Then the subject inhales some pure oxygen from a bag. The subject inhales and exhales several times from this bag of oxygen in order to wash out most of the nitrogen or carbon dioxide that was in the lungs during the underwater weighing. Then the bag is taken away, and the total amount of nitrogen (or carbon dioxide) is measured using a special machine. The total amount of nitrogen (or carbon dioxide) gives a direct indication of RV. The value of "100" in the preceding formula is an estimate of the amount of intestinal gases in a typical subject.

## Units in Equations Used for Densitometry

In examining any equation involving multiple terms, one should make sure that the units are consistent. To see if the units are consistent, one should erase each term and, in its place, plug in the name of the unit. Replace "density" by (kg/ml). Replace "weight in air" by kg, "weight in water" by kg, density of water by (kg/ml), and "RV" by ml. The result will give this equation: kg/ml = kg/ml. Such a result for this type of exercise indicates that the equation makes sense.

*Equations Used for Densitometry*

The value for the density is then placed in the following formula to give the percentage of body fat. This formula was developed by W. E. Siri in the early 1950s:

$$\text{Percent body fat} = [495/\text{density}] - 450$$

This equation rests on the assumption that the density of fat is 0.9 g/cm³, and that of fat-free mass is 1.1 g/cm³. Fat-free mass is composed of water, protein, mineral, and carbohydrate. Most of the fat-free mass in the human body is water, which has a density of 1.0 g/cm³. About 73% of the fat-free mass is water. About 20% of the fat-free mass is protein (1.34 g/cm³). About 7% of the fat-free mass is mineral (3.00 g/cm³).

The Siri equation is somewhat less accurate when applied to persons with a greater (African-Americans) or lesser (people with osteoporosis) proportion of fat-free mass occurring as bone. Please note that, of the various components of fat-free mass, bone has the greatest density. Adjustments of the Siri equation have been suggested for use with African-Americans (Zillikens and Conway, 1990; Conway *et al.*, 1995), muscular weight trainers (Modlesky *et al.*, 1996), elderly persons (Durnin and Womersley, 1974), and obese persons (Deurenberg *et al.*, 1989a,b). The amount of fat in the human body varies to a surprising extent, according to a study that employed densitometry and the Siri equation (Durnin and Womersley, 1974).

## Assessment of Body Fat by Computed Tomography

Computed tomography (CT), as applied to nutrition research, was described by Kvist *et al.* (1988). The technique involves exposure to X-rays for about 5 seconds, with a resulting cross-sectional picture of the abdomen called a "slice." The picture represents a "slice" of the body, where the slice is about 12 mm thick. The abdominal slice is taken at the level of one of the lumbar vertebra. The results take the form of a picture where areas of variable densities can be interpreted as being subcutaneous fat, visceral fat, muscle, or bone.

A study of 57 obese women illustrates the use of CT for measuring subcutaneous fat and visceral fat, and the use of the data for predicting adverse risks to health (Zamboni *et al.*, 1992). Forty of the women were premenopausal (24–46 years old), while 17 were postmenopausal (52–68 years old). The results (Table 7.1) revealed that the younger obese women tended to have more subcutaneous fat and less visceral fat than the older obese women. The units of area that are used in CT are cm². The data suggest a trend in the aging process of obese women, where there occurs a progressive increase in visceral fat and a reduction in subcutaneous fat.

EXERCISE

What percent of fat was found to be subcutaneous and visceral fat (Table 7.1) in the study of young and old women? (See Zamboni *et al.*, 1992.)

**TABLE 7.1** Shifts in Location of Body Fat in Pre- and Postmenopausal Women

|  | Premenopausal | Postmenopausal |
|---|---|---|
| Age (years) | 35 ± 11 | 60 ± 8 |
| Body weight (kg) | 103 ± 18 | 96 ± 15 |
| BMI (kg/m$^2$) | 39 ± 6 | 37 ± 5 |
| Subcutaneous adipose tissue (cm$^2$) | 602 ± 150 | 500 ±113 |
| Visceral adipose tissue (cm$^2$) | 150 ± 75 | 205 ± 60 |

Source: Zamboni *et al.* (1992).

The main goal in studies that employ computed tomography is to measure visceral fat and thus determine which persons are at a greater health risk. A secondary goal is related to methodology, and here the goal is to determine which inexpensive technique (BMI or waist/hip ratio) gives results that are most similar to those from CT. A trend seems to be emerging. The trend is that persons with a (visceral fat)/(subcutaneous fat) ratio of over 0.4 have a distinct, increased health risk.

The use of computerized tomography to measure subcutaneous and visceral fat is illustrated in a small but detailed study by Conway *et al.* (1995) (Table 7.2). The study compared obese black women and obese white women. The terms "black" and "African-American" generally are used to indicate persons of West African descent, i.e., Angola, Senegal, and Gambia. The term "white" refers to persons of European descent. The ranges of ages and weights of both groups of women were similar. These, and other parameters, are listed in Table 7.2.

**TABLE 7.2** Search for Differences in Body Fat Distribution in Black and White Women

|  | Black women | White women |
|---|---|---|
| Age (years) | 35 ± 7 | 39 ± 6 |
| Weight (kg) | 109 ± 17 | 103 ± 24 |
| BMI | 40 ± 5 | 38 ± 8 |
| Waist/hip ratio | 0.86 ± 0.1 | 0.84 ±0.1 |
| % body fat (bioelectric impedance) | 53 ± 7 | 52 ± 5 |
| Subcutaneous fat (cm$^2$) | 449 ± 170 | 446 ± 152 |
| Visceral fat (cm$^2$) | 105 ± 25 | 160 ± 70 |
| Total cholesterol (m$M$) | 4.73 | 5.17 |
| HDL-cholesterol (m$M$) | 1.47 | 1.14 |
| Triglycerides (m$M$) | 0.93 | 1.60 |

Source: Conway *et al.* (1995).

EXERCISE

According to the CT measurements (Table 7.2), what percent of body fat of the obese women was subcutaneous and what percent was visceral?

A goal of the Conway *et al.* (1995) study was to assess the amount of fat that was subcutaneous and visceral in the abdomens of obese women, because of the emerging picture suggesting that visceral abdominal fat is associated with risk factors for cardiovascular disease. Risk factors for cardiovascular disease include high plasma cholesterol, high LDLs, low HDLs, and high plasma triglycerides. The researchers examined the blood picture of the obese women. As indicated in Table 7.2, the obese black women had more healthy blood values than the obese white women. An emerging picture from this, and related studies, is that in comparing black and white women of *equal obesity*, the black woman may be at less health risk because of the more favorable blood picture. For example, a study of 101 black and white children also revealed the previous trend, i.e., that blacks tend to have a lower proportion of visceral fat than whites (Goran *et al.*, 1997).

From these results, one might guess that the black population, overall, is at lesser risk for cardiovascular disease than whites. This notion is false because of the greater prevalence of obesity among blacks. Epidemiological studies have revealed that about one-third of white women are obese, while over half of all black women are obese.

EXERCISE

Studies of *premature infants* (1.2 kg body weight) revealed that they accumulated and retained protein at the same rate as *fetuses of the same postconception age*. However, the premature infants accumulated fat at a rate three times that of fetuses of the same age. Both groups gained total body weight at the same rate. Which methods would you use to estimate the accumulation of protein, water, and fat, over the course of time, in a premature infant? (See Reichman *et al.*, 1981.)

# SIGNALING PATHWAYS FOR REGULATING ADIPOCYTE FORMATION

The pathway of differentiation of pre-adipocytes to adipocytes is only now beginning to be detailed. Attention has focused on the transcription factors PPAR-γ and C/EBP. Transcription factors are proteins that bind to specific regions of DNA called **response elements**. The binding event provokes the activation of many genes and results in an increase in the rate of transcription. PPAR-γ appears to be active when it is bound by prostaglandin $J_2$ (PGJ$_2$). Knowledge of the pre-adipocyte maturation path is only fragmentary to date, but includes the following six-step sequence. PPAR is further detailed in the section on RAR, VDR, and THR in the Vitamin chapter.

**Step 1**. An excess of dietary energy somehow provokes the production of prostaglandin $J_2$ within the pre-adipocyte (Forman *et al.*, 1995). The biosynthetic pathway of PGJ$_2$ from arachidonic acid is outlined in Figure 7.11. Prostaglandins are detailed in the Essential Fatty Acid section.

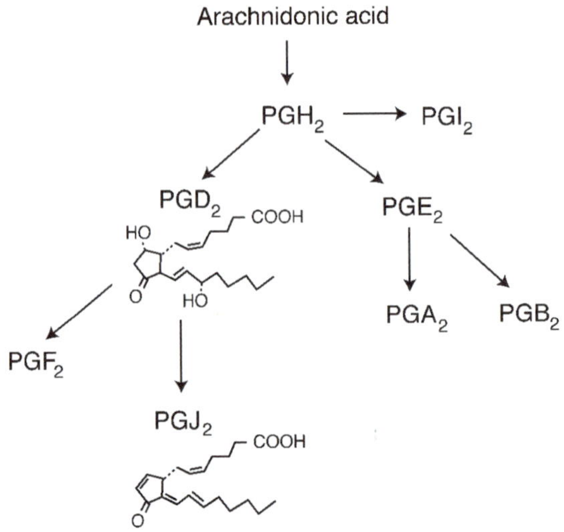

**FIGURE 7.11.**  The pathway for prostaglandin biosynthesis. Arachidonic acid is a fatty acid occurring in phospholipids in the plasma membrane of all cells. Hydrolysis of arachidonic acid allows it to be converted to prostaglandin $H_2$ and then to other prostaglandins, such as $PGD_2$ and $PGJ_2$. $PGJ_2$ can bind to a transcription factor called PPAR, and provoke the conversion of fibroblasts to adipocytes. It is not yet proven if $PGJ_2$ works this way in living animals. Fibroblasts are cells that are relatively undifferentiated. PPAR does not act alone as a transcription factor, but it must first bind to RXR. The PPAR/RXR complex of proteins is called a heterodimer. RXR can bind 9-*cis*-retinoic acid. It is likely that transcription activation by PPAR/RXR requires bound 9-*cis*-retinoic acid, for some genes, but not for other genes. (From Forman *et al.*, 1995.)

**Step 2**. $PGJ_2$ travels to the nucleus, where it binds to PPAR, followed by binding to RXR. The result is a complex composed of the hormone ($PGJ_2$) and the two proteins (PPAR and RXR). The particular family of receptor proteins that includes PPAR and RXR is water soluble and not membrane bound. Prostaglandin $J_2$ binds to an intracellular receptor (PPAR), in contrast to other types of prostaglandins that bind to extracellular receptors.

**Step 3**. The PPAR/RXR complex binds to a special response element with the sequence AGGTCANAGGTCA. The sequence AGGTCA is repeated twice here and is separated by a single nucleotide (A, T, G, or C). The response element always occurs as double-stranded DNA, but the sequence of only one strand is shown here. The AGGTCA at the left-hand side of AGGTCANAGGTCA binds PPAR (of the PPAR/RXR complex), while the AGGTCA to the right binds RXR (of the PPAR/RXR complex). PPAR alone cannot bind to the response element. PPAR must occur as the PPAR/RXR complex for binding to occur (Forman *et al.*, 1995; Spiegelman and Flier, 1996).

**Step 4**. AGGTCANAGGTCA resides near dozens of genes that code for proteins that are unique or necessary to adipocytes. The binding of the PPAR/RXR complex to these response elements in the cell activates the nearby genes — genes that are just "downstream" of the response element — and provokes the conversion of the pre-adipocyte to an adipocyte.

**Step 5.** C/EBP is a transcription factor of 42 kDa. It binds to a response element with the structure CCAAT. C/EBP forms dimers with itself to generate C/EBP–C/EBP. Activation by PPAR-γ is thought to be a primary event in adipocyte formation, with activation by C/EBP being secondary. Evidence suggests that both of these transcription factors act synergistically with each other.

**Step 6.** The changes occurring in adipocyte formation include the synthesis and deposition of collagen fibers at the outside surface of the cell, and the synthesis of GLUT-4 transporters, PEPCK, leptin, the insulin receptor, acetyl-CoA carboxylase, and citrate lyase (Smas and Sul, 1995). Adipocytes make a special protein called perilipin, which binds to the outside of lipid droplets. The existence of this protein might be compared with apolipoproteins, which also coat the outside of lipid droplets. A simple assay used to assess conversion of a pre-adipocyte, such as a fibroblast, to an adipocyte is staining with a red dye that binds to lipid droplets (Forman *et al.*, 1995).

EXERCISE

Genes include the regions of DNA that code for a particular protein or polypeptide, as well as regions of DNA used to control the rate of transcription. Please find the binding site for a transcription factor on this region of the leptin gene:

     5'-ACGTAGAAGCTTGAAATGAGGTAAAAGGTCAGAGTCCAAGCTT-3'

(See Hollenberg *et al.*, 1997.)

EXERCISE

A drug is available which binds to PPAR-γ and provokes the synthesis of adipocytes. Do you think that this drug would make diabetes better or worse? Hint: Ask yourself if you think the drug would result in a decline or increase in plasma glucose levels. (See Spiegelman and Flier, 1996.)

## GENETIC FACTORS IN OBESITY

### Genetic Factors in Human Obesity

The question whether some humans are genetically obese may be divided into two parts: (1) does genetics cause an obese person to deposit a greater amount of food energy into adipose tissue; and (2) can genetics cause an obese person to have an increased desire to eat food. The first question was addressed in careful studies by Bouchard and co-workers (Bouchard and Tremblay, 1997). These studies involved twelve pairs of twins. All of the subjects were males, and were fed an excess of energy (4.18 extra megajoules/day) for 100 days. The average gain in weight at the end of the study was 8 kg. The weight gain for one brother appears on the *y*-axis, while that for the twin brother is on the *x*-axis (Figure 7.12). The results show, for example, that Brothers W both gained 7 kg, that Brothers X each gained about 9 kg, and that Brothers Y both gained about 12 kg. The results suggest a genetic component in weight gain. On the other hand, Brothers Z gained dissimilar amounts of weight.

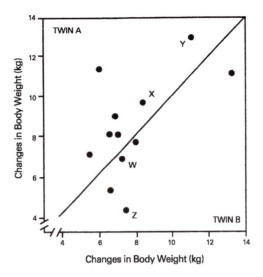

**FIGURE 7.12.** Changes in body weight after a 100-day overfeeding study. Each point represents two persons (twin brothers). Twelve pairs of twins were studied. The weight increase for one brother may be read from the Y-axis, while that for the other brother is on the X-axis. (Redrawn by permission from Bouchard and Tremblay, 1997.)

Bouchard and co-workers also addressed the question of weight loss. Seven pairs of twins exercised on cycle ergometers (exercycles) during a 93-day study. The exercise level was adjusted to an extra energy expenditure of 4.18 MJ/day for each subject. The subjects were fed a constant diet over the course of the study. The average loss in body weight, after the study, was 5 kg. The reduction in subcutaneous fat and visceral fat was measured by computed tomography. Thus, the amount of fat is measured in square centimeters (a cross-sectional picture). The results of the visceral fat measurements reveal, for most sets of twins, a striking similarity in fat loss (Figure 7.13). The results support the notion that some dieters have an unusual difficulty in losing weight because of genetic factors.

It should be noted that human genetics is exquisitely complex. Any person who is fat may be obese because of several, not just one, genetic changes in the body. In addition, the fact that two brothers are twins does not guarantee that they will be genetically identical. A variety of studies have shown that 40–70% of the features of obesity, in humans, are inherited (e.g., Comuzzie and Allison, 1998).

## Genetic Factors in Mouse Obesity

Several strains of mice have been discovered that are naturally genetically obese. The genetic defects have been traced to specific genes and to changes in the corresponding polypeptides. These polypeptides occur in various tissues in the body, but one thing they have in common is that they all act in the hypothalamus.

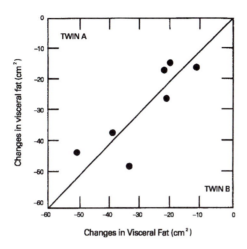

**FIGURE 7.13.** Changes in body weight after a 93-day negative energy study. Each point represents two persons (twin brothers). Exercise on a cycle ergometer was used to provoke negative energy balance. Seven pairs of twins were studied. Visceral fat was measured using computed tomography. The results showed a striking similarity of visceral fat loss among twins. Of interest to those interested in exercise nutrition is the finding that the subjects oxidized more lipids than carbohydrates after the 93-day study, as measured by RQ tests. Another observation is that some subjects (with exercise) tended to be high lipid oxidizers, while others were high carbohydrate oxidizers. (Redrawn by permission from Bouchard and Tremblay, 1997.)

Thus, an appreciation of the genetic factors of mouse obesity requires some background in the physiology of the hypothalamus.

## Physiology of the Hypothalamus

The hypothalamus has long been suspected as being an important regulator of feeding behavior. Work during the 1950s revealed that physical damage inflicted to parts of the hypothalamus result in increases or decreases in food intake and body weight. A cross-section of the hypothalamus of the rat is shown in Figure 7.14.

The hypothalamus contains regions called the **ventromedial hypothalamus** and the **ventrolateral hypothalamus**. The following commentary serves to emphasize the importance of the hypothalamus in the control of eating. Lesions (intentional damage) in the **ventromedial** region cause overeating and obesity, and provoke a decline in glucagon secretion. The **ventromedial** region is called the *satiety center* of the brain. The normal function of the ventromedial region is to inhibit the activity of the lateral region. Electrical stimulation of the **ventromedial** region provokes increased glucagon secretion and an increase in the activity of glycogen phosphorylase (and thus glycogen breakdown) in the liver. This effect is mediated by nerves leading to the liver. Lesions in the **ventrolateral** region cause less eating and loss in weight. The **ventrolateral** region is called the *feeding center*

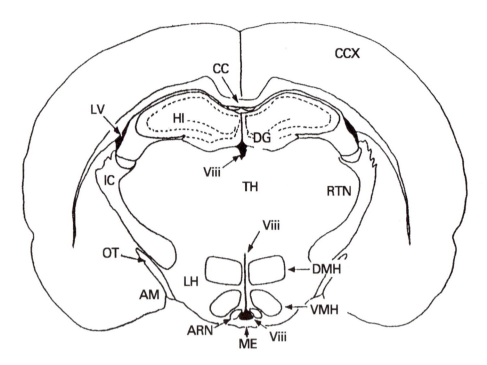

**FIGURE 7.14.** Cross-section of the hypothalamus of the rat and regions used to trigger hunger and satiety. The hypothalamus contains regions called the ventromedial hypothalamus (VMH) and the ventrolateral hypothalamus (LH). Lesions in the ventromedial region cause overeating and obesity. Lesions in the ventrolateral region cause less eating and loss in weight. The arcuate nucleus (ARN) of the hypothalamus is the site of synthesis of melanocortin hormones, where these hormones include POMC (pro-opiomelanocortin), ACTH, MSH, and LPH. POMC is a 265-amino-acid polypeptide that is the precursor of ACTH (39 amino acids), MSH (13 amino acids), and LPH (39 amino acids). The arcuate nucleus is also the site of synthesis of neuropeptide Y. Neurones extend from the arcuate nucleus (site of NPY synthesis) and convey NPY to the dorsomedial hypothalamic nucleus (DMH) and paraventricular nucleus (not shown). The DMH is also involved in melanocortin metabolism, since MC4 receptors (which bind melanocortin) occur in the ventromedial nucleus (VMH). The VMH and LH are connected via the DMH. Other structures are the corpus callosum (CC), cerebral cortex (CCX), hippocampus (HI), and optic tract (T). (Redrawn by permission from Schwartz *et al.*, 1992.)

of the brain. The satiety center acts by inhibiting the activity of the feeding center. Electric stimulation of the **ventrolateral** region of the hypothalamus provokes an increase in activity of liver glycogen synthase (Rohner-Jeanrenaud, 1995; Lynch *et al.*, 1997).

   The study of obesity has been aided by the occurrence of certain strains of mice that are genetically obese. These mice arose spontaneously in various animal colonies over the past few decades. Five separate strains have been maintained and bred. Their names and abbreviations are: (1) obese (ob), (2) diabetes (db), (3) fat (fat), (4) agouti yellow ($A^Y$), and (5) tubby (tub).

The names of these strains are used to refer to the strain of mouse, to the genetic makeup, to the specific gene that may be affected, and to the protein (if any) made by the gene. In recent years, the specific genes that are responsible for obesity have been identified for these strains. The *obese* gene (ob gene) codes for leptin, a hormone that circulates in the bloodstream. The hormone, in humans, is 116 amino acids long (Considine *et al.*, 1995). The *diabetes* gene (db gene) codes for the leptin receptor, i.e., for a membrane-bound protein. The protein is rather large and consists of 1178 amino acids (Caro *et al.*, 1996).

## Obese Gene and Leptin

The obese gene, or the "ob gene," codes for a hormone called **leptin**. A lack of leptin provokes obesity, as revealed by the existence of the obese strain of mouse. An increased amount of leptin in the bloodstream provokes obesity, as revealed by studies in rats (Figure 7.15) (Chen *et al.*, 1996). Plasma leptin levels were raised to the indicated concentrations by infecting rats with a virus that synthesizes leptin (Figure 7.15A; filled squares). Plasma leptin levels in control rats (Figure 7.15A; filled triangles) were also measured. The food intake (Figure 7.15B) and body weight (Figure 7.15C) in the two groups of rats was also measured. The results demonstrated that the high-leptin rats had a decreased food intake and decreased body weight, when compared with the control rats.

Separate studies were designed to test if the lack of weight gain in the high-leptin rats was solely due to the lower food intake, or was also due to other factors (Chen *et al.*, 1996). The food intake of the high-leptin rats was carefully measured on a daily basis. A separate group of normal rats was then fed the same amount of food (same amount as the high-leptin rats). These normal rats gained weight at the exact rate (lowered rate) as the high-leptin rats. This suggests that the effect of leptin is mainly on food intake — and possibly solely on food intake. This technique of food-weighing and feeding is called **pair feeding** (Appendix A). The researchers went a step further, and beyond the simple measurement of body weight. They performed surgery and discovered that the pair-fed group of normal rats still had substantial fat stores, while the high-leptin rats contained little or no adipose tissue. This means that the influence of leptin is on food intake, but also on other physiological events in the body.

Leptin is synthesized by adipocytes and is secreted into the bloodstream. Secretion is proportional to the amount of triglycerides stored in the adipocytes, that is, leptin secretion is proportional to the amount of body fat. Body fat (percentage of body weight that is fat) was estimated using bioelectric impedance. Serum leptin levels were measured in a study of 108 normal-weight subjects and 71 obese subjects (Figure 7.16) (Considine *et al.*, 1996). The average serum leptin concentration in the normal-weight subjects was about 8.0 ng/ml. The figure reveals that, once the percent body fat increases beyond 25%, the leptin levels tend to increase by some threefold to tenfold.

EXERCISE

The data in Figure 7.16 reveal an increase in serum leptin in obese persons. These data could mean an increase in leptin production (by fat tissue), or a decline in leptin uptake and degradation (by other tissues). Please devise an experiment for assessing changes in the rate of leptin synthesis. Please design a test for measuring changes in leptin uptake (uptake from blood).

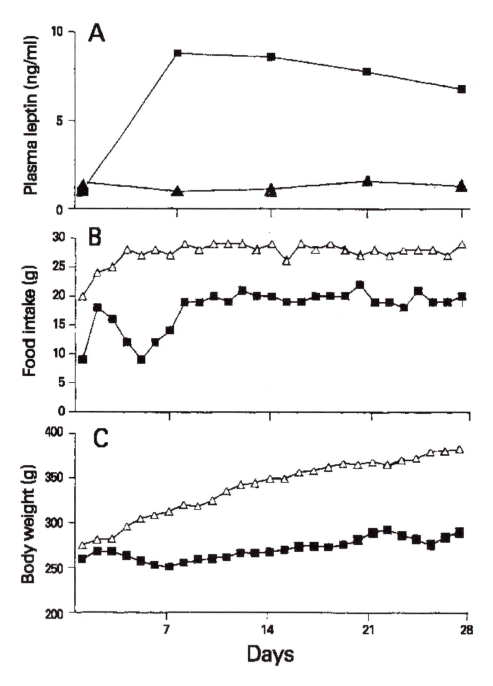

**FIGURE 7.15.** Measurements of rat biochemistry and physiology during 28 days of treatment with leptin. (A) Plasma leptin with administration of a virus that synthesizes leptin inside of rat liver cells (squares), or with no virus (triangles). (B) Food intake (grams/day) where rats were treated with the virus (squares) or with no virus (triangles). (C) Body weight where rats were treated with the virus (squares) or with no virus (triangles). The results showed that the leptin-treated rats ate less food and gained less weight than the nontreated rats. (Redrawn by permission from Chen *et al.*, 1996.)

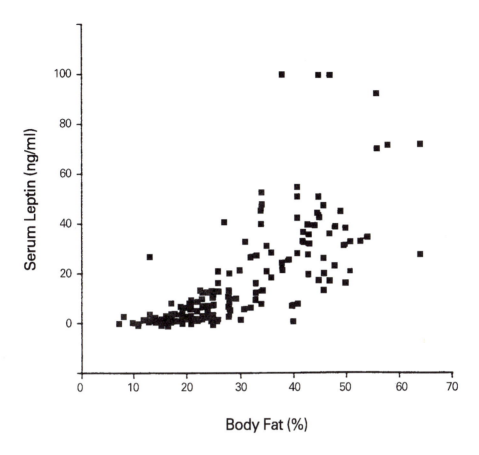

**FIGURE 7.16.** Plot of serum leptin versus percent body fat in 275 adults. The percentage of the body that consisted of fat was measured using bioelectric impedance. The results showed that fatter persons tended to have higher serum leptin levels. Separate work showed that obese humans had higher levels of mRNA coding for leptin than normal weight subjects, suggesting that higher serum leptin is caused by its higher rate of synthesis, rather than a lower rate of uptake by various tissues and degradation. (Drawn with permission from Considine *et al.*, (1996).)

## Summary of the Physiology of Adipose Tissue Regulation

An emerging picture suggests that the following sequence of events normally contributes to the regulation of food intake and storage:

1. Increase in size or number of adipocytes.
2. Increased expression of ob gene in adipocytes, and increased secretion of leptin by white adipose tissue into the bloodstream.
3. Increased levels of plasma leptin.
4. Increased binding of leptin to receptors (leptin receptor) in the hypothalamus.

5. Reduced synthesis of neuropeptide Y in nerves in one region of the hypothalamus (the arcuate nucleus of the hypothalamus). Note that nerves extend from the arcuate nucleus to another region of the hypothalamus (paraventricular and dorsomedial nuclei of the hypothalamus) (Wang *et al.*, 1997).

6. Decrease in eating and loss in body weight.

## Diabetes Gene and the Leptin Receptor

The term "diabetes gene," or "db gene," is only the name of a gene that is defective in a strain of genetically obese mice, and does not imply that human diabetics have problems with this gene. The db gene codes for a membrane-bound protein. This protein is the leptin receptor. The protein consists of 1178 amino acids where the extracellular region is 840 amino acids long, the transmembrane section is 34 amino acids, and the cytosolic part is 304 amino acids. In the normal mouse, the protein crosses through the plasma membrane and contains extra- and intracellular regions. But in the mutant mouse, the protein lacks most of the intracellular region. The protein is said to be "truncated." A specific change in the gene was traced to this defect. A guanine residue was mutated (changed) to a residue of thymine (Chen *et al.*, 1996). The hypothalamus is the main site of expression for the leptin receptor in the body, but the receptor also occurs elsewhere in the brain, and in the liver, lungs, heart, kidney, testes, and adipose tissue (Caro *et al.*, 1996). Interest is focusing on the possibility that obese humans have leptin receptors with an altered structure.

## Agouti Gene and Agouti Protein

The agouti gene codes for a 131-amino-acid protein that normally influences hair color in rats. Agouti mice are distinguished in that their fur is yellow, rather than white, and by the fact that they are obese. In the agouti mouse, the agouti protein is overproduced, and the overproduced protein greatly inhibits the effects of a hormone called **melanocortin** or **melanocyte-stimulating hormone** (MSH). Melanocortin is a polypeptide with the structure shown below (Adan *et al.*, 1994). Those with a background in endocrinology might recognize that this sequence is contained within the longer sequence of the hormone ACTH (ACTH is adrenocorticotropin, a 39-amino-acid hormone, which also serves as the precursor of melanocortin):

$$H_2N–Ser–Tyr–Ser–Met–Glu–His–Phe–Arg–Trp–Gly–Lys–Pro–Val–COOH$$

(Melanocortin)

The agouti protein is secreted by specialized cells and is bound by adjacent melanocytes. These specialized cells and the melanocytes reside near each other in the skin. Melanocytes are cells that create skin pigment, i.e., eumelanin, which is black, and pheomelanin, which is yellow. The agouti protein binds to a receptor on the melanocyte, called the **melanocortin receptor**, and instructs the melanocytes to switch from making the black to the yellow pigment.

## Melanocortin Receptors

The melanocortin receptor actually is a family of receptors with similar coding sequences in the DNA, and similar amino acid sequences. Five melanocortin receptors have been found: MC1 (melanocortin-1), MC2, MC3, MC4, and MC5. MC1 occurs in the skin, where it resides in the membrane of the melanocytes — cells used to lend color to the skin. MC2 occurs in the adrenal gland, a gland situated just above the kidney. MC3 occurs in the hypothalamus, a region of the brain used for regulating food intake. MC4 is distributed throughout the brain, while MC5 occurs throughout the body. One might be especially interested in the interactions between the agouti protein and its receptors in brain (MC3 and MC4), since the brain has a proven role in hunger.

### The Question of Which Melanocortin Receptors Are Bound by Agouti Protein

In the agouti strain of mouse, the agouti gene is changed in a way that causes it to be overexpressed in hair cells, and also expressed in cells throughout the body — not just in hair cells (Zemel *et al.*, 1995). This means that the agouti protein has acquired the ability to contact and bind melanocortin receptors throughout the body. It is believed that, in the agouti mouse, the agouti protein binds to the melanocortin receptor of the hypothalamus and of skeletal muscle. Since skeletal muscle is the major target of insulin, it is proposed that insulin resistance (diabetes) in the agouti mouse results from the agouti protein's action on muscle.

The following commentary concerns the obesity that occurs in the agouti mouse. MC1 and MC4 are unique among all the melanocortin receptors, as they bind the agouti protein very tightly. Using standard genetic techniques, a mouse lacking the melanocortin-4 receptor (MC4 receptor) was created (Huszar *et al.*, 1997). The genetic technique produced what is known as a "knock-out mouse," since the gene was knocked out. The mouse lived and thrived.

Would you expect the mouse to be obese or to be abnormally thin? The lack of the receptor would prevent any melanocortin from relaying signals to the cell. The scenario that results would be expected to be the same as that in the agouti mouse, where the agouti protein circulates throughout the entire body and actively inhibits all of the melanocortin receptors. The mutant mice lacking in MC4 proved to overeat and to become obese.

The interpretation arising from this is as follows. Normally, melanocortin is produced in the hypothalamus, specifically in the **arcuate nucleus** of the hypothalamus. The melanocortin travels to the **ventromedial** region of the hypothalamus, where it binds to MC4 receptors, which produce a signal that *reduces eating*. In the agouti mouse, the agouti protein travels from the skin, through the bloodstream, and to the hypothalamus, where it inhibits the action of melanocortin. The result is *increased eating* (Mountjoy and Wong, 1997).

Would you expect the knock-out mouse to be white or to be yellow? The MC4 receptor has absolutely no relevance to skin color. Only the MC1 receptor is relevant to skin color. The knock-out mice were white.

*Summary of Melanocortin and Agouti Protein and Their Receptors*

To summarize, melanocortin is the normal **agonist** of the MC1 to MC5 receptors. Normally, agouti protein is the **antagonist** of MC1, but, in the agouti strain of mouse, where agouti is distributed throughout the body, agouti protein is a potential **antagonist** of all the melanocortin receptors (receptors MC1–MC5). To review some terminology used by all endocrinologists, an **agonist** binds to a receptor and provokes it to relay a signal, while an **antagonist** binds to a receptor and prevents it from relaying its signal. A number of drug companies are developing agonists for MC4, with the hope that it can be used as a drug to prevent obesity (Campfield *et al.*, 1998).

EXERCISE

An agonist of melanocortin called "MTII" was injected into the brain of a mouse. Would you expect the mouse to eat more food or less food? (See Fan *et al.*, 1997.)

Let us pause to review the big picture. Normally, starvation results in low plasma leptin, while obesity results in increased plasma leptin. These two extremes in hormone concentrations provoke different responses by the hypothalamus. Starvation in the normal mouse results in low plasma leptin, where low levels of leptin binding to the leptin receptor (in the hypothalamus) provoke a signal in the hypothalamus that results in increases in neuropeptide Y. The hypothalamus is a major site of production of the hormone, neuropeptide Y. Neuropeptide Y contains 36 amino acids and has the following structure (the one-letter abbreviations for the amino acids are used):

```
H₂N-Tyr-Pro-Ser-Lys-Pro-Asp-Asn-Pro-Gly-Glu-Asp-
  Ala-Pro-Ala-Glu-Asp-Leu-Ala-Arg-Tyr-Tyr-Ser-Ala-Leu-
    Arg-His-Tyr-Ile-Asn-Leu-Ile-Thr-Arg-Gln-Arg-Tyr-COOH
```

The increase in neuropeptide Y results in an increase in food intake. Obesity in the normal mouse results in an increase in plasma leptin, where increased levels of leptin at the leptin receptor (in the hypothalamus) provokes an increase in melanocortin (MSH). The increase in MSH in the brain provokes an increase in stimulation of the melanocortin-4 receptor in the brain, which results in a decline in food intake. MSH is the hormone that binds to melanocortin receptors, where this interaction can be antagonized if agouti protein is allowed to contact the receptor.

Neuropeptide Y may be selectively increased in two different regions of the hypothalamus. In mutant strains of mice that eat too much (the agouti mouse and the MC4 receptor knock-out mouse) neuropeptide Y was found to be produced at *increased levels* in the **dorsal medial** hypothalamus. In another strain of mouse that eats too much (the ob mouse), neuropeptide Y levels were found to be *higher than normal* in the **arcuate nucleus** of the hypothalamus (Spiegelman and Flier, 1996).

## Proposed Feedback Loop

The following illustrates a feedback loop. This loop involves leptin, the leptin receptor, NPY, and insulin. It has been proposed to have a role in regulating hunger and satiety. The onset of starvation provokes a decline to low plasma leptin levels. The low leptin levels, in turn, provoke an *increase in NPY synthesis* in the hypothalamus, where the increased NPY provokes eating and weight gain. Insulin completes this particular feedback loop. Insulin levels rise in the bloodstream shortly after a meal. One of insulin's targets is the brain. Insulin acts at the hypothalamus to *decrease the synthesis of NPY*. More specifically, the insulin receptors occur in the arcuate nucleus of the hypothalamus. The result is a decrease in feeding after a meal of food has been consumed (Figlewicz *et al.*, 1996; Woods *et al.*, 1998).

## Fat Gene and Carboxypeptidase E

The fat gene is defective in the fat strain of genetically obese mice. The fat gene codes for an enzyme (carboxypeptidase E) that is used for the processing and maturation of hormones. The mutation of this gene in the fat strain of mice results in conversion of a serine residue to proline, and the resulting complete loss in carboxypeptidase E activity (Fricker *et al.*, 1996). The story of the fat gene seems more or less fragmentary at this time.

## Tubby Gene

The tubby gene codes for a protein that occurs in the brain, and not much in other organs. The tubby gene in normal mice codes for a protein that is 505 amino acids long. The tubby gene is also expressed in the mutant mouse, but here it codes for a protein that is only 485 amino acids long. The gene is expressed in the hypothalamus, a region of the brain, but also in the eye and testis. The purpose of the tubby gene is not clear at this time.

## SUMMARY

Obesity is a major health problem because of its prevalence and because it is a risk factor for cardiovascular disease and diabetes (NIDDM). The measurement and treatment obesity requires a knowledge of the various classifications of fat. Biochemically and genetically, the two types of fat that are most different are white fat and brown fat. However, only white fat is an obesity issue. White fat was shown to be classified according to its distribution or location in the body. These classes are lower body and upper body fat, and subcutaneous and visceral fat. Various techniques for measuring body fat were presented. The commentary was embellished with data acquired from studies of nurses, pregnant women, obese children, and African-Americans. A number of genes and proteins that may control body weight in people were discovered in the past few years. These developments were made possible because of the existence of mutant strains of mice that are genetically obese, and because of the availability of a continuously evolving recipe book of techniques in molecular biology. Most of the attention has focused on the genes coding for leptin and the leptin receptor.

## REFERENCES

Adan, R. A., Cone, R., Burbach, J., and Gispen, W. (1994). Differential effects of melanocortin peptides on neural melanocortin receptors. *Mol. Pharmacol.* **46**, 1182–1190.

Anderson, A. E. (1995). Eating disorders in males. *In* "Eating Disorders and Obesity" (K. D. Brownell and C. G. Fairburn, eds.), pp. 177–182. Guilford Press, New York.

Bjorntorp, P. (1985). Regional patterns of fat distribution. *Ann. Intern. Med.* **103**, 994–995.

Bouchard, C., and Tremblay, A. (1997). Genetic influences on the response of body fat and fat distribution to positive and negative energy balances in human identical twins. *J. Nutr.* **127**, 943S–947S.

Braun, B., Zimmermann, M. B., and Kretchmer, N. (1995). Effects of exercise intensity on insulin sensitivity in women with non-insulin-dependent diabetes mellitus. *J. Appl. Physiol.* **78**, 300–306.

Brodie, D. A. (1988). Techniques of measurement of body composition part I. *Sports Med.* **5**, 11–40.

Campbell, P. J., and Carlson, M. G. (1993). Impact of obesity on insulin action in NIDDM. *Diabetes* **42**, 405–410.

Campfield, L., Smith, F., and Burn, P. (1998). Strategies and potential molecular targets for obesity treatment. *Science* **280**, 1383–1387.

Caro, J. F., Sinha, M., Kolaczynski, J., Zhang, P., and Considine, R. V. (1996). Leptin: The tale of an obesity gene. *Diabetes* **45**, 1455–1462.

Chen, G., Koyama, K., Yuan, X., Lee, Y., Zhou, Y.-T., O'Doherty, R., Newgard, C. B., and Unger, R. H. (1996). Disappearance of body fat in normal rats induced by adenovirus-mediated leptin gene therapy. *Proc. Natl. Acad. Sci. U.S.A.* **93**, 14795–14799.

Comuzzie, A. G., and Allison, D. B. (1998). The search for human obesity genes. *Science* **280**, 1374–1377.

Considine, R. V., Considine, E. L., Williams, C., Nyce, M., Magosin, S., Bauer, T., Rosato, E., Colberg, J., and Caro, J. F. (1995). Evidence against either a premature stop codon or the absence of obese gene mRNA in human obesity. *J. Clin. Invest.* **95**, 2986–2988.

Considine, R. V., Sinha, M., Heiman, M., Kriauciunas, A., Stephens, T., Nyce, M., Ohannesian, J., Marco, C., McKee, L. J., Bauer, T., and Caro, J. F. (1996). Serum immunoreactive-leptin concentrations in normal-weight and obese humans. *N. Engl. J. Med.* **334**, 292–295.

Conway, J. M., Yanovski, S. Z., Avila, N. A., and Hubbard, V. (1995). Visceral adipose tissue differences in black and white women. *Am. J. Clin. Nutr.* **61**, 765–771.

Craig, L. S. (1969). Anthropometric determinants of obesity. *In* "Obesity" (N. L. Wilson, ed.). F. A. Davis, Philadelphia.

Craig, L. S, and Bayer, L. M. (1967). Androgynic phenotypes in obese women. *Am. J. Phys. Anthrop.* **26**, 23–34.

Deurenberg, P., Kooy, K., and Leenen, R. (1989a). Differences in body impedance when measured with different instruments. *Eur. J. Clin. Nutr.* **43**, 885–886.

Deurenberg, P., Leenen, R., Kooy, K., and Hautvast, J. (1989b). In obese subjects the body fat percentage calculated with Siri's formula is an overestimation. *Eur. J. Clin. Nutr.* **43**, 569–575.

Durnin, J. V., and Womersley, J. (1974). Body fat assessed from total body density and its estimation from skinfold thickness. *Brit. J. Nutr.* **32**, 77–97.

Fan, W., Boston, B., Kesterson, R., Hruby, V., and Cone, R. D. (1997). Role of melanocortin-ergic neurons in feeding and the agouti obesity syndrome. *Nature* **385**, 165–168.

Figlewicz, D. P., Schwarz, M., Seeley, R., Chavez, M., Baskin, D., Woods, S., and Porte, D. (1996). Endocrine regulation of food intake and body weight. *J. Lab. Clin. Med.* **127**, 328–332.

Flatt, J. P. (1995). Body composition, respiratory quotient, and weight maintenance. *Am. J. Clin. Nutr.* **62**, 1107S–1117S.

Forman, B. M., Tontonoz, P., Chen, J., Brun, R., Spiegelman, B. M., and Evans, R. M. (1995). 15-Deoxy-prostaglandin $J_2$ is a ligand for the adipocyte determination factor PPARγ. *Cell* **83**, 803–812.

Fricker, L. D., Berman, Y. L., Leiter, E. H., and Devi, L. A. (1996). Carboxypeptidase E activity is deficient in mice with the fat mutation. *J. Biol. Chem.* **271**, 30619–30624.

Goran, M. I., Naby, T., Treuth, M., Trowbridge, C., Dezenberg, C., McGloin, A., and Glover, B. A. (1997). Visceral fat in white and African American prepubertal children. *Am. J. Clin. Nutr.* **65**, 1703–1708.

Hervey, G. R. (1959). The effects of lesions in the hypothalamus in parabiotic rats. *J. Physiol. (London)* **145**, 336–352.

Hoek, H. W. (1995). The distribution of eating disorders. *In* "Eating Disorders and Obesity" (K. D. Brownell and C. G. Fairburn, eds.), pp. 207–211. Guilford Press, New York.

Hollenberg, A. N., Susulic, V. S., Madura, J. P., Ahang, B., Moller, D. E., Tontonoz, P., Sarraf, P., Spiegelman, B. M., and Lowell, B. B. (1997). Functional antagonism between CCAAT/enhancer binding protein-α and peroxisome proliferator-activated receptor-γ on the leptin promoter. *J. Biol. Chem.* **272**, 5283–5290.

Hughes, V. A., Fiatarone, M. A., Fielding, R. A., Kahn, B., Merrara, C., Shepard, P., Fisher, E., Wolfe, R. R., Elai, D., and Evans, W. J. (1993). Exercise increases muscle GLUT-4 levels and insulin action in subjects with impaired glucose tolerance. *Am. J. Physiol.* **264**, E855–E862.

Huszar, D., Lynch, C. A., Fairchild-Huntress, V., Dumore, J., Fang, Q., Berkemeier, L., Gu, W., Kesterson, R., Boston, B. A., Cone, R. D., Smith, F. J., Campfield, L., Burn, P., and Lee, F. (1997). Targeted disruption of the melanocortin-4 receptor results in obesity in mice. *Cell* **88**, 131–141.

Kennedy, G. R. (1950). The hypothalamic control of food intake in rats. *Proc. R. Soc. (London), Ser. B* **137**, 535–549.

Krotkiewski, M., Bjorntorp, P., Sjostrom, L., and Smith, U. (1983). Impact of obesity on metabolism in men and women. *J. Clin. Invest.* **72**, 1150–1162.

Kvist, H., Chowdhury, B., Grangard, U., Tylen, U., and Sjostrom, L. (1988). Total and visceral adipose-tissue volumes derived from measurements with computed tomography in adult men and women. *Am. J. Clin. Nutr.* **48**, 1351–1361.

Lee, R. D., and Nieman, D. C. (1996). "Nutritional Assessment," 2nd ed., pp. 223–288. Mosby Year Book, St. Louis.

Leibel, R. L., Rosenbaum, M., and Hirsch, J. (1995). Changes in energy expenditure resulting from altered body weight. *N. Engl. J. Med.* **332**, 621–628.

Lohman, T. G. (1987). The use of skinfolds to estimate total body fatness on children and youth. *J. Phys. Educ., Recr. Dance* **58**, 98–102.

Lynch, J., Woinar, M., and Lang, C. (1997). Ventromedial hypothalamic lesions impair glucoregulation in response to endotoxin. *Am. J. Physiol.* **272**, R1525–R1531.

Manson, J. E., Willett, W. C., Stampfer, M. J., Colditz, G. A., Hunter, D. J., Hankinson, S. E., Hennekens, C. H., and Speizer, F. E. (1995). Body weight and mortality among women. *N. Engl. J. Med.* **333**, 677–685.

Maurer, D., and Sobel, J. (1995). "The Medicalization and Demedicalization of Obesity." A. de Gruyter, New York.

Modlesky, C. M., Cureton, K., Lewis, R. D., Prior, B. M., Sloniger, M. A., and Rowe, D. A. (1996). Density of the fat-free mass and estimates of body composition in male weight trainers. *J. Appl. Physiol.* **80**, 2085–2096.

Mountjoy, K., and Wong, J. (1997). Obesity, diabetes and functions for proopiomelanocortin-derived peptides. *Mol. Cell. Endocrinol.* **128**, 171–177.

Reichman, B., Chessex, P., Putet, G., Verellen, G., Smith, J. M., Heim, T., and Swyer, P. R. (1981). Diet, fat accretion, and growth in premature infants. *N. Engl. J. Med.* **305**, 1495–1500.

Rohner-Jeanrenaud, F. (1995). A neuroendocrine reappraisal of the dual-centre hypothesis. *Int. J. Obesity* **19**, 517–534.

Schneider, N. B., and Ruderman, S. H. (1992). Diabetes, exercise, and atherosclerosis. *Diab. Care* **15**, 1787–1793.

Schwartz, M., Sipols, A. J., Marks, J., Sanacora, G., White, J., Scheurink, A., Kahn, S., Baskin, D., Woods, S. C., Figlewicz, D., and Porte, D. (1992). Inhibition of hypothalamic neuropeptide Y gene expression by insulin. *Endocrinology* **130**, 3608–3616.

Schwartz, M., Seeley, R. J., Campfield, L., Burn, P., and Baskin, D. G. (1996). Identification of targets of leptin action in rat hypothalamus. *J. Clin. Invest.* **98**, 1101–1106.

Sjostrom, L. (1980). Fat cells and body weight. *In* "Obesity" (A. J. Stunkard, ed.), pp. 72–100. Saunders, Philadelphia.

Smas, C. M., and Sul, H. S. (1995). Control of adipocyte differentiation. *Biochem. J.* **309**, 697–710.

Spiegelman, B. M., and Flier, J. S. (1996). Adipogenesis and obesity: Rounding out the big picture. *Cell* **87**, 377–389.

Taggart, N. R., Holliday, R. M., Billewicz, W. Z., Hytten, F. E., and Thomson, A. M. (1967). Changes in skinfolds during pregnancy. *Brit. J. Nutr.* **21**, 439–451.

Trayhurn, P. (1996). Uncoupling protein in brown adipose tissue. *Biochem. Soc. Trans.* **24**, 402–406.

Wadden, T. A., and Stunkard, A. J. (1985). Social and psychological consequences of obesity. *Ann. Intern. Med.* **103**, 1062–1067.

Walsh, B. T., and Devlin, M. J. (1998). Eating disorders: Progress and problems. *Science* **280**, 1387–1390.

Wang, Z., Bing, C., Al-Barazanji, K., Mossakowaska, D. E., Wang, X.-M., McBay, D. L., Neville, W. A., Taddayon, M., Pickavance, L., Dryden, S., Thomas, M., McHale, M. T., Gloyer, I., Wilson, S., Buckingham, R., Arch, J., Trayhurn, P., and Williams, G. (1997). Interactions between leptin and hypothalamic neuropeptide Y neurons in the control of food intake and energy homeostasis in the rat. *Diabetes* **46**, 335–341.

Weigle, D. S. (1994). Appetite and the regulation of body composition. *FASEB J.* **8**, 302–310.

Wilfley, D. E., and Rodin, J. (1995). Cultural influences on eating disorders. *In* "Eating Disorders and Obesity" (K. D. Brownell and C. G. Fairburn, eds.), pp. 78–82. Guilford Press, New York.

Wilmore, J. H., Vodak, P. A., Parr, R. B., Girandola, R. N., and Billing, J. (1980). Further simplification of a method for determination of residual lung volume. *Med. Sci. Sports. Exercise* **12**, 216–218.

Woods, S. C., Seeley, R., Porte, D., and Schwartz, M. (1998). Signals that regulate food intake and energy homeostasis. *Science* **280**, 1378–1383.

Zamboni, M., Armellini, F., Milani, M., Marchi, M., Todesco, T., Robbi, R., Bergamo-Andreis, I., and Bosello, O. (1992). Body fat distribution in pre- and post-menopausal women. *Int. J. Obesity* **16**, 495–504.

Zemel, M. B., Kim, J. H., Woychick, R. P., Michaud, E. J., Kadwell, S. H., Patel, I., and Wilkinson, W. O. (1995). Agouti regulation of intracellular calcium. *Proc. Natl. Acad. Sci.* **92**, 4733–4737.

Zillikens, M. C., and Conway, J. M. (1990). Anthropometry in blacks: Applicability of generalized skinfold equations and differences in fat patterning between blacks and whites. *Am. J. Clin. Nutr.* **52**, 45–51.

## BIBLIOGRAPHY

*Anthropometry — Methodology and Application*

Albu J. B., Murphy, L., Frager, D., Johnson, J., and Pi-Sunyer, F. X. (1997). Visceral fat and race-dependent health risks in obese nondiabetic premenopausal women. *Diabetes* **46**, 456–462.

Arner, P. (1995). Techniques for the measurement of white adipose tissue metabolism: A practical guide. *Int. J. Obesity* **19**, 435–442.

Brodie, D. A. (1988). Techniques of measurement of body composition part I. *Sports Med.* **5**, 11–40.

Brozek, J., Henschel, and Keys, A. (1949). Effect of submersion in water on the volume of residual air in man. *J. Appl. Physiol.* **2**, 240–246.

Conway, J. M., Yanovski, S. Z., Avila, N. A., and Hubbard, V. (1995). Visceral adipose tissue differences in black and white women. *Am. J. Clin. Nutr.* **61**, 765–771.

Crain, L. S. (1969). Anthropometric determinations of obesity. *In* "Obesity" (N. L. Wilson, ed.), pp. 13–23. F. A. Davis, Philadelphia.

Deurenberg, P., Kooy, K., and Leenen, R. (1989a). Differences in body impedance when measured with different instruments. *Eur. J. Clin. Nutr.* **43**, 885–886.

Deurenberg, P., Leenen, R., Kooy, K., and Hautvast, J. (1989b). In obese subjects the body fat percentage calculated with Siri's formula is an overestimation. *Eur. J. Clin. Nutr.* **43**, 569–575.

Deurenberg, P., Westerterp, K., and Wierik, E. (1994). Between-laboratory comparison of densitometry and bio-electrical impedance measurements. *Brit. J. Nutr.* **71**, 309–316.

Durnin, J. V., and Rahaman, M. (1967). The assessment of the amount of fat in the human body from measurements of skinfold thickness. *Brit. J. Nutr.* **21**, 681–689.

Durnin, J. V., and Womersley, J. (1974). Body fat assessed from total body density and its estimation from skinfold thickness. *Brit. J. Nutr.* **32**, 77–97.

Ellis, K. J. (1997). Visceral fat mass in childhood: A potential early marker for increased risk of cardiovascular disease. *Am. J. Clin. Nutr.* **65**, 1887–1888.

Goran, M. I., Naby, T., Treuth, M., Trowbridge, C., Dezenberg, C., McGloin, A., and Glover, B. A. (1997). Visceral fat in white and African American prepubertal children. *Am. J. Clin. Nutr.* **65**, 1703–1708.

Kekes-Szabo, T., Hunter, G. R., Nyikos, I., Williams, M., Blaudeau, T., and Snyder, S. (1996). Anthropometric equations for estimating abdominal adipose tissue distribution in women. *Int. J. Obesity* **20**, 753–758.

Krotkiewski, M., Bjorntorp, P., Sjostrom, L., and Smith, U. (1983). Impact of obesity on metabolism in men and women. *J. Clin. Invest.* **72**, 1150–1162.

Kvist, H., Chowdhury, B., Grangard, U., Tylen, U., and Sjostrom, L. (1988). Total and visceral adipose-tissue volumes derived from measurements with computed tomography in adult men and women. *Am. J. Clin. Nutr.* **48**, 1351–1361.

Lee, R. D., and Nieman, D. C. (1996). "Nutritional Assessment," 2nd ed., pp. 223–288. Mosby Year Book, St. Louis.

Manson, J. E., Willett, W. C., Stampfer, M. J., Colditz, G. A., Hunter, D. J., Hankinson, S. E., Hennekens, C. H., and Speizer, F. E. (1995). Body weight and mortality among women. *N. Engl. J. Med.* **333**, 677–685.

Modlesky, C. M., Cureton, K., Lewis, R. D., Prior, B. M., Sloniger, M. A., and Rowe, D. A. (1996). Density of the fat-free mass and estimates of body composition in male weight trainers. *J. Appl. Physiol.* **80**, 2085–2096.

Rice, T., Borecki, I. B., Bouchard, C., and Rao, D. C. (1993). Segregation analysis of fat mass and other body composition measures derived from underwater weighing. *Am. J. Hum. Genet.* **52**, 967–973.

Rolland-Cachera, M., Brambilla, P., Manzoni, P., Akrout, M., Sironi, S., Maschio, A., and Chiumello, G. (1997). Body composition assessed on the basis of arm circumference and triceps skinfold thickness. *Am. J. Clin. Nutr.* **65**, 1709–1713.

Yanovski, J. A., Yanovski, S. Z., Filmer, K., Hubard, V., Avila, N., Lewis, B., Reynolds, J., and Flood, M. (1996). Differences in body composition of black and white girls. *Am. J. Clin. Nutr.* **64**, 833–839.

Wilmore, J. H., Vodak, P. A., Parr, R. B., Girandola, R. N., and Billing, J. (1980). Further simplification of a method for determination of residual lung volume. *Med. Sci. Sports. Exercise* **12**, 216–218.

Zamboni, M., Armellini, F., Milani, M., Marchi, M., Todesco, T., Robbi, R., Bergamo-Andreis, I., and Bosello, O. (1992). Body fat distribution in pre- and post-menopausal women. *Int. J. Obesity* **16**, 495–504.

Zamboni, M., Armellini, F., Turcato, E., Todesco, T., Bissoli, L., Bergamo-Andreis, I., and Bosello, O. (1993). Effect of weight loss on regional body fat distribution in premenopausal women. *Am. J. Clin. Nutr.* **58**, 29–34.

Zillikens, M. C., and Conway, J. M. (1990). Anthropometry in blacks: Applicability of generalized skinfold equations and differences in fat patterning between blacks and whites. *Am. J. Clin. Nutr.* **52**, 45–51.

## Physiology and Psychology of Obesity

Anderson, A. E. (1995). Eating disorders in males. *In* "Eating Disorders and Obesity" (K. D. Brownell and C. G. Fairburn, eds.), pp. 177–182. Guilford Press, New York.

Bjorntorp, P. (1985). Regional patterns of fat distribution. *Ann. Intern. Med.* **103**, 994–995.

Bjorntorp, P. (1991). Adipose tissue distribution and function. *Int. J. Obesity* **15**, 67–81.

Bjorntorp, P. (1996). The regulation of adipose tissue distribution in humans. *Int. J. Obesity* **20**, 291–302.

Bouchard, C., and Tremblay, A. (1997). Genetic influences on the response of body fat and fat distribution to positive and negative energy balances in human identical twins. *J. Nutr.* **127**, 943S–947S.

Bouchard, C., Despres, J.-P., Mauriege, P., Marcotte, M., Chagnon, M., Dionne, F. T., and Belanger, A. (1991). The genes in the constellation of determinants of regional at distribution. *Int. J. Obesity* **15**, 9–18.

Braun, B., Zimmermann, M. B., and Kretchmer, N. (1995). Effects of exercise intensity on insulin sensitivity in women with non-insulin-dependent diabetes mellitus. *J. Appl. Physiol.* **78**, 300–306.

Bray, G. A. (1985). Complications of obesity. *Ann. Intern. Med.* **103**, 1052–1062.

Campbell, P. J., and Carlson, M. G. (1993). Impact of obesity on insulin action in NIDDM. *Diabetes* **42**, 405–410.

Flatt, J. P. (1995). Body composition, respiratory quotient, and weight maintenance. *Am. J. Clin. Nutr.* **62**, 1107S–1117S.

Golay, A., and Bobbioni, E. (1997). The role of dietary fat in obesity. *Int. J. Obesity* **21**, S2–S11.

Greenwood, M., and Pittman-Waller, V. A. (1988). Weight control: A complex, various, and controversial problem. *In* "Obesity and Weight Control" (R. T. Frankle, ed.). pp. 3–15. Aspen Publishers, Rockville, MD.

Hervey, G. R. (1959). The effects of lesions in the hypothalamus in parabiotic rats. *J. Physiol. (London)* **145**, 336–352.

Hoek, H. W. (1995). The distribution of eating disorders. *In* "Eating Disorders and Obesity" (K. D. Brownell and C. G. Fairburn, eds.), pp. 207–211. Guilford Press, New York.

Hughes, V. A., Fiatarone, M. A., Fielding, R. A., Kahn, B., Merrara, C., Shepard, P., Fisher, E., Wolfe, R. R., Elai, D., and Evans, W. J. (1993). Exercise increases muscle GLUT-4 levels and insulin action in subjects with impaired glucose tolerance. *Am. J. Physiol.* **264**, E855–E862.

Kennedy, G. R. (1950). The hypothalamic control of food intake in rats. *Proc. R. Soc. (London), Ser. B* **137**, 535–549.

Kral, J. G. (1985). Morbid obesity and related health risks. *Ann. Intern. Med.* **103**, 1043–1047.

Leibel, R. L., Rosenbaum, M., and Hirsch, J. (1995). Changes in energy expenditure resulting from altered body weight. *N. Engl. J. Med.* **332**, 621–628.

Maurer, D., and Sobel, J. (1995). "The Medicalization and Demedicalization of Obesity." A. de Gruyter, New York.

Reichman, B., Chessex, P., Putet, G., Verellen, G., Smith, J. M., Heim, T., and Swyer, P. R. (1981). Diet, fat accretion, and growth in premature infants. *N. Engl. J. Med.* **305**, 1495–1500.

Schoen, R. E., Evans, R. W., Sankey, S. S., Weissfeld, J. L., and Kuller, L. (1996). Does visceral adipose tissue differ from subcutaneous adipose tissue in fatty acid content? *Int. J. Obesity* **20**, 346–352.

Sjostrom, L. (1980). Fat cells and body weight. *In* "Obesity" (A. J. Stunkard, ed.), pp. 72–100. Saunders, Philadelphia.

Smas, C. M., and Sul, H. S. (1995). Control of adipocyte differentiation. *Biochem. J.* **309**, 697–710.

Trayhurn, P. (1996). Uncoupling protein in brown adipose tissue. *Biochem. Soc. Trans.* **24**, 402–406.

Wadden, T. A., and Stunkard, A. J. (1985). Social and psychological consequences of obesity. *Ann. Intern. Med.* **103**, 1062–1067.

Weigle, D. S. (1994). Appetite and the regulation of body composition. *FASEB J.* **8**, 302–310.

Wilfley, D. E., and Rodin, J. (1995). Cultural influences on eating disorders. *In* "Eating Disorders and Obesity" (K. D. Brownell and C. G. Fairburn, eds.), pp. 78–82. Guilford Press, New York.

Zamboni, M., Armellini, F., Turcato, E., Todisco, P., Gallagher, D., Grave, R., Heymsfield, S., and Bosello, O. (1997). Body fat distribution before and after weight gain in anorexia nervosa. *Int. J. Obesity* **21**, 33–36.

## Various Hormones and Neuroendocrinology of Appetite and Obesity

Adan, R. A., Cone, R., Burbach, J., and Gispen, W. (1994). Differential effects of melanocortin peptides on neural melanocortin receptors. *Mol. Pharmacol.* **46**, 1182–1190.

Fan, W., Boston, B., Kesterson, R., Hruby, V., and Cone, R. D. (1997). Role of melanocortinergic neurons in feeding and the agouti obesity syndrome. *Nature* **385**, 165–168.

Figlewicz, D. P., Schwarz, M., Seeley, R., Chavez, M., Baskin, D., Woods, S., and Porte, D. (1996). Endocrine regulation of food intake and body weight. *J. Lab. Clin. Med.* **127**, 328–332.

Forman, B. M., Tontonoz, P., Chen, J., Brun, R., Spiegelman, B. M., and Evans, R. M. (1995). 15-Deoxy-prostaglandin $J_2$ is a ligand for the adipocyte determination factor PPARγ. *Cell* **83**, 803–812.

Krotkiewski, M., Blohme, B., Lindholm, N., and Bjorntorp, P. (1976). The effects of adrenal corticosteroids on regional adipocyte size in man. *J. Clin. Endocrinol. Metab.* **42**, 91–97.

Leibowitz, S. F. (1992). Neurochemical–neuroendocrine systems in the brain controlling macronutrient intake and metabolism. *Trends Neurosci.* **15**, 491–497.

Morley, J. E. (1987). Neuropeptide regulation of appetite and weight. *Endocrine Rev.* **8**, 256–278.

Mountjoy, K. G. (1994). The human melanocyte stimulating hormone receptor has evolved to become "super-sensitive" to melanocortin peptides. *Mol. Cell. Endocrinol.* **102**, R7–R11.

Mountjoy, K., and Wong, J. (1997). Obesity, diabetes and functions for proopiomelanocortin-derived peptides. *Mol. Cell. Endocrinol.* **128**, 171–177.

Rohner-Jeanrenaud, F. (1995). A neuroendocrine reappraisal of the dual-centre hypothesis. *Int. J. Obesity* **19**, 517–534.

Schwartz, M. W., Marks, J. L., Sipols, A. J., Baskin, D., Woods, S., Kahn, S., and Porte, D. (1991). Central insulin administration reduces neuropeptide Y mRNA expression in the arcuate nucleus of food-deprived lean (Fa/Fa) but not obese (fa/fa) Zucker rats. *Endocrinology* **128**, 2645–2647.

Schwartz, M., Sipols, A. J., Marks, J., Sanacora, G., White, J., Scheurink, A., Kahn, S., Baskin, D., Woods, S. C., Figlewicz, D., and Porte, D. (1992). Inhibition of hypothalamic neuropeptide Y gene expression by insulin. *Endocrinology* **130**, 3608–3616.

Spiegelman, B. M., and Flier, J. S. (1996). Adipogenesis and obesity: Rounding out the big picture. *Cell* **87**, 377–389.

Stephens, T. W. (1996). Life without neuropeptide Y. *Nature* **381**, 377–418.

## The "Obesity Genes": Leptin, Leptin Receptor, Agouti, Tubby, and Fat

Caro, J. F., Sinha, M., Kolaczynski, J., Zhang, P., and Considine, R. V. (1996). Leptin: The tale of an obesity gene. *Diabetes* **45**, 1455–1462.

Chen, G., Koyama, K., Yuan, X., Lee, Y., Zhou, Y.-T., O'Doherty, R., Newgard, C. B., and Unger, R. H. (1996). Disappearance of body fat in normal rats induced by adenovirus-mediated leptin gene therapy. *Proc. Natl. Acad. Sci. U.S.A.* **93**, 14795–14799.

Chen, H., Charlat, O., Tartaglia, L. A., Woolf, E. A., Weng, X., Ellis, S. J., Lakey, N. D., Culpepper, J., Moore, K. J., Breitbart, R., Duyk, G. M., Tepper, R., and Morgenstern, J. P. (1996). Evidence that the diabetes gene encodes the leptin receptor. *Cell* **84**, 491–495.

Considine, R. V., Considine, E. L., Williams, C., Nyce, M., Magosin, S., Bauer, T., Rosato, E., Colberg, J., and Caro, J. F. (1995). Evidence against either a premature stop codon or the absence of obese gene mRNA in human obesity. *J. Clin. Invest.* **95**, 2986–2988.

Considine, R. V., Sinha, M., Heiman, M., Kriauciunas, A., Stephens, T., Nyce, M., Ohannesian, J., Marco, C., McKee, L. J., Bauer, T., and Caro, J. F. (1996). Serum immunoreactive-leptin concentrations in normal-weight and obese humans. *N. Engl. J. Med.* **334**, 292–295.

Cool, D. R., Norman, E., Shen, F.-S., Chen, H.-C., Pannell, L., Zhang, Y., and Loh, Y. P. (1997). Carboxypeptidase E is a regulated secretory pathway sorting receptor. *Cell* **88**, 73–83.

Fricker, L. D., Berman, Y. L., Leiter, E. H., and Devi, L. A. (1996). Carboxypeptidase E activity is deficient in mice with the fat mutation. *J. Biol. Chem.* **271**, 30619–30624.

Haffner, S. M., Gingerich, R. L., Miettinen, H., and Stern, M. P. (1996). Leptin concentrations in relation to overall adiposity and regional body fat distribution in Mexican Americans. *Int. J. Obesity* **20**, 904–908.

Hollenberg, A. N., Susulic, V. S., Madura, J. P., Ahang, B., Moller, D. E., Tontonoz, P., Sarraf, P., Spiegelman, B. M., and Lowell, B. B. (1997). Functional antagonism between CCAAT/enhancer binding protein-$\alpha$ and peroxisome proliferator-activated receptor-$\gamma$ on the leptin promoter. *J. Biol. Chem.* **272**, 5283–5290.

Huszar, D., Lynch, C. A., Fairchild-Huntress, V., Dumore, J., Fang, Q., Berkemeier, L., Gu, W., Kesterson, R., Boston, B. A., Cone, R. D., Smith, F. J., Campfield, L., Burn, P., and Lee, F. (1997). Targeted disruption of the melanocortin-4 receptor results in obesity in mice. *Cell* **88**, 131–141.

Klebig, M. L., Wilkinson, J. E., Geisler, J. G., and Woychik, R. P. (1995). Ectopic expression of the agouti gene in transgenic mice causes obesity, features of type II diabetes, and yellow fur. *Proc. Natl. Acad. Sci. U.S.A.* **92**, 4728–4732.

Kleyn, P. W., Fan, W., Kovats, S., Lee, J., Pulido, J., Wu, Y., Berkemeier, L., Misumi, D., Holmgren, L., Charlat, O., Woolf, E., Tayber, O., Brody, T., Shu, P., Hawkins, F., Kennedy, B., Baldini, L., Ebeling, C., Alperin, G. D., Deeds, J., Lakey, N. D., Culpepper, J., Chen, H., Glucksmann-Kuis, M., Carlson, G., Duyk, G., and Moore, K. J. (1996). Identification and characterization of the mouse obesity gene tubby. *Cell*, 85, 281–290.

Kling, J. (1997). Carboxypeptidase E. *Nature Biotechnol.* **15**, 128–129.

Lahlou, N., Landais, P., Boissieu, D., and Bougneres, P.-F. (1997). Circulating leptin in normal children and during the dynamic phase of juvenile obesity. *Diabetes* **46**, 989–993.

MacDougald, O. A., Hwang, C.-S., Fan, H., and Lane, M. D. (1995). Regulated expression of the obese gene product (leptin) in white adipose tissue and 3T3-L1 adipocytes. *Proc. Natl. Acad. Sci. U.S.A.* **92**, 9034–9037.

Manne, J., Argeson, A., and Siracusa, L. D. (1995). Mechanisms for the pleiotropic effects of the agouti gene. *Proc. Natl. Acad. Sci. U.S.A.* **92**, 4721–4724.

Muzzin, P., Eisensmith, R. C., Copeland, K. C., and Woo, S. L. (1996). Correction of obesity and diabetes in genetically obese mice by leptin gene therapy. *Proc. Natl. Acad. Sci. U.S.A.* **93**, 14804–14808.

Ollmann, M. M., Wilson, B. D., Yang, Y.-K., Kerns, J. A., Chen, Y., and Gantz, I., and Barsh, G. S. (1997). Antagonism of central melanocortin receptors *in vitro* and *in vivo* by agouti-related protein. *Science* **278**, 135–138.

Schwartz, M., Seeley, R. J., Campfield, L., Burn, P., and Baskin, D. G. (1996). Identification of targets of leptin action in rat hypothalamus. *J. Clin. Invest.* **98**, 1101–1106.

Shimabukuro, M., Koyama, K., Chen, G., Wang, M.-Y., Trieu, F., Lee, Y., Newgard, C. B., and Unger, R. H. (1997). Direct antidiabetic effect of leptin through triglyceride depletion of tissues. *Proc. Natl. Acad. Sci. U.S.A.* **94**, 4637–4641.

Sinha, M. K., Ohannesian, J. P., Heiman, M., Kriawclunas, A., Stephens, T., Magosin, S., Marco, C., and Caro, J. F. (1996). Nocturnal rise in leptin in lean, obese, and non-insulin-dependent diabetes mellitus subjects. *J. Clin. Invest.* **97**, 1344–1347.

Stehling, O., Doring, H., Ertl, J., Preibisch, G., and Schmidt, I. (1996). Leptin reduces juvenile fat stores by altering the circadian cycle of energy expenditure. *J. Biol. Chem.* **271**, R1770–R1774.

Wang, Z., Bing, C., Al-Barazanji, K., Mossakowaska, D. E., Wang, X.-M., McBay, D. L., Neville, W. A., Taddayon, M., Pickavance, L., Dryden, S., Thomas, M., McHale, M. T., Gloyer, I., Wilson, S., Buckingham, R., Arch, J., Trayhurn, P., and Williams, G. (1997). Interactions between leptin and hypothalamic neuropeptide Y neurons in the control of food intake and energy homeostasis in the rat. *Diabetes* **46**, 335–341.

Zemel, M. B., Kim, J. H., Woychick, R. P., Michaud, E. J., Kadwell, S. H., Patel, I., and Wilkinson, W. O. (1995). Agouti regulation of intracellular calcium. *Proc. Natl. Acad. Sci.* **92**, 4733–4737.

Zhou, Y.-T., Shimabukuro, M., Koyama, K., Lee, Y., Wang, M.-Y., Trieu, F., Newgard, C. B., and Unger, R. H. (1997). Induction of leptin uncoupling protein-2 and enzymes of fatty acid oxidation. *Proc. Natl. Acad. Sci. U.S.A.* **94**, 6386–6390.

# 8

# PROTEIN

## OVERVIEW

Protein is composed of a polymer of amino acids. The α-amino acids are molecules distinguished as having an amino group at the 2-carbon of carboxylic acids. This amino group is the α-amino group. The nitrogen atoms of the amino acids occur at the α-amino group, as well as at the amide nitrogens of glutamine and asparagine, the guanidino group of arginine, the ε-amino group of lysine, the indole nitrogen of tryptophan, and the imidazole nitrogens of histidine. The average ratio of protein/nitrogen, by weight, is 6.25, for the typical protein in the diet. This number is used as a conversion factor for expressing the amount of protein in the diet, as is the protein equivalent excreted in the urine in the form of nitrogenous waste products. For example, a rat consuming 1.0 g of nitrogen in the form of protein consumes about 6.25 g of protein.

Meat, fish, and dairy products are good sources of high-quality protein (Table 8.1). Salmon, for example, contains nearly 30 g of protein per 100 g of food. Skim milk is also a good source. Its protein content appears to be low (3.6 g/100 g food), but this is because

**TABLE 8.1** Protein Content and Protein Density of Various Foods[a]

| Food | Protein content (g/100 g food) | Energy content (kJ/100 g food) | "Protein density" (g/1000 kJ energy) |
|---|---|---|---|
| Baked salmon | 27 | 761 | 36 |
| Fried liver | 26 | 960 | 28 |
| Skim milk | 3.6 | 150 | 24 |
| Pork, canned ham | 18 | 810 | 23 |
| Cooked hamburger | 24 | 2000 | 20 |
| Hard-boiled eggs | 13 | 680 | 19 |
| Green peas | 5.1 | 290 | 18 |
| Cooked white beans | 7.8 | 490 | 16 |
| Snap beans | 1.6 | 110 | 15 |
| Cheddar cheese | 25 | 1670 | 15 |
| Whole milk | 3.5 | 270 | 13 |
| Peanuts | 26 | 2380 | 11 |
| Cooked oatmeal | 2.0 | 230 | 8.7 |
| Bread | 9.1 | 1210 | 7.5 |
| Canned corn, drained | 2.6 | 350 | 7.4 |
| Spaghetti | 3.4 | 460 | 7.3 |
| Baked potatoes | 2.6 | 390 | 6.7 |
| Raw carrots | 1.1 | 180 | 6.3 |
| Walnuts | 14.8 | 2720 | 5.4 |
| Raw orange | 1.0 | 210 | 4.9 |
| Cooked white rice | 2.0 | 460 | 4.4 |
| Orange juice | 0.8 | 200 | 4.0 |
| Banana | 1.1 | 360 | 3.1 |
| Beer | 0.3 | 180 | 1.7 |
| Raw apple | 0.2 | 240 | 0.8 |
| Soybean oil | 0.0 | 3700 | 0.0 |

*Source*: Pennington (1976).

[a]The protein contents of a variety of foods are listed. The energy content of the foods is also listed. The joule (J) is the unit of energy. For comparison, it might be useful to note that the energy requirement for a typical "70-kg man" is 11,300 kJ/day. The foods are listed in order of their protein density. Foods that are richer in protein have a higher content of protein and a lesser content of carbohydrate or fat. The importance of including a certain amount of carbohydrate in the diet is covered under Sparing of Protein by Carbohydrate (the total fast versus the glucose fast). The danger of consuming foods with very low protein densities is described under Starvation.

of the relatively high water content of milk. The protein contents of both skim and whole milk are nearly identical when expressed as grams per 100 g of food; however, the proportion of total energy value contributed by protein is higher for skim milk than for whole milk. This is because skim milk is very low in fat. The weight of protein per unit of energy in the food is called the **protein density**. The weight of any nutrient in food per unit of energy is called the **nutrient density**. Note that the terms *protein density* and *nutrient density* are not scientifically accurate. Density means the mass per unit of volume. It does not mean mass per unit of energy. Loose interpretations of the term *density* are in common use, however. For example, population density refers to the number of people per unit of area.

Beans and peas, which are legumes, are good sources of protein. Their protein densities are quite similar to that of eggs. The grains, including rice, wheat, and corn, are moderately good sources of protein. The protein densities of wheat products, such as bread and spaghetti, are similar to that of corn. The protein density of rice is a bit lower. Bananas, cassava, sweet potatoes, and taro are the major staples of the tropics. These foods are poor sources of protein. The protein density of cassava is well below that of corn, potatoes, wheat, or rice. From Table 8.1, it can be seen that apples, beer, and vegetable oil are extremely poor sources of protein. It is ironic that apples, which somehow have been associated with good nutrition, are junk food as far as protein nutrition is concerned.

The concerns of the nutritionist regarding protein extend well beyond protein content and protein density. A major concern in human and animal nutrition is the quality of the protein. The term **protein quality** refers to the ability of a particular protein to provide a balanced pattern of indispensable amino acids. A high-quality protein contains high concentrations of the indispensable amino acids. A low-quality protein is deficient or lacking in one or more of these amino acids. Although beans and eggs are similar, as far as protein density is concerned, the quality of egg protein is considerably better than that of protein from beans.

This chapter focuses initially on the catabolism of the amino acids. Aminotransferases can be used to catalyze the first step in the breakdown of nearly all of the amino acids. Lysine catabolism, in contrast, does not begin with an aminotransferase-catalyzed step. Some amino acids can be catabolized via more than one pathway. Glutamate catabolism, for example, can begin by reactions catalyzed by glutamate oxaloacetate aminotransferase or glutamate dehydrogenase.

Unique biochemical pathways are followed in the catabolism of each of the amino acids. Some amino acids are degraded to yield carbon skeletons that are intermediates of the Krebs cycle. Amino acids that produce such carbon skeletons are called glucogenic, as the carbon skeletons can be withdrawn from the Krebs cycle and used for gluconeogenesis. Other amino acids are catabolized to yield acetyl-CoA. Acetyl-CoA cannot be used for the net production of glucose, though it can be used for making ketone bodies. Hence, these amino acids are called ketogenic. The terms **glucogenic** and **ketogenic** do not imply that catabolism of the amino acid in question always results in the synthesis of glucose and ketone bodies. It simply indicates metabolic pathways that are potentially open to the amino acid.

Amino acid degradation results in the production of **waste nitrogen**. Most of the waste nitrogen produced in the body is packaged in the form of urea and excreted in the urine. A small fraction of the waste nitrogen is excreted as urinary ammonia, though this proportion can rise with prolonged fasting. Waste nitrogen may appear in the cell initially in the form of ammonia, as during the catabolism of glutamate by the action of glutamate dehydrogenase; however, the catabolism of most amino acids results initially in the direct transfer of waste nitrogen to a carrier. The carrier serves to prevent the accumulation of free ammonium ions in the body. Here, waste nitrogen may take the form of the α-amino groups of glutamate, aspartate, or alanine, or the amide group of glutamine. The production of urea and ammonium ions is illustrated in this chapter by studies involving cats, rats, and humans.

The second half of this chapter concerns material of a less biochemical and a more nutritional nature. Three types of protein nutrition are illustrated by studies with humans: (1) the total fast, (2) the carbohydrate-only fast, and (3) a diet intended to maintain nitrogen balance.

## MILK PROTEIN

An outline of the properties of a common food protein, namely milk protein, will be used to anchor the chapter in everyday dietary experience, prior to focusing on metabolic pathways. Human milk contains about 7.0% lactose, 3.7% fat, and 1.2% protein. The corresponding values for bovine milk are 4.8, 4.3, and 3.3%. The term percent (%) means grams of nutrient per 100 ml of fluid milk. One can see that human milk is higher in lactose but lower in protein than cow milk (Vorherr, 1974). Lactose is a disaccharide, composed of galactose and glucose.

The terms **curds** and **whey** are in the layperson's vocabulary and are used to refer to specific types of proteins in milk. When acid is added to milk, to give a mildly acid pH of 4.6, a precipitate forms. The precipitate, which consists of a protein called **casein**, is called curds. The material that does not precipitate in mild acid is called whey, and consists of a fluid that contains the whey proteins. The whey proteins are α-lactalbumin, antibodies, and albumin. α-Lactalbumin is a very small protein, having a molecular weight of 14 kDa. β-Lactoglobulin is the major protein of whey in *bovine milk*, but it does not occur in *human milk*. This protein is quite small and has a molecular weight of only 18 kDa or, to be exact, the molecular weight is 18,276 Da. To dwell on the characteristics of these small proteins, one might note that α-lactalbumin binds calcium, and other divalent metal ions. The protein serves as a cofactor for **lactose synthase**, allowing the enzyme to catalyze the synthesis of lactose. β-Lactoglobulin is distinguished in that it binds retinol, a form of vitamin A.

In the kitchen, one can acidify milk by adding bacteria that excrete lactic acid. The result is the formation of curds, and this is the first step in making cottage cheese, cream cheese, and other cheeses. The proteins of whey are distinguished from those of curds in that they remain soluble in mild acid, but precipitate (coagulate) when heated at 80°C or higher. When whole milk is heated, the scum that develops consists of precipitated *whey proteins*. With prolonged boiling, however, the *curd proteins* also precipitate.

Concentrated preparations of unheated whey protein are sometimes used to replace eggs in desserts, mayonnaise, and bakery items because they form a gel (coagulate) when cooked.

The following commentary applies to milk proteins, but also to most other proteins in nature as well. When a protein precipitates in mild acid it is because the added hydrogen ions bind to the carboxylic acid groups of residues of glutamate or aspartate. Many proteins tend to contain more negatively charged amino acids (glutamate or aspartate) than positively charged amino acids (lysine). This binding event neutralizes some of the negative charges that are contained on the surface of the protein. When this partial neutralization occurs, and where the result is a protein with an equal number of positive and negative charges on its surface, the result is a protein with a zero overall charge. With this sort of neutralization, the proteins clump together and form a clumped mass that is too large to remain suspended in solution. The exact pH where a protein has no overall positive or negative charge is called the **isoelectric point**. Every protein in nature has its own unique isoelectric point, i.e., pH 4.2, pH 5.7, or pH 8.2.

When proteins precipitate with **heating**, it is because the protein turns inside-out, exposing many of the lipophilic amino acid residues (residues previously facing the inner core of the protein) to the external aqueous environment. The process of turning inside-out is called **denaturation**. The lipophilic amino acids tend not to attract water molecules, and the protein loses its interactions with the aqueous environment and instead binds to other denatured proteins, thus creating a clump of proteins. The phrase *randomly oriented polypeptide* is more accurate than "inside-out protein." A generalization kept in mind by most food scientists and biochemists is as follows. Protein that is **precipitated by mild acidification** is most probably not denatured. Neutralization of the pH can easily bring the proteins back into solution. Protein that is **precipitated by heating** is generally in the denatured state, and usually cannot easily be made soluble again.

The casein proteins of milk are distinguished in that they occur as complexes of several many similar proteins. The casein proteins are assembled into clusters or micelles having a diameter of about 0.08 μm. The caseins consist of α-casein and β-casein, which form the core of the micelle, and κ-casein (kappa-casein), which forms a surface coating around the core. The caseins are all small proteins having a molecular weight of about 24 kDa, and consist of about 200 amino acids (Wong *et al.*, 1996; Dev *et al.*, 1994; Gastaldi *et al.*, 1996; Zhang *et al.*, 1996).

In making yogurt, milk is initially heated at 80°C for 15–30 min in order to denature the *whey proteins*. This heating provokes the whey proteins to denature, and consequently associate with the casein micelles, resulting in a yogurt with improved viscosity and texture (Law and Leaver, 1997; Boye *et al.*, 1997; Otte *et al.*, 1996).

In addition to the proteins of the curds and whey, milk contains small droplets of fat that are coated with a special protein called butyrophilin (Goldfarb, 1997). The fat droplets of milk are about 3–6 μm in diameter. The coat protein consists of lipophilic regions that face the fat and hydrophilic regions that face the water.

In reviewing any commentary on milk proteins, one might ask whether treatment of a particular protein with heat causes gel-formation or coagulation (clumping; aggregation) and whether exposure to acid results in gel-formation or coagulation.

The properties of other proteins — such as soybean protein, egg protein, and meat protein — as well as their food sources (matrices), are major topics in food science. Rather than dwell on this subject, we will turn to some of the features of dietary protein, after digestion and acquisition of their component amino acids by the body. The role of dietary protein in allowing growth and the maintenance of health will be detailed, though the biochemical machinery (ribosomes, tRNAs, aminoacyl tRNA synthases) used for the assembly of amino acids into proteins will not be described.

## AMINO ACID CATABOLISM AND AMINOTRANSFERASES

The first step in the catabolism of most amino acids involves the removal of the α-amino group. Removal of this group results in conversion of the amino acid to the corresponding keto acid. The structures of a number of amino acids and their corresponding keto acids are shown in Table 8.2. For example, this table lists pyruvate as the keto acid version of alanine and α-ketoglutarate as the keto acid of glutamate.

The conversion of amino acids to keto acids is usually catalyzed by enzymes called **aminotransferases**. Some aminotransferases can recognize a variety of different amino acids as substrates; others are more specific in their action. Aminotransferases use vitamin $B_6$ as a cofactor, and the "$B_6$" in the figures indicates that the enzyme is a vitamin $B_6$-requiring enzyme. Cofactors are small molecules that bind to specific enzymes and participate in the chemistry of catalysis. Some

**TABLE 8.2**  Amino Acids and Keto Acids

| Number of carbons | Amino form (amino acids) | Keto form (keto acids) |
|---|---|---|
| 3 | CH₃CHCOOH<br>&#124;<br>NH₂<br>Alanine | CH₃CCOOH<br>&#124;&#124;<br>O<br>Pyruvate |
| 4 | HOOC—CH₂CH—COOH<br>&#124;<br>NH₂<br>Aspartic acid | HOOC—CH₂C—COOH<br>&#124;&#124;<br>O<br>Oxaloacetic acid (OAA) |
| 5 | HOOC—CH₂CH₂CH—COOH<br>&#124;<br>NH₂<br>Glutamic acid | HOOC—CH₂CH₂C—COOH<br>&#124;&#124;<br>O<br>α-Ketoglutarate |
| 6 | H₃C<br>&#124;<br>CH—CH₂CH—COOH<br>&#124;  &#124;<br>H₃C  NH₂<br>Leucine<br>(branched-chain amino acid) | H₃C<br>&#124;<br>CH—CH₂C—COOH<br>&#124;  &#124;&#124;<br>H₃C  O<br>α-Ketoisocaproic acid<br>(branched-chain keto acid) |

cofactors, such as NAD, bind only momentarily to the enzyme; others, such as vitamin $B_6$, remain bound to the enzyme before, after, and during the event of catalysis. In the case of vitamin $B_6$-requiring enzymes, the cofactor occurs in the forms pyridoxal phosphate and pyridoxamine phosphate.

The removal of the α-amino group of an amino acid by an aminotransferase is illustrated in the case of glutamate–pyruvate aminotransferase (Figure 8.1). This enzyme is also called alanine aminotransferase. The reaction depicted in Figure 8.1 actually involves the interconversion of amino acids and keto acids. The enzyme catalyzes the removal of the amino group of alanine and its transfer to α-ketoglutarate, resulting in the production of pyruvate and glutamate. In the body, the pyruvate formed in this reaction may enter the Krebs cycle for oxidation to carbon dioxide. The reaction (Figure 8.1) is drawn with double-headed arrows. This means that the reaction is freely reversible. The reverse reaction involves conversion of glutamate to α-ketoglutarate, with the concomitant conversion of pyruvate to alanine. This reaction may be useful, for example, when the body needs more alanine for protein biosynthesis where the cell contains adequate quantities of pyruvate and glutamate.

The reactions catalyzed by aminotransferases are called transamination reactions. It might be noted that in these reactions the amino group being transferred initially is transferred to the cofactor pyridoxal phosphate, resulting in its conversion to pyridoxamine phosphate. In the second half of the reaction, the amino group residing on the cofactor is transferred to the keto acid cosubstrate, thus regenerating the cofactor in the pyridoxal phosphate form. As stated earlier, the cofactor remains bound to the enzyme when it occurs as the pyridoxal phosphate and pyridoxamine phosphate forms.

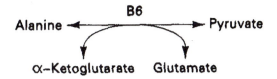

**FIGURE 8.1** Reaction catalyzed by glutamate–pyruvate aminotransferase.

A large number of aminotransferases have been discovered. They catalyze the reversible transfer of an amino group from specific amino acids to a cosubstrate. α-Ketoglutarate is the cosubstrate used by most aminotransferases. α-Ketoglutarate and glutamate are used in a cyclic manner during amino acid breakdown in the body (Figure 8.2). A specific amino acid, such as tyrosine, donates its amino group to α-ketoglutarate in a reaction catalyzed by a specific enzyme, in this case tyrosine aminotransferase. The keto form of the amino acid then undergoes further catabolism, as indicated in the figure. The glutamate product is converted back to α-ketoglutarate (see Figure 8.2) in a reaction catalyzed by glutamate–oxaloacetate aminotransferase. This enzyme is also called aspartate aminotransferase. The overall picture, in this example, is as follows. The amino group initially occurring in tyrosine eventually appears as part of aspartic acid. The molecule of aspartate, then, may deliver the amino group to the urea cycle, as detailed later in this chapter. The preceding scenario characterizes the catabolic pathways of most of the amino acids. The continuation of the tyrosine catabolic pathway is shown at the end of this chapter.

## Aminotransferases and the Alanine Cycle

Glutamate–pyruvate aminotransferase has received attention for its role in facilitating the breakdown of muscle amino acids, as during prolonged exercise or fasting. With the initial step in amino acid breakdown, as catalyzed by a specific aminotransferase, the amino group appears on glutamate. This glutamate tends not, however, to leave muscle for the bloodstream. Instead, glutamate–oxaloacetate aminotransferase catalyzes the transfer of glutamate's amino group to pyruvate, forming alanine. This alanine then leaves the muscle for the bloodstream (DeRosa and Swick, 1975).

The alanine released from muscle is taken up by the liver. Glutamate–pyruvate aminotransferase, in liver, catalyzes the conversion of alanine to pyruvate. This pyruvate can then be used for gluconeogenesis. The amino group initially dis-

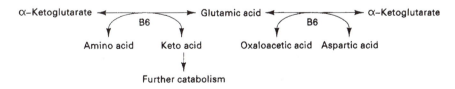

**FIGURE 8.2** Transfer of an amino group from an amino acid to α-ketoglutarate and then to oxaloacetic acid.

charged during amino acid catabolism in muscle eventually occurs as part of glutamic acid in the liver. The amino group of glutamate can, in turn, be transferred to oxaloacetic acid (see Figure 8.2). The preceding scenario illustrates some of the reactions that constitute the alanine cycle.

## Amino Acid Catabolic Pathways Starting with Oxidation

Glutamic acid can be catabolized by a mode other than transamination, namely by oxidation via glutamate dehydrogenase (Figure 8.3). Glutamate dehydrogenase is a mitochondrial enzyme. It catalyzes the reversible discharge of an ammonium ion from glutamate, yielding α-ketoglutarate. NAD is the cofactor of this enzyme. The ammonium ion released can be packaged as urea, as discussed later in this chapter. Specific NAD- or NADP-requiring enzymes catalyze the first step in the breakdown of proline (proline oxidase), lysine (saccharopine dehydrogenase), threonine (threonine dehydrogenase), and glycine (glycine cleavage complex), with the enzymes involved given in parentheses.

The first step in lysine catabolism is catalyzed by saccharopine dehydrogenase. (This enzyme is named for the reverse reaction.) The step involves the condensation of lysine with α-ketoglutarate to form a Schiff base. The Schiff base is then reduced to form saccharopine. In this step, the Schiff base involves lysine's side-chain amino group and α-ketoglutarate's keto group. In a subsequent enzymatic step, the α-ketoglutarate moiety is split off, taking with it the amino group and occurring as glutamic acid. At this point, the lysine catabolite occurs as α-**amino adipic acid-δ-semialdehyde**. The final products of the pathway consist of two molecules of $CO_2$ and one of acetoacetyl-CoA.

Catabolism of tyrosine and tryptophan begins with oxygen-requiring steps. The tyrosine catabolic pathway, shown at the end of this chapter, results in the formation of fumaric acid and acetoacetic acid. Tryptophan catabolism commences with the reaction catalyzed by tryptophan-2,3-dioxygenase. This enzyme catalyzes conversion of the amino acid to N-formyl-kynurenine. The enzyme requires iron and copper and thus is a metalloenzyme. The final products of the pathway are acetoacetyl-CoA, acetyl-CoA, formic acid, four molecules of carbon dioxide, and two ammonium ions. One of the intermediates of tryptophan catabolism, α-amino-β-carboxymuconic-δ-semialdehyde, can be diverted from complete oxidation, and used for the synthesis of NAD (see Niacin in Chapter 9).

L-Amino acid oxidase and D-amino acid oxidase are oxygen-requiring enzymes that catalyze the conversion of amino acids to the corresponding keto acids and an ammonium ion. The oxygen cosubstrate is converted to hydrogen peroxide. Flavin adenine dinucleotide (FAD) is required as a cofactor. Both enzymes are of minor importance in amino acid catabolism. Note that all of the amino acids described in this text are L-amino acids. D-Amino acids are minor components of

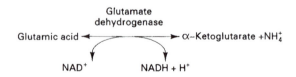

**FIGURE 8.3** Oxidation of glutamic acid by glutamate dehydrogenase.

the diet and can occur in invertebrate tissues and in certain bacteria, and with the cooking and processing of foods (Man and Bada, 1987). The L- and D-notation refers to the three-dimensional arrangement of the four bonds at the $\alpha$-carbon of the amino acid.

## Catabolism of Serine, Threonine, and Histidine

Serine catabolism can be catalyzed by three different enzymes, namely serine dehydratase (Figure 8.4), serine aminotransferase, and serine hydroxymethyltransferase (see Folate in Chapter 10). Serine can be converted to pyruvate in a reaction catalyzed in serine dehydratase. The molecule of pyruvate can then enter the Krebs cycle. The ammonium ion produced in the reaction can be packaged as urea, as detailed later in this chapter. Serine dehydratase, though not an aminotransferase, requires vitamin $B_6$ as a cofactor. Rat liver contains high levels of serine dehydratase, whereas human liver contains relatively low levels of the enzyme (Snell, 1986; Ogawa et al., 1989). In human liver, the first step in serine catabolism may proceed via transamination to yield hydroxypyruvate. This keto acid can then be reduced to glyceric acid by glycerate dehydrogenase. Glycerate can then be phosphorylated by glycerate kinase to generate 2-phosphoglycerate, an intermediate in glycolysis.

Threonine can be broken down by two separate pathways. Serine dehydratase catalyzes the conversion of threonine to 2-ketobutyrate plus an ammonium ion; 2-ketobutyrate is then converted by branched-chain keto acid (BCKA) dehydrogenase to propionyl-CoA plus carbon dioxide. Propionyl-CoA catabolism is described later in this chapter. Threonine can also be broken down by a complex that has been suggested to be composed of threonine dehydrogenase and acetoacetone synthase (Tressel et al., 1986). Here, threonine catabolism results in the production of acetyl-CoA plus glycine.

Histidine breakdown is initiated by histidase in a reaction somewhat similar to that of serine dehydratase (see Folate in Chapter 10).

## Catabolism of Glutamine and Asparagine

Glutamine catabolism occurs by hydrolysis of the amide group from the side chain. Glutaminase catalyzes the reaction, resulting in the production of glutamate and an ammonium ion. Asparagine catabolism occurs in an identical manner, and here the reaction is catalyzed by asparaginase.

## Catabolism of the Branched-Chain Amino Acids

The branched-chain amino acids (BCAAs) are leucine, isoleucine, and valine. Pathways for the breakdown of leucine and isoleucine appear in Figures 8.5 and

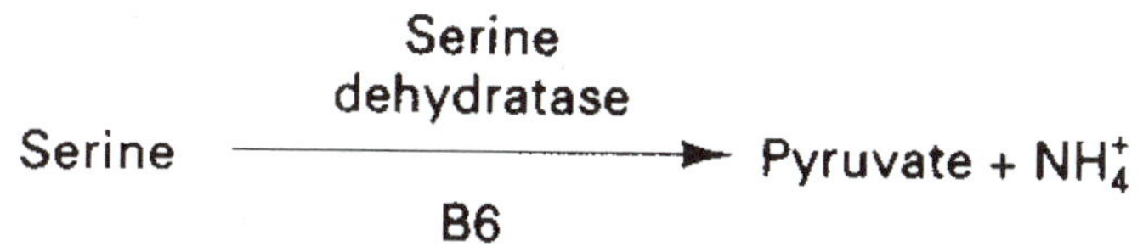

$$\text{Serine} \xrightarrow[\text{B6}]{\text{Serine dehydratase}} \text{Pyruvate} + NH_4^+$$

FIGURE 8.4  Serine catabolism via serine dehydratase.

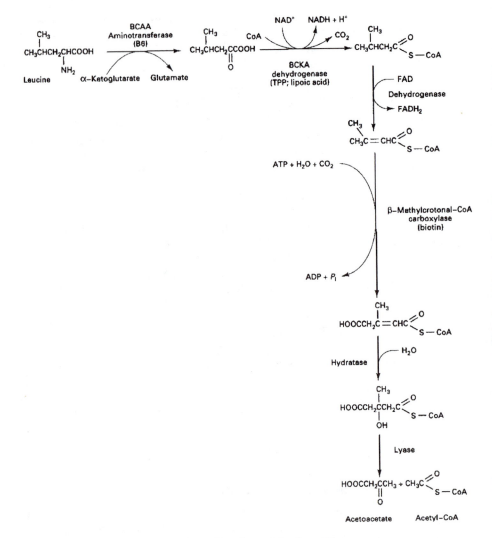

**FIGURE 8.5** Leucine catabolic pathway.

8.6, respectively. The first step in the breakdown of leucine, isoleucine, and valine is catalyzed by BCAA aminotransferase. The cosubstrate is α-ketoglutarate. The second step in the catabolism of all three BCAAs is catalyzed by BCKA dehydrogenase. This enzyme is similar to pyruvate dehydrogenase in its cofactor requirements, in its production of carbon dioxide, and in the appearance of the product as the CoA derivative. In addition, it might be noted that both enzymes are regulated by the covalent attachment of a phosphate group to the enzyme. The third step in BCAA catabolism is catalyzed by dehydrogenases that use FAD as the cofactor. The final products of the leucine catabolic pathway are acetoacetate, which is a ketone body, and acetyl-CoA.

Leucine catabolism also occurs by a second, minor pathway. The first step in this pathway is catalyzed by leucine aminomutase (Poston, 1984). Leucine aminomutase is one of the three vitamin $B_{12}$-requiring enzymes in mammalian tissues.

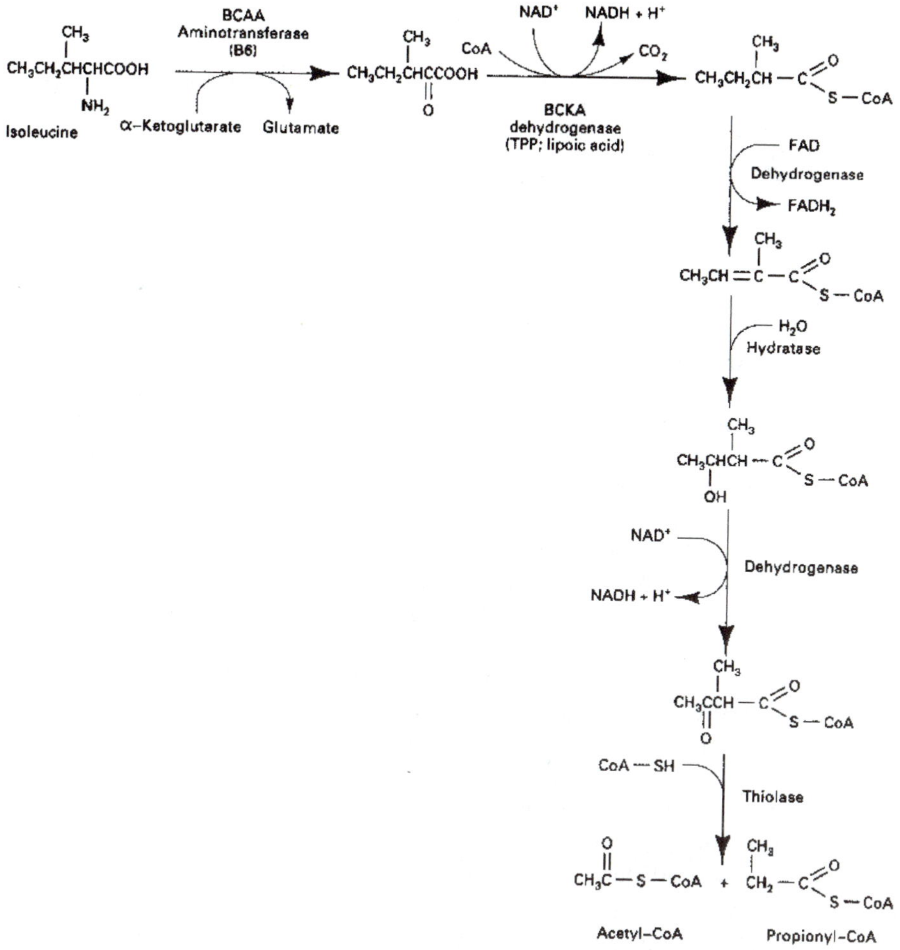

**FIGURE 8.6** Isoleucine catabolic pathway.

The products of the isoleucine catabolic pathway are propionyl-CoA and acetyl-CoA; valine catabolism produces one molecule of propionyl-CoA and two molecules of carbon dioxide. Propionyl-CoA is further catabolized to succinyl-CoA, an intermediate of the Krebs cycle (Figure 8.7). This pathway is also used for catabolism of the short-chain fatty acid propionic acid, after its conversion to the thiol ester form by thiokinase. The first step in propionyl-CoA breakdown is catalyzed by propionyl-CoA carboxylase, a biotin-requiring enzyme. The second step is catalyzed by methylmalonyl-CoA mutase, a vitamin $B_{12}$-requiring enzyme.

## Arginine Catabolism and Urea Synthesis

The catabolic pathway of arginine is of special interest because the first step of this pathway is catalyzed by arginase, an enzyme noted for its participation in the

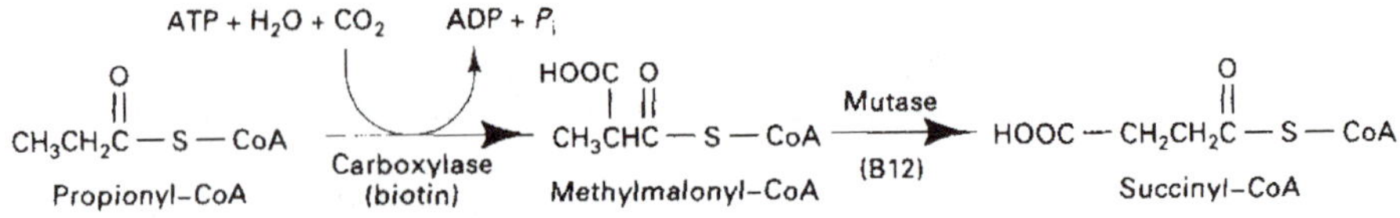

$$ATP + H_2O + CO_2 \qquad ADP + P_i$$

FIGURE 8.7  Conversion of propionyl-CoA to succinyl-CoA.

series of reactions known as the urea cycle. The arginine catabolic pathway results in the production of glutamate. Arginine is distinguished by the presence of a **guanidino group**, whose structure is shown in Figure 8.8.

The guanidino group has one of the strongest positive charges found in biology. This means that it retains its plus charge at pH extremes that result in the loss of a proton, and hence positive charge, of other protonated compounds. The guanidino group was previously encountered in the section on Creatine Phosphate (Chapter 4). To review, creatine synthesis involves transfer of the guanidino group to glycine, generating guanidinoacetic acid. In the reaction catalyzed by arginase, the guanidino group is not transferred to a molecule of glycine, but to water, resulting in the production of urea. Transfer of the guanidino group of arginine, whether during the synthesis of creatine or of urea, results in the production of ornithine, an amino acid. This amino acid is not incorporated into proteins.

The pathway resulting in the conversion of arginine to glutamate is shown in Figure 8.9. The carbon skeletons of all of the structures are five carbons long. The first step involves hydrolysis of the guanidino group from arginine, generating urea and ornithine. The second step is catalyzed by an aminotransferase and results in replacement of the terminal amino group of ornithine with an aldehyde group. The final step is catalyzed by a dehydrogenase and results in oxidation of the aldehyde group to a carboxylic acid group. The proline catabolic pathway, which also involves five-carbon skeletons, is also shown in Figure 8.9.

## The Three-Step Series

A series of three reactions — dehydrogenation, hydration, and dehydrogenation — is a recurring theme in biochemistry. Dehydrogenation results in the conversion of an alkane to an alkene group. Hydration results in the introduction of a hydroxyl group (and a hydrogen) at the double bond. Dehydrogenation results in oxidation of the hydroxyl group to a keto group. This sequence occurs in leucine catabolism (see Figure 8.5), as well as in the Krebs cycle and fatty acid oxidation.

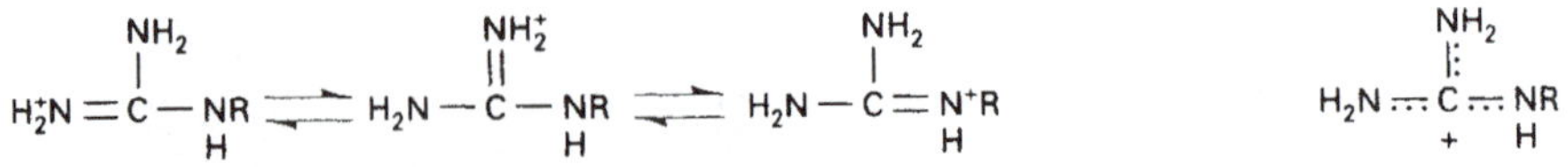

FIGURE 8.8  The guanidino group. The three resonating forms may be represented by the one form at the right.

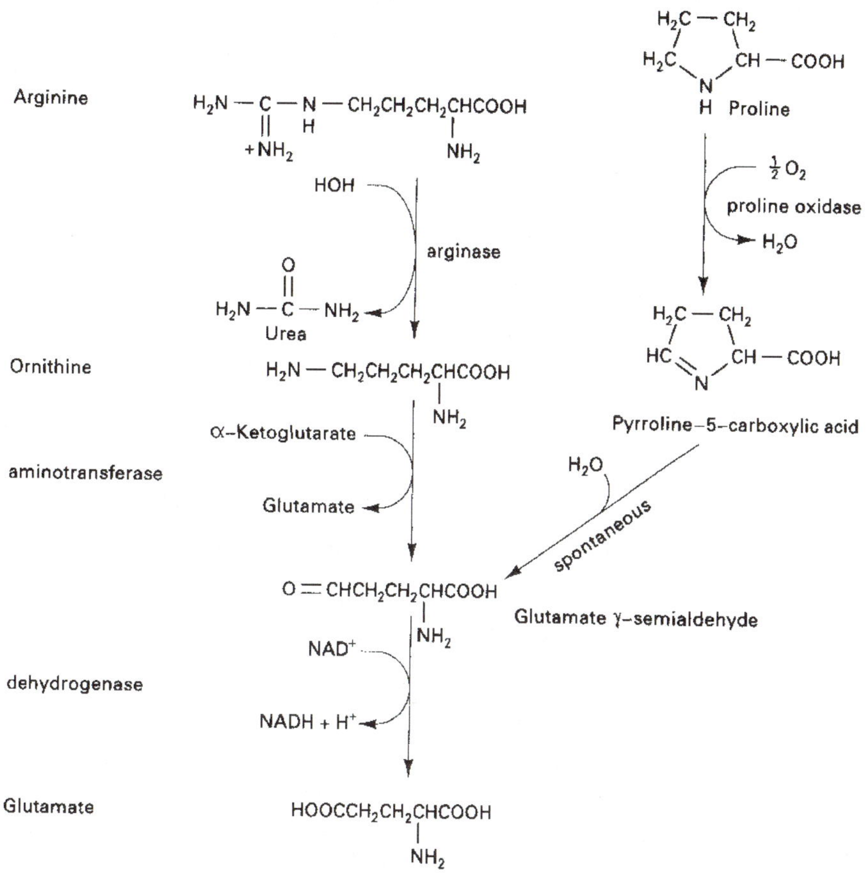

**FIGURE 8.9** Arginine and proline catabolism. Glutamate-γ-semialdehyde, an intermediate in the pathways of arginine and proline breakdown, received its name because its structure consists of glutamate, but with the γ-carboxyl group reduced to the aldehyde form.

## Enzyme Specificity

The concept that an enzyme recognizes, or shows a preference for, specific substrates is illustrated by the data in Table 8.3. Enzymes recognize specific substrates or classes of substrates. A very specific enzyme is glutamate dehydrogenase, which acts on glutamate, with little or no activity occurring with other amino acid substrates (Struck and Sizer, 1960). The fact that BCAA aminotransferase recognizes a particular class of substrates raises the question of how specific is the activity of the enzyme. Data are shown from an experiment with BCAA aminotransferase purified from rat brain (Table 8.3), conducted as follows. A series of test tubes contained identical quantities of the enzyme, as well as identical quantities of α-ketoglutarate. Different test tubes contained different cosubstrates, as indicated. The test tubes were incubated for a period to allow the enzyme to catalyze its reaction. The results demonstrate that transfer of the amino group from the cosubstrate to α-ketoglutarate occurred at or near the maximal rate with

**TABLE 8.3** Substrate Specificity of BCAA Aminotransferase

| Substrate | Percent maximal activity |
|---|---|
| Isoleucine | 100 |
| Leucine | 78 |
| Valine | 69 |
| Glutamate | 23 |
| Methionine | 5 |
| Phenylalanine | 2 |
| Tryptophan | 1 |
| Alanine | 1 |
| Aspartate | 1 |

*Source*: Hall *et al.* (1993).

isoleucine, leucine, and valine. Glutamate was fairly active, whereas methionine and the other amino acids tested were poor substrates.

### Methylmalonyl-CoA, a Metabolite Unique in Its Sensitivity to Vitamin B₁₂ Status

Breakdown of isoleucine, valine, threonine, and methionine results in the production of propionyl-CoA. Propionyl-CoA, in turn, is catabolized to succinyl-CoA via the intermediate methylmalonyl-CoA. Methylmalonyl-CoA is a compound of unusual interest to nutritional scientists. This compound accumulates in the cell during a vitamin $B_{12}$ deficiency. Vitamin $B_{12}$ deficiency is not a rare disease, as it appears in a common autoimmune disease called **pernicious anemia**. Vitamin $B_{12}$ deficiency also occurs in strict vegetarians who avoid meat, fish, poultry, and dairy products. Methylmalonyl-CoA can also build up with rare genetic diseases that involve the production of defective, mutant forms of methylmalonyl-CoA mutase. Most of the methylmalonyl-CoA that accumulates to abnormally high levels in the cell is hydrolyzed to **methylmalonic acid** (MMA), which leaves the cell for the bloodstream and eventual excretion in the urine. Some of the MMA is converted back to propionyl-CoA, resulting in the production and accumulation of propionic acid in the cell. The measurement of plasma and urinary MMA has proven to be a method of choice for the diagnosis of vitamin $B_{12}$ deficiency, whether induced by pernicious anemia or by dietary deficiency.

EXERCISE

Propionyl-CoA carboxylase may be defective in a rare genetic disease. Please devise a test for the detection of this disease.

EXERCISE

Propionate is used as a preservative to retard the growth of mold in bread and cake. Please devise an argument that this preservative is harmless to humans.

EXERCISE

A short-chain fatty acid thiokinase catalyzes the ATP-dependent conversion of propionic acid to propionyl-CoA. Which cells of the gastrointestinal tract might be expected to have high levels of short-chain thiokinase and of methylmalonyl-CoA mutase. Hint: see the section on Dietary Fiber in Chapter 3.

## BIOSYNTHESIS OF DISPENSABLE AMINO ACIDS

The dispensable amino acids can be synthesized in the body by pathways that are shared, in part, by amino acid catabolism. Glutamate, aspartate, and alanine are synthesized by reversal of the transamination reactions detailed earlier in this chapter. Serine is synthesized from an intermediate of the glycolysis pathway, 3-phosphoglycerate. 3-Phosphoglycerate is oxidized to 3-phosphohydroxypyruvate by an NAD-requiring enzyme. 3-Phosphohydroxypyruvate is then transaminated to yield 3-phosphoserine, which, via the action of a phosphatase, is hydrolyzed to produce serine. Glycine is synthesized from serine by the action of serine hydroxymethyltransferase (see Folate in Chapter 9). Cysteine is synthesized in the body from methionine and tyrosine from phenylalanine, as shown at the end of this chapter.

Proline is synthesized from glutamate. First, the γ-carboxyl group of glutamate is oxidized, yielding an aldehyde group. The aldehyde group condenses with the α-amino group, cyclizing and yielding a Schiff base. The Schiff base is then reduced to produce proline. The structures involved are the same as in the proline catabolic pathway, shown earlier in this chapter.

Arginine appears to be a dispensable amino acid in humans. Evidence suggests that this amino acid is synthesized in the body from glutamate. The terminal carboxyl group of glutamate is first phosphorylated, producing phosphoryl glutamate. This compound is then reduced to yield glutamate γ-semialdehyde plus phosphate. This step is followed by transamination to form ornithine. Ornithine is then converted to arginine via enzymes of the urea cycle, as detailed later. Arginine is only stimulatory for growth in rat and pig and is not an essential component of their diets. Arginine is indispensable for the growing cat and dog (Carey et al., 1987).

Glutamine synthase catalyzes the ATP-dependent synthesis of glutamine from glutamate and ammonia. In providing energy to drive this reaction, ATP is converted to ADP plus inorganic phosphate ($P_i$). Glutamine synthase is regulated in mammalian tissues, and the amount of enzyme protein can increase when the cell is deficient in glutamine (Milman et al., 1975).

Asparagine synthase catalyzes the ATP-dependent synthesis of asparagine from aspartate, using glutamine as the source of the amino group. In donating its amino group, glutamine is converted to glutamate. In providing energy for the reaction, ATP is converted to AMP plus inorganic pyrophosphate ($PP_i$).

## Glucogenic and Ketogenic Amino Acids

The amino acids can be classified according to whether they are glucogenic or ketogenic. This classification is based on the pathways of catabolism followed by

**TABLE 8.4**  Glucogenic and Ketogenic Amino Acids

| | |
|---|---|
| Purely ketogenic | Leucine, lysine |
| Both ketogenic and glucogenic | Isoleucine, phenylalanine, tyrosine, tryptophan |
| Purely glucogenic | Alanine, serine, glycine, cysteine, aspartic acid, asparagine, glutamic acid, glutamine, arginine, histidine, valine, threonine, methionine, praline |

the amino acid. Although all of the amino acids can be completely oxidized and their carbon skeletons broken down to $CO_2$, the carbon skeletons can also be conserved in liver metabolism and converted to glucose or ketone bodies. Amino acids that can be catabolized to produce glucose are called **glucogenic**. Amino acids that can be catabolized to produce acetyl-CoA or ketone bodies are called **ketogenic**. Some amino acids, when catabolized, yield two separate carbon skeletons, one of which is glucogenic and the other ketogenic. These amino acids are considered to be both glucogenic and ketogenic. Isoleucine, for example, is both glucogenic and ketogenic. Leucine and lysine are purely ketogenic. The classification is shown in Table 8.4.

### Leucine and Lysine — Purely Ketogenic

Leucine and lysine are purely ketogenic amino acids. The catabolism of these amino acids does not yield intermediates of the Krebs cycle. It does not yield pyruvate. It does not produce compounds that can result in the net synthesis of glucose. Leucine catabolism results in the production of a molecule of acetyl-CoA and a molecule of acetoacetate; lysine breakdown produces acetoacetyl-CoA.

### Entry of Glucogenic Amino Acids into the Krebs Cycle

The positions of entry of glucogenic components of the amino acids into the Krebs cycle are shown in Figure 8.10. The amino acids that can be oxidized to pyruvate can be converted to oxaloacetic acid (OAA), after which the carbon skeleton exits the mitochondrion in the form of malate and is converted to glucose. Amino acids that can be oxidized to α-ketoglutarate or to succinyl-CoA can be converted to OAA and then to glucose.

## WASTE NITROGEN

Many of the reactions described so far point to the concept of waste nitrogen. What happens to the ammonium ions discharged during catabolism of the various amino acids? What happens to the amino groups transferred from various amino acids to oxaloacetic acid, yielding aspartate? The ammonium ions and excess amino groups, carried in the form of aspartate, are handled in the following manner. They are incorporated into a small, water-soluble molecule known as

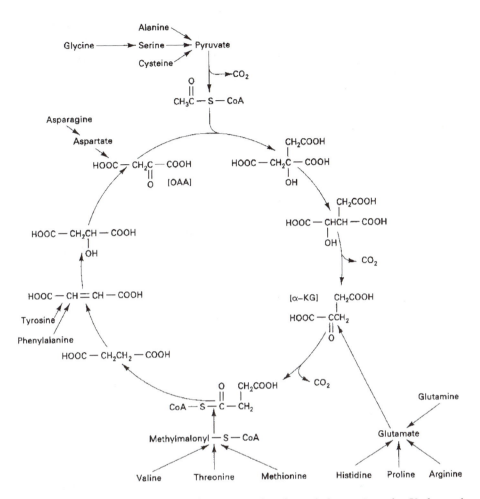

**FIGURE 8.10** Positions of entry of amino acid carbon skeletons into the Krebs cycle. Amino acids that are broken down to three-carbon skeletons may enter the Krebs cycle at the point of pyruvate. Glycine, after conversion to serine, can enter the Krebs cycle as pyruvate. Alternatively, glycine can be broken down by the glycine cleavage system (see Folate section). The products of this reaction are carbon dioxide and an ammonium ion. Five-carbon skeletons enter the Krebs cycle at the point of α-ketoglutarate, whereas four-carbon skeletons enter at the points of succinyl-CoA and oxaloacetic acid.

urea. Urea is excreted in the urine. Most of the waste nitrogen produced via amino acid catabolism is excreted in the form of urea. A small fraction of waste nitrogen is excreted in the form of free ammonium ions. The cycle of reactions in which waste nitrogen is packaged as urea is called the urea cycle.

## Glutamate and Aspartate as Carriers of Waste Nitrogen

The first step in the breakdown of a variety of amino acids results in transfer of the α-amino group to α-ketoglutarate, yielding glutamic acid. What happens to the glutamic acid formed? The amino group of glutamate can be transferred to oxaloacetate in a reaction catalyzed by glutamate–oxaloacetate aminotransferase. Glutamate may be seen as intermediate in the pathway for disposing of the α-amino groups during the breakdown of a variety of amino acids. Aspartic acid is produced from oxaloacetate. Aspartic acid is a carrier of the amino group discharged from the initial breakdown of the catabolized amino acid. Aspartate acts as the penultimate carrier of waste nitrogen in the body; urea is the ultimate carrier.

## Urea Cycle

The urea cycle, also called the ornithine cycle, was discovered by Hans Krebs at the University of Freiburg in Germany, in 1932 (Holmes, 1993). The cycle is depicted in Figure 8.11. A priming reaction is employed for introducing the nitrogen atom of the ammonium ion into the cycle. Carbamyl phosphate synthase catalyzes this priming, or activation reaction (Step 1). The reaction results in the packaging of one nitrogen atom in the form of carbamyl phosphate and requires

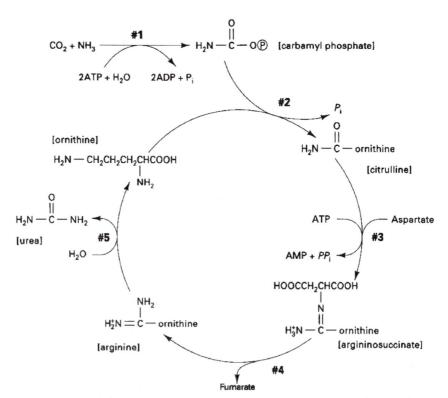

FIGURE 8.11 The urea cycle. The urea cycle is based on the ornithine molecule. Ornithine acts as a chemical scaffold, on which a moiety of urea is built. The finished moiety of urea is discharged by hydrolysis, producing free urea and regenerating the original molecule of ornithine.

the expenditure of two ATPs. This means that energy is required for the disposal of waste nitrogen. The nitrogen atom is introduced into the urea cycle at Step 2. Ornithine transcarbamylase catalyzes the reaction indicated in Step 2. Ornithine is converted to citrulline in this step. The structure of citrulline is depicted in the form of carbamyl-ornithine. This emphasizes the fact that the ornithine molecule is the basic structure used to carry the nitrogen atoms introduced into and discharged from the urea cycle.

A second nitrogen atom is introduced into the urea cycle in a set of reactions catalyzed by argininosuccinate synthase (Step 3) and argininosuccinate lyase (Step 4). The intermediate is argininosuccinate. Arginine is the product formed after introduction of the second nitrogen atom into the urea cycle. The structure of arginine is shown as a derivative of ornithine. Step 3 in Figure 8.11 shows that one molecule of ATP is used to drive the reaction. Cleavage of only one phosphate–phosphate bond occurs in the reaction. Hence, one might be tempted to say that the reaction requires the equivalent of only one molecule of ATP; however, the reaction is unusual in that the product is not $P_i$, but $PP_i$. Whenever $PP_i$ is produced in the body, it is promptly hydrolyzed to two molecules of $P_i$. Hence, Step 3 requires expenditure of the equivalent of two molecules of ATP. In other words, Step 3 requires the energy equivalent of the conversion of 2 ATP to 2 $P_i$ + 2 ADP. The total amount of energy used in Steps 1 to 3 is equivalent to 4 ATP, that is, the conversion of 4 ATP to 4 ADP + 4 $P_i$. Arginase catalyzes the final step in the urea cycle (Step 5). The reaction involves hydrolysis of the guanidino moiety from arginine, producing urea. For each nitrogen entering the urea cycle at Step 3 (see Figure 8.11), a nitrogen must enter at Step 2 for operation of the urea cycle.

Figure 8.12 illustrates the role of the urea cycle in accepting waste nitrogen from a variety of sources and in packaging it in the form of urea.

## Compartmentation of the Urea Cycle

The urea cycle may be considered to be a mitochondrial pathway, as carbamyl phosphate synthase and ornithine transcarbamylase are mitochondrial enzymes; however, the enzymes catalyzing subsequent steps of the pathway are cytosolic. The steps leading to conversion of citrulline to ornithine occur in the cytosol. Hence, the pathway is shared by the mitochondrial and cytosolic compartments. The fumarate produced by the urea cycle is converted to malate by a cytoplasmic form of fumarase. Mitochondrial fumarase is part of the Krebs cycle. Cytoplasmic malate can enter the mitochondrion by means of a transport system, such as the malate/phosphate exchanger or the malate/α-ketoglutarate exchanger. These transport systems are membrane-bound proteins.

## Ammonia

Ammonium ions are produced by the catabolism of a number of amino acids. Glutamate dehydrogenase is the major source of ammonium ions in the body. Ammonium ions are also produced from the catabolic pathways of serine, histidine, tryptophan, glycine, glutamine, and asparagine. L-Amino acid oxidase and

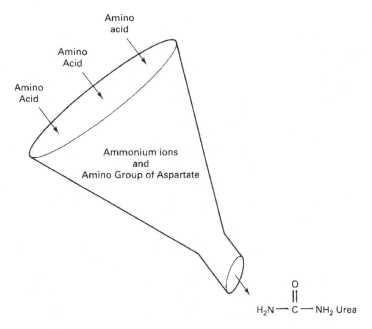

**FIGURE 8.12** Overall direction of flux of waste nitrogens derived from amino acid catabolism. The waste nitrogens are "funneled," via free ammonium ions and aspartate, to produce urea.

D-amino acid oxidase are minor sources of ammonium ions in the body. Apparently, most blood ammonia originates directly from the catabolism of dietary protein in the gut. The digestion of dietary protein and catabolism by enzymes in the enterocytes and gut microflora produce ammonium ions. The ammonia enters the circulation via the portal vein, where it travels to the liver for eventual packaging as urea. Much of the waste nitrogen produced in tissues such as skeletal muscle and brain is packaged as glutamine. Glutamine synthase catalyzes the conversion of ammonia and glutamate to glutamine (Figure 8.13). Glutamine serves as a temporary carrier of waste nitrogen in the bloodstream, acting to reduce levels of free plasma ammonia and to prevent neurotoxicity (Souba, 1987).

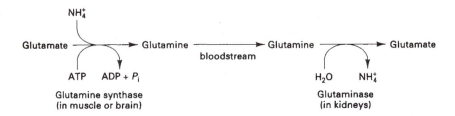

**FIGURE 8.13** The role of glutamine in transporting waste nitrogen through the bloodstream. Glutamine synthase catalyzes the synthesis of glutamine in an ATP-requiring reaction. Glutaminase catalyzes the hydrolysis of the amide group of glutamine, producing glutamate and an ammonium ion.

The glutamine travels through the bloodstream and is taken up by a number of organs. The glutamine that enters the cells of the intestines and kidney tends to be deaminated, releasing ammonium ions (see Figure 8.13). Glutaminase catalyzes the hydrolysis of the ammonium group from glutamine. The ammonia produced in the gut can travel to the liver for packaging as urea. The ammonia produced in the kidney can be excreted in the urine, as ammonium ions. Glutamine, as a carrier of waste nitrogen, works in conjunction with the alanine cycle, as shown in Figures 4.39 and 4.40.

Amino acids are not the only source of ammonium ions produced in the body. Much of the ammonia produced, especially in the brain, arises from the hydrolysis of purines. Adenylate deaminase catalyzes the hydrolysis of AMP, yielding IMP and ammonium ions (Cooper and Plum, 1987). IMP is inosine monophosphate (inosinic acid). GMP may also be hydrolyzed in this manner, yielding xanthosine and ammonium ions. Further details on purine metabolism occur at the end of this chapter and under Folate in Chapter 9.

## Toxicity and Packaging of Waste Nitrogen Illustrated by Experimental Nutrition

In considering the oxidation of dietary protein, or tissue protein, the fate of the nitrogen atoms in the protein is a major concern. Most of the nitrogen atoms are packaged in the form of urea, and excreted as such; the residual nitrogen is generally excreted as ammonium ions. It might prove interesting to view three nutritional studies on waste nitrogen. The first study is quite dramatic and involves cats. The second study concerns humans consuming various amounts of protein. The third experiment involves the intestines of rats.

Arginine is required in the diet of growing cats, but is not required by humans. The cats used in the study were fed diets deficient in arginine. One group of cats consumed the deficient diet, while a second group consumed the diet containing an ornithine supplement. Blood samples were taken at the indicated times following the meals (Table 8.5). The concentration of plasma ammonium ions was measured over time. The plasma ammonium concentration rose from a basal level of about 50 mM well over tenfold, in the hours following the consumption of the

**TABLE 8.5** Plasma Concentration of Ammonium Ions Following the Test Meal

| Time after meal (min) | Diet | |
|---|---|---|
| | – Arginine, – ornithine | – Arginine, + ornithine |
| | $\mu M\ NH_4^+$ | |
| 0 | 54 | 50 |
| 45 | 183 | 140 |
| 90 | 638 | 341 |
| 150 | 786 | 240 |

*Source*: Morris and Rogers (1978).

unsupplemented meal. At 2 to 3 hours after the meal, some of the cats suffered tetanic spasms and died. These effects are due to the toxic effect of ammonium ions on the nervous system. The increase in plasma ammonia was tempered in the cats whose diets were supplemented with ornithine. Here, the increase in ammonium levels was only about fivefold over the basal level. The cats consuming the protein meal with ornithine did not have spasms and did not die.

EXERCISE

How could ornithine have the remarkable effect just described?

The second study concerned humans. The subjects consumed various amounts of a meal consisting of milk and soy protein, at levels of 1, 2, or 4 g of protein/kg of lean body mass. (The term **lean body mass** refers generally to the mass of the metabolically active part of the body, i.e., of the cells. The lean body mass excludes the parts of the body that are relatively inactive metabolically, that is, fat, extracellular fluids, and mineral deposits.) Urine was collected over the course of several hours following the protein meal. The urinary urea was measured, as indicated in Figure 8.14. A steady flow of urine was induced by having the subjects drink 200 ml of water every half-hour. Without the water, a concentrated solution of urea would have remained in the bladder. The data in Figure 8.14 demonstrate that urinary urea concentrations may increase within an hour of consuming protein. The amount of urea excreted increases with the amount of protein consumed. In addition, the duration of elevated excretion is longer with larger amounts of dietary protein.

EXERCISE

What percentage of the nitrogen consumed was excreted within 8 or 9 hours of consuming the protein meals (see Figure 8.14)?

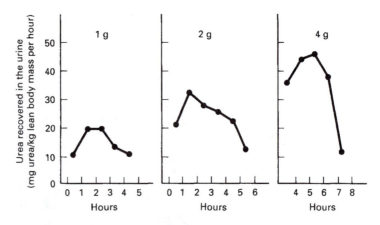

**FIGURE 8.14** Urea excreted following the consumption of 1, 2, or 4 g of protein. (Redrawn with permission from Rafoth and Onstad, 1975.)

EXERCISE

Generally, adults consume about 100 g of protein per day. Was the amount of protein consumed by the subjects (see Figure 8.14) higher or lower than an amount that might be eaten in a typical meal?

The third study involved rats. A solution containing an amino acid was dispensed into the lumen of a rat's intestines. The amino acid was glutamine. All of the blood leaving the portal vein was collected for analysis. The portal vein is the route followed by water-soluble metabolites absorbed from the diet, from the small intestines to the liver. To prevent a net loss of blood, the rat received a continual infusion of blood from donor rats. Various nitrogen-containing metabolites were detected in the portal blood. The data in Table 8.6 reveal the fate of the nitrogen of the glutamine instilled into the gut. The results reveal that most of the nitrogen appeared as ammonium ions, citrulline, and alanine.

The recovery of much of the nitrogen in the form of ammonium ions can be accounted for by the reaction of glutaminase. Glutaminase catalyzes the hydrolysis of the amide moiety of glutamine (see Figure 8.13). Some of the ammonium ions may have originated from glutamic acid via the action of glutamate dehydrogenase. The results also reveal that a substantial fraction of the nitrogen appeared in the form of alanine. The alanine probably arose via the action of glutamate–pyruvate aminotransferase. The data indicate that there was very little increase in glutamic acid, and no detectable glutamine, in the portal vein. One might wonder what happened to the carbon skeletons of the glutamine instilled into the lumen. Separate experiments showed that much of the glutamine was completely oxidized to $CO_2$. It is thought that the complete oxidation of glutamate and glutamine contributes about half to the energy requirement of the enterocyte. In this manner, the enterocyte is different from most cells. Most cells derive their energy primarily from fatty acids and/or glucose (Souba *et al.*, 1985; Windmueller and Spaeth, 1978).

EXERCISE

By which biochemical pathway did the rat gut cells produce citrulline in the study depicted in Table 8.6?

**TABLE 8.6**  Nitrogenous Metabolites from Gut Cells

| Metabolite appearing in portal vein after perfusing gut with glutamine | Percent of total nitrogen released into the portal vein |
|---|---|
| Ammonia | 40 |
| Citrulline | 30 |
| Alanine | 25 |
| Glutamate | 2 |

*Source*: Windmueller and Spaeth (1978).

## REGULATION OF PROTEIN CATABOLISM

The regulation of the catabolism of proteins and amino acids in the body is a complex topic. One goal of the regulatory mechanism might be to increase the rate of amino acid catabolism when dietary proteins are in excess of the body's needs. Another goal might be to allow the selective catabolism of nonvital proteins in the body, during starvation, to support gluconeogenesis. The less vital protein includes close to half of the body's skeletal muscle; the more vital protein includes that of the nervous system, for example. Mechanisms must exist during long-term starvation to allow the net degradation of nonvital proteins while conserving those that are more important for survival.

### Different Styles of Protein Catabolism

Proteins are degraded in cells by various systems, including the ubiquitin-dependent pathway, macroautophagy, and microautophagy. Ubiquitin, a small protein of 76 amino acids, is attached to damaged or abnormal proteins by means of an ATP-dependent reaction. The ubiquitin molecule serves as a tag that directs the faulty protein to be hydrolyzed by a particle called the proteasome.

Macroautophagy and microautophagy are processes involving small vesicles or vacuoles and occur in the cytoplasm. Macroautophagy involves capture of parts of the cytoplasm by a membrane, followed by hydrolysis of the captured proteins within a vesicle. Microautophagy involves capture of smaller portions of the cytoplasm by very small vesicles. These vesicles deliver their contents to the lysosomes. Lysosomes are organelles containing a large variety of hydrolytic enzymes. In addition, muscle contains Ca-dependent proteases that are thought to be used for the degradation of contractile proteins (see Calcium in Chapter 10).

The proteasome is a barrel-shaped particle in the cell that is used for the degradation of proteins to small peptides. In mammals, the proteasome consists of 28 polypeptides and has a molecular weight of 2,000,000. The proteins of this particle constitute about 1% of the all the protein in the cell (Goldberg, 1995).

The proteasome is used for the normal turnover of all proteins, and serves to degrade proteins that have sustained some sort of damage during the course of the day. Proteasomes are vital for degrading signaling proteins (cyclins and transcription factors), which, in some circumstances, need to be present in the cell for brief and limited periods of time. When a protein sustains some sort of damage, it becomes tagged with ubiquitin, in an ATP-requiring reaction catalyzed by ubiquitin-conjugating enzyme. This step occurs on the cytosolic surface of the endoplasmic reticulum (Biederer et al., 1997; Wilkinson, 1995). The proteasome contains a doorway, or gate, which contains special proteins that recognize and unfold the tagged protein (Stuart and Jones, 1997). The unfolded protein then passes through the gate and into the interior of the proteasome, where a variety of proteases catalyze the degradation of the target protein to peptides of 7–10 amino acids in length. Five types of proteases occur in the mammalian proteasome. During fasting, the ubiquitin-dependent pathway is activated, thus stimulating the degradation of proteins (in the proteasome) and supporting enhanced gluconeogenesis.

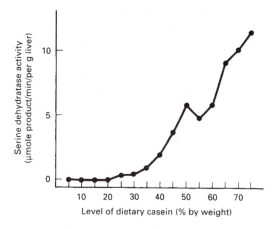

**FIGURE 8.15** Enzyme activity versus level of protein in the diet. Each point represents enzyme activity in livers from rats fed the indicated diet. (Redrawn with permission from Peters and Harper, 1985.)

### Regulation of Protein Catabolism as Illustrated by Studies with Rats

Increases in the activities of amino acid catabolic enzymes during the consumption of high-protein diets have been clearly documented. Examples of the adjustments occurring with high-protein diets are illustrated by two studies involving rats (Figures 8.15 and 8.16). In the first study, rats were fed diets containing different amounts of protein, as indicated in Figure 8.15. Each diet was fed for 1.5 weeks, after which the liver was removed and the activity of serine dehydratase determined. The data demonstrate that the activity of the enzyme, as measured *in vitro*, increased dramatically and progressively once the protein level increased beyond 30%, by weight, of the dry diet. This type of experiment has been used to help define the dietary requirement for protein. In short, a dietary protein level that

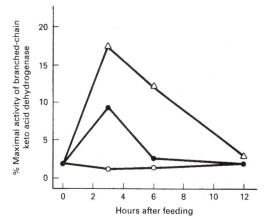

**FIGURE 8.16** Enzyme activity versus time after feeding. (Redrawn with permission from Block *et al.*, 1987.)

induces an increase in serine dehydratase activity might be expected to be above the minimal requirement, whereas a level that induces little or no increase might be expected to be at or below the requirement.

The second study also involved rats. All of the rats were raised on a diet containing 25% protein for 2 weeks. The protein was casein. On the day of the experiment, the rats were fed test diets containing 9% protein (O), 25% protein (●), or 50% protein (Δ). The rats were sacrificed at various times after the test meals, as indicated in Figure 8.16, and used for assays of an enzyme used for BCAA catabolism. Muscle was used as the source of enzyme. The results demonstrate that the enzyme activity was clearly higher in rats fed diets containing elevated amounts of protein.

In addition to regulation occurring at the levels of proteolysis and amino acid degradation, the regulation of protein catabolism involves the regulation of enzymes of the urea cycle. The urea cycle is used for packaging and disposing of most of the waste nitrogen formed in the body and for preventing plasma concentrations of ammonium ions from rising to toxic levels. Carbamyl phosphate synthetase requires **N-acetylglutamic acid** for activity. This compound binds to the enzyme, resulting in its activation. N-Acetylglutamate is synthesized in the mitochondria by acetylation of the amino group of glutamate. Studies with rats have revealed that injecting amino acids into the bloodstream can result, within minutes, in threefold increases in the concentration of N-acetylglutamate in the liver and in the activity of carbamyl phosphate synthetase. Apparently, the increase in concentration of N-acetylglutamate parallels the increase in the concentration of mitochondrial glutamate arising from the conversion of various amino acids to glutamate, as well as from glutamate itself (Stewart and Walser, 1980). Regulation by N-acetylglutamate has been proposed as a means for rapidly adjusting the rate of urea synthesis in response to high levels of dietary protein.

The activities of all five enzymes of the urea cycle increase with consumption of a high-protein diet (Kato *et al.*, 1978; Schimke, 1962). This increase occurs somewhat slowly. Maximal adjustment takes place over the course of several days of a high-protein diet. The stimulatory effect of the high-protein diet has been traced to specific amino acids. Methionine, glycine, and alanine appear to be more potent than the other amino acids in inducing increases in the activities of the urea cycle enzymes (Snodgrass and Lin, 1981). The mechanisms involved in controlling the rates of protein breakdown and synthesis are complex. Our understanding of these mechanisms is rather fragmentary. This is because several different pathways are used to hydrolyze proteins: macro- and microautophagy, the ubiquitin-dependent pathway, and the calcium-dependent proteases. In addition, the rate of protein breakdown is influenced by insulin and glucagon, as well as by specific amino acids.

Glucagon is often associated with the stimulation of protein breakdown in the liver, though it may reduce protein breakdown in muscle. Insulin is often associated with a decrease in protein degradation in liver and muscle. Studies with rats have revealed that brief fasting results in a large protein loss in the liver, but not in the muscle. More specifically, the rough endoplasmic reticulum of the liver is degraded with brief fasting. In muscle, the noncontractile proteins are readily degraded, but with prolonged fasting the contractile proteins are lost through proteolysis. Apparently, the process of macroautophagy is regulated more than the other pathways. It is stimulated by an amino acid deficiency. This effect has been traced to specific amino acids. A deficiency in leucine, in particular, may stimulate macroautophagy. Apparently, contractile proteins are not degraded by macroauto- phagy. They are catabolized by the calcium-activated proteases.

## NUTRITIONAL ASPECTS OF PROTEIN CATABOLISM

The remainder of this chapter takes on a more "nutritional" aura. It concerns the concepts of growth and maintenance. The material may be especially interesting to those involved in infant nutrition, animal science, and weight reduction by low-energy diets. The material reveals the techniques used for determining the protein requirement and the recommended daily allowance (RDA) for protein. Clinical issues involving protein nutrition, such as renal disease and gout, are also outlined.

### Nitrogen Balance

**Nitrogen balance** is the difference between the amount of nitrogen consumed per day and the amount of nitrogen excreted per day. This definition can be expressed by the formula

$$\text{Nitrogen balance} = (\text{grams of nitrogen eaten} - \text{grams of nitrogen lost})$$

Nitrogen balance experiments require measurement of the amount of nitrogen consumed per day for several days. This quantity is easy to measure for several reasons. All of the nitrogen consumed enters the body via only one opening, namely the mouth. The amount of N consumed can be determined from the weight of the amount of food consumed and a knowledge of the N content of the food. The N content of the food can be determined by chemical analysis. The amount of N consumed can be controlled in advance, at least in the case of humans. Nitrogen balance experiments also require measurement of the amount of N excreted per day for several days. This measurement requires the collection of urine and feces. Quantitative collection of these materials, homogenization of the feces, and analyses of N content of both urine and feces can be arduous, particularly when hundreds of samples are involved. In the hands of exacting researchers, the measurement of excreted nitrogen also involves the collection and analysis of hair, sweat, semen, menstrual loss, and tooth brushings.

A meaningful value for N balance cannot be determined by analyzing the food and excreta collected during 1 day. Because of the biological variability involved in experiments involving animals and humans, samples must be collected over several days. The study described next involved measurement of the daily N balance over 5-day periods.

## Growth

What is growth? Growth may mean an increase in height over a period. Is it possible to have growth but without an increase in height? Growth may also mean an increase in weight with or without an increase in height. An increase in weight can, however, be caused by accumulation of fat or water retention. Weight increases as a result of these factors are not considered growth. Growth may be considered to involve an increase in the lean body mass. Lean body mass is defined as the total mass of the bodily tissues, excluding the adipose tissue, mineral deposits, and extracellular fluids such as blood plasma, lymph, and gut lumen fluids. Roughly, lean body mass is the mass of the body that is metabolically active, to the exclusion of tissues that are relatively inactive.

Growth can be measured in several ways. It can be measured by *counting the increase in the number of cells* in an organism. This technique is practical only for animals that have a small number of cells, such as nematodes (small worms). Growth can be measured by *measuring the amount of DNA* in the organism. This technique has been used by nutritional scientists for measuring growth of specific tissues, such as the brain, with the consumption of different diets (Zamenhof *et al.*, 1971). The preceding techniques fail to take into account that some cells are larger than others. White blood cells are much larger than red blood cells. Skeletal muscle cells are much larger than smooth muscle cells. In measuring growth, the researcher is interested in the increase in cell numbers, but also in the contribution due to increase in cell size. The contribution to growth by increase in cell number and size can be estimated by measuring the body's level of potassium, as revealed in the coming paragraphs.

### Positive N Balance Indicates Growth

Nitrogen balance data are presented in Table 8.7. This study involved human subjects who consumed different diets containing different mixtures of amino acids. The amount of N consumed (12 g) was fairly constant in the different diets. The amounts of nitrogen excreted daily in the urine and feces are also listed. One might be surprised to see that the amount of nitrogen excreted in the urine is many times greater than that excreted in the feces. The N balance is also recorded for each of the dietary periods. A positive N balance indicates that the subject is increasing in lean body mass. Lean body mass increases during growth, pregnancy, muscle building, and recovery from starvation.

### Accumulation of Potassium Indicates Growth

Growth can be assessed by measuring the content of potassium ions in the body on the following basis. Potassium ions occur mainly in intracellular fluids, in which the K concentration is about 142 mM. Extracellular fluids, on the other hand, contain much lower levels of potassium, about 10 mM. Adipose tissue is largely water free and thus contains minimal amounts of fluids and salts. The potassium ion content may be determined in growing humans. How is it possible to measure the total content of K ions in a living person? A fraction of the potassium in the

**TABLE 8.7**  Nitrogen Balance Experiment

| Dietary period (days) | Daily nitrogen intake (g) | Daily nitrogen output (g) | | Average daily nitrogen balance (g) |
|---|---|---|---|---|
| | | Urine | Feces | |
| 5 | 12.1 | 10.4 | 0.55 | +1.2 |
| 3 | 12.1 | 12.1 | 0.66 | −0.63 |
| 5 | 12.2 | 11.2 | 0.45 | +0.57 |
| 5 | 12.2 | 12.0 | 0.51 | −0.25 |

*Source*: Nakagawa *et al.* (1961).

body is naturally radioactive. The natural abundance of radioactive potassium ($^{40}$K) is 0.0118%. $^{40}$K emits $\gamma$-rays. The $\gamma$ radiation released by the potassium in the human body can be measured with a machine called a whole-body counter. The adult body, for example, contains about 0.1 microcurie of $^{40}$K. The microcurie is a unit of radioactivity. The whole body counter has been used in a variety of studies involving children and adults; however, the smaller quantities of potassium ions in small animals and infants makes it difficult to measure accurately the amount of intracellular fluid and, thus, the lean body mass in these subjects.

Studies with pregnant teenagers revealed that the subjects stored or accumulated about 3.4 mmol of potassium per day (King *et al.*, 1973). The rate of accumulation of potassium was determined by measuring the increase in $^{40}$K with a whole-body counter. The ratio of potassium to total nitrogen in tissues is relatively constant. From the known, constant ratio it was calculated that the increase of 3.4 mmol of potassium per day was equivalent to an increase of 1.6 g of nitrogen per day (King *et al.*, 1973). The value of 1.6 g of N means that the subjects were in a state of positive N balance. Such experiments have been used to determine the increment in dietary protein required to support growth during various stages of life. Other work has revealed that the ratio of potassium to lean body mass differs in men and women (Lukaski, 1987). Once the total body K has been determined, the lean body mass can be estimated from this ratio. The lean body mass determined at two different times reveals the growth taking place between readings.

The $^{40}$K technique may produce data that are not easy to interpret. The method requires a long period between readings to detect growth. Shorter periods, that is, less than a few months, may not produce usable data. Another problem is that the ratio of K to total body N may differ in humans at different stages of life or with gender. Another problem is that the values for N retention may differ when determined by the $^{40}$K technique and by the N balance technique. The values for N retention in the study by King *et al.* (1973) were determined by both techniques. The results from the two methods were similar, though the factors leading to their slight disagreement remain unclear. The relationships between dietary protein and growth are illustrated next by studies involving rats and monkeys. The studies concern the measurement of N balance, responses to consumption of proteins of differing quality, responses to consumption of incomplete proteins supplemented with amino acids, and responses to consumption of complementary proteins.

*Protein Requirement during Growth*

The protein requirement can be determined from the amount of protein required in the diet to maintain nitrogen balance, in the case of nongrowing animals, and from the amount required to support a maximal positive N balance, in the case of growing animals. The following study involved growing animals, infant monkeys. Figure 8.17A depicts data from monkeys fed lactalbumin-based diets. The study involved 15 different monkeys. Each animal was fed a diet of milk, in which each milk bottle contained a formula with a different concentration of protein. All of the diets contained lactalbumin at a concentration of 3.8% or less, with carbohydrate making up the difference. This substitution is reasonable because carbohydrate and protein are isoenergetic; that is, one gram of either of these nutrients supplies about the same amount of energy. Each monkey received the diet during a 6-day trial. The amount of nitrogen intake was calculated from the volume of milk consumed by the monkey. The N balance was calculated by subtracting the amount of N excreted (urinary + fecal) from the amount of N consumed over the entire trial. The monkey drinking the milk lowest in protein, for example, received 0.1 g of nitrogen per day, as recorded on the x axis. The monkey drinking the milk highest in protein received 0.45 g N per day (see Figure 8.17A). The entire experi-

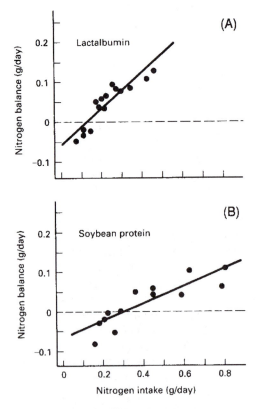

**FIGURE 8.17** (A, B) Nitrogen balance data obtained with different diets. (Redrawn with permission from Ausman *et al.*, 1986.)

ment was repeated using milk based on soybean protein. The results from this study appear in Figure 8.17B.

### EXERCISE

What intake of lactalbumin is sufficient to maintain an infant monkey in zero N balance?

### EXERCISE

What was the maximal rate of growth (rate of maximal accumulation of N) that could be supported by the lactalbumin-based diet?

### EXERCISE

What percentage of the N consumed was retained by the monkeys that consumed the lactalbumin-based milk? What percentage was retained with the soy-based milk?

### EXERCISE

Would you expect the line to intercept the same point on the *y* axis with the lactalbumin and soy groups, or a different point?

### EXERCISE

Is there an upper limit to the line in the N balance curves? Think of two or three factors that may impose an upper limit on the value for the daily N balance.

### EXERCISE

Imagine that the points in Figure 8.17A have been erased. Draw the results of an N balance study using *adult* monkeys.

## Soybean Protein versus Lactalbumin

The study in monkeys involved lactalbumin or soybean protein-based diets. Nitrogen balance was attained with the feeding of relatively low levels of **lactalbumin** (0.15 g N/day) but relatively *high levels of soy protein* (0.30 g N/day). Another point is that lactalbumin supported a positive N balance at lower concentrations than did soy protein. These findings indicate that lactalbumin is a higher-quality protein than soy protein. The concept of protein quality is explored next. Lactalbumin is a higher-quality protein than soy protein, as the soy protein is somewhat deficient in methionine. For this reason, animal diets based on soybean protein are usually supplemented with methionine.

## Why Do Adults Need to Eat Protein?

Growing animals need to eat protein because the newly formed tissues and fluids of the body contain protein. But why do adult, nongrowing animals need to eat protein? The answer is that all of the proteins in the body are in a state of flux. The amide groups of asparagine and glutamine of all proteins slowly and spontaneously hydrolyze over the course of several days or weeks, resulting in abnormal proteins. The proteins of the body are continuously being hydrolyzed to their component amino acids and then resynthesized. This phenomenon of hydrolysis and resynthesis is called **protein turnover**. During the hydrolysis of a protein in the body, a fraction of the free amino acids enter the pathways of amino acid catabolism, thus leading to their permanent loss. Small amounts of intact protein and intact amino acids are also lost in the urine and feces. Fecal protein and amino acids arise, in part, from the proteins of the pancreatic juices and from cells sloughed off from the gut mucosa. Other nitrogenous compounds, such as creatinine and uric acid, are lost in the urine. The amount of nitrogen lost from the body as a result of the breakdown of the body's own proteins and amino acids is called the **obligatory nitrogen loss**. Dietary protein must be consumed to replace these losses. The amount of protein required to replace the obligatory losses is called the **protein required for maintenance**. Adult animals, as well as growing animals, have maintenance requirements for protein, but only growing animals have the additional requirement of extra protein for growth.

A species difference has been found regarding the percentage of dietary protein used to support growth versus maintenance (during growth). In growing rats, about 95% of the dietary protein is used for growth. In the growing pig, about 90% of the dietary protein is used for growth, with 10% for maintenance. In the 6-month-old human infant, about 50% is used for growth (and 50% for maintenance), while in the growing 5 year old child, only 5% of dietary protein is used for growth (Young and El-Khoury, 1995).

The concept of protein turnover may seem a bit abstract. A concrete example of protein turnover follows. About 75 g of human protein enters the gut each day. Some of this protein is in the form of gastric and pancreatic enzymes. Some is in the form of the enterocytes that constantly slough off into the lumen of the gut. Eventually, most of the protein is degraded and hydrolyzed in the gut to amino acids. A fraction of these amino acids are broken down and then immediately oxidized in the Krebs cycle. The value of 75 g of human-derived protein is not very different from the typical daily intake of dietary protein — 100 g. The breakdown of protein to its constituent amino acids is part of protein turnover. The breakdown of the amino acids to simpler molecules, such as $CO_2$, is amino acid turnover. The turnover of the body's proteins affects not only the proteins released into the gut lumen, but all proteins of the body. Some proteins turn over relatively rapidly, as shown in Table 8.8, whereas others such as collagen and other structural proteins turn over quite slowly. Ornithine decarboxylase turns over with a half-life of about 0.2 hr and thus has a relatively rapid turnover rate. This means that any given molecule of the protein has a 50% chance of being degraded during a 12-min period. RNA polymerase II is relatively stable in the cell and has a half-life of about half a day.

What can be accomplished by protein turnover? One benefit is the removal of proteins that have sustained spontaneous hydrolytic damage or oxidative damage. Another benefit is that protein turnover makes possible the control of enzyme levels by means of genetic regulation. In a condition where an *increase* in the activity of a particular enzyme is needed, the rate of transcription of the mRNA coding for the enzyme can be increased. This increase results in an increase in translation, that is, the synthesis of the protein's polypeptide chain. In a condition

**TABLE 8.8** *In Vivo* Half-Life of Various Proteins

| Protein | Half-life (hr) |
|---|---|
| Ornithine decarboxylase | 0.2 |
| RNA polymerase I | 1.3 |
| Tyrosine aminotransferase | 2.0 |
| Serine dehydratase | 4.0 |
| PEPCK | 5.0 |
| RNA polymerase II | 12.0 |
| 3-Phosphoglycerate dehydrogenase | 15.0 |

*Source*: Goldberg and St. John (1976).

where a *decrease* in the activity of a particular enzyme is needed, the rate of transcription can be decreased. If protein turnover did not occur, a decrease in the rate of transcription would not result in a prompt drop in the enzyme's activity in the cell. Enzymes whose activity can be rapidly decreased by regulation at the genetic level might be expected to undergo a relatively rapid rate of turnover.

## Obligatory Nitrogen Losses of the Adult

The following study concerns nitrogen excretion data and involves humans. It was designed to determine the obligatory losses of nitrogen. Subjects consumed a diet that supplied 150 g of glucose per day. The diet contained no protein. During the first day, about 11 g of nitrogen was excreted in the urine and about 0.8 g in the feces. Over the course of several days on the glucose-only diet, urinary N stabilized at about 3.0 g/day and fecal N at about 0.8 g/day. With stabilization of the obligatory N losses, the total N losses amounted to about 3.8 g N/day. This value is equivalent to about 24 grams of the body's protein lost per day. The urinary losses of N, as urea and ammonia, during the course of the glucose-only diet are shown in Figure 8.18. The data demonstrate that most of the urinary N losses took the form of urea, not ammonium ions.

## Sparing of Protein by Carbohydrate

### The Total Fast versus the Glucose Fast

The subjects described in the preceding study (see Figure 8.18) were not strictly fasting. They consumed glucose as a source of energy. The goal of the work was to determine the obligatory nitrogen losses; however, the purpose of the glucose may seem obscure at this point. Why was glucose given to the subjects? In the real world, diets usually contain a certain amount of protein, carbohydrate, and fat. The fact that diets usually contain carbohydrates suggests that the researchers gave the subjects glucose to simulate a real-world situation. This notion, however, does not rigorously justify the inclusion of the glucose. Glucose was given to the subjects to reduce, or spare, the catabolism of the body's proteins and amino acids.

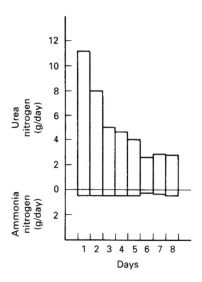

**FIGURE 8.18** Amount of waste nitrogen in urine over the course of a week. (Redrawn with permission from Aoki *et al.*, 1975.)

The concept of "protein sparing" has been known since the 1940s (Gamble, 1948; Holliday, 1993), when it was discovered during studies designed to create a better life raft ration. This work demonstrated that, with total fasting, a 70-kg man loses an amount of nitrogen equivalent to 80 g of protein per day. The work also revealed that about half of these losses could be prevented by feeding carbohydrates to the fasted man. The maximal effect of the glucose in preventing or sparing the N losses was about 50%. The glucose brought the losses of protein down to the equivalent of about 40 g of protein per day.

The studies depicted in Figure 8.19 illustrate changes taking place during the transition from a total fast to a carbohydrate-only diet. Subjects were fasted for 7 days. The N losses were fairly constant during this period (10 g N/day). Carbohydrate was introduced on the eighth day; the fasted subjects were fed fructose (100 g/day). The data demonstrate that the fructose induced a decrease in N losses to about 5.0 g N/day (Figures 8.19A and 8.19B). Figure 8.19B presents data that distinguish the amount of N lost as urea and as ammonia. The study reveals that the proportion of N excreted in the form of ammonia increases with prolonged fasting. The proportion of N lost as ammonia on the first days of fasting is quite small; however, it increases to about 25% of the total nitrogen by day 7 of fasting. The excretion of N as ammonia may serve two purposes. First, the synthesis of urea requires energy. The equivalent of four ATPs are expended in the synthesis of each molecule of urea. In contrast, the packaging of two nitrogens as glutamine in the liver requires only two ATPs. This glutamine then leaves the liver for the kidney, where it is hydrolyzed to produce ammonia (for excretion) and glutamate. Second, and more important, an increase in excretion of waste N as ammonium ions, rather than as urea, reduces the concentration of acid in the body. The increases in ammonium ion excretion and reduction in urea excretion help to counteract the acidifying effects of excessive ketone body production (ketoacidosis).

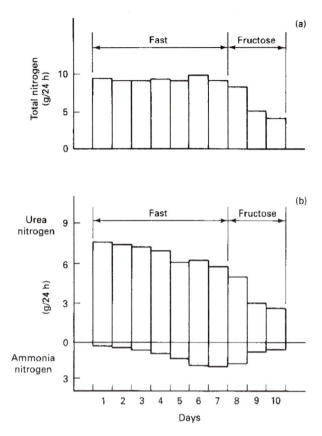

**FIGURE 8.19** (A)Total urinary nitrogen and (B) urinary nitrogen as urea and ammonia. Total urinary nitrogen decreases with transition from total fast to fructose-only fast. (Redrawn with permission from Gelfand and Sherwin, 1986.)

## Changing Styles of Waste Nitrogen Excretion with Fasting

The ammonium ion ($NH_4^+$) is a very weak acid. It tends to remain protonated, rather than dissociate, in the physiological pH range. It is excreted by the kidney in the protonated form, removing protons from the body. Urea, on the other hand, exists in the body as $H_2NCONH_2$, not as $H_3^+NCONH_2$ or $H_3^+NCONH_3^+$. The excretion of waste N in the form of urea does not result in the excretion of protons from the body. When waste nitrogen in the form of ammonium ion is excreted, it takes with it a proton; however, when the ammonium ion is packaged as urea, the proton is discharged within the body. A shift from urea excretion to ammonium ion excretion may thus aid in counteracting ketoacidosis during prolonged fasting, as summarized in Figure 8.20 (Atkinson and Bourke, 1984; Boon et al., 1996). Note that acidosis is controlled by a number of other mechanisms, such as the renal sodium/proton exchanger (Krapf et al., 1991). This exchanger acts to resorb a sodium ion from the glomerular filtrate in exchange for a proton in the renal tubule cell. The proton then is lost in the urine. Walser (1986) has discussed, in detail, the production of acid or alkali during the oxidation of protein and various amino acids.

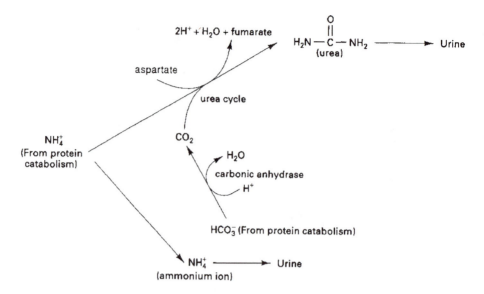

FIGURE 8.20 Elimination of ammonium ions via the urea cycle or via direct excretion. With protein catabolism, the excretion of waste nitrogen via the urea cycle results in net production of acid in the body; however, excretion of ammonium ions by the kidney into the urine does not result in this production of acid in the body.

### Attaining Zero Nitrogen Balance

The subjects in the preceding studies were in negative N balance. The next study concerns an attempt to produce N balance (zero N balance) in adults consuming a semistarvation diet. This diet contains carbohydrate and just enough protein to offset the daily obligatory losses of protein in the body. The material is directly related to the use of low-energy diets for weight reduction. Attainment and maintenance of zero N balance are critical to those interested in losing weight by consuming energy-deficient diets. The loss of lean body mass is an undesirable consequence when attempting to lose weight by total fasting or by consuming carbohydrate-only diets.

The study shown in Figure 8.21 concerns obese human subjects. The subjects consumed a semistarvation diet for 8 weeks. The diet supplied 60 g of protein (9.6 g N/day) and 54 g of carbohydrate per day. The energy value of these nutrients is 2100 kJ/day. This value is quite low, considering that the recommended energy intake for the 70-kg man is 11,300 kJ/day. The goal of the study was to restrict the energy intake of the subjects while providing enough carbohydrate to spare the body's losses of protein and providing enough protein to replace the nitrogen lost via the obligatory losses. The subjects weighed about 130 kg prior to the study. They lost about 20 kg by the end of the 8-week period. The data in Figure 8.21 show that the subjects were in negative N balance during the 8-week study, though the losses tended to decrease with time, resulting in a near-zero N balance by the eighth week.

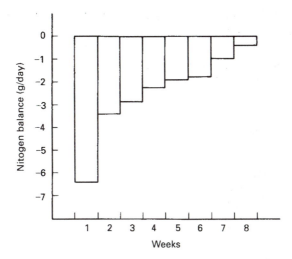

**FIGURE 8.21** Nitrogen balance of a group of subjects over time. (Redrawn with permission from Pasquali *et al.*, 1987.)

## Variability between Subjects

The semistarvation study just outlined involved six subjects. Data from this study are averages of the data collected from all six subjects. The histograms in Figure 8.22 show data for each of the subjects: CC, LV, PG, BT, AG, and AB. Subject LV was close to zero N balance from the third to the eighth weeks of the study. Subject PG was in negative N balance throughout the entire study. Subject BT seemed to be in positive N balance in the final weeks of the study. The positive value for the N balance suggests that the subject was gaining lean body mass during this time. This suggestion seems unlikely given the nature of the diet. The histograms

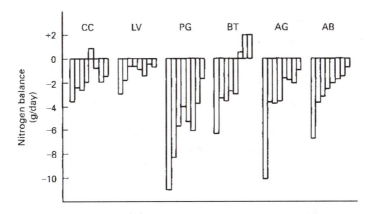

**FIGURE 8.22** Nitrogen balance of different subjects over time. (Redrawn with permission from Pasquali *et al.*, 1987.)

demonstrate the large variations in the ability of different humans to conserve nitrogen during fasting or semifasting conditions. These large variations have been noted in other studies as well (Fisler *et al.*, 1982). Errors in sample collection and analysis are a major concern in metabolic balance studies as they often lead to erroneous results.

## *Growth Assessment by Carcass Analysis*

The nutrient balance technique, whether used for measuring N balance, calcium balance, or balance of other nutrients, is not particularly accurate. Inaccuracies are introduced with feeding. It may not always be possible to measure exactly how much food is consumed by the animal or human subject. Inaccuracies are introduced in collecting the excreta, especially as collections must be made over the course of many days. An important source of error is the fact that the value for N balance is determined by subtraction, that is, subtracting the amount excreted from that consumed. The values for balance, whether positive or negative, are often small compared to the values for the amount of nutrient consumed.

The aforementioned problems are reduced by the carcass analysis technique. This technique, sometimes used with growing rats or chickens but not with higher animals, requires analysis of the amount of nitrogen in groups of animals. The group is measured before and after the feeding trial. Before and after the trial, animals are sacrificed, their carcasses dissolved in acid, and the total amounts of nitrogen measured in the resulting solutions. Unlike the N balance technique, the carcass analysis technique requires destruction of the animal.

## Summary of Issues in Protein Nutrition

To summarize, the studies of protein nutrition considered so far may be organized into those that represent decreasing levels of severity:

1. *Total fast*: As far as protein nutrition is concerned, the total fast is the most severe condition. During a total fast, the loss of urinary nitrogen is equivalent to about 80 g of the body's proteins.

2. *Carbohydrate-only diet*: A less severe condition is a carbohydrate-containing, protein-free diet. The obligatory losses of nitrogen may be reduced to the equivalent of about 40 g of bodily proteins per day with this diet. A diet that contains carbohydrate but little or no protein may be viewed as good or bad. The reduction of nitrogen losses, in itself, seems desirable. It is clearly desirable, from the standpoint of a subject interested in weight reduction, to prevent the bodily protein losses that occur with a low-energy diet; however, the losses of skeletal muscle proteins tend to be minimized, and losses of other proteins such as serum albumin and retinol-binding protein continue with the carbohydrate-containing, protein-free diet. Chronic consumption of this diet can result in kwashiorkor. Total fast may result in greater nitrogen losses from the body, but the breakdown of muscle protein occurring during the total fast liberates a quantity of amino acids that is readily used by the liver for the synthesis of more vital proteins.

3. *Protein-containing low-energy diet*: The least severe condition described, the protein-containing low-energy diet, supplies enough protein to replace the obligatory losses of nitrogen.

## Determination of the Daily Requirement and Recommended Daily Allowance for Protein

The protein requirement of the adult human was determined as follows. Ordinarily, the requirement might be assumed to be equivalent to the amount of protein excreted in the form of urea and ammonia during the total fast. In the real world, however, humans consume carbohydrates. These carbohydrates are usually present in abundance. Hence, calculation of the protein requirement uses, as a starting point, the value of the obligatory N losses with the carbohydrate-containing, protein-free diet.

Can the consumption of an amount of protein equivalent to the obligatory losses replace the N lost from the breakdown of the body's proteins? The answer to this question is "no," for the following reason. A fraction of the protein consumed in the diet is not incorporated into the body's proteins. This fraction of protein is catabolized and used for energy before it has a chance to be used for protein synthesis in the body. This value is thought to range from 30 to 50% of the protein consumed. Hence, the amount of protein consumed to replace the obligatory losses of N is greater than the equivalent of the obligatory losses.

Two other correction factors follow. Obligatory N losses, as well as the amounts of protein required to replace these losses, differ from person to person. These amounts may be expressed as a range for any population. The upper value of this range used for calculating the RDA for protein. By use of statistical methods, a correction factor of 25% was used in order that the RDA for protein would supply the protein requirement for 97.5% of the human population. The second correction factor concerns the variability in quality of the proteins in the American diet. This correction factor is rather small and has a value of 6–10%. The quality of dietary proteins may be lower than that of the reference protein (ovalbumin) for two reasons. The first is that ovalbumin contains a complement of amino acids that closely matches the dietary protein needs of the body, whereas the proteins in a typical mixed diet may contain amino acids in proportions that do not as perfectly match the needs of the body. The second factor contributing to slight impairments in protein quality is **digestibility**. Protein digestibility refers to the percentage of consumed protein that is absorbed by the gastrointestinal tract, that is, not recovered as fecal N. Egg protein is about 95% digestible, milk protein 95%, meat 94%, corn protein 85%, whole wheat protein 86%, refined wheat protein 96%, oatmeal 86%, and protein from beans 78% (Pellet, 1990).

The protein requirement and the RDA for protein were derived by the following steps:

1. *Obligatory losses*: The obligatory losses of protein with consumption of the carbohydrate-containing, protein-free diet are close to 54 mg N/kg body weight/day. This is equivalent to 0.34 g protein/kg or about 24 g of protein for the 70-kg man (FAO/WHO/UNU, 1985).

2. *Catabolism of amino acids from dietary protein prior to use by the body*: The efficiency of utilization of dietary protein for the replacement of obligatory losses has been estimated to be 54 to 75% (Young and Marchini, 1990). The catabolism of dietary amino acids for energy production, prior to their incorporation into the body's proteins, contributes to this factor. In correcting for this factor, a protein requirement of about 0.6 g protein/kg body weight is found. This value was directly determined by long-term N balance studies with human subjects consuming diets containing various levels of protein (Garza *et al.*, 1977).

3. *The mixed quality of proteins in the diet*: The values for protein, noted earlier, refer to high-quality proteins such as those found in eggs, fish, and meat. In the real world, the diet contains a mixture of proteins from plant and animal sources. The mixed quality of the proteins, as determined by their component amino acids as well as their variable digestibility, suggests a correction factor in the calculation of the RDA for protein. With this correction, a protein requirement of about 0.64 g protein/kg body weight is calculated.

4. *Biological variability*: A correction factor of 25%, which was based on statistical considerations, was used to increase the calculated protein requirement to yield the RDA. The 25% correction results in a value for the requirement that would be expected to meet the needs of 97.5% of a normal population. With this correction, an RDA of 0.8 g protein/kg body weight for the adult was calculated (Food and Nutrition Board, 1989).

Note that the value for obligatory N losses, as determined by several studies with humans, is fairly reproducible; however, the amount of dietary protein required to produce N balance has proven difficult to pinpoint, thus lending an air of approximation to the RDA for protein.

## Amino Acid Requirements

The requirements for the nine indispensable amino acids have been studied in detail. There is no requirement, however, for any particular dispensable amino acid. Maximal rates of growth are possible with diets lacking a few of the dispensable amino acids. Omission of all of the dispensable amino acids is, in fact, quite compatible with life. This omission results in the body's diversion of some of its indispensable amino acids for the synthesis of the missing dispensable amino acids. The presence of the dispensable amino acids in the diet ensures the efficient use of the indispensable amino acids for protein synthesis, rather than for conversion to other amino acids. Growth is greatly impaired when the diet is lacking in all of the dispensable amino acids. In some studies, the dispensable amino acids are supplied by glutamic acid. Glutamate has proven to be an efficient source of amino groups for the synthesis of the dispensable amino acids. The use of glutamate for this purpose is said to supply the requirements for **nonspecific nitrogen**. In other words, the glutamate is not used for the synthesis of a specific amino acid, but of a variety of other amino acids.

An interesting species difference exists regarding the nonspecific N used for growth. Glutamine alone is sufficient to support maximal rates of growth in pigs and chicks, while a combination of several dispensable amino acids is needed for growing rats, mice, and cats (Hiramatsu *et al.*, 1994). For adult humans, glutamine functions as well as a mixture of dispensable amino acids (Ala, Arg, Asp, Glu, Gly, Pro, Ser) as a source of nonspecific N (Hiramatsu *et al.*, 1994).

**TABLE 8.9**  Glutamate Plus Glutamine Content of Various Proteins

| Protein | Glutamate + glutamine (g/100 g protein) |
|---|---|
| Gliadin (wheat) | 46 |
| Zein (corn) | 27 |
| Hordenin (barley) | 38 |
| Arachin (peanut) | 21 |
| Glycinin (soybean) | 21 |
| Albumin (human serum) | 17 |
| Albumin (egg white) | 17 |
| α-Casein (milk) | 23 |
| β-Lactoglobulin (milk) | 20 |
| Actin (muscle) | 15 |
| Myosin (muscle) | 21 |
| Collagen (tendon) | 11 |

*Source*: Filer *et al.* (1979).

Some information regarding the content of glutamate in the diet follows. Glutamate is the amino acid present in greatest quantities in many proteins. The wheat protein **gliadin** is nearly 50% glutamate plus glutamine, by weight (Table 8.9). A major protein of corn, **zein**, is about 25% glutamate plus glutamine. Although many proteins contain more residues of glutamate than glutamine, both amino acids are sometimes reported as a sum. This is because the method of amino acid analysis involves acid hydrolysis of the protein in question, which results in conversion of glutamine to glutamate. Free glutamate also occurs in foods. The overall consumption of glutamate by various populations is shown in Table 8.10. The high value for Asian nations arises from the use of monosodium glutamate as a flavoring agent. Members of Western societies also use glutamate as a flavoring agent but without knowing it. Cheese contains 0.5 to 1.0% glutamate, by weight. The value for free glutamate in tomatoes is 0.14%. The free glutamate in these foods may contribute to the irresistible quality of pizza pies.

The requirements for the indispensable amino acids are useful for several reasons. Several proteins have been chosen as ideal, reference proteins. These reference proteins are used by food scientists for comparing the qualities of different proteins. The ideal proteins are egg protein, cow milk protein, and human milk protein. These proteins, although all of high nutritional quality, differ in their proportions of the various amino acids. Additionally, human milk may differ from source to source in its tryptophan content, leaving the optimal or ideal level of this amino acid unclear. Another reason for determining the requirements is that they may differ for animals or humans during different stages of life. More specifically, there is some evidence that the amino acid requirements may differ for premature and term infants. Still another reason for determining the requirements is that the requirements for specific indispensable amino acids can change depending on the available levels of specific dispensable amino acids. This statement applies to methionine and phenylalanine. The requirement for methionine is directly influenced by the amount of cysteine in the diet. The requirement for phenylalanine is directly influenced by the amount of tyrosine.

**TABLE 8.10** Daily Glutamate Consumption in Different Countries

| Country | Glutamate consumed per day per person (g) |
|---|---|
| Taiwan | 3.0 |
| Korea | 2.3 |
| Japan | 1.6 |
| Italy | 0.4 |
| United States | 0.35 |

*Source*: Filer *et al.* (1979).

### *Methods Available for Determining Amino Acid Requirements*

The first step in determining the requirement for a specific amino acid is to assemble a diet containing all of the amino acids except the one of interest. A series of diets containing different levels of the amino acid can be prepared and then fed to experimental subjects. A variety of biological parameters can be measured. In growing subjects, the growth rate can be used to assess the requirement. In adult subjects, the nitrogen balance technique can be used. The goal of the work is to determine the level of the amino acid in question that results in the highest-quality mixture of amino acids and to define the amount of the amino acid that might be considered excessive or deficient.

A study of the histidine requirement of rats provides an example (Figures 8.23–8.26). Growing rats were raised on diets that were chemically defined and nutritionally complete in all respects except that the content of histidine was varied. The diets contained 0 to 0.6% histidine; for example, a diet containing 0.6% histidine has 0.6 g of histidine per 100 g of diet.

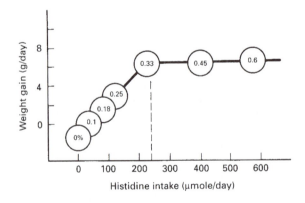

**FIGURE 8.23** Weight gain per day versus diet. The numbers within the circles indicate the composition of the diet fed to that particular group of rats, and refer to the percentage histidine, by weight, in the diet. (Redrawn with permission from Kang-Lee and Harper, 1977.)

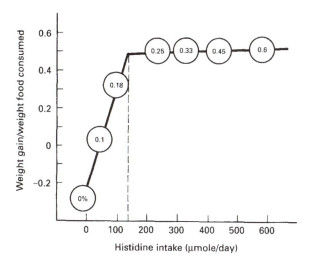

**FIGURE 8.24** Weight gain per weight of food consumed versus diet. (Redrawn with permission from Kang-Lee and Harper, 1977.)

The data in Figure 8.23 show that the maximal rate of growth was attained when the diet contained 0.33% histidine. Figure 8.24 shows that the weight gained/food consumed was maximal with the diet containing 0.25% histidine. Figure 8.25 reveals the concentrations of plasma histidine with the different diets. The data demonstrate that the concentrations are fairly stable with 0 to 0.2% histidine. Even with the histidine-free diet, the plasma level of this amino acid during the feeding trial remained at about 40 $\mu M$. With progressively higher levels of dietary histid-

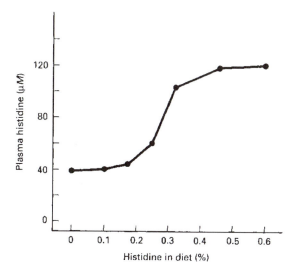

**FIGURE 8.25** Plasma histidine concentration versus diet. (Redrawn with permission from Kang-Lee and Harper, 1977.)

ine, the plasma concentration appears to plateau at about 120 μM. The plasma amino acid levels alone do not provide information on the histidine requirement; however, a knowledge of correlations between plasma histidine levels and growth can form the basis for defining the dietary requirement for this amino acid.

The histidine catabolic pathway is discussed under Folate in Chapter 9. The material reveals that histidine is catabolized to produce glutamate. Glutamate, in turn, can be converted to α-ketoglutarate and completely oxidized to $CO_2$ in the Krebs cycle. In the study depicted in Figure 8.26, the dietary histidine was spiked with [$^{14}C$]histidine. The term "spiked" means that only a very small proportion of the histidine contained carbon-14. The metabolic behavior of the radioactive histidine, which can be followed, mirrors the metabolic fate of nonradioactive histidine in the diet. All of the $CO_2$ exhaled by the rats can be easily collected. The $^{14}CO_2$ present in the rat's breath can be measured by use of a liquid scintillation counter. The amount of $^{14}CO_2$ produced directly mirrors the proportion of histidine, absorbed from the diet, that was degraded the rat's body.

The results in Figure 8.26 reveal that only a tiny proportion of histidine is catabolized with diets containing up to 0.25% histidine. With diets containing more than 0.25% histidine, there is a steady increase in the proportion of absorbed amino acid undergoing oxidation. For example, about 25% of the absorbed histidine (both radioactive and nonradioactive) is degraded in the rats fed diets containing 0.6% histidine.

Figure 8.23 reveals that the histidine requirement for the rat is about 0.33%. The data in Figure 8.26 show that dietary histidine is most effectively used at levels of up to 0.25% and that higher levels have an increasing tendency to be broken down. The amount of breakdown depends on the exact level of amino acid in the diet.

EXERCISE

Is there any indication that toxic, high levels of histidine were used (Figure 8.23)?

EXERCISE

What plasma histidine levels might be correlated with the intake of nutritionally adequate levels of histidine (Figure 8.25)?

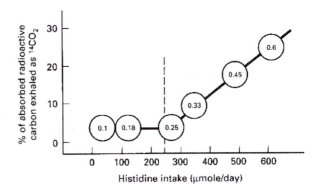

**FIGURE 8.26** Percentage of absorbed radioactive carbon (consumed as histidine) that is exhaled as $^{14}CO_2$ versus concentrations of histidine in the diet. (Redrawn with permission from Kang-Lee and Harper, 1977.)

*Plasma Concentrations of Amino Acids*

The concentration of amino acids in plasma is a recurring topic in the nutritional sciences. Plasma amino acids have received attention through research on the control of the appetite (Peters and Harper, 1981), the time course of protein digestion (Adibi and Mercer, 1973), glutamate (MSG) consumption and its possible mild side effects (Stegink *et al.*, 1985), the toxic effects of various amino acids during liver disease (Bernardini and Fischer, 1982), and the influence of insulin on amino acid metabolism during bed rest, diabetes, and pregnancy (Fukagawa *et al.*, 1986; Shangraw *et al.*, 1988; Fitch and King, 1987; Traxinger and Marshall, 1989). Plasma amino acids are also important for the diagnosis and treatment of genetic diseases, such as phenylketonuria (Scriver *et al.*, 1989), hyperglycinemia (Halton and Krieger, 1980), and maple syrup urine disease (Parsons *et al.*, 1987).

The plasma concentrations of the amino acids in normal human subjects are listed in Table 8.11. All of the concentrations are averages from a group of human subjects. There was some variation between subjects, with small variations occurring in individuals during the course of the day. The blood samples were taken from nonfasting subjects, that is, within an hour or so of a meal. One striking aspect of the results is the similarity of concentrations in different studies. Glutamine consistently occurs at the highest concentration, about 0.7 mM, whereas methionine is consistently low, at about 0.025 mM. Clearly, homeostatic mechanisms

**TABLE 8.11**  Amino Acid Concentrations in Human Plasma

| Amino acid | Concentration (μM) | | | | |
|---|---|---|---|---|---|
| Glutamine | 656[a] | 600–800[b] | 657[c] | 724[d] | – |
| Alanine | 360 | 344 | 340 | 228 | – |
| Cysteine | – | – | – | – | 270[e] |
| Valine | 264 | 212 | 258 | 266 | – |
| Glycine | 232 | 215 | 233 | 204 | – |
| Proline | – | 175 | 215 | 213 | – |
| Lysine | 192 | 164 | – | 181 | – |
| Threonine | 145 | 134 | 131 | 119 | – |
| Leucine | 133 | 112 | 143 | 134 | – |
| Serine | 113 | 109 | 106 | 116 | – |
| Histidine | 94 | 73 | – | 80 | – |
| Arginine | 94 | 69 | – | 74 | – |
| Isoleucine | 64 | 59 | 97 | 60 | – |
| Asparagine | 62 | – | – | 52 | – |
| Cystine | 61 | – | – | – | – |
| Tyrosine | 64 | 54 | 74 | 56 | – |
| Phenylalanine | 58 | 49 | 68 | 47 | – |
| Tryptophan | – | 39 | – | – | – |
| Glutamic acid | 34 | 30–70 | 44 | 66 | – |
| Methionine | 24 | 24 | 34 | 23 | 27 |
| Homocysteine | – | – | – | – | 12 |
| Aspartic acid | 7 | – | – | 5 | – |

Data in these columns are taken from the following sources: [a]Scriver *et al.* (1985); [b]Felig *et al.* (1970); [c]Haymond *et al.* (1982); [d]Albert *et al.* (1986); [e]Stabler *et al.* (1988).

actively regulate the levels of amino acids in the various compartments of the body.

## Sparing of Methionine by Cysteine and of Phenylalanine by Tyrosine

The concept of sparing of one nutrient by another was introduced earlier, where it was demonstrated that dietary carbohydrate can spare protein. Similarly, cysteine can spare methionine and tyrosine can spare phenylalanine. A certain proportion of dietary methionine is converted to cysteine. Methionine normally supplies part of the body's needs for cysteine. With cysteine-free diets, methionine can supply all of the body's needs for cysteine. The methionine catabolic pathway that leads to cysteine production is shown in Figure 8.27. Only the sulfur atom of methionine appears in the molecule of cysteine; serine supplies the carbon skeleton of cysteine. α-Ketobutyrate is a byproduct of the pathway. α-Ketobutyrate is further degraded to propionyl-CoA by BCKA dehydrogenase or pyruvate dehydrogenase. Propionyl-CoA is then converted to succinyl-CoA, an intermediate of the Krebs cycle.

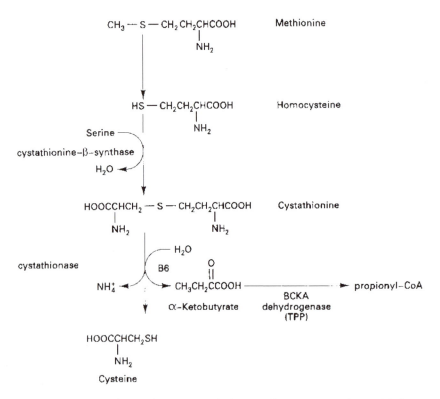

**FIGURE 8.27** Pathway for methionine catabolism and cysteine synthesis. Methionine is the source of the sulfur atom of cysteine. Serine is the source of the carbon skeleton of serine. In methionine catabolism, the carbon skeleton of methionine is converted to propionyl-CoA, which eventually enters the Krebs cycle at the point of succinyl-CoA. BCAA dehydrogenase catalyzes the oxidation of α-ketobutyrate to propionyl-CoA.

**TABLE 8.12**  Nitrogen Balance Experiment

| Daily nitrogen intake (g) | Urinary nitrogen (g) | Fecal nitrogen (g) | Daily nitrogen balance (g) | Supplement |
|---|---|---|---|---|
| 10.1 | 8.82 | 0.92 | +0.36 | 2.0 g methionine |
| 10.1 | 8.89 | 0.89 | +0.32 | 1.0 g methionine |
| 10.1 | 8.92 | 0.88 | +0.30 | 0.9 g methionine |
| 10.1 | 9.19 | 0.92 | −0.01 | 0.8 g methionine |
| 10.1 | 9.39 | 0.84 | −0.13 | 0.7 g methionine |
| 10.1 | 8.97 | 0.75 | +0.38 | 0.1 g methionine + 0.81 g cysteine |
| 10.1 | 9.46 | 0.79 | −0.15 | 0.0 g methionine + 0.81 g cysteine |

*Source*: Rose and Wixom (1955)

The influence of cysteine on the methionine requirement is demonstrated by the following study. The subject was a student who was awarded a Ph.D. after conducting the study on himself. The subject consumed diets that were complete except that the concentration of methionine was varied (Table 8.12). The dietary amino acids were supplied in the form of pure amino acids, rather than as protein, to allow full control over the levels of amino acids supplied by the diet. Cysteine was supplied as cystine. Cystine is a dimer of cysteine, in which the two cysteine residues are connected via a disulfide bond (R—S—S—R). Cystine is readily converted to cysteine in the body. The methionine requirement was assessed by determining the conditions that supported a zero or slightly positive nitrogen balance.

The results demonstrate that a near-zero N balance occurred with the diet containing 0.8 g of methionine. Higher levels of methionine resulted in a positive N balance. The N balance was clearly negative with the 0.7-g methionine diet. The following statement illustrates the ability of cysteine to spare methionine: An unusually small amount of methionine (0.1 g) was able to support a positive N balance when cystine was included in the diet. Cysteine alone, that is, with no methionine, resulted in a negative N balance. Cysteine (or cystine) can replace a large fraction of our requirement for methionine. For this reason, the requirement for methionine is sometimes stated as a combined requirement for methionine plus cysteine.

Tyrosine can spare the requirement for phenylalanine. In many species, about 50% of the phenylalanine requirement can be replaced by tyrosine. For this reason, the requirement for phenylalanine is sometimes expressed as the requirement for the sum of both amino acids. The biochemical relationship between phenylalanine and tyrosine is shown in Figure 8.28. Phenylalanine monooxygenase catalyzes the conversion of phenylalanine to tyrosine. This enzyme uses the cofactor tetrahydrobiopterin. Biopterin is not required in the diet. It is synthesized from GTP. Folate (in plants and bacteria) and molybdopterin are also synthesized from GTP. Figure 8.28 also depicts the catabolic pathway for tyrosine.

Phenylalanine monooxygenase is defective in persons with the genetic disease **phenylketonuria**. The name of this disease derives from the fact that some of the

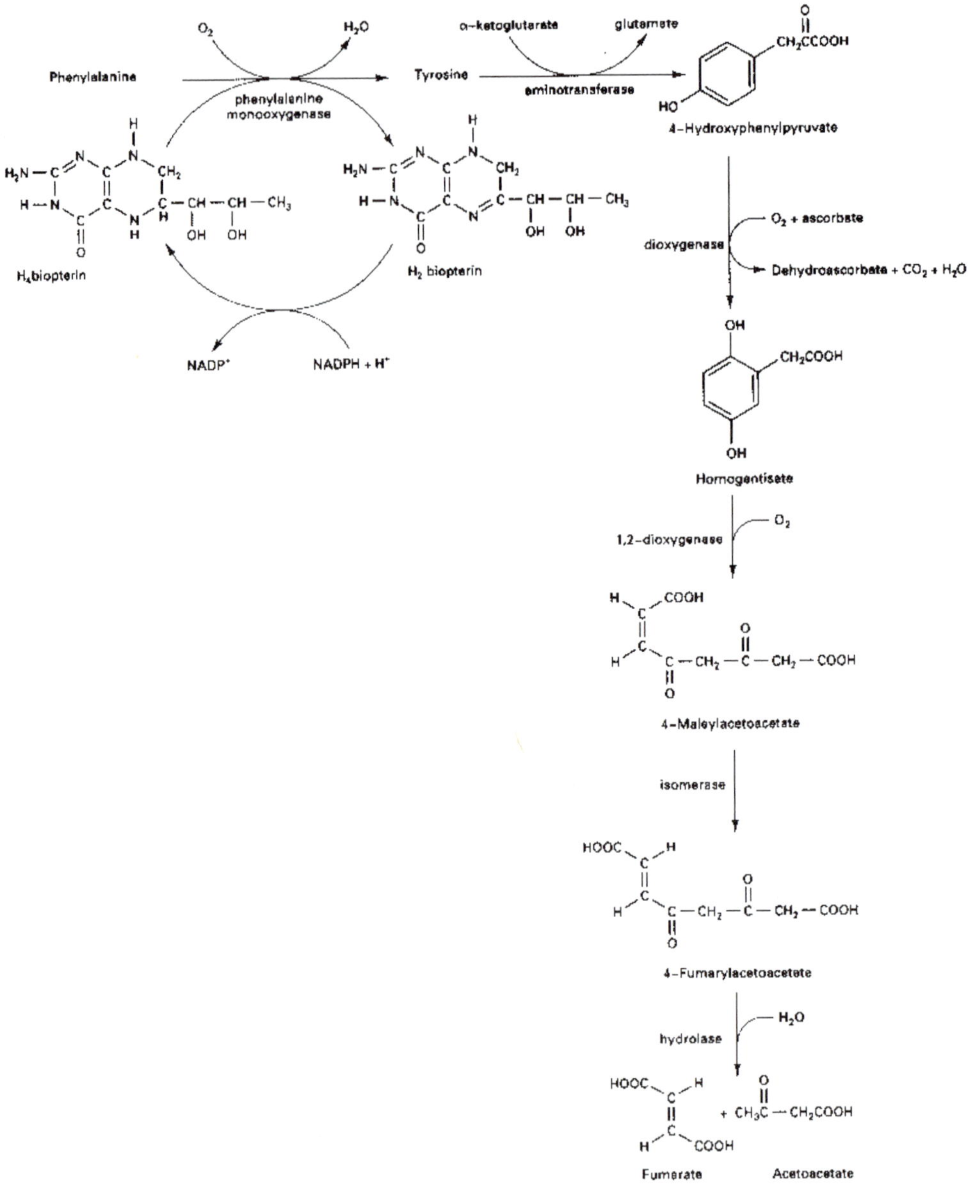

**FIGURE 8.28** Phenylalanine and tyrosine catabolism. Phenylalanine is converted to tyrosine by phenylalanine monooxygenase. This enzyme requires tetrahydrobiopterin as a cofactor. This cofactor is synthesized in the body from GTP and must be in the fully reduced, tetrahydro form to be active. The cofactor is converted to the dihydro form in the course of the reaction. A separate enzyme, which uses NADPH as a reducing agent, catalyzes the reduction of dihydroprotein back to tetrahydrobiopterin. Oxygen is the cosubstrate of phenylalanine monooxygenase, as well as of two other enzymes, in the pathway shown.

phenylalanine that builds up in the body is converted to the corresponding keto acid by the action of an aminotransferase. This keto acid, phenylpyruvic acid, appears in elevated amounts in the urine. The symptoms of the disease, including mental retardation, are caused by amounts of dietary phenylalanine that are in slight excess of the requirement. The symptoms can be minimized, or even prevented, by the consumption of a low-protein diet. This diet is generally consumed until the age of 6, at which point the disease becomes less severe. The diet may also involve avoidance of special sources of phenylalanine, such as aspartame. Aspartame, an artificial sweetener, is a dipeptide composed of aspartic acid and phenylalanine (Asp–Phe–methyl). The carboxyl group of the phenylalanine is esterified with a methyl group. Some think that the low-phenylalanine diet should be resumed, in adulthood, during pregnancy. This recommendation is intended to avoid the possible poisoning of the fetus by the high levels of phenylalanine in the bloodstream of the mother who has phenylketonuria. About 1 in 10,000 persons has phenylketonuria.

## PROTEIN QUALITY

Protein quality refers to the ability of a dietary protein to supply the amino acid needs of the body. Some foods contain high levels of protein; others contain low levels (see Table 8.1). The fact that a specific food is a rich source of protein does not indicate that the food has any particular value in supporting growth or maintenance. Gelatin, for example, is a protein that is sometimes used in cooking. This protein is available in a pure, powdered form; however, the use of gelatin as a food and as the sole source of protein cannot supply the body's amino acid needs. A diet based on gelatin and excluding other proteins cannot sustain life. This is because gelatin is a low-quality protein. It lacks the indispensable amino acid tryptophan.

The quality of different proteins can be expressed according to their chemical score, protein efficiency ratio (PER), biological value (BV), and net protein utilization (NPU). These parameters refer to the different tests used to define the protein's quality. The chemical score refers only to the property of the protein in question. The PER, BV, and NPU refer to relationships between the dietary protein and the consumer. The values of the PER, BV, and NPU depend on properties of both the protein in question and the needs of the animal. The chemical score, PER, BV, and NPU of proteins in various foods, including eggs, fish, rice, and maize, are listed in Table 8.13.

The value of the chemical score is determined by a chemical test. The chemical score for a particular food or protein would not be expected to vary much when determined by different researchers. The PER, BV, and NPU are biological tests and involve animal feeding trials. The BV and NPU require nitrogen balance measurements, and hence their determination requires more work than the other tests. The values for the PER, BV, and NPU might be expected to vary somewhat with different determinations; however, the relative order of the different foods in Table 8.13 might not be expected to change much when the PER, BV, and NPU are determined by different researchers.

**TABLE 8.13**  Protein Quality

| Protein | Chemical score | PER | BV | NPU |
|---|---|---|---|---|
| Egg | 100 | 3.92 | 94 | 94 |
| Fish | 71 | 3.55 | 76 | 80 |
| Beef | 69 | 2.30 | 74 | 67 |
| Sunflower seed | 61 | 2.10 | 70 | 58 |
| Casein | 58 | 2.86 | 80 | 72 |
| Oats | 57 | 2.19 | 65 | 66 |
| Rice | 56 | 2.18 | 64 | 57 |
| Peanuts | 55 | 1.65 | 55 | 43 |
| Soybeans | 47 | 2.32 | 73 | 61 |
| Wheat | 43 | 1.53 | 65 | 40 |
| Lima beans | 41 | 1.53 | 66 | 52 |
| Maize (corn) | 41 | 1.12 | 59 | 51 |
| Lentils | 31 | 0.93 | 45 | 30 |

*Source*: Committee on Amino Acids, Food and Nutrition Board (1974).

### The Chemical Score and the Concept of the Limiting Amino Acid

The chemical score is determined as follows. The contents of the indispensable amino acids of egg albumin (ovalbumin), which is used as a reference protein, and the food protein in question are determined. Egg albumin is considered to be ideal and nutritionally complete. The test requires several steps. The protein must be purified, hydrolyzed to the constituent amino acids, and then subjected to analysis using an amino acid analyzer. The values for each of the indispensable amino acids for ovalbumin and the test protein are then listed as shown in Table 8.14. The contents of the various amino acids in the two proteins are then compared. The amino acid in the test protein that is present in the lowest level, on a percentage basis, can be found by examining the table. This amino acid is called the **limiting amino acid** of the protein. The value of the percentage is the **chemical score**. Table 8.14 indicates that the quantity of lysine in oat protein is 51% that in egg protein. Hence, the chemical score of oat protein is 51.

### Protein Efficiency Ratio

Determination of the PER requires growing animals. The PER is defined by the formula

$$PER = \frac{\text{weight gained}}{\text{weight of protein consumed}}$$

The conditions for gathering PER data are standardized. The animals to be used are weanling rats; rats are generally weaned at 21 days of age. The protein is used at a concentration of 10% of the dry weight of the diet. The PER of the test protein should always be compared with that of ovalbumin, where ovalbumin is fed to a group of control rats. The weight gained and food consumed are recorded for a

**TABLE 8.14** Amino Acid Composition of Two Protein Sources

| Amino acid | Egg protein | Oat protein |
|---|---|---|
| | g of amino acid per 16 g of protein nitrogen | |
| Lysine | 6.9 | 3.5 |
| Histidine | 2.6 | 1.9 |
| Threonine | 4.9 | 4.7 |
| Methionine | 3.2 | 2.2 |
| Cysteine | 2.1 | 1.9 |
| Valine | 7.1 | 4.7 |
| Leucine | 8.8 | 5.8 |
| Isoleucine | 5.9 | 3.3 |
| Phenylalanine | 5.5 | 4.2 |
| Tyrosine | 3.8 | 2.6 |

*Source*: Maruyama *et al.* (1975).

3-week feeding trial. Table 8.13 reveals, for example, that the PER for egg protein is about twice that for soybean protein.

The PER is acceptable for comparing the food values of different proteins, though it does have shortcomings. Because the test can be conducted only with growing animals, it cannot be used to determine the protein requirement for maintenance, but only the combined requirement for growth and maintenance. Another problem is that the PER cannot distinguish between weight gained as fat and as lean body mass. An advantage of the test is that the only equipment needed are rat cages, food jars, water bottles, and a balance.

### Biological Value

The BV of protein is determined by measuring the amount of N consumed and that excreted. First, the obligatory losses of urinary and fecal N must be determined, which requires the feeding of nitrogen-free diets. This is followed by a determination of the amounts of urinary and fecal N with consumption of the test protein. The differences in N excreted between the two dietary conditions is expressed as [Δ fecal N] and [Δ urinary N]. The Greek capital delta conventionally means "the change in." The formula for the BV is

$$BV = \frac{\text{retained N}}{\text{absorbed N}} = \frac{[\text{N intake}] - [\Delta \text{ fecal N}] - [\Delta \text{ urinary N}]}{[\text{N intake}] - [\Delta \text{ fecal N}]}$$

The BV represents the fraction of amino acids absorbed by the gut that is retained by the body. The BV ignores that part of the dietary protein that is not absorbed and appears in the feces. The change in fecal N occurring with the two diets is subtracted from the values of N intake, in both the numerator and denominator. The equation seems to state, "Let's ignore that part of the dietary protein that is poorly digested and appears in the feces." The equation seems to ask, "As far as the absorbed amino acids are concerned, how suitable are they for the needs of the animal?" A high BV indicates that the amino acids occur in proportions that are highly compatible with the needs of the body. A protein that is completely lacking in one of the indispensable amino acids would be expected to have a BV close to or equal to zero. The formula for the BV yields a fraction. This fraction may be multiplied by 100 to express the BV as a percentage.

## Net Protein Utilization

The NPU involves the same measurements used in determination of the BV. These measurements are of the urinary and fecal N losses and of urinary and fecal N losses with consumption of the test protein. The formula for the NPU is

$$
\text{NPU} = \frac{\text{retained N}}{\text{consumed N}} = \frac{[\text{N intake}] - [\Delta \text{ fecal N}] - [\Delta \text{ urinary N}]}{[\text{N intake}]}
$$

The equation for the NPU seems to ask, "Of the protein that is consumed, how much is retained?" This contrasts with the BV, which asks, "Of the protein that is absorbed, how much is retained?" The NPU is of practical importance in agriculture. A poultry farmer may use NPU values to evaluate new feeds and to determine the cheapest way to produce the largest amount of poultry meat. The carcass analysis technique is often used to determine the NPU.

## Amino Acid Supplements

The quality of a dietary protein may be increased by supplementing the diet with appropriate amino acids. This concept is illustrated by the study shown in Figure 8.29, which involved supplementation of a wheat-based diet with lysine. The change in the quality of the diet was determined simply by measuring changes in the growth rate.

The study involved two groups of growing rats. The first group was fed diets containing various concentrations of wheat protein. The levels of wheat protein ranged from 10 to 80% of the diet, by weight (O, Figure 8.29). The increase in protein content was at the expense of glucose. In short, the 10% wheat protein diet contained 10 g of wheat protein, 80 g of glucose, and 10 g of oil, salts, and vitamins. The 40% wheat protein diet contained 40 g of wheat protein, 50 g of glucose, and 10 g of oil, salts, and vitamins. The second group of rats consumed the same diets with the exception that all of the diets were supplemented with 1% lysine (1.0 g lysine/100 g diet) (●, Figure 8.29). The feeding trial lasted 2 weeks. The results demonstrate that the maximal weight gain during the feeding period was about 85 g. Maximal growth was attained with the 70% wheat protein diet (O) and the 20% wheat protein, lysine-supplemented diet (●).

The results from the study can be interpreted as follows. Lysine appears to be the first or most limiting amino acid of wheat protein. Maximal growth occurred with reduced levels of dietary protein where the diet was supplemented with lysine. If lysine was the second most limiting amino acid of wheat, the lysine

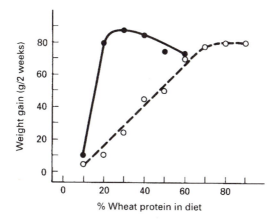

**FIGURE 8.29** Weight gain with different diets. (Redrawn with permission from Munaver and Harper, 1959.)

supplement would not be expected to overcome the lack of the most limiting amino acid of wheat protein. Consumption of the 70% wheat protein diet resulted in the same growth increases as incurred with the 20% wheat protein diet containing 1% lysine. This suggests that the high levels of protein supplied by the 70% protein diet provided a sufficient amount of lysine, as well as adequate quantities of other indispensable amino acids, as did the 20% protein lysine-supplemented diet.

The steadily increasing growth rates occurring with greater concentrations of wheat protein (○, Figure 8.29) may result from the increasingly ample supply of lysine. With increasing amounts of dietary protein, an increase in catabolism of amino acids other than lysine might be expected. Hence, progressively greater rates of urea excretion may be attained with the diets higher in protein. With the diets containing elevated levels of protein, regulatory changes may occur in metabolism that increase the catabolism of all of the amino acids except lysine. The existence of such regulatory mechanisms is not clear.

EXERCISE

Is it possible for two different amino acids to be equally limiting in a particular dietary protein?

EXERCISE

The PER test uses test and reference proteins at a level of 10%, by weight, of the diet. Please state the consequences of a test conducted with protein levels of 70%, by weight.

## Complementary Proteins

The most limiting amino acid of wheat protein is lysine. The study described earlier revealed the growth-promoting effects of supplementing wheat protein with lysine. The most limiting amino acid of soybeans is methionine. Soybean-based infant formulas are supplemented with methionine. Generally, people do not supplement their diets with amino acids. The dinner table does not bear shakers containing methionine or lysine, along with the salt and pepper shakers. On the other hand, people do supplement their moderate-quality proteins with the limiting amino acids. People also supplement their moderate-quality proteins with other moderate-quality proteins. What does this mean? The major staples of the world include sweet potatoes, cassava, beans, rice, corn, and wheat. The first two foods in this list contain relatively little protein; however, the legumes and grains are moderately good sources of protein. Legumes contain ample amounts of lysine, threonine, and tryptophan, but are limiting in methionine. Grains contain methionine, but are limiting in lysine and sometimes threonine or tryptophan. A diet containing both legumes and grains supplies a mixture of amino acids that is of higher quality than a diet containing legumes alone or grains alone. These mixtures are not as high in quality as the protein of meat, milk, or fish, but they are better than that in legumes or grain alone. The use of both legume and grain proteins in a diet supplies **complementary proteins**. Complementary proteins are supplied by such time-tested food combinations as the beans and rice consumed in Central and South America, the tofu and rice consumed in Asia, and the peanut butter sandwiches consumed by American children.

The following study illustrates the effect of using complementary proteins. Growing rats were fed diets containing various combinations of rice and bean protein (Figure 8.30). For example, one test diet consisted of 100% rice protein and no bean protein, whereas another diet contained 40% rice protein and 60% bean protein. The PER was determined for each combination. The results demonstrate that the most valuable sources of protein were those containing 50 to 80% rice protein, with the remainder consisting of bean protein.

An interesting sidelight concerning the use of complementary proteins is that they must be consumed together, or at about the same time, to be of maximal

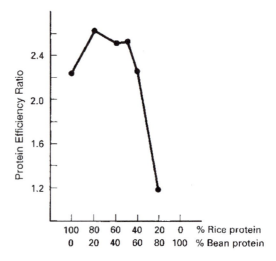

**FIGURE 8.30** Protein efficiency ratio versus type of diet. (Redrawn with permission from Hulse *et al.*, 1977.)

benefit as a protein source. For example, eating beans at breakfast and corn at dinner would fail to supply complementary proteins. The following description is paraphrased from a brief report concerning complementary mixtures of amino acids (Cannon *et al.*, 1947): Protein-deficient rats recovered weight steadily when fed a diet containing 10 indispensable amino acids. If, however, the diet is divided into two portions — one containing arginine, histidine, leucine, lysine, and threonine and the other containing isoleucine, methionine, phenylalanine, tryptophan, and valine — and these two incomplete rations are fed alternately, the recovery of weight is poor. Poor weight recovery occurs when the rations are offered to rats at alternate hours over a 14-hour period. The rats lose weight. On the other hand, the animals eat well and recover weight rapidly when the two incomplete rations are mixed together.

A fine point might be added concerning differences in human and animal nutrition. Some populations seem to thrive and grow on diets containing only one source of a moderate-quality protein. Persians, in particular, may consume diets where grain may be the sole source of protein (Vaghefi *et al.*, 1974). Although the growth of rats consuming diets containing only bread protein may be impaired, such effects are not readily detected in humans. Human children require a lower density of protein in the diet than do rats for supporting a maximal rate of growth. Rats require diets containing close to three times higher amounts of protein per total energy content than do children. A human infant is expected to grow at a near-maximal rate when its only source of protein is bread or beans. This situation is clearly not the case with rats. Rat growth is strikingly impaired when grain protein-based or bean protein-based diets are used.

## CLINICAL ISSUES IN PROTEIN NUTRITION

Protein nutrition is of special concern in several conditions. These include liver disease, protein-calorie malnutrition, diets limiting in indispensable amino acids, modified fasting (for the purpose of weight reduction), chronic renal disease, and hyperuricemia. The goal of protein nutrition in modified fasting is to supply an amount of protein for minimizing losses of endogenous proteins and for maintaining nitrogen balance. The goal of protein nutrition during the treatment of chronic renal disease is to reduce the buildup of urea and other waste products of protein metabolism in the bloodstream by limiting the amount of dietary protein. The progress of chronic renal disease may be slowed with low-protein diets. Chronic renal disease is usually fatal. The possibility of protein malnutrition during treatment is a concern. Uric acid is a catabolite of purines. Uric acid occurs in elevated concentrations in the plasma of certain populations. High plasma uric acid, or hyperuricemia, results primarily from genetic factors, but can be influenced by the diet as well. The danger of hyperuricemia is that it is a risk factor for the deposit of crystals of uric acid in the joints, leading to **gout**, and in the kidneys, leading to **uric acid kidney stones**. Hyperuricemia is treated mainly by drugs, though dietary intervention is also important. Dietary intervention involves the elimination of certain high-purine foods, consumption of the RDA of protein, and elimination of high-protein diets. Details of modified fasting, renal disease, and hyperuricemia are presented next.

## Modified Fasting

Modified fasting using low-energy diets is used to treat obesity. Total fasting should not be used for weight reduction. The main problem with total fasting, as far as protein nutrition is concerned, is catabolism of the lean body mass. Muscle accounts for most of the lean body mass. Catabolism of lean body mass accounts for about 50% of the weight loss in the first month of a total fast and for 25 to 33% thereafter (Seim and Rigden, 1990). Total fasting can provoke cardiac arrhythmias, destruction of cardiac muscle fibers, and hepatic and renal damage.

The modified fast supplies 1.2 to 1.4 g of protein per kilogram of ideal body weight. This amount of protein minimizes losses of the body's protein and maintains nitrogen balance. The protein can be supplied by commercially available powders consisting of protein, carbohydrate, vitamins, and minerals. Alternatively, the protein can be supplied by lean meat, fish, or poultry. Potassium losses are a vital concern during total fasting and during modified fasting as well. Low plasma levels of potassium (hypokalemia) can induce cardiac arrhythmias. Patients undergoing modified fasting are periodically monitored for cardiac arrhythmias, by use of an electrocardiogram, and for plasma potassium levels.

Sudden death has occurred among unsupervised fasting persons consuming commercially available "liquid protein" diets. The chronic consumption of such diets has resulted in about 60 deaths. The exact cause of death is not clear, though it might have resulted from cardiac arrest caused by hypokalemia. It should be noted that the term liquid protein is not scientifically accurate. Proteins can be dissolved in water, and they can be suspended in water, but they generally do not exist in the liquid state. Many of the deaths occurred when the liquid protein diet was discontinued and the subject consumed large amounts of food (binging). Apparently, the resultant increases in the levels of various metabolites in the bloodstream induced cardiac arrest.

The use of low-energy diets for weight reduction is safer for markedly obese individuals than for the mildly obese, because the weight lost with restricting energy to the mildly obese consists of a greater proportion of lean body mass and a lesser proportion of fat, than for the markedly obese (Wadden *et al.*, 1990).

Weight reduction programs may involve the participation of a dietician, physician, and psychologist. The dietician evaluates the choices of food of the subjects and makes sure that the diet contains sufficient amounts of protein, vitamins, minerals, and water and insufficient amounts of energy. The physician assesses cardiac function and evaluates status for a variety of nutrients. These include tests for plasma potassium, sodium, calcium, folate, and iron. The psychologist provides extra motivation for the weight reduction program and suggests changes in lifestyle. Without counseling, the majority of patients regain much of their lost weight in the year following treatment. With counseling, one-third or more of the patients regain much of their lost weight.

Exercise is used in weight reduction programs. Exercise helps reduce body fat and enhance muscle mass. The exercise program should be started when the modified fast begins. It involves changes in daily activities, such as using stairs instead of the elevator and substituting walking or bicycling for the automobile. A loss of about 50 pounds can be achieved in a 3-month weight reduction program.

## Chronic Renal Failure

Chronic renal failure is treated by dialysis, kidney transplants, and drugs, as well as by low-protein diets. For this reason, an outline of chronic renal failure occurs in this chapter. The normal **glomerular filtration rate** (GFR) is 80 to 120 ml/min. In severe renal disease, the GFR can be reduced to 10 ml/min or less. This represents a 90% loss of renal function. Diabetes mellitus and hypertension (high blood pressure) are the main causes of chronic renal failure. Sustained and chronic injury to the kidneys leads to the destruction of the nephrons, where this destruction is usually not reversible. The nephron, which is the smallest unit of kidney function, is detailed in the section on Sodium, Potassium, Chloride, and Water. The severe loss of nephrons results in alterations of functions of many other organs of the body. The collection of abnormalities that results is called **uremia**.

Moderate renal failure is defined as the situation where the GFR is 30–70 ml/min, while severe renal failure is where the GFR is under 30 ml/min. In chronic renal failure, the gradual destruction of nephrons is compensated by hypertrophy of the remaining nephrons. Hence, the kidney continues to excrete normal levels of salt and water even with a moderate reduction in the GFR (Martin, 1996). Eventually, however, urea and other nitrogenous compounds (ammonia, guanidine, urate) begin to accumulate in the bloodstream. The increase in some of these compounds produces the nausea, vomiting, and tiredness that makes the patient first aware of the renal failure. Uremia occurs when the GFR falls below 20–25% the normal value (Brenner and Lazarus, 1994).

With progressing renal failure, the kidney loses its ability to excrete acid into the urine. The result is acidosis. Acidosis can be treated with sodium bicarbonate or sodium citrate (Brenner and Larazus, 1994). The kidney performs several functions not directly related with urine production, and these include the synthesis of calcitriol (the active form of vitamin D) and erythropoietin (a hormone that stimulates red blood cell production). Thus, treatment of chronic renal failure may include injections of calcitriol and erythropoietin.

Diet modification is an important part of therapy for uremia. Dietary protein is restricted, and this technique reduces nausea, vomiting, and anorexia. Protein is restricted to no less than 0.6 g protein/kg body weight, in order to avoid negative N balance (Brenner and Lazarus, 1994). The normal plasma concentration of urea is 3.9 to 7.9 mM (0.24–0.49 mg/ml). The normal serum urea nitrogen level is 0.11 to 0.23 mg/ml. The goal of restricting dietary protein is to prevent serum urea nitrogen from rising above 0.8 mg/ml. Where the GFR is 20 to 25 ml/min, protein may be restricted to 60 to 90 g/day. Where the GFR is 10 ml/min or less, dietary protein may be restricted to 40 g/day (Maroni and Mitch, 1997).

End-stage renal disease is defined as the situation where the GFR falls below 20 ml/min (Martin, 1996). Each year, over 30,000 persons in the United States acquire end-stage renal disease. Hypertension is the most common complication of end-stage renal failure, where the hypertension can provoke heart failure and leakage of fluid into the lungs (pulmonary edema). This hypertension can be treated with dietary restriction of salt and water, and with drugs (inhibitor of angiotensin-converting enzyme).

EXERCISE

In treating the uremic patient with the protein-restricted diet, would you use a protein of high or low biological value?

## Hyperuricemia

Dietary purines are largely catabolized in the gut, rather than used by the body for the synthesis of nucleic acids. The end-product of purine catabolism in humans is **uric acid**. The diet accounts for less than half of the uric acid appearing in the bloodstream. Most of the plasma uric acid, or urate, originates from catabolism of the purines synthesized by the body (endogenous purines). The major purines are adenine and guanine. They occur mainly as nucleotides, such as adenosine triphosphate (ATP) and guanosine triphosphate (GTP), and as parts of nucleic acids. For example, the adenine in RNA occurs as adenosine monophosphate, and the adenine in DNA occurs as deoxyadenosine monophosphate.

Normal serum uric acid is 0.025 to 0.080 mg/ml in males and 0.015 to 0.060 mg/ml in females. A uric acid level greater than 0.070 mg/ml is associated with an increased risk for gout. Purine-restricted diets can result in plasma uric acid levels of 0.005 to 0.015 mg/ml. About two-thirds of the uric acid in the body is excreted via the urine; the rest is excreted in the feces. Uric acid accounts for only about 5% of urinary nitrogen.

Gout is a metabolic disease characterized by the painful inflammation of joints. Gout is a type of arthritis. Hyperuricemia is a risk factor for gout and for the development of uric acid stones in kidneys. Gout often presents as a violent attack of pain in the big toe. Less commonly, it involves the wrists, ankles, or knees. The pain spontaneously disappears after a few days, but recurs later, with an increasing frequency. Eventually, crystalline deposits of uric acid accumulate in the joints and appear as bumps on the skin, each with a diameter of a millimeter to a few centimeters. The deposits of urate lead to erosion of the bone and to destruction of the joints. Gout can be crippling.

The mechanism producing the pain and inflammation of the gouty attack involves the neutrophils. The neutrophil is a type of white blood cell that consumes foreign objects, such as bacteria, and attacks them with toxic forms of oxygen, such as HOOH and HO•. High plasma uric acid results in the formation and deposit of uric acid crystals in joints. The urate crystals are taken up by the neutrophils. The neutrophils release toxic forms of oxygen, as well as hydrolytic enzymes. This reaction is part of the inflammatory response of gout and contributes to the pain and fever of the gouty attack.

Uric acid stones account for about 10% of all kidney stones in the United States. Most kidney stones consist of calcium phosphate or calcium oxalate. The risk for developing urate stones increases with increased levels of plasma urate and with increased levels of urinary urate. Sometimes the stones that form in the kidney are excreted in the urine, where they appear as a fine gravel. A larger stone can obstruct the ureter, impairing the formation of the urine and leading to intense pain. Kidney stones are also called **renal calculi**.

Hyperuricemia and gout are treated with drugs, such as **allopurinol**. Allopurinol inhibits one of the enzymes of the purine catabolic pathway and decreases the formation of uric acid. Dietary intervention may be part of the treatment program. Dietary intervention involves the elimination of foods containing high levels of purines. This reduces exogenous purines as a source of uric acid. These foods include glandular and organ meats, dried legumes, shrimp, sardines, and mackerel. Other sources of meats and fish are restricted but not eliminated. High dietary protein can stimulate the formation of uric acid. The mechanism of this effect is not clear. It is thought that high levels of dietary protein can interfere with the normal regulation of the pathway of purine catabolism, resulting in an increase in its rate.

The following study demonstrates the effect of dietary protein on uric acid. In the study, human subjects consumed diets containing no protein, normal levels of protein, and extremely high levels of protein. The diets were chemically defined and did not contain purines. The protein-free diet supplied 0.9 g N/day. The normal-protein diet supplied 13 g N/day. The extremely high-protein diet supplied 62 g N/day. The feeding trials were 2 weeks in duration. The normal-protein diet contained 90 g of egg albumin, which was supplemented with 162 g of soy protein plus 156 g of casein in the high-protein diet.

Table 8.15 lists the values of the nitrogenous compounds in the plasma and urine. The plasma values represent the levels in the bloodstream after an overnight fast and are in the units of milligrams/liter. The urinary values represent the daily output and are in the units of grams/day. Uric acid levels responded to the diets in the following manner. Plasma uric acid levels were nearly identical with the three diets (second line). Urinary uric acid levels changed little with the protein-free and normal-protein diets; however, the fivefold increase in protein intake with the high-protein diet resulted in a 2.5-fold increase in urinary uric acid levels (last line). Although the mechanism by which high levels of protein increase the pro-

**TABLE 8.15**  Nitrogenous Metabolites in Human Plasma and Urine

| Nitrogen | Protein free | Normal protein | High protein |
|---|---|---|---|
| Plasma urea N (mg/liter) | 49 | 147 | 366 |
| Plasma uric acid N (mg/liter) | 19 | 17 | 18 |
| Total urinary N (g/day) | 2.1 | 9.6 | 52.0 |
| Urinary urea N (g/day) | 1.0 | 7.0 | 48.0 |
| Urinary ammonia N (g/day) | 0.04 | 0.18 | 1.11 |
| Urinary amino acid N (g/day) | 0.09 | 0.13 | 0.34 |
| Urinary creatinine N (g/day) | 0.55 | 0.56 | 0.59 |
| Urinary urate N (g/day) | 0.10 | 0.13 | 0.37 |

*Source*: Bowering *et al.* (1970).

duction of uric acid is not clear, the data demonstrate the value of avoiding high-protein diets during the treatment of patients with hyperuricemia and gout.

Studies by Clifford *et al.* (1976) revealed the effect of purine consumption on uric acid levels. The purines were supplied to human subjects in the form of ribonucleic acid (RNA) (4 g/day). Purine consumption resulted in a near doubling of plasma levels of uric acid and a 2.5-fold increase in urinary uric acid. This demonstrates the need for avoiding purine-rich foods in treating hyperuricemia and gout. It has been recommended that the maximal safe limit of RNA in the diet is 2.0 g/day (Clifford *et al.*, 1976). This amount of RNA can be supplied by 340 g of sardines, 415 g of dried lentils or pinto beans, or 500 g of chicken liver. As few people consume, or would be willing to consume, 500 g of liver per day the limitation of dietary RNA to safe levels would not be expected to be a common concern.

The purine catabolic pathway appears in Figure 8.31. The end-product of purine catabolism in primates, and in some other vertebrates, is uric acid. Purine catabolism differs in other species. Urate oxidase catalyzes the breakdown of uric acid to allantoin. Allantoin can be further broken down to produce urea and glyoxylate. Allantoin is the purine excretory product in some mammals and reptiles. Urea is the purine excretory product in fish. Guanine is the purine excretory product in pigs and spiders. Uric acid is used for the packaging and excretion of waste N from amino acids in birds and some reptiles.

Most of the free purines derived from the breakdown of DNA, RNA, and nucleotides in the diet are catabolized to xanthine and then to uric acid in the gut mucosa. The AMP and GMP biosynthesized in the body can also be broken down to free purines, such as adenine, guanine, and hypoxanthine. These purines, in contrast to those derived from the diet, are largely reused for the synthesis of ATP and GTP. They are first converted back to AMP or GMP in a pathway of reutilization called the purine salvage pathway. For example, adenine phosphoribosyltransferase (PRPP) catalyzes the conversion of adenine to AMP. Here, PRPP serves as the source of the phosphoribose group. Pyrophosphate is a product of the reaction.

## Wasting in Cancer and AIDS

Cancer results in anorexia (lack of interest in eating), malnutrition, and weight loss. Although weight loss may be due to loss of muscle and organ tissue (lean body mass), as well as of fat, the loss of fat is a minimal concern in cancer and other wasting diseases. Cachexia is a term that is often used to refer to the wasting of muscle and other tissues that occurs with various diseases.

Wasting is a common feature of cancer of the lung, prostate, head, and stomach, to give a few examples, but not breast cancer. Weight loss is often the first detected sign that a patient may have cancer. In cancer, wasting tends to occur once tumors comprise over 0.01% of the patient's body weight (Tisdale, 1996). In famine or anorexia nervosa, loss of fat accounts for about three quarters of the weight loss, with loss of lean body mass accounting for the remaining weight loss. In contrast, with cancer, weight loss occurs about equally in muscle and fat.

AIDS may be defined as the disease that occurs with infection by human immunodeficiency virus (HIV), where symptoms such as low white blood cell

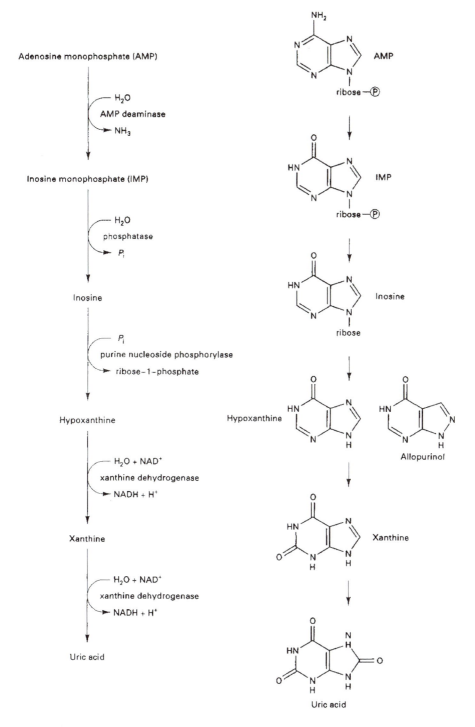

**FIGURE 8.31** Purine catabolic pathway. Purines occur in a number of forms in the body. The figure illustrates the pathway using the example of the degradation of AMP. AMP is degraded in five steps to produce uric acid. Allopurinol is a drug that inhibits the activity of xanthine dehydrogenase.

counts and opportunistic infections also occur. Wasting is common in AIDS. With AIDS, for example, the patient may lose 5% of body weight over the course of a month. The wasting in cancer and AIDS occurs to a greater extent in lean body mass, when compared with the wasting of healthy persons who are fasting for prolonged periods of time (Hellerstein, 1996). The loss of lean body mass, with cancer, tends to occur more in muscle than in visceral tissues (Tisdale, 1997). Wasting increases the risk of death in cancer patients and in AIDS patients.

Surprisingly, diet therapy, even with tube feeding directly into the gut (enteral nutrition) or circulatory system (parenteral nutrition), has little or no influence on cancer-induced loss of lean body mass (Chlebowski *et al.*, 1996). Hence, weight loss is not simply due to anorexia, but also due to the failure of the body to utilize protein and energy to maintain the lean body mass. Where tube feeding does lead to weight gain, the increased weight is due to increases in body fat, and not in muscle. The changes in energy metabolism are striking, but the exact mechanisms leading to these changes are not clear. Where the healthy person undergoes starvation, the plasma triglyceride levels decline to low levels. In contrast, in wasting in cancer and AIDS, plasma triglyceride levels increase, and there is an increased rate of fatty acid oxidation, and increases in the basal metabolic rate. In wasting in AIDS, for example, the basal metabolic rate can be 7–12% higher than expected. The anorexia and tissue wasting in cancer is thought to result from interactions between tumors and normal white blood cells, where the white blood cells are provoked to release certain hormones called cytokines.

A variety of drugs have been used to prevent wasting in AIDS patients. Drugs that have been shown to be successful include growth hormone and androgens. Methods unique to the science of nutrition have found use in assessing whether a drug that prevents loss of body weight actually prevents loss of lean body mass. These methods include skinfold measurements and bioelectrical impedance (Oster *et al.*, 1994). These methods are detailed in the Obesity chapter.

## REFERENCES

Adibi, S. A., and Mercer, D. W. (1973). Protein digestion in human intestine as reflected in luminal mucosal and plasma amino acid concentrations after meals. *J. Clin. Invest.* **52**, 1586–1594.

Albert, J. D., Legaspi, A., Horowitz, G. D., Tracey, K. J., Brennan, M. F., and Lowry, S. F. (1986). Extremity amino acid metabolism during starvation and intravenous feeding in humans. *Am. J. Physiol.* **251**, E604–E610.

Aoki, T. T., Muller, W. A., Brennan, M. F., and Cahill, G. F. (1975). Metabolic effects of glucose in brief and prolonged fasting man. *Am. J. Clin. Nutr.* **28**, 507–511.

Arunepanlop, B., Morr, C. V., Karleskind, D., and Laye, I. (1996). Partial replacement of egg white proteins with whey proteins in angel food cakes. *J. Food Sci.* **61**, 1085–1093.

Atkinson, D. E., and Bourke, E. (1984). The role of ureagenesis in pH homeostasis. *Trends Biol. Sci.* July, pp. 297–300.

Atkinson, D. E., and Bourke, E. (1988). Letters to the editors. *Am. J. Physiol.* **255**, F808–F809.

Ausman, L. M., Gallina, D., Hayes, K. C., and Hegsted, D. M. (1986). Comparative assessment of soy and milk protein quality in infant Cebus monkeys. *Am. J. Clin. Nutr.* **43**, 112–127.

Bernardini, P., and Fischer, J. E. (1982). Amino acid imbalance and hepatic encephalopathy. *Annu. Rev. Nutr.* **2**, 419–454.

Biederer, T., Volkwein, C., and Sommer, T. (1997). Role of cue1p in ubiquitination and degradation at the ER surface. *Science* **278**, 1806–1809.

Block, K. P., Aftering, R. P., Mehard, W. B., and Buse, M. G. (1987). Modulation of rat skeletal muscle branched-chain α-keto acid dehydrogenase *in vivo*. *J. Clin. Invest.* **79**, 1349–1358.

Boon, L., Blommaart, P., Meijer, A., Lamers, W., and Schoolwerth, A. C. (1996). Response of hepatic amino acid consumption to chronic metabolic acidosis. *Am. J. Physiol.* **271**, F198–F202.

Bowering, J., Calloway, D. H., Margen, S., and Kaufman, N. A. (1970). Dietary protein and uric acid metabolism in normal men. *J. Nutr.* **100**, 249–261.

Boye, J., Alli, I., Ramaswamy, H., and Raghavan, V. (1997). Interactive effects of factors affecting gelation of whey proteins. *J. Food Sci.* **62**, 57–65.

Brenner, B. M., and Lazarus, J. M. (1994). Chronic renal failure. *In* "Harrison's Principles of Internal Medicine," 13th ed. (K. J. Isselbacher, ed.). McGraw-Hill, New York.

Cannon, P. R., Steffee, C. H., Frazier, L. J., Rowley, D. A., and Stepto, R. C. (1947). The influence of time of ingestion of essential amino acids upon utilization in tissue synthesis. *Fed. Proc.* **6**, 390.

Carey, G. P., Kime, Z., Rogers, Q. R., Morris, J. G., Hargrove, D., Buffington, C. A., and Brusilow, S. W. (1987). An arginine-deficient diet in humans does not evoke hyperammonemia or orotic aciduria. *J. Nutr.* **117**, 1734–1739.

Chlebowski, R., Palomares, M., Lillington, L., and Grosvenor, M. (1996). Recent implications of weight loss in lung cancer management. *Nutrition* **12**, S43–S47.

Clifford, A. J., Riumallo, J. A., Young, V. R., and Scrimshaw, N. S. (1976). Effect of oral purines on serum and urinary uric acid of normal, hyperuricemic and gouty humans. *J. Nutr.* **106**, 428–450.

Cooper, A. J., and Plum, F. (1987). Biochemistry and physiology of brain ammonia. *Physiol. Rev.* **67**, 440–519.

DeRosa, G., and Swick, R. W. (1975). Metabolic implications of the distribution of the alanine aminotransferase isoenzymes. *J. Biol. Chem.* **250**, 7961–7967.

Dev, B. C., Sood, S. M., DeWind, S., and Slattery, C. W. (1994). κ-Casein and β-caseins in human milk micelles: Structural studies. *Arch. Biochem. Biophys.* **314**, 329–336.

FAO/WHO/UNO (1985). "Energy and Protein Requirements," World Health Organization Technical Report Series 724, p. 56. World Health Organization, Geneva.

Felig, P., Marliss, E., Pozefsky, T., and Cahill, G. F. (1970). Amino acid metabolism in the regulation of gluconeogenesis in man. *Am. J. Clin. Nutr.* **23**, 986–992.

Filer, L. J., Garrattini, S., Kare, M. R., Reynolds, W. A., and Wurtman, R. J. (1979). Glutamic acid. *In* "Advances in Biochemistry and Physiology," p. 28. Raven Press, New York.

Fisler, J. S., Drenick, E. J., Blumfield, D. E., and Swendseid, M. E. (1982). Nitrogen economy during very low calorie reducing diets. *Am. J. Clin. Nutr.* **35**, 471–486.

Fitch, W. L., and King, J. C. (1987). Plasma amino acid, glucose, and insulin responses to moderate-protein and high-protein test meals in pregnant, nonpregnant, and gestational diabetic women. *Am. J. Clin. Nutr.* **46**, 243–249.

Food and Nutrition Board (1989). "Recommended Dietary Allowances," 10th ed., p. 66. National Academy Press, Washington, DC.

Fukagawa, N. K., Minaker, K. L., Young, V. R., and Rowe, J. W. (1986). Insulin dose-dependent reactions in plasma amino acids in man. *Am. J. Physiol.* **250**, E13–E17.

Gamble, J. L. (1948). Physiological information gained from studies on the life raft ration. *Harvey Lect.* **42**, 247–273.

Garza, C., Scrimshaw, N. S., and Young, V. R. (1977). Human protein requirements: A long term metabolic nitrogen balance study in young men to evaluate the 1973 FAO/WHO safe level of egg protein intake. *J. Nutr.* **107**, 335–352.

Gastaldi, E., Lagaude, A., and Fuente, B. (1996). Micellar transition state in casein between pH 5.5 and 5.0. *J. Food Sci.* **61**, 59–68.

Gelfand, R. A., and Sherwin, R. S. (1986). Nitrogen conservation in starvation revisited: Protein sparing with intravenous fructose. *Metabolism* **35**, 37–44.

Goldberg, A. L. (1995). Functions of the proteasome: The lysis at the end of the tunnel. *Science* **268**, 522–524.

Goldberg, A. L., and St. John, A. C. (1976). Intracellular protein degradation in mammalian and bacterial cells, part 2. *Annu. Rev. Biochem.* **45**, 747–803.

Goldfarb, M. (1997). Two-dimensional electrophoretic analysis of human milk-fat globule membrane proteins with attention to apolipoprotein E patterns. *Electrophoresis* **18**, 511–515.

Goodman, M. W., Zieve, L., Konstantides, F. N., and Cerra, F. B. (1984). Mechanism of arginine protection against ammonia intoxication in the rat. *Am. J. Physiol.* **247**, G290–G295.

Hall, T. R., Wallin, R., Reinhart, G. D., and Hutson, S. M. (1993). Branched-chain amino-transferase isoenzymes. *J. Biol. Chem.* **268**, 3092–3098.

Halton, D. M., and Krieger, I. (1980). Studies of glycine metabolism and transport in fibroblasts from patients with nonketotic hyperglycinemia. *Pediatr. Res.* **14**, 932–934.

Hayase, K., Yokogoshi, H., and Yoshida, A. (1980). Effect of dietary proteins and amino acid deficiencies on urinary excretion of nitrogen and the urea synthesizing system in rats. *J. Nutr.* **110**, 1327–1337.

Haymond, M. W., Karl, I. E., Clarke, W. L., Pagliara, A. S., and Santiago, J. V. (1982). Differences in circulating gluconeogenic substrates during short-term fasting in men, women, and children. *Metabolism* **31**, 33–42.

Hellerstein, M. K. (1996). Management of nutritional and endocrinologic consequences of HIV infection. *In* "Management of the HIV-Infected Patient" (S. Crowe, J. Hoy, and J. Mills, eds.). Cambridge Univ. Press, Cambridge.

Hiramatsu, T., Fukagawa, N., Marchini, J., Cortiella, J., Yong-Ming, Y., Chapman, T., and Young, V. (1994). Methionine and cysteine kinetics at different intakes of cystine in healthy adult men. *Am. J. Clin. Nutr.* **60**, 525–533.

Holmes, F. L. (1993). "Hans Krebs: Architect of Intermediary Metabolism, 1933–1937," Vol. 2. Oxford Univ. Press, New York.

Hulse, J. H., Rachie, K. O., and Billingsly, L. W. (1977). "Nutritional Standards and Methods of Evaluation for Food Legume Breeders." International Development Research Centre, Ottawa, Canada.

Kang-Lee, Y., and Harper, A. E. (1977). Effect of histidine intake and hepatic histidase activity on the metabolism of histidine *in vivo*. *J. Nutr.* **107**, 1427–1443.

Kato, H., Mizutani-Funahashi, M., Shiosaka, S., and Nakagawa, H. (1978). Circadian rhythms of urea formation and argininosuccinate synthetase activity in rat liver. *J. Nutr.* **108**, 1071–1077.

King, J. C., Calloway, D. H., and Margen, S. (1973). Nitrogen retention. Total body [40]K and weight gain in teenaged pregnant girls. *J. Nutr.* **103**, 772–785.

Klahr, S., Schreiner, G., and Ichikawa, I. (1988). The progression of renal disease. *N. Engl. J. Med.* **318**, 1657–1666.

Krapf, R., Pearce, D., Lynch, C., Xi, X., Reudelhuber, T. L., Pouyssegur, J., and Rector, F. C. (1991). Expression of rat renal Na/H antiporter mRNA levels in response to respiratory and metabolic acidosis. *J. Clin. Invest.* **87**, 747–750.

Kushner, R. F. (1986). Total parenteral nutrition-associated metabolic acidosis. *J. Parenteral Enteral. Nutr.* **10**, 306–310.

Law, A. J., and Leaver, J. (1997). Effect of protein concentration on rates of thermal denaturation of whey proteins in milk. *J. Agric. Food Chem.* **45**, 4255–4261.

Lockwood, A. H., McDonald, J. M., Reiman, R. E., Gelbard, A. S., Laughlin, J. S., Duffy, T. E., and Plum, F. (1979). The dynamics of ammonia metabolism in man. *J. Clin. Invest.* **63**, 449–460.

Lukaski, H. C. (1987). Methods for the assessment of human body composition: Traditional and new. *Am. J. Clin. Nutr.* **46**, 537–556.

Man, E. H., and Bada, J. L. (1987). Dietary D-amino acids. *Annu. Rev. Nutr.* **7**, 209–225.

Maroni, B. J., and Mitch, W. E. (1997). Role of nutrition in prevention of progress of renal disease. *Annu. Rev. Nutr.* **17**, 435–455.

Martin, K. J. (1996). Renal failure. *In* "Textbook of Primary Care Medicine," 2nd ed. (J. Noble, ed.). Mosby, St. Louis.

Maruyama, K., Shands, H. L., Harper, A. E., and Sunde, M. L. (1975). An evaluation of the nutritive value of new high protein oat varieties (cultivars). *J. Nutr.* **105**, 1048–1054.

Milman, G., Portnoff, L. S., and Tiemeier, D. C. (1975). Immunochemical evidence for glutamine-mediated degradation of glutamine synthetase in cultured Chinese hamster cells. *J. Biol. Chem.* **250**, 1393–1399.

Morris, J. G., and Rogers, Q. R. (1978). Arginine: An essential amino acid for the cat. *J. Nutr.* **108**, 1944–1953.

Munaver, S. M., and Harper, A. E. (1959). Amino acid balance and imbalance. *J. Nutr.* **69**, 58–64.

Nakagawa, I., Takahashi, T., and Suzuki, T. (1961). Amino acid requirements of children: Minimal needs of lysine and methionine based on the nitrogen balance method. *J. Nutr.* **74**, 401–407.

Ogawa, H., Gomi, T., Konishi, K., Date, T., Nakashima, H., Nose, K., Matsuda, Y., Peraino, C., Pitot, H. C., and Fujioka, M. (1989). Human liver serine dehydratase. *J. Biol. Chem.* **264**, 15818–15823.

Otte, J., Ju, Z. Y., Faergemand, M., Lomholt, S. B., and Qvist, K. B. (1996). Protease-induced aggregation and gelation of whey proteins. *J. Food Sci.* **61**, 911–923.

Parsons, H., Fung, E., and Snyder, F. F. (1987). Branched-chain α-keto acids for the diagnosis of maple-syrup-urine disease. *N. Engl. J. Med.* **316**, 951.

Pasquali, R., Casimirri, F., and Melchionda, N. (1987). Protein metabolism in obese patients during very-low-calorie mixed diets containing different amounts of proteins and carbohydrates. *Metabolism* **36**, 1141–1148.

Pellet, P. L. (1990). Protein requirements of humans. *Am. J. Clin. Nutr.* **51**, 723–737.

Pennington, J. A. (1976). "Dietary Nutrient Guide." Avi, Westport, CT.

Peters, J. C., and Harper, A. E. (1981). Protein and energy consumption, plasma amino acid ratios, and brain neurotransmitter concentrations. *Physiol. Behav.* **27**, 287–298.

Peters, J. C., and Harper, A. E. (1985). Adaption of rats to diets containing different levels of protein. *J. Nutr.* **115**, 383–398.

Poston, J. M. (1984). The relative carbon flux through the α- and the β-keto pathways of leucine metabolism. *J. Biol. Chem.* **259**, 2059–2061.

Rafoth, R. J., and Onstad, G. R. (1975). Urea synthesis after oral protein ingestion in man. *J. Clin. Invest.* **56**, 1170–1174.

Rose, W. C., and Wixom, R. L. (1955). The amino acid requirements of man. *J. Biol. Chem.* **216**, 763–773.

Schimke, R. T. (1962). Adaptive characteristics of urea cycle enzymes in the rat. *J. Biol. Chem.* **237**, 459–468.

Scriver, C. R., Gregory, D. M., Sovetts, D., and Tissenbaum, G. (1985). Normal plasma amino acid values in adults: The influence of some common physiological variables. *Metabolism* **34**, 868–873.

Scriver, C. R., Beaudet, A. L., Sly, W. S., and Valle, D. (1989). "The Metabolic Basis of Inherited Disease," 6th ed. McGraw-Hill Information Services, New York.

Seim, H. C., and Rigden, S. R. (1990). Approaching the protein-sparing modified fast. *Am. J. Fam. Pract.* **52**, 51S–56S.

Shangraw, R. E., Stuart, C. A., Prince, M. J., Peters, E. J., and Wolfe, R. R. (1988). Insulin responsiveness of protein metabolism *in vivo* following bed rest in humans. *Am. J. Physiol.* **255**, E548–E558.

Snell, K. (1986). The duality of pathways for serine biosynthesis is a fallacy. *Trends. Biol. Sci.* **11**, 241–243.

Snodgrass, P. J., and Lin, R. C. (1981). Induction of urea cycle enzymes of rat liver by amino acids. *J. Nutr.* **111**, 586–601.

Souba, W. W. (1987). Interorgan ammonia metabolism in health and disease: A surgeon's view. *J. Parenteral Enteral Nutr.* **11**, 569–579.

Souba, W. W., Smith, R. J., and Wilmore, D. W. (1985). Review — Glutamine metabolism by the intestinal tract. *J. Parenteral Enteral Nutr.* **9**, 608–617.

Stabler, S. P., Marcell, P. D., Podell, E. R., Allen, R. H., Savage, D. G., and Lindenbaum, J. (1988). Elevation of total homocysteine in the serum of patients with cobalamin or folate deficiency detected by capillary gas chromatography–mass spectrometry. *J. Clin. Invest.* **81**, 466–474.

Steginck, L. D., Filer, L. J., and Beker, G. L. (1985). Plasma glutamate concentrations in adult subjects ingesting monosodium L-glutamate in consomme. *Am. J. Clin. Nutr.* **42**, 220–225.

Stewart, P. M., and Walser, M. (1980). Short-term regulation of ureagenesis. *J. Biol. Chem.* **255**, 5270–5280.

Struck, J., and Sizer, I. W. (1960). Oxidation of L-$\alpha$-amino acids by liver microsomes. *Arch. Biochem. Biophys.* **90**, 22–30.

Stuart, D. I., and Jones, E. Y. (1997). Cutting complexity down to size. *Nature* **386**, 437–438.

Takiguchi, M., and Mori, M. (1995). Transcriptional regulation of genes for ornithine cycle enzymes. *Biochem. J.* **312**, 195–205.

Tisdale, M. J. (1996). Inhibition of lipolysis and muscle protein degradation by EPA in cancer cachexia. *Nutrition* **12**, S31–S33.

Tisdale, M. J. (1997). Cancer cachexia: Metabolic alterations and clinical manifestations. *Nutrition* **13**, 1–7.

Traxinger, R. R., and Marshall, S. (1989). Role of amino acids in modulating glucose-induced desensitization of the glucose transport system. *J. Biol. Chem.* **264**, 20910–20916.

Tressel, T., Thompson, R., Zieske, L. R., Menendez, M., and Davis, L. (1986). Interaction between L-threonine dehydrogenase and aminoacetone synthetase and mechanisms of aminoacetone production. *J. Biol. Chem.* **261**, 16428–16437.

Vaghefi, S. B., Makdani, D. D., and Mickelsen, O. (1974). Lysine supplementation of wheat proteins: A review. *Am. J. Clin. Nutr.* **27**, 1231–1246.

Vorherr, H. (1974). "The Breast Morphology, Physiology, and Lactation." Academic Press, New York.

Wadden, T. A., Van Itallie, T. B., and Blackburn, G. L. (1990). Responsible and irresponsible use of very-low-calorie diets in the treatment of obesity. *JAMA* **263**, 83–85.

Walser, M. (1986). Roles of urea production, ammonium excretion, and amino acid oxidation in acid–base balance. *Am. J. Physiol.* **250**, F181–F188.

Wilkinson, K. D. (1995). Roles of ubiquitylation in proteolysis and cellular regulation. *Annu. Rev. Nutr.* **15**, 161–189.

Windmueller, H. G. (1982). Glutamine utilization by the small intestine. *Adv. Enzymol.* **53**, 201–237.

Windmueller, H. G., and Spaeth, A. E. (1978). Identification of ketone bodies and glutamine as the major respiratory fuels *in vivo* for postabsorptive rat small intestine. *J. Biol. Chem.* **253**, 69–76.

Wong, D. W., Camirand, W. M., and Pavlath, A. E. (1996). Structures and functionalities of milk proteins. *CRC Crit. Rev. Food Sci. Nutr.* **36**, 807–844.

Young, V. R., and El-Khoury, A. (1995). Can amino acid requirements for nutritional maintenance in adult humans be approximated from the amino acid composition of body mixed proteins? *Proc. Natl. Acad. Sci. U.S.A.* **92**, 300–304.

Young, V. R., and Marchini, J. S. (1990). Mechanisms and nutritional significance of metabolic responses to altered intakes of protein and amino acids, with reference to nutritional adaption in humans. *Am. J. Clin. Nutr.* **51**, 270–289.

Zamenhof, S., Martheus, E. V., and Grauel, L. (1971). DNA (cell number) and protein in neonatal rat brain: Alteration by timing of maternal dietary protein restriction. *J. Nutr.* **101**, 1265–1270.

Zhang, Z. P., Fujii, M., and Aoki, T. (1996). Behavior of calcium and phosphate in artificial casein micelles. *J. Dairy Sci.* **79**, 1722–1727.

## BIBLIOGRAPHY

### *Regulation of Protein Catabolism*

Dice, J. F. (1987). Molecular determinants of protein half-lives in eukaryotic cells. *FASEB J.* **1**, 349–357.

Driscoll, J., and Goldberg, A. L. (1990). The proteasome (multicatalytic protease) is a component of the 1500-kDa proteolytic complex which degrades ubiquitin-conjugated proteins. *J. Biol. Chem.* **265**, 4789–4792.

Dunn, W. A. (1990). Studies on the mechanisms of autophagy: Formation of the autophagic vacuole. *J. Cell Biol.* **110**, 1923–1933.

Hershko, A. (1988). *Ubiquitin-mediated protein degradation. J. Biol. Chem.* **263**, 15237–15240.

Holliday, M. A. (1993). Biography of James L. Gamble. *J. Pediatr.* **122**, 156–161.

Furuno, K., Goodman, M. N., and Goldberg, A. L. (1990). Role of different proteolytic systems in the degradation of muscle proteins during denervation atrophy. *J. Biol. Chem.* **265**, 8550–8557.

Kato, H., Mizutani-Funahashi, M., Shiosaka, S., and Nakagawa, H. (1978). Circadian rhythms of urea formation and argininosuccinate synthetase activity in rat liver. *J. Nutr.* **108**, 1071–1077.

Lam, Y. A., Xu, W., DeMartino, G., and Cohen, R. (1997). Editing of ubiquitin conjugates by an isopeptidase in the 26S proteasome. *Nature* **385**, 737–740.

Li, J. B., and Wassner, S. J. (1984). Effects of food deprivation and refeeding on total protein and actomyosin degradation. *Am. J. Physiol.* **246**, E32–E37.

Lund, P., and Wiggins, D. (1984). Is *N*-acetyl-glutamate a short-term regulator of urea synthesis? *Biochem J.* **218**, 991–994.

Mortimore, G. E., Lardeux, B. R., and Adams, C. E. (1988). Regulation of microautophagy and basal protein turnover in rat liver. *J. Biol. Chem.* **263**, 2506–2512.

Mortimore, G. E., and Poso, A. R. (1987). Intracellular protein catabolism and its control during nutrient deprivation and supply. *Annu. Rev. Nutr.* **7**, 539–564.

Schimke, R. T. (1962). Adaptive characteristics of urea cycle enzymes in the rat. *J. Biol. Chem.* **237**, 459–468.

Snodgrass, P. J., and Lin, R. C. (1981). Induction of urea cycle enzymes of rat liver by amino acids. *J. Nutr.* **111**, 586–601.

Stewart, P. M., and Walser, M. (1980). Short-term regulation of ureagenesis. *J. Biol. Chem.* **255**, 5270–5280.

Takiguchi, M., and Mori, M. (1995). Transcriptional regulation of genes for ornithine cycle enzymes. *Biochem. J.* **312**, 195–205.

Turinsky, J., and Long, C. L. (1990). Free amino acids in muscle: Effect of muscle fiber population and denervation. *Am. J. Physiol.* **258**, E485–E491.

Young, V. R., and Marchini, J. S. (1990). Mechanisms and nutritional significance of metabolic responses to altered intakes of protein and amino acids, with reference to nutritional adaption in humans. *Am. J. Clin. Nutr.* **51**, 270–289.

## Protein Sparing and Growth

FAO/WHO/UNO (1985). "Energy and Protein Requirements," World Health Organization Technical Report Series 724, p. 56. World Health Organization, Geneva.

Fisler, J. S., Drenick, E. J., Blumfield, D. E., and Swendseid, M. E. (1982). Nitrogen economy during very low calorie reducing diets. *Am. J. Clin. Nutr.* **35**, 471–486.

Food and Nutrition Board (1989). "Recommended Dietary Allowances," 10th ed., p. 66. National Academy Press, Washington, DC.

Forbes, G. B. (1987). "Human Body Composition." Springer-Verlag, New York.

Gamble, J. L. (1948). Physiological information gained from studies on the life raft ration. *Harvey Lect.* **42**, 247–273.

Garza, C., Scrimshaw, N. S., and Young, V. R. (1977). Human protein requirements: A long term metabolic nitrogen balance study in young men to evaluate the 1973 FAO/WHO safe level of egg protein intake. *J. Nutr.* **107**, 335–352.

King, J. C., Calloway, D. H., and Margen, S. (1973). Nitrogen retention. Total body $^{40}K$ and weight gain in teenaged pregnant girls. *J. Nutr.* **103**, 772–785.

Lukaski, H. C. (1987). Methods for the assessment of human body composition: Traditional and new. *Am. J. Clin. Nutr.* **46**, 537–556.

Moore, F. D., and Sastry, K. S. (1982). Intracellular potassium: $^{40}K$ as a primordial gene irradiatory. *Proc. Natl. Acad. Sci. U.S.A.* **79**, 3556–3559.

Pellet, P. L. (1990). Protein requirements of humans. *Am. J. Clin. Nutr.* **51**, 723–737.

Zamenhof, S., Martheus, E. V., and Grauel, L. (1971). DNA (cell number) and protein in neonatal rat brain: Alteration by timing of maternal dietary protein restriction. *J. Nutr.* **101**, 1265–1270.

Zeman, F. J., and Stanbrough, E. C. (1969). Effect of maternal protein deficiency on cellular development in the fetal rat. *J. Nutr.* **99**, 274–282.

## Amino Acids

Adibi, S. A., and Mercer, D. W. (1973). Protein digestion in human intestine as reflected in luminal mucosal and plasma amino acid concentrations after meals. *J. Clin. Invest.* **52**, 1586–1594.

Benevenga, N. J., and Steele, R. D. (1984). Adverse effects of excessive consumption of amino acids. *Annu. Rev. Nutr.* **4**, 157–181.

Bernardini, P., and Fischer, J. E. (1982). Amino acid imbalance and hepatic encephalopathy. *Annu. Rev. Nutr.* **2**, 419–454.

Cannon, P. R., Steffee, C. H., Frazier, L. J., Rowley, D. A., and Stepto, R. C. (1947). The influence of time of ingestion of essential amino acids upon utilization in tissue synthesis. *Fed. Proc.* **6**, 390.

Carey, G. P., Kime, Z., Rogers, Q. R., Morris, J. G., Hargrove, D., Buffington, C. A., and Brusilow, S. W. (1987). An arginine-deficient diet in humans does not evoke hyperammonemia or orotic aciduria. *J. Nutr.* **117**, 1734–1739.

DeRosa, G., and Swick, R. W. (1975). Metabolic implications of the distribution of the alanine aminotransferase isoenzymes. *J. Biol. Chem.* **250**, 7961–7967.

Fitch, W. L., and King, J. C. (1987). Plasma amino acid, glucose, and insulin responses to moderate-protein and high-protein test meals in pregnant, nonpregnant, and gestational diabetic women. *Am. J. Clin. Nutr.* **46**, 243–249.

Fukagawa, N. K., Minaker, K. L., Young, V. R., and Rowe, J. W. (1986). Insulin dose-dependent reactions in plasma amino acids in man. *Am. J. Physiol.* **250**, E13–E17.

Halton, D. M., and Krieger, I. (1980). Studies of glycine metabolism and transport in fibroblasts from patients with nonketotic hyperglycinemia. *Pediatr. Res.* **14**, 932–934.

Laidlaw, S. A., and Kopple, J. D. (1987). Newer concepts of the indispensable amino acids. *Am. J. Clin. Nutr.* **46**, 593–605.

Milman, G., Portnoff, L. S., and Tiemeier, D. C. (1975). Immunochemical evidence for glutamine-mediated degradation of glutamine synthetase in cultured Chinese hamster cells. *J. Biol. Chem.* **250**, 1393–1399.

Milner, J. A., Garton, R. L., and Burns, R. A. (1984). Phenylalanine and tyrosine requirement of immature beagle dogs. *J. Nutr.* **114**, 2212–2216.

Parsons, H., Fung, E., and Snyder, F. F. (1987). Branched-chain α-keto acids for the diagnosis of maple-syrup-urine disease. *N. Engl. J. Med.* **316**, 951.

Peters, J. C., and Harper, A. E. (1981). Protein and energy consumption, plasma amino acid ratios, and brain neurotransmitter concentrations. *Physiol. Behav.* **27**, 287–298.

Poston, J. M. (1984). The relative carbon flux through the α- and the β-keto pathways of leucine metabolism. *J. Biol. Chem.* **259**, 2059–2061.

Scriver, C. R., Beaudet, A. L., Sly, W. S., and Valle, D. (1989). "The Metabolic Basis of Inherited Disease," 6th ed. McGraw-Hill Information Services, New York.

Shangraw, R. E., Stuart, C. A., Prince, M. J., Peters, E. J., and Wolfe, R. R. (1988). Insulin responsiveness of protein metabolism *in vivo* following bed rest in humans. *Am. J. Physiol.* **255**, E548–E558.

Stegink, L. D., Filer, L. J., and Beker, G. L. (1985). Plasma glutamate concentrations in adult subjects ingesting monosodium L-glutamate in consomme. *Am. J. Clin. Nutr.* **42**, 220–225.

Traxinger, R. R., and Marshall, S. (1989). Role of amino acids in modulating glucose-induced desensitization of the glucose transport system. *J. Biol. Chem.* **264**, 20910–20916.

Tressel, T., Thompson, R., Zieske, L. R., Menendez, M., and Davis, L. (1986). Interaction between L-threonine dehydrogenase and aminoacetone synthetase and mechanisms of aminoacetone production. *J. Biol. Chem.* **261**, 16428–16437.

Vaghefi, S. B., Makdani, D. D., and Mickelsen, O. (1974). Lysine supplementation of wheat proteins: A review. *Am. J. Clin. Nutr.* **27**, 1231–1246.

Woo, S. L. (1989). Molecular basis and population genetics of phenylketonuria. *Biochemistry* **28**, 1–6.

## Clinical Issues in Protein Nutrition

Bowering, J., Calloway, D. H., Margen, S., and Kaufman, N. A. (1970). Dietary protein and uric acid metabolism in normal men. *J. Nutr.* **100**, 249–261.

Clifford, A. J., Riumallo, J. A., Young, V. R., and Scrimshaw, N. S. (1976). Effect of oral purines on serum and urinary uric acid of normal, hyperuricemic and gouty humans. *J. Nutr.* **106**, 428–450.

Hoffer, L. J., and Forse, R. A. (1990). Protein metabolic effects of a prolonged fast and hypocaloric refeeding. *Am. J. Physiol.* **258**, E832–E840.

Ihle, B. E., Becker, G. J., Whitworth, J. A., Charlwood, R. A., and Kincaid-Smith, P. S. (1989). The effect of protein restriction on the progression of renal insufficiency. *N. Engl. J. Med.* **321**, 1773–1777.

Klahr, S., Schreiner, G., and Ichikawa, I. (1988). The progression of renal disease. *N. Engl. J. Med.* **318**, 1657–1666.

Puig, J. G., Anton, F. M. Jiminez, M. J., and Gutierrez, P. C. (1986). Renal handling of uric acid in gout. *Metabolism* **35**, 1147–1153.

Seim, H. C., and Rigden, S. R. (1990). Approaching the protein-sparing modified fast. *Am. J. Fam. Pract.* **52**, 51S–56S.

Wadden, T. A., Van Itallie, T. B., and Blackburn, G. L. (1990). Responsible and irresponsible use of very-low-calorie diets in the treatment of obesity. *JAMA* **263**, 83–85.

# 9

# VITAMINS

## OVERVIEW

Vitamins are required in the diet for growth, maintenance, and reproduction. These qualities are shared by other nutrients, such as the amino acids and the minerals. Why are amino acids and minerals not called vitamins? The vitamins share a family of characteristics. They are organic compounds required in the diet in only small amounts. They are not catabolized to satisfy part of the energy requirement and are not used for structural purposes. Many vitamins are used as cofactors for enzymes. Pyruvate dehydrogenase is one striking example. This enzyme uses five cofactors, where four of these cofactors are derived from vitamins.

Not all cofactors are derived from vitamins. Coenzyme Q, lipoic acid, dolichol phosphate, biopterin, heme, and molybdopterin are cofactors that are synthesized in the body from simple organic compounds. Heme and molybdopterin are relatively complex, from a nutritional point of view, because they require metal ions as part of their structure.

Two of the vitamins are converted to hormones. These are vitamin A and vitamin D. Vitamin A also functions as a cofactor in the visual cycle.

One quality shared by most of the vitamins is their inability to be used in the form in which they are absorbed. Most of the vitamins absorbed from the diet must be converted or modified to the active form. For example, vitamin A, in the form of retinyl ester, must be converted to retinoic acid to perform some of its functions. Thiamin must be converted to thiamin pyrophosphate to be coenzymatically active. Vitamin K must be reduced to dihydrovitamin K, folate must be polyglutamated, and niacin must be converted to NAD to be active in the cell.

Some vitamins undergo a rather unique transformation prior to becoming functional. They are covalently attached to specific enzymes. Biotin, for example, is covalently bound to the biotin-requiring enzymes. Pantothenic acid, in a modified form, is covalently bound to fatty acid synthase. Riboflavin, following conversion to the cofactor form, is bound to succinate dehydrogenase, as well as to a few other enzymes requiring riboflavin-based cofactors.

Some of the vitamins in the coenzyme form associate tightly with specific enzymes, but not via a covalent linkage. Immediately after biosynthesis on the ribosome, enzymes do not contain their cofactor, and these are called apoenzymes. An enzyme containing its required cofactor is called a **holoenzyme**. With removal of the cofactor, the enzyme is also called an apoenzyme. The enzymes that exist in apoenzyme and holoenzyme forms include those that use vitamin $B_{12}$, vitamin $B_6$, thiamin, and riboflavin-based cofactors. Enzymes that use niacin-based cofactors, folate, ascorbate, and vitamin K are not said to exist in apoenzyme and holoenzyme forms. These enzymes bind their cofactors relatively weakly, and the cofactors behave in a manner similar to substrates.

The existence of apoenzyme and holoenzyme forms of various enzymes is of use to the clinician. The proportion of a specific enzyme occurring in apoenzyme and holoenzyme forms is used to assess vitamin status in the cases of vitamin $B_6$, thiamin, and riboflavin. Vitamin status is determined by measuring the percentage stimulation of enzyme activity that occurs after adding the appropriate cofactor to a biological sample (blood) containing the enzyme of interest.

A **cofactor** can be any chemical required by an enzyme, that is, a metal ion, coenzyme, lipid, or accessory protein. A **coenzyme** is a small organic molecule required by an enzyme that participates in the chemistry of catalysis. Most of the coenzymes work according to the following principle. They shuttle back and forth between two or more different forms. Here, one of the forms may be considered to be the coenzymatically active form, and the other may be seen as requiring regeneration to the active form. This is the case for folate, vitamin $B_6$, riboflavin- and niacin-based cofactors, ascorbic acid, and vitamin K. The coenzymatically active (inactive) forms of three of these coenzymes are tetrahydrofolate (dihydrofolate), ascorbic acid (dehydroascorbic acid), and vitamin $KH_2$ (vitamin K).

The vitamins are discussed in the following order. Folate is presented first because deficiencies in this vitamin may be more widespread than those of other vitamins. Folate, vitamin $B_{12}$, vitamin K, and biotin are presented consecutively because they all mediate the transfer of 1-carbon units. Vitamin $B_6$ also mediates the transfer of small groups, including carboxyl groups, and thus is presented afterward. The relationship between folate and vitamin $B_{12}$ is especially close, as a deficiency in vitamin $B_{12}$ can induce impairments in folate metabolism.

Vitamin A and vitamin D, which can both function as hormones, are presented next. Vitamin A deficiency is widespread throughout poorer regions of the world; the consequences of this deficiency are especially severe. Niacin, thiamin, and riboflavin are presented in sequence, as these vitamins are very important for energy metabolism. Deficiencies in these three vitamins tend to occur with poor dietary practices. Pantothenic acid is used for the synthesis of coenzyme A, and thus is of paramount importance in energy metabolism and many biosynthetic and catabolic reactions. Pantothenic acid deficiency is rare. Ascorbic acid and vitamin E are presented consecutively, as they are used as reductants and serve to protect the cell from damage caused by toxic forms of oxygen. Ascorbic acid is used in the biosynthesis of collagen and specific hormones. The essential fatty acids are presented after vitamin E, as a major role for vitamin E in the body is the protection of essential fatty acids, and other unsaturated fatty acids, from toxic oxygen damage. The essential fatty acids are not considered vitamins. They are used for structural purposes in the cell and can contribute to a substantial proportion of our energy intake. Essential fatty acids, however, are required in the diet in only small amounts, when compared with the overall fatty acids in a typical diet. For these reasons, they are included in this chapter.

Some vitamins are water soluble, while others are fat soluble. This classification is valuable as it indicates whether the vitamin is likely to be absorbed similarly to lipids or like other water-soluble nutrients. The fat-soluble vitamins are A, D, E, and K. The water-soluble vitamins are ascorbic acid, biotin, folate, niacin, pantothenic acid, riboflavin, thiamin, vitamin $B_6$, and vitamin $B_{12}$. The classification is also valuable, as it helps chemists decide on the best way to extract and analyze a particular vitamin in foods and biological tissues. Aside from having some bearing on the pathways of absorption and distribution throughout the body, the question of whether a particular vitamin is fat soluble or water soluble has little or no relevance to its function in the body.

Vitamin biochemistry is a fascinating topic of knowledge. Many of the functions of "the vitamins," as revealed in this chapter, also take place in plants, bacteria, and archae. However, those interesting in delving deeper will be delighted to learn about further functions of "the vitamins" that take place in these life forms. One of the stranger functions, for example, is the role of folates as a structural component in some bacterial viruses.

# FOLATE

Folate is a water-soluble vitamin. The recommended daily allowances (RDAs) of folate for the adult man and woman are 200 and 180 µg, respectively. The RDA is defined in the Preface. Folate is critically important for growth, and for this reason it is required in increased amounts during pregnancy (400 µg). The need for dietary folate remains elevated after pregnancy and during lactation because of the trans-

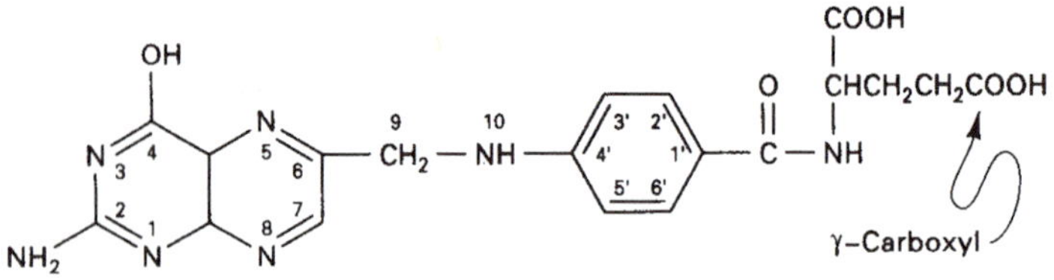

**FIGURE 9.1** Structure of folic acid (pteroylglutamic acid, PteGlu).

fer of the mother's vitamin to the milk. The RDA for infants is 3.6 µg/kg per day. Liver, egg yolk, orange juice, and green vegetables such as broccoli, spinach, and bell peppers are good sources of folate.

Folate is a generic term referring to a family of related compounds. All of these compounds represent modifications of the simplest form of the vitamin, folic acid (pteroylglutamic acid, PteGlu). Folic acid does not occur in nature in appreciable amounts, though it is readily assimilated by the body and converted to the active cofactor forms of the vitamin. Folic acid is the form of the vitamin used in folate supplements. Folates are modified by reduction and by a polyglutamyl chain or tail. The reduced folates include dihydrofolate and tetrahydrofolate.

Folates containing the polyglutamate chain are called folylpolyglutamates. The folylpolyglutamates in mammalian tissues consist mainly of folylpentaglutamates and folylhexaglutamates (Brody *et al.*, 1976; Brody and Stokstad, 1990, 1991). These contain five residues of glutamic acid (folylpentaglutamates) and six residues of glutamic acid (folylhexaglutamates). The structure of folic acid appears in Figure 9.1. The γ-carboxyl group, which is the point of attachment of the amino group of glutamate (in folylpolyglutamates), is indicated.

In addition to modification by reduction and by the polyglutamyl chain, the vitamin is modified by 1-carbon units, such as the 5-methyl- and 10-formyl-groups. The cofactor forms of the vitamin present in the greatest concentrations in mammalian tissues are tetrahydrofolylpentaglutamate ($H_4PG5$) (which does not bear a 1-carbon unit), 5-methyl-tetrahydrofolylpentaglutamate (5-methyl-$H_4PG5$), and 10-formyl-tetrahydrofolylpentaglutamate (10-formyl-$H_4PG5$). The structure of 5-methyl-$H_4PG5$ is shown in Figure 9.2. The residues of glutamate in the polyglutamyl chain are connected via the γ-carboxyl group and the α-amino group of adjacent glutamate residues to form a peptide bond. This peptide bond differs from that involving glutamic acid in proteins. The glutamate in proteins is connected to adjacent amino acids via its α-carboxyl and α-amino groups.

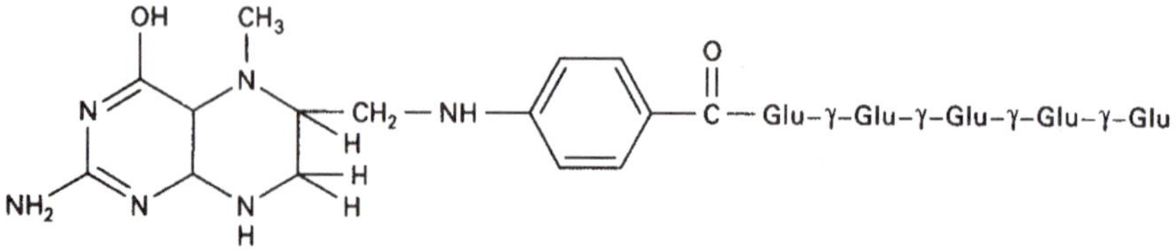

**FIGURE 9.2** 5-Methyl-tetrahydrofolylpentaglutamate (5-methyl-$H_4$PteGlu$_5$, 5-methyl-$H_4PG_5$).

## Biochemistry of Folate

Dietary folates consist mainly of folylpolyglutamates. The folylpolyglutamates of foods contain from four to seven residues of glutamic acid. The folates may bear a formyl group, a methyl group, or no 1-carbon group. The pteridine ring of folates in foods occurs mainly in the reduced tetrahydro- form. Other forms of the pteridine ring are the dihydro- and the nonreduced forms. Folylpolyglutamates do not readily cross cell membranes. They are poorly transported by gut, liver, and other cells. The folylpolyglutamates also tend to be poorly transported by folate-requiring microorganisms, including *Lactobacillus casei* and *Streptococcus faecalis*.

### γ-Glutamyl Hydrolase and Digestion of Folylpolyglutamates

The enterocyte contains a membrane-bound enzyme, γ-glutamyl hydrolase, which catalyzes the hydrolysis of the folylpolyglutamates to folylmonoglutamates. For example, it hydrolyzes 5-methyl-$H_4$PG7 to 5-methyl-$H_4$PG1, and $H_4$PG5 to $H_4$PG1. γ-Glutamyl hydrolase is required for the efficient absorption of dietary folates; however, the enzyme is not required for the absorption of medicinal forms of the vitamin, such as folic acid and 5-formyl-$H_4$PteGlu. The enzyme recognizes the γ-glutamyl bond and is not able to catalyze the hydrolysis of the more typical α-peptide bonds occurring in proteins and other polypeptides.

### Folates in the Bloodstream

Once absorbed by the gut, dietary folates appear in the bloodstream in the mono-glutamyl form. The main form of serum folate is 5-methyl-$H_4$folylmonoglutamate. The absorbed folates may be converted to the 5-methyl-form during passage through the enterocyte. They may be converted to the 5-methyl-form in the liver, with the release of 5-methyl-$H_4$folylmonoglutamate back into the bloodstream. A small amount of hydrolysis of food folates to folylmonoglutamates can occur in the food itself. This is because most animal and plant foods appear to contain γ-glutamyl hydrolase. Cooking destroys the activity of the enzyme.

### Polyglutamation of Folates in the Cell

Plasma folylmonoglutamates are transported into cells, where they are converted to the polyglutamyl form, an event catalyzed by folylpolyglutamate synthase. This enzyme uses tetrahydrofolates as a substrate. It seems not to recognize or use 5-methyl-$H_4$folates. Hence, the 5-methyl-$H_4$folate absorbed from the bloodstream must be converted to $H_4$folate prior to polyglutamation. The reaction occurs in a

$$H_4PteGlu \longrightarrow H_4PteGlu_2 \longrightarrow H_4PteGlu_3 \longrightarrow H_4PteGlu_4 \longrightarrow H_4PteGlu_5 \longrightarrow H_4PteGlu_6$$

**FIGURE 9.3**  Polyglutamation of tetrahydrofolic acid.

series of steps (Figure 9.3). Addition of each glutamate residue requires the energy of a molecule of ATP, that is, the conversion of ATP to ADP + $P_i$.

## Functions of the Polyglutamyl Group of Tissue Folates

Essentially all cellular folates occur in the polyglutamyl form. The function of the polyglutamyl chain is threefold. It is required for the retention of the vitamin in the cell. The long, polyanionic chain tends to prevent the folate from passing through cell membranes. The chain may also enhance the binding of the folate cofactor to folate-requiring enzymes. In some cases, the enhanced binding dramatically increases the catalytic rate of the enzyme. The most dramatic and unusual function of the polyglutamate tail is to allow channeling, a phenomenon where the folate coenzyme shuttles from one active site to another (of a bifunctional enzyme) without dissociating from the surface of the enzyme (Paquin *et al.*, 1985).

## 1-Carbon Metabolism

Folates are used as cofactors in a series of reactions known as **1-carbon metabolism**. 1-Carbon metabolism is presented in three diagrams of increasing complexity. The simplest diagram (Figure 9.4) depicts a cycle involving three cofactor forms: $H_4$folate, 5,10-methylene-$H_4$folate, and 5-methyl-$H_4$folate. All of these forms occur mainly as the pentaglutamate in mammalian tissues. Tetrahydrofolate ($H_4$folate) may be thought of as the starting and ending point of the cycle of reactions shown. The diagram also illustrates the **methylation cycle**.

Folates function by mediating the transfer of 1-carbon units. Most 1-carbon units in the cell originate from the 3-carbon of serine. The point of entry of this carbon is indicated by an asterisk. Serine donates a carbon to $H_4$folate, generating 5,10-methylene-$H_4$folate, which can then be reduced to 5-methyl-$H_4$folate. 5-

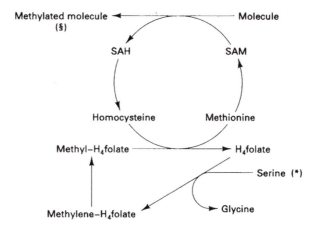

**FIGURE 9.4** The methylation cycle and its relationship to folate metabolism.

Methyl-H$_4$folate is a cofactor of methionine synthase. Methionine synthase cata-lyzes the transfer of the 1-carbon unit from 5-methyl-H$_4$folate to homocysteine, generating methionine. Methionine synthase also uses vitamin B$_{12}$ as a cofactor.

The methylation cycle proceeds as follows. Methionine can be converted to S-adenosylmethionine (SAM). SAM is a universal methyl donor and is required in most or all methylation events occurring in the body. For example, SAM is used in the synthesis of creatine and carnitine and in the methylation of nucleic acids and proteins. With the donation of the methyl group, SAM is converted to S-ade-nosylhomocysteine (SAH), as shown in Figure 9.4. SAH is finally broken down to homocysteine, completing the methylation cycle. The point of departure of the 1-carbon unit, derived from serine, from the methylation cycle is indicated by the section symbol (§).

EXERCISE

The data in Table 9.1 are from an experiment involving folylpolyglutamate synthase. The enzyme was obtained in purified form from liver and was mixed with glutamic acid, ATP, and the indicated form of folate. The mixture was incubated to permit the enzyme to catalyze the attachment of glutamate residues to the folate. What do you think would be the effect of a vitamin B$_{12}$ deficiency on the rate of polyglutamation of dietary folates. Hint: Dietary folates occur in the bloodstream initially as 5-methyl-H$_4$PteGlu, and enter the cell in this form.

EXERCISE

Please estimate the folate requirement of the human infant from the following data. A baby may consume 800 ml of milk/day. Human milk may contain folate at 40 ng/ml.

A more complete version of folate metabolism is given in Figure 9.5. The cycle of reactions used to regenerate methionine is featured in the center of the diagram. H$_4$folate may be considered the starting and ending point in the cycles depicted. The cycles involve the introduction of a 1-carbon unit, derived from serine, into tissue folates. This is followed by use of the 1-carbon unit for the synthesis of methionine, purines, and thymidylate. Conditions that result in a decline in tissue

TABLE 9.1  Activity of Folylpolyglutamate Synthase

| Folate substrate | Relative polyglutamation activity (%) |
|---|---|
| H$_4$folic acid | 100 |
| H$_2$folic acid | 156 |
| 10-Formyl-H$_4$folic acid | 95 |
| Folic acid | 6.4 |
| 5-Methyl-H$_4$folic acid | 5.6 |
| 5,10-Methylene-H$_4$folic acid | 5.2 |

Source: Cichowicz and Shane (1987).

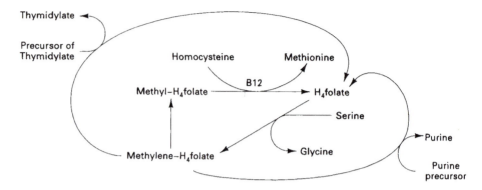

**FIGURE 9.5** Simplified version of folate metabolism. Serine is the source of most of the 1-carbon units carried by folates. These 1-carbon units are used in the synthesis of thymidylate (a pyrimidine nucleotide), purine nucleotides, and methionine.

levels of $H_4$folate might be expected to result in a decline in the rates of synthesis of these compounds. These conditions include dietary folate deficiency, vitamin $B_{12}$ deficiency, and treatment with methotrexate.

### Impairment of 1-Carbon Metabolism by Vitamin $B_{12}$ Deficiency

Methionine synthase is a vitamin $B_{12}$-dependent enzyme. It is a component of the cycle of reactions involved in regeneration of methionine from homocysteine. This regeneration is impaired by vitamin $B_{12}$ deficiency, leading to an accumulation of homocysteine and 5-methyl-$H_4$folate in the cell. The accumulation of 5-methyl-$H_4$folate occurs at the expense of $H_4$folate, with the consequent impairment in the rate of reactions that require or depend on $H_4$folate. Vitamin $B_{12}$ deficiency can result in an impairment in synthesis of thymidylate and purine. It can impair growth, as growth requires the synthesis of new nucleic acids. A decline in growth, whether induced by a deficiency in folate or in vitamin $B_{12}$, can most easily be detected by examining the bone marrow. Such an examination can reveal the consequent impairment in division in immature red blood cells.

### Refined Diagram of 1-Carbon Metabolism

A more complete and refined diagram of 1-carbon metabolism is given by Figure 9.6. The cycle of reactions that regenerate methionine is shown at the center, as in the previous diagrams. Thymidylate synthase (TS), at the left, catalyzes the conversion of deoxyuridylic acid (dUMP) to thymidylic acid (dTMP).

Thymidylate synthase is the target of **5-fluorouracil** (5FU), an anticancer drug. The active form of the drug, 5-fluorodeoxyuridylate, which is synthesized in the cell, inhibits the enzyme by forming a covalent complex with it. This inhibition results in a buildup of dUMP in the cell. Some of the dUMP is converted to dUTP. Normally, DNA polymerase recognizes dATP, dTTP, dGTP, and dCTP as substrates

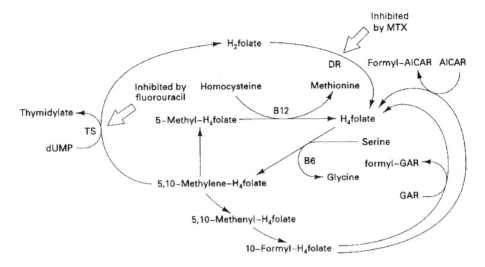

FIGURE 9.6 Detailed diagram of folate metabolism. Serine hydroxymethyltransferase catalyzes the transfer of 1-carbon units to folylpolyglutmates. The enzyme requires vitamin $B_6$ as the cofactor, as indicated by the $B_6$. The arrows depict the flow of 1-carbon units toward the synthesis of methionine, thymidylate, and purine nucleotides. A unique form of serine hydroxymethyltransferase occurs in the mitochondria. The interconversion of 5,10-methylene-$H_4$folate, 5,10-methenyl-$H_4$folate, and 10-formyl-$H_4$folate are catalyzed by one polypeptide chain. The protein is therefore called a multifunctional enzyme (Paukert et al., 1976). The protein also catalyzes the ATP-dependent addition of formic acid to $H_4$folate, producing 10-formyl-$H_4$folate. In bacteria, all of the above reactions are catalyzed by separate enzymes, and not by multifunctional enzymes.

for the synthesis of DNA. However, with an increase in cellular levels of dUTP, DNA polymerase tends to accept dUTP as a substrate, and increasing amounts of uracil become incorporated into the DNA (in place of thymidine). The abnormal base (uracil) in the strand of DNA is recognized by a special DNA repair enzyme, which immediately proceeds to repair the abnormal base (uracil). The enzyme is called **uracil DNA glcosylase**.

The event of DNA repair is a multistep process, and one of the steps involves the temporary breakage of the strand of DNA that contains the offending residue of uracil. After removal of the incorrect nucleotide, the strands of DNA are normally stitched back together again. But with the unusually high accumulation of dUTP in the cell (which occurs with 5FU treatment) opposite strands of DNA accumulate nearby strand breaks, with the consequent splitting of the double-stranded DNA helix. Where this event occurs simultaneously at many points in the genome, death results. The preceding mechanism has not been proven, but it can account for the killing properties of 5-fluorouracil.

Dihydrofolate reductase (DR) is shown near the top of Figure 9.6. This enzyme catalyzes the reduction of $H_2$folate to $H_4$folate. DR is part of the cycle of reactions in the synthesis of thymidylic acid. The enzyme also catalyzes the reduction of folic acid to dihydrofolic acid and then to tetrahydrofolic acid. DR is the target of the anticancer drug methotrexate (MTX). MTX exerts its toxic effects more on rapidly growing cells than on slowly growing cells. In addition to killing some types of cancer cells, it inhibits the growth of immature red blood cells and of the rapidly growing epithelial cells that line the gastrointestinal tract. MTX can also result in damage to the liver.

## Purine and Pyrimidine Biosynthesis

The purine biosynthetic pathway requires 10-formyl-H$_4$folate. 5,10-Methylene-H$_4$folate is oxidized to 5,10-methenyl-H$_4$folate, as shown in Figure 9.6. 5,10-Methenyl-H$_4$folate is then hydrated to produce 10-formyl-H$_4$folate. 10-Formyl-H$_4$folate is used at two steps of the purine biosynthetic pathway, namely for conversion of glycinamide ribonucleotide (GAR) to formyl-GAR and of aminoimi-dazolecarboxamide ribonucleotide (AICAR) to formyl-AICAR.

5,10-Methenyl-H$_4$folate can also be generated during histidine catabolism, though histidine is a minor source of the 1-carbon units of folate metabolism. Brody and Stokstad (1991) clearly demonstrated that the ring-carbon of histidine is transferred to the folate cofactor *in vivo*, in a study that took advantage of the commercial availability of [$^{14}$C]histidine bearing a radioactive [$^{14}$C]ring-carbon.

The exact points at which folate is used in pyrimidine metabolism are shown in Figures 9.7 to 9.9. Figure 9.7 shows the utilization of the carbon skeleton of aspartic acid for the synthesis of the pyrimidine ring. Carbamyl phosphate synthase catalyzes the synthesis of carbamyl phosphate. Carbamyl phosphate is the source of one of the carbon atoms and one of the nitrogen atoms of the pyrimidine ring. Carbamyl phosphate synthase is cytosolic, is not activated by acetylgluta-mate, and is different from the mitochondrial enzyme used in the urea cycle. UMP is the product of the pathway. It can be converted to UTP in an ATP-dependent reaction. The conversion of UTP to CTP is shown in Figure 9.8. The reduction of ribonucleotides, such as UDP, to deoxyribonucleotides, such as dUDP, is described under Iron in Chapter 10. Figure 9.9 shows the conversion of dUMP to dTMP. The reaction is catalyzed by thymidylate synthase and is folate dependent. The reaction is unusual, as far as folate metabolism is concerned, because folate serves as both the source of a 1-carbon unit (as usual) and a reductant (unusual).

Figures 9.10 and 9.11 show the biosynthetic pathways for AMP and GMP. Phosphoribosylpyrophosphate (PRPP) is the source of the sugar group of the pyrimidine and purine nucleotides. A nucleotide consists of the purine or pyrimidine base connected to ribose phosphate. PRPP is synthesized as shown in Figure 9.12.

The purine biosynthetic pathway begins with glutamine as the source of an amino group. Glycine supplies two of the carbons and one of the nitrogens of the purine ring. Folate provides another carbon atom, and glutamine yet another amino group. The remainder of the purine ring is derived from carbon dioxide, the amino group of aspartate, and another 1-carbon unit donated by folate. The product, IMP, represents the completed purine ring. Figure 9.11 depicts the conversion of IMP to AMP and of IMP to GMP. Aspartate and glutamine, respectively, are the sources of amino groups in AMP and GMP.

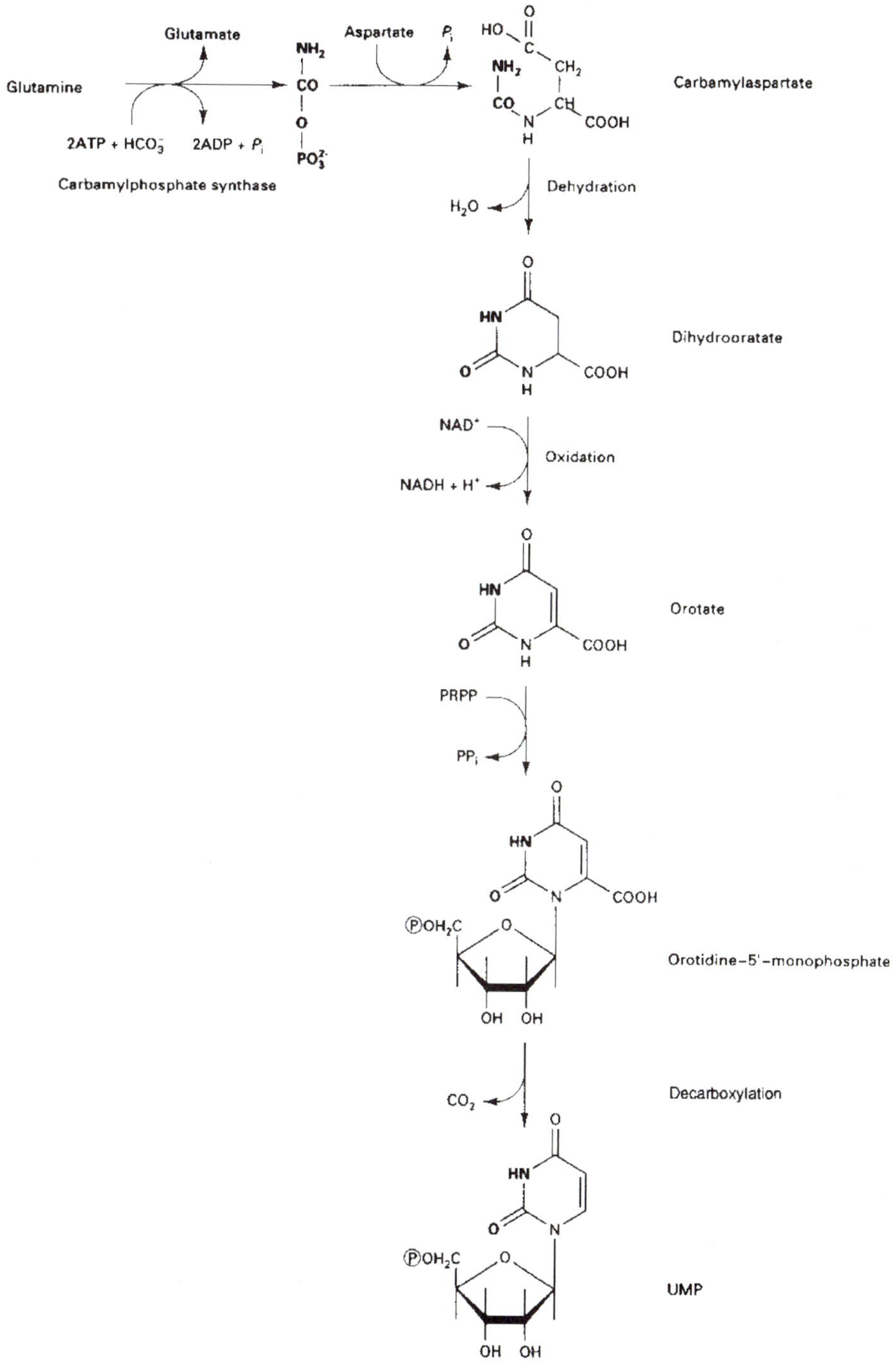

**FIGURE 9.7** Pyrimidine biosynthetic pathway. The pathway of pyrimidine biosynthesis involves six steps and results in the production of uridine 58-monophosphate. Folate is not used in this pathway. The pathway commences with the transfer of the amide nitrogen of glutamine to bicarbonate to produce carbamyl phosphate. This molecule then reacts with aspartate to form the beginnings of the six-membered pyrimidine ring.

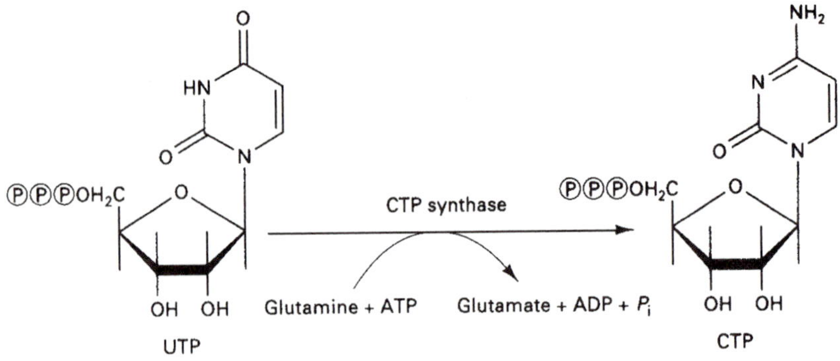

FIGURE 9.8 CTP synthase. CTP synthase catalyzes the transfer of an amino group to UTP, generating CTP. Glutamine serves as the donor of the amino nitrogen. Folate is not involved in the synthesis of CTP. Cytidine 5'-diphosphate (CDP) is distinguished in that it is used in lipid metabolism, where it takes the forms of CDP-choline, CDP-ethanolamine, and CDP-diacylglycerol.

## Mitochondrial Folate Metabolism

The catabolism of choline and glycine occurs in the mitochondria of the liver. The choline that is not needed for the synthesis of phospholipids and other compounds can be catabolized by the series of reactions depicted in Figure 9.13. Choline is oxidized to N,N,N-(trimethyl)glycine (trimethylglycine for short). Trimethylglycine then donates one of its methyl groups to homocysteine, generating methionine, in a reaction catalyzed by **betaine-homocysteine methyltransferase**. The enzyme accounts for nearly 1% of the protein of the liver (Garrow, 1996). The reaction of betaine-homocysteine methyltransferase should not be confused with that of methionine synthase, a cytosolic enzyme.

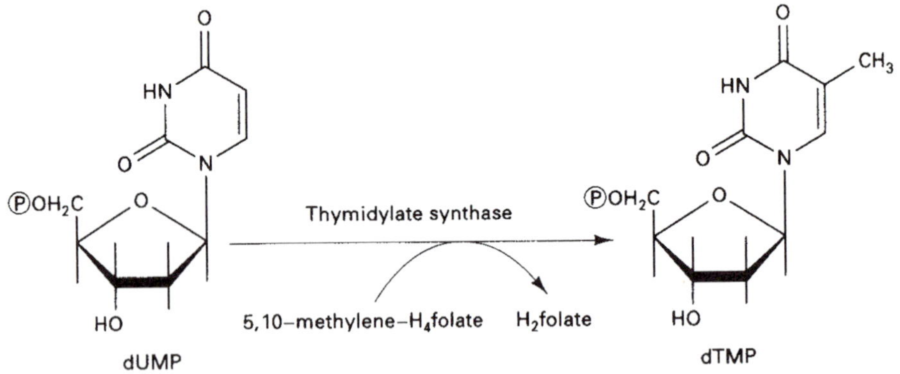

FIGURE 9.9 Thymidylate synthase. Thymine is the pyrimidine base that occurs in thymidylic acid (thymidylate). Thymidylic acid (TMP) is the same as deoxythymidylic acid (dTMP), and both terms are in use. Thymine occurs in DNA but not in RNA. Conversely, uracil occurs in RNA but usually not in DNA. Thymidylate synthase is a well-known folate-requiring enzyme.

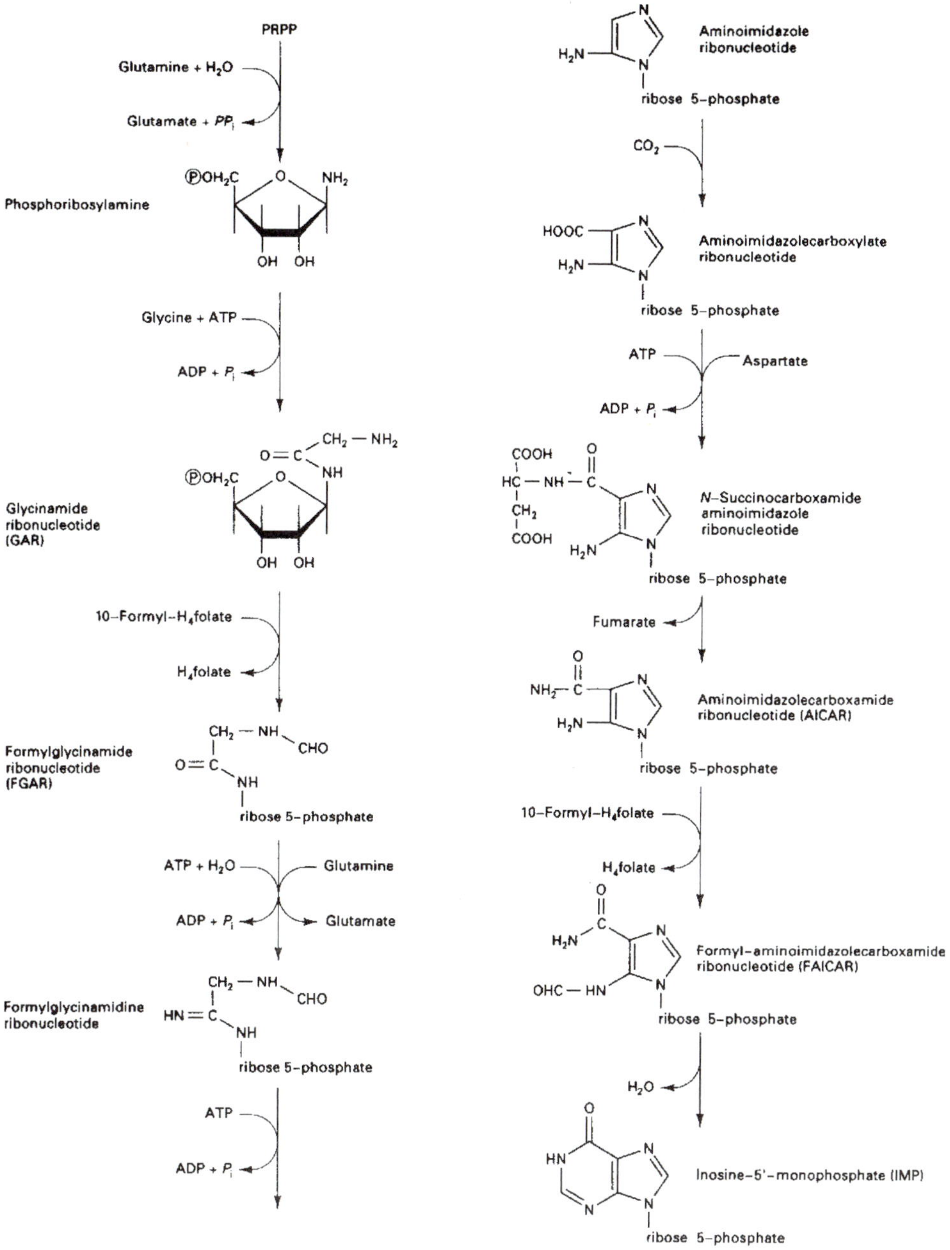

**FIGURE 9.10** Purine biosynthetic pathway. The purine biosynthetic pathway is a 10-step pathway. The folate cofactor is used at Steps 3 and 9. Glutamine is used at Steps 1 and 4 in the donation of an amino group. Aspartate is used at Steps 7 and 8 to donate an amino group. The product of the pathway is IMP, which is subsequently converted to AMP or GMP.

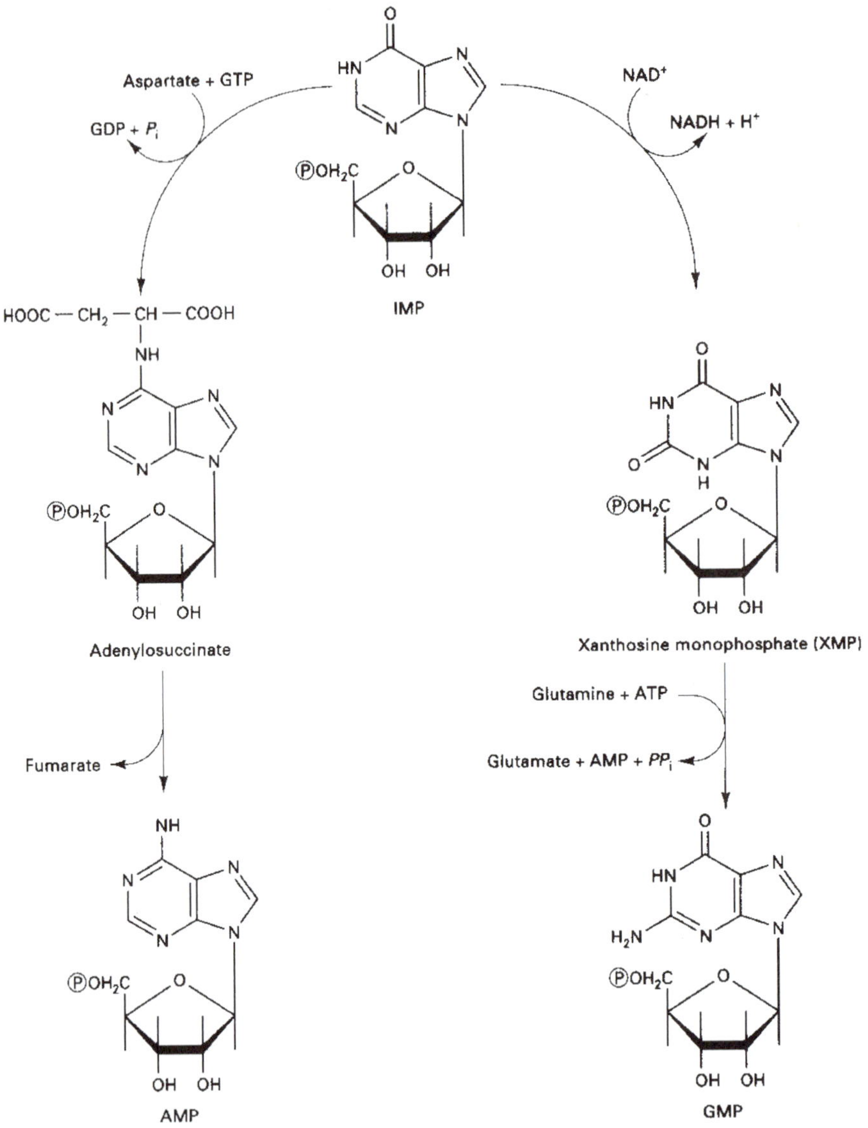

**FIGURE 9.11** Conversion of inosine monophosphate to adenosine monophosphate and guanosine monophosphate. Aspartate donates an amino group in the synthesis of AMP. Glutamine donates an amino group in the synthesis of GMP. Folate is not used in these steps of purine metabolism.

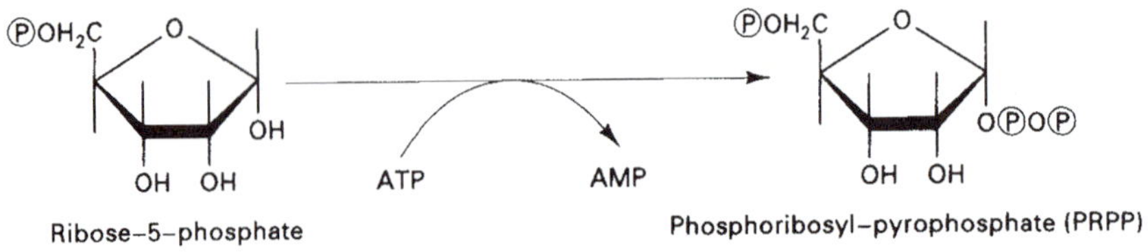

**FIGURE 9.12** Synthesis of phosphoribosylpyrophosphate (PRPP).

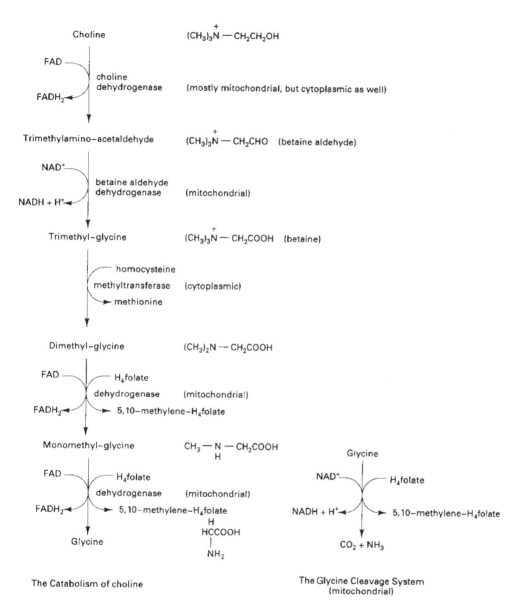

The Catabolism of choline

The Glycine Cleavage System
(mitochondrial)

**FIGURE 9.13** Complete oxidation of choline to carbon dioxide. The oxidation of choline to $CO_2$ requires the participation of enzymes of the mitochondria and cytoplasm. This means that some of the intermediates are expected to leave and reenter the mitochondria in the course of the pathway. The pathway involves the participation of three folate-requiring enzymes. Betaine-homocysteine methyltransferase is not a folate-requiring enzyme.

The methyl groups of *N,N*-(dimethyl)glycine are then removed in two steps, yielding glycine. The 1-carbon units of dimethylglycine occur at the oxidation level of methanol. They are oxidized to the oxidation level of formaldehyde. The formaldehyde groups generated apparently are not released into the surrounding medium, but seem to be bound by tetrahydrofolate, generating 5,10-methylene-

$H_4$folate, as shown in Figure 9.13. The 1-carbon unit bound to 5,10-methylene-$H_4$ folate was at one time called "active formaldehyde." The FAD cofactor is covalently bound to dimethyl- and monomethylglycine dehydrogenases. Trimethylglycine is also called **betaine**. Monomethylglycine is also called **sarcosine**.

Glycine catabolism is catalyzed by the glycine cleavage system. The glycine cleavage system is an enzyme that requires pyridoxyl phosphate as a cofactor. Thus, its reaction is reminiscent of that of serine hydroxymethyltransferase. The glycine cleavage system catalyzes the conversion of glycine to $CO_2$, ammonia, and 5,10-methylene-$H_4$folate, as shown in Figure 9.13. Note that under physiological conditions ammonia is protonated and occurs as the ammonium ion. Evidence suggests that the serine hydroxymethyltransferase occurring in the mitochondrion, rather than the version of this enzyme occurring in the cytoplasm, plays a major role in supplying the cell's requirement for glycine (Stover *et al.*, 1997).

Support of purine and thymidylate synthesis also requires that the folate cofactor exist in the triglutamate form, or as longer derivatives. However, support of glycine and methionine synthesis by folate cofactors does not occur with the triglutamate, and requires longer-chain-length folates, i.e., folylpentaglutamates (Lowe *et al.*, 1993).

EXERCISE

Studies with rats revealed that it was possible to omit methionine from the diet and replace it with two other chemicals that are not classical amino acids. Describe this diet after consulting Figure 9.13. (See Du Vigneaud *et al.*, 1939.)

## Risk Factors for Folate Deficiency

*Poor Diets*

Chronic alcoholism is the major cause of folate deficiency in the United States. Alcoholics generally have poor diets — for example, one liter of whiskey per day. It is not clear if the alcohol induces metabolic defects that interfere with the metabolism and function of folate. Beer does contain folate, as this product is brewed with yeast, an organism containing high levels of the vitamin. Wine and hard liquors, on the other hand, contain little or no folate. The elderly may also be at risk for folate deficiency. It is thought that in the elderly the deficiency is due to poor diets rather than age-related defects in the absorption and utilization of folate.

Food preparation is a concern because losses of the vitamin are due to destruction on exposure to heat and dilution in cooking water. Boiling vegetables in large amounts of water can result in loss of more than half of the food folate in the water. Folate, in particular the tetrahydrofolate and 10-formyl-tetrahydrofolate forms, is unstable. The vitamin is destroyed by storage in water, by heating, and even with prolonged storage in the frozen state. Fresh and frozen foods are good sources of folate; canned foods are poor sources.

*Pregnancy*

Folate deficiency occurs in pregnancy on a wide scale. The increased utilization of the vitamin by the fetus and related tissues, as well as the secretion in milk during lactation, can place an increased demand on the dietary folates consumed by the mother. Severe folate deficiency leads to **megaloblastic anemia**. This disease, in which the synthesis of red blood cells is impaired, tends to occur with pregnancy in underdeveloped countries, but usually not in North America or Europe. Goat's milk is a poor source of folate and vitamin $B_{12}$. Overreliance on goat's milk as a source of food for infants can result in a deficiency in these vitamins and in anemia.

*Vitamin $B_{12}$ Deficiency*

Vitamin $B_{12}$ deficiency induces deficiency in folate, because vitamin $B_{12}$ is required for the conversion of folate from a form that has limited use (5-methyl-$H_4$folate) to a form ($H_4$folate) that can be readily assimilated and used in a variety of reactions. Vitamin $B_{12}$ is the cofactor of methionine synthase, the enzyme required for conversion of 5-methyl-$H_4$folate to $H_4$folate. The metabolism of folate and that of vitamin $B_{12}$ are, in part, intimately related. Because of this relationship, the symptoms of deficiencies in both of these vitamins are shared.

*Epilepsy*

A small fraction of epileptics treated with anticonvulsants (dilantin, phenytoin, diphenylhydantoin) develop folate deficiency. Epilepsy is not a rare disease. Hence, there is an awareness of the possibility of the occurrence of megaloblastic anemia in epileptics treated with the aforementioned anticonvulsant. Supplementing epileptics with folate can alleviate the deficiency; however, the supplements may also result in an increase in the seizure rate. Thus, physicians must be prepared to halt folate supplementation of epileptics being treated for anemia.

## Signs of Folate Deficiency

The hematological signs that present during a folate deficiency may also be considered functional tests for vitamin status (see Assessment of Folate Status), as they reflect the tissue content of the vitamin. These hematological signs are not specific to folate status, as they can present in diseases other than folate deficiency. The hematological signs can only suggest, but do not prove, that the subject is folate deficient.

The hematological signs that present during the course of progressive folate deficiency are depicted in the histogram in Figure 9.14, which reveals the course of events during consumption of a folate-deficient diet by a healthy human subject. The subject was the author of the study. Low serum folate levels occur with 2 weeks of consumption of the diet. About 8 to 9 weeks are required for detection of the first hematological sign, **hypersegmentation of the neutrophils**. Continued consumption of the diet for about 12 weeks can result in elevated levels of urinary

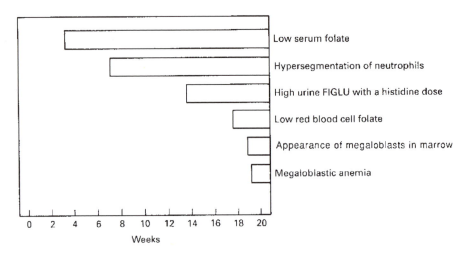

FIGURE 9.14 Development of symptoms of a folate deficiency with consumption of a folate-deficient diet. (Redrawn with permission from Herbert, 1967.)

formiminoglutamic acid (FIGLU), following the test dose of histidine. A prolonged deficiency can result in low levels of folate in red blood cells, as well as two more hematological signs, megaloblastic marrow and megaloblastic anemia. Megaloblastic marrow presents prior to megaloblastic anemia.

## Assessment of Folate Status

Folate status can be assessed by the measurement of serum or red blood cell folate levels, by the histidine load test, and by hematological methods.

### Measure of Serum and Red Blood Cell Folate

Serum folates can be determined by the competitive binding assay (as detailed under Vitamin $B_{12}$) or by microbiological assays. Microbiological assays can be used to measure most vitamins, including folate, vitamins $B_6$ and $B_{12}$, thiamin, and biotin. The concept behind the microbiological assay is the same in all cases. A test organism, such as a lactic acid bacterium, is used. Lactic acid bacteria are similar to animals in that they require a wide variety of nutrients, including a number of vitamins. Most other species of bacteria, in contrast, are able to synthesize all of the vitamins from glucose and other simple nutrients. A special type of growth medium is required. In the case of the assay for folate, the synthetic growth medium contains all of the required nutrients except folate. The lactic acid bacterium cannot grow on this medium; however, addition of known quantities of the vitamin, or of a biological sample containing an undetermined amount of folate, does support growth. The growth of the bacterium can reflect in a very sensitive and precise manner the exact amount of folate included in the growth medium.

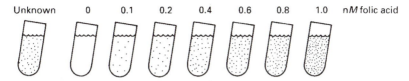

Unknown    0    0.1    0.2    0.4    0.6    0.8    1.0    n*M* folic acid

**FIGURE 9.15** Growth of *Lactobacillus casei*. Growth of *L. casei* in a sample containing an unknown quantity of folate and in a series of tubes containing known quantities of folic acid is shown. The growth of the cells can be easily assessed by measuring the absorbance of the cell suspension in a spectrophotometer. The growth data from the tubes containing folic acid in the concentration range 0–1.0 nM is used to form the standard curve. The concentration of folate in the unknown is determined by measuring the absorbance of the bacterial suspension in the tube and then reading the folate level from the standard curve.

Lactobacillus casei is the bacterium used in microbiological assays for folate. A series of tubes containing increasing amounts of folic acid are used to construct a standard curve. The tubes in the standard curve contain folic acid in concentrations ranging from 0 to 1.0 nM (Figure 9.15). The tubes are inoculated with a very small number of bacterial cells and incubated 20 hours. In the tubes containing little or no vitamin, there is little or no growth. Where the tubes contain larger amounts of vitamin, there are corresponding increased densities of cell growth. The results from the standard curve are shown in Figure 9.15. Cell growth densities in tubes containing unknown quantities of folate were also measured and compared with the standard curve to calculate the amount of folate in the unknown sample.

Serum contains folate monoglutamates. The normal level of serum folate is 6.0 ng/ml or greater. A value of 3.0 ng/ml or less indicates a folate deficiency. Serum folate levels are sensitive to recent dietary practices and hence may not accurately reflect the amount of folate in tissues. True folate status is closely related to the tissue or cellular level of folate, rather than the serum level.

Red blood cell folate levels closely reflect the true folate status of the subject, as these folates are an intracellular form. The folates of red blood cells, as well as of all other cells, occur mainly as folylpentaglutamates. Folylpentaglutamates, as well as other folylpolyglutamates, are poorly absorbed by bacterial cells. Hence, microbiological assays might not be expected to be useful in determining tissue folate levels. The problem can be avoided by treating tissue folates with γ-glutamyl hydrolase. This enzyme catalyzes the hydrolysis of folylpolyglutamates, producing folylmonoglutamate. The various folylmonoglutamates, such as 5-methyl-$H_4$PteGlu, $H_4$PteGlu, 10-formyl-$H_4$PteGlu, and PteGlu, are readily used by *L. casei* for growth. A red blood cell folate level greater than 160 ng/ml of packed cells indicates normal folate status. Values of 140 ng/ml or less indicate a deficiency.

### Histidine Load Test

The histidine load test is not used in the clinical setting and is only sometimes used by researchers; however, a description of this test provides a clear-cut example of how folates behave in the mediation of 1-carbon metabolism. Histidine catabolism takes place in the liver according to the pathway shown (Figures 9.16 and 9.17). The intermediates, formiminoglutamic acid and 5-formimino-$H_4$folate, bear the formimino group: —CH=NH.

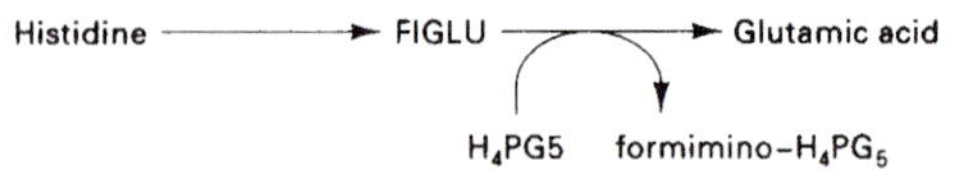

**FIGURE 9.16** Simplified diagram of histidine catabolism. FIGLU is an intermediate in the pathway of histidine catabolism that may appear in elevated amounts in the urine during folate deficiency.

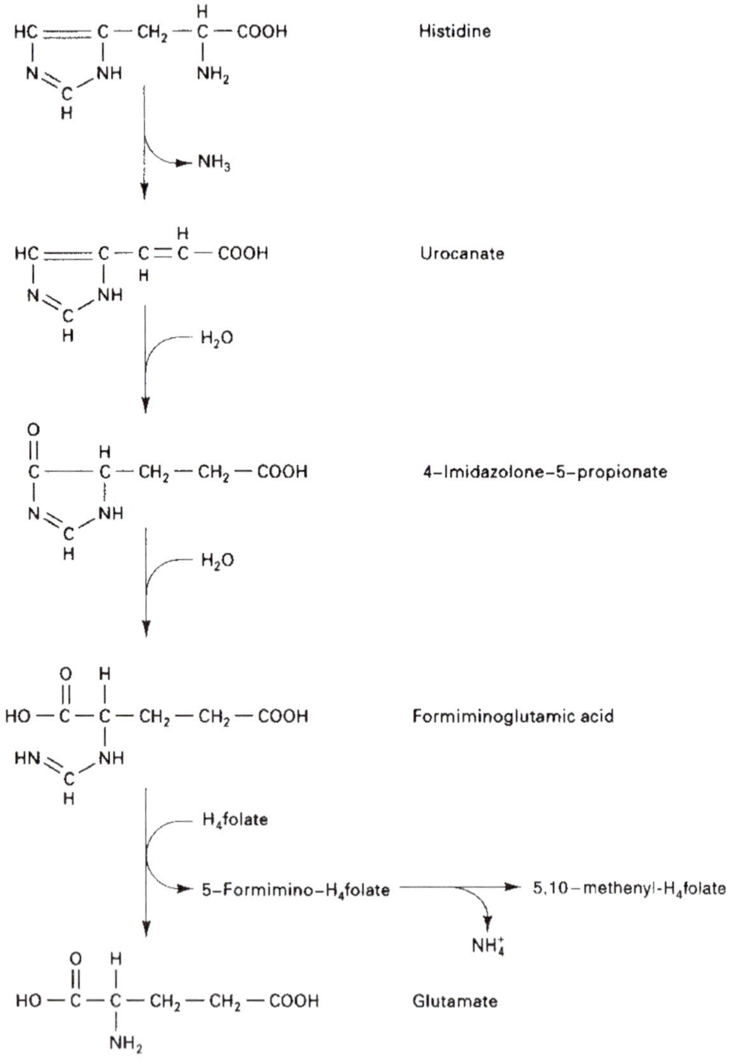

**FIGURE 9.17** Histidine catabolic pathway. The final two steps of the pathway are catalyzed by a folate-dependent enzyme, formiminotransferase/cyclodeaminase. This enzyme contains two active sites. The first of these sites catalyzes the transfer of the formimino group from FIGLU to H₄folate. The second site catalyzes the conversion of 5-formimino-H₄folate to 5,10-methenyl-H₄folate. The reactions are distinguished by the fact that the polyglutamyl chain of the folate cofactor enables the cofactor to remain bound to the enzyme during the course of the reaction. H₄folylmonoglutamate can be used by the enzyme, but this form of the vitamin tends to dissociate from the active site after the first step and only inefficiently rebinds to the second active site, resulting in a low rate for the overall two-step reaction. The folylpolyglutamate substrate tends to remain bound to the enzyme during both steps, thus facilitating the flow of the ring-carbon of histidine into 1-carbon metabolism.

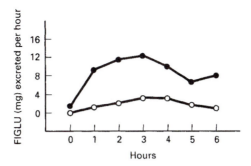

**FIGURE 9.18** FIGLU excretion in a healthy woman (○) and a folate-deficient pregnant woman (●). (Redrawn with permission from Chanarin *et al.*, 1963.)

Folate deficiency can result in an overall decline in the concentrations of all the cofactor forms of folate in the cell. Vitamin $B_{12}$ deficiency can result in a *decline* in the concentration of tetrahydrofolate and an *increase* in that of 5-methyl-$H_4$folate. Both deficiencies would be expected to result in impairment of the catabolism of histidine.

Histidine catabolism was assessed in studies of folate deficiency in pregnancy and in experimental vitamin $B_{12}$ deficiency, as revealed further on. Figure 9.18 depicts a study of a healthy woman and a folate-deficient pregnant woman. In humans, urinary FIGLU is difficult to detect unless the body is challenged with an oral dose of histidine (10 g). The histidine load is consumed in a few small doses to avoid nausea. Both women consumed the dose. Urine was collected over the course of several hours, in increments. The amount of FIGLU excreted per hour is recorded in the figure. The data demonstrate that urinary FIGLU levels were elevated in the folate-deficient subject. The histidine load test is not used in the clinic because the results are influenced by other factors, such as liver disease, and because the urine collection is inconvenient.

The data in Table 9.2 concern a study in rats. The animals were raised on nutritionally complete diets or vitamin $B_{12}$-deficient diets for 3 months. Then urine was collected over the course of a day and used for analysis of metabolites related to folate and vitamin $B_{12}$ (FIGLU and methylmalonic acid). The data show that $B_{12}$ deficiency induces a dramatic increase in urinary FIGLU. Rats do not require a

**TABLE 9.2** Urinary Metabolites from Rats

| | $\mu$mol/100 g body weight per day | |
| --- | --- | --- |
| Diet | FIGLU | Methylmalonic acid |
| $+B_{12}$ | 0.05 | 7.4 |
| $-B_{12}$ | 3.9 | 172.0 |

*Source*: Shin *et al.* (1975).

challenge dose of histidine to induce an increase in urinary FIGLU, as do humans. The relationship between methylmalonic acid and vitamin $B_{12}$ is discussed under Vitamin $B_{12}$.

## Hematologic Methods

The circulatory system contains a number of types of cells. Nearly half of the volume of whole blood consists of red blood cells, whereas only about 1% of the blood consists of white blood cells. The function of the red blood cells is to transport oxygen to the tissues. White blood cells are part of the immune system. There are several types of white blood cells: neutrophils, macrophages, and lymphocytes. Neutrophils are also called **polymorphonuclear leukocytes**. The smallest type of cell in the bloodstream is the platelet. Platelets are used in the clotting of blood, as outlined under Vitamin K.

The various blood cells arise and develop from one type of cell, the **stem cell**, as shown in Figure 9.19. Stem cells are located in the marrow of the bone. They are not often seen in the bloodstream. The stem cell can differentiate to produce neutrophils. The neutrophil is distinguished by its nucleus, which appears to have multiple lobes or segments. The stem cell can also differentiate to form macrophages. One type of macrophage, the Kupfer cell, constitutes a small proportion of the cells of the liver. They are located in the walls of the capillaries passing through the liver. The stem cell gives rise to megakaryocytes. The megakaryocytes synthesize platelets. On discharge from the megakaryocyte, platelets initially appear as a thread of small cells peeling off of the megakaryocyte spool.

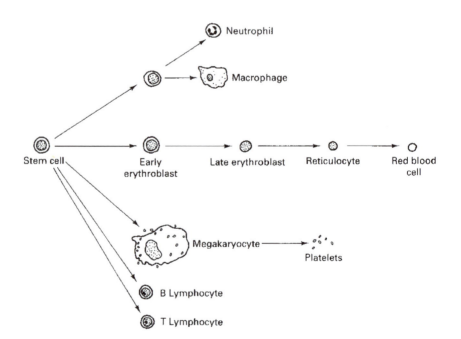

**FIGURE 9.19** Differentiation of the stem cell to various types of blood cells. (Redrawn with permission from Hood *et al.*, 1984.)

Normal Neutrophils                Hypersegmented Neutrophils

**FIGURE 9.20** Neutrophils. The examination of neutrophils is used to diagnose vitamin status.

Several intermediates occur with differentiation of the stem cell to a mature red blood cell. These intermediates include the early erythroblast, late erythroblast, and reticulocyte (see Figure 9.19). The early erythroblast is relatively large and has a diameter of 15 μm. This cell has a nucleus and is a dividing cell. It does not contain much hemoglobin, the oxygen-carrying protein of the red blood cell. The early erythroblast matures into the smaller late erythroblast (8–10 μm). The late erythroblast contains much hemoglobin and is red. It has lost its nucleus and cannot divide. The late erythroblast further matures, losing more internal structures, such as the mitochondria, and becomes a reticulocyte. Reticulocytes are found in the bone marrow, as well as in the bloodstream. The reticulocyte matures to form the red blood cell.

The hematological signs of folate deficiency manifest first in the neutrophils and then in red blood cells. Normally, close to half of the neutrophils contain three lobes: 20 to 40% have two lobes and 15 to 25% have four lobes. In folate deficiency, segmentation of the neutrophils increases; cells with 6 to 10 lobes may be detected. The detection of three five-lobed neutrophils in a count of 100 neutrophils indicates the possibility of folate deficiency. Diagrams of normal and hypersegmented neutrophils are shown in Figure 9.20.

The second hematological sign of folate deficiency involves the pathway of red blood cell synthesis. Early erythroblasts accumulate in folate deficiency, initially in the bone marrow, where they can be easily detected by the clinician, and then in the bloodstream. Because of the large size of early erythroblasts, the marrow containing the accumulated cells is called a **megaloblastic marrow**. Generally, marrow samples are taken from the hip (pelvis), though they can also be taken from the sternum.

A fraction of the early erythroblasts accumulating in the marrow escape to the bloodstream. The folate deficiency induces an overall decrease in the concentration of red blood cells in the bloodstream, resulting in anemia. Because of the appearance of the early erythroblasts in the blood, and because of the anemia, the condition is called **megaloblastic anemia**. Normally, the mean volume of the red blood cell is 82 to 92 fl (μm³). In slight or moderate megaloblastic anemia, the mean volume is 95 to 110 fl, and in severe megaloblastic anemia, 110 to 130 fl or greater. The clinical signs of anemia are weakness and shortness of breath. These symptoms occur only when the hematocrit (see Iron in Chapter 10) is markedly depressed. One important point concerning the hematological signs should be made: Hypersegmentation of the neutrophils and megaloblastic anemia can be induced by both folate deficiency and vitamin $B_{12}$ deficiency. The clinician who detects these two signs in a patient requires further information, in the form of blood vitamin levels, to determine if the patient is deficient in folate, vitamin $B_{12}$, or both vitamins.

Treatment of the folate-deficient patient with folic acid permits the early eryth-roblast to divide, producing late erythroblasts and, eventually, reticulocytes and red blood cells. These effects can very easily be detected by examining blood samples taken before and after the injection and determining the percentage of reticulocytes. Normally, the concentration of reticulocytes in the bloodstream is quite low. A burst in the number can be induced by injecting folic acid (0.1 mg) into a folate-deficient patient. It should be pointed out that injecting vitamin $B_{12}$ into a $B_{12}$-deficient patient, or folate into some $B_{12}$-deficient patients, can also cause this burst. Thus, the hematological response to vitamin injections is not a reliable indicator of which vitamin deficiency was present.

The coexistence of deficiencies in iron and folate (or $B_{12}$) can produce an am-biguous hematological picture, as iron deficiency produces microcytic anemia. The combination of microcytes (small red blood cells) and megaloblasts (large red blood cells) that may occur can make the diagnosis of megaloblastic anemia difficult; however, the occurrence of hypersegmented neutrophils during the com-bined deficiency clearly reveals the existence of megaloblastic anemia. If the mixed anemia is treated only with iron, and not with folate or $B_{12}$, the megaloblastic features may become more clear.

## Folic Acid and Neural Tube Defects

**Neural tube defects** (NTDs) are a family of birth defects, involving the brain or spinal cord, which arise from mistakes made early in the development of the embryo. The neural tube is a part of the developing human embryo. It begins to form at 18 days after conception. The neural tube seals off and closes, thus forming a tube, by the 30th day after conception. To reveal more detail, during the 2nd and 3rd week after conception, when the human embryo is 2–3 mm long, a flat structure called the **neural plate** forms two parallel ridges. These ridges fold over and move towards each other and form a tube (the neural tube). Any drug, nutrient imbalance, or genetic defect that interferes with this stage of development may prevent normal closure of the tube, resulting in defects in the newborn. At its least severe, an NTD may be undetectable except by examination of the spine by the fingers of a physician. At its most severe, an NTD involves the absence of the brain. The most common NTDs are **spina bifida** (spinal cord at the lumbar vertebra not covered with bone), **anencephaly** (no brain), and **encephalocele** (tissue protruding through a hole in the skull).

Early studies by R. W. Smithells and K. M. Laurence in Great Britain suggested that the rate of NTD births could be reduced by supplements of multivitamins or by folic acid. These results sparked a number of studies, for example:

1. Intervention studies involving about 1800 women *with a prior NTD birth* (Wald, 1991), and 7540 women *with no prior NTD birth* (Czeizel and Dudas, 1992) provided dramatic evidence that folic acid supplements reduce the rate of NTD births. These studies were carried out in Great Britain and Hungary.

2. John Scott and co-workers (Daly *et al.*, 1995) collected about 56,000 blood samples from women attending prenatal clinics in Ireland, and measured the level of folate in red blood cells. Of these women, 81 produced an NTD infant. The results indicated a dramatic association between an NTD birth and low maternal red cell folate (under 0.15 µg folate/ml cells).

These studies are suggestive of a protective effect of folic acid, but the final word regarding folic acid and NTDs is not yet available. Studies conducted with a general population in the United States have been less clear than those already mentioned. Epidemiological studies conducted in Boston and Ontario (Werler *et al.*, 1996) and in California (Shaw *et al.*, 1996) seemed not to support the notion that dietary folate intake influences NTD outcome, but did suggest that maternal obesity is a risk factor. A survey of about 600 mothers with an NTD birth revealed no reduction in risk factor with vitamin supplements (Mills *et al.*, 1989). A scenario that seems to be emerging is as follows:

1. Some populations, such as those found in parts of Great Britain and Hungary, have greater rates of NTDs than other groups. In Ireland, the rate of NTD births is 1.9 per 1000 births (Daly *et al.*, 1995). The rates in the United States are lower, where the frequency of NTD births in California is 0.9/1000 births (Cunningham, 1995), and that in Georgia is 1.2/1000 births (Mills *et al.*, 1989).

2. Mothers having one NTD infant are high-risk mothers. Subsequent births may be about 10 times more likely to be NTD births than the typical birth.

3. Supplements of folic acid (0.4 mg/day), or possibly less, can reduce the rate of NTDs. To achieve an extra 0.4 mg folic acid per day, using food, one would have to drink eight glasses of orange juice or eat three servings of Brussels sprouts per day (Wald and Bower, 1995). Since it is not likely that any population, as a whole, would be expected to consume folic acid supplements on a consistent, daily basis, food fortification has become an attractive option. Because neural tube formation occurs early in pregnancy, it is probably necessary for folic acid supplements (when taken) to be consumed for 2–3 weeks (every day) prior to the day of conception, and for 2–3 weeks (every day) after conception, in order for the vitamin to produce its apparent effect.

4. Folic acid seems to exert its anti-NTD effect in some mothers but not in other mothers.

A consequence of these studies is that the Food and Drug Administration (FDA) has recommended that commercially available flour and cereal products be fortified with folic acid (1.4 mg folic acid/kg flour) for preventing NTDs (Tucker *et al.*, 1996). The consumer interested in the folate level in any particular food can view the label on the package. Folic acid supplements have the effect of reducing the level of homocysteine in the blood. This homocysteine effect appears directly relevant to the prevention of atherosclerosis, but may also be relevant to neural tube defects. The reader interested in continuing developments regarding neural tube defects and folate should take note of the relationship between folate and homocysteine, presented in the Vitamin $B_6$ section.

A concern with this government policy is that folic acid supplementation may mask the symptoms of disease in persons with **pernicious anemia** (PA) (see Vitamin $B_{12}$ section). In short, PA is an autoimmune disease that involves anemia (low concentrations of red blood cells in the blood), nerve damage, or both anemia and nerve damage. A potential problem with taking folic acid pills is that the *anemia* of PA can be reversed or prevented with folic acid, resulting in an apparent cure. However, this supplement only masks the signs of PA, and allows the unsuspected *nerve damage* to continue. There has been some general agreement, however, that the amount of folic acid recommended to prevent neural tube defects is not sufficient to mask the anemia of pernicious anemia.

Animals bearing genetic defects carrying various diseases have arisen spontaneously in various animal colonies. Where possible, these ill animals have been maintained for many generations. In some cases, these mutant animals are commercially available. Several strains of mice that have NTDs have been discovered. Quite striking is the fact the spina bifida that occurs in offspring of one such mouse can be prevented by feeding extra methionine to the pregnant mother (Essien and Wannberg, 1993). The genes that control neural tube formation, as well as all embryological development, is a field of intense research. One exciting result is that the genetic defect, in one strain of NTD mouse, occurs in the gene coding for a transcription factor. This transcription factor is called PAX3 (Baldwin *et al.*, 1992). More recent studies with mice bearing the mutant PAX3 gene have revealed that, under certain conditions (growing the embryos in a dish), adding folic acid can prevent NTDs (Fleming and Copp, 1998).

## VITAMIN $B_{12}$

Vitamin $B_{12}$ is a water-soluble vitamin. *Cobalamin* is the term used to refer to compounds having vitamin $B_{12}$ activity, as well as to related compounds. The RDA for vitamin $B_{12}$ for the adult is 2.0 µg. The vitamin is present in animal products, such as meat, poultry, milk, and fish, but is not present in plant products or in yeast. All of the vitamin $B_{12}$ in our environment originated with synthesis in bacteria, fungi, and algae.

### Biochemistry of Vitamin $B_{12}$

Vitamin $B_{12}$ is unique among all the vitamins in that it is the largest and most complex and because it contains a metal ion. This metal ion is cobalt. Cobalt occurs in three oxidation states: $Co^+$, $Co^{2+}$, and $Co^{3+}$. The medicinal forms of the vitamin are cyanocobalamin and hydroxocobalamin. In cyanocobalamin, a molecule of cyanide is complexed to the $Co^{3+}$ atom. Cyanocobalamin is readily converted in the body to the cofactor forms methylcobalamin and 5-deoxyadenosylcobalamin. Methylcobalamin contains cobalt in the $Co^+$ state, where it acts as a cofactor for methionine synthase. 5-Deoxyadenosylcobalamin, which contains cobalt in the $Co^{2+}$ state, is the cofactor for methylmalonyl-CoA mutase. Vitamin $B_{12}$ is also a cofactor for leucine aminomutase, an enzyme used in leucine metabolism (Poston, 1984). This enzyme appears not to have a vital function in metabolism. No more than three cobalamin-requiring enzymes are known to occur in mammalian tissues.

The structure of vitamin B$_{12}$ appears in Figure 9.21. The X indicates the point of attachment of the cyanide group in cyanocobalamin, the hydroxyl group in hydroxocobalamin, and the methyl and deoxyadenosyl groups in the cofactor forms of the vitamin.

## Vitamin B$_{12}$ Deficiency

Vitamin B$_{12}$ deficiency results in impairment in the activities of the B$_{12}$-requiring enzymes. This impairment prevents synthesis of the enzyme's products and forces the accumulation of reactants in the cell. Inhibition of methionine synthase prevents the synthesis of methionine and the regeneration of tetrahydrofolate. This inhibition results in interruption of the methylation cycle, which involves S-adenosylmethionine. The inhibition also results in an impairment of folate-mediated metabolism, because of the failure to regenerate H$_4$folate from 5-methyl-H$_4$folate. The major effect of B$_{12}$ deficiency is an impairment of growth, particularly of rapidly growing cells such as immature red blood cells. B$_{12}$ deficiency also results in the buildup of homocysteine in the cell and bloodstream.

Inhibition of methylmalonyl-CoA mutase can prevent the synthesis of succinyl-CoA from such precursors as propionic acid, valine, and isoleucine. This effect might be expected to have no adverse consequences in metabolism, as the supply of succinyl-CoA derived from fats and carbohydrates is plentiful; however,

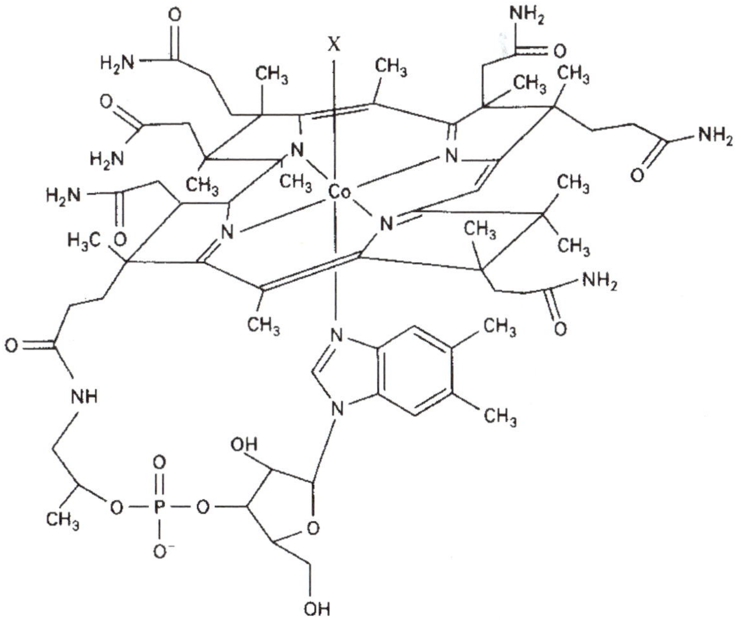

FIGURE 9.21 Structure of vitamin B$_{12}$. Vitamin B$_{12}$, also called cobalamin, contains an atom of cobalt.

inhibition of the enzyme also results in buildup of methylmalonyl-CoA and of intermediates prior to this compound. A fraction of the methylmalonyl-CoA is broken down in the cell, yielding methylmalonic acid. The accumulation of methylmalonic acid in the cell, which is reflected by increases in the plasma and urine, is of interest to the clinician. The methylmalonic acid produced as a consequence of hydrolysis of methylmalonyl-CoA is used to diagnose $B_{12}$ status.

The buildup of methylmalonyl-CoA in the cell may lead to reversal of the reaction of propionyl-CoA carboxylase and, as a consequence, an increase in the levels of propionyl-CoA. Increased levels of propionyl-CoA can lead to its use, in place of acetyl-CoA, by fatty acid synthase. Use of the 3-carbon propionyl group, rather than the 2-carbon acetyl group, by this enzyme can result in the production of small amounts of odd-chain fatty acids. These fatty acids contain an odd number of carbons, that is, 15, 17, or 19 carbons. The addition of large amounts of propionic acid to the diet during $B_{12}$ deficiency can be used to artificially enhance the production of odd-chain fatty acids.

Increased concentrations of methylmalonyl-CoA can lead to its recognition by fatty acid synthase in place of malonyl-CoA. The use of this abnormal substrate leads to the production of branched-chain fatty acids. Here, the branch consists of a methyl group. $B_{12}$ deficiency induces the synthesis of very small amounts of methylated fatty acids. There is some thought that the odd-chain fatty acids and branched-chain fatty acids that increase in the deficiency, and become incorporated into nerve cells membranes, may contribute to the neurological problems associated with the deficiency.

The first signs of $B_{12}$ deficiency include low levels of liver $B_{12}$, increasing levels of serum and urinary methylmalonic acid (MMA), and, often, low serum $B_{12}$. The signs that appear later in the course of the deficiency include megaloblastic marrow, megaloblastic red blood cells, and anemia. Anemia means an abnormally low concentration of red blood cells in the blood or low hemoglobin concentrations in the red blood cells. Neurological symptoms present with severe deficiency. The neurological signs may occur, in some cases, without the presentation of anemia.

## Causes of Vitamin $B_{12}$ Deficiency

### Pernicious Anemia

Vitamin $B_{12}$ deficiency is commonly caused by **pernicious anemia** (PA). PA is an autoimmune disease resulting from the body's production of antibodies that recognize intrinsic factor or other proteins of the parietal cell. The binding of antibodies to these proteins results in loss of their function. The parietal cells may be destroyed and be undetectable in patients with PA. The major defect in PA is gastric atrophy. There may be a lack of all gastric secretions, including intrinsic factor, gastric acid, and pepsin.

PA is characterized by megaloblastic anemia, megaloblastic marrow, and, occasionally, neurological symptoms. The hematological characteristics of the disease are *indistinguishable from those of folate deficiency*. The biochemical signs of the disease include low serum $B_{12}$ and elevated MMA. The neurological problems affect about 25% of those with the disease and include numbness of the feet and hands. Left untreated, the B$_{12}$ deficiency results in irreversible nerve damage and eventually death, hence the name pernicious anemia.

PA is a risk factor for certain cancers of the stomach. How could the body's destruction of its own parietal cells be connected, somehow, to cancer in nearby cells? The consequent lack of stomach acid interrupts the normal feedback loop for gastrin secretion (see Figure 2.14). With the feedback loop no longer operating, the result is a continual secretion of gastrin by the G-cells of the stomach (Scully *et al.*, 1997). The resulting continual, elevated concentrations of gastrin in the lining of the stomach constitutes an environment that allows precancerous cells to become cancer cells. In addition to its usual function, gastrin is a growth hormone, and thus can be a contributing factor in supporting the growth of cancer cells (Varro *et al.*, 1997).

Generally, PA first presents among older populations. It tends to be first detected between the ages of 40 and 80. It affects about 0.1% of all persons in Great Britain and Scandinavia. A recent survey in Los Angeles, California, indicated that PA occurs in about 3% of elderly blacks and 3% of elderly whites, with a lower prevalence among Latinos and Asians (Carmel, 1996). Blacks and Latinos seem to have a tendency to acquire PA at a younger age, i.e., during their 30s (Carmel *et al.*, 1987; Carmel and Johnson, 1978). A Latino is a person originating from Mexico, Central America, or Cuba.

In previous decades, PA was diagnosed in the doctor's office, after the patient complained of feelings of weakness and tiredness. At the present time, the disease tends more to be detected earlier and diagnosed when the patient complains about gastrointestinal symptoms (dyspepsia, loss of appetite, and flatulence).

PA can easily be treated with large oral doses of hydroxocobalamin (1 mg/day) or with periodic injections of the vitamin (1 mg/month). Absorption of the small fraction of the large oral dose occurs by non-intrinsic factor-dependent mechanisms and can satisfy the vitamin requirement. Treatment must be continued throughout life.

EXERCISE

MMA accumulates in the bloodstream and urine in genetic diseases involving mutations in methylmalonyl-CoA mutase. Plasma levels of MMA may rise to very high levels, resulting in acidification of the blood. The consequent acidosis may be fatal. Why are these genetic diseases treated with low-protein diets? (See Ampola *et al.*, 1975.)

## Vegetarian Diets and B$_{12}$ Deficiency

Rare genetic diseases affecting intrinsic factor may occur in infants. A child will not show symptoms of anemia until the age of 2 years, indicating that the amount of vitamin present at birth can fulfill the child's needs for this period. The B$_{12}$ deficiency can be induced in normal infants by feeding the pregnant mother and the breast-feeding infant a B$_{12}$-deficient diet. Unfortunately, such an "experiment" has been performed, as described next.

A strict vegetarian woman consumed a plant-based diet for 8 years. Strict vegetarianism involves the exclusion of meat, fish, and dairy products from the diet. The woman gave birth to a child who was exclusively breast-fed by the vegetarian mother. Symptoms of $B_{12}$ deficiency presented in the infant within a few months of birth. The symptoms included anemia and neurological problems, such as flaccidity, poor muscular control, twitching, and an abnormal electroencephalogram. The infant's serum $B_{12}$ levels were well below normal (0.02 ng/ml); the urinary MMA levels were quite high, about 2000-fold greater than normal (Higginbottom *et al.*, 1978).

The second case history involves an adult vegetarian who complained of a sore tongue and fatigue that increased over 6 to 8 months. The patient was extremely pale. There were no neurological symptoms. The hematocrit was only 19%; thus, the patient was anemic. The hematocrit is the percentage of the blood volume consisting of red blood cells. Normal hematocrits are 47% (men) and 42% (women). The patient was studied for 45 days, as indicated in Figure 9.22. Treatment with vitamin $B_{12}$ injections was initiated on day 28.

Vitamin $B_{12}$ deficiency due to vegetarianism may be common in some Hindu communities, because of their strict vegetarian practices (Chanarin *et al.*, 1985). The deficiency is common in rural Mexico because of the infrequency of dietary meat (once a week) and because of the practice of heating cow's milk, which destroys the vitamin (Allen *et al.*, 1995).

EXERCISE

Did the vitamin $B_{12}$ injections succeed in bringing the serum $B_{12}$ values into the normal range (Figure 9.22)?

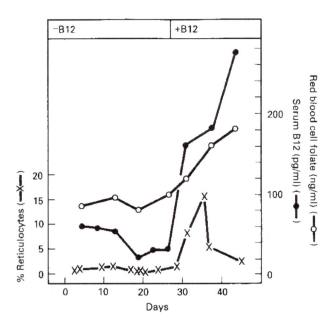

**FIGURE 9.22** Effect of vitamin $B_{12}$ injections on hematocrit of vegetarian woman with anemia. (Redrawn with permission from Hines, 1966.)

EXERCISE

A vegetarian diet would be expected to supply folate in ample amounts. B$_{12}$ deficiency can provoke decreases in tissue folate levels, even with a normal intake of folate. What was the effect of the B$_{12}$ injections on red blood cell folate levels (Figure 9.22)?

## Vitamin B$_{12}$ Deficiency in the Elderly

Diagnosis of B$_{12}$ deficiency in the elderly may be difficult. Often serum B$_{12}$ concentrations occur in the normal range, but plasma levels of MMA and of homocysteine may occur at elevated levels, indicating that B$_{12}$-requiring enzymes are not acquiring enough of their cofactor (Stabler *et al.*, 1997). According to surveys using tests for serum MMA, about 15% of the elderly are deficient in the vitamin. Homocysteine appears to be less reliable as an indicator of vitamin status (it sometimes fails to increase with B$_{12}$ deficiency; and it is sensitive to folate status as well as to B$_{12}$ status).

Gastric atrophy (atrophic gastritis) is the usual cause of B$_{12}$ deficiency in the elderly, while lack of intrinsic factors is only occasionally the cause (Carmel, 1997). Gastric atrophy becomes especially frequent with each decade after the age of 60 years. It results in the lack of stomach acid (achlorhydria), and the failure of gastric enzymes to provoke the release of cobalamin cofactors from their enzymes (Herbert, 1994). The problem may be corrected by feeding acid and pepsin to the patient. An easier way to solve the problem is to eat B$_{12}$ supplements. However, as gastric atrophy progresses to a more severe stage, the stomach may fail to produce intrinsic factor, and here B$_{12}$ injections become a necessity. A difference between gastric atrophy and PA is that the former involves the *disappearance* of parietal cells, while the latter involves the *destruction* of parietal cells.

## Tapeworms

Vitamin B$_{12}$ deficiency may occur with a parasitic infection by tapeworms. Tapeworms, which can be many meters long, can live in the gut. They can take up dietary vitamin B$_{12}$, thus limiting its availability to the human host. Tapeworms can arise from the consumption of raw fish. Although raw fish is consumed in several countries, the vitamin B$_{12}$ deficiency arising from tapeworm infections occurs mainly in Finland. The infection results in megaloblastic anemia as well as neurological symptoms, such as paraesthesia, decreased vibration sense, and ataxia.

## A Warning Regarding the Use of Folic Acid

The symptoms of B$_{12}$ deficiency may be divided into two classes: (1) impairment of DNA synthesis and cell growth, and (2) neurological problems. Administration of folic acid can alleviate the hematological signs, probably by increasing the concentration of H$_4$folate in the cell and its consequent use in the synthesis of thymidylate and purines. Folate treatment does not, however, reverse the neurological problems. Because of this, cases of megaloblastic anemia must not be treated by the indiscriminate use of folic acid. Treatment with folic acid may provide temporary relief from the hematological problems, while allowing the neurological situation to worsen. This warning originally appeared in 1947 (see "A warning regarding the use of folic acid," 1947).

There is some concern that the use of high doses of folic acid supplements, such as 4.0 mg/day, with the goal of reducing the risk for neural tube defect birth, may prevent the detection of pernicious anemia (by preventing the development of the hematological signs). However, one might note that the population at risk for PA is somewhat different (older adults) than the population planning to become pregnant (younger female adults).

## Assessment of Vitamin B$_{12}$ Status

A number of assay methods for vitamin B$_{12}$ status, and for testing defects in the pathway of vitamin B$_{12}$ absorption, are listed here:

1. Test for the concentration of the vitamin in serum (direct test).
2. Test for serum levels of methylmalonic acid (functional test).
3. Test for vitamin B$_{12}$ absorption (Schilling test).
4. Test for vitamin B$_{12}$ absorption (TC II test).

The assay for serum B$_{12}$ levels is a direct test, as it measures the concentration of the vitamin itself. The assay of MMA levels is a functional test of B$_{12}$ status, as it measures a compound whose metabolism is dependent on the correct functioning of vitamin B$_{12}$. The results of the MMA test reflect the activity of methylmalonyl-CoA mutase in the liver. It is thought that the functional test is more valuable than the direct test. Serum vitamin B$_{12}$ levels may not reflect the functioning of the B$_{12}$-requiring enzymes of the cell. Serum B$_{12}$ levels may sometimes be within the normal range despite an increased excretion of MMA. Normal serum B$_{12}$ values range from 0.2 to 1.0 ng/ml. Normal serum MMA levels range from 20 to 75 ng/ml, and normal urinary MMA levels from 0.8 to 3.0 µg/ml.

Serum B$_{12}$ levels can be measured by two methods: microbiological assays using *Lactobacillus leichmannii* and the competitive binding method. The competitive binding assay is detailed in Figure 9.23. A serum sample is mixed with a known amount of commercially available, radioactively labeled vitamin B$_{12}$, [$^{57}$Co]cobalamin. The mixture is then added to a suspension of synthetic microscopic beads containing covalently bound intrinsic factor (IF), a protein that binds vitamin B$_{12}$ extremely tightly. The B$_{12}$ originating from the blood serum and the synthetic [$^{57}$Co]cobalamin then compete with each other to bind to the IF.

The amount of vitamin B$_{12}$ in the serum sample is determined as follows. The amount of radioactive [$^{57}$Co]cobalamin bound to the beads is measured. The purpose of using the beads is to facilitate the separation of bound [$^{57}$Co]cobalamin from the nonbound [$^{57}$Co]cobalamin remaining in solution (floating around versus sinking to the bottom of the test tube). High levels of bound radioactivity indicate that the serum sample contains low vitamin levels. Low levels of bound radioactivity indicate that the serum contains high levels of vitamin. Plasma folate is

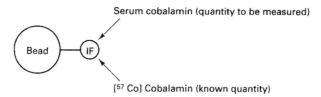

Serum cobalamin (quantity to be measured)

Bead — IF

[$^{57}$ Co] Cobalamin (known quantity)

**FIGURE 9.23** Competitive binding assay.

conventionally measured using competitive binding assays. The agents used in this case are synthetic radioactive folic acid and a folate-binding protein derived from milk.

PA is essentially a disease affecting the transport of vitamin B$_{12}$ as a result of the lack of intrinsic factor. Diagnosis of the disease involves two components: (1) an abnormal Schilling test and (2) correction of the disease with exogenous intrinsic factor. The detection of anti-intrinsic factor is also used in the diagnosis. The Schilling test is a functional test in which an oral dose of radioactive B$_{12}$ is administered. In normal persons, about one-third of an oral test dose of vitamin B$_{12}$ is excreted in the urine. Urinary excretion is dependent on the prior absorption by the gut. In persons with PA, less than 1.0% of the test dose appears in the urine, reflecting an impairment in the absorptive process. The diagnosis of PA requires repetition of the test dose, but with the consumption of 60 mg of intrinsic factor. Intrinsic factor is commercially available.

Following absorption of dietary vitamin B$_{12}$, the vitamin is packaged in transcobalamin II (TCII), which circulates in the bloodstream, and then taken up by various cells in the body. The B$_{12}$ bound to TCII generally accounts for only 20% of the total vitamin in serum, while the rest of the vitamin is bound to other proteins. With the onset of B$_{12}$ deficiency, the levels of vitamin bound to TCII drop first, and holotranscobalamin II accumulates as apotranscobalamin II. Eventually, the levels of B$_{12}$ bound to the other proteins decline (Herbert et al., 1990). Plasma holotranscobalamin II levels of under 30 pM have been used to indicate B$_{12}$ deficiency (Allen et al., 1995). Holotranscobalamin II levels reflect recent intake and absorption of B$_{12}$, while serum vitamin B$_{12}$ levels tend more to reflect the long-term picture of diet and absorption.

When testing for B$_{12}$ absorption by the gut, a test for holo-TCII and apo-TCII has an advantage over the Schilling test. The Schilling test uses pure vitamin B$_{12}$, as a component of the test protocol. Because of this, the Schilling test cannot detect defects in B$_{12}$ absorption due to the lack of stomach acid or the lack of activity of gastric enzymes. The TCII test is sensitive to the these problems, but also to the lack of intrinsic factor (Herbert et al., 1990).

The biochemical "cutoff" values indicating deficiency may vary from laboratory to laboratory, or when using reagents supplied by different manufacturers. In general, it is important for each clinical laboratory to conduct its own survey among a population perceived as healthy and normal, in order to establish the reliability of its tests for vitamin B$_{12}$ status.

*Inconsistent Nature of Signs and Symptoms of B₁₂ Deficiency*

The patient suffering from $B_{12}$ deficiency may feel sick because of the anemia, the neurological symptoms, or both. Variability is a characteristic of $B_{12}$ nutrition. Often, serum $B_{12}$ levels are normal with dietary or functional $B_{12}$ deficiency; often neurological problems occur without megaloblastic anemia. In children fed vegetarian diets, the consequent low serum $B_{12}$, high serum MMA, and growth impairment all may occur without development of megaloblastic anemia (Stabler *et al.*, 1997).

EXERCISE

During the course of a day, one might consume about 100 mmol of isoleucine and valine, the two major precursors of methylmalonyl-CoA. Why is it that only about 4.0 mmol of MMA is normally excreted per day? (See Rasmussen, 1989.)

## VITAMIN K

Vitamin K is a fat-soluble vitamin. It is essential for the synthesis of several proteins involved in blood clotting. Another word for *clotting* is *coagulation*. The abbreviation "K" is derived from the German word *koagulation*. The vitamin K-dependent proteins that are involved in regulating blood clotting are factor II, factor VII, factor IX, factor X, protein C, and protein S. Other vitamin K-dependent proteins are uniquely involved in regulating bone growth (osteocalcin) and in regulating cell growth (Gas6 protein) (Varnum *et al.*, 1995). Other vitamin K-dependent proteins of unclear function exist (protein Z) (Hogg and Stenflo, 1991). Vitamin K serves as a cofactor for **vitamin K-dependent carboxylase**. This enzyme catalyzes the conversion of certain residues of glutamic acid to γ-carboxy-glutamic acid, during the process of translation.

There exist two naturally occurring forms of vitamin K: phylloquinone and menaquinone. Both forms are used with only slight modification occurring in the body, that is, reduction to dihydrovitamin K. Phylloquinone, also called vitamin $K_1$, is synthesized in plants. Menaquinone, also known as vitamin $K_2$, is synthesized by bacteria. These two forms differ only in the structure of the side chain. The side chain is a lipophilic chain built of **isoprenoid** units, as shown in Figure 9.24. Obviously, the vitamin is not used for blood clotting in plants and bacteria, as these forms of life do not have blood. Instead, the vitamin is used for electron transport and energy production. A third form of the vitamin is the synthetic compound, menadione (vitamin $K_3$). Menadione can be alkylated in the liver by enzymes that attach the polyisoprenoid chain (see Figure 9.24). Phylloquinone is used in medicine, whereas water-soluble analogues of menadione are sometimes used in experimental nutrition where they are included in the drinking water consumed by rats.

The RDA for vitamin K is 80 μg for the adult and 5 to 10 μg for the infant. Spinach, lettuce, broccoli, brussels sprouts, and cabbage are good sources of vitamin K. These plants may contain up to 8.0 mg vitamin/kg of food. Cow's milk contains about 20 μg/liter; human milk contains a lower, perhaps tenfold lower,

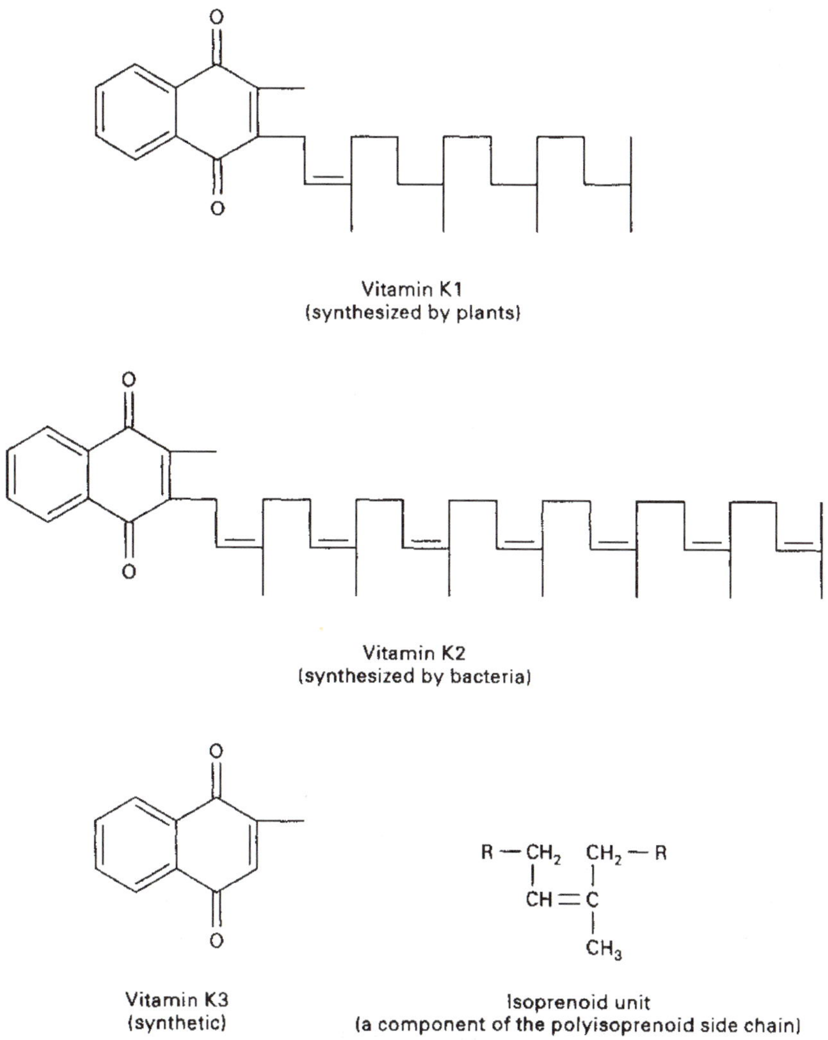

FIGURE 9.24 Vitamin K and related compounds. Vitamin $K_1$, also called phylloquinone, is 2-methyl-3-phytyl-1,4-naphthoquinone. Vitamin $K_2$, also called menaquinone, occurs naturally with a polyisoprenoid tail containing from 5 to 13 isoprenoid units. The figure shows menaquinone-7 (MK-7), which contains a seven-isoprenoid-unit tail. Studies of human liver have revealed that MK-7, MK-8, MK-10, MK-11, and MK-12 account for most of the vitamin K in this tissue, while phylloquinone accounts for about 10% of the vitamin K (Suttie, 1995). Vitamin $K_3$ is a synthetic version of the vitamin.

concentrations of the vitamin. It is thought that human breast milk can supply only 20% of the infant's requirement for vitamin K.

Vitamin K is absorbed by the small intestines, where it enters the lymph packaged in chylomicrons. A deficiency in vitamin K can occur in adults and children suffering from fat malabsorption syndromes, such as cystic fibrosis.

## The Vitamin K Cycle

Vitamin K-dependent carboxylase is a membrane-bound enzyme of the endoplasmic reticulum. It catalyzes the incorporation of $CO_2$ into specific residues of glutamate in the polypeptide chain of specific blood clotting proteins, resulting in their conversion to residues of γ-carboxyglutamate. Dihydrovitamin K is the cofactor of the enzyme. Oxygen is required as the co-substrate. With each catalytic event, dihydrovitamin K is converted to vitamin K epoxide. Vitamin K epoxide is reduced back to the active cofactor by epoxide reductase (Figure 9.25) (Preusch and Suttie, 1983). The anticoagulant warfarin blocks the normal functioning of the cycle by inhibiting epoxide reductase. This inhibition results in the buildup of non- or undercarboxylated clotting factors within the liver or in the bloodstream. The structure of γ-carboxyglutamate (GLA) is shown in Figure 9.25. Apparently, epoxide reductase also catalyzes the reduction of vitamin K to dihydrovitamin K.

## History of Vitamin K Research

Work on vitamin K had three independent beginnings. They were toxicological, nutritional, and biochemical in nature.

### Toxicology

A hemorrhagic (bleeding) disease was noted in cattle that had been eating spoiled sweet clover. Clover generally contains a bitter chemical called coumarin, though sweet clover contains less coumarin than other forms of the plant. During spoilage, microbial action converts the coumarin to **dicoumarol**. The dicoumarol in the spoiled clover caused the death of many animals, especially after procedures such as dehorning and castration. The structure of dicoumarol was identified by K. P. Link at the University of Wisconsin. His laboratory synthesized a simplified analogue of the compound called **warfarin**. Warfarin first found use as a rat poison

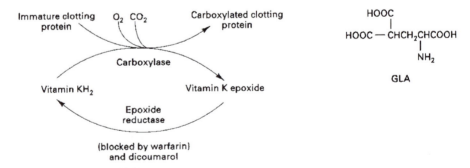

**FIGURE 9.25** Vitamin K cycle. The cofactor probably remains associated with the membranes of the endoplasmic reticulum during the reactions depicted in the figure.

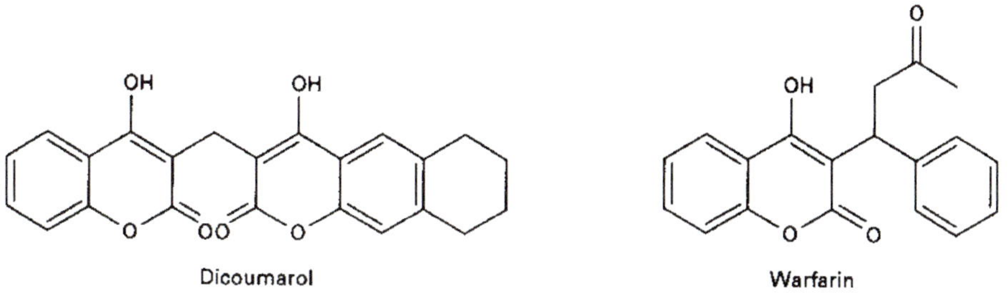

**FIGURE 9.26** Structures of anti-vitamin K compounds.

because of its delayed effect. The rat willingly consumes the poison and introduces it to its friends well before death occurs from internal bleeding. The structures of dicoumarol and warfarin are shown in Figure 9.26. Both of these compounds inhibit the activity of vitamin K epoxide reductase. This enzyme is required for the production of the active form of vitamin K, dihydrovitamin K.

## Nutrition

Vitamin K research also has nutritional beginnings. Dam (1934) in Denmark, as well as Almquist (1935) at the University of California at Berkeley, fed rats diets deficient in lipids. These research groups discovered that the diets induced a previously unrecognized bleeding disease. The disease could be cured by treatment with an extract of alfalfa. It was later recognized that alfalfa contained vitamin $K_1$. The disease could also be cured by treatment with spoiled fish meal; it was later found that the meal contained the bacterial product vitamin $K_2$. The finding that phthiocol could cure the bleeding disease provided an early clue to the structure of the vitamin. Phthiocol has a structure that is quite similar to that of vitamin $K_3$, except that it contains a hydroxyl group.

## Biochemistry

Work during the early 1900s led to the discovery of a variety of proteins required for blood clotting. Most of these proteins are proteases. These proteases are not general proteases. They catalyze the cleavage of specific peptide bonds of specific proteins. The substrates of these proteases are other blood clotting proteins. The proteolytic event involves the cleavage of one or more peptide bonds of the substrate, resulting in its conversion to an active blood clotting protein. The blood clotting proteins include factors XII, XI, IX, VII, X, II, and XIII and fibrinogen. Factor II is also known as prothrombin. The activated versions of these proteins are factors $XII_a$, $XI_a$, $IX_a$, $VII_a$, $X_a$, $II_a$, and $XIII_a$, and fibrin, respectively.

The relationship between vitamin K and blood clotting is as follows. Generation of the blood clot, which is composed of cross-linked platelets and cross-linked fibrin, is dependent on the activity of a number of proteins. Some of these proteins are vitamin K-dependent proteins. This means that a vitamin K-dependent reaction was involved in the biosynthesis of the protein in the liver prior to its secretion into the bloodstream. In the presence of an adequate vitamin K status, the vitamin K-dependent proteins are modified by the formation of residues of γ-carboxyglutamic acid. With vitamin deficiency, the proteins are synthesized in the liver, but fail to be modified in the normal manner, and appear in the bloodstream in an immature form. The deficiency tends to affect clotting proteins II and VII more than the other vitamin K-dependent proteins.

The time course of blood clotting, *in vitro*, is illustrated in Figure 9.27. The study involved a plasma sample contained in a test tube. Transmission of light through the tube was measured with a spectrophotometer. At time zero, the transmission of light was low and was arbitrarily set to zero. The transmission was low because of the dense suspension of platelets in the sample.

The clotting pathway, *in vivo*, is initiated by a protein called **tissue factor**. However, other macromolecules, such as collagen, can initiate blood clotting in the laboratory. Collagen was used in this study and was added at time zero (Figure 9.27). The results demonstrate an increase in light transmission through the blood sample after addition of the collagen. This increase was due to the aggregation of platelets into a relatively small mass. This mass is the blood clot.

The term **plasma** refers to whole blood with the red blood cells removed. Usually, plasma is prepared and stored with an added artificial anticoagulant, in order to prevent spontaneous formation of the blood clot inside of the blood-collection tube. **Serum** means plasma after formation of the blood clot, and with removal of the clump composed of coagulated proteins.

## The Blood Clotting Cascade

Injury to a tissue results in exposure of tissue factor from the torn blood vessel. Within the wall of the artery, tissue factor occurs in the **adventitia**, and not in the

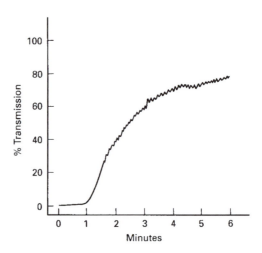

**FIGURE 9.27** Illustration of blood clotting in a test tube. (Redrawn with permission from Wilner *et al.*, 1968.)

**intima** or **media** (Schecter *et al.*, 1997). These layers are detailed in the Lipids chapter (Figure 6.34). When tissue factor is exposed to the blood clotting proteins that circulate within the bloodstream, the result is the activation of the blood clotting cascade. The blood clotting cascade consists of two branches. One branch results in the cross-linking of fibrin, while the other provokes the cross-linking of platelets. The resulting three-dimensional network is called the blood clot. Some of the proteins involved in the clotting cascade are as follows. Some of these terms are listed as pairs, in cases where the two proteins occur as a sequence in time, or where the two proteins work together at the same time:

- Tissue factor
- Factor X and Factor $X_a$
- Factor II (prothrombin) and Factor $II_a$ (thrombin)
- Docking proteins
- Fibrinogen and fibrin
- von Willebrand factor
- Integrin $\alpha_{IIb}\beta_3$
- Platelets and microparticles
- Anticoagulation
- Thrombomodulin and protein C

Figure 9.28 shows a simplified diagram of the clotting cascade. The figure shows that the exposure of tissue factor to the bloodstream provokes the conversion of prothrombin to thrombin, and that thrombin catalyzes the cross-linking of fibrin and the cross-linking of platelets. The clotting cascade is shown in greater detail in Figure 9.29. In viewing the figure, one might consider thrombin to be a focal point, and ask, "Does this particular step occur prior to the activation of thrombin or after?" and "Is this reaction catalyzed by thrombin?" A goal of the blood clotting cascade is to provoke an explosive activation of thrombin. Thrombin is a protease, and once its active form is generated, it catalyzes the cleavage of three target polypeptides: (1) fibrinogen, (2) factor XIII, and (3) the thrombin receptor, a membrane-bound protein of the platelet.

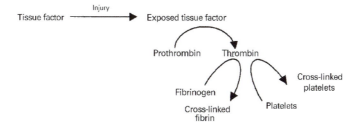

**FIGURE 9.28** Simple diagram of the blood clotting cascade.

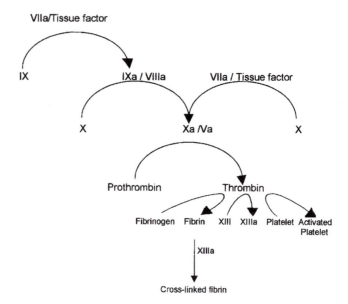

**FIGURE 9.29** Detailed diagram of the blood clotting cascade. The initiating action is the exposure of tissue factor to the bloodstream. With exposure, tissue factor combines with factor VII$_a$, resulting in the generation of a catalytically active complex of factor VII$_a$ plus tissue factor. Please note that factor VII is not active, and that factor VII$_a$ without tissue factor is also not active. The factor VII$_a$/tissue factor complex provokes the activation of two different factors, namely factor IX and factor X. Factor IX$_a$ binds to factor VIII$_a$ (a docking protein)to form the factor IX$_a$/factor VIII$_a$ complex. This resides on a phospholipid surface because of the adhesive qualities of factor VIII$_a$. Perhaps the most significant phospholipid surface, in the bloodstream, which contributes to blood coagulation are the microparticles. The factor IX$_a$/factor VIII$_a$ complex cleaves specific peptide bonds in factor X, generating factor X$_a$. Factor X$_a$ spontaneously complexes with factor V$_a$, another docking protein, and the active factor X$_a$/factor V$_a$ complex cleaves specific peptide bonds of prothrombin, generating thrombin. Thrombin cleaves a number of target polypeptides, resulting in activation events. These target polypeptides include fibrinogen, factor XIII, and the thrombin receptor. The end-result of the clotting cascade is a cross-linked network consisting of platelets and fibrin.

## Xymogens and Active Proteases

Tissue factor is a membrane-bound protein of the epithelial cells that line the blood vessels. The protein is small and consists of 263 amino acids. A tear in the blood vessel results in the exposure of tissue factor to the bloodstream, and thus to all of the proteins within the plasma. Exposure of tissue factor to factor VII results in the formation of the factor VII/tissue factor complex. This complex catalyzes the cleavage of factor IX, thus generating factor IX$_a$. The **factor IX$_a$/factor VIII$_a$ complex**, in turn, catalyzes the activation of factor X. Factor IX is a xymogen, meaning that it is a nonactivated protein. The subscript "a" in factor IX$_a$ means that the factor has been activated, and is no longer a xymogen. The **factor VII$_a$/tissue factor complex** also catalyzes the cleavage of factor X, generating factor X$_a$. To summarize, factor X can be activated by two separate complexes: (1) the **factor VII$_a$/tissue**

**factor complex**, and (2) the **factor IX$_a$/factor VIII$_a$ complex**. The reason for these two separate pathways for activating factor X is not clear, but evidence suggests that they tend to act at different points in time after an injury.

## Three of the Proteases Work While Complexed with Docking Proteins

Figure 9.29 shows factor X$_a$ in a complex with factor V$_a$. Factor V$_a$ is a docking protein; this factor does not have enzymatic activity. It serves as a biochemical glue that mediates the binding of factor X$_a$ to the phospholipid surface of the platelet. Factor X$_a$, even though activated, does not have much enzymatic activity. However, once complexed with its docking protein, and bound to a phospholipid surface, factor X$_a$ is highly active. Other docking proteins used in regulating blood clotting are factor VIII$_a$ and thrombomodulin. Each of the docking proteins may also be called a cofactor.

## Factor X$_a$/Factor V$_a$ Catalyzes Thrombin Formation

Prothrombin (factor II) is a 581-amino-acid protein that contains one disulfide bond (Figure 9.30). Prothrombin is cleaved by factor X$_a$ in between two specific

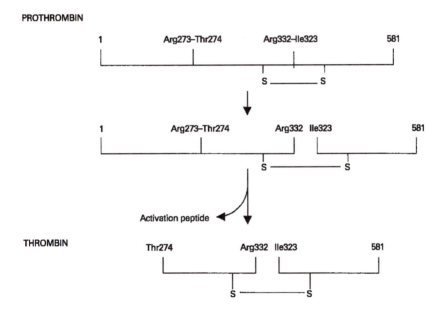

**FIGURE 9.30** Structures of prothrombin and thrombin. Prothrombin is a protein consisting of 581 amino acids, while thrombin consists of 308 amino acids. Prothrombin is cleaved within the bloodstream, at the indicated peptide bonds, by the factor X$_a$/factor V$_a$ complex. The intermediary form of thrombin is called **meizothrombin**. The **activation peptide** might be considered to be a waste product of the activation scheme. Please note that thrombin consists of two separate polypeptide chains, and that these are held together by a disulfide bond. Studies by researchers have revealed that prothrombin can also be activated by factor X$_a$ alone, but here the pathway of peptide bond cleavage is somewhat different than that produced by the factor X$_a$/factor V$_a$ complex.

peptide bonds that involve arginine. These bonds, which occur at Arg(273)–Thr(274) and at Arg(332)–Ile(323) are indicated. The first of the two cleavages results in the generation of two polypeptide chains, which are held together by the disulfide bond. The second cleavage results in the generation of the thrombin (factor II$_a$), and a byproduct called the **activation peptide** (Figure 9.30).

### *Thrombin Provokes Formation of the Fibrin Soft Clot, and then the Fibrin Firm Clot*

Thrombin itself does not constitute a blood clot. Once thrombin is produced, there is still quite a way to go before bleeding can be halted. Thrombin catalyzes the cleavage of fibrinogen to give fibrin. Please note the "gen" suffix on fibrinogen, which indicates that it is a xymogen. Fibrinogen is composed of three kinds of polypeptide chains, i.e., the α-chain, β-chain, and γ-chain. Actually, fibrinogen contains two copies of each of these chains, and thus the entire protein is more accurately represented as $\alpha_2\beta_2\gamma_2$.

Thrombin cleaves at Arg(16)–Gly(17) of the α-chain, and at Arg(14)–Gly(15) of the β-chain. This results in the release of activation peptides (byproducts) from fibrinogen. The resulting fibrin spontaneously forms a soft clot. Thrombin also catalyzes the activation of factor XIII, and factor XIII$_a$ catalyzes the covalent cross-linking of different fibrins with each other. The result is a firm clot.

Once the fibrin clot has been formed, fibrin may also stimulate healing. Evidence suggests that part of the β-chain of fibrin acts as a hormone to stimulate the spreading of cells and in the formation of new capillaries. Amino acids 15–42 of the β-chain of fibrin, for example, seem especially critical for this function. To summarize, the blood clotting cascade functions to seal up the wound and stop bleeding, but also to stimulate healing of the wounded area.

### *Summary*

The material detailed in this section may be summarized by the following diagram, which indicates the succession of events that occur during the course of several seconds:

### *Thrombin Catalyzes the Activation of Platelets*

Thrombin catalyzes the formation of the fibrin clot and, at the same time, the activation of platelets. Are fibrin clot formation and platelet activation totally separate biochemical pathways? The answer to this question is yes and no. To some extent, fibrin clot formation and platelet activation occur separately and independently of each other. However, these two pathways do converge. Platelet activation results in the formation of links between the platelets and the fibrin clot.

Thrombin initially acts by binding to the thrombin receptor, a membrane-bound protein in the plasma membrane of the platelet. Thrombin binds to the thrombin receptor, and then catalyzes the cleavage of its N-terminal region. Thrombin cleaves in between the residues of arginine and serine (Arg–Ser), as indicated herein. The following diagram depicts the first 18 amino acid residues of the thrombin receptor. The various features of the drawing are not to scale:

The thrombin receptor is distinguished in that its polypeptide chain weaves seven times in or out of the plasma membrane (phospholipid bilayer), as shown in the drawing. Several other hormone-receptors weave seven times through the plasma membrane and, in common with all of these receptors, the activation of the thrombin receptor provokes the activation of G protein. Activated G protein, in turn, provokes a number of events within the platelet, as revealed later on. Activation of G protein, by many types of hormones, relays a signal within the plasma membrane that provokes the activation of phospholipase $A_2$ and of phospholipase C. The first of these two enzymes acts on membrane phospholipids and cleaves arachidonic acid, allowing this fatty acid to be converted to a number of hormones (prostaglandins and thromboxanes). The second of these two enzymes acts on phosphatidylinositol (PI), cleaves the inositol group, resulting in the release of inositol-1,4,5-trisphosphate (IP3). IP3, in turn, is a signal that provokes an increase in concentration of calcium ions within the cell (calcium signaling). The actions of these two enzymes are universal events in many, or most, types of cells. Further details are in the Essential Fatty Acid section and the Calcium and Phosphate section.

## Consequences of Activation of Signaling Pathways within the Platelet

Cleavage of the thrombin receptor provokes several events in the platelet, including:

1. Filopodia formation (tiny feet or spikes).
2. Microparticle formation (tiny spheres).
3. Activation of nearby platelets in the bloodstream.
4. Activation of integrins in the platelet's membrane.
5. Release of contents of secretory vesicles into the bloodstream.

TABLE 4.16  Masses of Utilizable Energy Stores of a Standard 70-kg Man

| Energy store or source | Mass (kg) | Energy (kJ) |
| --- | --- | --- |
| Fat | 15 | 590 |
| Protein (muscle) | 6 | 100 |
| Glycogen (muscle) | 0.15 | 2.51 |
| Glycogen (liver) | 0.075 | 1.25 |
| Plasma glucose | 0.020 | 0.33 |
| Plasma triglycerides | 0.003 | 0.125 |
| Plasma free fatty acids | 0.0003 | 0.012 |

*Source*: Cahill (1970).

## Body Fuels Available during Fasting

The preceding discussion of ketone bodies leads to the topic of fasting. Prolonged fasting produces symptoms of deficiencies in many nutrients, most notably water (if water is omitted), energy, protein, and potassium. An evaluation of energy supplies available during fasting requires information on body composition. Tables 4.16 and 4.17 present data on the body composition of the standard 70-kg man. Masses of the different energy stores or usable sources appear in Table 4.16; Table 4.17 lists the weights of the organs. Fat (15 kg) constitutes 80 to 85% of the energy stores in the nonobese man. During starvation, this energy source can sustain life for 50 to 60 days. An obese person can live over 100 days using internal fuels as the sole energy source. At rest, skeletal muscle accounts for 35 to 40% of the body's total consumption of oxygen. Resting muscle depends mainly on fat, not carbohydrates, as an energy fuel.

About 20% of the body's muscle mass (28 kg) consists of protein (6 kg). Muscle protein accounts for about 17% of the body's energy reserves; however, not all muscle protein can be used for energy production. It is thought that the use of more than 50% of the 6 kg available is fatal. Muscle mass decreases during fasting.

TABLE 4.17  Organ Masses of a Standard 70-kg Human

| Organ | Mass (kg) |
| --- | --- |
| Skeletal muscle | 28.0 |
| Adipose tissue | 15.0 |
| Skeleton | 10.0 |
| Skin | 4.9 |
| Liver | 1.8 |
| Brain | 1.4 |
| Heart | 0.33 |
| Kidneys | 0.31 |

*Source*: Forbes (1987).

Noncontractile proteins are catabolized first, because contractile fibers resist proteolysis and breakdown. With prolonged starvation, however, contractile fibers are broken down, and it is not clear what mechanisms control the catabolism of muscle proteins.

Glycogen is a minor energy store. The amount stored in the body is sufficient to supply the energy needs of the brain for only 3 days of fasting. Clearly, the energy value of the nutrients in the bloodstream itself is minimal.

## STARVATION

Severe malnutrition is a complex problem. Factors involved include the overall availability of food; the availability only of starchy, protein-deficient foods; and contaminated water supplies. Infants and young children are special targets because they are rapidly growing, are unusually susceptible to infections, and may not be able to find food for themselves.

There are two general types of severe malnutrition. The first, **marasmus**, is an overall lack of energy and protein and results in overall wasting of the body. The second type occurs with chronic consumption of a diet supplying ample amounts of carbohydrate but little or no protein. This diet results in preservation of fat stores but, unfortunately, also in a tendency to preserve the stores of amino acids, the proteins of skeletal muscle. This failure to mobilize muscle protein results in continued turnover of proteins that are vital to survival, such as serum albumin and immunoglobulins (antibodies). Continued failure to consume protein leads to gradual depletion of these essential proteins, a condition called **kwashiorkor**. Malnutrition can occur as marasmus, kwashiorkor, or a range of intermediate forms. All forms, both mild and severe, are called **protein-energy malnutrition**.

Severe protein-energy malnutrition often occurs after weaning, the transition from nursing to the consumption of foods from other sources (i.e., solid foods). The marasmus common in Latin America is caused by early weaning, followed by use of overdiluted commercial milk formulas. Kwashiorkor occurs in Africa, where babies are fed starchy roots, such as cassava, that are low in protein. It also occurs in the Caribbean, where babies are fed sugar cane. The major symptom of kwashiorkor is edema, mainly of the feet and legs. **Edema** is the condition produced when water normally held in the bloodstream by osmotic pressure leaks into other extracellular spaces. It can result from reduced osmotic pressure in the bloodstream caused by catabolism and depletion of serum albumin.

Kwashiorkor is thought to involve "confused hormones." An adequate supply of carbohydrates, in the form of starchy roots or sugar cane, can result in maintenance of high levels of plasma insulin, which can prevent the net proteolysis that otherwise would supply the liver and immune system with amino acids. The mechanism by which insulin controls muscle turnover is not well understood; however, in marasmus the body's stores of fat and muscle are broken down and mobilized in a more coordinated fashion, resulting in a relatively gradual decline in overall health.

All proteins in the body, whether occurring in the plasma, as membrane-bound proteins, or inside the various compartments of the cell, are degraded or hydrolyzed. The rate of hydrolysis of any specific protein is characteristic of and unique for that protein. Albumin, transferrin, thyroxine-binding protein, and retinol-bind-

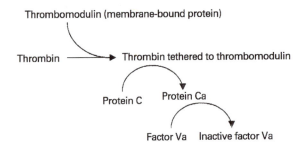

FIGURE 9.31 The anticlotting pathway. The anticoagulation pathway is used to catalyze the inactivation of factor $V_a$, and thus halt continued activation of prothrombin in regions of the blood downstream of the newly formed blood clot. The anticlotting pathway, along with antithrombin III, both serve to prevent an injury from provoking the conversion of the entire bloodstream into a massive clump of coagulated blood. When thrombin becomes tethered to thrombomodulin, it no longer recognizes its procoagulation substrates, and acquires the ability to activate protein C. Activated protein C, in turn, catalyzes the inactivation of factor $V_a$.

conversion of arachidonic acid to thromboxane. Further details are revealed in the Essential Fatty Acid section.

## Difference between Warfarin and Heparin

Warfarin and heparin are both anticoagulants. They are both used to prevent the formation of potentially dangerous blood clots. Warfarin may be administered orally; heparin requires injection. Warfarin inhibits the activity of epoxide reductase, an enzyme of the vitamin K cycle. This inhibition results in a reduction in the rate of synthesis of the GLA-containing blood clotting proteins and, thus, in a lesser tendency to form blood clots. Heparin is a naturally occurring polysaccharide, though it does not naturally circulate in the bloodstream. Warfarin and related drugs influence events in the liver. Heparin does not interfere with the vitamin K cycle, but acts only in the bloodstream. Heparin enhances the rate of inactivation of thrombin in the plasma. This type of inactivation normally occurs throughout the day, and heparin enhances the rate. Stray molecules of thrombin are continuously mopped up by the protein antithrombin. Antithrombin forms a tight complex with thrombin, resulting in its inactivation. Heparin enhances the rate of formation of this complex.

A functional deficiency in vitamin K occurs during treatment with warfarin or related drugs. The moderate functional deficiency provoked by the drug treatment is intentional and is used to "thin the blood" and reduce the tendency of those with cardiovascular disease to form undesirable blood clots. The drugs inhibit the normal cycling of vitamin K during synthesis of the vitamin K-dependent blood clotting proteins. The phrase *thin the blood* is not scientifically accurate. Warfarin treatment does not reduce the viscosity of the blood. The drug acts in the liver, where its effect is to reduce the production of mature, fully carboxylated clotting proteins. The immature clotting proteins may accumulate in the liver, as the lack of carboxylation appears to impair their secretion from the hepatocytes. A fraction of the immature clotting proteins are secreted into the bloodstream.

## Hemophilias (Bleeding Disorders Not Related to Vitamin K Status)

The physician confronted with a bleeding disorder must distinguish between a deficiency in vitamin K and one of a number of rare genetic disorders that result in defects in one of the blood clotting proteins. Vitamin K therapy can restore normal clotting in cases of vitamin deficiency but not in genetic diseases. These genetic diseases are called the **hemophilias**. Common types of hemophilias are those affecting factor VIII (hemophilia A) and factor IX (hemophilia B). Hemophilia A occurs in 1 in 10,000 males and involves bleeding in soft tissues, muscles, and joints that bear weight. When factor VIII levels are under 5% the normal level, the symptoms of hemophilia A become moderate or severe. Bleeding can continue for hours or days after trauma. Hemophilia B occurs in 1 in 100,000 males. Its symptoms are the same as in hemophilia A.

**Von Willebrand factor** (vWF) is a protein of the clotting cascade that was not mentioned earlier. vWF is synthesized by endothelial cells of the blood vessels, and released into the bloodstream. The protein is large (260 kDa) and consists of 2050 amino acids. It is also found in platelets, where it resides inside secretory vesicles. vWF serves two functions. It tightly binds to factor VIII, and stabilizes it, and maintains the usual and normal levels of factor VIII in the bloodstream. Here, vWF and factor VIII occur in a 1:1 complex. In the absence of vWF, plasma levels of factor VIII decline to dangerously low levels. The second function of vWF is as follows. vWF is stored in special vesicles in platelets, and secreted from platelets with platelet activation. The secreted vWF helps the platelets to bind to the extracellular matrix residing under torn endothelial cells. Specifically, vWF binds to integrin $\alpha_{IIb}\beta_3$ (of platelets) and to collagen (of the extracellular matrix). Genetic defects in vWF that are serious enough to impair blood clotting occur in about 1 in 1000 persons. The disease (von Willebrand disease) involves spontaneous bleeding in joints and mucosal surfaces.

## Vitamin K Deficiency

Newborns are prone to vitamin K deficiency for three reasons: (1) they are born with low stores of the vitamin, (2) mother's milk is low in the vitamin, and (3) newborns have a sterile gastrointestinal tract. Vitamin K deficiency can lead to **hemorrhagic disease of the newborn**. The disease can present as bleeding beneath the skin or elsewhere in the body. It is thought to affect up to 1.0% of all newborns. In very rare instances, bleeding can occur in the brain with fatal consequences. One can imagine the confusion arising when parents are suspected of traumatizing their newborn when the real problem was a vitamin deficiency. It is common practice for all infants to be injected with phylloquinone (1.0 mg) *on the day of birth*.

Vitamin K deficiency can occur during treatment with certain antibiotics that wipe out the gut microflora. Vitamin K synthesized and released in the large

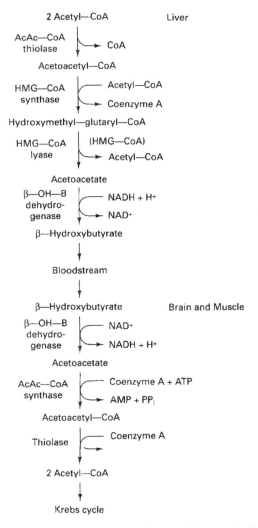

FIGURE 4.65 Synthesis, circulation, and degradation of ketone bodies. Ketone body metabolism involves synthesis in the mitochondria of the liver, distribution via the bloodstream, and oxidation by the Krebs cycle in various organs, such as the brain and muscle.

A comparison of fatty acid and ketone body levels in the bloodstream may overemphasize the importance of the latter compounds as energy sources for the following reasons. First, fatty acid molecules contain 16 to 18 carbons, whereas ketone bodies contain only four carbons. Therefore, the weight of 1 mmol of fatty acid is several times that of 1 mmol of β-hydroxybutyrate. Second, fatty acids are highly reduced, alkane-like structures, whereas ketone bodies are partially oxidized. This means that on a per-weight basis complete oxidation of fatty acids yields more energy than complete oxidation of ketone bodies.

Under normal conditions, glucose is the exclusive fuel of the brain; however, when ketone bodies accumulate in the bloodstream, they are also used by the

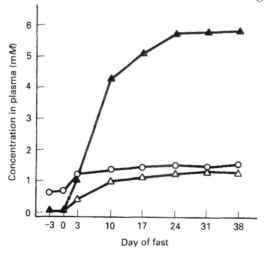

**FIGURE 4.66** Formation and reaction of the intermediate, hydroxymethylglutaryl-CoA (HMG-CoA). The arrow pointing from the 2-carbon of acetyl-CoA to the carbon-3 of acetoacetyl-CoA depicts the position of attack of the 2-carbon on the carbonyl carbon during the reaction.

brain. The study reported in Table 4.15 involved obese human subjects fasting for a lengthy period. After 1 month, extraction of fuels by the brain was measured by the AV difference technique. Samples of blood entering and leaving the brain were taken and AV differences of the nutrients determined. The concentration of glucose in venous blood was 0.26 m$M$ less than that in arterial blood, indicating its removal and catabolism by the brain. The concentration of lactate was higher in the venous

**FIGURE 4.67** Plasma concentrations of ketone bodies in fasting humans. Plasma concentrations of FFAs (○), β-hydroxybutyrate (▲), and acetoacetate (△) are shown before and during fasting. (Redrawn with permission from Owen *et al.*, 1969.)

## Biotin Deficiency

Biotin deficiency in humans is rare. The first indicator of human biotin deficiency appears to be a drop in urinary biotin, with a maintenance of plasma levels of the vitamin. The deficiency can result from a rare genetic deficiency in biotinidase. The disease occurs in about 1 in 40,000 infants and results in a rash about the eyebrows and cheeks and neurological symptoms, such as muscle pain, extreme tiredness, and numbness. Biotin deficiency may occur in persons who consume raw eggs (six per day) over many months. Egg white contains a protein called **avidin**, which binds to biotin extremely tightly, though not via a covalent linkage. The physiological function of the avidin in egg white is not clear. The protein tightly binds the biotin released during the digestion of dietary protein and prevents its absorption. Avidin is destroyed during cooking. The protein has proven useful in inducing experimental biotin deficiency in studies in humans and animals. Biotin deficiency has occurred during prolonged total parenteral nutrition, where biotin had been inadvertently omitted from the liquid formula. Persons who are taking antibiotics or who have had parts of their intestines surgically removed may be at risk for a deficiency.

Biotin deficiency in animals results in alopecia, a scaly dermatitis, anorexia, and eventually death. Some of the biochemical changes occurring in rats consuming a biotin-deficient diet containing raw egg white are illustrated in Figure 9.33. The animals consumed the diet for up to 30 days. The activities of three biotin-requiring enzymes were determined in the livers of rats killed at the indicated times. The enzymes measured were acetyl-CoA carboxylase (●), propionyl-CoA carboxylase (○), and pyruvate carboxylase (Δ).

The decrease in propionyl-CoA carboxylase activity during biotin deficiency induces the accumulation of propionyl-CoA in tissues. This compound, in elevated concentrations, is recognized by fatty acid synthase, resulting in the production of

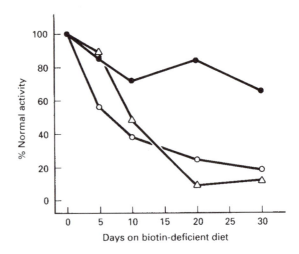

**FIGURE 9.33** Biochemical changes occurring in rats consuming a biotin-deficient diet containing raw egg white. (Redrawn with permission from Arinze and Mistry, 1971.)

small amounts of fatty acids containing an odd number of carbons, that is, 15 or 17 carbons.

Studies with pregnant animals have revealed that biotin deficiency tends to leave the mothers in a healthy state, while producing birth defects in the fetuses (Mock *et al.*, 1997a).

## Assay of Biotin

Biotin can be precisely measured in the laboratory using microbiological assays, where the test organism is *Lactobacillus plantarum*. The key to utilizing any type of microbiological assay is having a growth medium that contains all nutrients required by the test organism, but is totally lacking in the nutrient of interest, i.e., biotin. Biotin can also be measured by high-pressure liquid chromatography (HPLC). The measurement of biotin in biological fluids may be complicated by the presence of certain breakdown products of the vitamin, such as **bisnorbiotin** (Berg, 1997; Mock *et al.*, 1997c). Breakdown products tend to interfere with the interpretation of results from HPLC, but not with microbiological assays.

EXERCISE

Would you expect a biotin deficiency to result in greater impairment of the Cori cycle or in fatty acid synthesis, on the basis of the data in Figure 9.33?

## VITAMIN B$_6$

Vitamin B$_6$ is a water-soluble vitamin. The RDA for adults is 2.0 mg; that for young infants is 0.3 mg. The vitamin has several forms: pyridoxine, pyridoxal, pyridox-amine, and the phosphorylated versions of these forms (Figure 9.34). Pyridoxine is the form used in vitamin supplements. The coenzymatically active forms of the vitamin are pyridoxal phosphate (PLP) and pyridoxamine phosphate (PMP). The cofactor remains tightly bound to the enzyme before, during, and after catalysis of the reaction. PLP is bound more tightly than PMP, as the aldehyde group of PLP forms a Schiff base with a lysine residue of the enzyme, lending stability to the coenzyme–apoenzyme complex (Figure 9.35). This Schiff base dissociates when the cofactor participates in the chemistry of catalysis. The major catabolite and urinary metabolite of vitamin B$_6$ is pyridoxic acid.

Poultry, fish, liver, and eggs are good sources of the vitamin; meat and milk contain lesser amounts. The vitamin in these foods is almost completely available to the body. Plants contain a unique form of vitamin B$_6$, in addition to the forms found in animals. This form is pyridoxine glucoside (Figure 9.36). About 50 to 75% of the vitamin B$_6$ in plant foods, such as beans, carrots, orange juice, and broccoli, occurs in this form, whereas only a very small proportion of the vitamin in grains and nuts occurs in this form. Studies with rats have indicated that pyridoxine glucoside in the diet may have an availability of less than 50%, though the situation with humans is not clear.

fatal auto accidents occur in the United States each year. More than half involve a driver who had recently been drinking alcohol. Disease problems of alcoholism include alcoholic hepatitis, cirrhosis of the liver, **fetal alcohol syndrome** (FAS), and pancreatitis. Alcohol drinking is also associated with an increased risk of cancer of the mouth and esophagus, and breast cancer in women. For example, the rate of breast cancer may be about 30% greater than normal in women who consume one drink per day (Thun *et al.*, 1997).

Alcoholic liver disease has three stages. In the first stage, fat accumulates in the liver, producing a fatty liver. Fatty liver and the associated enlargement of the liver are completely reversible by discontinuing alcohol consumption. Histological examination of an enlarged liver may reveal a ballooning effect, in which the hepatocytes increase in diameter two- to threefold. Cellular disorganization and mitochondrial deformities are other features of the livers of alcoholics. The fatty liver condition is illustrated by the experiment in Table 4.20. For 18 days, human subjects consumed a nonalcoholic diet from which about 25% of the energy was derived from protein, 25% from fat, and 50% from carbohydrate. The protein foods were ground beef, fish, eggs, and turkey. Then, for about 2 weeks the same diet was consumed, except that alcohol replaced the carbohydrates; that is, pure alcohol replaced sugar, bread, cereals, and fruits. Liver biopsies taken after each dietary period were used for measurement of liver TGs. The results demonstrated a striking increase in liver TGs when 50% of the energy had been supplied by ethanol.

The second stage of alcoholic liver disease, **alcoholic hepatitis**, is characterized by the death of a number of liver cells and inflammation of the affected areas. The damage in this stage sometimes proves fatal.

In the third stage, **cirrhosis** (fibrous structures and scar tissue) forms in the liver. In humans, development of cirrhosis requires 5 to 20 years of steady consumption of a high-alcohol diet. It is not clear whether the hepatitis stage is required for formation of cirrhotic tissue. About 75% of the cases of cirrhosis of the liver result from alcoholism. Alcoholism does not always cause cirrhosis. Only about one-quarter of heavy drinkers develop cirrhosis.

TABLE 4.20  Effect of an Alcohol Diet on the Fat Content of the Liver

| Subject | Triglyceride (mg TG/100 mg protein) | |
| --- | --- | --- |
| | Nonalcoholic diet | Alcohol diet |
| 1 | 3.2 | 48.3 |
| 2 | 2.5 | 13.8 |
| 3 | 3.1 | 10.4 |
| 4 | 6.7 | 54.4 |
| 5 | 1.5 | 10.7 |

*Source*: Lieber and Rubin (1968).

The fatty liver condition can be described as follows. With alcohol consumption, the liver's main source of energy (fatty acids) is replaced by alcohol. This results in accumulation of unused fatty acids, in the form of triglycerides, with the consequent deposition of these TGs as fat. This mechanism is different from the fatty liver provoked by kwashiorkor, where a fatty liver results from the failure to synthesize the apolipoproteins needed for exporting TGs from the liver. The fatty liver of kwashiorkor is not associated with cirrhosis, but that of alcoholism is associated with conversion of stellate cells to cells resembling those of connective tissue, i.e., fibroblasts. Stellate cells are fat-storing cells that occur in the interstitial space between the capillaries and hepatocytes, called the **space of Disse**.

**Fetal alcohol syndrome** (FAS) occurs in the fetuses of chronic alcoholic mothers who drink heavily during pregnancy. The frequency of FAS births ranges from 2.5 to 10% of alcoholic mothers. For the pregnant woman, heavy drinking is defined as occasional consumption of 5 drinks, or 45 drinks per month. Consumption of one drink per day has not yet been associated with FAS or with milder forms of the syndrome. FAS is characterized by growth retardation, characteristic facial abnormalities, and mental retardation. It is a major cause of mental retardation, and its effects do not correct themselves as the child grows older.

A minority of alcoholics develop nutrient deficiencies. In Western countries, alcoholics represent the largest population segment that can benefit from dietary intervention. Alcoholics are at risk for deficiencies in folate, thiamin, riboflavin, vitamin $B_6$, vitamin A, and magnesium, particularly when the intake of these substances is low. In some cases, absorption of the nutrient is impaired; in others, catabolism of the nutrient is increased. Thiamin deficiency is a firmly established consequence of alcoholism, as discussed in the Thiamin section.

The case of vitamin A is particularly interesting. Alcoholism induces an increase in the level of cytochrome P450 in the liver, which in turn catalyzes the conversion of vitamin A to inactive products. Drugs such as phenobarbital can also induce increased production of cytochrome P450. Thus, an alcoholic who chronically takes this drug may be at risk of developing a vitamin A deficiency.

Alcohol can stimulate the absorption of iron, probably because it increases gastric acid secretion. Gastric acid enhances the solubility of ferric iron, thereby enhancing its absorbability. Some alcoholics suffer from iron-deficiency anemia; others absorb too much iron, resulting in excess iron stores.

EXERCISE

Why might one expect the activity of the malate–aspartate shuttle to be changed by alcohol consumption? (See Sugano *et al.*, 1990.)

EXERCISE

The action of cytochrome P450 can result in conversion of components of cigarette smoke to cancer-producing chemicals, called **carcinogens**. Explain how alcoholism can increase one's risk of developing cancer. (See Lieber, 1988.)

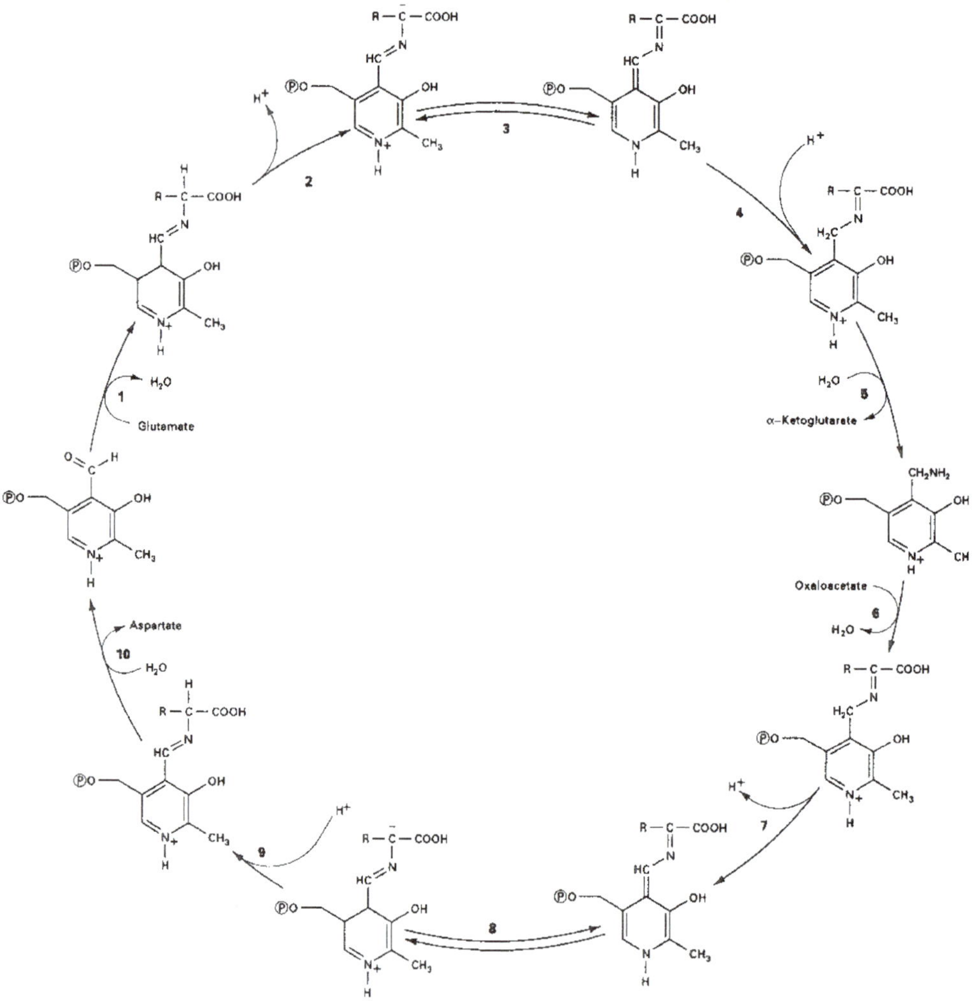

**FIGURE 9.37** Detailed mechanism of aminotransferase reaction. All of the steps are reversible. Double arrows are shown only at Steps 3 and 8 to emphasize the importance of the resonance that stabilizes the carbanion, that is, allows a negative charge to develop on the α-carbon of the substrate.

PLP is also the cofactor of glycogen phosphorylase, an enzyme featured promi-nently in Chapter 4. The PLP functions in a rather unorthodox manner in this enzyme. The catalytically active part of the cofactor is the phosphate group, rather than the aldehyde group and aromatic ring, as in all of the enzymes listed in Table 9.3. The phosphate group of PLP participates in the transfer of inorganic phos-phate to the glucose units of glycogen, resulting in the release of glucose 1-phos-phate. Muscle contains most of the body's glycogen and thus most of its glycogen phosphorylase. Most of the vitamin $B_6$ in muscle occurs bound to this enzyme.

Vitamin $B_6$ is released from enzymes in the stomach and gut by the action of proteolytic enzymes. The phosphate groups of PLP and PMP are hydrolyzed by

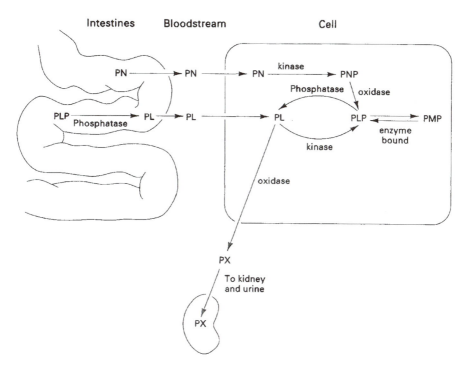

**FIGURE 9.38** Transport and metabolism of vitamin B$_6$.

phosphatases during the process of absorption. The nonphosphorylated versions of the vitamin are transported in the bloodstream to the various tissues, where they are converted to PLP, as shown in Figure 9.38. The hydroxyl group of PNP is oxidized to an aldehyde group, generating PLP. Some of the PLP in the liver may reenter the bloodstream, where it occurs largely bound to serum albumin. About half of the plasma vitamin is in the form of PLP. The vitamin may also be oxidized in various tissues to pyridoxic acid, an event that represents the degradation of the vitamin. Pyridoxic acid is excreted in the urine.

## Vitamin B$_6$ Deficiency

The population thought to be most at risk for B$_6$ deficiency comprises chronic alcoholics. The deficiency arises from a low intake of the vitamin as well as from alcohol-induced impairments in the metabolism of the vitamin. An alcoholic deriving 80% of his or her energy requirement from whiskey might be expected to be consuming only 20% of the RDA for vitamin B$_6$, as well as for other nutrients such as protein, folate, and thiamin. The symptoms of B$_6$ deficiency are not specific for this nutrient. They include depression, confusion, and sometimes convulsions.

During the 1950s, some infants fed canned formulas suffered from convulsions. This effect was traced to the very low levels of B$_6$ in the formulas. Infants receiving under 0.1 mg of B$_6$ per day are at risk for developing seizures. It is believed that

lum (ER). Alcohol dehydrogenase is a zinc metalloenzyme containing two subunits; each subunit has a molecular weight of 40,000 and contains two zinc atoms. One zinc atom, bound to the enzyme via residues of histidine and cysteine, is required for catalytic activity. The function of the second zinc atom, bound to residues of cysteine only, is structural.

Alcohol dehydrogenase, a cytosolic enzyme, catalyzes the oxidation of ethanol to acetaldehyde with the generation of NADH. The energy from this NADH is transferred to mitochondria, primarily by means of the malate–aspartate shuttle. When an alcoholic consumes over 50% of his or her energy in the form of alcohol, this shuttle becomes vastly important. Most of the acetaldehyde is oxidized in the mitochondria by mitochondrial aldehyde dehydrogenase, though a cytosolic version of the enzyme also exists. These reactions are shown in Figure 4.69.

An interesting sidelight, Asians are often deficient in aldehyde dehydrogenase. About 50% of Japanese and Chinese people lack the mitochondrial enzyme. About 40% of native South Americans also lack the enzyme. This results in a number of adverse reactions to drinking alcohol, including facial flushing from dilation of peripheral blood vessels and increased heart rate. Apparently, these reactions are provoked by an unusual buildup of plasma acetaldehyde.

## Family of Alcohol Dehydrogenases

Alcohol dehydrogenase contains two atoms of zinc. One atom participates in the catalytic mechanism, while the other serves a structural role. The enzyme actually is a dimer, and is composed of two subunits. Several types of alcohol dehydrogenase occur in the body. The enzymes have closely related amino acid sequences, and thus consist of a family of related enzymes.

Class I alcohol dehydrogenase catalyzes the metabolism of most dietary ethanol. This enzyme occurs in the liver. The class IV enzyme occurs in the mucosa of the stomach, and thus may begin catabolism of dietary ethanol even before the alcohol is absorbed by the gut. However, in comparing the weight of the human stomach mucosa (50 g), the weight of the human liver (1500 g), and the levels of enzyme activity in these two organs, it appears that the liver has 300 times the capacity for ethanol oxidation than does the stomach (Yin *et al.*, 1996). Class III alcohol dehydrogenase occurs in all tissues and tends not to oxidize ethanol, but instead is used in vitamin A metabolism. Specifically, this enzyme catalyzes the conversion of retinol (an alcohol) to retinaldehyde. The class IV enzyme requires glutathione as a co-substrate, and catalyzes the oxidation of formaldehyde to formic acid. Please note that formaldehyde is not an alcohol, but an aldehyde. This activity serves the useful purpose of eliminating a naturally occurring toxic com-

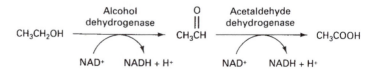

**FIGURE 4.69** Formation and oxidation of acetaldehyde.

pound (formaldehyde) from the body. The enzyme is also called **formaldehyde dehydrogenase** (Uotila and Koivusalo, 1996).

## Alcohol Catabolism in the Endoplasmic Reticulum

A second pathway of alcohol oxidation becomes important with consumption of large amounts of alcohol or with chronic alcoholism. This pathway, called the **ethanol-oxidizing system of the ER**, involves the heme enzyme cytochrome P450 and one other protein. Cytochrome P450 alone cannot catalyze the oxidation. As homogenization of the liver results in disruption of the ER to form tiny vesicles called microsomes, the system has also been called the **microsomal ethanol-oxidizing system**.

At moderate concentrations of alcohol, oxidation is mostly catalyzed by alcohol dehydrogenase. At high alcohol concentrations, this enzyme becomes saturated and cannot function at a greater rate. The ethanol-oxidizing system becomes increasingly important at these high concentrations, because it is not saturated by moderate concentrations of alcohol and is induced by chronic alcohol consumption. Electron microscopy shows that this induction involves an increase in the amount of ER present in the cell. Induction of the ethanol-oxidizing system in the regular drinker can result in more than a 50% increase in the rate of ethanol metabolism, which may account, in part, for the greater tolerance of alcohol by regular drinkers. Of greater importance is the adaption of the nervous system to alcohol, which allows the chronic alcoholic to remain sober despite blood alcohol levels (220 mM) that would ordinarily prove fatal.

The conversion of ethanol to acetic acid, by sequential action of the ethanol-oxidizing system of the ER and acetaldehyde dehydrogenase, is shown in Figure 4.70. The conversion of acetic acid to acetyl-CoA, as catalyzed by thiokinase, is shown in Figure 4.71. Thiokinase accepts short-chain fatty acids, as well as acetic acid, as its substrate. Acetyl-CoA is a substrate of the Krebs cycle.

## Human Subject Study with Alcohol Feeding

The experiment reported in Figure 4.72 illustrates production of acetaldehyde and acetate from ethanol. Human subjects received a dose of ethanol via a stomach tube (0.15 g of ethanol/kg of body weight). Blood samples indicated that plasma ethanol reached a peak concentration of 1.5 mM after a half-hour. Acetate reached

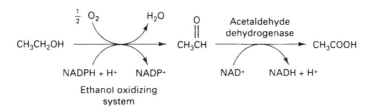

**FIGURE 4.70**  Reaction catalyzed by the ethanol-oxidizing system.

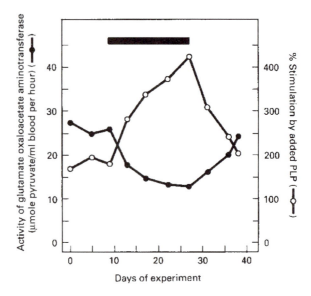

**FIGURE 9.39** Results of an apoenzyme stimulation test. (Reprinted with permission from Linkswiler, 1967.)

ment for vitamin $B_6$. The subject consumed a $B_6$-sufficient diet (4 mg/day) for the first 10 days, followed by a vitamin $B_6$-free diet for 15 days, as indicated by the black bar in Figure 9.39. Finally, a diet supplying 5 mg of vitamin per day was consumed. Blood samples were taken at the indicated times and used for the assay of basal aminotransaminase activity (no added PLP) (●) and for determining the percentage stimulation over basal activity occurring after addition of PLP to the enzyme assay mixtures (○).

The basal activity of aminotransferase fell by 50% during consumption of the $B_6$-deficient diet. The stimulation occurring with addition of PLP to the enzyme incubation mixtures rose in this period from 200% (twofold stimulation) to 400% (fourfold). The basal activity of the aminotransferase, and probably of all enzymes of the body, varies from subject to subject. It may even vary with repeated enzyme assays using the same sample of red blood cells. Thus, the basal activity is not used to assess vitamin $B_6$ status. The percentage stimulation is relatively constant in normal subjects and is thus a more useful indicator of $B_6$ status. It should be noted that the *storage of red blood cells* can lead to gradual release of the cofactor from the enzyme; thus, an artefactual diagnosis of $B_6$ deficiency is possible.

### Pyridoxic Acid Excretion Test

Most of the vitamin in the body is eventually degraded to pyridoxic acid (PX) and excreted in the urine. Vitamin $B_6$ deficiency can result in a decrease in the amount of PX excreted, as illustrated by the data in Table 9.4. Human subjects who had consumed a $B_6$-sufficient diet were fed a $B_6$-deficient diet for 45 days. The results

**TABLE 9.4**  Vitamin B$_6$ Metabolite and Urine with B$_6$ Deficiency

| Days on B$_6$-deficient diet | Urinary pyridoxic acid (mg/day) |
|:---:|:---:|
| 0 | 1.0 |
| 5 | 0.20 |
| 10 | 0.14 |
| 15 | 0.10 |
| 20 | 0.10 |
| 25 | 0.05 |
| 45 | 0.05 |

*Source*: Linkswiler (1967).

demonstrate an initial rapid fall in urinary PX followed by more gradual declines in PX production.

## Tryptophan Load Test

The tryptophan catabolic pathway, in simplified form, is shown in Figure 9.40. The PLP-dependent enzymes in this multibranched pathway are indicated. The tryptophan load test involves the consumption of a challenge dose of the amino acid (2–5 g) and the measurement of tryptophan catabolites in the urine for 1 day. The levels of excreted kynurenine, kynurenic acid, 3-OH-kynurenine, and xanthurenic acid may change little with the challenge dose in persons with normal vitamin B$_6$ status; however, the dose can induce dramatic increases in the excretion of these compounds under B$_6$-deficient conditions. The increases found during an experimentally induced deficiency involving human subjects are recorded in Figure 9.40.

EXERCISE

If the production of xanthurenic acid requires a B$_6$-dependent enzyme, then why may the urinary levels of this compound *increase* with B$_6$ deficiency?

At one time it was thought that women taking oral contraceptives were at risk for B$_6$ deficiency. This notion seems to have been in error. The error was due to a misinterpretation of the tryptophan load test. As mentioned earlier, a deficiency in vitamin B$_6$ can induce the accumulation of specific intermediates of the tryptophan catabolic pathway and enhanced excretion in the urine. Oral contraceptives can also induce an increase in the formation and excretion of specific intermediates by stimulating the activity of specific enzymes of the tryptophan catabolic pathway. This stimulation was responsible for the false indications of B$_6$ deficiency. Independently of the tryptophan load test, there continues to be some evidence for risk associated with the use of oral contraceptives. Oral contraceptive use may result in lowered levels of plasma vitamin B$_6$. These lowered levels may result in a vitamin B$_6$ deficiency when coupled with pregnancy and lactation.

## REFERENCES

Aguilar-Parada, E., Eisentraut, A. M., and Unger, R. H. (1969). Effects of starvation on plasma pancreatic glucagon in normal man. *Diabetes* **18**, 717–723.

Ahlborg, G., and Felig, P. (1982). Lactate and glucose exchange across the forearm, legs, and splanchnic bed during and after prolonged leg exercise. *J. Clin. Invest.* **69**, 45–54.

Ahlborg, G., Felig, P., Hagenfeldt, S., Hendler, R., and Wahren, J. (1974). Substrate turnover during prolonged exercise in man. *J. Clin. Invest.* **53**, 1080–1090.

Ajayi, O. A., and James, O. A. (1984). Effect of riboflavin supplement on riboflavin nutriture of a secondary school population in Nigeria. *Am. J. Clin. Nutr.* **39**, 787–791.

Bach, J.-F. (1994). Insulin-dependent diabetes mellitus as an autoimmune disease. *Endocrine Rev.* **15**, 516–542.

Bach, J.-F. (1997). Autoimmunity and type I diabetes. *Trends Endocrinol. Metab.* **8**, 71–74.

Bai, D. H., Pape, M. E., Lopez-Casillas, F., Luo, X. C., Dixon, J. E., and Kim, K.-H. (1986). Molecular cloning of cDNA for acetyl-coenzyme A carboxylase. *J. Biol. Chem.* **261**, 12395–12399.

Barford, D., Hu, S., and Johnson, L. N. (1991). Structural mechanism for glycogen phosphorylase control by phosphorylation and AMP. *J. Mol. Biol.* **218**, 233–260.

Bass, A., Brdiczka, D., Eyer, P., Hofer, S., and Pette, D. (1969). Metabolic differentiation of distinct muscle types at the level of enzymatic organization. *Eur. J. Biochem.* **10**, 198–206.

Beckman, J. S., and Koppenol, W. H. (1996). Nitric oxide, superoxide, and peroxynitrite: The good, the bad, and the ugly. *Am. J. Physiol.* **271**, C1424–C1437.

Bernlohr, D., Simpson, M., Hertzel, A., and Banaszak, L. J. (1997). Intracellular lipid-binding proteins and their genes. *Annu. Rev. Nutr.* **17**, 277–303.

Bergman, E. N. (1990). Energy contributions of volatile fatty adds from the gastrointestinal tract in various species. *Physiol. Rev.* **70**, 567–590.

Blom, P. C., Hostmark, A. T., Vaage, O., Kardel, K. R., and Maehlum, S. (1987). Effect of different post-exercise sugar diets on the rate of muscle glycogen synthesis. *Med. Sci. Sports Exercise* **19**, 491–496.

Bolli, G., DeFeo, P., Perriello, G., Cosmo, S. D., Compagnucci, P., Santeusanio, F., Brunetti, P., and Unger, R. H. (1984). Mechanisms of glucagon secretion during insulin-induced hypoglycemia in man. *J. Clin. Invest.* **73**, 917–922.

Cahill, G. F. (1970). Starvation in man. *N. Engl. J. Med.* **282**, 668–675.

Carruthers, C., Unson, C., Kim, H., and Sakmar, T. P. (1994). Synthesis and expression of a gene for the rat glucagon receptor. *J. Biol. Chem.* **269**, 29321–29328.

Casteno, J. G., Nieto, A., and Feliu, J. E. (1979). Inactivation of phosphofructokinase by glucagon. *J. Biol. Chem.* **254**, 5576–5579.

Castillo, L., Beaumier, L., Ajami, A. M., and Young, V. R. (1996). Whole body nitric oxide synthesis in healthy men determined from [$^{15}$N]arginine-to-[$^{15}$N]citrulline labeling. *Proc. Natl. Acad. Sci. U.S.A.* **93**, 11460–11465.

Cook, G. (1984). Differences in the sensitivity of carnitine palmitoyltransferase to inhibition by malonyl-CoA are due to differences in $K_i$ values. *J. Biol. Chem.* **259**, 12030–12033.

Cooper, D. M., Mons, N., and Karpen, J. (1995). Adenylyl cyclases and the interaction between calcium and cAMP signalling. *Nature* **374**, 421–424.

Cross, R. L. (1981). The mechanism and regulation of ATP synthesis by F1-ATPases. *Annu. Rev. Biochem.* **50**, 681–714.

Diamond, I., and Gordon, A. S. (1997). Cellular and molecular neuroscience of alcoholism. *Physiol. Rev.* **77**, 1–20.

DiPadova, C., Worner, T. M., Julkunen, R. J., and Lieber, C. S. (1987). Effects of fasting and chronic alcohol consumption on the first-pass metabolism of ethanol. *Gastroenterology* **92**, 1169–1173.

Dohm, G. L., Kasperek, G. J., and Barakat, H. A. (1985). Time course of changes in gluconeogenic enzyme activities during exercise and recovery. *Am. J. Physiol.* **249**, E6–E11.

Efrat, S., Tal, M., and Lodish, H. F. (1994). The pancreatic β-cell glucose sensor. *Trends Biol. Sci.* **19**, 535–538.

Ekdahl, E. N., and Ekman, P. (1984). The effect of fructose-2,6-bisphosphate and AMP on the activity of phosphorylated and unphosphorylated fructose-1,6-bisphosphatase from rat liver. *FEBS Lett.* **167**, 203–209.

Elmendorf, J. S., Damrau-Abney, A., Smith, T., David, T., and Turinsky, J. (1997). Phosphatidylinositol 3-kinase and dynamics of insulin resistance in denervated slow and fast muscles *in vivo. Am. J. Physiol.* **272**, E661–E670.

Felig, P., and Wahren, J. (1975). Fuel homeostasis in exercise. *N. Engl. J. Med.* **293**, 1078–1084.

Ferguson, S. J. (1986). The ups and downs of P/O ratios. *Trends Biochem. Sci.* **11**, 351–353.

Foley, J. E., Kashiwagi, A., Verso, M. A., Beaven, G., and Andrews, J. (1983). Improvement in *in vitro* insulin action after one month of insulin therapy in obese noninsulin-dependent diabetes. *J. Clin. Invest.* **72**, 1901–1909.

Forbes, G. B. (1987). "Human Body Composition," p. 171. Springer-Verlag, New York.

Fritz, I. B. (1959). Action of carnitine on long chain fatty acid oxidation by liver. *Am. J. Physiol.* **197**, 297–304.

Franz, M. J., Horton, E. S., Bantle, J. P., Beebe, C., Brunzell, J., Coulston, A., Henry, R. R., Hoogwerf, B., and Stacpoole, P. W. (1994). Nutrition principles for the management of diabetes and related complications. *Diabetes Care* **17**, 490–522.

Gerbitz, K., Gempel, K., and Brdiczka, D. (1996). Mitochondria and diabetes. *Diabetes* **45**, 113–126.

Gollnick, P. D., and Hodgson, D. R. (1986). The identification of fiber types in skeletal muscle: A continual dilemma. *Exercise Sports Sci. Rev.* **14**, 81–104.

Gopher, A., Vaisman, N., Mandel, H., and Lapidot, A. (1990). Determination of fructose metabolism pathways in normal and fructose-intolerant children. *Proc. Natl. Acad. Sci. U.S.A.* **87**, 5449–5453.

Guyton, A. C. (1971). "Textbook of Medical Physiology," 4th ed., p. 915. Saunders, Philadelphia.

Halliwell, B. (1988). Albumin — an important extracellular antioxidant? *Biochem. Pharmacol.* **37**, 569–57.

Harris, R. C., Edwards, R. H., Hultman, E., Nordesjo, L., Nylind, B., and Sahlin, K. (1976). The time course of phosphorylcreatine resynthesis during recovery of the quadriceps muscle in man. *Pfluegers Arch.* 367, 137-142.

Hatefi, Y. (1985). The mitochondrial electron transport and oxidative phosphorylation system. *Annu. Rev. Biochem.* **54**, 1015–1069.

Haussinger, D. (1986). Regulation of hepatic ammonia metabolism: The intercellular glutamine cycle. *In* "Advances in Enzyme Regulation" (G. Weber, ed.), Vol. 25, pp. 159–180. Pergamon, Oxford.

Hickner, R. C., Fisher, J. C., Ehsani, A., and Kohrt, W. (1997). Role of nitric oxide in skeletal muscle blood flow at rest and during exercise in humans. *Am. J. Physiol.* **273**, H405–H410.

Holloszy, J., and Kohrt, W. (1996). Regulation of carbohydrate and fat metabolism during and after exercise. *Annu. Rev. Nutr.* **16**, 121–138.

Holmes, F. L. (1993). "Hans Krebs: Architect of Intermediary Metabolism, 1933–1937," Vol. 2. Oxford Univ. Press, New York.

Jensen, M. D., Haymond, M. W., Gerich, J. E., Cryer, P. E., and Miles, J. M. (1987). Lipolysis during fasting. *J. Clin. Invest.* **79**, 207–213.

Jensen, M. D., Haymond, M. W., Rizza, R. A., Cryer, P. E., and Miles, J. M. (1989). Influence of body fat distribution on free fatty acid metabolism in obesity. *J. Clin. Invest.* **83**, 1168–1173.

Jungermann, K., and Kietzmann, T. (1996). Zonation of parenchymal and nonparenchymal metabolism in liver. *Annu. Rev. Nutr.* **16**, 179–203.

elevated levels of homocysteine (Boushey *et al.*, 1995; Molloy *et al.*, 1997). This particular mutation occurs in 5–15% of the population.

The striking connection between premature cardiovascular disease and *high levels* of homocysteine, in homocysteinuria, prompted a look at possible relationships between cardiovascular disease and *moderate levels* of homocysteine. This close look has included genetic analysis, the biochemical analysis of plasma metabolite levels, and epidemiological studies. Clearly, exploration by researchers into the effects of moderate elevations of homocysteine has been a multidisciplinary effort. The studies have taken these forms:

1.  Studies of plasma homocysteine and the thickness of atherosclerotic lesions or of the diameter of the coronary artery or the carotid artery.

2.  Studies of plasma homocysteine and the rate of spontaneous blood clot formation.

3.  Studies of plasma homocysteine (in living patients with severe cardiovascular disease) and the number of subsequent years between medical examination and death.

4.  Studies of plasma homocysteine and supplements with vitamins that are relevant to enzymes of the pathway of homocysteine breakdown. These enzymes are cystathionine β-synthase (vitamin $B_6$), methionine synthase (vitamin $B_{12}$), and 5,10-methylene-$H_4$folate reductase (folic acid). The associated vitamins are listed.

5.  Surveys designed to test the possible connection between a history of eating foods rich (or poor) in folate, rich (or poor) in vitamin $B_6$, and the eventual diagnosis of cardiovascular disease.

These studies are detailed in what follows. Although the **severe** increases in plasma homocysteine that occur with homozygous cystathionine β-synthase deficiency can be clearly treated with vitamin $B_6$, the **moderate** increases of homocysteine (which are widespread in the population and do not lead to mental retardation) can be treated best with folate.

A study of 1401 subjects involved measuring plasma homocysteine and measuring the lesions in the carotid artery, one of the arteries in the body that tends to acquire atherosclerotic lesions. The thickness of the lesions were measured by **ultrasonography**. Plasma folate, vitamin $B_6$, and vitamin $B_{12}$ were also measured. A correlation was found between narrowing of the artery and homocysteine levels at above 15 $\mu M$ homocysteine. A correlation with narrowing of the artery was also found with folate levels below 2.5 ng/ml, and with vitamin $B_6$ levels below 30 nM (Selhub *et al.*, 1995). A study of 231 normal subjects and 304 patients with atherosclerosis (coronary artery under 30% normal diameter) revealed a striking correlation between plasma homocysteine of 12 $\mu M$, or greater, and atherosclerosis (Robinson *et al.*, 1995). These studies help define an upper limit of acceptable plasma homocysteine levels.

Spontaneous blood clot formation was dramatically correlated with plasma homocysteine levels of over 22 $\mu M$, as revealed by a study of 269 normal subjects and 269 patients with episodes of spontaneous clots (Heijer *et al.*, 1996). The spontaneous clots were detected by ultrasonography, and other techniques. This study helps define the upper limit of acceptable homocysteine.

A study of 587 patients with atherosclerosis, and with a diameter of the coronary artery that was half the normal value, monitored the time from initial diagnosis until death. The disease revealed that premature death was correlated with increased homocysteine, where the correlation was especially strong with plasma homocysteine levels of over 15 $\mu M$ (Nygard *et al.*, 1997). Again, this study helps define acceptable and undesirable plasma homocysteine concentrations.

An epidemiological study supports the connection between low intakes of folate and of vitamin B$_6$ with cardiovascular disease. The study involved 80,082 women who were followed for 14 years. Vitamin intake was estimated by questionnaires, while the eventual acquisition of disease was determined by physicians who reviewed medical records. Although no biochemical tests were performed, it should be noted that this and other epidemiological studies acquire their power by their large number of subjects, as well as by small pilot studies that validate the accuracy of the questionnaires (Rimm *et al.*, 1998).

A study of 285 human subjects revealed that treatment with three vitamins together (folate, vitamin B$_6$, and vitamin B$_{12}$) can provoke a decline of plasma homocysteine from an initial level of about 12 $\mu M$ to the lower level of 8 $\mu M$ (Naurath *et al.*, 1995). A study of 100 men with moderate levels of plasma homocysteine (18–40 $\mu M$) involved separate supplements of placebo, folic acid, vitamin B$_{12}$, or vitamin B$_6$. Folic acid alone resulted in a 40% decline in plasma homocysteine. Vitamin B$_{12}$ alone provoked a 15% decrease in the amino acid, while vitamin B$_6$ alone produced no consistent change (Ubbink *et al.*, 1994).

Other studies have involved measuring plasma homocysteine levels after a challenge dose of methionine. It was found that vitamin B$_6$ can reduce plasma homocysteine, where testing involved the methionine dose. It was found that vitamin B$_6$ tends not to influence the amino acid with fasting human subjects who did not receive any methionine dose (Boushey *et al.*, 1995). The methionine challenge test involves an oral dose of 100 mg methionine/kg body weight, with blood samples taken at intervals over the course of 8 hours. Measurements of plasma homocysteine, with the methionine dose, may be a more sensitive test for metabolic defects than measurements of plasma homocysteine in the fasting subject (Mudd, 1985; Clarke *et al.*, 1991).

Vitamin B$_{12}$ deficiency is a clearly defined condition that tends to be somewhat common in the older population, due to gastric atrophy and pernicious anemia. Moderate increases in plasma homocysteine can occur with B$_{12}$ deficiency, and these levels can be cut in half by supplements of this vitamin (Brattstrom *et al.*, 1988; Sumner *et al.*, 1996). Researchers interested in the mechanisms by which plasma homocysteine can influence metabolism need to consider the possible effects of other sulfhydryl compounds (R—SH) that occur in plasma. These sulfhydryl compounds include cysteine (10–100 $\mu M$) (Hiramatsu *et al.*, 1994; Fukagawa, 1996), glutathione (2 $\mu M$ in humans; 20 $\mu M$ in rats) (Halliwell and Gutteridge, 1990), and the free sulfhydryl in albumin (640 $\mu M$) (DeMaster *et al.*, 1995). The micromolar levels of cysteine, glutathione, and the albumin in plasma are indicated.

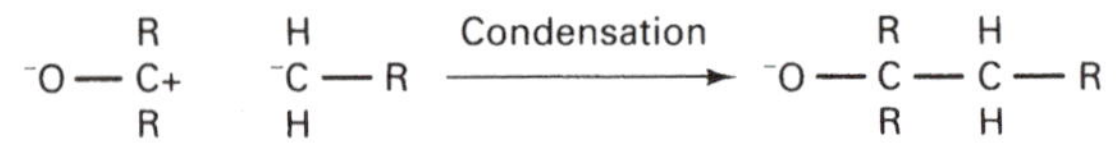

$$^-O - \overset{R}{\underset{R}{C}}+ \quad \overset{H}{\underset{H}{C}} - R \quad \xrightarrow{\text{Condensation}} \quad ^-O - \overset{R}{\underset{R}{C}} - \overset{H}{\underset{H}{C}} - R$$

**FIGURE 4.73** Condensation of positive and negative carbons to form a covalent bond. The carbanion does not normally occur in solution. It is formed at the active site of enzymes where it exists only momentarily.

Acetyl-CoA carboxylase and propionyl-CoA carboxylase catalyze the condensation of $CO_2$ with the 2-carbon of the thiol ester substrate. Citrate synthase catalyzes the condensation of OAA at the 2-carbon of the thiol ester substrate. These condensation reactions involve attack of an atom with a partial positive charge (the carbon of $CO_2$ or the keto group of OAA) on an atom with a negatively charged character (the 2-carbon of the substrate). Condensation of positively and negatively charged carbons to form a stable, covalent bond is shown in Figure 4.73.

The central role of thiol ester compounds in nutritional biochemistry is illustrated by considering some features of the mechanisms of the reactions catalyzed by citrate synthase and acetyl-CoA carboxylase. A brief background on the properties of the thiol ester bond is first presented. This material has been simplified and is intended only to clarify why the reactions employ thiol esters of carboxylic acids, rather than oxygen esters or the free, unesterified carboxylic acid.

Carboxylic acids resonate in the manner shown in Figure 4.74, allowing the negative charge to be shared by both oxygen atoms. Oxygen esters (Figure 4.75) tend to resonate in a similar manner. The oxygen atom of the ester linkage contains a pair of electrons not involved in covalent bond formation. These electrons can participate in the resonance shown, though the degree of resonance is much less than that of a free carboxylic acid. Thiol esters (Figure 4.76), however, do not tend to resonate, as indicated by the large X. Although the sulfur atom has an electronic structure similar to that of the oxygen atom, its atomic weight is twice that of oxygen. It is thought that sulfur's larger size prevents it from resonating with the keto group. This lack of resonance is useful, because it permits the keto group of the thiol ester to participate in electron shifts involving the R group of the molecule.

The reactions catalyzed by citrate synthase and acetyl-CoA carboxylase are thought to begin with dissociation of a proton from the 2-carbon of the acetyl moiety of acetyl-CoA (Figure 4.77). This produces a carbanion (a carbon atom carrying a negative charge) centered on the 2-carbon. This carbanion is stabilized by the resonance with the keto group also shown in Figure 4.77. Stabilization of the carbanion facilitates removal of the proton. Occurrence of the acetyl group as a free acid or oxygen ester would be expected to prevent resonance and thus deter

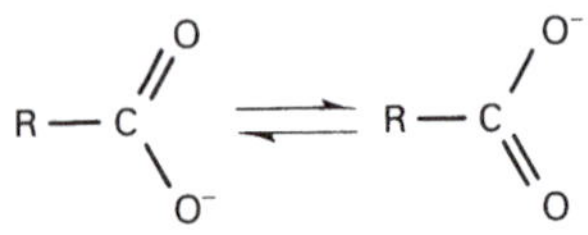

$$R - C \overset{\displaystyle O}{\underset{\displaystyle O^-}{\Big\backslash}} \quad \rightleftharpoons \quad R - C \overset{\displaystyle O^-}{\underset{\displaystyle O}{\Big\backslash}}$$

**FIGURE 4.74** Resonance of a carboxylic acid.

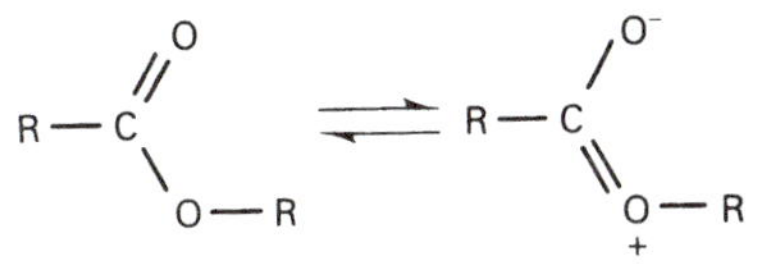

FIGURE 4.75  Resonance of an oxygen ester.

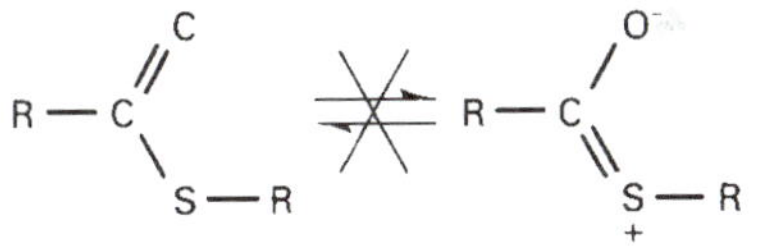

FIGURE 4.76  No resonance in a thiol ester.

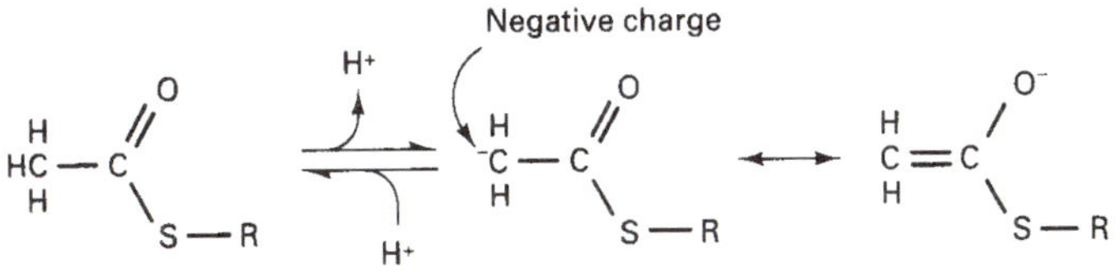

FIGURE 4.77  Development of a negative charge at the α-carbon of a thiol ester.

proton removal. Resonance involving the 2-carbon of a keto group is called **eno-lization**.

Stabilization of the 2-carbanion by enolization is thought to be a critical step in the reactions catalyzed by citrate synthase and acetyl-CoA carboxylase. With the former enzyme, the carbanion attacks an incoming molecule of OAA. As shown in Figure 4.78, condensation of acetyl-CoA with OAA forms citryl-CoA. Hydrolysis of citryl-CoA then produces citric acid and acetyl-CoA. In the case of acetyl-CoA carboxylase, the carbanion attacks an incoming molecule of $CO_2$, resulting in formation of malonyl-CoA (Figure 4.79).

## SUMMARY

The glucagon/insulin ratio was used to introduce the chapter and to provide a theme or reference point for the more complex matters to come. This ratio is of central importance to those interested in energy regulation during working, fasting, feeding, and diabetes. Simple diagrams were used to introduce the major types of energy metabolism, and these were followed by diagrams of biochemical pathways and some of the loci where regulation occurs.

Glucagon and insulin are secreted by endocrine cells of the pancreas into the bloodstream in response to fluctuations in plasma glucose. These hormones, in turn, act at various target organs to influence energy metabolism. Glucose metabolism is regulated at the points of glucose transport and at various points in the pathways of glycolysis and gluconeogenesis, including glycogen synthesis and breakdown. Triglyceride (TG) metabolism is regulated at the points of TG hydroly-

cleavage can also occur asymmetrically (not at the exact center), resulting in a version of retinal (8′-apocarotenal) bearing an extra long isoprenoid chain plus a useless byproduct. The apocarotenal is then enzymatically trimmed to produce a single molecule of retinal (Furr and Clark, 1997; Nagao *et al.*, 1996; Duszka *et al.*, 1996).

The cleavage of β-carotene to form retinal, followed by the reduction of retinal to retinol, is shown in Figure 9.41. The retinol is converted to the retinyl ester, packaged in chylomicrons, and exported in the lymphatic system.

## Absorption and Storage of Vitamin A

Retinyl esters, as well as the carotenoids not converted to vitamin A in the gut, first appear in the chylomicrons. Following conversion to the chylomicron remnants, the material is taken up by the liver. The β-carotene tends to be exported to other tissues, where it accumulates mainly in fat deposits. β-Carotene and other carotenoids lend an orange-yellow color to the fat. Animals that do not readily absorb carotenoids, such as rodents, rabbits, and pigs, are known as "white fat" animals. What happens to the retinyl esters? The retinyl esters are released and rapidly hydrolyzed to retinol. The retinol binds to a protein, retinol-binding protein (RBP), and is transported to the stellate cells. **Stellate cells** occur in the liver, in the space between the capillaries and the hepatocytes. This space is called the **space of Disse**. Stellate cells are also called Ito cells. They store fat. The vitamin A delivered to the stellate cells is converted back to retinyl esters for storage. Fat droplets in these cells may contain about 40% retinyl esters, 13% cholesterol, 28% triglycerides, and 4% phospholipids. The following studies illustrate the role of the liver, and of stellate cells in particular, in storing vitamin A. The studies involved rats.

Rats were fed a control diet (2.4 mg retinol/kg diet) or a deficient diet (0.6 mg retinol/kg diet). Following the feeding trial, the stellate cells of the liver were analyzed for their content of vitamin A and other components, as shown in Table 9.5. The results demonstrate that the low-vitamin A diet resulted in a decrease in vitamin A content with little effect on the contents of the other lipids.

Figure 9.42 illustrates a series of events that may occur over the course of prolonged deficiency. The study involved measurement of the vitamin A content

**TABLE 9.5**  Vitamin A Stellate Cells

|  | $\mu g/10^6$ Cells | |
| --- | --- | --- |
|  | Control diet | Deficient diet |
| Vitamin A | 33 | 6.6 |
| Triglycerides | 28 | 25 |
| Cholesterol esters | 17 | 15 |
| Cholesterol | 15 | 15 |

*Source*: Moriwaki *et al.* (1988).

FIGURE 9.42  A series of events that may occur over the course of a prolonged vitamin A deficiency. (Redrawn with permission from Diplock, 1985.)

of the liver (●), the vitamin A concentration in the blood (——), and the content of rhodopsin in the eye (○). Rhodopsin is a vitamin A-containing protein of the eye. It is required for mediating vision in dim light. Note that the defect most closely associated with a mild to moderate deficiency in vitamin A is **nightblindness**. Rats were fed a deficient diet for 8 weeks, and the aforementioned parameters were measured at the indicated intervals. The results demonstrate that blood levels of vitamin A and rhodopsin levels were normal until the 4th week. After 4 weeks, there occurred a drastic decrease in plasma vitamin levels and a more gradual drop in rhodopsin levels. These results indicate that the initial maintenance of plasma vitamin A and rhodopsin was due to mobilization of the vitamin stored in the liver. One definition of a "storage form" of a nutrient implies that it can be selectively mobilized and used in the presence of a dietary deficiency of the nutrient.

## Biochemistry of Vitamin A

### Retinol-Binding Protein

Vitamin A is stored in the stellate cells of the liver in the form of retinyl esters. With mobilization of the vitamin, the retinyl ester is converted to retinol and released into the bloodstream as a complex with RBP. The complex contains one molecule of retinol and one molecule of protein. RBP is not released into the circulation unless it contains a molecule of retinol. Various cells of the body contain RBP receptors. The RBP receptor is a membrane-bound protein. It is required for the uptake of the vitamin. Apparently, RBP does not enter the cell when it discharges retinol. The following experiment, which involved the epithelial cells of bovine eyes, illustrates the role of RBP. Vitamin A is very important for the health of many epithelial cells, including those of the eyes. The cells were maintained in culture in a petri dish. Retinol was added to the cells in the form of a retinol–RBP complex

Wolfe, R. R. (1987). Does exercise stimulate protein breakdown in humans? *Med. Sci. Sports Exercise* **19**, S172–S178.

## Arteriovenous Differences

Ahlborg, G., and Felig, P. (1982). Lactate and glucose exchange across the forearm, legs, and splanchnic bed during and after prolonged leg exercise. *J. Clin. Invest.* **69**, 45–54.

Ahlborg, G., Felig, P., Hagenfeldt, L., Hendler, R., and Wahren, J. (1974). Substrate turnover during prolonged exercise in man. *J. Clin. Invest.* **53**, 1080–1090.

Hagenfeldt, L., and Wahren, J. (1971). Human forearm muscle metabolism during exercise. *Scand. J. Clin. Lab. Invest.* **27**, 299–306.

Nestel, P. J., Ishikawa, T., and Goldrick, R. B. (1978). Diminished plasma free fatty acid clearance in obese subjects. *Metabolism* **27**, 589–596.

Owen, O. E., Morgan, A. P., Kemp, H. G., Sullivan, J. M., Herrara, M. G., and Cahill, G. F. (1967). Brain metabolism during fasting. *J. Clin. Invest.* **46**, 1589–1595.

## Arginine, Nitric Oxide, and Vasodilation

Beckman, J. S., and Koppenol, W. H. (1996). Nitric oxide, superoxide, and peroxynitrite: The good, the bad, and the ugly. *Am. J. Physiol.* **271**, C1424–C1437.

Clarkson, P., Adams, M. R., Powe, A., Donald, A., McCredle, R., Robinson, J., McCarthy, S., Keech, A., Celermajer, D., and Deanfield, J. (1996). Oral L-arginine improves endothelium-dependent dilation in hypercholesterolemic young adults. *J. Clin. Invest.* **97**, 1989–1994.

Drewett, J. G., and Garbers, D. L. (1994). The family of guanylyl cyclase receptors and their ligands. *Endocrine Rev.* **15**, 135–162.

Hickner, R., Fisher, J., Ehsani, A., and Kohrt, W. (1997). Role of nitric oxide in skeletal muscle blood flow at rest and during dynamic exercise in humans. *Am. J. Physiol.* **273**, H405–H410.

Huang, P. L., Huang, Z., Mashimo, H., Bloch, K., Moskowitz, M., Bevan, J., and Fishman, M. (1995). Hypertension in mice lacking the gene for endothelial nitric oxide synthase. *Nature* **377**, 239–242.

Komalavilas, P., and Lincoln, T. M. (1996). Phosphorylation of the inositol 1,2,5-trisphosphate receptor. *J. Biol. Chem.* **271**, 21933–21938.

Smith, J. A., Francis, S., Walsh, K., Kumar, S., and Corbin, J. (1996). Autophosphorylation of type Ib cGMP-dependent protein kinase increases basal catalytic activity and enhances allosteric activation by cGMP or cAMP. *J. Biol. Chem.* **271**, 20756–20762.

Wickelgren, I. (1997). Biologists catch their first detailed look at NO enzyme. *Science* **278**, 389.

## Creatine Phosphate

Bessman, S. P., and Geiger, P. J. (1981). Transport of energy in muscle: The phosphorylcreatine shuttle. *Science* **211**, 448–452.

Lente, F. V., and Suit, P. (1989). Assessment of renal function by serum creatinine and creatinine clearance: Glomerular filtration rate estimated by four procedures. *Clin. Chem.* **35**, 2326–2330.

Levey, A. S., Perrone, R. D., and Madias, N. E. (1988). Serum creatinine and renal function. *Annu. Rev. Med.* **39**, 465–490.

Sahlin, K., Harris, R. C., and Hultman, E. (1975). Creatine kinase equilibrium and lactate content compared with muscle pH in tissue samples obtained after isometric exercise. *Biochem. J.* **152**, 173–180.

Walser, M. (1987). Creatinine excretion as a measure of protein nutrition in adults of varying age. *J. Parenteral Enteral Nutr.* **11**, 73S–78S.

*Glucose Transport*

Cartee, G. D., Young, D. A., Sleeper, M. D., Zierath, J., Wallberg-Hendriksson, H., and Holloszy, J. O. (1989). Prolonged increase in insulin-stimulated glucose transport in muscle after exercise. *Am. J. Physiol.* **256**, E494–E499.

Chasiotis, D., Edstrom, L., Sahlin, K., and Sjoholm, H. (1986). Activation of glycogen phosphorylase by electrical stimulation of isolated fast-twitch and slow-twitch muscles from rats. *Acta Physiol. Scand.* **123**, 43–47.

Fushiki, T., Wells, J. A., Tapscott, E. B., and Dohm, G. L. (1989). Changes in glucose transporters in muscle in response to glucose. *Am. J. Physiol.* **256**, E580–E587.

Gardemann, A., Strulik, H., and Jungermann, K. (1987). Nervous control of glycogenolysis and blood flow in arterially and portally perfused liver. *Am. J. Physiol.* **253**, E238–E245.

Heller-Harrison, R., Morin, M., Guilherme, A., and Czech, M. P. (1996). Insulin-mediated targeting of phosphatidylinositol 3-kinase to GLUT4-containing vesicles. *J. Biol. Chem.* **271**, 10200–10204.

Hendriksson, J., Salmons, S., Chi, M., Hintz, C. S., and Lowry, O. H. (1988). Chronic stimulation of mammalian muscle: Changes in metabolite concentration in individual fibers. *Am. J. Physiol.* **255**, C543–C551.

Hughes, V. A., Fiatarone, M., Fielding, R., Kahn, B. B., Ferrara, C., Shepherd, P., Fisher, E. C., Wolfe, R. R., Elahi, D., and Evans, W. J. (1993). Exercise increases muscle GLUT-4 levels and insulin action in subjects with impaired glucose tolerance. *Am. J. Physiol.* **264**, E855–E862.

Wallberg-Hendriksson, H. (1986). Repeated exercise regulates glucose transport capacity in skeletal muscle. *Acta Physiol. Scand.* **127**, 39–43.

*Fructose*

Gelfand, R. A., and Sherwin, R. S. (1986). Nitrogen conservation in starvation revisited: Protein sparing with intravenous fructose. *Metabolism* **35**, 37–44.

Gopher, A., Vaisman, N., Mandel, H., and Lapidot, A. (1990). Determination of fructose metabolism pathways in normal and fructose-intolerant children. *Proc. Natl. Acad. Sci. U.S.A.* **87**, 5449–5453.

Hallfrisch, J. (1990). Metabolic effects of dietary fructose. *FASEB J.* **4**, 2652–2660.

Kneepkens, C. M. (1989). What happens to fructose in the gut? *Scand. J. Gastroenterol., Suppl.* **24**(171), 1–8.

Pozza, G., Galansino, G., Hoffeld, H., and Foa, P. P. (1958). Stimulation of insulin output by monosaccharides and monosaccharide derivatives. *Am. J. Physiol.* **192**, 497–500.

Tappy, L., Randin, J.-P., Felber, J.-P., Chiolero, R., Simonson, D. C., Jequier, E., and DeFronzo, R. A. (1986). Comparison of thermogenic effect of fructose and glucose in normal humans. *Am. J. Physiol.* **250**, E718–E724.

Truswell, A. S., Seach, J. M., Diet, D. N., and Thorburn, A. W. (1988). Incomplete absorption of pure fructose in healthy subjects and the facilitating effects of glucose. *Am. J. Clin. Nutr.* **48**, 1424–1430.

Youn, J. H., Kaslow, H. R., and Bergman, R. N. (1987). Fructose effect to suppress hepatic glycogen degradation. *J. Biol. Chem.* **262**, 11470–11477.

*Medium-Chain Fatty Acids*

Cook, G. A., and Gamble, M. S. (1987). Regulation of carnitine palmitoyltransferase insulin results in decreased activity and decreased apparent $K_i$ values for malonyl-CoA. *J. Biol. Chem.* **262**, 2050–2055.

Delafosse, B., Viale, J., Pachiaudi, C., Normand, S., Goudable, J., Bouffard, Y., Annat, G., and Bertrand, O. (1997). Long- and medium-chain triglycerides during parenteral nutrition in critically ill patients. *Am. J. Physiol.* **272**, E550–E555.

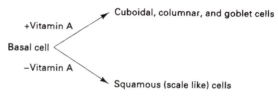

FIGURE 9.46 Differentiation of basal cells.

## Hormonal Effects of Vitamin A

After entering the cell, retinol and retinoic acid (RA) are bound by separate cytoplasmic proteins. These cytosolic proteins are called **cytoplasmic-retinol binding protein** (CRBP) and **cytoplasmic retinoic acid-binding protein** (CRABP). Two forms of CRBP exist, and two forms of CRABP exist. One form of CRBP occurs mainly in the villi of the small intestines, while the other form occurs in the liver and other organs. One form of CRABP occurs in the skin, while the other form occurs in many tissues.

CRBP is thought to assure that retinol is metabolized by specific enzymes, while preventing metabolism by other enzymes. For example, by binding retinol, CRBP prevents excessive rates of conversion of retinol to retinoic acid (Napoli, 1996).

Both forms of CRABP increase dramatically during fetal development, then decline again before birth, as revealed by studies of mice. Using standard genetic techniques, knock-out mice were created that lack either or both forms of CRABP. Some of the mice lacking both types of CRABP had extra fingers, while about 10% died prematurely (Li and Norris, 1996). The research community has expressed surprise that the defects in these knock-out mice were relatively mild, and not more severe. The functions of the preceding cytosolic proteins remain unclear. CRBP and CRABP resemble cytosolic fatty acid-binding proteins, as they consist of about 130 amino acids and have similar amino acid sequences.

The nucleus contains several retinoic-binding proteins that are totally different and unrelated to the cytosolic proteins. The nuclear proteins are used for genetic control, and here the vitamin asserts its hormonal activity. The complex of vitamin with its receptor protein binds to specific regions of chromatin and provokes changes in the rate of transcription of specific genes. The complex of hormone with receptor protein binds to specific regions of the chromatin and regulates gene expression. Several types of RA-binding proteins occur in the nucleus. These proteins bind either all-*trans*-RA or 9-*cis*-RA. The protein–RA complex does not act alone when it binds to our DNA, with the consequent activation of specific genes. Instead, the complex forms a dimer, where this dimer binds to regulatory regions of the chromosome.

The discovery of this dimer has provoked the interest of a number of disciplines. Why is this? Because the second protein in the dimer can bind vitamin D, thyroid hormone, prostaglandin, or a second molecule of RA, depending on the exact nature of the second protein. This means that, when vitamin A performs its hormonal role in the body, it behaves in much the same way as vitamin D, thyroid hormone, and certain prostaglandins. (Prostaglandins are special hormones that are made from fatty acids.) Further details appear in the section entitled Vitamin

A, Vitamin D, and Thyroid Hormone at the Genome, and in the Essential Fatty Acid section.

### 9-Cis-*Retinoic Acid and How It Is Acquired*

9-*Cis*-retinoic acid is a version of vitamin A that, along with all-*trans*-retinoic acid, is used as a hormone. 9-*Cis*-RA binds to a protein called RXR, while all-*trans*-RA binds to RAR. 9-*Cis*-RA can be formed by cleavage of dietary 9-*cis*-β-carotene, or from the enzymatic isomerization of all-*trans*-RA, at a single double bond. 9-*Cis*-β-carotene is typically present in the diet, but since it is not a required nutrient and is not a vitamin, it is apparent that isomerization of all-*trans*-RA may supply all the 9-*cis*-RA that is needed.

## Vitamin A and the Visual Cycle

Vision under low-light conditions requires the vitamin A-dependent visual cycle. The light-sensitive molecule of this cycle is 11-*cis*-retinal. 11-*Cis*-retinal is bound to a protein called **opsin**. The complex formed by 11-*cis*-retinal and opsin is called **rhodopsin**. Light energy induces conversion of the protein-bound 11-*cis*-retinal to all-*trans*-retinal. This event results in dissociation of the vitamin from the protein and in discharge of an electrical impulse to the brain. The remaining events of the visual cycle involve regeneration of the complex of 11-*cis*-retinal and opsin. The aldehyde group of retinal in rhodopsin occurs in a Schiff base with a lysine residue of opsin.

The visual cycle is shown in Figure 9.47. The regeneration of rhodopsin involves several steps. All-*trans*-retinal is reduced to the alcohol form, all-*trans*-retinol. Under bright light conditions, this retinol is converted to retinyl esters and stored until dim light prevails. All-*trans*-retinol is then isomerized to yield 11-*cis*-retinol. This isomerization involves conversion of the *trans* double bond at carbon-11 to a *cis* double bond. The vitamin is then oxidized to 11-*cis*-retinal, followed by reassociation with opsin. The visual cycle takes place in structures of the eye called **rods** (Figure 9.48). Rods are used for vision in dim light. Other structures of the eye, known as **cones**, are used for the perception of bright light and colored light. Rhodopsin accounts for nearly all of the protein content of the outer segment of the rod (see Figure 9.48). The end of the rod opposite the outer segment features a synapse. This synapse mediates the transmission of an impulse to a nerve.

Naturally occurring mutations in the polypeptide sequence of rhodopsin, in cGMP phosphodiesterase (an enzyme used in the rods and cones), and other proteins have been shown to be responsible for various types of vision defects and blindness (Nathans *et al.*, 1992).

## Assessment of Vitamin A Status and Consequences of a Deficiency

Vitamin A deficiency occurs with the chronic consumption of a vitamin A-deficient diet or during a prolonged starvation. The first symptom is nightblindness. This is followed by damage to the cornea. The nightblindness is reversible, but corneal

Uyeda, K., Furuya, E., and Luby, L. J. (1981). The effect of natural and synthetic D-fructose-2,6-bisphosphate on the regulatory kinetic properties of liver and muscle phosphofructokinase. *J. Biol. Chem.* **256**, 8394–8399.

Wallberg-Hendriksson, H., and Holloszy, J. O. (1985). Activation of glucose transport in diabetic muscle: Responses to contraction and insulin. *Am. J. Physiol.* **249**, C233–C237.

Weber, J., and Senior, A. E. (1997). Catalytic mechanism of $F_1$-ATPase. *Biochim. Biophys. Acta* **1319**, 19–58.

Westergaard, H., and Dietschy, J. M. (1976). The mechanism whereby bile acid micelles increase the rate of fatty acid and cholesterol uptake into the intestinal mucosal cell. *J. Clin. Invest.* **58**, 97–108.

Yin, S.-J., Han, C.-L., Liao, C.-S., and Wu, C.-W. (1996). Expression, activities, and kinetic mechanisms of human stomach alcohol dehydrogenase. *In* "Enzymology and Molecular Biology of Carbonyl Metabolism," Vol. 6 (H. Weiner, R. Lindahl, D. Crabb, and T. Flynn, eds.), pp. 347–355. Plenum, New York.

Yudkoff, M., Daikhin, Y., Nissim, I., Grunstein, R., and Nissim, I. (1997). Effects of ketone bodies on astrocyte amino acid metabolism. *J. Neurochem.* **69**, 682–692.

## BIBLIOGRAPHY

*Brain Energy*

Bray, G. A., Teague, J., and Lee, C. K. (1987). Brain uptake of ketones in rats with differing susceptibility to dietary obesity. *Metabolism* **36**, 27–30.

Ferre, P., Satabin, P., Decaus, J.-F., Escriva, F., and Girard, J. (1983). Development and regulation of ketogenesis in hepatocytes isolated from newborn rats. *Biochem. J.* **214**, 937–942.

Hawkins, R. A., Mans, A. M., Davis, D. W., Vina, J. R., and Hibbard, L. S. (1985). Cerebral glucose use measured with [$^{14}$C]glucose labeled in the 1, 2, or 6 position. *Am. J. Physiol.* **248**, C170–C176.

Levitsky, L. L., Fisher, D. E., Paton, J. B., and Delannoy, C. W. (1977). Fasting plasma levels of glucose, acetoacetate, D-β-hydroxybutyrate, glycerol, and lactate in the baboon infant. *Pediatr. Res.* **11**, 298–302.

Mehta, S., Kalsi, H. K., Nain, C. K., and Menkes, J. H. (1977). Energy metabolism brain in human protein-calorie malnutrition. *Pediatr. Res.* **11**, 290–293.

Mellerup, E. T., and Rafaelsen, O. J. (1969). Brain glycogen after intracisternal insulin injection. *J. Neurochem.* **16**, 777–781.

Siegel, G. J., Agranoff, B. W., Albers, R. W., and Molinoff, P. B. (1989). "Basic Neurochemistry," 4th ed., p. 570. Raven Press, New York.

*Regulation of Glycolysis and Gluconeogenesis*

Acheson, K. J., Flatt, J. P., and Jequier, E. (1982). Glycogen synthesis versus lipogenesis after a 500 gram carbohydrate meal in man. *Metabolism* **31**, 1234–1240.

Acheson, K. J., Schutz, Y., Bessard, T., Ravussin, E., Jequier, E., and Flatt, J. P. (1984). Nutritional influences on lipogenesis and thermogenesis after a carbohydrate meal. *Am. J. Physiol.* **246**, E62–E70.

Argaud, D., Kirby, T. L., Newgard, C., and Lange, A. (1997). Stimulation of glucose-6-phosphatase gene expression by glucose and fructose-2,6-bisphosphate. *J. Biol. Chem.* **272**, 12854–12861.

Bergman, E. N. (1990). Energy contributions of volatile fatty acids from the gastrointestinal tract in various species. *Physiol. Rev.* **70**, 567–590.

Blom, P. C., Hostmark, A. T., Vaage, O., Kardel, K. R., and Maehlum, S. (1987). Effect of different post-exercise sugar diets on the rate of muscle glycogen synthesis. *Med. Sci. Sports Exercise* **19**, 491–496.

Blom, P. C., Vollestad, N. K., and Costill, D. L. (1986). Factors affecting changes in muscle glycogen concentration during and after prolonged exercise. *Acta Physiol., Suppl.* **128**(556), 67–74.

Casteno, J. G., Nieto, A., and Feliu, J. E. (1979). Inactivation of phosphofructokinase by glucagon. *J. Biol. Chem.* **254**, 5576–5579.

Dohm, G. L., Kasperek, G. J., and Barakat, H. A. (1985). Time course of changes in gluconeogenic enzyme activities during exercise and recovery. *Am. J. Physiol.* **248**, E6–E11.

Ekdahl, E. N., and Ekman, P. (1984). The effect of fructose-2,6-bisphosphate and AMP on the activity of phosphorylated and unphosphorylated fructose-1,6-bisphosphatase from rat liver. *FEBS Lett.* **167**, 203–209.

Felig, P., and Wahren, J. (1975). Fuel homeostasis in exercise. *N. Engl. J. Med.* **293**, 1078–1084.

Ferrannini, E., Bjorkman, O., Reichard, G. A., Pilo, A., Olsson, M., Wahren, J., and DeFronzo, R. A. (1985). The disposal of an oral glucose load in healthy subjects. *Diabetes* **34**, 580–588.

Hargreaves, M., Costill, D. L., Fink, W. J., King, D. S., and Fielding, R. A. (1987). Effect of pre-exercise carbohydrate feeding on endurance cycling performance. *Med. Sci. Sports Exercise* **19**, 33–36.

Haussinger, D. (1986). Regulation of hepatic ammonia metabolism: The intercellular glutamine cycle. *In* "Advances in Enzyme Regulation" (G. Weber, ed.), Vol. 25, pp. 159–180. Pergamon, Oxford.

Hers, H. G., and Hue, L. (1983). Gluconeogenesis and related aspects of glycolysis. *Annu. Rev. Biochem.* **52**, 617–653.

Hultman, E., and Bergstrom, J. (1967). A study of the glycogen metabolism during exercise in man. *Scand. J. Clin. Lab. Invest.* **19**, 218–228.

Katz, J., and McGarry, J. D. (1984). The glucose paradox. *J. Clin. Invest.* **74**, 1901–1909.

Louveau, I., Chaudhuri, S., and Etherton, T. D. (1991). An improved method for isolating RNA from porcine adipose tissue. *Anal. Biochem.* **196**, 308–310.

Magnusson, I., Chandramouli, V., Schumann, W. C., Kumaran, K., Wahren, J., and Landau, B. R. (1987). Quantitation of the pathways of hepatic glycogen formation on ingesting a glucose load. *J. Clin. Invest.* **80**, 1748–1754.

Mair, T., and Muller, S. C. (1996). Traveling NADH and proton waves during oscillatory glycolysis *in vitro*. *J. Biol. Chem.* **271**, 627–630.

Mitchell, J. B., Costill, D. L., Houmard, J. A., Fink, W. J., Pascoe, D. D., and Pearson, D. R. (1989). Influence of carbohydrate dosage on exercise performance and glycogen metabolism. *J. Appl. Physiol.* **67**, 1843–1849.

Neufer, P. D., Costill, D. L., Flynn, M. G., Kirwan, J. P., Mitchell, J. B., and Houmard, J. (1987). Improvements in exercise performance: Effects of carbohydrate feedings and diet. *J. Appl. Physiol.* **62**, 983–988.

Pilkis, S. J., El-Maghrabi, M. R., and Claus, T. H. (1988). Hormonal regulation of hepatic gluconeogenesis and glycolysis. *Annu. Rev. Biochem.* **57**, 755–783.

Richter, E. A., Ploug, T., and Galbo, H. (1985). Increased muscle glucose uptake after exercise. *Diabetes* **34**, 1041–1048.

Scheele, G., Adler, G., and Kern, H. (1987). Exocytosis occurs at the lateral phase membrane of the pancreatic acinar cell during supramaximal secretatogogue stimulation. *Gastroenterology* **92**, 345–353.

Schacter, E., Chock, P. B., Rhee, S. G., and Stadtman, E. (1986). Cyclic cascades and metabolic regulation. *In* "The Enzymes" (P. D. Boyer and E. G. Krebs, eds.), 3rd ed., Vol. 17, Part A, pp. 23–25. Academic Press, San Diego.

Smith, M. (1987). Position of the American Dietetic Association: Nutrition for physical fitness and athletic performance for adults. *J. Am. Diet. Assoc.* **87**, 933–939.

damage is not. The corneal damage involves drying of the conjunctiva, followed by the appearance of opaque white spots called Bitot's spots. The tendency toward drying of the eye is called xerosis. The eventual irreversible damage to the cornea, lens, and consequent total blindness is part of the disease called **xerophthalmia**.

Total blindness has occurred in infants in Denmark, during the early 1900s, due to vitamin A deficiency. Here deficiency arose from the practice of using milk fat to make butter, and exporting the butter for commercial purposes, while using the byproduct (skimmed milk) to feed infants. Total blindness has occurred on a greater scale in infants of impoverished populations, as in India, where the only foods given to infants were lentils, eggplant, and bitter gourd (Moore, 1957). These foods are low in β-carotene.

Experimentally induced vitamin A deficiency in animals results in anorexia (loss of appetite), lack of growth, infections, and xerophthalmia. Often death from infections precedes the onset of xerophthalmia. Vitamin A deficiency results in an impairment of antibody production, as well as damage to the epithelial linings of the respiratory and gastrointestinal tracts. Both results allow invasion by pathogenic organisms.

Studies in underdeveloped countries have shown that xerophthalmia results in the blindness of a quarter million children per year in Asia alone. Attempts have been made to detect a relationship between the occurrences of xerophthalmia and respiratory or diarrheal infections. Some studies have found an association; others have not. Generally, the populations suffering from malnutrition and chronic diarrheal infections are those afflicted with xerophthalmia.

Vitamin A supplementation programs have been used in Indonesia, the Philippines, and Africa, where they have resulted in moderate reductions in child mortality and dramatic decreases in the incidence of nightblindness and xerophthalmia. Public health workers have used the following schedule for preventing xerophthalmia and increasing the liver's reserves of vitamin A. Oral retinyl palmitate (110 mg), retinyl acetate (66 mg), or injected retinyl palmitate (55 mg) are administered on each of two successive days, and once a few weeks later if symptoms are not relieved. One approach that seems to be especially workable is to give high-dose capsules at the time of a scheduled immunization. One clever

vision. The terms "rods" and "cones" were derived from the shape of the **outer segment** of the rods and cones. The outer segment of the cone is pointed. Each outer segment contains several hundred tiny disks, where each disk is a flat membranous structure. Photoreceptive proteins, such as the rhodopsin in the rods, resides in the tiny disks. The disks are continuously being created and destroyed (they turn over), as is the case with most or all other structures in the body. New disks are created at the proximal end of the outer segment. They are gradually pushed to the distal end (the tip). Eventually the disk is destroyed at the distal end. About 100 new disks are created in each rod (or cone) per day. The reader interested in comparing the disks in the rods and cones might like to know that the disks in rods are free-floating, while those in the cones maintain a connection to the plasma membrane of the photoreceptor (Geneser, 1986). The **inner segment** of the rod or cone contains mitochondria and glycogen. The structure called the inner fiber has the properties of a nerve. Below the inner fiber occurs synaptic vesicles, and these contain neurotransmitters. When light falls upon the rod or cone, the synaptic vesicles are provoked to fuse with the base, thus relaying a signal to nerves.

approach is to give new mothers high-dose capsules in order to boost the levels of retinol in the milk (Filteau and Tomkins, 1995).

Vitamin A, carotenoids with previtamin A activity, and carotenoids with no previtamin A activity have been associated with anticancer effects. These protective effects have been demonstrated in cell culture systems as well as in animals. The studies have involved cancer induced by ultraviolet radiation or with carcinogens. In some cases, vitamin A or carotenoids have been shown to have a protective effect; in other cases no effect or even a cancer-promoting effect has been shown.

Epidemiological evidence has linked the consumption of carotenoid-containing foods with lower incidences of certain types of cancers, notably cancer of the lung, stomach, and bladder. For example, the regular consumption of dark green and yellow vegetables, cruciferous vegetables, and tomatoes has been associated with reduced rates of lung cancer (Marchand et al., 1989). The proposed anticancer effect of most of these plants is due to β-carotene, rather than the vitamin A produced in the body from the carotenoids. The proposed anticancer effect of tomatoes is associated with lycopene, and that of dark green vegetables is associated with lutein. In other studies low incidences of cancer were associated with vitamin A intakes of 5000 IU per day, and high incidences were associated with intakes of 1700 to 2500 IU per day.

## Hazards of High Intakes

The chronic consumption of high levels of vitamin A is dangerous. An intake of vitamin A 10 times the RDA or higher by pregnant women can cause brain damage in their infants. These levels of the vitamin can lead to neurological symptoms and damage to the eyes when consumed by children or adults. It is the responsibility of the nutritionist to discourage the intake of "high-potency" preparations of vitamin A and to limit consumption to 6000 IU per day or less.

High dietary levels of β-carotene, in contrast, appear not to be toxic. The chronic consumption of high levels of β-carotene can result in elevated levels in the plasma, though the response differs markedly in different individuals. High levels of plasma carotenoids have little or no effect on plasma vitamin A levels. The mechanism of conversion of β-carotene to vitamin A, though poorly understood, seems to be carefully regulated. Chronic intake of 30 to 300 mg of β-carotene per day results in increased storage of the compound in fat to the point where the skin (subcutaneous fat) appears orange.

## VITAMIN D

Vitamin D is a fat-soluble vitamin. The vitamin is closely associated with sunlight, as periodic exposure to sunlight can eliminate the dietary requirement for the vitamin. The pathway for vitamin synthesis in animals requires a light-dependent step: the opening of an aromatic ring, resulting in conversion of a precursor of the vitamin to vitamin D. Failure to obtain exposure to sunshine can result in the materialization of a requirement for the preformed vitamin. For this reason, RDAs for vitamin D have been established to help ensure that persons with indoor occupations, or those confined indoors, receive sufficient vitamin D.

The RDA for vitamin D for adults is 5 µg. The RDA during pregnancy and lactation has been set at 10 µg, though it is not certain that there is an increased need during these times. The usual dietary intake of vitamin D in the United States is 1 to 2 µg/day.

Foods may contain precursors of vitamin D as well as vitamin D. The precursors are converted to the vitamin following their absorption from the diet. The body is a source of one of the precursors, 7-dehydrocholesterol. Conversion of the precursor to the vitamin form occurs in the skin. The major function of vitamin D is to maintain the concentration of calcium ions in the plasma. Maintenance of these concentrations is vital for normal functioning of the nervous system and for the growth of bones and the maintenance of bone mass. Although vitamin D is closely associated with bone growth, it is thought that the relationship is only indirect. The vitamin stimulates the absorption of dietary calcium by the gut, but it seems not to stimulate the subsequent deposit of calcium in the bone.

A vitamin-deficient diet can result in lowered levels of plasma calcium; however, the use of high-calcium diets during vitamin deficiency can result in the maintenance of normal plasma calcium levels. The requirement for the vitamin may be overridden, providing that the diet contains high levels of calcium. The physiologically important form of vitamin D, 1,25-dihydroxyvitamin $D_3$, is a hormone. Another hormone, parathyroid hormone (PTH), is used in the control of vitamin D metabolism and calcium levels, as detailed in the following section.

## Biochemistry of Vitamin D

*The Provitamin and the Vitamin*

Plants contain the steroid ergosterol. Ergosterol is a provitamin of vitamin D. Exposure of ergosterol to ultraviolet light results in its conversion to ergocalciferol (vitamin $D_2$). Farmers, for example, leave their hay out in the sun to dry. This process results in the conversion of ergosterol to vitamin $D_2$. Cows benefit from the increased levels of vitamin $D_2$ produced in their feed.

Animals can convert cholesterol to 7-dehydrocholesterol. 7-Dehydrocholesterol is a provitamin of vitamin D. The exposure of one's body to sunlight results in the conversion of 7-dehydrocholesterol in the skin to cholecalciferol (vitamin $D_3$). The provitamin in both plants and animals is converted to vitamin D by ultraviolet light via identical chemical reactions that involve the opening of the steroid ring (Figures 9.49 and 9.50).

Vitamin $D_2$ and vitamin $D_3$ as is do not have any direct function, but must be hydroxylated to be of use to the body. The events of 1-hydroxylation and 25-hydroxylation proceed in an identical manner for both vitamins $D_2$ and $D_3$. Both vitamins have equal potency for humans, but vitamin $D_2$ is more potent in rats, while vitamin $D_3$ is much more potent for chickens. Vitamin $D_2$ does not much occur in our diet, in contrast to vitamin $D_3$, but vitamin $D_2$ is the form used most for food supplementation and for vitamins.

**FIGURE 9.49** Vitamin D metabolism. The 7-dehydrocholesterol synthesized in the skin is converted to vitamin $D_3$ in a light-catalyzed reaction. Vitamin $D_3$ is hydroxylated in the liver, producing 25-hydroxy-vitamin $D_3$. This is followed by a second hydroxylation in the kidney, yielding 1,25-dihydroxyvitamin $D_3$.

Ergosterol
(provitamin D)

7-Dehydrocholesterol
(provitamin D)

UV light

UV light

Vitamin $D_2$

Vitamin $D_3$

Hydroxylations

1,25-Dihydroxyvitamin $D_3$

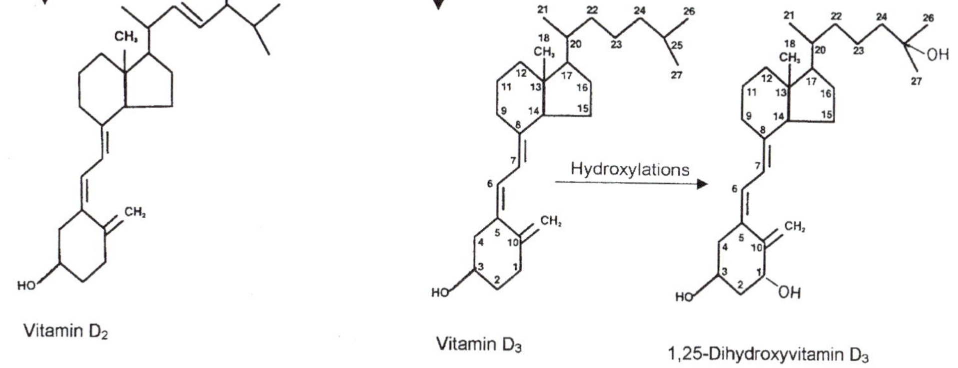

**FIGURE 9.50** UV light is required for conversion of plant (ergesterol) and animal (7-dehydrocholesterol) forms of the provitamin to the vitamin. The plant form differs from the animal form, in that the side-chain contains a double bond and an extra methyl carbon. UV light results in conversion of both forms of provitamin to vitamin $D_2$ and $D_3$, respectively. (The side-chain double bond of $D_2$ is not reduced in the body, but it can be artificially reduced in the laboratory, where the product is called vitamin $D_4$.) Hydroxylation is shown for $D_3$, but not for $D_2$.

Vitamin $D_3$ is converted to 1,25-dihydroxyvitamin $D_3$ by a series of two hydroxylation reactions. The 25-hydroxylation, which occurs first, is catalyzed by an enzyme of the liver. The subsequent conversion of 25-hydroxyvitamin $D_3$ to 1,25-dihydroxyvitamin $D_3$ is catalyzed by an enzyme of the kidney. The enzyme that catalyzes this latter step is called "1-hydroxylase" in this text. Figure 9.50 indicates the positions of the 1-carbon and 25-carbon of the steroid ring. Vitamin D can also be converted to 24,25-dihydroxyvitamin $D_3$, as well as a number of other metabolites. Although there is much interest in the possible physiological value of these compounds, they seem generally to be inactive catabolites of the vitamin. 1,25-Dihydroxyvitamin $D_3$ is also called **calcitrol**. 25-Hydroxyvitamin $D_3$ is also called **calcidiol**.

## *Detection of Light-Catalyzed Changes in Vitamin Structure*

The following study illustrates the effect of light on the conversion of 7-dehydrocholesterol to vitamin D *in vivo* (Figure 9.51). Human subjects were exposed to ultraviolet light for periods ranging from 0 to 60 minutes. Patches of skin were cut from the subjects at the indicated times. The lipid-soluble compounds of the skin were then extracted from the skin with a solvent, the compounds analyzed by high-pressure liquid chromatography (also high-performance liquid chromatography, HPLC). In HPLC, the sample is run through a solid matrix, where the different metabolites separate. The various metabolites of vitamin D have different affinities (tendencies to bind to) for the solid matrix. During passage through the matrix, each metabolite migrates through the tube or column at a different rate. The differences in migration rate result in different times required to pass through the entire chromatograph column. This time is called the elution time. The elution time of 7-dehydrocholesterol was determined to be about 11 minutes, whereas that of cholecalciferol was 5 minutes. The breakdown products of vitamin $D_3$ include lumisterol and tachysterol. These elute at 6 and 8 minutes, respectively.

The results in Figure 9.51 demonstrate that the major vitamin D-related compound in skin not exposed (0 minutes) to ultraviolet (UV) light was 7-dehydro-

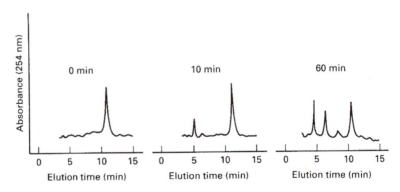

**FIGURE 9.51** Effect of light on the conversion of 7-dehydrocholesterol to vitamin D *in vivo*. (Redrawn with permission from Holick *et al.*, 1981.)

cholesterol. Brief exposure to UV light resulted in some conversion of this compound to cholecalciferol (10 minutes); prolonged exposure to UV light yielded a variety of compounds (60 minutes).

EXERCISE

Which compound accumulated to greater concentrations in the skin, tachysterol or lumisterol (see Figure 9.51)?

## Assessment of Vitamin D Status

### Direct Assay for Vitamin D

The normal range of plasma concentrations of vitamins $D_2$ and $D_3$ is 1 to 2 ng/ml. 25-Hydroxyvitamin $D_3$ is the form of the vitamin present at highest concentrations, with normal values ranging from 8 to 40 ng/ml. A value of 12 ng/ml has been used as a cutoff point to indicate vitamin D deficiency, though some clinicians have used lower levels to indicate vitamin D deficiency. The hormonally active form of the vitamin, $1,25\text{-}(OH)_2D_3$, occurs at much lower levels (15–45 pg/ml).

Vitamin status can be assessed by a direct test for the levels of $25\text{-}(OH)D_3$ in the serum. This competitive binding test involves three components: (1) a serum sample, (2) radioactive 25-hydroxy[$^3$H]vitamin $D_3$, and (3) vitamin D-binding protein. The source of vitamin D-binding protein may be sheep serum. The functional tests for determination of vitamin D include the diagnostic tests for rickets and osteomalacia. A test for osteomalacia, for example, may include measurement of the width of the osteoid in a bone biopsy. The osteoid is described later.

### Vitamin D and the Regulation of Plasma Calcium

The role of vitamin D in the regulation of plasma calcium levels involves complex mechanisms. Our discussion of these mechanisms begins with the concrete and vivid example of "milk fever." This example might be used as a reference point while reading through the studies on chicks, rats, and humans.

Milk fever is a problem on the dairy farm. The disease occurs in cows whose plasma calcium levels drop below 5.0 mg/100 ml. Milk fever often results in coma and death. The disease presents when a cow has been fed a diet high in calcium prior to calving and lactation. ("Calving" means giving birth to a calf.) Alfalfa is an example of a high-calcium food. The mother cow starts to produce milk at the time of calving. The milk production places a sudden drain on the calcium present in the cow's plasma. This calcium must originate either from the diet or from the bones of the cow. Animals are able to dissolve or "mobilize" calcium from their own bones in times of need to maintain plasma calcium levels.

With calving, the mother cow tends to stop eating. Hence, the diet may no longer be a reliable source of calcium at the onset of lactation. This leaves the bones as a source of calcium. The body requires a period of adjustment to activate the calcium-mobilizing mechanisms. The prior consumption of high-calcium foods leaves these mechanisms in the nonactivated state. The sudden change from a feeding lifestyle to a nonfeeding, lactating lifestyle may be too sudden to allow the cow to adjust and mobilize bone calcium at an adequate rate. The consequence is a drop in plasma Ca levels, coma, and sometimes death. Milk fever can be prevented by feeding the pregnant cow a low-calcium diet (100 g/day). The disease can easily be treated by injecting calcium or 1,25-$(OH)_2D_3$ into the sick mother cow.

The following highlights the major functions of 1,25-$(OH)_2D_3$ and the regulation of plasma levels of this hormone:

1. *Calcium absorption*: Calcium in the diet can be absorbed by the intestines by mechanisms that are dependent on and independent of 1,25-dihydroxyvitamin $D_3$. The vitamin D-independent mechanism predominates during the consumption of high-calcium diets. The vitamin D-dependent mechanism is activated and predominates with diets low in calcium. The biochemical mechanisms by which increased levels of 1,25-$(OH)_2D_3$ in the plasma activate the calcium transport system in the enterocyte are not clear. It is thought that the hormone binds to the cell and travels to the nucleus, where it activates genes coding for components of the calcium transport system.

2. *Parathyroid gland*: The parathyroid gland senses the calcium levels in the bloodstream and responds to maintain these levels. Hence, the role of the parathyroid in controlling plasma calcium may be thought of as analogous to that of the pancreas in maintaining plasma glucose levels. When plasma calcium levels drop, the parathyroid gland is stimulated to secrete parathyroid hormone (PTH) into the bloodstream. PTH travels to the kidney, where it stimulates the activity of 1-hydroxylase. 1-Hydroxylase catalyzes the conversion of 25-$(OH)D_3$ to 1,25-$(OH)D_3$.

3. *Three effects of 1,25-$(OH)_2D_3$*: 1,25-Dihydroxyvitamin $D_3$, in increased concentrations, has three effects that act to restore plasma calcium to the normal value of about 1.5 m$M$: (1) activation of the vitamin D-dependent calcium transport system of the enterocyte, (2) activation of the osteoclasts, and (3) enhancement of calcium resorption by the kidney. Osteoclasts are cells that reside on the outside and inside surfaces of bones. They dissolve the bone, releasing the constituent calcium and phosphate ions into the circulation. 1,25-$(OH)_2D_3$ can induce an increase in the activity of the osteoclasts.

4. *Three effects of parathyroid hormone*: The presence of parathyroid hormone in the plasma is required for the maintenance of 1-hydroxylase activity in the kidney. The increase in PTH levels induces an increase in the activity of the hydroxylase, resulting in an increase in the production of 1,25-$(OH)D_3$ and an increase in the plasma levels of 1,25-$(OH)D_3$. In addition, PTH works in tandem with 1,25-$(OH)_2D_3$ in stimulating the osteoclasts to dissolve bones and in increasing the efficiency of calcium resorption by the kidneys. Both PTH and 1,25-$(OH)_2D_3$ are required for these latter two effects. PTH is not required for the effect of the vitamin on the intestines.

The targets of action of 1,25-$(OH)_2D_3$ and PTH + 1,25-$(OH)_2D_3$ are summarized in Figure 9.52.

The study shown in Figure 9.53 illustrates the role of 1,25-dihydroxyvitamin $D_3$ in stimulating the uptake of calcium ions by the intestines. Chick duodena were incubated in a salt solution containing radioactive calcium ions and vitamin D at the indicated levels. Each point in the figure represents the uptake of calcium occurring during a separate incubation period. The forms of the vitamin used included 1,25-$(OH)_2D_3$ (●), 1-$(OH)D_3$ (Δ), 25-$(OH)D_3$ (▲), 24,25-$(OH)_2D_3$ (○), and

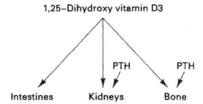

1,25–Dihydroxy vitamin D3

PTH          PTH

Intestines    Kidneys    Bone

FIGURE 9.52  Targets of action of hormonally active forms of vitamin D.

vitamin $D_3$ ($\square$). Of all the forms of the vitamin tested, 1,25-$(OH)_2D_3$ stimulated the uptake of calcium ions most efficiently, that is, when the vitamin was added at very low concentrations. Maximal stimulation of uptake occurred at about 10 nanomolar 1,25-$(OH)_2D_3$; equivalent stimulation by other forms of the vitamin occurred only at concentrations 10- to 1000-fold greater.

EXERCISE

Please compare the approximate concentrations of vitamin required to provoke half-maximal stimulation of calcium transport. How much more active was 1,25-$(OH)_2D_3$ than vitamin $D_3$ (see Figure 9.53)?

The importance of maintaining plasma calcium levels is illustrated by the following study, as well as by material under Calcium in Chapter 10. The ability of vitamin-deficient and -sufficient rats to maintain their plasma calcium levels with the consumption of diets containing low, medium, and high calcium levels (0.02, 0.5, and 1.4%, by weight) was tested. Lactating rats were used because milk production places an extra strain on the rat's supply of calcium and, hence, highlights the importance of the vitamin D-dependent regulatory mechanisms. Figure 9.54A depicts the plasma calcium levels of rats fed the vitamin-deficient

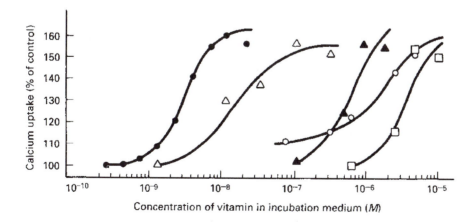

FIGURE 9.53  Role of 1,25-dihydroxyvitamin $D_3$ in stimulating uptake of calcium ions by intestines. (Redrawn with permission from Franceschi and DeLuca, 1981.)

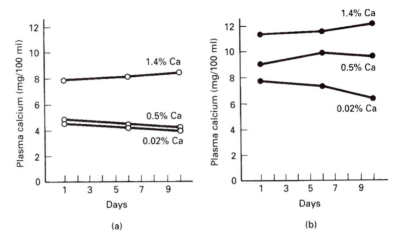

**FIGURE 9.54** Plasma calcium levels of rats fed (A) vitamin-deficient diets and (B) vitamin-sufficient diets. (Redrawn with permission from Brommage and DeLuca, 1985.)

diet. Figure 9.54B shows the plasma calcium levels of rats fed the vitamin-sufficient diets. The normal plasma calcium level is considered to be about 9 mg/100 ml.

The data demonstrate that the vitamin-deficient rats (see Figure 9.54A) maintained near-normal levels of plasma calcium with the high-calcium diet, though *not* with the moderate- or low-calcium diets. The vitamin-sufficient rats (Figure 9.54B) maintained near-normal plasma calcium, even with the low-calcium diet. The study shows that normal levels of plasma calcium can be maintained with the high-calcium diet in the absence of the vitamin, whereas vitamin D was required to maintain plasma calcium with diets containing lower levels of calcium.

The next study illustrates a relationship between different plasma levels of vitamin D and dietary calcium. Human subjects consumed vitamin D-sufficient diets containing moderate (0.3 g/day) or high (2.0 g/day) levels of calcium for 4 weeks. Blood samples were taken at the conclusion of the trial. As the human subjects were not fed a vitamin D-deficient diet and were in good health, their plasma calcium levels might be expected to be normal in all cases; however, because the groups were fed low- and high-calcium diets, the difference in the plasma levels of hormones used for calcium regulations might be expected (Table 9.6).

EXERCISE

Please find evidence that PTH and 1,25-(OH)$_2$-D$_3$ are involved in regulating plasma calcium levels from the data in Table 9.6.

EXERCISE

The rate of calcium absorption by the intestines was measured in human subjects by releasing calcium into the gut using long tubes and measuring the rate of disappearance (absorption) of the calcium by withdrawing small samples of the fluids of the gut lumen and analyzing the fluids. State evidence that the body's response to a low-calcium diet includes a change in the efficiency of absorption by the gut (see Table 9.6).

**TABLE 9.6**  Calcium Metabolism with High- and Low-Ca Diets

|  | Low-calcium diet | High-calcium diet |
|---|---|---|
| Plasma calcium (mg/dl) | 9.4 | 9.4 |
| Plasma PTH (units/ml) | 17 | 15 |
| 1,25-(OH)$_2$D$_3$ (pg/ml) | 44 | 27 |
| Calcium absorption (µmol Ca/30 min) | 200 | 110 |

*Source*: Norman *et al*. (1981).

## Parathyroid Hormone

The parathyroid gland plays a central role in the regulation of plasma 1,25-dihydroxyvitamin D$_3$. This gland is able to sense changes in the concentration of calcium ions in the bloodstream. It secretes PTH at a higher rate when it senses low concentrations. It releases PTH at a lower rate when plasma calcium rises above the normal concentration. The PTH enters the circulation and travels to the kidneys, where it regulates the activity of 1-hydroxylase. Low plasma calcium levels, for example, induce an increased release of PTH, activation of 1-hydroxylase, an increase in conversion of 25-(OH)D$_3$ to 1,25(OH)$_2$D$_3$, and finally an increase in the rate of delivery of calcium into the circulation (Figure 9.55). Calcium ions can enter the bloodstream from three sources — the diet, skeleton, and glomerular filtrate.

The data in Table 9.6 indicate an association between calcium metabolism and PTH levels, but do not support a strong connection or conclusion. The study depicted in Figure 9.56 provides a firm connection between PTH and calcium regulation. Parathyroid glands were removed from rats at the time indicated in the figure. The kidneys were removed from different rats at the indicated times after parathyroidectomy and used for assay of 1-hydroxylase. In short, the kidneys were homogenized and the extract was placed in test tubes and used for assaying 1-hydroxylase activity. Some of the parathyroidectomized rats were treated with PTH at the indicated times. This treatment is a form of replacement therapy. The kidneys of these rats were removed at the indicated times and used for 1-hydroxylase assays.

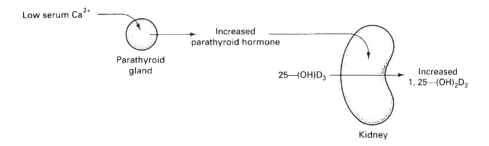

**FIGURE 9.55**  Relationship between the parathyroid gland and vitamin D.

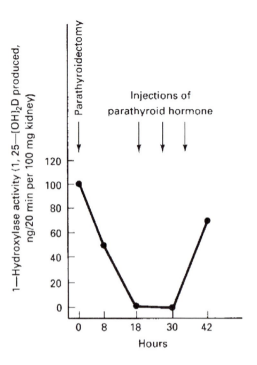

**FIGURE 9.56** Connection between PTH and calcium regulation. (Redrawn with permission from Tanaka and DeLuca, 1984.)

The results demonstrate that 1-hydroxylase activity dramatically decreases with removal of the parathyroid gland and that activity can be restored by PTH. The experiment demonstrates the role of PTH in maintaining plasma calcium concentrations.

EXERCISE

Imagine that you are examining a patient with severe kidney failure. The patient has abnormally low plasma calcium levels. Discuss which of the following two treatments would be better: injections of PTH or injections of 25-(OH)$_2$D$_3$.

### Calcium Resorption by the Kidneys

Calcium ions are resorbed by the kidneys. In the adult man, about 11 g of calcium ions passes from the plasma into Bowman's space of the glomeruli and into the renal tubules each day. Only 1.0% or less of this calcium is lost to the urine. Most of the calcium is resorbed through the tubules to the bloodstream. This resorption process is controlled by 1,25-(OH)$_2$D$_3$ and PTH. Increased urinary losses of calcium may occur with vitamin D deficiency or after damage to the parathyroid gland.

## Action of 1,25-Dihydroxyvitamin D3 in Regulating Transcription

1,25-Dihydroxyvitamin $D_3$ exerts its effects at the nuclear and nonnuclear events. The nonnuclear effects may occur within a few minutes of exposure of the cell to 1,25-$(OH)_2D_3$. In participating in its nonnuclear role, 1,25-$(OH)_2D_3$ may bind to a receptor in the plasma membrane and provoke the liberation of inositol-1,2,5-trisphosphate (IP3) into the cytoplasm. IP3 is a modified sugar molecule that is attached to diglycerides in cell membranes, but readily available for cleavage and liberation into the cytoplasm. After liberation, the IP3 molecule provokes a burst in levels of intracellular calcium ions and the consequent activation of protein kinases. The end-result may be a change in the rate of growth, in differentiation of the cell, or in migration of the cell to a different location in the body. The nonnuclear action of vitamin D is a relatively recent field of research (Mellay *et al.*, 1997; Khare *et al.*, 1997).

The nuclear effects of 1,25-$(OH)_2D_3$ requires the participation of a 1,25-$(OH)_2D_3$ receptor protein and chromatin. Chromatin consists of the complex of DNA and protein that resides in the nucleus. These proteins include regulatory proteins and histones. Histones have mainly a structural role. They maintain the proper folding of our DNA into an orderly, compact structure. The regulatory proteins include the transcription factors, a group of several hundred different proteins. One of these transcription factors is **vitamin D receptor** (VDR).

Maximal expression of the nuclear effects of the vitamin requires an hour or so after exposure of the cell to 1,25-$(OH)_2D_3$. The 1,25-$(OH)_2D_3$ entering the cell binds to VDR to form a hormone/receptor complex. The hormone/receptor complex binds to small stretches of DNA called **hormone response elements** and provokes a change in the rate of transcription of a nearby gene. The term "nearby" means that the hormone response element and the gene may be separated from each other by a stretch of from 20 to 1000 nucleotides away from each other. Further details on the nuclear action of vitamin D occur in the section entitled Vitamin A, Vitamin D, and Thyroid Hormone at the Genome.

## Vitamin D Deficiency

The definition of vitamin D deficiency is a condition where the concentration of 25-hydroxyvitamin $D_3$ in the blood serum occurs at 12 ng/ml or less. Prolonged deficiency can result in two diseases, namely rickets (in children) and osteomalacia (in adults). Vitamin D deficiency tends to occur in those who do not get enough exposure to sunlight and who also fail to eat foods that are rich in vitamin D. In an ideal world, where everyone was regularly exposed to sunshine, vitamin D would never be classified as a vitamin.

Vitamin D deficiency, rickets, and osteomalacia tend to arise in several conditions or environments, as listed hereunder. Generally, the combined lack of sunlight *and* dietary deficiency must exist before any sign of the deficiency arises.

1. *Infancy*: Infants who are never brought outside, or who are totally protected from the sunshine during excursions out of doors, are at risk for rickets. The risk is increased for infants born shortly before wintertime at northern latitudes, and who are never given fortified milk formula.

2. *Being elderly and unable or unwilling to go outside, especially during frigid winter weather*: The elderly are also at increased risk for osteomalacia because of a reduced ability to synthesize vitamin D, even with exposure to sunlight (Dawson-Hughes *et al.*, 1997; Kinyamu *et al.*, 1997; Chapuy *et al.*, 1992; Wielen *et al.*, 1995).

3. *Living in northern latitudes*: Vitamin D deficiency continues to be documented in Canada (Binet and Kooh, 1996).

4. *Having dark skin, as in those originating from Africa or India*: A typical victim of rickets in the year 2000 may be a dark-skinned infant born of Asian-Indian parents in Canada, who is raised exclusively on breast milk and never given fortified milk formula.

5. *Scrupulously covering the skin, e.g., for religious reasons, whenever going outside*: Osteomalacia has been documented in young Arab women in Kuwait and Israel (Lowenthal and Shany, 1994; El-Sonbaty and Abdul-Ghaffar, 1996). The women cover their faces with veils, and their hands and feet with black gloves and black socks. They acquire osteomalacia, even though they live in a sunny climate, and suffer from bone fractures, bone pain, muscle weakness, and a waddling walk.

6. *Consuming vegetarian diets and avoiding fortified foods*: Vegetarian diets that contain no milk, animal fat, or meat contain little vitamin D. This risk factor can be revealed by noting that the RDA for vitamin D can be supplied by 1.5 kg of beef, 2.0 kg of corn oil, or 100 kg of cabbage. Few persons would be willing or able to consume these quantities of food on a daily or even weekly basis. On the other hand, saltwater fish, such as salmon, sardines, and herring, are rich sources of vitamin D. Oils produced from these foods contain very high levels of the vitamin. The RDA can be supplied by eating 50 g of salmon or 2.0 g of cod liver oil. Fortified milk contains 400 IU per quart. A half-quart of fortified milk provides the RDA. For comparison, human breast milk is relatively low in the vitamin and contains only 4–60 IU per quart.

7. *Calcium deficiency*: Although rare throughout the world, rickets has been found to occur in various parts of Africa due to dietary deficiency in calcium, but with sufficient vitamin D (Oginni *et al.*, 1996). No amount of vitamin D can prevent the rickets that may develop with calcium deficiency.

8. *Fat malabsorption syndromes*: Fat malabsorption syndromes such as cystic fibrosis and cholestatic liver diseases (lack of bile salts) can impair the absorption of vitamin D.

9. *Kidney failure*: The kidney plays a vital role in the conversion of dietary vitamin D and skin-synthesized vitamin D to the hormonally active form of the vitamin. Severe renal disease can impair the activity of the enzyme required for the catalysis of this conversion.

10. *Epilepsy*: Signs of vitamin D deficiency may occur in epileptics treated with anticonvulsants such as dilantin. The drugs can stimulate the activity of enzymes of the endoplasmic reticulum that catabolize and inactivate the vitamin.

11. *Genetic disease*: Vitamin D metabolism is adversely affected in a rare genetic disease that results in impairment of the conversion of vitamin D3 to the hormonally active form. The disease affects 1-hydroxylase, an enzyme of the kidney.

## Structure and Synthesis of Bone

Vitamin D is used in the maintenance of plasma calcium ion concentrations. The normal level of free calcium ions in the plasma ranges from 1.0 to 1.5 m$M$. This concentration is needed to support a normal rate of deposit of calcium in bone during growth and during bone turnover. Apparently, vitamin D has no direct effect on the deposit of calcium ions in bone. It seems to act only indirectly and in maintaining plasma calcium at a level required to support bone mineralization. Note, however, that there remains interest in the possibility that vitamin D does have a direct effect on the cells that synthesize bone. A few details on bone formation and structure and on the vitamin D-dependent process of bone resorption are presented here.

## Cartilage and the Growth Plate

Cartilage and bone consist of living cells sparsely distributed in a matrix of extracellular protein and polysaccharides. Cartilage contains a large proportion of matrix. Bone contains a small proportion of matrix and is mineralized. Cartilage does not contain capillaries. It receives the nutrients that diffuse slowly through the matrix. Bone contains a network of canals through which course nutrient-carrying blood vessels. Cartilage growth occurs by the action of cells called **chondrocytes**. Bone growth occurs by the action of cells called **osteoblasts**. Both of these cells synthesize and secrete collagen, which is a protein of connective tissue. They also secrete a variety of polysaccharides, such as chondroitin sulfate. The structures of the collagen and the polysaccharides found in cartilage and bone are similar but not identical. Further information on collagen is found under Ascorbic Acid (Vitamin C).

To summarize the first steps in bone mineralization, the chondrocytes produce and release small vesicles via budding process. Then, the chondrocytes synthesize and release collagen and proteoglycan. Finally, mineralization occurs. The initial site of mineralization, in bone formation, is thought to be on the phosphatidylserine that exists in the small vesicles (Wu *et al.*, 1993).

The osteoblasts, unlike chondrocytes, are used in mineralization of the organic matrix. Osteoblasts synthesize bone at two different regions of the bone. The first is at the **epiphyseal plate**, also called the **growth plate**. The growth plate consists of cartilage and is located at the ends of long bones in the growing animal or human. It disappears at the onset of adulthood. This disappearance is called "the fusion of the epiphysis." The second region of action of the osteoblast is in the osteon, the structural unit of bone. The osteon looks like a canal surrounded by layers of mineral. Hard bone consists of an array of osteons.

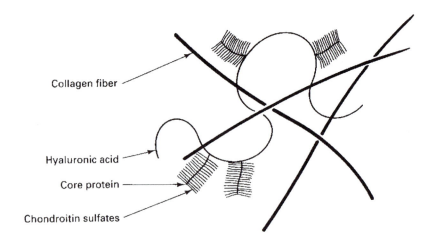

**FIGURE 9.57** Diagram of cartilage.

Cartilage is a firm, gellike substance. It is capable of bearing weight, but is flexible and not rigid. Cartilage contains a matrix of proteins and sulfated polysaccharides. This matrix, which is also called the "ground substance," consists of a backbone of hyaluronic acid from which branch core proteins at various intervals. Each core protein is coated with about 100 molecules of sulfated polysaccharide. The structure of the ground substance, which is interwoven with collagen fibers, is shown in Figure 9.57.

**Hyaluronic acid** is a linear polymer of alternating units of two sugars, glucuronic acid and N-acetylglucosamine. The sulfated polysaccharides connected to the core protein include chondroitin sulfate and keratin sulfate. Chondroitin sulfate is a polymer of two alternating sugar units. The two sugars are glucuronic acid and N-acetylgalactosamine. The sulfate groups are connected to the hydroxyl groups of the residues of N-acetylgalactosamine.

**Glycosaminoglycan** and **proteoglycan** are two terms used to refer to some of the complex macromolecular structures in cartilage. The glycosaminoglycans include hyaluronic acid and chondroitin sulfate. Hyaluronic acid is the only glycosaminoglycan without a covalently attached core protein. The proteoglycans are the large aggregates of protein and oligosaccharides found in cartilage, bone, and other types of connective tissue. The proteoglycan of cartilage has an overall molecular weight of up to 4 million. The term *ground substance* is used by histologists and refers to the same structure known as proteoglycan, the term used by biochemists.

### The Mineralization of Cartilage

The location of the growth plate is shown in Figure 9.58. The growth plate is surrounded by hard bone and porous bone. The side of the growth plate facing

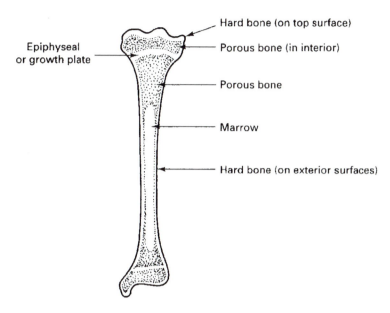

**FIGURE 9.58** Diagram of a cross-section of a long bone.

the bone shaft is gradually invaded by osteoblasts. These osteoblasts deposit mineral at one face of the growth plate. This deposit might be expected to lead to thinning of the growth plate in the growing animal; however, chondrocytes in the growth plate continue to synthesize cartilage, thus maintaining the mass of the growth plate. The overall effect is a lengthening of the bone in the growing animal. Collagen and other proteins, as well as the polysaccharides at the mineralized face of the growth plate, are eventually catabolized and replaced by bone-specific proteins and polysaccharides by the osteoblasts. When adulthood is reached, the growth plate is replaced by porous bone.

### The Osteon

The osteon is a cylindrical structure of the long bones. The cylinder runs roughly parallel to the length of the bone. The osteon has a diameter of about 0.1 to 0.3 mm. A cross-section of the hard bone of the long bones, when examined under the microscope, reveals an array of these cylindrical structures or osteons. Each osteon is composed of a number (4–20) of layers, or concentric rings, resembling the cross-section of a tree. A central channel of a diameter of 0.03 to 0.07 mm contains blood vessels and nerves. Between the mass of concentric rings, which consist of mineralized bone, and the channel is a soft layer called the osteoid. The **osteoid** is composed of a matrix of proteins and polysaccharides. The osteoid might be compared with the matrix synthesized by the osteoblasts that invade the growth

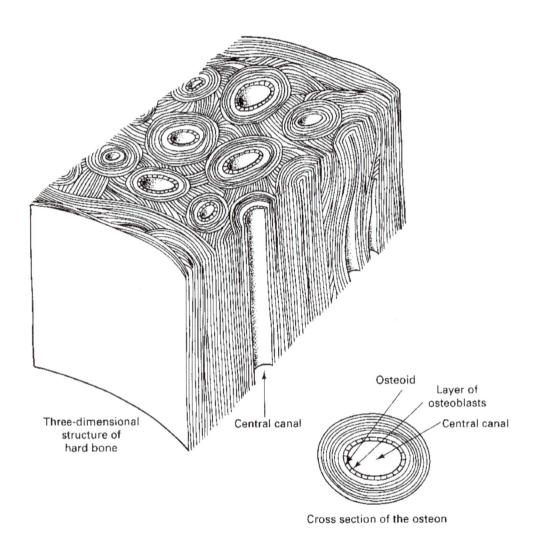

Three-dimensional
structure of
hard bone

Central canal

Osteoid

Layer of
osteoblasts

Central canal

Cross section of the osteon

**FIGURE 9.59**  Three-dimensional structure of hard bone and a cross-section of the osteon.

plate. The osteoid is synthesized by a layer of osteoblasts (Figure 9.59). The osteon is also called the **haversian system**.

*Bone Biochemistry*

Bone is considered to be a mineralized connective tissue. About one-third of the dry weight of bone is organic matter, consisting mainly (90% by weight) of collagen. The remaining organic material consists of other extracellular proteins, such as **osteonectin, osteocalcin (bone GLA protein), bone sialoprotein,** and complexes of protein and chondroitin sulfate. Collagen is used as scaffolding for the deposit of minerals.

**Osteonectin,** which has a molecular weight about 32,000, is a phosphorylated glycoprotein. It has a high affinity for collagen and bone mineral. It is thought that osteonectin coats the collagen fibers and facilitates the deposit of mineral on the collagen.

**Bone sialoprotein** has a molecular weight of about 25,000. About 20% of the weight of the protein is sialic acid, an acidic sugar. The protein may aid in the formation of collagen fibers prior to the deposit of minerals. It is thought that osteonectin promotes the adhesion of osteoblasts to unmineralized surfaces. The protein component of the protein–chondroitin sulfate complex of bone has a molecular weight of 45,000, whereas the molecular weight of each of the associated chondroitin sulfate chains is 40,000. The complex, in bone, contains only one or two chains of chondroitin sulfate.

**Osteocalcin** (bone Gla protein) has a molecular weight of only 5700. It is distinguished by the fact that it contains residues of γ-carboxyglutamic acid (Gla) and thus requires vitamin K-dependent carboxylase for its synthesis (see Vitamin K). Bone Gla protein contains 2–3 residues of Gla. Bone Gla protein adheres to the calcium ions of bone via these Gla residues. It is thought that the function of this protein is to prevent or limit the rate of mineralization of the growth plate. Osteocalcin is not needed for life, as revealed by studies of knock-out mice that lacked the gene coding for the protein (Ducy *et al.,* 1996). The knock-out mice appeared normal, and were able to reproduce. The bones of the mice were somewhat different from normal bones. Their thickness and density was greater than normal. The exact molecular interactions involving osteocalcin and bone development remain unclear.

Bone mineral consists of very small crystalline particles, about 30 nm in diameter. For comparison, the diameters of calcium ion and hemoglobin are about 0.2 and 6.4 nm, respectively. Bone mineral consists of an imperfect form of **hydroxyapatite.** Pure hydroxyapatite has the formula $Ca_{10}(PO_4)_6(OH)_2$. The hydroxyapatite of bone incorporates anions, such as carbonate, and cations, such as sodium, into its structure. About 5% of the weight of bone mineral is carbonate $(HCO_3^{2-})$, with 1% being citrate and 0.5 to 1.0% being sodium and magnesium. The minerals that are part of the crystalline structure of bone are said to be "poorly exchangeable." These minerals can be released from the bone at appreciable rates only when bone is mobilized. Bone mobilization, or dissolution, is stimulated when plasma levels of 1,25-$(OH)_2D_3$ and PTH increase. The surfaces of bones also contain loosely bound minerals. These minerals are freely exchangeable, are in rapid equilibrium with the fluids of the body, and can be used by the various tissues of the body during a dietary deficiency of the mineral in question. The net loss of freely exchangeable minerals from bone does not require the participation of hormones. The freely exchangeable minerals of bone represent a small and perhaps insignificant pool of minerals.

*Turnover and Remodeling of Bone*

The material composing the weight-bearing, outer shell of bone is hard, compact, and dense. This bone is called **cortical bone**. The bone in the interior regions is porous and spongy and contains a latticework of rods, plates, and arches. This bone is called **trabecular bone**. The term *trabecula* refers to a small bar or ridge. Both cortical and trabecular bone normally undergo turnover. Bone turnover occurs in growing and adult bones and is carried out by two types of cells, osteoblasts and osteoclasts. Both types of cells cooperate in the turnover process. The osteoclasts appear on the surface of the bone. Over a period of about 2 weeks they bore tunnels (in cortical bone) or pits (in trabecular bone). Then the osteoclasts are replaced by osteoblasts, which fill in the cavities over a period of 3 to 4 weeks to create new bone. The remineralization of the cavity involves the initial formation by osteoblasts of the osteoid, which takes the form of a thin layer between the cells and the mineralized bone. The osteoid then "disappears" as bone mineral is deposited in it.

Osteoblasts and osteoclasts are involved in the bone remodeling process. One can easily see that certain parts of bone must be eroded and resorbed in the growing infant or child. The skull, for example, must be resorbed to make room for the growing brain. The osteoclast adheres to the bone and secretes acid in its zone of contact with the bone. A pH of about 3 may be produced in this zone of contact. Normally, bone mineral is very insoluble in water. An attempt to dissolve hydroxyapatite in water at pH 7 results in a solution of calcium ions with a concentration of about 0.1 m$M$. Bone mineral can, however, be dissolved in acid. Apparently, the bone mineral dissolved in the acid secreted in the region of contact can result in a solution of calcium reaching about 40 m$M$. The osteoclast is distinguished by its ruffled border.

Bone resorption involves removal of the organic constituents of bone, as well as of the mineral. Apparently, the osteoclast secretes a number of proteases, which are catalytically active in an acidic environment, into the region of contact. The mechanisms that control osteoclast activity during normal bone turnover and during vitamin D- and PTH-induced bone resorption are unknown. Vitamin D deficiency and rare diseases that impair the synthesis of $1,25\text{-}(OH)_2D_3$ result in a failure to mineralize bone. This failure is apparent in the epiphyseal plate of growing persons and results in a disease called rickets, in which the bones are deformed. Rapidly growing bones are the most severely affected. The bones become soft and bendable. Rickets in infants results in a delay in the closure of the fontanelle (soft spot) of the skull. It can also induce the formation of a series of bumps along the infant's ribs called the "rachitic rosary." The classical signs of rickets in older infants — those able to stand up — is bowed legs. The femur is so soft that it bends under the weight of the child. X-ray radiography can be used to diagnose the disease; it reveals the continued growth of the epiphyseal plate and its failure to mineralize on the face contacting the shaft of the bone. The overall effect is an increase in the thickness of the growth plate.

Vitamin D deficiency in adults cannot affect the epiphyseal plate, as it has disappeared, but it can prevent normal mineralization of the osteoid layer in bone that turns over. In vitamin D deficiency the osteoclasts continue to create tunnels and pits in the bone. The osteoblasts continue to synthesize the protein matrix; however, complete mineralization of the osteoid may not occur. The result is osteomalacia. This disease may present as bone pain about the hips. Osteomalacia can be diagnosed using a bone biopsy. A sample is taken from the iliac crest — the hip bone. An abnormally wide osteoid is indicative of the disease. X-rays can also be used to diagnose osteomalacia, which is characterized by arrays or zones of tiny fractures in such bones as the pelvis and femur.

Rickets and osteomalacia can be treated with daily oral doses of ergocalciferol (0.01–15.0 mg/day). Treatment is continued for 1 to 3 months and is coupled with adequate levels of dietary calcium and phosphate. Where the disease is due to a defect in 1-hydroxylase, treatment is with $1,25\text{-}(OH)_2D_3$.

## Bone and Atherosclerosis

Bone proteins play a role (unfortunately) in the process of atherosclerosis (Schwartz et al., 1995). Although atherosclerosis is not a disease of bone, it is a disease that involves bone matrix proteins, such as osteopontin, bone morphogenic protein, and osteocalcin. The mineralization of the atherosclerotic plaques that can develop in the coronary artery, and other arteries, contributes to the danger of this pathological structure. Specifically, the intimal layer of the artery becomes calcified.

## Difference between Osteomalacia and Osteoporosis

Osteomalacia and osteoporosis are two different diseases of the bone, though both result in an increase in the fragility of bone. Osteomalacia involves a decrease in the mineral content of the bone, with an increase in the content of the osteoid matrix. Here, the ratio of unmineralized/mineralized bone increases. Osteoporosis results in a decrease in the bone mass but no change in its histological appearance. Here, the ratio of unmineralized/mineralized bone is normal, representing more of a quantitative change than a qualitative change in the bone.

Osteoporosis is a widespread bone disease; osteomalacia is relatively rare. The former disease occurs in old age and most commonly affects postmenopausal Caucasian women. By the age of 65, about half of all persons show signs of osteoporosis. The disease results in fractures of the vertebra, hip, and wrist that occur either spontaneously or with minimal trauma, such as getting out of bed or opening a window. Osteoporosis involves the thinning of bone, enlargement of the cavities and canals in bone, and gradual loss of bone at a rate of 5 to 10% per decade. It results from the continued action of the osteoclasts in forming cavities in bone and the failure of the osteoblasts to fill in the cavities with osteoid and mineral.

The biochemical mechanisms that lead to osteoporosis are not clear. There is some thought that the disease can be aggravated by a decrease in the activity of

1-hydroxylase in elderly people and a consequent impairment in the absorption of dietary calcium. In addition, the ability of PTH to stimulate 1-hydroxylase may be impaired in those with the disease. The possible benefit of calcium supplements at levels higher than the RDA is controversial. The possible benefit of regular exercise, in the form of walking, dancing, and tennis, in preventing osteoporotic bone losses is not clear; it seems to preserve mineral in some bones but not in others.

## Phosphate Metabolism

Phosphate metabolism is regulated in a manner similar, but not identical, to calcium metabolism. The close association of these two nutrients is reasonable as they are both components of the bone mineral hydroxyapatite. Hydroxyapatite has the structure $Ca_{10}(PO_4)_6(OH)_2$. About 99% of the body's calcium and 85% of its phosphate is stored in bone. 1,25-Dihydroxyvitamin $D_3$ stimulates the absorption of dietary phosphate. This stimulation seems not to be dependent on the presence of dietary calcium. Plasma levels of phosphate are not as closely regulated as those of calcium, though both are regulated by 1,25-$(OH)_2D_3$ and PTH. The kidney is an important site for the regulation of plasma phosphate levels. This control is exerted by varying the extent of resorption of the phosphate filtered through the glomerulus.

## Treatment of Vitamin D Deficiency and Hazards of High Intake

Rickets, which is diagnosed by X-rays of leg bones, heals promptly with 4000 IU of oral vitamin D per day, with treatment for a month. In performing this treatment, the physician needs to monitor plasma 25-hydroxyvitamin D to make certain that they are raised to the normal range. The bone abnormalities (visible by X-ray) disappear gradually over the course of 3–9 months. Parents are instructed to take their infants outdoors for about 20 minutes per day with their faces exposed in order to prevent deficiency. Osteomalacia is treated by eating 2500 IU/day for about three months. Measurements of 25-hydroxyvitamin D, calcium, and parathyroid hormone are also part of the treatment process.

Food fortification has almost completely eliminated rickets in the United States. For those who cannot drink fortified milk and cannot go outside, supplements of vitamin D pills should be considered. In some elderly persons, a 400 IU supplement may not be enough to support normal calcium absorption by the gut, and daily doses of 10,000 IU per day may be needed.

One should realize that high doses of vitamin D are dangerous and can result in the permanent deposit of minerals in the heart, lungs, and kidneys. Symptoms of toxicity include nausea, vomiting, pain in joints, and loss of interest in eating food. Toxicity occurs in adults with eating 50,000 IU/day for an extended period of time. In infants, toxicity occurs with 1000 IU/day. Continued eating of toxic doses can lead to death. Ergocalciferol, rather than hormonally active forms of the vitamin, is used in vitamin D therapy, in order to reduce the chance of hypercalcemia (high plasma $Ca^{2+}$).

Vitamin D has recently found an association with psoriasis, a common disease of the skin. Psoriasis is not curable, but a variety of skin ointments can reduce the severity of the skin lesions. One of these ointments is a chemical analogue of vitamin D called calcipotriene. Calcipotriene was developed after initial observations that oral or topical calcitriol was effective against the disease. The drug results in improvement in 60% of patients (Greaves and Weinstein, 1995).

# VITAMIN A, VITAMIN D, AND THYROID HORMONE AT THE GENOME

The hormonal forms of vitamin A are all-*trans*-retinoic acid and 9-*cis*-retinoic acid. The hormonal form of vitamin D is 1,25-dihydroxyvitamin $D_3$, and that of thyroid hormone is T3. These hormones act within the nucleus, where they bind to special proteins. These proteins are classed as **transcription factors**. Various transcription factors bind to the regulatory regions of all genes and influence the rate of transcription (Figure 9.60). Many genes are continuously transcribed, and here the term "basal level of transcription" is used to describe the rate of transcription. In cases where the gene is regulated, special transcription factors are used to enhance or inhibit the basal level of transcription.

Usually, transcription factors bind to special regions just upstream of a specific gene, but in some instances they may bind somewhat downstream of a gene. The term "downstream" always means the same direction taken by RNA polymerase when it makes mRNA from the gene. The term "downstream" also means traveling from the 5'-end towards the 3'-end of the DNA. The term "genome" simply means all the DNA in the cell, i.e., the collection of all the coding and noncoding sequences in all of the chromosomes.

All-*trans*-retinoic acid binds a transcription factor called RAR, 9-*cis*-retinoic acid binds to RXR, 1,25-$(OH)_2D_3$ binds to VDR, and thyroid hormone binds to THR. Another transcription factor belonging to the presently discussed group is PPAR. PPAR is used in the control of fat cell formation. PPAR binds prostaglandins. The abbreviation "RAR" stands for retinoic acid receptor, "VDR" for vitamin D receptor, and "THR" for thyroid hormone receptor. "PPAR" is a trivial name that is not relevant to most of the work done on this transcription factor. A **ligand** is the general term for any hormone or activator that binds to any transcription factor. The ligand remains bound to the transcription factor when the transcription factor binds to the hormone response element, and the ligand remains bound when factor transcription provokes the activation of a gene.

RAR, RXR, VDR, THR, and PPAR are distinguished in that they usually act, not as a monomers, but as dimers. RXR is generally used as a partner in these dimers. These dimers, which are complexes of two proteins, include PPAR/RXR, RAR/RXR, RXR/VDR, and RXR/THR. When the dimers bind to DNA, they bind to regions that contain this sequence: GGGTCA (guanine–guanine–guanine–thymine–cytosine–adenine). This sequence is called the "half-core sequence." GGGTCA is only the most often occurring half-core sequence, and slight variations in this sequence are common. The half-core sequence actually occurs twice, where the first to occur usually binds RXR and the second binds the other member of the complex, i.e., VDR or THR. The entire sequence of DNA that binds PPAR/RXR is GGGTCANGGGTCA, where N means any nucleotide (A, T, G, or C). The entire sequence of DNA that binds the RXR/VDR complex is GGGTCANNNGGGTCA. The complete sequence of DNA that binds the RXR/THR complex is GGGTCANNNNGGGTCA. The full sequence of DNA that binds the RXR/RAR complex is GGGTCANNNNNGGGTCA. A major distinguishing difference between the sequences of DNA which bind these four heterodimer complexes is the number of bases occurring between the half-core sequences. Only one strand of the DNA helix was shown in these examples.

Table 9.7 summarizes the ligands, transcription factors, and response elements detailed in this section. The typical sequence of events occurring with gene activation is as follows (Figure 9.60):

1.  The ligand binds to RAR (or PPAR, THR, or VDR) to form a ligand/protein complex.

2.  The ligand/protein binds to RXR to form a heterodimer composed of two proteins. Usually, RXR does not contain its ligand, 9-*cis*-retinoic acid, during the scenario of gene activation. On the other hand, if 9-*cis*-retinoic acid does bind to RXR, the RXR may be prevented from participating in the heterodimer complex. In this way, the action of binding of 9-*cis*-retinoic acid to RXR may impair the activation of a specific gene.

3.  The heterodimer binds tightly to the regulatory sequence of DNA. The regulatory sequence containing the pair of GGGTCA sequences is called the *hormone response element*. The half-core sequences that bind to the RXR/RAR complex, for example, are separated by four nucleotides, where the exact identity of these nucleotides is usually not vital to the functioning of the

**TABLE 9.7**  List of Ligands, Transcription Factors, and Hormone Binding Elements

| Ligand | Transcription factor complex | Hormone binding element (both strands of DNA are shown) |
|---|---|---|
| Prostaglandin (proposed ligand) | PPAR/RXR | GGGTCANGGGTCA<br>CCCAGTNCCCAGT |
| 1,25-(HO)$_2$-Vit.D$_3$ | RXR/VDR | GGGTCANNNGGGTCA<br>CCCAGTNNNCCCAGT |
| T3 | RXR/THR | GGGTCANNNNGGGTCA<br>CCCAGTNNNNCCCAGT |
| all-*trans*-retinoic acid | RXR/RAR | GGGTCANNNNNGGGTCA<br>CCCAGTNNNNNCCCAGT |

The first protein listed, in the complex, binds to the upstream half-core, while the second protein listed binds to the downstream half-core. RXR binds to the upstream half-core for all of the complexes, except for the PPAR/RXR complex, where RXR binds to the downstream half-core.

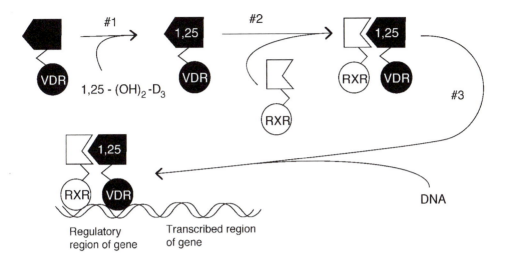

**Figure 9.60.** Sequence of events in gene regulation by hormone receptors, as illustrated by the vitamin D receptor and its binding to a vitamin D response element. VDR is shown some of its domains: a DNA binding domain (round shape), a ligand binding domain (pentagon), and a dimerization domain (point on the pentagon). *Step 1.* The ligand binds to VDR. *Step 2.* VDR binds RXR, forming a heterodimer. *Step 3.* The heterodimer binds to its hormone response element, located in the regulatory region of the gene. The binding of the heterodimer provokes an increase (or decrease) in the rate of transcription of a nearby gene.

hormone response element. Sometimes, the exact identities of the nucleotides just upstream or downstream of the hormone response element are important and may influence the event of gene activation.

4.  The heterodimer stimulates RNA polymerase to begin transcribing the gene. A typical stimulation is a 10- to 50-fold stimulation over the basal rate. As long as the transcription factor complex (with bound ligand) remains associated with the hormone response element, RNA polymerase acts over and over again, at the gene to make more and more copies of the mRNA.

Which genes are activated by the binding of the heterodimer complexes? Insight into these genes was supplied by a number of nutritional, pharmacological, and genetic studies. Techniques, such as dietary vitamin deficiency, vitamin overdose, and use of vitamin analogues and antagonists have proven useful in the past. In recent years, the functions of genes in mammals has been probed by use of the "knock-out" technique, where specific genes are deleted or "knocked out" in the fetal animal, usually a mouse. In the case of some genes, the resulting knock-out mice develop normally and continue to thrive, and it may take much effort before any consequent change in metabolism is discovered, and thus much effort to find the use of the gene, if any. In other cases, severe anatomical defects occur in the fetal mouse and the animal dies shortly before or after birth, again making it difficult to determine the use of the gene.

## Response Elements and the Activation of Genes

### Adipocyte Development

The PPAR/RXR complex, apparently, can activate transcription when either PPAR or RXR contains its ligand (Mangelsdorf and Evans, 1995). The PPAR/RXR complex provokes the differentiation of precursor cells to adipocytes. Specifically, this complex activates the genes coding for malic enzyme, acyl-CoA synthase, enoyl-CoA hydratase, PEPCK, and other enzymes (Ijpenberg et al., 1997). Malic enzyme catalyzes the conversion of malic acid to pyruvate, resulting in the production of NADPH. NADPH is needed for fatty acid synthesis. The hormone response element in the malic enzyme gene is: <u>GGGTCA</u>AA<u>GTTGA</u> (Ijpenberg et al., 1997). The half-core sequences are underlined. Please note that the downstream half-core does not perfectly match the typical half-core. Further details of PPAR are revealed in the Obesity chapter.

### Bone Remodeling

Vitamin D regulates the genes for several proteins, including osteocalcin, osteopontin, $\beta 3$-integrin, vitamin D 24-hydroxylase, and parathyroid hormone. In all cases, the ligand for VDR is 1,25-$(OH)_2$-$D_3$ and in all cases the partner protein RXR does not bind its ligand. An increase in plasma 1,25-$(OH)_2$-$D_3$ can trigger bone remodeling by stimulating various genes in osteoblasts and osteoclasts. Osteocalcin, for example, functions to halt bone matrix formation, where this activity promotes the resorption of bone and an increased delivery of calcium ions to the bloodstream. At the same time, an increase in plasma 1,25-$(OH)_2$-$D_3$ can trigger its feedback inactivation by stimulating the synthesis of vitamin D 24-hydroxylase. This enzyme catalyzes the inactivation of 1,25-$(OH)_2$-$D_3$ by converting it to 1,24,25-$(OH)_3$-$D_3$. 1,25-Dihydroxyvitamin $D_3$ decreases the expression of the gene coding for parathyroid hormone. This constitutes a type of feedback inhibition because parathyroid hormone functions to stimulate the synthesis of 1,25-$(OH)_2$-$D_3$. The hormone response element that resides in the gene for parathyroid hormone is <u>GGTTCA</u>AAG<u>CAGACA</u> (Darwish and DeLuca, 1996; Mackey et al., 1996). The half-core sequences are underlined, and these sequences are separated by three nucleotides.

### Metabolic Rate

Thyroid hormone, when complexed with RXR/THR, regulates the number of mitochondria in the cell as well as the function of mitochondria. Several mitochondrial proteins, including cytochrome c oxidase, NADH dehydrogenase, and the $\beta$-subunit of $F_0F_1$-ATPase (the enzyme that makes most of the ATP in the body) are induced by thyroid hormone (Koiduchi et al., 1996; Almeida et al., 1997). Thyroid hormone regulates the activity of an important protein of the plasma membrane, Na,K-ATPase. This enzyme is used for transporting sodium out of the cell, and indirectly for driving other transport systems which are forced to operate by the return of sodium back into the cell. Thyroid hormone provokes an increase in

Na,K-ATPase activity. The hormone response element for Na,K-ATPase or, more accurately one of its subunits, is <u>AGGTCA</u>CTCC<u>GGGACG</u> (Feng *et al.*, 1993). The half-core sequences are underlined. Note that four nucleotides separate the half-core sequences, and that one of these sequences deviates from the typical AG-GTCA sequence. Thyroid hormone's stimulation of mitochondrial proteins and Na,K-ATPase is consistent with this hormone's action in raising the body's metabolic rate.

In the developing brain, TH regulates nerve growth, nerve migration, and the formation of connections between nerves. These events are probably the result of thyroid hormone's regulation of the genes coding for various proteins of the nervous system, including myelin basic protein, PCP-2 protein, and proteolipid protein (Oppenheimer and Schwartz, 1997).

## *Embryonic Development*

RXR/RAR activates the HOX gene. Here, RAR contains all-*trans*-retinoic acid, while RXR is unoccupied. The HOX gene actually consists of a family of some 40 related genes. The hormone response element for the HOXA-1 gene is <u>GGTTCAC</u>-CGAA<u>AGTTCA</u> (Giguere, 1994). All of the HOX genes code for transcription factors. Thus, the overall scenario is that one transcription factor (RXR/RAR) controls the synthesis of other transcription factors (HOX proteins). The HOX family of transcription factors is used in the developing embryo, i.e., for regulating the development of the skeleton and other tissues. The HOX gene is used for controlling the positioning of the limbs and brain of the developing embryo (Marshall *et al.*, 1996, Roy *et al.*, 1995; Morrison *et al.*, 1996). Using standard genetic techniques, knock-out mice lacking RAR were produced. As the mouse fetuses developed, they had skeletal abnormalities. Defects in the mouth, eye, spinal cord and feet also occurred (Morriss-Kay and Sokolova, 1996). Some, but not all, of these abnormalities also occur when pregnant mother mice are fed vitamin A deficient diets. This picture represents an exciting meeting of classical nutritional science and contemporary mouse genetics.

## *Four Domains in the Hormone Receptors (Transcription Factors)*

The described transcription factors consist of a polypeptide of about 400 amino acids, and contain special regions, or domains, that are used for four purposes (Liu *et al.*, 1997):

1. To bind double-stranded DNA at the site of the response element.

2. To bind the partner protein in the heterodimer.

3. To bind the ligand.

4. To transmit a signal to RNA polymerase, thus provoking an increase in the frequency of repeated transcription.

The VDR for example, consists of 427 amino acids. A number of naturally occurring mutations have been found to occur in the human gene coding for the vitamin D binding protein (VDR). Gly-33-Asp and Arg-73-Gln are but two examples of these mutations. These mutations result in permanent impairment in the functioning of the DNA-binding domain, and result in rickets in the person bearing the mutation (Haussler *et al.*, 1997). The abbreviation "Gly-33-Asp" means that the glycine, which normally occurs at the 33rd amino acid of the polypeptide chain, has been mutated to aspartate.

The structure typical of hormone receptors, and its constituent domains, is as follows:

This discussion specifically applies to RAR and VDR, but the overall scheme may apply to all hormone-binding proteins that bind to response elements. The **A domain** and **B domain** are used for the activation process, and may contact the basal transcription factors. The **C domain** is the DNA binding domain, and contains zinc fingers. The C domain also contains the nuclear localization sequence, i.e., a stretch of amino acids that is used to bring the entire protein into the nucleus, shortly after it as been created on the ribosome. The **D domain** contains a hinge region, and may maintain specific orientations between the C domain and E domain. The **E domain** is the ligand binding region. Regions of the D domain and E domain are used to maintain dimerization, i.e., for binding to the RXR protein. The **F domain** plays a part in activating transcription (Petkovich, 1992; Chambon, 1996; Haussler *et al.*, 1997).

EXERCISE

One type of genetic disease, resulting in rickets, can be cured by large doses of vitamin D. This disease involves a mutation in one of the four domains mentioned earlier. In which domain does this mutation occur?

## *The Zinc Finger — A Structure in DNA Binding Proteins*

The domain of the hormone receptor that is directly involved in binding to DNA contains one or more short stretches of amino acids known as a **zinc finger** (Figure 9.61). The zinc finger sequence is about 30 amino acids long, and binds a zinc atom with specific residues of cysteine and histidine. The anion of sulfur (on Cys) and the lone pair electrons (on the N of histidine's ring) interact with and bind the positively charged zinc atom. RAR, RXR, VDR, THR, and PPAR all contain zinc fingers, and all probably require zinc atoms for binding to DNA. The zinc-binding region occurs in the C domain. Other transcription factors, such as those that bind steroid hormones, also contain zinc fingers.

## Vitamin A and Cancer

Vitamin A deficiency in animals leads to a greater frequency of various types of cancer. Most of these animal studies involved the induction of cancer using chemical carcinogens, that is, by feeding, injection, or application to the skin.

Epidemiological studies on human populations have suggested that humans deficient in vitamin A may be at a greater risk for certain types of cancer. The converse situation, namely the treatment of cancer patients with vitamin A, has

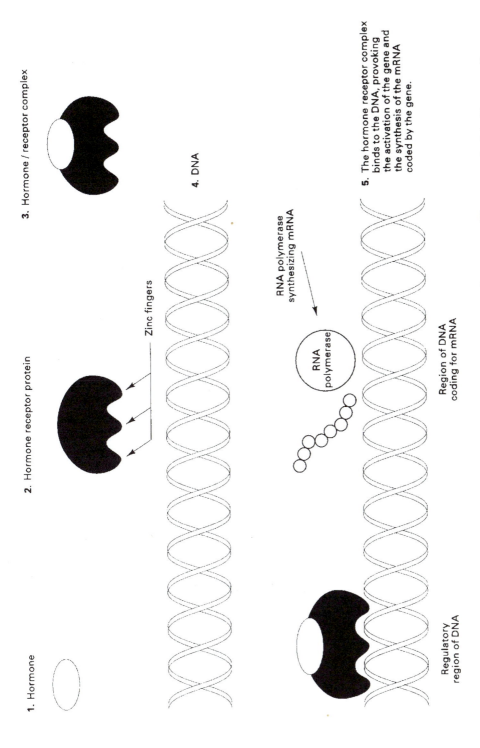

1. Hormone

2. Hormone receptor protein

Zinc fingers

3. Hormone / receptor complex

4. DNA

RNA polymerase
synthesizing mRNA

RNA
polymerase

Regulatory
region of DNA

Region of DNA
coding for mRNA

5. The hormone receptor complex
binds to the DNA, provoking
the activation of the gene and
the synthesis of the mRNA
coded by the gene.

**FIGURE 9.61** Diagram of hormone receptor and its binding to a regulatory region on the genome. The events depicted here are the same as those depicted in Figure 9.60, but with an added schematic detail showing the location of zinc fingers in a hormone-binding protein.

yielded some provocative results. Certain cancers of the skin, mouth, larynx, liver, cervix, lung, and one type of leukemia (promyelocytic leukemia) may respond somewhat in some cases, or dramatically in others, to treatment with retinoic acid (Lotan, 1996; Shafritz, 1996). Clear and definite data have revealed that some types of cancer cells synthesize reduced levels of one of the retinoic acid receptors (RARs) (Minna and Mangelsdorf, 1997). Evidence suggests that the curative effect of retinoic acid, in this case, is due to its restoration of normal levels of RARs. Synthetic retinoic acid compounds, which selectively bind to only one of the three types of RAR (or to one of the three types of RXR), are expected to be more powerful and more selective as anticancer agents than naturally occurring forms of retinoic acid.

## Relation of Vitamin Binding Proteins to Other Transcription Factors

Several families of transcription factors exist. These include basal (or general) transcription factors, activators, and repressors. The basal transcription factors include transcription factor IIA (TFIIA), TFIIB, TFIID, TFIIE, TFIIF, and TFIIH. Most of these transcription factors exist as multiprotein complexes. These transcription factors must be assembled just upstream of the transcribed part of the gene before RNA polymerase begins its catalytic activity. When assembled, the entire "mega-complex" consists of about 50 proteins (Jacobson and Tjian, 1996).

TFIID is involved at the beginning of the assembly of the mega-complex. TFIID binds to the sequence thymine–adenine–thymine–adenine (TATA), which occurs about 30 base pairs upstream of the transcription start site. TFIIB and TFIIF recruit RNA polymerase and incorporates this enzyme into the mega-complex. The correct term for this mega-complex is an *initiation complex*. TFIIA is necessary for regulating the rate of formation of the initiation complex (Geiger *et al.*, 1996). Once RNA polymerase is stabilized at the transcription start site, TFIIE and TFIIH act by provoking RNA polymerase to escape from the initiation complex, and to begin polymerizing ribonucleotides to create mRNA (Zawel and Reinberg, 1995).

These events are illustrated in the simple diagram that follows. The region called the promoter consists of about 20 base pairs. The promoter binds RNA polymerase. The transcription start site begins a few base pairs downstream of the promoter. The TATA sequence occurs about 30 base pairs upstream of the transcription start site. Response elements can reside a bit further, or quite a bit further, upstream of the TATA sequence. The sizes of the various regions in the diagram are not to scale:

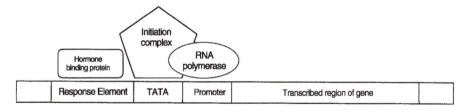

A variety of special transcription factors regulate the behavior of the initiation complex. These proteins include JUN, FOS, and CREBP. CREBP (cAMP response element binding protein) is a protein that specially responds to cyclic AMP. JUN and FOS are used in the regulation of cell division. Further details on these two transcription factors appear in the Diet and Cancer chapter. Other special transcription factors require binding by a hormone. These include the **RAR/VDR/THR/PPAR family**, which usually occur as heterodimers, and the **steroid hormone binding proteins**, which occur as homodimers. The steroid hormone binding proteins occur as homodimers and bind aldosterone, glucocorticoids, estrogen, androgen, or testosterone.

How do the special transcription factors regulate the activities of the basal transcription factors? Information on this issue is only beginning to become available. One might expect, for example, to find a scenario where RXR/VDR directly contacts TFIIA, to stimulate the activity of the initiation complex.

The researcher always needs to take care in the quest for discovering new response elements and new transcription factors. Just because a sequence in DNA resembles the order of bases in an established response element does not mean that it actually binds any transcription factor. Just because a specific DNA sequence binds a transcription factor does not mean that it actually functions to regulate any gene in the living cell.

## NIACIN

Niacin is a water-soluble vitamin. The RDA of niacin for the adult man is 19 mg. Niacin is converted in the body to the cofactor **nicotinamide adenine dinucleotide** (NAD). NAD also exists in a phosphorylated form, NADP. The phosphate group occurs on the 2-hydroxyl group of the AMP half of the coenzyme. NAD and NADP are used in the catalysis of oxidation and reduction reactions. These reactions are called redox reactions. NAD cycles between the oxidized form, NAD, and the reduced form, NADH + H$^+$. The coenzyme functions to accept and donate electrons. NADP behaves in a similar fashion. It occurs as NADP$^+$ and NADPH + H$^+$. The utilization of NAD is illustrated in the sections on glycolysis, the malate–aspartate shuttle, ketone body metabolism, and fatty acid oxidation. The utilization of NADP is illustrated in the sections concerning fatty acid synthesis and the pentose phosphate pathway.

The term *niacin* refers to both nicotinic acid and to nicotinamide. Niacin in foods occurs mainly in the cofactor form, NAD and NADP, and their reduced versions. NAD is hydrolyzed by enzymes of the gut mucosa to yield nicotinamide. Dietary NAD can also be hydrolyzed in the gut mucosa at the pyrophosphate bond to yield nicotinamide nucleotide. Nicotinamide and nicotinamide nucleotide are then broken down, possibly by enzymes in the gut and liver, to yield nicotinic acid. The conversion of nicotinic acid to NAD (nicotinamide adenine dinucleotide) is shown in Figure 9.62. The first step involves the transfer of a ribose phosphate group from PRPP to nicotinic acid, forming nicotinic acid nucleotide. The second step involves the transfer of an ADP group from ATP, forming nicotinic acid adenine dinucleotide. The final step is an amidation reaction. Here, glutamine donates its amide group to a carboxyl group, forming NAD. The asterisk indicates the point of attachment of the phosphate group of NADP.

Conversion of nicotinic acid to NAD is illustrated by the following experiment involving mice (Figure 9.63). The animals were injected with carbon-14-labeled nicotinic acid. The livers were removed at the indicated times — 0.33, 1.0, 3.0, and 10 minutes — and used for analysis of the radioactive metabolites. At 20 seconds, unchanged nicotinic acid (○) was the major metabolite. At 1 to 3 minutes, there was a temporary accumulation of nicotinic acid ribonucleotide (∇) and deamido-NAD (△). NAD (●) was the major metabolite after 3 minutes.

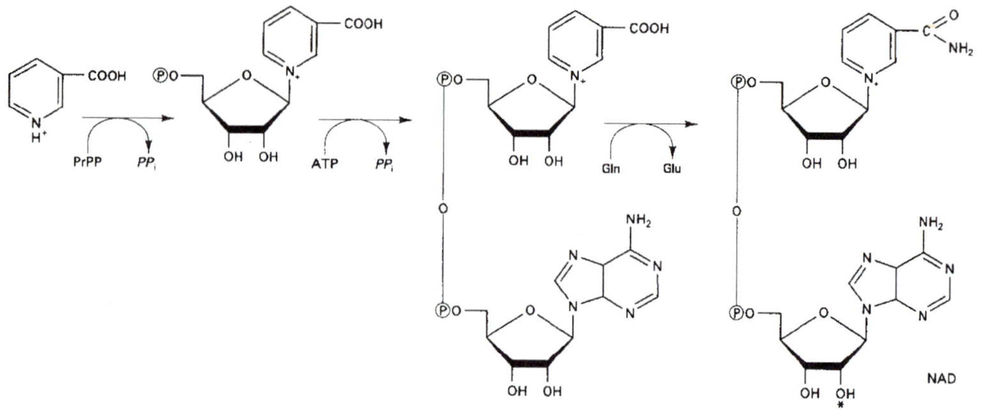

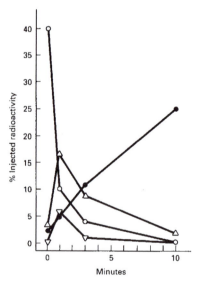

**FIGURE 9.62** Conversion of nicotinic acid to NAD.

## Biochemistry of NAD

NAD tends to be an electron acceptor in *catabolic reactions* involving the degradation of carbohydrates, fatty acids, ketone bodies, amino acids, and alcohol. NAD is used in energy-producing reactions. NADP, which is cytosolic, tends to be involved in *biosynthetic reactions*. Reduced NADP is generated by the pentose phosphate pathway (cytosolic) and used by cytosolic pathways, such as fatty acid biosynthesis and cholesterol synthesis, and by ribonucleotide reductase. The niacin coenzymes are used for two-electron transfer reactions. The oxidized form of NAD is $NAD^+$. There is a positive charge on the cofactor because the aromatic amino group is a **quaternary amine**. A quaternary amine participates in four

**FIGURE 9.63** Illustration of the conversion of nicotinic acid to NAD using radioactively labeled nicotinic acid. (Redrawn with permission from Ijichi *et al.*, 1966.)

covalent bonds. NAD-dependent reactions involve the transfer of two electrons and two protons. NAD accepts two of the electrons and one of the protons; the remaining proton remains in solution. Hence, the reduced form of NAD is not written as $NADH_2$, but as $NADH + H^+$ (Figure 9.64).

The niacin coenzymes might be compared with the riboflavin coenzymes. Niacin coenzymes are used by enzymes for the transfer of **two electrons at a time**, where both electrons are transferred without the accumulation of a one-electron reduced intermediate. The riboflavin coenzymes, in accepting two electrons, can accept **one electron at a time**, with a detectable free radical intermediate. Another difference is that niacin coenzymes do not readily react with molecular oxygen, whereas riboflavin coenzymes can form covalent bonds with oxygen. Hence, flavoenzymes are used for introducing oxygen, from $O_2$, into various metabolites. Another difference is that the reducing power of $NADH + H^+$ is greater than that of $FADH_2$. Electrons from reduced NAD are often transferred to flavoproteins, resulting in a reduced flavoprotein; that is, the FAD cofactor is converted to $FADH_2$. The reverse event, namely the reduction of NAD by $FADH_2$, does not tend to occur. This point is illustrated by the fact that $NADH + H^+$ can generate more ATPs in the respiratory chain than can $FADH_2$.

## NAD Used for Nonredox Purposes

*Posttranslational Modification of Proteins*

NAD is used in posttranslational modification of a variety of proteins, notably some of the proteins of the chromosomes. The chromosomes are composed of DNA, histones, and nonhistone proteins. The histones, which are distinguished by their high content of basic amino acids, serve as a scaffold and maintain the coiled and folded structure of the DNA. The other proteins are used in regulating the expression of specific genes. Poly(ADP-ribose) polymerase catalyzes the attachment of ADP-ribose to various chromosomal proteins. This modification, shown in Figure 9.65A, is more dramatic than a simple methylation or phosphorylation. The enzyme uses NAD as a substrate. Here, NAD does not serve its usual role as an oxidant or reductant. The ADP-ribosyl moiety of NAD is donated to the acceptor protein. A molecule of nicotinamide is discharged with each event of

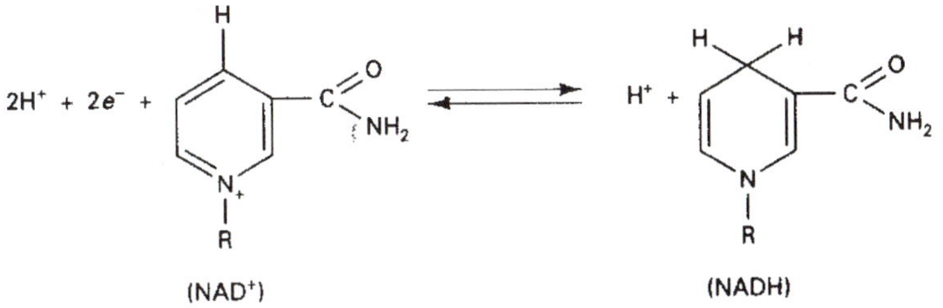

**FIGURE 9.64** NAD accepts two electrons and one proton. When used as a substrate by enzymes, NAD is a 2-electron acceptor and a 2-electron donor. However, when acting purely as a chemical, in the absence of enzymes, NAD (or $NADH + H^+$) can accept or donate a single electron.

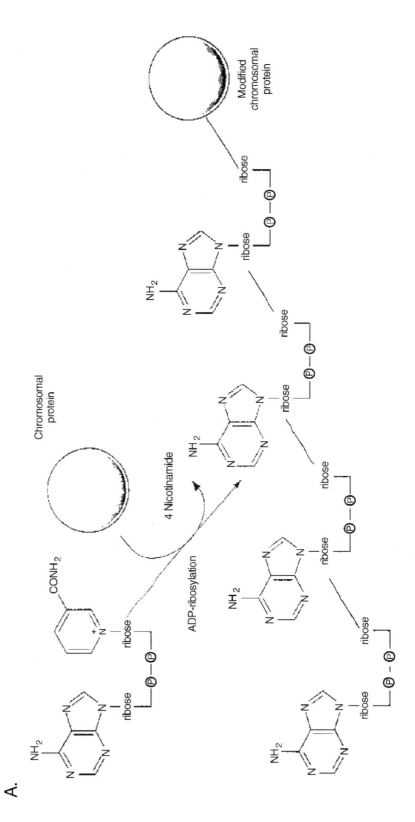

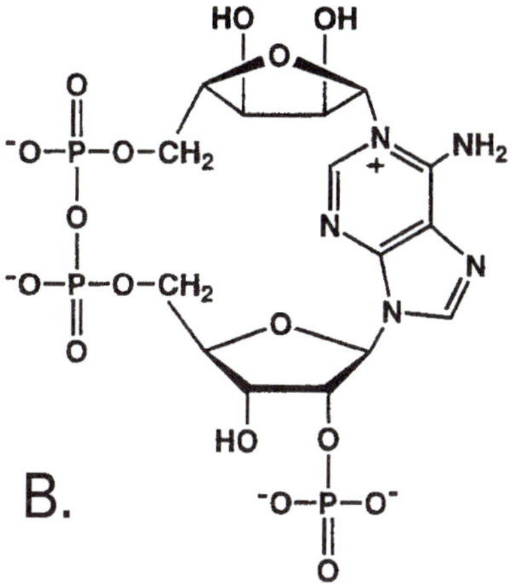

**FIGURE 9.65  A** (facing page). Involvement of NAD for ADP-ribosylation reactions. Four molecules of NAD are shown for the synthesis of a polymer consisting of four residues of ADP-ribose. **B** (above). Cyclic ADP-ribose, a molecule thought to be used for cell-signaling, is made from NAD.

ADP-ribosylation. In the case of some proteins, multiple ADP-ribose units are transferred to the protein, forming a poly(ADP-ribose) group. The activity of poly (ADP-ribose) polymerase is increased during cell growth, cell differentiation, and recovery from damage to the DNA. The enzyme is involved in controlling the repair of DNA (Satoh *et al.*, 1993).

### Cyclic ADP-Ribose, a Molecule Used in Cell-Signaling

NAD serves as a donor of an ADP-ribose group for the synthesis of **cyclic ADP-ribose**. ADP-ribosyl cyclase catalyzes the synthesis of cyclic ADP-ribose (Figure 9.65B). This molecule works within the cell and provokes the release of calcium ions from an internal storage site (Jacobson *et al.*, 1997). The overall concentration of cyclic ADP-ribose in various organs is about 2.0 micromolar (Dousa *et al.*, 1996). The action of cyclic ADP-ribose in the cell is thought to be similar to that of another signaling molecule, IP3 (see Calcium section).

NADP, the phosphorylated version of NAD, is converted in the cell to nicotinic acid adenine dinucleotide phosphate (NAADP). NAD cannot be used directly for this purpose. NAADP seems also to be used as a cell-signaling molecule, and to provoke a burst of intracellular calcium ions (Aarhus *et al.*, 1996; Dousa *et al.*, 1996).

## History of Niacin Nutrition

The nutritional history of niacin is unusually colorful and involves Italy, Mexico, and the American South. The documentation of niacin deficiency seems to have started a long time ago, when corn was brought to Spain by Columbus. Corn is distinguished as being low in niacin. The introduction of corn to Europe resulted in its widespread cultivation; it replaced rye as a major crop. Corn became popular in Spain, southern France, Italy, Rumania, southern Russia, and Egypt. **Pellagra**, a disease resulting from severe niacin deficiency, followed the cultivation of corn. This disease is characterized by severe dermatitis and fissured scabs, diarrhea, and mental depression. The disease is associated with "the four Ds": dermatitis, diarrhea, dementia, and death. Because of the third "D," mental hospitals proved to be a convenient place to find and study the disease.

Pellagra was noted in the 1760s in Spain and Italy. A special hospital was employed for pellagrins in Legano, Italy, in 1784. Later, in 1917, a special hospital was set up in Spartanburg, South Carolina, for those suffering from the disease. In America, during the depression of the 1930s, about a quarter of a million cases of pellagra occurred, with 1 case in 35 being fatal.

A variety of theories were devised to explain the disease. These theories involved toxic reactions to moldy corn or to the consumption of rancid vegetable oil and bacterial infections. It was realized that pellagra was associated with poverty, the corn-eating practices of the poor, and the fact that corn was sometimes moldy. This realization resulted in an attempt to reduce the spoilage of corn. The French government reduced the consumption of corn, and the disease disappeared from France by 1900. Alternate approaches or ideas arose in Italy and America. The Italian government constructed municipal corn driers to reduce the growth of mold. In America, it was believed that pellagra was caused by eating vegetable oils rather than animal fat, because vegetable oils, which contain polyunsaturated fatty acids, spoil relatively easily. Approaches based on either of these two would not be expected to have any effect on the disease.

Joseph Goldberger, in America, provided definitive results regarding the cause of pellagra. Goldberger was interested in determining if pellagra was an infectious disease. He took blood samples from pellagra patients and injected them into himself. He consumed their scaly skin, urine, and feces (not directly, but mixed with food). Goldberger and his co-workers did not become ill, though the pellagrins continued to die. In later studies in South Carolina, Goldberger found that the disease could be cured by milk. He concluded that pellagra was not infectious but resulted from the elimination of meat and milk from the diet.

After experimenting with rats and other animals, Goldberger found that dogs consuming a pellagric diet developed a condition he called "black tongue disease." This disease included signs found in humans, such as diarrhea and an inflamed mouth. In the 1930s, after niacin was found to have a biochemical role as a coenzyme, researchers at the University of Wisconsin discovered that niacin could cure black tongue disease. From there, it was demonstrated that niacin could cure pellagra.

The list in Figure 9.66 was translated from a publication appearing in Italy in 1905. The list contains some sound advice, such as "consume milk" and "do not be ashamed to go to the doctor," as well as advice later found to be irrelevant.

How can milk and alkali-treated corn prevent pellagra? The history of pellagra contains some apparent contradictions. A description of these apparent contradictions may bring into vivid view some interesting aspects of the nutrition of niacin. One of the contradictions involves Central Americans and the other concerns milk. It was realized that Mexicans resembled Italians in being extremely poor and in eating corn; however, it was known that Mexicans did not suffer from pellagra. There was some thought that the resistance of Mexicans to pellagra resulted from

## ADVICE AND RULES FOR AVOIDING PELLAGRA

SPOILED CORN IS THE CAUSE OF PEL-LAGRA, and corn readily becomes spoiled, moldy or poisonous when harvested too early, before it is ripe, and stored in places which are damp or poorly ventilated; it may also be of poor quality when imported from some other place and may contain a large percentage of damaged grains.

CINQUANTINA is a variety of corn which should be used only for feeding animals.

SPOILED CORN MAY BE RECOGNIZED by its pale or greenish color, by the shriveled and cracked surface of the grains which are also covered with greenish, bluish or brownish spots, by its musty odor and its bitterish, disgusting taste.

The damaged corn also weighs less than sound corn; and the surface of spoiled grain lacks the shining appearance of the sound article.

KEEP your corn then in places WELL DRIED AND AIRED.

Distrust WHITE CORN because it is more likely to spoil than other kinds.

IMPORTED CORN IS FREQUENTLY DAMAGED.

\* \* \*

Keep watch over your corn while it is being ground in the MILLS of the country.

If you have carried good corn to the mill, see to it that you receive meal ground from that corn, and do not allow the miller to substitute meal ground from inferior grain.

\* \* \*

If your corn is GROUND BY A ROLLER MILL (a cilindro), the spoiled grains are not likely to be ground into the meal.

\* \* \*

The establishment of a COMMUNAL, CO-OPERATIVE OR INDEPENDENT BAKERY is a great benefit in a section where pellagra is very severe.

The utmost care should be used when one undertakes the PREPARATION AND BAK-ING OF BREAD AT HOME.

\* \* \*

Instead of SPENDING YOUR MONEY ON WINES AND LIQUORS, buy wheat bread; limit your use of polenta. If you have milk, eggs, cheese, limit the sale of these articles to others, and use at least a part of such products for your own home food.

\* \* \*

REDUCE the cultivation of corn; EXTEND cultivation on your low lands and do not grow corn in mountainous regions or in very moist soils.

\* \* \*

DO NOT BE ASHAMED to go to the doctor if you are a pellagrin; and HAVE YOURSELF ENTERED AT THE LOCANDA SANITARIA or at the ECONOMIC KITCHEN.

Get cured in time and so avoid the HOSPI-TAL or the INSANE ASYLUM.

\* \* \*

Remember that the LAW AGAINST PEL-LAGRA requires a CURATIVE DIET for pellagrins. It is your RIGHT to demand it, and your DUTY to procure it.

\* \* \*

The CLEANLINESS and HEALTHFUL-NESS Of YOUR HOMES are necessary conditions for preventing the moulding of corn which is kept in your houses.

Never keep your corn in BED ROOMS, and see to it that you have proper places for the STORING and SEASONING OF YOUR GRAIN.

KEEP the corn DRY.

\* \* \*

PROFIT BY YOUR INSTRUCTION IN AG-RICULTURE, and better your crops.

\* \* \*

EXERT YOURSELVES TO CO-OPERATE WITH OTHERS FOR YOUR OWN SALVA-TION by acquiring knowledge of, and interest in the application of the LAW AGAINST PELLAGRA.

**FIGURE 9.66** Advice and rules for avoiding pellagra. (Redrawn with permission from Carpenter, 1981.)

their custom of cooking corn in hot alkali, as it might be expected that this treatment would kill the mold on the corn. The solution to the problem is as follows. Corn is low in niacin; however, treatment of corn with hot alkali (calcium hydroxide) results in the degradation of a compound in the plant called "bound niacin." This degradation produces a number of products, including free niacin. The treatment of corn with calcium hydroxide is a step used in the manufacture of corn tortillas. It might also be noted that the Hopi of Arizona harvest corn in the immature, milky stage. Corn at this stage contains ample amounts of niacin and hence would be expected to protect against pellagra.

The following study illustrates the effect of alkali treatment on the growth-promoting properties of corn. The growth-promoting effect was attributed to the release of bound niacin. Rats consumed diets in which the only source of niacin was corn meal (O) or alkali-treated corn meal (●). Other rats consumed corn meal diets (△) or alkali-treated corn meal diets (▲) supplemented with niacin (Figure 9.67).

EXERCISE

Was the quantity of bound niacin released during alkali treatment sufficient to support maximal growth of the rats (Figure 9.67)?

As mentioned earlier, milk was often used to treat pellagra; however, neither milk nor eggs contain very much niacin. The question arises: How can milk reverse the symptoms of a disease known to result mainly from niacin deficiency? The answer lies in a consideration of the pathway of catabolism of one of the amino acids, tryptophan. The breakdown of tryptophan may follow a number of different routes, including that shown in Figure 9.68. The final product of this pathway is nicotinic acid ribonucleotide, which can be converted to NAD. A small fraction of

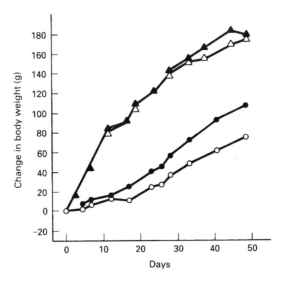

**FIGURE 9.67** Effect of alkali treatment on the growth-promoting properties of corn. (Redrawn with permission from Squibb *et al.*, 1959.)

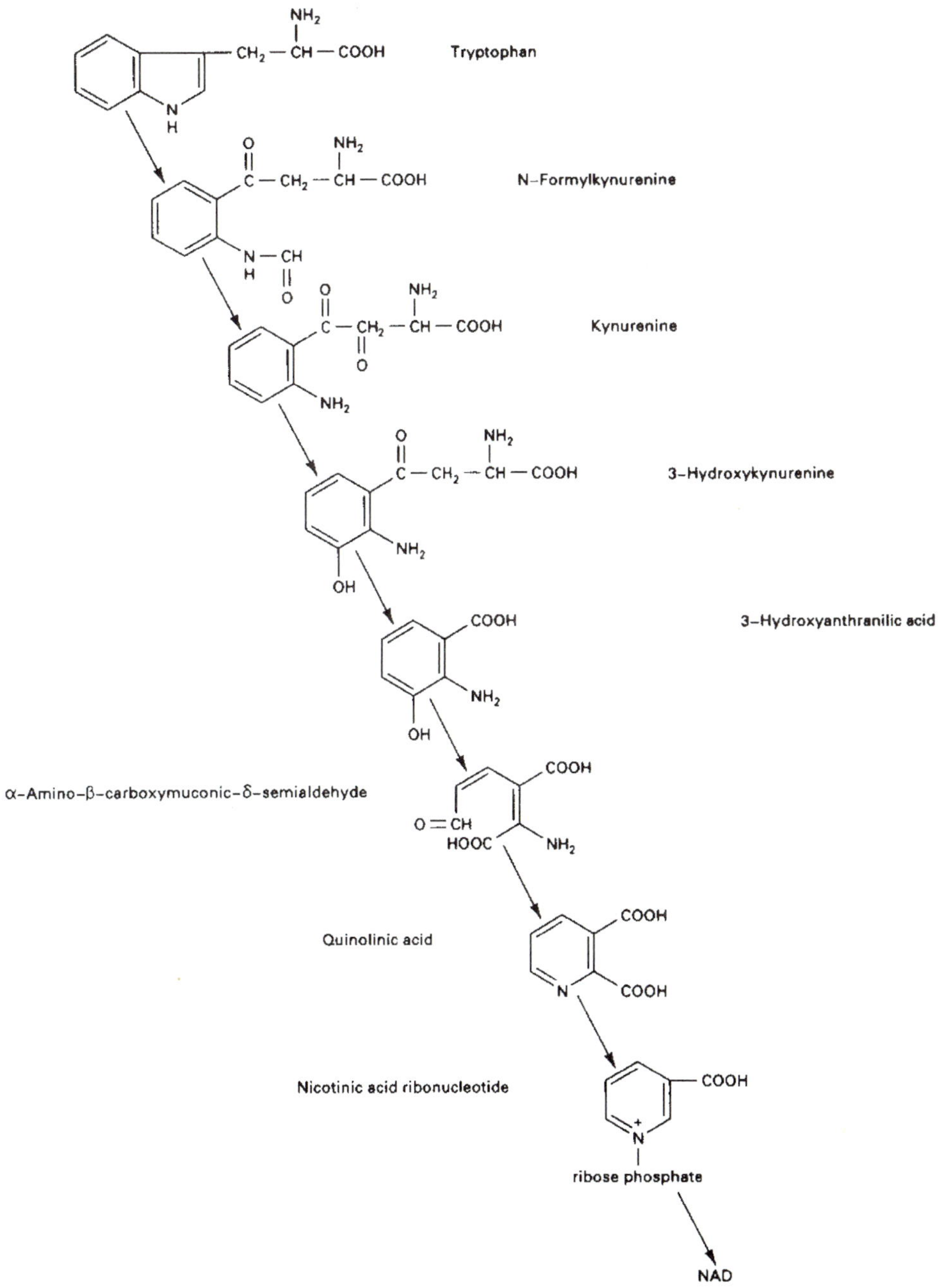

**FIGURE 9.68** Biosynthesis of NAD.

the tryptophan consumed each day, in the form of dietary proteins, is degraded to form NAD. It has generally been agreed that 60 mg of dietary tryptophan is equivalent to 1.0 mg of niacin. For this reason, the RDA for niacin is expressed as "mg of niacin equivalents" rather than "mg of niacin." Milk protein is high in tryptophan and thus is, in effect, a good source of niacin. Milk and egg protein are about 1.4% tryptophan, by weight, whereas corn protein contains about 0.6% tryptophan. Corn that is not treated with alkali is a poor source of niacin and a relatively poor source of tryptophan.

EXERCISE

The average diet in the United States may supply 0.7 to 1.1 g of tryptophan per day. What percentage of the RDA of niacin might be expected to be supplied by this amino acid?

## Assessment of Niacin Status

That nongrowing animals require niacin implies that it is lost from the body either as intact niacin or as a modified or breakdown product of the vitamin. An amount of niacin equivalent to nearly 90% of our daily intake is excreted in the forms of N-methyl-2-pyridone-5-carboxamide (pyridone for short) and N-methyl-nicotinamide. The structures of these compounds are shown in Figure 9.69. Measurement of these urinary metabolites can be used to assess niacin status. Loss of the normal quantity in the urine each day indicates that the supply in the diet is adequate. In humans, the healthy adult excretes 4 to 6 mg of N-methyl-nicotinamide per day. An abnormally low level indicates that the dietary intake is not adequate. Measurement of urinary niacin metabolites has proven useful in determining the amount of niacin available in a variety of foods. The body's ability to use niacin in different foods may vary even if the foods contain identical quantities of the vitamin. One contributing factor to the low availability of niacin is the occurrence of the vitamin in the "bound form," as mentioned earlier. Excretion of normal levels of pyridone, for example, depends not only on normal absorption of the vitamin from the diet, but also on its conversion to NAD or NADP, followed by catabolism to the metabolite.

There is some interest in using the ratio of NAD/NADP in red blood cells as a convenient measure of niacin status (Jacobson and Jacobson, 1997). How could this assay work? Data from human studies has shown that niacin deficiency allows a decline in intracellular NAD, but a maintenance of the levels of NADP.

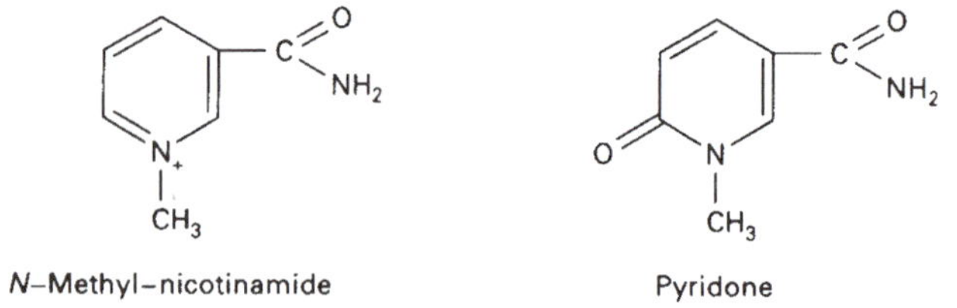

N–Methyl–nicotinamide                Pyridone

FIGURE 9.69  Excretory forms of niacin.

## Pharmacological Use of Nicotinic Acid

Large doses of nicotinic acid are often used for the treatment of atherosclerosis and to lower blood cholesterol. In detail, nicotinic acid results in a lowering of LDL-cholesterol ("bad cholesterol"), in increase in HDL-cholesterol ("good cholesterol"), and a decrease in plasma triglycerides (Blankenhorn *et al.*, 1993; Holvoet and Collen, 1995; Illingworth *et al.*, 1994; McKenney *et al.*, 1994). Therapy involves daily doses of 1.5–4.0 grams of niacin per day. The drug may produce troublesome side effects, such as liver damage, nausea, and diarrhea. Flushing of the skin (due to dilation of arteries) occurs as a side effect when nicotinic acid therapy is started, but may disappear with continued therapy. Still, along with the statin drugs and resins that bind bile salts, niacin continues to be a mainstream drug for treating or preventing cardiovascular disease.

## THIAMIN

Thiamin is a water-soluble vitamin. The RDA of thiamin for the adult man is 1.5 mg. The vitamin is present in a variety of foods of plant and animal origin, as well as in yeast. The populations most at risk of developing a deficiency are chronic alcoholics in Western countries and those with an overdependence on polished rice as a staple in underdeveloped nations. The consumption of large amounts of raw seafood can also induce the deficiency. Thiamin deficiency in humans is called **beriberi**. The disease occurs in a variety of forms and causes different problems in infants, adults, and alcoholics.

## Absorption and Metabolism of Thiamin

The cofactor form of thiamin is thiamin pyrophosphate (TPP). TPP is released from dietary proteins during hydrolysis in the gastrointestinal tract and then hydrolyzed to thiamin. The thiamin is absorbed and transported into various tissues, where it is converted back to TPP by the action of thiaminokinase (Figure 9.70). A small proportion of the body's thiamin occurs as thiamin monophosphate (TMP) and thiamin trisphosphate (TTP).

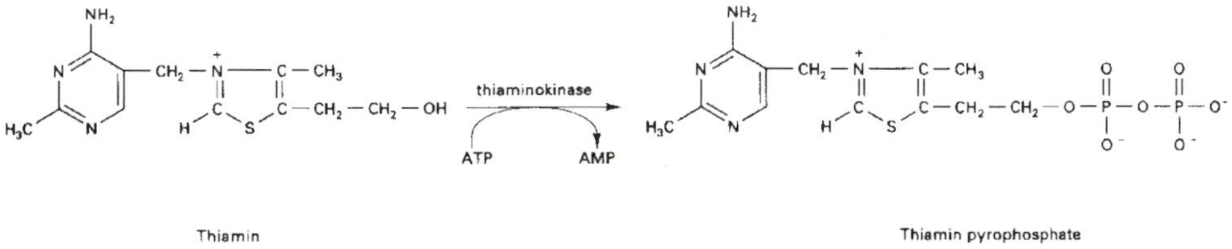

**FIGURE 9.70** Conversion of thiamin to thiamin triphosphate (TPP).

## Biochemistry of Thiamin

Thiamin pyrophosphate is a cofactor for only a small number of enzymes, namely pyruvate dehydrogenase, α-ketoglutarate dehydrogenase, BCKA dehydrogenase, and transketolase. The three dehydrogenases catalyze the reduction of NAD and the discharge of a molecule of $CO_2$. Transketolase catalyzes the transfer of 2-carbon units at two positions of the pentose phosphate pathway. The dehydrogenases are described at earlier points in the text (Chapters 4 and 8). The regulatory aspects of the pentose phosphate pathway are detailed with the material on fatty acid synthetase (Chapter 5). This pathway (Figure 9.71) comprises a series of related reactions involving the interconversion of carbohydrates containing three to seven carbons. The most important intermediate of the pathway is ribose 5-phosphate, which is required for the synthesis of ribonucleotides such as ATP and GTP, deoxyribonucleotides such as dATP and dGTP, and nucleic acids such as RNA and DNA. The pentose phosphate pathway is distinguished by its involvement in the synthesis of ribose 5-phosphate, as well as in the reduction of NADP. This reduction occurs in the first two steps of the pathway. In the first step, the aldehyde group of glucose is oxidized to a carboxylic acid group. The second step involves the oxidative decarboxylation of 6-phosphogluconate. The ribose 5-phosphate may be used for nucleic acid synthesis, or it may remain in the pathway for eventual catabolism to $CO_2$. The final products, fructose-6-phosphate and glyceraldehyde-3-phosphate, can enter glycolysis for conversion back to glucose-6-phosphate for reentry into the pentose phosphate pathway. Continued operation of the entire pathway requires the conversion of ribulose 5-phosphate to xylulose-5-phosphate, ribose-5-phosphate, and xylulose-5-phosphate (and, with recycling of products, to glucose-6-phosphate), and results in the complete oxidation to carbon dioxide. The steps of the pathway catalyzed by transketolase are indicated by TPP. The 2-carbon units transferred in these steps are circled (see Figure 9.71).

## Thiamin Deficiency

Thiamin deficiency produces anorexia and weight loss, heart problems, and neurological symptoms. The wasting of tissues and edema may occur. Where edema is present, the wasting may not be apparent to the eye. The symptoms involving the heart include tachycardia, an enlarged heart, and cardiac failure. Tachycardia is inappropriately rapid beating of the heart. The symptoms involving the nervous system include mental confusion, anorexia, ataxia, and nystagmus. Ataxia is quivering of the hands or limbs. Nystagmus is impaired movement of the eyes involving paralysis of the muscles of the eyes.

Beriberi associated with the consumption of polished rice diets was once widespread in the East. It still occurs in rural parts of Southeast Asia. The thiamin deficiency arises from an overdependence on polished rice. Polished rice is the result of removal, during milling, of the rice bran which contains thiamin.

Three types of beriberi are recognized: dry beriberi, wet beriberi, and infantile beriberi. **Dry beriberi** occurs mainly in older adults and involves the wasting

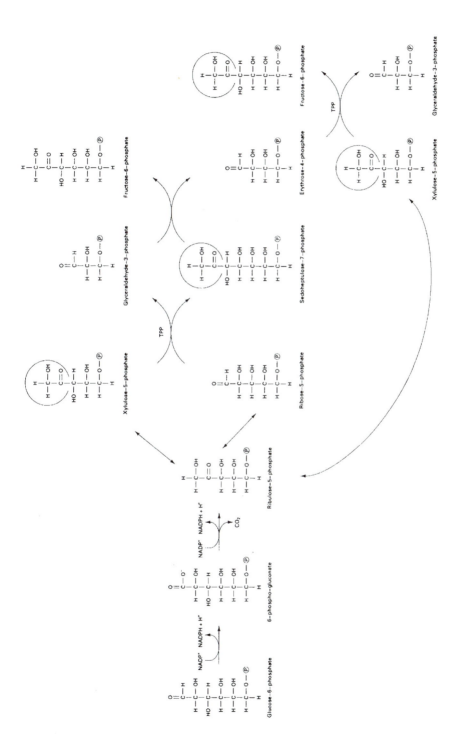

**FIGURE 9.71** Pentose phosphate pathway. The pentose phosphate pathway is used for the metabolism of various sugars. It is required for the biosynthesis of ribose 5-phosphate (a component of ATP, GTP, CTP, UTP, TTP, and the nucleic acids). The pentose phosphate pathway is used for the reduction of NADP. Thiamin pyrophosphate is a cofactor for two enzymes of the pathway, as indicated by TPP. The circled groups are the two-carbon units transferred by TPP.

away and weakness of the legs. Neuropathies result from the degeneration of sensory and motor nerves. This type of beriberi does not often involve the heart. **Wet beriberi** is characterized by edema and heart failure. Leakage through the capillaries results in edema. The heart increases its output to maintain blood pressure. The skin is unusually warm; however, when the heart begins to fail, the skin becomes cool and bluish. **Infantile beriberi** occurs in breast-fed infants of mothers who are thiamin deficient. The disease may occur even if the nursing mother appears to be healthy. Infantile beriberi presents rapidly and death from heart failure can result within a few hours. This disease can result in death before edema or loss of motor function occurs. Convulsions and coma may also occur.

The thiamin deficiency that occurs with chronic consumption of raw fish results from the activity of thiaminase. Thiaminase catalyzes the cleavage and thus destruction of the vitamin. The thiaminase content of various types of raw seafood has been measured (Hilker and Peters, 1966). Unfortunately, a reliable study of thiaminase in raw seafood is not yet available. Bracken ferns also contain thiaminase. Grazing animals such as sheep consume these ferns and develop thiamin deficiency. The deficiency results in brain lesions in the animal and in the bending backward of the neck. Australians call this phenomenon "stargazing." The animal falls to the ground and pedals its feet in the air.

Thiamin trisphosphate accounts for a small proportion of the thiamin in the brain. The function of this compound is not clear, nor is it known if it plays a part in the neurological signs that occur with thiamin deficiency. Recent studies have shown that exposure of baboons to a flickering light can induce changes in brain levels of thiamin triphosphate. Flickering lights are often used in the study of convulsions. The work available on the compound is not extensive.

Thiamin deficiency in alcoholics may be caused by decreased intake, reduced absorption, and impaired ability to use the absorbed vitamin. The ataxia and ocular symptoms associated with the deficiency in alcoholics are known as **Wernicke's disease**. Vitamin therapy can provide relief from nystagmus within a few hours of treatment and from ataxia within several weeks. The treatment of alcoholics also involves the supply of other nutrients lacking in the diet, such as folate, vitamin $B_{12}$, and protein. Left untreated, patients suffering from Wernicke's disease continue to develop **Korsakoff's psychosis**, which involves amnesia and confusion. Only about 25% of patients with Korsakoff's psychosis can be completely cured by thiamin treatment, which must be continued for a few weeks or months. The two conditions just described constitute the Wernicke–Korsakoff syndrome. The syndrome was named after two researchers. Karl Wernicke, a German, noted impaired or paralyzed eye movements and unstable walking and disorientation in his patients, most of whom were alcoholics. Polyneuropathy, a weakness of the hands, calves, and feet, was also noted. Sergei Korsakoff, a Russian, observed amnesia and confusion and an inability to learn new names or tasks in alcoholic patients.

Thiamin therapy for alcoholics may involve a single injection of 10 mg thiamin or 50 mg oral thiamin propyl disulfide. The latter compound is a fat-soluble version of thiamin that permits absorption of the vitamin where alcohol-induced damage would prevent efficient absorption of thiamin itself. Thiamin propyldisulfide is converted in the body to thiamin.

## Assessment of Thiamin Status

Thiamin status has been assessed by direct tests involving the measurement of thiamin levels in the blood or urine. The vitamin can be assayed by the thiochrome method or by microbiological assays. The disadvantage of these methods is that thiamin levels in normal individuals can vary greatly. The test organism used for microbiological assays may be *Lactobacillus viridescens* or *Lactobacillus fermenti*.

The thiochrome method involves the addition of alkaline ferricyanide to the biological sample containing thiamin. This treatment results in the oxidative conversion of thiamin to thiochrome, which can be measured by fluorescence.

Thiamin compounds can also be measured by HPLC. A clever technique has been used to facilitate the detection of thiamin compounds immediately after they have completed their chromatographic separation by HPLC. The mixture of unseparated compounds is exposed to conditions expected to provoke the conversion of thiamin to its thiochrome derivative, and then applied to the HPLC column (Tallakesen *et al.*, 1997; Rindi and Laforenza, 1997). This approach, called "pre-column derivatization," is used widely by chemists and biochemists during the separation and detection of many compounds.

A few aspects of the chemistry of the vitamin should be noted. Thiamin is stable at mildly acidic pH but unstable and easily destroyed at pH 8.0 or higher, especially with heating. Thiamin is also destroyed by sulfite, especially at pH 6.0 or higher.

The most reliable method for assessing thiamin status involves the measurement of red blood cell transketolase. This enzyme is measured with and without the addition of TPP to the enzyme assay mixtures. In dietary thiamin deficiency, synthesis of transketolase continues, but conversion of the apoenzyme to the holoenzyme in the cell is inhibited, resulting in the accumulation of the enzyme in the apoenzyme form. Addition of TPP to cell homogenates results in the conversion of apoenzyme to holoenzyme. This conversion can easily be detected by enzyme assays. The amount of stimulation of enzyme activity by the added TPP is used to assess thiamin status. A deficiency is indicated by a stimulation of over 20%. The TPP-dependent stimulation, using red blood cells from normal subjects, ranges from 0 to 15%.

## Determination of the Thiamin Requirement

The thiamin requirement was estimated by first inducing a thiamin deficiency and then attempting to reverse the deficiency with various levels of the vitamin. In one study, human subjects consumed a thiamin-deficient diet for 2 weeks. The diet contained protein (100 g casein/day), carbohydrate (350 g glucose/day) oil (134 g/day), as well as minerals and vitamins other than thiamin. After 2 weeks, the same diet was consumed, but with the indicated amounts of thiamin. Diets containing different levels of thiamin (0–1.1 mg/day) were consumed for a 2-week period before switching to a higher level. The stimulatory effect of adding TPP during assays of red blood cell transketolase, using blood samples taken during consumption of each of the diets, is shown in Figure 9.72. The thiamin status of the subjects, as assessed by transketolase assays, indicates a 20% stimulatory effect during the first 2-week period. The stimulatory effect increased to about 45% with

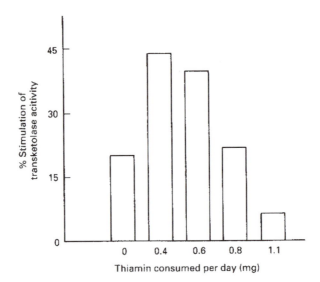

**FIGURE 9.72** Stimulatory effect of adding thiamin triphosphate (TPP) during assays of red blood cell transketolase, using blood samples taken during the consumption of different diets. (Redrawn with permission from Ariaey-Nejad *et al.*, 1970.)

consumption of 0.4 mg thiamin per day, indicating a continuing trend toward deficiency, rather than a reversal. The supplementation of 0.6 mg thiamin did little to reverse the deficiency, whereas 1.1 mg thiamin/day reduced the stimulatory effect, successfully reversing the deficiency.

The severity of the symptoms of thiamin deficiency has been associated with energy intake. The consumption of large doses of glucose has been found to induce an unusual rise in plasma pyruvate and lactate, as well as neurological symptoms, in thiamin deficient humans. Because of this association, the thiamin requirement is sometimes expressed on a per energy intake basis.

EXERCISE

At one time it was thought that plasma pyruvate levels could be used to assess thiamin status. Please rationalize this test. State how it would work. Then explain how the test could be influenced by factors other than thiamin status. (See Chong, 1970.)

EXERCISE

The half-life of thiamin in humans is about 2 weeks. This means that any given molecule of TPP in the body has a 50% chance of being excreted within a 2-week period. If one consumes a thiamin-free diet, how long would it take tissue TPP levels to drop to one-eighth the normal level? (See Ariaey-Nejad *et al.*, 1970.)

## Use of Thiamin in Maple Syrup Urine Disease

Maple syrup urine disease is a genetic disease involving a defect in BCKA dehydrogenase. The disease affects one in 100,000 births, manifests in infants as lethargy and seizures, and can result in mental retardation. It might be emphasized that it is the responsibility of the clinician to insist on the prompt diagnosis of an infant presenting with these symptoms. The genetic defect results in increases in the plasma and urinary levels of BCKAs and BCAAs. The disease derives its name from the maple syrup odor of the keto acid of leucine. In some cases, the disease responds to high doses of thiamin (10–200 mg/day). The thiamin-responsive version of the disease is thought to result from a defect in the ability of the apoenzyme of BCKA dehydrogenase to bind the cofactor. The high doses of vitamin result in elevated levels of TPP in cells, which forces the binding of cofactor to the active site of the defective enzyme. The usual treatment of maple syrup urine disease is a low-protein diet.

## RIBOFLAVIN

Riboflavin is a water-soluble vitamin. The RDA for the adult man is 1.7 mg. Free riboflavin rarely accumulates in the cell, though it is found in cow's milk, blood plasma, and urine. The vitamin in food and in the body occurs mainly in the cofactor forms. Liver is an excellent source of riboflavin; considerable amounts also occur in meat and dairy products and in dark green vegetables. Broccoli and spinach are good sources of the vitamin. Grains and legumes also are sources of dietary riboflavin. Riboflavin is susceptible to destruction by light. For example, exposure of milk to sunlight for several hours can lead to a substantial destruction of its riboflavin. For this reason, milk should be protected from bright light during storage.

Riboflavin is absorbed by the gut and enters the bloodstream, where close to half of it is loosely bound to serum albumin. The main site of absorption is the ileum. As might be expected, when large doses (20–60 mg) of riboflavin are eaten, much of the dose is promptly excreted in the urine (Zempleni *et al.*, 1996). The FAD cofactors based on riboflavin are called **flavins**; enzymes using flavins as a cofactor are called **flavoproteins**.

The conversion of riboflavin to flavin mononucleotide (FMN) is catalyzed by flavokinase (Figure 9.73). This conversion may occur during absorption through the gut mucosa or in other organs. The subsequent conversion of FMN to flavin adenine dinucleotide (FAD) is catalyzed by **FAD synthase**. FAD synthase uses ATP as a source of an adenylyl group, in this conversion (McCormick *et al.*, 1997). Various phosphatases, including those of the gut mucosa, can catalyze the breakdown of FAD to FMN and of FMN to free riboflavin. Dietary flavins that are covalently bound to proteins are thought to be unavailable and not to contribute to our dietary needs (Bates *et al.*, 1997).

## Biochemistry of Riboflavin

Flavins are used as cofactors by about 50 enzymes in mammals. The most well known of these enzymes are those used in mainstream energy metabolism, namely dihydrolipoyl dehydrogenase, fatty acyl-CoA dehydrogenase, succinate dehydrogenase, and NADH dehydrogenase. Dihydrolipoyl dehydrogenase is a component of pyruvate dehydrogenase and $\alpha$-ketoglutarate dehydrogenase. Lipoic acid

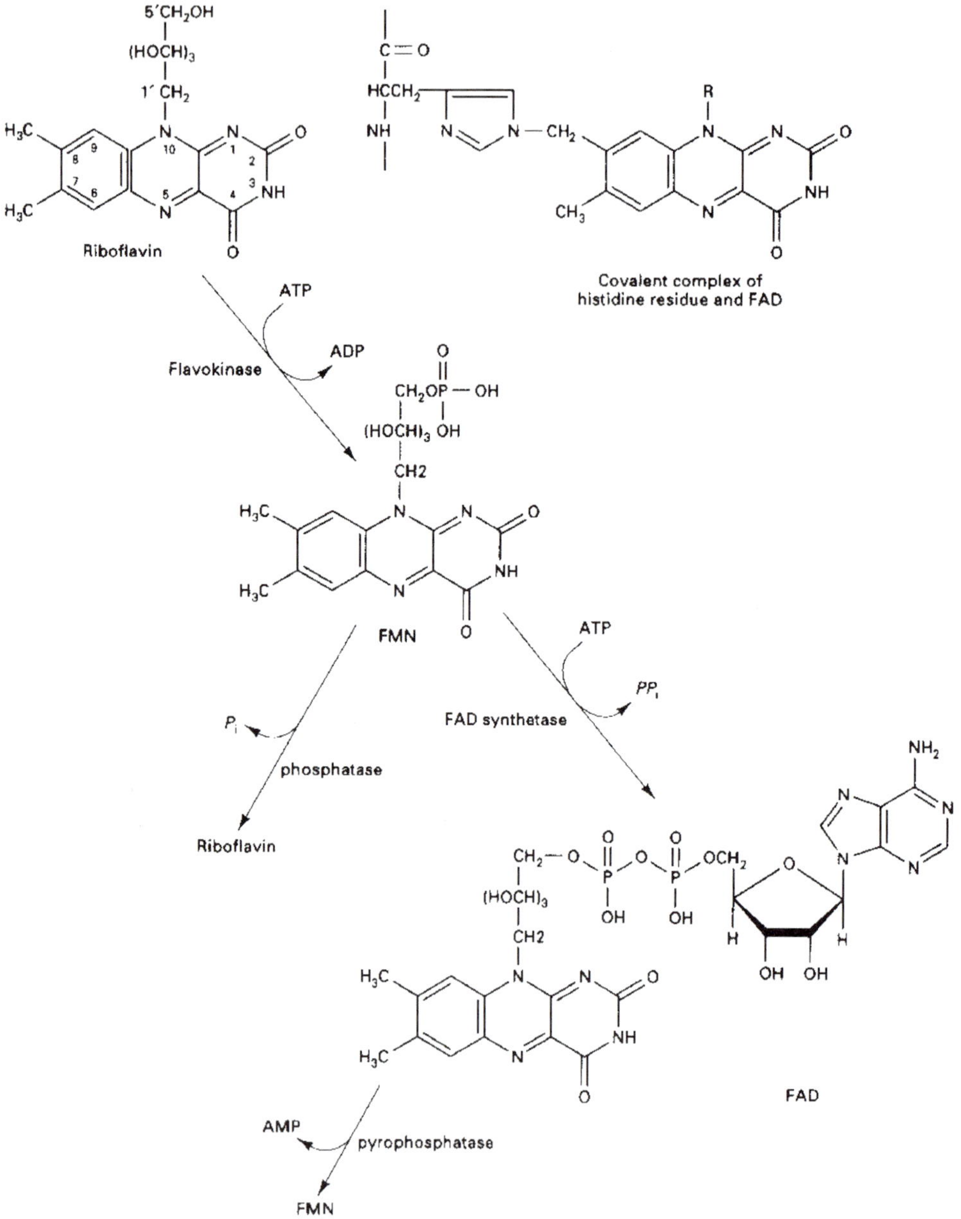

**FIGURE 9.73** Riboflavin metabolism and FAD covalently bound to an enzyme via a residue of histidine.

is a cofactor that is synthesized in the body. It occurs covalently bound to a residue of lysine of a protein that is part of a multiprotein complex. A number of flavoproteins are listed in Table 9.8.

The cofactor forms are flavin mononucleotide (FMN) and flavin adenine dinucleotide (FAD). These cofactors are tightly, but not covalently, bound to enzymes. They accept and transfer electrons. They may accept electrons from substrates or from another cofactor, namely NADH + $H^+$. The flavin-containing enzymes transfer electrons to a number of substrates, including $O_2$ and NAD. About 10% of the FAD in the cell is covalently bound to enzymes. Enzymes containing covalently bound flavin cofactors include succinate dehydrogenase, monoamine oxidase, and monomethylglycine dehydrogenase. FAD is bound to succinate dehydrogenase via a residue of histidine, as shown in Figure 9.74. The FAD of monoamine oxidase is bound to the sulfur atom of a cysteine residue. Dietary FAD and FMN are hydrolyzed by phosphatases or pyrophosphatases of the gut mucosa, which liberate free riboflavin (see Figure 9.73).

A brief review of the complex of proteins constituting pyruvate dehydrogenase is a must for anyone interested in vitamin metabolism, as the complex requires four different vitamins for activity: riboflavin (FAD), thiamin (TPP), niacin (NAD), and pantothenic acid (coenzyme A). The reactions catalyzed by the complex are shown in Figure 9.74. Five steps are involved. In Step 1, pyruvate is decarboxylated, resulting in the transfer of a hydroxyethyl group to TPP. In Step 2, the hydroxyethyl group is oxidized to an acetyl group and transferred to lipoic acid. In Step 3, the acetyl group is transferred to coenzyme A, generating acetyl-CoA; dihydrolipoic acid is also generated in this step. In Step 4, dihydrolipoic acid

**TABLE 9.8** Flavoproteins

| Enzyme | Cofactor | Function |
|---|---|---|
| Dihydrolipoyl dehydrogenase | FAD | Energy metabolism |
| Fatty acyl-CoA dehydrogenase | FAD | Fatty acid oxidation |
| Succinate dehydrogenase | FAD | Krebs cycle |
| NADH dehydrogenase | FMN | Respiratory chain |
| Xanthine dehydrogenase | FAD | Purine catabolism |
| Glutathione reductase | FAD | Reduction of GSSG to 2 GSH |
| Methylene-$H_4$folate reductase | FAD | Production of 5-methyl-$H_4$folate |
| Sphinganine oxidase | FAD | Sphingosine synthesis |
| Pyridoxine phosphate oxidase | FMN | Vitamin $B_6$ metabolism |
| Monoamine oxidase | FAD | Metabolism of neurotransmitters |
| D-Amino acid oxidase | FAD | Catabolism of D-amino acids to keto acids |

$$\text{D-amino acid} \xrightarrow{\text{FAD}} \text{imino acid} \xrightarrow{\text{spontaneous}} \text{keto acid} + NH_3$$
$$O_2 \quad HOOH$$

| | | |
|---|---|---|
| L-amino acid oxidase | FMN | Catabolism of L-amino acids to keto acids |
| Choline dehydrogenase | FAD | Choline catabolism |
| Dimethylglycine dehydrogenase | FAD | Choline catabolism |
| Monomethylglycine dehydrogenase | FAD | Choline catabolism |

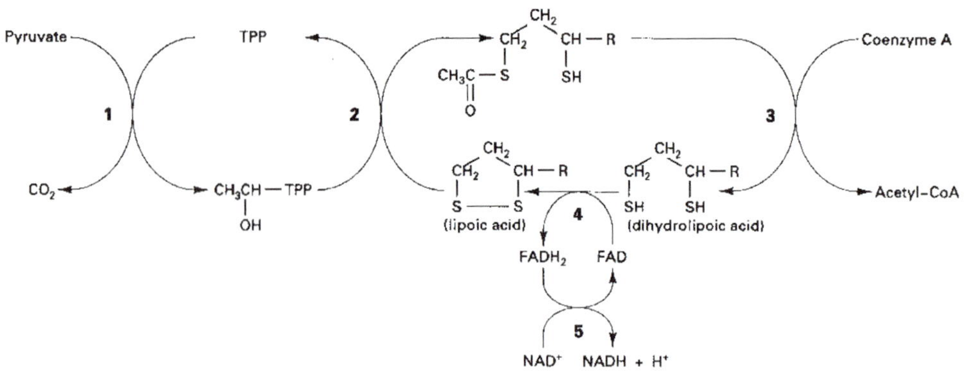

FIGURE 9.74 Cycle of reactions catalyzed by pyruvate dehydrogenase.

transfers its electrons to FAD, which in Step 5 are transferred to NAD. Note that lipoic acid does not occur as a free acid. Its carboxyl group occurs in an amide bond with a lysine residue.

## Riboflavin Deficiency

The signs of riboflavin deficiency include lesions of the mouth known as cheilosis and angular stomatitis. Cheilosis is swelling and fissuring of the lips. It is painful and results in bleeding. Angular stomatitis is fissuring and ulceration at the angles of the mouth. Other symptoms include dermatitis and a rash on the scrotum or vulva. A naturally occurring deficiency of riboflavin only, and not of other vitamins, is unknown. Poorer populations in the United States may be deficient in riboflavin among other nutrients. The deficiency is widespread in underdeveloped countries. Surprisingly, riboflavin deficiency has commanded little attention. This is because its symptoms are not incapacitating, providing that the deficiency is not too severe. A severe deficiency, which can be induced experimentally in animals, results in a failure to grow and reproduce, dermatitis, and nerve degeneration.

## Assessment of Riboflavin Status

Flavins are lost from the body as intact riboflavin, rather than as a breakdown product of riboflavin. Hence, vitamin status may be assessed by measuring the level of urinary riboflavin. Generally, the loss of 30 µg of riboflavin/g creatinine or less per day indicates a deficiency. This method of assessment is not preferred because it is influenced by a number of factors unrelated to vitamin status. Another problem with this method is its great sensitivity to a short-term deficiency; thus, it does not necessarily reflect the true concentrations of FAD and FMN in tissues.

The most reliable way to assess riboflavin status is by a functional test. The test involves the assay of **glutathione reductase**, using red blood cells as the source of enzyme. A clear illustration of the use of urinary riboflavin and glutathione reductase activity to assess vitamin status was provided by Boisvert *et al.* (1993). Glutathione reductase, which requires FAD as a cofactor, catalyzes the reduction of the disulfide bond of the glutathione dimer (GSSG) to produce two molecules of glutathione (GSH):

$$
\begin{array}{c}
\text{(Glutathione reductase)} \\
\text{Glutathione dimer} \xrightarrow{\hspace{2cm}} \text{2 Glutathione} \\
\text{(GS-SG)} \qquad \text{(FAD)} \qquad \text{(2 GSH)} \\
\text{NADPH + H}^+ \qquad \text{NADP}^+
\end{array}
\tag{9.5}
$$

Glutathione is discussed further in the section on selenium and glutathione in Chapter 10. The enzyme assay is conducted using glutathione reductase extracted from red blood cells with and without added FAD. Chronic consumption of a diet deficient in riboflavin allows the continued synthesis of a variety of flavoproteins, but results in the accumulation of apoenzyme without its conversion to holoenzyme. Addition of chemically pure FAD to a biological fluid containing apoenzyme results in the stimulation of enzyme activity because of the formation of the holoenzyme. It is this stimulation of enzyme activity that is used to determine vitamin status in humans.

EXERCISE

State two reasons why the FAD stimulation test, using assays of succinate dehydrogenase and red blood cells as a possible source of enzyme, would probably not work.

The following study illustrates the use of the FAD stimulation test to assess vitamin status in high school students. Blood samples were withdrawn from 184 Caucasian girls (open bars) and 34 Afro-American girls (shaded bars). The red blood cells were broken by mixing them with distilled water. Reduced NAD and GSSG were added to the cell extracts, with and without FAD. The mixtures were incubated for 10 minutes to allow the holoenzyme to catalyze its reaction and then the product formed was measured. The results of the test are shown in Figure 9.75. A stimulation of 20% or less is considered to be normal and indicative of adequate riboflavin status. A stimulation of 30 to 40% was found in 7 Caucasian girls and 6 Afro-American girls, indicating a deficiency in these students. Adding FAD resulted in no stimulation of enzyme activity in 36 of the students, as shown. The subsequent supplementation of the diets of the girls with 5.0 mg riboflavin per day resulted in a decline in the stimulation to only 0 to 5%, demonstrating recovery from the riboflavin deficiency.

# PANTOTHENIC ACID

Pantothenic acid is a water-soluble vitamin. The vitamin has two functions, in the biosynthesis of coenzyme A and in the synthesis of the cofactor of fatty acid

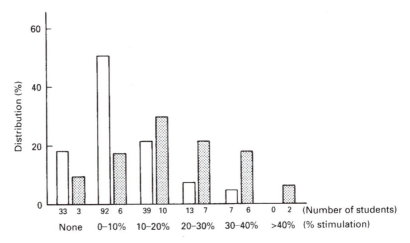

| | | | | | (Number of students) |
|---|---|---|---|---|---|
| 33  3 | 92  6 | 39  10 | 13  7 | 7  6 | 0  2 |
| None | 0–10% | 10–20% | 20–30% | 30–40% | >40%    (% stimulation) |

FIGURE 9.75  Results of the FAD stimulation test used to assess vitamin status in high school students. (Redrawn with permission from Sauberlich *et al.*, 1972.)

synthase. This cofactor is 4′-phosphopantetheine. The immediate precursor of this cofactor is coenzyme A. 4′-Phosphopantetheine is covalently bound to fatty acid synthase and thus is called a prosthetic group. Coenzyme A is used in a variety of reactions, for example, the Krebs cycle, fatty acid synthesis and oxidation, amino acid metabolism, ketone body metabolism, cholesterol synthesis, and conjugation of bile salts. The coenzyme received its name when its use in transferring the acetyl group was recognized. Coenzyme A, in its acetylated form, is used in many vital reactions such as those catalyzed by citrate synthase (Krebs cycle) and choline acyltransferase (acetylcholine synthesis) and the priming reaction of fatty acid synthase. In addition, acetyl-CoA is a substrate of acetyl-CoA carboxylase, the first enzyme of the fatty acid biosynthetic pathway. Apparently, 4′-phosphopantetheine is a cofactor only of one enzyme, fatty acid synthase.

The average intake of pantothenic acid, as free pantothenic acid and as coenzyme A, acetyl-coenzyme A, and long-chain fatty acyl-coenzyme A, is 5 to 10 mg/day. An RDA for the vitamin has not been established because the vitamin is plentiful in a variety of foods. Pantothenic acid is present in all plant and animal foods. The richest sources of the vitamin are liver, yeast, egg yolk, and vegetables. In foods, the vitamin occurs mainly as coenzyme A.

Coenzyme A does not readily cross cell membranes, including those of the gut. Dietary coenzyme A is hydrolyzed in the gut lumen to yield pantothenic acid, which is readily absorbed. Studies with rats have shown that coenzyme A is hydrolyzed in the gut lumen according to the pathway in Figure 9.76.

The serum level of pantothenic acid is about 1 to 5 $\mu M$ (Lopaschuk *et al.*, 1987). The vitamin in the bloodstream is transported into various tissues, where it is then converted to coenzyme A. Coenzyme A is synthesized from pantothenic acid, ATP, and cysteine. The pathway of coenzyme A synthesis is shown in Figure 9.77. The cofactor of fatty acid synthase is synthesized from coenzyme A and does not involve the direct participation of pantothenic acid. A specific enzyme catalyzes

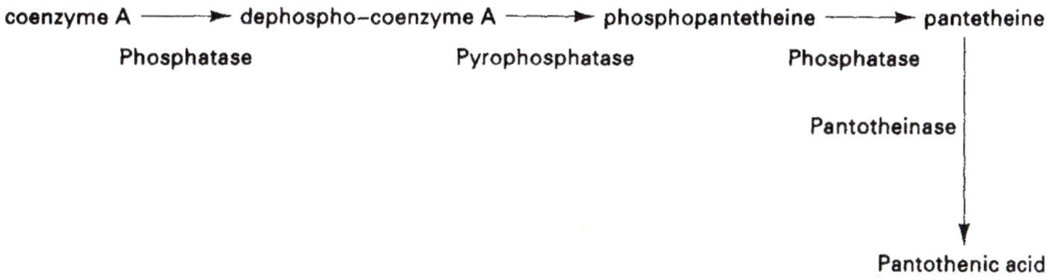

**FIGURE 9.76** Digestion of dietary coenzyme A with the release of pantothenic acid.

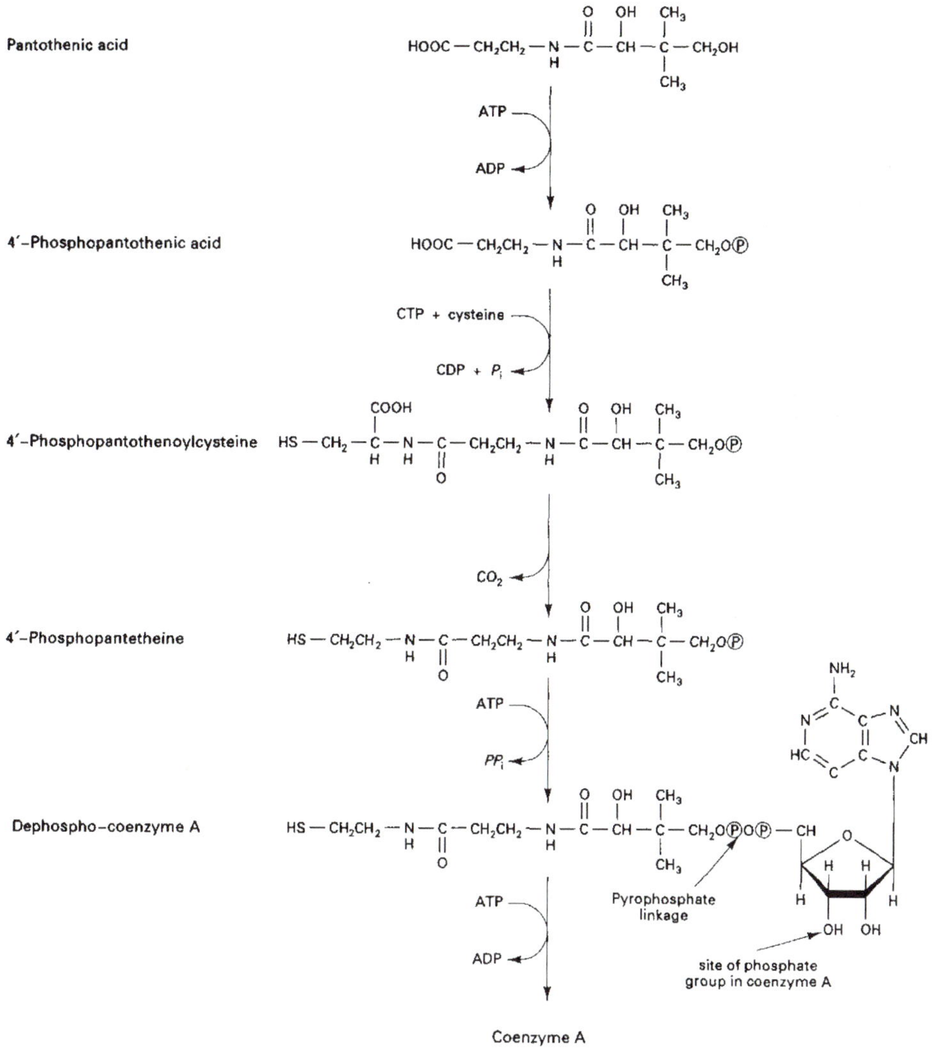

**FIGURE 9.77** Biosynthesis of coenzyme A.

FIGURE 9.78 Conversion of fatty acid synthase to the holoenzyme form.

the covalent attachment of the 4′-phosphopantetheine group to the apoenzyme, resulting in the formation of the holoenzyme (Figure 9.78).

## 4′-Phosphopantetheine and Fatty Acid Synthase

Fatty acid synthase is a large enzyme with a molecular weight of 540,000. It consists of two identical subunits. To each subunit is attached a molecule of 4′-phosphopantetheine. The cofactor is attached to the enzyme via a residue of serine. The sulfhydryl group of the cofactor is used for the covalent attachment of the growing fatty acid. The bond used in this attachment is the thiol ester linkage. The cofactor acts as a swinging arm, enabling the growing fatty acid chain to reach the different catalytic sites located in different regions on the surface of the enzyme. For example, the swinging arm enables the growing fatty acid moiety to reach the sites used in the catalysis of reduction and dehydration reactions. Each subunit contains seven different catalytic sites.

The cofactor of fatty acid synthase is bound to the enzyme at a point near a specific residue of cysteine. This cysteine residue is important in the catalytic mechanism. The sulfhydryl group of this cysteine is used for temporarily holding the fatty acid moiety each time a new molecule of malonic acid is transferred to the 4-phosphopantetheine group. One might refer to the diagram of the enzyme in Chapter 5, where the sulfhydryl groups of the cysteine residue and of 4-phosphopantetheine are shown.

## Coenzyme Concentrations in the Cell

The concentration of free pantothenic acid in the liver is about 15 $\mu M$; that in the heart is about tenfold greater (Robishaw and Neely, 1985). The concentration of the cofactor form of the vitamin, coenzyme A, is higher in the mitochondrion than in the cytosol. In the liver, cytosolic coenzyme A is about 0.06 mM, and mitochondrial coenzyme A, about 2.6 mM. In the liver, about 70% of coenzyme A is mitochondrial, whereas in the heart about 95% is mitochondrial (Tahiliani and Neely, 1987). These values might be compared with that for carnitine, another molecule used in the handling of fatty acids. Please consult the Carnitine section in Chapter Four. About half of the coenzyme A in liver occurs as the long-chain fatty acyl-coenzyme A derivative. The concentration of fatty acid synthase in the cytoplasm is quite low, about 0.01 $\mu M$. Hence, the concentration of the 4′-phosphopantetheine cofactor is much lower than that of coenzyme A. The pantothenic acid bound to this enzyme does not make a significant contribution to our dietary vitamin.

## Measurement of Pantothenic Acid

Pantothenic acid levels in foods and body fluids can easily be measured by microbiological assays. Lactic acid bacteria are used as the test organism. Where measurement of the vitamin occurring as coenzyme A is desired, the coenzyme must first be treated with hydrolytic enzymes to liberate the pantothenic acid prior to the microbiological assay.

## Pantothenic Acid Deficiency

A deficiency purely in pantothenic acid has probably never occurred, except in controlled studies. Persons suffering from severe malnutrition would be expected to be deficient in the vitamin. Studies with animals have shown that consumption of a diet deficient in the vitamin results in a loss of appetite, slow growth, skin lesions, ulceration of the intestines, weakness, and eventually death. Pantothenic acid deficiency also results in the production of gray fur in animals whose fur is colored. Biochemical studies with deficient animals have revealed severe decreases in pantothenic acid levels in a variety of tissues, but only moderate declines in the levels of coenzyme A in liver and kidney and maintenance of coenzyme A levels in the brain (Smith *et al.*, 1987). Some striking defects in glycogen and ketone body metabolism have been noted in pantothenic acid-deficient animals.

Plasma and urinary levels of pantothenic acid have been measured in dietary surveys as well as in controlled studies of the vitamin deficiency. One fairly recent study with human subjects involved the feeding of a pantothenic acid-free diet for 9 weeks. The urinary pantothenic acid levels (4–6 mg/day) in vitamin-sufficient subjects were roughly half that of the intake (10 mg/day). With consumption of the vitamin-free diet, urinary pantothenic acid levels gradually declined to about 0.8 mg/day over the 9-week period (Fry *et al.*, 1976). Both urinary and blood serum levels of pantothenate have been used to assess dietary status. Values from urinary measurements seem to be somewhat better correlated with intake of this vitamin, than blood measurements data (Berg, 1997).

## ASCORBIC ACID (VITAMIN C)

Ascorbic acid, also known as vitamin C, is a water-soluble vitamin. The RDA for the adult is 60 mg. Good sources of ascorbic acid are bell peppers, broccoli, citrus fruit, spinach, tomatoes, and potatoes. Animal products contain some vitamin C while grains contain essentially none.

Ascorbic acid is an unusual vitamin in that it can be synthesized by most mammals; however, it cannot be made by humans, primates, guinea pigs, and fruit bats. Ascorbic acid is synthesized from glucose in a six-step pathway. Mammals that cannot make the vitamin lack the last enzyme of this pathway, gulonolactone oxidase. This enzyme occurs in the kidney of chickens, amphibians, and reptiles, and in the liver of most mammals (Banhegyi *et al.*, 1997). Gulonolactone oxidase is a flavoprotein. Each catalytic event of the enzyme results in the conversion of $O_2$ to HOOH. It might strike one as a paradox that a vitamin (vitamin C) that helps protect the body from toxic oxygen is synthesized by a pathway that produces toxic oxygen (HOOH).

### Transport of Vitamin C in the Body

Vitamin C occurs in two forms, ascorbic acid and dehydroascorbic acid. Ascorbic acid is absorbed from the diet by a special $Na^+$-dependent transporter. Once inside the bloodstream, ascorbic acid is absorbed into the adrenal glands, and other organs, by the same transporter. However, certain cells, such as neutrophils and red blood cells, cannot take up ascorbic acid, but instead acquire dehydroascorbic acid via glucose transporters (Guaiquil *et al.*, 1997). Vitamin C is eliminated from the body via the urine in the forms dehydroascorbate, ketogulonate, ascorbate 2-sulfate, and oxalic acid. When consumed in large doses (2 g/day), the vitamin is excreted mainly as ascorbic acid. The major pathway of breakdown and excretion in rats and guinea pigs is via oxidation to $CO_2$. This route seems to be only of slight importance in humans.

## Vitamin C Deficiency

Deficiency in vitamin C, unlike the case with most vitamins, is associated with a specific disease. This disease is **scurvy**. The symptoms of scurvy include swollen or bleeding gums and hemorrhages under the skin. These symptoms occur when the body's ascorbate is depleted to the point where plasma ascorbate levels are under 0.2 mg/100 ml. Controlled studies with human subjects revealed that symptoms of the disease may develop within 4 weeks with the consumption of an ascorbate-free diet. Scurvy is rarely encountered in developed countries, though it may occur in chronic alcoholism. The disease in humans may be prevented by consuming 10 to 15 mg ascorbic acid per day.

### Deficiency in Humans

Vitamin C status is often assessed by measuring the plasma levels of ascorbic acid. A plasma concentration of 0.2 mg/100 ml or greater indicates normal vitamin status. A concentration of 0.1 mg/100 ml or less indicates a deficiency. The following study of human subjects illustrates the progression of a deficiency (Figure 9.79). The subjects consumed a diet supplying 75 mg ascorbic acid per day from days 1 to 17. They consumed an ascorbate-free diet from days 18 to 117. During this period, plasma vitamin levels fell below 0.1 mg/100 ml, indicating a deficiency. The resumption of a vitamin-sufficient diet (66 mg/day) at day 117 restored plasma vitamin concentrations to normal levels. The first signs of scurvy in this study occurred after 1 month of eating the ascorbate-free diet. These signs included petechial hemorrhages. (Petechial refers to small purple spots on the skin.) After another month on the diet, other signs materialized, including bleeding gums and joint pain. Bleeding due to vitamin C deficiency means scurvy.

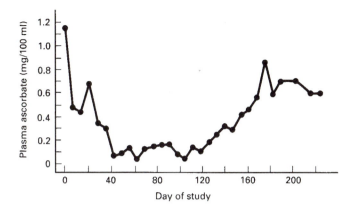

**FIGURE 9.79** Progression of a deficiency with consumption of an ascorbate-free diet. (Redrawn with permission from Hodges *et al.*, 1971.)

## Deficiency in Pigs

Let us now view some data acquired from pigs. The ascorbic acid concentration in various tissues of the normal pig are approximately as follows: plasma (0.034 m$M$), aqueous humor (0.5 m$M$), cerebrum (0.9 m$M$), liver (1.3 m$M$), and adrenal gland (10 m$M$). Studies with a special strain of vitamin C-requiring pigs revealed the effects of vitamin C deficiency. Dietary deficiency provoked bleeding in various parts of the body. The pigs used in the study were pregnant, and the deficiency produced defects in bone and the associated cartilage in the fetuses (Wegger and Palludan, 1994).

## Biochemistry of Ascorbic Acid

The term *vitamin C* refers to ascorbic acid (the fully reduced form of the vitamin) and to dehydroascorbic acid. Removal of one electron from ascorbic acid yields semidehydroascorbic acid (ascorbate radical). This form of the vitamin is a free radical; it contains an unpaired electron. The structures of free radicals are written with large dots. The removal of a second electron yields dehydroascorbic acid. Conversion of ascorbate to dehydroascorbate, via the removal of two electrons, can occur under two conditions: (1) with use of ascorbic acid by ascorbate-dependent enzymes; and (2) with the spontaneous reaction of ascorbate with oxygen. Semidehydroascorbate is an intermediate in this conversion pathway.

**Dehydroascorbate reductase** catalyzes the regeneration of ascorbic acid from dehydroascorbate. The enzyme requires glutathione (GSH) as a source of reducing power. GSH and ascorbic acid are both biological reductants. The levels of GSH in the cell are maintained by glutathione reductase, as discussed under Riboflavin. Because of the regeneration, both ascorbate and dehydroascorbate have biological activity. The latter compound may break down to form diketogulonic acid (Figure 9.80). Diketogulonic acid is an orange compound that has no biological activity. Its formation represents a loss of vitamin C.

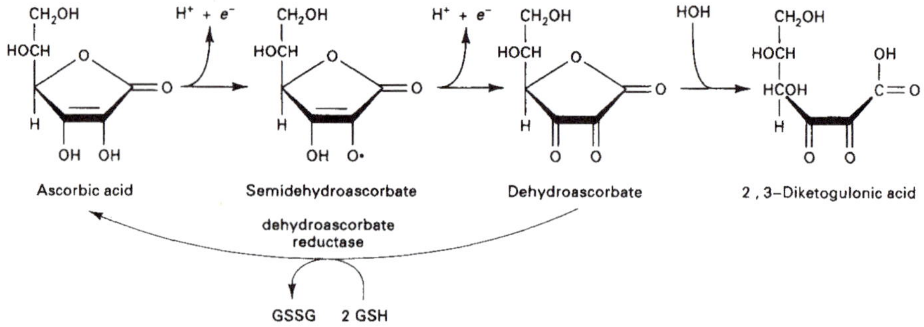

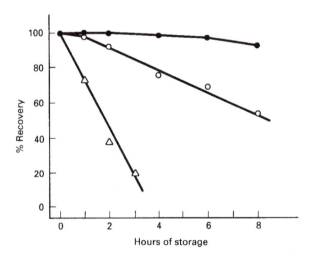

Ascorbic acid     Semidehydroascorbate     Dehydroascorbate     2,3–Diketogulonic acid

**FIGURE 9.80** Ascorbate metabolism.

The instability of ascorbic acid in solution is a concern to nutritionists and food scientists. This instability is illustrated by a study of ascorbate in blood plasma (Figure 9.81). The data in the figure depict the time course for the loss of various forms of the vitamin with storage. The figure shows the levels of ascorbate + dehydroascorbate with storage at 4°C (●), the levels of ascorbate only with storage at 4°C (○), and the levels of ascorbate only with storage of the plasma at room temperature (δ). The rate of loss of ascorbate is greater at room temperature than in the cold. The remarkable stability of the vitamin, when expressed as the sum of ascorbate plus dehydroascorbate, shows that the primary pathway of deterioration of the vitamin was via conversion to dehydroascorbate, rather than by breakdown to ketogulonic acid.

Ascorbic acid is a cofactor in various hydroxylation reactions. These reactions include the hydroxylation of proline residues in a variety of proteins, such as the

**FIGURE 9.81** Instability of ascorbic acid in solution as illustrated by a study of ascorbate in blood plasma. (Redrawn with permission from Lee *et al.*, 1988.)

connective tissue proteins collagen and elastin, and the hydroxylation of catecholamines. Ascorbic acid is also a cofactor in a step in the maturation of certain polypeptide hormones. This step is an amidation reaction. The bleeding that occurs in scurvy is probably a result of increased capillary fragility caused by a failure to produce normal levels of the connective tissue proteins.

## Ascorbic Acid and Collagen Formation

Ascorbic acid plays an important role in collagen formation. Connective tissues such as skin, tendons, ligaments, and cartilage contain a dense network of collagen fibers. Collagen is synthesized in special cells called fibroblasts and chondrocytes. The protein is synthesized in the endoplasmic reticulum and packaged inside secretory vesicles that can fuse with the cell membrane, releasing their contents from the cell. The secreted collagen fibers form a deposit close to the cell surface. The collagen in tendons occurs in long, thick bundles. The collagen in blood vessels is wrapped around the vessels. The collagen in bone is used as a matrix for the deposit of calcium crystals. Cartilage is present in walls of respiratory passages, in the friction-free surfaces of bone, and at the points of attachment of tendons and ligaments to bone. Cartilage is produced by chondrocytes.

**Proline monooxygenase** participates in the maturation of collagen. It catalyzes the conversion of specific proline residues to hydroxyproline. The enzyme does not hydroxylate free proline, though it can act on small peptides containing proline. Molecular oxygen is the source of the oxygen atom in the hydroxyl group of hydroxyproline. In common with many other oxygen-using enzymes, proline monooxygenase contains an iron atom. Hence, the enzyme is an iron metalloenzyme. The iron must be in the reduced state (ferrous iron, $Fe^{2+}$) rather than in the oxidized state (ferric iron, $Fe^{3+}$) to support catalytic activity. A general property of ferrous iron is that it is not particularly stable. It can spontaneously oxidize to the ferric state. Ascorbate plays a vital role in maintaining the enzyme's iron in the reduced state.

The conversion of ferrous to ferric iron in the enzyme is not coupled to each event of hydroxylation. In other words, the conversion is not an obligatory component of the hydroxylation reaction. The oxidation of iron appears to occur after every 10 to 20 hydroxylation events. Then the iron must be re-reduced for enzyme catalysis to continue. Ascorbic acid reduces the iron again and, in turn, is converted to semidehydroascorbate (Figure 9.82). The further metabolism of semidehydroascorbate is not entirely clear. It is believed that one molecule of semidehydroascorbate may donate one electron to another semidehydroascorbate, generating one molecule of ascorbate plus one of dehydroascorbate. This type of transfer, which is generally common in the chemistry of various types of free radicals, is called **disproportionation**.

Vitamin C deficiency results in an impairment in the hydroxylation of collagen. Properly hydroxylated collagen molecules self-associate to form a triple-helix structure within the cell (Figure 9.83). Collagen that is not hydroxylated and does not form the triple helix is not readily secreted from the cell. Its secretion is impaired. The underhydroxylated collagen tends not to build up in the cell but instead is rapidly degraded. Historically, one confusing aspect of vitamin C research was that the deficiency seemed not to result in an accumulation of abnormal

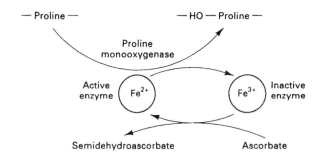

FIGURE 9.82 Ascorbic acid reduces the iron atom of proline monooxygenase.

collagen. This failure did not seem to support the contention that ascorbate was required for the formation of hydroxyproline. More recent work, however, demonstrated that there is a slight buildup of underhydroxylated collagen in deficiency and that the administration of ascorbic acid results in the prompt hydroxylation of the protein and its secretion from the cell. Most of the underhydroxylated collagen is rapidly degraded before it has a chance to build up.

Elastin is a hydroxyproline containing protein of connective tissue. Unlike collagen, elastin does not form a triple helix. With a vitamin deficiency, elastin continues to be produced and secreted from the cell, but in an underhydroxylated state. The function of the hydroxyl group in elastin is not clear.

Collagen is a major protein of connective tissue and is the major protein, by weight, of the body. The maturation of this protein requires the hydroxylation of proline residues and the subsequent association of three collagen molecules to form a triple helix. An additional step is also required, namely the formation of cross-links between adjacent triple helices. These cross-links involve residues of lysine (Figure 9.83). **Lysyl oxidase** catalyzes the removal of the terminal amino group of lysine and the oxidation of the terminal carbon to an aldehyde group. Hence, the lysine residue is converted to an α-aminoadipic-δ-semialdehyde residue. The resultant aldehyde group than condenses with the amino group of a

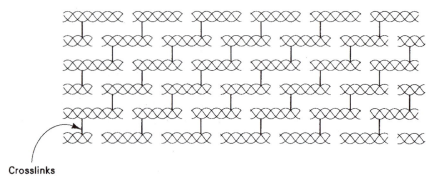

Crosslinks

FIGURE 9.83 Collagen consists of a triple helix that is cross-linked at intervals.

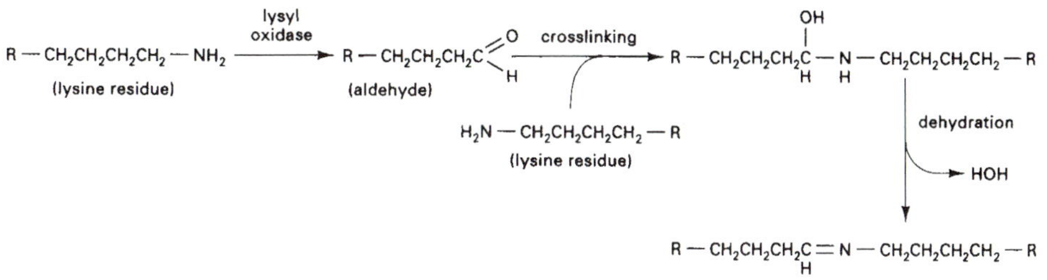

**FIGURE 9.84** Cross-linking of collagen triple helices. The linkage, once formed, is a Schiff base.

lysine residue of a nearby triple helix. The condensation product results in the formation of a Schiff base linkage. Polymers of elastin are cross-linked in a manner similar to those of collagen. The reaction catalyzed by lysyl oxidase and the subsequent event of cross-linking are shown in Figure 9.84.

## Ascorbic Acid and Hormone Synthesis

Ascorbic acid is required for the synthesis of catecholamine hormones and amidated hormones. Metabolic studies conducted with cells of the adrenal gland, which uses ascorbic acid for the synthesis of catecholamines and amidated hormones, revealed that this gland contains high levels of vitamin C (Hornig, 1975). To view the numbers, human adrenal gland contains about 40 mg ascorbate/100 g tissue, whereas skeletal muscle contains only 4 mg/100 g. Metabolic studies conducted with cells of the pituitary gland (Glembotski, 1986), for example, revealed the role of ascorbate in the production of α-melanotropin, an amidated hormone of the pituitary gland.

Norepinephrine and epinephrine are hormones in the class called **catecholamines**. The catecholamines are synthesized and stored in the adrenal gland. With exercise, nerve impulses stimulate the adrenal gland to release the hormones into the bloodstream. Elevated levels of the plasma catecholamines, in turn, induce the contraction or dilation of specific arteries, and the synthesis of cAMP in various cells. Norepinephrine and epinephrine are stored in and released by nerve endings and, for this reason, these hormones are also classed as **neurotransmitters**. The catecholamine biosynthetic pathway begins with tyrosine (Figure 9.85).

Tyrosine monooxygenase uses **biopterin** as a cofactor. Biopterin is made in the body and is not a vitamin. Its structure resembles that of folic acid. **Dopa decarboxylase** is a vitamin $B_6$-requiring enzyme. Dopamine hydroxylase is a copper metalloenzyme. The active form of the enzyme contains copper in the reduced state (cuprous, $Cu^+$). With each catalytic event, the copper is oxidized to the cupric state ($Cu^{2+}$). The enzyme uses ascorbic acid as a cofactor for converting the cupric copper back to cuprous copper. Thus, each catalytic event also results in the conversion of ascorbic acid to semidehydroascorbate. The semidehydroascorbate, perhaps by disproportionation, is converted to ascorbate and dehydroascorbate. The catalytic cycle of dopamine hydroxylase is shown in Figure 9.86. Dopamine hydroxylase, as well as the stored catecholamines, are located in special vesicles

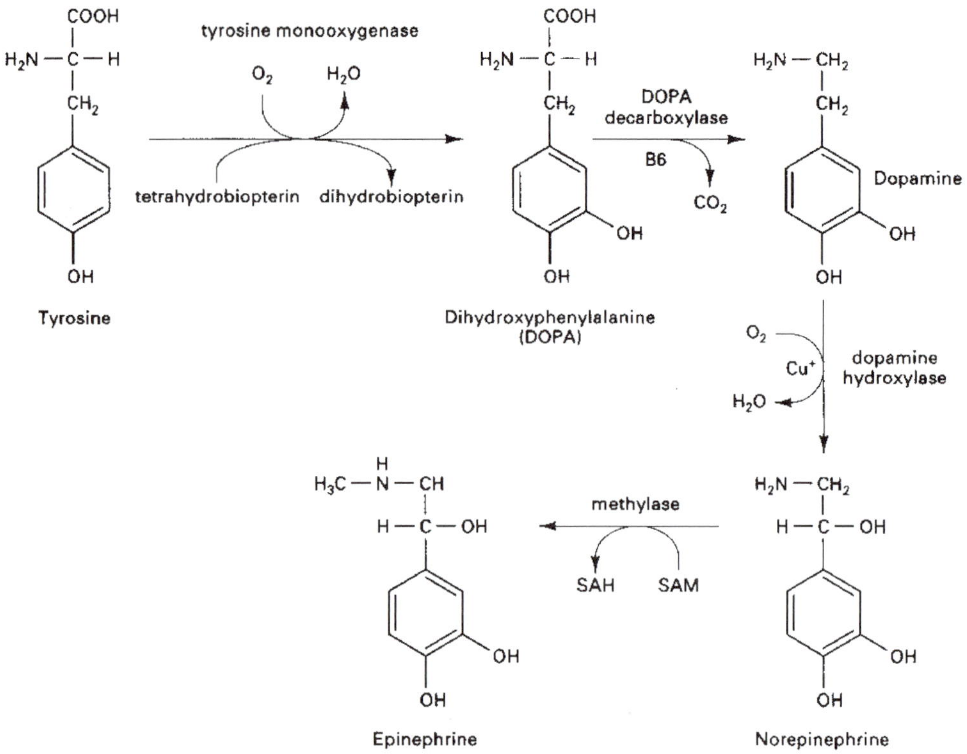

COOH
|
$H_2N - C - H$
|
$CH_2$

tyrosine monooxygenase

$O_2$    $H_2O$

tetrahydrobiopterin    dihydrobiopterin

OH

Tyrosine

COOH
|
$H_2N - C - H$
|
$CH_2$

OH

OH

Dihydroxyphenylalanine
(DOPA)

DOPA
decarboxylase

B6

$CO_2$

$H_2N - CH_2$
|
$CH_2$

Dopamine

OH

OH

$O_2$

$Cu^+$

$H_2O$

dopamine
hydroxylase

H
|
$H_3C - N - CH$
|
$H - C - OH$

methylase

OH

OH

Epinephrine

SAH    SAM

$H_2N - CH_2$
|
$H - C - OH$

OH

OH

Norepinephrine

**FIGURE 9.85** Biosynthesis of catecholamines. Tyrosine is used for the synthesis of various small molecules, which are used as hormones and neurotransmitters. The nutritional biochemist might be especially interested in the pathway of epinephrine biosynthesis, as it requires the participation of four separate cofactors. These are: (1) biopterin; (2) pyridoxal phosphate; (3) ascorbic acid; and (4) S-adenosyl-methionine.

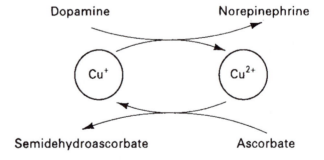

**FIGURE 9.86** Ascorbate is required for the activity of dopamine hydroxylase, also called dopamine-β-monooxygenase. The mechanism of the reaction is quite similar to that of amidating enzyme, an ascorbate-requiring enzyme that catalyzes the hydroxylation of polypeptides, during the course of a two-step sequence.

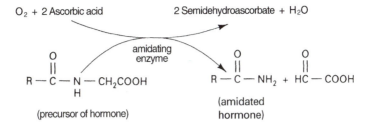

FIGURE 9.87 Ascorbate is used for the synthesis of amidated hormones. Two molecules of ascorbate are used for the production of one molecule of amidated hormone, where each ascorbate is converted to semidehydroascorbic acid. Glyoxylate is a byproduct of the reaction. The reaction occurs in two steps, which are catalyzed by a bifunctional enzyme, as revealed in the text. The chemistry of the first reaction, which results in the hydroxylation of the substrate, is quite similar to that catalyzed by dopamine-β-monooxygenase.

in nerve endings and in vesicles in the adrenal gland. The vesicles in the adrenal are called chromaffin granules.

Ascorbate is required for the synthesis of amidated hormones. Many polypeptide hormones and neurotransmitters have a C-terminal amide group. In other words, the C-terminal carboxyl carbon is covalently attached to ammonia. The hormones containing C-terminal amides are called **amidated hormones**. In most cases, the C-terminal amide group is required for the biological activity of the hormone. **Peptidylglycine α-amidating monooxygenase** catalyzes the amidation reaction (Figure 9.87). For short, this enzyme may be called "amidating enzyme." The enzyme acts on polypeptides containing glycine at the C terminus.

Amidating enzyme is a copper metalloenzyme that requires both oxygen and ascorbic acid. One copper atom is bound to three residues of histidine (his 107, 108, and 172). The other copper atom is bound to two residues of histidine (his 242 and 244), and to a residue of methionine (met 314). The enzyme is actually bifunctional, that is, it consists of two separate enzymes occurring in a single polypeptide chain. The two enzymes work, one after the other, to create the amidated polypeptide (Prigge *et al.*, 1997).

The N-terminal half of the enzyme, which catalyzes the first reaction, contains two atoms of copper. Here, $O_2$ is split in half, where one of the oxygen atoms is used to create water, and the other oxygen atom is used as the source of the hydroxyl group on the α-carbon of the glycine residue (reaction #1):

$$
\begin{array}{c}
\text{CH}_2\text{–COOH} \qquad \xrightarrow{\text{Reaction \#1}} \qquad \overset{\text{OH}}{\text{CH–COOH}} \\
\text{Polypeptide–NH} \qquad\qquad\qquad \text{Polypeptide–NH} \\
\text{(hydroxylated intermediate)} \\
\text{2 Ascorbate} + O_2 \qquad\quad \text{2 Semi-dehydroascorbate} + H_2O
\end{array}
$$

The C-terminal part of the bifunctional enzyme catalyzes the cleavage of the hydroxylated intermediate, generating the amidated polypeptide plus glyoxylate, as shown here (reaction #2):

$$
\begin{array}{c}
\text{OH} \\
\text{CH–COOH} \\
\underset{\text{(hydroxylated intermediate)}}{\text{Polypeptide–NH}}
\end{array}
\xrightarrow{\text{Reaction \#2}}
\text{Polypeptide–NH}_2 \; + \; \text{glyoxylate}
$$

To summarize the properties of the amidating reaction, the two copper ions are initially in the cupric state. Two separate molecules of ascorbate donate a total of two electrons, reducing both copper atoms to the cuprous state. Then $O_2$ binds to one of the copper atoms, while the other copper atom donates an electron to the bound oxygen, resulting in an enzyme-bound peroxide group (enzyme–copper–OOH). The peroxide group splits in half — where one half reacts with a hydrogen atom to form *water*, and the other half attacks the polypeptide, forming the *hydroxylated polypeptide*. The hydroxylated polypeptide goes on to be processed at the second active site, producing the amidated hormone (Prigge *et al.*, 1997).

## Ascorbic Acid and Damage from Radicals

Ascorbic acid seems to be involved in reducing damage to the body from radicals. Covalent bonds generally involve the sharing of a pair of electrons between two atoms. A radical is an atom with an electron that occurs by itself and is unpaired. Radicals occur temporarily at the active site of certain enzymes during the desired reaction and are a necessary part of the catalytic mechanism; however, some radicals are not desirable because they can produce indiscriminant damage to various components of the cell. Cell membranes, soluble metabolites, and the chromosomes may all be targets of this damage.

Metabolites of oxygen are considered to be a major source of the damaging radicals. There is some thought that toxic oxygen damage contributes to the natural aging process and is a factor in some diseases, such as cancer, cardiovascular disease, arthritis, and cataracts (Frei *et al.*, 1989). Oxygen radicals are one type of toxic oxygen, and these can originate from oxygen-utilizing enzymes, such as cytochrome *c* oxidase and some flavoproteins, as well as from radiation. One type of oxygen radical is superoxide ($O_2^-$). Superoxide may be produced by the imperfect and incomplete reduction of oxygen by the respiratory chain, where $O_2$ is only reduced to $O_2^-$, rather than completely reduced to water (Imlay and Fridovich, 1991; Halliwell and Gutteridge, 1990). Superoxide dismutase catalyzes the removal of superoxide by a dismutation reaction, resulting in the formation of $O_2$ and HOOH:

$$
2\,O_2^- + 2\,H^+ \quad \underset{\text{(Dismutase)}}{\longrightarrow} \quad \text{HOOH} + O_2 \tag{9.6}
$$

Superoxide is not particularly reactive, compared with another oxygen radical (the hydroxyl radical), though it is thought to contribute to cellular damage.

Hydrogen peroxide (HOOH) is produced by superoxide dismutase. HOOH is also a byproduct of the activity of specific flavoproteins. HOOH is a toxic com-

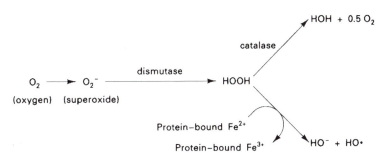

FIGURE 9.88 General pathway of formation and removal of toxic forms of oxygen.

pound that, fortunately, is efficiently destroyed by catalase. Catalase catalyzes the dismutation of two HOOH molecules to form two molecules of water and one of $O_2$ (Figure 9.88). Both of these products, obviously, are harmless to the cell.

Another pathway of metabolism available to HOOH is reaction with iron in the cell. The reaction is apparently not catalyzed by any particular enzyme, but occurs in a nonspecific manner by iron bound to protein or to DNA. Free iron occurs in vanishingly low levels in the cell. This reaction, which requires iron to be in its reduced form ($Fe^{2+}$), is called the **Fenton reaction**. The hydroxyl radical (•OH) is highly reactive and has a fleeting lifetime in the cell. It reacts with a great variety of molecules. The reaction occurs at or near the site of generation of the hydroxyl radical.

Ascorbate may be involved in reducing damage to the cell from radicals. A simplified mechanism (Figure 9.89) shows the hydroxyl radical reacting with a component in the cell, abstracting (pulling off) a hydrogen radical. The product is a radical, but it is one that is more stable than •OH. Ascorbate may donate a hydrogen radical (H•) to this product, thus repairing it before further deterioration can occur. Here, the ascorbate is converted to semidehydroascorbate, a relatively stable radical (Buettner, 1993). Whether this scenario occurs to a significant extent in the body remains somewhat speculative. Vitamin E protects membranes from damage by radicals and, in doing so, is converted to the vitamin E radical. Evidence suggests that there is an ascorbate–vitamin E interaction. Ascorbate can react with the vitamin E radical and regenerate vitamin E in its original, chemically active form (Gey, 1998; Thomas *et al.*, 1995).

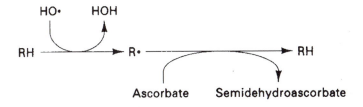

FIGURE 9.89 Possible use of ascorbate in reducing damage from radicals.

## VITAMIN E

Vitamin E is a fat-soluble vitamin. The RDA for the adult man is 10 mg of α-tocopherol, or its biological equivalent. The RDA for infants should also be mentioned, as vitamin E deficiency, when it occurs, tends to strike this population. The RDA for the newborn is 3 mg of α-tocopherol, or its equivalent. The vitamin needs of the infant have been expressed in terms of the amount of polyunsaturated fatty acids (as fats and oils) in the diet, for example, 0.7 mg of α-tocopherol per gram of linoleic acid. A common level of dietary intake is about 10 mg per day. A deficiency in the vitamin is quite rare.

Good sources of vitamin E are vegetable oils, such as corn, soy, and peanut oil. Animal fats, such as butter and lard, contain lower amounts of the vitamin. The content of the most important form of vitamin E, α-tocopherol, in various foods is as follows. Corn oil contains about 16 mg of α-tocopherol per 100 g; sunflower oil 50 mg/100 g; wheat germ oil 120 mg/100 g; and fish, eggs, and beef 0.5 to 2.0 mg/100 g. In plants, α-tocopherol resides in chloroplasts, while other forms of tocopherol (β-, γ-, and δ-tocopherol) occur elsewhere in the plant cell. Tocotrienols, which also have vitamin E activity, are not found in the green parts of plants, but in the bran and germ of seeds.

The term *vitamin E* refers to two groups of compounds, the tocopherols and the tocotrienols. The structures of these compounds appear in Figure 9.90. All forms of the vitamin contain two parts, a "head" and a "tail." The head consists of an aromatic ring structure, called **chroman** or **chromanol**, and is the site of antioxidant action. The tail of tocopherols is a **phytyl group**, while the tail of tocotrienols is a **polyisoprenoid group**. The tail of vitamin E serves to anchor the vitamin in lipid membranes, in the lipids of adipose tissue, and in the lipid surface and core of the lipoproteins.

The term **antioxidant** is defined as any substance that, when present at low concentrations compared to those of oxidizable compounds, can delay or prevent the oxidation of that compound (Halliwell, 1996). The term can be used to refer to vitamin E, ascorbate, catalase, peroxidase, superoxide dismutase, and other compounds. The term "antioxidant" should only be used as a rhetorical term to make it easier to refer to these types of compounds. This is because, under some conditions, "antioxidants" such as vitamin E or ascorbate can promote the oxidative destruction of lipids, nucleic acids, and other macromolecules.

### Chemistry of Antioxidant Action

The chroman ring of vitamin E bears a hydroxyl group that can donate a hydrogen radical (H•) to other radicals in the immediate environment. Since the immediate environment of vitamin E in the cell consists of lipids, vitamin E usually donates hydrogen radicals to free radicals occurring on polyunsaturated fatty acids. In calling vitamin E an antioxidant, one means that vitamin E can bring a halt to a damaging chain reaction that had earlier been caused by toxic oxygen.

A number of aromatic hydroxyl compounds, such as the food additive butylated hydroxytoluene (BHT), have the same sort of antioxidant activity as vitamin

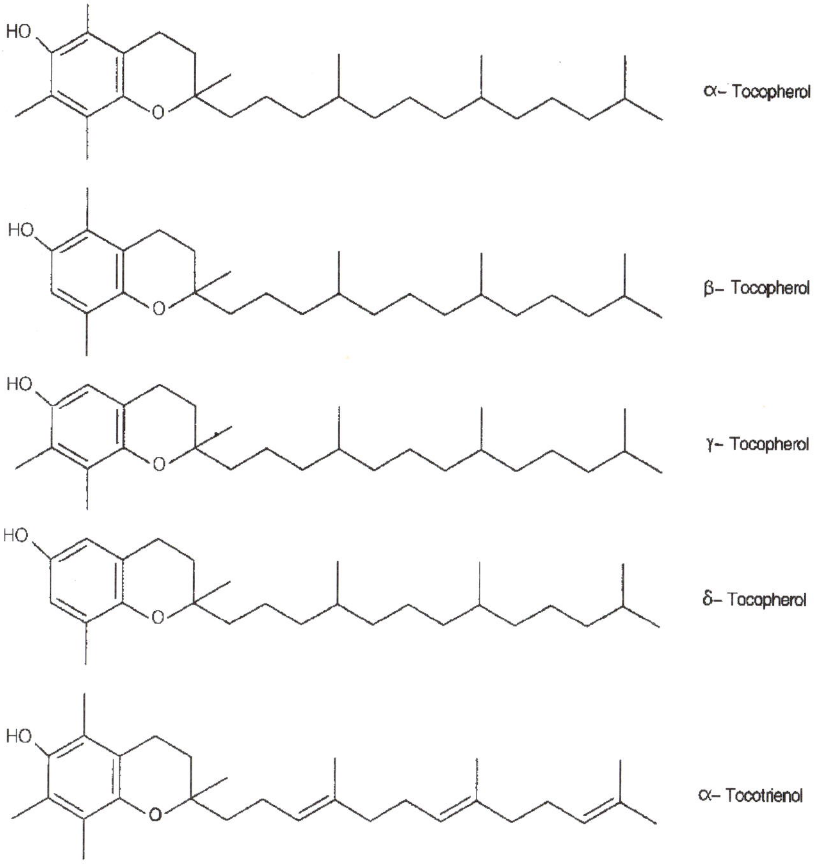

**FIGURE 9.90**  Various forms of vitamin E.

E. However, vitamin E is hundreds of times more potent as an H• donor than BHT and other aromatic hydroxyls. Why is this? Directly opposite the aromatic ring from the hydroxyl group is an oxygen atom (occurring as a member of the six-sided ring). This oxygen atom contains a "lone pair" of electrons (see Chapter 1). This lone pair of electrons may be thought of as an antenna, which hovers above the chroman group, waiting to interact with other atoms. When vitamin E donates an H•, vitamin E's hydroxyl group occurs as an oxygen radical (R—O•). The lone pair electrons and the oxygen radical electron are situated in a way that encourages all three electrons to interact. The oxygen radical is stabilized by resonance about the aromatic ring, and the end-result is the stabilization of the vitamin E radical (Kamal-Eldin and Appleqvist, 1996; Liebler and Burr, 1992).

The differing H• donating potencies of various forms of vitamin E depend, in part, on how well the lone pair electrons can stabilize the vitamin E radical. The vitamin E radical may also be called a **chromanoxyl radical**.

Vitamin E is not very stable to storage. The stability can be greatly improved by esterifying the vitamin via the hydroxyl group to acetic acid. The resulting molecule, α-tocopheryl acetate, is used as a commercial form of vitamin E. The ester linkage is hydrolyzed in the body, liberating the vitamin in its active form.

## Biological Potency of Various Forms of Vitamin E

The predominant form of vitamin E in food is α-tocopherol. This form of the vitamin is also the most biologically potent form (100%), as determined by the rat fertility test. Other forms (and their relative potencies) are β-tocopherol (40%), γ-tocopherol (10%), δ-tocopherol (1%), and α-tocotrienol (25%). The rat fertility test is performed as follows. Female rats are fed diets deficient in vitamin E, sufficient in α-tocopherol, or containing a known amount of the test compound. The rats are then mated with male rats. The number of living fetuses in the uterus of the female rat is then used to assess the potency of the test compound, relative to α-tocopherol. The deficient state results in dead fetuses, spontaneous abortions, and fetal resorptions.

## Absorption and Storage of Vitamin E

Vitamin E is absorbed from the gut with the aid of bile salts. The vitamin is not esterified to a fatty acid during absorption, as is the case with cholesterol and retinol. Vitamin E is transported to the bloodstream in chylomicrons and distributed to the various tissues via the lipoproteins.

The various forms of vitamin E are all absorbed by the gut in proportion to their abundance in the diet, where they appear in the chylomicrons. As the chylomicrons travel through the blood, the various forms of vitamin E are partly transferred to all the other types of lipoprotein particles. However, once vitamin E enters the liver, α-tocopherol is the only form that is preferentially packaged into the VLDLs (Traber, 1997). This preference seems to be a result of the activity of **α-tocopherol transfer protein**. This is a small protein that exists only in the cytoplasm of hepatocytes.

EXERCISE

A rare genetic disease in humans results in mutations in the gene coding for α-tocopherol transfer protein. Naturally occurring mutations occur in different parts of the gene, in different patients. Where the mutation results in a severe defect in the structure of the protein, the result is extremely low levels of plasma vitamin E and in neurological damage. The neurological damage presents by ataxia (unsteady limbs) and paresthesia (numbness in limbs). How would you treat this disease? (Consult Gotoda *et al.*, 1995).

Although adipose tissue contains high levels of vitamin E, the tissue might not be considered to function efficiently as a storage site of the vitamin. Studies with animals have demonstrated that a deficiency in vitamin E results in the rapid depletion of the vitamin in plasma and various organs of the body with little effect on that in adipose tissue. In contrast, a dietary deficiency in vitamin A results in the maintenance of the plasma concentration at the expense of the vitamin A stored in the liver.

## Vitamin E Deficiency

The signs of a vitamin E deficiency may be quite different in animals and humans. The signs occurring in experimental animals include impaired reproduction and muscle weakness (muscular dystrophy). Reproductive defects in female animals involve a failure of the fetus to thrive. In males, the deficiency results in an inhibition of sperm production. One feature common to humans and animals is the formation of lesions (pathological structures) in nerves and muscles. The deficiency can cause the degeneration of nerves and the accumulation of a compound called lipofuscin in various tissues, such as muscle. Lipofuscin has an amorphous structure and is thought to be composed of lipid degradation products and cross-linked proteins.

It is likely that all of the biological problems and lesions that occur during the deficiency arise from a failure to halt the rise in oxidized lipids. This function is summarized in Figure 9.91. All of the membranes of the cell contain polyunsaturated fatty acids (PUFAs). PUFAs are readily damaged by toxic forms of oxygen. One particularly toxic form of oxygen is the hydroxyl radical, •OH. The hydroxyl radical is highly reactive and never accumulates in the cell in quantities that are directly measurable. The fleeting presence of •OH is detected only by analyzing for the damage that is has inflicted.

Damage to cell membranes may be initiated by the reaction of the hydroxyl radical with a PUFA, generating a PUFA radical. The hydroxyl radical is converted to water (Figure 9.91). The PUFA radical, in turn, can be repaired by vitamin E (Figure 9.92). Involvement of vitamin E in this repair results in its conversion to a vitamin E radical. The vitamin E radical is relatively unreactive and does not cause further damage to the cell. In the absence of vitamin E, the PUFA radical may induce a chain reaction, resulting in widespread damage to cell membranes.

### Deficiency in Animals

Figure 9.93 depicts an experiment with guinea pigs. The animals consumed a vitamin E-deficient diet for up to 8 weeks, as indicated. The results demonstrate that the levels of vitamin in plasma and liver decreased rapidly, whereas that in adipose tissue tended to be maintained and to decrease at a slower rate. Data on the heart are also shown. The declines in vitamin levels (see Figure 9.93) were induced by three different factors: (1) consumption of the vitamin-free diet; (2) the continued, normal daily turnover and loss of the vitamin; and (3) the growth of young animals. Growth results in a decreased concentration of vitamin E in tissues as a result of dilution of the existing pool of the vitamin by new tissue. Lesions of

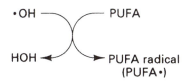

**FIGURE 9.91** Production of damage to cell membranes by radicals.

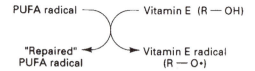

**FIGURE 9.92**  Repair or halt of damage by radicals of membranes with vitamin E.

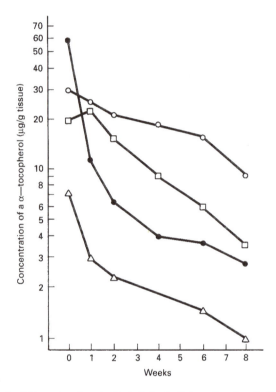

**FIGURE 9.93**  Effect of 8-week vitamin E-deficient diet on levels of vitamin E in plasma (Δ), liver (●), adipose tissue (○), and heart (□). (Redrawn with permission from Machlin *et al.*, 1979.)

muscle were observed in the guinea pigs after 8 weeks. The lesions occurred at a time when substantial quantities of the vitamin remained in the adipose tissue. Hence, it is concluded that the vitamin in fat cannot be efficiently mobilized in times of need.

### Deficiency in Humans

Normal serum values for vitamin E range from 8.0 to 16.0 μg/ml. The marked influence of the lipid levels in the bloodstream on plasma tocopherol makes it preferable to express the plasma vitamin levels as a ratio to total lipid. The use of this ratio can correct for conditions that result in increases in plasma lipid levels, such as hyperlipidemia. A ratio of serum α-tocopherol/total lipid of under 0.8 mg/g indicates a deficiency in vitamin E. The concentration of vitamin E in the blood of newborns is generally less than half that of adults. These levels can increase to adult levels within a few days of birth during breast-feeding.

The chemical test used for measuring vitamin E in serum is as follows. The vitamin is extracted from the water-soluble components of serum by mixing the serum with the solvent hexane. The vitamin, along with other lipids, leaves the aqueous layer and enters the hexane layer. The hexane layer may be called the solvent phase. The amount of vitamin in the hexane layer can be measured by fluorescence spectrometry. Exposure (or excitation) of the vitamin to light results in the production of light by the vitamin. The wavelength of emission is always longer (lower energy) than the wavelength of excitation, in all cases of fluorescence. In the case of vitamin E, the wavelength of light used for excitation is 295 nm, whereas that of the light emitted is 325 nm. The amplitude (intensity) of the light fluorescing from the sample is proportional to the concentration of the vitamin in the sample.

Two functional tests have been used in studies of humans: the red blood cell hemolysis test and the tissue biopsy test. Both tests reflect properties of biological membranes. The red cell test measures the ability of the cell to resist peroxide-induced damage. Cells taken from normal, vitamin E-sufficient subjects can withstand the oxidizing effects of hydrogen peroxide (HOOH); however, red blood cells taken from vitamin E-deficient subjects may be broken (hemolyzed) in response to the HOOH, resulting in the release of hemoglobin. In detail, red cells are suspended in sodium chloride (0.85%) and incubated in the presence or absence of HOOH (1.2%) for 3 hours. The extent of hemolysis is then measured using a spectrophotometer, which measures the red-colored material released into solution. Note that the susceptibility of red cells to mechanical damage, as opposed to oxidative damage, seems to be the same for cells from vitamin-deficient and vitamin-sufficient subjects.

Vitamin E deficiency is quite rare. Persons at risk include those with fat malabsorption syndromes, including cystic fibrosis, cholestatic liver disease, and diseases that prevent the functioning of the lipoproteins. Infants suffering from fat malabsorption diseases can develop symptoms of vitamin E deficiency by the age of 1.5 to 2.0 years. In adults, the onset of a disease preventing fat absorption may provoke the vitamin deficiency after a longer period, that is, 10 years or more. The deficiency in humans can induce a variety of neurological symptoms such as ataxia, lack of reflexes, decreased vibratory sensation, and paralysis of the eye muscles. These neurological symptoms cannot be reversed rapidly. When possible, reversal may require months of treatment with vitamin E. It is thought, but not proven, that the neurological problems arise from the failure of the vitamin to protect nerves from toxic forms of oxygen.

Cholestatic liver disease may occur in infancy. One particularly severe symptom is the inability to walk. The neurological symptoms can be treated with weekly injections of 100 mg α-tocopherol over half a year. Vitamin E deficiency in newborns has been associated with hemolytic anemia. Anemia is a decreased concentration of red blood cells in whole blood as well as a drop in the hemoglobin level below the normal range. Hemolytic anemia can be treated relatively promptly, that is, with a week or two of vitamin E therapy.

The premature infant may be at risk for vitamin E deficiency, because infants may be born with low supplies of the vitamin and because the premature infant in particular has a poorly developed capacity for absorbing lipids.

## Vitamin E, Antioxidants, and Atherosclerosis

It has been proposed that the vitamin E occurring in low-density lipoprotein (LDL) particles in the bloodstream, along with other antioxidants, normally acts to reduce or minimize the trend to atherosclerosis that occurs in most or all people. The natural rate of progression of atherosclerosis, as measured during its mature phase, involves the reduction in diameter of arteries by about 1.5% per year (Superko and Krauss, 1994).

One scenario that has been proposed to account for atherosclerosis is summarized by the following diagram. LDLs in the bloodstream are exposed to toxic oxygen, which results in damage to its protein component, apo B. The LDL bearing the damaged apo B is then taken up by cells bearing **scavenger receptor A** (monocytes and macrophages (phagocytes)):

Continued uptake by phagocytes that reside by the endothelial cells of the coronary artery results in their conversion to foam cells, in the invasion of smooth muscle cells, and in eventual narrowing of the lumen of the artery. Evidence suggests that the vitamin E ($\alpha$-tocopherol) present in LDLs serves to delay the onset of damage to apo B, and give it time to be eventually taken up by the LDL receptor in hepatocytes of the liver. This part of the scenario is shown in the following diagram:

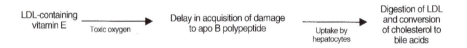

LDL particles that have undamaged apo B generally circulate in the bloodstream for 3 days, with eventual uptake by the LDL receptor. An LDL-bearing modified or damaged polypeptide may no longer be recognized by the LDL receptor and may promptly (within a few minutes) be taken up by scavenger receptor A of various phagocytes.

A typical LDL particle contains about 700 molecules of phospholipid, 600 free cholesterol molecules, 1600 cholesteryl esters, 185 triglycerides, and one polypeptide of apo B (Steinberg, 1997; Noguchi et al., 1998). A typical LDL particle also contains 3–16 molecules of α-tocopherol. α-Tocopherol is the major antioxidant of the LDL. An LDL particle may also contain 0–2 molecules of β-carotene, with similar low levels of ubiquinone, γ-tocopherol, lycopene, cryptoxanthin, canthaxanthin, and lutein (Ziouzenkova *et al.*, 1996; Tribble *et al.*, 1994). These latter compounds may also serve as antioxidants in the LDL.

## Indirect Pathway of Damage to Apo A

Oxidative damage to the lipids in the LDL does not, in itself, result in recognition and uptake by scavenger receptor A. But where lipid breakdown products (mainly aldehydes) react with apolipoprotein A (on the surface of the LDL), the LDL particle will be taken up by scavenger receptor A. In addition, where toxic oxygen directly damages apo A (without damage to lipids), the LDL particle may also be taken up by scavenger receptor A.

Some forms of toxic oxygen, such as HOOH, may react with contaminating metal ions binding to the surface of the LDL and, via the Fenton reaction, produce hydroxyl radicals with the consequent formation of damaged lipids (Lynch and Frei, 1995). To be specific, the target lipids include the polyunsaturated fatty acids occurring in phospholipids and in cholesteryl esters. The damaged PUFAs may then break down to produce aldehydes (malondialdehyde; 4-hydroxynonenal), which may then react with free amino groups of the apo A polypeptide (Hanzell *et al.*, 1996). The structure of the condensation product is the Schiff base.

## Direct Pathway of Damage to Polypeptides

Hypochlorite (HOCl) can react directly with amino acid residues of apo A. Hypochlorite, which is produced by the catalytic action of **myeloperoxidase**, can react with and damage residues of tyrosine, methionine, and cysteine (Yang *et al.*, 1997). Myeloperoxidase is secreted by phagocytes into the extracellular fluid. Intracellular myeloperoxidase also produces hypochlorite, which may leak out into the extracellular fluids during phagocytosis. In either case, the result can be damage to any nearby protein, including apo A.

White blood cells (monocytes) tend to migrate through the endothelial cells of arteries and reside in and infiltrate the "subintimal space." The **intima** is the layer of smooth muscle cells that coats all arteries (Schwartz *et al.*, 1995). Any LDL particle that happens to diffuse between the epithelial cells and into the subintimal space may find itself confronted with a monocyte. Toxic oxygen released by the monocyte can damage the LDL. The damaged LDL, in turn, attracts more monocytes into the subintimal space, and induces the monocytes to differentiate into macrophages (Steinberg, 1997). The phagocytes (monocytes; macrophages) continue to produce toxic oxygen. The macrophages continue to take up damaged LDL particles, and to accumulate cholesterol, with their eventual conversion to **foam cells** and the generation of the **fatty streak**. The coronary artery is distinguished, apart from most other arteries in the body, in that its intimal layer tends to have a high tendency to develop into an atherosclerotic lesion (Schwartz *et al.*, 1995).

The preceding commentary, regarding the role of oxidized LDLs in the formation of the fatty streak, represents a coherent proposal, but only partial evidence exists to support it. The proposed scenario is summarized in Figure 9.94.

*Antioxidants May Act as a Free-Radical Buffer and Delay Onset of Damage to the Apo B Polypeptide*

Where might α-tocopherol and other "antioxidants" fit into the aforementioned scenario? Where oxidative damage to lipids in the LDL particle occurs at a low rate, any α-tocopherol within the LDL particle might be expected to terminate free radical damage, according to mechanisms detailed in the present section and in the section on Essential Fatty Acids. Evidence suggests that when all molecules of α-tocopherol become used up and depleted (converted to the α-tocopherol free radical), other antioxidants within the LDL particle may take over, and thus further delay the onset of damage to apo A. These secondary antioxidants include β-carotene and ubiquinone. Evidence also suggests that plasma ascorbate also serves the important role of converting α-tocopherol free radical back to α-tocopherol, thus delaying or preventing its depletion.

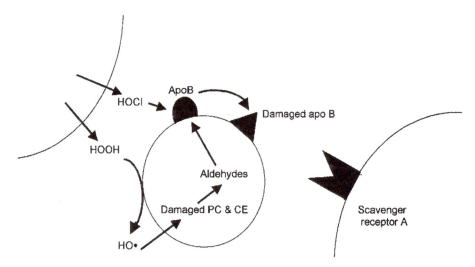

**FIGURE 9.94** Sources of toxic oxygen and pathways leading to damage to apo B. Certain white blood cells function in the body to kill and eat invading microorganisms. This type of white blood cell is called a phagocyte. There exist three types of phagocytes in our bloodstream: monocytes, macrophages, and neutrophils. These phagocytes produce toxic oxygen (HOOH and HOCl) and use it to kill unwanted organisms. Exposure of LDL particles to these forms of toxic oxygen can result in damage to the lipid component of the LDL, as well as to the protein component of the LDL (apo B-100). The figure depicts HOCl directly reacting with apo B on the surface of the LDL. The resulting damaged apo B facilitates uptake of the entire LDL particle by monocytes and macrophages. The figure also depicts hydrogen peroxide, reacting with reduced metal ions contaminating the surface of the LDL, to produce hydroxyl radicals. Evidence suggests that copper atoms, but not iron atoms, can be reduced at the surface of the LDL. However, there is very little evidence, to date, that metal ions naturally occur as contaminants on the surface of our LDLs. The figure depicts the hydroxyl radical provoking damage to lipids (phosphatidylcholine and cholesteryl ester) in the LDL particle, the production of aldehyde breakdown products, and the condensation of these byproducts with apo B. The resulting altered apo B facilitates uptake of the entire LDL particle by monocytes and macrophages.

## Antioxidants May Be Overwhelmed

In the event that the LDL particle is exposed to high levels of toxic oxygen, the α-tocopherol within the particle may be overwhelmed, allowing damage to apo A. In the event that ascorbic levels are reduced or lowered in the immediate environment of the LDL particle, α-tocopherol may be similarly be overwhelmed. This situation may be more serious than it sounds, as the α-tocopherol free radical (if given the time) may facilitate and catalyze damage to lipids at the core of the LDL particle. In this way, vitamin E may facilitate the transfer of free radicals from the aqueous phase at the surface of the LDL particle to the fatty phase at its core (Upston *et al.*, 1996; Thomas *et al.*, 1995; Bowry *et al.*, 1992).

There has been some interest in raising the concentration of α-tocopherol, β-carotene, and ubiquinol (reduced coenzyme Q) in the LDL particle by dietary supplements, with the hope of reducing oxidative damage to its lipids and to apo A. There has also been some interest in raising plasma ascorbate for the same purpose. Studies with humans have revealed that the levels of α-tocopherol and ubiquinol inside the LDL particle can be increased by two- to fourfold, by dietary means (Kayden and Traber, 1993; Jialal and Grundy, 1992, Thomas *et al.*, 1995, Furr and Clark, 1997).

Studies with humans have shown that a diet low in β-carotene (0.025 mg/day) results in plasma levels of about 0.5 μM, while large doses of β-carotene (90 mg/day) can result in a tenfold increase in plasma β-carotene, to give about 5 μM. Most of the β-carotene (70%) in plasma resides in the LDL particles (Johnson *et al.*, 1995; Johnson and Russell, 1992). Again, one should recall that some people are "responders," so that a dose of dietary β-carotene provokes a rise in plasma β-carotene, whereas other people are "nonresponders."

Studies with humans have demonstrated that large doses of ascorbic acid (1 g/day) result in increases in plasma levels, but the increases are only modest, and the levels rise to a plateau level of about 80 μM (Levine *et al.*, 1997).

A number of studies have suggested an association between elevated plasma vitamin E levels and reduced risk for cardiovascular disease (Bonithon-Kopp *et al.*, 1997; Gey, 1998; Rimm *et al.*, 1993; Stampfer *et al.*, 1993). The task of the nutritionist attempting to correlate vitamin intake with cardiovascular disease is complicated by the fact that different food oils contain different amounts of the various tocopherols. Olive oil contains about 120 mg α-tocopherol/kg oil; soybean oil (70 mg α-tocopherol and 900 mg γ-tocopherol/kg); safflower oil (340 mg α-tocopherol and 35 mg γ-tocopherol/kg); and wheat germ oil (1500 mg α-tocopherol and 800 mg γ-tocopherol/kg) (Chase *et al.*, 1994; McLaughlin and Weihrauch, 1979). Most of the vitamin E present in blood plasma is α-tocopherol (rather than γ-tocopherol, for example) because of the influence of α-tocopherol transfer protein. Most of the plasma vitamin E resides in the LDLs. A paradox seems to present itself where people who have elevated LDLs (and who are more at risk for cardiovascular disease) should also have elevated α-tocopherol (and possibly be at lesser risk for cardiovascular disease). This paradox can be avoided by expressing plasma α-tocopherol levels as α-tocopherol/cholesterol (Gey, 1998).

The question of whether deficiencies (or large doses) of any of the aforementioned antioxidants can reduce atherosclerotic lesions has provoked considerable interest among the research community. Atherosclerosis is a complex, multistep event, and any detected influence of an "antioxidant" should not automatically be assumed to be due to a reduction of damage to the protein component of lipoprotein particles. For example, one might ask whether the antioxidants can influence the rate of colonization and adhesion of phagocytes in the intima, or if they influence the infiltration of smooth muscle cells into this region.

## ESSENTIAL FATTY ACIDS

The essential fatty acids (EFAs) are linoleic acid and linolenic acid. These fatty acids occur in the diet as parts of phospholipids and triglycerides. Both fatty acids are used in the body for structural purposes, where they are incorporated into the phospholipids of cell membranes. They can also be oxidized and used to produce energy. The most interesting use of EFAs is for hormone synthesis. These hormones comprise three classes of compounds having similar structures: prostaglandins, thromboxanes, and leukotrienes. **Linoleic acid** is converted to arachidonic acid, which is a precursor of certain prostaglandins, thromboxanes, and leukotrienes. Arachidonic acid is a long-chain fatty acid. It is a relatively minor component of dietary fats and oils and has sometimes been called an essential fatty acid, even though it is readily synthesized from linoleate. **Linolenic acid** is converted to eicosapentanoic acid (EPA), which in turn is also used for the synthesis of a number of prostaglandins, thromboxanes, and leukotrienes (Dajani, 1993).

The essential fatty acids are also converted in the body to the 22-carbon fatty acids **docopentaenoic acid** (DPA) and **docohexaenoic acid** (DHA). DPA is made from linoleic acid; DHA is made from linoleic acid. The functions of these 22-carbon fatty acids are not clear, but they may be important for vision and for other functions of the nervous system. DPA and DHA can be further elongated, in the body, to the "very-long-chain fatty acids." The very-long-chain fatty acids contain 24 to 34 carbons, and occur in the brain, rods of the retina, and in the testes (Suh *et al.*, 1996). Their functions are not clear.

The information presented so far is summarized in the following diagram. Arachidonic acid could be considered to be the most important metabolite in the diagram:

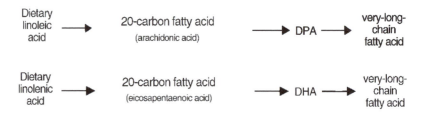

The RDA for the essential fatty acids is 1 to 2% of total energy intake. Generally, between 5 and 10% of our energy intake consists of EFAs. Because of our ample intake of fat, a deficiency in EFAs is quite rare. The biochemical steps in the modification of nonessential fatty acids, such as oleic acid, and essential fatty acids (linoleic acid and linolenic acid) are generally the same. These steps include **elongation** and **desaturation**. Modification of fatty acids by their repeated desaturation, elongation, desaturation, and elongation leads to a variety of interesting fatty acids. These fatty acids have functions that are only beginning to be understood.

Two systems are used for identifying fatty acids. One system employs the Greek capital Delta ($\Delta$), where numbering of the carbons starts from the carboxylic acid carbon. In other words, the carboxylic acid carbon is carbon number 1. The other system employs the Greek lowercase omega ($\omega$) and is attuned to the biochemical origins of the fatty acid. The omega numbering system is based on the properties of metabolism, rather than on organic chemistry. Here, numbering starts with the tail-end carbon, i.e., the "methyl" carbon.

## Metabolism and Nomenclature

Palmitic acid and stearic acid are the major products of fatty acid synthase. Both of these fatty acids are **saturated**; that is, they contain no double bonds between adjacent carbons. "Saturated" means that they contain a maximal content of hydrogen atoms. Although palmitate is a major fatty acid in cell membranes, most or all membranes contain longer-chain fatty acids as well. These fatty acids, having from 18 to 22 carbon atoms, are synthesized by the **fatty acyl chain elongation system**. The chain elongation system is located primarily in the endoplasmic reticulum, in contrast to fatty acid synthase, which is cytosolic. The chain elongation system catalyzes the addition of 2-carbon units to the growing fatty acid.

### Mammalian Desaturases

In addition to chain elongation, fatty acids are modified by the introduction of double bonds (desaturation). Enzymes, called **desaturases**, catalyze the synthesis of unsaturated fatty acids. They can use saturated or partially unsaturated fatty acids as substrates. $\Delta^9$-Desaturase, for example, catalyzes the introduction of a double bond between carbons 9 and 10 of a fatty acid (counting from the carboxylic acid end). Three examples of reactions of $\Delta$-desaturases are shown in Figure 9.95.

Mammalian desaturases can introduce double bonds into fatty acids *only between carbon-10 and the carboxyl carbon*. In contrast, plants can produce linoleic acid, which has double bonds between carbons 9 and 10, and between carbons 12 and 13. Plants also synthesize linolenic acid, which has double bonds between *carbons 9 and 10, carbons 12 and 13*, and *carbons 15 and 16*.

### Delta Nomenclature and Omega Nomenclature

Linolenic acid is abbreviated as $18:3\Delta^{9,12,15}$ or $18:3\omega3$. The number *before* the colon indicates the total number of carbon atoms in the fatty acid. The number *after* the colon indicates the number of double bonds. According to the delta terminology, the superscripts indicate the carbon numbers at which a double bond occurs (counting from the —COOH end). According to the omega terminology, the num-

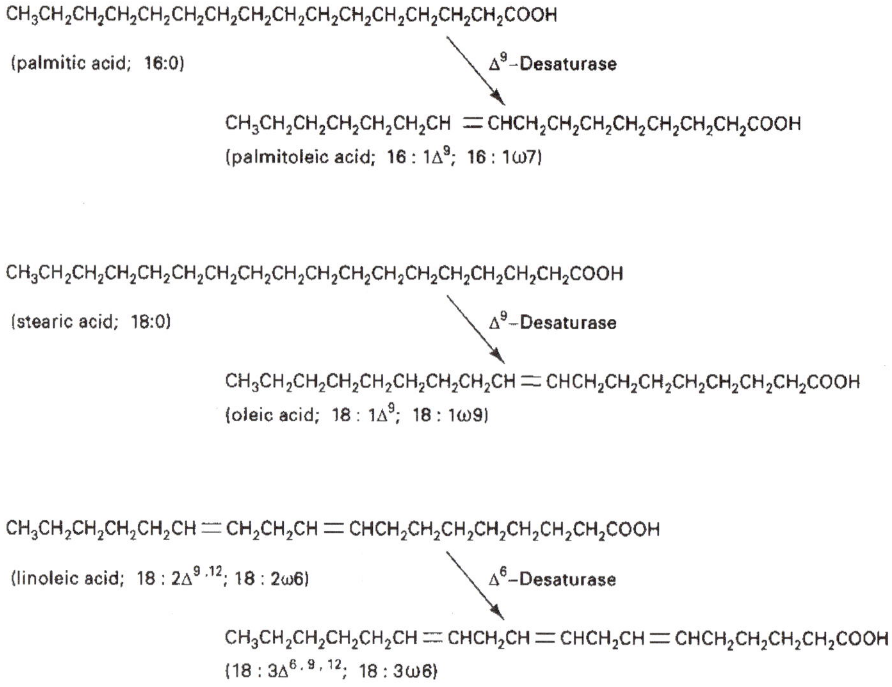

$CH_3CH_2CH_2CH_2CH_2CH_2CH_2CH_2CH_2CH_2CH_2CH_2CH_2CH_2CH_2COOH$

(palmitic acid; 16:0)                                                                $\Delta^9$-Desaturase

$CH_3CH_2CH_2CH_2CH_2CH_2CH = CHCH_2CH_2CH_2CH_2CH_2CH_2CH_2COOH$

(palmitoleic acid; 16 : 1$\Delta^9$; 16 : 1$\omega$7)

$CH_3CH_2CH_2CH_2CH_2CH_2CH_2CH_2CH_2CH_2CH_2CH_2CH_2CH_2CH_2CH_2CH_2COOH$

(stearic acid; 18:0)                                                                $\Delta^9$-Desaturase

$CH_3CH_2CH_2CH_2CH_2CH_2CH_2CH_2CH = CHCH_2CH_2CH_2CH_2CH_2CH_2CH_2COOH$

(oleic acid; 18 : 1$\Delta^9$; 18 : 1$\omega$9)

$CH_3CH_2CH_2CH_2CH_2CH = CH_2CH_2CH = CHCH_2CH_2CH_2CH_2CH_2CH_2CH_2COOH$

(linoleic acid; 18 : 2$\Delta^{9,12}$; 18 : 2$\omega$6)                                $\Delta^6$-Desaturase

$CH_3CH_2CH_2CH_2CH_2CH = CHCH_2CH = CHCH_2CH = CHCH_2CH_2CH_2CH_2COOH$

(18 : 3$\Delta^{6,9,12}$; 18 : 3$\omega$6)

FIGURE 9.95  Activities of desaturases. The desaturases are enzymes of the endoplasmic reticulum that recognize the coenzyme A derivatives of the fatty acids. The enzymes do not act on the free fatty acids. The desaturases require oxygen and NADH for activity. Two other proteins are required as well, and these are cytochrome $b_5$ and cytochrome $b_5$ reductase (Sprecher *et al.*, 1995). Once a fatty acid has received a desaturation, the fatty acid may (or may not) be lengthened by a 2-carbon unit. This step is catalyzed in the endoplasmic reticulum by chain elongation enzymes. These enzymes utilize malonyl-CoA as the source of the 2-carbon unit, as does fatty acid synthase, a cytosolic enzyme. Once the 2-carbon elongation has occurred, the fatty acid may be the substrate for another desaturase enzyme.

ber at the right refers to the carbon at which the first double bond occurs (counting from the methyl end of the fatty acid).

The "n minus" nomenclature is sometimes used in place of the omega nomenclature. Instead of writing 18:3$\omega$3, for example, one can use the n minus nomenclature to form the abbreviation, 18:3n-9,n-12,n-15.

### Nonessential Fatty Acids Can Be Desaturated and Elongated

Oleic acid (18:1$\omega$9), a monounsaturated fatty acid, is the major fatty acid in human milk. Oleic acid, as well as the products resulting from its further desaturation and elongation, are called the *omega-9 fatty acids*. The omega-9 fatty acid 24:1$\omega$9 is a major fatty acid in the membranes of nerves. The omega-9 fatty acid 20:3$\omega$9 is called the **Mead acid**. Mead acid accumulates in the body in response to a deficiency in the essential fatty acids, and there is some thought that Mead acid may compensate, with the decline in linoleic acid, in serving structural purposes in membranes. Mead acid is probably not used for the synthesis of any hormone.

## Essential Fatty Acids Can Be Desaturated and Elongated

The pathway of conversion of **linoleic acid**, in mammals, to arachidonic acid and to longer-chain-length fatty acids, is shown in Figure 9.96. The conversion of **linolenic acid**, in mammals, to longer-chain length fatty acids is shown in Figure 9.97. The structures of the fatty acids have been simplified by omitting the hydrogen atoms. The asterisks indicate the two carbons added during the elongation step. The fatty acids occur as the coenzyme A derivative while undergoing desaturation and elongation. For simplicity, coenzyme A was omitted from the diagrams in Figure 9.96 and 9.97.

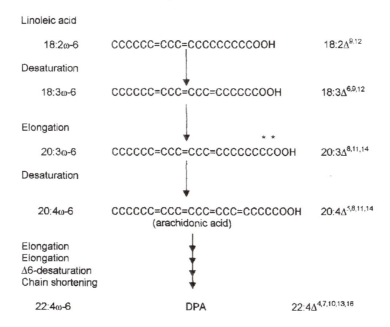

**FIGURE 9.96** Omega-6 fatty acid biosynthetic pathway. Linoleic acid (18:2ω6) is the essential fatty acid that is the starting material for the biosynthesis of arachidonic acid. Arachidonic acid, in turn, can be converted to DPA. The reactions occur with the fatty acids esterified with coenzyme A but, for simplicity, this cofactor, as well as the hydrogen atoms, have been left out of the diagram. The enzymes $\Delta^6$-desaturase and $\Delta^5$-desaturase are used, respectively, in the two desaturation steps used for arachidonic acid biosynthesis. The steps leading from arachidonic acid (20:4ω6) to DPA(22:5ω6) consists of a four-step sequence, involving: (1) elongation; (2) elongation; (3) 6-desaturation; and (4) chain-shortening. This round-about method of achieving an elongation by 2-carbons may seem surprising. Even more surprising is that all steps occur on the surface of the endoplasmic reticulum, except for the chain-shortening step, which occurs in an entirely different organelle, the peroxisome (Sprecher *et al.*, 1995).

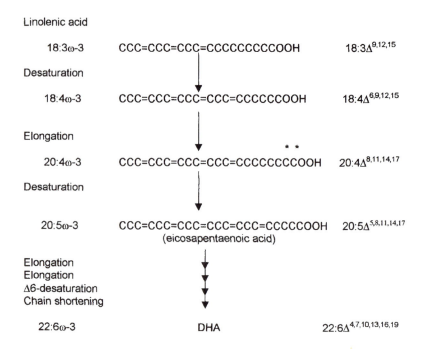

Linolenic acid

| | | |
|---|---|---|
| 18:3ω-3 | CCC=CCC=CCC=CCCCCCCCCCOOH | 18:3Δ$^{9,12,15}$ |

Desaturation

| | | |
|---|---|---|
| 18:4ω-3 | CCC=CCC=CCC=CCC=CCCCCCOOH | 18:4Δ$^{6,9,12,15}$ |

Elongation

| | | |
|---|---|---|
| 20:4ω-3 | CCC=CCC=CCC=CCC=CCCCCCCCOOH | 20:4Δ$^{8,11,14,17}$ |

Desaturation

| | | |
|---|---|---|
| 20:5ω-3 | CCC=CCC=CCC=CCC=CCC=CCCCCOOH | 20:5Δ$^{5,8,11,14,17}$ |
| | (eicosapentaenoic acid) | |

Elongation
Elongation
Δ6-desaturation
Chain shortening

| | | |
|---|---|---|
| 22:6ω-3 | DHA | 22:6Δ$^{4,7,10,13,16,19}$ |

**FIGURE 9.97** Omega-3 fatty acid biosynthetic pathway. Linolenic acid (18:3ω3) is the essential fatty acid that is the starting material for the biosynthesis of EPA and DHA. The reactions occur with the fatty acids esterified with coenzyme A, but, for simplicity, this cofactor, as well as the hydrogen atoms, have been left out of the diagram. The enzymes Δ$^6$-desaturase and Δ$^5$-desaturase are used, respectively, in the two desaturation steps. The steps leading from EPA (20:5ω3) to DHA (22:6ω3) consist of a four-step sequence, involving: (1) elongation; (2) elongation; 3) 6-desaturation; and (4) chain-shortening. This roundabout method of achieving an elongation by 2-carbons may seem surprising. All steps occur on the surface of the endoplasmic reticulum, except for the chain-shortening step, which occurs in an entirely different organelle, the peroxisome (Sprecher *et al.*, 1995). DHA occurs at especially high concentrations in the retina and in nerve endings.

### Desaturase Enzymes

Δ$^9$-Desaturase, Δ$^6$-desaturase, and Δ$^5$-desaturase catalyze the fatty acid desaturation reactions in mammals. When Δ$^6$-desaturase acts on an omega-6 fatty acid, it remains an omega-6 fatty acid. When Δ$^6$-desaturase acts on an omega-3 fatty acid, it remains an omega-3 fatty acid. The desaturases received their names according to where their site of action is relative to the COOH-carbon.

### Linoleic Acid and Other Omega-6 Fatty Acids

Linoleic acid and the family of fatty acids derived from it are called the omega-6 fatty acids (Figure 9.96). This is because the number of carbons from the methyl end to the first double bond is six. Linoleic acid is used to make arachidonic acid (20:4ω6), a fatty acid essential for the synthesis of various hormones. These hormones are the prostaglandins, thromboxanes, and leukotrienes. These three classes of hormones are used for the regulation of many physiological processes. DPA ($22:5\omega6$) is a 22-carbon fatty acid synthesized from arachidonic acid. DPA is found in high concentrations in the central nervous system and other tissues.

## Linolenic Acid and Other Omega-3 Fatty Acids

Linolenic acid tends to occur at much lower levels in the diet and in the tissues of the body than does linoleic acid. Linolenic acid can undergo successive desaturations and elongations to yield EPA and DHA (Figure 9.97). The relationship between linolenic acid ($18:3\omega3$), EPA ($20:5\omega3$), and DHA ($22:6\omega3$) is revealed by their nomenclature, which shows they are all omega-3 fatty acids. EPA is a precursor for the synthesis of prostaglandin $E_3$, thromboxane $A_3$, and leukotriene $A_5$ (Dajani, 1993). DHA is a precursor to a series of prostaglandins that includes $PGI_3$, $PGD_3$, and $PGE_3$. There is some evidence that DHA can alter the excitability and activity of certain nerves (Hamano *et al.*, 1996).

## Fish Oils (Marine Oils)

Fish oils are especially rich in the omega-3 fatty acids EPA and DHA (Soyland *et al.*, 1993). Dietary omega-3 fatty acids can reduce plasma triglyceride levels in humans, and may reduce any tendency for spontaneous blood clot formation. This combination of effects might be expected to reduce the risk for cardiovascular disease. (Note that omega-3 fatty acids do not reduce LDL-cholesterol). A recent study with pigs fed milk containing vegetable oil or fish oil revealed that the plasma triglycerides were higher (1.2 m$M$) with the vegetable oil, but lower (0.8 m$M$) with fish oil (Arbuckle and Innis, 1993).

The long-term influence of omega-3 fatty acid supplements is not known, and, for this reason, the American Heart Association has not recommended that supplements be taken (Krauss *et al.*, 1996). There is a very real concern that dietary PUFAs of all types may enhance the accumulation of lipid oxidation products in lipoprotein particles, thus allowing the attack of lipid oxidation products on the protein component of the LDL, the enhanced uptake of the damaged LDL by macrophages in the artery wall, and increased atherosclerosis (Allard *et al.*, 1997; Felton *et al.*, 1994). (Monounsaturated fatty acids are thought not to be involved in the oxidation scenario leading to atherosclerosis.)

Psoriasis is a disease of the skin that involves inflammation. There has been some interest in using dietary fish oils to treat psoriasis, especially since Eskimos (fish-eaters) tend not to acquire psoriasis. Both the omega-6 and omega-3 fatty acids are converted to hormones that activate white blood cells and provoke inflammatory reactions. However, the omega-3 fatty acids are generally less potent, and the associated inflammatory reaction is smaller. Hence, there is some thought that an increase in dietary omega-3 fatty acids may inhibit the processing of omega-6 fatty acids to the inflammatory hormones.

Unfortunately, careful studies have shown that fish oils do not have this effect (Soyland *et al.*, 1993). Psoriasis patients might find comfort with the recent demonstration that vitamin D-based compounds, when applied to the skin, can result in a remarkable decline in skin lesions.

EXERCISE

Name four things that are wrong with the notation $16:1\Delta^{1,5,6,17}$.

## Prostaglandins and Thromboxanes

*Conversion of Arachidonic Acid to Prostaglandins and Thromboxanes*

The arachidonic acid present in cell membranes is esterified to the 2-position of such phospholipids as phosphatidylcholine, phosphatidylethanolamine, and phosphatidylinositol. It is thought that arachidonic acid present only in phosphatidylinositol (PI), but not in the other phospholipids, is involved in the synthesis of prostaglandins and related hormones. The inositol group may occur in PI as unmodified inositol, inositol-4-phosphate, and inositol-4,5-bisphosphate.

Hydrolysis of IP3 from membrane lipids yields 1-acyl-2-arachidonyl-glycerol, which remains in the plasma membrane. The arachidonate esterified at the 2-position may be hydrolyzed by phospholipase $A_2$, to yield free arachidonate. The pathway leading to the release of IP3 and arachidonate is shown in Figure 9.98. The events depicted occur in a burst. They occur mainly within a time frame of a minute or so of stimulation of the cell. Stimulation of the cell can induce an increase in the concentration of free arachidonic acid in the cell. This arachidonate can be used by cyclooxygenase. **Cyclooxygenase** is a membrane-bound enzyme of the endoplasmic reticulum. The active site faces the cytoplasm. The enzyme is bifunctional. It catalyzes the attachment of oxygen molecules to arachidonic acid,

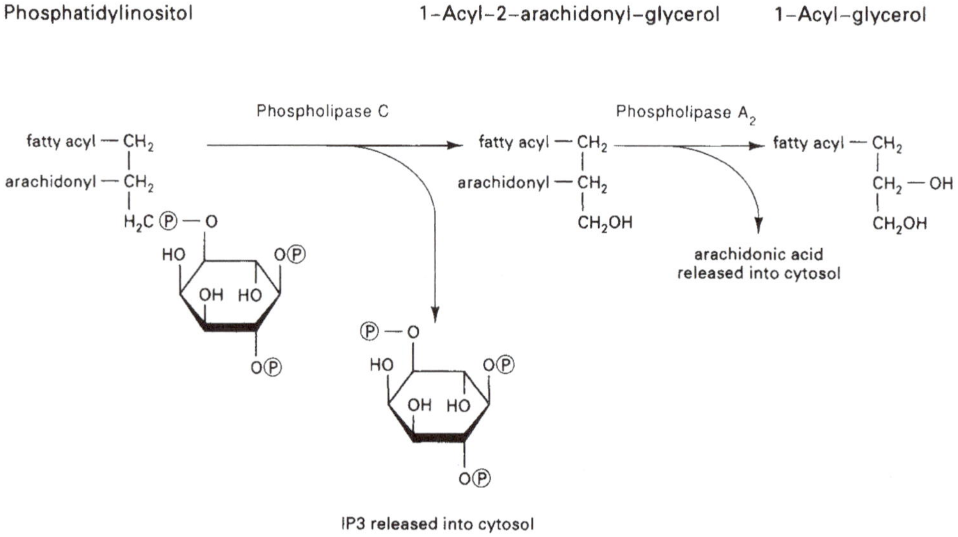

FIGURE 9.98 Events taking place in the plasma membrane on stimulation of a cell. Phosphatidylinositol (PI) and more highly phosphorylated versions of this lipid account for 2 to 8% of the lipids of the plasma membrane of eukaryotic cells. The inositol 1,4,5-triphosphate (IP3) moiety of phosphatidylinositol- 4,5-diphosphate may be hydrolyzed from this lipid immediately after the cell is stimulated. For example, the stimulation of platelets by thrombin or the islets of the pancreas by glucose is followed by the release of IP3 into the cytoplasm. In some cells, arachidonic acid is hydrolyzed from 1-acyl-2-arachidonyl-glycerol, which can support a burst of prostaglandin synthesis.

resulting in its modification by a hydroperoxide group (R—OOH) and an endoperoxide group (R—OO—R). The metabolite containing both groups is PGG$_2$ (Figure 9.99). The enzyme also catalyzes the further conversion of PGG$_2$ to PGH$_2$. Other enzymes can convert prostaglandins to thromboxanes. The conversion of PGH$_2$ to thromboxane A$_2$, for example, is shown in Figure 9.99.

## Classification of Prostaglandins

Any event that provokes the activation of phospholipase A$_2$ may result in liberation of arachidonic acid from diglycerides in the plasma membrane. Cyclooxygenase catalyzes the conversion of arachidonic acid to prostaglandin H$_2$ (PGH$_2$). PGH$_2$ is the parent compound of other prostaglandins and thromboxanes. Four groups of prostaglandins are formed from the PGH parent. These are PGD, PGE, PGF, and PGI. The subscript (as in PGE$_2$) represents the number of double bonds in the side chain, and this subscript can be 1, 2, or 3 (Kobayashi *et al.*, 1997). The physiological activity of any prostaglandin depends more on the structure of the head group, and not much on the number of double bonds in the side chain.

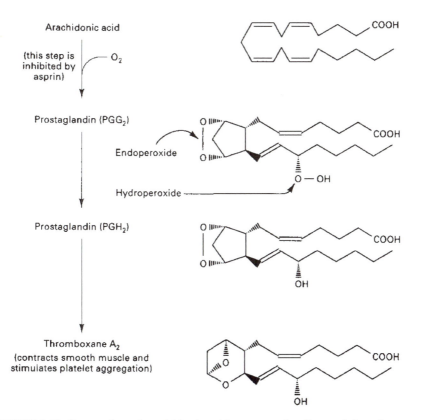

**FIGURE 9.99**  Conversion of arachidonic acid to prostaglandins and thromboxanes.

*Prostaglandin Receptors*

The various prostaglandins and their receptors represent a bewildering array of functions. However, these have been conveniently divided into three functions: (1) relaxation; (2) contraction; and (3) inhibition. The **relaxation receptors** are those that bind PGI, PGD, $PGE_2$, and $PGE_4$. The stimulation of these receptors results in an increase in cytosolic cAMP, and muscle relaxation.

The **contractile receptors** are those that bind PGF, $PGE_1$, and thromboxane. Stimulation of these receptors provokes a burst of calcium ions, and the contraction of smooth muscle. The **inhibitory receptor** is the $PGE_3$ receptor. Its stimulation results in a decline in cAMP levels, and the consequent inhibition of nerve activity, inhibition of gastric acid secretion, and decline in water resorption (Kobayashi *et al.*, 1997). These three functions provide a working guideline, but are not strict rules.

The following effects take place because the receptors are coupled to adenylyl cyclase or to phospholipase C. Adenylyl cyclase catalyzes the synthesis of cAMP, while phospholipase C catalyzes the hydrolysis of the phosphatidyl-4,5-bisphosphate, releasing inositol-1,4,5-trisphosphate (IP3). This IP3, in turn, travels to the endoplasmic reticulum, where it provokes the momentary release of calcium ions. The increase in $Ca^{2+}$ ion levels, in turn, provokes activation of a number of protein kinases, as discussed in the Calcium and Phosphate section. All of these receptors are polypeptides that weave seven times in or out of the plasma membrane. All of these receptors are directly linked to G protein, and require G protein for transmitting their message within the membrane to various enzymes.

*Details of Events Provoked by the Stimulation of Prostaglandin Receptors, as Revealed by the Examples of Prostaglandin F and Thromboxane A*

PGF appears not to be necessary for life or for embryological development. However, this prostaglandin is required for the delivery of a fully developed baby. Knock-out mice that lack the PGF receptor were prepared by standard genetic techniques. These mice appeared normal, but when pregnancy came to the end of its term the baby mice were not ejected from the uterus; the uterus did not contract. If a cesarean birth was performed, the babies grew and developed normally; otherwise, they died inside their mothers (Sugimoto *et al.*, 1997). PGF has also been found to regulate the pressure of the eye (Abramovitz *et al.*, 1994).

The stimulation of platelets to form a blood clot involves thromboxane production. The sequence of events resulting in the aggregation of platelets is summarized in Figure 9.100. The initial event of cell stimulation is the reaction of thrombin with the platelet surface, which provokes the release of arachidonic acid, as mentioned in the Vitamin K section. Arachidonic acid is converted to thromboxane $A_2$. The $TXA_2$, in turn, is released from the platelet. It travels a short distance, within the bloodstream, and binds to its receptor on the surface of nearby platelets.

Thromboxane $A_2$ also causes contraction of the smooth muscles in the artery, thus minimizing blood loss in the artery in the vicinity of the injury. As mentioned earlier, $TXA_2$ binds to a "contractile receptor."

The prostaglandins mediate inflammation and pain. Aspirin is a universally used drug. The pain-relieving properties of the drug are a consequence of its inhibition of cyclooxygenase. Aspirin covalently reacts with the enzyme, inhibiting the first reaction catalyzed by the enzyme, oxygenation. Daily doses of aspirin

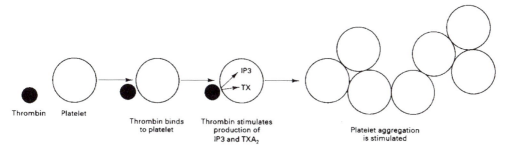

**FIGURE 9.100** Thrombin activation of platelets. Thrombin binds to the thrombin receptor on the platelet membrane, resulting in the activation of pathways that provoke platelet aggregation. The occupancy of the receptors by these agents relays a signal, via G protein, to activate phospholipase C. Phospholipase C catalyzes a cleavage reaction on phosphatidylinositol-4,5-bisphosphate, releasing IP3, which in turn travels to the endoplasmic reticulum and provokes the release of a burst of intracellular calcium ions. The $Ca^{2+}$ activates protein kinase C, which catalyzes the phosphorylation of a number of target proteins. One result of the above events is the activation of the $\alpha_{IIb}\beta_3$ integrin. In its activated form, this integrin takes a new conformational change and is able to bind extracellular matrix proteins, such as fibrin and von Willebrand factor. The binding of thrombin to its receptor also provokes the activation of phospholipase $A_2$, which catalyzes the cleavage of arachidonic acid from the plasma membrane. Arachidonic acid, in turn, is then converted to thromboxane $A_2$ ($TXA_2$), which diffuses out of the platelet and provokes the activation of nearby platelets. $TXA_2$ binds to the $TXA_2$-receptor on its target platelets. This receptor weaves seven times in or out of the plasma membrane, and is linked via G protein, to phospholipase C. Stimulation of target platelets by $TXA_2$ thus provokes the release of IP3 and a burst of intracellular calcium ions (D'Angelo *et al.*, 1996). The activation of a few platelets by thrombin results in a feedforward activation pathway that activates many platelets by thromboxane. The amplified signal that results provokes the activation of a great multitude of platelets, and the formation of a visible blood clot.

are used to lower the tendency of blood clotting, especially in persons with cardiovascular disease who have a tendency to develop spontaneous blood clots. Aspirin lessens the tendency toward platelet aggregation and, hence, can reduce the risk of embolisms.

*Leukotrienes*

Two pathways of metabolism are available for arachidonic acid. The first is the **cyclooxygenase pathway**, which leads to the formation of prostaglandins, thromboxanes, and prostacyclin. The second is the **lipoxygenase pathway**. Lipoxygenase catalyzes the first step in the conversion of arachidonic acid to a number of noncyclized metabolites. These metabolites include 5-hydroperoxyeicosatetraenoic acid (HPETE) and the leukotrienes (Figure 9.101).

The first step in the formation of the leukotrienes is cleavage of arachidonic acid from membrane phospholipids. 5-Lipoxygenase, which is a calcium-dependent enzyme, catalyzes the introduction of $O_2$ into the arachidonic acid and the conversion of the resulting hydroperoxide to an epoxide. Leukotriene $A_4$ contains an epoxide group. Leukotriene $A_4$ ($LTA_4$) can be converted to leukotriene $B_4$ ($LTB_4$) by enzymatic hydrolysis. $LTB_4$ is a dihydroxy fatty acid; it regulates the functions

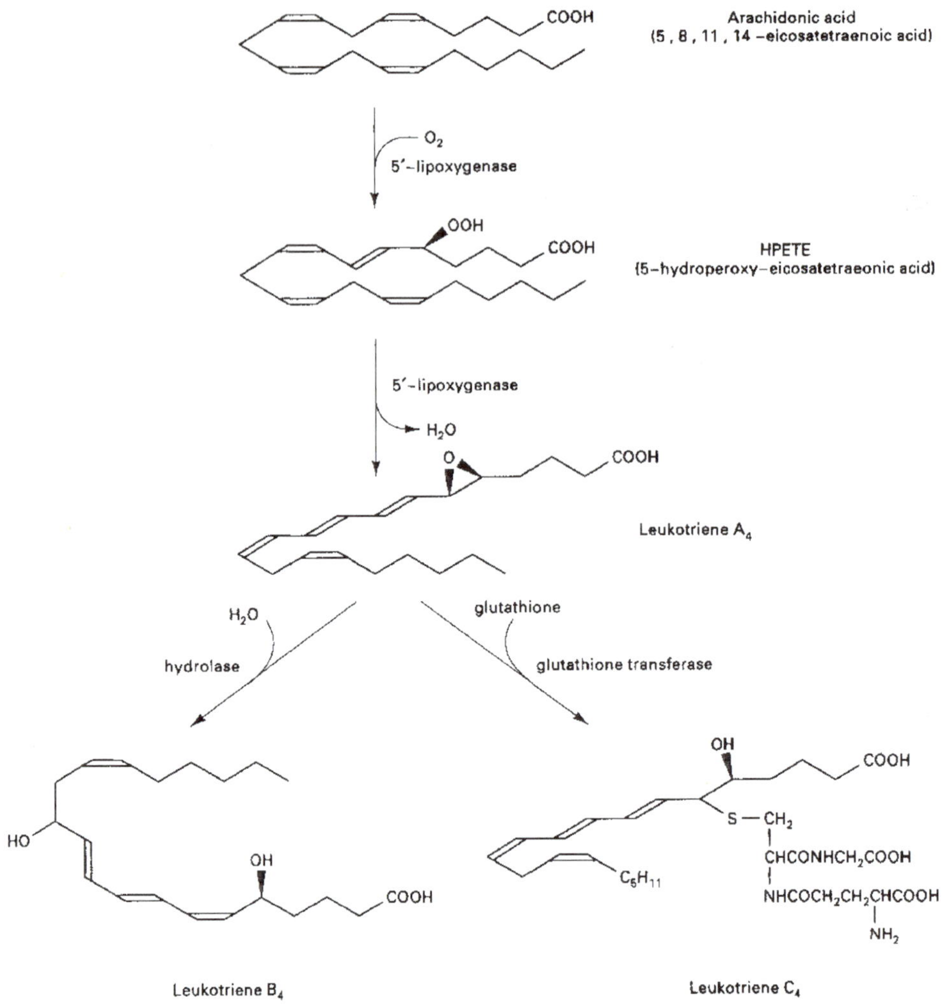

**FIGURE 9.101** Pathways of leukotriene metabolism. The arachidonic acid present in the plasma membrane is esterified to the 2-position of various fatty acids. A burst in intracellular calcium ions, as provoked by various hormones or immunological stimuli, causes phospholipase $A_2$ to bind to the inside of the plasma membrane. This enzyme is a cytosolic protein. Phospholipase $A_2$ then catalyzes the hydrolysis of arachidonate. 5'-Lipoxygenase catalyzes the conversion of the free arachidonate to form $LTA_4$, in a two-step reaction. 5'-Lipoxygenase occurs only in white blood cells, i.e., neutrophils, eosinophils, monocytes, macrophages, mast cells, basophils, and B lymphocytes (Henderson, 1994). Further metabolism of $LTA_4$ may occur within the white blood cell, or it may be released and then undergo secondary metabolism by an unrelated tissue. Enzymes used for the conversion of $LTA_4$ to leukotriene $B_4$ and leukotriene $C_4$ (as well as $LTD_4$ and $LTE_4$) are widely distributed in the body. $LTC_4$ contains a molecule of glutathione. Hydrolysis of the glutamate residue from this glutathione generates $LTD_4$. Further trimming of the $LTD_4$ results in $LTE_4$, which contains only the cysteine residue of glutathione. To give a specific example of this trimming pathway, it has been found that after release of $LTC_4$ by white blood cells the $LTC_4$ travels to its target tissue, where the glutamate is removed to give $LTD_4$ (Carter et al., 1997). The question of whether the $LTC_4$ family ($LTC_4$, $LTD_4$, and $LTE_4$) of leukotrienes all bind to different receptors and have different biological activities is currently under study.

of certain white blood cells (neutrophils, eosinophils, macrophages, lymphocytes). Neutrophils have receptor sites for $LTB_4$. For example, $LTB_4$ can provoke the movement and migration of the neutrophils.

$LTA_4$ has a second possible fate, namely conversion to leukotriene $C_4$. Glutathionine transferase catalyzes this conversion, which involves the covalent attachment of glutathione to the epoxide oxygen of $LTA_4$. Hydrolysis of the glutamate residue results in the conversion of $LTC_4$ to $LTD_4$. The subsequent hydrolysis of the glycine residue results in conversion of $LTD_4$ to $LTE_4$.

## Biological Properties of LTB$_4$ and LTC$_4$

$LTB_4$ acts to attract the migration of white blood cells into various tissues. Hence, where a tissue contains white blood cells that synthesize and release this leukotriene, more white blood cells will migrate into the tissue. $LTC_4$, and the related $LTD_4$ and $LTE_4$, act to increase the permeability of blood vessels and to provoke the contraction of smooth muscle (Henderson, 1994). The precursor to these leukotrienes, which is $LTA_4$, is synthesized only in white blood cells, i.e., the 5'-lipoxygenase occurs only in white blood cells. However, the downstream enzymes used to convert $LTA_4$ to $LTB_4$ or $LTC_4$ occurs in a wide variety of tissues. In addition, one might keep in mind that a complete understanding of the actions of the leukotrienes, as well as of all other hormones, also requires a knowledge of the target cells that contain the corresponding hormone receptors.

The physiological effects of the leukotrienes can be illustrated by conditions where the leukotrienes are involved in **inflammatory diseases**. These inflammatory diseases include asthma, arthritis, glomerulonephritis, psoriasis, inflammatory bowel syndrome (Crohn's disease), adult respiratory distress syndrome, and ulcerative colitis. In *rheumatoid arthritis*, for example, neutrophils are attracted into the joints between bones, resulting in pain and inflammation. The accumulated neutrophils also produce leukotrienes, which attract still more neutrophils into the joint (Griffiths *et al.*, 1995). To use the example of *asthma*, the leukotrienes produced by white blood cells in the lung provoke spasms of the smooth muscles in the airways and cause an inability to breath. **Edema** is a problem in a variety of diseases. Edema, which involves the net leakage of fluids from the bloodstream into extracellular spaces, is easily diagnosed because it leads to puffy skin. It is thought that the influence of leukotriene $C_4$ in increasing permeability of veins serves the useful purpose of allowing more white blood cells to migrate from the bloodstream, across the veins, and into tissues that contain infections. Edema can be dangerous when it involves the lungs, as in *adult respiratory distress syndrome*, because it results in leakage of extracellular fluids into the lungs, resulting in an inability to breath (Henderson, 1994).

A variety of drugs have been devised to treat these diseases. Some of these drugs bind to 5'-lipoxygenase and inhibit its activity, while other drugs prevent the binding of the leukotriene to its receptor. Drugs have been devised that prevent $LTB_4$ from binding to the $LTB_4$ receptor in the endothelial cells of veins, and to prevent $LTC_4$, $LTD_4$, or $LTE_4$ from binding to their receptors in the membranes of smooth muscles (Griffiths *et al.*, 1995).

Many of the effects of leukotrienes are reduced in animals raised on diets deficient in essential fatty acids. Such diets lead to a reduction in all the functions of the essential fatty acids. The advantage of drugs is that they can be designed to affect specific functions of the EFAs rather than all functions.

## Essential Fatty Acid Deficiency

Persons at risk for EFA deficiency tend to be the same as those at risk for vitamin E deficiency. Some signs are shared by both deficiencies. Premature infants may be at risk for EFA deficiency because of their low stores of lipids and their rapid growth, especially when they are fed diets that do not contain EFAs. For example, fats have been omitted from diets used to feed preterm infants (to avoid a variety of complications). EFA deficiency may develop later in life with fat malabsorption syndromes. EFA deficiency has presented in adults fed by total parenteral nutrition for longer periods, where EFAs had not been included in the liquid diet.

Signs of an EFA deficiency can appear in humans after a few weeks on an EFA-deficient diet. The most obvious sign of deficiency is eczematous dermatitis, or scaly skin. Hair loss and impaired wound healing may also result. EFA deficiency in the rat results in dermatitis, a rough coat of fur, scaly paws and tail, fragility of the capillaries, impaired growth, and impaired reproductive ability. These signs have been identified specifically with a *deficiency in linoleic acid, not linolenic acid*.

EFAs are plentiful in the diet. Wide variations in the intake of EFAs may have little or no effect on the physiological functions of the body because of the stores of linoleic acid in adipose tissue and because of possible metabolic adjustments by the body. A functional test for assessing EFA status has not been adopted for routine use. One functional test, for example, might be the measurement of platelet aggregating ability; a complicating factor is that some prostaglandins inhibit platelet aggregation whereas others stimulate aggregation.

### Accumulation of Mead Acid Indicates EFA Deficiency

The following material concerns a test for assessing EFA status. The test is a direct test, not a functional one. Under normal conditions, a proportion of the linoleic acid in the diet is converted to form arachidonic acid and other long-chain fatty acids. During EFA deficiency, oleic acid (not an EFA) is increasingly used for desaturation and elongation. Oleic acid is an omega-9 fatty acid. Increased desaturation and elongation of this fatty acid occur in EFA deficiency and result in the production of another omega-9 fatty acid, 20:3ω9, also called Mead acid. Mead acid can also be abbreviated $20:3\Delta^{5,8,11}$. The ratio of the levels of Mead acid/linoleic acid in plasma membrane lipids is used to assess essential fatty acid status. The ratio increases with an EFA deficiency. Generally, a ratio of 0.4 or greater is considered to indicate deficiency. The ratio may increase to 4.0 or more during severe deficiency in humans or experimental animals.

EXERCISE

How many separate steps of desaturation and elongation are required to convert oleic acid to the Mead acid?

EXERCISE

Feeding infants a liquid formulation based on safflower oil can result in linolenic acid deficiency. Explain this effect. (See Heird and Gomez, 1996.)

The accumulation of Mead acid in humans is shown in Figure 9.102. The study involved human subjects recovering from surgery of the gastrointestinal tract. They were fed by parenteral nutrition. The diet infused contained no lipids and induced the development of essential fatty acid deficiency. Plasma lipids were analyzed over the course of several weeks, as indicated in the figure. The results demonstrate that the levels of linoleic acid (●) and arachidonic acid (Δ) in plasma lipids decreased over time. Concomitantly, the concentration of Mead acid increased (○). The changes recorded demonstrate a biochemical abnormality, but not necessarily a change that is physiological harmful.

With a deficiency in linolenic acid (18:3ω3) (but not in linoleic acid), there is a decline in DHA (22:6ω3) in the retina and brain, as revealed in studies with rodents and monkeys. With deficiency in only linolenic acid (omega-3), there is also an increase in conversion of the available linoleic acid (omega-6) to produce elevated levels of 22-carbon omega-6 fatty acids (Craig-Schmidt et al., 1996; Innis et al., 1994). Deficiency in only linolenic acid results in altered learning behaviors and a decreased visual sense.

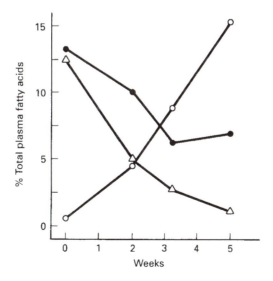

**FIGURE 9.102** Accumulation of Mead acid (20:3ω9) in humans (○) during the development of a deficiency in essential fatty acids. (Redrawn with permission from Richardson and Sgoutas, 1975.)

The question of which dietary practices lead to EFA deficiency is difficult to predict. A study of women suffering from anorexia nervosa, a disease that provokes pathological fasting behavior and marked weight loss, indicated that the women were not deficient in EFAs. Surprisingly, their plasma levels of Mead acid were lower, rather than higher, than those of control human subjects (Holman *et al.*, 1995).

## Does Dietary DPA or DHA Influence Brain Development?

DHA normally accounts for 0.2–0.6% of the fatty acids in human milk lipids. Milk DHA levels can range from 0.1 to 1.9%, depending on the mother's dietary intake of preformed DHA. DPA may be undetectable in human milk. This information, in itself, does not suggest that infants can benefit materially from the DHA in milk (Craig-Schmidt *et al.*, 1996; Arbuckle and Innis, 1993; Innis *et al.*, 1994).

A study with newborn pigs asked, "Can dietary manipulation provoke changes in the DHA or DPA content of the brain of pig?" Baby pigs were fed diets containing linoleic acid plus 0.7% linolenic acid or plus 3.9% linolenic acid. The percentage values indicate grams of the fatty acid/100 grams of total fatty acid. The diets contained no DPA or DHA. At the age of 15 days, the baby pigs were sacrificed, their brains removed and the lipids extracted and analyzed. Let us compare the results acquired with the two formulas. The high-linolenic-acid formula resulted in higher levels of omega-3 fatty acids ($22:5\omega3$ (EPA); $22:6\omega3$ (DHA)) in the brain, and lower levels of an omega-6 fatty acid ($22:5\omega6$ (DPA)) (Arbuckle *et al.*, 1994). Similar results were found with examination of the retinas of the baby pigs. A similar study involving baby pigs revealed that feeding dietary supplements of DHA (rather than its precursor) can provoke an increase in the DHA present in the phospholipids of the retina (Craig-Schmidt *et al.*, 1996).

This information in itself does not suggest that infants can benefit from the DHA in milk. The information does not suggest that DPA or DHA serves any function in the body. The data only reveal that certain fatty acids do occur in the body and that their levels can be manipulated by the diet.

There does exist evidence that human infants fed formula lacking in the longer polyunsaturated fatty acids have lower levels of these metabolites in their brains than infants fed human milk. There is some emerging concern that there may be some impairment in development of the human infant's nervous system with feeding of formula containing only vegetable oils, and hence lacking in the longer-chain polyunsaturated fatty acids (Heird and Gomez, 1996).

## Genetic Diseases that Disrupt Peroxisomal Steps of Fatty Acid Metabolism

Several genetic diseases involving the peroxisomes have been discovered and characterized. These diseases result in the accumulation of very-long-chain fatty acids (26:0) in certain tissues of the body (adrenal gland and brain). This fatty acid may increase by some 20-fold in the brain. The peroxisomal diseases can result in loss of the myelin layer of nerves (demyelination) and in seizures, spasticity, mental retardation, and death.

The most common genetic disease of the peroxisomes is **X-linked adrenoleuk-odystrophy**. This disease involves a defect in a membrane-bound protein of the peroxisome. Adrenoleukodystrophy is distinguished in that there is much interest in preventing its symptoms by low-fat diets that are supplemented with oil containing erucic acid. This oil is popularly known as Lorenzo's oil. Erucic acid is a 22-carbon fatty acid (22:1ω9). It is thought to exert its effect by reducing the synthesis of long-chain fatty acids in the body. Lorenzo's oil may be effective prior to the development of symptoms, but not if symptoms have already occurred (Moser and Moser, 1996; Rasmussen *et al.*, 1994).

The peroxisomes are organelles that, like the lysosomes, do not contain DNA. The contents and membrane are made by proteins that are coded for by the nuclear DNA. The peroxisome contains about 50 different enzymes, including those that are involved in the metabolism of fatty acids, phospholipids, prostaglandins, cholesterol, bile salts, and purines. The organelle received its name because many oxidase enzymes, which generate HOOH as a useless byproduct, reside within. The peroxisomes also contain catalase, the enzyme that converts HOOH to water and oxygen. How do the enzymes within this organelle acquire their location? All of these enzymes contain a special sequence of amino acids at their C terminus. This sequence, which is serine–lysine–leucine, is a targeting sequence, and contains enough information to make sure that the entire protein is placed inside the peroxisome shortly after synthesis on ribosomes. In diseases that result in defective peroxisomes, these enzymes may continue to be biosynthesized, but may also be degraded at an abnormally high rate within the cytosol. The transcription factor, **peroxisome proliferation activated receptor** (PPAR), is used to regulate the rate of transcription of peroxisomal enzymes, and other enzymes as well (Subramani, 1993).

## Relationship between Vitamin E and the PUFAs

The relationship between vitamin E and the polyunsaturated fatty acids, including the essential fatty acids, is quite close. It has been said that, if the body did not need or use PUFAs, there would also be no need for vitamin E. A dietary relationship exists between PUFAs and vitamin E. Greater intakes of PUFAs tend to increase the requirement for vitamin E. An increase in intake of PUFAs, however, does not necessarily mean that vitamin E supplements are needed, because the foods highest in PUFAs (vegetable oils) are also good sources of vitamin E.

The polyunsaturated fatty acids are somewhat unstable in the presence of oxygen. Although a variety of mechanisms have been proposed for the oxygen-induced damage to PUFAs, only one is described here. This mechanism involves the attack of the hydroxyl radical on a double bond of the PUFA. Damage to PUFAs, and the prevention of damage by vitamin E, may occur as shown in Figure 9.103. The first step (#1) may involve the transfer of a hydrogen radical from the PUFA to the hydroxyl radical (HO•). This transfer generates a PUFA radical and water. The PUFA radical may rearrange (Step 2). The PUFA radical may react with a molecule of oxygen, generating a peroxy radical (Step 3). The damage inflicted on the PUFA is quite clear at this point; however, the potential for *far greater damage* remains, as shown in Step 4, a propagation step. The peroxy radical may abstract

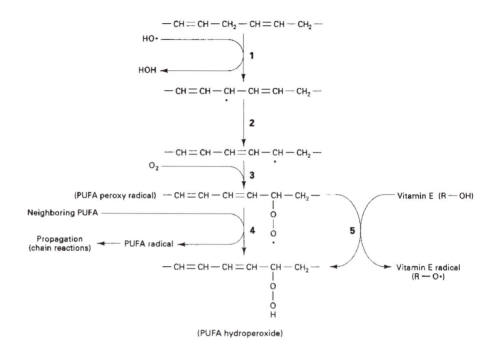

**FIGURE 9.103** Pathway leading to damage to polyunsaturated fatty acids and the protective effect of vitamin E.

a hydrogen atom from a neighboring PUFA, converting this PUFA to a PUFA radical. This new PUFA radical may, in turn, cause damage to a neighboring polyunsaturated fatty acid. This represents a chain reaction. The chain reaction, initiated by single hydroxyl radical, may result in damage to a large number of PUFAs in the cell membrane (Figure 9.103).

Vitamin E can interrupt or prevent the chain reaction. Vitamin E can act to convert the PUFA peroxy radical to a PUFA hydroperoxide (Step 5). This prevents the PUFA peroxy radical from inflicting damage on a neighboring PUFA. The vitamin, in turn, is converted to a vitamin E radical. The vitamin E radical is relatively stable and unreactive and is thought not to cause further damage to the cell membrane.

## Food Science and Polyunsaturated Fatty Acids

Damaged PUFAs may generate a number of breakdown products. These include a variety of short-chain aldehydes, such as malondialdehyde, and small alkanes, such as ethane and pentane (Figure 9.104). Concrete examples of oxygen-induced damage to PUFAs can be found in food science. The prevention of lipid peroxidation is a major concern in the food industry. Damage to lipids in poultry may occur during the mechanical deboning process, leading to off-flavors. Damage to the PUFAs in fish may occur during storage, leading to toughness. Damage to the lipids in meat may occur with prolonged heating, leading to "warmed-over"

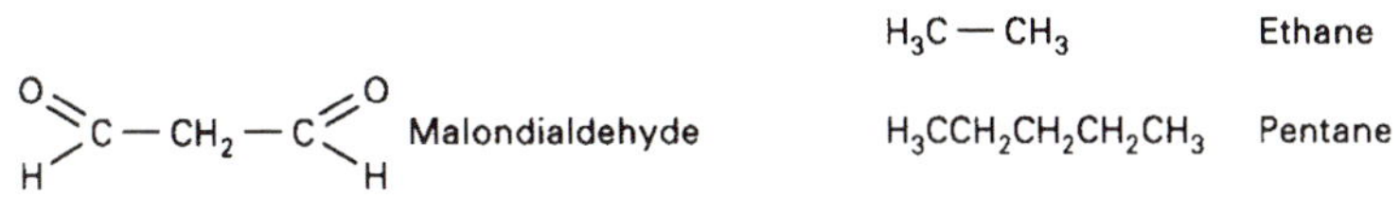

$$H_3C - CH_3 \qquad \text{Ethane}$$

$$\overset{O}{\underset{H}{\diagdown}}C - CH_2 - C\overset{O}{\underset{H}{\diagup}} \quad \text{Malondialdehyde} \qquad H_3CCH_2CH_2CH_2CH_3 \quad \text{Pentane}$$

**FIGURE 9.104** Small products of breakdown of polyunsaturated acids.

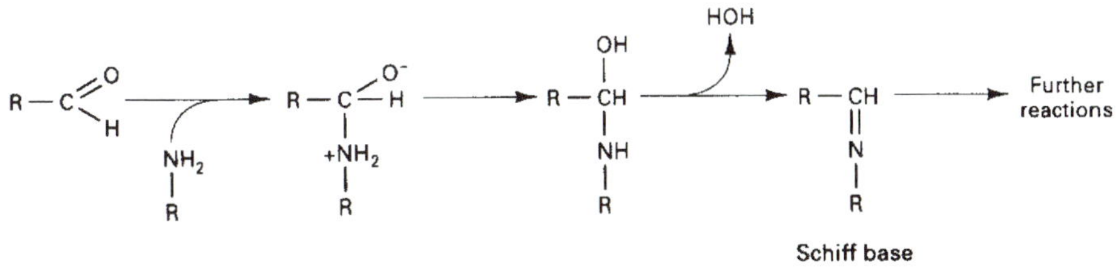

$$R - C\overset{O}{\underset{\underset{R}{\overset{|}{NH_2}}}{\diagup}}{\diagdown}_H \quad \longrightarrow \quad R - C\overset{O^-}{\underset{\underset{R}{\overset{|}{+NH_2}}}{\diagdown}}H \quad \longrightarrow \quad R - \overset{OH}{\underset{\underset{R}{\overset{|}{NH}}}{\overset{|}{C}H}} \quad \overset{HOH}{\longrightarrow} \quad R - \overset{\phantom{O}}{\underset{\underset{R}{\overset{|}{N}}}{\overset{||}{C}H}} \quad \longrightarrow \quad \begin{array}{l}\text{Further} \\ \text{reactions}\end{array}$$

Schiff base

**FIGURE 9.105** Condensation of aldehydes with amino groups.

flavors. The off-flavors are due to the production of volatile aldehydes and acids. The toughness is due to the production of malondialdehyde, which can cross-link the proteins in the food. The aldehyde groups of malondialdehyde may react with the amino groups of different proteins, resulting in their cross-linkage (Figure 9.105).

It is probable that some of the reactions that lead to the off-flavors and tough-ness in foods described occur in tissues of the vitamin E-deficient animal. The study shown in Figure 9.106 concerns the production of ethane by vitamin E-de-ficient rats. The animals were raised on complete (●) or vitamin E-deficient (○) diets. They were placed in a metabolic chamber to collect ethane released in the

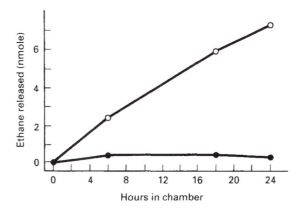

**FIGURE 9.106** Production of ethane by vitamin E-deficient rats. (Redrawn with permission from Hafeman and Hoekstra, 1977.)

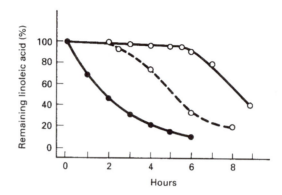

**FIGURE 9.107**  Relationship between vitamin E and polyunsaturated fatty acids (PUFAs). (Redrawn with permission from Mead and Wu, 1976.)

breath. The results demonstrate the steady release of this PUFA breakdown product by the deficient rats, but not by the control animals.

The next study is purely chemical in nature and concerns the relationship between vitamin E and PUFAs. The study traces the breakdown of PUFAs over time. Layers of linoleic acid were placed in dishes. One dish contained a layer consisting only of linoleic acid (●, Figure 9.107). Another dish contained a mixture of linoleic acid and vitamin E, with one molecule of vitamin E per 2000 molecules of the fatty acid (--O--). A third dish contained a greater proportion of vitamin E, one molecule of vitamin E per 400 molecules of linoleate (—O—). The dishes were stored in air for 9 hours and samples taken at the times indicated in Figure 9.107. The results show the protective effect of vitamin E. Damage to the PUFA commenced immediately where the layer did not contain the vitamin. The damage was delayed where increasing concentrations of the vitamin had been included.

The preceding material revealed the roles of superoxide and hydrogen peroxide in initiating damage to components of the cell. $O_2^-$ can arise from a number of sources. These include NADH dehydrogenase (an enzyme of the respiratory chain), as well as dihydroorotate dehydrogenase, xanthine oxidase, aldehyde oxidase, and the autooxidation of reduced flavins and reduced biopterin. HOOH can arise from uric oxidase, α-hydroxy acid oxidase, aldehyde oxidase, and xanthine oxidase. The major source of HOOH appears to be a number of enzymes in the peroxisomes. The peroxisome is an organelle in the cell. Xanthine oxidase represents a damaged or altered version of xanthine dehydrogenase. Normally, about 20% of the xanthine dehydrogenase in the cell occurs as xanthine oxidase. Xanthine dehydrogenase, in its unaltered form, does not produce toxic oxygen.

The food scientist might wonder how oxidative damage can occur in vegetable oil or in stored cooked meats. These foods do not contain active enzymes and thus would not be expected to produce superoxide or HOOH. What initiates oxidative damage to the PUFAs in foods during storage? The answer is that all foods, including fats and oils, contain trace amounts of iron and copper. These metals, in their oxidized forms ($Fe^{3+}$, $Cu^{2+}$), may react directly with PUFAs. The metal ions can remove a hydrogen radical from the PUFAs, resulting in formation of a PUFA radical and the chain reaction described earlier.

# REFERENCES

Abramovitz, M., Boie, Y., Nguyen, T., Rushmore, T. H., Bayne, M. A., Metters, K. M., Slipetz, D., and Grygorczyk, R. (1994). Cloning and expression of a cDNA for the human prostanoid FP receptor. *J. Biol. Chem.* **269**, 2632–2636.

Allard, J. P., Kurian, R., Aghdassi, E., Muggli, R., and Royall, D. (1997). Lipid peroxidation during n-3 fatty acid and vitamin E supplementation in humans. *Lipids* **32**, 535–541.

Arbuckle, L. D., and Innis, S. M. (1993). Docosahexaenoic acid is transferred through maternal diet to milk and to tissues of natural milk-fed piglets. *J. Nutr.* **123**, 1668–1675.

Arbuckle, L. D., MacKinnon, M. J., and Innis, S. M. (1994). Formula 18:2(n-6) and 18:3(n-3) content and ratio influence long-chain polyunsaturated fatty acids in the developing piglet liver and central nervous system. *J. Nutr.* **124**, 289–298.

Aarhus, R., Dickey, D., Graeff, R. M., Gee, K. R., Walseth, T. F., and Lee, H. C. (1996). Activation and inactivation of $Ca^{2+}$ release by $NAADP^+$. *J. Biol. Chem.* **271**, 8513–8516.

Allen, L. H., Rosado, J. L., Casterline, J. E., Martinez, H., Lopez, P., Munoz, E., and Black, A. K. (1995). Vitamin B-12 deficiency and malabsorption are highly prevalent in rural Mexican communities. *Am. J. Clin. Nutr.* **62**, 1013–1019.

Almeida, A., Loopez-Mediavilla, C., and Medina, J. M. (1997). Thyroid hormones regulate the onset of osmotic activity of rat liver mitochondria after birth. *Endocrinology* **138**, 764–770.

Ampola, M. G., Mahoney, M. J., Nakamura, E., and Tanaka, K. (1975). Prenatal therapy of a patient with vitamin $B_{12}$-responsive methylmalonic acidemia. *N. Engl. J. Med.* **293**, 313–317.

Ariaey-Nejad, M. R., Balaghi, M., Baker, E. M., and Sauberlich, H. E. (1970). Thiamin metabolism in man. *Am. J. Clin. Nutr.* **23**, 764–778.

Arinze, J. C., and Mistry, S. P. (1971). Activities of some biotin enzymes and certain aspects of gluconeogenesis during biotin deficiency. *Comp. Biochem. Physiol. B* **38**, 285–294.

A warning regarding the use of folic acid. (Editorial) (1947). *N. Engl. J. Med.* **237**, 713–714.

Baldwin, C. T., Hoth, C. F., Amos, J. A., da-Silva, E. O., and Milunsky, A. (1992). An exonic mutation in hte HuP2 paired domain gene causes Waardenburg's syndrome. *Nature* **355**, 637–638.

Banhegyi, G., Braun, L., Csala, M., Puskas, F., and Mandl, J. (1997). Ascorbate metabolism and its regulation in animals. *Free Radical Biol. Med.* **23**, 793–803.

Bankier, A., Turner, M., and Hopkins, I. J. (1983). Pyridoxine-dependent seizures — A wider clinical spectrum. *Arch. Dis. Child.* **58**, 415–418.

Bates, C. J. (1997). Bioavailability of riboflavin. *Eur. J. Clin. Nutr., Suppl.* **51**, S38–S42.

Berg, H. (1997). Bioavailability of biotin. *Eur. J. Clin. Nutr., Suppl.* **51**(1), S60–S61.

Binet, A., and Kooh, S. (1996). Persistence of vitamin D-deficiency rickets in Toronto in the 1990s. *Can. J. Public Health* July-Aug., pp. 227–230.

Blankenhorn, D. H., Selzer, R. H., Crawford, D., Barth, J., Liu, C., Mack, M., and Alaupovic, P. (1993). Beneficial effects of colestipol–niacin therapy on the common carotid artery. *Circulation* **88**, 20–28.

Boisvert, W. A., Mendoza, I., Castaneda, C., Portocarrero, L., Solomons, N., Gershoff, S. N., and Russell, R. M. (1993). Riboflavin requirements of healthy elderly humans and its relationship to macronutrient composition of the diet. *J. Nutr.* **123**, 915–925.

Bonithon-Kopp, C., Coudray, C., Berr, C., Touboul, P., Feve, A., Favier, J., and Ducimetiere, P. (1997). Combined effects of lipid peroxidation and antioxidant status on carotid atherosclerosis in a population aged 59–71: The EVA study. *Am. J. Clin. Nutr.* **65**, 121–127.

Boushey, C., Beresford, S., Omenn, G., and Motulsky, A. G. (1995). A quantitative assessment of plasma homocysteine as a risk factor for vascular disease. *JAMA* **274**, 1049–1057.

Bowry, V. W., Ingold, K., and Stocker, R. (1992). Vitamin E in human low-density lipoprotein. *Biochem. J.* **288**, 341–344.

Brattsrom, L., Israelsson, B., Lindgarde, F., and Hultberg, B. (1988). Higher total plasma homocyteine in vitamin $B_{12}$ deficiency than in heterozygosity for homocyteinuria due to cystathione β-synthase deficiency. *Metabolism* **37**, 175–178.

Brody, T., and Stokstad, E. L. R. (1990). Nitrous oxide provokes changes in folypenta- and hexaglutamates. *J. Nutr.* **120**, 71–80.

Brody, T., and Stokstad, E. L. R. (1991). Incorporation of the 2-ring carbon of histidine into folylpolyglutamate coenzymes. *J. Nutr. Biochem.* **2**, 492–498.

Brody, T., Shin, Y. S., and Stokstad, E. L. R. (1976). Rat brain folate identification. *J. Neurochem.* **27**, 409–413.

Brommage, R., and DeLuca, H. F. (1985). Regulation of bone mineral loss during lactation. *Am. J. Physiol.* **248**, E182–E187.

Buczylko, J., Saari, J. C., Crouch, R. K., and Palczewski, K. (1996). Mechanisms of opsin activation. *J. Biol. Chem.* **271**, 20621–20630.

Buettner, G. R. (1993). The pecking order of free radicals and antioxidants. *Arch. Biochem. Biophys.* **300**, 535–543.

Carlson, A., and Bok, D. (1996). Promotion of the release of 11-*cis*-retinal from cultured retinal pigment epithelium by interphotoreceptor retinoid-binding protein. *Biochemistry* **31**, 9056–9062.

Carmel, R. (1996). Prevalence of undiagnosed pernicious anemia in the elderly. *Arch. Intern. Med.* **156**, 1097–1100.

Carmel, R. (1997). Cobalamin, the stomach, and aging. *Am. J. Clin. Nutr.* **66**, 750–759.

Carmel, R., and Johnson, C. S. (1978). Racial patterns in pernicious anemia. *N. Engl. J. Med.* **298**, 647–650.

Carmel, R., Johnson, C., and Weiner, J. (1987). Pernicious anemia in Latin Americans is not a disease of the elderly. *Arch. Intern. Med.* **147**, 1995–1996.

Carter, B. Z., Wiseman, A. L., Orkiszewski, R., Ballard, K., Ou, C.-N., and Lieberman, M. W. (1997). Metabolism of leukotriene $C_4$ in γ-glutamyl transpeptidase-deficient mice. *J. Biol. Chem.* **272**, 12305–12310.

Carpenter, K. J. (1981). "Pellagra." Hutchison Ross, Stroudsburg, PA.

Chambon, P. (1996). A decade of molecular biology of retinoic acid receptors. *FASEB J.* **10**, 940–954.

Chanarin, I., Rothman, D., and Watson-Williams, E. J. (1963). Normal formiminoglutamic acid excretion in megaloblastic anemia in pregnancy. *Lancet* **1**, 1068–1072.

Chanarin, I., Malkowska, V., O'Hea, A.-M., Rinsler, M. G., and Price, A. B. (1985). Megaloblastic anaemia in a vegetarian Hindu community. *Lancet*, November, pp. 1168–1172.

Chapuy, M., Arlot, M., Duboeuf, F., Brun, J., Crouzet, B., Arnaud, S., Delmas, P., and Meunier, P. (1992). Vitamin $D_3$ and calcium to prevent hip fractures in elderly women. *N. Engl. J. Med.* **327**, 1637–1642.

Chase, G., Akoh, C., and Eitenmiller, R. (1994). Analysis of tocopherols in vegetable oils by high-performance liquid chromatography. *J. Am. Oil Chem. Soc.* **71**, 877–880.

Chen, C.-C., and Heller, J. (1977). Uptake of retinol and retinoic acid from serum retinol-binding protein by retinal pigment epithelial cells. *J. Biol. Chem.* **252**, 5216–5221.

Chong, Y. H. (1970). Erythrocyte transketolase activity. *Am. J. Clin. Nutr.* **23**, 261–266.

Cichowicz, D. J., and Shane, B. (1987). Mammalian folylpoly-γ-glutamates synthetse, 2: Substrate specificity and kinetic properties. *Biochemistry* **26**, 513–521.

Clarke, R., Daly, L., Robinson, K., Naughten, E., Cahalane, S., Fowler, B., and Graham, I. (1991). Hyperhomocysteinemia: An independent risk factor for vascular disease. *N. Engl. J. Med.* **324**, 1149–1155.

Coburn, S. P., Ziegler, P., Costill, D. L., Mahuren, J., Fink, W. J., Schaltenbrand, W., Pauly, T. A., Pearson, D. R., Conn, P. S., and Guilarte, T. R. (1991). Response of vitamin B-6 content of muscle to changes in vitamin B-6 intake in men. *Am. J. Clin. Nutr.* **53**, 1436–1442.

Cormack, D. H. (1984). "Introduction to Histology." J. B. Lippincott, Philadelphia.

Craig-Schmidt, M. C., Stieh, K., and Lien, E. L. (1996). Retinal fatty acids of piglets fed docohexaenoic and arachidonic acids from microbial sources. *Lipids* **31**, 53–59.

Cui, J., O'Shea, S., Purkayastha, A., Saunders, T. L., and Ginsburg, D. (1996). Fatal haemorrhage and incomplete block to embryogenesis in mice lacking coagulation factor V. *Nature* **384**, 66–68.

Cunningham, G. C. (1995). California's public health policy in preventing neural tube defects by folate supplementation. *West. J. Med.* **162**, 265–267.

Czeizel, A. E., and Dudas, I. (1992). Prevention of the first occurrence of neural-tube defects by periconceptual vitamin supplementation. *N. Engl. J. Med.* **327**, 1832–1835.

Dajani, E. Z. (1993). ω-3 fatty acids and bowel cancer. *Gastroenterology* **104**, 1239–1240.

Daly, L. E., Kirke, P. N., Molloy, A., Weir, D. G., and Scott, J. M. (1995). Folate levels and neural tube defects. *JAMA* **274**, 1698–1702.

Darwish, H., and DeLuca, H. F. (1996). Analysis of binding of the 1,25-dihydroxyvitamin $D_3$ receptor to positive and negative vitamin D response elements. *Arch. Biochem. Biophys.* **334**, 223–234.

D'Angelo, D., Eubank, J. J., Davis, M. G., and Dorn, G. W. (1996). Mutagenic analysis of platelet thromboxane receptor cysteines. *J. Biol. Chem.* **271**, 6233–6240.

Dawson-Hughes, B., Harris, S. S., Krall, E. A., and Dallal, G. (1997). Effect of calcium and vitamin D supplementation on bone density in men and women 65 years of age or older. *N. Engl. J. Med.* **337**, 670–676.

DeMaster, E. G., Quast, B. J., Redfern, B., and Nagasawa, H. T. (1995). Reaction of nitric oxide with the free sulfhydryl group of human serum albumin yields a sulfenic acid and nitrous oxide. *Biochemistry* **34**, 11494–11499.

Demming-Adams, B., Gilmore, A. M., and Adams, W. W. (1996). In vivo functions of carotenoids in higher plants. *FASEB J.* **10**, 403–412.

Diplock, A. T. (1985). "The Fat-Soluble Vitamins." Technomic, Lancaster, PA.

Dousa, T. P., Chini, E. N., and Beers, K. W. (1996). Adenine nucleotide diphosphates: Emerging second messengers acting via intracellular $Ca^{2+}$ release. *Am. J. Physiol.* **271**, C1007–C1024.

Ducy, P., Desbois, C., Boyce, B., Pinero, G., Story, B., Dunstan, C., Smith, E., Bonadio, J., Goldstein, S., Gundberg, C., Bradley, A., and Karsenty, G. (1996). Increased bone formation in osteocalcin-deficient mice. *Nature* **382**, 448–452.

Duszka, C., Grolier, P., Azim, E., Alexandre-Gouabau, M.-C., Borel, P., and Azais-Braesco, V. (1996). Rat intestinal β-carotene dioxygenase activity is located primarily in the cytosol of mature jejunal enterocytes. *J. Nutr.* **126**, 2550–2556.

Du Vigneaud, V., Chandler, J. P., Moyer, A. W., and Keppel, D. M. (1939). The effect of choline on the ability of homocystine to replace methionine in the diet. *J. Biol. Chem.* **131**, 57–76.

El-Sonbaty, M., and Abdul-Ghaffer, N. (1996). Vitamin D deficiency in veiled Kuwaiti women. *Eur. J. Clin. Nutr.* **50**, 315–318.

Essien, F. B., and Wannberg, S. L. (1993). Methionine but not folinic acid or vitamin B-12 alters the frequency of neural tube defects in axd mutant mice. *J. Nutr.* **123**, 27–34.

Felton, C. V., Crook, D., Davies, M. J., and Oliver, M. F. (1994). Dietary polyunsaturated fatty acids and composition of human aortic plaques. *Lancet* **344**, 1195–1196.

Feng, J., Orlowski, J., and Lingrel, J. B. (1993). Identification of a functional thyroid hormone response element in the upstream flanking region of the human Na,K-ATPase α1 gene. *Nucl. Acids Res.* **21**, 2619–2626.

Filteau, S. M., and Tomkins, A. M. (1995). Vitamin A supplementation in developing countries. *Arch. Dis. Child.* **72**, 106–109.

Franceschi, R. T., and DeLuca, H. F. (1981). Characterization of 1,25-dihydroxyvitamin $D_3$-dependent calcium uptake in cultured embryonic chick duodenum. *J. Biol. Chem.* **256**, 3840–3847.

Frei, B., Stocker, R., and Ames, B. (1988). Antioxidant defenses and lipid peroxidation in human blood plasma. *Proc. Natl. Acad. Sci. U.S.A.* **85**, 9748–9752.

Fry, P. C., Fox, H. M., and Tao, H. G. (1976). Metabolic response to a pantothenic acid deficient diet in humans. *J. Nutr. Sci. Vitaminol.* **22**, 339–346.

Fukagawa, N., Ajami, A., and Young, V. R. (1996). Plasma methionine and cysteine kinetics in response to an intravenous glutathione infusion in adult humans. *Am. J. Physiol.* **270**, E209–E214.

Furr, H. C., and Clark, R. M. (1997). Intestinal absorption and tissue distribution of carotenoids. *Nutr. Biochem.* **8**, 364–377.

Garrow, T. A. (1996). Purification, kinetic properties, and cDNA cloning of mammalian betaine-homocysteine methyltransferase. *J. Biol. Chem.* **271**, 22831–22838.

Geiger, J. H., Hahn, S., Lee, S., and Sigler, P. B. (1996). Crystal structure of the yeast TFIIA/TBP/DNA complex. *Science* **272**, 830–836.

Geneser, F. (1986). "Textbook of Histology," pp. 696–707. Lea and Febiger, Philadelphia.

Gey, K. F. (1998). Vitamins E plus C and interacting conutrients required for optimal health. *Biofactors* **7**, 113–174.

Giguere, V. (1994). Retinoic acid receptors and cellular retinoid binding proteins. *Endocrine Revs.* **15**, 61–79.

Glembotski, C. C. (1986). The characterization of the ascorbic acid-mediated $\alpha$-amidation of $\alpha$-melanotropin in cultured intermediate pituitary lobe cells. *Endocrinology* **118**, 1461–1468.

Gotoda, T., Arita, M., Arai, H., Inoue, K., Yokota, T., Fukuo, Y., Yazaki, Y., and Yamada, N. (1995). Adult-onset spinocerebellar dysfunction caused by a mutation in the gene for the $\alpha$-tocopherol-transfer protein. *N. Engl. J. Med.* **333**, 1313–1318.

Greaves, M., and Weinstein, G. (1995). Treatment of psoriasis. *N. Engl. J. Med.* **332**, 581–588.

Griffiths, R. J., Pettipher, E., Koch, K., Farrell, C., Breslow, J., Conklyn, M., Smith, M., Hackman, B., Wimberly, D., Milici, A., Scampoli, D., Cheng, J., Pillar, J., Pazoles, C., Doherty, N., Melvin, L., Reiter, L., Biggars, M., Falkner, F., Mitchell, D., Liston, T., and Showell, H. (1995). Leukotriene $B_4$ plays a critical role in the progression of collagen-induced arthritis. *Proc. Natl. Acad. Sci. U.S.A.* **92**, 517–521.

Guaiquil, V., Farber, C., Golde, D., and Vera, J. C. (1997). Efficient transport and accumulation of vitamin C in HL-60 cells depleted of glutathione. *J. Biol. Chem.* **272**, 9915–9921.

Hafeman, D. G., and Hoekstra, W. G. (1977). Lipid peroxidation *in vivo* during vitamin E deficiency and selenium deficiency in the rat as monitored by ethane evolution. *J. Nutr.* **107**, 666–672.

Halliwell, B. (1996). Vitamin C: Antioxidant or pro-oxidant *in vivo*? *Free Radical Res.* **25**, 439–454.

Halliwell, B., and Gutteridge, J. M. (1990). The antioxidants of human extracellular fluids. *Arch. Biochem. Biophys.* **280**, 1–8.

Hamano, H., Nabekura, J., Nishikawa, M., and Ogawa, T. (1996). Docohexaenoic acid reduces GABA response in substantia nigra neuron of rat. *J. Neurophysiol.* **75**, 1264–1270.

Hanzell, L. J., Arnold, L., Flowers, D., Waeg, G., Malle, E., and Stocker, R. (1996). Presence of hypochlorite-modified proteins in human atherosclerotic lesions. *J. Clin. Invest.* **97**, 1535–1544.

Haussler, M. R., Haussler, C., Jurutka, P., Thompson, P., Hsieh, J.-C., Remus, L. S., Selznick, S., and Whitfield, G. K. (1997). The vitamin D hormone and its nuclear receptor: Molecular actions and disease states. *J. Endocrinol.* **154**, S57–S73.

Heijer, M., Koster, T., Blom, H. J., Bos, G. M., Briet, E., Reitsma, P., Vandenbroucke, J., Rosendaal, F. R. (1996). Hyperhomocysteinemia as a risk factor for deep-vein thrombosis. *N. Engl. J. Med.* **334**, 759–762.

Heird, W. C., and Gomez, M. R. (1996). Parenteral nutrition in low-birth-weight infants. *Annu. Rev. Nutr.* **16**, 471–499.

Henderson, W. R. (1994). The role of leukotrienes in inflammation. *Ann. Intern. Med.* **121**, 684–697.

Herbert, V. (1967). Biochemical and hematological lesions in folic acid deficiency. *Am. J. Clin. Nutr.* **20**, 562–569.

Herbert, V. (1994). Vitamin B-12 and elderly people. *Am. J. Clin. Nutr.* **59**, 1093–1094.

Herbert, V., Fong, W., Gulle, V., and Stopler, T. (1990). Low holotranscobalamin II is the earliest serum marker for subnormal vitamin $B_{12}$ (cobalamin) absorption in patients with AIDS. *Am. J. Hematol.* **34**, 132–139.

Higginbottom, M. C., Sweetman, L., and Nyhan, W. L. (1978). A syndrome of methylmalonic aciduria, homocysteinuria, megaloblastic anemia, and neurological abnormalities in a vitamin $B_{12}$-deficient breast-fed infant of a strict vegetarian. *N. Engl. J. Med.* **299**, 317–323.

Hilker, D. M., and Peters, O. F. (1966). Anti-thiamin activity in Hawaii fish. *J. Nutr.* **89**, 419–421.

Hines, J. D. (1966). Megaloblastic anemia in an adult vegan. *Am. J. Clin. Nutr.* **19**, 260–268.

Hiramatsu, T., Fukagawa, N. K., Marchini, J., Cortiella, J., Yu, Y.-M., Chapman, T., and Young, V. R. (1994). Methionine and cysteine kinetics at different intakes of cystine in healthy adult men. *Am. J. Clin. Nutr.* **60**, 525–533.

Hodges, R. E., Hood, J., Canham, J. E., Sauberlich, H. E., and Baker, E. M. (1971). Clinical manifestations of ascorbic acid deficiency in man. *Am. Clin. Nutr.* **24**, 432–443.

Hogg, P. J., and Stenflo, J. (1991). Interaction of vitamin K-dependent protein Z with thrombin. *J. Biol. Chem.* **266**, 10953–10958.

Holick, M. F., MacLaughlin, J. A., and Doppelt, S. H. (1981). Regulation of cutaneous previtamin $D_3$ photosynthesis in man. *Science* **211**, 590–593.

Holman, R. T., Adams, C. E., Nelson, R. A., Grater, S. J., Jaskiewicz, J. A., Johnson, S. B., and Erdman, J. W. (1995). Patients with anorexia nervosa demonstrate deficiencies of selected essential fatty acids, compensatory changes in nonessential fatty acids and decreased fluidity of plasma lipids. *J. Nutr.* **125**, 901–907.

Holvoet, P., and Collen, D. (1995). Lipid lowering and enhancement of fibrinolysis with niacin. *Circulation* **92**, 698–699.

Hood, L. E., Weissman, I. L., Wood, W. B., and Wilson, J. H. (1984). "Immunology," 2nd ed. Benjamin/Cummings, Menlo Park, CA.

Hornig, D. (1975). Distribution of ascorbic acid, metabolites and analogues in man and animals. *Ann. N. Y. Acad. Sci.* **258**, 103–118.

Ijichi, H., Ichiyama, A., and Hayaishi, S. (1966). Studies on the biosynthesis of nicotinamide adenine dinucleotide. *J. Biol. Chem.* **241**, 3701–3707.

Ijpenberg, A., Jeannin, E., Wahli, W., and Desvergne, B. (1997). Polarity and specific sequence requirements of peroxisome proliferator activated receptor (PPAR)/retinoid X receptor heterodimer binding to DNA. *J. Biol. Chem.* **272**, 20108–20117.

Illingworth, D., Stein, E., Mitchel, Y., Dujovne, C., Frost, P., Knopp, R., Tun, P., Zupkis, R., and Greguski, R. (1994). Comparative effects of lovastatin and niacin in primary hypercholesterolemia. *Arch. Intern. Med.* **154**, 1586–1595.

Imlay, J. A., and Fridovich, I. (1991). Assay of metabolic superoxide production in *E. coli. J. Biol. Chem.* **266**, 6957–6965.

Innis, S. M., Nelson, C. M., Rioux, M. F., and King, D. J. (1994). Development of visual acuity in relation to plasma and erythrocyte ω-6 and ω-6 fatty acids in healthy term gestation infants. *Am. J. Clin. Nutr.* **60**, 347–352.

Jacobson, E. L., and Jacobson, M. K. (1997). Tissue NAD as a biochemical measure of niacin status in humans. *Methods Enzymol.* **280**, 221–230.

Jacobson, M. K., Coyle, D. L., Vu, C. Q., Kim, H., and Jacobson, E. L. (1997). Preparation of cyclic ADP-ribose, 2′-phospho-cyclic ADP-ribose, and nicotinate adenine dinucleotide phosphate: Possible second messengers of calcium signaling. *Methods Enzymol.* **280**, Part J, 265–275.

Jacobson, R. H., and Tjian, R. (1996). Transcription factor IIA: A structure with multiple functions. *Science* **272**, 827–828.

Jialal, I., and Grundy, S. (1992). Effect of dietary supplementation with α-tocopherol on the oxidative modification of low-density lipoprotein. *J. Lipid Res.* **33**, 899–906.

Johnson, E. J., and Russell, R. M. (1992). Distribution of orally administered β-carotene among lipoproteins in healthy men. *Am. J. Clin. Nutr.* **56**, 128–135.

Johnson, E. J., Suter, P. M., Sahyoun, N., Ribaya-Mercado, J., and Russell, R. M. (1995). Relation between β-carotene intake and plasma and adipose tissue concentrations of carotenoids and retinoids. *Am. J. Clin. Nutr.* **62**, 598–603.

Kamal-Eldin, A., and Appleqvist, L. (1996). The chemistry and antioxidant properties of tocopherols and tocotrienols. *Lipids* **31**, 671–701.

Kang-Yoon, S. A., Kirksey, A., Giacoia, G. P., and West, K. D. (1995). Vitamin B-6 adequacy in neonatal nutrition: Associations with preterm delivery, type of feeding, and vitamin B-6 supplementation. *Am. J. Clin. Nutr.* **62**, 932–942.

Kayden, H. J., and Traber, M. G. (1993). Absorption, lipoprotein transport, and regulation of plasma concentrations of vitamin E in humans. *J. Lipid Res.* **34**, 343–358.

Khare, S., Bolt, M., Wali, R., Skarosi, S., Roy, H., Niedziela, S., Scaglione-Sewell, B., Aquino, B., Abraham, C., Sitrin, M., Brasitus, T., and Bissonnette, M. (1997). 1,25-Dihydroxyvitamin $D_3$ stimulates phospholipase C in rat colonocytes. *J. Clin. Invest.* **99**, 1831–1841.

Kinyamu, H., Gallagher, J., Balhorn, K., Petranick, K., and Rafferty, K. (1997). Serum vitamin D metabolites and calcium absorption in normal young and elderly free-living women and in women living in nursing homes. *Am. J. Clin. Nutr.* **65**, 790–797.

Kobayashi, K., Haradi, Y., and Hayashi, K. (1991). Kinetic behavior of monodehydroascorbate radical studied by pulse radiolysis. *Biochemistry* **30**, 8310–8315.

Koibuchi, N., Matsuzaki, S., Ichimura, K., Ohtake, H., and Yamaoka, S. (1996). Ontogenic changes in the expression of cytochrome *c* oxidase subunit I gene in the cerebellar cortex of the perinatal hypothyroid rat. *Endocrinology* **137**, 5096–5108.

Krauss, R. M., Deckelbaum, R. J., Ernst, N., Fisher, E., Howard, B. V., Knopp, R. H., Kotchen, T., Lichtenstein, A. H., McGill, H. C., Pearson, T. A., Prewitt, T., Stone, N. J., Horn, L. V., and Weinberg, R. (1996). Dietary guidelines for healthy American adults. *Circulation* **94**, 1795–1800.

Lee, W., Davis, K. A., Rettmer, R. L., and Labbe, R. F. (1988). Ascorbic acid status. *Am. J. Clin. Nutr.* **48**, 286–290.

Levine, M., Rumsey, S., Wang, Y., Park, J., Kwon, O., and Amano, N. (1997). In situ kinetics: An approach to recommended intake of vitamin C. *Methods Enzymol.* **281**, Part K, 425–437.

Li, E., and Norris, A. W. (1996). Structure/function of cytoplasmic vitamin A-binding proteins. *Annu. Rev. Nutr.* **16**, 205–234.

Liebler, D., and Burr, J. (1992). Oxidation of vitamin E during iron-catalyzed lipid peroxidation. *Biochemistry* **31**, 8278–8284.

Linkswiler, H. (1967). Biochemical and physiological changes in vitamin $B_6$ deficiency. *Am. J. Clin. Nutr.* **20**, 547–557.

Liu, Y.-Y., Collins, E. D., Norman, A. W., and Peleg, S. (1997). Differential interaction of 1,25-dihydroxyvitamin $D_3$ analogues and their 20-epi homologues with the vitamin D receptor. *J. Biol. Chem.* **272**, 3336–3345.

Lopaschuk, G. D., Michalak, M., and Tsang, H. (1987). Regulation of pantothenic acid transport in the heart. *J. Biol. Chem.* **262**, 3615–3619.

Lotan, R. (1996). Retinoids in cancer chemoprevention. *FASEB J.* **10**, 1031–1039.

Lowe, K. E., Osborne, C., Lin, B.-F., Kim, J.-S., Hsu, J.-C., and Shane, B. (1993). Regulation of folate and one-carbon metabolism in mammalian cells. *J. Biol. Chem.* **268**, 21665–21673.

Lowenthal, M., and Shany, S. (1994). Osteomalacia in Bedouin women of the Negev. *Isr. J. Med. Sci.* **30**, 520–523.

Lynch, S. M., and Frei, B. (1995). Reduction of copper, but not iron, by low density lipoprotein (LDL). *J. Biol. Chem.* **270**, 5158–5163.

Machlin, L. J., Keating, J., Nelson, J., Brin, J., Filipski, R., and Miller, O. N. (1979). Availability of adipose tissue tocopherol in the guinea pig. *J. Nutr.* **109**, 105–109.

Mackey, S. L., Heymont, J. L., Kronenberg, H. M., and Demay, M. B. (1996). Vitamin D receptor binding to the negative human parathyroid hormone vitamin D response element does not require the retinoid X receptor. *Mol. Endocrinol.* **10**, 298–305.

Mangelsdorf, D. J., and Evans, R. M. (1995). The RXR heterodimers and orphan receptors. *Cell* **83**, 841–850.

Marchand, L. L., Yoshizawa, C. N., Kolonel, L. N., Hankin, J. H., and Goodman, M. T. (1989). Vegetable consumption and lung cancer risk: A population-based case-control study in Hawaii. *J. Natl. Cancer Inst.* **81**, 1158–1164.

Marshall, H., Morrison, A., Studer, M., Popperi, H., and Krumlauf, R. (1996). Retinoids and hox genes. *FASEB J.* **10**, 969–978.

McCormick, D. B., Oka, M., Bowers-Komro, D., Yamada, Y., and Hartman, H. A. (1997). Purification and properties of FAD synthetase from liver. *Methods Enzymol.* **280**, Part J, 407–413.

McKenney, J. M., Proctor, J., Harris, S., and Chinchili, V. (1994). A comparison of the efficacy and toxic effects of sustained- vs. immediate-release niacin in hypercholesterolemic patients. *JAMA* **271**, 672–677.

Mead, J. F., and Wu, G.-S. (1976). Protective effect of vitamin E against PUFA oxidation. *In* "Lipids" (R. Paoletti, G. Porcellati, and G. Jacini, eds.), Vol. 1, pp. 197–201. Raven Press, New York.

Mellay, V., Grosse, B., and Lieberherr, M. (1997). Phospholipase Cβ and membrane action of calcitriol and estradiol. *J. Biol. Chem.* **272**, 11902–11907.

Micozzi, M. S., Beecher, G. R., Taylor, P. R., and Khachick, F. (1990). Carotenoid analysis of selected raw and cooked foods associated with a lower risk for cancer. *J. Natl. Cancer Inst.* **82**, 282–285.

Mills, J. L., Rhoads, G. G., Simpson, J. L., Cunningham, G. C., Conley, M. R., Lassman, M. R., Walden, M., Depp, O. R., and Hoffman, H. J. (1989). The absence of a relation between the periconceptual use of vitamins and neural-tube defects. *N. Engl. J. Med.* **321**, 430–435.

Minna, J. D., and Mangelsdorf, D. J. (1997). Retinoic acid receptor expression abnormalities in lung cancer. *J. Natl. Cancer Inst.* **89**, 602–604.

Mock, D. M., Mock, N. I., and Stratton, S. L. (1997a). Concentrations of biotin metabolites in human milk. *J. Pediatr.* **131**, 456–458.

Mock, D. M., Stadler, D. D., Stratton, S. L., and Mock, N. I. (1997b). Biotin status assessed longitudinally in pregnant women. *J. Nutr.* **127**, 710–716.

Mock, D. M., Wang, K.-S., and Kearns, G. L. (1997c). The pig is an appropriate model for human catabolism as judged by the urinary metabolite profile of radioisotope-labeled biotin. *J. Nutr.* **127**, 365–369.

Molloy, A. M., Daly, S., Mills, J. L., Kirke, P., Whitehead, A., Ramsbottom, D., Conley, M., Weir, D. G., and Scott, J. M. (1997). Thermolabile variant of 5,10-methylenetetrahydrofolate reductase associated with low red-cell folates: Implications for folate intake recommendations. *Lancet* **349**, 1591–1593.

Moore, T. (1957). "Vitamin A." Elsevier, Amsterdam.

Morrison, A., Moroni, M., Ariza-McNaughton, L., Krumlauf, R., and Mavilio, F. (1996). In vitro and transgenic analysis of a human HOXD4 retinoid-responsive enhancer. *Development* **122**, 1895–1907.

Morriss-Kay, G. M., and Sokolova, N. (1996). Embryonic development and pattern development and pattern formation. *FASEB J.* **10**, 961–968.

Moser, H., and Moser, A. (1996). Very-long-chain fatty acids in diagnosis, pathogenesis, and therapy of peroxisomal disorders. *Lipids* **31**, S141–S145.

Mudd, S. H., Skovby, F., Levy, H. L., Pettigrew, K., Wilcken, B., Pyeritz, R., Andria, G., Boers, G., Bromberg, I., Cerone, R., Fowler, B., Grobe, H., Schmidt, H., and Schweitzer, L. (1985). The natural history of homocystinuria due to cystathionine β-synthase deficiency. *Am. J. Hum. Genet.* **37**, 1–31.

Nagao, A., During, A., Hoshino, C., Terao, J., and Olson, J. A. (1996). Stoichometric conversion of all-*trans*-β-carotene to retinal by pig intestinal extract. *Arch. Biochem. Biophys.* **328**, 57–63.

Napoli, J. L. (1996). Retinoic acid biosynthesis and metabolism. *FASEB J.* **10**, 993–1001.

Nathans, J., Merbs, S. L., Sung, C.-H., Weitz, C., and Wang, Y. (1992). Molecular genetics of human visual pigments. *Annu. Rev. Genet.* **26**, 403–424.

Naurath, H. J., Joosten, E., Riezler, R., Stabler, S. P., Allen, R. H., and Lindenbaum, J. (1995). Effects of vitamin $B_{12}$, folate, and vitamin $B_6$ supplements in elderly people with normal serum vitamin concentrations. *Lancet* **346**, 85–89.

Noguchi, N., Gotoh, N., and Niki, E. (1998). Action of vitamin E as an antioxidant against oxidative modification of low-density lipoprotein. *Biofactors* **7**, 41–50.

Norman, D A., Fordtran, J. S., Brinkely, L. J., Zerwekh, J. E., Nicar, M. J., Strowig, S. M., and Pak, C. Y. (1981). Jejunal and ileal adaption to alterations in dietary calcium. *J. Clin. Invest.* **67**, 1599–1603.

Nygard, O., Nordrehaug, J., Refsum, H., Ueland, P. M., Farstad, M., and Vollset, S. (1997). Plasma homocysteine levels and mortality in patients with coronary artery disease. *N. Engl. J. Med.* **337**, 230–236.

Oginni, L., Worsfold, M., Oyelami, O., Sharp, C., Powell, D., and Davie, M. (1996). Etiology of rickets in Nigerian children. *J. Pediatr.* **128**, 692–694.

Oppenheimer, J. H., and Schwartz, H. L. (1997). Molecular basis of thyroid hormone-dependent brain development. *Endocrine Rev.* **18**, 462–475.

Paquin, J., Baugh, C. M., and MacKenzie, R. E. (1985). Channeling between the active sites of formiminotransferase-cyclodeaminase. *J. Biol. Chem.* **260**, 14925–14931.

Paukert, J. L., Straus, L., and Rabinowitz, J. C. (1976). Formyl-methenyl-methylene tetrahydrofolate synthetase. *J. Biol. Chem.* **251**, 5104–5111.

Petkovich, M. (1992). Regulation of gene expression by vitamin A. *Annu. Rev. Nutr.* **12**, 443–471.

Poston, J. M. (1984). The relative carbon flux through the α- and β-keto pathways of leucine metabolism. *J. Biol. Chem.* **259**, 2059–2061.

Preusch, P. C., and Suttie, J. W. (1983). Stereospecificity of vitamin K-epoxide reductase. *J. Biol. Chem.* **258**, 714–716.

Prigge, S. T., Kolhekar, A. S., Eipper, B. A., Mains, R. E., and Amzel, L. M. (1997). Amidation of bioactive peptides: The structure of peptidyl-glycine α-hydroxylating monooxygenase. *Science* **278**, 1300–1305.

Raiten, D. J., Reynolds, R. D., Andon, M. B., Robbins, S. T., and Fletcher, A. B. (1991). Vitamin B-6 metabolism in premature infants. *Am. J. Clin. Nutr.* **53**, 78–83.

Rasmussen, K. (1989). Studies on methylmalonic acid in humans. *Clin. Chem.* **35**, 2271–2280.

Rasmussen, M., Moser, A. B., Borel, J., Khangoora, S., and Moser, H. W. (1994). Brain, liver, and adipose tissue erucic and very long chain fatty acid levels in adrenoleukodystrophy patients treated with glyceryl trierucate and trioleate oils (Lorenzo's oil). *Neurochem. Res.* **19**, 1073–1082.

Richardson, T. J., and Sgoutas, D. (1975). Essential fatty acid deficiency in four adult patients during total parenteral nutrition. *Am. J. Clin. Nutr.* **28**, 258–263.

Rimm, E., Stampfer, M., Ascherio, A., Giovannucci, E., Colditz, G., and Willett, W. (1993). Vitamin E consumption and the risk for coronary heart disease in men. *N. Engl. J. Med.* **328**, 1450–1456.

Rindi, G., and Gastaldi, G. (1997). Measurements and characteristics of intestinal riboflavin transport. *Methods Enzymol.* **280**, Part J, 399–407.

Robinson, K., Mayer, E. L., Miller, D., Green, R., Lente, F., Gupta, A., Kottke-Marchant, K., Savon, S., Selhub, J., Nissen, S., Kutner, M., Topol, E., and Jacobsen, D. W. (1995). Hyperhomocysteinemia and low pyridoxal phosphate. *Circulation* **92**, 2825–2830.

Robishaw, J. D., and Neely, J. R. (1985). Coenzyme A metabolism. *Am. J. Physiol.* **248**, E1–E9.

Rosenberg, L. E. (1991). Inherited disorders of amino acid metabolism. *In* "Principles of Internal Medicine" (J. Wilson, E. Braunwald, K. J. Isselbacher, R. G. Petersdorf, J. B. Martin, A. Fauci, and R. K. Root, eds.), pp. 1868–1875. McGraw-Hill, New York.

Roy, B., Taneja, R., and Chambon, P. (1995). Synergistic activation of retinoic acid (RA)-responsive genes and induction of embryonal carcinoma cell differentiation by an RA receptor a (RARα)-, RARβ-, or RARγ-selective ligand in combination with a retinoid X receptor-specific ligand. *Mol. Cell. Biol.* **15**, 6481–6487.

Satoh, M. S., Poirier, G. G., and Lindahl, T. (1993). NAD$^+$-dependent repair damaged DNA by human cell extracts. *J. Biol. Chem.* **268**, 5480–5487.

Sauberlich, H. E., Judd, J. H., Nichoalds, G. E., Broquist, H. P., and Darby, W. J. (1972). Application of the erythrocyte glutathione reductase assay in evaluating riboflavin status in a high school student population. *Am. J. Clin. Nutr.* **25**, 756–762.

Schaumburg, H., Kaplan, J., Windebank, A., Vick, N., Rasmus, S., Pleasure, D., and Brown, M. (1983). Sensory neuropathy from pyridoxine abuse. *N. Engl. J. Med.* **309**, 445–448.

Schecter, A., Rollins, B., Zhang, Y., Charo, I. F., Fallon, J. T., Rossikhina, M., Giesen, P., Nemerson, Y., and Taubman, M. B. (1997). Tissue factor is induced by monocyte chemoattractant protein-1 in human aorta smooth muscle and THP-1 cells. *J. Biol. Chem.* **272**, 28568–28573.

Schwartz, S., deBlois, D., and O'Brien, E. (1995). The intima soil for atherosclerosis and restenosis. *Circ. Res.* **77**, 445–465.

Scully, R. E., Mark, E., McNeely, W. F., Ebeling, S. H., and Pillips, L. (1997). Presentation of a case. *N. Engl. J. Med.* **336**, 861–867.

Seawright, A., English, P., and Gartner, R. (1965). Hypervitaminosis A and hyperostosis of the cat. *Nature* **206**, 1171–1172.

Selhub, J., Jacques, P. F., Bostom, A. G., D'Agostino, R. B., Wilson, P., Belanger, A. J., O'Leary, D., Wolf, P., Schaefer, E. J., and Rosenberg, I. H. (1995). Association between plasma homocysteine concentrations and extracranial carotid-artery stenosis. *N. Engl. J. Med.* **332**, 286–291.

Shafritz, D. A. (1996). Synthetic retinoids for the secondary prevention of hepatocellular carcinoma. *N. Engl. J. Med.* **334**, 1600–1601.

Shaw, G. M., Velie, E. M., and Schaffer, D. (1996). Risk of neural tube defect-affected pregnancies among obese women. *JAMA* **275**, 1093–1096.

Shin, Y. S., Buehring, K. U., and Stokstad, E. L. R. (1975). The relationship between vitamin B$_{12}$ and folic acid and the effect of methionine on folate metabolism. *Mol. Cell. Biochem.* **9**, 97–108.

Smith, C. M., Narrow, C. M., Kendrick, Z. V., and Steffan, C. (1987). The effect of pantothenate deficiency in mice on their metabolic response to fast and exercise. *Metabolism* **36**, 115–121.

Smolin, L. A., and Benevenga, N. J. (1984). The use of cyst(e)ine in the removal of protein-bound homocysteine. *Am. J. Clin. Nutr.* **39**, 730–737.

Sokoll, L., Booth, S., O'Brien, M., Davidson, K. W., Tsaioun, K. I., and Sadowski, J. A. (1997). Changes in serum osteocalcin, plasma phylloquinone, and urinary γ-carboxyglutamic acid in response to altered intakes of dietary phylloquinone in human subjects. *Am. J. Clin. Nutr.* **65**, 779–784.

Soyland, E., Funk, J., Rajka, G., Sandberg, M., Thune, P., Rustad, L., Helland, S., Middelfart, K., Odu, S., Falk, E., Solvoll, K., Bjorneboe, G., and Drevon, C. A. (1993). Effect of dietary supplementation with very-long chain n-3 fatty cids in patients with psoriasis. *N. Engl. J. Med.* **328**, 1812–1816.

Sprecher, H., Luthria, D. L., Mohammed, B. S., and Baykousheva, S. P. (1995). Reevaluation of the pathways for the biosynthesis of polyunsaturated fatty acids. *J. Lipid Res.* **36**, 2471–2477.

Squibb, R. L., Braham, J. E., Arroyave, G., and Scrimshaw, N. S. (1959). A comparison of the effect of raw corn and tortillas (lime-treated corn) with niacin, tryptophan or beans on the growth and muscle niacin of rats. *J. Nutr.* **67**, 351–361.

Stabler, S., Lindenbaum, J., and Allen, R. H. (1997). Vitamin B-12 deficiency in the elderly: Current dilemmas. *Am. J. Clin. Nutr.* **66**, 741–749.

Stampfer, M., Hennekens, C., Manson, J., Colditz, G., Rosner, B., and Willet, W. (1993). Vitamin E consumption and the risk for coronary disease in women. *N. Engl. J. Med.* **328**, 1444–1449.

Steinberg, D. (1997). Low-density lipoprotein oxidation and its pathological significance. *J. Biol. Chem.* **272**, 20963–20966.

Stover, P. J., Chen, L., Suh, J., Stover, D., Keyomarsi, K., and Shane, B. (1997). Molecular cloning, characterization, and regulation of the human mitochondrial serine hydroxymethyltransferase gene. *J. Biol. Chem.* **272**, 1842–1848.

Subramani, S. (1993). Protein import into peroxisomes and biogenesis of the organelle. *Annu. Rev. Cell. Biol.* **9**, 445–478.

Sugimoto, Y., Yamasaki, A., Segi, E., Tsuboi, K., Aze, Y., Nishimura, T., Oida, H., Yoshida, N., Tanaka, T., Katsuyama, M., Hasumoto, K., Murata, T., Hirata, M., Ushikubi, F., Negishi, M., Ichikawa, A., and Narumiya, S. (1997). Failure of parturition in mice lacking the prostaglandin F receptor. *Science* **277**, 681–683.

Suh, M., Wierzbicki, A. A., Lien, E., and Clandinin, M. T. (1996). Relationship between dietary supply of long-chain fatty acids and membrane composition of long- and very-long-chain essential fatty acids in developing rat photoreceptors. *Lipids* **31**, 61–64.

Sumner, A. E., Chin, M. M., Abrahm, J. L., Berry, G. T., Gracely, E. J., Allen, R. H., and Stabler, S. P. (1996). Elevated methylmalonic acid and total homocysteine levels show high prevalence of vitamin $B_{12}$ deficiency after gastric surgery. *Ann. Intern. Med.* **124**, 469–476.

Superko, H. R., and Krauss, R. M. (1994). Coronary artery disease regression. *Circulation* **90**, 1056–1069.

Surya, A., Foster, K. W., and Knox, B. E. (1995). Transducin activation by the bovine opsin apoprotein. *J. Biol. Chem.* **270**, 5024–5031.

Suttie, J. W. (1995). The importance of menaquinones in human nutrition. *Annu. Rev. Nutr.* **15**, 399–417.

Svardal, A., Refsum, H., and Ueland, P. M. (1996). Determination of *in vivo* protein binding of homocysteine and its relation to free homocysteine in the liver and other tissues of the rat. *J. Biol. Chem.* **261**, 3156–3163.

Tahiliani, A. G., and Neely, J. R. (1987). A transport system for coenzyme A in isolated heart mitochondria. *J. Biol. Chem.* **262**, 11607–11610.

Tallaksen, C. M., Bohmer, T., Karlsen, J., and Bell, H. (1997). Determination of thiamin and its phosphate esters in human blood, plasma, and urine. *Methods Enzymol.* **279**, Part I, 67–74.

Tanaka, T., and DeLuca, H. F. (1984). Rat renal 25-hydroxyvitamin $D_3$ 1- and 24-hydroxy-lases: Their *in vivo* regulation. *Am. J. Physiol.* **246**, E168–E173.

Thomas, S. R., Neuzil, J., Mohr, D., and Stocker, R. (1995). Coantioxidants make α-tocopherol an efficient antioxidant for low-density lipoprotein. *Am. J. Clin. Nutr.* **62**, 1357S–1364S.

Traber, M. G. (1997). Regulation of human plasma vitamin E. *Adv. Pharmacol.* **38**, 49–63.

Tribble, D. L., and Krauss, R. M. (1994). HDL and coronary artery disease. *Adv. Intern. Med.* **38**, 1–29.

Tsang, S. H., Gouras, P., Yamashita, C. K., Kjeldbye, H., Fisher, J., Farber, D. B., and Goff, S. P. (1996). Retinal degeneration in mice lacking the γ subunit of the rod cGMP phosphodiesterase. *Science* **272**, 1026–1029.

Tsuge, H. (1997). Determination of vitamin $B_6$ vitamers and metabolites in a biological sample. *Methods Enzymol.* **280**, Part J, 3–12.

Tucker, K. L., Mahnken, B., Wilson, P. W., Jacques, P., and Selhub, J. (1996). Folic acid fortification of the food supply. *JAMA* **276**, 1879–1885.

Ubbink, J. B., Vermaak, W., Merwe, A., Becker, P., Delport, R., and Potgieter, H. (1994). Vitamin requirements for the treatment of hyperhomocysteinemia in humans. *J. Nutr.* **124**, 1927–1933.

Upston, J. M., Neuzil, J., and Stocker, R. (1996). Oxidation of LDL by recombinant human 15-lipoxygenase: Evidence for α-tocopherol-dependent oxidation of esterified core and surface lipids. *J. Lipid Res.* **37**, 2650–2661.

Varnum, B. C., Young, C., Elliott, G., Garcia, A., Bartley, T., Fridell, Y.-W., Hunt, R., Trall, G., Clogston, C., Toso, R. J., Yanagihara, D., Bennett, L., Sylber, M., Merewether, L., Tseng, A., Escobar, E., Liu, E., and Yamane, H. K. (1995). Axl receptor tyrosine kinase stimulated by the vitamin K-dependent protein encoded by growth-arrest-specific gene 6. *Nature* **373**, 623–626.

Varro, A., Dockray, G. J., Bate, G., Vaillant, C., Higham, A., Armitage, E., and Thompson, D. G. (1997). Gastrin biosynthesis in the antrum of patients with pernicious anemia. *Gastroenterology* **112**, 733–741.

Velazquez, A., Teran, M., Baez, A., Gutierrez, J., and Rodriguez, R. (1995). Biotin supplementation affects lymphocyte carboxylases and plasma biotin in severe protein-energy malnutrition. *Am. J. Clin. Nutr.* **61**, 385–391.

Wald, N. (1991). Prevention of neural tube defects: Results of the medical research council vitamin study. *Lancet* **338**, 131–137.

Wald, N., and Bower, C. (1995). Folic acid and the prevention of neural tube defects. *Brit. Med. J.* **310**, 1019–1020.

Wegger, I., and Palludan, B. (1994). Vitamin C deficiency causes hematological and skeletal abnormalities during fetal development in swine. *J. Nutr.* **124**, 241–248.

Werler, M. M., Louik, C., Shapiro, S., and Mitchell, A. A. (1996). Prepregnant weight in relation to risk of neural tube defects. *JAMA* **275**, 1089–1092.

Wielen, R. P., Lowik, M., Berg, H., Groot, L., Haller, J., Moreiras, O., and Staveren, W. (1995). Serum vitamin D concentrations among elderly in Europe. *Lancet* **346**, 207–210.

Wilner, G. D., Nossel, H. L., and LeRoy, E. C. (1968). Aggregation of platelets by collagen. *J. Clin. Invest.* **47**, 2616–2621.

Wu, L. N., Yoshimori, T., Genge, B., Sauer, G., Kirsch, T., Ishikawa, Y., and Wuthier, R. (1993). Characterization of the nucleational core complex responsible for mineral induction by growth plate cartilage matrix vesicles. *J. Biol. Chem.* **268**, 25084–25094.

Yang, C.-Y., Gu, Z.-W., Yang, H.-X., Yang, M., Wiseman, W., Rogers, L., Welty, S., Kata, V., Rohde, M., and Smith, C. V. (1997). Oxidation of bovine β-casein by hypochlorite. *Free Radical Biol. Med.* **22**, 1235–1240.

Zawel, L., and Reinberg, D. (1995). Common themes in assembly and function of eukaryotic transcription complexes. *Annu. Rev. Biochem.* **64**, 533–561.

Zempleni, J., Galloway, J. R., and McCormick, D. B. (1996). Pharmacokinetics of orally and intravenously administered riboflavin in healthy humans. *Am. J. Clin. Nutr.* **63**, 54–66.

Ziouzenkova, O., Winklhofer-Roob, B., Puhl, H., Roob, J., and Esterbauer, H. (1996). Lack of correlation between the α-tocopherol content of plasma and LDL, but high correlations for γ-tocopherol and carotenoids. *J. Lipid Res.* **37**, 1936–1946.

## BIBLIOGRAPHY

*Folate*

Bailey, L. B. (1988). Factors affecting folate bioavailability. *Food Technol.* (Oct.) **42**, 206–238.

Baker, S. J., and DeMeyer, E. M. (1979). Nutritional anemia: Its understanding and control with special reference to the work of the World Health Organization. *Am. J. Clin. Nutr.* **32**, 368–417.

Blakely, R. L. (1969). "Frontiers of Biology," Vol. 13: "The Biochemistry of Folic Acid and Related Pteridines" (A. Neuberger and E. L. Tatum, eds.). North-Holland, Amsterdam.

Brody, T., Shin, Y. S., and Stokstad, E. L. R. (1976). Rat brain folate identification. *J. Neurochem.* **27**, 409–413.

Brody, T., Shane, B., and Stokstad, E. L. R. (1979). Separation and identification of pteroylpolyglutamates by polyacrylamide gel chromatography. *Anal. Biochem.* **92**, 501–509.

Brody, T., and Stokstad, E. L. R. (1990). Nitrous oxide provokes changes in folypenta- and hexaglutamates. *J. Nutr.* **120**, 71–80.

Brody, T., and Stokstad, E. L. R. (1991). Incorporation of the 2-ring carbon of histidine into folylpolyglutamate coenzymes. *J. Nutr. Biochem.* **2**, 492–498.

Case, G. L., Kaisaki, P. J., and Steele, R. D. (1988). Resolution of rat liver 10-formyltetrahydrofolate dehydrogenase/hydrolase activities. *J. Biol. Chem.* **263**, 10204–10207.

Dansky, L. V., Andermann, E., Rosenblatt, D., Sherwin, A. L., and Andermann, F. (1987). Anticonvulsants, folate levels, and pregnancy outcome: A prospective study. *Ann. Neurol.* **21**, 176–182.

Du Vigneaud, V., Chandler, J. P., Moyer, A. W., and Keppel, D. M. (1939). The effect of choline on the ability of homocystine to replace methionine in the diet. *J. Biol. Chem.* **131**, 57–76.

Garrow, T. A. (1996). Purification, kinetic properties, and cDNA cloning of mammalian betaine-homocysteine methyltransferase. *J. Biol. Chem.* **271**, 22831–22838.

Halsted, C. H. (1980). Folate deficiency in alcoholism. *Am. J. Clin. Nutr.* **33**, 2736–2740.

Jencks, D. A., and Matthews, R. G. (1987). Allosteric inhibition of methylenetetrahydrofolate reductase by adenosylmethionine. *J. Biol. Chem.* **262**, 2485–2493.

Jolivet, J., Cowan, K. H., Curt, G. A., Clendeninn, N. J., and Chabner, B. A. (1983). The pharmacology and clinical use of methotrexate. *N. Engl. J. Med.* **309**, 1094–1104.

Lowe, K. E., Osborne, C., Lin, B.-F., Kim, J.-S., Hsu, J.-C., and Shane, B. (1993). Regulation of folate and one-carbon metabolism in mammalian cells. *J. Biol. Chem.* **268**, 21665–21673.

Paquin, J., Baugh, C. M., and MacKenzie, R. E. (1985). Channeling between the active sites of formiminotransferase-cyclodeaminase. *J. Biol. Chem.* **260**, 14925–14931.

Paukert, J. L., Straus, L., and Rabinowitz, J. C. (1976). Formyl-methenyl-methylene tetrahydrofolate synthetase. *J. Biol. Chem.* **251**, 5104–5111.

Rabinowitz, J. C., and Tabor, H. (1958). The urinary excretion of formic acid and formiminoglutamic acid in folic acid deficiency. *J. Biol. Chem.* **233**, 252–255.

Reisenauer, A. M., Chandler, C. J., and Halsted, C. H. (1986). Folate binding and hydrolysis by pig intestinal brush-border membranes. *Am. J. Physiol.* **251**, G481–G486.

Shane, B., and Stokstad, E. L. R. (1976). Transport and utilization of methyltetrahydrolate by *Lactobacillus casei. J. Biol. Chem.* **251**, 3405–3410.

Stokstad, E. L. R. (1968). Experimental anemias in animals resulting from folic acid and vitamin $B_{12}$ deficiencies. *Vit. Horm.* **26**, 443–463.

Stover, P., and Schirch, V. (1993). The metabolic role of leucovorin. *Trends Biochem. Sci.* **18**, 102–106.

Stover, P. J., Chen, L., Suh, J., Stover, D., Keyomarsi, K., and Shane, B. (1997). Molecular cloning, characterization, and regulation of the human mitochondrial serine hydroxymethyltransferase gene. *J. Biol. Chem.* **272**, 1842–1848.

Tamura, T., and Stokstad, E. L. R. (1973). The availability of food folate in man. *Br. J. Haematol.* **25**, 513–532.

## Folate and Neural Tube Defects

Baldwin, C. T., Hoth, C. F., Amos, J. A., da-Silva, E. O., and Milunsky, A. (1992). An exonic mutation in hte HuP2 paired domain gene causes Waardenburg's syndrome. *Nature* **355**, 637–638.

Cunningham, G. C. (1995). California's public health policy in preventing neural tube defects by folate supplementation. *West. J. Med.* **162**, 265–267.

Czeizel, A. E., and Dudas, I. (1992). Prevention of the first occurrence of neural-tube defects by periconceptual vitamin supplementation. *N. Engl. J. Med.* **327**, 1832–1835.

Daly, L. E., Kirke, P. N., Molloy, A., Weir, D. G., and Scott, J. M. (1995). Folate levels and neural tube defects. *JAMA* **274**, 1698–1702.

Essien, F. B., and Wannberg, S. L. (1993). Methionine but not folinic acid or vitamin B-12 alters the frequency of neural tube defects in axd mutant mice. *J. Nutr.* **123**, 27–34.

Habibzadeh, N., Wild, J., Lucock, M. D., and Schorah, C. J. (1995). One-carbon metabolism in pregnancies complicated by neural tube defects. Lancet **345**, 791.

Mills, J. L., Rhoads, G. G., Simpson, J. L., Cunningham, G. C., Conley, M. R., Lassman, M. R., Walden, M., Depp, O. R., and Hoffman, H. J. (1989). The absence of a relation between the periconceptual use of vitamins and neural-tube defects. *N. Engl. J. Med.* **321**, 430–435.

Mills, J. L., McPartlin, J., Kirke, P. N., Lee, Y. J., Conley, M. R., Weir, D. G., and Scott, J. M. (1995). Homocysteine metabolism in pregnancies complicated by neural-tube defects. *Lancet* **345**, 149–151.

Mills, J. L., Scott, J. M., Kirke, P. N., McPartlin, J. M., Conley, M. R., Weir, D. G., Molloy, A. M., and Lee, Y. J. (1996). Homocysteine and neural tube defects. *J. Nutr.* **126**, 756S–760S.

Moephuli, S., Klein, N., Baldwin, M., and Krider, H. M. (1997). Effects of methionine on the cytoplasmic distribution of actin and tubulin during neural tube closure in rat embryos. *Proc. Natl. Acad. Sci. U.S.A.* **94**, 543–548.

Motulsky, A. G. (1996). Nutritional ecogenetics: Homocysteine-related arteriosclerotic vascular disease, neural tube defects, and folic acid. *Am. J. Hum. Genet.* **58**, 17–20.

Shaw, G. M., Velie, E. M., and Schaffer, D. (1996). Risk of neural tube defect-affected pregnancies among obese women. *JAMA* **275**, 1093–1096.

Stanier, P., Henson, J. N., Eddleston, J., Moore, G., and Copp, A. J. (1995). Genetic basis of neural tube defects: The mouse gene loop-tail maps to a region of chromosome 1 syntenic with human 1q21-q23. *Genomics* **26**, 473–478.

Wald, N. (1991). Prevention of neural tube defects: Results of the medical research council vitamin study. *Lancet* **338**, 131–137.

Wald, N., and Bower, C. (1995). Folic acid and the prevention of neural tube defects. *Brit. Med. J.* **310**, 1019–1020.

Werler, M. M., Louik, C., Shapiro, S., and Mitchell, A. A. (1996). Prepregnant weight in relation to risk of neural tube defects. *JAMA* **275**, 1089–1092.

*Vitamin B₁₂*

A warning regarding the use of folic acid. (Editorial) (1947). *N. Engl. J. Med.* **237**, 713–714.

Akesson, B., Fehling, C., and Jagerstad, M. (1979). Lipid composition and metabolism in liver and brain of vitamin $B_{12}$-deficient rat sucklings. *Br. J. Nutr.* **41**, 263–274.

Allen, L. H., Stabler, S. P., Savage, D. G., and Lindenbaum, J. (1993). Metabolic abnormalities in cobalamin (vitamin $B_{12}$) and folate deficiency. *FASEB J.* **7**, 1344–1353.

Allen, L. H., Rosado, J. L., Casterline, J. E., Martinez, H., Lopez, P., Munoz, E., and Black, A. K. (1995). Vitamin B-12 deficiency and malabsorption are highly prevalent in rural Mexican communities. *Am. J. Clin. Nutr.* **62**, 1013–1019.

Allen, L. H., and Casterline, J. (1994). Vitamin B-12 deficiency in elderly individuals: Diagnosis and requirements. *Am. J. Clin. Nutr.* **60**, 12–14.

Ampola, M. G., Mahoney, M. J., Nakamura, E., and Tanaka, K. (1975). Prenatal therapy of a patient with vitamin $B_{12}$-responsive methylmalonic acidemia. *N. Engl. J. Med.* **293**, 313–317.

Birn, H., Verrous, P. J., Nexo, E., Hager, H., Jacobsen, C., Christiansen, E. I., and Moestrup, S. K. (1997). Characterization of an epithelial 460-kDa protein that facilitates endocytosis of intrinsic factor–vitamin $B_{12}$ and binds receptor-associated protein. *J. Biol. Chem.* **272**, 26497–26504.

Carmel, R. (1996). Prevalence of undiagnosed pernicious anemia in the elderly. *Arch. Intern. Med.* **156**, 1097–1100.

Carmel, R. (1997). Caobalamin, the stomach, and aging. *Am. J. Clin. Nutr.* **66**, 750–759.

Carmel, R., and Johnson, C. S. (1978). Racial patterns in pernicious anemia. *N. Engl. J. Med.* **298**, 647–650.

Carmel, R., Johnson, C. S., and Weiner, J. M. (1987). Pernicious anemia in Latin Americans is not a disease of the elderly. *Arch. Intern. Med.* **147**, 1995–1996.

Chanarin, I. (1969). "The Megaloblastic Anemias." Blackwell Scientific, Oxford.

Chanarin, I., Malkowska, V., O'Hea, A.-M., Rinsler, M. G., and Price, A. B. (1985). Megaloblastic anaemia in a vegetarian Hindu community. *Lancet*, November, pp. 1168–1172.

Cooper, B. A., and Lowenstein, L. (1966). Vitamin $B_{12}$–folate interrelationships in megaloblastic anemia. *Br. J. Haematol.* **12**, 283–296.

Garton, G. A., Scaife, J. R., and Smith, A. (1975). Effect of vitamin $B_{12}$ status on the occurrence of branched-chain and odd-numbered fatty acids in the liver lipids of the baboon. *Lipids* **10**, 855–857.

Herbert, V. (1994). Vitamin B-12 and elderly people. *Am. J. Clin. Nutr.* **59**, 1093–1094.

Herbert, V., Fong, W., Gulle, V., and Stopler, T. (1990). Low holotranscobalamin II is the earliest serum marker for subnormal vitamin $B_{12}$ (cobalamin) absorption in patients with AIDS. *Am. J. Hematol.* **34**, 132–139.

Higginbottom, M. C., Sweetman, L., and Nyhan, W. L. (1978). A syndrome of methylmalonic aciduria, homocysteinuria, megaloblastic anemia, and neurological abnormalities in a vitamin $B_{12}$-deficient breast-fed infant of a strict vegetarian. *N. Engl. J. Med.* **299**, 317–323.

Hines, J. D. (1966). Megaloblastic anemia in an adult vegan. *Am. J. Clin. Nutr.* **19**, 260–268.

Lee, E. Y., Seetharam, B., Alpers, D. H., and DeSchryver-Kecskemeti, K. (1989). Immunohistochemical survey of cobalamin-binding proteins. *Gastroenterology* **97**, 1171–1180.

Poston, J. M. (1984). The relative carbon flux through the α- and β-keto pathways of leucine metabolism. *J. Biol. Chem.* **259**, 2059–2061.

Pounder, R. E. (1996). *Helicobacter pylori* and gastroduodenal secretory function. *Gastroenterology* **110**, 947–950.

Rasmussen, K. (1989). Studies on methylmalonic acid in humans. *Clin. Chem.* **35**, 2271–2280.

Stabler, S., Lindenbaum, J., and Allen, R. H. (1997). Vitamin B-12 deficiency in the elderly: Current dilemmas. *Am. J. Clin. Nutr.* **66**, 741–749.

Scully, R. E., Mark, E., McNeely, W. F., Ebeling, S. H., and Pillips, L. (1997). Presentation of a case. *N. Engl. J. Med.* **336**, 861–867.

Stokstad, E. L. R. Webb, R. E., and Shah, E. (1966). Effect of vitamin $B_{12}$ and folic acid on the metabolism of formiminoglutamate, formate, and propionate in the rat. *J. Nutr.* **88**, 225–232.

Varro, A., Dockray, G. J., Bate, G., Vaillant, C., Higham, A., Armitage, E., and Thompson, D. G. (1997). Gastrin biosynthesis in the antrum of patients with pernicious anemia. *Gastroenterology* **112**, 733–741.

Vu, T., Amin, J., Ramos, M., Flener, V., Vanyo, L., and Tisman, G. (1993). New assay for the rapid determination of plasma holotranscobalamin II levels. *Am. J. Hematol.* **42**, 202–211.

Zimran, A., and Hershko C. (1983). The changing pattern of megaloblastic anemia in Israel. *Am. J. Clin. Nutr.* **37**, 855–861.

## Vitamin K

Almquist, H. J., and Stokstad, E. L. R. (1935). Hemorrhagic chick disease of dietary origin. *J. Biol. Chem.* **111**, 105–113.

Bale, M. D., and Mosher, D. F. (1986). Effects of thrombospondin on fibrin polymerization and structure. *J. Biol. Chem.* **261**, 862–868.

Banner, D. W., D'Arcy, A., Chene, C., Winkler, F., Guha, A., Konigsberg, W., Nemerson, Y., and Kirchhofer, D. (1996). The crystal structure of the complex of blood coagulation factor VIIa with soluble tissue factor. *Nature* **380**, 41–46.

Barry, O. P., Pratico, D., Lawson, J. A., and Fitzgerald, G. A. (1997). Transcellular activation of platelets and endothelial cells by bioactive lipids in platelet microparticles. *J. Clin. Invest.* **99**, 2118–2127.

Canfield, L. M., Hopkinson, J., Lima, A. F., Silva, B., and Garza, C. (1991). Vitamin K in colostrum and mature human milk over the lactation period — a cross-sectional study. *Am. J. Clin. Nutr.* **53**, 730–735.

Clouse, L. H., and Comp, P. C. (1986). The regulation of hemostasis: The protein C system. *N. Engl. J. Med.* **314**, 1298–1304.

Cote, H., Bajzar, L., Stevens, W. K., Samis, J. A., Morser, J., MacGillivray, R. T., and Nesheim, M. E. (1997). Functional characterization of recombinant human meizothrombin and meizothrombin (desF1). *J. Biol. Chem.* **272**, 6194–6200.

Cui, J., O'Shea, S., Purkayastha, A., Saunders, T. L., and Ginsburg, D. (1996). Fatal haemorrhage and incomplete block to embryogenesis in mice lacking coagulation factor V. *Nature* **384**, 66–68.

Dahlback, B. (1985). Ultrastructure of human coagulation factor V. *J. Biol. Chem.* **260**, 1347–1349.

Dam, H., and Schonheyder, F. (1934). A deficiency disease in chicks resembling scurvy. *Biochem. J.* **28**, 1355–1360.

Desbois, C., Hogue, D., and Karsenty, G. (1994). The mouse osteocalcin gene cluster contains three genes with two separate spatial and temporal patterns of expression. *J. Biol. Chem.* **269**, 1183–1190.

Doellgast, G. J., Triscott, M. X., Buss, D. H., and West, J. (1988). Extrinsic pathway enzyme-linked coagulation assay (EP-ELCA). *Clin. Chem.* **34**, 294–299.

Esmon, C. T. (1989). The roles of protein C and thrombomodulin in the regulation of blood coagulation. *J. Biol. Chem.* **264**, 4743–4746.

Filer, L. J. (1971). Vitamin K supplementation for infants receiving milk substitute infant formulas and for those with fat malabsorption. *Pediatrics* **48**, 483–487.

Fleming, A., and Copp, A. J. (1998). Embryonic folate metabolism and mouse neural tube defects. *Science* **280**, 2107–2109.

Glader, B. E., and Buchanan, G. R. (1976). The bleeding neonate. *Pediatrics* **58**, 548–555.

Gilbert, G. E., and Arena, A. A. (1995). Phosphatidylethanolamine induces high affinity binding sites for factor VIII on membranes containing phosphatidyl-L-serine. *J. Biol. Chem.* **270**, 18500–18505.

Gilbert, G. E., Sims, P. H., Wiedmer, T., Furie, B., Furie, B. C., and Shattil, S. H. (1991). Platelet-derived microparticles express high affinity receptors for factor VIII. *J. Biol. Chem.* **266**, 17261–17268.

Grinnell, B. W. (1997). Tipping the balance of blood coagulation. *Nature Biotechnol.* **15**, 124–125.

Hildebrandt, E. F., and Suttie, J. W. (1982). Mechanisms of coumarin action: Sensitivity of vitamin K metabolizing enzymes of normal and warfarin-resistant rat liver. *Biochemistry* **21**, 2406–2411.

Isbell, D. T., Du, S., Schroering, A., Colombo, G., and Shelling, J. G. (1993). Metal ion binding to dog osteocalcin studied by $^1$H NMR spectroscopy. *Biochemistry* **32**, 11352–11362.

Krishnaswamy, S. (1992). The interaction of human factor VIIa with tissue factor. *J. Biol. Chem.* **267**, 23696–23706.

Kuliopulos, A., Hubbard, B. R., Lam, Z., Koski, I., Furie, B., Furie, B. C., and Walsh, C. T. (1992). Dioxygen transfer during vitamin K dependent carboxylase catalysis. *Biochemistry* **31**, 7722–7728.

Lane, P. A., Hathaway, W. E., Githens, J. H., Krugman, R. D., and Rosenberg, D. A. (1983). Fatal intracranial hemorrhage in a normal infant secondary to vitamin K deficiency. *Pediatrics* **72**, 562–564.

Lollar, P., and Parker, C. G. (1990). pH-dependent denaturation of thrombin-activated porcine factor VII. *J. Biol. Chem.* **265**, 1688–1692.

Mazzucato, M., Marco, L. D., Masotti, A., Pradella, P., Bahou, W., and Ruggeri, Z. (1998). Characterization of the initial α–thrombin interaction with glycoprotein Ibα in relation to platelet activation. *J. Biol. Chem.* **273**, 1880–1887.

Molino, M., Bainton, D., Hoxie, J., Coughlin, S., and Brass, L. F. (1997). Thrombin receptors on human platelets. *J. Biol. Chem.* **272**, 6011–6017.

Naski, M. C., and Shafer, J. A. (1990). α-Thrombin-catalyzed hydrolysis of fibrin I. *J. Biol. Chem.* **265**, 1401–1407.

Nesheim, M., Pittman, D. D., Giles, A., Fass, D. N., Wang, J. H., Slonosky, D., and Kaufman, R. J. (1991). The effect of plasma von Willebrand factor on the binding of human factor VIII to thrombin-activated human platelets. *J. Biol. Chem.* **266**, 17815–17820.

Olson, J. A. (1987). Recommended dietary intakes (RDI) of vitamin K in humans. *Am. J. Clin. Nutr.* **45**, 687–692.

Overman, R. S., Stahmann, M. A., Huebner, C. F., Sullivan, W. R., Spero, L., Doherty, D. G., Ikawa, M., Graf, L., Roseman, S., and Link, K. P. (1944). Studies on hemorrhagic sweet clover disease. *J. Biol. Chem.* **153**, 5–24.

Preusch, P. C., and Suttie, J. W. (1983). Stereospecificity of vitamin K-epoxide reductase. *J. Biol. Chem.* **258**, 714–716.

Regan, L. M., Mollica, J. S., Rezaie, A. R., and Esmon, C. T. (1997). The interaction between the endothelial cell protein C receptor and protein C is dictated by the γ-carboxyglutamic acid domain of protein C. *J. Biol. Chem.* **272**, 26279–26284.

Rosing, J., Zwaal, R. F., and Tans, G. (1986). Formation of meizothrombin as intermediates in factor Xa-catalyzed prothrombin activation. *J. Biol. Chem.* **261**, 4224–4228.

Sadler, J. E. (1991). Von Willebrand factor. *J. Biol. Chem.* **206**, 22777–22780.

Schecter, A., Rollins, B., Zhang, Y., Charo, I. F., Fallon, J. T., Rossikhina, M., Giesen, P., Nemerson, Y., and Taubman, M. B. (1997). Tissue factor is induced by monocyte chemoattractant protein-1 in human aorta smooth muscle and THP-1 cells. *J. Biol. Chem.* **272**, 28568–28573.

Schneider, D. L., Fluckiger, H. B., and Manes, J. D. (1974). Vitamin K1 content of infant formula products. *Pediatrics* **53**, 273–275.

Sekiya, F., Yoshida, M., Yamashita, T., and Morita, T. (1996). Magnesium (II) is a crucial constituent of the blood coagulation cascade. *J. Biol. Chem.* **271**, 8541–8544.

Shapiro, M., Trejo, J., Zeng, D., and Coughlin, S. R. (1996). Role of the thrombin receptor's cytoplasmic tail in intracellular trafficking. *J. Biol. Chem.* **271**, 32874–32880.

Shearer, M., Bach, A., and Kohlmeier, M. (1996). Chemistry, nutritional sources, tissue distribution and metabolism of vitamin K with special reference to bone health. *J. Nutr.* **126**, 1181S–1186S.

Silverburg, S. A., and Nemerson, Y. (1975). The control of prothrombin conversion. *Biochemistry* **14**, 2636–2644.

Sims, P. J., Wiedmer, T., Esmon, C. T., Weiss, H. J., and Shattil, S. J. (1989). Assembly of the platelet prothrombinase complex is linked to vesiculation of the platelet membrane. *J. Biol. Chem.* **264**, 17049–17057.

Sokoll, L., Booth, S., O'Brien, M., Davidson, K. W., Tsaioun, K. I., and Sadowski, J. A. (1997). Changes in serum osteocalcin, plasma phylloquinone, and urinary γ-carboxyglutamic acid in response to altered intakes of dietary phylloquinone in human subjects. *Am. J. Clin. Nutr.* **65**, 779–784.

Suttie, J. W. (1995). The importance of menaquinones in human nutrition. *Annu. Rev. Nutr.* **15**, 399–417.

Suttie, J. W., Mummah-Schendel, L. L., Shah, D. V., Lyle, B. J., and Gregor, J. L. (1988). Vitamin K deficiency from dietary vitamin K restriction in humans. *Am. J. Clin. Nutr.* **47**, 475–480.

Svensson, P., and Dahlback, B. (1994). Resistance to activated protein C as a basis for venous thrombosis. *N. Engl. J. Med.* **330**, 517–522.

Tijburg, P., Heerde, W., Leenhouts, H., Hessing, M., Bouma, B., and de Groot, P. (1991). Formation of meizothrombin as intermediate in factor Xa-catalyzed prothrombin activation on endothelial cells. *J. Biol. Chem.* **266**, 4017–4022.

Varnum, B. C., Young, C., Elliott, G., Garcia, A., Bartley, T., Fridell, Y.-W., Hunt, R., Trall, G., Clogston, C., Toso, R. J., Yanagihara, D., Bennett, L., Sylber, M., Merewether, L., Tseng, A., Escobar, E., Liu, E., and Yamane, H. K. (1995). Axl receptor tyrosine kinase stimulated by the vitamin K-dependent protein encoded by growth-arrest-specific gene 6. *Nature* **373**, 623–626.

## Biotin

Balnave, D. (1977). Clinical symptoms of biotin deficiency in animals. *Am. J. Clin. Nutr.* **30**, 1408–1413.

Baugh, C. M., Malone, J. H., and Butterworth, C. E. (1968). Human biotin deficiency. *Am. J. Clin. Nutr.* **21**, 173–182.

Berg, H. (1997). Bioavailability of biotin. *Eur. J. Clin. Nutr., Suppl.* **51**(1), S60–S61.

Bowman, B. B., and Rosenberg, I. H. (1987). Biotin absorption by distal rat intestine. *J. Nutr.* **117**, 2121–2126.

Carey, C. J., and Morris, J. G. (1977). Biotin deficiency in the cat and the effect on hepatic propionyl CoA carboxylase. *J. Nutr.* **107**, 330–334.

Dakshinamurti, K., and Chauhan, J. (1988). Regulation of biotin enzymes. *Annu. Rev. Nutr.* **8**, 211–233.

Dakshinamurti, K., and Chauhan, J. (1989). Biotin. *Vit. Horm.* **45**, 337–384.

Hyman, J., Fleischhauer, K., and Wolf, B. (1997). Biotinidase in serum and tissues. *Methods Enzymol.* **279**, Part I, 422–434.

Kramer, T. R., Briske-Anderson, M., Johnson, S. B., and Holman, R. T. (1988). Effects of biotin deficiency on serum fatty acid composition: Evidence for abnormalities in humans. *J. Nutr.* **118**, 342–348.

Mock, D. M. (1997). Determinations of biotin in biological fluids. *Methods Enzymol.* **279**, Part I, 265–275.

Mock, D. M., Mock, N. I., and Stratton, S. L. (1997a). Concentrations of biotin metabolites in human milk. *J. Pediatr.* **131**, 456–458.

Mock, D. M., Stadler, D. D., Stratton, S. L., and Mock, N. I. (1997b). Biotin status assessed longitudinally in pregnant women. *J. Nutr.* **127**, 710–716.

Mock, D. M., Wang, K.-S., and Kearns, G. L. (1997c). The pig is an appropriate model for human catabolism as judged by the urinary metabolite profile of radioisotope-labeled biotin. *J. Nutr.* **127**, 365–369.

Oizumi, J., and Hayakawa, K. (1988). Biotinidase in human breast milk. *Am. J. Clin. Nutr.* **48**, 295–297.

Pomponio, R. J., Reynolds, T. R., Mandel, H., Admoni, O., Melone, P., Buck, G. A., and Wolf, B. (1997). Profound biotinidase deficiency caused by point mutation that creates a downstream cryptic 3′ splice acceptor site within an exon of the human biotinidase gene. *Human Mol. Genet.* **6**, 739–745.

Said, H. M., and Redah, R. (1988). Ontogenesis of the intestinal transport of biotin in the rat. *Gastroenterology* **94**, 68–72.

Stubbe, J., and Abeles, R. H. (1977). Biotin carboxylations — Concerted or not concerted? That is the question! *J. Biol. Chem.* **252**, 8338–8340.

Sweetman, L., and Nyhan, W. L. (1986). Inheritable biotin-treatable disorders and associated phenomena. *Annu. Rev. Nutr.* **6**, 317–343.

Velazquez, A., Teran, M., Baez, A., Gutierrez, J., and Rodriguez, R. (1995). Biotin supplementation affects lymphocyte carboxylases and plasma biotin in severe protein-energy malnutrition. *Am. J. Clin. Nutr.* **61**, 385–391.

Wang, K.-S., Mock, N. I., and Mock, D. M. (1997). Biotin biotransformation to bisnorbiotin is accelerated by several peroxisome proliferators and steroid hormones in rats. *J. Nutr.* **127**, 2212–2216.

Wolf, B., Heard, G. S., Jefferson, L. G., Proud, V. K., Nance, W. E., and Weissbecker, K. A. (1985). Clinical findings in four children with biotinidase deficiency through a state-wide neonatal screening programs. *N. Engl. J. Med.* **313**, 16–19.

## Vitamin B$_6$

Bankier, A., Turner, M., and Hopkins, I. J. (1983). Pyridoxine-dependent seizures — A wider clinical spectrum. *Arch. Dis. Child.* **58**, 415–418.

Brown, R. R., Yess, N., Price, J. M., Linkswiler, H., Swan, P., and Henkes, L. V. (1965). Vitamin B$_6$ depletion in man: Urinary excretion of quinolinic acid and niacin metabolites. *J. Nutr.* **87**, 419–423.

Choi, S.-Y., Churchich, J. E., Zaiden, E., and Kwok, F. (1987). Brain pyridoxine-5-phosphate oxidase. *J. Biol. Chem.* **262**, 12013–12017.

Coburn, S. P., Ziegler, P., Costill, D. L., Mahuren, J., Fink, W. J., Schaltenbrand, W., Pauly, T. A., Pearson, D. R., Conn, P. S., and Guilarte, T. R. (1991). Response of vitamin B-6 content of muscle to changes in vitamin B-6 intake in men. *Am. J. Clin. Nutr.* **53**, 1436–1442.

Contractor, S. F., and Shane, B. (1970). Blood and urine levels of vitamin B$_6$ in the mother and fetus before and after loading of the mother with vitamin B$_6$. *Am. J. Obstet. Gynecol.* **107**, 635–640.

Dakshinamurti, K. (1990). Vitamin B$_6$. *Ann. N. Y. Acad. Sci.* **585**, 1–567.

Ferroli, C. E., and Trumbo, P. R. (1994). Bioavailability of vitamin B-6 in young and older men. *Am. J. Clin. Nutr.* **60**, 68–71.

Hanna, M. C., Turner, A. J., and Kirkness, E. F. (1997). Human pyridoxal kinase. *J. Biol. Chem.* **272**, 10756–10760.

Hansen, C. M., Leklem, J. E., and Miller, L. T. (1996). Vitamin B-6 status indicators decrease in women consuming a diet high in pyridoxine glucoside. *J. Nutr.* **126**, 2512–2518.

Hansen, C. M., Leklem, J. E., and Miller, L. T. (1996). Vitamin B-6 status of women with a constant intake of vitamin B-6 changes with three levels of dietary protein. *J. Nutr.* **126**, 1891–1901.

Hansen, C. M., Leklem, J. E., and Miller, L. T. (1997). Changes in vitamin B-6 status indicators of women fed a constant protein diet with varying levels of vitamin B-6. *Am. J. Clin. Nutr.* **66**, 1379–1387.

Hunter, J. E., and Harper, A. E. (1976). Stability of some pyridoxal phosphate-dependent enzymes in vitamin B-6 deficient rats. *J. Nutr.* **106**, 653–664.

Ink, S. L., and Henderson, L. M. (1984). Vitamin $B_6$ metabolism. *Annu. Rev. Nutr.* **4**, 455–470.

Kang-Yoon, S. A., Kirksey, A., Giacoia, G. P., and West, K. D. (1995). Vitamin B-6 adequacy in neonatal nutrition: Associations with preterm delivery, type of feeding, and vitamin B-6 supplementation. *Am. J. Clin. Nutr.* **62**, 932–942.

Linkswiler, H. (1967). Biochemical and physiological changes in vitamin $B_6$ deficiency. *Am. J. Clin. Nutr.* **20**, 547–557.

Manore, M. M., Vaughan, L. A., Carroll, S. S., and Leklem, J. E. (1989). Plasma pyridoxal 5'-phosphate concentration and dietary vitamin B-6 intake in free-living, low-income elderly people. *Am. J. Clin. Nutr.* **50**, 339–345.

Mascher, H. J. (1997). High-performance liquid chromatography determination of total pyridoxal in human plasma. *Methods Enzymol.* **280**, Part J, 12–21.

McMahon, L., Nakano, H., Levy, M.-D., and Gregory, J. F. (1997). Cytosolic pyridoxine-β-D-glucoside hydrolase from porcine jejunal mucosa. *J. Biol. Chem.* **272**, 32025–32033.

Middleton, H. M. (1986). Intestinal hydrolysis of pyridoxal 5'-phosphate *in vitro* and *in vivo* in the rat. *Gastroenterology* **91**, 343–350.

Raiten, D. J., Reynolds, R. D., Andon, M. B., Robbins, S. T., and Fletcher, A. B. (1991). Vitamin B-6 metabolism in premature infants. *Am. J. Clin. Nutr.* **53**, 78–83.

Reynolds, R. D. (1988). Bioavailability of vitamin B-6 from plant foods. *Am. J. Clin. Nutr.* **48**, 863–867.

Shane, B., and Contractor, S. F. (1975). Assessment of vitamin $B_6$ status. Studies on pregnant women and oral contraceptive users. *Am. J. Clin. Nutr.* **28**, 739–747.

Snell, E. E. (1970). Analogues of pyridoxal or pyridoxal phosphate: Relation of structure to binding with apoenzymes and to catalytic activity. *Vit. Horm.* **28**, 265–290.

Snell, E. E., and Di Mari, S. J. (1970). Schiff base intermediates in enzyme catalysis. *In* "The Enzymes," Vol. 2, 3rd ed., pp. 335–370. Academic Press, New York.

Tagaya, M., Horinishi, M., and Fukui, T. (1987). Flexibility in the phosphorylase catalytic reaction. *J. Biol. Chem.* **262**, 3092–3097.

Tsuge, H. (1997). Determination of vitamin $B_6$ vitamers and metabolites in a biological sample. *Methods Enzymol.* **280**, Part J, 3–12.

Williams, M. A., McIntosh, D. J., and Hincenbergs, I. (1966). Changes in fatty acid composition in liver lipid fractions of pyridoxine-deficient rats fed cholesterol. *J. Nutr.* **88**, 193–201.

## *Vitamin B6, Folate, and Homocysteine*

Boushey, C., Beresford, S., Omenn, G., and Motulsky, A. G. (1995). A quantitative assessment of plasma homocysteine as a risk factor for vascular disease. *JAMA* **274**, 1049–1057.

Brattsrom, L., Israelsson, B., Lindgarde, F., and Hultberg, B. (1988). Higher total plasma homocyteine in vitamin $B_{12}$ deficiency than in heterozygosity for homocyteinuria due to cystathione β-synthase deficiency. *Metabolism* **37**, 175–178.

Clarke, R., Daly, L., Robinson, K., Naughten, E., Cahalane, S., Fowler, B., and Graham, I. (1991). Hyperhomocysteinemia: An independent risk factor for vascular disease. *N. Engl. J. Med.* **324**, 1149–1155.

DeMaster, E. G., Quast, B. J., Redfern, B., and Nagasawa, H. T. (1995). Reaction of nitric oxide with the free sulfhydryl group of human serum albumin yields a sulfenic acid and nitrous oxide. *Biochemistry* **34**, 11494–11499.

Engbersen, A., Franken, D., Boers, G., Stevens, E., Trijbels, F., and Blom, H. (1995). Thermolabile 5,10-methylenetetrahydrofoalte reductase as a cause of mild hyperhomocysteinemia. *Am. J. Hum. Genet.* **56**, 142–150.

Finkelstein, J. D., and Mudd, S. H. (1964). Homocysteinuria due to cystathionine synthetase deficiency: Mode of inheritance. *Science* **146**, 785–787.

Fukagawa, N., Ajami, A., and Young, V. R. (1996). Plasma methionine and cysteine kinetics in response to an intravenous glutathione infusion in adult humans. *Am. J. Physiol.* **270**, E209–E214.

Halliwell, B., and Gutteridge, J. M. (1990). The antioxidants of human extracellular fluids. *Arch. Biochem. Biophys.* **280**, 1–8.

Heijer, M., Koster, T., Blom, H. J., Bos, G. M., Briet, E., Reitsma, P., Vandenbroucke, J., Rosendaal, F. R. (1996). Hyperhomocysteinemia as a risk factor for deep-vein thrombosis. *N. Engl. J. Med.* **334**, 759–762.

Hiramatsu, T., Fukagawa, N. K., Marchini, J., Cortiella, J., Yu, Y.-M., Chapman, T., and Young, V. R. (1994). Methionine and cysteine kinetics at different intakes of cystine in healthy adult men. *Am. J. Clin. Nutr.* **60**, 525–533.

Jakubowski, H. (1997). Metabolism of homocyteine thiolactone in human cell cultures. *J. Biol. Chem.* **272**, 1935–1942.

Kang, S.-S., Wong, P., Cook, H., Norusis, M., and Messer, J. V. (1986). Protein-bound homocyst(e)ine. *J. Clin. Invest.* **77**, 1482–1486.

Kluijtmans, L., Heuvel, L., Boers, G., Frosst, P., Stevens, E., Oost, B., Heijer, M., Trijbels, F., Rozen, R., and Blom, G. (1996). Molecular genetic analysis in mild hyperhomocysteinemia: A common mutation in the methylenetetrahydrofolate reductase gene is a genetic risk factor for cardiovascular disease. *Am. J. Hum. Genet.* **58**, 35–41.

Moghadasian, M., McManus, B. M., and Frohlich, J. J. (1997). Homocyst(e)ine and coronary artery disease. *Arch. Intern. Med.* **157**, 2299–2308.

Molloy, A. M., Daly, S., Mills, J. L., Kirke, P., Whitehead, A., Ramsbottom, D., Conley, M., Weir, D. G., and Scott, J. M. (1997). Thermolabile variant of 5,10-methylenetetrahydrofolate reductase associated with low red-cell folates: Implications for folate intake recommendations. *Lancet* **349**, 1591–1593.

Mudd, S. H., Skovby, F., Levy, H. L., Pettigrew, K., Wilcken, B., Pyeritz, R., Andria, G., Boers, G., Bromberg, I., Cerone, R., Fowler, B., Grobe, H., Schmidt, H., and Schweitzer, L. (1985). The natural history of homocystinuria due to cystathionine β-synthase deficiency. *Am. J. Hum. Genet.* **37**, 1–31.

Naurath, H. J., Joosten, E., Riezler, R., Stabler, S. P., Allen, R. H., and Lindenbaum, J. (1995). Effects of vitamin $B_{12}$, folate, and vitamin $B_6$ supplements in elderly people with normal serum vitamin concentrations. *Lancet* **346**, 85–89.

Nygard, O., Nordrehaug, J., Refsum, H., Ueland, P. M., Farstad, M., and Vollset, S. (1997). Plasma homocysteine levels and mortality in patients with coronary artery disease. *N. Engl. J. Med.* **337**, 230–236.

Pancharuniti, N., Lewis, C. A., Sauberlich, H. E., Perkins, L., Go, R. C., Alvarez, J. O., Macaluso, M., Acton, R. T., Copeland, R. B., Cousins, A. L., Gore, T. B., Cornwell, P., and Roseman, J. M. (1994). Plasma homocyst(e)ine, folate, and vitamin B-12 concentrations and risk for early-onset coronary artery disease. *Am. J. Clin. Nutr.* **59**, 940–948.

Rimm, E. B., Willett, W. C., Hu, F. B., Sampson, L., Colditz, G. A., Manson, J. E., Hennekens, C., and Stampfer, M. J. (1998). Folate and vitamin $B_6$ from diet and supplements in relation to risk of coronary heart disease among women. *JAMA* **279**, 359–364.

Robinson, K., Mayer, E. L., Miller, D., Green, R., Lente, F., Gupta, A., Kottke-Marchant, K., Savon, S., Selhub, J., Nissen, S., Kutner, M., Topol, E., and Jacobsen, D. W. (1995). Hyperhomocysteinemia and low pyridoxal phosphate. *Circulation* **92**, 2825–2830.

Rosenberg, L. E. (1991). Inherited disorders of amino acid metabolism. *In* "Principles of Internal Medicine" (J. Wilson, E. Braunwald, K. J. Isselbacher, R. G. Petersdorf, J. B. Martin, A. Fauci, and R. K. Root, eds.), pp. 1868–1875. McGraw-Hill, New York.

Selhub, J., and Miller, J. W. (1992). The pathogenesis of homocyteinemia: Interruption of the coordinate regulation by *S*-adenosylmethionine of the remethylaton and transsulfuration of homocysteine. *Am. J. Clin. Nutr.* **55**, 131–138.

Selhub, J., Jacques, P. F., Bostom, A. G., D'Agostino, R. B., Wilson, P., Belanger, A. J., O'Leary, D., Wolf, P., Schaefer, E. J., and Rosenberg, I. H. (1995). Association between plasma homocysteine concentrations and extracranial carotid-artery stenois. *N. Engl. J. Med.* **332**, 286–291.

Smolin, L. A., and Benevenga, N. J. (1984). The use of cyst(e)ine in the removal of protein-bound homocysteine. *Am. J. Clin. Nutr.* **39**, 730–737.

Stampfer, M., and Malinow, M. R. (1995). Can lowering homocyteine levels reduce cardiovascular risk? *N. Engl. J. Med.* **332**, 328–329.

Sumner, A. E., Chin, M. M., Abrahm, J. L., Berry, G. T., Gracely, E. J., Allen, R. H., and Stabler, S. P. (1996). Elevated methylmalonic acid and total homocysteine levels show high prevalence of vitamin $B_{12}$ deficiency after gastric surgery. *Ann. Intern. Med.* **124**, 469–476.

Ubbink, J. B., Vermaak, W., Merwe, A., Becker, P., Delport, R., and Potgieter, H. (1994). Vitamin requirements for the treatment of hyperhomocysteinemia in humans. *J. Nutr.* **124**, 1927–1933.

*Vitamin A*

Batres, R. O., and Olson, J. A. (1987). A marginal vitamin A status alters the distribution of vitamin A among parenchymal and stellate cells in rat liver. *J. Nutr.* **117**, 874–879.

Bendich, A., and Olson, J. A. (1989). Biological actions of carotenoids. *FASEB J.* **3**, 1927–1932.

Blomhoff, R., Green, M. H., Berg, T., and Norum, K. R. (1990). Transport and storage of vitamin A. *Science* **250**, 399–404.

Buczylko, J., Saari, J. C., Crouch, R. K., and Palczewski, K. (1996). Mechanisms of opsin activation. *J. Biol. Chem.* **271**, 20621–20630.

Cullum, M. E., and Zile, M. H. (1985). Metabolism of all-*trans*-retinoic acid and all-*trans*-retinyl acetate. *J. Biol. Chem.* **260**, 10590–10596.

Demming-Adams, B., Gilmore, A. M., and Adams, W. W. (1996). In vivo functions of carotenoids in higher plants. *FASEB J.* **10**, 403–412.

Duszka, C., Grolier, P., Azim, E., Alexandre-Gouabau, K., Borel, P., and Azais-Braesco, V. (1996). Rat intestinal β-carotene dioxygenase activity is located primarily in the cytosol of mature jejunal enterocytes. *J. Nutr.* **126**, 2550–2556.

Farahbakhsh, Z., Ridge, K. D., Khorana, H. G., and Hubbell, W. L. (1995). Mapping light-dependent structural changes in the cytoplasmic loop connecting helices C and D in rhodopsin: A site-directed spin labeling study. *Biochemistry* **34**, 8812–8819.

Filteau, S. M., and Tomkins, A. M. (1995). Vitamin A supplementation in developing countries. *Arch. Dis. Child.* **72**, 106–109.

Geneser, F. (1986). "Textbook of Histology," pp. 696–707. Lea and Febiger, Philadelphia.

Green, S. (1993). Promiscuous liasons. *Nature* **361**, 590–591.

Krinsky, N. 1. (1988). The evidence for the role of carotenes in preventive health. *Clin. Nutr.* **7**, 107–112.

Li, E., and Norris, A. W. (1996). Structure/function of cytoplasmic vitamin A-binding proteins. *Annu. Rev. Nutr.* **16**, 205–234.

Marchand, L. L., Yoshizawa, C. N., Kolonel, L. N., Hankin, J. H., and Goodman, M. T. (1989). Vegetable consumption and lung cancer risk: A population-based case-control study in Hawaii. *J. Natl. Cancer Inst.* **81**, 1158–1164.

Micozzi, M. S., Beecher, G. R., Taylor, P. R., and Khachick, F. (1990). Carotenoid analysis of selected raw and cooked foods associated with a lower risk for cancer. *J. Natl. Cancer Inst.* **82**, 282–285.

Nagao, A., During, A., Hoshino, C., Terao, J., and Olson, J. A. (1996). Stoichometric conversion of all-*trans*-β-carotene to retinal by pig intestinal extract. *Arch. Biochem. Biophys.* **328**, 57–63.

Napoli, J. L. (1996). Retinoic acid biosynthesis and metabolism. *FASEB J.* **10**, 993–1001.

Nathans, J., Merbs, S. L., Sung, C.-H., Weitz, C., and Wang, Y. (1992). Molecular genetics of human visual pigments. *Annu. Rev. Genet.* **26**, 403–424.

O'Toole, B. A., Fradkin, R., Warkany, J., Wilson, J. G., and Mann, G. V. (1974). Vitamin A deficiency and reproduction in rhesus monkeys. *J. Nutr.* **104**, 1513–1524.

Ramadori, G., Rieder, H., Theiss, F., and Bushenfelde, K. (1989). Fat-storing (Ito) cells of rat liver synthesize and secrete apolipoproteins: Comparison with hepatocytes. *Gastroenterology* **97**, 163–172.

Rao-Mirotznik, R., Harkins, A. B., Buchsbaum, G., and Sterling, P. (1995). Mammalian rod terminal: Architecture of a binary synapse. *Neuron* **14**, 561–569.

Ridge, K. D., Lee, S. S., and Abdulaev, N. (1996). Examining rhodopsin folding and assembly through expresson of polypeptide fragments. *J. Biol. Chem.* **271**, 7860–7867.

Simon, A., Hellman, U., Wernstedt, C., and Eriksson, U. (1995). The retinal pigment epithelial-specific 11-*cis* retinol dehydrogenase belongs to the family of short chain alcohol dehydrogenases. *J. Biol. Chem.* **270**, 1107–1112.

Solon, F. S., Latham, M. C., Guirriec, R., Florentino, R., Williamson, D. F., and Aguilar, J. (1985). Fortification of MSG with vitamin A: The Philippine experience. *Food Technol.* (Nov.), pp. 71–77.

Stump, D. G., Lloyd, R. S., and Chytil, F. (1991). Site-directed mutagenesis of rat cellular retinol-binding protein. *J. Biol. Chem.* **266**, 4622–4630.

Surya, A., Foster, K. W., and Knox, B. E. (1995). Transducin activation by the bovine opsin apoprotein. *J. Biol. Chem.* **270**, 5024–5031.

Tsang, S. H., Gouras, P., Yamashita, C. K., Kjeldbye, H., Fisher, J., Farber, D. B., and Goff, S. P. (1996). Retinal degeneration in mice lacking the γ subunit of the rod cGMP phosphodiesterase. *Science* **272**, 1026–1029.

Vliet, T., Schreurs, W., and Berg, H. (1995). Instestinal β-carotene absorption and cleavage in men: Response of β-carotene and retinyl esters in the triglyceride-rich lipoprotein fraction after a single oral dose of β-carotene. *Am. J. Clin. Nutr.* **62**, 110–116.

West, K. P., Howard, G. R., and Sommer, A. (1989). Vitamin A and infection: Public health implications. *Annu. Rev. Nutr.* **9**, 63–86.

Wolf, G., Lane, M. D., and Johnson, B. C. (1957). Studies on the function of vitamin A metabolism. *J. Biol. Chem.* **225**, 995–1008.

Wolf, G. (1995). The enzymatic cleavage of β-carotene: Still controversial. *Nutr. Revs.* **53**, 134–138.

*Vitamin D*

Bell, N. H. (1985). Vitamin D-endocrine system. *J. Clin. Invest.* **76**, 1–6.

Brommage, R., and DeLuca, H. F. (1984). A maternal defect is responsible for growth failure in vitamin D-deficient rat pups. *Am. J. Physiol.* **246**, E216–E220.

Binet, A., and Kooh, S. (1996). Persistence of vitamin D-deficiency rickets in Toronto in the 1990s. *Can. J. Public Health* July-Aug., pp. 227–230.

Chapuy, M., Arlot, M., Duboeuf, F., Brun, J., Crouzet, B., Arnaud, S., Delmas, P., and Meunier, P. (1992). Vitamin $D_3$ and calcium to prevent hip fractures in elderly women. *N. Engl. J. Med.* **327**, 1637–1642.

Darwish, H. M., Krisinger, J., Strom, M., and DeLuca, H. F. (1987). Molecular cloning of the cDNA and chromosomal gene for vitamin D-dependent calcium-binding protein of rat intestine. *Proc. Natl. Acad. Sci. U.S.A.* **84**, 6108–6111.

Dawson-Hughes, B., Harris, S. S., Krall, E. A., and Dallal, G. (1997). Effect of calcium and vitamin D supplementation on bone density in men and women 65 years of age or older. *N. Engl. J. Med.* **337**, 670–676.

DeLuca, H. (1988). The vitamin D story: A collaborative effort of basic science and clinical medicine. *FASEB J.* **2**, 224–236.

Ducy, P., Desbois, C., Boyce, B., Pinero, G., Story, B., Dunstan, C., Smith, E., Bonadio, J., Goldstein, S., Gundberg, C., Bradley, A., and Karsenty, G. (1996). Increased bone formation in osteocalcin-deficient mice. *Nature* **382**, 448–452.

Elmer, L. K. (1984). Switch dry cows to grass to prevent milk fever. *Hoard's Dairyman* **129**, 415–416.

El-Sonbaty, M., and Abdul-Ghaffer, N. (1996). Vitamin D deficiency in veiled Kuwaiti women. *Eur. J. Clin. Nutr.* **50**, 315–318.

Greaves, M., and Weinstein, G. (1995). Treatment of psoriasis. *N. Engl. J. Med.* **332**, 581–588.

Haddad, J. G. (1996). Seasonal diminution of vitamin D stores in the United States. *Trends Endocrinol. Metab.* **7**, 209–212.

Haussler, M. R., and McCain, T. A. (1977). Basic and clincal concepts related to vitamin D metabolism and action. *N. Engl. J. Med.* **297**, 1041–1050.

Holick, M. F. (1994). McCollum Award Lecture, 1994: Vitamin D — new horizons for the 21st century. *Am. J. Clin. Nutr.* **60**, 619–630.

Holick, M. F. (1996). Vitamin D and bone health. *J. Nutr.* **126**, 1159S–1164S.

Jacques, P. F., Felson, D., Tucker, K., Mahnken, B., Wilson, P. W., Rosenberg, I. H., and Rush, D. (1997). Plasma 25-hydroxyvitamin D and its determinants in an elderly population sample. *Am. J. Clin. Nutr.* **66**, 929–936.

Khare, S., Bolt, M., Wali, R., Skarosi, S., Roy, H., Niedziela, S., Scaglione-Sewell, B., Aquino, B., Abraham, C., Sitrin, M., Brasitus, T., and Bissonnette, M. (1997). 1,25-Dihydroxyvitamin $D_3$ stimulates phospholipase C in rat colonocytes. *J. Clin. Invest.* **99**, 1831–1841.

Kinyamu, H., Gallagher, J., Balhorn, K., Petranick, K., and Rafferty, K. (1997). Serum vitamin D metabolites and calcium absorption in normal young and elderly free-living women and in women living in nursing homes. *Am. J. Clin. Nutr.* **65**, 790–797.

Lowenthal, M., and Shany, S. (1994). Osteomalacia in Bedouin women of the Negev. *Isr. J. Med. Sci.* **30**, 520–523.

Mellay, V., Grosse, B., and Lieberherr, M. (1997). Phospholipase Cβ and membrane action of calcitriol and estradiol. *J. Biol. Chem.* **272**, 11902–11907.

Muhe, L., Lulseged, S., Mason, K. E., and Simoes, E. A. (1997). Case-control study of the role of nutritional rickets in the risk of developing pneumonia in Ethiopian children. *Lancet* **349**, 1801–1804.

Nemere, I., Yoshimoto, Y., and Norman, A. W. (1984). Calcium transport in perfused duodena from normal chicks: Enhancement within fourteen minutes of exposure to 1,25-dihydroxyvitamin $D_3$. *Endocrinology* **115**, 1476–1483.

Norman, A. W. (1980). "Vitamin D Molecular Biology and Clinical Nutrition." Dekker, New York.

Oginni, L., Worsfold, M., Oyelami, O., Sharp, C., Powell, D., and Davie, M. (1996). Etiology of rickets in Nigerian children. *J. Pediatr.* **128**, 692–694.

Okuda, K., Usui, E., and Ohyama, Y. (1995). Recent progress in enzymology and molecular biology of enzymes involved in vitamin D metabolism. *J. Lipid Res.* **36**, 1641–1652.

Reichel, H., Koeffler, H. P., and Norman, A. W. (1989). The role of the vitamin D endocrine system in health and disease. *N. Engl. J. Med.* **320**, 980–991.

Steenbock, H., and Herting, D. C. (1955). Vitamin D and growth. *J. Biol. Chem.* **57**, 449–468.

Takeyama, K., Kitanaka, S., Sato, T., Kobori, M., Yanagisawa, J., and Kato, S. (1997). 25-Hydroxyvitamin $D_3$ 1α-hydroxylase and vitamin D synthesis. *Science* **277**, 1827–1830.

Underwood, J. L., and DeLuca, H. F. (1984). Vitamin D is not directly necessary for bone growth and mineralization. *Am. J. Physiol.* **246**, E493–E498.

Wielen, R. P., Lowik, M., Berg, H., Groot, L., Haller, J., Moreiras, O., and Staveren, W. (1995). Serum vitamin D concentrations among elderly in Europe. *Lancet* **346**, 207–210.

## Cells and Protein of Bone

Casser-Bette, M., Murray, A. B., Closs, E. I., and Schmidt, J. (1990). Bone formation by osteoblast-like cells in a three dimensional cell culture. *Calcif. Tissue Int.* **46**, 46–56.

Escarot-Charrier, B., Bouchard, F., and Delloye, C. (1989). Bone sialoprotein II synthesized by cultured osteoblasts contains tyrosine sulfate. *J. Biol. Chem.* **264**, 20049–20053.

Goldberg, H. A., Domenicucci, C., Pringle, G. A., and Sodek, J. (1988). Mineral-binding proteoglycans of fetal porcine calvarial bone. *J. Biol. Chem.* **263**, 12092–12101.

McSheehy, P. M., and Chambers, T. J. (1986). Osteoblastic cells mediate osteoclastic responsiveness to parathyroid hormone. *Endocrinology* **118**, 824–828.

Pacifici, M., Oshima, O., Fisher, L. W., Young, M. F., Shapiro, I. M., and Leboy, P. S. (1990). Changes in osteonectin distribution and levels are associated with mineralization of the chicken tibial growth cartilage. *Calcif. Tissue Int.* **47**, 51–61.

Ruoslahti, E. (1989). Proteoglycans in cell regulation. *J. Biol. Chem.* **264**, 13369–13372.

Silver, I. A., Murrills, R. J., and Etherington, D. J. (1988). Microelectrode studies on the acid microenvironment beneath adherent macrophages and osteoclasts. *Exp. Cell. Res.* **175**, 266–276.

## Vitamin A, Vitamin D, and Thyroid Hormone at the Genome Vitamin A and General References

Allenby, G., Janocha, R., Kazmer, S., Speck, J., Grippo, J. F., and Levin, A. A. (1994). Binding of 9-*cis*-retinoic acid and all-*trans*-retinoic acid to retinoic acid receptors α, β, and γ. *J. Biol. Chem.* **269**, 16689–16695.

Chambon, P. (1996). A decade of molecular biology of retinoic acid receptors. *FASEB J.* **10**, 940–954.

Feng, X., Peng, Z.-H., Wen, D., Li, X.-Y., Rochette-Egly, C., Chambon, P., Voorhees, J. J., and Xiao, J.-H. (1997). Suprabasal expression of a dominant-negative RXR mutant in transgenic mouse epidermis. *Genes Dev.* **11**, 59–71.

Forman, B. M., Umesono, K., Chen, J., and Evans, R. M. (1995). Unique response pathways are established by allosteric interactions among nuclear hormone receptors. *Cell* **81**, 541–550.

Geiger, J. H., Hahn, S., Lee, S., and Sigler, P. B. (1996). Crystal structure of the yeast TFIIA/TBP/DNA complex. *Science* **272**, 830–836.

Giguere, V. (1994). Retinoic acid receptors and cellular retinoid binding proteins. *Endocrine Revs.* **15**, 61–79.

Goodrich, J. A., and Tjian, R. (1994). Transcription factors IIE and IIH and ATP hydrolysis direct promoter clearance by RNA polymerase II. *Cell* **77**, 145–156.

Hanna-Rose, W., and Hansen, U. (1996). Active repression mechanisms of eukaryotic transcription repressors. *Trends Genet.* **12**, 229–234.

Ijpenberg, A., Jeannin, E., Wahli, W., and Desvergne, B. (1997). Polarity and specific sequence requirements of peroxisome proliferator activated receptor (PPAR)/retinoid X receptor heterodimer binding to DNA. *J. Biol. Chem.* **272**, 20108–20117.

Jacobson, R. H., and Tjian, R. (1996). Transcription factor IIA: A structure with multiple functions. *Science* **272**, 827–828.

Lala, D. S., Mukherjee, R., Schulman, I. G., Koch, S. S., Dardashti, L. J., Nadzan, A. M., Croston, G., Evans, R. M., and Heyman, R. A. (1996). Activation of specific RXR heterodimers by an antagonist of RXR homodimers. *Nature* **383**, 450–453.

Locker, J. (1996). "Transcription Factors." Wiley, New York.

Lotan, R. (1996). Retinoids in cancer chemoprevention. *FASEB J.* **10**, 1031–1039.

Mangelsdorf, D. J., and Evans, R. M. (1995). The RXR heterodimers and orphan receptors. *Cell* **83**, 841–850.

Mangelsdorf, D. J., Thummel, C., Beato, M., Herrlich, P., Schutz, G., Umesono, K., Blumberg, B., Kastner, P., Manuel, M., Chambon, P., and Evans, R. M. (1995). The nuclear receptor subfamily: The second decade. *Cell* **83**, 835–839.

Marshall, H., Morrison, A., Studer, M., Popperi, H., and Krumlauf, R. (1996). Retinoids and hox genes. *FASEB J.* **10**, 969–978.

Minna, J. D., and Mangelsdorf, D. J. (1997). Retinoic acid receptor expression abnormalities in lung cancer. *J. Natl. Cancer Inst.* **89**, 602–604.

Morrison, A., Moroni, M., Ariza-McNaughton, L., Krumlauf, R., and Mavilio, F. (1996). In vitro and transgenic analysis of a human HOXD4 retinoid-responsive enhancer. *Development* **122**, 1895–1907.

Morriss-Kay, G. M., and Sokolova, N. (1996). Embryonic development and pattern development and pattern formation. *FASEB J.* **10**, 961–968.

Petkovich, M. (1992). Regulation of gene expression by vitamin A. *Annu. Rev. Nutr.* **12**, 443–471.

Pugh, B. F., and Tjian, R. (1992). Diverse transcriptional functions of the multisubunit eukaryotic TFIID complex. *J. Biol. Chem.* **267**, 679–682.

Ross, A. C. (1993). Introduction. *J. Nutr.* **123**, 344–345.

Roy, B., Taneja, R., and Chambon, P. (1995). Synergistic activation of retinoic acid (RA)-responsive genes and induction of embryonal carcinoma cell differentiation by an RA receptor a (RARα)-, RARβ-, or RARγ-selective ligand in combination with a retinoid X receptor-specific ligand. *Mol. Cell. Biol.* **15**, 6481–6487.

Shafritz, D. A. (1996). Synthetic retinoids for the secondary prevention of hepatocellular carcinoma. *N. Engl. J. Med.* **334**, 1600–1601.

Taneja, R., Roy, B., Plassat, J.-L., Zusi, C. F., Ostrowski, J., Reczek, P. R., and Chambon, P. (1996). Cell-type and promoter-context dependent retinoic acid receptor (RAR) redundancies for RAR and Hoxa-1 activation in F9 and P19 cells can be artefactually generated by gene knockouts. *Proc. Natl. Acad. Sci. U.S.A.* **93**, 6197–6202.

Tate, B. F., Levin, A. A., and Grippo, J. F. (1994). The discovery of 9-*cis* retinoic acid. *Trends Endocrinol. Metab.* **5**, 189–194.

Tjian, R., and Maniatis, T. (1994). Transcriptional activation: A complex puzzle with few easy pieces. *Cell* **77**, 5–8.

Zawel, L., and Reinberg, D. (1995). Common themes in assembly and function of eukaryotic transcription complexes. *Annu. Rev. Biochem.* **64**, 533–561.

Zhang, X.-K., Lehmann, J., Hoffmann, B., Dawson, M. I., Cameron, J., Graupner, G., Hermann, T., Tran, P., and Pfahl, M. (1992). Homodimer formation of retinoid X receptor induced by 9-*cis* retinoic acid. *Nature* **358**, 587–591.

*Thyroid Hormone Receptor*

Almeida, A., Loopez-Mediavilla, C., and Medina, J. M. (1997). Thyroid hormones regulate the onset of osmotic activity of rat liver mitochondria after birth. *Endocrinology* **138**, 764–770.

Collingwood, T. N., Butler, A., Tone, Y., Clifton-Bligh, R., Parker, M. G., and Chatterjee, V. (1997). Thyroid hormone-mediated enhancement of heterodimer formation between thyroid hormone receptor and retinoid X receptor. *J. Biol. Chem.* **272**, 13060–13065.

Feng, J., Orlowski, J., and Lingrel, J. B. (1993). Identification of a functional thyroid hormone response element in the upstream flanking region of the human Na,K-ATPase α1 gene. *Nucl. Acids Res.* **21**, 2619–2626.

Fondell, J. D., Ge, H., and Roeder, R. G. (1996). Ligand induction of a transcriptionally active thyroid hormone receptor coactivator complex. *Proc. Natl. Acad. Sci. U.S.A.* **93**, 8329–8333.

Glass, C. K. (1996). Some new twists in the regulation of gene expression by thyroid hormone and retinoic acid receptors. *J. Endocrinol.* **150**, 349–357.

Ikeda, M., Wilcox, E. C., and Chin, W. W. (1996). Different DNA elements can modulate the conformation of thyroid hormone receptor heterodimer and its transcription activity. *J. Biol. Chem.* **271**, 23096–23104.

Koibuchi, N., Matsuzaki, S., Ichimura, K., Ohtake, H., and Yamaoka, S. (1996). Ontogenic changes in the expression of cytochrome *c* oxidase subunit I gene in the cerebellar cortex of the perinatal hypothyroid rat. *Endocrinology* **137**, 5096–5108.

Leng, X., Blanco, J., Tsai, S., Ozato, K., O'Malley, B. W., and Tsai, M.-J. (1994). Mechanisms for synergistic activation of thyroid hormone receptor and retinoid X receptor on different response elements. *J. Biol. Chem.* **269**, 31436–31442.

Olson, D. P., and Koenig, R. J. (1997). 5'-Flanking sequences in thyroid hormone response element half-sites determine the requirement of retinoid X receptor for receptor-mediate gene expression. *J. Biol. Chem.* **272**, 9907–9914.

Oppenheimer, J. H., and Schwartz, H. L. (1997). Molecular basis of thyroid hormone-dependent brain development. *Endocrine Rev.* **18**, 462–475.

Park, E. A., Song, S., Olive, M., and Roesler, W. J. (1997). CCAAT-enhancer-binding protein α (C/EBPα) is required for the thyroid hormone but not the retinoic acid induction of phosphoenolpyruvate carboxykinase (PEPCK) gene transcription. *Biochem. J.* **322**, 343–349.

Zhu, X.-G., McPhie, P., Lin, K.-H., and Cheng, S.-Y. (1997). The differential hormone-dependent transcriptional activation of thyroid hormone receptor isoforms is mediated by interplay of their domains. *J. Biol. Chem.* **272**, 9048–9054.

## *Vitamin D Receptor*

Darwish, H., and DeLuca, H. F. (1996). Analysis of binding of the 1,25-dihydroxyvitamin D₃ receptor to positive and negative vitamin D response elements. *Arch. Biochem. Biophys.* **334**, 223–234.

Haussler, M. R., Haussler, C., Jurutka, P., Thompson, P., Hsieh, J.-C., Remus, L. S., Selznick, S., and Whitfield, G. K. (1997). The vitamin D hormone and its nuclear receptor: Molecular actions and disease states. *J. Endocrinol.* **154**, S57–S73.

Jurutka, P. W., Hsieh, J.-C., Remus, L. S., Whitfield, G. K., Thompson, P. D., Haussler, C., Blanco, J. C., Ozato, K., and Haussler, M. R. (1997). Mutations in the 1,25-dihydroxyvitamin D₃ receptor identifying C-terminal amino acids required for transcriptional activation that are functionally dissociated from hormone binding, heterodimeric DNA binding, and interaction with basal transcription factor IIB, *in vitro*. *J. Biol. Chem.* **272**, 14592–14599.

Liu, Y.-Y., Collins, E. D., Norman, A. W., and Peleg, S. (1997). Differential interaction of 1α,25-dihydroxyvitamin D₃ analogues and their 20-epi homologues with the vitamin D receptor. *J. Biol. Chem.* **272**, 3336–3345.

Mackey, S. L., Heymont, J. L., Kronenberg, H. M., and Demay, M. B. (1996). Vitamin D receptor binding to the negative human parathyroid hormone vitamin D response element does not require the retinoid X receptor. *Mol. Endocrinol.* **10**, 298–305.

Whitfield, G. K., Hsieh, J.-C., Jurutka, P. W., Selznick, S. H., Haussler, C., Macdonald, P., and Haussler, M. R. (1995). Genomic actions of 1,25-dihydroxyvitamin D₃. *J. Nutr.* **125**, 1690S–1694S.

Yen, P. M., Liu, Y., Sugawara, A., and Chin, W. W. (1996). Vitamin D receptors repress basal transcription and exert dominant negative activity on triiodothyronine-mediated transcriptional activity. *J. Biol. Chem.* **271**, 10910–10916.

Zou, A., Elgort, M. G., and Allegretto, E. A. (1997). Retinoid X receptor (RXR) ligands activate the human 25-hydroxyvitamin $D_3$ 24 hydroxylase promoter via RXR heterodimer binding to two vitamin D-responsive elements and elicit additive effects with 1,25-dihydroxyvitamin $D_3$. *J. Biol. Chem.* **272**, 19027–19034.

*Niacin*

Aarhus, R., Dickey, D., Graeff, R. M., Gee, K. R., Walseth, T. F., and Lee, H. C. (1996). Activation and inactivation of $Ca^{2+}$ release by $NAADP^+$. *J. Biol. Chem.* **271**, 8513–8516.

Baum, C. L., Selhub, J., and Rosenberg, I. H. (1982). The hydrolysis of nicotinamide adenine dinucleotide by brush border membranes of rat intestines. *Biochem. J.* **204**, 203–207.

Berg, H. (1997). Bioavailability of niacin. *Eur. J. Clin. Nutr., Suppl.* **51**(1), S64–S65.

Cardenas-Corona, M. E., Jacobson, E. L., and Jacobson, M. K. (1987). Endogenous polymers of ADP-ribose are associated with the nuclear matrix. *J. Biol. Chem.* **262**, 14863–14866.

Carpenter, K. J. (1981). "Pellagra." Hutchison Ross, Stroudsburg, PA.

Carpenter, K. J., Schelstraete, M., Vilicich, V. C., and Wall, J. S. (1988). Immature corn as a source of niacin for rats. *J. Nutr.* **118**, 165–169.

Carter, E. G., and Carpenter, K. J. (1982). The bioavailability for humans of bound niacin from wheat bran. *Am. J. Clin. Nutr.* **36**, 855–861.

Cervantes-Laurean, D., Jacobson, E. L., and Jacobson, M. K. (1997). Preparation of low-molecular-weight model conjugates for ADP–ribose linkages to protein. *Methods Enzymol.* **280**, 275–287.

Dousa, T. P., Chini, E. N., and Beers, K. W. (1996). Adenine nucleotide diphosphates: Emerging second messengers acting via intracellular $Ca^{2+}$ release. *Am. J. Physiol.* **271**, C1007–C1024.

Everse, J., Anderson, B., and You, K.-S. (1982). "The Pyridine Nucleotide Coenzymes." Academic Press, New York.

Henderson, L. M. (1983). Niacin. *Annu. Rev. Nutr.* **3**, 289–307.

Jacob, R. A., Swendseid, M. E., McKee, R. W., Fu, C. S., and Clemens, R. A. (1989). Biochemical markers for assessment of niacin status in young men: Urinary and blood levels of niacin metabolites. *J. Nutr.* **119**, 591–598.

Jacobson, E. L., and Jacobson, M. K. (1997). Tissue NAD as a biochemical measure of niacin status in humans. *Methods Enzymol.* **280**, 221–230.

Jacobson, M. K., Coyle, D. L., Vu, C. Q., Kim, H., and Jacobson, E. L. (1997). Preparation of cyclic ADP-ribose, 2′-phospho-cyclic ADP-ribose, and nicotinate adenine dinucleotide phosphate: Possible second messengers of calcium signaling. *Methods Enzymol.* **280**, Part J, 265–275.

Krehl, W. A. (1981). Discovery of the effect of tryptophan on niacin deficiency. *Fed. Proc.* **40**, 1527–1530.

Lee, H. C., and Aarhus, R. (1995). A derivative of NADP mobilizes calcium stores insensitive to inositol trisphosphate and cyclic ADP-ribose. *J. Biol. Chem.* **270**, 2152–2157.

Loetscher, P., Alvarez-Gonzalez, R., and Althaus, F. R. (1987). Poly(ADP-ribose) may signal changing metabolic conditions to the chromatin of mammalian cells. *Proc. Natl. Acad. Sci. U.S.A.* **84**, 1286–1289.

Shambaugh, G. E., Koehler, R. R., and Radosevich, J. A. (1988). Developmental pattern of poly(ADP-ribose) synthetase and NAD glycohydrolase in the brain of the fetal and neonatal rat. *Neurochem. Res.* **13**, 973–981.

Satoh, M. S., Poirier, G. G., and Lindahl, T. (1993). $NAD^+$-dependent repair damaged DNA by human cell extracts. *J. Biol. Chem.* **268**, 5480–5487.

Ueda, K., and Hayaishi, O. (1985). ADP-ribosylation. *Annu. Rev. Biochem.* **54**, 73–100.

Vu, C. Q., Lu, P.-J., Chen, C.-S., and Jacobson, M. K. (1996). 2'-Phospho-cyclic ADP-ribose, a calcium-mobilizing agent derived from NADP. *J. Biol. Chem.* **271**, 4747–4754.

Walseth, T. F., Wong, L., Graaeff, R. M., and Lee, H. C. (1997). Bioassay for determining endogenous levels of cyclic ADP-ribose. *Methods Enzymol.* **280**, Part J, 287–294.


## Thiamin

Ariaey-Nejad, M. R., Balaghi, M., Baker, E. M., and Sauberlich, H. E. (1970). Thiamin metabolism in man. *Am. J. Clin. Nutr.* **23**, 764–778.

Bettendorff, L., Schoffeniels, E., Naquet, R., Silva-Barrat, S., Riche, D., and Menini, C. (1989). Phosphorylated thiamine derivatives and cortical activity in the baboon *Papio papio*: Effect of intermittent light stimulation. *J. Neurochem.* **53**, 80–87.

Brin, M. (1962). Effects of thiaminase deficiency and of oxythiamine on rat tissue transketolase. *J. Nutr.* **78**, 179–183.

Brin, M. (1963). Thiamine deficiency and erythrocyte metabolism. *Am. J. Clin. Nutr.* **12**, 107–116.

Chong, Y. H. (1970). Erythrocyte transketolase activity. *Am. J. Clin. Nutr.* **23**, 261–266.

Costello, C. A., Kelleher, N. L., Abe, M., McLafferty, F. W., and Begley, T. P. (1996). Mechanistic studies on thiaminase I. *J. Biol. Chem.* **271**, 3445–3452.

Dryefus, P. M. (1979). Nutritional disorders of the nervous system. *In* "Human Nutrition" (R. B. Alfin-Slater and D. Kritchevsky, eds.), Vol. 4, pp. 53–81. Plenum, New York.

Fayol, V. (1997). High-performance liquid chromatography determination of total thiamin in biological and food products. *Methods Enzymol.* **279**, Part I, 57–66.

Gibson, G., Nielsen, P., Mykytyn, V., Carlson, K., and Blass, J. (1989). Regionally selective alterations in enzymatic activities and metabolic fluxes during thiamin deficiency. *Neurochem. Res.* **14**, 17–24.

Elnageh, K. M., and Gaitonde, M. K. (1988). Effect of a deficiency of thiamine on brain pyruvate dehydrogenase: Enzyme assay by three different methods. *J. Neurochem.* **51**, 1482–1489.

Giguere, J.-F., and Butterworth, R. F. (1987). Activities of thiamine-dependent enzymes in two experimental models of thiamine deficiency encephalopathy. *Neurochem. Res.* **12**, 305–310.

Hilker, D. M., and Peters, O. F. (1966). Anti-thiamin activity in Hawaii fish. *J. Nutr.* **89**, 419–421.

Laforenza, U., Patrini, C., Alvisi, C., Faelli, A., Licandro, A., and Rindi, G. (1997). Thiamine uptake in human intestinal biopsy specimens, including observations from a patient with acute thiamine deficiency. *Am. J. Clin. Nutr.* **66**, 320–326.

Pekovich, S. R., Martin, P. R., and Singleton, C. K. (1996). Thiamine pyrophosphate-requiring enzymes are altered during pyrithiamine-induced thiamine deficiency in cultured human lymphocytes. *J. Nutr.* **126**, 1791–1798.

Sax, C. M., Salamon, C., Kays, W. T., Guo, J., Yu, F., Cuthbertson, R., and Piatigorsky, J. (1996). Transketolase is a major protein in the mouse cornea. *J. Biol. Chem.* **271**, 33568–33574.

Tallaksen, C. M., Bohmer, T., Karlsen, J., and Bell, H. (1997). Determination of thiamin and its phosphate esters in human blood, plasma, and urine. *Methods Enzymol.* **279**, Part I, 67–74.

Thornber, E. J., Dunlop, R. H., and Gawthorne, J. M. (1980). Thiamin deficiency in the lamb: Changes in thiamin phosphate esters in the brain. *J. Neurochem.* **35**, 713–717.

Vimokesant, S. L., Hilker, D. M., Nakornchai, S., Rungruangsak, K., and Dhanamitta, S. (1975). Effects of betel nut and fermented fish on the thiamin status of northeastern Thais. *Am. J. Clin. Nutr.* **28**, 1458–1463.

Wilkinson, T. J., Hanger, H., Elmslie, J., George, P., and Sainsbury, R. (1997). The response to treatment of subclinical thiamine deficiency in the elderly. *Am. J. Clin. Nutr.* **66**, 925–928.

Wood, B., Gijsbers, B., Goode, A., Davis, S., Mulholland, J., and Breen, K. (1980). A study of partial thiamin restriction in human volunteers. *Am. J. Clin. Nutr.* **33**, 848–861.

Ziporin, A. A., Nunes, W. T., Powell, R. C., Waring, P. P., and Sauberlich, H. E. (1965). Excretion of thiamin and its metabolites in the urine of young adult males receiving intakes of the vitamin. *J. Nutr.* **85**, 287–296.

## Riboflavin

Ajayi, O. A., and James, O. A. (1984). Effect of riboflavin supplementation on riboflavin nutriture of a secondary population in Nigeria. *Am. J. Clin. Nutr.* **39**, 787–791.

Akiyama, T., Selhub, J., and Rosenberg, I. H. (1982). FMN phosphatase and FAD pyrophosphates in rat intestinal brush borders: Role in intestinal absorption of dietary riboflavin. *J. Nutr.* **112**, 263–268.

Bates, C. J. (1997). Bioavailability of riboflavin. *Eur. J. Clin. Nutr., Suppl.* **51**, S38–S42.

Boisvert, W. A., Mendoza, I., Castaneda, C., Portocarrero, L., Solomons, N., Gershoff, S. N., and Russell, R. M. (1993). Riboflavin requirements of healthy elderly humans and its relationship to macronutrient composition of the diet. *J. Nutr.* **123**, 915–925.

Bruice, T. C. (1980). Mechanisms of flavin catalysis. *Acc. Chem. Res.* **13**, 256–262.

Burch, H. B., Lowry, O. H., Padilla, A. M., and Coombs, A. M. (1956). Effects of a riboflavin deficiency and realimentation on flavin enzymes of tissues. *J. Biol. Chem.* **223**, 29–45.

Chastain, J. L., and McCormick, D. B. (1987). Flavin catabolites: Identification and quantitation in human urine. *Am. J. Clin. Nutr.* **46**, 830–834.

Cline, J. L., Odle, J., and Easter, R. A. (1996). The riboflavin requirement of adult dogs at maintenance is greater than previous estimates. *J. Nutr.* **126**, 984–988.

Hoppel, C., DiMarco, J. P., and Tandler, B. (1979). Riboflavin and rat hepatic cell structure and function. *J. Biol. Chem.* **254**, 4164–4170.

McCormick, D. B., Oka, M., Bowers-Komro, D., Yamada, Y., and Hartman, H. A. (1997). Purification and properties of FAD synthetase from liver. *Methods Enzymol.* **280**, Part J, 407–413.

Merrill, A. H., Froehlich, J. A., and McCormick, D. D. (1979). Purification of riboflavin-binding proteins from bovine plasma and discovery of a pregnancy-specific riboflavin-binding protein. *J. Biol. Chem.* **254**, 9362–9364.

Merrill, A. H., Lambeth, J. D., Edmondson, D. E., and McCormick, D. B. (1981). Formation and mode of action of flavoproteins. *Annu. Rev. Nutr.* **1**, 281–317.

Ohkawa, H., Ohishi, N., and Yagi, K. (1983). New metabolites of riboflavin appear in human urine. *J. Biol. Chem.* **258**, 5623–5628.

Rindi, G., and Gastaldi, G. (1997). Measurements and characteristics of intestinal riboflavin transport. *Methods Enzymol.* **280**, Part J, 399–407.

Rivlin, R. S. (1975). "Riboflavin." Plenum, New York.

Walsh, C. (1980). Flavin coenzymes: At the crossroads of biological redox chemistry. *Acc. Chem. Res.* **13**, 148–155.

Zempleni, J., Galloway, J. R., and McCormick, D. B. (1996). Pharmacokinetics of orally and intravenously administered riboflavin in healthy humans. *Am. J. Clin. Nutr.* **63**, 54–66.

## Pantothenic Acid

Banno, K. (1997). Measurement of pantothenic acid and hopantenic acid by gas chromatography-mass spectroscopy. *Methods Enzymol.* **279**, Part I, 213–219.

Barbarat, B., and Podevin, R.-A. (1986). Pantothenate-sodium cotransport in renal brush-border membranes. *J. Biol. Chem.* **261**, 14455–14460.

Berg, H. (1997). Bioavailability of pantothenic acid. *Eur. J. Clin. Nutr., Suppl.* **51**(1), S62–S63.

Fox, H. M., and Linkswiler, H. (1961). Pantothenic acid excretion on three levels of intake. *J. Nutr.* **75**, 451–454.

Fry, P. C., Fox, H. M., and Tao, H. G. (1976). Metabolic response to a pantothenic acid deficient diet in humans. *J. Nutr. Sci. Vitaminol.* **22**, 339–346.

Lopaschuk, G. D., Michalak, M., and Tsang, H. (1987). Regulation of pantothenic acid transport in the heart. *J. Biol. Chem.* **262**, 3615–3619.

Mattick, J. S., Tsukamoto, Y., Nickless, J., and Wakil, S. J. (1983). The architecture of animal fatty acid synthetase. *J. Biol. Chem.* **258**, 15291–15299.

Risher, M. N., Robishaw, J. D., and Neely, J. R. (1985). The properties and regulation of pantothenate kinase from rat heart. *J. Biol. Chem.* **260**, 15745–15751.

Robishaw, J. D., and Neely, J. R. (1985). Coenzyme A metabolism. *Am. J. Physiol.* **248**, E1–E9.

Shibata, K., Gross, C. J., and Henderson, L. M. (1983). Hydrolysis and absorption of pantothenate and its coenzymes in the small intestines. *J. Nutr.* **113**, 2207–2215.

Smith, C. M., Narrow, C. M., Kendrick, Z. V., and Steffan, C. (1987). The effect of pantothenate deficiency in mice on their metabolic response to fast and exercise. *Metabolism* **36**, 115–121.

Sohby, C. (1979). Regulation of fatty acid synthetase activity. *J. Biol. Chem.* **254**, 8561–8566.

Tahiliani, A. G., and Neely, J. R. (1987). A transport system for coenzyme A in isolated heart mitochondria. *J. Biol. Chem.* **262**, 11607–11610.

Williams, M. A., Chu, L.-C., McIntosh, D. J., and Hincenbergs, I. (1968). Effects of dietary fat level on pantothenate depletion and liver fatty acid composition in the rat. *J. Nutr.* **94**, 377–382.

Wittwer, C. T., Beck, S., Peterson, M., Davidson, R., Wilson, D. E., and Hansen, R. G. (1990). Mild pantothenate deficiency in rats elevates serum triglyceride and free fatty acid levels. *J. Nutr.* **120**, 719–725.

*Ascorbic Acid (Vitamin C)*

Banhegyi, G., Braun, L., Csala, M., Puskas, F., and Mandl, J. (1997). Ascorbate metabolism and its regulation in animals. *Free Radical Biol. Med.* **23**, 793–803.

Barnes, M. J., Constable, B. J., Morton, L. F., and Kodicek, E. (1970). Studies *in vivo* on the biosynthesis of collagen and elastin in ascorbic acid-deficient guinea pigs. *Biochem. J.* **119**, 575–585.

Beers, M. F., Johnson, R. G., and Scarpa, A. (1986). Evidence for an ascorbate shuttle for the transfer of reducing equivalents across chromaffin granule membranes. *J. Biol. Chem.* **261**, 2529–2535.

Buettner, G. R. (1993). The pecking order of free radicals and antioxidants. *Arch. Biochem. Biophys.* **300**, 535–543.

Carpenter, K. J. (1986). "The History of Scurvy and Vitamin C." Cambridge Univ. Press, Cambridge.

Chou, P.-T., and Khan, A. U. (1983). L-Ascorbic acid quenching of singlet delta molecular oxygen in aqueous media: Generalized antioxidant property of vitamin C. *Biochem. Biophys. Res. Commun.* **115**, 932–936.

Crandon, J. H., Lund, C. C., and Dill, D. B. (1940). Experimental human scurvy. *N. Engl. J. Med.* **223**, 353–369.

Davidson, J. M., LuValle, P., Zoia, O., Quaglino, D., and Giro, M. (1997). Ascorbate differentially regulates elastic and collagen biosynthesis in vascular smooth muscle cells and skin fibroblasts by pretranslational mechanisms. *J. Biol. Chem.* **272**, 345–352.

Eipper, B. A., Park, L., Keutmann, H. T., and Mains, R. E. (1986). Amidation of joining peptide, a major pro-ACTH/endorphin-derived product peptide. *J. Biol. Chem.* **261**, 8686–8694.

England, S., and Seifter, S. (1986). The biochemical functions of ascorbic acid. *Annu. Rev. Nutr.* **6**, 365–406.

Frei, B., Stocker, R., and Ames, B. (1988). Antioxidant defenses and lipid peroxidation in human blood plasma. *Proc. Natl. Acad. Sci. U.S.A.* **85**, 9748–9752.

Glembotski, C. C. (1986). The characterization of the ascorbic acid-mediated α-amidation of α-melanotropin in cultured intermediate pituitary lobe cells. *Endocrinology* **118**, 1461–1468.

Guaiquil, V., Farber, C., Golde, D., and Vera, J. C. (1997). Efficient transport and accumulation of vitamin C in HL-60 cells depleted of glutathione. *J. Biol. Chem.* **272**, 9915–9921.

Halliwell, B., and Gutteridge, J. M. (1990). The antioxidants of human extracellular fluids. *Arch. Biochem. Biophys.* **280**, 1–8.

Hornig, D. (1975). Distribution of ascorbic acid, metabolites and analogues in man and animals. *Ann. N. Y. Acad. Sci.* **258**, 103–118.

Imlay, J. A., and Fridovich, I. (1991). Assay of metabolic superoxide production in *E. coli. J. Biol. Chem.* **266**, 6957–6965.

Imlay, J. A., and Linn, S. (1988). DNA damage and oxygen radical toxicity. *Science* **240**, 1302–1309.

Kobayashi, K., Haradi, Y., and Hayashi, K. (1991). Kinetic behavior of monodehydroascorbate radical studied by pulse radiolysis. *Biochemistry* **30**, 8310–8315.

Jacob, R. A., Skala, J. H., and Omaye, S. T. (1987). Biochemical indices of human vitamin C status. *Am. J. Clin. Nutr.* **46**, 818–826.

Janes, S. M., Palcic, M. M., Scaman, C. H., Smith, S. J., Brown, D. E., Dooley, D. M., Mure, M., and Klinman, J. P. (1992). Identification of topaquinone and its consensus sequence in copper amine oxidases. *Biochemistry* **31**, 12147–12154.

Johnston, C. S., Cartee, G. D., and Haskell, B. E. (1985). Effect of ascorbic acid nurtiture on protein-bound hydroxyproline in guinea pig plasma. *J. Nutr.* **115**, 1089–1093.

Kolhekar, A., Mains, R. W., and Eipper, B. A. (1997). Peptidylglycine α-amidating monoooxygenase: An ascorbate-requiring enzyme. *Methods Enzymol.* **279**, Part I, 35–43.

Levine, M. (1986). New concepts in the biology and biochemistry of ascorbic acid. *N. Engl. J. Med.* **314**, 892–902.

Levine, M., and Morita, K. (1985). Ascorbic acid in endocrine systems. *Vit. Horm.* **42**, 1–64.

May, S. W., Phillips, R. S., Mueller, P. W., and Herman, H. H. (1981). Dopamine-β-hydroxylase. *J. Biol. Chem.* **256**, 8470–8475.

Kenyon, J., and Munro, N. (1948). The isolation and some properties of dehydro-L-ascorbic acid. *J. Chem. Soc.* Part 1, pp. 158–161.

Prigge, S. T., Kolhekar, A. S., Eipper, B. A., Mains, R. E., and Amzel, L. M. (1997). Amidation of bioactive peptides: The structure of peptidyl-glycine α-hydroxylating monooxygenase. *Science* **278**, 1300–1305.

Rose, R. C., Choi, J.-L., and Koch, M. J. (1988). Intestinal transport and metabolism of oxidized ascorbic acid (dehydroascorbic acid). *Am. J. Physiol.* **254**, G824–G828.

Sauberlich, H. E., Tamura, T., Craig, C. B., Freeberg, L., and Liu, T. (1996). Effects of erythorbic acid on vitamin C metabolism in young women. *Am. J. Clin. Nutr.* **64**, 336–346.

Sawyer, D. T., Chiericato, G., and Tsuchiya, T. (1982). Oxidation of ascorbic acid and dehydroascorbic acid by superoxide ion in aprotic media. *J. Am. Chem. Soc.* **104**, 6273–6278.

Snook, J. T., London, C. P., and DeLany, J. P. (1983). Supplementation frequency and ascorbic acid status in adult males. *Am. J. Clin. Nutr.* **37**, 532–539.

Tuderman, L., Myllyla, R., and Kivirikko, K. I. (1977). Mechanism of the prolyl hydroxylase reaction. *Eur. J. Biochem.* **80**, 341–348.

Wegger, I., and Palludan, B. (1994). Vitamin C deficiency causes hematological and skeletal abnormalities during fetal development in swine. *J. Nutr.* **124**, 241–248.

Wimalasena, K., Herman, H. H., and May, S. W. (1989). Effects of dopamine β-monooxygenase substrate analogues on ascorbate levels and norepinephrine synthesis in adrenal chromaffin granule ghosts. *J. Biol. Chem.* **264**, 124–130.

*Vitamin E*

Berger, H. M., Ouden, A. L., and Calame, J. J. (1985). Pathogenesis of liver damage during parenteral nutrition: Is lipofuscin a clue? *Arch. Dis. Child.* **60**, 774–776.

Bieri, J. G., and McKenna, M. C. (1981). Expressing dietary values for fat-soluble vitamins: Changes in concepts and terminology. *Am. J. Clin. Nutr.* **34**, 289–295.

Bowry, V. W., Ingold, K., and Stocker, R. (1992). Vitamin E in human low-density lipoprotein. *Biochem. J.* **288**, 341–344.

Chio, K. S., and Tappel, A. L. (1969). Inactivation of ribonuclease and other enzymes by peroxidizing lipids and by malonaldehyde. *Biochemistry* **8**, 2827–2832.

Dawson, L. E., and Schierholz, K. (1976). Influence of grinding, cooking and refrigerated storage on lipid stability in turkey. *Poult. Sci.* **55**, 618–622.

Dhanakoti, S. N., and Draper, H. H. (1987). Response of urinary malondialdehyde to factors that stimulate lipid peroxidation *in vivo*. *Lipids* **22**, 643–646.

Draper, H. H., McGirr, L. G., and Hadley, M. (1986). The metabolism of malondialdehyde. *Lipids* **21**, 305–307.

Fitch, C. D. (1968). The red blood cell in the vitamin E-deficient monkeys. *Am. J. Clin. Nutr.* **21**, 51–56.

Fuller, C., Chandalia, M., Garg, A., Grundy, S., and Jialal, I. (1996). RRR-α-tocopheryl acetate supplementation at pharmacologic doses decreases low-density-lipoprotein oxidative susceptibility but not protein glycation in patients with diabetes mellitus. *Am. J. Clin. Nutr.* **63**, 753–759.

Furr, H. C., and Clark, R. M. (1997). Intestinal absorption and tissue distribution of carotenoids. *Nutr. Biochem.* **8**, 364–377.

Gordon, H. H., Nitowsky, H. M., and Cornblath, M. (1955). Studies of tocopherol deficiency in infants and children. *Am. J. Dis. Child.* **90**, 669–681.

Gotoda, T., Arita, M., Arai, H., Inoue, K., Yokota, T., Fukuo, Y., Yazaki, Y., and Yamada, N. (1995). Adult-onset spinocerebellar dysfunction caused by a mutation in the gene for the α-tocopherol-transfer protein. *N. Engl. J. Med.* **333**, 1313–1318.

Guggenheim, M. A., Ringel, S. P., Silverman, A., and Grabert, B. E. (1982). Progressive neuromuscular disease in children with chronic cholestasis and vitamin E deficiency. *J. Pediatr.* **100**, 51–58.

Halliwell, B. (1996). Vitamin C: Antioxidant or pro-oxidant *in vivo*? *Free Radical Res.* **25**, 439–454.

Hanzell, L. J., Arnold, L., Flowers, D., Waeg, G., Malle, E., and Stocker, R. (1996). Presence of hypochlorite-modified proteins in human atherosclerotic lesions. *J. Clin. Invest.* **97**, 1535–1544.

Harding, A. E., Matthews, S., Jones, S., Ellis, C. J., Booth, I. W., and Muller, D. P. (1985). Spinocerebellar degeneration associated with a selective defect of vitamin E absorption. *N. Engl. J. Med.* **313**, 32–35.

Kamal-Eldin, A., and Appelqvist, L.-A. (1996). The chemistry and antioxidant properties of tocopherols and tocotrienols. *Lipids* **31**, 671–701.

Kanner, J., Harel, S., and Hazan, B. (1986). Muscle membranal lipid peroxidation by an "iron redox cycle." *J. Agric. Food Chem.* **34**, 506–510.

Katz, M. L., Groome, A. B., and Robison, W. G. (1985). Localization of lipofuscin in the duodenums of vitamin E-deficient rats. *J. Nutr.* **115**, 1355–1365.

Kayden, H. J., and Traber, M. G. (1993). Absorption, lipoprotein transport, and regulation of plasma concentrations of vitamin E in humans. *J. Lipid Res.* **34**, 343–358.

Khayat, A., and Schwall, D. (1983). Lipid oxidation in seafood. *Food Technol.* (July) **37**, 130–140.

Kontush, A., Finckh, B., Karten, B., Kohlschutter, A., and Beisiegel, U. (1996). Antioxidant and prooxidant activity of α-tocopherol in human plasma and low density lipoprotein. *J. Lipid Res.* **37**, 1436–1448.

Kontush, A., Meyer, S., Finckh, B., Kohlschutter, A., and Beisiegel, U. (1996). α-Tocopherol as a reductant for Cu(II) in human lipoproteins. *J. Biol. Chem.* **271**, 11106–11112.

Levine, M., Rumsey, S., Wang, Y., Park, J., Kwon, O., and Amano, N. (1997). In situ kinetics: An approach to recommended intake of vitamin C. *Methods Enzymol.* **281**, Part K, 425–437.

Liebler, D., and Burr, J. (1992). Oxidation of vitamin E during iron-catalyzed lipid peroxidation: Evidence for electron-transfer reactions of the tocopheroxyl radical. *Biochemistry* **31**, 8278–8284.

Lynch, S. M., and Frei, B. (1995). Reduction of copper, but not iron, by low density lipoprotein (LDL). *J. Biol. Chem.* **270**, 5158–5163.

Oski, F. A., and Barness, L. A. (1968). Hemolytic anemia in vitamin E deficiency. *Am. J. Clin. Nutr.* **21**, 45–50.

Rhee, K. S. (1988). Enzymic and nonenzymic catalysis of lipid oxidation in muscle foods. *Food Technol.* (June), pp. 127–132.

Roher, A., Wolfe, D., Palutke, M., and Kuruga, D. (1986). Purification, ultrastructure, and chemical analysis of Alzheimer disease amyloid plaque core protein. *Proc. Natl. Acad. Sci. U.S.A.* **83**, 2662–2666.

Schwartz, S., deBlois, D., and O'Brien, E. (1995). The intima soil for atherosclerosis and restenosis. *Circ. Res.* **77**, 445–465.

Sevanian, A., and Hochstein, P. (1985). Mechanisms and consequences of lipid peroxidation in biological systems. *Annu. Rev. Nutr.* **5**, 365–390.

Sokol, R. J., Guggenheim, M., Iannaccone, S. T., Barkhaus, P. E., Miller, C., Silverman, A., Balistreri, W. F., and Heubi, J. E. (1985). Improved neurologic function after long-term correction of vitamin E deficiency in children with chronic cholestasis. *N. Engl. J. Med.* **313**, 1580–1586.

Steinberg, D. (1997). Low-density lipoprotein oxidation and its pathological significance. *J. Biol. Chem.* **272**, 20963–20966.

Superko, H. R., and Krauss, R. M. (1994). Coronary artery disease regression. *Circulation* **90**, 1056–1069.

Thomas, S. R., Neuzil, J., Mohr, D., and Stocker, R. (1995). Coantioxidants make α-tocopherol an efficient antioxidant for low-density lipoprotein. *Am. J. Clin. Nutr.* **62**, 1357S–1364S.

Traber, M. G. (1997). Regulation of human plasma vitamin E. *Adv. Pharmacol.* **38**, 49–63.

Traber, M. G., and Kayden, H. J. (1989). Preferential incorporation of α-tocopherol vs. γ-tocopherol in human lipoproteins. *Am. J. Clin. Nutr.* **49**, 517–526.

Traber, M. G., and Sies, H. (1996). Vitamin E in humans: Demand and deliver. *Annu. Rev. Nutr.* **16**, 321–347.

Traber, M. G., Sokol, R. J., Ringel, S. P., Neville, H. E., Thellman, C. A., and Kayden, H. J. (1987). Lack of tocopherol in peripheral nerves of vitamin E-deficient patients with peripheral neuropathy. *N. Engl. J. Med.* **317**, 262–265.

Ulshen, M. H. (1985). Vitamin E in cholestasis. *Gastroenterology* **88**, 588–592.

Upston, J. M., Neuzil, J., and Stocker, R. (1996). Oxidation of LDL by recombinant human 15-lipoxygenase: Evidence for α-tocopherol-dependent oxidation of esterified core and surface lipids. *J. Lipid Res.* **37**, 2650–2661.

Yang, C.-Y., Gu, Z.-W., Yang, H.-X., Yang, M., Wiseman, W., Rogers, L., Welty, S., Kata, V., Rohde, M., and Smith, C. V. (1997). Oxidation of bovine β-casein by hypochlorite. *Free Radical Biol. Med.* **22**, 1235–1240.

Ziouzenkova, O., Winklhofer-Roob, B., Puhl, H., Roob, J., and Esterbauer, H. (1996). Lack of correlation between the α-tocopherol content of plasma and LDL, but high correlations for γ-tocopherol and carotenoids. *J. Lipid Res.* **37**, 1936–1946.

*Essential Fatty Acids*

Abramovitz, M., Boie, Y., Nguyen, T., Rushmore, T. H., Bayne, M. A., Metters, K. M., Slipetz, D., and Grygorczyk, R. (1994). Cloning and expression of a cDNA for the human prostanoid FP receptor. *J. Biol. Chem.* **269**, 2632–2636.

Allard, J. P., Kurian, R., Aghdassi, E., Muggli, R., and Royall, D. (1997). Lipid peroxidation during n-3 fatty acid and vitamin E supplementation in humans. *Lipids* **32**, 535–541.

Arbuckle, L. D., and Innis, S. M. (1993). Docosahexaenoic acid is transferred through maternal diet to milk and to tissues of natural milk-fed piglets. *J. Nutr.* **123**, 1668–1675.

Arbuckle, L. D., MacKinnon, M. J., and Innis, S. M. (1994). Formula 18:2(n-6) and 18:3(n-3) content and ratio influence long-chain polyunsaturated fatty acids in the developing piglet liver and central nervous system. *J. Nutr.* **124**, 289–298.

Belury, M. A., Patrick, K. E., Lockniskar, M., and Fisher, S. M. (1989). Eicosapentaenoic acid and arachidonic acid: Comparison of metabolism and activity in murine epidermal cells. *Lipids* **24**, 423–429.

Boie, Y., Rushmore, T., Darmon-Goodwin, A., Grygorczyk, R., Slipetz, D. M., Metters, K. M., and Abramovitz, M. (1994). Cloning and expression of a cDNA for the human prostanoid IP receptor. *J. Biol. Chem.* **269**, 12173–12178.

Bourre, J. M., Youyou, A., Durand, G., and Pascal, G. (1987). Slow recovery of the fatty acid composition of sciatic nerve in rats fed a diet initially low in n-3 fatty acids. *Lipids* **22**, 535–538.

Craig-Schmidt, M. C., Stieh, K., and Lien, E. L. (1996). Retinal fatty acids of piglets fed docohexaenoic and arachidonic acids from microbial sources. *Lipids* **31**, 53–59.

D'Angelo, D., Eubank, J. J., Davis, M. G., and Dorn, G. W. (1996). Mutagenic analysis of platelet thromboxane receptor cysteines. *J. Biol. Chem.* **271**, 6233–6240.

Dajani, E. Z. (1993). ω-3 fatty acids and bowel cancer. *Gastroenterology* **104**, 1239–1240.

Farrell, P. M., Gutcher, G. R., Palta, M., and DeMets, D. (1988). Essential fatty acid deficiency in premature infants. *Am. J. Clin. Nutr.* **48**, 220–229.

Felton, C. V., Crook, D., Davies, M. J., and Oliver, M. F. (1994). Dietary polyunsaturated fatty acids and composition of human aortic plaques. *Lancet* **344**, 1195–1196.

Funk, C. D., Furci, L., Fitzgerald, G., Grygorczyk, R., Rochette, C., Bayne, M. A., Abramovitz, M., Adam, M., and Metters, K. M. (1993). Cloning and expression of a cDNA for the human prostaglandin E receptor EP1 subtype. *J. Biol. Chem.* **268**, 26767–26772.

Griffiths, R. J., Pettipher, E., Koch, K., Farrell, C., Breslow, J., Conklyn, M., Smith, M., Hackman, B., Wimberly, D., Milici, A., Scampoli, D., Cheng, J., Pillar, J., Pazoles, C., Doherty, N., Melvin, L., Reiter, L., Biggars, M., Falkner, F., Mitchell, D., Liston, T., and Showell, H. (1995). Leukotriene $B_4$ plays a critical role in the progression of collagen-induced arthritis. *Proc. Natl. Acad. Sci. U.S.A.* **92**, 517–521.

Guarnieri, M., and Johnson, R. M. (1970). The essential fatty acids. *Adv. Lipid Res.* **8**, 115–174.

Hamano, H., Nabekura, J., Nishikawa, M., and Ogawa, T. (1996). Docohexaenoic acid reduces GABA response in substantia nigra neuron of rat. *J. Neurophysiol.* **75**, 1264–1270.

Heird, W. C., and Gomez, M. R. (1996). Parenteral nutrition in low-birth-weight infants. *Annu. Rev. Nutr.* **16**, 471–499.

Henderson, W. R. (1994). The role of leukotrienes in inflammation. *Ann. Intern. Med.* **121**, 684–697.

Holman, R. T. (1988). George O. Burr and the discovery of essential fatty acids. *J. Nutr.* **118**, 535–540.

Holman, R. T., Adams, C. E., Nelson, R. A., Grater, S. J., Jaskiewicz, J. A., Johnson, S. B., and Erdman, J. W. (1995). Patients with anorexia nervosa demonstrate deficiencies of selected essential fatty acids, compensatory changes in nonessential fatty acids and decreased fluidity of plasma lipids. *J. Nutr.* **125**, 901–907.

Homayoun, P., Durand, G., Pascal, G., and Bourre, J. M. (1988). Alteration in fatty acid composition of adult rat brain capillaries and choroid plexus induced by a diet deficient in n-3 fatty acids. *J. Neurochem.* **51**, 45–48.

Innis, S. M., Nelson, C. M., Rioux, M. F., and King, D. J. (1994). Development of visual acuity in relation to plasma and erythrocyte ω-6 and ω-6 fatty acids in healthy term gestation infants. *Am. J. Clin. Nutr.* **60**, 347–352.

Ivanetich, K. M., Bradshaw, J. J., and Ziman, M. R. (1996). Δ6-Desaturase: Improved methodology and analysis of the kinetics in a multi-enzyme system. *Biochem. Biophys. Acta* **1292**, 120–132.

Kobayashi, T., Kiriyama, M., Hirata, T., Hirata, M., Ushikubi, F., and Narumiya, S. (1997). Identification of domains conferring ligand binding specificity to the prostanoid receptor. *J. Biol. Chem.* **272**, 15154–15160.

Krauss, R. M., Deckelbaum, R. J., Ernst, N., Fisher, E., Howard, B. V., Knopp, R. H., Kotchen, T., Lichtenstein, A. H., McGill, H. C., Pearson, T. A., Prewitt, T., Stone, N. J., Horn, L. V., and Weinberg, R. (1996). Dietary guidelines for healthy American adults. *Circulation* **94**, 1795–1800.

Kromhout, D., Bosschieter, E. B., and Coulander, C. (1985). The inverse relation between fish consumption and 20-year mortality from coronary heart disease. *N. Engl. J. Med.* **312**, 1205–1216.

Leaf, A., and Weber, P. C. (1988). Cardiovascular effects of n-3 fatty acids. *N. Engl. J. Med.* **318**, 549–557.

Leyton, J., Drury, P. J., and Crawford, M. A. (1987). In vivo incorporation of labeled fatty acids in rat liver lipids after oral administration. *Lipids* **22**, 553–558.

Makrides, M., Neumann, M. A., and Gibson, R. A. (1996). Is dietary docohexaenoic acid essential for term infants? *Lipids* **31**, 115–119.

Makrides, M., Neumann, M., Simmer, K., Pater, J., and Gibson, R. (1995). Are long-chain polyunsaturated fatty acids essential nutrients in infancy? *Lancet* **345**, 1463–1468.

Mantzioris, E., James, M. J., Gibson, R. A., and Cleland, L. G. (1995). Differences exist in the relationships between dietary linoleic acid and α-linolenic acids and their respective long-chain metabolites. *Am. J. Clin. Nutr.* **61**, 320–324.

Martinez, M., and Ballabriga, A. (1987). Effects of parenteral nutrition with high doses of linoleate on the developing human liver and brain. *Lipids* **22**, 133–138.

Marzo, I., Alava, M. A., Pineiro, A., and Naval, J. (1996). Biosynthesis of docohexaenoic acid in human cells: Evidence that two different Δ6-desaturase activities may exist. *Biochim. Biophys. Acta* **1301**, 263–272.

Moser, H. W., and Borel, J. (1995). Dietary management of X-linked adrenoleukodystrophy. *Annu. Rev. Nutr.* **15**, 379–397.

Negishi, M., Irie, A., Sugimoto, Y., Namba, T., and Ichikawa, A. (1995). Selective coupling of prostaglanding E receptor EP3D to Gi and Gs through interaction of α-carboxylic acid agonist and arginine residue of seventh transmembrane domain. *J. Biol. Chem.* **270**, 16122–16127.

Neuringer, M., Anderson, B., and Connor, W. E. (1988). The essentiality of n-3 fatty acids for the development and function of the retina and brain. *Annu. Rev. Nutr.* **8**, 517–541.

Neuringer, M., Connor, W. E., Lin, D. S., Barstad, L., and Luck, S. (1986). Biochemical and functional effects of prenatal and postnatal ω3 fatty acid deficiency on retina and brain in rhesus monkeys. *Proc. Natl. Acad. Sci. U.S.A.* **83**, 4021–4025.

Pawlosky, R. J., Ward, G., and Salem, N. (1996). Essential fatty acid uptake and metabolism in the developing rodent brain. *Lipids* **31**, S103–S107.

Rasmussen, M., Moser, A. B., Borel, J., Khangoora, S., and Moser, H. W. (1994). Brain, liver, and adipose tissue erucic and very long chain fatty acid levels in adrenoleukodystrophy patients treated with glyceryl trierucate and trioleate oils (Lorenzo's oil). *Neurochem. Res.* **19**, 1073–1082.

Reddy, J., and Mannaerts, G. P. (1994). Peroxisomal lipid metabolism. *Annu. Rev. Nutr.* **14**, 343–370.

Richardson, T. J., and Sgoutas, D. (1975). Essential fatty acid deficiency in four adult patients during total parenteral nutrition. *Am. J. Clin. Nutr.* **28**, 258–263.

Smith, W. L., Garavito, R., and DeWitt, D. L. (1996). Prostaglandin endoperoxidase H synthases (cyclooxygenases)-1 and -2. *J. Biol. Chem.* **271**, 33157–33160.

Soyland, E., Funk, J., Rajka, G., Sandberg, M., Thune, P., Rustad, L., Helland, S., Middelfart, K., Odu, S., Falk, E., Solvoll, K., Bjorneboe, G., and Drevon, C. A. (1993). Effect of dietary supplementation with very-long chain n-3 fatty cids in patients with psoriasis. *N. Engl. J. Med.* **328**, 1812–1816.

Sprecher, H., Luthria, D. L., Mohammed, B. S., and Baykousheva, S. P. (1995). Reevaluation of the pathways for the biosynthesis of polyunsaturated fatty acids. *J. Lipid Res.* **36**, 2471–2477.

Subramani, S. (1993). Protein import into peroxisomes and biogenesis of the organelle. *Annu. Rev. Cell. Biol.* **9**, 445–478.

Sugimoto, Y., Yamasaki, A., Segi, E., Tsuboi, K., Aze, Y., Nishimura, T., Oida, H., Yoshida, N., Tanaka, T., Katsuyama, M., Hasumoto, K., Murata, T., Hirata, M., Ushikubi, F., Negishi, M., Ichikawa, A., and Narumiya, S. (1997). Failure of parturition in mice lacking the prostaglandin F receptor. *Science* **277**, 681–683.

Suh, M., Wierzbicki, A. A., Lien, E., and Clandinin, M. T. (1996). Relationship between dietary supply of long-chain fatty acids and membrane composition of long- and very-long-chain essential fatty acids in developing rat photoreceptors. *Lipids* **31**, 61–64.

Tinoco, J., Babcock, R., Hincenbergs, I., Medwadowski, B., and Miljanich, P. (1978). Linolenic acid deficiency: Changes in fatty acid patterns in female and male rats raised on a linolenic acid-deficient diet for two generations. *Lipids* **13**, 6–17.

Tinoco, J., Williams, M. A., Hincenbergs, I., and Lyman, R. L. (1971). Evidence for nonessentiality of linolenic acid in the diet of the rat. *J. Nutr.* **101**, 937–946.

# 10

# INORGANIC NUTRIENTS

## OVERVIEW

The number of inorganic nutrients required for human and animal life is impressive, when compared with the number of required organic nutrients. Some inorganic nutrients, such as iron, calcium, and zinc, are used for a large number of functions, while others, such as selenium and molybdenum, have only a few uses. The differences in the personalities of the various inorganic nutrients are relatively great. Sodium, for example, is highly water soluble and occurs in high concentrations in the extracellular fluids of the body. This nutrient is not tightly bound to macromolecules; one of its functions is to pass through cell membranes. Sodium functions in this manner to convey nerve impulses and to drive nutrient transport systems. Potassium, chloride, and calcium also perform some of their functions by passing through biological membranes. In contrast, iron is relatively insoluble. It occurs as free iron in biological fluids only in very small concentrations. Iron is bound very tightly by various proteins, for example, proteins involved in oxygen metabolism.

Some inorganic nutrients are distinguished as being stored in the body. They are stored in a reservoir that can be tapped during periods of a dietary deficiency. These nutrients are iron, calcium, and phosphate. Most of the other inorganic nutrients tend not to be stored in usable, mobilizable forms in the body.

Water, sodium, potassium, and chloride are discussed first in this chapter because disturbances in the metabolism of these nutrients are more serious and more widespread than disturbances involving the other inorganic nutrients. Iodine is discussed next because deficiencies in this nutrient, widespread in some areas of the world, lead to serious and irreversible problems. Iron follows, as moderate deficiencies in this nutrient occur in all regions of the world. Iron deficiency is often associated with increased losses of the nutrient rather than with the consumption of poor diets.

Calcium and phosphate are presented together, as both of these nutrients are used to make bone. Calcium has commanded the attention of many cell biologists, as $Ca^{2+}$ ions are used for transmitting signals within the cell. Attention to plasma calcium levels is a vital concern to the clinician, as are plasma levels of sodium and potassium. To avoid ambiguity, the phosphate content of foods is often expressed in terms of the quantity of *phosphorus atoms*, rather than as *phosphate groups*. One might keep in mind, however, the fact that phosphorus never exists as a free atom in foods or biological tissues. It usually occurs as phosphate.

The remaining inorganic nutrients are fascinating from a biochemical standpoint, but tend not to be of routine importance in clinical nutrition or medicine. Assessment of nutritional status for a number of the inorganic nutrients involves the direct analysis for the nutrient, that is, in serum, rather than a functional test. For example, clearly defined levels of potassium and calcium are closely tied with human health. But suitable functional tests have not been devised.

## TABLES OF INORGANIC NUTRIENTS

Tables 10.1 to 10.5 may be useful as an orientation and a source of specific information for sections in this chapter. Table 10.1 lists the inorganic nutrients in various foods. One of the more striking aspects of these data is the fact that potassium concentrations in plant foods are much higher than those of sodium. Another point, raised under calcium and phosphate, is that green leafy vegetables (broccoli) and dairy products (cottage cheese) are high in calcium, whereas meats have relatively low levels of this nutrient. The calcium/phosphate ratios of various foods are also discussed in this section. These ratios can be easily calculated from the data in the table. Food iron data from two sources are listed. Milk and milk products, which contain high levels of many nutrients, are very low in iron.

A convenient list of most of the required inorganic nutrients in foods, along with interpretive commentary, has appeared (Pennington *et al.*, 1995a–c).

Table 10.2 lists the inorganic components in various milks. Milk was discussed at earlier points in the text, that is, with regard to its antibody (Chapter 3) and casein (Chapter 8) content. Table 10.2 reveals that the levels of most inorganic nutrients are similar in early and late milks.

**TABLE 10.1**  Inorganic Nutrients in Foods (mg element/100 g raw food)

|  | Na | K | Ca | Mg | P[a] | Fe | Fe[c] | Cu | Zn | Cl |
|---|---|---|---|---|---|---|---|---|---|---|
| Fresh peas | 1 | 340 | 15 | 30 | 100 | 1.9 | 1.49 | 0.23 | 0.7 | 38 |
| Broccoli | 12 | 340 | 100 | 18 | 67 | 1.5 | 0.56 | 0.07 | 0.6 | 55 |
| Potatoes | 7 | 1570 | 8 | 24 | 40 | 0.5 | 1.36 | 0.15 | 0.3 | 79 |
| Grapes | 1 | 270 | 4 | 3 | 14 | 0.3 | 0.38 | 0.07 | 0.1 | – |
| Orange juice | 2 | 180 | 12 | 12 | 22 | 0.3 | 0.16 | 0.05 | 0.2 | 1 |
| White flour[b] | 3 | 130 | 15 | 36 | 130 | 1.5 | – | 0.22 | 0.9 | 62 |
| Polished rice | 6 | 110 | 4 | 13 | 100 | 0.5 | 1.2 | 0.06 | 1.3 | 27 |
| Chicken meat | 81 | 320 | 10 | 25 | 200 | 0.7 | 1.1 | 0.19 | 1.1 | 78 |
| Lean beef | 59 | 350 | 10 | 22 | 190 | 2.1 | 2.9 | 0.19 | 5.5 | 65 |
| Eggs | 140 | 140 | 52 | 12 | 220 | 2.0 | 2.1 | 0.10 | 1.5 | 160 |
| Cottage cheese | 450 | 54 | 60 | 6 | 140 | 0.1 | 0.08 | 0.02 | 0.47 | 670 |
| Whole cow milk | 50 | 150 | 120 | 12 | 95 | 0.05 | 0.02 | 0.02 | 0.35 | 95 |

*Source*: Paul and Southgate (1978).
[a]The value of phosphorus can be converted to that of phosphate ($PO_4$) by multiplying by 3.06.
[b]The values refer to unfortified white flour.
[c]Pennington *et al*. (1995).


**TABLE 10.2**  Inorganic Nutrients in Milks

| | mg element/liter milk | | | |
|---|---|---|---|---|
| | Human colostrum[a] (1–5 days) | Mature human milk[a] (over 30 days) | Rat milk[b] (Day 1) | Rat milk[b] (Day 25) |
| Calcium | 230 | 280 | 792 | 683 |
| Chloride | 91 | 40 | –[c] | – |
| Potassium | 14 | 15 | – | – |
| Sodium | 48 | 15 | – | – |
| Magnesium | 34 | 30 | 174 | 122 |
| Phosphorus[d,e] | 14 | 15 | – | – |
| Zinc | 5.4 | 1.7 | 13.8 | 5.2 |
| Iron | 0.45 | 0.40 | 8.6 | 5.5 |
| Copper | 0.46 | 0.35 | 9.2 | 1.8 |
| Iodine | 0.12 | 0.07 | – | – |
| Manganese | 0.007[f] | 0.004[f] | 0.33 | 0.29 |

[a]Casey and Hambidge (1983).
[b]Keen *et al*. (1981).
[c]Data not reported.
[d]About half of the phosphorus occurs as inorganic phosphate and half as phosphate esters.
[e]Jenness (1985).
[f]Stastny *et al*. (1984).

Table 10.3 lists the approximate concentrations of various inorganic nutrients in the biological fluids. In most cases, the data refer to the total concentration of the ion, including free ion and ion complexed with organic compounds. The data for intracellular calcium and magnesium refer to the approximate concentration of the free, uncomplexed ions. A vital relationship revealed by the table is that Na⁺ levels are high in the plasma, and low in the cells, while K⁺ levels are low in plasma, and high in cells.

Inorganic nutrients lost in the urine represent nutrients that must be consumed in the diet to maintain nutrient balance. The metal ions that are secreted in gastric juices and intestinal juices are not necessarily lost from the body. They are efficiently reabsorbed by the intestinal tract. Vomiting, however, leads to loss of both gastric and intestinal juices. Diarrhea leads to loss of intestinal juices.

The inorganic components of bone appear in Table 10.4. For comparison, those of liver are also tabulated. The major inorganic components of bone are calcium and phosphate. Bone physiology is discussed under the sections Vitamin D and Calcium and Phosphate. One useful aspect of bone is that it can be dissolved at an increased rate during periods when the dietary intake of calcium is low. The rate of solubilization is under the control of 1,25-dihydroxyvitamin $D_3$ and parathyroid hormone. The rate of hormone production responds directly to changes in plasma calcium levels. Dissolution of bone crystal results in the liberation of all of the organic and inorganic components of bone, including sodium, not just calcium. During a deficiency of only sodium or potassium, for example, the deficiency cannot be corrected by increasing the rate of bone dissolution. The aforementioned

**TABLE 10.3** Millimolar Concentrations of Ions in Biological Fluids

| Ion | Intra-cellular fluid[a,b] | Plasma[c] | Glo-merular filtrate[c] | Urine[a] | Gastric juice[d] | Small intestinal juice[d] | Secretory diarrhea[d] | Sweat[d,e] |
|---|---|---|---|---|---|---|---|---|
| Na | 10 | 136–145 | 136 | 128 | 20–100 | 72–120 | 50–60 | 10–60 |
| K | 140 | 3.5–5.0 | 3.8 | 60 | 5–10 | 3.5–6.8 | 30–50 | 4–5 |
| Ca | 0.0001[f] | 4.5–5.5 | 1.3 | 2.4 | —[g] | – | – | 4–9 |
| Mg | 1–3[f] | 1.5–2.5 | 0.8 | 7.5 | – | – | – | 1.5–5 |
| Cl | 4 | 98–106 | 108 | 134 | 120–160 | 69–127 | 40–45 | 5–40 |
| HCO₃ | 10 | 23–28 | 28 | 14 | 0 | 30 | 45 | – |
| Phosphate | 75 | 1.2–3.0 | 2 | 50 | – | – | – | – |
| Sulfate | 10[h] | 0.3–1.5 | 0.5 | 1.6 | – | – | – | – |

[a]Guyton (1971).

[b]The concentrations of intracellular ions are only approximate and would be expected to vary somewhat in different types of cells. The technology for measuring intracellular levels of free calcium ions is well developed; that used for measuring the levels of other ions is still under development.

[c]Koushanpour and Kirz (1986).

[d]Zeman and Ney (1988).

[e]Costill (1977).

[f]Tsien (1983).

[g]Data not reported.

[h]Bunning and Riordan (1987).

**TABLE 10.4**  Inorganic Components of Bone

| | Rat bone (mg/g dry weight) | | Rat liver[a] (µg/g wet weight) |
| --- | --- | --- | --- |
| | Femur[a] | Tibia[b] | |
| Calcium | 151.0 | 329 | 35.3 |
| Potassium | 10.1 | 21 | 3376 |
| Sodium | 7.1 | 19 | 789 |
| Magnesium | 4.1 | – | 218 |
| Phosphorus | 80.2 | 166 | 3123 |
| Zinc | 0.176 | – | 37.1 |
| Iron | 0.089 | – | 50.0 |
| Copper | 0.005 | – | 10.6 |
| Manganese | 0.002 | – | 1.9 |

[a]Rader *et al.* (1984).
[b]Fine *et al.* (1987).

hormones are responsive to plasma calcium levels, but not to levels of sodium, potassium, zinc, or the other inorganic nutrients.

Interactions between some of the metal ions and various proteins and metabolites are detailed throughout this chapter. Prothrombin and metallothionein are examples of proteins that bind metal ions. Citrate, cysteine, ATP, and pyrophosphate ($PP_i$) are metabolites that can bind metal ions. Table 10.5 is intended to bring some overlying order to the many examples of interactions between metal ions and organic constituents of the body. The table reveals, for example, that magnesium tends to bind to oxygen. In the body, magnesium ions bind to the charged oxygen atoms of ATP. Iron tends to bind to sulfur. In fact, an entire class of metalloenzymes exists, called **iron-sulfur proteins**, where iron is bound directly to sulfur atoms, and not to a heme prosthetic group. Mercury tends to bind to sulfur. In the body, mercury can produce toxicity by forming strong complexes with cysteine residues of vital enzymes. The mercury binds to the charged atom of sulfur (R—S⁻).

# SODIUM, POTASSIUM, CHLORIDE, AND WATER

Sodium and potassium are monovalent cations. Chloride is a monovalent anion. Na and K generally do not function by supporting the activity of specific enzymes,

**TABLE 10.5**  Binding Preferences of Metal Ions

| | | | | | | |
| --- | --- | --- | --- | --- | --- | --- |
| Metal ions that tend to bind to oxygen | $Mg^{2+}$ | $Ca^{2+}$ | $Mn^{2+}$ | $Fe^{3+}$ | $Co^{3+}$ | |
| Metal ions that tend to bind to oxygen or sulfur | $Fe^{2+}$ | $Cu^{2+}$ | $Zn^{2+}$ | $Co^{2+}$ | $Pb^{2+}$ | $Ni^{2+}$ |
| Metal ions that tend to bind to sulfur | $Cu^+$ | $Cd^{2+}$ | $Hg^{2+}$ | | | |

*Source*: Pearson (1966).

in striking contrast to all of the other metal ions. Na and K are unusual in that they function by changing their location, that is, by passing from one side of the plasma membrane to the other. This passage may be actively driven by ATP-using mechanisms (ion pumps). In other cases, the passage occurs in a passive manner through channels, without the use of ATP.

Na and K are distributed on both sides of the plasma membrane in such a way that the fluid inside the cells has a slightly lower concentration of positively charged ions than the extracellular fluid. This results in a potential difference (voltage difference) across the plasma membrane. In the event that a small hole is poked in the cell membrane, the potential difference would provoke the flow of negatively charged ions out of the cell, with the consequence of neutralizing the charge difference. In living cells, however, ion flow is controlled by channels that mediate the passage only of specific ions. The Na channel, for example, allows the passage of $Na^+$, but not of $K^+$ or of $Cl^-$.

An understanding of ion pumps and channels, especially those of the kidney, may aid in understanding the issues of Na and K nutrition. These issues, discussed along with the material on kidney physiology, include the use of diuretics, sweating, blood pressure, and life-threatening conditions involving abnormal levels of plasma Na and K.

Most diets, even poor ones, provide sufficient amounts of Na, K, Cl, and water. Deficiencies in these nutrients tend to occur in a variety of pathological situations. These situations often involve excessive losses of these nutrients, rather than the lack of intake or impaired absorption. Plasma levels of Na or K can be low even with sufficient intakes of these nutrients and with a normal body content. Low plasma Na and K conditions are called **hyponatremia** and **hypokalemia**, respectively.

Na and K are sometimes called **electrolytes**. This term arose because these ions are used for generating an electric charge difference across the plasma membrane of most or all cells. These ions, however, have many functions unrelated to this use. Na, for example, is used to maintain blood pressure and K is used as a cofactor by several enzymes.

## Water

### Dietary Requirements

The requirement for water varies from individual to individual. It can be influenced by the composition of the diet, by climate and temperature, and by physical activity. The Food and Nutrition Board recommends an intake of 1000 ml of water per 1000 kcal (4184 kJ) of energy expended for adults. The recommendation for infants is 1500 ml/1000 kcal. The water requirement increases slightly during pregnancy, by about 30 ml/day, and markedly during lactation, by about 750 ml/day. Infants may be especially prone to a water deficiency because their daily water loss is equivalent to about 15% of their body weight. This figure is greater than that for adults. Other risk factors for infants include their inability to find water on their own and their susceptibility to diarrheal infections.

The nonexercising adult loses the equivalent of about 4% of his or her body weight in water per day. About 1400 ml is lost in the urine, 100 to 200 ml with the feces, and about 1250 ml by insensible losses. Insensible losses include the water lost by evaporation through the lungs and mild perspiration through the skin. The fluid lost through the skin by insensible losses is essentially water. Sweat, which is not considered an insensible loss, has a marked salt content. Despite their quality of being "insensible," the losses through the lungs and skin each range from 400 to 600 ml per day. The exact value for an individual is correlated with the body surface area. The obligatory loss of urine for the young adult is 500 to 600 ml per day. This is the minimal volume required for the removal of urea, uric acid, and other solutes from the body.

The failure to consume water can lead to death within several days. Water deficiency is most often provoked by diarrheal infections that are severe and prolonged. As body water is intimately associated with sodium, death might also be considered to result, in part, from sodium deficiency. Diarrhea-provoked fluid deficiency is probably the most common type of nutrient deficiency in the world.

The following material outlines the adverse effects of water loss. Depletion of body water equivalent to about 2% of body weight may impair athletic ability. Water loss resulting in a 5% reduction in body weight can lead to cramps and heat exhaustion. A 7 to 10% loss in body weight as water can result in hallucinations or a life-threatening heat stroke. Severe water losses can be fatal for two reasons. First, the loss of water can result in a decrease in the volume of the blood plasma to the point where oxygen and other nutrients cannot be efficiently delivered to the brain and other tissues. Second, the evaporation of water from the skin is used to maintain body temperature. With water loss, the body temperature may rise, resulting in heat stroke, coma, and death. Heat stroke can be a problem during prolonged exercise where the athlete fails to drink enough water. Heat stroke is a concern to military personnel stationed in hot areas as well as to football players training in the summer. Heat stroke also strikes the elderly during hot summer periods. This problem may be accentuated by the decline in the thirst sensation in the elderly and a general decline in the agility of the mechanisms that control homeostasis. High humidity is a contributing factor to heat stoke. It results in sweat rolling off the body, rather than evaporating and cooling the body.

## Compartmentation of Water in the Body

Definition of the different fluid compartments of the body may bring a more precise understanding to the material described here. Water constitutes about 60% of the body weight of the young adult male and 50% of that of the young adult female. The lesser value for the female arises from her greater proportion of adipose tissue. The fat stored in adipose tissue is relatively anhydrous (free of water). About 55% of the total body water is intracellular in the adult, with the remainder being extracellular. The **extracellular fluids** consist of plasma, interstitial fluid, fluids within the intestinal lumen, cerebrospinal fluid, and the water contained within the skeleton, connective tissue, and cartilage. As a shortcut, the extracellular fluid (ECF) is sometimes considered to include only the plasma and interstitial fluid. The plasma and interstitial fluid together constitute about 23% of the body weight. Further information on the compartmentation of the fluid volumes is given in Table 10.6.

**TABLE 10.6**  Distribution of Water in the Body

| Body compartment | Approximate volume of fluid in 70-kg man (liters) |
|---|---|
| Total body water | $42.0^{a,b}$ |
| Blood plasma | $3.2,^d$  $2.8^b$ |
| Interstitial fluid (including lymph) | $8.4^a$ |
| Cartilage and connective tissue | $3.2^a$ |
| Contents of lumen of gastrointestinal tract | $0.5^a$ |
| Cerebrospinal fluid | $0.2^a$ |
| Bile | $0.15^a$ |
| Intracellular fluid | $23,^a$  $28^b$ |

[a]Koushanpour and Kirz (1986).
[b]Goldberger (1986).

The ionic composition of the plasma and interstitial fluid is closely regulated by the kidneys. This regulation is essential for normal functioning of red and white blood cells, the nerves and brain, and other organs. The cells exposed to the **plasma** are the blood cells and epithelial cells that line the lumen of the blood vessels, that is, arteries, capillaries, and veins. All other cells are exposed to the **interstitial fluids**. The ionic composition of the interstitial fluids is similar to that of the plasma. Water, ions, $O_2$, $CO_2$, and small molecules of the plasma and interstitial fluid are in rapid equilibrium with each other because of the pores in the capillaries. Plasma proteins and blood cells tend to be excluded from the interstitial fluid because they are too large to pass through the pores. White blood cells have the ability to migrate through blood vessels and to enter any part of the body where invading organisms need to be destroyed.

### Water, Salt, and the Blood Volume

Water, salt, and blood pressure are related. The blood volume is closely related to the blood pressure. A loss in blood volume can occur with water deficiency or because of extensive bleeding. The lack of enough blood to fill up the vessels of the circulatory system leads to a drop in blood pressure. A severe drop in blood pressure results in the inability of the heart to pump vital nutrients to the brain and other tissues. A loss in blood volume can also result from sodium deficiency. The concentrations of sodium and its counterion chloride must be maintained to maintain the osmotic strength of the blood plasma. Osmotic strength is expressed by the term **osmolality**. Osmolality is equal to the sum of the molarities of the separate particles (ions or molecules) in a liquid. For example, a solution of 1 mole of NaCl in 1 liter has an osmolality of 2.0 osmol/liter. Na and Cl ions dissociate completely in solution. Osmotic pressure develops when two solutions of differing osmolalities are placed in contact with each other but separated by a semipermeable membrane. The walls of capillaries are semipermeable membranes. The renal tubules are, to some extent, composed of semipermeable membranes. The difference in osmotic strength of the solutions located on either side of a semipermeable membrane results in the movement of water from the solution of lower osmotic strength to that of greater osmotic strength. In short, water moves from the more dilute solution to the more concentrated solution. A drop in the osmotic strength of blood plasma, when due to a drop in NaCl concentration, results in an increase in the net flow of water through the capillary wall and out into the interstitial fluid.

The pathological accumulation of water in the interstitial fluid is called **edema**. Edema sometimes occurs in the legs. It may be unaesthetic as it results in a puffy appearance, but it can also result in infections. Edema in the air spaces in the lungs can impair breathing. Blood pressure is regulated by hormones that control the resorption of salt. An increased rate of salt resorption results in an increase in water reabsorption by the renal tubule. This is because the water follows the passage of salt by virtue of osmotic forces. The resorption of water is also under the control of vasopressin, as discussed later.

## Resorption of Water by the Renal Tubule

Nearly 200 liters of water passes through the glomerulus into the renal tubule per day. Essentially all of this water is resorbed through the renal tubule back into the body. The water is resorbed by osmotic forces. Resorption of water is induced by the resorption of Na and other ions. About 80% of the water appearing in the glomerular filtrate is resorbed by the proximal tubule, 6% at the loop of Henle, 9% by the distal tubule, and 4% by the collecting duct. Only about 1% of the filtered water appears in the urine. One powerful mechanism that induces the osmotic reabsorption of water is the countercurrent multiplier system. This effect is described in most college-level physiology texts.

## Biochemistry of Water

Water is required for the activity of all enzymes and for the functioning of all components of cell membranes, cytoskeleton, nucleic acids, and chromatin. (Chromatin is the complex of DNA, histones (structural proteins), and regulatory proteins that occurs in the nucleus.) Water is required for supporting the movements of ions that constitute the electrical impulses of nerves and muscle cells. Water is used as a medium for dispersing or dissolving nutrients and nutrient carriers in the body. Nutrient carriers include albumin, lipoproteins, and red blood cells. Water is also required for dissolving and eliminating wastes, such as $CO_2$, urea, and uric acid. Water is required in the extracellular fluids for maintaining the volume of the circulatory system and for maintaining blood pressure. Blood pressure is closely identified with the plasma concentration of sodium.

The activities of all enzymes and the three-dimensional conformations of all proteins and nucleic acids are sensitive to the overall concentration of ionic charges in the surrounding fluids. In extracellular fluids, the major cation is sodium. In intracellular fluids, the major cation is potassium. The major anions of biological fluids are chloride and phosphate. The general requirement of macromolecules for anions and cations might be called a requirement for a medium having a specific ionic strength. Large changes in ionic strength influence the catalytic and regulatory properties of all enzymes. Dramatic changes in the ionic strength of plasma and intracellular fluids rarely or never occur in mammalian tissues and would not be expected to be involved in regulating the behavior of cells.

The ionic strength of any fluid can be calculated from the following formula. This calculation requires knowledge of the concentration of each positively and negatively charged ion. The number of charges (no. charges) indicated in the formula refers to the number of plus or minus charges of the ion in question:

$$\text{Ionic Strength} = 0.5 \, [(\text{molarity of ion A}) \, (\text{no. charges of ion A})$$
$$+ \, (\text{molarity of ion B}) \, (\text{no. charges of ion B})$$
$$+ \, (\text{molarity of ion C}) \, (\text{no. charges of ion C}) + \ldots]$$

Only a few environments in the body may be considered water free or nearly water free. These include areas where triglycerides are stored, the interior recesses of many enzymes, and bone crystals.

## Sodium, Potassium, and Chloride

### Dietary Requirements

The RDA for sodium is expressed as a range (0.5–2.4 g Na/day). The minimal possible requirement for Na of the adult, in a cool climate, who is not engaging in activities that cause sweating, is about 0.115 g. The RDA for Na takes into account the range of Na needs for a population having a wide range of lifestyles. The slight increases in Na requirement during pregnancy and lactation are easily met by the usual intakes. The usual intake of Na is 1.8 to 5.0 g/day. About one-third of our Na is supplied by bread and cereals, 20% by meats, and 14% by dairy products. Human milk contains about 7 m$M$ Na, though the levels may range from 3 to 19 m$M$.

The amount of chloride in the diet tends to parallel the amount of sodium. Losses of chloride from the body also parallel those of sodium. The minimal requirement of Cl for the adult is about 0.75 g/day. The minimal requirement for sodium chloride for the adult is thus about 1.25 g per day.

The minimal requirement for potassium is about 1.6 to 2.0 g/day. Fruits and vegetables contain high levels of potassium. Persons who consume large amounts of these foods may have a potassium intake of 8 to 11 g/day. A more typical intake is 2.5 g/day. Human milk contains about 12 m$M$ potassium.

Nutrient levels of foods are usually expressed in terms of weight, that is, milligrams or grams; however, the levels of nutrients and other metabolites in the bloodstream and other biological fluids are more meaningfully expressed in terms of molarity, that is, nanomolar (n$M$), micromolar ($\mu M$), or millimolar (m$M$). The molar levels of metal ions, amino acids, and enzymes in the bloodstream can provoke interesting discussions and conclusions. Na and K intakes by adults in the United States range from 80 to 220 mmol Na and from 50 to 145 mmol K per day. Fecal Na ranges from 0.4 to 5.0 mmol per day and seems to vary little with sodium intakes from 20 to 180 mmol. This means that essentially all of the sodium of the diet is absorbed. The amount of fecal Na is equivalent to about 1 to 2% of the Na intake. Fecal K ranges from 3 to 22 mmol/day, a range somewhat greater than that of sodium. The amount of fecal K is equivalent to about 15% of the dietary K intake. Fecal Na and K can arise from that which is not absorbed from the diet, as well as from that secreted or lost from the gut mucosa. Fecal Na can arise from the fluids secreted by the cells of the crypts. Potassium ions are closely associated with cells and tissues. Thus, fecal K can arise from the dead cells that are sloughed off from the tips of the villi. Most of the remaining Na and K is lost in the urine. Na losses in the sweat may be substantial with prolonged exercise in a warm climate.

## Biochemistry of Sodium and Potassium

Sodium ions are not known to be required by many enzymes. Although the normal functioning of many enzymes can be supported by $Na^+$, when supplied as sodium chloride or sodium phosphate, potassium ions can be equally as effective. There is one known example of a sodium-requiring enzyme: sucrase. Potassium ions do not effectively activate this enzyme (Alvarado and Mahmood, 1979).

Sodium is also required in many nutrient transport systems. The transport of glucose, amino acids, and various ions across membranes may require the cotransport of sodium ions. The transport proteins that mediate the passage of these nutrients across the membrane recognize and bind both sodium and the nutrient. Na and K are both used by Na,K-ATPase. Na,K-ATPase is a membrane-bound protein whose function is intimately involved with the functioning of the Na-dependent nutrient transport systems, as described in Chapter 2. Sodium ions have proven effective in supporting the activity of thrombin, a protein of the blood clotting cascade (Cera *et al.*, 1995).

Potassium ions are required by a variety of enzymes (Suelter, 1970). These enzymes are activated by K to a greater extent than by Na. Pyruvate kinase, an enzyme of the glycolytic pathway, is the most well known K-requiring enzyme. It is activated by a variety of monovalent cations, as indicated in Table 10.7. Ammonium ions at 100 m$M$ support catalytic activity; however, the concentration of ammonium ions in the cell is under 1.0 m$M$. Sodium at 100 m$M$ weakly supports activity. The concentration of Na in the cell is only about 10 m$M$, indicating that the importance of Na in supporting enzyme activity *in vivo* is nil. Rubidium, which is chemically similar to K, supports activity; however, Rb is not a physiological cation.

Pyruvate carboxylase also requires a monovalent cation for activity. The activity of the purified enzyme was measured in the presence of various monovalent cations, as indicated in Table 10.7. Similar patterns of stimulation have been found for acetyl-CoA synthetase, an enzyme used in acetate metabolism (Webster, 1966), propionyl-CoA carboxylase (Giorgio and Plaut, 1967), and several other enzymes (Suelter, 1970). Maximal activity of the aforementioned enzymes usually occurs at a wide range of potassium concentrations, that is, from 50 to 150 m$M$. There is therefore little reason to believe that the slight changes in intracellular K concentrations that can occur under normal conditions or during K deficiency result in an impairment in the activities of these enzymes or in some type of regulation of the activities.

The enzyme most closely associated with Na and K metabolism is Na,K-ATPase (the sodium pump). This enzyme is a membrane-bound protein. It catalyzes the transport of sodium ions out of the cell in exchange for potassium ions. Three Na

**TABLE 10.7**  Stimulation of Enzyme Activities by Monovalent Cations

|              | Relative activity (%) | |
| --- | --- | --- |
|              | Pyruvate kinase[a] | Pyruvate carboxylase[b] |
| $K^+$        | 100 | 100 |
| $NH_4^+$     | 57  | 105 |
| $Rb^+$       | 51  | 110 |
| $Cs^+$       | 13  | 90  |
| $Na^+$       | 7   | 35  |
| $Li^+$       | 1   | 5   |

[a]Kwan *et al.* (1975).
[b]McClure *et al.* (1971).

ions are driven out for every two K ions entering. This mechanism results in a negative charge in the cell, when compared with the outside. The relatively high concentration of extracellular Na is used to drive various nutrient transport systems. These transport mechanisms involve the cotransport of Na and the nutrient. The Na that enters the cell via this mechanism is driven back out by Na,K-ATPAse. The sodium gradient — higher outside, lower inside — drives the transport of the nutrient into the cell.

Na,K-ATPase is required for the functioning of nerves and muscle cells. The electrical impulses that occur in these cells are dependent on the relatively high concentrations of extracellular Na and intracellular K. These ion gradients are maintained during the resting state, as well as during excitation of the cell. An electrical impulse in a nerve or muscle cell may be initiated by the deposit of a neurotransmitter in the region of a receptor site of the cell. A number of different neurotransmitters exist, each with specific functions. In the case of muscle cells, the neurotransmitter is deposited near the center of the cell, rather than at one end of it. The skeletal muscle cell, also called the muscle fiber, is 10 to 100 μm wide and often is as long as the entire muscle. Nerve cells may be 20 μm wide and 1 m long.

The neurotransmitter initiates the momentary opening of Na channels in the area of the neurotransmitter receptor. Na ions flow inward until there is no charge difference across the plasma membrane. This event is called **depolarization of the membrane**. The amount of Na that flows across the plasma membrane results in only a very small increase in the cytosolic level of sodium. The inflow of Na is self-limiting. Na channels close again after a fraction of a second; however, depolarization induces the opening of Na-channels a bit further along the nerve cell or muscle fiber. Sodium ions then enter the cell at this new region of the cell for a fraction of a second, leading to depolarization. The process continues, in a self-generating manner, from the point of initial stimulation to the far end(s) of the cell. The pattern of Na inflow along the length of the cell resembles that of a wave traveling from one end of a long trough to the other.

Potassium ions are also used in conduction of impulses along nerves and muscles. The brief inflow of Na ions results in neutralization of the electrical charge across the membrane. The charge difference is restored by potassium ions. Immediately after the Na channels have opened and closed, K channels open and close. This latter event results in momentary flow of potassium out of the cell, which restores the relative negative charge within the cell. The nerve or muscle cell is then able to transmit another impulse.

About 100 impulses may be transmitted by a nerve or muscle cell before the gradient difference between Na and K ions deteriorates to the point where impulses can no longer be transmitted. The continued activity of Na,K-ATPase ensures that the gradients of Na and K are maintained and that intracellular Na and extracellular K do not rise to abnormally high concentrations.

## Biochemistry of Chloride

Chloride, fluoride, bromide, and iodide are collectively known as the **halides**. The nonionized elemental forms of the halides are called halogens. The term halogen comes from a Greek word meaning "salt-producing." The halogens combine with the alkali metals to produce salts, such as sodium chloride. Chloride is only occasionally at the center of scientific attention. This is because dietary deficiencies of specifically chloride ions are quite rare. Also, the biology of a salt is sometimes viewed in terms of the cation ($Na^+$ or $K^+$) rather than the chloride counterion. Often, the movement of chloride ions across a membrane is seen only as a passive movement where it serves to balance a charge difference caused by the movement of sodium or potassium. This situation is changing. There is accumulating evidence that active transport mechanisms exist for the chloride anion.

A problem of nutritional interest related to chloride metabolism is **cystic fibrosis**. This disease is the most common lethal genetic disease in Caucasians. In cystic fibrosis a thick, dehydrating mucus develops in the lungs. This mucus impairs breathing and invites recurrent infections that eventually destroy the lungs, usually resulting in death by the age of 25. The disease results from a genetic mutation leading to a defect in chloride transport. The failure to transport $Cl^-$ from the lung epithelium into the air passage results in the failure of the counterion, sodium, to follow passively. Hence, the disease causes a problem in sodium chloride transport. Generally, the transport of salt through a membrane is accompanied by the flow of water. This flow of water is passive and occurs in response to osmotic pressure. Thus, in cystic fibrosis, water transport into the airway is impaired, resulting in mucus dehydration.

The biochemical functions of chloride are less diverse than those of other anions. Apparently, chloride does not occur covalently bound to mammalian metabolites, unlike other anions, such as iodide, sulfate, and phosphate. Chloride does not accept protons in the physiological pH range, making it an excellent counterion for gastric acid. The formation of a gastric acid consisting solely of protons might be possible in the laboratory but could not be accomplished in live organisms. A gastric acid consisting of acetic acid or phosphoric acid would be biologically possible, but these acids are not as strong as hydrochloric acid. Hence, chloride is recognized as being important as a constituent of gastric acid.

Chloride ions are known to be required for the activity of only a few enzymes — certain peptidases. These peptidases include angiotensin II (Bunning and Riordan, 1983, 1987), an enzyme that participates in the regulation of salt metabolism, and the cathepsins. The cathepsins are located in lysosomes, organelles used for the hydrolysis of nutrients recently transported into the cell.

## Regulation of Sodium, Potassium, and Water Balance

A central issue in Na and K metabolism is the mechanism by which plasma and tissue levels of these ions are maintained with variations in intake and excretion. The main site of regulation of Na and K balance is the kidney; the intestines comprise a minor site. With the consumption of a diet deficient in sodium, the kidneys respond by lowering urinary sodium and thereby conserving this nutrient. With a low-potassium diet, the kidneys also respond by decreasing the loss of this ion in the urine. Urinary sodium losses are controlled by varying the rate of **sodium reabsorption** by the renal tubules. Urinary potassium losses are controlled by varying the rate of **potassium secretion** by the renal tubules. The biochemical mechanisms that regulate Na and K balance are shared, in part. In some cases, changes in Na status can influence K balance. In other cases, Na balance and K balance appear to be independently regulated.

Evidence suggests that the skeleton may play a part in maintaining plasma Na levels. Bone is considered to be a reservoir of Na ions that may be used in times of need. It is thought that about one-third of the sodium in bone is in equilibrium with that in the extracellular fluids and is freely exchangeable. This freely exchangeable Na may be used for maintaining normal levels of plasma Na. The Na may dissociate from the bone, thus counteracting any trend toward a decline in plasma Na levels that may occur with excessive losses or with a dietary deficiency (Alcantra *et al.*, 1980).

The following study illustrates the conservation of the body's Na and K with the consumption of diets low in these nutrients. All of the subjects consumed a diet containing adequate amounts of Na (130 mmol/day) and K (90 mmol/day) for a week. They then consumed diets deficient in Na only, in K only, or in both ions. The daily urinary losses of Na and K are recorded in Figure 10.1. In Figure 10.1A are results for subjects who consumed a low-sodium (14 mmol/day) diet containing normal or low levels of potassium. The data demonstrate that daily losses of sodium progressively decreased to very low levels. In both diets, urinary Na losses fell to levels lower than the sodium intake. This indicates that, despite the highly restricted Na diet, the subjects may have been in positive sodium balance. In Figure 10.1B are results for subjects who consumed a low-potassium (14 mmol/day) diet containing normal or low levels of sodium. The data demonstrate that the daily losses of potassium progressively decreased to about 15 to 20 mmol.

The body's ability to conserve sodium by restricting its loss in the urine is more efficient than its ability to conserve potassium. As reviewed by Ornt *et al.* (1987), the body can restrict its K losses to 5 to 15 mmol/day in the urine; the mechanisms for Na conservation can restrict urinary Na losses to less than 1 mmol per day. As detailed later, the intestine's ability to absorb Na is also more efficient than its ability to absorb K. Thus, a starving person may experience a K deficiency prior to the onset of an Na deficiency. It is thought that our ability to better conserve Na arises from the fact that plant foods are rich in potassium but highly deficient in sodium. As indicated in Table 10.1, peas contain over 300 times more K than Na, on a per-weight basis. The difference is somewhat less extreme for potatoes and other plant foods.

### *Renin Increases Angiotensin II; Angiotensin II Increases Aldosterone*

Three molecules that participate in the regulation of Na balance are renin, angiotensin II, and aldosterone. These three molecules occur in the plasma and originate in the kidney, liver, and adrenal gland, respectively.

**Renin,** which has a molecular weight of about 40,000, is a peptidase. Renin is synthesized and secreted by cells in the walls of the arterioles entering the

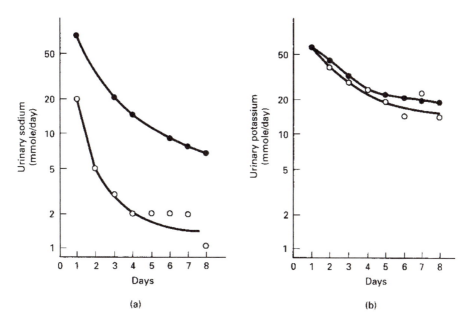

**FIGURE 10.1**  Daily urinary losses of Na and K of subjects who consumed (a) a low-Na diet containing normal (●) or low (○) K or (b) a low-K diet containing normal (●) or low (○) Na. (Redrawn with permission from Squires and Huth, 1959.)

glomerulus. These cells secrete renin into the bloodstream, where it catalyzes the cleavage of a peptide bond of the prohormone, angiotensinogen. Angiotensinogen, a protein with a molecular weight of about 50,000, is secreted by the liver. The cleavage produces angiotensin I, which is biologically inactive and is converted to **angiotensin II** in a reaction catalyzed by the "converting enzyme" (Figure 10.2). The converting enzyme is a peptidase present in the blood vessels of the lungs.

**Aldosterone** is a steroid hormone secreted by the zona glomerulosa of the adrenal gland (Figure 10.3). The steroid hormones are biosynthesized from cholesterol. Aldosterone is called a mineralocorticoid because it regulates the transport of minerals, Na and K. Aldosterone received its name because it contains an aldehyde group: —CH=O.

The adrenal gland appears to be one organ, but functionally it is two different organs. The inner core of the gland is the adrenal medulla. It is under nervous control and secretes epinephrine (adrenaline) and norepinephrine (noradrenaline). The outer layer of the adrenal gland is the adrenal cortex. The cells of the adrenal cortex closest to the surface of the gland constitute the zona glomerulosa. These cells synthesize aldosterone and release it into the bloodstream.

Angiotensin II induces the adrenal gland to synthesize and secrete aldosterone. The adrenal glands rest on the anterior side (the top side) of the kidneys. Aldosterone travels through the bloodstream to the vicinity of the renal tubule, where it enters the renal tubule cell through the basolateral membrane. Aldosterone enhances the body's conservation of Na by inducing the tubule cell to increase its efficiency of reabsorption of sodium ions. Figure 10.4 summarizes the renin/angiotensin II/aldosterone pathway that is used to enhance sodium conser-

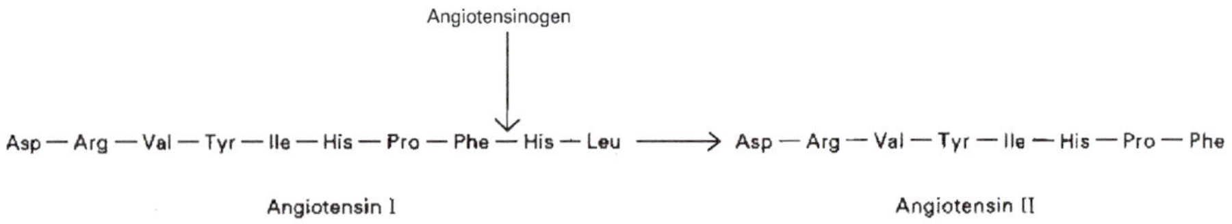

**FIGURE 10.2** The conversion of angiotensinogen to angiotensin II occurs in two steps that are catalyzed by renin and converting enzyme.

vation during the consumption of a low-Na diet. Low plasma Na stimulates the release of renin. Renin promotes the generation of angiotensin II in the plasma. Angiotensin II stimulates the adrenal gland to secrete aldosterone, which in turn stimulates the renal tubules to reabsorb or reclaim the sodium ions appearing in the glomerular filtrate.

In the following study, subjects initially consumed a diet containing typical levels of Na and K (200 mmol Na and 100 mmol K per day). They then consumed for 6 days a low-sodium diet containing 10 mmol Na and 100 mmol of K per day. The urinary losses of Na and K are shown, along with the levels of aldosterone, renin, angiotensin II, Na, and K. Table 10.8 shows a striking trend toward sodium conservation, as reflected in the urinary Na, with the maintenance of serum Na levels at a constant value. The data suggest that the subjects were in negative sodium balance during the initial 2 days or so with the low-Na diet, but reached Na balance by the sixth day on the low-Na diet. The data also demonstrate that increases in plasma renin and angiotensin II had occurred by the first day on the low-Na diet, whereas aldosterone levels did not increase until the second day on the low-Na diet. This indicates that the effects of aldosterone are somewhat de-layed, after a change from an Na-sufficient diet to an Na-deficient diet. This raises a fine point. The major regulator of Na conservation is aldosterone, which exerts its effect on the cells of the distal tubule. Angiotensin II, however, also directly promotes conservation of Na; its effect, which is minor compared to that of aldosterone, is exerted on the proximal tubule. Research studies have revealed that angiotensin II seems to stimulate the synthesis of aldosterone by the adrenal gland only with consumption of a low-sodium diet. With normal levels of plasma Na, the effect of angiotensin II on the adrenal may be small.

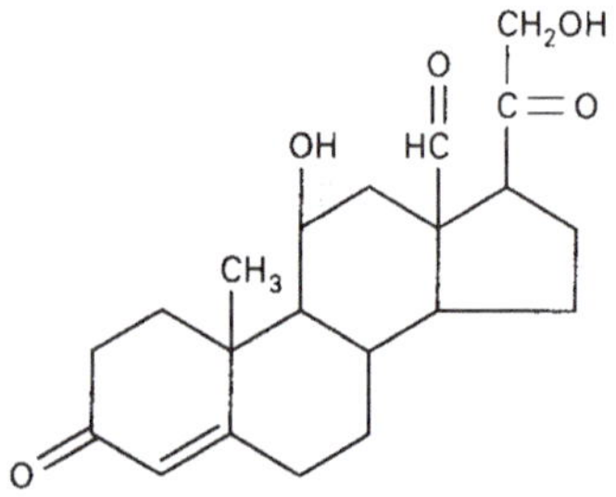

**FIGURE 10.3** Structure of aldosterone.

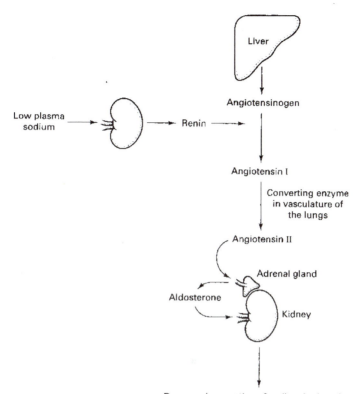

**FIGURE 10.4** Regulation of plasma sodium levels. Low plasma Na levels induce activation of Na conservation mechanisms in the kidney, resulting in a drop in urinary Na losses. High plasma Na levels induce inactivation of the mechanisms, resulting in increased urinary losses of sodium. Renin, angiotensinogen, angiotensin I, angiotensin II, and aldosterone are present in the general circulation and enter and exit the organs as indicated by the arrows. Renin is released from cells in the arteriole supplying the glomerulus. Each human kidney contains a million glomeruli. Each glomerulus is associated with one renal tubule. Aldosterone acts not at the proximal end of the tubule near the glomerulus, but at the distal end of the tubule, where the more mature urine occurs en route to the ureter.

The next study (Figure 10.5) illustrates an intermediate step in the control of sodium balance, namely the influence of angiotensin II on the adrenal gland. Rat zona glomerulosa cells were incubated in a flask containing a solution of the typical nutrients (glucose, amino acids, and salts). Angiotensin II, at the levels indicated in the figure, was added to the cells, which were then incubated for 1 hour. After this hour, the amount of aldosterone released into the medium was measured. The results demonstrate that the hormone was released from the cultured cells in response to very low concentrations of angiotensin.

*Renal Tubule*

The **nephron** is the functional unit of the kidney. There are 1 million nephrons per human kidney. Each nephron contains elements shown in Figure 10.6. Different

**TABLE 10.8** Changes in Metabolites Occurring with Low-Sodium Diets

| | | Low Na | | |
|---|---|---|---|---|
| | Normal Na | 1 day | 2 days | 6 days |
| Urinary Na (mmol/day) | 217 | 105 | 59 | 9.9 |
| Urinary K (mmol/day) | 83 | 74 | 85 | 92 |
| Aldosterone (ng/100 ml plasma) | 10.4 | 11.7 | 22.5 | 37 |
| Plasma renin (units/ml) | 1.0 | 2.4 | 2.4 | 4.3 |
| Plasma angiotensin II (pg/ml) | 22.0 | 29.5 | 30.8 | 38.8 |
| Serum Na (m$M$) | 139 | 139 | 139 | 138 |
| Serum K (m$M$) | 4.2 | 4.2 | 4.2 | 4.2 |

*Source*: Rogacz *et al.* (1990).

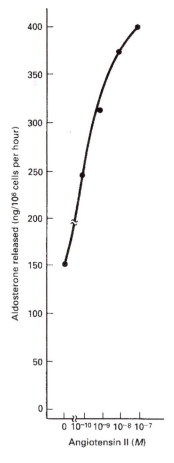

**FIGURE 10.5** Results of a study on rat zona glomerulosa cells. The effect of angiotensin II was measured as the amount of aldosterone released into the medium by the cells. (Redrawn with permission from Braley *et al.*, 1989.)

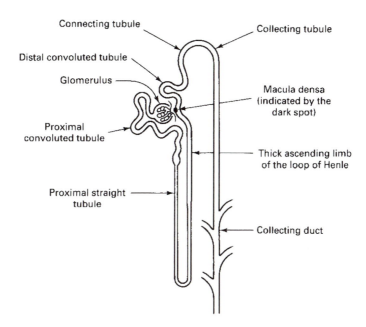

**FIGURE 10.6** The nephron. The glomerular filtrate is formed at the glomerulus. Its composition is essentially the same as that of blood plasma except that it is nearly free of proteins. Smaller proteins tend to pass through the small pores of the glomerulus and enter the filtrate, however. Membrane-bound peptidases, as well as transport systems for amino acids and short peptides, exist to allow resorption of the proteins. The water, minerals, and bicarbonate moving through the renal tubule are reabsorbed at specific sites of the tubule to enter the interstitial spaces and then the bloodstream. The water transporter AQP-1, for example, occurs in the proximal straight tubule, while AQP-2 occurs in the principal cells of the collecting duct. AQP-2 is regulated by vasopressin. The water and solutes that are not reabsorbed constitute the mature urine. The renal tubule might be considered to be a microscopic version of the intestinal tract, in form as well as in function. The **proximal tubule** contains a brush border, as does the small intestines. The **distal tubule** does not contain a brush border, in analogy to the large intestines.

types of nephrons contain the various elements in different proportions. The nephron consists of Bowman's capsule, the network or tuft of capillaries inside it, and the tubule leading from Bowman's capsule to ducts leading to the bladder. The tubule consists of the proximal tubule, the loop of Henle, and the distal tubule. The proximal tubule consists of the proximal convoluted tubule and the proximal straight tubule, as indicated in the figure. In this region, most of the filtered bicarbonate and potassium are reabsorbed into the bloodstream. The distal tubule — which consists of the distal convoluted tubule, the connecting tubule, and the initial collecting tubule — participates in regulation of Na and K balance. This region contains the principal cells and is sensitive to aldosterone. The distal ends of several tubules empty into the collecting duct. The **macula densa** is a specialized region of the tubule occurring between the loop of Henle and the distal convoluted tubules. It senses the Na or Cl level in the lumen of the tubule and relays the information to nearby granulated cells located in the wall of the arteriole entering Bowman's capsule.

Renin is released from granulated cells located in the arteriole entering the glomerulus. The release of renin is controlled at two points. One point is directly sensitive to the blood pressure; the other is sensitive to the concentration of NaCl. The arteriole entering the glomerulus contains stretch receptors, called **barorecep-tors**. The cells that constitute these receptors are subjected to stretch when the blood pressure increases. The stretch is relaxed with a drop in blood pressure. A decrease in stretch triggers the granule cells to secrete renin into the bloodstream. Recent evidence suggests that the baroreceptors are identical to granulated cells. The release of renin is also controlled by a signal from the macula densa. (*Macula densa* is Latin for "dark spot.") The macula densa is part of the renal tubule. It is in close contact with the arteriole entering the glomerulus. Apparently, the macula densa can sense the concentration of Na or Cl passing into the distal tubule and respond by sending a signal to the granulated cells. A drop in salt concentration results in a signal that induces secretion of renin. It is thought that the signal arising from the baroreceptors is more important than that from the macula densa.

Classically, the renin/angiotensin system has been viewed as requiring the dispersal of these polypeptides in the general circulation, that is, systemically; however, recent work has revealed that angiotensinogen and renin are made in a variety of specialized regions of the body. The dot blot technique has revealed messenger RNAs coding for angiotensinogen and renin in these tissues. The dot blot technique has an advantage over techniques used for the direct detection of polypeptides or proteins. Techniques sensitive to polypeptides cannot distinguish whether the polypeptide found in a tissue originated in the tissue under scrutiny or arrived there from some other tissue via the bloodstream. Locally acting renin/angiotensin systems appear to be responsive to the diet. Ingelfinger *et al.* (1986) found that the amount of mRNA coding for angiotensinogen in the kidney was 3.5-fold greater with a low-salt diet than with a high-salt diet. This result, however, should not minimize the importance of the liver as the major source of the prohormone in the body.

## Principal Cells

The major sites of regulation of Na balance are the **principal cells** of the renal tubule. The proteins involved in regulating salt balance include Na,K-ATPase, the sodium channel, and the potassium channel. These proteins are membrane-bound proteins. They are used for transporting Na and K across the membranes of the tubule cell. Sodium reabsorption involves the transport of Na appearing in the lumen of the renal tubule into the tubule cell and on through the cell to the interstitial space. The Na appearing in the interstitial space can then pass into the capillaries to enter the bloodstream.

In the initial step, Na is resorbed by the principal cells (in the collecting duct) through the Na channel (Figure 10.7(1)). The second step in Na resorption is the active transport of Na out of the cell by Na,K-ATPase (the sodium pump, (2)). Na,K-ATPase is composed of two subunits and has an overall molecular weight of 147,000. It catalyzes the ATP-dependent exchange of Na ions for K ions. Three Na ions leave the cell for every two K ions entering the cell. In other words, the transport of the two ions is coupled. The exit of three Na ions is coupled with the entry of two K ions. K returns to the interstitial fluid by the K channel (3).

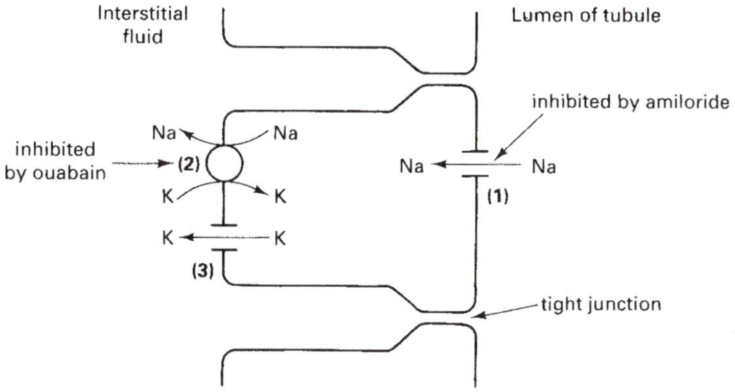

FIGURE 10.7  Sodium resorption by the principal cells of the renal tubule.

The apical side of the cell faces the lumen; this side is also the mucosal side in many tissues. The basal side faces the interstitial fluid and the capillaries; this side is also called the serosal side. The sides of the cell facing adjacent cells are called the lateral sides. Nutrients entering and exiting the basal and lateral membranes (basolateral membrane) can freely enter and exit the permeable membranes of nearby capillaries. Thus, the Na secreted by the basolateral membrane can enter the circulatory system. Potassium leaking from the capillaries can be taken up by the Na,K-ATPase of the basolateral membrane. The cells are connected to each other by impermeable junctions called **tight junctions**.

**Amiloride** (Figure 10.8) is used clinically as a diuretic. It blocks the passage of sodium through the Na channel of the distal tubule. This blockage results in a failure of Na resorption at this point of the tubule. It can also result in a failure in urinary potassium excretion, as K excretion requires the concomitant resorption of sodium. **Ouabain** (pronounced "wha-bane") (Figure 10.9) is an inhibitor of Na,K-ATPase. This chemical is purified from the leaves of *Digitalis lanata*. A related compound, digoxin, is used clinically to augment the force of contraction of the heart.

The Na channel and Na,K-ATPase are regulated as follows. Normally, the distal tubule resorbs 5 to 10% of the total amount of filtered sodium. Increased levels of plasma aldosterone appear to stimulate the entry of Na through the Na channel

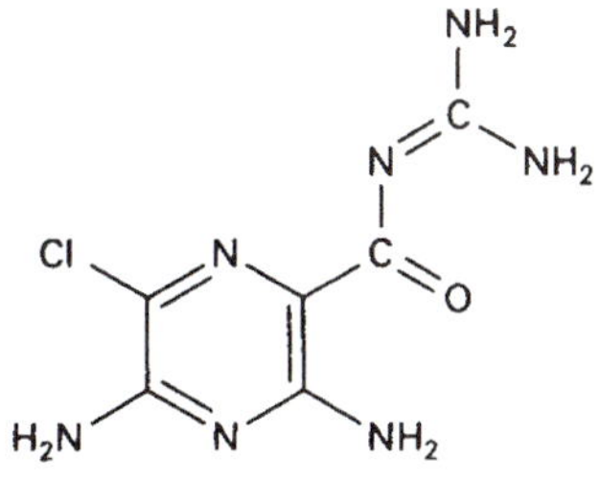

FIGURE 10.8  Structure of amiloride.

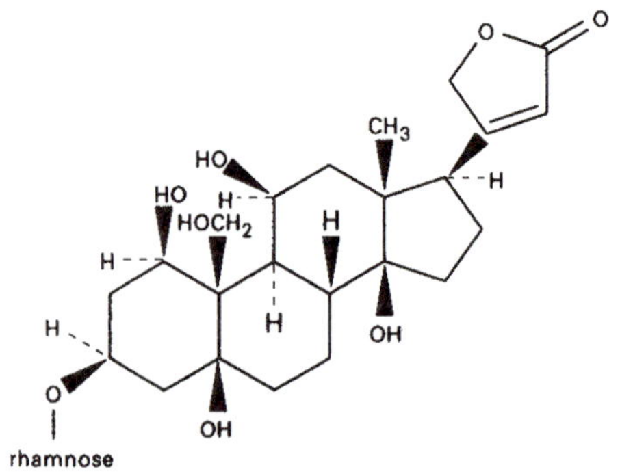

**FIGURE 10.9** Structure of ouabain.

by increasing the number of Na channels in the apical membrane. Aldosterone also plays a part in enhancing the passage of Na out of the cell through the basolateral membrane by stimulating the synthesis of new Na,K-ATPase proteins, which can then be inserted into the plasma membrane. Activation of Na,K-ATPase actually comprises two steps. First, the increased concentration of Na in the tubule cell induces insertion of preexisting molecules of Na,K-ATPase into the membrane. Second, aldosterone acts at the cell nucleus to stimulate the synthesis of more Na,K-ATPase molecules. The hormone binds a receptor protein in the principal cell. This binding is antagonized by the diuretic drug spironolactone.

The following study illustrates the influence of aldosterone on Na,K-ATPase in the collecting tubule. The experiment involved rabbits (Figure 10.10). The animals were prepared in three ways. The first group were sham operated, that is, anesthetized, cut open, and then sutured. The second group were adrenalectomized, that is, anesthetized, cut open, adrenal glands removed, and sutured. The third group were adrenalectomized and then treated with aldosterone replacement therapy. The rabbits in the third group were injected with 10 µg aldosterone/kg body weight. Three hours after injection, the kidneys were removed and a kidney tubule was dissected out and then separated into the proximal convoluted tubule, the ascending limb of the loop of Henle, and the collecting tubule. The dissected segments were used for assays of Na,K-ATPase. Assays involved the addition of ATP to the segment and measurement of the rate of conversion of the ATP to ADP + $P_i$. All of the segments contained Na,K-ATPase activity. The activity in the loop of Henle and collecting tubule were decreased with removal of the adrenal gland. This may have been in response to the lack of one or more of the many hormones secreted by this gland. Aldosterone replacement therapy restored Na,K-ATPase activity only in the collecting tubule. It was concluded that only Na,K-ATPase in the collecting tubule (which contains principal cells) was regulated by aldosterone.

Angiotensin II controls Na reabsorption by the proximal tubule. More specifically, it acts on the proximal convoluted tubule. This effect is independent of aldosterone. Angiotensin II binds to receptors on the membranes of the tubule cell, resulting in stimulation of the Na–H exchanger. The overall effect is an increase in the reabsorption of $NaHCO_3$. The flow of ions is accompanied by a flow of water. Thus, the action of angiotensin II at the proximal tubule can result in an increase in plasma volume and a rise in blood pressure.

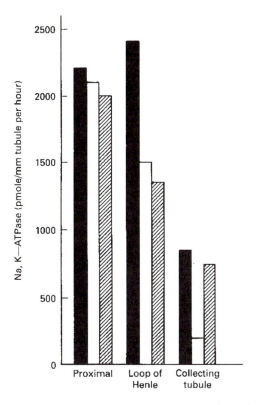

**FIGURE 10.10** Influence of aldosterone on Na,K-ATPase in the collecting tubule of three groups of rabbits: (black bars) sham operated, (open bars) adrenalectomized, (hatched bars) adrenalectomized, and then treated with aldosterone replacement theory. (Redrawn with permission from El Mernissi and Doucet, 1983.)

## Intestines

Intestinal absorption of salt and water is stimulated by angiotensin II and aldosterone. Angiotensin II stimulates the absorption of Na by the jejunum and ileum. Nerves leading from the brain to the gut may also control sodium absorption by the gut. Angiotensin II in the brain may play a part in this mechanism. Aldosterone seems to have no effect on salt absorption by the small intestines. The kidney is controlled by the nervous system. The nervous control of salt absorption by the kidney seems to play only a minor role in salt balance. Aldosterone stimulates the colon to absorb Na. Plasma angiotensin II is not thought to have a direct effect on salt absorption by the colon. The material described here is outlined in Figure 10.11.

## Analogy between Nutrient Transport Systems of the Renal Tubule and Intestines

An analogy may be drawn between the *proximal* kidney tubule and the *proximal* intestines. Here, angiotensin II directly stimulates Na reabsorption. An analogy may also be made between the *distal* tubule and the *distal* intestines. Here, aldosterone

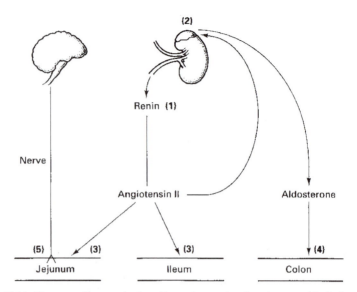

**FIGURE 10.11** Hormonal control of sodium absorption by the gut. (1) A drop in plasma Na levels and in blood pressure stimulates the release of renin into the bloodstream, resulting in an increase in the rate of generation of angiotensin II. (2) Angiotensin II stimulates the synthesis of aldosterone by the adrenal cortex. (3) Angiotensin II directly stimulates the small intestines to absorb Na. (4) Aldosterone stimulates the large intestines to absorb Na. (5) The nervous system, possibly by angiotensin II-dependent mechanisms, stimulates the small intestines to absorb Na.

directly stimulates Na absorption. In addition, the proximal kidney tubule and the small intestines both absorb potassium. Continuing the analogy, the distal kidney tubule and the large intestines can secrete K to regulate the levels of plasma K.

### Resorption of Other Nutrients by the Renal Tubule

Most of the filtered sodium is resorbed, not by the distal tubule, but by more proximal segments of the renal tubule. Sodium resorption in these regions can be regulated, though the adjustments that occur are not as important to maintaining Na balance as those occurring in the distal tubule. The transport systems to be described involve the resorption of glucose, amino acids, bicarbonate, potassium, and chloride. Sodium ions are resorbed, or cotransported, with all of these nutrients.

About 70% of the filtered Na and K is resorbed in the proximal tubule. Sodium is cotransported with glucose and the amino acids, resulting in the resorption of over 99% of these nutrients. Cotransport of Na with bicarbonate results in resorption of 80 to 90% of the filtered bicarbonate. The uptake of these nutrients at the apical membrane occurs by Na–amino acid and Na–glucose cotransport systems. One of two sodium ions are cotransported with each glucose. Seven Na–amino acid cotransport systems have been found in the renal tubule. These include systems specific for acidic amino acids, basic amino acids, glycine, neutral amino acids, cysteine, and cystine. One or more sodium ions are cotransported with each amino acid.

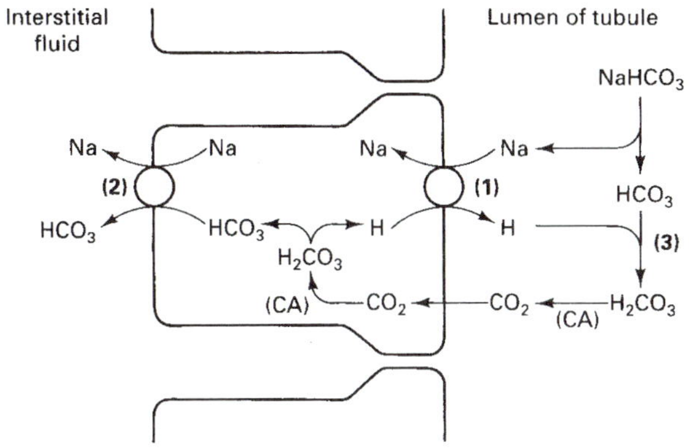

**FIGURE 10.12** Sodium bicarbonate resorption by cells of the proximal renal tubule. (1) Sodium enters through the apical membrane via the Na–H exchanger. (2) The resorbed sodium passes out of the cell into the interstitial space by way of the Na–HCO$_3$ cotransporter. (3) The bicarbonate in the filtrate accepts a proton to become carbonic acid. Steps that are catalyzed by carbonic anhydrase are indicated by (CA).

The pathway for bicarbonate resorption is described in some detail in Figure 10.12. Sodium enters through the apical membrane via the Na–H exchanger, and the resorbed Na passes out of the cell into the interstitial space by way of the Na–HCO$_3$ cotransporter. The bicarbonate in the filtrate accepts a proton to become carbonic acid. Carbonic anhydrase (CA) catalyzes the dehydration of carbonic acid to yield CO$_2$, which readily diffuses through the plasma membrane. Cystolic carbonic anhydrase generates a proton and bicarbonate. The luminal enzyme is membrane bound, whereas the cytosolic enzyme is free. The proton leaves the cell by way of the Na–H exchanger. The proton, in effect, is recycled and does not result in the accumulation of acid in any compartment. Bicarbonate leaves the cell for the bloodstream by way of the Na–HCO$_3$ cotransporter. This transport system participates not only in the resorption of nutrients, but also in the regulation of plasma pH and blood pressure. In chronic acidosis (acidic plasma), the activity of the bicarbonate transport system increases to restore the plasma pH to normal. Here, the increased resorption of NaHCO$_3$ leads to a rise in plasma pH and corrects the acidosis. Apparently, the activity of the bicarbonate transport system is controlled at the site of the Na–H exchanger. The protonation of a regulatory site on this membrane-bound protein results in its activation.

Potassium balance is regulated by an aldosterone-dependent mechanism. An increase in plasma K levels stimulates the adrenal cortex to synthesize and release aldosterone. Aldosterone stimulates the activity of Na,K-ATPase in the collecting tubule. Apparently, aldosterone also stimulates an increase in the number of K channels in the apical membrane of the tubule cell. High dietary intakes of K result in an increase in activity of Na,K-ATPase and an increase in the number of K channels. Most of the K appearing in the glomerular filtrate is resorbed by the proximal convoluted tubule. Here, K is absorbed through special channels in the tight junctions between the cells. This pathway is called a *paracellular pathway*. Some of the K in the glomerular filtrate is resorbed in the thick ascending loop of Henle. The transporter used here is the Na–K–2Cl cotransporter, located in the apical membrane and thus in contact with the lumen of the renal tubule ((1) in

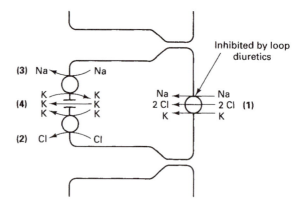

**FIGURE 10.13** Resorption of sodium, potassium, and chloride by cells of the thick ascending tube of Henle.

Figure 10.13). Potassium continues through the cell and leaves the basolateral membrane, possibly through a K–Cl cotransporter (2). The cotransport of Na, Cl, and K through the apical membrane is driven by an ATP-using pump in the basolateral membrane, namely Na,K-ATPase (3). The ions that are resorbed in the ascending limb of the loop of Henle are not followed by water, as this part of the renal tubule is not permeable to water; however, these ions are in close proximity to the descending limb and thus enhance the passage of water out of the descending limb. Figure 10.13 also includes a K channel, which may be used in K resorption (4).

Aldosterone alone cannot induce the secretion of K. An increased concentration of plasma K is needed for aldosterone to take effect. Studies have revealed that artificially elevated levels of aldosterone do not stimulate urinary losses of K in rats consuming a normal-K diet; however, an artificially raised aldosterone level did stimulate K excretion in rats consuming a high-K diet (Stanton, 1987).

The pathway of potassium secretion by the principal cell is diagrammed in Figure 10.14). The pumps and channels are those involved in sodium resorption

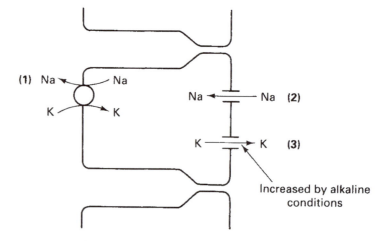

**FIGURE 10.14** Potassium secretion by the principal cells.

by the principal cells. The K channel occurs in both apical and basolateral membranes. For clarity, only the K channel in the apical membrane is shown in the figure. Continued operation of the Na,K-ATPase (1) pump is made possible by the Na channel in the apical membrane (2). The exit of potassium into the renal lumen, and thus excretion in the urine, is made possible by the K channel in the apical membrane (3). Apparently, the number of K channels in the apical membrane can be increased in certain pathological conditions (alkalosis), leading to increases in K secretion and K depletion.

## Na–K–2Cl Cotransporter and Potassium Resorption

The Na–K–2Cl cotransporter (Figure 10.13) has a molecular weight of about 86,000. Transport of Na, K, or Cl is dependent on the cotransport of the other two ions. The transport protein is the target of certain diuretics, including furosemide and bumetanide, known as **loop diuretics** because their target resides in the loop of Henle. Furosemide is shown in Figure 10.15. Use of the loop diuretics results in inhibition of transport of Na, K, and Cl, and in an increase in the urinary excretion of these ions and of water. This loss of water into the urine is the desired effect of the drug. The increase in fluid flow rate beyond the ascending limb of the loop of Henle can also induce increases in K secretion in parts of the distal tubule. A side effect of the use of loop diuretics is potassium depletion, as a result of increased secretion in the distal tubule. This loss of potassium is not desired. Persons taking these diuretics may need to take K supplements to make up for the increased urinary losses.

More than 90% of the filtered K is resorbed by the time the filtrate reaches the distal tubule. The quantity of K excreted in the urine is regulated by the principal cells in the collecting tubule. This regulation is not affected by changing the rate of K *reabsorption*, but by changing the *excretion* rate. Here, the rate of secretion from the principal cells can be varied from a value near zero to a rate equivalent to 1.5 times that of the initial appearance of potassium in the filtrate of the glomerulus.

## Vasopressin

Water balance is controlled by vasopressin in addition to mechanisms that control Na and K balance. Vasopressin (Figure 10.16) stimulates the resorption of water by the kidney tubule. **Vasopressin**, also called **antidiuretic hormone**, is synthesized by the hypothalamus. The hormone is a nonapeptide. The two cysteine

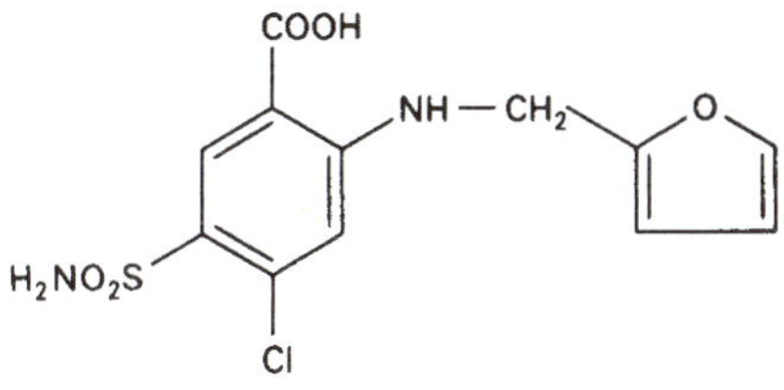

**FIGURE 10.15** Structure of furosemide.

$$\text{Cys} - \text{Tyr} - \text{Phe} - \text{Gln} - \text{Asn} - \text{Cys} - \text{Pro} - \text{Arg} - \text{Gly} - \text{NH}_2$$

FIGURE 10.16  Structure of human vasopressin. The hormone is amidated at the C terminus. The cysteine residues are connected via a disulfide bond.

residues are linked by a disulfide bond. Vasopressin contains a C-terminal amide group. The half-life of vasopressin in the bloodstream is about 6.5 minutes.

Vasopressin is packaged into secretory vesicles. The vesicles are transported from the hypothalamus to the nearby pituitary gland. These two glands are located at the base of the brain and are connected by a short stalk. The hormone is stored in the pituitary for release into the bloodstream. The hypothalamus contains osmoreceptors, which are sensitive to the concentration of solute in the bloodstream. An increase in osmolality induces an increase in the rate of secretion of vasopressin. The hormone travels to the kidney, where it binds to receptors on the basolateral membrane of the nephron. It induces an increase in the permeability of the apical membrane to water, allowing the passage of water from the tubule lumen into the cell. This is followed by increased passage of water from the cell to the extracellular fluids. Vasopressin acts at the distal part of the tubule, on the cells of the collecting duct. Although most of the water resorbed by the renal tubule is reabsorbed in the proximal tubule prior to the ascending loop of Henle, vasopressin influences only resorption by the collecting duct.

## Atrial Natriuretic Peptide Pathway of Blood Pressure Regulation

The heart contains special receptors that respond to stretching. Activation of the stretch receptors results in the release of atrial natriuretic peptide (ANP) from the heart (Reis *et al.*, 1997). The role of stretch in provoking ANP release was shown in the laboratory by inflating a balloon within the heart (Christensen and Leistad, 1997). ANP travels through the bloodstream to the kidneys, where it reduces the rate of sodium resorption. The result is a lowering of blood pressure and the appearance more salt in the urine (Lopez *et al.*, 1995). ANP is a 28-amino-acid polypeptide with the following sequence (Drewett and Garbers, 1994):

H₂N–Ser–Leu–Arg–Arg–Ser–Ser–Cys–Phe–Gly–Gly–Arg–Ile–Asp–Arg–Ile–Gly–Ala–Gln–Ser–Gly–
Leu–Gly–Cys–Asn–Ser–Phe–Arg–Tyr–COOH

When ANP binds to its target cell in the renal tubule, it provokes the activation of a protein kinase. The kinase, in turn, catalyzes the phosphorylation and inactivation of Na,K-ATPase. Na,K-ATPase is a vital membrane-bound protein in most, or all, cells of the body, as mentioned earlier. Its inhibition in the renal tubule results in decline in sodium resorption (Wang and Robinson, 1997). The exact kinase that is involved is not certain, but it appears to be protein kinase A (PKA) (Ledoux *et al.*, 1997; Belusa *et al.*, 1997).

Mice have been created in the laboratory (knock-out mice) that lack ANP, and these mice acquire salt-sensitive hypertension. The mouse's blood pressure increases upon feeding a high-salt diet (John *et al.*, 1995). The question of whether salt-sensitive people have altered levels or structures of ANP is currently under study. Of the people with hypertension, about half do not respond to low-salt diets (they are salt-insensitive) while the other half are salt-sensitive.

It is interesting to point out that the actions of angiotensin II and ANP appear to have a common meeting point, namely at the site of Na,K-ATPase in the renal tubule. Angiotensin II provokes the *dephosphorylation* of Na,K-ATPase and an increase in its activity, whereas ANP provokes its *phosphorylation* and a decline in activity (Aperia *et al.*, 1994).

## Nutritional Problems

Problems in water and salt nutrition can occur with severe and prolonged diarrhea or vomiting and with prolonged exercise. Chronic changes in salt nutrition can change blood pressure and influence the course of cardiovascular disease. These nutritional concerns are addressed after first discussing normal plasma levels of sodium and potassium.

Normal serum sodium ranges from 136 to 145 m$M$. **Hyponatremia** is defined as an Na level under 130 m$M$. Where plasma Na falls below 120 m$M$, such symptoms as headaches, confusion, seizures, and coma can occur. These symptoms result from increased flow of water into the cells of the brain, due to osmotic forces, causing swelling of the brain cells. Hyponatremia most often occurs (as an undesirable side-effect) where a low-sodium diet is combined with diuretic drugs, such as thiazide and furosemide. These drugs provoke the excretion of sodium by the kidneys and, in this way, reduce high blood pressure. Severe diarrhea involving losses of 8–10 liters of fluid per day can result in hyponatremia, where the particular nature of the diarrhea is such that losses of sodium are greater and out of proportion to the losses of water. A rare cause of hyponatremia, which occurs in some psychotic patients, is the drinking of more than 20 liters of water per day. Similarly, hyponatremia can result from excess beer-drinking when combined with a poor diet. Hyponatremia can result from disorders that alter our body's handling of sodium and water (lack of aldosterone in Addison's disease; excess vasopressin in certain cancers (Fried and Palevksy, 1997).

**Hypernatremia** (high plasma sodium levels) occurs less commonly than hyponatremia. Hypernatremia is defined as a serum sodium level above 145 m$M$. The first signs of hypernatremia are restlessness, lethargy, and irritability. Seizures and death may occur when plasma Na levels rise above 160 m$M$. These problems occur because high plasma levels draw water from the brain cells, resulting in their shrinkage. Hypernatremia occurs with a loss of water that is disproportionately greater than the loss of sodium. This loss may be associated with excessive hyperventilation or sweating. A rise in the concentration of plasma solutes, such as sodium, stimulates the thirst mechanism. The consequent drinking of water rapidly corrects the hypernatremia. Hypernatremia occurs where there are uncorrected losses of water. The problem may remain uncorrected where the thirst mechanism is impaired, as occurs in the elderly; with damage to the hypothalamus; or with the excessive consumption of salt (e.g., large amounts of broth or seawater).

Normal serum potassium ranges from 3.5 to 5.0 m$M$. **Hypokalemia** occurs at under 3.5 m$M$ potassium. Mild hypokalemia can result in weakness and muscle cramps and can cause cardiac arrhythmias in patients with heart disease. Severe hypokalemia (under 2.5 mM K) can result in paralysis; the consequent paralysis of the lungs can be fatal. Hypokalemia is most commonly caused by diuretics, such as thiazide or furosemide (Wingo and Wingo, 1997). The second most common cause of hypokalemia is excessive diarrhea or vomiting, where laxative abuse and enema abuse are related causes. Up to 20% of persons complaining of chronic diarrhea practice laxative abuse. A third cause of hypokalemia is prolonged fasting and starvation. With 3 weeks of fasting, plasma potassium levels may decline to below 3.0 mM. However, in some persons plasma potassium may be naturally maintained at about 3.0 mM, even with 100 days of fasting. About half of hospitalized alcoholics, who are undergoing withdrawal symptoms, experience hypokalemia, where the causes are poor nutrition, vomiting, and diarrhea. It is interesting to point out that potassium loss, with vomiting, is only partly due to loss of potassium through the mouth. Vomiting, which involves loss of acid from the mouth, results in alkalosis. Alkalosis, in turn, provokes the kidneys to release more potassium in the urine. In this way, severe and continual vomiting can result in hypokalemia.

**Hyperkalemia**, or high plasma potassium levels (5.0 mM or higher), results in cardiac arrhythmias. A K level of 8.0 mM can provoke cardiac arrest and death. Hyperkalemia may also result in neurological signs such as tingling, paresthesia (numbness of hands and feet), and paralysis. Hyperkalemia can arise from consuming too much potassium salt (but vomiting usually prevents toxicity), from failure of the kidneys to perform their normal task of excreting potassium into the urine, and from loss of potassium from tissues into the bloodstream (as with acidosis, severe trauma, or burns). Renal failure accounts for three quarters of all cases of hyperkalemia, and here hyperkalemia occurs when the GFR declines to only 5 ml/min, especially where high-potassium foods are eaten. A GFR of only 5 ml/min occurs with severe renal failure. Another kidney-related source of hyperkalemia is use of spironolactone or triamterene. These are diuretic drugs that are used to treat hypertension (they provoke sodium loss), but these particular drugs also promote potassium retention, and thus the drugs can result in hyperkalemia. Acidosis, which occurs in a number of diseases, can provoke a shift in location of potassium from inside muscle to the bloodstream, resulting in hyperkalemia (Greenberg, 1998).

Hypokalemia or hypernatremia occur with the overproduction of aldosterone (aldosteronism). Aldosteronism occurs after the age of 40 and results in increased Na resorption, increased blood volume, and increased urinary losses. In rare cases, hypokalemia occurs with the habitual consumption of large amounts of black licorice. Licorice contains **glycyrrhizic acid**. This compound has a structure and an effect similar to those of aldosterone.

## Diarrhea and Vomiting

Diarrheal diseases are a major world health problem. They are responsible for about a quarter of the 10 million infant deaths that occur each year. Although 99% of these deaths occur in poor parts of Asia and Africa, diarrheal diseases are among the leading sources of infant mortality in the United States. Severe and prolonged diarrhea results in losses of water, sodium, potassium, and bicarbonate. Diarrheal diseases can result from contamination of the food and water supply with the fecal matter of humans and animals, as may occur when proper sanitation is lacking or when farm animals are allowed to roam near water wells, inside houses, and in areas intended for the preparation of food. Another provoking factor is an early halt in breast-feeding, which prevents infants from receiving the antibodies in breast milk and exposes them to the public water supply.

Some extra perspective on diarrhea might be acquired, with a few words regarding the time-course and various causes of the diarrheas. Most diarrheas due to infectious organisms spontaneously disappear within 4 weeks. Chronic diarrhea can be due to infections with giardia or amoeba, to fat malabsorption (steatorrhea), to inflammatory bowel disease, or to laxative abuse. From 4 to 20% of persons complaining of chronic diarrhea practice laxative abuse. Laxative abuse, which is nearly always denied by patients, is a component of various eating disorders. One such eating disorder is anorexia nervosa. This disease is considered to be a mental illness. Metabolic derangements, such as hypokalemia and kidney damage, can occur with laxative abuse. It is the responsibility of the physician to order screening tests for laxatives that may be present in the urine or feces of suspected abusers (Donowitz *et al.*, 1995).

The primary nutritional concern in treating diarrheal diseases is correction of the dehydration. Dehydration is identified with a *loss of blood volume* and the consequent condition called **shock**. Dehydration can be reversed by oral fluids or, in emergency cases, intravenous fluids. The term **fluid** means water-containing solutes, such as ions and organic compounds. Severe diarrhea of infancy is defined as a fecal volume greater than 60 ml/kg body weight per day. An infant who has lost 5% of its body weight because of an acute diarrheal episode is at risk for shock. A more gradual depletion of body water may not result in shock, because the water losses are shared by the extra- and intracellular fluids, rather than being derived mainly from the extracellular fluid. Shock invariably occurs when dehydration produces a 10% loss in body weight. Although diarrhea contains $Na^+$, as well as $HCO_3^-$ and $K^+$, the primary concerns in treating severe diarrhea are to replace sodium and water and to restore the blood volume.

## Therapy for Diarrhea

Maintenance of the volume of the circulatory system is dependent on sodium ions and water. Oral rehydration therapy may involve drinking a solution of 90 m$M$ NaCl. Replacing K and $HCO_3$ may be a vital concern as well, though they are secondary to water and sodium. Diarrhea can result in acidosis because of the bicarbonate secreted into the lumen and the concomitant delivery of protons to the plasma. Hence, bicarbonate may be included during rehydration therapy to reverse acidosis. **Acidosis** is defined as a plasma pH less than 7.10. Generally, serious and life-threatening acidosis does not occur with mild or moderate cases of diarrhea.

An oral formula suitable for the correction of a variety of diarrheal fluid losses has been recommended by the World Health Organization (WHO). It consists of 90 m$M$ Na, 20 m$M$ K, 80 m$M$ Cl, 30 m$M$ $HCO_3$, and 110 m$M$ glucose. The glucose aids in intestinal absorption of sodium ions, because glucose transport is mediated by the Na-glucose cotransporter. Only low levels of sugar are used in oral replacement therapy, as higher levels can enhance diarrhea in children suffering from malnutrition. The aforementioned formula containing 90 mM Na is generally effective in the treatment of all diarrheas. Solutions containing lower levels of Na (50–60 mM) are considered preferable for noncholera diarrheas. Cholera diarrheas have a higher Na content than others. In noncholera diarrheas, it is recommended that consumption of the 90 mM oral formula be followed by breast-feeding or by drinking water.

Not all diarrheas are the same. The Na, Cl, and K concentrations may differ in different diarrheas. Moderate-volume diarrheas (25 g stool/kg per day) may result in losses of 50 mmol each of Na, Cl, and K per kilogram of stool. High-volume diarrheas tend to contain higher levels of Na and Cl, but lower levels of K. High-volume diarrheas (40–100 g/kg per day) have been associated with losses of about 70 mmol of Na, 70 mmol of Cl, and 30 mmol of K per kilogram of stool per day.

Potassium may be lost via three routes. Potassium can be lost in the diarrhea as well with vomiting. The vomiting of fluids secreted by the upper small intestines results in K losses by the oral route. The vomiting of gastric juices, however, does not lead to significant losses of potassium by the oral route. Extensive vomiting induces *losses of potassium in the urine* and can lead to serious hypokalemia. How can vomiting gastric acid (HCl) cause the loss of potassium in the urine? This situation seems to be a paradox. Referring to Figure 10.14, one can visualize the scenario that has been proposed to account for K losses with prolonged vomiting. Alkalinization of the plasma may induce losses of potassium ions via the K channel of the principal cell.

Potassium deficiency caused by a lack of dietary K is relatively rare outside of cases of prolonged fasting. The deficiency is more commonly caused by vomiting and diarrhea. The basal rate of gastric acid production is under 50 ml per hour in the adult. With persistent vomiting, 1 to 2 liters may be lost per day. The production of gastric acid (HCl), in effect, results in the replacement of some of the sodium chloride in the plasma by sodium bicarbonate. This results in alkalinization of the plasma, as shown in Figure 10.17. Figures 10.18 and 10.19 contrast the pH changes occurring in the plasma during the normal production of gastric acid (Figure 10.18) and during the vomiting of gastric acid (Figure 10.19). Normally, the production of gastric acid results in temporary alkalinization of plasma after a meal. This slight rise in pH is called the *alkaline tide*. Persistent alkalinization is prevented by the body's secretion of $NaHCO_3$ into the small intestines. $NaHCO_3$ is secreted by the pancreas as well as by enterocytes. This secretion is accompanied by the release

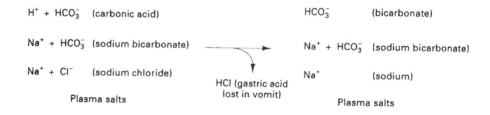

$H^+ + HCO_3^-$ (carbonic acid)

$Na^+ + HCO_3^-$ (sodium bicarbonate)

$Na^+ + Cl^-$ (sodium chloride)

Plasma salts

HCl (gastric acid lost in vomit)

$HCO_3^-$ (bicarbonate)

$Na^+ + HCO_3^-$ (sodium bicarbonate)

$Na^+$ (sodium)

Plasma salts

**FIGURE 10.17** Changes in plasma salts with vomiting.

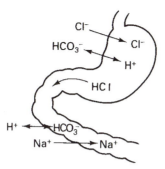

**FIGURE 10.18** Eating. With the consumption of a meal, the parietal cells produce $H_2CO_3$. The proton is released into the lumen while the bicarbonate enters the bloodstream. The chloride of HCl comes from the bloodstream. The enterocyte produces $H_2CO_3$. The bicarbonate enters the lumen while the proton enters the bloodstream. The gastric acid is neutralized by the bicarbonate produced by the enterocytes.

into the bloodstream of protons derived from the dissociation of $H_2CO_3$ in the cell, which results in neutralization of the alkaline tide. With vomiting, however, the alkalosis is not neutralized by the protons, leading to a persistent alkalosis.

## Renal Proton Pumps and the Regulation of Blood pH

The pH of the bloodstream and interstitial fluids is regulated by a number of means, including changing the rates of bicarbonate fixation via the urea cycle, and changing ammonium ion excretion (see the Protein chapter). The renal tubules can regulate blood pH by changing the rate of bicarbonate resorption (Figure 10.12), as well as by changing their rate of proton excretion.

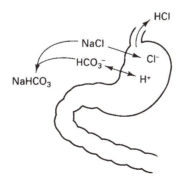

**FIGURE 10.19** Vomiting. With the vomiting of gastric acid, HCl is produced as shown; however, $H_2CO_3$ is not generated for the purpose of neutralizing gastric acid. Thus the proton of carbonic acid, which would otherwise be produced in the enterocyte, is not available for counteracting the alkalosis.

During acidosis, the cells of the renal tubule can respond by inserting two proteins into the apical region of the plasma membrane. (The apical part is that region that is exposed to the developing urine.) The two proteins are H,K-**ATPase** and H⁺-**ATPase**. H,K-ATPase, and the enzymes that act in concert with it, is better known as a component of the parietal cell where it creates stomach acid. The other proton pump of the renal tubule, which is H⁺-ATPase, is closely related to $F_0F_1H^+$-ATPase of the mitochondrial membrane. Hence, anyone who understands how protons are pumped out of the mitochondrion and how stomach acid is made will clearly understand how the renal tubule can shuttle protons to the lumen of the renal tubule and into the developing urine.

Most of the H,K-ATPase and H⁺-ATPase of the kidney occurs in the cells of the collecting duct (Tsuruoka and Schwartz, 1997; Cougnon *et al.*, 1996; Gluck *et al.*, 1996).

## Long-Duration Exercise

The most vital health concern during prolonged exercise is maintenance of the body's temperature. The contraction of muscles during exercise or work results in an increase in heat production. This heat is rapidly carried from the muscles, via the blood, near the surfaces of the body. The heat is then lost to the atmosphere, where the air is cooler than the body. Another source of heat loss is the evaporation of sweat water. The purpose of sweating is to cool the body during exercise. A typical rate of sweat loss, during a marathon race, for example, is 1 liter per hour. Where exercise takes place in a warm humid climate sweating can result in large losses of fluid from the body but little cooling. Here, the sweat rolls off the body without evaporating. A marathon taking place in a warm humid climate can evoke fluid losses of 2.8 liters per hour. The American Dietetic Association recommends that fluid losses occurring with exercise of moderate duration (1 hour or less) be replaced with plain cool water (Smith, 1987). In hot and humid environments, the fluid should be replaced by drinking cool water prior to endurance exercise and at 15-minute intervals during exercise.

A topic of periodic concern is the risk from sodium and potassium losses in sweat. The concentrations of Na and K in sweat are influenced by the rate of sweating. Lower sweat rates (200 ml/hr) result in the loss of fluids containing about 12 m$M$ Na and 4 to 5 m$M$ K. Higher sweat rates (1000 ml/hr), such as those provoked by fast running races, produce a fluid containing about 40 m$M$ Na and 4 to 5 m$M$ K. The concentration of Na in sweat is about half that in plasma, whereas the concentrations of K in sweat and plasma are about the same. A sweat rate of 1 liter per hour would lead to the loss of about 4 mmol of K per hour. A 5-hour marathon might thus cause the loss of about 20 mmol of potassium. The volume of extracellular fluid (ECF) in the 70-kg man is about 14 liters. The ECF has a K level of about 4 m$M$ and thus contains a total of about 56 mmol of potassium. Therefore, the marathon might be expected to cause a severe decrease in the concentration of K in the ECF, resulting in hypokalemia. Studies with exercising subjects have, however, demonstrated that plasma K levels do not decrease with exercise and hypokalemia does not result. This is because muscle and other tissues serve as a vast reservoir of potassium. The intracellular fluids contain about 140 m$M$ K and have a total volume of about 28 liters. Thus, the 70-kg man contains about 3920 mmol of potassium. The potassium losses during a marathon are negligible compared with the amount of K present in the tissues of the body.

Research to date indicates that Na and K supplements are not required during running races of short (1 hour or less) and moderate (5 hours) duration. The salts provided by the diet are sufficient to maintain Na and K balance during repeated days of competition at moderate levels. With moderate-duration exercise, the kidneys enhance their efficiency of Na and K reabsorption. Because of this adaption, any Na and K supplements taken during the course of exercise would be superfluous and would result in the excretion of amounts of Na and K equivalent to those in the supplements. Sodium-restricted diets are commonly used to control high blood pressure. Low-Na diets should not be consumed with repeated days of moderate-duration exercise.

EXERCISE

How many hours of running and sweating (without eating K) might be expected to deplete tissue potassium levels by 5.0%? Ignore the obligatory urinary K losses for this calculation.

A sweat rate of one liter per hour, as occurs with continuous moderate exercise, would lead to the loss of about 40 mmol of Na per hour. A 5-hour running marathon would result in the loss of 200 mmol of sodium. This loss represents depletion of about 12% of the sodium in the ECF. This loss does not, however, result in a drop in plasma sodium concentrations. Plasma sodium is maintained during prolonged exercise by the loss of plasma water, resulting in a drop in plasma volume. Experiments involving human subjects exercising on a stationary bicycle for 3 hours in a warm humid room suggested that about 10% of the water lost in sweat comes from plasma, 38% from interstitial fluid, and about 52% from intracellular fluids (Table 10.9).

An interesting effect may be observed during ultramarathons. Ultramarathons are races of 50 miles or longer, lasting about 8 hours. The continued loss of fluids in sweat and the continued consumption of plain water can result in hyponatremia. Sweating results in the loss of both water and sodium. Continued drinking of plain water during prolonged exercise results in the replacement of lost water, which is lost at a relatively rapid rate, but not in the replacement of lost sodium,

**TABLE 10.9**  Changes in Electrolytes with Exercise

|  | Sweat losses | Urinary losses | Preexercise | | Postexercise | |
|---|---|---|---|---|---|---|
|  |  |  | Concentration in plasma | Total amount in plasma | Concentration in plasma | Total amount in plasma |
| Volume | 4.11 liters | 0.24 liter | – | – | – | – |
| Sodium | 155 mmol | 32 mmol | 135 m$M$ | 462 mmol | 141 m$M$ | 419 mmol |
| Potassium | 16 mmol | 29 mmol | 4.0 m$M$ | 13.7 mmol | 4.6 m$M$ | 13.7 mmol |
| Chloride | 137 mmol | 39 mmol | 104 m$M$ | 358 mmol | 106 m$M$ | 320 mmol |

Source: Costill et al. (1976).

which is lost at a relatively slower rate. The consumption of plain water at a rate greater than that lost by sweating may produce a dangerous hyponatremia (125 m$M$ plasma Na or less) that causes disorientation, confusion, and seizures.

The results of the following study illustrate the losses of Na and K in sweat and the changes in plasma Na and K levels during a period of exercise. Subjects exercised on a stationary bicycle in a warm humid room, without drinking water, until they lost 6% of their body weight. They experienced extreme fatigue during the final stages of the exercise-induced sweat loss. The data in Table 10.9 demonstrate that the amount of Na lost in sweat was equivalent to about half the amount of sodium consumed in a typical diet. The kidneys can easily adapt, over the course of a few days, to maintain sodium balance with the consumption of half the amount of sodium in a typical American diet. The amount of potassium lost in the sweat was small compared with that consumed in the diet per day. The amount of K lost in the sweat was about the same as that in a medium-sized potato. The kidneys also adapt with the consumption of low K levels. Note that the body may become partially depleted of Na and K between the onset of heavy exercise and the time when maximal adaption has occurred. This adaption may take a few days for maximal Na conservation and about a week for maximal K conservation. A negative ion balance might be expected with the sudden onset of a period of prolonged exercise, exposure to hot and humid conditions, and fasting.

## Blood Pressure

A major concern in sodium nutrition is hypertension. Hypertension, which is abnormally high blood pressure, is a significant risk factor for cardiovascular disease and renal failure. About a quarter of the adults in the United States have hypertension. The highest incidence of hypertension is found in northern Japan, where the Na intake may be above 400 mmol per day (20–30 g of NaCl/day). The source of the Na is the salt added in pickled vegetables, preserved marine products, soy sauce, and miso (broth). Moderate incidences of hypertension occur in the United States, where the Na intake generally ranges from 100 to 200 mmol or more per day (100 mmol of Na is supplied by 6.4 g of NaCl). The moderately restricted Na diet used in hospitals and other institutions provides about 44 mmol of Na per day. A strictly restricted Na diet contains about 22 mmol of Na per day. High levels of sodium are also found in catsup, potato chips, pretzels, canned food, frozen convenience foods, and bread made with baking soda.

Very low incidences of hypertension have been found in "primitive" societies, where the Na intake is very low (10–60 mmol/day). For example, certain groups in Papua New Guinea, do not add salt to their diet. The diet, which may consist of sweet potatoes, plantains, vegetable leaves, and corn, supplies 20 to 40 mmol of Na per day. Eskimos also consume a low-Na diet. Their diet, which consists of fish and bear meat, may supply only 60 mmol of Na per day.

A role for Na intake in hypertension was suggested by the finding that people who consume low-Na diets, such as those just mentioned, tend to acquire hypertension when they move to an industrialized society. Apparently, hypertension can be caused by the diet more easily than it can be reversed. When high levels of Na have been in the diet for several years, the associated hypertension may be only partially reversed by restricting the Na intake. Sodium intake represents only part of the etiology of hypertension. Apparently, genetics plays a role. Some people are predisposed to developing hypertension and acquire high blood pressure with the consumption of excess sodium. These persons are called "salt sensitive." Others are "salt-resistant" or "salt-insensitive" and do not develop high blood pressure with the excessive consumption of sodium, within certain limits. In salt-resistant persons, the blood pressure remains constant and unchanged even with a low-salt diet (Lopez *et al.*, 1995).

Salt-sensitive people tend to develop hypertension with an intake of 125 to 250 mmol of Na per day, whereas salt-resistant people may not have a rise in blood pressure with up to 300 mmol of Na per day. Very large amounts of Na (800 mmol/day) lead to hypertension in both salt-sensitive and -insensitive people. It has been estimated that close to 20% of adult Americans are salt sensitive. African-Americans have a greater incidence of Na sensitivity and hypertension than Americans of European descent. Generally, blood pressure tends to increase with age in salt-sensitive individuals consuming moderate or high levels of sodium. The prevalence of hypertension is very low in children and adolescents. The interracial differences in blood pressure that are found in adults are not found among children (Sinaiko, 1996).

Increased levels of potassium in the diet have been associated with a decrease in blood pressure. The exact populations that might benefit from increased levels of dietary K are currently being studied. In one study, K supplements given to hypertensive patients resulted in decreases in blood pressure. These supplements resulted in a doubling of the usual K intake to about 160 mmol K per day. The blood pressure was only partially reduced and did not reach normal values (Treasure and Ploth, 1983). Increasing the dietary K levels of normotensive (normal blood pressure) persons consuming typical Na- and K-sufficient diets had little or no effect on blood pressure (Miller *et al.*, 1987). The association between K intake and blood pressure is stronger in African-Americans than in Americans of European descent. This statement is based on studies showing that African-Americans have relatively lower intakes of K and greater blood pressures.

Hypertension is often treated with diuretics. Diuretics are drugs that promote the loss of sodium from the body, though some diuretics can cause the loss of potassium, resulting in hypokalemia. The use of dietary supplements of K to correct this hypokalemia has been shown to be of benefit and to result in decreases in blood pressure. The best source of potassium is plant food, as is strikingly apparent from the data in Table 10.1; however, the interest in nonfood supplements of potassium continues. One problem with potassium salts is that they taste bad and can produce nausea. Hence, there has been some interest in the manufacture of aesthetically acceptable forms of potassium salts. One form, a mixture of NaCl and KCl, is useful for those who feel compelled to add sodium chloride to their food. One study revealed that NaCl–KCl mixtures are accepted and their use can result in a reduction of sodium intake. The study, which involved normotensive subjects, did not lead to any consistent change in blood pressure (Mickelsen *et al.*, 1977).

The hypertension of salt-resistant persons is often treated with drugs that inhibit angiotensin converting enzyme (ACE). When left untreated, hypertension can provoke damage to kidneys (nephropathy) (Chowdhury *et al.*, 1996).

*Pregnancy and High Blood Pressure*

Hypertension is a common disorder during pregnancy, affecting 1–5% of pregnant mothers, and results in increased release of protein in the urine and sometimes in death of the fetus. In pregnancy, the combination of hypertension with increased urinary protein is called **preeclampsia**. Preeclampsia is a concern during pregnancies, as it may lead to eclampsia. **Eclampsia** involves convulsions and possibly death to the mother. The hypertension of preeclampsia does not respond to a low-salt diet — it is treated with bed rest and a drug (methyl-dopa) (Sibai, 1996). Among women with preeclampsia, the incidence of progress to eclampsia is only 0.2%. Although this incidence is rather low, it is the responsibility of the physician to monitor urinary protein levels during preeclampsia. Surprisingly, a common nutrient is the drug of choice for preventing the convulsions of eclampsia. This nutrient is magnesium ($MgSO_4$) (Sibai, 1996). An especially dangerous combination is pregnancy in women with IDDM (insulin-dependent diabetes). Close to 15% of these diabetic women acquire hypertension during the course of their pregnancy. When a diabetic woman becomes hypertensive during pregnancy the fetal death rate is relatively high (Garner, 1995). When the diabetes is left untreated, the rate of birth defects such as neural tube defects and heart abnormalities increases.

*The Dahl Rat*

The availability of animals bearing various maladies — such as cancer, diabetes, and obesity — has proven vital to our understanding of the corresponding diseases in humans. The discovery of a salt-sensitive strain of rat, called the **Dahl rat**, has led to insights into human hypertension. With feeding of a high-salt diet, normal rats do not acquire high blood pressure, but the Dahl rat does acquire high blood pressure. The mechanism seems to be due to an increased rate of NaCl resorption by the renal tubule. Specifically, there is increased NaCl resorption in the thick ascending loop. With prolonged salt-feeding to the Dahl rat, and with prolonged hypertension, the glomeruli become damaged (glomerular sclerosis). This further reduces salt excretion, which leads to even higher blood pressure. This pathological scenario in the Dahl rat closely resembles that in salt-sensitive humans (Cowley, 1997).

In addition to the Dahl rat, several other strains of rats have been found to be salt sensitive. Naturally occurring mutations in the rat genome have been found to occur in the genes coding for angiotensin (Kim *et al.*, 1995), 11β-steroid hydroxylase, renin, angiotensin converting enzyme (ACE), and other polypeptides (Cowley, 1997). The results from rats have given researchers a number of places to begin their hunt for variations in the genes in salt-sensitive humans. In humans, the only defect found, to date, to result in hypertension is a mutation in the gene for angiotensinogen. This mutation results in the replacement of Met 235 by Thr 235. This particular mutation has been found in woman who acquire preeclampsia, a conditions that occasionally arises during pregnancy (Jeunemaitre *et al.*, 1997).

# IODINE

Iodide is a monovalent anion. Its only known occurrence in mammals is as a component of the thyroid hormones. These hormones are essential during embryological development and for regulating metabolic rate and heat production throughout life. The RDA for iodine is 0.15 mg. Iodine occurs in foods mainly as iodide and, to a lesser extent, covalently bound to amino acids. Iodide is rapidly absorbed by the gut and assimilated by the thyroid gland for use in producing the thyroid hormones. The iodine occurring as part of amino acids, that is, as part of tyrosine, is somewhat less well absorbed. The main excretory route of iodine is the urine. Urinary excretion is a reliable indicator of iodine intake and status. A urinary level greater than 50 µg I/g creatinine is considered to indicate adequate iodine status. Lower levels (25–50 µg I/g creatinine) indicate a risk for deficiency, and even lower levels indicate a serious risk.

Iodine deficiency is closely associated with the iodine content of the soils used in agriculture in the affected areas. Although iodine supplementation programs have eradicated iodine deficiency in many areas of the world, serious deficiencies still occur in certain regions.

## Metabolism of Inorganic Iodine

Iodine supplements may take the form of potassium iodide (KI) or potassium iodate (KIO$_3$). The iodate anion is rapidly reduced to iodide in the body. The reduction is nonenzymatic and involves sulfhydryl-containing compounds such as glutathione. Glutathione is a tripeptide containing glutamate, cysteine, and glycine (Glu–Cys–Gly). Glutathione is present in the plasma and inside cells. The conversion of iodate to iodide that uses the reducing power of sulfhydryl groups requires the oxidation of six sulfhydryl groups (R—SH) to three disulfide groups (R—S—S—R):

$$6\,\text{R–SH} \;+\; \text{IO}_3^- \longrightarrow 3\,\text{R–S–S–R} \;+\; \text{I}^- \;+\; 3\,\text{H}_2\text{O}$$

(Sulfhydryl compound, such as glutathione)    (Iodate)    (Disulfide)    (Iodide)    (10.1)

## Conversion of Inorganic Iodine into Organic Iodine, and Thyroid Hormone Synthesis

Iodide in the bloodstream enters the thyroid gland via the Na–I cotransporter. This cotransporter facilitates the coupled entry of sodium and iodide into the cell. The first step in thyroid hormone synthesis is the incorporation of iodide into a large protein. The iodide is incorporated into tyrosine residues of **thyroglobulin**, a protein of molecular weight 660,000. The protein is composed of two identical subunits. It contains 140 residues of tyrosine, but does not have a higher proportion of tyrosine than the average protein. Normally, thyroglobulin contains 10 to 50 iodine atoms. Less than one-third of the tyrosines contain iodine in the form of monoiodinated and diiodinated tyrosine (Figure 10.20).

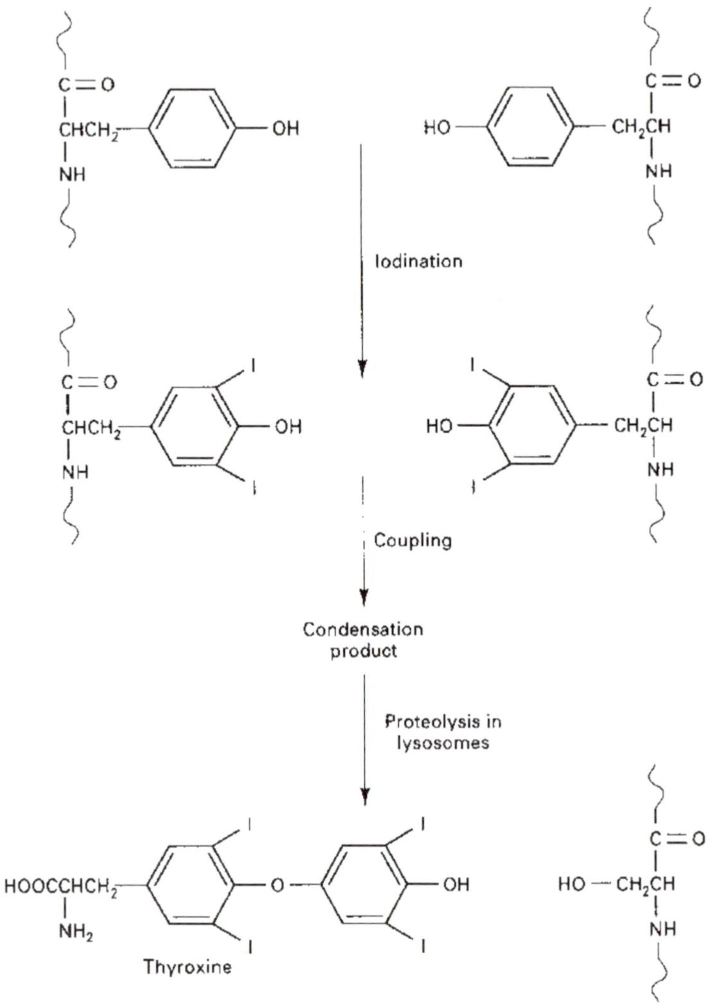

**FIGURE 10.20** Synthesis of T4 from thyroglobulin.

The second step in thyroid hormone synthesis is the covalent bridging of two different residues of iodinated tyrosine. A dimer is formed, reminiscent of the dimers of cysteine in proteins (the cysteine dimer is called cystine). Only a small fraction of the iodinated tyrosines is bridged in this way. More specifically, only four of the iodinated tyrosines, located at positions 5, 2555, 2569, and 2748, participate in the reaction. The numbers refer to the amino acid, counting from the amino terminus of the protein. Thyroglobulin has 2748 amino acids. The first and second steps are catalyzed by **thyroperoxidase**, a heme protein. It requires hydrogen peroxide for activity. To summarize, thyroperoxidase catalyzes the attachment of iodine atoms to residues of tyrosine as well as the subsequent cross-linking of the iodinated tyrosine residues.

The third step in thyroid hormone synthesis is cleavage of the aromatic R group from one of the iodinated tyrosines in the dimer. As a result, one of the iodotyrosines contains an extra aromatic ring, and the other iodotyrosine residue loses its aromatic ring (Figure 10.20). Then the amino acids at the N-terminal and C-terminal ends are cleaved, liberating the coupled iodotyrosine residues. Coupled iodotyrosines that contain four atoms of iodine are called 3,5,3′,5′-**tetraiodothyronine** (thyroxine, T4). Coupled iodotyrosines that contain only three residues of iodine are called 3,5,3′-**triiodothyronine** (T3). Most of the thyroid hormone released from the thyroid gland takes the form of T4, with T3 making up only about 10% of the hormone released. After liberation of T4 and T3, the remainder of thyroglobulin is proteolyzed, releasing the large quantity of iodotyrosines that did not participate in thyroid hormone synthesis. These iodotyrosines are deiodinated, resulting in highly efficient recycling of the iodine atoms in the thyroid gland.

Iodinated thyroglobulin is stored in a viscous gel in the thyroid gland called **colloid**. The store of iodinated thyroglobulin is sufficient for about 100 days of normal thyroid hormone secretion with the consumption of an iodine-free diet. Thyroglobulin accounts for about 75% of the weight of the entire thyroid gland (see Figure 10.21).

The concentrations of T4 (80 ng/ml) and T3 (1.2 ng/ml) in normal human serum are indicated. Most of the T3 in the bloodstream is not directly produced by the thyroid, but results from deiodination of T4 in the liver and kidney, a reaction catalyzed by 5′-**deiodinase** (Figure 10.22). The enzyme is bound to the membrane of the endoplasmic reticulum. This means that most of the T3 entering target organs such as the skeletal muscle is produced with the participation of the

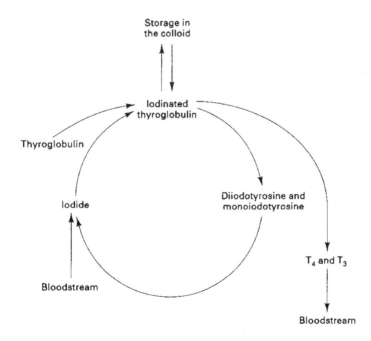

**FIGURE 10.21** Iodine uptake and thyroid hormone synthesis by the thyroid gland. Thyroglobulin is broken down to yield T4 and T3 as well as iodotyrosine byproducts. These byproducts are further broken down in the thyroid to yield iodide.

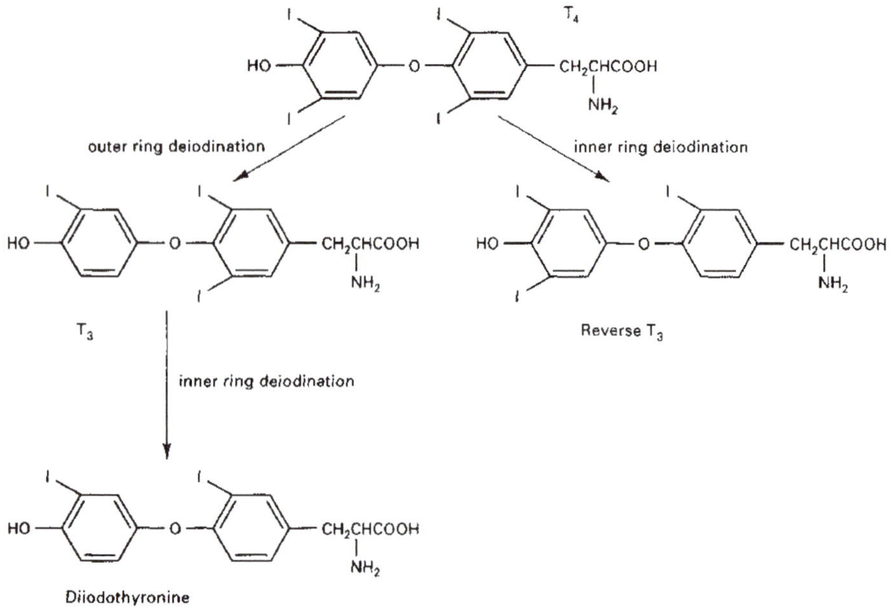

**FIGURE 10.22** Conversion of T4 to T3 and to inactive catabolites.

thyroid and the liver. The brain, on the other hand, produces its own T3 from T4. About 40% of T4 is converted to a biologically inactive metabolite, **reverse T3**. Removal of one of the iodine atoms from the inner ring of T4 produces reverse T3. Both T3 and reverse T3 are further deiodinated to produce other inactive compounds, mono- and diiodothyronine, which are excreted in the urine.

Thyroid hormones have a lengthy life-span in the bloodstream — several days — probably because they are bound to proteins in the circulation. More than 99% of T4 and T3 is bound to plasma proteins. These proteins are thyroid hormone-binding protein, transthyretin, and albumin. Most of the hormone is carried by thyroid hormone-binding protein. Transthyretin (from thyroid and retinol) occurs in a 1:1 complex with retinol-binding protein in the bloodstream. This complex serves to prevent the loss of retinol-binding protein, which is a small protein, in the urine. Transthyretin has also been called prealbumin. It binds T4 and not T3.

## The Deiodinases

The deiodinases are selenium-containing enzymes that are used for the synthesis of the active form of thyroid hormone, T3. The deiodinases also catalyze the inactivation of the various forms of thyroid hormone. Three types of deiodinase exist, and these are called type I, type II, and type III deiodinase. Further details appear in the Selenium section.

Type I deiodinase appears to be responsible for most of the T3 in the bloodstream. The type I enzyme catalyzes the conversion of T4 to T3 within the thyroid gland. This is a 5′-deiodination reaction. The dominant role of this enzyme is revealed by the fact that only type I deiodinase is inhibited by the propylthiouracil, a drug used to produce thyroid hormone deficiency. Propylthiouracil does not inhibit the type II or III enzymes (St. Germain, 1994; Croteau *et al.*, 1996).

Type II deiodinase occurs in the brain and brown adipose tissue of rats (but not in muscle of rats), and in the brain, skeletal muscle, heart, and thyroid gland in humans (Pallud *et al.*, 1997). This enzyme catalyzes the conversion of T4 to T3. When the thyroid gland is stimulated, the type II deiodinase takes on an increased importance in the conversion of T4 to T3 (Salvatore *et al.*, 1996). Type II deiodinase is unique among the deiodinases in that it appears to contain two selenium atoms, rather than just one. The physiological role of the enzyme is to utilize T4 acquired from the bloodstream and to convert it to T3 within the target tissue.

Type III deiodinase catalyzes the conversion of T4 to reverse T3, and the conversion of T3 to T2. These steps constitute 5'-deiodination reactions. The enzyme occurs in the brain and skin of rats. The physiological role of the type III enzyme is thought to be to protect the brain from possible toxic effects of active thyroid hormone (T3). The placenta is distinguished in that it contains both type II and type III deiodinases (Glinoer, 1997).

## Regulation of the Thyroid Gland

The activity of the thyroid gland, which is located in the neck, is regulated by hormones secreted by two glands in the brain, namely the pituitary gland and the hypothalamus. The activity of the thyroid gland is controlled via its stimulation by **thyroid-stimulating hormone** (TSH), secreted by the pituitary. TSH, also called thyrotropin, is a protein of molecular weight 28,000. It is a glycoprotein and contains oligosaccharides bound to residues of asparagine. Some of the sugar residues contain sulfate groups. The thyroid is able to change its sensitivity, depending on the availability of dietary iodine. With iodine deficiency, the sensitivity to TSH increases, resulting in stimulation of the thyroid gland. This stimulation takes the form of an increase in iodide transport, an increase in thyroperoxidase activity, and enlargement of the thyroid gland.

The pituitary gland is involved in feedback regulation of thyroid activity. High levels of T4 in the bloodstream result in inhibition of TSH secretion. Low levels of T4 result in an increase in TSH secretion. These effects are dependent on the conversion of T4 to T3 within the pituitary. The activity of the pituitary is controlled by **thyrotropin-releasing hormone** (TRH), a hormone synthesized in the hypothalamus. TRH is a tripeptide with the structure pyroglutamate–histidine–proline–$NH_2$. Note the C-terminal amide group, which is required for the activity of many peptide hormones. TRH stimulates the synthesis and secretion of TSH. Apparently, TRH is involved in regulating the sensitivity of the pituitary to the inhibitory feedback control mechanism mentioned earlier.

## Hormonal Effects of Thyroid Hormone

An increase in thyroid hormone levels results in an increase in the basal metabolic rate (BMR). BMR measurements can be used to assess thyroid status, as discussed in detail in Chapter 5. This method for the diagnosis of hypo- or hyperthyroidism is not in common use because it is cumbersome. The increase in BMR has been associated with increases in various reactions that use ATP. The increased use of ATP is matched by an increase in activity of the respiratory chain and in $O_2$

reduction. The two reactions most closely associated with the increase in BMR with higher levels of plasma thyroid hormones are that of Na,K-ATPase (the sodium pump) and those of fatty acid synthesis. Na,K-ATPase is present in the membranes of most or all cells of the body. The increase in fatty acid synthesis with the rise in thyroid activity is associated with diversion of fatty acids in the liver away from triglyceride synthesis and toward oxidation. A rise in thyroid activity also induces an increase in the rate of fatty acid synthesis. The overall effect is a futile increase in fatty acid oxidation and fatty acid synthesis that results in excess heat production.

Changes in thyroid hormone levels often occur in diseases unrelated to iodine status. These diseases are known as hyperthyroidism and hypothyroidism. They may arise from defects in the thyroid gland, pituitary, or hypothalamus. **Hyperthyroidism** results in weight loss, despite a normal or increased energy intake. It may result in an exaggerated release of fatty acids from the adipose tissue with fasting. Hyperthyroidism can be treated with drugs that inhibit 5'-deiodinase. One such drug is **propylthiouracil**. Propylthiouracil has a structure similar to those of the antithyroid compounds present in tiny amounts in cabbage. In **hypothyroidism**, the opposite trends in metabolism can occur. There may be decreases in BMR and body temperature, as well as a mild gain in weight. Thyroid hormone plays a vital role in the normal growth of the fetus. A deficiency in this hormone has especially severe effects on brain development.

Changes in thyroid hormone levels in the body provoke consequent changes in metabolism by forming a complex with a special transcription factor, the thyroid hormone receptor. The hormone/receptor complex then binds to special regions of the chromosome and provoke changes in the rate of transcription of nearby genes. Details of this scenario were presented in the vitamin chapter in the section on Vitamin A, Vitamin D, and Thyroid Hormone at the Genome.

## Iodine Deficiency

An iodine intake of 100 to 150 µg/day is considered nutritionally adequate. Urinary iodine levels are similar to intake levels and can be used to estimate iodine consumption. Iodine deficiency occurs with an intake below 50 µg/day. Persons consuming this level are at risk for developing **goiter**. Goiter is almost always induced by intakes under 10 µg/day. Goiter is enlargement or hypertrophy of the thyroid gland. The thyroid in the iodine-sufficient subject is barely visible in the neck. Goiter, of increasing severity, is categorized in three grades: (1) a small enlargement detectable by palpitation, (2) a thick neck, and (3) a large swelling that may be asymmetrical and is visible from a distance. A grade 3 goiter compresses the trachea and can result in shortness of breath during heavy work. The highest incidences of goiter are found in less developed areas of the world such as the Czech Republic, the former Yugoslavia, parts of India, Paraguay, Peru, Argentina, Pakistan, most of Africa, Southeast Asia, and New Guinea. The disease does not frequently strike more developed areas because of the practice of fortifying salt with iodine. Goiter was once common in the United States in areas near the Great Lakes and westward to Washington and the Pacific. Goiter nearly disappeared by the 1950s because of salt fortification. Table salt is fortified with 100 mg KI/kg NaCl. Milk and bread may be inadvertently fortified with iodine. The iodine in milk originates from disinfectants used in the dairy industry. The iodide in bread (1 mg iodine/kg bread) originates from its use as a dough oxidizer by the bread industry. Goiter is reversible.

The most serious complication of iodine deficiency is cretinism. In societies suffering from widespread goiter, about 2% of the population may be cretins. Cretins suffer from mental retardation and have a characteristic facial appearance and a large tongue. Some are deaf and mute. Dwarfism, diplegia, and quadriplegia may also occur. Cretinism results from maternal iodine deficiency and thus is a diet-related birth defect. The mental and physical impairments of cretinism are not reversible. They can be prevented by treating the iodine-deficient mother with iodine early in pregnancy. Goiter is easy to diagnose because it presents by a swelling in the throat. Cretinism may be difficult to diagnose because it presents in a variety of different ways. The defect illustrates the importance of the thyroid hormones in fetal development.

Iodine deficiency and the associated goiter and cretinism are treated by public health programs involving salt fortification and injections with iodized oil. The salt may be fortified with iodide (KI) or iodate ($KIO_3$). Iodate is more stable to moisture and sunlight and is used as a supplement in underdeveloped countries. The iodine administered in oil is covalently bound to fatty acids and is released with the catabolism of the oil. Oil injections may be more acceptable in areas of the world where food is not salted, such as in parts of New Guinea. The efficacy of oil was shown in a study of iodine-deficient schoolchildren (Furnee *et al.*, 1995). A cutoff point of urinary iodine, which was considered indicative of iodine deficiency, was considered to be 0.4 μmol iodine per liter of urine. At the beginning of the study, the children were excreting about 0.17 μmol I/liter. A single oral dose of triglycerides containing 675 mg iodine resulted in a concentration of urinary iodine above the cutoff point for just over one full year.

The effect of iodine deficiency on thyroid weight and plasma T4 is briefly illustrated by the following experiment. Rats were fed diets containing normal (0.2 mg iodine/kg diet) or low (0.1 mg iodine/kg diet) levels of iodine for 4 months. Table 10.10 demonstrates a marked increase in the weight of the thyroid gland, a decrease in plasma T4, and an increase in blood flow through the gland with the deficient diet. The increase in blood flow is thought to represent an adaptive mechanism that enhances the rate of delivery of plasma iodine to the thyroid gland in the deficient animal. Enlargement of the thyroid may be attributed to stimulation by increased levels of TSH. TSH levels increased by a very small extent in the study with rats. As discussed earlier, the thyroid can increase its sensitivity to TSH and thus be more greatly stimulated by TSH even with normal plasma TSH levels. Note that the normal human thyroid gland weighs 20 to 25 g.

**TABLE 10.10** Effects of Iodine Deficiency on Organs and Hormones

|  | Normal | Deficient |
|---|---|---|
| Thyroid gland weight (mg) | 13 | 23 |
| Plasma T4 (ng/ml) | 40 | 20 |
| Thyroid blood flow (ml/min per g of tissue) | 23 | 68 |
| Thyroid-stimulating hormone (ng/ml) | 2.4 | 2.9 |

*Source*: Michalkiewicz *et al.* (1989).

## Goitrogens

The risk of developing goiter during iodine deficiency is increased with the consumption of foods containing **goitrogens**. Goitrogens are chemicals that are toxic to the thyroid or that break down to produce toxic chemicals. Goitrogens are present in various foods, such as cassava, cabbage, turnips, and rutabagas. Cassava, for example, is a staple in Africa and tropical areas in the Pacific Basin. Cassava contains cyanogenic glycosides, which are a source of cyanide. In cassava intended for animal food, the cyanide is removed by warming the chopped cassava in the sun. When cassava is intended for human consumption, the cyanide is removed by boiling it in water. If not adequately removed, the cyanide is released from the goitrogen either in the plant or in the body. In the body, it is converted to thiocyanate. Thiocyanate induces goiter by inhibiting uptake of iodine by the thyroid. Thiocyanate also inhibits the activity of thyroperoxidase. Goitrogens in cabbages and related plants inhibit thyroperoxidase, especially its activity in catalyzing the coupling reaction. The goitrogen in cabbage is called goitrin. A similar chemical, propylthiouracil, is used as a drug in the treatment of hyperthyroidism. The structures of the compounds mentioned are shown in Figure 10.23.

High iodine intake can also have an antithyroid effect. An intake of 2.0 mg of iodide per day can impair the synthesis of thyroid hormone, resulting in lower levels of plasma T4 and T3. The thyroid gland enlarges in response, producing goiter. This problem occurs in Japan, where large amounts of iodine are consumed in the form of seaweed. The daily consumption of seaweed soup can supply 80 to 200 mg iodine per day, causing "iodine goiter" in children and adults.

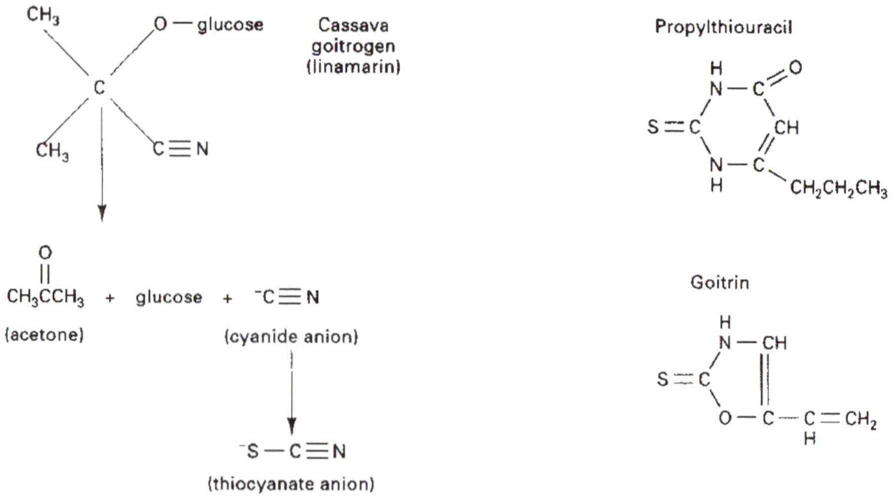

FIGURE 10.23 Structures of goitrogens. The breakdown of linamarin is shown.

# IRON

The RDA for iron for the adult man is 10 mg; that for the adult woman is 15 mg. The RDA during pregnancy is 30 mg. The RDAs are markedly influenced by the relatively low absorbability of dietary iron and by the losses of iron occurring during menstruation. Iron is present in a variety of foods. In the United States, the daily intake of iron is about 10 mg, with about one-third of this being supplied by meat, poultry, and fish. The need for iron as a nutrient is universally appreciated. Its role as a component of the blood and as a nutrient needed to prevent anemia is common knowledge; however, iron is a rather atypical nutrient. First, a deficiency in this nutrient is not associated with symptoms that are striking or devastating, in contrast to the blindness caused by vitamin A deficiency and the bleeding caused by vitamin C deficiency. Another unique feature of iron is that it is stored in large amounts in the body. It is stored in a protein called **ferritin**. The iron in ferritin enables the newborn infant to grow while consuming foods marginal in iron content, such as breast milk. A further striking feature of iron is that it occurs as free iron in concentrations that are extremely low in the body. This is in contrast to free calcium ions, which occur at about 2 m$M$ in the plasma, and free potassium ions, which are about 140 m$M$ in the cell. There are two reasons that nearly all of the iron in the body is protein bound, rather than free. The first is that iron (ferric iron) is not soluble in water. Another reason is that free iron (ferrous iron) can be toxic to the cell. Ferrous iron can react with the hydrogen peroxide in biological fluids to form the hydroxyl radical, via the Fenton reaction. The Fenton reaction is shown in Figure 10.24. The complexation of iron with proteins may prevent its participation in generation of hydroxyl radicals.

## Issues in Iron Solubility

A discussion of the solubilities of iron salts may prove useful in understanding the role of chelating agents in promoting or impairing the absorption of dietary iron. It may also contribute to an appreciation of the role of iron transport and storage proteins in facilitating the handling of iron in an aqueous environment. A brief outline of the locations of iron in the body introduces some of the iron transport and storage proteins and allows a comparison between the amount of iron stored in the body and the amount depleted by obligatory losses. A close look at iron biochemistry reveals a few details about hemoglobin, the most well known iron metalloprotein.

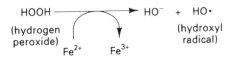

FIGURE 10.24 One-electron oxidation of ferrous iron to ferric iron. The oxidation of ferrous to ferric iron can take place under a variety of conditions. One condition, shown here, involves the reduction of HOOH, producing a hydroxide anion and a hydroxyl radical. This reaction is the Fenton reaction.

A description of iron absorption and the effects of various foods on absorption represents topics of a more nutritional nature. Dietary iron consists of iron bound to the heme prosthetic group of heme-iron proteins and of iron bound to cysteine residues of nonheme iron proteins. The life cycle of iron involves absorption from the diet and insertion into various iron metalloproteins. It also involves the catabolism of iron metalloproteins, such as hemoglobin, liberation of the iron atom, and reutilization of the iron. The life cycle of the iron atom is quite eventful. An understanding of this life cycle requires descriptions of the proteins used for the transport and storage of iron and of the birth and death of the red blood cell. A number of mechanisms are used to dismantle the senescent red blood cell. Tests for iron status involve an examination of red blood cells, as one might expect; however, it might come as a surprise that iron status is often assessed by an examination of iron transport and storage proteins, rather than by functional tests involving iron metalloenzymes. This is in contrast to the various functional tests used for assessing status for the various vitamins, as stressed in Chapter 9. There are several problems in iron nutrition that affect persons of various age groups.

The following descriptive material concerns the solubility of iron in the reduced form (ferrous iron, $Fe^{2+}$) and the oxidized form (ferric iron, $Fe^{3+}$). In a solution of strong acid, ferrous iron exists as a complex with water, $Fe(H_2O)_6^{2+}$. The complex contains six molecules of water. At higher pH, some of the protons of the complex are released, generating $Fe(OH)_2$. $Fe(OH)_2$ is pale green and, if present at sufficiently high concentrations, can form a gelatinous precipitate. Its maximal solubility at pH 7 is about 0.1 M. Hence, this form of iron is fairly soluble at neutral pH.

In a solution of strong acid, ferric iron exists as a complex with water, $Fe(H_2O)_6^{3+}$. At higher pH, some of the protons are released, generating $Fe(OH)_3$. $Fe(OH)_3$ is somewhat soluble in mild acid (pH 2.0), where it can be dissolved at a concentration of about 1.0 mM. At higher pH, such as at neutrality, $Fe(OH)_3$ is not water soluble. At pH 7.0 $Fe(OH)_3$ can be dissolved to a concentration of only $10^{-18}$ M. This is one-billionth of a billionth molar. In solutions of mildly acidic or neutral pH, $Fe(OH)_3$ forms a brown polymer or aggregate.

The levels of iron that can be maintained in solution can be enhanced by chelating agents. A chelating agent, or chelator, is usually a small organic compound that forms a complex with a metal ion. Ascorbate and apo-heme can form chelates with iron. Sugars, such as fructose, glucose, and sucrose, can also chelate iron. Multivalent anions, such as citrate, are good chelators. A level of 5 µM ferric iron can be kept in soluble form in the presence of 5 mM citrate. The iron present in these chelators is nutritionally available. The iron present in other chelating agents, in contrast, is not readily available to people or animals. These chelators include phosphate, phytic acid, and certain clays.

## Iron in the Human Body

The adult man contains 40 to 50 mg of iron per kilogram body weight. The adult woman contains 35 to 50 mg iron per kilogram body weight. The newborn infant contains relatively high levels of iron, about 70 mg/kg. These high levels reflect the high levels of iron stored in ferritin and the relatively high concentration of red blood cells in the bloodstream of the neonate. The premature infant may have lower stores of iron.

The obligatory losses of iron in the adult man occur by way of the skin (0.2 mg/day), gastrointestinal tract (0.6 mg), and urinary tract (0.1 mg). Thus, the overall losses amount to about 0.9 mg per day. Women have additional losses as a result of menstruation. The normal menstrual flow is about 35 ml per period. This is equivalent to about 18 mg iron, as blood contains 0.5 mg of iron per milliliter. Excessive menstrual blood loss is the most common cause of iron deficiency in women. The upper limit of the normal period is about 80 ml; excessive losses may rise over 200 ml per period. The populations at risk for an iron deficiency are those that are rapidly growing and those that are losing blood. These populations include infants between the ages of 0.5 and 4 years, young adolescents, and women during pregnancy and during the reproductive period.

Hemoglobin, which represents more than 95% of the protein of the red blood cell, contains about 60% of the body's iron. Myoglobin is an oxygen storage protein and represents under 1.0% of the protein of muscle. Myoglobin contains about 4% of the body's iron. Thus, about 64% of the body's iron occurs in proteins that transport or store oxygen.

The iron occurring in the iron storage protein ferritin can represent between 5 and 30% of the body's iron. The exact value depends on a number of factors, including the dietary history of the individual. Ferritin has a molecular weight of 450,000 and contains 24 subunits, that is, 24 separate polypeptides. Each subunit forms a shell around a core of iron. The iron occurs as ferric iron (ferric hydroxyphosphate). Iron enters ferritin as reduced iron. The protein shell has 14 channels. It has been proposed that some of these channels are used for entry and exit of iron atoms. The iron is oxidized within the interior of the protein shell to ferric iron. The mechanism responsible for this oxidation as iron enters ferritin, and for the reduction that occurs as iron leaves ferritin, is not clear.

The iron needs of the body are controlled by changing the content of iron in the storage protein ferritin. The iron needs of the body are not controlled by varying the rate of iron excretion, in striking contrast to the mechanisms used to control the body's sodium and potassium. Hemosiderin is an additional iron storage protein. This protein occurs in lysosomes and is thought to represent a partially degraded form of ferritin. Iron tends to be stored in hemosiderin with an iron overload, that is, a high dose of iron.

Transferrin, a protein with a molecular weight of 74,000, has the capacity to bind two atoms of iron. The iron is bound in the oxidized form. Transferrin is an iron transport protein. The iron bound to this protein accounts for only 0.1% of the body's iron. This protein does not reside in a storage facility in the body, but occurs in transit from one organ to another. The role of transferrin in delivering iron to the cell is as follows. First, holotransferrin (containing iron) binds to the transferrin receptor on the cell surface (Step 1). This is followed by invagination of a region of the plasma membrane containing a number of bound holotransferrins to form a vesicle, which buds off within the cell (Step 2). This vesicle fuses with a second vesicle that contains an acid pump. The acid pump is used to lower the pH within the vesicle. The acid environment promotes the release of ferric iron from transferrin. Transferrin remains bound to its receptor. The iron remains bound within the vesicle (Step 3). The iron is reduced to ferrous iron, possibly by a membrane-bound enzyme. $Fe^{2+}$ readily crosses the membrane into the cytoplasm. The source of electrons for this reduction is not clear; it may be NADH, ascorbic acid, or glutathione. Glutathione contains a sulfhydryl group, which is a biological reductant (Step 4). The vesicle, which contains the apotransferrin–transferrin receptor complex in its membrane, fuses with the plasma membrane (Step 5). In the neutral environment of the bloodstream, apotransferrin dissociates from the receptor and enters the circulation (Step 6). The steps described are illustrated in Figure 10.25.

The levels of transferrin receptor and ferritin are regulated in a coordinated manner. When a cell needs more iron, the transferrin receptor increases in number in the plasma membrane of the cell, thus promoting uptake of Fe. At the same time, ferritin synthesis decreases, promoting the use of iron by Fe-requiring proteins in the cell. The levels of transferritin and ferritin are controlled by changes in the mRNA coding for these proteins, as detailed later in this section.

A second iron-transport system may be used by various cells. This system is thought to involve ceruloplasmin, a small peptide, and a membrane-bound transporter. A fraction of the iron in the bloodstream occurs, in the ferrous form, bound to a small peptide. Ceruloplasmin is a plasma protein that can catalyze the oxidation of ferrous iron to ferric iron. Evidence suggests that the ceruloplasmin in the bloodstream catalyzes the oxidation of iron, and the coincident transfer of the iron

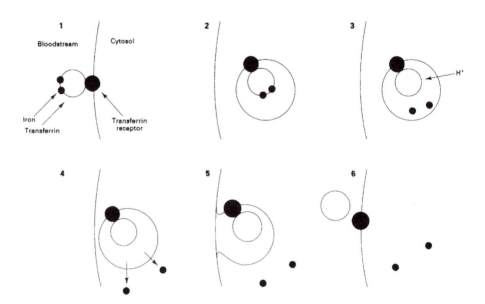

**FIGURE 10.25** Utilization of transferrin and its receptor for delivering iron into the cell. The iron/transferrin complex circulates throughout the bloodstream. Eventually, the iron/transferrin complex binds to the transferrin receptor, a membrane-bound protein (Step 1). Part of the plasma membrane pinches off, creating an endocytotic vesicle, which resides in the cytoplasm (Step 2). The interior of the vesicles becomes acidified (Step 3), and the iron atoms leave the vesicle (Step 4). Mobilferrin is a cytosolic protein that is thought to bind the released iron atoms, and to shuttle them to newly synthesized iron metalloproteins (Conrad *et al.*, 1996). Finally, the transferrin receptor is inserted back into the plasma membrane (Steps 5 and 6).

to the membrane-bound transporter (Mukhopadhyay *et al.*, 1998). Ceruloplasmin is not an iron-transport protein; it is an iron-oxidizing enzyme, and remains outside the cell during the proposed pathway of transport. The aforementioned small peptide is unique, in that its levels in plasma increase markedly during hemochromatosis, a disease of iron metabolism.

## Biochemistry of Iron

About 7% of the body's iron occurs in iron metalloenzymes. These enzymes are classified into the heme enzymes and the nonheme enzymes. Heme is a coenzyme that contains one atom of tightly bound iron. The iron in nonheme enzymes generally is tightly bound to the sulfur atoms of cysteine residues. Hence, these enzymes are called "iron-sulfur proteins." The heme proteins include hemoglobin, myoglobin, and the cytochromes. The cytochromes of the mitochondria are used in the respiratory chain. Cytochromes also occur in the endoplasmic reticulum, and these include cytochrome P450 and cytochrome $b_5$. Cytochrome P450 catalyzes the hydroxylation of drugs, pesticides, carcinogens, and nonnutritive compounds found in plants. The enzyme also participates in alcohol metabolism. Cytochrome $b_5$ is discussed in this section. The heme proteins also include catalase and peroxidase. These two enzymes are used for removing forms of oxygen that are potentially toxic to the cell, that is, hydrogen peroxide (HOOH) and organic peroxides (R—OOH), respectively. Another heme enzyme is endoperoxide synthase, discussed under Essential Fatty Acids in Chapter 9. Some iron enzymes are used in the immune system, as discussed later. These include myeloperoxidase, a heme enzyme, and ribonucleotide reductase, a nonheme enzyme. Table 10.11 lists some of the nonheme iron enzymes of the body.

### *Hemoglobin*

**Hemoglobin** (Hb) has the following properties. It is the major protein of the red blood cells. Hemoglobin, along with its cofactor heme, is synthesized in immature red blood cells. Hemoglobin functions as an oxygen carrier. It has a molecular weight of 64,500 and is composed of four subunits. Two of the subunits are called α-globin, and the other two are β-globin. Each subunit contains a heme group, which can bind one molecule of $O_2$. The iron atom present in the heme group must be in the reduced ferrous form to bind $O_2$. The affinity of Hb for oxygen varies depending on the concentration of $O_2$ in the environment, that is, in the bloodstream. The binding is strong in the presence of high concentrations of oxygen, as in the capillaries of the lungs. Here, the red blood cell replenishes its supply of $O_2$. As the red blood cell encounters environments of lower oxygen levels, the affinity for Hb for oxygen adjusts itself, resulting in a lower binding strength. Here, the oxygen dissociates from the Hb and enters the tissues.

Another feature of Hb is that it responds to the pH of the bloodstream. Acidic pH induces a decrease in the affinity of Hb for oxygen — the Bohr effect. This effect is useful in promoting the release of $O_2$ in exercising tissues, where there is increased production of $CO_2$. The $CO_2$ enters the red blood cell, where it is converted to carbonic acid by the action of carbonic anhydrase. The acid dissoci-

**TABLE 10.11**  Nonheme Iron Enzymes

| Nonheme iron enzymes | Function |
|---|---|
| Aconitase | Krebs cycle |
| Succinate dehydrogenase | Krebs cycle |
| Ribonucleotide reductase | DNA synthesis. The enzyme catalyzes the reduction of ribonucleotide diphosphates (ADP, UDP, GDP, CDP) to the respective deoxyribonucleotide diphosphates (dADP, dUDP, dGDP, dCDP). These latter compounds are then converted to dATP, dTTP, dGTP, and dCTP, respectively, by other enzymes and used for DNA synthesis. |
| Xanthine dehydrogenase | The enzyme is used in the catabolism of the purine ring. It catalyzes the NAD-dependent oxidation of xanthine to uric acid. The enzyme also contains FAD and molybdenum. A fraction of the enzyme normally occurs in the body as xanthine oxidase, which represents an altered form of the enzyme. Xanthine oxidase uses $O_2$ as an oxidant, rather than $NAD^+$. Xanthine oxidase converts xanthine to uric acid, and $O_2$ to HOOH and the hydroxyl radical. |
| Adrenodoxin | The enzyme is used in the synthesis of steroid hormones from cholesterol. It is used for the introduction of oxygen atoms into cholesterol in the biosynthetic pathways for aldosterone, the glucocorticoids, and the sex hormones. |
| $\Delta^9$-Desaturase | The enzyme is used in the synthesis of unsaturated fatty acids. |
| NADH dehydrogenase | Respiratory chain |
| Coenzyme Q reductase | Respiratory chain |

ates in the cell, inducing release of $O_2$ to the tissues. The Bohr effect is dependent on the presence of both forms of globin. Artificially produced Hb containing four subunits of β-globin, for example, does not display the Bohr effect.

A very small fraction of Hb spontaneously oxidizes per day, producing ferric Hb. Ferric Hb is called **methemoglobin**. Normally, the amount of Hb in this form constitutes less than 1% of the total Hb. Methemoglobin is reduced back to Hb within the red blood cell by a system composed of three proteins: cytochrome $b_5$ reductase, cytochrome $b_5$, and methemoglobin reductase. These proteins are not bound to the endoplasmic reticulum because the mature red blood cell does not contain organelles. NADH, derived from red blood glycolysis, supplies electrons to cytochrome $b_5$ reductase, which are transferred to cytochrome $b_5$, to methemoglobin reductase, and finally to methemoglobin. Methemoglobin can accumulate in the blood with nitrite poisoning. Infants are especially susceptible to poisoning by nitrite or nitrate. Ingested nitrate is converted to nitrite by the gut microflora and absorbed into the bloodstream. Poisoning by these chemicals manifests as a blue color of the skin.

## Nitrite Poisoning

Ingested nitrate is converted to nitrite by the gut microflora, or by the oral microflora, and then absorbed into the bloodstream. One estimate suggests that about 8% of oral nitrate is converted to nitrite by the bacteria of the mouth, where this conversion is promoted by the recirculation of nitrate from the gut, to the blood, out into the saliva, and back to the gut (de Vries, 1997). Nitrite poisoning has occurred in several hundred infants, with over a dozen infant deaths, due to feeding the infants water from wells in Minnesota, South Dakota, and Iowa (Johnson *et al.*, 1987). Here nitrate is consumed, but once consumed it is converted into nitrate within the infant's body. The well water in question contained about 150 mg nitrate/liter. Infants tend to have low levels of cytochrome $b_5$ reductase in their red blood cells, possibly accounting for their increased susceptibility to poisoning by nitrate or nitrite. Nitrite poisoning, with the typical symptoms of nausea, vomiting, and blue skin color, has occurred with consumption of Polish blood sausages (Bakshi *et al.*, 1967). Although preserved meats usually contain added nitrite, in order to prevent the production of botulinum toxin, nitrite is also added to lend a red color to the meat. Nitrite poisoning from sausages has not occurred when the correct levels of nitrite were used, but only when the manufacturer accidentally added too much nitrite or failed to disperse the concentrated nitrite additive.

## Other Iron Metalloproteins

Myoglobin is a monomeric protein with a molecular weight of 16,900. It is a minor protein of muscle and is used for short-term storage of oxygen. Myoglobin binds oxygen more tightly than does Hb. The rate of $O_2$ release increases when the environment contains very low levels of $O_2$. In diving mammals, such as dolphins and seals, the myoglobin content of muscle can reach 3 to 8% of the muscle protein. The enzymes of the respiratory chain include NADH dehydrogenase, which is a nonheme iron protein, and the cytochromes, which are heme proteins. NADH dehydrogenase is bound to the mitochondrial membrane. It mediates the transfer of electrons from NADH to coenzyme Q reductase. Coenzyme Q reductase receives electrons from a second source, via direct transfer from succinate dehydrogenase. Succinate dehydrogenase, an enzyme of the Krebs cycle, is bound to the mitochondrial membrane and in this way is unique among the enzymes of the Krebs cycle. Coenzyme Q reductase is a nonheme iron protein. It mediates the transfer of electrons from the preceding two sources (NADH and succinate dehydrogenase) to coenzyme Q. These electrons are then transferred to cytochrome $b$, which then transfers them to cytochrome $c$. Finally, cytochrome $c$ oxidase catalyzes the transfer of the electrons from cytochrome $c$ to oxygen, forming water.

If one were required to name the most important enzyme in the body, one might name cytochrome $c$ oxidase, because of its vital interfacing with oxidative metabolism and $O_2$. At this point, electrons derived from components of the diet meet the oxygen absorbed by the lungs. Cytochrome $c$ oxidase is a multisubunit protein containing about a dozen proteins. These include two cytochromes ($a$ and $a_3$) and two copper proteins. A simplified mechanism of action for cytochrome $c$ oxidase is outlined in four steps in Figure 10.26.

The cytochromes of the respiratory chain are usually described in elementary biology courses; however, a number of other heme-containing enzymes also called cytochromes exist. These include cytochrome P450 and cytochrome $b_5$. Cytochrome P450 received its name because, under certain conditions, it absorbs light having a wavelength of 450 nm. Cytochrome P450 is not one protein but a group

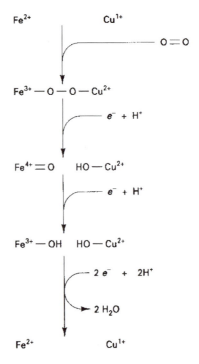

**FIGURE 10.26** Simplified mechanism of action for cytochrome c oxidase. (1) A molecule of oxygen binds to the iron and copper of the protein. Both of these metals occur in the reduced form at this point. Oxygen contains a double bond. The electrons of one of these bonds is shared between the iron and copper atoms. The metal atoms also share one of their electrons with the oxygen, resulting in their conversion to ferric and cupric ions. (2) Cytochrome c donates an electron to cytochrome c oxidase. This electron originated from food, that is, the oxidation of food. The electron is accepted at the copper side of the oxygen complex. A proton is also accepted at this side, generating a hydroxyl group bound at the copper. Iron is generated in the 4+ state. (3) A second electron arrives from cytochrome c and is accepted from the iron side of the oxygen. A proton is also accepted. (4) Two more electrons arrive from cytochrome c. This regenerates cytochrome c oxidase in its original reduced state. Two protons bind to the oxygen, generating water.

of closely related proteins involved in electron transfer reactions. One function of cytochrome P450 was mentioned under Alcohol in Chapter 4. Its function in reducing methemoglobin was mentioned earlier. Cytochromes P450 and $b_5$ also participate in the metabolism of lipids. The proteins are part of electron chains that are small compared to the elaborate apparatus of the respiratory chain.

Cytochromes are used in the conversion of cholesterol to the steroid hormones. These hormones include aldosterone, cortisol, and the sex hormones. Synthesis of aldosterone, for example, occurs in the mitochondria of the adrenal cortex. One of the steps in aldosterone synthesis is a hydroxylation that is catalyzed by a complex of adrenodoxin reductase (FAD-containing protein), adrenodoxin (nonheme iron protein), and cytochrome P450 (heme protein).

The synthesis of unsaturated fatty acids is described under Essential Fatty Acids in Chapter 9. One of the steps involves the $\Delta^9$-desaturase, also called stearyl-CoA

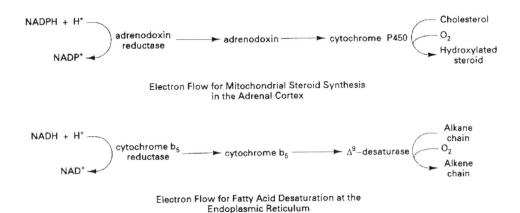

Electron Flow for Mitochondrial Steroid Synthesis
in the Adrenal Cortex

Electron Flow for Fatty Acid Desaturation at the
Endoplasmic Reticulum

**FIGURE 10.27** Role of cytochromes in steroid hormone synthesis and fatty acid desaturation.

desaturase. The desaturation reaction requires a complex of membrane-bound proteins of the endoplasmic reticulum. The active sites of the proteins face the cytoplasm rather than the lumen of the endoplasmic reticulum. Thus, the event of long-chain fatty acid desaturation might be considered a cytoplasmic event. Desaturation requires the participation of three proteins: cytochrome $b_5$ reductase (FAD-containing proteins), cytochrome $b_5$ (heme protein), and $\Delta^9$-desaturase (non-heme iron protein). The pathway of electron transport, from NADH to molecular oxygen, involving $\Delta^9$-desaturase is shown in Figure 10.27. NADPH can also serve as the electron donor.

Ribonucleotide reductase is a cytoplasmic enzyme required by all growing cells. This enzyme converts ribonucleotides to the corresponding deoxyribonucleotides, the building blocks of DNA synthesis. The enzyme uses ribonucleotides in the diphosphate form, rather than the more familiar triphosphate form:

$$\text{ADP} \xrightarrow[\substack{\text{Ribonucleotide reductase}\\ \text{(non-heme iron)}}]{} \text{dADP} \tag{10.2}$$

## Iron-Containing Enzymes of the Immune System

Severe iron deficiency has been associated with an increased incidence of infections. There is some thought that this is due to an impairment of the activity of the iron-requiring enzymes of cells of the immune system. These cells are the neutrophils, which are phagocytic cells, and the lymphocytes, which are antibody-producing cells. Neutrophils phagocytize bacteria and kill them. The neutrophil engulfs the bacterium, surrounds it with a vesicle, and then exposes it to toxic forms of oxygen such as hydrogen peroxide and the hydroxyl radical. In addition, the bacterium is exposed to hypochlorous acid (HOCl), which is a strong oxidant and highly toxic. Hypochlorous acid is synthesized from HOOH and chloride ions by the heme metalloenzyme myeloperoxidase:

$$\text{HOOH} + \text{Cl}^- \xrightarrow[\text{Myeloperoxidase}]{} \text{H}_2\text{O} + \text{HOCl} \tag{10.3}$$

Myeloperoxidase resides in a vesicle in the neutrophil called the azurophilic granule. This vesicle fuses with the phagocytic vesicle, which contains the bacterium, exposing the bacterium to the enzyme as well as to its toxic product.

Lymphocytes synthesize antibodies. Antibodies are proteins that recognize and bind to structures that are foreign to the host organism. Each lymphocyte biosynthesizes only one type of antibody. The lymphocyte is stimulated to divide and multiply and to produce its one type of antibody when the body is exposed to foreign proteins. The iron enzyme most vital to the process of cell multiplication is **ribonucleotide reductase**.

Attempts to establish a connection between iron deficiency, impairment of neutrophil and lymphocyte function, and increased rate of infections have not been conclusive to date. One complicating issue in making the connection between iron deficiency and resultant infections is that some infections can induce iron deficiency. One example is the hookworm infection, described later in this section. Among the most severely iron-deficient people in the world are children in Africa in areas where malaria is prevalent. The children die of malaria rather than of physiological defects resulting from the iron deficiency.

## Regulation of Iron-Requiring Enzymes and the Iron Response Element

Response elements are short stretches of DNA that are used for regulating the rate of transcription of nearby genes. The response elements that bind hormone binding proteins were detailed in the section on Vitamin A, Vitamin D, and Thyroid Hormone at the Genome. These response elements occur in DNA. The **iron response element** (IRE) occurs in mRNA, and is used to regulate the level of certain iron metalloproteins. An iron-binding protein that is used in conjunction with the IRE is called the **iron regulatory protein** (IRP). The iron response element is coded for by our DNA, but the iron regulatory protein only binds to the IRE which occurs in messenger RNA.

Iron response elements consist of short stretches of ribonucleic acid, and they occur in the mRNA coding for ferritin, transferrin receptor, and δ-aminolevulinic acid synthase. Each IRE can be bound by an IRP. Several different IRPs exist. Surprisingly, one of the IREs is quite similar to aconitase, the Krebs cycle enzyme (Kim *et al.*, 1996; Toth and Bridges, 1995; Henderson *et al.*, 1996). All IRPs contain an iron binding site. This site consists of several residues of cysteine. To be more specific, the sulfhydryl groups of cysteine residues function to bind iron atoms.

When iron is low in the cell, the mRNA for ferritin is *translated at a lesser rate* (resulting in lesser amounts of this protein in the cell). The goal here is to cut down on the excessive synthesis of our major iron storage protein, if no excess iron is available for storing. When iron is low in the cell, the mRNA for the transferrin receptor is *translated more* (creating more transferrin). When iron is low in the cell, the mRNA for δ-aminolevulinate synthase, an enzyme in the heme biosynthetic pathway, is *translated less*. The overall goal here is to cut down on the wasteful synthesis of heme, if no iron is available for completing this cofactor.

The following scenario is proposed, but not proven. When the supply of iron is high, more iron binds to each IRP, and the IRP no longer binds to iron response element. When dietary iron is low, less iron binds to the IRPs, resulting in more of the IRP binding to the iron response element. IRP binding to the iron response element reduces translation, while no binding of the IRP increase translation.

The structure of the IRE that occurs in the sequence of ferritin mRNA appears in Figure 10.34. The iron response element is a small region of RNA, and it is distinguished in that it spontaneously folds upon itself to form a hairpin shape. The ribonucleotides of RNA follow similar base-pairing rules as in DNA. In RNA, guanine binds to cytosine, and adenine binds to uracil. It is accurate to state that within the hairpin the RNA occurs as double-stranded RNA. To repeat, one might note that DNA contains a sequence of DNA bases that is used to code for the iron response element, but these DNA bases do not bind the IRP.

In the case of **transferrin**, the rate of degradation of the transferrin mRNA is altered, with a consequent change in the amount of transferritin in the cell. In the case of **ferritin**, the rate of translation of ferritin mRNA is directly altered, with a consequent change in the amount of ferritin protein in the cell. The hairpin, or stem-and-loop, of the iron response element interacts with cellular iron in a way that changes the rate of degradation of the entire mRNA (transferrin) or rate of translation (ferritin). Each hairpin consists of only 30 to 40 ribonucleotides and hence accounts for a small portion of the entire mRNA (Figure 10.34). The entire mRNA used for making ferritin, for example, consists of 1100 ribonucleotides. The first 92 ribonucleotides do not code for any part of ferritin, and it is here, in the upstream untranslated region (5'-UTR), where the iron-responsive element occurs. The coding region of the ferritin's mRNA occurs at ribonucleotides 92 to 662. The function of the remaining ribonucleotides (the downstream untranslated region; 3'-UTR), if any, is not clear. Much of the work on iron regulation was performed in the laboratory of Hamish Munro.

## Absorption and Recycling of Iron

Iron absorption is an issue of continuing interest in the nutritional sciences because of the relatively high frequency of iron-deficiency anemia and the remarkably poor efficiency of absorption of most forms of dietary iron. The biochemistry of iron absorption by the gut is not much understood. Hence, the material that follows consists mainly of valuable descriptive information. The availability of the iron in plant foods such as beans, peas, corn, bread, and rice is quite poor. It ranges from less than 1% to 10%. The availability of iron in meat is considerably higher than that in plant products. The nonheme iron in meat, fish, chicken and liver may be about 20% available. The heme iron of meat may be close to 30% available. Nearly all of the iron in plants is nonheme iron. Much of the iron in meat is nonheme iron as well. The most available source of iron is human milk (50%). The term "availability" describes the percentage of the iron in the food that is absorbed and used for physiological purposes, such as red blood cell formation. The availability of iron in various foods can be measured by preparing a food containing radioactive iron, feeding it to a subject, and measuring the incorporation of radioactivity into the hemoglobin of the red blood cells. Radioactive food can be prepared by adding the radioactive iron to the food or by growing the food (plant or animal) in the presence of radioactive iron.

The interactions between different foods have sparked some interest. For example, if rice is consumed with orange juice, the orange juice can enhance the absorption of the iron in the rice. This effect results from the chelation of the iron by the ascorbate in the juice and the increased absorbability of the iron from the complex. On the other hand, if rice is consumed with tea, the tannins in the tea can reduce the absorption of the iron in the rice because the iron in the iron–tannin complex is not readily available. In general, including meat in the diet can increase the availability of iron from other foods. The mechanism of this effect is not clear. The availability of heme iron is not much influenced by other components of the food known to influence availability of nonheme iron. These components include ascorbate, tannins, phytate, phosphates, and fibers.

The following study illustrates the range in availability of iron when consumed with various foods. The subjects consumed a dose of radioactive iron ($^{55}$Fe or $^{59}$Fe) that had been dispersed in the test food. The test foods included starch (Experiment 1), corn meal mush (Expt. 2), corn meal mush with meat (Expt. 3), bread (Expt. 4), and bread and meat (Expt. 5). The availability of the iron was determined by measuring the amount of radioactivity incorporated into the red blood cells 2 weeks after consumption of the test meal. The results, listed in Table 10.12, show that iron's availability was lowest when mixed with the corn meal and that availability was improved when meat was consumed with the corn meal. The meat seemed not to affect the availability of the iron in the bread. It is not clear why the effect of the meat differed in Experiments 3 and 5.

## Phytic Acid Impairs Iron Absorption

Phytic acid has been identified as a major inhibitor of iron absorption in plant foods. Phytic acid is inositol hexaphosphoric acid (Figure 10.28). Phytate is a constituent of plants and constitutes from 1 to 5% of the weight of legumes, cereals, and nuts. About half of the phosphate in grain may be in the form of phytate. Cereal grains are the seeds of plants such as wheat, rice, maize, millet, barley, and oats. The seeds of cereals are covered with a husk containing indigestible fiber. The outer layers of the seed also contain most of the phytate of the seed. When removed by milling, the outer layer is called bran. Phytate also binds calcium and zinc ions, limiting their availability as well. In the United States, bread is enriched

**TABLE 10.12** Effects of Diet on Iron Availability

| Experiment | Test meal | Percent availability |
|:---:|:---|:---:|
| 1 | 4 mg radioactive Fe$^{3+}$ + 67 g hydrolyzed starch in water + 35 g corn oil | 11 |
| 2 | 4 mg radioactive Fe$^{3+}$ + 60 g boiled corn meal | 0.5 |
| 3 | 4 mg radioactive Fe$^{3+}$ + 60 g boiled corn meal + 92 g cooked ground beef | 1.6 |
| 4 | 4 mg radioactive Fe$^{3+}$ + 60 g wheat bread | 3.0 |
| 5 | 4 mg radioactive Fe$^{3+}$ + 60 g wheat bread + 92 g cooked ground beef | 3.2 |

*Source*: Hurrell et al. (1989).

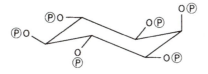

**FIGURE 10.28** Structure of phytic acid.

with iron at a level of 12.5 mg iron per pound. It is difficult to enrich bread or flour in underdeveloped countries, where iron deficiency is a widespread public health problem, because of the practice of milling wheat near the home, rather than at a central location, making iron fortification programs difficult to implement.

## Iron Supplements and Food Fortification

Iron deficiency anemia occurs mainly in infants, children, and fertile women. For this reason, a variety of foods, including infant formula and infant cereals, is fortified with iron. Ferrous sulfate is a form of iron that is most readily absorbed by the gut, but when added to dry cereals it can promote their spoilage and rancidity. For this reason, dry cereals are fortified with elemental iron particles, ferric pyrophosphate, or ferrous fumarate (Davidsson *et al.*, 1997). Ascorbic acid may also be added to the cereal to enhance iron absorption. To view some of the numbers, infant cereals may contain 75 mg iron/kg cereal (1.3 mmol iron/kg), 1 mmol phytic acid/kg, and 2.6 mmol ascorbic acid/kg (Davidsson *et al.*, 1997). Although phytic acid impairs iron absorption, the added ascorbate serves to prevent this effect. An alternate method for preventing phytate from impairing iron absorption is to treat the food with the enzyme phytase. A parent interested in enhancing a child's iron absorption can easily feed a child some orange juice, but it would not be practical to pretreat the child's cereal with phytase. A typical availability of ferrous sulfate in infants is about 3–5% (with no ascorbate), and 6–10% (with ascorbate). Ascorbate is effective when present in a twofold molar excess over the iron.

A large proportion of women in underdeveloped countries are iron-deficient before becoming pregnant. In this case, iron-deficiency anemia tends to develop by mid-pregnancy. One problem with some community programs intended to correct iron-deficiency anemia is that *daily* doses of 60–120 mg iron (as ferrous sulfate) are given. These doses tend to produce toxic effects (pain, nausea) and have the end-result of poor compliance. A solution to this problem is to begin *weekly* supplements of 60–120 mg iron (not daily) at the start of pregnancy (well before anemia occurs) or for all women of childbearing age (Viteri, 1998). Evidence suggests that infants with iron-deficiency anemia may have reduced mental development, where assessment of intelligence was conducted at the age of 5 years (Lozoff *et al.*, 1991).

## Regulation of Iron Absorption

To some extent, the body's needs for iron can be controlled at the point of absorption. Human subjects can vary their absorption of heme iron between 20 and 50%, and of nonheme iron between 1 and 40%, according to the body's needs. With iron deficiency, about 40% of both types of iron may be absorbed. With normal iron status, the percentage absorption of heme iron may decrease somewhat (to 20–30%), while the absorption of nonheme iron may decline more strikingly (to 2.5–10%). The mechanisms involved in controlling the rate of iron absorption by the gut are not known (Lynch *et al.*, 1989; Hallberg *et al.*, 1997).

### Catabolism of Dietary Heme

In the cases of dietary heme and nonheme iron, the iron appears in the bloodstream bound to the transport protein transferrin. After its dissociation from dietary proteins by proteases, the heme is absorbed intact by the enterocyte. The heme is then degraded by heme oxidase. Heme oxidase catalyzes the $O_2$-dependent degradation of heme to **biliverdin**. Biliverdin is further degraded to **bilirubin**, which is excreted from the body in the bile. Heme absorption, as well as heme oxidase activity, is somewhat higher in the duodenum than in the jejunum and ileum, as determined in studies with rats. The heme catabolic pathway is shown in Figure 10.29. Most of the bilirubin in the body is not produced by the catabolism of dietary heme, but by the catabolism of the heme present in old, or senescent, red blood cells. Between 75 and 80% of the bilirubin formed in the body is derived from senescent red blood cells; most of the remainder is derived from the normal turnover of the heme proteins in the liver.

### Catabolism of Red Blood Cells

The human red blood cell has a life-span in the circulatory system of about 120 days. This life-span is due to the natural senescence of the cell. Apparently, mechanisms exist that trigger destruction of the red blood cell at about 120 days. The senescent red blood cell is removed from the circulation by special phagocytic cells called **macrophages**. Most senescent red blood cells are consumed by macrophages in the spleen. The liver, lungs, and connective tissues also contain macrophages. The macrophages that reside in the liver are called **Kupffer cells**. Kupffer cells reside in the walls of the capillaries in the liver, where they have access to large particles in the blood. The macrophages, as well as a few other types of cells, are part of the **mononuclear phagocytic system** (MPS). The cells of the MPS are all derived from the stem cells of the bone marrow. Red blood cells, neutrophils, and lymphocytes are not part of the MPS.

About 90% of senescent red blood cells are taken up by the macrophages; only 10% hemolyze (rupture) while in the bloodstream, releasing hemoglobin into the plasma. The free hemoglobin binds immediately to haptoglobin to form a com-

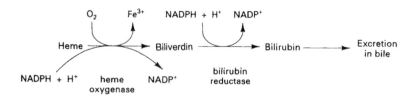

**FIGURE 10.29** Heme degradation pathway.

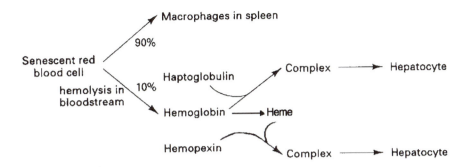

FIGURE 10.30 Breakdown of red blood cells to give hemoglobin and free heme.

plex. Any heme released binds to hemopexin to form a complex as well. These complexes are taken up by the hepatocyte, where the iron is stored or used for other purposes (Figure 10.30).

The body's iron is extensively recycled. This recycling involves its packaging in the red blood cells as part of the heme prosthetic group of hemoglobin, a period of 4 months in the circulating red blood cell, dismantling in the macrophage, packaging in transferrin, and delivery back to a developing red blood cell. The temporary residence of iron in ferritin may be part of this cycle. Erythropoiesis is the production of red blood cells. The pathway of erythropoiesis and the names of the immature forms of the red blood cell are discussed in the Folic Acid section.

## Biosynthesis of Heme

Heme synthesis is described in eight steps. Glycine and succinyl-CoA are the starting materials; succinyl-CoA is an intermediate of the Krebs cycle, and glycine can be derived from the diet or from serine. The first step in heme synthesis is catalyzed by aminolevulinic acid synthase, a vitamin $B_6$-requiring enzyme. The second step is catalyzed by aminolevulinic acid dehydratase, a zinc metalloenzyme. The third step results in polymerization of four pyrrole units. The structure of the pyrrole ring is shown in Figure 10.31. The fourth step involves cyclization of the tetrapyrrole structure to form a large ring, the **porphyrin ring**. This is followed by modification of the ring by decarboxylations and oxidations. The final step involves insertion of ferrous iron into protoporphyrin IX to form heme. Protoporphyrin IX is apo-heme. All of the steps are shown in Figure 10.32. The heme biosynthetic pathway is distinguished by the fact that the first and final three

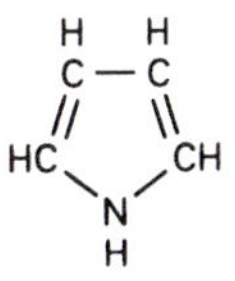

FIGURE 10.31 Structure of pyrrole.

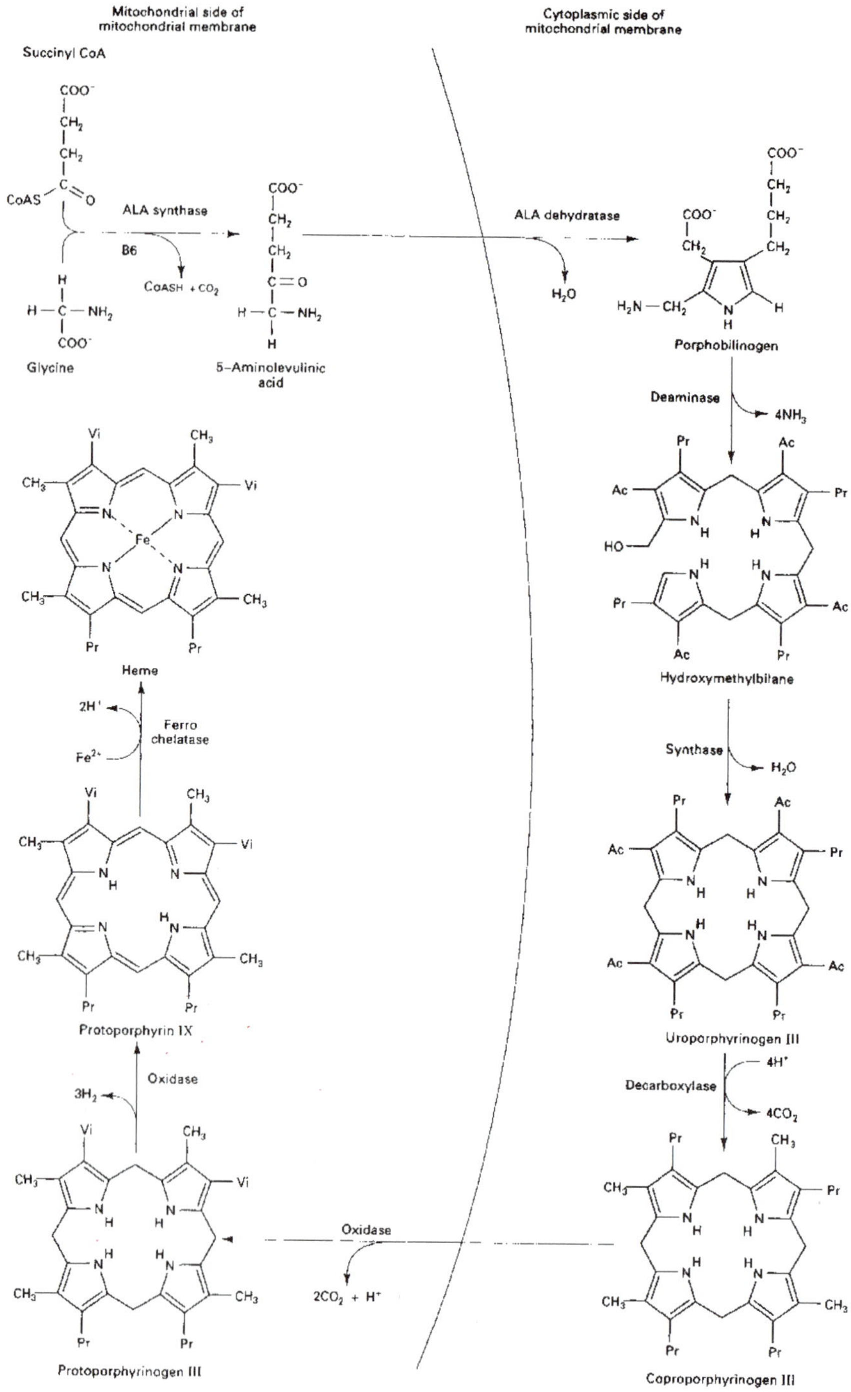

**FIGURE 10.32** Biosynthesis of heme. Heme is the iron-containing prosthetic group of hemoglobin, myoglobin, the cytochromes, and other heme enzymes. Heme synthesis is catabolized by enzymes in the mitochondria and the cytoplasm. Pr, propionic acid group; Ac, acetic acid group; Vi, vinyl group. The vinyl group is R—CH=CH₂.

steps take place in the mitochondrion. The intermediate steps take place in the cytosol.

Aminolevulinic acid dehydratase has the dubious distinction of being extremely sensitive to lead ions, $Pb^{2+}$. Lead can combine with the sulfhydryl groups of the enzyme, resulting in its inactivation. Lead poisoning occurs in children who consume paint with a high lead concentration and in workers at lead smelting plants. It should be noted that people who die of lead poisoning do not die because of heme deficiency. They die because of damage to the nervous system.

The iron inserted into apo-heme is acquired from the transferrin circulating in the bloodstream. Transferrin is taken up by cells according to their need for iron. This transport is mediated by a membrane-bound protein of the plasma membrane called the **transferrin receptor**. The erythropoietic cell contains a large number of transferrin receptors in its plasma membrane. Stem cells contain very few transferrin receptors because of their lack of hemoglobin synthesis. The normoblast contains 0.3 to 0.8 million transferrin receptors, whereas the more mature reticulocyte contains 0.1 million receptors. After release of iron, transferrin returns to the bloodstream. Most of the transferrin contains iron derived from recently catabolized red blood cells, as shown in Figure 10.33. About 99% of the transferrin contains iron acquired from macrophages, where old red blood cells are dismantled; only 1% is derived from recently absorbed dietary iron.

Erythropoiesis is controlled by the hormone erythropoietin. A protein of molecular weight 40,000, erythropoietin is fairly large for a hormone. Erythropoietin is also called a growth factor. The hormone is produced in the kidney, by cells in the walls of the capillaries in the region of the glomerulus. The concentration of erythropoietin in the bloodstream is normally 0.02 n$M$. Increases in the hormone stimulate differentiation of the stem cells as well as the rate of division of the red blood cell precursors. Increases in plasma erythropoietin are most closely associated with exposure to low oxygen conditions, but can also occur following losses of large amounts of blood.

## Iron Deficiency

Those at risk for iron deficiency include infants and children between the ages of 0.5 and 4.0 years, because of the rapid rate of growth at this time and because the

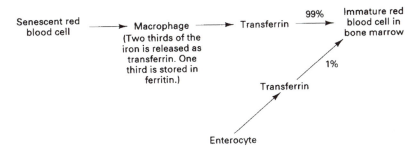

**FIGURE 10.33** Sources of iron for red blood cell synthesis.

infant's iron stores are not sufficient to last beyond the age of 6 months. Children in early adolescence are also at risk because of their rapid growth. In addition, menstruation is a risk factor for a deficiency. Pregnancy is another risk factor because of the mother's expanding blood volume, the demands of the fetus and placenta, and the blood losses during childbirth.

The earliest signs of iron deficiency involve the storage forms of iron. Iron is stored intracellularly in ferritin. A small proportion of tissue ferritin leaves the cells by mechanisms that are not clear, resulting in a very low serum concentration of ferritin. The serum ferritin level accurately mirrors the tissue stores of iron. Iron status can be assessed during the early stages of the deficiency by measuring serum ferritin. This test is easy to perform, in contrast to the more direct test of measuring the ferritin content in a bone marrow sample. A serum ferritin level under 12 ng/ml provides a firm diagnosis of iron deficiency. In patients with serum ferritin levels below 12 ng/ml, marrow iron cannot be detected by the histological staining test. The correlation between low serum ferritin and low marrow iron stores is quite strong. On the other hand, a serum ferritin level in the normal range does not always indicate normal iron status. This is because a variety of disorders, such as infections and liver disease, can induce an elevation of serum ferritin, even during iron deficiency. Here, an examination of bone marrow iron can lead to a conclusive diagnosis of the deficiency.

In one other instance, plasma ferritin levels may be elevated even when the body's iron stores are low. Plasma ferritin may be elevated in the first few days of iron therapy in anemia, particularly where the doses of iron are high. Anemia in infants, for example, can be treated with 6 mg of iron/kg body weight per day. The iron can be supplied as oral ferrous sulfate. Iron deficiency anemia in adults can be treated with 50 mg of iron three times a day. The iron can be supplied as ferrous sulfate. Early rises in serum ferritin may not occur at these doses, but can occur at higher doses. With the use of standard doses, serum ferritin may rise into the normal range only after the anemia has been corrected.

A more prolonged iron deficiency affects transport iron, as well as the population of red blood cells most recently released into the bloodstream. A deficiency of intermediate duration can be revealed by three tests:

1. *Saturation of serum transferrin*: This test measures the proportion of transferrin occurring in the apo-transferrin form. Apo-transferrin is transferrin lacking iron.

2. *Apo-heme in the red blood cell*: This test measures the accumulation of apo-heme in red blood cells. Apo-heme is the same thing as protoporphyrin IX.

3. *Mean corpuscular volume (MCV)*: This is a test for the average volume of the red blood cell. The MCV is calculated by taking a known volume of packed, settled cells and determining the number of cells in the sample. The number of cells in any given suspension of red blood cells can be determined by counting the cells under a light microscope. The MCV is calculated from the ratio of volume of the cell sample/number of cells in the sample.

Once an iron deficiency has been detected by low serum ferritin levels, the preceding three tests can be used to estimate its severity.

Depletion of the iron stores is followed by a drop in serum iron, with a deficiency of intermediate duration. Almost all of the serum iron occurs as transferrin-bound iron. Hence, a measurement of serum iron may be thought of as a measure of transferrin iron. A drop in serum iron involves an increase in the apotransferrin/holotransferrin ratio. Normally, about 20 to 25% of transferrin occurs in the saturated, holotransferrin form. A saturation of 20 to 25% strongly indicates normal iron status. Where the saturation drops below 16%, there may be impairment in the supply of iron to the developing red blood cells in the marrow. This is because holotransferrin is required to deliver iron to these cells. In short, erythropoiesis may be impaired at a transferrin saturation below 16%. The transferrin saturation test is in common use, though the results are somewhat variable. The results vary in daily cycles and can vary from day to day in an individual. According to Cook (1982), the transferrin saturation test is more applicable for surveying populations and less so for diagnosing patients.

Increased levels of apo-heme can be found in red blood cells during iron deficiency. Normally, the red blood cell contains about 350 ng of apo-heme per milliliter of packed red blood cells. Levels greater than 1000 ng/ml cells indicate iron deficiency. Hematological tests are useful for assessing iron status when the iron stores have been depleted for a relatively long period. The first hematological signs are lightly colored red blood cells (hypochromic cells) and small red blood cells (microcytic cells) in the circulation. The microcytic anemia of the iron deficiency should be contrasted with the megaloblastic anemia of folate and vitamin $B_{12}$ deficiencies.

Anemia is the most severe sign of the iron deficiency that affects the red blood cells. Anemia is indicated by an MCV under 70 fl, a hemoglobin level under 130 mg/ml of blood, and a hematocrit under 38%. The hematocrit is the proportion of whole blood, by volume, composed of red blood cells. The hematocrit is sometimes called the "packed cell volume." The hematological signs of anemia are listed in more detail in Table 10.13.

Severe anemia is indicated by hemoglobin levels under 70 mg Hb/ml whole blood. The anemia is characterized by weakness and shortness of breath, and may not be suspected with a sedentary style of life. A decrease in the Hb concentration has been directly related to decreases in maximal work capacity, as determined by a standard stairstepping test, and in maximal aerobic capacity. The consequences of anemia are serious among populations that are dependent on physically de-

**TABLE 10.13**  Standard Values for Hemoglobin and Hematocrit

|  | Hemoglobin (mg/ml) | Hematocrit (%) |
|---|---|---|
| Men |  |  |
| Normal | 130–160 | >44 |
| Anemia | <130 | <38 |
| Women |  |  |
| Normal | 120–160 | >33 |
| Anemia | <120 | <32 |
| Pregnant women |  |  |
| Anemia | <110 | <31 |

manding tasks, such as the tropical rural poor. These populations may also suffer from malnutrition and high rates of infection, which can complicate the diagnosis of iron deficiency.

Surprisingly, individuals with hemoglobin levels as low as 20 to 30 mg Hb/ml can continue to be physically active. This is due to adaptive mechanisms involving an increase in the heart rate and an increase in efficiency of extraction of oxygen from the blood. Experiments with rats have revealed that the decreased work capacity of the iron-deficient animals, as determined by their ability to run, was due to a deficiency in the enzymes of the respiratory chain of muscle, as well as to anemia. In short, transfusion with normal blood failed to restore fully the running capacity of the deficient animals.

Iron is rapidly mobilized from tissue stores (ferritin) during early pregnancy. This mobilization is reflected by decreases in serum ferritin levels. Ferritin levels may drop from a normal value of 60 ng/ml to about 15 ng/ml during the first 2 months of pregnancy. The iron is being mobilized to expand the blood volume of the mother and to produce placental and fetal tissues. Apparently, the drop in serum ferritin in early pregnancy cannot be diminished by dietary iron supplements.

Both human and bovine milk contain rather low levels of iron. Human milk iron gradually falls from about 0.5 mg/liter during the first month of lactation to about 0.3 mg/liter by 4 to 6 months. Cow milk contains about 0.5 to 1.0 mg iron/liter. Breast milk, however, is a vastly better source of iron than cow milk; the iron in breast milk is 50% absorbed by the gut, whereas cow milk is only 10% absorbed. Breast-feeding until the age of about 6 months supplies the iron needed by the infant to avoid iron deficiency. During this time the infant's needs are supplied by the milk as well as by the infant's stores. Fomon and Strauss (1978) have recommended supplementing young infants with about 7 mg of iron per day (ferrous sulfate) to maintain the infant's stores during the first 6 months of life and to ensure growth thereafter.

Preterm infants may be at a risk for iron deficiency because of their relatively rapid rate of growth and low iron stores. The marginal supply of breast milk iron dictates that preterm infants be supplemented with iron at an early age. Dallman et al. (1980) recommend that preterm infants receive supplements beginning at 2 months.

Cow milk-based infant formulas are supplemented with iron at levels of about 12 mg/liter. Here, the iron is added as ferrous sulfate or ferrous gluconate. Commercial infant cereals are supplemented with small iron particles (powdered iron) at levels close to 50 mg iron per 100 g dry cereal. The particles have a diameter of less than 10 mm. This amount of iron is tenfold greater than the iron naturally present in the cereal. The absorption of the iron supplied in particulate form is about 4%.

Anemia in infants is defined as an Hb level below 109 mg/ml of whole blood and a hematocrit less than 33%. Extensive surveys conducted in the United States have revealed that the incidence of anemia in lower-income families is about 6%. The lower-income family was defined as that participating in the WIC (Women, Infants, Children) program. The decline in infant anemia in the decade prior to 1985 is attributed to an increase in the use of iron-fortified formulas and cereals.

As mentioned earlier, blood losses from menstruation can result in iron deficiency and anemia. Other sources of blood loss are important as well. The normal rate of blood loss in the feces is 0.5 to 1.0 ml/day. Intestinal blood losses can increase in certain cancers. About 60% of the cancers of the colon and rectum lead to further blood losses of 2 to 10 ml/day. The fecal blood test is universally used to screen for the presence of cancer of the gastrointestinal tract. In the absence of periodic testing, cancer may first present as iron deficiency anemia. Duodenal and gastric ulcers, as well as hemorrhoids, are extremely common sources of blood loss and can result in iron depletion. Aspirin, when consumed in large amounts, can provoke bleeding of the gastrointestinal tract, resulting in blood losses of 1 to 5 ml/day.

EXERCISE

How much chicken must one eat per day to replace the iron lost in the feces as a result of blood losses and the normal exfoliation of intestinal cells? Hint: Chicken contains about 9.0 mg iron/100 g meat.

The hookworm is a parasitic worm. It is a common source of anemia in warm climates, including the southern United States. The parasitic worms are classed as flukes (trematodes), tapeworms (cestodes), and roundworms (nematodes). Tapeworms, for example, are mentioned under Vitamin $B_{12}$ in Chapter 9. The hookworm is a roundworm. It enters the body via the skin, that is, bare feet. The hookworm resides in the lumen of the small intestines where it attaches itself to the villi. This results in damage to the villi, blood losses, secondary infections by other microorganisms, and inflammation. The roundworm secretes anticoagulants that promote continued bleeding. Each worm may be responsible for the loss of up to 0.25 ml of blood per day. Hb levels as low as 20 mg/ml have been associated with hookworm infections. The disease can be prevented by the sanitary disposal of feces and by wearing shoes.

# Iron Overload and Hemochromatosis

Hemochromatosis is a group of disorders involving the progressive deposit of iron in the hepatocytes and in cells of the heart, pancreas, and joints. These deposits can result from the chronic consumption of excessive levels of iron, repeated blood transfusions, or the genetic disease hematochromatosis. This disease is relatively rare and appears to result from an inappropriately high absorption of iron. The symptoms of hemochromatosis include weakness and weight loss, joint and abdominal pain, enlargement of the liver, and eventually cirrhosis of the liver. Liver damage with iron overload is especially problematic in alcoholics. Hemochromatosis can be screened for by a test for serum ferritin and definitively diagnosed by a biopsy of the liver. The biopsy is examined by a staining method sensitive to the stored iron. Hemochromatosis is treated by the weekly removal of 500 ml of blood for about 2 years. This procedure is called **phlebotomy**. Because of the possibility of provoking hemochromatosis in the small population having the genetic disease or in chronic alcoholics, an upper limit has been set on the level of iron added to fortified foods.

In an iron overload, the excessive amounts of iron are stored in ferritin, as might be expected, but to an increasing extent in hemosiderin. Hemosiderin is a protein in the lysosomes. With increased iron storage, the lysosomes may be damaged, releasing iron. It is thought that the iron released may cause damage to the membrane lipids and to proteins of the cell, leading to cirrhosis. This damage is thought to result from the hydroxyl radicals produced by the Fenton reaction (Figure 10.24; see also Figure 10.34).

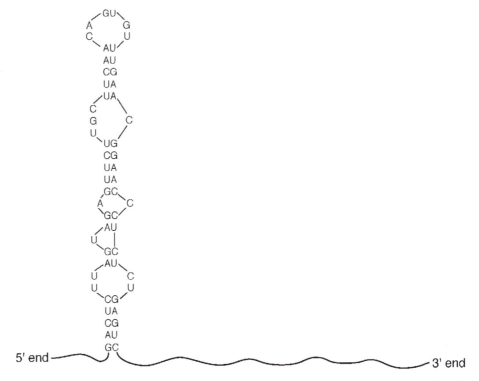

FIGURE 10.34.  Iron response element. The IRE occurring in the mRNA that codes for the H chain of ferritin is shown. IREs are short regions occurring in messenger RNA that form hairpin structures and are used for regulating the availability of the mRNA for use in protein synthesis. The IRE occurring in the ferritin's mRNA (shown) resides in the 5'-UTR, and serves to regulate the rate of binding of the mRNA to the ribosome. The IREs in the mRNA for aminolevulinic acid synthase works in the same manner. In contrast, the IRE in the mRNA coding for the transferrin receptor occurs in the 3'-UTR, and is used to control the rate of degradation of the mRNA molecule by nucleases. The binding of the iron regulatory protein to the IRE has the effect of reducing translation, in the case of ferritin and aminolevulinic acid synthase, and of increasing translation, in the case of transferrin receptor. The tip of the hairpin is a loop consisting of CAGUGU. The stem structure is maintained, because A bonds non-covalently with U, and G bonds non-covalently with C. The stem of ferritin's IRE is interrupted by UCG, on the 5'-side of the hairpin, and by C, on the 3'-side of the hairpin, as shown.

# CALCIUM AND PHOSPHATE

Calcium is a divalent cation. The most obvious function of calcium is as a component of the skeleton, where it occurs as a salt with phosphate. A 70-kg man contains about 1.3 kg of calcium, with about 99% occurring in the skeleton. This calcium is not metabolically inert. About 0.5 g of calcium leaves the bone and is deposited back into the bone each day. About 85% of the body's phosphate occurs in the skeleton. Calcium is used in the activation of a number of hydrolytic enzymes. The calcium in these enzymes is associated with residues of glutamate or aspartate. The enzymes include those that hydrolyze polysaccharides, phospholipids, and proteins. Calcium is associated with a number of calcium-binding proteins. The calcium, in these cases, supports interactions between macromolecules, that is, the binding of one protein to another and the binding of a protein to a phospholipid membrane. In the case of some of the blood-clotting proteins, calcium is bound to residues of γ-carboxyglutamic acid, where it is required to support the functions of the proteins. The calcium-binding proteins include components of the cytoskeleton. The cytoskeleton consists of two networks. One network hugs the inside of the plasma membrane of the cell. It maintains the shape of the cell, controls locomotion (in migrating cells), and controls phagocytosis (by white blood cells). The other network crisscrosses through the interior of the cell. It controls the movements of secretory vesicles and organelles, and is involved in moving the chromosomes during mitosis. The roles of calcium ions in supporting the preceding events are very complex.

## Hormones and Neurotransmitters Bind to Cells and the Cells Respond

Calcium ions participate in the transmissions of signals to the cell. Different cells respond to different signals. The nerve ending responds to a nerve impulse. In brief, a nerve impulse is a traveling wave consisting of a brief influx of sodium ions. Muscle cells respond to neurotransmitters, such as acetylcholine and norepinephrine. Pancreatic acinar cells respond to cholecystokinin. Adrenal glomerulosa cells respond to angiotensin II. The adrenal medulla responds to nerve impulses. (Cells of the adrenal medulla respond by secreting norepinephrine and epinephrine into the bloodstream.) White blood cells respond to the binding of foreign proteins and particles to their surface. In these and other cases, the response of the cell to the signal is mediated by calcium ions. The signal induces an increase in the concentration of $Ca^{2+}$ in the cytoplasm. In the case of some cells, such as the endings of nerve cells, the $Ca^{2+}$ is acquired from the extracellular fluid. In the case of other cells, such as skeletal muscle cells, the immediate source of the $Ca^{2+}$ is a calcium storage organelle.

What is the effect of the cytosolic $Ca^{2+}$ in elevated concentrations? The immediate effects are clear in some cases and ill-defined in others. $Ca^{2+}$ ions bind to various target proteins, causing changes in their behavior. They can bind to proteins of the cytoskeleton, inducing their polymerization or dissociation. They can bind to kinases, inducing their activation and resulting in phosphorylation of a number of proteins. In some cases, calcium binds directly to the target protein. In other cases, calcium binds to **calmodulin**, a Ca-binding protein, to form a complex. The Ca–calmodulin complex, in turn, binds to a target protein, producing the final effect. The final effect in calcium signaling is muscle contraction (in the case of muscle cells), hormone secretion (in the instances of nerves and cells of the adrenal gland), or phagocytosis and cell division (in the case of certain white blood cells).

Calcium signaling plays a part in activating cells to engage in cell division, as well as in transforming normal cells to cancer cells.

## Phosphate and Energy Storage and Transfer

Phosphate consists of a central atom of phosphorus, four atoms of oxygen, and zero to three atoms of hydrogen. Phosphate resonates among several forms. A few forms of phosphate are shown in Figure 10.35A. Phosphate occurs in equilibrium with $H_3PO_4$, $H_2PO_4^-$, $HPO_4^{2-}$, and $PO_4^{3-}$ (Figure 10.35B). The predominant form at neutral pH is $HPO_4^{2-}$. The fully protonated form, which is the predominant form in an environment of low pH, is phosphoric acid ($H_3PO_4$).

Free phosphate is also called **inorganic phosphate**. It is abbreviated by $P_i$. The phosphate covalently bound to sugars, proteins, and other components of the cell is called **organic phosphate**. The bond is via a hydroxyl group of the parent molecule. The ability of the phosphate group to resonate is impaired when it occurs as organic phosphate. The number of possible resonating forms is decreased. In addition, the ability of the phosphate group to rotate and move in solution is impaired when it occurs as organic phosphate. The enzyme-catalyzed cleavage of organic phosphate results in an increase in the number of movements available for the free phosphate group. Hence, the cleavage of the phosphate group from organic phosphate results in an increase in entropy, and therefore in the liberation of energy. ATP-dependent enzymes have the property that they can capture this release of energy, and couple it with an otherwise energetically unfavorable reaction, such as the formation of a peptide bond.

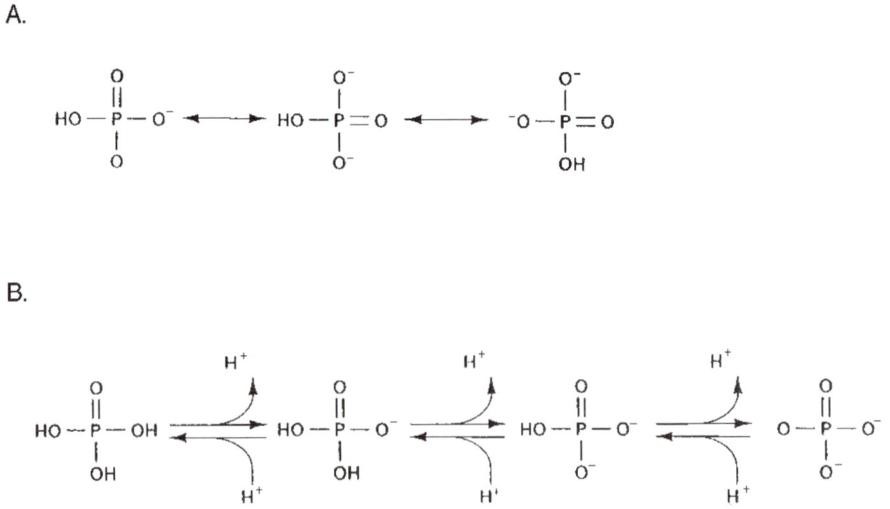

A.

B.

FIGURE 10.35  A. Resonance of inorganic phosphate. B. Release of protons by phosphoric acid and acceptance of protons by phosphate. Phosphoric acid is shown at the far left, while fully ionized phosphate is at the far right.

The phosphate group most commonly associated with temporary energy storage, and energy transfer, is the terminal phosphate group of ATP. However, ATP-dependent reactions that produce AMP with the release of the terminal two phosphate groups, bound together as **pyrophosphate** ($PP_i$), also occur in the cell. Generally, ATP-dependent reactions that result in pyrophosphate production are further driven to completion (in the forward direction) because of the action of phosphatase enzymes in splitting the pyrophosphate into two molecules of inorganic phosphate. If one were required to name the biggest user of ATP in the body, one could mention ion transport systems. It is thought that about 25% of the ATP synthesized per day is used by the sodium pump (Na,K-ATPase).

The amount of energy discharged with the hydrolysis of various organic phosphate bonds varies from a low of 13.8 kJ/mol for glucose-6-phosphate to a high of 61.9 kJ/mol for phosphoenolpyruvate (PEP). The amount of energy released with the hydrolysis of ATP to ADP + $P_i$ is 31.8 kJ/mol (Bridger and Henderson, 1983). Hydrolysis of creatine phosphate releases 43.1 kJ/mol. The relatively large amount of energy released with the hydrolysis of molecules such as ATP and PEP is due to the separation of the negative charges as the $P_i$ is released from the parent molecule. Hydrolysis of ATP, for example, relieves the internal strain caused by the negative charges of adjacent phosphate groups. Cleavage of phosphate from glucose-6-phosphate does not result in any such charge separation.

## Phosphate in Nucleic Acids

A universal function of phosphate is as a component of DNA and RNA. DNA is a polymer of deoxyribonucleosides. RNA is a polymer of ribonucleosides. The molecular glue in these polymers is the phosphate group. The phosphate group links two adjacent nucleosides. The phosphate group remains negatively charged in this linkage. The negative charge stabilizes the polymer against spontaneous, nonenzymatic hydrolysis. The advantage of the phosphate group as a linking molecule is that it can bind to two organic molecules, via their hydroxyl groups, and yet remain negatively charged (Westheimer, 1987). Accidental hydrolysis of a DNA polymer may involve the attack of a free hydroxyl anion. The negative charge of the phosphate linkages of DNA repels such an attack (dR stands for deoxyribose) (Figure 10.36).

One might envision a type of DNA or RNA held together by sulfate groups, as shown in Figure 10.37. Westheimer (1987) has pointed out that the sulfate bond is relatively unstable. It is able to link two organic molecules, but it lacks a negative charge in this linkage. It is susceptible to attack by a hydroxyl group. The imaginary DNA shown would undergo spontaneous hydrolysis by the attack of hydroxyl anions, resulting in an unacceptably high rate of damage to the genetic material.

## Phosphate Groups Prevent Phosphorylated Compounds from Leaking Out

In addition to its functions in bone, energy transfer, and nucleic acids, phosphate serves to prevent the leakage of biochemicals from the cell. The phosphate groups of nucleotides, intermediates of glycolysis, and vitamin $B_6$ greatly impair the

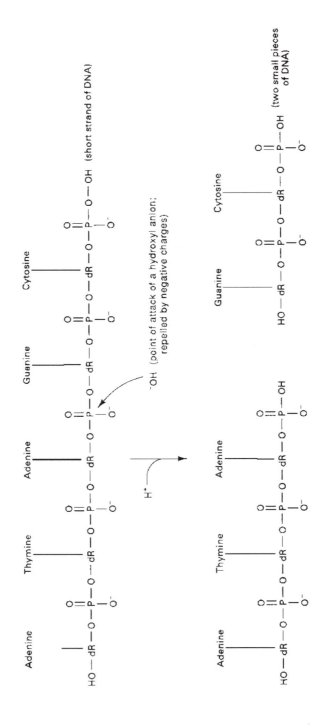

**FIGURE 10.36** Phosphate groups are used to link together DNA polymers. The polymer may be hydrolyzed by attack with hydroxide followed by a proton. RNA is also linked by phosphodiester bonds. These bonds are fairly stable to spontaneous hydrolysis but, as indicated by Westheimer (1987), are about 100 times less stable than those of DNA.

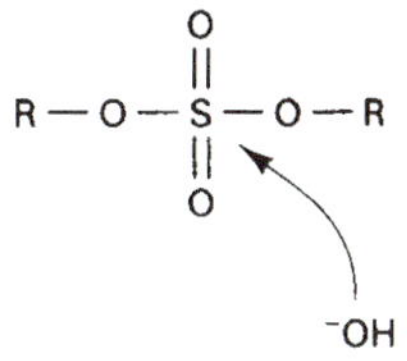

FIGURE 10.37  Imaginary version of nucleic acid using sulfate links and point of attack of hydroxide during hydrolysis.

passage of these molecules through membranes. The phosphate group increases the hydrophilicity of the compound, reducing its tendency to enter into the lipophilic environment of the membrane's interior. In cases where phosphorylated molecules rapidly pass through membranes, the passage is mediated by a specific transport system, as in the example of the **ATP transporter** in the mitochondrial membrane. Other roles of phosphate groups occur in phospholipids and in the covalent modification of proteins.

## Dietary Requirements

The RDA for calcium in adults over the age of 24 is 0.8 g. The RDA for women during pregnancy and lactation is 1.2 g. The increased level of 1.2 g is required to supply the fetus with the 30 g of calcium present in the newborn and to provide the 0.24 g of calcium secreted in the milk each day. The RDA for persons from 11 to 24 years of age is 1.2 g; the RDAs for children (0.8 g) and infants (0.6 g) are lower. Eggs supply about 30% of dietary calcium and 30% of dietary phosphate for the overall population in the United States. Meat, poultry, and fish supply 20–25% of our phosphate, but only 10% of our calcium. Milk and dairy products supply 20–25% of our phosphate, and 50% of our calcium (Calvo and Park, 1996). A dietary deficiency in calcium is quite rare, though calcium nutrition receives much attention because of mainstream health concerns related to calcium, such as osteoporosis, hypercalcemia, and hypocalcemia.

Studies involving the food content of phosphate and dietary phosphate often express the amount of phosphate in terms of the amount of phosphorus. Phosphorus never occurs as free phosphorus in biological systems, just as nitrogen never occurs as free nitrogen atoms; however, use of phosphorus content avoids ambiguities that might arise with the use of the phosphate content, as different people might consider phosphate to be $NaH_2PO_4$, $Na_2HPO_4$, $H_3PO_4$, or $PO_4^{3-}$. In addition, the term *phosphate* may sometimes be thought to mean total phosphate or just free inorganic phosphate. The use of molar units would also avoid these ambiguities.

## Ratio of Calcium/Phosphorus in Various Foods

Phosphate requirements have not been determined for humans. The Food and Nutrition Board recommends a 1/1 ratio of calcium to phosphorus, by weight, in the diet and states that the precise ratio of Ca/P is not important in human nutrition. The ratio of Ca/P, by weight, in typical diets in the United States is about 1/2. Milk and green leafy vegetables are rich in calcium and contain more calcium than phosphorus. Dairy products contribute up to two-thirds of the dietary calcium, with vegetables, fruits, and grains supplying most of the remainder. Meat, poultry, and fish supply only a small amount of the dietary Ca. Human milk and bovine milk contain Ca and P at ratios of about 2/1 and 1/1, by weight, respectively. Hence, human milk is relatively poor in phosphorus. Generally, meat, poultry, and fish contain 15 to 20 times more phosphorus than calcium, by weight. Liver and heart and foods made from these organs, such as salami and liverwurst, provide some 25 to 50 times as much phosphorus as calcium. Carbonated soft drinks also contain much more phosphorus than calcium because of the polyphosphate additives.

## Calcium in Milk

The fact that milk and milk products are rich in calcium is known by most people. Casein is the main protein of milk, accounting for about 80% of milk protein. Casein has a molecular weight of about 23,000 and, thus, is a relatively small protein; however, the casein in milk occurs as a large complex of casein molecules having an overall molecular weight of about $100 \times 10^6$. Casein is phosphorylated on residues of serine. These phosphate groups bind calcium ions. The calcium ions help maintain the stability of the large complex.

About 30% of the phosphate of bovine milk occurs as free inorganic phosphate. About 20% occurs as phosphoserine residues of casein. About 40% occurs as inorganic phosphate tightly associated with the casein complex. The remaining phosphate occurs as phospholipids. The binding constant involving calcium and casein is about $10^3$ $M^{-1}$ (Berrocal et al., 1989). This indicates fairly weak binding. Further details on milk protein are revealed at the beginning of the Protein chapter.

## Absorption and Excretion

With the typical dietary intake of calcium (800 mg/day), about 20% of the calcium is absorbed. Fecal Ca is about 640 mg/day and urinary Ca is about 160 mg/day. Urinary calcium increases somewhat with an increase in intake. For any given increase, urinary Ca increases by an amount equivalent to about 6.0% of the increase in intake. Fecal Ca consists of the unabsorbed mineral and a small amount of Ca secreted into the gastrointestinal tract (100–150 mg/day). An amount of Ca equivalent to 1 to 6% of the total fecal Ca is excreted in the bile (DeGrazia and Rich, 1964). Urinary Ca generally ranges from 100 to 250 mg/day. The Ca lost via the skin is about 15 mg/day, with increased losses occurring in the sweat during work in warm climates.

## Routes of Calcium Loss with Moderate and High Calcium Intake

Controlled studies with human subjects consuming moderate levels of Ca (700 mg/day) revealed that an amount of Ca equivalent to 26% of the intake was excreted in the urine, with 74% excreted in the feces. With high-calcium diets (1600 mg/day), urinary Ca was equivalent to 15 to 18% of intake, with the remainder being fecal, indicating a decrease in the proportion of dietary Ca absorbed by the gut. In the high-Ca diets, the extra mineral was supplied by milk, $CaCO_3$ pills, or a $CaCl_2$ solution (Lewis *et al.*, 1989).

Studies with baby pigs revealed that urinary Ca levels were low and equivalent to about 2% of the intake. Fecal Ca was equivalent to about 21% of the intake. The remaining calcium was retained and used for growth. The excreted phosphate was more equally distributed in the urine and feces, where the urinary phosphate was equivalent to 12 to 27% of the intake and the fecal phosphate to about 13% of the intake. The remaining phosphate was used for growth (Miller *et al.*, 1964).

## Routes of Calcium Loss during Fasting

Studies with human subjects on a total fast revealed the obligatory losses of calcium. The total fast was continued for 40 days and involved obese subjects (Fisler and Drenick, 1984). Subjects drank 2 liters of water per day. The urinary losses during the first 3 weeks of fasting averaged about 150 mg/day, whereas fecal Ca was about 24 mg/day. With more prolonged fasting, urinary Ca decreased to about 90 mg/day, and fecal Ca decreased slightly. The obligatory losses of phosphate were also studied. Urinary phosphate decreased from about 1.0 g/day (0.33 g phosphorus/day) in the initial weeks of fasting to about 0.35 g/day later. Fecal phosphate fluctuated at about 0.3 g/day throughout the fast. The results from the total fasting study demonstrate that the ability to conserve Ca is less than that for sodium or potassium. With fasting, urinary calcium continued at a rate of 100 mg/day or higher, a significant fraction of the RDA for this mineral (Fisler and Drenick, 1984).

## Influence on the Counterion on Calcium Absorption

Calcium absorption may be influenced by the solubility of the specific calcium salt. Rates of absorption of calcium from calcium salts, such as Ca acetate, Ca lactate, Ca gluconate, Ca citrate, and Ca carbonate, seem to be similar and to fall in the range 25 to 40%. These values were determined in studies with humans in the absence of a concurrent meal (Sheikh *et al.*, 1987). The test dose contained 500 mg of Ca. The absorbability of the calcium in whole milk falls in this range and is about 30%.

Calcium carbonate is about 30% absorbed and is the preferred form of calcium when used as supplements, because of the relatively low molecular weight of the counterion, which means the pills are small. Highly soluble calcium salts, such as those with citrate, malate, or glycine as the counterion, have been used as supplements, though they make for large-sized pills. Calcium citrate is more rapidly dissolved than calcium carbonate, though this difference has been found to make little or no difference on the availability of the mineral. Commercially available calcium carbonate supplements are available in a number of physical forms, some of which, unfortunately, have been found to be extremely resistant to dissolution in water (Sheikh and Fordtran, 1990). The solubilities of a number of calcium salts in water are listed in Table 10.14.

**TABLE 10.14**  Solubility of Various Forms of Chlorine

|  | Maximal solubility in water at pH 7 (mM) |
|---|---|
| Calcium oxalate | 0.04 |
| Hydroxyapatite | 0.08 |
| Calcium carbonate | 0.14 |
| Tricalcium phosphate | 1.0 |
| Calcium citrate | 7.3 |
| Calcium citrate malate | 80 |
| Calcium (glycine)$_2$ complex | 1500 |

*Source*: Heaney *et al.* (1990).

Calcium oxalate is relatively insoluble and is poorly absorbed by the gut. Calcium oxalate can be dissolved to concentrations attainable by other "insoluble" nutrients, such as the long-chain fatty acids. Only about 10% of the calcium supplied as Ca oxalate is absorbed by the human gut. The calcium of spinach is only 5% absorbable. This calcium occurs largely as calcium oxalate. It is not clear why the mineral in spinach appears to be less absorbable than pure calcium oxalate. Hydroxyapatite, a pure crystal resembling that of bone, has a solubility and absorbability similar to those of Ca oxalate.

## Calcium Absorption with Gastric Atrophy

The issue of calcium absorption involves the possible influence of gastric acid. Calcium salts are more soluble in acid than at neutral pH. This fact can be illustrated by soaking an eggshell or oyster shell (CaCO$_3$) in vinegar and noting the disappearance of hard material over the course of a week. The influence of gastric acid might be expected to be of interest to those who are not able to produce gastric acid (achlorhydria). This condition occurs with gastric atrophy, and with pernicious anemia, and is called **achlorhydria**. Gastric atrophy occurs in 10 to 20% of the elderly. Supplements of calcium in the form of calcium carbonate may be poorly absorbed (4% of dose absorbed) by those with achlorhydria, whereas highly soluble forms such as Ca citrate are well absorbed (Recker, 1985). If those with achlorhydria eat food with the calcium carbonate, the absorption of calcium may be increased (20% of dose absorbed). Food has been found to increase the absorption of both the soluble and less soluble forms of calcium supplements in normal persons. A typical absorption of 20% on an empty stomach may be increased to 30 to 35% by food. Apparently, the food results in a more gradual entry of the calcium into the intestines, resulting in its more complete absorption.

A variety of dietary sugars may enhance the rate of calcium absorption, though the mechanism of this effect remains unclear (Schuette *et al.*, 1989). This effect has brought attention to the sugar in milk (lactose) as a nutrient that can enhance calcium absorption. This effect of lactose is not unique; other sugars can produce the same effect.

## Methods for Measuring Calcium Absorption

The measurement of calcium absorption is of interest to those studying the availability of the calcium in various foods or supplements. The quantity and availability of food calcium are concerns in the prevention of osteoporosis. Calcium absorption is also of interest to those studying the interactions between different components of food and their influence on calcium absorption. Measurement of either fecal or urinary Ca can indicate the extent of absorption, but some fine-tuning is needed to yield an accurate picture, as described in what follows. The radioactive calcium can be supplied as calcium-45 or calcium-47. Where the absorption of food Ca is of interest, the study can be run most accurately by using food containing naturally labeled calcium, that is, foods grown or animals raised in the presence of radioactive calcium. A less accurate picture can be attained by mixing radioactive calcium with the food, that is, with the plant or animal product.

Calcium absorption can be measured by the **double-isotope technique**. In this technique a meal containing calcium-45 is consumed and the radioactivity in the urine measured. The measurement of urinary $^{45}Ca$ alone cannot provide the fractional absorption, because some of the $^{45}Ca$ absorbed is taken up by cells, deposited in bone, or excreted in the bile. A second isotope of calcium, $^{47}Ca$, is used to correct for the fates of absorbed calcium, other than excretion in the urine. The use of the $^{47}Ca$ is intended to eliminate cell uptake, bone deposit, and biliary losses as variables in the study of the absorption of the dose of calcium-45.

The double-isotope technique is carried out as follows (DeGrazia *et al.*, 1965). The subject consumes a test food containing calcium-45. About halfway through the absorptive process, calcium-47 is injected into a vein, in the form of a solution of $^{47}CaCl_2$. The absorptive process is about halfway complete at about 2 hours after a meal. The injected $^{47}Ca$ is calcium that is 100% absorbed. A urine sample is collected 1 day after the oral and intravenous doses of radioactive calcium. The fraction of the oral dose absorbed is calculated by the formula:

$$\text{fraction absorbed} = \frac{\text{fraction of oral dose in urine sample}}{\text{fraction of injected dose in urine sample}}$$

The use of the second, injected isotope is based on the assumption that once inside the bloodstream it behaves identically to the first isotope entering the bloodstream from the gut.

Calcium absorption can also be measured by determining the fraction of the mineral in a test dose that is not absorbed. Here, a test meal containing calcium-47 is consumed and the feces collected for 12 days to measure the isotope excreted. The calcium lost in biliary and intestinal secretions may be corrected for by an intravenous injection of radioactive calcium (Spencer *et al.*, 1978).

## Calcium/Phosphate Ratios

The relationship between calcium and phosphate metabolism is very complex. The question of the ratio of calcium/phosphate in the diet may be raised when discussing diets needed to support maximal growth or when discussing pathological phenomena, such as hypocalcemia, osteoporosis, kidney stone formation, and the calcification of soft tissues. A firm grounding in one or two relationships in calcium

and phosphate metabolism is presented here in a study involving rats. Consideration of this study may facilitate the understanding of material on more subtle or
complex issues of calcium and phosphate metabolism.

### Ratio of Calcium/Phosphate in Bone

A special relationship between calcium and phosphate is suggested by the fact that
the calcium/phosphorus ratio in bone is about 2.2/1.0, by weight. As almost all of
the body's calcium and most of the body's phosphate occurs in bone, one might
expect that the ratio of calcium/phosphorus in the diet that supports maximal
growth would be similar to the ratio of these elements in bone. This ratio would
be expected to be influenced by the fact that dietary P is more readily absorbed
than Ca and by the use of P for soft tissues.

### Animal Studies with Variable Calcium, and Constant Ca/P Ratios

The following study concerns the dietary requirement for calcium. Different
groups of rats consumed diets containing different levels of calcium, ranging from
about 0.8 to 6.0 g Ca/kg diet. All of the diets contained calcium/phosphorus in a
ratio of about 1/1, by weight. Daily rates of food consumption by the rats consuming the different diets were similar or identical. The calcium was supplied as
calcium carbonate, the phosphorus as ammonium phosphate. The feeding trial
lasted 3 weeks. The amount of food consumed per day was recorded for each
group of rats and used to calculate the amount of calcium consumed. Weight gain
was also recorded. The amount of carcass calcium was determined at the end of
the feeding trial. The results of the study are depicted in Figure 10.38. The total

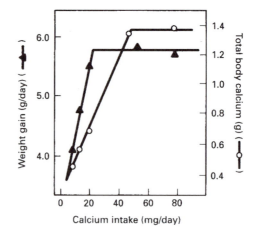

**FIGURE 10.38**  Rate of weight gain (▲) and amount of total body calcium (○) for rats fed
diets containing different amounts of calcium for 3 weeks. (Redrawn with permission from
Bernhart *et al.*, 1969.)

body calcium is essentially identical to the total bone calcium. The data show that near-maximal rates of weight gain occurred with diets supplying 20, 50, and 80 mg Ca/day. Maximal rates of bone calcium deposition occurred with diets supplying 50 or 80 mg Ca/day. The data suggest that the minimal amounts of Ca needed to support maximal rate of growth and maximal rate of increase in skeletal mass are different. Once a sufficient amount of calcium is consumed to support maximal growth, a larger amount of calcium is needed to support a maximal increase in bone mass.

### Animal Studies with Constant Calcium and Variable Phosphate

A second study focused on the phosphate requirement. Here, all diets contained the same amount of calcium, but the amount of phosphate was varied. All of the diets contained 3.6 g Ca/kg diet. This amount was shown, in the preceding study, to support maximal growth and bone mass increase where the ratio of calcium/phosphorus was about 1/1. The amount of phosphate in the diet was varied to supply calcium/phosphorus in ratios of 1/0.2, 1/0.3, 1/0.4, 1/0.6, 1/0.9, and 1/1.4. Food consumption rates for the groups of rats were nearly identical. The data in Figure 10.39 demonstrate that near-maximal weight gains occurred when the calcium/phosphorus ratio was 1/0.6, 1/0.9, and 1/1.4. In short, dietary phosphorus could support near-maximal growth when present in the diet in an amount slightly lower than, about equal to, and slightly greater than that of calcium. The effect of the different diets on bone calcium was also determined. Diets containing calcium/phosphorus at ratios of 1/0.6, 1/0.9, and 1/1.4 also seemed to result in near-maximal bone calcium levels. The results in Figures 10.38 and 10.39 may be summarized as follows. Where the diet supplies sufficient amounts of calcium, an amount of phosphorus slightly lower or greater than the amount of calcium could support near-maximal rates of growth and bone formation.

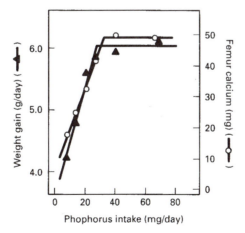

**FIGURE 10.39** Rate of weight gain (▲) and amount of femur calcium (○) for rats fed diets containing different amounts of phosphorus for 3 weeks. (Redrawn with permission from Bernhart *et al.*, 1969.)

Studies with dogs revealed that high dietary phosphate results in bone loss and a decline in bone mass. The diets contained calcium/phosphorus (mg/mg) at a 1/10 ratio, and resulted in a loss of trabecular bone at a greater rate than cortical bone (Calvo and Park, 1996).

The dietary calcium/phosphorus ratio is important for the maintenance of kidney health. Where there is too much phosphate, mineral deposits may develop in the kidneys, in a process called **nephrocalcinosis**. Female rats are especially prone to nephrocalcinosis, which occurs when the calcium/phosphorus ratio is under 1/0.77. For this reason, a recommended Ca/P ratio is 1.66 (the same as 1/0.6) (Reeves, 1997).

## Human Studies with Various Ca/P Ratios

Dietary levels of calcium and phosphate are relevant to such issues as hypocalcemic tetany in infants, growth and maintenance, and maximal bone mass and the consequent protection against osteoporosis. Studies with human subjects have revealed that a negative calcium balance is attained with the consumption of 200 to 400 mg Ca/day, whereas a near-neutral or positive calcium balance is provided by 500 to 850 mg Ca/day (Spencer *et al.*, 1978, 1984; Zemel and Linkswiler, 1981; Hegsted *et al.*, 1981). The experiments compared the calcium balances attained with diets containing Ca and P in ratios ranging from 1/0.4 to 1/10. In many cases, increasing the amount of dietary phosphate resulted in an increase in the amount of fecal calcium and a decrease in urinary calcium, but no change in calcium balance. In other cases, increasing the amount of dietary phosphate had a slight pro effect on the calcium balance. The data indicate that varying the Ca/P ratio has little effect on calcium balance. In contrast, elevated levels of dietary phosphate may have a striking impact on the newborn, as detailed later in this section. The optimal amounts of calcium and phosphate leading to the maintenance of bone mass over the long term, that is, several decades, are not clear and remain a subject of some controversy.

## Influence of High Protein Diets on Calcium

A variety of studies have shown that increasing dietary protein can induce excessive losses of calcium in the urine and a negative calcium balance. This effect is called the **calciuric effect of protein**. The effect may occur in some persons but not in others. It is most marked with low intakes of calcium and high intakes of protein. Many persons in the United States consume such diets. The calciuric effect of proteins has been demonstrated in a number of controlled studies with human subjects. The effect can be reduced, and perhaps minimized, by phosphate. A simultaneous increase in phosphate intake with protein intake may result in only a small increase in urinary calcium and maintenance of calcium balance. Foods that are high in protein, such as meat and eggs, also contain high levels of phosphate. Thus, the potential hypercalcemic effect of the meat or egg protein may be largely reduced by the phosphate in the same food.

The mechanism by which dietary protein induces an increase in urinary calcium is not clear. The effect has been attributed, in part, to the catabolism of sulfur-containing amino acids to yield sulfate. Elevated levels of plasma sulfate can form a complex with calcium. The complex passes into the renal tubule, where it is poorly reabsorbed, resulting in its excretion in the urine. The mechanism by which phosphate reverses the hypercalciuric effect of protein is also not clear.

## Phosphate Turnover in the Body

About 85% of the body's phosphate occurs in bones, with 14% in soft tissues and about 1.0% in the extracellular fluids. The normal range of phosphate intake is 20 to 50 mmol/day (0.6–1.5 g phosphorus/day). Phosphate is absorbed throughout the small intestines. In the duodenum, it is absorbed by an Na-dependent transport mechanism. Here, transport of the phosphate is coupled with the cotransport of a sodium ion. The rate of Na-dependent transport of phosphate is enhanced by $1,25\text{-}(OH)_2D_3$. Phosphate transport in the jejunum and ileum occurs by a passive mechanism. The rate of phosphate transport in this case is dependent mainly on the concentration of phosphate in the lumen and is independent of the levels of other nutrients and independent of energy-using processes. About 200 mg of phosphorus is excreted per day in fluids of the gastrointestinal tract. About two-thirds of this phosphorus is reabsorbed by the gut.

Phosphate is better used when supplied by food than when supplied by phosphate salts, such as potassium phosphate. The food phosphate occurs largely as phosphate esters and, with gradual hydrolysis, enters the body relatively slowly and is used efficiently in the metabolic processes. Phosphate supplied as inorganic phosphate is rapidly absorbed, with a tendency to be excreted in the urine rather than to be used by tissues (Schuette and Linkswiler, 1982). The phosphate in bran occurs largely in the form of phytic acid (inositol hexaphosphate). The phosphate groups of this compound may be only partially hydrolyzed in the gut. Phytic acid tends to be excreted via the fecal route as a complex with iron or calcium.

### Phosphate Deficiency

Dietary phosphate deficiency is relatively rare because the phosphate content in plant and animal foods is well above the requirement and because of the efficient absorption of phosphate (50–90%). Phosphate deficiency can occur in a number of situations. It can occur with the chronic intake of aluminum-based antacids, particularly with a low-phosphate diet. These antacids form a complex with dietary phosphate, preventing its absorption and resulting in the deficiency. Deficiency can occur with increased urinary excretion of phosphate that occurs with starvation and in diabetics experiencing ketoacidosis. Chronic alcoholics may be phosphate deficient because of decreased dietary intake, impaired absorption, and increased urinary excretion (Berner and Shike, 1988). Phosphate deficiency has been observed in the small, premature infant. The premature infant has a higher requirement for phosphate than the term infant, as it grows at a relatively greater rate. Phosphate is required for the synthesis of soft tissues and bone. The small, premature infant's requirement for phosphate cannot be fully supplied by human milk.

### Phosphate Toxicity and Hyperphosphatemia

The normal range of free inorganic phosphate, in plasma, is expressed as 2.0–4.3 mg phosphorus/100 ml. Hyperphosphatemia is defined as the condition where plasma phosphate levels rise above 5.0 mg phosphorus/100 ml. Hyperphosphatemia can result from overuse of laxatives or enemas that contain phosphate. Phosphate enemas are used in the hospital prior to examinations of the intestines for disease, such as colon cancer. However, hyperphosphatemia has occurred in the home situation with accidental drinking of enema formulas. Enema drinking has been found to result in serum phosphate levels of 40–60 mg phosphorus/100 ml. Death from hyperphosphatemia has occurred with administering of enemas to infants (Kirschbaum, 1998; Barcia *et al.*, 1997). With enema poisoning, the phosphate enters the bloodstream, complexes with calcium ions, and results in hypocalcemia. Hypocalcemia, whether due to phosphate poisoning or other causes, can result in tetany or even death.

Hyperphosphatemia is much more commonly produced by chronic renal failure, and by other problems of kidney function, than by phosphate poisoning. Chronic renal failure, hypoparathyroidism (low parathyroid hormone), and pseudoparathyroidism (failure of kidneys to respond to parathyroid hormone) all involve the failure of the kidneys to excrete phosphate into the urine at a normal rate. The danger of long-term hyperphosphatemia is the deposit of calcium phosphate crystals into the soft tissues of the body (Knochel, 1994; Holick, 1996).

## Osteoporosis

It is estimated that at least 15 million persons in the United States have osteoporosis. The sites of osteoporotic fractures most commonly are the vertebra and hip, though the forearm and wrist are also involved. A bone fracture in an elderly person is undesirable because it leads to periods of disability. In addition, about 10% of the fractures of osteoporosis are fatal. Osteoporosis results from the continued activity of osteoclasts but reduced activity of osteoblasts. Hence, the disease involves a lack of coordination in the rates of bone resorption and formation. The biochemical basis for this imbalance is not clear because of the complexity of the mechanisms controlling the activities of bone cells (and of all other cells). At least a dozen growth factors are used to regulate bone cells.

Persons at greatest risk for osteoporosis are Caucasian and Asian women, though males also acquire the disease. Peak levels in bone mass are reached by ages 30 to 35. Bone losses begin before the age of 40 and progress at a more rapid rate thereafter. Bone losses are especially rapid with the onset of menopause and in the following 5 years, after which the rate of bone loss declines. Caucasian and Asian women tend to have a smaller bone mass than other groups, and thus have less of a built-in safeguard against osteoporotic disease. African-Americans have a greater bone density in young adulthood than Caucasians or Asians and have a lower prevalence of osteoporosis. African-Americans also seem to lose bone with aging at a lesser rate than other groups. The benefit of attaining maximal bone mass by young adulthood, as a guard against osteoporosis, is a good rationale for maintaining proper calcium nutrition during adolescence and early adulthood.

Calcium nutrition and physical activity are contributing factors in maintaining skeletal mass. It is firmly established that the loss of bone and a negative calcium balance tend to occur throughout the population after the age of 40. It is also obvious that a dietary deficiency in calcium leads to progressive bone losses. However, when a diet of an adult already supplies the RDA for calcium, further calcium supplements are not expected to have a significant influence on the trend to bone loss. It is firmly established that complete immobilization results in bone losses. These losses occur with prolonged bed rest, as in those suffering from spinal cord injuries, as well as when the skeleton bears no weight, as during prolonged spaceflight. Evidence suggests that programs of physical activity involving walking, dancing, or sports may have no influence on reducing age-related bone losses and the advent to osteoporosis (NIH Consensus Development Panel, 1994).

The study of osteoporosis is complicated by a number of factors:

1. The rate of bone loss is relatively slow, thus requiring long periods of observation (1–4 years) to determine the influence of preventive measures. The rate of bone loss may be only 1% per year. This may be too small to measure reliably by the available technology.

2. Bone loss occurs at different rates at different ages, thus requiring attention to age and to the time of onset of menopause.

3. Bone losses occur at different rates in different bones, and in different regions of the bone (cortical versus trabecular). For example, from ages 40 to 80, women lose about 35% of **cortical bone** and 50% of **trabecular bone**. Losses in men amount to about 66% of these values. Thus, any study of osteoporosis must take into account the locale of the bone loss.

4. Determination of dietary calcium intake is important to those studying osteoporosis; however, an accurate value for Ca intake is often elusive.

Dietary habits and calcium intake during a previous decade of life usually cannot be determined reliably by means of an interview or questionnaire. These techniques yield data, but the data are likely to be worthless. Such techniques as 1-week food diaries, recorded at the beginning and end of a year-long study period, have been used to gain reliable data.

Age-related bone loss and the trend toward osteoporosis occur at a greater frequency in postmenopausal women than in other groups. Menopause is the permanent cessation of menstruation resulting from loss of cyclic ovarian function. It occurs spontaneously at an average age of 50 years. It also occurs immediately after surgical removal of the ovaries. This operation is fairly common. The loss in ovarian function results in a decrease in plasma estrogen levels. Estrogen plays a major role in maintaining bone mass in women. The rate of bone loss during the first 5 years of menopause is relatively rapid. The rate is twice that of men the same age. The rate then decreases and plateaus. After 10 to 15 years of estrogen deficiency, the bone mass is about one-third of its peak mass. The danger of bone loss is the eventual high risk for fractures of the vertebra, hip, and other bones.

Estrogen replacement therapy is started in many women immediately after the onset of menopause, when the rate of bone loss is greatest, and may be continued until the age of 75. It might be noted that all women are estrogen deficient after menopause but that osteoporosis does not strike all women. Thus, other factors contribute to the disease. Men usually do not undergo striking changes in levels of steroid sex hormones, such as testosterone, with aging. The possible role of male sex hormones and bone metabolism is not clear.

Calcitonin therapy is also used in the treatment of osteoporosis in men and women. Calcitonin is a hormone composed of 32 amino acids. It has a C-terminal amide group. Calcitonin is synthesized and secreted mainly by the thyroid gland and in much smaller amounts by other organs. Calcitonin occurs in the plasma at about 20 pg/ml. Osteoclasts contain calcitonin receptors. The hormone appears to impair the metabolic activity of the osteoclasts. The physiological function of calcitonin is not clear; however, treatment of people with the hormone does result in inhibition of bone resorption and maintenance of the bone mass. The treatment preserves the bone mass in the spine, femur, and radius. There is little or no evidence that low plasma levels of calcitonin are associated with an increased risk for osteoporosis.

Calcium supplements, along with estrogen and calcitonin therapy, are commonly used in the prevention and treatment of osteoporosis. Bone loss clearly occurs with diets supplying under 400 mg Ca/day. Bone loss can be minimized with the consumption of the RDA for calcium. There is some thought that all postmenopausal women should consume 1000 to 1500 mg of calcium per day, an amount higher than the RDA. There is some evidence that such supplementation can reduce bone losses of the ulna but have no effect on spinal bone losses (Smith and Gilligan, 1990).

Calcium absorption by the intestines decreases with aging, especially after the age of 70. This decrease is probably a consequence of the 50% decrease in production of 1,25-$(OH)_2D_3$ that also occurs with aging. Many homeostatic mechanisms of the body tend to become less agile with aging. The adaption of the gut to absorbing adequate calcium from borderline diets (500 mg Ca/day or less) may be impaired in the elderly. Because of these changes, there is much interest in increasing the RDA for calcium for postmenopausal women.

Although it might seem reasonable to treat osteoporosis with vitamin D, it must be realized that the primary function of vitamin D is to maintain plasma calcium levels, not to promote bone formation. An end-effect of vitamin D supplementation is an increase in bone resorption and increased excretion of calcium in the urine. Calcium supplements should not be used indiscriminately. Two types of persons should not receive calcium supplements: persons with hypercalcemia and persons with kidney stones or a family history of kidney stones. These two issues are discussed in the following pages.

In summary, persons with mild osteoporosis are treated with calcium supplements to bring the calcium intake to the RDA. Those with more severe bone losses receive estrogen (women only) or calcitonin. Estrogen and calcitonin decrease the rate of bone resorption. The abandonment of a sedentary lifestyle is encouraged. Chronic alcoholism is a risk factor, as is cigarette smoking. These practices are discouraged. Drugs such as fluoride and bisphosphonates are gaining attention in the treatment of osteoporosis (Sahni *et al.*, 1993; Delmas, 1996; Torgerson *et al.*, 1997). Bisphosphonate has a structure similar to that of diphosphate (pyrophosphate) (Figure 10.40). The bisphosphonate compounds absorb to the bone crystal. Bone containing the bisphosphonate resists resorption, resulting in maintenance of the bone mass.

Diphosphonate                    Pyrophosphate

FIGURE 10.40 Bisphosphonate, a drug, and pyrophosphate.

## Osteoporosis: An Issue in Child and Adolescent Nutrition

*Calcium Nutrition in Girls*

Adequate calcium nutrition during childhood and adolescence is thought to be an important deterrent to osteoporosis in later life. A one-year study of 149 girls (8 years old) revealed the effects of consuming a typical diet (900 mg Ca/day) or a diet supplemented with an additional 850 mg Ca/day (Bonjour *et al.*, 1997). The Ca supplement was supplied by specially prepared biscuits, juices, and candy bars (rather than pills) to assure compliance. To view some of the data, the density at the femoral diaphysis increased by 5% (unsupplemented) and by 6% (plus calcium) during the 1-year period. The density at the lumbar spine increased by 4% (unsupplemented) and by 4% (plus calcium). The data demonstrated a clear and definite increase in bone density at some sites, but not at others. This study, as well as others (Johnston *et al.*, 1992), tend to show that calcium supplements to children can benefit bones of the **appendicular skeleton**, but not the axial skeleton. The appendicular skeleton includes the pelvis, arm bones, and leg bones. The **axial skeleton** includes the skull, vertebra, ribs, and sternum.

Boys and men tend to consume an amount of calcium that meets the RDA; however, girls and women tend to consume only half the RDA level (Bronner, 1994; NIH Consensus Development Panel, 1994). Once the peak bone mass has been achieved, evidence suggests that Ca supplements have no influence on further increasing bone mass. Once menopause has been reached, calcium supplements cannot slow the loss of bone, unless estrogen is also taken (there is some evidence that this statement applies only to trabecular bone, and that cortical bone loss can be reduced by Ca supplements alone (Bronner, 1994)). In viewing the foregoing considerations, it seems reasonable that all parents of girls should be persuaded to take the RDA for calcium seriously for the duration of their childhood and adolescence.

*Prolonged and Intense Exercise May Provoke Lower Estrogen Levels and Bone Loss*

Intense and prolonged athletic training in girls and women sometimes leads to **amenorrhea**, i.e., the lack of menstrual periods over the course of an entire year. Although amenorrhea is easy to detect by the layperson, further focus on this issue revealed that intense and prolonged exercise can lead to endocrine disturbances and in bone loss. The endocrine disturbances include reduced estrogen, reduced progesterone, and reduced luteinizing hormone. Apparently, the reduced estrogen level is the provoking factor in this bone loss. Women who engage in competitive sports that require intense exercise should not expect to acquire any protection from osteoporosis (Prior *et al.*, 1990; Winters *et al.*, 1996).

*Bone Loss during Lactation*

During breast-feeding, bone mineral at specific sites (lumbar spine, neck of the femur) can decrease by 3–5%. In new mothers, these changes do not occur if the mother does not breast-feed. One might be alarmed that this loss might contribute to eventual osteoporosis. However, the losses have been documented to be reversed when lactation is concluded. Furthermore, attempts to prevent bone loss during lactation, by means of Ca supplements, have failed to prevent the bone loss. Thus, there is little or no reason to increase dietary Ca during the lactation period (Prentice, 1997).

## Method for Assessing Bone Density

In studies of bone growth and of osteoporosis, bone density is measured most precisely by **dual-energy X-ray absorptiometry**. The unit of bone density is $mg/cm^2$, and this unit applies to data acquired from a cross-section. The technique uses sodium iodide crystals, not X-ray film, to detect X-rays after passing through bone. With repeated measurements over the course of a day, for example, the variation is only 1%. Although this figure is impressive, one should take care in using the method since bone mass or density may change by only 1% during the course of a research study lasting a year or so (Epstein and Miller, 1997; Lukaski, 1993).

## Genetics of Osteoporosis

Osteoporosis appears to result from defects in one of several different genes, as is the case for diabetes, cancer, and atherosclerosis. In the case of osteoporosis, none of these genes have been identified with certainty. Candidate genes include those that code for collagen, collagen-degrading enzymes, certain hormones (cytokines; polypeptide growth factors), and the receptors coding for cytokines, polypeptide growth factors, estrogen, and vitamin D (Grant and Ralston, 1997).

## Fluoride and Bone

Fluoride is a monovalent anion. It appears not to be an essential nutrient, as raising animals on a fluoride-deficient diet over three generations has been found not to impair growth and reproduction. Fluoride has proven to be of value in reducing the rate of dental caries (tooth decay). There has been much interest in the possibility that fluoride can prevent and reverse osteoporosis, though such effects have not been proven.

The effect of fluoride in preventing dental caries was recognized when it was found that levels of tooth decay were low in regions of the world where the water supply naturally contained high levels of fluoride. In some regions, the levels of fluoride are so high that they cause a harmless mottling of the teeth called "fluorosis." Both the first set and the permanent set of teeth benefit from dietary fluoride. Water and other fluids account for most of our fluoride intake.

The optimal range of water fluoride is 0.7 to 1.2 mg/liter. This level of fluoride, whether provided by naturally occurring fluoride or by the artificial fluoridation of the community water supply, results in about a 50% reduction in the rate of dental caries. The water used for making coffee, tea, and soups is a significant source of fluoride. The water in soft drinks and alcoholic beverages can also be a significant source when the bottling plant is located in an area using fluoridated water. Commercial infant formulas are significant sources of fluoride for babies.

The fluoride content of human milk is slightly under 0.01 mg/liter (0.5 $\mu M$). Milk fluoride levels are not much influenced by changes in dietary intake, even though plasma levels may vary widely. The fluoride content of cow milk is usually under 0.05 mg/liter. Commercially available milk-based and soy-based formulas contain 0.13 to 0.30 mg F/liter. The recommended upper limit in infant formulas is 0.4 mg F/liter.

The American Dental Association recommends that fluoride supplements be taken in areas that do not have fluoridated water. Nonfluoridated water is defined as that containing less than 0.3 mg F/liter. The recommendation is 0.25 mg F/day from ages 0 to 2 years, 0.5 mg F/day from ages 2 to 3 years, and 1.0 mg F/day from ages 3 to 13 years. Fluoride levels greater than 2 mg F/liter may produce dental fluorosis.

Fluoride is incorporated into bone and teeth, resulting in the formation of fluoroapatite. Fluoroapatite is like hydroxyapatite, except that the fluoride anion substitutes for the hydroxyl anion. Fluoroapatite has a more orderly structure and is less soluble than hydroxyapatite.

There is much interest in the lifelong use of fluoridated water for preventing osteoporosis and in using pharmacological quantities of fluoride for reversing osteoporosis. Some epidemiological studies have shown that the incidence of osteoporosis in areas containing high fluoride levels (1–8 mg F/liter) is lower than in nonfluoridated areas (under 0.1 mg F/liter). These effects require many years of residence in the fluoridated area and are thought to require the concomitant consumption of the RDA of calcium. Pharmacological levels of fluoride (30–80 mg F/day) have been tested for their possible effect on osteoporosis. These levels are equivalent to that in 50 liters of drinking water. Normally, plasma fluoride is about 0.5 $\mu M$. A temporary rise to about 5.0 $\mu M$ occurs with the consumption of a liquid meal containing 5.0 mg F. The beneficial effects of fluoride in the treatment of osteoporosis are thought to require the maintenance of plasma fluoride at 5 to 10 $\mu M$ throughout the day. Higher levels of plasma F are associated with side effects, such as joint pain.

The use of fluoride as a drug for periods of 2 to 5 years has been shown to lead to increases in bone density in the spine, but little or no effect elsewhere. The increase in bone density may be equivalent to 35% of the initial bone density. One serious problem remains with this use of fluoride. Some studies have shown that treatment results in a decrease in the bone fracture rate (the desired effect); other studies have shown that the increase in bone density is associated not with a reduction in the fracture rate but in the formation of abnormally structured and unusually fragile bone.

The effect of fluoride might be compared with the effects of estrogen and calcitonin. Fluoride treatment results in an increase in bone mass; estrogen and calcitonin tend not to result in a gain in bone mass, but only in minimization of further bone losses.

## Calcium and Oxalate

Calcium and oxalate are closely associated with the formation of stones in the urinary tract. Kidney stones (renal calculi) and bladder stones are mineral deposits containing protein. They can have a diameter of a centimeter or greater. Most kidney stones (75%) are composed mainly of calcium oxalate or calcium oxalate with hydroxyapatite. Uric acid stones account for about 10% of stones; xanthine stones are rare. Calcium-containing kidney stones occur in Western nations and affect about one person in 1000. The disease may occur in children, but typically occurs after the age of 30 and in men. Calcium bladder stones occur mainly in the children of underdeveloped countries, such as Thailand, and occur rarely in Western nations. Some kidney stones do not result in symptoms. Others may cause blood loss in the urine. Stones that obstruct the flow of urine from the kidney into the ureter result in violent pain, nausea, and vomiting.

Small crystals of calcium oxalate are a normal component of the urine. They form in the glomerular filtrate as water is reabsorbed and the urine is concentrated. The urine of most persons contains compounds that inhibit the growth of crystals. These inhibitors include magnesium, citrate, pyrophosphate, and mucopolysaccharides. Apparently, persons who tend to form renal and bladder stones have reduced levels of these inhibitors. Stone formation has a genetic component. The disease may "run in the family."

Stone formation may occur in those with elevated levels of urinary calcium. Normally, adults excrete less than 200 mg of calcium in the urine per day, even with relatively high intakes of calcium. A fraction of the population absorbs more calcium than normal and excretes more calcium in the urine, resulting in hypercalciuria. Hypercalciuria is defined as urinary excretion of calcium of more than 300 mg/day. About half of patients with calcium stones have hypercalciuria and may be calcium hyperabsorbers. Persons with hypercalciuria are advised to limit their calcium intake to one serving of milk or cheese per day. They are also advised to limit their protein intake to the RDA. Their protein intake should be limited to minimize the calciuric effect of protein. They are also advised to *increase their water intake* to produce 2 liters of urine per day and to avoid oxalate-containing foods. Persons with hypercalciuria and with a familial history of stones should not take calcium supplements to raise their intake above the RDA.

Most of the oxalate in the body arises from endogenous sources, rather than the diet. About 40% of the oxalate formed in the body arises from ascorbic acid. About 60% arises from 2-carbon precursors, such as glycine and ethanolamine. Only about 0.1% of the body's glycine is catabolized via oxalate formation. Some people increase their intake of ascorbic acid to levels above the RDA by taking supplements. Consumption of large amounts of ascorbate results in increases in urinary oxalate in some persons but not in others. The normal, basal level of urinary oxalate is about 50 mg/day. The daily intake of 3.0 g of ascorbic acid may double the normal levels of urinary oxalate and thus increase the risk for calcium oxalate stones.

The pathways for the formation of oxalate from glycine, ethanolamine, and ascorbic acid, are shown in Figure 10.41. Apparently, oxalate has no function in the body, nor is it catabolized to carbon dioxide. Oxalate occurs in plants as sodium

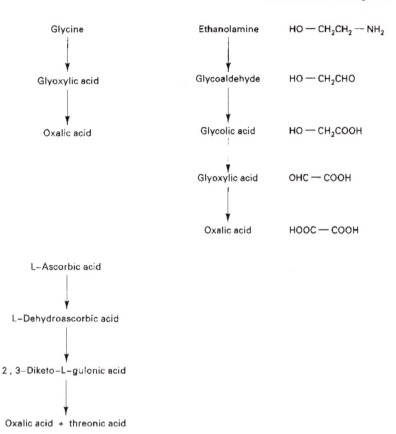

**FIGURE 10.41** Degradative pathways resulting in the formation of oxalic acid. The degradation of glycine and ethanolamine to yield oxalic acid is catalyzed by oxidases. The breakdown of ascorbic acid to produce oxalic and threonic acids occurs spontaneously over time.

oxalate or potassium oxalate. These compounds are water soluble. Oxalate also occurs in plants as calcium oxalate, which is insoluble at neutral and alkaline pH. Calcium oxalate is soluble at pH 3.0 or less, and thus may be solubilized by gastric acid; however, calcium oxalate re-forms with passage into the small intestines. Oxalate occurs in high concentrations in cocoa, chocolate, tea, rhubarb, bamboo shoots, spinach, chard, parsley, and beet tops. Only about 5% of dietary oxalate is absorbed. There is some evidence that eating large amounts of the preceding foods can result in the urinary excretion of elevated levels of oxalate, posing a risk for stone formation.

## Plasma Calcium Levels

The regulation of plasma calcium levels is detailed under Vitamin D in Chapter 9. Further details on plasma calcium levels in normal and pathological situations are revealed in the present section. Plasma calcium levels generally do not reflect calcium status. Plasma calcium is maintained in the normal range during periods of negative calcium balance because of the actions of PTH and 1,25-$(OH)_2D_3$. Our ability to increase the rate of bone resorption during a dietary deficiency in Ca makes it difficult to use plasma Ca levels as an indicator of dietary status. There has been some interest in using plasma PTH or 1,25-$(OH)_2D_3$ levels to assess calcium status, though attempts to use these methods have been met with difficulties. The use of bone density measurements to assess Ca status has also posed a number of problems. Bone density can vary by 10% in the normal population. A decrease in bone density can be detected only after prolonged calcium deficiency (Nordin *et al.*, 1987).

About 40% of serum calcium is protein bound, with most of it (80%) being bound to albumin. Generally, one or two calcium ions are associated with serum albumin. Albumin serves as a calcium buffer. It can bind more calcium ions when excessive concentrations of calcium appear in the bloodstream. About 13% of the calcium in serum is weakly complexed with phosphate, citrate, and sulfate. About half (47%) of serum calcium occurs as the free calcium ion. The level of free serum $Ca^{2+}$ is maintained within narrow limits, 1.0 to 1.25 m$M$ (40 to 50 µg/ml). The normal concentration of total serum calcium (bound plus free) is 85 to 105 µg/ml. Conditions in which the level of free serum calcium falls below and rises above the normal range are called **hypocalcemia** and **hypercalcemia**, respectively. The term *ionized calcium* is often used to refer to the concentrations of free calcium. This term is not scientifically accurate, because all of the calcium in the body is ionized. Calcium does not engage in the formation of covalent bonds (Cotton and Wilkinson, 1966).

## Calcium Sensor Protein Is Used in the Regulation of Plasma $Ca^{2+}$ Levels

Cells of the parathyroid gland are unique among most other cells in that they are sensitive to changes in *extracellular* $Ca^{2+}$ levels. Parathyroid gland cells contain a protein in the plasma membrane that senses extracellular $Ca^{2+}$, called **calcium sensor protein**. Calcium sensor protein is 1078 amino acids long. The N-terminal half (613 amino acids) is extracellular and contains regions that bind and respond to calcium. With varying amounts of bound calcium, the protein transmits a signal to the cell. The transmembrane part of the protein weaves seven times in or out of the plasma membrane. The C-terminal region (222 amino acids) is entirely intracellular. Signaling proteins that weave seven times in and out of the plasma membrane are common in biology, and all of these proteins communicate directly to another signaling protein, called **G protein**. G protein, in turn, relays a signal to phospholipase C. Activated phospholipase C catalyzes the cleavage of the IP3 moiety from the membrane phospholipid, phosphatidylinositol bisphosphate. This cleavage event releases the IP3 into the cytoplasm. Once free in the cytoplasm, the IP3 provokes a momentary increase in calcium levels within the cytoplasm. This burst, in turn, provokes the secretion of parathyroid hormone from the cell. The sequence of events just described can be represented as follows:

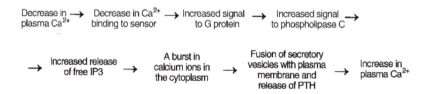

To summarize the overall scenario, as extracellular calcium ion levels decline, the parathyroid gland increases its release of parathyroid hormone, thus increasing plasma $Ca^{2+}$, and restoring it to proper levels. Most cells of the body contain some elements of the preceding signaling scheme, as revealed at a later point in this section.

Calcium sensor protein also occurs in the kidney. Specifically, the calcium sensor occurs in the thick ascending loop, where the N-terminal half juts out into the extracellular fluid (not into the lumen of the tubule, but in the space between tubules). When extracellular $Ca^{2+}$ ions increase in concentration, the thick ascending loop decreases its rate of calcium resorption (Chattopadhyay et al., 1996; Pearce and Thakkar, 1997; Pearce et al., 1996). The regulatory scenario in the kidney represents a mode of regulating the body's calcium balance that appears completely independent of vitamin D. The overall scenario is that, as extracellular $Ca^{2+}$ increases, the parathyroid gland decreases its release of PTH, and the kidney reduces its reabsorption of calcium ions from the developing urine.

## Bone and the Regulation of Plasma Calcium

It is thought that bone can act as a calcium buffer. The readily exchangeable calcium of bone accounts for about 5% of total bone Ca. The stable calcium of bone requires the resorptive action of osteoclasts (Pinto et al., 1988). The osteoclasts secrete acid in their region of contact with the bone, resulting in its dissolution. The freely exchangeable calcium occurs as calcium phosphate rather than hydroxyapatite. It is thought that sudden increases in plasma Ca levels can be dampened by the binding of this Ca to freely exchangeable sites on bone and that sudden decreases in plasma Ca can be dampened by the release of Ca. The hormonal control of bone resorption requires a period of at least 30 minutes to respond to changes in plasma Ca levels (Brindley et al., 1988).

## Causes and Symptoms of Abnormal Concentrations of Plasma Calcium

Hypocalcemia and hypercalcemia occur when the calcium regulatory hormones fail to respond in a normal fashion. Hypocalcemia can also occur with a severe deficiency in dietary calcium, though such a situation is relatively uncommon. Some of the events taking place during dietary Ca deficiency are illustrated by an example involving growing pigs. Baby pigs were raised on Ca-free and Ca-sufficient (10.0 g Ca/kg) diets. The feeding trial was 6 weeks long. The deficient pigs displayed a number of problems, including impaired growth, serum Ca levels that were 50% of normal, and prolonged blood clotting times. The deficient animals moved with difficulty and suffered from occasional tetanic convulsions. A variety of bone problems presented, including rickets and rib fractures. The signs of rickets included wide zones of cartilage and poor mineralization of the epiphyseal plates (Miller et al., 1962). Similar experiments have not been carried out with human subjects. Impaired blood clotting seems not to be a problem for humans with low serum calcium levels.

Hypocalcemia can result from hypoparathyroidism, chronic renal failure, vitamin D deficiency, and hypomagnesemia. Hypomagnesemia occurs mainly in those with alcoholism so severe as to require hospitalization. Magnesium deficiency results in a decline in the responsiveness of osteoclasts to PTH, resulting in interruption of the normal process of bone turnover. In this case, hypocalcemia cannot be effectively corrected unless magnesium therapy is used.

Hypoparathyroidism can result from decreased production of PTH or failure of target organs to respond to PTH. In rare cases, it involves the production of genetically defective PTH. PTH provokes the kidney to conserve calcium and to excrete phosphate. Thus, hypoparathyroidism results in low plasma calcium and high plasma phosphate levels. The disease may result in the calcification of soft tissues because of the high plasma phosphate level. Elevated phosphate levels result in an increased rate of precipitation of calcium and phosphate as the calcium phosphate salt. The disease is treated with oral calcium supplements and phosphate-binding antacids to minimize the absorption of dietary phosphate.

Chronic renal failure can result in the failure to synthesize $1,25(OH)_2D_3$ and the consequent decrease in intestinal absorption of calcium (Pak, 1990). Genetic diseases can also result in the synthesis of defective 1-hydroxylase, as well as the synthesis of defective $1,25(OH)_2D_3$-binding proteins. These proteins are involved in the regulation of specific genes. Each of these effects interrupts the normal action of vitamin D in maintaining plasma calcium levels.

Many sick persons have low serum albumin levels. The hypocalcemia occurring with low serum albumin results in a decrease in the total concentration of plasma calcium, but not in a decrease in that of free calcium ions. This type of hypocalcemia, which can occur with cirrhosis of the liver, does not result in the clinical signs of hypocalcemia, because free plasma Ca levels are maintained.

Hypocalcemia commonly occurs during the first 1 or 2 days of life in premature, low-birth-weight infants. The exact mechanism is not clear. Hypocalcemia can also present in newborns fed cow milk, because cow milk contains calcium and phosphorus in a ratio of about 1.34/1.0, by weight. Mother's milk contains relatively less phosphate; the calcium/phosphorus ratio is 2.25/1.0. The excess phosphate in cow milk promotes hyperphosphatemia in the newborn. It is thought that plasma phosphate, in elevated concentrations, forms a complex with plasma calcium. Formation of this complex reduces the levels of free calcium, resulting in symptoms of hypocalcemia. The newborn is not as able to make hormonal adjustments to maintain plasma calcium levels as is the older infant (Mizrachi *et al.*, 1968).

The symptoms of hypocalcemia present when free plasma Ca levels fall below 1.0 mM. The first symptoms are tingling sensations and paresthesia. (Paresthesia is a numbness of the hands and feet.) The most characteristic sign of hypocalcemia is **tetany**. Tetany includes spasms of various kinds, including those of the face, air passages, hands, and feet. Convulsions may also occur. Tetany represents an emergency situation. It can be relieved by oral calcium or by an intravenous injection of calcium gluconate (10 ml of 10% calcium gluconate). Premature infants may receive about 5.0 ml of the solution. The injections are halted at the first signs of recovery to avoid hypercalcemia. Tetany does not necessarily indicate hypocalcemia. It can be caused by hypomagnesemia as well as by hypokalemia and alkalosis (high blood pH). Acidosis (low blood pH) can prevent hypocalcemia-induced tetany. Tetany that does not respond to calcium may be the result of a low plasma magnesium level and can be corrected by injection of magnesium sulfate.

Hypercalcemia occurs with hyperparathyroidism, a disease involving the excessive production of PTH by the parathyroid gland. Hypercalcemia can result from cancers that produce $1,25-(OH)_2D_3$. Cancer cells produce a variety of other molecules that stimulate osteoclasts; this condition is called oncogenic hypercalcemia.

Hypercalcemia can result from an excessive intake of vitamin D. Prolonged immobilization can also result in hypercalcemia, as bone resorption increases with this immobilization, especially where there is concurrent renal failure (where the kidneys cannot excrete the excess calcium). Sudden, severe hypercalcemia results in vomiting, coma, and possibly death. Prolonged hypercalcemia can result in the formation of kidney stones and in the calcification of soft tissues, such as the eye. Stone formation and calcification are more likely to occur with concurrent hyperphosphatemia.

EXERCISE

Why might the consumption of a huge dose of oxalic acid lead to tetany?

EXERCISE

Why is furosemide plus large amounts of intravenous salt and water used to treat sudden hypercalcemia?

How does hypocalcemia result in tetany? Calcium ions are required at a number of points in the transmission of nerve impulses and in the contraction of skeletal muscles. Apparently, reduction of the $Ca^{2+}$ concentration in the extracellular fluids results in an increase in sensitivity of some nerves to the potential difference across the plasma membrane. When bathed in a medium of lowered $Ca^{2+}$ levels, the nerve "thinks" that it is somewhat depolarized and releases acetylcholine at the point of contact with muscle fibers.

## Calcium Signaling

Calcium ions constitute only one of several participants in the pathway of "calcium signaling." The role of cytosolic $Ca^{2+}$ ions, in this pathway, is to rise suddenly in concentration over a course of a few seconds or minutes, and then to fall back to basal levels. The following commentary concerns events that occur before, during, and after the burst of calcium ions. An extracellular stimulant, such as a hormone, binds to a receptor in the cell membrane. The event of binding provokes the activation of a membrane-bound protein, **phospholipase C.**

Phospholipase C catalyzes the cleavage of a phospholipid in the cell membrane, phosphatidylinositol-4,5-bisphosphate, generating diacylglycerol plus **inositol-**

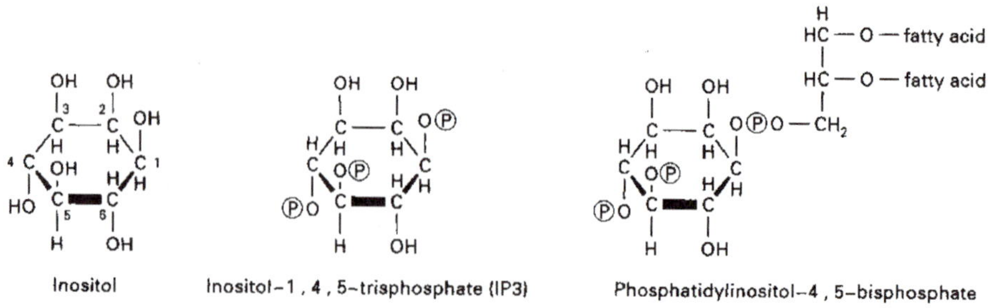

Inositol

Inositol–1 , 4 , 5–trisphosphate (IP3)

Phosphatidylinositol–4 , 5–bisphosphate

FIGURE 10.42 Structures of components of the calcium signaling pathway. Inositol hexaphosphate (phytic acid) appears not to be synthesized by mammalian cells. The phytic acid in the diet can, to some extent, be hydrolyzed to give inositol. Inositol is required in the diet of rodents but not of humans.

**1,4,5-trisphosphate** (IP3) (Figure 10.42). At this point, a branch occurs in the signaling pathway: (1) the protein kinase C branch, and (2) the calcium branch.

### The Protein Kinase C Branch

Diacylglycerol binds to **protein kinase C** and activates it. Activated protein kinase C catalyzes the phosphorylation of a number of target proteins, as discussed in the Diet and Cancer chapter.

### The Calcium Branch

IP3 diffuses to specialized regions of the endoplasmic reticulum and induces it to release a small amount of stored $Ca^{2+}$ (Golovina and Blaustein, 1997). The consequent increase in cytoplasmic $Ca^{2+}$ levels stimulates a number of events in the cytoplasm, including the further activation of protein kinase C. Calcium ions directly bind to and activate calmodulin, protein kinase C, phospholipase A2, proteins of muscle fibers (troponin, caldesmon), and proteins of the cytoskeleton (gelsolin, villin). One might hesitate to call these proteins Ca-metalloenzymes. It is more accurate to say that these enzymes are regulated by calcium.

### Example of Angiotensin

The release of IP3 from lipids of the plasma membrane is illustrated here by a study involving adrenal glomerulosa cells. Angiotensin was added to the cells at 0 seconds (Figure 10.43). Samples of cells were collected at the indicated times and immediately mixed with acid to halt further metabolism; the soluble metabolites were then analyzed. The data illustrate an increase in free intracellular IP3 immediately after addition of the hormone.

The release of $Ca^{2+}$ from the endoplasmic reticulum and into the cytoplasm is illustrated in another study of adrenal glomerulosa cells. The cells were incubated

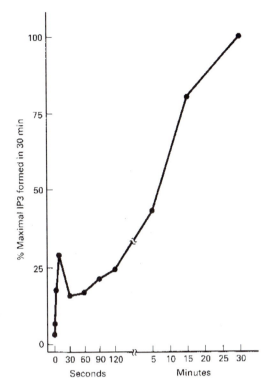

**FIGURE 10.43** Adrenal glomerulosa cells were incubated in a medium of salts and other nutrients. Inositol was added to the medium. The inositol was radioactive to allow the researcher to determine if metabolism resulted in its incorporation into phospholipids or its modification by phosphorylation. Incubation of [$^3$H]inositol with the cells resulted in its incorporation as phosphatidyl[$^3$H]inositol diphosphate ($^3$H is called tritium). (Redrawn with permission from Balla *et al.*, 1988.)

in a medium containing salts and other nutrients. A Ca-sensitive organic compound was added to the medium. Cells were allowed to take up the compound. Then the remaining extracellular compound was washed away. The compound was fura-2. Fura-2 is a synthetic molecule that has binding sites for $Ca^{2+}$. It exhibits fluorescence in the presence of a specific level of $Ca^{2+}$, but not at lower levels. The percentage of fura-2 molecules in the cell that bind calcium is proportional to the level of intracellular $Ca^{2+}$. Thus, the fluorescent response is proportional to intracellular $Ca^{2+}$. More specifically, in the presence of $Ca^{2+}$, fura-2 absorbs light of 340 nm and emits light (the fluorescence) of 480 nm. Fura-2 does not absorb much light of 340 nm in the absence of $Ca^{2+}$ and hence does not fluoresce under these conditions. Angiotensin was added to the cells at the indicated time (Figure 10.44). The basal level of free $Ca^{2+}$ was about 114 n*M*. The level rose to about 145 n*M* immediately after adding the hormone and then fell slowly back to the basal level. The increase in cytosolic $Ca^{2+}$ occurred in response to the IP3 liberated from the plasma membrane. An increase in aldosterone synthesis followed the increase in calcium, as shown by separate experiments. Some of these events are outlined in Figure 10.45.

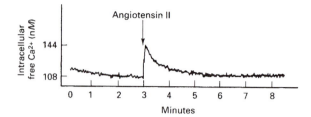

FIGURE 10.44 Adrenal glomerulosa cells were incubated in a medium containing salts and other nutrients. Angiotensin was added to the cells at the indicated time and the intracellular Ca²⁺ level was determined as the fluorescent response of fura-2. (Redrawn with permission from Hausdorff and Catt, 1988.)

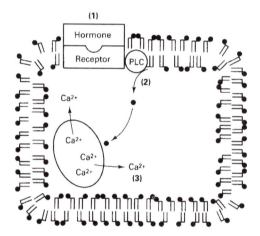

FIGURE 10.45 Calcium signaling. The tiny dots indicate the polar groups of membrane phospholipids. These polar groups include phosphocholine (PC), phosphoethanolamine (PE), and phosphoinositol-4,5-bisphosphate (PI). PC tends to face the extracellular fluid. PE tends to face the cytoplasm. Phospholipids that contain inositol are minor lipids, and account for only 0.8 to 5.0% of the total phospholipids of the plasma membrane; however, these phospholipids are vital for life. Most hormones act by binding to a receptor in the cell membrane (Step 1). One class of receptors, once stimulated, relay a signal to an enzyme that hydrolyzes the polar group from PI. This enzyme is **phospholipase C** (Step 2). The receptor does not directly contact phospholipase C; it relays its signal to the enzyme via **G protein**. The cleavage generates free IP3, but also generates diacylglycerol, which binds to protein kinase C, resulting in the activation of protein kinase C. IP3 induces the release of Ca²⁺ stored in an organelle in the cell (Step 3), resulting in an increase in the levels of cytosolic calcium ions. The cytosolic calcium ion levels then return to their basal level. This decrease is mediated by calcium pumps in the plasma membrane, which pump Ca²⁺ out of the cell, and by calcium pumps in the organelle, which pump Ca²⁺ back into the organelle. The continued stimulation of the cell would be expected eventually to deplete the store of Ca²⁺ in the organelle. Thus, the cell is ultimately dependent on a supply of extracellular calcium. G protein received its name because it binds GDP. This protein contains three different subunits, as will be apparent from the following discussion. The binding of a hormone (or thrombin) to its receptor provokes the release of GDP and its replacement by GTP. The complex of GTP and the α subunit then dissociates from the β/γ complex. The GTP/α subunit complex then activates phospholipase C. Eventually, the α subunit catalyzes the hydrolysis of GTP to GDP, producing an inactive complex of GDP and the α subunit. This is followed by reassociation to form an inactive complex composed of GDP and the three subunits. G proteins are also used in mediating the activation of phospholipase C, adenylyl cyclase, phospholipase A2, and certain ion channels, depending on the cell.

## Calcium Signaling Provokes Secretion

The Ca signal in some cells results in the fusion of vesicles with the plasma membrane; as a result the membrane-bound proteins of the vesicle are incorporated into the plasma membrane. One example is the acid pump in the stomach. The event of fusion also causes the contents of the vesicle to spill out into the extracellular fluid. For example, the acetylcholine released by nerve endings stimulates the adrenal medulla to secrete catecholamines into the bloodstream. Another example is that of cholecystokinin, which induces cells of the pancreas to secrete enzymes into the pancreatic duct. In these cases, the fusion of vesicles with the membrane is thought to be dependent on annexin or a related protein. Calcium binds to annexin. The Ca–annexin complex is thought to insert itself into two adjacent membranes, with the consequence of facilitating the fusion of the membranes with each other. It is thought that the Ca–annexin complex induces a localized disruption of the arrangement of the phospholipids in the two adjacent membranes, facilitating their intermixing and eventual fusion to form one membrane.

## Calmodulin is a Mediator of Calcium Signaling

Ca signaling involves a brief increase in the concentration of cytosolic calcium. In some cases, it is the Ca alone at elevated levels that binds to a target protein, thus changing its function. In other cases, the effect of the calcium is mediated by a complex of Ca and calmodulin. **Calmodulin** is a protein with four calcium-binding sites. $Ca^{2+}$ binds to these sites with association constants ranging from $10^5 \, M^{-1}$ (weaker) to $10^9 \, M^{-1}$ (stronger). Calmodulin is used in the processing of the Ca signal. The binding of Ca to calmodulin results in a change in the overall shape or conformation of the protein, enabling it to bind to specific target proteins, that is, kinases or phosphatases. For example, activated calmodulin binds to **calmodulin protein kinase II**. This kinase, in turn, catalyzes the phosphorylation of many target proteins (Konick and Schulman, 1998; Putney, 1998).

## Sarcoplasmic Reticulum and Muscle Contraction

A rise in cytosolic calcium levels during Ca signaling is provoked by IP3. IP3 seems to be involved in all events of Ca signaling, except perhaps in nerves and cardiac and skeletal muscle cells. The source of Ca in muscle cells is the **sarcoplasmic reticulum** (SR), a specialized type of endoplasmic reticulum. The SR contains ribosomes and many of the enzymes associated with the ER of other cells, but uniquely contains large quantities of the enzyme Ca-ATPase. **Ca-ATPase** accounts for about 75% of the protein of the SR. The SR takes up and accumulates calcium from the cytoplasm via Ca-ATPase, which is a membrane-bound protein. Ca-ATPase, also called *the calcium pump*, requires magnesium ions for activity. It uses the energy of ATP to drive calcium transport into the SR. The Ca release channel is used to release calcium from the SR during muscle contraction. The rise in cytosolic calcium levels during Ca signaling in cardiac and skeletal muscle cells is provoked by the wave of depolarization as it passes down the cell fiber. Depolarization involves the momentary inflow of sodium ions and induces a change in a specific macromolecular complex associated with the plasma membrane. This complex seems to be in contact with the membrane of the SR. A change in this complex causes the Ca channels of the SR to open, releasing $Ca^{2+}$ into the cytoplasm.

Cardiac muscle cells are dependent on both extra- and intracellular stores of calcium. The wave of depolarization traveling along the cell fiber appears to induce the flow of extracellular $Ca^{2+}$ into the cell. The rise in cytosolic calcium that results is not sufficient to cause contraction of the cardiac muscle. Instead, it seems to induce a more substantial release of calcium from the SR of the cardiac muscle cell, which then stimulates contraction.

In muscle, the main target of calcium ions in the Ca signaling pathway is troponin. **Troponin** is a protein of skeletal and cardiac muscle. Troponin is part of the contractile filament. The contractile filaments are huge, rodlike structures inside the muscle cell. They are aligned with the long dimension of the cell and are responsible for the contraction of each muscle cell, and hence the entire muscle. Binding of $Ca^{2+}$ to troponin results in a change in its conformation, which causes contraction. Troponin has a structure similar to that of calmodulin, another Ca-binding protein.

### Skeletal Muscles

The skeletal muscle cell contains two types of filaments, thin and thick. The basic contractile unit of the muscle might be considered to be a thick filament surrounded by thin filaments, as shown in Figure 10.46. The **sarcomere** is the name of this unit. This basic unit is 2.5 µm long. The skeletal muscle cell may be about 500 mm long and 50 µm wide. During contraction, the thin and thick filaments slide past each other at about 15 µm/sec.

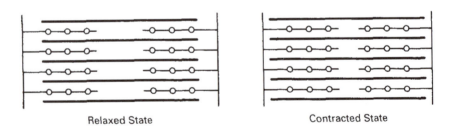

| | |
|---|---|
| Relaxed State | Contracted State |

**FIGURE 10.46** The contractile unit of skeletal muscle. The contractile unit (sarcomere) is shown in the relaxed and contracted states. The thin filaments are anchored to the cytoplasm and contain the Ca-sensitive element. The Ca-sensitive element is a protein called **troponin**, as indicated by the circles. The thick filaments lie between the thin filaments. The figure shows that myosin spans the central part of the sarcomere, while actin extends from the right and left edges of the sarcomere. Each vertical edge is called a Z line. Not shown is a protein called **titin**, which spans the entire sarcomere, and is attached to opposite Z lines. Titin, a huge protein of about 30,000 amino acids, acts as a molecular bungee cord, and gives muscles their springy quality (Keller, 1997).

The **thin filament** consists of a double-stranded helix of **actin** molecules. The filament appears to be relatively thin, as determined by electron microscopy, accounting for the name. Troponin is part of the thin filament. It consists of a polymer of actin subunits. Each subunit has a molecular weight of 42,000. Troponin occurs as a complex with tropomyosin, another protein of the thin filament. There is one molecule of troponin for every seven actin molecules in the thin filament. The **thick filament** is composed of a network of **myosin** molecules. Myosin has a molecular weight of 460,000. It consists of two identical polypeptide chains. Myosin is long rather than globular or spherical. Each myosin heavy chain is associated with two myosin light chains.

Both thin and thick filaments are present in the cytoplasm of the muscle cell. They are aligned with the long dimension of the muscle cell. In the resting state, troponin inhibits the interaction between actin and myosin filaments. When the concentration of $Ca^{2+}$ rises, the conformation of troponin changes, permitting the filaments to interact with each other. The skeletal muscle relaxes when cytosolic $Ca^{2+}$ levels return toward the basal level. This drop results in dissociation of the Ca-troponin complex.

Much of food we consume is used to provide energy to move our skeletal muscles. Hence, it might be of interest to focus on the role of ATP in skeletal muscle contraction. The relevant events occur in the contractile cycle, which consists of these steps:

1. ATP binds to a pocket that is formed by the folding of myosin's polypeptide chain. This pocket can bind ATP or ADP plus $P_i$.

2. When ATP is bound, the myosin and actin dissociate from each other (Rayment, 1996).

3. Myosin catalyzes the hydrolysis of ATP, generating ADP plus inorganic phosphate ($P_i$).

4. In the absence of ATP, myosin rebinds actin, provoking the release of ADP and $P_i$, and leading to muscle contraction.

Contraction of skeletal muscle does not occur at the same time as ATP hydrolysis, but shortly thereafter when ADP is released. The total cycle time of these steps is 50 milliseconds (Schnapp, 1995). Myosin and actin are detached from each other during most of the 50 msec cycle time. The strongly bound, actively pulling state lasts for only 5% of the cycle time. Each thick filament consists of a great number of myosin molecules, and each muscle contains a great number of thick and thin filaments. Thus, the many myosin molecules in a muscle can easily share the task of maintaining the contraction of a muscle, whether the contraction lasts for a fraction of a second or for many minutes.

## Smooth Muscles

Smooth muscles are present in tissues requiring sustained contraction, rather than rapid contraction and relaxation. Smooth muscles regulate the flow of blood through arteries, arterioles, and veins, where they control the size of the lumen of the vessel. They occur in the gastrointestinal tract, where they are responsible for movements of the stomach and the peristaltic waves of the intestines. Smooth muscles are present in sphincters, where they control the flow of fluids through ducts. These muscles are involved in the contraction of the uterus. Smooth muscles are stimulated to contract by catecholamines released by nerves in the vicinity of the muscle, as well as by a number of other hormones. Smooth muscles are also stimulated by the catecholamines in the bloodstream that originate in the secretions of the adrenal medulla. These hormones diffuse over the entire smooth muscle cell. Smooth muscle cells are generally quite small, about 200 to 300 μm long and 5 μm wide. The hormone binds to hormone receptors and activates phospholipase C, resulting in the release of calcium from the sarcoplasmic reticulum and contraction. Smooth muscles are different from skeletal muscles in that they tend to be less dependent on depolarization of the plasma membrane. The contraction of smooth muscles may involve both IP3-dependent release of intracellular calcium and depolarization-dependent entry of extracellular calcium through Ca channels in the plasma membrane.

### Phosphatidylinositol Is Used for Two Different Types of Signaling Pathways

Phosphatidylinositol-4,5-bisphosphate is cleaved by phospholipase C to produce IP3, an essential component in the Ca-signaling pathway detailed earlier. However, this phospholipid is used for a totally separate signaling pathway, one that does not involve calcium, and that involves **PI 3-kinase**. PI 3-kinase was first mentioned in Chapter 4 in the commentary on insulin signaling. PI 3-kinase catalyzes phosphorylation of PI-4,5-bisphosphate to give PI-3,4,5-trisphosphate. The latter phospholipid, while maintaining its residence in the membrane, provokes the activation of **protein kinase B**. Protein kinase B contains a special region that is used for binding to PI-3,4,5-trisphosphate. Various polypeptide growth hormones act by provoking the activation of PI 3-kinase, which results in the synthesis of PI-3,4,5-trisphosphate, and the eventual activation of protein kinase B (Hemmings, 1997). Protein kinase B is important to the survival of the cell, and its targets include proteins used for translation (protein synthesis) and for glycogen metabolism (Stokoe et al., 1997).

## Calcium-Requiring Enzymes and Proteins

Established calcium-requiring enzymes and proteins are listed in Table 10.15. The calcium metalloenzymes include various hydrolytic enzymes. These hydrolases act on polysaccharides, phospholipids, and proteins. Phospholipase C was discussed earlier. Even though phospholipase C requires calcium and is involved in the Ca signaling pathway, it is believed that the changes in cytosolic Ca levels that occur with Ca signaling do not influence the activity of this enzyme.

Possibly the best known Ca-requiring enzymes are those involved in the blood clotting cascade. These enzymes might better be described as Ca-binding proteins, as the calcium ion is not required for catalytic activity, but for regulation and proper physiological function. $Ca^{2+}$ ions are required for the binding of the clotting factors to phospholipid surfaces. The clotting factors bind $Ca^{2+}$ via residues of γ-carboxyglutamate (GLA). Binding of calcium to these sites results in a conforma-

**TABLE 10.15** Calcium Metalloenzymes and Binding Proteins

| | Function |
|---|---|
| Pancreatic α-amylase | Hydrolysis of dietary starches. |
| Pancreatic phospholipase $A_2$ | Hydrolysis of dietary phospholipids. The enzyme cleaves the fatty acid at the 2-position of the glycerol backbone. |
| Trypsinogen | The calcium ion promotes the activation of trypsinogen to trypsin, and stabilizes trypsin under harsh conditions. |
| Calcium-activated protease in muscle | The enzyme catalyzes the hydrolysis of troponin to tropomyosin. The enzyme is thought to be used in the degradation and turnover of muscle fibers. |
| Phospholipase C | The enzyme is loosely associated with the cytoplasmic side of the plasma membrane. It catalyzes the hydrolysis of the inositol phosphate group of phosphatidyl-inositol-4,5-diphosphate. |
| Protein kinase C | The enzyme is loosely associated with the cytoplasmic side of the plasma membrane. It is used in cell signaling. |
| Phosphorylase kinase | Glycolysis |
| GLA-containing blood clotting factors | See Vitamin K in Chapter 9. |
| Calmodulin | Calcium-dependent cell signaling. |
| Troponin | Calcium signaling in muscle fibers. |
| Annexin and related proteins | Secretion in response to the calcium signal. |
| Villin and gelsolin | Proteins of the cytoskeleton. |

tional change of the clotting factor, exposing lipophilic sites that facilitate the factor's avid binding to the phospholipid surface. In addition, the $Ca^{2+}$ ion may form a bridge between the negatively charged GLA residues and the negatively charged phospholipids, supporting the binding of the enzyme to the membrane. The clotting factors need to bind to a phospholipid surface to interact with each other in a normal fashion. This means, for example, that calcium ions and a phospholipid surface are required for clotting factor $X_a$ to catalyze the conversion of factor II to factor $II_a$. If the concentration of $Ca^{2+}$ in the bloodstream was severely reduced, blood clotting could not occur.

The study illustrated in Figures 10.47 and 10.48 reveals the nature of the calcium requirement by enzymes of the clotting cascade. Factor $X_a$, factor II, phospholipid, and the indicated cation were mixed together in a test tube. A small, synthetic substrate was also added. The substrate is hydrolyzed by factor $II_a$, but not by factor II, factor X, or factor $X_a$. The sequence of hydrolytic cleavages that occurs with the mixing of all of the required components is shown in Figure 10.47. The results show that calcium ions support maximal activity.

EXERCISE

Please explain the effect occurring with the addition of both $Mg^{2+}$ and $Ba^{2+}$ (Figure 10.48).

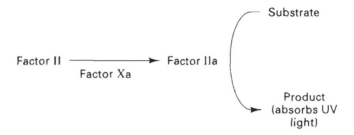

FIGURE 10.47 Diagram of the overall sequence of reactions measured in the experiment depicted in Figure 10.48.

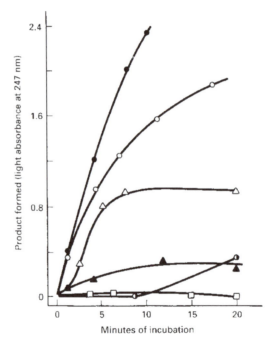

FIGURE 10.48 Results of a study illustrating the nature of the calcium requirement by enzymes of the clotting cascade. Factor $X_a$, factor II, phospholipid, and the indicated cation were mixed together in a test tube. A small, synthetic substrate was also added. Cations: (half-black circle) no addition, (●) calcium, (○) strontium, (▲) barium, (□) magnesium, (△) magnesium + barium. (Redrawn with permission from Nelsestuen *et al.*, 1976.)

## MAGNESIUM

The RDA for magnesium is 4.5 mg/kg body weight. Thus, the RDA for a 70-kg man is 315 mg. The typical intake of Mg in the United States is about 300 mg/day. Plants and meats contain about 0.2 to 0.3 mg Mg/kg of food. Milk is a relatively poor source of Mg. Human breast milk contains about 1.4 m$M$ magnesium. Thus, a typical rate of milk production (750 ml/day) results in the loss of about 1.0 mmol of Mg/day in the milk. The absorption of Mg from food ranges from 20 to 70%. Absorption of dietary Mg is high when low levels of Mg (50 mg) are present in the daily diet. Absorption is low when the diet contains high levels of Mg (1000 mg). Mg is absorbed through the gut by specific transport systems. There is some evidence that the transport systems include the vitamin D-sensitive transport system that functions to absorb calcium. Fecal Mg represents unabsorbed Mg. The Mg released in bile and in pancreatic and intestinal fluids is almost completely reabsorbed, resulting in efficient enterohepatic circulation of the mineral. Gastric acid, pancreatic fluid, and bile contain about 0.5, 0.05, and 0.7 m$M$ Mg, respectively. Most of the magnesium absorbed by the gut is eventually excreted via the urine rather than in the feces. Studies with human subjects revealed that, with a normal diet supplying 9 mmol Mg/day, about 50% of the Mg was absorbed. In other words, with this level of dietary Mg, about half the Mg was excreted in the feces (Dengel *et al.*, 1994).

Magnesium is distinguished by the fact that it is required by most ATP-using enzymes. Here, Mg$^{2+}$ occurs as a complex with ATP, as shown in Figure 10.49. In other words, the true substrate for most ATP–requiring enzymes is not ATP, but the Mg-ATP complex. A deficiency in magnesium is uncommon. When it does occur, the physiological function that is most sensitive is neuromuscular activity. In molecular terms, the enzymes involved in neuromuscular activity that appear to be sensitive to Mg deficiency are those involved in the transport of sodium, potassium, and calcium. These enzymes are Na,K-ATPase and the calcium pump (Ca-ATPase).

## Biochemistry of Magnesium

Most or all ATP-requiring enzymes use ATP in the form of the Mg–ATP complex. ATP chelates the magnesium ion, which is a divalent metal ion. The glycolytic pathway features a number of Mg-requiring enzymes. One of these enzymes is phosphofructokinase. The Mg requirement for this enzyme is illustrated by the data in Figure 10.50. The study involved phosphofructokinase purified from rabbit muscle. Each point in Figure 10.50 represents the catalytic activity of the enzyme that was expressed during incubation in separate test tubes. All of the test tubes

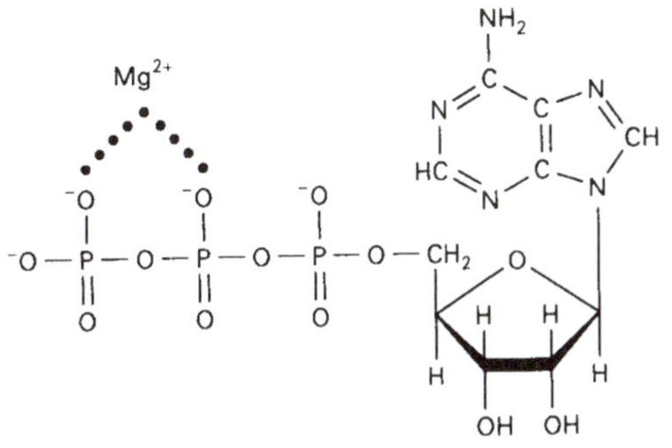

**FIGURE 10.49** Structure of the Mg–ATP complex.

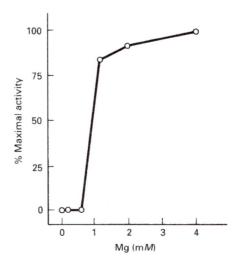

**FIGURE 10.50** Magnesium requirement of phosphofructokinase. (Redrawn with permission from Paetkau and Lardy, 1967.)

contained identical amounts of ATP (1 m$M$), substrate (fructose-6-phosphate), and enzyme. Different test tubes contained different concentrations of Mg, as indicated in the figure. All of the test tubes were incubated for 10 minutes to permit catalytic activity. After the 10-minute period, the amount of product (fructose-1,6-bisphosphate) formed in this interval was measured. The data demonstrate that enzyme activity did not occur until the concentration of Mg in the incubation mixture was equal to or higher than that of ATP. The results indicate the vital relationship between Mg ions and ATP.

EXERCISE

In the study of phosphofructokinase (Figure 10.50), is there any evidence that free ATP inhibits the activity of the enzyme? Is there any evidence that free magnesium inhibits the enzyme?

Pyruvate kinase is an Mg-requiring enzyme of glycolysis. The enzyme requires both potassium and magnesium ions for activity. These ions are weakly bound to the enzyme. Apparently, K aids in the binding of the carboxyl group of the substrate to the enzyme, and Mg aids in the binding of the phosphate group during its transfer from PEP to ADP. The following study involved pyruvate kinase purified from rabbit muscle. Enzyme activity was measured in the presence of a variety of divalent metal ions (Figure 10.51). The greatest activity occurred in the presence of 1.0 to 10.0 mM Mg ions. This range of concentrations is similar to that found in the cell and is therefore in the physiological range. Significant activity was also supported by manganese and cobalt; however, the concentrations of these ions required to support activity are greater than the concentrations that occur in the cell.

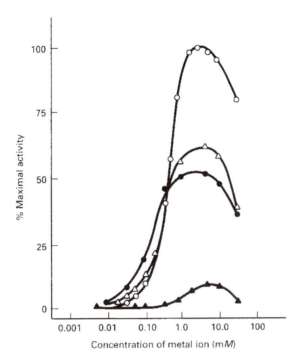

**FIGURE 10.51** Pyruvate kinase activity in the presence of a variety of divalent metal ions: (O) Mg, (Δ) Co, (●) Mn, (▲) Ni. (Redrawn with permission from Kwan *et al.*, 1975.)

The preceding experiments illustrate the function of Mg in enzymes where the metal ion occurs as the Mg–ATP complex and where it occurs in a complex with the enzyme. Some enzymes require Mg–ATP as well as free magnesium ions. One such enzyme is carbamyl phosphate synthetase, an enzyme of the urea cycle. These experiments also raise another issue, namely the magnesium requirement of enzymes that are also stimulated by other metal ions. Many Mg-requiring enzymes can use manganese in place of Mg. Manganese can form a complex with ATP and it can bind directly to enzymes. Because of the fact that magnesium occurs in the cell at concentrations much greater than those of manganese, magnesium would be expected to be the activator *in vivo*. In some cases, however, manganese may activate the enzyme at concentrations much lower than the required concentration of magnesium; the binding of manganese to the enzyme may occur with a strength more than 10,000 times that of magnesium. Here, it might be suspected that Mn is the activator *in vivo*, rather than Mg. Enzymes that tightly bind manganese are discussed under Manganese.

The documentation, by example, of all of the major Mg-requiring enzymes of the body would lead to a very long discussion. Magnesium is required at various steps in the synthesis of DNA, RNA, and protein. The metal ion is required by various enzymes participating in the synthesis of carbohydrates and lipids. $Mg^{2+}$ ions play a structural role, where they are bound to phospholipid membranes, ribosomes, and chromatin. $Mg^{2+}$ ions are required by enzymes that are used in transmitting signals within cells. These enzymes include adenylate cyclase, which catalyzes the synthesis of cyclic AMP, and protein kinase C, which catalyzes the phosphorylation of a number of proteins. $Mg^{2+}$ is required by other protein kinases as well. $Mg^{2+}$ is used by a number of proteins of the muscle, including creatine kinase, actin, Na,K-ATPase, and calcium pumps. Magnesium is vital for energy production, as it is required by the $F_1$ protein, the ATP-synthesizing protein in the mitochondrion.

The question for the nutritionist and clinician is: Which Mg-dependent function is most sensitive to depletion of the body's magnesium and to hypomagnesemia? The answer is probably ion transport systems, such as the calcium pump and Na,K-ATPase. The impaired activity of these ion pumps is likely to be responsible for the neuromuscular problems that present with an Mg deficiency. The defects would involve a difficulty in maintaining the normal movements of calcium, sodium, and potassium ions required for nerve conduction and muscle contraction.

## Magnesium in the Body

Normal plasma Mg ranges from about 1.6 to 2.1 m$M$. Tetany may occur when plasma Mg levels drop below 1.0 m$M$. About half of the plasma Mg occurs as the free ion, with about one-third bound to albumin and the remainder occurring in complexes with phosphate, citrate, and other anions. The nonprotein-bound Mg (70% of total Mg) in the plasma can enter the glomerular filtrate. Nearly all of the Mg entering the filtrate is reabsorbed by the renal tubule and is prevented from being lost in the urine.

The free cytosolic Mg concentration has been estimated as 0.4 to 1.0 m$M$. Total cellular Mg, which includes ions bound to metabolites, membranes, and proteins, is equivalent to about 10 m$M$. Mitochondria contain 25 to 40 m$M$ Mg, whereas free Mg in this organelle is about 0.5 m$M$ (Fahien et al., 1985).

About 60 to 65% of the body's Mg is in bone, 27% in muscle, 6% in other cells, and about 1% in extracellular fluids. About 20 to 30% of bone magnesium is freely exchangeable and is in a rapid equilibrium with serum. The remaining bone magnesium is intimately associated with the apatite crystal of the bone and is called *nonexchangeable magnesium*. Nonexchangeable bone Mg remains constant in bone in growing and adult animals fed an Mg-deficient diet.

What do *loosely bound* and *freely exchangeable* magnesium mean? The reversible interaction between two metabolites, such as a metal ion (M) and a protein (P), can be expressed as

$$M + P \rightleftharpoons MP$$

P can also represent a small molecule, such as ATP, phosphate, or citrate. MP represents a complex. The stability of the complex can be expressed by the formula

$$K = \frac{[MP]}{[M]\,[P]}$$

The brackets indicate "molar concentration of." $K$ is the binding constant, also called the association constant. The value of the binding constant reflects the chemical properties of M and P, as well as other factors such as temperature and solvent. For example, the binding constants for complexes of Mg and ATP, Mg and citrate, and Mg and pyruvate kinase are different.

The last formula can be rearranged to give:

$$[MP] = K\,[M]\,[P]$$

This formula indicates that the concentration of the MP complex is dependent on those of M and P. The formula can be used to estimate the concentration of MP, with a knowledge of the total Mg and total ATP. The cytosolic level of total Mg is about 10 m$M$ and that of ATP is about 2 m$M$. The value of $K$ is taken from Table 10.16. The concentration [M] of free magnesium is 10 – [MP]. The concentration [P] of free ATP is 2.0 – [MP]. The value for the concentration [MP] of the Mg–ATP complex can easily be found from the following formula using a computer:

$$[MP] = (73{,}000)\,(10 - [MP])\,(2.0 - [MP])$$

This formula can provide only a rough estimate of the amount of Mg–ATP in the cell because it ignores the binding constants and concentrations of metabolites other than ATP that bind Mg and thus compete with ATP for binding Mg. The calculation excludes the concentration of Mg tightly bound to proteins. This concentration should not be included in the calculation because the binding of Mg to ATP is only of moderate strength. The contribution of protein-bound Mg to formation of the Mg–ATP complex would be insignificant. In short, ATP cannot effectively compete for the binding of protein-associated Mg. This pool of Mg may reside deep within the structure of the protein, rather than on the surface.

The stability constants for ATP and a variety of cations can be used in the following way. One might be concerned that the high levels of potassium ions in the cell could form a K–ATP complex and thus prevent formation of the Mg–ATP complex required by a variety of enzymes. This concern may arise from the fact that the level of K in the cell is about tenfold greater than that of Mg. From Table

**TABLE 10.16**  Association Constants for Complexes Involving ATP or Related Chemicals

| Binding | Constants |
|---|---|
| Mg-ATP | 73,000 $M^{-1}$ |
| Ca-ATP | 35,000 |
| Mn-ATP | 100,000 |
| Na-ATP | 14 |
| K-ATP | 15 |
| Mg-ADP | 4,000 |
| Mg-AMP | 70 |
| Mg phosphate | 500 |
| Mg citrate | 4,500 |

*Source*: O'Sullivan and Smithers (1979).

10.16, the association constant for K–ATP formation, 15 $M^{-1}$, is much smaller than that for Mg–ATP formation, 73,000 $M^{-1}$. Using the formulas given earlier, one finds that the concentration of K–ATP in the cell is small compared with that of Mg–ATP, despite the higher concentration of cellular potassium.

Magnesium ions bind to inorganic phosphate ($P_i$) and to citrate. The levels of these anions, respectively, are about 3.3 and 1.2 m$M$ in the cytosol and 17.0 and 5.2 m$M$ in the mitochondria. The association constants for Mg-$P_i$ and Mg-citrate formation are small, compared with that for Mg–ATP formation (see Table 10.16). $P_i$ and citrate might be expected to only slightly impair the formation of the Mg–ATP complex in the cell.

The binding constant for Mg–ATP formation may seem large; however, the binding is only of moderate strength. The Mg bound to ATP is described as "freely exchangeable." Metals that are tightly bound to proteins and not readily exchangeable have binding constants ranging from $10^9$ to $10^{30}$ $M^{-1}$ or greater. The association constant for Mg–ATP might be compared with that for hormone-binding proteins. The 1,25-dihydroxyvitamin $D_3$ binding protein of the nucleus, for example, binds the hormone with a binding constant of about $10^{10}$ $M^{-1}$. The binding of free fatty acids by serum albumin was discussed earlier in the text. The binding constants for long-chain fatty acids to albumin range from $10^5$ to $10^8$ $M^{-1}$.

## Magnesium Deficiency

Magnesium deficiency is rare for two reasons. First, the mineral is plentiful in plant and animal foods. Second, the kidney is able to adjust its reabsorption of filtered magnesium to attain a high efficiency of tubular reabsorption. Persons at risk for Mg deficiency include those suffering from prolonged diarrhea and those taking loop diuretics. Both of these conditions are the same as those that can cause potassium and sodium deficiencies. The effects of Mg deficiency include low plasma magnesium, tetany, and cardiac arrhythmias. Although Mg deficiency and hypomagnesemia are not common, the clinician should evaluate plasma Mg levels when tetany presents during such conditions as severe diarrhea. Mg deficiency can also occur with malabsorption syndromes, such as sprue, steatorrhea, and that occurring with surgical removal of the small intestines. The Mg that is not absorbed in steatorrhea is lost in a complex with fatty acids. Mg occurs in a complex with the carboxyl group of the fatty acids. Mg deficiency may also occur during the malabsorption that accompanies chronic alcoholism, especially where there is also malnutrition.

It is doubtful that a deficiency purely in magnesium and not in any other nutrient ever occurs outside the laboratory. Magnesium deficiency has been induced and studied in humans and animals. The results of these studies have proven useful in the diagnosis of Mg deficiency, as it occurs in the real world, and in the study of interactions between Mg and other nutrients.

The following study involved an extended Mg deficiency induced in humans (Shils, 1969). Adult human subjects were fed an Mg-deficient diet for 40 to 100 days. Within a day or so of starting this diet, urinary Mg levels decreased from a normal value of 8 to 16 mmol/day to less than 0.5 mmol/day. Plasma Mg levels decreased gradually over the course of a month and then plateaued at the lower level. Normal plasma Mg values were 1.2 to 2.0 m$M$. The final low values were 10 to 30% of the initial values. The wide fluctuations in plasma Mg that can be tolerated would prove to be fatal in the case of potassium, calcium, or glucose. Red blood cell Mg levels were only moderately affected during the deficiency. They declined to about 60% of the initial value.

After the plasma Mg levels had decreased, plasma calcium levels began to drop. The drop in plasma Ca was moderate compared with that in Mg. Plasma Ca levels decreased even though the subjects were receiving adequate amounts of calcium and vitamin D. Neuromuscular symptoms developed after about 4 weeks of the diet. These symptoms included tetany, lack of reflexes, tremors, and muscle weakness. Tetany may be mild, as in mouth twitches, or more severe, as in painful spasms. Hypocalcemia produces a tetany similar to that produced by Mg deficiency; however, Mg deficiency-induced tetany can occur with normal serum calcium levels. The tetany of Mg deficiency can be reversed by administration of Mg but not calcium. Tetany can result from low concentrations of Ca or Mg in the extracellular fluids (which surround the nerves) or from an alkaline pH. Low potassium levels in the extracellular fluids, however, do not result in tetany.

Experimentally induced Mg deficiency in animals (rats, pigs, dogs) generally results in a progressive drop in plasma Mg from a normal level of about 0.8 to 0.3 m$M$ or lower. A diet that is completely lacking in Mg can produce death involving a seizure within 2 weeks. Diets that contain Mg, but at inadequate levels, can result in the formation of a variety of defects that may take a month or so to manifest. In a number of animals, Mg deficiency results in impaired growth, muscle weakness, tetany, and structural abnormalities in the mitochondria and sarcoplasmic reticulum. The deficiency can provoke a slight decrease in muscle potassium levels and increases in muscle sodium. It is thought that these effects result from impairment of the activity of Na,K-ATPase in the plasma membranes of muscle cells and the consequent deterioration in the normal gradients of Na and K.

Mg deficiency induces a drop in plasma calcium levels in a variety of animals (mice, dogs, monkeys) but not in the rat. Apparently, the hypocalcemia that occurs in humans and in some animals results from impairment of the process of bone reabsorption. Mg deficiency seems to result in a decrease in the sensitivity of the osteoclasts to parathyroid hormone. In addition, Mg deficiency may result in a decrease in the rate of secretion of this hormone, leading to low plasma parathyroid hormone levels.

Another calcium-related effect occurs with Mg deficiency. The diet induces the deposit of calcium in soft tissues, including the kidneys, heart, and aorta. The calcification can result in increases in tissue calcium of 30 to 160 times the normal levels.

## Magnesium and High Blood Pressure

Magnesium supplements may reduce hypertension in some persons. The issue of magnesium and the control of blood pressure has been difficult to establish conclusively. One study involving people with moderate hypertension revealed that daily doses of 20 mmol (485 mg) Mg could produce a decline in blood pressure, where at least three continuous months of supplementation were required for the effect (Witteman *et al.*, 1994). The 20-mmol supplement contained about double the usual dietary intake from food. Magnesium sulfate is commonly used as a drug to prevent convulsions that may occur during episodes of hypertension during pregnancy (see Sodium, Potassium, Chloride, and Water section).

## MANGANESE

An RDA for manganese has not been established. A provisional dietary Mn intake for adults of 2 to 5 mg/day is recommended by the National Research Council. Manganese deficiency rarely occurs and has not yet been documented in any specific population. Mn is found in grains, fruits, and vegetables. Meat, fish, and dairy products are relatively poor sources of the mineral.

### Biochemistry of Manganese

Manganese is required by mitochondrial superoxide dismutase. This enzyme catalyzes the same reaction catalyzed by the cytosolic form of the enzyme, namely the conversion of superoxide to hydrogen peroxide. Superoxide is a molecule of oxygen containing one additional electron. It is produced as a byproduct of various reactions in which molecular oxygen ($O_2$) is involved. These reactions may include those of cytochrome $c$ oxidase and various flavoproteins. The HOOH formed by superoxide is decomposed by catalase:

$$2O_2^- + 2H^+ \xrightarrow[\text{O}_2]{\overset{\text{Superoxide dismutase}}{}} \text{HOOH} \xrightarrow{\text{Catalase}} H_2O + \tfrac{1}{2}O_2 \qquad (10.4)$$

There is some thought that the cellular damage occurring in Mn deficiency is caused by the toxic effects of accumulating superoxide, though such a connection has been difficult to prove.

Pyruvate carboxylase is a well-known Mn metalloenzyme. The enzyme is a tetramer and contains one biotin cofactor per subunit and one divalent cation per subunit. The enzyme from calf liver, for example, contains four tightly bound Mn atoms. The enzyme from chicken liver contains two Mn atoms and two Mg atoms. Raising chickens on an Mn-deficient diet results in the production of an Mn-free enzyme, where magnesium ions replace the usually occurring manganese ions. The Mg-containing enzyme is catalytically active, leaving the requirement of the enzyme for Mn in question (Scrutton et al., 1972).

Phosphoenolpyruvate carboxylase may require manganese (Brinkworth et al., 1981; Columbo and Lardy, 1981). Manganese also appears to be necessary for the activity of acetyl-CoA carboxylase, the first enzyme of the fatty acid biosynthetic pathway. More explicitly, a regulatory enzyme that catalyzes the activation of acetyl-CoA carboxylase appears to require Mn (Thampy and Wakil, 1985). Tyrosine sulfotransferase, an enzyme involved in the metabolism of catecholamine hormones, requires manganese (Sakakibara et al., 1997). Arginase, an important enzyme of the urea cycle, has been reported to be an Mn-metalloenzyme (Brock et al., 1994). The biochemistry of manganese function represents an incomplete story.

The range of defects occurring during experimentally induced Mn deficiency suggests that a variety of Mn-dependent functions remain to be discovered.

## Manganese in the Body

The absorption of dietary Mn by humans is low, about 6.0%, and can range from 1 to 16%. About 99% of the body's losses of Mn are fecal, with about 0.7% lost via the skin and 0.1% via the urine. The primary route of excretion of the body's Mn is the bile, which contributes to the fecal manganese. Plasma Mn levels vary from 0.015 to 0.030 $\mu M$. Plasma Mn can vary within this range from day to day in an individual. Red blood cell manganese is about 20 ng/ml of packed cells. Urinary Mn is fairly constant at about 7.0 nmol/g of creatinine. Plasma and urinary Mn levels seem not to be closely correlated with Mn intake.

The total concentration of Mn in the liver cell is about 35 $\mu M$, about 1/1000th that of magnesium. Free Mn occurs at about 0.2 to 1.0 $\mu M$ in the liver cell. Loosely bound and freely exchangeable Mn occur at about 12 $\mu M$, with the rest of the ion occurring tightly bound to proteins. Mitochondria contain a concentration of Mn equivalent to 300 $\mu M$, about 1/100th that of mitochondrial magnesium (Senior *et al.*, 1980).

The body loses Mn at a rate of about 50% every 3 to 10 weeks (Davidsson *et al.*, 1989). This rate, which is called the biological half-life of Mn, was measured as follows. Human subjects consumed a meal containing radioactive manganese-54. Measurements of total-body [54]Mn were then made using a whole-body counter. A delay of 1 week was permitted before the first measurement to allow for excretion of the nonabsorbed [54]Mn. Then daily measurements of the total amount of radioactive Mn in the body were made over the course of several weeks. The biological half-life of magnesium is longer than that of manganese and has been reported to be 60 to 180 days (Corkey *et al.*, 1986).

## Manganese Deficiency

In animals Mn deficiency causes poor growth, impairs the ability to stand from a supine position, and causes swelling and disorganization of the endoplasmic reticulum and defects in the mitochondrial membrane. Mn deficiency in mother rats produces offspring with a variety of skeletal abnormalities, such as leg bones that are disproportionately short in comparison with the rest of the body. The offspring also have tremors and ataxia. In humans, Mn deficiency results in a scaly, red rash on the skin of the upper torso. Plasma levels of calcium alkaline phosphatase increase, suggesting that Mn deficiency provokes the net dissolution of bone (bone resorption) (Freeland-Graves and Turnlund, 1996).

## ZINC AND COPPER

The RDA for zinc is 15 mg. With mixed diets containing phytate and fiber, about 30% of dietary zinc is absorbed. The efficiency of absorption increases with diets containing very little zinc. Dietary zinc must replace the obligatory losses, which, at the minimum, are about 0.7 mg/day. The RDA for copper has not been determined. The National Research Council recommends 1.5 to 3.0 mg per day as a safe and adequate range of dietary copper intake for adults. The usual dietary intake of copper in the United States is about 1 mg/day. This amount is sufficient to maintain copper balance. About one-third of dietary copper is absorbed. Meat, poultry, and seafood are the best sources of zinc. Oysters are distinguished as a food containing very high levels of zinc and copper (Brody and Mathews, 1989). The metals are contained in vesicles or granules in the animal. The vesicles are thought to be used for metal detoxification, though their function is not really clear.

Zinc and copper are described together because of their nutritional and similarities. These metal ions differ biochemically, however. Copper is involved in oxidation–reduction reactions and occurs in the cuprous ($Cu^+$) and cupric ($Cu^{2+}$) states. Zinc does not donate and accept electrons; it occurs as $Zn^{2+}$. Health-threatening deficiencies in zinc and copper are fairly rare. There is some evidence that zinc deficiency occasionally occurs in certain poor populations of the Middle East and results in impaired growth and sexual development. Zinc and copper occur tightly bound to proteins in the cell, rather than as free ions. Protein-bound zinc serves, in some cases, a catalytic function and, in other cases, a structural function. The most famous of the zinc-requiring enzymes is carbonic anhydrase. Free zinc ions play a part in nerve transmission within the brain, though the exact nature of this role is not clear (Koh et al., 1996). A large number of regulatory proteins that bind to DNA contain zinc. The zinc-containing DNA-binding proteins belong to a class of proteins called **transcription factors**, i.e., they are used to regulate the rate of transcription. If one had to state an overall role of zinc in the body, one might say genetic regulation.

The copper metalloenzymes are involved in oxygen-using reactions. These enzymes include cytochrome $c$ oxidase (respiratory chain), lysyl oxidase (collagen synthesis), and dopamine $\beta$-hydroxylase (neurotransmitter synthesis). Lysyl oxidase is a small protein with a molecular weight of 32 kDa. This enzyme contains an unusual modification, namely cross-linking between two different parts of its polypeptide chain. The cross-linked region consists of a structure called **lysine tyrosylquinone** (Klinman, 1996). Two amino acids are involved in this cross-linked structure, and these are Lys 314 and Tyr 349. Lysine tyrosylquinone is used as a cofactor and is necessary for the catalytic activity of the enzyme. Other copper metalloenzymes contain a related cofactor, namely 2,4,5-trihydroxyphenylalanine (topaquinone, TPQ). **Serum amino oxidase** is a copper metalloenzyme that contains TPQ. TPQ consists of a modified residue of phenylalanine. The copper in the active site of the enzyme occurs immediately adjacent to the TPQ cofactor.

If one had to state an overall role of copper in the body, one might say oxygen metabolism. One major factor shared by both zinc and copper is that both metal ions occur bound to **metallothionein**. The function of metallothionein is not firmly established. Copper is bound to another protein, ceruloplasmin, which occurs in the cell and plasma. The function of this protein is not clear either. Zinc absorption, as iron absorption, is impaired by high levels of phytic acid. Copper absorption is not inhibited by phytic acid. The major route of excretion of both metal ions is fecal, rather than urinary.

## Biochemistry of Zinc and Copper

Zinc is used by a great number of enzymes and proteins, whereas copper seems to be limited to only a few functions. The most thoroughly studied zinc metalloenzymes of mammals are carbonic anhydrase, carboxypeptidase A and related peptidases, alkaline phosphatase, alcohol dehydrogenase, and cytosolic superoxide dismutase. The superoxide dismutase of the cytosol also contains copper. It catalyzes the same reaction as the mitochondrial enzyme (see Manganese). Table 10.17 provides a more complete list of the zinc metalloenzymes and metalloproteins. Copper metalloenzymes and metalloproteins are also listed. The functions of some of these proteins are described elsewhere in the text, in the Vitamin C and Alcohol sections.

Zinc usually binds to proteins via residues of cysteine and histidine. Sometimes, zinc is bound to residues of glutamate or aspartate. The zinc ion sometimes plays a catalytic role and sometimes a structural role. In the latter case, it helps maintain the three-dimensional structure or conformation of the protein. For example, carboxypeptidase A contains two atoms of zinc. One is required for catalytic activity and is bound to cysteine and histidine. The other, which plays a structural role, is bound only to cysteine. Cytoplasmic superoxide dismutase is a dimer. It contains one atom of $Cu^{2+}$ and one of $Zn^{2+}$ per subunit. The zinc is bound via three residues of histidine and one residue of aspartate. It is buried deep within the enzyme and serves a structural role. The copper atom is bound via four residues of histidine. It resides close to the surface of the protein and participates in the chemistry of catalysis.

## Zinc Fingers

A special sequence of amino acids that contains residues of cysteine and histidine has been identified in a number of DNA-binding proteins (transcription factors) and in some enzymes involved in DNA metabolism. This sequence assumes a coiled conformation at the surface of the protein and has been imagined to resemble a sausage or a finger. The sequence of amino acids, which is about 30 amino acids long, has been shown to tightly bind an atom of zinc. The structure is called a **zinc finger**. The zinc atom is required for the binding of the protein to DNA. The role of DNA-binding proteins that contain zinc fingers in regulating genetic expression is described in the section on Vitamin A, Vitamin D, and Thyroid Hormone at the Genome.

The zinc finger contains the following general sequence of amino acids (AAs):

$$-Phe-AA-Cys-AA-AA-Cys-(AA)_3-Phe-(AA)_5-Leu-AA-His-(AA)_3-His-$$

This sequence is part of the linear polypeptide chain of the entire protein. The sequence may be repeated one or more times in the DNA-binding protein, and the exact order of the amino acids differs in zinc fingers found in different types of DNA-binding proteins. DNA-binding proteins that contain zinc fingers include the transcription factors that bind the respective hormones: RAR, RXR, VDR, and THR (Zhu et al., 1997). Other transcription factors that contain zinc fingers include TFIIIA, Sp1 (Sp1 binds to the GC box) (Narayan et al., 1997), and steroid hormone receptors.

The sex steroid hormones (androgen and estrogen) function by binding to transcription factors that contain zinc fingers. The complex of the hormone and transcription factor binds to special response elements along the genome, and regulates hormone-responsive genes. It is conceivable that the reproductive de-

**TABLE 10.17** Zinc and Copper Metalloenzymes and Binding Proteins

| | Function |
|---|---|
| **Zinc** | |
| Carbonic anhydrase | Interconverts $CO_2$ and bicarbonate |
| Alcohol dehydrogenase | Alcohol catabolism |
| Alkaline phosphatase | Hydrolyzes phosphate groups |
| Carboxypeptidase A | Digestion of dietary protein |
| Carboxypeptidase B | Digestion of dietary protein |
| Angiotensin-converting enzyme | Regulation of blood pressure and salt balance |
| Cytoplasmic superoxide dismutase | Toxic oxygen removal |
| Aminolevulinic acid dehydratase | Biosynthesis of heme |
| Insulin in secretory vesicles | Used in packaging of insulin |
| Catecholamine containing vesicles | Used for strengthening the protein matrix used in packaging catecholamines |
| Sphingomyelinase | Mammals have several different enzymes that catalyze the hydrolysis of sphingomyelin; one of these enzymes requires zinc |
| Metallothionein | Storage or detoxification of zinc |
| Lactalbumin/galactosyltransferase | Synthesis of lactose: UDP–galactose complex + glucose → lactose + UDP |
| Component 9 of complement | Immune system |
| Protein kinase C | Transmission of signals within the cell |
| Fructose-1,6-diphosphatase | Gluconeogenesis |
| Thymulin | Hormone of the immune system |
| 5′-Nucleotidase | Cleavage of phosphate from nucleoside 5′-monophosphates |
| Glyoxalase | Detoxification of aldehydes |
| mRNA editing enzyme | Changes the mRNA coding for apolipoprotein B100 (by deamidating a cytosine residue), so that it codes for apo B48 |
| *Transcription factor Sp1 | DNA binding and basal transcription |
| *Transcription factor TFIIIA | DNA binding and basal transcription |
| *Glucocorticoid receptor | DNA binding and genetic regulation |
| * Androgen receptor | DNA binding and genetic regulation |
| *Steroid hormone receptor | DNA binding and genetic regulation |
| *Retinoic acid receptor | DNA binding and genetic regulation |
| *Vitamin D receptor | DNA binding and genetic regulation |
| *Poly(ADP-ribose) polymerase | Response to damage to DNA |
| *RAF | Used for cell signaling; RAF binds to RAS, a membrane-bound protein that is used in cell signaling |
| *HRS | Used for endocytosis, a process occurring at the plasma membrane that results in the creation of small vesicles, and in the uptake of chemicals from the extracellular fluid |

*Indicates presence of zinc finger structures.                    continued

TABLE 10.17  continued

| | Function |
|---|---|

**Copper**

| | |
|---|---|
| Cytochrome *c* oxidase | Respiratory chain |
| Lysyl oxidase | Collagen synthesis |
| Dopamine β-hydroxylase | Neurotransmitter synthesis |
| Tyrosine oxidase | Melanin synthesis |
| Cytoplasmic superoxide dismutase | Toxic oxygen removal |
| Amine oxidase | Catabolism of histamine and related endocrinologically active molecules: $R-CH_2NH_2 + O_2 + H_2O \rightarrow R-CHO + NH_3 + HOOH$ |
| Metallothionein | Copper storage or detoxification |
| Ceruloplasmin | Copper storage or transport |
| Uricase (not in humans) | Catabolism of uric acid in nonprimates |

fects that occur with zinc deficiency result from the inactivity of the binding protein when deprived of zinc. In short, the hormone is functionally ineffective unless it can form a complex with the protein and unless the complex can bind to the chromatin.

Cadmium is a divalent metal ion. It is a toxic metal ion that can induce damage to the reproductive system. In general, cadmium can bind to sites of proteins that are normally used to bind zinc. It is possible that the cadmium-induced damage to the reproductive system results from the binding to the zinc fingers of the steroid hormone-binding proteins.

## *Be Careful before Claiming that a Protein Requires Zinc*

The existence of a zinc finger sequence in a protein does not necessarily mean that it binds zinc ions *in vivo*. The researcher must determine that any protein bearing this sequence actually binds zinc before claiming that the sequence is a real zinc finger. One might expect, for example, to find a stoichiometry of one zinc atom per zinc finger sequence. In the actual practice of studying metalloproteins, one may find a stoichiometry of 0.7–0.9 metal ions per metal-binding sequence, because of the losses of metal ions during the purification of the proteins. The quantity of metal atoms bound to any protein can be determined by **atomic absorption spectrometry**.

The detection of one zinc atom per zinc finger sequence (or to a related zinc-binding sequence) does not necessarily mean that the protein actually requires zinc for activity. The researcher needs to determine that the zinc actually functions to support activity, to support a regulatory property, or to stabilize the structure of the protein before claiming that the protein is a zinc metalloprotein. The simplest method used to assess if a protein requires zinc for activity is to measure the activity in the presence and absence of a chelator, such as ethyle-nediaminetetraacetic acid (EDTA). However, one should be cautioned that the results from this chelator test can only be used to guide the design of more convincing tests. The chelator test can lead to misleading results, as shown by the example of zinc and DNA polymerase (Lehman, 1981). The concepts in the foregoing discussion apply to the study of most metalloproteins.

## Zinc in the β-Cells of the Pancreas

Insulin is stored in the β-cells of the pancreas, where it resides in secretory vesicles. The hormone occurs in an orderly, crystalline structure involving zinc ions. Each molecule of insulin is associated with two to four atoms of zinc in a complex that can only be slowly dissolved in water. The C-peptide is also contained in the secretory vesicle, but it is not included in the crystalline structure. The zinc/insulin complex is thought to allow the slow "timed release" of insulin after it is secreted into the bloodstream. Once in the circulation, insulin slowly dissolves out of the complex. The interior of the vesicle is kept slightly acidic (about pH 6.0) by the activity of a proton pump residing in the membrane of the vesicle (Aspinwall *et al.*, 1997). The zinc/insulin complex is insoluble at pH 7.0 or below. It might be noted that insulin is not complexed with zinc in the β-cells of some animals, for example, guinea pigs, indicating that the complex is not vital. Zinc deficiency seems not to have a profound effect on insulin secretion in rats.

Zinc also seems to serves to maintain a timed release of adrenaline from the adrenal gland. Here, zinc acts to cross-link and strengthen the matrix of proteins that exists within the secretory vesicle, thus impeding the dispersion of the hormone from the matrix, in the seconds immediately following secretion (Aspinwall *et al.*, 1997).

## Absorption of Zinc and Copper

Absorption of dietary zinc ranges from about 15 to 60%, depending on the nature of the diet. Red meat, liver, raisins, and molluscs are the foods highest in zinc, with beans and nuts also containing high levels. Milk, fruits, and vegetables are low in zinc (Sandstead and Smith, 1996). The zinc of whole grains is less well absorbed than that of milled flour, probably because of the phytic acid that chelates zinc ions; however, the higher zinc content of whole grains makes the foods comparable as a source of the mineral. The zinc in bread made with yeast has a higher availability than that in unleavened bread, probably because of the hydrolysis of the phosphate groups of phytic acid in the flour by enzymes produced by the yeast. The phytic acid content of the typical diet in the United States is not high enough to impair absorption of dietary zinc; however, the phytic acid in the whole-grain diets consumed by populations in the Middle East can reduce the absorption of zinc from the normal range to well under 20%. Zinc is absorbed by the small intestines, either as free zinc ions or possibly in a complex with amino acids.

The following study illustrates the effects of different foods on the absorption of zinc. Subjects ate a test diet consisting of a roll made with whole wheat flour or white flour, along with other foods, as indicated in Table 10.18. The roll contained 1.0 mg of radioactive zinc ($^{65}$Zn). The total amount of zinc in the test meal was about 3.5 mg. Two weeks later, the amount of radioactive zinc remaining in the

**TABLE 10.18** Absorption of Zinc from Various Meals

| | Absorption (%) | |
| --- | --- | --- |
| Test diet | Mean | Range |
| White flour roll | 13 | 9–24 |
| Whole meal roll | 8 | 6–11 |
| Whole meal roll + milk | 10 | 6–14 |
| Whole meal roll + beef | 11 | 6–16 |
| Whole meal roll + egg | 12 | 6–22 |

*Source*: Sandstrom *et al.* (1980).

body was measured with a whole-body counter. The amounts of $^{65}$Zn remaining in the body are listed in the table. The results might be compared with those from iron absorption tests, described under Iron. Zinc absorption, unlike iron absorption, was not drastically changed by including meat in the test meal (8% absorption, no meat; 11% absorption, plus meat). Including an egg with the test meal appeared to enhance slightly the absorption of the zinc. Absorption from the whole meal bread was slightly impaired compared with that from the white flour roll. The range of absorption of zinc varied from subject to subject, making it difficult to formulate strong and definitive statements concerning the effects of various foods on zinc absorption.

Studies on mineral absorption may raise the question, *What is absorption*? It might be easier to ask, What is *not* absorption? A nutrient that is not absorbed is one that does not enter the enterocyte and is excreted in the feces. Complete excretion of a nonabsorbed substance may take 12 days in humans (Turnlund *et al.*, 1983). A nutrient that enters the enterocyte, or gut mucosa, to be temporarily stored in the enterocyte and then released back into the gut lumen might also be considered to be not absorbed. A nutrient that passes through the gut mucosa into the bloodstream and is excreted into the urine or bile within 1 or 2 days is definitely absorbed; however, studies that measure the amount of nutrient in the body 1 week after consumption of the test food would fail to detect nutrients that are absorbed into the bloodstream and then rapidly excreted. The life cycle of metal ions tends not to follow the latter scenario. Generally, metal ions that are absorbed are incorporated into proteins, released during protein turnover (degradation and resynthesis), and promptly reused within the cell. The metal ion described by this scenario would be detected in the experiment with the radioactive bread rolls.

The absorption of zinc can change with different dietary intakes. Wada *et al.* (1985) found, in studies with human subjects, that the extent of absorption of dietary zinc is relatively high (50%) when low levels of zinc (5.5 mg/day) are consumed. The extent of absorption decreases to 25% when higher levels of zinc (16.5 mg/day) are consumed. In this study, the absorption of zinc was measured after consumption of 2.0 mg of $^{70}$Zn mixed in a fruit drink. Zinc-70 is not radioactive. It is a "stable isotope" of zinc. Samples of biological materials containing stable isotopes can be measured by the technique of thermal ionization mass spectrometry. In the fruit drink experiment, fecal samples were collected for 12 days following the consumption of $^{70}$Zn. The value for the absorption of zinc from the drink was calculated by subtracting the total amount of fecal isotope from the amount consumed.

From 30 to 50% of dietary copper is absorbed by humans. As with zinc, the extent of copper absorption changes with nutrient status. Studies with humans have revealed that a large proportion of dietary copper (56%) is absorbed by the gut with the consumption of diets low in copper (0.8 mg Cu/day). A low proportion (12%) of the dietary copper is absorbed with high-copper diets (8.0 mg Cu/day) (Turnlund *et al.*, 1989). Absorption of copper was determined by feeding diets containing a stable isotope of copper, copper-65, and measuring the fecal losses.

The zinc and copper ions that are absorbed from the diet enter the portal vein, where they are weakly bound to plasma albumin. This situation is reminiscent of that with dietary medium-chain fatty acids. The metal ions enter various tissues of the body, where they are assimilated into various metalloproteins.

## Plasma Zinc and Copper Levels

About two-thirds of the zinc in plasma is loosely bound to albumin. Most of the remaining zinc is tightly bound to other plasma proteins. A small fraction (2–3%) of plasma zinc is weakly bound to amino acids. The amino acids that most avidly bind zinc are histidine and cysteine. Amino acid-associated zinc enters the glomerular filtrate and thus is a source of the zinc ions destined for excretion in the urine. Most of the filtered zinc is reabsorbed and is prevented from immediate excretion. The zinc in red blood cells is bound to carbonic anhydrase. Carbonic anhydrase is present at a level about 0.1% that of hemoglobin in the red blood cell on a per-weight basis. The zinc content of mitochondria is about 1 nmol/mg protein (Link and Jagow, 1995).

Much of the copper in the plasma (60–95%) is bound to ceruloplasmin. The complex of copper and ceruloplasmin is assembled and secreted by the liver. A small fraction of plasma copper, under 7%, is weakly bound to albumin and to free amino acids, especially histidine, threonine, and glutamine. The copper bound to serum albumin is associated with a histidine residue near the amino terminus of the protein. The copper in red blood cells is bound to superoxide dismutase.

## Metallothionein and Ceruloplasmin

The metal-binding proteins metallothionein and ceruloplasmin are recurring themes in the study of zinc and copper. **Metallothionein** is a small protein with a protein binds zinc and copper ions, as well as nonnutritive heavy metals. The protein consists of about 60 amino acids, and has a molecular weight of about 7000. One third of these amino acids (20 of them) are cysteine. The 20 sulfhydryl groups of these cysteine residues can bind a total of 7 bivalent metal ions, i.e., 7 zinc atoms or 7 copper atoms (Kagi and Schaffer, 1988).

Knock-out mice lacking the gene coding for metallothionein were produced using standard genetic techniques. The mice thrived and reproduced. However, when exposed to cadmium, the knock-out mice showed an increased sensitivity to cadmium toxicity, as one might have expected (Lazo *et al.*, 1995). A similar type of experiment involved higher or lower levels of metallothionein, rather than its presence or complete absence. Experiments with rats have revealed a dramatic protective effect of metallothionein against cadmium toxicity. Rat liver cells were maintained in cell culture. The ability of liver cells to function normally and synthesize protein was inhibited when cadmium was added to the cells. Where the levels of metallothionein in the rat had been induced by injecting the animals with zinc one day before removing the liver, cadmium toxicity was prevented (Din and Frazier, 1985). The study with the knock-out mice also revealed that metallothionein is not absolutely required for life.

High levels of zinc stimulate the synthesis of metallothionein in the small intestines. The elevated levels of metallothionein then serve as a depot for the binding of high levels of zinc consumed in subsequent meals. The induced protein has been shown to limit the amount of zinc entering the bloodstream with consumption of a high-zinc diet (Menard *et al.*, 1981). High doses of copper can induce metallothionein synthesis to the same extent as can zinc. At levels near those found in the diet, zinc is a potent inducer while copper is only a weak inducer. Normally, hepatic metallothionein contains mainly zinc, whereas kidney metallothionein contains copper and, when present in the diet, cadmium. The copper entering the liver may be stored in hepatic metallothionein and released into the plasma in **ceruloplasmin** or secreted in the bile later.

Evidence suggests that the regulation of metallothionein levels by metal ions results from the binding of zinc (or other metal ions) to a special transcription factor (molecular weight = 105 kDa), with the subsequent binding of the zinc/transcription factor complex to a promoter that resides near the metallothionein gene. The zinc/transcription factor complex actually binds to the **metal response element** that resides in the promoter. The sequence of the metal response element is:

TGCACNC
ACGTGNG

The N indicates A, T, G, or C (Samson *et al.*, 1995). The responsiveness of metallothionein's mRNA to zinc levels has prompted some researchers to develop mRNA assays as a way to assess zinc status in humans (Sullivan and Cousins, 1997).

Metallothionein synthesis is stimulated by heavy metals, such as cadmium, zinc, copper, and mercury. This protein tightly binds these metals and prevents them from inflicting damage on more vital cysteine-containing proteins of the body. Cadmium toxicity, in particular, is a problem in certain parts of the world. Cadmium toxicity has occurred in factories producing alkaline batteries and in zinc refineries. Long-term exposure to cadmium produces kidney and lung damage and induces the release of proteins into the urine (proteinuria). Cadmium toxicity has been widespread in Japan as a result of the irrigation of rice fields with water contaminated with industrial cadmium. The toxicity manifests as fragility and deformity of bones. These signs have been found in a few hundred persons in Japan. In Japan, the disease is called *itai-itai disease*, which means "ouch-ouch," and refers to the pain occurring in the bones.

## Ceruloplasmin

**Ceruloplasmin** consists of a single polypeptide chain with a molecular weight of 132,000. About half of the copper atoms bound to ceruloplasmin are in the cupric state ($Cu^{2+}$) and half are in the cuprous state ($Cu^+$). Ceruloplasmin is synthesized in the liver, where it receives six copper atoms, and is then secreted into the plasma. The protein is also made in white blood cells (Juan *et al.*, 1997). Ceruloplasmin is larger than albumin, and is about the same size as antibodies. The large size prevents loss of ceruloplasmin (and the bound copper) through the glomerulus of the kidney, and into the developing urine. Most plasma copper does not originate from recently consumed meals, but from the copper consumed in prior weeks or months. This copper is directly incorporated into hepatic ceruloplasmin from the diet or is slowly released during the normal turnover of copper-requiring enzymes and then bound to ceruloplasmin.

Ceruloplasmin seems not absolutely necessary for life, as several people have been found to lack this protein. However, these people suffer from diabetes and retinal degeneration, and iron deposits occur in their brain, liver, and pancreas (Harris, 1995). The exact physiological role of ceruloplasmin remains unclear, but it is related somehow to the transfer of iron in and out of ferritin. Ferritin is a huge multisubunit iron storage protein with an overall molecular weight of 450 kDa. It can hold up to 2500 iron atoms. Studies have shown that iron can spontaneously be incorporated into ferritin, and also that ceruloplasmin can load iron into ferritin.

There is some evidence that ceruloplasmin is involved in iron metabolism. It has been suggested that ceruloplasmin and ferritin in plasma work together to reduce the levels of free ferrous ions in plasma. Here, the ceruloplasmin catalyzes the oxidation of $Fe^{2+}$ (ferrous) to $Fe^{3+}$ (ferric), the form of the metal that binds to ferritin. Ceruloplasmin acts as an oxidant in this process. This proposed function may reduce damage to membrane lipids possibly inflicted by the small amount of $Fe^{2+}$ in the circulation. It is thought that ceruloplasmin may be used in the mobilization of iron from intracellular stores. Here, the protein may facilitate the transfer of iron from ferritin to transferrin (Frieden and Hsieh, 1976). A relationship between copper and iron is suggested by the fact that copper-deficient rats may develop iron deficiency anemia, as revealed by measurements of hemoglobin and hematocrit (Johnson and Dufault, 1989; Cohen *et al.*, 1985). Ceruloplasmin may also function in a unique iron transport mechanism, as mentioned in the Iron section.

## Zinc Excretion and Zinc Deficiency

The normal range of serum zinc in the adult is 11 to 18 μM. Urinary zinc is 300 to 700 μg/day, which is equivalent to 2 to 5% of zinc intake. Most of the absorbed zinc is excreted in the bile and is eventually lost in the feces. Fecal zinc originates from nonabsorbed zinc and from pancreatic secretions and exfoliated enterocytes. The zinc released by the pancreas can be reabsorbed and thus participates in enteropancreatic circulation. Nearly all of the zinc in pancreatic juices is bound to carboxypeptidase A and B. About 1 mg of fecal zinc per day is derived from pancreatic secretions. In any accounting of the nutrients absorbed each day, one needs to realize that the enzymes and salts released in the digestive juices each day (during a meal) actually represent a second meal. The digestive enzymes, for

**TABLE 10.19**  Approximate Normal Values for Zinc Levels and Losses

| | |
|---|---|
| Plasma | 0.7–1.25 µg/ml |
| Red blood cells | 10–14 µg/ml |
| Urine | 0.3–0.7 mg/day |
| Feces | 5–15 mg/day |
| Saliva | 0.5 µg/day |
| Hair | 90–250 µg/g |
| Semen | 80–1500 µg/day (250 µg) |
| Sweat | 700 µg/day |

*Source*: Baer and King (1984).

example, are eventually hydrolyzed to amino acids and absorbed. With the onset of consumption of a zinc-deficient diet, zinc continues to be secreted in the pancreatic fluids. This continued release contributes to the negative zinc balance that occurs during the deficiency, since zinc is not efficiently absorbed from the gut lumen. Zinc levels in various tissues of the normal adult are listed in Table 10.19.

## Zinc Deficiency in Rats

Zinc deficiency in the rat is very well characterized. The symptoms of the deficiency in this animal may be used as a reference point when considering the effects of a severe deficiency in humans. In the rat, deficiency produces the following symptoms. Within 3 to 4 days of consumption of the deficient diet, the rats become **anorexic** (loss of appetite). They ate about half as much food per day as rats fed a complete diet. (In general, feeding animals diets that are totally lacking in one essential nutrient leads to a drop in food consumption. This effect seems to be not so pronounced in humans, in the few cases where it has been studied.) The efficiency of food utilization is poor with the zinc-free diet. The deficient rats gained only 0.2 g of body weight per gram of food consumed, which is about half that of the rats that ate a complete diet. Thus, the deficient rats gain weight at about one-quarter the rate of rats fed a normal diet. Interestingly, force-feeding the zinc-free diet in an attempt to overcome the anorexia results in a decline in health of the animals and can result in death within a few days.

Zinc-deficient rats develop fissures at the corners of the mouth, a poor hair coat, scaly feet, and a kangaroo-like posture. Reproductive defects are common. The female rats fail to have an estrus cycle, and the males show atrophy of the testicles and do not produce sperm.

Bone contains a small amount of zinc. Studies have shown that feeding growing rats a zinc-free diet results in low levels of bone zinc. Normal bone contains 0.4 mg zinc per gram of bone salts (bone ash), whereas deficient bone contains one-quarter of that level (Swenerton and Hurley, 1968). The zinc that is present in bone cannot be readily tapped as a reservoir during a dietary deficiency in the mineral. Only under one condition can this zinc be used: when bone mobilization is stimulated (i.e., by feeding a low-calcium diet) as shown in a clever experiment by Hurley and Tao (1972). Zinc deficiency during pregnancy produced birth defects in the neonatal rats. These defects strike a number of organ systems. An attempt was made to release skeletal zinc by feeding the pregnant rats a calcium-deficient diet. The calcium deficiency induced the mobilization of bone salts, resulting in the liberation of some zinc. Consequently, induction of calcium deficiency in rats on a zinc-free diet resulted in an increase in litter size and prevention of some birth defects.

## Zinc Deficiency in Monkeys

Raising young monkeys on adequate (50 mg Zn/kg diet) or deficient (2 mg Zn/kg diet) diets produced differences in plasma zinc, though these differences were small (12 $\mu M$ with normal diet; 9 $\mu M$ with deficient diet) (Golub *et al.*, 1996). The Zn-deficient diet did not produce anorexia. The monkeys continued to survive during a 2-year feeding trial. At the end of the trial the deficient animals weighed 20% less than the sufficient animals.

Bone density was measured by X-ray analysis, and revealed that the bone of the deficient animals was about 10% less dense than that of the control animals. Unexpectedly, the Zn-deficient diets seemed not to prevent the young monkeys from attaining sexual maturity (puberty) (Golub *et al.*, 1996), in contrast to an effect seen in some humans living in the Middle East. It is quite possible that, if the researchers had used a diet containing less zinc, or had added phytate to the diet, then impairment of sexual development would have resulted.

## Zinc Deficiency in Humans

The onset of dietary zinc deficiency induces a decline in urinary zinc. Studies with humans have revealed that consumption of normal zinc diets (16 mg/day) followed by low-zinc diets (0.3 mg/day) results in a decline in urinary zinc levels. Within 1 week of initiating the low-zinc diet, urinary zinc may fall from an initial value of about 450 µg/day to less than 150 µg/day (Baer and King, 1984). Plasma zinc levels tend to be maintained during a dietary deficiency in zinc. This relationship indicates a role for the kidney in conserving the body's zinc and in reabsorbing it from the glomerular filtrate. Zinc losses in the sweat alone change depending on the diet. Sweat losses may average about 600 µg/day with a high-zinc diet (34 mg/day) and 200 µg/day with a low-zinc diet (4.0 mg/day).

Zinc deficiency in humans is difficult to diagnose because a suitable functional test, or a test for zinc reservoirs, has been lacking. Milne *et al.* (1987) documented that a variety of zinc-requiring enzymes of the red blood cell seem to resist the effects of dietary zinc deficiency, but demonstrated that liver alcohol dehydrogenase may be sensitive to the deficiency. The activity of this enzyme was assessed by a functional test — the ethanol tolerance test. The ethanol tolerance test would be expected to be influenced by factors unrelated to zinc status, such as the state of induction of the ethanol oxidizing system, and hence may not be the functional test of choice for assessing zinc status.

Golub *et al.* (1984) pointed out a problem in using plasma zinc levels to assess zinc status in children. Periods of rapid growth are associated with decreased levels of plasma zinc. Therefore, nongrowing individuals may have relatively high plasma zinc levels even with a low-zinc diet. In short, plasma zinc levels are influenced by intake as well as by the needs of growing tissues. Illness may also influence plasma zinc levels. Plasma zinc levels may temporarily fall to less than half the normal levels in severe illness or inflammation. It is thought that this drop is due to increased synthesis of zinc-requiring enzymes. There has been some interest in using hair zinc content to assess zinc status. This technique has proven to be less useful than had been hoped.

Zinc deficiency is relatively uncommon in the United States, but may occur in adults with alcoholism or intestinal malabsorption problems. Low plasma zinc has been found in patients with alcoholic cirrhosis, Crohn's disease, and coeliac disease. Zinc deficiency, as well as deficiencies in a variety of other trace minerals, has occurred in humans fed for prolonged periods with intravenous diets from which the mineral had been inadvertently omitted. Controlled studies with humans have shown that the signs of zinc deficiency present after 2 to 5 weeks of consumption of a zinc-free diet. The signs include a facial rash, diarrhea, and alopecia. The rash occurs on the face, groins, hands, and feet. The symptoms can be reversed by administering zinc.

There seems to be little reason to suspect that the elderly have a tendency to be deficient, or even marginally deficient, in zinc (Wood et al., 1995). On the other hand, an emerging concern is that increased calcium intake can interfere with zinc absorption or retention (Sandstead and Smith, 1996). A study of postmenopausal women revealed that supplements of zinc plus calcium seemed to prevent bone loss, while supplements of calcium only permitted bone loss (Strause et al., 1994). Hence, there is some interest in the question of whether persons taking calcium, with the possible effect of preventing osteoporosis, should also take zinc supplements.

A variety of symptoms, not mentioned earlier, have also been associated with zinc deficiency. Zinc deficiency is thought to delay wound healing, impair the senses of taste and smell, and impair the functions of the immune system. (The impaired sense of taste is called *hypogeusia*.) Severe zinc deficiency results in impairment of the immune system and increased infections. These problems have been found in the severe deficiencies that occur in starvation (marasmus) and acrodermatitis enteropathica and in studies with animals. The immune system involves a large array of hormones, hormone receptor proteins, signaling proteins, and DNA-binding proteins, some of which require zinc for activity. It is not clear which of these components is most sensitive to zinc deficiency.

## Zinc Deficiency in the Middle East

Zinc deficiency has been found among peasant populations in rural areas of the Middle East. *Unleavened whole wheat bread* can account for 75% of the energy intake in these areas, that is, Iran, Egypt, and Turkey. This diet, which does not contain meat, does contain zinc, but it also contains phytic acid at a level of about 3 g/day. The phytic acid inhibits zinc absorption. The yeast used to *leaven bread* produces phosphatases that hydrolyze the phosphate groups from phytic acid.

The deficiency results from general malnutrition (lack of meat), as well as from the wheat-based diet. The symptoms include lack of sexual maturation and small stature. Supplementation of affected young adults with zinc provokes the growth of pubic hair and enlargement of genitalia to a normal size within a few months (Prasad, 1991). The zinc deficiency that results is thought to be a contributing factor in a type of dwarfism found among these populations. The victims are very short, show marked retardation in sexual development, and have levels of plasma zinc somewhat below normal. It is not clear if any lack of sexual development, due to malnutrition, which continues far beyond the age of 20, can be successfully treated with supplements of zinc or other nutrients. Ethical concerns have prevented collecting any clear data on this issue.

## Acrodermatitis Enteropathica

Zinc deficiency occurs in a rare genetic disease, **acrodermatitis enteropathica**. The disease develops in the early months of life, produces pigmented areas on the elbows, knees, face, and buttocks, and is fatal if left untreated. The disease results from an impairment in zinc absorption. Intestinal disturbances, diarrhea, and growth failure also occur. The disease can be treated successfully by administering oral zinc at two times the RDA (Grider and Young, 1996).

## Copper Excretion and Copper Deficiency

The normal range of serum copper in the adult is 11 to 24 $\mu M$. Urinary copper is normally about 20 $\mu g$/day. This level is equivalent to 0.5 to 3.0% of copper intake. Most of the copper absorbed into the body is excreted by way of the bile and lost via the feces. About 1.7 mg of copper is excreted in bile per day; this amount varies with the amount absorbed from the diet. This copper occurs complexed with protein and bilirubin. Bilirubin is a catabolite of heme. The copper is excreted in the bile and tends not to be absorbed back into the body. There is little or no enterohepatic circulation of copper. The concentration of bile copper drops markedly with a copper deficiency, contributing to the conservation of this mineral by the body.

## Copper Deficiency in Animals

Experimentally induced copper deficiency produces anemia in a variety of animals. It produces an anemia. As one might expect, this type of anemia cannot be cured by iron. The heart may enlarge and the mitochondria of the heart may enlarge dramatically. The heart and arteries may rupture with deficiency, probably because of a drop in the activity of lysyl oxidase, a copper metalloenzyme involved in the synthesis of connective tissue proteins. Skeletal defects, such as osteoporosis, occur in experimental copper deficiency. Infections appear to affect copper metabolism. Infections can induce a threefold increase in plasma ceruloplasmin levels. Copper-deficient animals have a reduced ability to exercise, as measured by running on treadmills (Klevay and Medeiros, 1996).

*Copper Deficiency in Humans*

Dietary copper deficiency has not been identified in any specific human population. It is probably very rare, but might be expected to occur with extreme malnutrition. Evidence suggests that the elderly do not have impaired copper metabolism and do not tend to be copper deficient (Wood *et al.*, 1995). In carefully controlled studies of human subjects, men were fed diets containing various amounts of copper (0.38 mg Cu/day; 0.66 mg Cu/day; 2.5 mg Cu/day). The associated plasma copper levels were similar in all cases, i.e., 12.6–13.8 μM (Turnlund *et al.*, 1997). Salivary copper levels ranged from 0.6 to 1.0 μM, and were considered not to change much in response to the diet. A clearer response was found with assays of copper in white blood cells. The diet supplying 0.38 mg Cu/day resulted in a level of 54 pmol Cu/mg protein (with the white blood cell assay), and 80–90 pmol Cu/mg protein with the higher copper diets (Turnlund *et al.*, 1997).

A response similar to that noted in the foregoing has been found with experimental deficiency of another mineral — zinc. Zinc deficiency in humans also leaves plasma zinc levels in the normal range, but provokes a decline in white blood cell zinc (Prasad, 1991).

## Genetic Diseases of Copper Metabolism

*Wilson's Disease and Menkes Disease*

Severe alterations in copper metabolism occur in two genetic disorders, **Wilson's disease** and **Menkes disease**. Both of these diseases are rare and occur in about one in 100,000 births. Both diseases involve naturally occurring mutations in copper transport proteins, i.e., membrane-bound proteins that mediate the passage of copper ions through cell membranes. The copper transporters that are defective in these two diseases are not the same protein, but they are related. To express this relation in numbers, over half (57%) of the sequence of amino acids, as they occur in the polypeptide chains, are identical. Both proteins are thought to utilize ATP to drive copper ions through membranes.

To highlight one contrast between these two proteins, the "Wilson's disease protein" occurs in several tissues but is most abundant in liver, while the "Menkes disease protein" appears to occur in a variety of tissues but to be absent in the liver. This statement applies to normal humans.

*Wilson's Disease and Copper Transporters*

Wilson's disease presents in teenagers and in young adults, but very rarely in early childhood. The disease results from defects in copper transport. The excretion of copper in the bile is impaired. Copper accumulates in the liver, kidney, and brain, resulting in hepatic and neurological damage. Copper also deposits in the cornea of the eyes, giving rise to characteristic yellow-brown rings. The symptoms of liver damage tend to occur in children over 8 years of age, whereas the neurological problems occur in middle age.

Wilson's disease often involves low plasma levels of ceruloplasmin, increased plasma nonceruloplasmin copper, and increased urinary copper. The disease can involve a tenfold increase in liver copper levels. The normal value for hepatic copper is 20 to 50 µg/g of liver (dry weight). Outward signs of the disease include episodes of jaundice, vomiting, and tiredness. Bone disorders such as osteoporosis can also occur. The neurological damage includes a loss in coordination. Wilson's disease does not result in mental retardation. The rate of incorporation of copper into ceruloplasmin is reduced and biliary excretion decreases to 20 to 40% the normal rate.

Wilson's disease can be diagnosed by increases in urinary copper and elevated levels of nonceruloplasmin copper in the plasma. The disease is fatal if left untreated. It can be successfully controlled by lifelong treatment with chelating agents. The copper chelating agent of choice is D-penicillamine. Treatment also involves avoiding foods that are high in copper, such as liver, nuts, chocolate, and mollusks. After an initial period of treatment with penicillamine, Wilson's disease may be treated with zinc (150 mg oral Zn/day). The zinc inhibits the absorption of dietary copper. The levels of zinc present in normal diets are not sufficient to impair the absorption of copper. The mechanism of the zinc effect is as follows. High levels of dietary zinc induce the synthesis of metallothionein in the cells of the gut, as shown in studies with rats. Metallothionein, at increased levels, tightly binds dietary copper. Most of this copper is then lost in the feces when the cells are shed from the intestinal villi.

The defect in Wilson's disease has been traced to a mutation in a copper transport protein. For clarity and brevity, the term "Wilson's protein" may be used to refer to normal and mutated versions of this copper transport protein. Mutations in the Wilson's protein result in a defect in the normal passage and elimination of copper from the liver cell via the bile. Another defect is the failure to load copper ions into ceruloplasmin. The most common occurring mutation in the gene coding for Wilson's protein results in a change of His 1069 to Gln 1069 (Shah *et al.*, 1997). The numbers refer to the sequence of amino acids in the polypeptide chain.

Studies with yeast cells have revealed that several proteins are used to pick up copper ions, as they are transported from the outside environment through the inside of the cell, and to shuttle them to appropriate copper metalloproteins. These shuttles, or "taxicabs," are proteins having names such as Atx1, Lys7, Cox17, and CCC. Each of these proteins have a human equivalent. For example, the human equivalent of the yeast's CCC protein is the human Wilson's protein (Valentine and Gralla, 1997).

One should take interest in attempts to predict human physiology from studies with yeast. Although one might hesitate to call yeasts "mini-mammals," the field of yeast genetics has provided us with much of our knowledge of how human cells work.

## Menkes Disease and Copper Transporters

Menkes disease is a genetic disease involving mental retardation and death before the age of 3 years. The disease is also called Menkes *steely hair syndrome*. This term came from the abnormal, steely or kinky hair that results. The hair is tangled, grayish, and easily broken.

The disease involves a *decrease* in copper levels in the serum, liver, and brain and *increases* in the gut mucosa cells, kidney, and connective tissue. Some of the consequences of Menkes disease result from a defect in the absorption of copper through the intestines. Instead of passing into the bloodstream, copper accumulates in the cells of the gut mucosa, the absorptive layer of the small intestines. More specifically, copper accumulates in the metallothionein in the gut cells. Growth failure, skeletal defects, and neuronal degeneration result.

Menkes disease results in low serum levels of copper (2–6 $\mu M$). Normal levels in the infant are 11 to 24 $\mu M$. Arterial weakness in Menkes syndrome results from the defective cross-linking of elastin and collagen due to the impaired activity of **lysyl oxidase**. Osteoporosis occurs, probably because of problems in the cross-linking of collagen. The neurological signs of Menkes disease may possibly result from the impaired activity of dopamine β-hydroxylase. The disease cannot easily be treated, though therapy has involved large oral doses of copper (0.6 g of copper in a complex with histidine).

The protein defective in Menkes disease is a membrane-bound protein consisting of 1500 amino acids. Evidence suggests that this protein occurs in the membrane of the endoplasmic reticulum, not in the plasma membrane. The mutations in the gene coding for Menkes protein that are responsible for the disease take a number of forms. The gene, as studied in hundreds of human subjects, may contain insertions (an extra nucleotide), deletions (one less nucleotide), conversions of an amino acid's codon to a stop codon (resulting in a truncated protein), and other types of mutations.

## MOLYBDENUM, SULFITE, AND SULFATE

### Molybdenum

Molybdenum is required in the diet. It is required by three enzymes in mammals: sulfite oxidase, xanthine dehydrogenase, and aldehyde oxidase. Molybdenum occurs in these enzymes as part of the molybdenum cofactor (Figure 10.52). This cofactor is biosynthesized in the body with GTP as the starting material. All known Mo metalloenzymes, with the exception of nitrogenase (a plant enzyme), use Mo in the form of the molybdenum cofactor.

An RDA for Mo has not been set, though the intake range of 75 to 250 $\mu$g/day has been recommended by the Food and Nutrition Board. This range is based on the typical intake of the mineral (0.1–0.2 mg/day) (Freeland-Graves and Turnlund, 1996). Apparently, Mo deficiency rarely occurs. Molybdenum deficiency cannot easily be induced by feeding animals an Mo-deficient diet. This is because it is difficult to remove or eliminate the molybdenum present as a trace contaminant

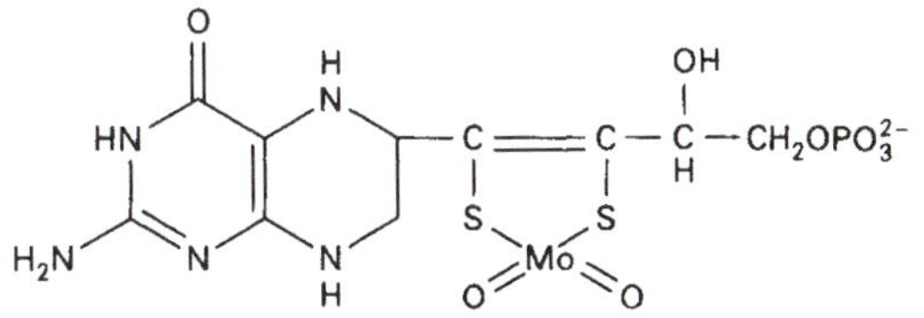

**FIGURE 10.52** Molybdenum cofactor.

in chemically defined diets. The effects of Mo deficiency have been studied by using an Mo antagonist in conjunction with the low-Mo diet. This antagonist is tungsten, which is similar in size and electronic structure to molybdenum. There has been one report of an Mo-deficient human. The mineral had been inadvertently omitted from a solution used for total parenteral feeding (Abumrad *et al.*, 1981). Conclusive evidence for a dietary requirement for Mo was provided by certain genetic diseases involving a failure to synthesize sulfite oxidase or a failure to synthesize the molybdenum cofactor. A failure to synthesize this cofactor leads to a lack of activity of the three Mo metalloenzymes. The genetic diseases lead to neurological damage, mental retardation, dislocation of the lens of the eye, and death. It is thought that these problems result from the lack of sulfite oxidase, rather than the lack of xanthine dehydrogenase or aldehyde oxidase.

## Absorption of Molybdenum

Molybdenum is very efficiently absorbed, in contrast, for example, to copper and iron. With doses of 0.025–1.4 mg Mb, only 1% of the mineral is recovered in the feces. With a single dose of a low level of the mineral (<0.025 mg), about 20% of the mineral is recovered in the urine during a 6-day collection period. With a large test dose (1.4 mg), about 90% of the dose is recovered in a six day period (Turnlund *et al.*, 1995).

## Molybdenum Biochemistry

The molybdenum of the Mo metalloenzyme is directly involved in the chemistry of catalysis. It changes its oxidation state and accepts and donates electrons during the catalytic cycle. During the event of catalysis, molybdenum is thought to cycle between the different ionic forms $Mo^{6+}$, $Mo^{4+}$, and $Mo^{5+}$. Hence, molybdenum may be compared with iron and copper, which also change their oxidation state during catalysis. Mo metalloenzymes are also iron metalloenzymes. The iron in sulfite oxidase occurs as heme. The iron in xanthine dehydrogenase and aldehyde oxidase occurs bound to sulfur. The latter two enzymes also contain FAD.

Sulfite oxidase catalyzes one of the final steps in the oxidation of the sulfur amino acids. The catabolism of methionine can result in the appearance of its sulfur atom in cysteine, as shown in Chapter 8. Cysteine can be oxidized to cysteine sulfonate, as shown in the section on taurine in Chapter 2, and then degraded to pyruvate. Daily, an average of 25 mmol of sulfite is produced in the body. This amount is large compared with the daily intake of food sulfite, which is about 2.5 mmol. The point at which sulfite oxidase occurs in the cysteine catabolic pathway is shown in Figure 10.53. Sulfate ($SO_4^{2-}$) is required for the synthesis of sulfated polypeptides and polysaccharides. It is thought that sulfate is not required in the diet.

EXERCISE

Please calculate the fraction of dietary sulfur amino acids that is degraded to generate sulfite.

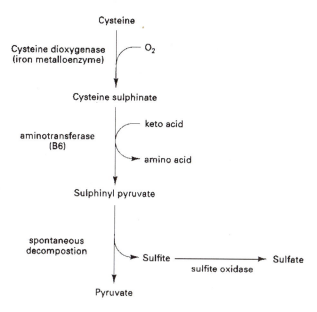

**FIGURE 10.53** Cysteine catabolic pathway and sulfate formation.

Sulfite oxidase is a vital enzyme. Its absence leads to the accumulation of sulfite in the body. Sulfite is a toxic substance. In addition, sulfite oxidase is required for the generation of sulfate in the body. If it were not for this enzyme, we would probably require sulfate in the diet.

## Sulfite

The sulfite in the diet occurs mainly as a food additive. The use of dietary sulfite has been a matter of concern to asthmatics, because a small percentage of persons with asthma respond to sulfite with an asthmatic attack. Apparently these asthmatics have low levels of sulfite oxidase, leading to impairment of sulfite catabolism and removal. The sulfite threshold associated with provoking an asthmatic attack is 3.0 mg. The concentration of sulfite is normally very low in the plasma, under 3.0 $\mu M$. The level may be about 15 $\mu M$ in persons deficient in sulfite oxidase. The level in these persons can increase to over 100 $\mu M$ with consumption of a diet high in cysteine.

Sulfite is in an equilibrium with various forms:

$$SO_2 \text{ in air} \updownarrow$$

$$SO_2 + H_2O \underset{\text{Dehydration}}{\overset{\text{Hydration}}{\rightleftharpoons}} H_2SO_3 \underset{H^+}{\overset{H^+}{\rightleftharpoons}} HSO_3^- \underset{H^+}{\overset{H^+}{\rightleftharpoons}} SO_3^{2-} \qquad (10.5)$$

| Sulfur dioxide | Sulfurous acid | Sulfite | Bisulfite |

In an acidic environment, it is protonated, and occurs mainly as sulfurous acid. In an alkaline environment, the protons dissociate, and it occurs mainly as bisulfite. Sulfurous acid is in an equilibrium with sulfur dioxide, which can leave a solution of water to enter atmosphere. The toxic effects of sulfite arise from its reactions with sulfhydryl groups, aldehyde groups, and ketones. Sulfite can also react with enzyme-bound NAD and FAD. It is well known that the sulfite added to foods can react with the thiamin in the food, destroying this vitamin. The reaction of sulfite with sulfhydryl groups (R—SH) results in its conversion to an $S$-sulfonate group (R—S—SO$_3^-$).

Sulfite is used by the food industry for the following applications. Sulfite (about 5 m$M$) is added to wine, where it acts as an antimicrobial agent. Sulfite is used to preserve the color of various products, such as canned beans, instant mashed potatoes, orange juice, and dehydrated apricots and peaches. The sulfite acts by bleaching away degradation products that would otherwise discolor the food. Sulfite is used for bleaching cherries before adding artificial colorings, in the production of maraschino cherries. These cherries contain small amounts of residual sulfite (50 mg SO$_2$/kg cherry). The source of the sulfite additive is usually sodium metabisulfite (Na$_2$S$_2$O$_5$). This compound is hydrolyzed to yield bisulfite when it is dissolved in water. Dried fruits receive sulfite by another technique. They are sulfited by exposure to fumes of burning sulfur. The fumes contain SO$_2$, which dissolves into the fruit. Dehydrated fruits contain up to 2.0 g of sulfite per kilogram of food.

## Sulfate

Sulfate is a component of major structural molecules of the body, such as chondroitin sulfate, heparan sulfate, and the sulfated gangliosides, as well as cholesterol sulfate. Sulfate is also a component of mucus. Mucus is secreted by epithelial cells as a protective agent. Gastric mucus, for example, is a glycoprotein containing about 500 chains of carbohydrate. The sulfate groups are connected to residues of the sugars $N$-acetylglucosamine and galactose. Sulfate is also a component of various proteins and hormones. The sulfate group is bound to residues of tyrosine. The proteins containing tyrosyl sulfate tend to be secretory proteins, that is, those occurring in the plasma and interstitial spaces.

The donor of the sulfate group in the compounds just mentioned is **3-phosphoadenosine-5′-phosphosulfate** (PAPS). The structure of PAPS and its synthesis from ATP and free sulfate appear in Figure 10.54. PAPS is used as a substrate by **sulfotransferases**. PAPS is synthesized in the cytoplasm and then transported into the Golgi, where it participates in the sulfation of macromolecules destined for secretion from the cell.

Free sulfate occurs in the plasma at concentrations of 1 to 2 m$M$. The sulfate in the plasma and glomerular filtrate has been a concern for those interested in calcium status. Consumption of high-protein diets leads to increases in urinary calcium levels. This effect has been attributed, in part, to the catabolism of sulfur amino acids to yield free sulfate. The sulfate forms a complex with the calcium in

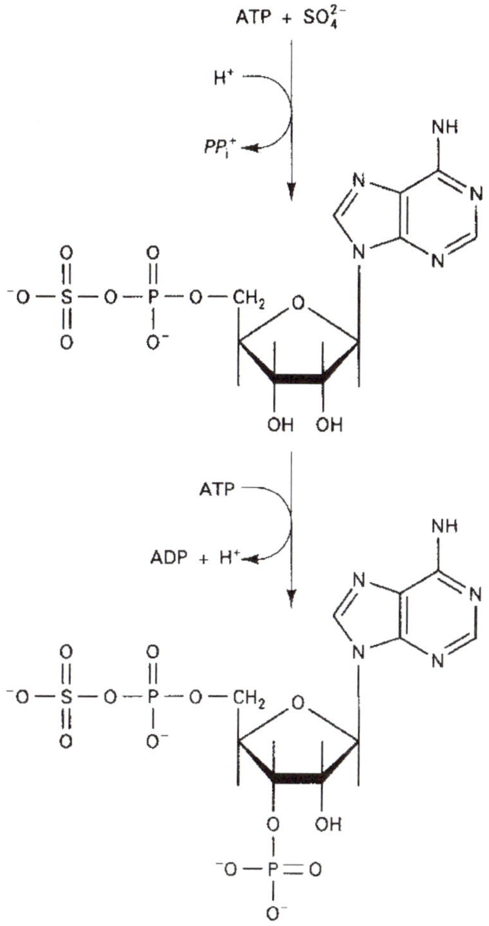

**FIGURE 10.54** Biosynthesis of PAPS from sulfate and ATP. PAPS is the biological donor of sulfate groups.

the glomerular filtrate and prevents its reabsorption by the epithelial cells of the renal tubules. Whiting and Draper (1981) have shown that feeding animals diets high in free sulfate can induce hypercalciuria. The continued feeding of sulfate can eventually lead to losses in bone mass in the animals.

## Xanthine Dehydrogenase

Xanthine dehydrogenase and aldehyde oxidase catalyze reactions involving hydroxylation. The enzymes recognize a range of substrates having similar structures. These structures include those with purine and pyrimidine rings. Both enzymes can hydroxylate hypoxanthine and xanthine, though the reaction catalyzed by xanthine dehydrogenase occurs at a much greater rate. Both enzymes can catalyze the conversion of acetaldehyde to acetic acid, where the rate of catalysis is slightly greater with aldehyde oxidase. This reaction, as catalyzed by these two enzymes, does not occur at a physiologically significant rate. Another enzyme, the NADH-dependent aldehyde oxidase, catalyzes the conversion of acetaldehyde to acetic acid at a rapid rate.

Xanthine dehydrogenase appears not to be a vital enzyme. Its absence is compatible with human life. One problem can arise where the activity of the enzyme is reduced or absent: formation of kidney stones composed of xanthine. This problem may be caused in rare instances by a genetic lack of enzyme activity, as well as by treatment with the drug allopurinol. Xanthine stone formation occurs because the compound is relatively insoluble and, thus, tends to crystallize in the renal tubule, as water is resorbed and the urine is concentrated. The tendency of any molecule to crystallize is enhanced when present at a greater concentration. Dilution reduces this tendency. The major reactions catalyzed by xanthine dehydrogenase are shown at the end of Chapter 8. The physiological function of aldehyde oxidase is not clear.

Xanthine dehydrogenase has received much attention because of its potential in the production of superoxide. Superoxide formation is often associated with the production of other toxic forms of oxygen, such as HOOH and the hydroxyl radical. Normally, xanthine dehydrogenase catalyzes an NAD-dependent reaction:

$$\text{Hypoxanthine} \xrightarrow[\text{H}_2\text{O} + \text{NAD}^+ \quad \text{NADH} + \text{H}^+]{\text{Xanthine dehydrogenase}} \text{Xanthine} \qquad (10.6)$$

Hypoxanthine (xanthine) → Xanthine (uric acid); H$_2$O + NAD$^+$ → NADH + H$^+$

Under certain conditions, xanthine dehydrogenase is converted to xanthine oxidase. These conditions involve the oxidation of a cysteine residue of the enzyme or the proteolytic cleavage of a specific peptide bond of the enzyme. In short, the conversion occurs with damage to the enzyme. Damage of this sort occurs with the purification of the enzyme from biological tissues. It can also occur under *in vivo* conditions, such as during a heart attack. The result is generation of an enzyme that catalyzes the conversion of molecular oxygen to toxic forms of oxygen. Xanthine oxidase (damaged xanthine dehydrogenase) is often used by researchers who require a convenient method for generating superoxide in a test tube. The *in vivo* production of xanthine oxidase is thought to contribute to the damage inflicted during a heart attack, because of its potential for producing superoxide. Xanthine oxidase catalyzes the oxygen-dependent reaction:

$$\text{Hypoxanthine} \xrightarrow[\text{H}_2\text{O} + 2\text{O}_2 \quad 2\text{O}_2^- + 2\text{H}^+]{\text{Xanthine oxidase}} \text{Xanthine} \qquad (10.7)$$

Hypoxanthine (xanthine) → Xanthine (uric acid); H$_2$O + 2O$_2$ → 2O$_2^-$ + 2H$^+$

# SELENIUM AND GLUTATHIONE

The Food and Nutrition Board has defined the recommended dietary allowance for selenium as 70 mg for adult males, 55 mg for adult females, and 10–15 mg for infants. An RDA for selenium has not been established. Seafoods and muscle and organ meats are high in selenium, whereas most fruits and vegetables are low in the element. Selenomethionine is the major form of selenium in plant foods. Selenomethionine is identical to methionine, except that selenium replaces the sulfur atom. The selenium consumed as selenocysteine is broken down to form alanine and hydrogen selenide ($H_2Se$). The breakdown pathway of selenomethionine is not clear. Although selenium does not occur to a great extent as selenite in foods, selenite is readily used as a source of the element by humans and animals. Selenium also occurs as selenate in foods such as beet leaves, garlic, and cabbage.

The chemistry of selenium resembles that of sulfur. Inorganic forms of sulfur include hydrogen sulfide ($H_2S$), sulfite ($SO_3^{2-}$), and sulfate ($SO_4^{2-}$). The corresponding forms of selenium are hydrogen selenide ($H_2Se$), selenite ($SeO_3^{2-}$), and selenate ($SeO_4^{2-}$). Cysteine is the most well known organic form of sulfur. The corresponding selenium derivative is selenocysteine.

A few differences in the chemistry of sulfur and selenium might be noted. The protonated forms of two selenium-based compounds, $H_2Se$ and $H_2SeO_3$, are more acidic than the corresponding sulfur-based compounds (Combs and Combs, 1984). In other words, when present in an aqueous solution at neutral pH, the selenium-based compounds tend to be dissociated and ionized, whereas the sulfur-based compounds remain protonated and nonionized. Sulfur-containing compounds tend to be oxidized in the body, resulting in the production of sulfate as the final product. In contrast, selenium-containing compounds tend to be reduced to produce selenide compounds, such as $H_2S$ and dimethyl selenide ($CH_3$—$Se$—$CH_3$). Elevated levels of dimethyl selenide are formed after a toxic dose of selenide, for example. Some of the dimethyl selenide is excreted in the breath, producing a garlic odor. A larger proportion of selenide may be trimethylated to produce $^+Se(CH_3)_3$, which is excreted in the urine.

## Selenium-Requiring Enzymes

Selenium is incorporated into Se-requiring enzymes by the modification of serine. This serine is not modified when it is in the free state or when it occurs in a polypeptide chain. The serine residue in question is modified when it occurs bound to transfer RNA, that is, as the aminoacyl-tRNA derivative. Seryl-tRNA is converted to selenocysteinyl-tRNA by the action of selenocysteine synthase (Sturchler et al., 1993). The codon for selenocysteine is UGA (TGA in DNA; UGA in mRNA). The fact that this particular triplet of bases codes for an amino acid is very unusual, as UGA normally is a stop codon. Stop codons occur in mRNA and signal the termination of synthesis of the protein; however, in the case of the UGA codons that code for selenocysteine residues, regions of the mRNA that lie beyond the coding sequence somehow convert the UGA from a codon that halts translation to one that codes for selenocysteine (Figure 10.55). The structure of selenocysteine is shown in Figure 10.56.

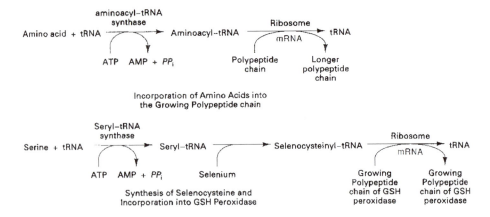

Incorporation of Amino Acids into
the Growing Polypeptide chain

Synthesis of Selenocysteine and
Incorporation into GSH Peroxidase

**FIGURE 10.55** Typical pathway for incorporation of an amino acid into a growing polypeptide chain (*top*). Pathway used for the incorporation of selenocysteine into a growing polypeptide chain (*bottom*).

Selenium-requiring proteins include GSH peroxidase, deiodinase, selenoprotein P, and selenoprotein W. GSH peroxidase is vital for the removal of certain forms of toxic oxygen, i.e., lipid peroxidides and hydrogen peroxide. 5'-Deiodinase is used for the synthesis of thyroid hormone. The functions of selenoproteins P and W are not known.

### Deiodinase

The deiodinase enzymes are used for the removal of iodine atoms from various thyroid hormones. Three types of deiodinase occur in mammals, and these are called type I, type II, and type III deiodinase (St. Germain, 1994; Pallud *et al.*, 1997). Type I enzyme is used for deiodinating T4 and T3. It is thought to act only on a modified form of thyroid hormones, namely on T4-sulfate and T3-sulfate. When type I enzyme acts on sulfated hormones, it catalyzes the deiodination at the 5'-position (on the phenolic ring). Type II deiodinase acts at the 5'-position (on the phenolic ring), and is used for converting T4 to T3. Type II enzyme occurs in the central nervous system, pituitary gland, and in brown adipose tissue. Type III deiodinase catalyzes the conversion of T4 to reverse T3, and the conversion of T3 to T2. Since these products (reverse T3 and T2) are hormonally inactive, type III deiodinase may be called the *inactivating deiodinase*. Type III enzyme acts only on the tyrosyl ring.

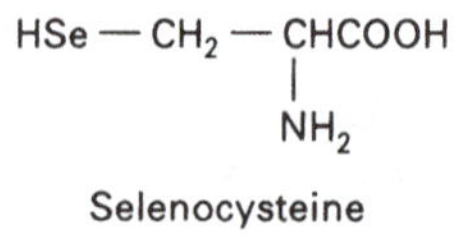

$$HSe - CH_2 - CHCOOH$$
$$|$$
$$NH_2$$

Selenocysteine

**FIGURE 10.56** Structure of selenocysteine.

The various deiodinases respond in different ways to selenium deficiency. Studies with rats revealed that nutritional Se deficiency results in a decline in activity of the type I enzyme in the liver and kidney, with little effect on the type I activity in the thyroid gland, and with little effect on the type II and type III enzymes in the brain (Pallud, 1997).

The genes for all three deiodinases contain the TGA triplet in the coding region of the DNA, while the mRNA contains a UGA triplet at the corresponding position. Part of the coding sequence of type II deiodinase is shown in what follows. The entire protein consists of 266 amino acids. The codon resulting in selenocysteine is indicated by "SeC," where this amino acids resides in the context –glycine–serine–alanine–threonine–selenocysteine–proline–proline–phenylalanine–threonine– (Pallud *et al.*, 1997):

$$- Gly - Ser - Ala - Thr - \textbf{SeC} - Pro - Pro - Phe - Thr -$$
$$-GGT–TCA–GCG–ACC–\textbf{TGA} –CCA–CCT– TTT – ACT–$$

In the cases of all three deiodinases, the mRNA contains a special sequence in the region of ribonucleotides following the coding region, that is, in the 3′-untranslated region (3′-UTR). In the mRNA coding for selenoproteins, the 3′-UTR spontaneously folds upon itself to form a structure that has a little stem topped with a loop. This structure, which is called a **stem-loop**, is required for the recognition, during translation, of the UGA as a codon for selenocysteine.

## Glutathione Peroxidase

Glutathione peroxidase is a selenium-containing enzyme that catalyzes the reaction

$$\begin{array}{ccc}
& \text{Glutathione} & \\
& \text{peroxidase} & \\
\text{HOOH} & \longrightarrow & 2H_2O \\
\text{(ROOH)} & & \text{(ROH + HOH)} \\
& 2GSH \quad GSSG &
\end{array} \tag{10.8}$$

The enzyme is used for the reduction of organic peroxides, including the peroxides of free fatty acids and other lipids. Apparently, esterified fatty acid peroxides are not recognized by the enzyme. The fatty acid peroxides that are part of membrane phospholipids (esterified to glycerol backbone) do not seem to be recognized by GSH peroxidase. GSH peroxidase activity results in the detoxification of peroxides, with the conversion of two molecules of glutathione (GSH) to the glutathione dimer (GSSG). Here, two GSHs are connected via a disulfide linkage, just as two residues of cysteine are connected via a disulfide linkage in many proteins. **Glutathione reductase**, a flavoprotein, catalyzes the NADPH-dependent reduction of GSSG back to 2 GSH.

Glutathione is a tripeptide composed of glutamate, cysteine, and glycine (Glu–Cys–Gly). The tripeptide is unusual in that the bond between the glutamate and

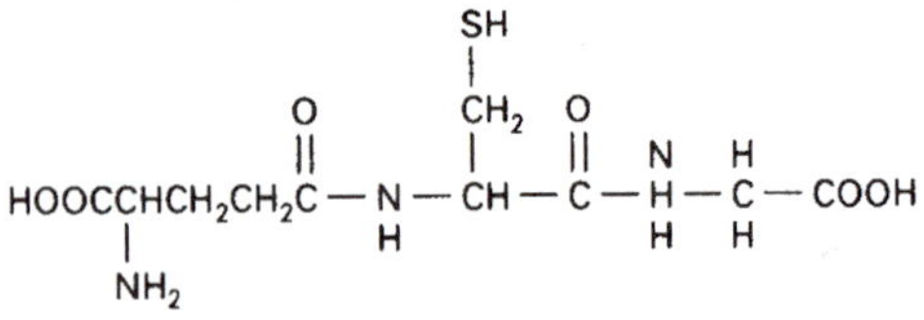

$$\begin{array}{c} \text{SH} \\ | \\ \text{CH}_2 \end{array}$$

HOOCCHCH$_2$CH$_2$C—N—CH—C—H—C—COOH

NH$_2$

**FIGURE 10.57** Structure of glutathione, a tripeptide with a variety of functions. The residue of glutamic acid is linked directly to the cysteine residue via the γ-carboxyl group of glutamate, not the α-carboxyl group of glutamate. The α-carboxyl group of cysteine is bound directly to glycine.

cysteine involves the γ-carboxyl group, rather than the α-carboxyl group, of glutamate (Figure 10.57) Glutathione is synthesized in the cytosol in two steps:

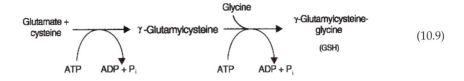

(10.9)

The concentration of GSH in the liver is 5 to 10 m*M*, whereas that of GSSG is about a tenth of this value. For comparison, note that cytosolic cysteine and coenzyme A occur at about 0.2 and 0.06 m*M*, respectively. Glutathione is thus a major sulfhydryl compound of the cell. In human plasma, GSH occurs at about 2 μ*M*, while the level of GSH in rat plasma is tenfold higher, 10–20 μ*M* (Halliwell and Gutteridge, 1990; Leeuwenburgh *et al.*, 1997).

Glutathione participates in a variety of reactions and has a number of functions in the body. These include the detoxification of hydrogen peroxide (HOOH) and organic peroxides (ROOH), the maintenance of vital sulfhydryl groups in the reduced form, the synthesis of certain hormones derived from arachidonic acid (leukotrienes), and the metabolism of a vast number of foreign compounds. These foreign compounds include the aromatic organic molecules responsible for the color, flavor, and fragrance of plants; the aromatic organic molecules produced by the immune systems of plants; the polycyclic aromatic hydrocarbons produced in the charring and smoking of foods; and a number of drugs and pesticides. Glutathione is also a cofactor in the metabolism of certain aldehydes (methylglyoxal, formaldehyde) and, perhaps to a small extent, in the resorption of amino acids by the kidney. The reactive part of GSH in all of the preceding functions is the sulfhydryl group of the cysteine residue.

Glutathione metabolism is depicted as a number of cycles and blind-end pathways (Figure 10.58). The synthesis of GSH (1), its use by GSH peroxidase (2), and the regeneration of GSH (3) have already been described. The function of GSH in maintaining cellular sulfhydryl groups (4), scavenging free radicals (5), conjugating with foreign compounds and various metabolites (6), and possibly in transporting amino acids (7) are outlined in Figure 10.58.

Important thiols (sulfhydryls) of the cell, such as the sulfhydryl groups of proteins and coenzyme A, are maintained in the reduced form by thioltransferase

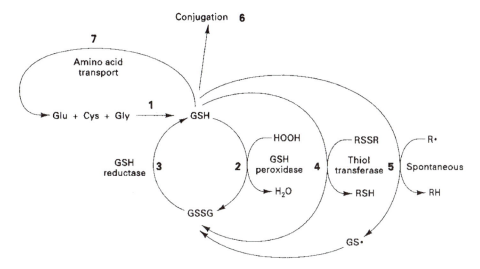

**FIGURE 10.58** Outline of glutathione metabolism. Glutathione is used as a component of an amino acid transport system (7), as a cofactor in peroxide elimination (2), as a reductant (4), and as a scavenger of radicals (5).

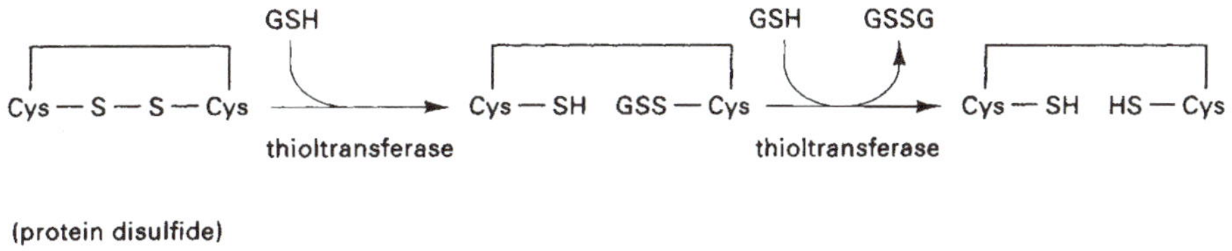

**FIGURE 10.59** Utilization of glutathione by thiol transferase.

(Figure 10.59). A sulfhydryl group can spontaneously condense with another sulfhydryl group, forming a disulfide bond. In many cases, the formation of a disulfide bond involving two residues of cysteine in a protein is required for the normal functioning of the protein and is desirable. In other cases, the disulfide bond represents an abnormal structure and may be reversed by thioltransferase.

Radicals can form in the cell with exposure to high-energy radiation, such as the ultraviolet rays of the sun and radiation used in cancer chemotherapy. Radicals can also be produced in the course of HOOH metabolism, where the HOOH is cleaved to produce a hydroxyl radical followed by reaction of this short-lived chemical with organic compounds in the cell. The direct, nonenzymatic reaction of GSH with a radical compound is

$$GSH + R\bullet \rightarrow GS\bullet + RH \tag{10.10}$$

This reaction results in production of a glutathione radical, which is relatively stable and unreactive. GSH can also react with superoxide, and here the products are the glutathione radical plus HOOH (Wefers and Sies, 1983):

$$GSH + O_2^{\overline{\cdot}} \xrightarrow{\text{H}^+} GS\cdot + HOOH \tag{10.11}$$

Two glutathione radicals can condense with each other to form GSSG:

$$GS\cdot + GS\cdot \rightarrow GSSG \tag{10.12}$$

It has been proposed that ascorbic acid can aid in the removal of the GSH radical (Figure 10.60) (Wefers and Sies, 1988). The reactions are nonenzymatic and occur in the aqueous phase.

## Selenium and Viruses

Selenium is an issue in the biology of one of the poxviruses, one of the largest viruses in existence, as well as in the biology of coxsackievirus, one of the smallest of viruses. In the case of MCV, a poxvirus, the virus's genome codes for the well-known selenoprotein *glutathione peroxidase*. Apparently, this enzyme protects the virus from the defense system of the host. In the case of coxsackievirus, selenium deficiency in the host, whether a mouse or human, provokes activation of the virus within the host cell, resulting in disease to the host (damage to the heart).

A type of poxvirus, called MCV, can live in the skin of people and cause tumors in children. The genome of this virus contains about 150 genes, where one of the genes codes for GSH peroxidase, a selenoprotein. The mRNA sequence of this enzyme, in the virus's genome, contains one UGA triplet in the coding region, which codes for selenocysteine, as well as a stem-loop in the 3'-UTR. The stem-loop consists of 50 consecutive ribonucleotides, i.e., it is 50 bases long, and allows the UGA to be recognized as a selenocysteine, rather than as a stop codon (McFadden *et al.*, 1998; Shisler *et al.*, 1998). Evidence suggests that the virus-coded GSH peroxidase serves to protect its home, that is, to protect itself (and the host cell) from the barrage of hydrogen peroxide produced by vigilant white blood cells, and thus allow the generation of large amounts of virus. One way that cells of the immune system (white blood cells) kill invading viruses is to release hydrogen peroxide and other forms of toxic oxygen in the immediate vicinity of the infected cells. Apparently, the poxvirus's own GSH peroxidase is used to protect its home against attack.

Coxsackievirus is a small virus that can infect mammals. The genome of this virus consists of RNA (not DNA), and it contains only 7400 ribonucleotides. The

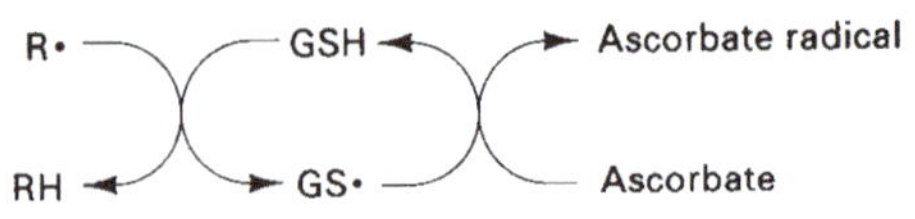

FIGURE 10.60 Detoxification of a free radical by glutathione, and regeneration of glutathione by ascorbate.

RNA genome codes for four proteins. Three of these proteins are virus capsule proteins (they coat the RNA), and one of these proteins is a viral protease. The virus can cause temporary fever and sore throats in children; however, about 5% of infections in adults and children result in damage to the heart. It is thought that Keshan disease in China, which is associated with selenium-deficient populations, is due to some influence of the selenium deficiency on the activation of cox-sackievirus. Studies with mice have revealed an interesting phenomenon. One particular strain of coxsackievirus can infect mice, but does not result in much harm to the mice. However, when the mice are raised on an Se-deficient diet, the virus acquires mutations in its genome and produces damage to the mouse's heart (Beck *et al.*, 1995).

An interpretation of these phenomena is that the Se-deficiency allows toxic oxygen to build up in the host cell. The increased levels of toxic oxygen result in an increased rate of damage to the virus's DNA. The increased DNA damage results in the acquisition of mutations in the virus's genome. Mutations have been detected in several different regions of the virus's genome (Beck *et al.*, 1995).

## Glutathione and Cancer

Plant foods contain a great variety of compounds that are not useful to mammals. Many of these compounds are aromatic and tend to be stored in the body's fat or excreted in breast milk, rather than immediately excreted in the urine or feces. The body attempts to make these compounds more readily excretable by introducing hydroxyl groups into the compounds. The hydroxylation reaction is catalyzed by one or more of the cytochrome P450 enzymes of the endoplasmic reticulum. In addition, the compound may be modified by conjugation with a sugar (glucuronic acid), sulfate, methyl group, acetyl group, or GSH. The substrate used in the donation of the sugar group is uridine-5'-diphosphate-glucuronic acid. PAPS, SAM, and acetyl-CoA are used in the enzymatic transfer of the sulfate, methyl, and acetyl groups, respectively. A number of glutathione *S*-transferase enzymes are used in the transfer of GSH to various compounds. Chemicals formed during the heating of foods (pyrolysis products), as well as drugs and other compounds, are also metabolized by hydroxylation and conjugation. The following material concerns the role of GSH in the detoxification of polycyclic aromatic hydrocarbons, aflatoxins, and anticancer drugs.

The structures of some of the polycyclic aromatic hydrocarbons are shown in Figure 10.61. They occur in smoked fish and meats, grilled and roasted foods, and vegetables and grains. They can originate from smoke, from the charring of foods, and from the soil used to grow plants. The most well known of these compounds is 3,4-benzpyrene. Benzpyrene can occur, for example, in smoked fish and charcoal-broiled steaks at levels reaching 10 to 50 µg/kg food. Benzpyrene is one of the more notorious compounds in food, as it is converted to a potent carcinogen in the body. Benzpyrene is converted to a 7,8-epoxide compound, followed by hydrolysis of the epoxide group to form two hydroxyl groups. The resulting 7,8-dihydroxide is again epoxidized, resulting in the formation of 7,8-dihydroxy-9,10-epoxy-benzpyrene. The series of reactions leading to the formation

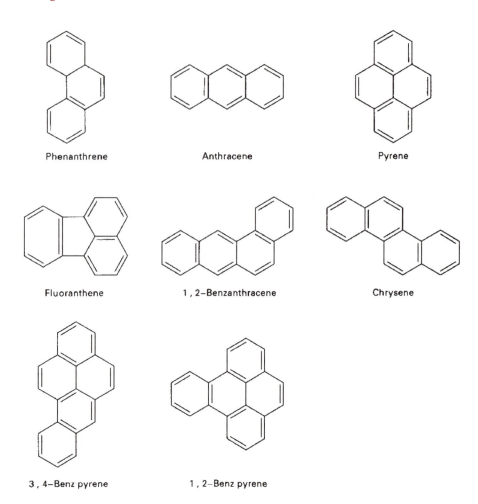

Phenanthrene          Anthracene          Pyrene

Fluoranthene       1 , 2–Benzanthracene          Chrysene

3 , 4–Benz pyrene          1 , 2–Benz pyrene

**FIGURE 10.61**  Polycyclic aromatic hydrocarbons.

of 7,8-dihydroxy-9,10-epoxy-benzpyrene appears in Figure 10.62. Epoxides are fairly reactive and are able to condense with a variety of target compounds including DNA. The reaction of benzpyrene epoxide with DNA to form an adduct is shown in Figure 10.62. The adduct consists of benzpyrene bound to a residue of guanine of the DNA. The significance of the adduct is that it is believed to be an early event in the conversion of a normal cell to a cancer cell, in the case of chemically induced cancer.

The most well-established form of chemically induced cancer is lung cancer caused by smoking. Tobacco smoking is the main cause of lung cancer. It has also been associated with cancer of the mouth, esophagus, bladder, and pancreas. It is thought that formation of an adduct between the polycyclic hydrocarbons of cigarette smoke and DNA is an early event in the development of lung cancer. Studies with extracts of lung tissue from deceased smokers have revealed that adducts occur at a frequency of about 100 adducts per $10^9$ DNA nucleotides (Randerath et al., 1989).

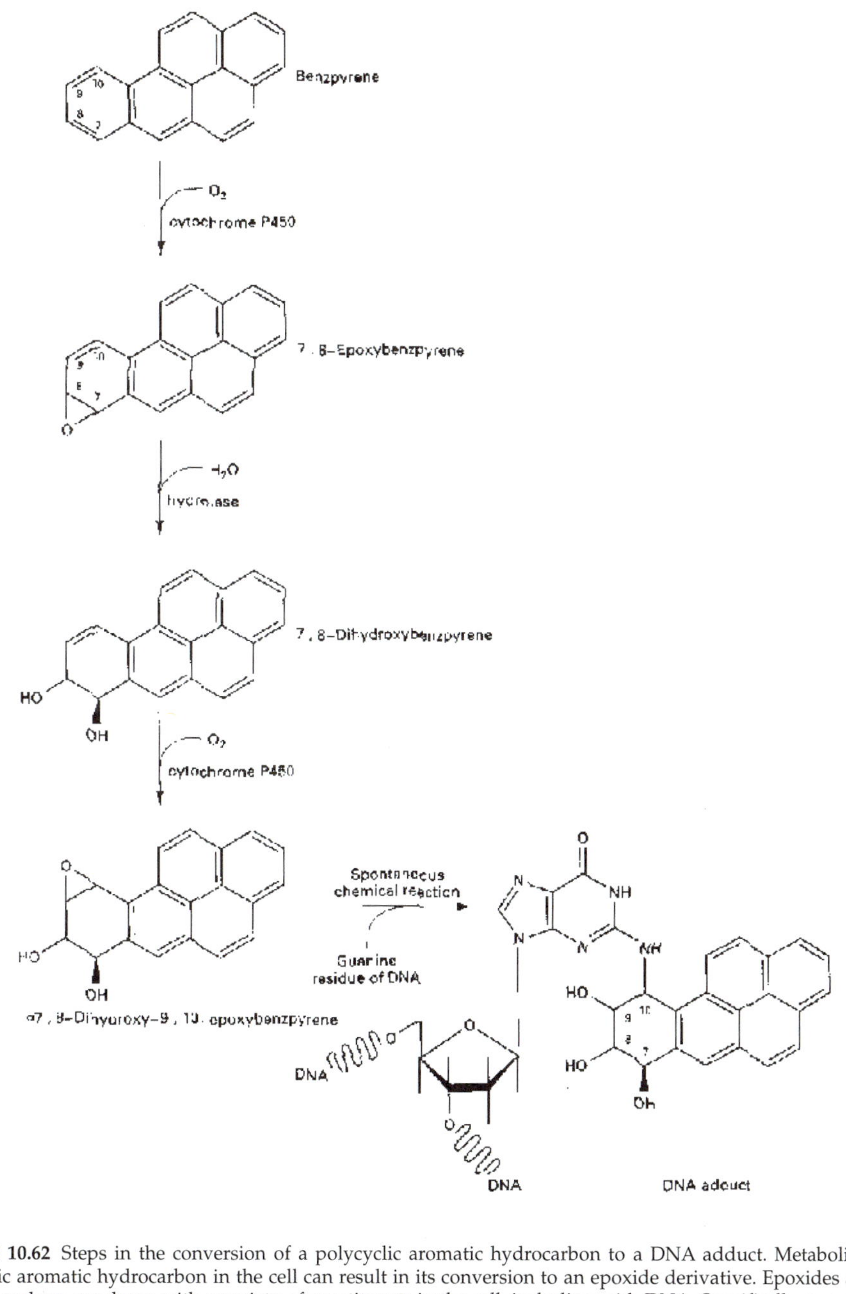

**FIGURE 10.62** Steps in the conversion of a polycyclic aromatic hydrocarbon to a DNA adduct. Metabolism of a polycyclic aromatic hydrocarbon in the cell can result in its conversion to an epoxide derivative. Epoxides are very reactive, and can condense with a variety of constituents in the cell, including with DNA. Specifically, epoxides can react with the bases of DNA. One of the functions of DNA is to support transcription. Another activity of DNA is replication. As DNA polymerase moves along a strand of DNA, during replication, it encounters many types of covalent damage. Most of this damage is repaired by a class of enzymes called DNA repair enzymes. However, if this damage is not repaired before the cell decides to divide, a mutation can occur. The type of mutation that results depends on the exact nature of the chemical modification of the DNA. These chemical modifications, and the mutations that result, are usually harmless to the cell, but sometimes they can cause lethal damage to the cell, or worse, result in the conversion of a normal gene to a "cancer gene" (oncogene). The conversion of covalent damage to a mutation is totally dependent on the process of replication, where the DNA polymerase encounters the damage, makes a mistake, and catalyzes the polymerization of an incorrect base (in the newly made DNA) across from the damage (residing in the old, template DNA).

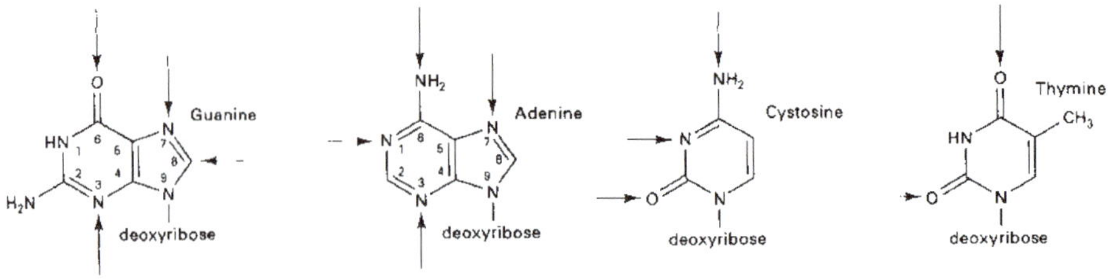

FIGURE 10.63 Possible positions of attack of chemical carcinogens on DNA bases.

Guanosine is not the only nucleotide base that can be modified. Adenine, cytosine, and thymine may also be modified. Figure 10.63 illustrates the positions of attack and complex formation by various chemical carcinogens.

7,8-Dihydroxy-9,10-epoxy-benzpyrene, as well as other epoxide compounds, can be detoxified by glutathione. A number of different GSH S-transferases can catalyze the transfer of GSH to the epoxide. This transfer results in the formation of a GSH conjugate. An example of the reaction is shown in Figure 10.64. The resulting conjugate is not carcinogenic. It can be readily excreted in the bile.

One of the most potent chemical carcinogens, as determined in studies with rats, is aflatoxin. Aflatoxin consists of a family of related compounds. They are natural products and are synthesized by the fungus Aspergillus flavus. Aflatoxin is most closely identified with peanuts and peanut butter; however, it is also present in dusts from grain and may contribute to cancer of the respiratory system in grain processors. Alfatoxins in grain and peanuts consumed in Africa and China are thought to contribute to the liver cancer found in these regions of the globe. Liver cancer is common in China and Africa but rare in Europe and America. Liver cancer is thought to be dependent on exposure to both aflatoxin and hepatitis virus. Conversion of aflatoxin $B_1$, the most potent of the aflatoxins, to the carcinogenic epoxide is shown in Figure 10.65. Also shown is detoxification of the epoxide, as catalyzed by GSH S-transferase. An alternate pathway of detoxification, involving hydrolysis of the epoxide group, is shown as well.

A few fine points concerning GSH S-transferase might be of interest. There is evidence that a large proportion of the population lacks one of the GSH S-trans-

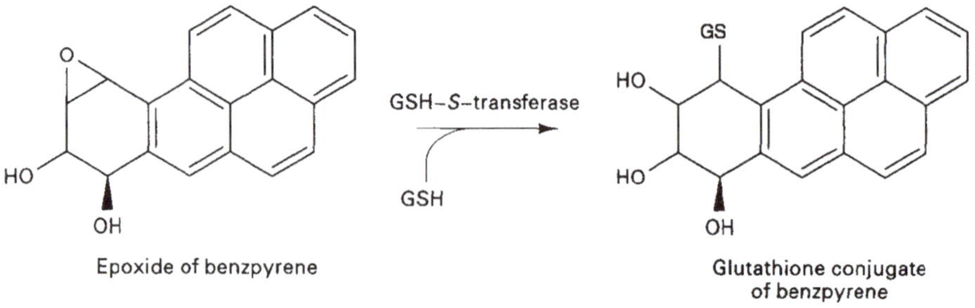

FIGURE 10.64 Formation of the glutathione conjugate of benzpyrene.

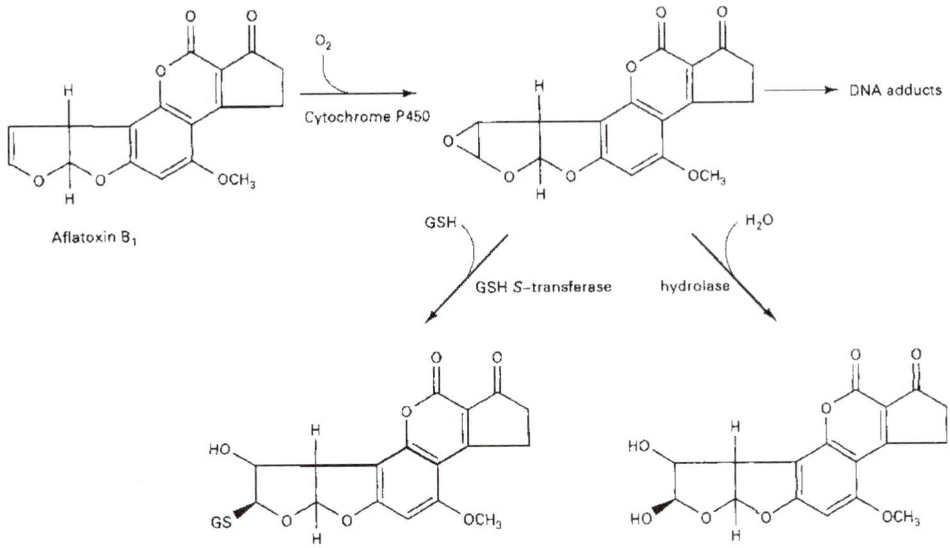

**FIGURE 10.65** Conversion of aflatoxin B$_1$ to the epoxide form and pathways for detoxification of the epoxide.

ferases. Hence, there is some thought that these persons are more susceptible to chemically induced cancers than might otherwise be prevented by the enzyme. There is also strong interest in designing and using inhibitors of GSH S-transferase. GSH S-transferase can catalyze the inactivation of certain anticancer drugs. Such drugs as nitrogen mustard, nitrosoureas, and melphalan can be inactivated by the action of GSH S-transferase. Inactivation results in insensitivity of the cancer cell to the drug. The use of drugs that inhibit GSH S-transferase or that induce depletion of the levels of GSH in the cancer cell would be expected to restore the effect of the anticancer drug (Waxman, 1990).

The modification of foreign compounds by hydroxylation and conjugation with glutathione or other small molecules does not always lead to detoxification of the compound. In many cases the modification results in the conversion of the compound to a more toxic substance. In this case, the process is called toxification.

## Glutathione and Aldehyde Metabolism

Glutathione is required as a cofactor by two enzymes in aldehyde metabolism: the glyoxylase system:

$$\tag{10.13}$$

and formaldehyde dehydrogenase:

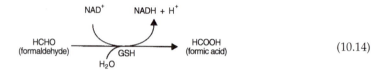

$$\text{HCHO (formaldehyde)} \quad \xrightarrow[\substack{\text{GSH} \\ H_2O}]{\substack{NAD^+ \quad NADH + H^+}} \quad \text{HCOOH (formic acid)} \tag{10.14}$$

The first step in the catabolism of an aldehyde by these enzymes is its condensation with the sulfur atom of GSH.

The glyoxalase system is active with a variety of 2-keto-aldehydes, including methylglyoxal. Methylglyoxal is formed from an intermediate of glycolysis, glyceraldehyde-3-phosphate. The concentration of glyceraldehyde-3-phosphate in the liver cell is about 30 $\mu M$ (Sarau et al., 1977), whereas that of methylglyoxal is 0.1 to 2.0 $\mu M$ (Thornalley, 1990). Apparently, methylglyoxal is formed with the slow spontaneous breakdown of glyceraldehyde-3-phosphate in the cell. Methylglyoxal is 2-keto-propionaldehyde. The glyoxalase system converts it to lactic acid.

Formaldehyde dehydrogenase catalyzes the oxidation of a number of aldehydes, including formaldehyde, to the corresponding acid. The enzyme is important as it catalyzes the detoxification of formaldehyde, a chemical present in small concentrations in most or all biological tissues. Formaldehyde, as well as other aldehydes, spontaneously condense with amino groups — via a Schiff base linkage — to form a condensation product. This type of condensation product is not desirable, and contributes to a small extent to the various types of damage inflicted upon the proteins of the body.

Formic acid, a product of the aforementioned reaction, is not toxic at low concentrations. Formic acid can be converted to $CO_2$ by catalase or by folate-dependent enzymes.

## Glutathione in the Glomerular Filtrate

The main purpose of the kidneys is to purify the blood and remove waste products. The kidney does this by two methods: (1) by means of a filter that retains the blood cells and most of the proteins, and produces a filtrate that resembles blood serum; and (2) by reclaiming all the nutrients in the filtrate (salts, amino acids, sugars, water) by transport systems in the renal tubule. In this way, the renal tubule behaves as a microscopic gut. One of the biochemicals that passes through the filtrate is glutathione. To retain this tripeptide, or its constituent amino acids, glutathione is broken down in the lumen of the renal tubule. This breakdown process is initiated by γ-**glutamyltranspeptidase**. The enzyme is membrane-bound, and its active site resides in the lumen of the renal tubule. The enzyme catalyzes the exchange of the cysteine–glycine moiety of GSH for an amino acid occurring in the glomerular filtrate. The result is the formation of two dipeptides. One of the dipeptides contains glutamic acid, where the glutamate residue is attached to its partner via its γ-carboxyl group (the α-carboxyl group remains free):

$$\text{Glutathione} \quad \xrightarrow[\substack{\text{Amino acid} \quad \text{Cys–Gly}}]{\text{γ-Glutamyltranspeptidase}} \quad \text{γ-Glutamyl-amino acid} \tag{10.15}$$

The amino acid substrates that are most recognized by γ-glutamyltranspeptidase are cysteine, glutamine, methionine, alanine, and serine. The products of the reaction, apparently, are efficiently transported into the epithelial cell, and then broken down into their constituent amino acids in the cell. The elucidation of the series of reactions involving formation of the γ-glutamyl-amino acid, absorption of the products, hydrolysis of the products, and re-formation of glutathione was mainly the work of Alton Meister (Meister, 1989).

The major function of γ-glutamyltranspeptidase appears to be to initiate the breakdown of glutathione, thus allowing the reclamation of its constituent amino acids. Mutant mice lacking in the enzyme were created by standard genetic techniques. The urine produced by these mice contained 600-fold greater levels of excreted glutathionine than that of normal mice. The mice became ill and failed to reproduce. Supplementing the mutant mice with cysteine prevented some of their health problems (Lieberman *et al.*, 1996; Harding *et al.*, 1997).

γ-Glutamyltranspeptidase appears to contribute only very slightly to the general amino acid resorption by the kidney. Still, there continues to unfold evidence that this enzyme, and related enzymes, play a part in some types of amino acid transport (Lee *et al.*, 1996).

## Selenium Nutrition

Selenium status in humans is a concern in certain parts of the world where the soils, and thus the foods, have very high or low contents of the element. Selenium status may also be a concern during total parenteral nutrition where Se is not included in the feeding solution. Assays of GSH peroxidase appear to be the most promising mode of assessing Se status. Se status may be assessed by measuring the Se content of liver or muscle. Because of the difficulty in obtaining biopsies, red blood cell Se levels have often been used for measuring intracellular Se levels. Red blood cell and tissue Se levels reflect long-term Se status. Plasma and urinary Se levels can also be measured, but these are very sensitive to the amount of Se consumed in foods the previous day or two and would not be expected to accurately reflect tissue levels. Assays of platelet GSH peroxidase have been proposed as a convenient method for assessing Se status. Assays of the enzyme in platelet extracts may be more accurate than those using red blood cell extracts because the hemoglobin in the red cell extracts can interfere with the enzyme assay. Note that the GSH peroxidase in red blood cells protects hemoglobin and red blood cell membranes from HOOH-induced damage.

Whole blood selenium levels can vary between 10 and 3000 ng Se/ml. These levels reflect dietary intake. With an Se intake of about 50 µg/day, plasma Se levels in humans are about 70 ng/ml, and red blood cell Se is about 90 ng/ml (Levander *et al.*, 1983). Maximal activity of GSH peroxidase, as determined by assays of the platelet enzyme, is supported where the plasma selenium level is about 100 ng/ml.

Selenium status may be a concern for premature infants, since this population may have lower levels of plasma Se and of GSH peroxidase. Full–term infants have 0.6-1.8 µM Se, while premature infants have been found to have 0.3–0.5 µM Se, though levels as high as 1.1–1.6 µM have also been found (Tyrala *et al.*, 1996).

Studies with infants have clearly shown that plasma Se levels respond to dietary Se levels. A very rough correlation may be made from the available data. A 2.5-fold increase in selenate intake provoked a 66% rise in plasma selenium levels (Tyrala *et al.*, 1996).

Selenium deficiency has occurred in patients fed by total parenteral nutrition. The signs of deficiency include muscle pain. The deficiency can be reversed by feeding 100 µg Se/day for a week in the form of selenomethionine. Se deficiency may also occur during severe malnutrition.

Selenium intake varies widely in different parts of the world. High intakes (750–5000 µg Se/day) occur in regions with high-Se soils. These regions include Hubei Province in China, where the mining and burning of high-selenium coal introduce Se-containing dust into the environment. Selenium poisoning (selenosis) occurs in areas of high-selenium soils, where the daily intake of Se is about 0.5 mg or greater. Selenium toxicity results in changes in fingernails and in hair loss. Neurological symptoms may also occur. The breath may have a garlic odor as a result of the increased production of dimethyl selenide. Low intakes (10–30 µg Se/day) are found in areas with low-selenium soils, such as Keshan County in Heilongjiang Province in China, New Zealand, and Finland. Keshan disease, named after the region, occurs in rural and mountainous areas in a belt extending from northeast to southwest China. The populations most affected are children and young women. The disease involves necrosis of areas of the heart and replacement of these areas by fibrous structures. The disease, which can prove fatal, is thought to be induced by two concomitant factors: Se deficiency and a virus. Apparently, the animal equivalent of Keshan disease cannot be produced by feeding Se-deficient diets. The damage to the heart may be caused by the decline in GSH peroxidase activity, the accumulation of toxic oxygen, and damage to the mitochondria. The incidence of Keshan disease has been reduced by supplementing children in China with selenium (1.0 mg sodium selenite per week).

Topics of continuing concern in Se nutrition include the availability of Se when supplied as selenomethionine or sodium selenite and the definition of normal levels of tissue selenium and the normal activity of GSH peroxidase. In many tissues, the Se associated with GSH peroxidase accounts for only a small proportion of tissue Se. This situation raises the question of whether values of tissue Se or of enzyme activity can more accurately define adequate or optimal Se status.

## Mild and Severe Selenium Deficiency in Animals

A mild deficiency in Se can be induced by feeding animals a diet deficient in selenium. A more severe deficiency can be induced by feeding the diet to the pregnant mother and to the offspring (multigenerational deficiency). Se deficiency in rats results in anorexia, a slow growth rate, lack of sperm production, loss of hair, cataracts, and death. Neurological problems and damage to the kidney and heart also occur.

Studies with animals have revealed an interesting pattern of relationships involving Se and vitamin E nutrition. In chicks, experimentally induced deficiencies in both Se and vitamin E result in exudative diathesis, muscular dystrophy, and pancreatic atrophy. The first two problems can be prevented by vitamin E alone. Pancreatic atrophy, on the other hand, cannot easily be prevented by vitamin E. Exudative diathesis involves edema resulting from increased permeability of capillaries. The muscular dystrophy in chicks involves the degeneration of skeletal muscles; the muscles acquire a pattern of white lines or striations. Selenium alone can reduce the muscular dystrophy but cannot completely prevent it. Pancreatic atrophy involves the degeneration of the acinar cells and a loss in exocrine function. With continued feeding of the Se-deficient diet the acinar cells are infiltrated by fibroblasts and macrophages. Eventually the pancreas becomes fibrotic. With continued feeding of the Se- and vitamin E-deficient diet, hemorrhages develop under the skin.

A combined deficiency in Se and vitamin E has occurred naturally in farm animals (sheep, cattle) in regions of the world where the soil is low in selenium, such as in western Oregon in the United States and in New Zealand, where forages low in vitamin E were used. This deficiency results in muscular dystrophy. The heart and skeletal muscle appear white. The disease has been called white muscle disease. Death may occur. The disease can be prevented by supplementing the animals with either Se or vitamin E.

The selenium in soils occurring as selenate is readily taken up by plants, whereas that occurring as elemental Se, selenide, or selenite is poorly accumulated by plants. The plant converts the selenium to organic forms, such as selenomethionine, Se-methyl-selenomethionine, selenocysteine, and Se-methyl-selenocysteine. The structures of two of these compounds are shown in Figure 10.66. In some plants these amino acids tend to be free; in other plants they tend to be incorporated into proteins. Although selenium is readily incorporated into the plant, Se has not been found to be required for plant life, despite attempts to address the possibility. Selenium deficiency can be prevented in farm animals by feeding them alfalfa grown on local soils treated with sodium selenite. There has been some hesitance in supplementing soils with selenate because of the possibility of the plant accumulating high levels of selenium, which may cause selenium toxicity in animals consuming the plants.

The close relationship between Se and vitamin E has a biochemical basis. Vitamin E acts to minimize damage to membranes inflicted by free radicals. Selenium (GSH peroxidase) acts to prevent the accumulation of HOOH, which is a source of hydroxyl radicals. Hydroxyl radicals can damage cell membranes, as well as other components of the cell. Damage that is more closely identified with Se than with vitamin E, such as pancreatic atrophy, may be caused by HOOH-inflicted damage to soluble components of the cell, rather than damage to cell membranes. The converse situation also holds. For example, vitamin E deficiency in rats produces fetal resorptions. This problem cannot be prevented by Se.

$$CH_3 - \overset{+}{Se} - CH_2CH_2CHCOOH \qquad CH_3 - Se - CH_2CH_2CHCOOH$$
$$\quad\;\; |\qquad\qquad | \qquad\qquad\qquad\qquad\qquad\;\; |$$
$$\quad\;\; CH_3 \qquad\;\; NH_2 \qquad\qquad\qquad\qquad\qquad NH_2$$

Se–methyl–selenomethionine                Selenomethionine

**FIGURE 10.66**  Structures of Se-methyl-selenomethionine and selenomethionine.

*Possible Anticancer Effect of Selenium Supplements*

The possibility that selenium supplements can reduce the risk for certain cancers arose from controlled studies with animals and from epidemiological studies on human populations. One might assume this effect to be related to the action of GSH peroxidase and the role of this enzyme in maintaining the body's defenses against toxic oxygen (Rayman, 1997). On the other hand, evidence suggests that any anticancer effect can occur by other mechanisms (Lanfear *et al.*, 1994; Reddy *et al.*, 1997). Some attention has focused on the notion that high doses of selenium can provoke **apoptosis**. Apoptosis is a normal event in the lifetime of many cells. Apoptosis, which results in the cell committing suicide, is also called *programmed cell death*. This process is closely identified with a protein called p53 and with the fragmentation of the entire genome catalyzed by special nucleases. In short, activation of the apoptosis pathway within a cancer cell, whether provoked by natural means, by selenium, or by a drug, would be expected to be beneficial to human health.

Epidemiological studies have suggested that selenium can prevent certain types of cancers, though conclusive evidence is not yet at hand (Colditz, 1996; Clark *et al.*, 1996). Epidemiological studies generally involve large numbers of people (1000–100,000) and attempt to identify risk factors for disease, by acquiring information on diet, genetic make-up, gender, blood biochemistry, and occupation. Cancer is a multistep disease, where the disease often cannot be detected until after all the steps have been taken. In contrast, diabetes can be detected by high plasma glucose well before the patient dies by the disease.

Similarly, risk for atherosclerosis can be detected by high plasma LDL-cholesterol long before a heart attack occurs. The goal of making correlations between any component of the diet, and risk for cancer, is a difficult one.

## OTHER INORGANIC NUTRIENTS

There is some evidence that chromium ($Cr^{3+}$), silicic acid ($Si(OH)_4$) or related silicon-based compounds, borate ($B_4O_7^{2-}$), nickel ($Ni^{2+}$), and arsenate ($AsO_4^{3-}$) are required inorganic nutrients. However, the available evidence tends to be not convincing to all researchers (Reeves, 1997). The functions of arsenate and nickel are not clear. There is some indication that arsenate is used in the metabolism of sulfur-containing amino acids. An experimentally induced deficiency in arsenate or nickel results in impaired growth.

*Chromium*

Chromium is a trivalent cation that occurs as $Cr^{3+}$. A typical daily intake is 0.5–3.8 μmol (25–200 μg/day). After absorption from the diet, chromium occurs bound to transferrin. A safe and adequate intake of 50 to 200 μg Cr/day has been established. Chromium appears to participate in glucose metabolism. The ion may play a part in mediating the hormonal effects of insulin. Chromium deficiency results in abnormally high glucose tolerance curves and impaired clearance of plasma glucose. Chromium deficiency can be induced in animals. There is evidence that this deficiency has occurred in some humans fed Cr-free liquid diets by total parenteral nutrition.

## Borate

The usual dietary intake of borate is about 0.1 mmol/day. Borate occurs as $B(OH)_3$ and $B(OH)_4^-$, and also occurs as weak, covalent complexes with sugars. Borate never occurs in biology as free boron atoms. Dietary borate is readily absorbed by the gut, and most of this chemical is excreted in the urine. Studies with humans have revealed that 88–97% of dietary borate is excreted in the urine (Hunt *et al.*, 1997). There is some evidence that borate deficiency provokes a further decline in health during vitamin D deficiency.

## Silicon

Silicon appears to be required for the normal synthesis of the organic matrix of bone and for its normal calcification. The silicon in the blood serum occurs entirely as silicic acid. Osteoblasts may contain the highest concentrations of silicon of all the cells of the body. Within these cells, silicon occurs mainly in the mitochondria. Silicon deficiency in animals results in bones with abnormal structures. These abnormalities include thinner cortical bone, reduced bone flexibility, and flattened cranial bones. Silicon deficiency also affects cartilage, and results in a dramatic reduction in the width of the epiphyseal cartilage (Carlisle, 1985).

## Cobalt

Cobalt is a required mineral since it is a vital component of vitamin $B_{12}$. There is no reason to believe that extra cobalt needs to be supplied, in addition to that occurring in dietary cobalamin.

## REFERENCES

Abumrad, N. N., Schneider, A. J., Steel, D., and Rogers, L. S. (1981). Amino acid intolerance during prolonged total parenteral nutrition reversed by molybdate therapy. *Am. J. Clin. Nutr.* **34**, 2551–2559.

Alcantra, P. F., Hanson, L. E., and Smith, J. D. (1980). Sodium requirements, balance, and tissue composition of growing pigs. *J. Anim. Sci.* **50**, 1092–1101.

Alvarado, F., and Mahmood, A. (1979). pH-dependent effects of the alkali-metal ions on intestinal brush-border sucrase. *J. Biol. Chem.* **254**, 9534–9541.

Aperia, A., Holtback, U., Syren, M.-L., Svensson, L.-B., Fryckstedt, J., and Greengard, P. (1994). Activation/deactivation of renal $Na^+,K^+$-ATPase: A final common pathway for regulation of natriuresis. *FASEB J.* **8**, 436–439.

Aspinwall, C., Brooks, S., Kennedy, R. T., and Lakey, J. (1997). Effects of intravesicular $H^+$ and extracellular $H^+$ and $Zn^{2+}$ on insulin secretion in pancreatic beta cells. *J. Biol. Chem.* **272**, 31308–31314.

Baer, M. T., and King, J. C. (1984). Tissue zinc levels and zinc excretion during experimental zinc depletion in young men. *Am. J. Clin. Nutr.* **39**, 556–570.

Bakshi, S., Fahey, J., and Pierce, L. (1967). Sausage cyanosis: Acquired methemoglobinemic nitrite poisoning. *N. Engl. J. Med.* **277**, 1072.

Balla, T., Baukal, A. J., Guillemette, G., and Catt, K. J. (1988). Multiple pathways of inositol polyphosphate metabolism in angiotensin-stimulated adrenal glomerulosa cells. *J. Biol. Chem.* **263**, 4083–4091.

Barcia, J. P., Strife, C. F., and Langman, C. (1997). Infantile hypophosphatemia: Treatment options to control hypercalcemia, hypercalciuria, and chronic bone demineralization. *J. Pediatr.* **130**, 825–828.

Beck, M. A., Shi, Q., Morris, V., and Levander, O. (1995). Rapid genomic evolution of a non-virulent Coxsackievirus B3 in selenium-deficient mice results in selection of identical virulent isolates. *Nature Med.* **1**, 433–436.

Belusa, R., Wang, Z.-W., Matsubara, T., Sahlgren, B., Dulubova, I., Nairn, A. C., Ruoslahti, E., Greengard, P., and Aperia, A. (1997). Mutation of the protein kinase C phosphorylation site on rat $\alpha$1 $Na^+,K^+$-ATPase alters regulation of intracellular $Na^+$ and pH and influences cell shape and adhesiveness. *J. Biol. Chem.* **272**, 20179–20184.

Berner, Y. N., and Shike, M. (1988). Consequences of phosphate imbalance. *Annu. Rev. Nutr.* **8**, 121–148.

Bernhart, F. W., Savini, S., and Tomarelli, R. M. (1969). Calcium and phosphorus requirements for maximal growth and mineralization of the rat. *J. Nutr.* **98**, 443–448.

Berrocal, R., Chanton, S., Juillerat, M. A., Pavillard, B., Scherz, J.-C., and Jost, R. (1989). Tryptic phosphopeptides from whole casein. *J. Dairy Res.* **56**, 335–341.

Bonjour, J.-P., Carrie, A.-L., Ferrari, S., Clavien, H., Slosman, D., Theintz, G., and Rizzoli, R. (1997). Calcium-enriched foods and bone mass growth in prepubertal girls: A randomized, double-blind, placebo-controlled trial. *J. Clin. Invest.* **99**, 1287–1294.

Braley L. M., Menachery, A., Rystedt, L., and Williams, G. H. (1989). Sodium-mediated modulation of aldosterone secretion. *Endocrinology* **125**, 1310–1314.

Bridger, W. A., and Henderson, J. F. (1983). "Cell ATP," pp. 9–19. Wiley, New York.

Brindley, G. W., Williams, E. A., Bronk, J. T., Meadows, T. H., Montgomery, R. J., Smith, S. R., and Kelly, P. J. (1988). Parathyroid hormone effects on skeleton exchangeable calcium and bone blood flow. *Am. J. Physiol.* **255**, H94–H100.

Brinkworth, R. I., Hanson, R. W., Fullin, F. A., and Schramm, V. L. (1981). $Mn^{2+}$-sensitive and insensitive forms of phosphoenolpyruvate carboxykinase (GTP). *J. Biol. Chem.* **256**, 10795–10802.

Brock, A., Chapman, S., Ulman, E., and Wu, G. (1994). Dietary manganese deficiency decreases rat hepatic arginase activity. *J. Nutr.* **124**, 340–344.

Brody, T., and Mathews, T. D. (1989). The release of zinc from leukocytes provoked by A23187 and EDTA is associated with the release of enzymes. *Comp. Biochem. Physiol. A* **94**, 693–697.

Bronner, F. (1994). Calcium and osteoporosis. *Am. J. Clin. Nutr.* **60**, 831–836.

Bunning, P., and Riordan, J. F. (1983). Activation of angiotensin converting enzyme by monovalent anions. *Biochemistry* **22**, 110–116.

Bunning, P., and Riordan, J. F. (1987). Sulfate potentiation of the chloride activation of angiotensin converting enzyme. *Biochemistry* **26**, 3374–3377.

Calvo, M. S., and Park, Y. K. (1996). Changing phosphorus content of the U.S. diet: Potential for adverse effects on bone. *J. Nutr.* **126**, 1168S–1180S.

Carlisle, E. M. (1985). Silicon as an essential trace element in animal nutrition. *In* "Silicon Biochemistry," Ciba Foundation Symposium 21, pp. 123–139. Wiley, New York.

Casey, E., and Hambidge, K. M. (1983). Nutritional aspects of human lactation. *In* "Lactation" (M. C. Neville and M. R. Neifert, eds.), p. 204. Plenum, New York.

Cera, E. D., Guinto, E., Vindigni, A., Dang, Q., Ayala, Y. L., Wuyi, M., and Tulinsky, A. (1995). The $Na^+$ binding site of thrombin. *J. Biol. Chem.* **270**, 22089–22092.

Chattopadhyay, N., Mithal, A., and Brown, E. M. (1996). The calcium-sensing receptor: A window into the physiology and pathophysiology of mineral ion metabolism. *Endocr. Rev.* **17**, 289–304.

Chowdhury, T. A., Barnett, A. H., and Bain, S. C. (1996). Pathogenesis of diabetic nephropathy. *Trends Endocrinol. Metab.* **7**, 320–323.

Christensen, G., and Leistad, E. (1997). Atrial systolic pressure, as well as stretch, is a principal stimulus for release of ANF. *Am. J. Physiol.* **272**, H820–H826.

Clark, L., Combs, G., Turnbull, B., Slate, E., Chalker, D., Chow, J., Davis, L., Glover, R., Graham, G., Gross, E., Krongrad, A., Lesher, J., Park, H., Sanders, B., Smith, C., Taylor, J.R. (1996). Effects of selenium supplementation for cancer prevention in patients with carcinoma of the skin. *JAMA* **276**, 1957–1963.

Cohen, N. L., Keen, C. L., Lonnerdal, B., and Hurley, L. S. (1985). Effects of varying dietary iron on the expression of copper deficiency in the growing rat. *J. Nutr.* **115**, 633–649.

Colditz, G. A. (1996). Selenium and cancer prevention. *JAMA* **276**, 1984–1985.

Columbo, G., and Lardy, H. (1981). Phosphoenolpyruvate carboxykinase (guanosine 5′-triphosphate) from rat liver cytosol. Divalent cation involvement in the decarboxylation reactins. *Biochemistry* **20**, 2758–2767.

Combs, G. F., and Combs, S. B. (1984). "The Role of Selenium in Nutrition." Academic Press, San Diego.

Conrad, M. E., Umbreit, J., Moore, E., and Heiman, D. (1996). Mobilferrin is an intermediate in iron transport between transferrin and hemoglobin in K562 cells. *J. Clin. Invest.* **98**, 1449–1454.

Cook, J. D. (1982). Clinical evaluation of iron deficiency. *Semin. Hematol.* **19**, 6–18.

Corkey, B. E., Duszynski, J., Rich, T. L., Matschinsky, B., and Williamson, J. R. (1986). Regulation of free and bound magnesium in rat hepatocytes and isolated mitochondria. *J. Biol. Chem.* **261**, 2567–2574.

Costill, D. L. (1977). Sweating: Its composition and effects on body fluids. *Ann. N. Y. Acad. Sci.* **301**, 160–174.

Costill, D. L., Cote, R., and Fink, W. (1976). Muscle water and electrolytes following varied levels of dehydration in man. *J. Appl. Physiol.* **40**, 6–11.

Cotton, F. A., and Wilkinson, G. (1966). "Advanced Inorganic Chemistry," 2nd ed., p. 431. Interscience, New York.

Cougnon, M., Planelles, G., Crowson, M., Shull, G., Rossier, B., and Jaisser, F. (1996) The rat distal colon P-ATPase α subunit encodes a ouabain-sensitive H+,K+-ATPase. *J. Biol. Chem.* **271**, 7277–7280.

Cowley, A. W. (1997). Genetic and nongenetic determinants of salt sensitivity and blood pressure. *Am. J. Clin. Nutr.* **65**, 587S–593S.

Croteau, W., Davey, J. C., Galton, V. A., and St. Germain, D. L. (1996). Cloning of the mammalian type II iodothyronine deiodinase. *J. Clin. Invest.* **98**, 405–417.

Dallman, P. R., Siimes, M. A., and Stekel, A. (1980). Iron deficiency in infancy and childhood. *Am. J. Clin. Nutr.* **33**, 86–118.

Davidsson, L., Cederblad, A., Lonnerdal, B., and Sandstrom, B. (1989). Manganese retention in man: A method for estimating manganese absorption in man. *Am. J. Clin. Nutr.* **49**, 170–179.

Davidsson, L., Galan, P., Cherouvrier, F., Kastenmayer, P., Juillerat, M., Hercberg, S., and Hurrell, R. F. (1997). Bioavailability in infants of iron from infant cereals: Effect of dephytinization. *Am. J. Clin. Nutr.* **65**, 916–920.

DeGrazia, J. A., and Rich, C. (1964). Studies of intestinal absorption of calcium in man. *Metabolism* **13**, 650–660.

DeGrazia, J. A., Ivanovich, P., Fellows, H., and Rich, C. (1965). A double isotope method for measurement of intestinal absorption of calcium in man. *J. Lab. Clin. Med.* **66**, 822–829.

Delmas, P. D. (1996). Bisphosphonates in the treatment of bone diseases. *N. Engl. J. Med.* **335**, 1836–1837.

Dengel, J. L., Mangels, A. R., and Moser-Veillon, P. B. (1994). Magnesium homeostasis: Conservation mechanism in lactating women consuming a controlled-magnesium diet. *Am. J. Clin. Nutr.* **59**, 990–994.

Din, W. S., and Frazier, J. M. (1985). Protective effect of metallothionein on cadmium toxicity in isolated rat hepatocytes. *Biochem. J.* **230**, 395–402.

Donowitz, M., Kokke, F. T., and Saidi, R. (1995). Evaluation of patients with chronic diarrhea. *N. Engl. J. Med.* **332**, 725–729.

Drewett, J. G., and Garbers, D. L. (1994). The family of guanylyl cyclase receptors and their ligands. *Endocr. Rev.* **15**, 135–162.

El Mernissi, G., and Doucet, A. (1983). Short-term effects of aldosterone and dexamethasone on Na,K-ATPase along the rabbit nephron. *Pfluegers Arch.* **399**, 147–151.

Epstein, S., and Miller, P. (1997). Bone mass measurements. *Trends Endocrinol. Metab.* **8**, 157–160.

Fahien, L. A., Kmiotek, E. H., Woldegiorgis, G., Evenson, M., Shrago, E., and Marshall, M. (1985). Regulation of aminotransferase-glutamate dehydrogenase interactions by carbamyl phosphate synthase. I. $Mg^{2+}$ plus leucine versus citrate and malate. *J. Biol. Chem.* **260**, 6069–6079.

Fine, B. P., Ty, A., Lestrange, N., and Levine, O. R. (1987). Sodium deprivation growth failure in the rat: Alterations in tissue composition and fluid spaces. *J. Nutr.* **117**, 1l623–1628.

Fisler, J. S., and Drenick, E. J. (1984). Calcium, magnesium, and phosphate balances during very low calorie diets of soy or collagen protein in obese man: Comparison to total fasting. *Am. J. Clin. Nutr.* **40**, 14–25.

Fomon, S. J., and Strauss, R. G. (1978). Nutrient deficiencies in breast-fed infants. *N. Engl. J. Med.* **299**, 355–356.

Freeland-Graves, J.H., and Turnlund, J. R. (1996). Deliberations and evaluations of the approaches, endpoints and paradigms for manganese and molybdenum dietary recommendations. *J. Nutr.* **126**, 2435S–2440S.

Fried, L., and Palevsky, P. (1977). Hyponatremia and hypernatremia. *Med. Clinics N. Am.* **81**, 585–609.

Frieden, E., and Hsieh, H. S. (1976). Ceruloplasmin: The copper protein with essential oxidase activity. *Adv. Enzymol.* **44**, 187–236.

Furnee, C. A., Pfann, G. A., West, C., Haar, F., Heide, D., and Hautvast, J. (1995). New model for describing urinary iodine excretion: Its use for comparing different oral preparations of iodized oil. *Am. J. Clin. Nutr.* **61**, 1257–1262.

Garner, P. (1995). Type I diabetes mellitus and pregnancy. *Lancet* **346**, 157–161.

Giorgio, A. J., and Plaut, G. W. (1967). The effect of univalent cations on activities catalyzed by bovine liver propionyl-CoA carboxylase. *Biochim. Biophys. Acta* **139**, 487–501.

Glinoer, D. (1997). The regulation of thyroid function in pregnancy: Pathways of endocrine adaptation from physiology to pathology. *Endocr. Rev.* **18**, 404–433.

Gluck, S. L., Underhill, D. M., Iyori, M., Holliday, L., Kostrominova, T., and Lee, B. S. (1996). Physiology and biochemistry of the kidney vacuolar $N^{+}$-ATPase. *Annu. Rev. Physiol.* **58**, 427–425.

Goldberger, E. (1986). "A Primer of Water, Electrolyte and Acid-Base Syndromes," 7th ed. Lea & Febiger, Philadelphia.

Golovina, V. A., and Blaustein, M. P. (1997). Spatially and functionally distinct $Ca^{2+}$ stores in sarcoplasmic and endoplasmic reticulum. *Science* **275**, 1643–1648.

Golub, J. S., Gershwin, M. E., Hurley, L. S., Saito, W. Y., and Hendrickx, A. G. (1984). Studies of marginal zinc deprivation in rhesus monkeys. IV. Growth of infants in the first year. *Am. J. Clin. Nutr.* **40**, 1192–1202.

Golub, M. S., Keen, C., Gershwin, M., Styne, D., Takeuchi, P., Ontell, F., Walter, R. M., and Hendrickx, A. G. (1996). Adolescent growth and maturation in zinc-deprived rhesus monkeys. *Am. J. Clin. Nutr.* **64**, 274–282.

Grant, S. F., and Ralston, S. H. (1997). Genes and osteoporosis. *Trends Endocrinol. Metab.* **8**, 232–236.

Greenberg, A. (1998). Hyperkalemia: Treatment options. *Semin. Nephrol.* **18**, 46–57.

Grider, A., and Young, E. M. (1996). The acrodermatitis enteropathica mutation transiently affects zinc metabolism in human fibroblasts. *J. Nutr.* **126**, 219–224.

Guyton, A. C. (1971). "Textbook of Medical Physiology," 4th ed., p. 404. Saunders, Philadelphia.

Hallberg, L., Hulten, L., and Gramatkovski, E. (1997). Iron absorption from the whole diet in men: How effective is the regulation of iron absorption? *Am. J. Clin. Nutr.* **66**, 347–356.

Halliwell, B., and Gutteridge, J. M. (1990). The antioxidants of human extracellular fluids. *Arch. Biochem. Biophys.* **280**, 1–8.

Harding, C. W., Williams, P., Wagner, E., Chang, D., Wild, K., Colwell, R., and Wolff, J. (1997). Mice with genetic $\gamma$-glutamyl transpeptidase deficiency exhibit glutathionuria, severe growth failure, reduced life spans, and infertility. *J. Biol. Chem.* **272**, 12560–12567.

Harris, E. D. (1995). The iron-copper connection: The link to ceruloplasmin grows stronger. *Nutr. Rev.* **53**, 170–173.

Hausdorff, W. P., and Catt, K. J. (1988). Activation of dihydropyridine-sensitive calcium channels and biphasic cytosolic calcium responses by angiotensin II in rat adrenal glomerulosa cells. *Endocrinology* **123**, 2818–2826.

Heaney, R. P., Recker, R. R., and Weaver, C. M. (1990). Absorbability of calcium sources: The limited role of solubility. *Calcif. Tissue Int.* **46**, 300–304.

Hegsted, M., Schuette, S. A., Zemel, M. B., and Linkswiler, H. M. (1981). Urinary calcium and calcium balance in young men as affected by level of protein and phosphorus intake. *J. Nutr.* **111**, 553–562.

Hemmings, B. A. (1997). PtdIns(3,4,5)P$_3$ gets its message across. *Science* **277**, 534.

Henderson, B., Menotti, E., and Kuhn, L. C. (1996). Iron regulatory proteins 1 and 2 bind distinct sites of RNA target sequences. *J. Biol. Chem.* **271**, 4900–4908.

Holick, M. F. (1996). Evaluation and treatment of disorders in calcium, phosphorus, and magnesium metabolism. *In* "Textbook of Primary Care Medicine" (J. Noble, ed.). Mosby, St. Louis.

Hunt, C. D., Herbel, J., and Nielsen, F. H. (1997). Metabolic responses of postmenopausal women to supplemental dietary boron and aluminum during usual and low magnesium intake: Boron, calcium, and magnesium absorption and retenton and blood mineral concentrations. *Am. J. Clin. Nutr.* **65**, 803–813.

Hurley, L. S., and Tao, S.-H. (1972). Alleviation of teratogenic effects of zinc deficiency by simultaneous lack of calcium. *Am. J. Physiol.* **222**, 322–325.

Hurrell, R. F., Lynch, S. R., Trinidad, T. P., Dassenko, S. A., and Cook, J. D. (1989). Iron absorption in humans: Bovine serum albumin with beef muscle and egg white. *Am. J. Clin. Nutr.* **47**, 102–107.

Ingelfinger, J. R., Pratt, R. E., Ellison, K., and Dzau, V. J. (1986). Sodium regulation of antiotensinogen mRNA expression in rat kidney cortex and medulla. *J. Clin. Invest.* **78**, 1311–1315.

Jenness, R. (1985). Biochemical and nutriltional aspects of milk and colostrum. *In* "Lactation" (B. L. Larson, ed.), p. 186. Iowa State Univ. Press, Ames.

Jeunemaitre, X., Inoue, I., Williams, C., Charru, A., Tichet, J., Powers, M., Sharma, A., Gimenez-Roqueplo, A., Hata, A., Corvol, P., and Lalouel, J. (1997). Haplotypes of angiotensinogen in essential hypertension. *Am. J. Hum. Genet.* **60**, 1448–1460.

John, S., Krege, J., Oliver, P., Hagaman, J., Hodgin, J., Pang, S., Flynn, T., and Smithies, O. (1995). Genetic diseases in atrial natriuretic peptide and salt-sensitive hypertension. *Science* **267**, 679–681.

Johnson, C. J., Bonrud, P., Dosch, T., Kilness, A., Senger, K., Busch, D., and Meyer, M. (1987). Fatal outcome of methemoglobinemia in an infnt. *JAMA* **257**, 2796–2797.

Johnson, W. T., and Dufault, S. N. (1989). Altered cytoskeletal organization and secretory response of thrombin-activated platelets from copper-deficient rats. *J. Nutr.* **119**, 1404–1410.

Johnston, C., Miller, J. Z., Slemenda, C., Reister, T., Hui, S., Christian, J., and Peacock, M. (1992). Calcium supplementation and increases in bone mineral density in children. *N. Engl. J. Med.* **327**, 82–87.

Juan, S.-H., Guo, J.-H., and Aust, S. D. (1997). Loading of iron into recombinant rat liver ferritin heteropolymers by ceruloplasmin. *Arch. Biochem. Biophys.* **341**, 280–286.

Kagi, J., and Schaffer, A. (1988). Biochemistry of metallothionein. *Biochemistry* **27**, 8509–8515.

Keen, C. L., Lonnerdal, B., Clegg, M., and Hurley, L. S. (1981). Developmental changes in composition of rat milk: Trace elements, minerals, protein, carbohydrate and fat. *J. Nutr.* **111**, 226–230.

Keller, T. C. (1997). Molecular bungees. *Nature* **387**, 233–235.

Kim, H.-S., Krege, J. H., Kluckman, K. D., Hagaman, J., Hodgin, J., Best, C., Jennette, J., Coffman, T., Maeda, N., and Smithies, O. (1995). Genetic control of blood pressure and the angiotensinogen locus. *Proc. Natl. Acad. Sci. U.S.A.* **92**, 2735–2739.

Kim, H.-Y., LaVaute, T., Iwai, K., Klausner, R. D., and Rouault, T. A. (1996). Identification of a conserved and functional iron-responsive element in the 5'-untranslated region of mammalian mitochondrial aconitase. *J. Biol. Chem.* **271**, 24226–24230.

Kirschbaum, B. (1998). The acidosis of exogenous phosphate intoxication. *Arch. Intern. Med.* **158**, 405–408.

Klevay, L. M., and Medeiros, D. M. (1996). Deliberations and evaluations of the approaches, endpoints and paradigms for dietary recommendations about copper. *J. Nutr.* **126**, 2419S–2426S.

Klinman, J. (1996). New quinocofactors in eukaryotes. *J. Biol. Chem.* **271**, 27189–27192.

Knochel, J. P. (1994). Disorders of phosphorus metabolism. *In* "Harrison's Principles of Internal Medicine" (K. Isselbacher, ed.), Vol. 2. McGraw-Hill, New York.

Koh, J.-Y., Suh, S., Gwag, B., He, Y., Hsu, C., and Choi, D. W. (1996). The role of zinc in selective neuronal death after transient global cerebral ischemia. *Science* **272**, 1013–1016.

Koninck, P., and Schulman, H. (1998). Sensitivity of CaM kinase II to the frequency of $Ca^{2+}$ oscillations. *Science* **279**, 227–230

Koushanpour, E., and Kirz, W. (1986). "Renal Physiology," 2nd ed., pp. 11–13. Springer-Verlag, New York.

Kwan, C.-Y., Erhard, K., and Davis, R. C. (1975). Spectral properties of Co(II)- and Ni(II)-activated rabbit muscle pyruvate kinase. *J. Biol. Chem.* **250**, 5951–5959.

Lanfear, J., Fleming, J., Wu, L., Webster, G., and Harrison, P. R. (1994). The selenium metabolite selenodiglutathione induces p53 and apoptosis: Relevance to the chemoprevention effects of selenium? *Carcinogenesis* **15**, 1387–1392.

Lazo, J., Kondo, Y., Dellapiazza, D., Michalska, A., Choo, K., and Pitt, B. R. (1995). Enhanced sensitivity to oxidative stress in cultured embryonic cells from transgenic mice different in metallothionein I and II genes. *J. Biol. Chem.* **27**, 5506–5510.

Ledoux, S., Dussaule, J.-C., Chatziantoniou, C., Ardaillou, N., Vandermeersch, S., and Ardaillou, R. (1997). Protein kinase A activity modulates natriuretic peptide-dependent cGMP accumulation in renal cells. *Am. J. Physiol.* **272**, C82–C89.

Lee, W., Hawkins, R., Peterson, D., and Vina, J. R. (1996). Role of oxoproline in the regulation of neutral amino acid transport across the blood-brain barrier. *J. Biol. Chem.* **271**, 19129–19133.

Leeuwenburgh, C., Hollander, J., Leichtweis, S., Griffiths, M., Gore, M., and Ji, L. L. (1997). Adaptations of glutathione antioxidant system to endurance training are tissue and muscle fiber specific. *Am. J. Physiol.* **272**, R363–R369.

Lehman, I. R. (1981). DNA Polymerase I of *Eschericia coli*. *In* "The Enzymes," (P. D. Boyer, ed.), Vol. 14, Part A, 3rd ed., pp. 15–37. Academic Press, San Diego.

Levander, O. A., Alfthan, G., Arvilommi, H., Gref, C. G., Huttunen, J. L., Kataja, M., Koivistoinen, P., and Pikkarainen, J. (1983). Bioavailability of selenium to Finnish men as assessed by platelet glutathione peroxidase activity and other blood parameters. *Am. J. Clin. Nutr.* **37**, 887–897.

Lewis, N. M., Marcus, M. S., Behling, A. R., and Greger, J. L. (1989). Calcium supplements and milk: Effects on acid–base balance and on retention of calcium, magnesium, and phosphorus. *Am. J. Clin. Nutr.* **49**, 527–533.

Lieberman, M. W., Wiseman, A., Shi. Z.,, Carter, B., Barrios, R., Ou, C., Chavez-Barrios, P., Wang, Y., Habib, G., Goodman, J., Huang, S., Lebovitz, R., and Matzuk, M. (1996). Growth retardation and cysteine deficiency in γ-glutamyl transpeptidase-deficient mice. *Proc. Natl. Acad. Sci. U.S.A.* **93**, 7923–7926.

Link, T. A., and Jagow, G. (1995). Zinc ions inhibit the Qp center of bovine heart mitochondrial $bc_1$ complex by blocking a protonable group. *J. Biol. Chem.* **270**, 25001–25006.

Lopez, M., Wong, S., Kishimoto, I., Dubois, S., Mach, V., Friesen, J., Garbers, D. L., and Beuvo, A. (1995). Salt-resistant hypertension in mice lacking the guanylyl cyclase-A receptor for atrial natriuretic peptide. *Nature* **378**, 65–68.

Lukashi, H. C. (1993). Soft tissue composition and bone mineral status: Evaluation by dual-energy X-ray absorptiometry. *J. Nutr.* **123**, 438–443.

Lynch, S. R., Skikne, B. S., and Cook, J. D. (1989). Food iron absorption in ideopathic hemochromatosis. *Blood* **74**, 2187–2193.

McClure, W. R., Lardy, H. A., and Kneifel, H. P. (1971). Rat liver pyruvate carboxylase. *J. Biol. Chem.* **246**, 3569–3578.

McFadden, G. (1998). Even viruses can learn to cope with stress. *Science* **279**, 40–41.

Meister, A. (1989). Metabolism and function of glutathione. *In* "Glutathione," Part A (D. Dolphin, O. Avramovic, and R. Poulson, eds.), pp. 367–474. Wiley, New York.

Menard, M., McCormick, C. C., and Cousins, R. J. (1981). Regulation of intestinal metallothionein biosynthesis in rats by dietary zinc. *J. Nutr.* **111**, 1353–1361.

Michalkiewicz, M., Huffman, L. J., Connors, J. M., and Hedge, G. A. (1989). Alterations in thyroid blood flow induced by varying levels of iodide intake in the rat. *Endocrinology* **125**, 54–60.

Mickelsen, O., Makdani, D., Gill, J. L., and Frank, R. L. (1977). Sodium and potassium intakes and excretions of normal men consuming sodium chloride or a 1:1 mixture of sodium and potassium chlorides. *Am. J. Clin. Nutr.* **30**, 2033–2040.

Miller, E. R., Ullrey, D. E., Zutaut, C. L., Baltzer, B. V., Schmidt, D. A. Hoefer, J. A., and Luecke, R. W. (1962). Calcium requirement of the baby pig. *J. Nutr.* **77**, 7–17.

Miller, E. R., Ullrey, D. E., Zutaut, C. L., Hoefer, J. A., and Luecke, R. W. (1964). Mineral balance studies with the baby pig: Effects of dietary phosphorus level upon calcium and phosphorus balance. *J. Nutr.* **82**, 111–114.

Miller, J. Z., Weinberger, M. H., and Christian, J. C. (1987). Blood pressure response to potassium supplementation in normotensive adults and children. *Hypertension* **10**, 437–442.

Milne, D. B., Canfield, W. K., Gallagher, S. K., Hunt, J. R., and Klevay, L. M. (1987). Ethanol metabolism in postmenopausal women fed a diet marginal in zinc. *Am. J. Clin. Nutr.* **46**, 688–693.

Mizrachi, A., London, R. D., and Gribetz, D. (1968). Neonatal hypocalcemia — Its causes and treatment. *N. Engl. J. Med.* **21**, 1163–1165.

Mukhopadyay, C., Attieh, Z., and Fox, P. L. (1998). Role of ceruloplasmin in cellular iron uptake. *Science* **279**, 714–717.

Narayan, V., Kriwacki, R., and Caradonna, J. P. (1997). Structures of zinc finger domains from transcription factor Sp1. *J. Biol. Chem.* **272**, 7801–7809.

Nelsestuen, G. L., Broderius, M., and Martin, G. (1976). Role of γ-carboxyglutamic acid. *J. Biol. Chem.* **251**, 6886–6893.

NIH Consensus Development Panel on Optimal Calcium Intake (1994). Optimal calcium intake. *JAMA* **272**, 1942–1948.

Nordin, B. E., Polley, K. J., Need, A. G., Morris, H. A., and Marshall, D. (1987). The problem of calcium requirement. *Am. J. Clin. Nutr.* **45**, 1295–1304.

Ornt, D. B., Scandling, J. D., and Tannen, R. L. (1987). Adaptation for potassium conservation during dietary potassium deprivation. *Semin. Nephrol.* **7**, 193–205.

O'Sullivan, W., and Smithers, G. W. (1979). Stability constants for biologically important metal–ligand complexes. *In* "Methods in Enzymology" (D. L. Lynch, ed.), Vol. 63, pp. 294–336. Academic Press, San Diego.

Paetkau, V., and Lardy, H. A. (1967). Phosphofructokinase. *J. Biol. Chem.* **242**, 2035–2042.

Pak, C. Y. (1990). Calcium disorders: Hypercalcemia and hypocalcemia. *In* "Fluids and Electrolytes" (J. P. Kokko and R. L. Tannen, eds.), 2nd ed., pp. 596–630. Saunders, Philadelphia.

Pallud, S., Lennon, A.-M., Ramauge, M., Gaveret, J.-M., Croteau, W., Pierre, M., Courtin, F., and St. Germain, D. L. (1997). Expression of the type II iodothyronine deiodinase in cultured rat astrocytes is selenium-dependent. *J. Biol. Chem.* **272**, 18104–18110.

Paul, A. A., and Southgate, D. A. T. (1978). "The Composition of Foods." Elsevier/North-Holland, Amsterdam.

Pearce, S., and Thakker, R. V. (1997). The calcium-sensing receptor: Insights into extracellular calcium homeostasis in health and disease. *J. Endocrinol.* **154**, 371–378.

Pearce, S., Williamson, C., Kifor, O., Bai, M., Coulthard, M., Davies, M., Lewis-Barned, N., McCredie, D., Powell, H., Kendall-Taylor, P., Brown, E. M., and Thakker, R. V. (1996). A familial syndrome of hypocalcemia with hypercalcemia due to mutations in the calcium-sensing receptor. *N. Engl. J. Med.* **335**, 1115–1122.

Pearson, R. G. (1966). Acids and bases. *Science* **151**, 172–177.

Pennington, J., Schoen, S., Salmon, G., Young, B., Johnson, R., and Marts, R. (1995a). Composition of core foods of the U.S. food supply, I. Sodium, phosphorus, and potassium. 1982–1991. *J. Food Comp. Anal.* **8**, 91–128.

Pennington, J., Schoen, S., Salmon, G., Young, B., Johnson, R., and Marts, R. (1995b). Composition of core foods of the U.S. food supply, 1982–1991, II. Calcium, magnesium, iron, and zinc. *J. Food Comp. Anal.* **8**, 129–169.

Pennington, J., Schoen, S., Salmon, G., Young, B., Johnson, R., and Marts, R. (1995c). Composition of core foods of the U.S. food supply, III. Copper, manganese, selenium, and iodine. 1982–1991. *J. Food Comp. Anal.* **8**, 171–217.

Pinto, M. R., Gorski, J. P., Penniston, J. T., and Kelly, P. J. (1988). Age-related changes in composition and $Ca^{2+}$-binding capacity of canine cortical bone extracts. *Am. J. Physiol.* **255**, H101–H110.

Prasad, A. S. (1991). Discovery of zinc deficiency and studies in an experimental human model. *Am. J. Clin. Nutr.* **53**, 403–412.

Prentice, A. (1997). Calcium supplementation during breast-feeding. *N. Engl. J. Med.* **337**, 558–559.

Prior, J., Vigna, Y., Schechter, M., and Burgess, A. E. (1990). Spinal bone loss and ovulatory disturbances. *N. Engl. J. Med.* **323**, 1221–1227.

Putney, J. W. (1998). Calcium signaling: Up, down, up, down ... What's the point? *Science* **279**, 191–192.

Rader, J. L., Wolnik, K. A., Gaston, C. M., Celesk, E. M., Peeler, J. T., Spivey Fox, M. R., and Fricke, F. L. (1984). Trace element studies in weanling rats: Maternal diets and baseline tissue values. *J. Nutr.* **114**, 1946–1954.

Randerath, E., Miller, R. H., Mittal, D., Aritts, T. A., Dunsford, H. A., and Randerath, K. (1989). Covalent DNA damage in tissues of cigarette smokers as determined by $^{32}$P-postlabeling assay. *J. Natl. Cancer Inst.* **81**, 341–347.

Rayman, M. P. (1997). Dietary selenium: Time to act. *Brit. Med. J.* **314**, 387–388.

Rayment, I. (1996). The structural basis of the myosin ATPase activity. *J. Biol. Chem.* **271**, 15850–15853.

Recker, R. R. (1985). Calcium absorption and achlorohydria. *N. Engl. J. Med.* **313**, 70–73.

Reddy, B., Rivenson, A., El-Bayoumy, K., Upadhyaya, P., Pittman, B., and Rao, C. V. (1997). Chemoprevention of colon cancer by organoselenium compounds and impact of high- or low-fat diets. *J. Nat. Cancer Inst.* **89**, 506–512.

Reeves, P. G. (1997). Components of the AIN-93 diets as improvements in the AIN-76A diet. *J. Nutr.* **127**, 838S–841S.

Reis, A. M., Jankowski, M., Mukaddam-Daher, S., Tremblay, J., Dam, T., and Gutkowska, J. (1997). Regulation of the natriuretic peptide system in rat uterus during the estrous cycle. *J. Endocrinol.* **153**, 345–355.

Rogacz, S., Williams, G., and Hollenberg, N. K. (1990). Time course of enhanced adrenal responsiveness to angiotensin on a low-salt diet. *Hypertension* **15**, 376–380.

Sahni, M., Guenther, H. L., Fleisch, H., Collin, P., and Martin, T. J. (1993). Bisphosphonates act on bone resorption through the mediation of osteoblasts. *J. Clin. Invest.* **91**, 2004–2011.

Sakakibara, Y., Katafuchi, J., Takami, Y., Nakayama, T., Suiko, M., Nakajima, H., and Liu, M.-C. (1997). Manganese-depedent DOPA/tyrosine sulfation in HepG2 human hepatoma cells. *Biochim. Biophys. Acts* **1335**, 102–106.

Salvatore, D., Tu, H., Harney, J. W., and Larsen, P. R. (1996). Type 2 iodothyronine deiodinase is highly expressed in human thyroid. *J. Clin. Invest.* **98**, 962–968.

Samson, S., Paramchuk, W., Shworak, N., and Gedamu, L. (1995). Functional analysis of the human metallotionein-IG gene. *J. Biol. Chem.* **270**, 25194–25199.

Sandstead, H. H., and Smith, J. C. (1996). Deliberations and evaluations of approaches, endpoints and paradigms for determining zinc dietary recommendations. *J. Nutr.* **126**, 2410S–2418S.

Sandstrom, B., Arvidsson, B., Cederblad, A., and Bjorn-Rasmussen, E. (1980). Zinc absorption from composite meals. *Am. J. Clin. Nutr.* **33**, 739–745.

Sarau, H. M., Foley, J. J., Moonsammy, G., and Sachs, G. (1977). Metabolism of dog gastric mucosa. *J. Biol. Chem.* **252**, 8572–8581.

Schnapp, B. J. (1995). Two heads are better than one. *Nature* **373**, 655–656.

Schuette, S. A., and Linkswiler, H. M. (1982). Effects on Ca and P metabolism in humans by adding meat, meat plus milk, or purified proteins plus Ca and P to a low protein diet. *Am. J. Clin. Nutr.* **112**, 338–349.

Schuette, S. A., Knowles, J. B., and Ford, H. E. (1989). Effect of lactose or its component sugars on jejunal calcium absorption in adult man. *Am. J. Clin. Nutr.* **50**, 1084–1087.

Scrutton, M. C., Griminger, P., and Wallace, J. C. (1972). Pyruvate carboxylase. *J. Biol. Chem.* **247**, 3305–3313.

Senior, A. E., Richardson, L. V., Baker, K., and Wise, J. G. (1980). Tight divalent cation-binding sites of soluble adenosine triphosphatase (F1) from beef heart mitochondria and *Escherichia coli*. *J. Biol. Chem.* **255**, 7211–7217.

Shah, A., Chernov, I., Zhang, H., Ross, B., Das, K., Lutsenko, S., Parano, E., Pavone, L., Evgrafo, O., Ivanova-Smolenskaya, I., Anneren, G., Westmark, K., Urrutia, F., Penchaszadeh, G., Sternlieb, I., Scheinberg, I., Gilliam, T., and Perukhin, K. (1997). Identification and analysis in the Wilson disease gene (ATP7B). *Am. J. Hum. Genet.* **61**, 317–328.

Sheikh, M. S., and Fordtran, J. S. (1990). Calcium bioavailability from two calcium carbonate preparations. *N. Engl. J. Med.* **323**, 921.

Sheikh, M. S., Santa Ana, C., Nicar, M. J., Schiller, L. R., and Fordtran, J. S. (1987). Gastrointestinal absorption of calcium from milk and calcium salts. *N. Engl. J. Med.* **317**, 532–536.

Shils, M. E. (1969). Experimental human magnesium depletion. *Medicine* **48**, 61–85.

Shisler, J. L., Senkevich, T. G., Berry, M., and Moss, B. (1998). Ultraviolet-induced cell death blocked by a selenoprotein from a human dermatotropic poxvirus. *Science* **279**, 102–105.

Sibai, B. M. (1996). Treatment of hypertension in pregnant women. *N. Engl. J. Med.* **335**, 257–265.

Sinaiko, A. R. (1996). Hypertension in children. *N. Engl. J. Med.* **335**, 1968–1973.

Smith, E. L., and Gilligan, C. (1990). Calcium and exercise in prevention of bone loss with age. *Clin. Nutr.* **9**, 17–20.

Smith, M. (1987). Position of the American Dietetic Association: Nutrition for physical fitness and athletic performance for adults. *J. Am. Diet. Assoc.* **87**, 933–939.

Spencer, H., Kramer, L., Osis, D., and Norris, C. (1978). Effect of phosphorus on the absorption of calcium and on the calcium balance in man. *J. Nutr.* **108**, 447–457.

Spencer, H., Kramer, L., and Osis, D. (1984). Effect of calcium on phosphorus metabolism in man. *Am. J. Clin. Nutr.* **40**, 219–225.

Squires, R. D., and Huth, E. J. (1959). Experimental potassium depletion in normal human subjects. *J. Clin Invest.* **38**, 1134–1148.

Stanton, B. A. (1987). Regulation of Na and K transport by mineralcorticoids. *Semin. Nephrol.* **7**, 82–90.

Stastny, D., Vogel, R. S., and Picciano, M. F. (1984). Manganese intake and serum manganese concentration of human milk-fed and formula-fed infants. *Am. J. Clin. Nutr.* **39**, 872–878.

St. Germain, D. L. (1994). Iodothyronine deiodinases. *Trends Endocrinol. Metab.* **5**, 36–42.

Stokoe, D., Stephens, L., Copeland, T., Gaffney, P., Reese, C. B., Painter, G., Holmes, A. B., McCormick, F., and Hawkins, P. T. (1997). Dual role of phosphatidyl-3,4,5-trisphosphate in the activation of protein kinase B. *Science* **277**, 567–570.

Strause, L. G., Hegenauer, J., Saltman, P., Cone, R., and Resnick, D. (1986). Effects of long-term dietary manganese and copper deficiency on rat skeleton. *J. Nutr.* **116**, 135–141.

Sturchler, C., Westhof, E., Carbon, P., and Krol, A. (1993). Unique secondary and tertiary structural features of the eucaryotic selenocysteine tRNA. *Nucl. Acids Res.* **21**, 1073–1079.

Suelter, C. H. (1970). Enzymes activated by monovalent cations. *Science* **168**, 789–795.

Sullivan, V. K., and Cousins, R. J. (1997). Competitive reverse transcriptase-polymerase chain reaction shows that dietary zinc supplementation in humans increases monocyte metallothionein mRNA levels. *J. Nutr.* **127**, 694–698.

Swenerton, H., and Hurley, L. S. (1968). Severe zinc deficiency in male and female rats. *J. Nutr.* **95**, 8–18.

Thampy, K. G., and Wakil, S. J. (1985). Activation of acetyl-CoA carboxylase. *J. Biol. Chem.* **260**, 6318–6323.

Thornalley, P. J. (1990). The glyoxalase system: New developments towards functional characterization of a metabolic pathway fundamental to biological life. *Biochem. J.* **269**, 1–11.

Torgerson, D. J., Gosden, T., and Reid, D. M. (1997). The economics of osteoporosis prevention. *Trends Endocrinol. Metab.* **8**, 236–239.

Toth, I., and Bridges, K. R. (1995). Ascorbic acid enhances ferritin mRNA translation by an IRP/aconitase switch. *J. Biol. Chem.* **270**, 19540–19544.

Treasure, J., and Ploth, D. (1983). Role of dietary potassium in the treatment of hypertension. *Hypertension* **5**, 864–872.

Tsien, R. Y. (1983). Intracellular measurements of ion activities. *Annu. Rev. Biophys. Bioeng.* **12**, 91–116.

Tsuruoka, S., and Schwarz, G. J. (1997). Metabolic acidosis stimulates $H^+$ secretion in the rabbit outer medullary collecting duct (inner stripe) of the kidney. *J. Clin. Invest.* **99**, 1420–1431.

Turnlund, J. R., Swanson, C. A., and King, J. C. (1983). Copper absorption and retention in pregnant women fed diets based on animal and plant proteins. *J. Nutr.* **113**, 2346–2352.

Turnlund, J. R., Keyes, W. R., Anderson, H. L., and Acord, L. L. (1989). Copper absorption and retention in young men at three levels of dietary copper by use of the stable isotope $^{65}$Cu. *J. Nutr.* **49**, 870–878.

Turnlund, J. R., Keyes, W. R., and Peiffer, G. L. (1995). Molybdenum absorption, excretion, and retention studied with stable isotopes in young men at five intakes of dietary molybdenum. *Am. J. Clin. Nutr.* **62**, 790–796.

Turnlund, J., Scott, K., Peiffer, G., Jang, A., Keyes, W., Keen, C., and Sakanashi, T. M. (1997). Copper status of young men consuming a low-copper diet. *Am. J. Clin. Nutr.* **65**, 72–78.

Tyrala, E., Borschel, M. W., and Jacobs, J. R. (1996). Selenate forification of infant formulas improves the selenium status of preterm infants. *Am. J. Clin. Nutr.* **64**, 860–865.

Valentine, J. S., and Gralla, E. B. (1997). Delivering copper inside yeast and human cells. *Science* **278**, 817–819.

Wada, L., Turnlund, J. R., and King, J. C. (1985). Zinc utilization in young men fed adequate and low zinc intakes. *J. Nutr.* **115**, 1345–1354.

Wang, X., and Robinson, P. J. (1997). Cyclic GMP-dependent protein kinase and cellular signaling in the nervous system. *J. Neurochem.* **68**, 443–456.

Waxman, D. J. (1990). Glutathione *S*-transferases: Role in alkalating agent resistance and possible target for modulation chemotherapy — a review. *Cancer Res.* **50**, 6449–6454.

Webster, L. T. (1966). Studies on the acetyl-CoA synthetase reaction. *J. Biol. Chem.* **241**, 5504–5510.

Wefers, H., and Sies, H. (1983). Oxidation of glutathione by the superoxide radical to the disulfide and the sulfonate yielding singlet oxygen. *Eur. J. Biochem.* **137**, 29–36.

Wefers, H., and Sies, H. (1988). The protection by ascorbate and glutathione against microsomal lipid peroxidation is dependent on vitamin E. *Eur. J. Biochem.* **174**, 353–357.

Westheimer, F. H. (1987). Why nature chose phosphates. *Science* **235**, 1173–1178.

Whiting, S. J., and Draper, H. H. (1981). Effect of chronic acid load as sulfate or sulfur amino acids on bone metabolism in adult rats. *J. Nutr.* **111**, 1721–1726.

Wingo, I., and Wingo, C. (1997). Hypokalemia: Consequences, causes, and correction. *J. Am. Soc. Nephrol.* **8**, 1179–1188.

Winters, K. M., Adams, W. C., Meredith, C. N., Loan, M., and Lasley, B. L. (1996). Bone density and cyclic ovarian function in trained runners and active controls. *Med. Sci. Sports Exerc.* **28**, 776–785.

Witteman, J. C., Grobbee, D. E., Derkx, F., Bouillon, R., Bruijn, A., and Hofman, A. (1994). Reduction of blood pressure with oral magnesium supplementation in women with mild to moderate hypertension. *Am. J. Clin. Nutr.* **60**, 129–135.

Wood, R. J., Suter, P. M., and Russell, R. M. (1995). Mineral requirements of elderly people. *Am. J. Clin. Nutr.* **62**, 493–505.

Zeman, F. J., and Ney, D. M. (1988). "Applications of Clinical Nutrition," p. 200. Prentice-Hall, Englewood Cliffs, NJ.

Zemel, M. B., and Linkswiler, H. M. (1981). Calcium metabolism in the young adult male as affected by level and form of phosphorus intake and level of calcium intake. *J. Nutr.* **111**, 315–324.

Zhu, X.-G., McPhie, P., Lin, K.-H., and Cheng, S.-Y. (1997). The differential hormone-dependent transcriptional activation of thyroid hormone receptor isoforms is mediated by interplay of their domains. *J. Biol. Chem.* **272**, 9048–9054.

## BIBLIOGRAPHY

*Food Content of Inorganic Nutrients*

Pennington, J., Schoen, S., Salmon, G., Young, B., Johnson, R., and Marts, R. (1995a). Composition of core foods of the U.S. food supply, I. Sodium, phosphorus, and potassium. 1982–1991. *J. Food Comp. Anal.* **8**, 91–128.

Pennington, J., Schoen, S., Salmon, G., Young, B., Johnson, R., and Marts, R. (1995b). Composition of core foods of the U.S. food supply, 1982–1991, II. Calcium, magnesium, iron, and zinc. *J. Food Comp. Anal.* **8**, 129–169.

Pennington, J., Schoen, S., Salmon, G., Young, B., Johnson, R., and Marts, R. (1995c). Composition of core foods of the U.S. food supply, III. Copper, manganese, selenium, and iodine. 1982–1991. *J. Food Comp. Anal.* **8**, 171–217.

Paul, A. A., and Southgate, D. A. T. (1978). "The Composition of Foods." Elsevier/North-Holland, Amsterdam.

*Review Articles on Sodium and Potassium*

Berry, C. A., and Rector, F. C. (1989). Electroneutral NaCl absorption in the proximal tubule: Mechanisms of apical Na-coupled transport. *Kidney Int.* **36**, 403–411.

Boron, W. F., and Boulpaep, E. L. (1989). The electrogenic $Na/HCO_3$ cotransporter. *Kidney Int.* **36**, 392–402.

Breyer, J., and Jacobson, H. R. (1990). Molecular mechanisms of diuretic agents. *Annu. Rev. Med.* **41**, 265–275.

Campbell, D. J. (1987). Circulating and tissue angiotensin systems. *J. Clin. Invest.* **79**, 1–6.

Cogan, M. G. (1990). Angiotensin II: A powerful controller of sodium transport in the early proximal tubule. *Hypertension* **15**, 451–458.

Cowley, A. W. (1997). Genetic and nongenetic determinants of salt sensitivity and blood pressure. *Am. J. Clin. Nutr.* **65**, 587S–593S.

Dahl, L. K. (1958). Salt intake and salt need. *N. Engl. J. Med.* **258**, 1152–1157.

Ehlers, R. W., and Riordan, J. F. (1989). Angiotensin-converting enzyme: New concepts concerning its biological role. *Biochemistry* **28**, 5311–5318.

Ely, D. L. (1997). Overview of dietary sodium effects on and interactions with cardiovascular and neuroendocrine functions. *Am. J. Clin. Nutr.* **65**, 594S–605S.

Food and Nutrition Board (1989). "Recommended Dietary Allowances," 10th ed., p. 66. National Academy Press, Washington, DC.

Field, M., Rao, M. C., and Chang, E. B. (1989). Intestinal electrolyte transport and diarrheal disease. *N. Engl. J. Med.* **321**, 800–806.

Fuller, P., and Lim-Tio, S. (1996). Aldosterone action, sodium channels and inherited disease. *J. Endocrinol.* **148**, 387–390.

Frazier, H. S. (1968). Regulation of sodium balance. *N. Engl. J. Med.* **279**, 868–975.

Garty, H. (1986). Mechanisms of aldosterone action in tight epithelia. *J. Membr. Biol.* **90**, 193–205.

Gluck, S. L., Underhill, D. M., Iyori, M., Holliday, L., Kostrominova, T., and Lee, B. S. (1996). Physiology and biochemistry of the kidney vacuolar $N^+$-ATPase. *Annu. Rev. Physiol.* **58**, 427–425.

Goldberger, E. (1986). "A Primer of Water, Electrolyte and Acid-Base Syndromes," 7th ed. Lea & Febiger, Philadelphia.

Guthrie, G. M. (1988). Six to eighteen — the perilous months. Nutr. Today 23, 4-11.

Hayslett, J. P., and Binder, H. J. (1982). Mechanisms of potassium adaptation. *Am. J. Physiol.* **12**, F103–F112.

Hirschhorn, N. (1980). The treatment of acute diarrhea in children: An historical and physiological perspective. *Am. J. Clin. Nutr.* **33**, 637–663.

Jacobson, H. R. (1981). Functional segmentation of the mammalian nephron. *Am. J. Physiol.* **241**, F203–F218.

Kimura, S., Yokomukai, Y., and Komai, M. (1987). Salt consumption and nutritional state especially dietary protein level. *Am. J. Clin. Nutr.* **45**, 1271–1276.

Kleyman, T. R. (1988). Amiloride and its analogues as tools in the study of ion transport. *J. Membr. Biol.* **105**, 1–21.

Koushanpour, E., and Kirz, W. (1986). "Renal Physiology," 2nd ed., pp. 11–13. Springer-Verlag, New York.

Levens, N. R. (1985). Control of intestinal absorption by the renin–angiotensin system. *Am. J. Physiol.* **249**, G3–G15.

Morel, F., and Doucet, A. (1986). Hormonal control of kidney functions at the cell level. *Physiol. Rev.* **66**, 377–468.

Nichols, B. L., and Soriano, H. A. (1977). A critique of oral therapy of dehydration due to diarrheal syndromes. *Am. J. Clin. Nutr.* **30**, 1457–1472.

Ornt, D. B., Scandling, J. D., and Tannen, R. L. (1987). Adaptation for potassium conservation during dietary potassium deprivation. *Semin. Nephrol.* **7**, 193–205.

Schwartz, W. B., Strihou, C., and Kassirer, J. P. (1968). Role of anions in metabolic alkalosis and potassium deficiency. *N. Engl. J. Med.* **279**, 630–639.

Shapiro, A. P., and Jacob, R. G. (1983). Nonpharmacologic approaches to the treatment of hypertension. *Annu. Rev. Public Health* **4**, 285–310.

Stanton, B. A. (1987). Regulation of Na and K transport by mineralcorticoids. *Semin. Nephrol.* **7**, 82–90.

Taylor, C. E., and Greenough, W. B. (1989). Control of diarrheal diseases. *Annu. Rev. Public Health* **10**, 221–244.

Quinn, S. J., and Williams, G. H. (1988). Regulation of aldosterone secretion. *Annu. Rev. Physiol.* **50**, 409–426.

Velazquez, H., and Giebisch, G. (1988). Effect of diuretics on specific transport systems. *Semin. Nephrol.* **8**, 295–304.

Wright, F. S. (1987). Renal potassium handling. *Semin. Nephrol.* **7**, 174–184.

Young, D. B. (1988). Quantitative analysis of aldosterone's role in potassium regulation. *Am. J. Physiol.* **255**, F811–F822.

*Research Articles on Sodium, Potassium, Angiotensin, and Vasopressin*

Aoki, Y., Albrecht, F., Bergman, K. R., and Jose, P. (1996). Stimulation of $Na^+$–$K^+$–$2Cl^-$ cotransport in rat medullary thick ascending limb by dopamine. *Am. J. Physiol.* **271**, R1561–R1567.

Aperia, A., Holtback, U., Syren, M.-L., Svensson, L.-B., Fryckstedt, J., and Greengard, P. (1994). Activation/deactivation of renal $Na^+$,$K^+$-ATPase: A final common pathway for regulation of natriuresis. *FASEB J.* **8**, 436–439.

Barlet-Bas, C., Khadouri, C., Marsy, S., and Doucet, A. (1990). Enhanced intracellular sodium concentration in kidney cells recruits a latent pool of Na-K-ATPase whose size is modulated by corticosteroids. *J. Biol. Chem.* **265**, 7799–7803.

Belusa, R., Wang, Z.-W., Matsubara, T., Sahlgren, B., Dulubova, I., Nairn, A. C., Ruoslahti, E., Greengard, P., and Aperia, A. (1997). Mutation of the protein kinase C phosphorylation site on rat $\alpha$1 $Na^+$,$K^+$-ATPase alters regulation of intracellular $Na^+$ and pH and influences cell shape and adhesiveness. *J. Biol. Chem.* **272**, 20179–20184.

Benos, D. J., Saccomani, G., and Sariban-Sohraby, S. (1987). The epithelial sodium channel. *J. Biol. Chem.* **262**, 10613–10618.

Blot-Chabaud, M., Wanstok, F., Bonvalet, J.-P., and Farman, N. (1990). Cell sodium induced recruitment of $Na^+$–$K^+$-ATPase pumps in rabbit cortical collecting tubules is aldosterone-dependent. *J. Biol. Chem.* **265**, 11676–11681.

Campbell, D. J., and Habener, J. F. (1986). Angiotensinogen gene is expressed and differentially regulated in multiple tissues of the rat. *J. Clin. Invest.* **78**, 31–39.

Chen, X., Li, W., Yoshida, H., Tsuchida, S., Nishimura, H., Takemoto, F., Okubo, S, Fogo, A., Matsusaka, T., and Ichikawa, I. (1997). Targeting deletion of antiotensin type 1B receptor gene in the mouse. *Am. J. Physiol.* **272**, F299–F304.

Ciampolillo, F., McCoy, D., Green, R. B., Karlson, K. H., Dagenais, A., Molday, R. S., and Stanton, B. A. (1996). Cell-specific expression of amiloride-sensitive Na$^+$-conducting ion channels in the kidney. *Am. J. Physiol.* **271**, C1303–C1315.

Clark, A. F., Sharp, M., Morley, S. D., Fleming, S., Peters, J., and Mullins, J. J. (1997). Renin-1 is essential for normal renal juxtaglomerluar cell granulation and macula densa morphology. *J. Biol. Chem.* **272**, 18185–18190.

Connor, J. A., Cornwall, M. C., and Williams, G. H. (1987). Spatially resolved cytosolic calcium response to angiotensin II and potassium in rat glomerulosa cells measured by digital imaging techniques. *J. Biol. Chem.* **262**, 2919–2927.

Cougnon, M., Planelles, G., Crowson, M., Shull, G., Rossier, B., and Jaisser, F. (1996) The rat distal colon P-ATPase α subunit encodes a ouabain-sensitive H$^+$,K$^+$-ATPase. *J. Biol. Chem.* **271**, 7277–7280.

Dai, L.-J., Raymond, L., Friedman, P., and Quamme, G. A. (1997). Mechanisms of amiloride stimulation of Mg$^{2+}$ uptake in immortalized mouse distal convoluted tubule cells. *Am. J. Physiol.* **272**, F249–F256.

Dawson-Hughes, B., Fowler, S. E., Dalsky, G., and Gallagher, C. (1996). Sodium excretion influences calcium homeostasis in elderly men and women. *J. Nutr.* **126**, 2107–2112.

Dluhy, R. G., Greenfield, M., and Williams, G. H. (1977). Effect of simultaneous potassium and saline loading on plasma aldosterone levels. *J. Clin. Endocrinol. Metab.* **45**, 141–146.

Ellison, D. H., Velazquez, H., and Wright, F. S. (1989). Adaptation of the distal convoluted tubule of the rat. *J. Clin. Invest.* **83**, 113–126.

Fisch, C. (1973). Relation of electrolyte disturbances to cardiac arrhythmias. *Circulation* **47**, 408–419.

Fujii, Y., Mujais, S. K., and Katz, A. I. (1989). Renal potassium adaptation: Role of the Na$^+$–K$^+$ pump in rat cortical collecting tubules. *Am. J. Physiol.* **256**, F279–F284.

Garg, L. C., and Narang, N. (1985). Renal adaptation to potassium in the adrenalectomized rabbit. *J. Clin. Invest.* **76**, 1065–1070.

Hein, L., Barsh, G. S., Pratt, R. E., Dzau, V., and Koblika, B. K. (1995). Behavioral and cardiovascular effects of disrupting the angiotensin II type-2 receptor gene in mice. *Nature* **377**, 744–750.

Higashimori, K., Mizuno, K., Nakajo, S., Boehm, F. H., Marcotte, P. A., Egan, D. A., Holleman, W. H., Heusser, C., Poisner, A. M., and Inagami, T. (1989). Pure human inactive renin. *J. Biol. Chem.* **264**, 14662–14667.

Ingelfinger, J. R., Pratt, R. E., Ellison, K., and Dzau, V. J. (1986). Sodium regulation of antiotensinogen mRNA expression in rat kidney cortex and medulla. *J. Clin. Invest.* **78**, 1311–1315.

Kim, W.-S., Hatsuzawa, K., Ishizuka, Y., Hashiba, K., Murakami, K., and Nakayama, K. (1990). A processing enzyme for prorenin in mouse submandibular gland. *J. Biol. Chem.* **265**, 5930–5933.

Kim, H.-S., Krege, J. H., Kluckman, K. D., Hagaman, J., Hodgin, J., Best, C., Jennette, J., Coffman, T., Maeda, N., and Smithies, O. (1995). Genetic control of blood pressure and the angiotensinogen locus. *Proc. Natl. Acad. Sci. U.S.A.* **92**, 2735–2739.

Kojima, I., Kojima, K., Kreutter, D., and Rasmussen, H. (1984). The temporal integration of the aldosterone secretory response to angiotensin occurs via two intracellular pathways. *J. Biol. Chem.* **259**, 14448–14457.

Lang, U., and Vallotton, M. B. (1987). Angiotensin II but not potassium induces subcellular redistribution of protein kinase C in bovine adrenal glomerulosa cells. *J. Biol. Chem.* **262**, 8047–8050.

Liu, F.-Y., and Cogan, M. G. (1988). Angiotensin II stimulation of hydrogen ion secretion in the rat early proximal tubule. *J. Clin. Invest.* **82**, 601–607.

Niimura, F., Okubo, S., Fogo, A., and Ichikawa, I. (1997). Temporal and spatial expression pattern of the angiotensinogen gene in mice and rats. *Am. J. Physiol.* **272**, R142–R147.

Northrup, T. E., Garella, S., Perticucci, E., and Cohen, J. J. (1988). Acidemia alone does not stimulate rat renal Na⁺–H⁺ antiporter activity. *Am. J. Physiol.* **259**, F237–F243.

Oliverio, M. I., Best, C. F., Hung-Suk, K., Arendshorst, W., Smithies, O., and Coffman, T. M. (1997). Angiotensin II responses in AT1A receptor-deficient mice: A role for AT1B receptors in blood pressure regulation. *Am. J. Physiol.* **272**, F515–F520.

Pratt, J. H. (1982). Role of angiotensin II in potassium-mediated stimulation of aldosterone secretion in the dog. *J. Clin. Invest.* **70**, 667–672.

Rechkemmer, G., and Halm, D. R. (1989). Aldosterone stimulates K secretion across mammalian colon independent of Na absorption. *Proc. Natl. Acad. Sci. U.S.A.* **86**, 397–401.

Satlin, L. M., and Palmer, L. G. (1997). Apical K⁺-conductance in maturing rabbit principal cell. *Am. J. Physiol.* **272**, F397–F404.

Stanton, B. A., Pan, L., Deetjen, H., Guckian, V., and Giebisch, G. (1987). Independent effects of aldosterone and potassium on induction of potassium adaptation in rat kidney. *J. Clin. Invest.* **79**, 198–206.

Thompson, C. B., and McDonough, A. A. (1996). Skeletal muscle Na,K-ATPase α and β subunit protein levels respond to hypokalemic challenge with isoform and muscle type specificity. *J. Biol. Chem.* **271**, 32653–32658.

Tsuruoka, S., and Schwarz, G. J. (1997). Metabolic acidosis stimulates H⁺ secretion in the rabbit outer medullary collecting duct (inner stripe) of the kidney. *J. Clin. Invest.* **99**, 1420–1431.

Velazquez, H., Good, D. W., and Wright, F. S. (1984). Mutual dependence of sodium and chloride absorption by renal distal tubule. *Am. J. Physiol.* **247**, F904–F911.

Winkler, C., Kittelberger, A., and Schwartz, G. J. (1997). Expression of carbonic anhydrase IV mRNA in rabbit kidney: Stimulation by metabolic acidosis. *Am. J. Physiol.* **272**, F551–F560.

Wright, F. S., Striedner, N., Fowler, N. B., and Giebisch, G. (1971). Potassium secretion by distal tubule after potassium adaptation. *Am. J. Physiol.* **221**, 437–448.

Yanagibashi, K., Haniu, M., Shively, J. E., Shen, W. H., and Hall, P. (1986). The synthesis of aldosterone by the adrenal cortex. *J. Biol. Chem.* **261**, 3556–3562.

Young, D. B., Jackson, T. E., Tipayamontri, U., and Scott, R. C. (1984). Effects of sodium intake on steady-state potassium excretion. *Am. J. Physiol.* **246**, F772–F778.

*Salt and Water Losses with Exercise and Illness*

Alcantra, P. F., Hanson, L. E., and Smith, J. D. (1980). Sodium requirements, balance, and tissue composition of growing pigs. *J. Anim. Sci.* **50**, 1092–1101.

Armstrong, L. E., Hubbard, R. W., Szlyk, P. C., Matthew, W. T., and Sils, I. V. (1985). Voluntary dehydration and electrolyte losses during prolonged exercise in the heart. *Aviat. Space Environ. Med.* **56**, 765–770.

Barr, S. I., and Costill, D. L. (1989). Water: Can the endurance athlete get too much of a good thing? *J. Am. Diet. Assoc.* **89**, 1629–1635.

Convertino, V. A., Keil, L. C., Bernauer, E. M., and Greenleaf, J. E. (1981). Plasma volume, osmolarity, vasopressin, and renin activity during graded exercise in man. *J. Appl. Physiol.* **50**, 123–128.

Costill, D. L. (1977). Sweating: Its composition and effects on body fluids. *Ann. N. Y. Acad. Sci.* **301**, 160–174.

Costill, D. L., Cote, R., Miller, E., Miller, T., and Wynder, S. (1975). Water and electrolyte replacement during repeated days of work in the heat. *Aviat. Space Environ. Med.* **46**, 795–800.

Donowitz, M., Kokke, F. T., and Saidi, R. (1995). Evaluation of patients with chronic diarrhea. *N. Engl. J. Med.* **332**, 725–729.

Drewett, J. G., and Garbers, D. L. (1994). The family of guanylyl cyclase receptors and their ligands. *Endocr. Rev.* **15**, 135–162.

Freund, B. J., Claybaugh, J. R., Dice, M. S., and Hashiro, G. (1987). Hormonal and vascular fluid responses to maximal exercise in trained and untrained males. *J. Appl. Physiol.* **63**, 669–675.

Frizzell, R. T., Lang, G. H., Lowance, D. C., and Lathan, S. R. (1986). Hyponatremia and ultramarathon running. *JAMA* **255**, 772–774.

Garner, P. (1995). Type I diabetes mellitus and pregnancy. *Lancet* **346**, 157–161.

Jeunemaitre, X., Inoue, I., Williams, C., Charru, A., Tichet, J., Powers, M., Sharma, A., Gimenez-Roqueplo, A., Hata, A., Corvol, P., and Lalouel, J. (1997). Haplotypes of angiotensinogen in essential hypertension. *Am. J. Hum. Genet.* **60**, 1448–1460.

Jensen, B. L., Ellekvist, P., and Skott, O. (1997). Chloride is essential for contraction of afferent arterioles after agonists and potassium. *Am. J. Clin. Nutr.* **272**, F389–F396.

John, S., Krege, J., Oliver, P., Hagaman, J., Hodgin, J., Pang, S., Flynn, T., and Smithies, O. (1995). Genetic diseases in atrial natriuretic peptide and salt-sensitive hypertension. *Science* **267**, 679–681.

Keren, G., Epstein, Y., and Magazanik, A. (1981). Temporary heat intolerance in a heatstroke patient. *Aviat. Space Environ. Med.* **52**, 116–117.

Knochel, J. P., Dotin, L. N., and Hamburger, R. J. (1972). Pathophysiology of intense physical conditioning in a hot climate. *J. Clin. Invest.* **51**, 242–255.

Ledoux, S., Dussaule, J.-C., Chatziantoniou, C., Ardaillou, N., Vandermeersch, S., and Ardaillou, R. (1997). Protein kinase A activity modulates natriuretic peptide-dependent cGMP accumulation in renal cells. *Am. J. Physiol.* **272**, C82–C89.

Lopez, M., Wong, S., Kishimoto, I., Dubois, S., Mach, V., Friesen, J., Garbers, D. L., and Beuvo, A. (1995). Salt-resistant hypertension in mice lacking the guanylyl cyclase-A receptor for atrial natriuretic peptide. *Nature* **378**, 65–68.

Needle, M. A., Kaloyanides, G. J., and Schwartz, W. B. (1964). The effects of selective depletion of hydrochloric acid on acid–base and electrolyte equilibrium. *J. Clin. Invest.* **43**, 1836–1846.

Reis, A. M., Jankowski, M., Mukaddam-Daher, S., Tremblay, J., Dam, T., and Gutkowska, J. (1997). Regulation of the natriuretic peptide system in rat uterus during the estrous cycle. *J. Endocrinol.* **153**, 345–355.

Sibai, B. M. (1996). Treatment of hypertension in pregnant women. *N. Engl. J. Med.* **335**, 257–265.

Sinaiko, A. R. (1996). Hypertension in children. *N. Engl. J. Med.* **335**, 1968–1973.

Smith, M. (1987). Position of the American Dietetic Association: Nutrition for physical fitness and athletic performance for adults. *J. Am. Diet. Assoc.* **87**, 933–939.

Sperotto, G., Carrazza, F. R., and Marcondes, E. (1977). Treatment of diarrheal dehydration. *Am. J. Clin. Nutr.* **30**, 1447–1456.

Wade, C. E., and Claybaugh, J. R. (1980). Plasma renin activity, vasopressin concentration, and urinary excretory responses to exercise in men. *J. Appl. Physiol.* **49**, 930–936.

Wang, X., and Robinson, P. J. (1997). Cyclic GMP-dependent protein kinase and cellular signaling in the nervous system. *J. Neurochem.* **68**, 443–456.

Welsh, M. J., and Frick, R. B. (1987). Cystic fibrosis. *J. Clin. Invest.* **80**, 1523–1526.

*Hypertension and Atrial Natriuretic Peptide*

Boegehold, M. A., and Kotchen, T. A. (1989). Relative contributions of dietary $Na^+$ and $Cl^-$ to salt-sensitive hypertension. *Hypertension* **14**, 579–583.

Chowdhury, T. A., Barnett, A. H., and Bain, S. C. (1996). Pathogenesis of diabetic nephropathy. *Trends Endocrinol. Metab.* **7**, 320–323.

Christensen, G., and Leistad, E. (1997). Atrial systolic pressure, as well as stretch, is a principal stimulus for release of ANF. *Am. J. Physiol.* **272**, H820–H826.

Grim, R. H., Neaton, J. D., Elmer, P. J., Svendsen, K. H., Levin, J., Segal, M., Holland, L., Witte, L. J., Clearman, D. R., Kofron, P., Labounty, R. K., Crow, R., and Prineas, R. J. (1990). The influence of oral potassium chloride on blood pressure in hypertensive men on a low-sodium diet. *N. Engl. J. Med.* **322**, 569–574.

Holbrook, J. T., Patterson, K. Y., Bodner, J. E., Douglas, L. W., Veillon, C., Kelsey, J. L., Mertz, W., and Smith, J. C. (1984). Sodium and potassium intake and balance in adults consuming self-selected diets. *Am. J. Clin. Nutr.* **40**, 786–793.

Joosens, J. V., and Begoers, J. (1987). Dietary salt and risks to health. *Am. J. Clin. Nutr.* **45**, 1277–1288.

Kaplan, N. M. (1987). Dietary aspects of the treatment of hypertension. *Annu. Rev. Public Health* **7**, 503–519.

Kaplan, N. M., Carnegie, A., Raskin, P., Heller, J. A., and Simmons, M. (1985). Potassium supplementation in hypertensive patients with diuretic-induced hypokalemia. *N. Engl. J. Med.* **312**, 746–749.

Krishna, G. G., Miller, E., and Kapoor, S. (1989). Increased blood pressure during potassium depletion in normotensive men. *N. Engl. J. Med.* **320**, 1177–1182.

Luft, F. C., and Weinberger, M. H. (1987). Potassium and blood pressure regulation. *Am. J. Clin. Nutr.* **45**, 1289–1294.

Mickelsen, O., Makdani, D., Gill, J. L., and Frank, R. L. (1977). Sodium and potassium intakes and excretions of normal men consuming sodium chloride or a 1:1 mixture of sodium and potassium chlorides. *Am. J. Clin. Nutr.* **30**, 2033–2040.

Miller, J. Z., Weinberger, M. H., and Christian, J. C. (1987). Blood pressure response to potassium supplementation in normotensive adults and children. *Hypertension* **10**, 437–442.

Rikimaru, T., Fujita, Y., Okuda, T., Kajiwara, N., Miyatani, S., Alpers, M., and Koishi, H. (1988). Responses of sodium balance, blood pressure, and other variables to sodium handling in Papua New Guinea highlanders. *Am. J. Clin. Nutr.* **47**, 502–508.

Rocchini, A. P., Key, J., Bondie, D., Chico, R., Moorehead, C., Katch, V., and Martin, M. (1989). The effect of weight loss on the sensitivity of blood pressure to sodium in obese adolescents. *N. Engl. J. Med.* **321**, 580–585.

Shah, M., Jeffery, R. W., Laing, B., Savre, S. G., Natta, M. V., and Strickland, D. (1990). Hypertension prevention trial (HPT): Food pattern changes resulting from intervention on sodium, potassium, and energy intake. *J. Am. Diet. Assoc.* **90**, 69–76.

Shapiro, A. P., and Jacob, R. G. (1983). Nonpharmacologic approaches to the treatment of hypertension. *Annu. Rev. Public Health* **4**, 285–310.

Suki, W. N. (1988). Dietary potassium and blood pressure. *Kidney Int., Suppl.* **34**(25), S175–S176.

Tobian, L. (1979). Dietary salt (sodium) and hypertension. *Am. J. Clin. Nutr.* **32**, 2659–2662.

Tobian, L., and Hanlon, S. (1990). High sodium chloride diets injure arteries and raise mortality without changing blood pressure. *Hypertension* **15**, 900–903.

Treasure, J., and Ploth, D. (1983). Role of dietary potassium in the treatment of hypertension. *Hypertension* **5**, 864–872.

Wilber, J. A. (1982). The role of diet in the treatment of high blood pressure. *J. Am. Diet. Assoc.* **80**, 25–29.

## Enzymology of Sodium and Potassium (Non-Renal)

Alvarado, F., and Mahmood, A. (1979). pH-dependent effects of the alkali-metal ions on intestinal brush-border sucrase. *J. Biol. Chem.* **254**, 9534–9541.

Bunning, P., and Riordan, J. F. (1983). Activation of angiotensin converting enzyme by monovalent anions. *Biochemistry* **22**, 110–116.

Bunning, P., and Riordan, J. F. (1987). Sulfate potentiation of the chloride activation of angiotensin converting enzyme. *Biochemistry* **26**, 3374–3377.

Cera, E. D., Guinto, E., Vindigni, A., Dang, Q., Ayala, Y. L., Wuyi, M., and Tulinsky, A. (1995). The Na$^+$ binding site of thrombin. *J. Biol. Chem.* **270**, 22089–22092.

Giorgio, A. J., and Plaut, G. W. (1967). The effect of univalent cations on activities catalyzed by bovine liver propionyl-CoA carboxylase. *Biochim. Biophys. Acta* **139**, 487–501.

Suelter, C. H. (1970). Enzymes activated by monovalent cations. *Science* **168**, 789–795.

Webster, L. T. (1966). Studies on the acetyl-CoA synthetase reaction. *J. Biol. Chem.* **241**, 5504–5510.

## *Iodine*

Bagchi, M., Palaniswami, N., Desai, H., Felicetta, J., and Brown, T. R. (1988). Decreased thyroidal response to thyrotropin in type II diabetes mellitus. *Metabolism* **37**, 669–671.

Croteau, W., Davey, J. C., Galton, V. A., and St. Germain, D. L. (1996). Cloning of the mammalian type II iodothyronine deiodinase. *J. Clin. Invest.* **98**, 405–417.

Cruz, I. P., and Nillni, E. (1996). Intracellular sites of prothyrotropin-releasing hormone processing. *J. Biol. Chem.* **271**, 22736–22745.

Dai, G., Levy, O., and Carrasco, N. (1996). Cloning and characterization of the thyroid iodide transporter. *Nature* **379**, 458–460.

DePalo, D., Kinlaw, W. B., Zhao, C., Engerlberg-Kulka, H., and St. Germain, D. L. (1994). Effect of selenium deficiency on type I 5′-deiodinase. *J. Biol. Chem.* **269**, 16223–16228.

Eskandari, S., Loo, D., Dai, D., Dai, G., Levy, O., Wright, E. M., and Carrasco, N. (1997). Thyroid Na$^+$/I$^-$ symporter. *J. Biol. Chem.* **272**, 27230–27236.

Evans, R. M. (1988). The steroid and thyroid hormone response superfamily. *Science* **240**, 889–895.

Fisher, D. A. (1989). Upper limit of iodine in infant formulas. *J. Nutr.* **119**, 1865–1868.

Furnee, C. A., Pfann, G. A., West, C., Haar, F., Heide, D., and Hautvast, J. (1995). New model for describing urinary iodine excretion: Its use for comparing different oral preparations of iodized oil. *Am. J. Clin. Nutr.* **61**, 1257–1262.

Glass, C. K., Lipkin, S. M., Devary, O. V., and Rosenfeld, M. G. (1989). Positive and negative regulation of gene transcription by a retinoic acid-thyroid hormone receptor heterodimer. *Cell* **59**, 697–708.

Glinoer, D. (1997). The regulation of thyroid function in pregnancy: Pathways of endocrine adaptation from physiology to pathology. *Endocr. Rev.* **18**, 404–433.

Hetzel, B. S., and Dunn, J. Y. (1989). The iodine deficiency disorders. *Annu. Rev. Nutr.* **9**, 21–38.

Hetzel, B. S., and Mano, M. T. (1989). A review of experimental studies of iodine deficiency during fetal development. *J. Nutr.* **119**, 145–151.

Hoppner, W., and Sietz, H.-J. (1989). Effect of thyroid hormones on glucokinase gene transcription in rat liver. *J. Biol. Chem.* **264**, 20643–20647.

Kaptein, E. M. (1996). Thyroid hormone metabolism and thyroid diseases in chronic renal failure. *Endocr. Rev.* **17**, 45–63.

Magner, J. A. (1990). Thyroid-stimulating hormone: Biosynthesis, cell biology, and bioactivity. *Endocr. Rev.* **11**, 354–385.

Matovinovic, J. (1983). Endemic goiter and cretinism at the dawn of the third millennium. *Annu. Rev. Nutr.* **3**, 341–412.

Muller, M. J., Acheson, K. J., Jequier, E., and Burger, A. G. (1990). Thyroid hormone action on lipid metabolism in humans: A role for endogenous insulin. *Metabolism* **39**, 480–485.

Ohtaki, S., Nekagawa, H., Nakamura, S., Nakamura, M., and Yamazaki, I. (1985). Characterization of hog thyroid peroxidase. *J. Biol. Chem.* **260**, 441–448.

Paire, A., Bernier-Valentin, F., Selmi-Ruby, S., and Rousset, B. (1997). Characterization of the rat thyroid iodide transporter using anti-peptide antibodies. *J. Biol. Chem.* **272**, 18245–18249.

Pallud, S., Lennon, A.-M., Ramauge, M., Gaveret, J.-M., Croteau, W., Pierre, M., Courtin, F., and St. Germain, D. L. (1997). Expression of the type II iodothyronine deiodinase in cultured rat astrocytes is selenium-dependent. *J. Biol. Chem.* **272**, 18104–18110.

Palumbo, G., Gentile, F., Condorelli, G. L., and Salvatore, G. (1990). The earliest site of iodination in thyroglobulin is residue number 5. *J. Biol. Chem.* **265**, 8887–8892.

Paschke, R., and Ludgate, M. (1997). The thyrotropin receptor in thyroid diseases. *N. Engl. J. Med.* **337**, 1675–1681.

Salvatore, D., Tu, H., Harney, J. W., and Larsen, P. R. (1996). Type 2 iodothyronine deiodinase is highly expressed in human thyroid. *J. Clin. Invest.* **98**, 962–968.

Samuels, H. H., Forman, B. M., Horowitz, Z., and Ye, Z.-S. (1989). Regulation of gene expression by thyroid hormone. *Annu. Rev. Physiol.* **51**, 623–639.

St. Germain, D. L. (1994). Iodothyronine deiodinases. *Trends Endocrinol. Metab.* **5**, 36–42.

Suda, A. K., Pittman, C. S., Shimizu, T., and Chambers, J. B. (1978). The production and metabolism of 3,5,38-triiodothyronine and 3,38,58-triiodothyronine in normal and fasting subjects. *J. Clin. Endocrinol. Metab.* **47**, 1311–1319.

Umesono, K., and Evans, R. M. (1989). Determinants of target gene specificity for steroid/thyroid hormone receptors. *Cell* **57**, 1139–1146.

Unger, J. (1988). Fasting induces a decrease in serum thyroglobulin in normal subjects. *J. Clin. Endocrinol. Metab.* **67**, 1309–1311.

Vilijn, F., and Carrasco, N. (1989). Expression of the thyroid sodium/iodide symporter in *Xenopus laevis* oocytes. *J. Biol. Chem.* **264**, 11901–11903.

Werner, S. C., and Ingbar, S. H. (1978). "The Thyroid," 4th ed. Harper and Row, New York.

*Review Articles on Iron*

Aisen, P., and Listowsky, I. (1980). Iron transport and storage proteins. *Annu. Rev. Biochem.* **49**, 357–393.

Benesch, R., and Benesch, R. E. (1974). Homos and heteros among the hemos. *Science* **185**, 905–908.

Beutler, E. (1997). How little we know about the absorption of iron. *Am. J. Clin. Nutr.* **66**, 419–420.

Bottmonley, S. S., and Muller-Eberhard, U. (1988). Pathophysiology of heme synthesis. *Semin. Hematol.* **25**, 282–302.

Carpenter, K. J. (1990). The history of a controversy over the role of inorganic iron in the treatment of anemia. *J. Nutr.* **120**, 141–147.

Charlton, R. W., and Bothwell, T. H. (1983). Iron absorption. *Annu. Rev. Med.* **34**, 55–68.

Cook, J. D. (1977). Absorption of food iron. *Fed. Proc.* **36**, 2028–2016.

Cook, J. D. (1982). Clinical evaluation of iron deficiency. *Semin. Hematol.* **19**, 6–18.

Cook, J. D. (1995). Iron supplementation: Is less better? *Lancet* **346**, 587.

Craig, W. J. (1994). Iron status in vegetarians. *Am. J. Clin. Nutr.* **59**, 1233S–1237S.

Dallman, P. R. (1987). Iron deficiency and the immune response. *Am. J. Clin. Nutr.* **46**, 329–334.

Dallman, P. R. (1989). Upper limit of iron in infant formulas. *J. Nutr.* **119**, 1852–1855.

Dallman, P. R., Siimes, M. A., and Stekel, A. (1980). Iron deficiency in infancy and childhood. *Am. J. Clin. Nutr.* **33**, 86–118.

Finch, C. A., and Cook, J. D. (1984). Iron deficiency. *Am. J. Clin. Nutr.* **39**, 471–477.

Fomon, S. J., and Strauss, R. G. (1978). Nutrient deficiencies in breast-fed infants. *N. Engl. J. Med.* **299**, 355–356.

Huebers, H. A., and Finch, C. A. (1987). The physiology of transferrin and transferrin receptors. *Physiol. Rev.* **67**, 520–582.

Jagow, G., and Sebald, W. (1980). β-Type cytochromes. *Annu. Rev. Biochem.* **49**, 281–314.

Lee, K., and Clydesdale, F. M. (1979). Iron sources used in food fortification and their changes due to food processing. *CRC Crit. Rev. Food Sci. Nutr.* **11**, 117–152.

Lombard, M., Chua, E., and O'Toole, P. (1997). Regulation of intestinal non-haem iron absorption. *Gut* **40**, 435–439.

Powell, L., Bassett, M. L., and Halliday, J. W. (1980). Hemochromatosis: 1980 update. *Gastroenterology* **78**, 374–381.

Spiro, T. G., and Saltman, P. (1969). Polynuclear complexes of iron and their biological implications. *Struct. Bonding* **6**, 116–156.

Theil, E. C. (1987). Ferritin: Structure, gene regulation, and cellular function in animals, plants, and microorganisms. *Annu. Rev. Biochem.* **56**, 289–315.

Traugh, J. A. (1989). Heme regulation of hemoglobin synthesis. *Semin. Hematol.* **26**, 54–62.

Viteri, F. E. (1989). Influence of iron nutrition on work capacity and performance. *In* "Dietary Iron: Birth to Two Years" (L. J. Filer, ed.), pp. 141–160. Raven Press, New York.

Whitehead, M. W., Thompson, R., and Powell, J. J. (1996). Regulation of metal absorption in the gastrointestinal tract. *Gut* **39**, 625–628.

Winzerling, J., and Law, J. H. (1997). Comparative nutrition of iron and copper. *Annu. Rev. Nutr.* **17**, 501–526.

*Research Articles on Iron*

Azevedo, J. L., Willis, W. T., Turcotte, L. P., Rovner, A. S., Dallman, P. R., and Brooks, G. A. (1989). Reciprocal changes of muscle oxidases and liver enzymes with recovery from iron deficiency. *Am. J. Physiol.* **256**, E401–E405.

Banerjee, D., Flanagan, P. R., Cluett, J., and Valberg, L. S. (1986). Transferrin receptors in the human gastrointestinal tract. *Gastroenterology* **91**, 861–869.

Berner, L. A., Miller, D. D., and Campen, D. V. (1985). Availability to rats of iron in ferric hydroxide polymers. *J. Nutr.* **115**, 1042–1049.

Braughler, J. M., Duncan, L. A., and Chase, R. L. (1986). The involvement of iron in lipid peroxidation. *J. Biol. Chem.* **261**, 10282–10289.

Conrad, M. E., and Schade, S. G. (1968). Ascorbic acid chelates in iron absorption: A role for hydrochloric acid and bile. *Gastroenterology* **55**, 35–45.

Conrad, M. E., Umbreit, J., Moore, E., and Heiman, D. (1996). Mobilferrin is an intermediate in iron transport between transferrin and hemoglobin in K562 cells. *J. Clin. Invest.* **98**, 1449–1454.

Dallman, P. R., Refino, C., and Yland, M. J. (1982). Sequence of development of iron deficiency in the rat. *Am. J. Clin. Nutr.* **35**, 671–677.

Davidsson, L., Galan, P., Cherouvrier, F., Kastenmayer, P., Juillerat, M., Hercberg, S., and Hurrell, R. F. (1997). Bioavailability in infants of iron from infant cereals: Effect of dephytinization. *Am. J. Clin. Nutr.* **65**, 916–920.

Fairweather-Tait, S. J., Balmer, S. E., Scott, P. H., and Minski, M. J. (1987). Lactoferrin and iron absorption in newborn infants. *Pediatr. Res.* **22**, 651–654.

Fenton, V., Cavill, I., and Fisher, J. (1977). Iron stores in pregnancy. *Br. J. Haematol.* **37**, 145–149.

Garcia-Lopez, J. S., Erdman, J. W., and Sherman, A. R. (1990). Iron retention by rats from casein-legume test meals: Effect of tannin level and previous diet. *J. Nutr.* **120**, 760–766.

Graf, E., Empson, K. L., and Eaton, J. W. (1987). Phytic acid. *J. Biol. Chem.* **262**, 11647–11650.

Grisham, M. B., Ritter, C. V., Smith, B. F., Lamont, J. T., and Granger, D. N. (1987). Interaction between oxygen radicals and gastric mucin. *Am. J. Physiol.* **253**, G93–G96.

Hallberg, L., Hulten, L., and Gramatkovski, E. (1997). Iron absorption from the whole diet in men: How effective is the regulation of iron absorption? *Am. J. Clin. Nutr.* **66**, 347–356.

Henderson, B., Menotti, E., and Kuhn, L. C. (1996). Iron regulatory proteins 1 and 2 bind distinct sites of RNA target sequences. *J. Biol. Chem.* **271**, 4900–4908.

Hurrell, R. F., Lynch, S. R., Trinidad, T. P., Dassenko, S. A., and Cook, J. D. (1989). Iron absorption in humans as influenced by bovine milk proteins. *Am. J. Clin. Nutr.* **49**, 546–552.

Juan, S.-H., Guo, J.-H., and Aust, S. D. (1997). Loading of iron into recombinant rat liver ferritin heteropolymers by ceruloplasmin. *Arch. Biochem. Biophys.* **341**, 280–286.

Kim, H.-Y., LaVaute, T., Iwai, K., Klausner, R. D., and Rouault, T. A. (1996). Identification of a conserved and functional iron-responsive element in the 5′-untranslated region of mammalian mitochondrial aconitase. *J. Biol. Chem.* **271**, 24226–24230.

Klausner, R. D., Rouault, T. A., and Harford, J. B. (1993). Regulating the fate of mRNA: The control of cellular iron metabolism. *Cell* **72**, 19–28.

Koerper, M. A., and Dallman, P. R. (1977). Serum iron concentration and transferrin saturation in the diagnosis of iron deficiency in children. *J. Pediatr.* **91**, 870–874.

Lipschitz, D. A., Simpson, K. M., Cook, J. D., and Morris, E. R. (1979). Absorption of monoferric phytate by dogs. *J. Nutr.* **109**, 1154–1160.

Lynch, S. R., Skikne, B. S., and Cook, J. D. (1989). Food iron absorption in ideopathic hemochromatosis. *Blood* **74**, 2187–2193.

McKie, A. T., Raja, K. B., Peters, T. J., Farzaneh, F., and Simpson, R. J. (1996). Expression of genes involved in iron metabolism in mouse intestine. *J. Biol. Chem.* **271**, G772–G779.

Mukhopadyay, C., Attieh, Z., and Fox, P. L. (1998). Role of ceruloplasmin in cellular iron uptake. *Science* **279**, 714–717.

Olivieri, N. F., Brittenham, G. M., Matsui, D., Berkovitch, M., Blendis, L. M., Cameron, R. G., McClelland, R. A., Liu, P. P., Templeton, D. M., and Koren, G. (1995). Iron-chelation therapy with oral deferiprone in patients with thalassemia major. *N. Engl. J. Med.* **332**, 918–922.

Saarinen, U. M., and Siimes, M. A. (1977). Developmental changes in serum iron, total iron-binding capacity, and transferrin saturation in infancy. *J. Pediatr.* **91**, 875–877.

Senozan, N. M. (1985). Methemoglobinemia: An illness caused by the ferric state. *J. Chem. Educ.* **62**, 181–185.

Surinya, K. H., Cox, T. C., and May, B. K. (1997). Transcriptional regulation of the human erythroid 5-aminolevulinate synthase gene. *J. Biol. Chem.* **272**, 26585–26594.

Taketani, S., Kohno, H., Sawamura, T., and Tokunaga, R. (1990). Hemopexin-dependent downregulation of expression of the human transferrin receptor. *J. Biol. Chem.* **265**, 13981–13985.

Toth, I., and Bridges, K. R. (1995). Ascrobic acid enhances ferritin mRNA translation by an IRP/aconitase switch. *J. Biol. Chem.* **270**, 19540–19544.

Wjebu, M. S. (1980). Effect of iron therapy on serum ferritin levels in iron-deficiency anemia. *Blood* **56**, 138–140.

Yip, R., Binkin, N. J., Fleshood, L., and Trowbridge, F. L. (1987). Declining prevalence of anemia among low-income children in the United States. *JAMA* **258**, 1619–1623.

Yip, R., Johnson, C., and Dallman, P. R. (1984). Age-related changes in laboratory values used in the diagnosis of anemia and iron deficiency. *Am. J. Clin. Nutr.* **39**, 427–436.

*Calcium and Phosphate*

Anderson, M. P., Hunt, R. D., Griffiths, H. J., McIntyre, K. W., and Zimmerman, R. E. (1977). Long-term effect of low dietary calcium:phosphate ratio on the skeleton of Cebus albifrons monkeys. *J. Nutr.* **107**, 834–839.

Bell, R. R., Draper, H. H., Tzeng, D. Y., Shin, H. K., and Schmidt, G. R. (1977). Physiological responses of human adults to foods containing phosphate additives. *J. Nutr.* **107**, 42–50.

Berner, Y. N., and Shike, M. (1988). Consequences of phosphate imbalance. *Annu. Rev. Nutr.* **8**, 121–148.

Berrocal, R., Chanton, S., Juillerat, M. A., Pavillard, B., Scherz, J.-C., and Jost, R. (1989). Tryptic phosphopeptides from whole casein. *J. Dairy Res.* **56**, 335–341.

Bridger, W. A., and Henderson, J. F. (1983). "Cell ATP," pp. 9–19. Wiley, New York.

Brindley, G. W., Williams, E. A., Bronk, J. T., Meadows, T. H., Montgomery, R. J., Smith, S. R., and Kelly, P. J. (1988). Parathyroid hormone effects on skeleton exchangeable calcium and bone blood flow. *Am. J. Physiol.* **255**, H94–H100.

Calvo, M. S., and Park, Y. K. (1996). Changing phosphorus content of the U.S. diet: Potential for adverse effects on bone. *J. Nutr.* **126**, 1168S–1180S.

Chu, J.-Y., Marge, S., and Costa, F. M. (1975). Studies in calcium metabolism. *Am. J. Clin. Nutr.* **28**, 1028–1035.

Cotton, F. A., and Wilkinson, G. (1966). "Advanced Inorganic Chemistry," 2nd ed., p. 431. Interscience, New York.

DeGrazia, J. A., and Rich, C. (1964). Studies of intestinal absorption of calcium in man. *Metabolism* **13**, 650–660.

DeGrazia, J. A., Ivanovich, P., Fellows, H., and Rich, C. (1965). A double isotope method for measurement of intestinal absorption of calcium in man. *J. Lab. Clin. Med.* **66**, 822–829.

Fisler, J. S., and Drenick, E. J. (1984). Calcium, magnesium, and phosphate balances during very low calorie diets of soy or collagen protein in obese man: Comparison to total fasting. *Am. J. Clin. Nutr.* **40**, 14–25.

Goldsmith, R. S., Killian, P., Ingbar, S. H., and Bass, D. E. (1969). Effect of phosphate supplementation during immobilization of normal men. *Metabolism* **18**, 349–368.

Greer, F. R. (1989). Calcium, phosphorus, and magnesium: How much is too much for infant formulas? *J. Nutr.* **119**, 1846–1851.

Hagler, L., and Herman, R. H. (1973). Oxalate metabolism, IV. *Am. J. Clin. Nutr.* **26**, 1073–1079.

Heany, R. P., Recker, R. R., and Hinders, S. M. (1988). Variability of calcium absorption. *Am. J. Clin. Nutr.* **47**, 262–264.

Heany, R. P., Smith, K. T., Recker, R. R., and Hinders, S. M. (1989). Meal effects on calcium absorption. *Am. J. Clin. Nutr.* **49**, 372–376.

Hegsted, M., Schuette, S. A., Zemel, M. B., and Linkswiler, H. M. (1981). Urinary calcium and calcium balance in young men as affected by level of protein and phosphorus intake. *J. Nutr.* **111**, 553–562.

Kerstetter, J. E., and Allen, L. H. (1989). Dietary protein increases urinary calcium. *J. Nutr.* **120**, 134–136.

Knochel, J. P. (1977). The pathophysiology and clinical characteristics of severe hypophosphatemia. *Arch. Intern. Med.* **137**, 203–220.

Lehmann, J., Adams, N. D., and Gray, R. W. (1979). Urinary calcium excretion in human beings. *N. Engl. J. Med.* **301**, 535–541.

Lehmann, J., and Worcester, E. M. (1989). Nephrolithiasis. *In* "Textbook of Nephrology" (S. G. Massry and R. J. Glasscock, eds.), 2nd ed., Vol. 1, pp. 920–941. Williams & Wilkins, Baltimore.

Lewis, N. M., Marcus, M. S., Behling, A. R., and Greger, J. L. (1989). Calcium supplements and milk: Effects on acid–base balance and on retention of calcium, magnesium, and phosphorus. *Am. J. Clin. Nutr.* **49**, 527–533.

Miller, E. R., Ullrey, D. E., Zutaut, C. L., Baltzer, B. V., Schmidt, D. A. Hoefer, J. A., and Luecke, R. W. (1962). Calcium requirement of the baby pig. *J. Nutr.* **77**, 7–17.

Miller, E. R., Ullrey, D. E., Zutaut, C. L., Hoefer, J. A., and Luecke, R. W. (1964). Mineral balance studies with the baby pig: Effects of dietary phosphorus level upon calcium and phosphorus balance. *J. Nutr.* **82**, 111–114.

Mizrachi, A., London, R. D., and Gribetz, D. (1968). Neonatal hypocalcemia — Its causes and treatment. *N. Engl. J. Med.* **21**, 1163–1165.

NIH Consensus Development Panel on Optimal Calcium Intake (1994). Optimal calcium intake. *JAMA* **272**, 1942–1948.

Nordin, B. E., Polley, K. J., Need, A. G., Morris, H. A., and Marshall, D. (1987). The problem of calcium requirement. *Am. J. Clin. Nutr.* **45**, 1295–1304.

Pak, C. Y. (1990). Calcium disorders: Hypercalcemia and hypocalcemia. *In* "Fluids and Electrolytes" (J. P. Kokko and R. L. Tannen, eds.), 2nd ed., pp. 596–630. Saunders, Philadelphia.

Pinto, M. R., Gorski, J. P., Penniston, J. T., and Kelly, P. J. (1988). Age-related changes in composition and $Ca^{2+}$-binding capacity of canine cortical bone extracts. *Am. J. Physiol.* **255**, H101–H110.

Recker, R. R. (1985). Calcium absorption and achlorohydria. *N. Engl. J. Med.* **313**, 70–73.

Reeves, P. G. (1997). Components of the AIN-93 diets as improvements in the AIN-76A diet. *J. Nutr.* **127**, 838S–841S.

Schuette, S. A., and Linkswiler, H. M. (1982). Effects on Ca and P metabolism in humans by adding meat, meat plus milk, or purified proteins plus Ca and P to a low protein diet. *Am. J. Clin. Nutr.* **112**, 338–349.

Schuette, S. A., Knowles, J. B., and Ford, H. E. (1989). Effect of lactose or its component sugars on jejunal calcium absorption in adult man. *Am. J. Clin. Nutr.* **50**, 1084–1087.

Sheikh, M. S., and Fordtran, J. S. (1990). Calcium bioavailability from two calcium carbonate preparations. *N. Engl. J. Med.* **323**, 921.

Sheikh, M. S., Santa Ana, C., Nicar, M. J., Schiller, L. R., and Fordtran, J. S. (1987). Gastrointestinal absorption of calcium from milk and calcium salts. *N. Engl. J. Med.* **317**, 532–536.

Sie, T.-L., Draper, H. H., and Bell. R. R. (1974). Hypocalcemia, hyperparathyroidism and bone resorption in rats induced by dietary phosphate. *J. Nutr.* **104**, 1195–1201.

Spencer, H., Kramer, L., Osis, D., and Norris, C. (1978). Effect of phosphorus on the absorption of calcium and on the calcium balance in man. *J. Nutr.* **108**, 447–457.

Spencer, H., Kramer, L., DeBartolo, M., Norris, C., and Osis, D. (1983). Further studies of the effect of a high protein diet as meat on calcium metabolism. *Am. J. Clin. Nutr.* **37**, 924–929.

Spencer, H., Kramer, L., and Osis, D. (1984). Effect of calcium on phosphorus metabolism in man. *Am. J. Clin. Nutr.* **40**, 219–225.

Spencer, H., Kramer, L., and Osis, D. (1988). Do protein and phosphorus cause calcium loss? *J. Nutr.* **118**, 657–660.

Vipperman, P. E., Peo, E. R., and Cunningham, P. J. (1974). Effect of dietary calcium and phosphorus level upon calcium, phosphorus and nitrogen balance in swine. *J. Anim. Sci.* **38**, 758–765.

Westheimer, F. H. (1987). Why nature chose phosphates. *Science* **235**, 1173–1178.

Wood, R. J., Gerhardt, A., and Rosenberg, I. H. (1987). Effects of glucose and glucose polymers on calcium absorption in healthy subjects. *Am. J. Clin. Nutr.* **46**, 699–701.

Wood, R. J., Suter, P. M., and Russell, R. M. (1995). Mineral requirements of elderly people. *Am. J. Clin. Nutr.* **62**, 493–505.

Zemel, M. B., and Linkswiler, H. M. (1981). Calcium metabolism in the young adult male as affected by level and form of phosphorus intake and level of calcium intake. *J. Nutr.* **111**, 315–324.

## Calcium Signaling

Abdel-Latif, A. A. (1986). Calcium-mobilizing receptors, polyphosphoinositides, and the generation of second messengers. *Pharm. Rev.* **38**, 227–272.

Berridge, M. J. (1984). Inositol triphosphate and diacylglycerol as second messengers. *Biochem. J.* **220**,345–360.

Chattopadhyay, N., Mithal, A., and Brown, E. M. (1996). The calcium-sensing receptor: A window into the physiology and pathophysiology of mineral ion metabolism. *Endocr. Rev.* **17**, 289–304.

Clapham, D. E. (1995). Calcium signaling. *Cell* **80**, 259–268.

Golovina, V. A., and Blaustein, M. P. (1997). Spatially and functionally distinct $Ca^{2+}$ stores in sarcoplasmic and endoplasmic reticulum. *Science* **275**, 1643–1648.

Goode, N., Hughes, K., Woodgett, J. R., and Parker, P. J. (1992). Differential regulation of glycogen synthase kinase-3b by protein kinase C isotypes. *J. Biol. Chem.* **267**, 16878–16882.

Hata, A., Akita, Y., Suzuki, K., and Ohno, S. (1993). Functional divergence of protein kinase C (PKC) family members. *J. Biol. Chem.* **268**, 9122–9129.

Hemmings, B. A. (1997). PtdIns(3,4,5)$P_3$ gets its message across. *Science* **277**, 534.

Kawanishi, T., Blank, L. M., Harootunian, A. T., Smith, M. T., and Tsien, R. Y. (1989). $Ca^{2+}$ oscillations induced by hormonal stimulation of individual fura-2-loaded hepatocytes. *J. Biol. Chem.* **264**, 12859–12866.

Keller, T. C. (1997). Molecular bungees. *Nature* **387**, 233–235.

Koninck, P., and Schulman, H. (1998). Sensitivity of CaM kinase II to the frequency of $Ca^{2+}$ oscillations. *Science* **279**, 227–230

Muallem, S., Pandol, S. J., and Beeker, T. G. (1989). Hormone-evoked calcium release from intracellular stores is a quantal process. *J. Biol. Chem.* **264**, 205–212.

Pearce, S., and Thakker, R. V. (1997). The calcium-sensing receptor: Insights into extracellular calcium homeostasis in health and disease. *J. Endocrinol.* **154**, 371–378.

Pearce, S., Williamson, C., Kifor, O., Bai, M., Coulthard, M., Davies, M., Lewis-Barned, N., McCredie, D., Powell, H., Kendall-Taylor, P., Brown, E. M., and Thakker, R. V. (1996). A familial syndrome of hypocalcemia with hypercalcemia due to mutations in the calcium-sensing receptor. *N. Engl. J. Med.* **335**, 1115–1122.

Phung, T. L., Roncone, A., Jensen, K., Sparks, C., and Sparks, J. D. (1997). Phosphoinositide 3-kinase activity is necessary for insulin-dependent inhibition of apolipoprotein B secretion by rat hepatocytes and localizes to the endoplasmic reticulum. *J. Biol. Chem.* **272**, 30693–30702.

Putney, J. W. (1998). Calcium signaling: Up, down, up, down ... What's the point? *Science* **279**, 191–192.

Rayment, I. (1996). The structural basis of the myosin ATPase activity. *J. Biol. Chem.* **271**, 15850–15853.

Rock, C. O., and Jackowski, S. (1987). Thrombin- and nucleotide-activated phosphatidylinositol 4,5-bisphosphate phospholipase C in human platelet membranes. *J. Biol. Chem.* **262**, 5492–5498.

Schnapp, B. J. (1995). Two heads are better than one. *Nature* **373**, 655–656.

Somlyo, A. P., Walker, J. W., Goldman, Y. E., Trentham, D. R., Kobayashi, S., Kitazawa, T., and Somlyo, A. V. (1988). Inositol triphosphate, calcium and muscle contraction. *Philos. Trans. R. Soc. (London), Ser. B* **320**, 399–414.

Stokoe, D., Stephens, L., Copeland, T., Gaffney, P., Reese, C. B., Painter, G., Holmes, A. B., McCormick, F., and Hawkins, P. T. (1997). Dual role of phosphatidyl-3,4,5-trisphosphate in the activation of protein kinase B. *Science* **277**, 567–570.

Streb, H., Heslop, J. P., Irvine, R. F., Schulz, I., and Berridge, M. J. (1985). Relationship between secretagogue-induced $Ca^{2+}$ release and inositol polyphosphate production in permeabilized pancreatic acinar cells. *J. Biol. Chem.* **260**, 7309–7315.

Subramanian, K., and Meyer, T. (1997). Calcium-induced restructuring of nuclear envelope and endoplasmic reticulum calcium stores. *Cell* **89**, 963–971.

*Bone Calcium and Osteoporosis*

Avioli, L. V. (1997). Salmon calcitonin in the prevention and treatment of osteoporosis. *Trends Endocrinol. Metab.* **8**, 89–92.

Bonjour, J.-P., Carrie, A.-L., Ferrari, S., Clavien, H., Slosman, D., Theintz, G., and Rizzoli, R. (1997). Calcium-enriched foods and bone mass growth in prepubertal girls: A randomized, double-blind, placebo-controlled trial. *J. Clin. Invest.* **99**, 1287–1294.

Bronner, F. (1994). Calcium and osteoporosis. *Am. J. Clin. Nutr.* **60**, 831–836.

Dawson-Hughes, B., Dallal, G. E., Krall, E. A., Sadowski, L., Sahyoun, N., and Tannenbaum, S. (1990). A controlled trial of the effect of calcium supplementation on bone density in postmenopausal women. *N. Engl. J. Med.* **323**,878–883.

Delmas, P. D. (1996). Bisphosphonates in the treatment of bone diseases. *N. Engl. J. Med.* **335**, 1836–1837.

Delmas, P. D., and Meunier, P. J. (1997). The management of Paget's disease of bone. *N. Engl. J. Med.* **336**, 558–566.

Engelman, V., Nickols, G., Ross, F. P., Horton, M., Griggs, D., Settle, S. L., Ruminski, P., and Teitelbaum, S. L. (1997). A peptidomimetic antagonist of $\alpha_v\beta_3$ integrin inhibits bone resorption *in vitro* and prevents osteoporosis *in vivo*. *J. Clin. Invest.* **99**, 2284–2292.

Epstein, S., and Miller, P. (1997). Bone mass measurements. *Trends Endocrinol. Metab.* **8**, 157–160.

Gallagher, J. C., Riggs, B. L., Eisman, J., Hamstra, A., Arnaud, S. B., and DeLuca, H. F. (1979). Intestinal calcium absorption and serum vitamin D metabolites in normal subjects and osteoporotic patients. *J. Clin. Invest.* **64**, 729–736.

Grant, S. F., and Ralston, S. H. (1997). Genes and osteoporosis. *Trends Endocrinol. Metab.* **8**, 232–236.

Halioua, L., and Anderson, J. J. (1989). Lifetime calcium intake and physical activity habits: Independent and combined effects on the radial bone of healthy premenopausal Caucasian women. *Am. J. Clin. Nutr.* **49**, 534–541.

Hegsted, D. M. (1986). Calcium and osteoporosis. *J. Nutr.* **116**, 2316–2319.

Jackman, L. A., Millane, S., Martin, B., Wood, O. B., McCabe, G., Peacock, M., and Weaver, C. M. (1997). Calcium retention in relation to calcium intake and postmenarcheal age in adolescent females. *Am. J. Clin. Nutr.* **66**, 327–333.

Johnston, C., Miller, J. Z., Slemenda, C., Reister, T., Hui, S., Christian, J., and Peacock, M. (1992). Calcium supplementation and increases in bone mineral density in children. *N. Engl. J. Med.* **327**, 82–87.

Lindsay, R., Nieves, J., Formica, C., Nenneman, E., Woelfert, L., Shen, V., Dempster, D., and Cosman, F. (1997). Randomised controlled study of effect of parathyroid hormone on vertebral-bone mass and fracture incidence among postmenapausal women on oestrogen with osteoporosis. *Lancet* **350**, 550–555.

Lukashi, H. C. (1993). Soft tissue composition and bone mineral status: Evaluation by dual-energy X-ray absorptiometry. *J. Nutr.* **123**, 438–443.

Marcus, R. (1987). Calcium intake and skeletal integrity: Is there a critical relationship? *J. Nutr.* **117**, 631–635.

McDermott, M. T., and Kidd, G. S. (1987). The role of calcitonin in the development and treatment of osteoporosis. *Endocr. Rev.* **8**, 377–390.

McKeon, V. A. (1990). Estrogen replacement therapy. *J. Gerontol. Nurs.* **16**, 6–11.

Mostov, K., and Werb, Z. (1997). Journey across the osteoclast. *Science* **276**, 219–266.

Riggs, B. L. (1990). A new option for treating osteoporosis. *N. Engl. J. Med.* **323**, 124–125.

Riggs, B. L., and Melton, L. J. (1986). Involutional osteoporosis. *N. Engl. J. Med.* **314**, 1676–1684.

Roodman, G. D. (1996). Advances in bone biology: The osteoclast. *Endocr. Rev.* **17**, 308–332.

Sahni, M., Guenther, H. L., Fleisch, H., Collin, P., and Martin, T. J. (1993). Bisphosphonates act on bone resorption through the mediation of osteoblasts. *J. Clin. Invest.* **91**, 2004–2011.

Smith, E. L., and Gilligan, C. (1990). Calcium and exercise in prevention of bone loss with age. *Clin. Nutr.* **9**, 17–20.

Torgerson, D. J., Gosden, T., and Reid, D. M. (1997). The economics of osteoporosis prevention. *Trends Endocrinol. Metab.* **8**, 236–239.

Winters, K. M., Adams, W. C., Meredith, C. N., Loan, M., and Lasley, B. L. (1996). Bone density and cyclic ovarian function in trained runners and active controls. *Med. Sci. Sports Exerc.* **28**, 776–785.

*Fluoride*

Ekstrand, J. (1989). Fluoride intake in early infancy. *J. Nutr.* **119**, 1856–1860.
Forsyth, D. M., Pond, W. G., Wasserman, R. H., and Krook, L. (1972). Dietary calcium and fluoride interactions in swine. *J. Nutr.* **102**, 1623–1638.
Harrison, J. E. (1990). Fluoride treatment for osteoporosis. *Calcif. Tissue Int.* **46**, 287–288.
Lindsay, R. (1990). Fluoride and bone — Quantity versus quality. *N. Engl. J. Med.* **322**, 845–846.
Navia, J. M., Lopez, H., and Bradley, E. L. (1981). Biological rat assay for total fluoride availability in foods. *J. Nutr.* **111**, 755–761.
Pak, C. Y. (1989). Fluoride and osteoporosis. *Proc. Soc. Exp. Biol. Med.* **184**, 278–286.
Riggs, B. L., Hodgson, S. F., O'Fallon, W. M., Chao, E. Y., Wahner, H. W., Muhs, J. M., Cedel, S. L., and Melton, L. J. (1990). Effect of fluoride treatment on the fracture rate in post-menopausal women with osteoporosis. *N. Engl. J. Med.* **322**, 802–809.
Schamschula, R. G., and Barmes, D. E. (1981). Fluoride and health: Dental caries, osteoporosis, and cardiovascular disease. *Annu. Rev. Nutr.* **1**, 427–435.
Subba Rao, G. (1984). Dietary intake and bioavailability of fluoride. *Annu. Rev. Nutr.* **4**, 115–136.
Tao, S., and Suttie, J. W. (1976). Evidence for a lack of an effect of dietary fluoride level on reproduction in mice. *J. Nutr.* **106**, 1115–1122.

*Magnesium Nutrition*

Alfrey, A. C., Miller, N. L., and Trow, R. (1974). Effect of age and magnesium depletion on bone magnesium pools in rats. *J. Clin. Invest.* **54**, 1074–1081.
Britton, W. M., and Stokstad, E. L. R. (1970). Aorta and other soft tissue calcification in the magnesium-deficient rat. *J. Nutr.* **100**, 1501–1506.
Bunce, G. E., Jenkins, K. J., and Phillips, P. H. (1962). The mineral requirements of the dog III. The magnesium requirement. *J. Nutr.* **76**, 17–29.
Day, H. G. (1982). E. V. McCollum and magnesium in nutrition. *Trends Biol. Sci.* **7**, 112–113.
Dengel, J. L., Mangels, A. R., and Moser-Veillon, P. B. (1994). Magnesium homeostasis: Conservation mechanism in lactating women consuming a controlled-magnesium diet. *Am. J. Clin. Nutr.* **59**, 990–994.
Fischer, P. W., and Giroux, A. (1987). Effects of dietary magnesium on sodium-potassium pump action in the heart of rats. *J. Nutr.* **117**, 2091–2095.
Karbach, U., and Rummel, W. (1990). Cellular and paracellular magnesium transport across the terminal ileum of the rat and its interaction with the calcium transport. *Gastroenterology* **98**, 985–992.
Koh, E. T., Reiser, S., and Fields, M. (1989). Dietary fructose as compared to glucose and starch increases the calcium content of kidney of magnesium-deficient rats. *J. Nutr.* **119**, 1173–1178.
Miller, E. R., Ullrey, D. E., Zutaut, C. L., Baltzer, B. V., Schmidt, D. A., Hoefer, J. A., and Luecke, R. W. (1965). Magnesium requirement of the baby pig. *J. Nutr.* **85**, 13–20.
Potter, J. D., Robertson, S. P., and Johnson, J. D. (1981). Magnesium and the regulation of muscle contraction. *Fed. Proc.* **40**, 2653–2656.
Shafik, I. M., and Quamme, G. A. (1989). Early adaptation of renal magnesium reabsorption in response to magnesium restriction. *Am. J. Physiol.* **257**, F974–F977.
Shils, M. E. (1969). Experimental human magnesium depletion. *Medicine* **48**, 61–85.
Sperotto, G., Carrazza, F. R., and Marcondes, E. (1977). Treatment of diarrheal dehydration. *Am. J. Clin. Nutr.* **30**, 1447–1456.

Turlapaty, P., and Altura, B. M. (1980). Magnesium depletion produces spasms of coronary arteries: Relationship to etiology of sudden death ischemic heart disease. *Science* **208**, 198–200.

Vink, R., McIntosh, T. K., Demediuk, P., Weiner, M. W., and Faden, A. I. (1988). Decline in intracellular free $Mg^{2+}$ is associated with irreversible tissue injury after brain trauma. *J. Biol. Chem.* **263**, 757–761.

Wacker, W. E., and Parisi, A. F. (1968). Magnesium metabolism. *N. Engl. J. Med.* **278**, 772–776.

Wester, P. O. (1987). Magnesium. *Am. J. Clin. Nutr.* **45**, 1305–1312.

Witteman, J. C., Grobbee, D. E., Derkx, F., Bouillon, R., Bruijn, A., and Hofman, A. (1994). Reduction of blood pressure with oral magnesium supplementation in women with mild to moderate hypertension. *Am. J. Clin. Nutr.* **60**, 129–135.

## Enzymology of Magnesium

Blair, J. (1969). Magnesium and the aconitase equilibrium: Determination of apparent stability constants of magnesium substrate complexes from equilibrium data. *Eur. J. Biochem.* **8**, 287–291.

Cohen, N. S., and Raijman, L. (1980). The apparent $K_m$ of ATPMg for carbamyl phosphate synthetase (ammonia) *in situ*. *J. Biol. Chem.* **255**, 3352–3357.

Cohn, M., and Hughes, T. R. (1962). Nuclear magnetic resonance spectra of adenosine di- and triphosphate. *J. Biol. Chem.* **237**, 176–181.

Ebel, H., and Gunther, T. (1980). Magnesium metabolism: A review. *J. Clin. Chem. Clin. Biochem.* **18**, 257–270.

Fahien, L. A., Kmiotek, E. H., Woldegiorgis, G., Evenson, M., Shrago, E., and Marshall, M. (1985). Regulation of aminotransferase-glutamate dehydrogenase interactions by carbamyl phosphate synthase. I. $Mg^{2+}$ plus leucine versus citrate and malate. *J. Biol. Chem.* **260**, 6069–6079.

Frieden, C. (1982). The $Mg^{2+}$-induced conformational change in rabbit skeletal muscle G-actin. *J. Biol. Chem.* **257**, 2882–2886.

Hannum, Y. A., and Bell, R. M. (1990). Rat brain protein kinase C. *J. Biol. Chem.* **265**, 2962–2972.

Laughlin, M. R., and Thompson, D. (1996). The regulatory role for magnesium in glycolytic flux of the human erythrocyte. *J. Biol. Chem.* **271**, 28977–28983.

Londos, C., and Preston, M. S. (1977). Activation of the heart adenylate cyclase system by divalent cations. *J. Biol. Chem.* **252**, 5957–5961.

Mendlein, J., and Sachs, G. (1989). The substitution of calcium for magnesium in $H^+,K^+$-ATPase catalytic cycle. *J. Biol. Chem.* **264**, 18512–18519.

Romani, A., and Scarpa, A. (1992). Regulation of cell magnesium. *Arch. Biochem. Biophys.* **298**, 1–12.

Shertzer, H. G., and Racker, E. (1976). Reconstitution and characterization of the adenine nucleotide transporter derived from bovine heart mitochondria. *J. Biol. Chem.* **251**, 2446–2452.

Wacker, W. E. C. (1969). The biochemistry of magnesium. *Ann. N. Y. Acad. Sci.* **162**, 717–726.

## Manganese

Bell, L. T., and Hurley, L. S. (1973). Ultrastructural effects of manganese deficiency in liver, heart, kidney, and pancreas of mice. *Lab. Invest.* **29**, 723–736.

Brinkworth, R. I., Hanson, R. W., Fullin, F. A., and Schramm, V. L. (1981). $Mn^{2+}$-sensitive and insensitive forms of phosphoenolpyruvate carboxykinase (GTP). *J. Biol. Chem.* **256**, 10795–10802.

Brock, A., Chapman, S., Ulman, E., and Wu, G. (1994). Dietary manganese deficiency decreases rat hepatic arginase activity. *J. Nutr.* **124**, 340–344.

Columbo, G., and Lardy, H. (1981). Phosphoenolpyruvate carboxykinase (guanosine 5'-triphosphate) from rat liver cytosol. Divalent cation involvement in the decarboxylation reactins. *Biochemistry* **20**, 2758–2767.

Corkey, B. E., Duszynski, J., Rich, T. L., Matschinsky, B., and Williamson, J. R. (1986). Regulation of free and bound magnesium in rat hepatocytes and isolated mitochondria. *J. Biol. Chem.* **261**, 2567–2574.

Davidsson, L. A., and Lonnerdal, B. (1989). Fe-saturation and proteolysis of human lactoferrin: Effect on brush-border receptor mediated uptake of Fe and Mn. *Am. J. Physiol.* **257**, G930–G934.

Davidsson, L., Cederblad, A., Lonnerdal, B., and Sandstrom, B. (1989). Manganese retention in man: A method for estimating manganese absorption in man. *Am. J. Clin. Nutr.* **49**, 170–179.

Finley, J. W., Johnson, P. E., and Johnson, L. K. (1994). Sex affects manganese absorption and retention by humans from a diet adequate in manganese. *Am. J. Clin. Nutr.* **60**, 949–955.

Freeland-Graves, J.H., and Turnlund, J. R. (1996). Deliberations and evaluations of the approaches, endpoints and paradigms for manganese and molybdenum dietary recommendations. *J. Nutr.* **126**, 2435S–2440S.

Freeland-Graves, J. H., Behmardi, F., Bales, C. W., Bougherty, V., Lin, P.-H., Crosby, J. B., and Trickett, P. C. (1988). Metabolic balance of manganese in young men consuming diets containing five levels of dietary manganese. *J. Nutr.* **118**, 764–773.

Gregor, J. L., Davis, C. D., Suttie, J. W., and Lyle, B. J. (1990). Intake, serum concentrations, and urinary excretion of manganese by adult males. *Am. J. Clin. Nutr.* **51**, 457–461.

Hatano, S., Nishi, Y., and Usui, T. (1983). Erythrocyte manganese concentration in healthy Japanese children, adults, and the elderly, and in cord blood. *Am. J. Clin. Nutr.* **37**, 457–460.

Hebda, C. A., and Nowak, T. (1982). Phosphoenolpyruvate carboxykinase. *J. Biol. Chem.* **257**, 5515–5522.

Krieger, D., Krieger, S., Jansen, O., Gass, P., Theilmann, L., and Lichtnecker, H. (1995). Manganese and chronic hepatic encephaloopathy. *Lancet* **346**, 270–274.

Libby, C. B., Frey, W. A., Villafranca, J. J., and Benkovic, S. J. (1975). Kinetic and binding studies of Mn(II) and fructose 1,6-bisphosphate with rabbit liver hexosebisphosphatase. *J. Biol. Chem.* **250**, 7564–7573.

Matte, A., Tari, L., Goldi, H., and Delbaere, L. (1997). Structure and mechanism of phosphoenolpyruvate carboxylase. *J. Biol. Chem.* **272**, 8105–8108.

Mildvan, A. S., Scrutton, M. C., and Utter, M. F. (1966). Pyruvate carboxylase. *J. Biol. Chem.* **241**, 3488–3498.

Sakakibara, Y., Katafuchi, J., Takami, Y., Nakayama, T., Suiko, M., Nakajima, H., and Liu, M.-C. (1997). Manganese-depedent DOPA/tyrosine sulfation in HepG2 human hepatoma cells. *Biochim. Biophys. Acts* **1335**, 102–106.

Scrutton, M. C., Griminger, P., and Wallace, J. C. (1972). Pyruvate carboxylase. *J. Biol. Chem.* **247**, 3305–3313.

Senior, A. E., Richardson, L. V., Baker, K., and Wise, J. G. (1980). Tight divalent cation-binding sites of soluble adenosine triphosphatase (F1) from beef heart mitochondria and *Escherichia coli*. *J. Biol. Chem.* **255**, 7211–7217.

Stastny, D., Vogel, R. S., and Picciano, M. F. (1984). Manganese intake and serum manganese concentration of human milk-fed and formula-fed infants. *Am. J. Clin. Nutr.* **39**, 872–878.

Thampy, K. G., and Wakil, S. J. (1985). Activation of acetyl-CoA carboxylase. *J. Biol. Chem.* **260**, 6318–6323.

Zidenberg-Cherr, S., Keen, C. L., Lonnerdal, B., and Hurley, L. S. (1983). Superoxide dismutase activity and lipid peroxidation in the rat: Developmental correlations affected by manganese deficiency. *J. Nutr.* **113**, 2498–2504.

*Zinc*

Apgar, J. (1985). Zinc and reproduction. *Annu. Rev. Nutr.* **5**, 43–68.

Aspinwall, C., Brooks, S., Kennedy, R. T., and Lakey, J. (1997). Effects of intravesicular $H^+$ and extracellular $H^+$ and $Zn^{2+}$ on insulin secretion in pancreatic beta cells. *J. Biol. Chem.* **272**, 31308–31314.

Baer, M. T., and King, J. C. (1984). Tissue zinc levels and zinc excretion during experimental zinc depletion in young men. *Am. J. Clin. Nutr.* **39**, 556–570.

Berg, J. M. (1990). Zinc fingers and other metal-binding domains. *J. Biol. Chem.* **265**, 6513–6516.

Blanchard, R. K., and Cousins, R. J. (1996). Differential display of intestinal mRNAs regulated by dietary zinc. *Proc. Natl. Acad. Sci.* **93**, 6863–6868.

Brand, I. A., and Kleineke, J. (1996). Intracellular zinc movement and its effect on the carbohydrate metabolism of isolated rat hepatocytes. *J. Biol. Chem.* **271**, 1941–1949.

Bremner, I., and Beattie, J. H. (1990). Metallothionein and the trace minerals. *Annu. Rev. Nutr.* **10**, 63–83.

Brody, T., and Mathews, T. D. (1989). The release of zinc from leukocytes provoked by A23187 and EDTA is associated with the release of enzymes. *Comp. Biochem. Physiol. A* **94**, 693–697.

Bruggenwirth, H., Boehmer, A., Ramnarain, S., Verleun-Mooijman, M., Satijn, D., Trapman, J., Grootegoed, J., and Brinkmann, A. (1997). Molecular analysis of the androgen-receptor gene in a family with receptor-positive partial androgen insensitivity. *Am. J. Hum. Genet.* **61**, 1067–1077.

Chesters, J. K., and Quarterman, J. (1970). Effects of zinc deficiency on food intake and feeding patterns of rats. *Br. J. Nutr.* **24**, 1061–1069.

Cohen, N. L., Keen, C. L., Lonnerdal, B., and Hurley, L. S. (1985). Effects of varying dietary iron on the expression of copper deficiency in the growing rat. *J. Nutr.* **115**, 633–649.

Cossack, Z. T. (1988). The efficacy of oral zinc therapy as an alternative to penicillamine for Wilson's disease. *N. Engl. J. Med.* **318**, 322–323.

Cousins, R. J. (1985). Absorption, transport, and hepatic metabolism of copper and zinc: Special reference to metallothionein and ceruloplasmin. *Physiol. Rev.* **65**, 238–309.

Din, W. S., and Frazier, J. M. (1985). Protective effect of metallothionein on cadmium toxicity in isolated rat hepatocytes. *Biochem. J.* **230**, 395–402.

DiSilvestro, R. A., and Cousins, R. J. (1983). Physiological ligands for copper and zinc. *Annu. Rev. Nutr.* **3**, 261–288.

Egan, C., Smith, F., Houk, R., and Serfass, R. (1991). Zinc absorption in women: Comparison of intrinsic and extrinsic stable-isotope labels. *Am. J. Clin. Nutr.* **53**, 547–553.

Figlewicz, D. P., Formby, B., Hodgson, A. T., Schmid, F. G., and Grodsky, G. M. (1980). Kinetics of $^{65}Zn$ uptake and distribution in fractions from cultured rat islets of langerhans. *Diabetes* **29**, 767–773.

Fosmire, G. J. (1990). Zinc toxicity. *Am. J. Clin. Nutr.* **51**, 225–227.

Freeland-Graves, J. (1988). Mineral adequacy of vegetarian diets. *Am. J. Clin. Nutr.* **48**, 859–862.

Frieden, E., and Hsieh, H. S. (1976). Ceruloplasmin: The copper protein with essential oxidase activity. *Adv. Enzymol.* **44**, 187–236.

Golub, J. S., Gershwin, M. E., Hurley, L. S., Saito, W. Y., and Hendrickx, A. G. (1984). Studies of marginal zinc deprivation in rhesus monkeys. IV. Growth of infants in the first year. *Am. J. Clin. Nutr.* **40**, 1192–1202.

Golub, M. S., Keen, C., Gershwin, M., Styne, D., Takeuchi, P., Ontell, F., Walter, R. M., and Hendrickx, A. G. (1996). Adolescent growth and maturation in zinc-deprived rhesus monkeys. *Am. J. Clin. Nutr.* **64**, 274–282.

Grider, A., and Young, E. M. (1996). The acrodermatitis enteropathica mutation transiently affects zinc metabolism in human fibroblasts. *J. Nutr.* **126**, 219–224.

Hamer, D. H. (1986). Metallothionein. *Annu. Rev. Biochem.* **55**, 913–951.

Hempe, J. M., Carlson, J. M., and Cousins, R. J. (1991). Intestinal metallothionein gene expression and zinc absorption in rats are zinc-responsive but refractory to dexamethasone and interleukin 1a. *J. Nutr.* **121**, 1389–1396.

Hurley, L. S., and Tao, S.-H. (1972). Alleviation of teratogenic effects of zinc deficiency by simultaneous lack of calcium. *Am. J. Physiol.* **222**, 322–325.

Johnson, W. T., and Dufault, S. N. (1989). Altered cytoskeletal organization and secretory response of thrombin-activated platelets from copper-deficient rats. *J. Nutr.* **119**, 1404–1410.

Kadonaga, J. T., Carner, K. R., Fasiarz, F. R., and Tjian, R. (1987). Isolation of cDNA encoding transcription factor Sp1 and functional analysis of the DNA binding domain. *Cell* **51**, 1079–1090.

Kagi, J., and Schaffer, A. (1988). Biochemistry of metallothionein. *Biochemistry* **27**, 8509–8515.

King, J. C. (1996). Does poor zinc nutriture retard skeletal growth and mineralization in adolescents? *Am. J. Clin. Nutr.* **64**, 375–376.

Koh, J.-Y., Suh, S., Gwag, B., He, Y., Hsu, C., and Choi, D. W. (1996). The role of zinc in selective neuronal death after transient global cerebral ischemia. *Science* **272**, 1013–1016.

Lazo, J., Kondo, Y., Dellapiazza, D., Michalska, A., Choo, K., and Pitt, B. R. (1995). Enhanced sensitivity to oxidative stress in cultured embryonic cells from transgenic mice different in metallothionein I and II genes. *J. Biol. Chem.* **27**, 5506–5510.

Lehman, I. R. (1981). DNA Polymerase I of *Eschericia coli. In* "The Enzymes," (P. D. Boyer, ed.), Vol. 14, Part A, 3rd ed., pp. 15–37. Academic Press, San Diego.

Liden, J., Delaunay, F., Rafter, I., Gustafsson, J., and Okret, S. (1997). A new function for the C-terminal zinc finger of the glucocorticoid receptor. *J. Biol. Chem.* **272**, 21467–21472.

Link, T. A., and Jagow, G. (1995). Zinc ions inhibit the $Q_p$ center of bovine heart mitochondrial $bc_1$ complex by blocking a protonable group. *J. Biol. Chem.* **270**, 25001–25006.

Lisle, R. C., Sarras, M., Hidalgo, J., and Andrews, G. K. (1996). Metallothionein is a component of exocrine pancreas secretion: Implications for zinc homeostasis. *Am. J. Physiol.* **271**, C1103–C1110.

Menard, M., McCormick, C. C., and Cousins, R. J. (1981). Regulation of intestinal metallothionein biosynthesis in rats by dietary zinc. *J. Nutr.* **111**, 1353–1361.

Michael, J., Carroll, R., Swift, H. H., and Steiner, D. F. (1987). Studies on the molecular organization of rat insulin secretory granules. *J. Biol. Chem.* **262**, 16531–16535.

Milne, D. B., Canfield, W. K., Gallagher, S. K., Hunt, J. R., and Klevay, L. M. (1987). Ethanol metabolism in postmenopausal women fed a diet marginal in zinc. *Am. J. Clin. Nutr.* **46**, 688–693.

Narayan, V., Kriwacki, R., and Caradonna, J. P. (1997). Structures of zinc finger domains from transcription factor Sp1. *J. Biol. Chem.* **272**, 7801–7809.

Prasad, A. S. (1985). Clinical manifestations of zinc deficiency. *Annu. Rev. Nutr.* **5**, 341–363.

Prasad, A. S. (1991). Discovery of zinc deficiency and studies in an experimental human model. *Am. J. Clin. Nutr.* **53**, 403–412.

Reyes, J. G. (1996). Zinc transport in mammalian cells. *Am. J. Physiol.* **270**, C401–C410.

Samson, S., Paramchuk, W., Shworak, N., and Gedamu, L. (1995). Functional analysis of the human metallotionein-IG gene. *J. Biol. Chem.* **270**, 25194–25199.

Sandstead, H. H. (1995). Requirements and toxicity of essential trace elements, illustrated by zinc and copper. *Am. J. Clin. Nutr.* **61**, 621S–624S.

Sandstead, H. H., and Smith, J. C. (1996). Deliberations and evaluations of approaches, endpoints and paradigms for determining zinc dietary recommendations. *J. Nutr.* **126**, 2410S–2418S.

Sazawal, S., Black, R. E., Bhan, M., Bhandari, N., Sinha, A., and Jalla, S. (1995). Zinc supplementation in young children with acute diarrhea in India. *N. Engl. J. Med.* **333**, 839–844.

Schissel, S., Schuchman, E. H., Williams, K., and Tabas, I. (1996). $Zn^{2+}$-stimulated sphingomyelinase is secreted by many cell types and is a product of the acid sphingomyelinase gene. *J. Biol. Chem.* **271**, 18431–18436.

Sullivan, V. K., and Cousins, R. J. (1997). Competitive reverse transcriptase-polymerase chain reaction shows that dietary zinc supplementation in humans increases monocyte metallothionein mRNA levels. *J. Nutr.* **127**, 694–698.

Swenerton, H., and Hurley, L. S. (1968). Severe zinc deficiency in male and female rats. *J. Nutr.* **95**, 8–18.

Swenerton, H., and Hurley, L. S. (1980). Zinc deficiency in rhesus and bonnet monkeys, including effects on reproduction. *J. Nutr.* **110**, 575–583.

Vallee, B. L., and Galdes, A. (1984). The metallobiochemistry of zinc enzymes. *Adv. Enzymol.* **56**, 283–430.

Wada, L., and King, J. C. (1986). Effect of low zinc intakes on basal metabolic rate, thyroid hormones, and protein utilization in adult men. *J. Nutr.* **116**, 1045–1053.

Wada, L., Turnlund, J. R., and King, J. C. (1985). Zinc utilization in young men fed adequate and low zinc intakes. *J. Nutr.* **115**, 1345–1354.

Wallwork, J. C., Fosmire, G. J., and Sandstead, H. H. (1981). Effect of zinc deficiency on appetite and plasma amino acid concentrations in the rat. *Br. J. Nutr.* **45**, 127–136.

Wood, R. J., Suter, P. M., and Russell, R. M. (1995). Mineral requirements of elderly people. *Am. J. Clin. Nutr.* **62**, 493–505.

Zhu, X.-G., McPhie, P., Lin, K.-H., and Cheng, S.-Y. (1997). The differential hormone-dependent transcriptional activation of thyroid hormone receptor isoforms is mediated by interplay of their domains. *J. Biol. Chem.* **272**, 9048–9054.

## Copper

Aldred, A. R., Grimes, A., Schreiber, G., and Mercer, J. F. (1987). Rat ceruloplasmin. *J. Biol. Chem.* **262**, 2875–2878.

Bremner, I. (1987). Involvement of metallothionein in the hepatic metabolism of copper. *J. Nutr.* **117**, 19–29.

Bull, P. C., Thomas, G. R., Rommens, J. M., Forbes, J. R., and Cox, D. W. (1993). The Wilson disease gene is a putative copper transporting P-type ATPase similar to the Menkes gene. *Nature Genet.* **5**, 327–337.

Castillo-Duran, C., and Uauy, R. (1988). Copper deficiency impairs growth of infants recovering from malnutrition. *Am. J. Clin. Nutr.* **47**, 710–714.

Chung, K., Romero, N., Tinker, D., Keen, C. L., Amemiya, K., and Rucker, R. (1988). Role of copper in the regulation and accumulation of superoxide dismutase and metallothionein in rat liver. *J. Nutr.* **118**, 859–864.

Chelly, J., Tumer, Z., Tonnesen, T., Petterson, A., Ishikawa-Brush, Y., Tommerup, N., Horn, N., and Monaco, A. P. (1993). Isolation of a candidate gene for Menkes disease that encodes a potential heavy metal binding protein. *Nature Genet.* **3**, 14–19.

Dubick, M. A., Yu, G. S., and Majumdar, A. P. (1989). Morphological and biochemical changes in the pancreas of the copper-deficient female rat. *J. Nutr.* **119**, 1165–1172.

Fleming, R. E., and Gitlin, J. D. (1990). Primary structure of rat ceruloplasmin and analysis of tissue-specific gene expression during development. *J. Biol. Chem.* **265**, 7701–7707.

Freedman, J. H., Weiner, R. J., and Peisach, J. (1986). Resistance to copper toxicity of cultured hepatoma cells. *J. Biol. Chem.* **261**, 11840–11848.

Gitlin, J. D. (1988). Transcriptional regulation of ceruloplasmin gene expression during inflammation. *J. Biol. Chem.* **263**, 6281–6287.

Harris, E. D. (1995). The iron-copper connection: The link to ceruloplasmin grows stronger. *Nutr. Rev.* **53**, 170–173.

Klevay, L. M., and Medeiros, D. M. (1996). Deliberations and evaluations of the approaches, endpoints and paradigms for dietary recommendations about copper. *J. Nutr.* **126**, 2419S–2426S.

Klinman, J. (1996). New quinocofactors in eukaryotes. *J. Biol. Chem.* **271**, 27189–27192.

Lutsenko, S., Petrukhin, K., Cooper, M. J., Gilliam, C., and Kaplan, J. H. (1997). N-terminus domains of human copper-transporting adenosine triphosphatases (the Wilson's and Menkes disease proteins) bind copper selectively *in vivo* and *in vitro* with stoichiometry of one copper per metal-binding repeat. *J. Biol. Chem.* **272**, 18939–18944.

Prohaska, J. R., and Lukasewycz, O. A. (1989). Copper deficiency during perinatal development. *J. Nutr.* **119**, 922–931.

Rayton, J. K., and Harris, E. D. (1979). Induction of lysyl oxidase with copper. *J. Biol. Chem.* **254**, 621–626.

Riodan, J. R., and Jolicoeur-Paquet, L. (1982). Metallothionein accumulation may account for intracellular copper retention in Menkes disease. *J. Biol. Chem.* **257**, 4639–4645.

Rucker, R. B., Romero-Chapman, N., Wong, T., Lee, J., Steinberg, F. M., McGee, C., Clegg, M. S., Reiser, K., Kosonen, T., Uriu-Hare, J., Murphy, J., and Keen, C. L. (1996). Modulation of lysyl oxidase by dietary copper in rats. *J. Nutr.* **126**, 51–60.

Shah, A., Chernov, I., Zhang, H., Ross, B., Das, K., Lutsenko, S., Parano, E., Pavone, L., Evgrafo, O., Ivanova-Smolenskaya, I., Anneren, G., Westmark, K., Urrutia, F., Penchaszadeh, G., Sternlieb, I., Scheinberg, I., Gilliam, T., and Perukhin, K. (1997). Identification and analysis in the Wilson disease gene (ATP7B). *Am. J. Hum. Genet.* **61**, 317–328.

Strause, L. G., Hegenauer, J., Saltman, P., Cone, R., and Resnick, D. (1986). Effects of long-term dietary manganese and copper deficiency on rat skeleton. *J. Nutr.* **116**, 135–141.

Tagliavacca, L., Moon, N., Dunham, W., and Kaufman, R. (1997). Identication and functional requirement of Cu(I) and its ligands within coagulation factor VIII. *J. Biol. Chem.* **272**, 27428–27434.

Tumer, Z., Lund, C., Tolshave, J., Vural, B., Tonnesen, T., and Horn, N. (1997). Identification of point mutations in 41 unrelated patients affected with Menkes disease. *Am. J. Hum. Genet.* **60**, 63–71.

Turnlund, J. R., Swanson, C. A., and King, J. C. (1983). Copper absorption and retention in pregnant women fed diets based on animal and plant proteins. *J. Nutr.* **113**, 2346–2352.

Turnlund, J., Scott, K., Peiffer, G., Jang, A., Keyes, W., Keen, C., and Sakanashi, T. M. (1997). Copper status of young men consuming a low-copper diet. *Am. J. Clin. Nutr.* **65**, 72–78.

Turnlund, J. R., Keyes, W. R., Anderson, H. L., and Acord, L. L. (1989). Copper absorption and retention in young men at three levels of dietary copper by use of the stable isotope $^{65}$Cu. *J. Nutr.* **49**, 870–878.

Valentine, J. S., and Gralla, E. B. (1997). Delivering copper inside yeast and human cells. *Science* **278**, 817–819.

Vulpe, C., Levinson, B., Whitney, S., Packman, S., and Gitschier, J. (1993). Isolation of a candidate gene for Menkes disease and evidence that it encodes a copper-transporting ATPase. *Nature Genet.* **3**, 7–13.

Wood, R. J., Suter, P. M., and Russell, R. M. (1995). Mineral requirements of elderly people. *Am. J. Clin. Nutr.* **62**, 493–505.

## *Molybdenum, Sulfite, and Sulfate*

Abdelhaq, E. H., and Labuza, T. P. (1987). Air drying characteristics of apricots. *J. Food Sci.* **52**, 342–345.

Abumrad, N. N., Schneider, A. J., Steel, D., and Rogers, L. S. (1981). Amino acid intolerance during prolonged total parenteral nutrition reversed by molybdate therapy. *Am. J. Clin. Nutr.* **34**, 2551–2559.

Amy, N. K. (1988). Effect of dietary protein and methionine on sulfite oxidase activity in rats. *J. Nutr.* **118**, 941–944.

Bolin, H. R., and Stafford, A. E. (1976). Sulfite absorption increase in peaches for drying. *J. Food Sci.* **41**, 1484–1485.

Bougle, D., Bureau, F., Foucault, P., Duhamel, J.-F., Muller, G., and Drosdowsky, M. (1988). Molybdenum content of term and preterm human milk during the first 2 months of lactation. *Am. J. Clin. Nutr.* **48**, 652–654.

Carter, S. R., Slomiany, A., Gwozdzinski, K., Liaau, Y. H., and Slomiany, B. L. (1988). Enzymatic sulfation of mucus glycoprotein in gastric mucosa. *J. Biol. Chem.* **263**, 11977–11984.

Cohen, H. J., Drew, R. T., Johnson, J. L., and Rajagopalan, K. V. (1973). Molecular basis of the biological function of molybdenum. The relationship between sulfite oxidase and the acute toxicity of bisulfite and $SO_2$. *Proc. Natl. Acad. Sci. U.S.A.* **70**, 3655–3659.

Esko, J. D., Elgavish, A., Prasthofer, T., Taylor, W. H., and Weinke, J. L. (1986). Sulfate transport-deficient mutants of Chinese hamster ovary cells. *J. Biol. Chem.* **261**, 15725–15733.

Freeland-Graves, J., and Turnlund, J. R. (1996). Deliberations and evaluations of the approaches, endpoints and paradigms for manganese and molybdenum dietary recommendations. *J. Nutr.* **126**, 2435S–2440S.

Gardlik, S., and Rajagopalan, K. V. (1990). The state of reduction of molybdopterin in xanthine oxidase and sulfite oxidase. *J. Biol. Chem.* **265**, 13047–13054.

Gunnison, A. F. (1981). Sulphite toxicity: A critical review of *in vitro* and *in vivo* data. *Food Cosmet. Toxicol.* **19**, 667–682.

Gunnison, A. F., Dulak, L., Zaccardi, J., and Farruggella, T. J. (1981). A sulfphite oxidase deficient rat model: Subchronic toxicity. *Food Cosmet. Toxicol.* **19**, 221–232.

Huttner, W. B. (1987). Protein tyrosine sulfation. *Trends Biol. Sci.* **12**, 361–363.

Inoue, H., Otsu, K., Yoneda, M., Kimata, K., Suzuki, S., and Nakanishi, Y. (1986). Glycosaminoglycan sulfotransferases in human and animal sera. *J. Biol. Chem.* **261**, 4460–4469.

Johnson, J. L., Wuebbens, M. M., Mandell, R., and Shih, V. E. (1989). Molybdenum cofactor biosynthesis in humans. *J. Clin. Invest.* **83**, 897–903.

Kramer, S. P., Johnson, J. L., Ribeiro, A. A., Mullington, D. S., and Rajagopalan, K. V. (1987). The structure of molybdenum cofactor. *J. Biol. Chem.* **262**, 16357–16363.

Krenitsky, T. A., Neil, S. M., Elion, G. B., and Hitchings, G. H. (1972). A comparison of the specificities of xanthine oxidase and aldehyde oxidase. *Arch. Biochem. Biophys.* **150**, 585–599.

Niehrs, C., Kraft, M., Lee, R. W., and Huttner, W. B. (1990). Analysis of the substrate specificity of tyrosylprotein sulfotransferase using synthetic peptides. *J. Biol. Chem.* **265**, 8525–8532.

Nordlee, J. A., Martin, L. B., and Taylor, S. L. (1985). Sulfite residues in marachino cherries. *J. Food Sci.* **50**, 256–257.

Rajagopalan, K. V. (1988). Molybdenum: An essential trace element in human nutrition. *Annu. Rev. Nutr.* **8**, 401–427.

Smith, J. T. (1973). An optimal level of inorganic sulfate for the diet of a rat. *J. Nutr.* **103**, 1008–1011.

Stipanuk, M. H. (1986). Metabolism of sulfur-containing amino acids. *Annu. Rev. Nutr.* **6**, 179–209.

Tsongas, T. A., Meglen, R. R., Walravens, P. A., and Chappell, W. R. (1980). Molybdenum in the diet: An estimate of average daily intake in the United States. *Am. J. Clin. Nutr.* **33**, 1103–1107.

Turnlund, J. R., Keyes, W. R., and Peiffer, G. L. (1995). Molybdenum absorption, excretion, and retention studied with stable isotopes in young men at five intakes of dietary molybdenum. *Am. J. Clin. Nutr.* **62**, 790–796.

Whiting, S. J., and Draper, H. H. (1981). Effect of chronic acid load as sulfate or sulfur amino acids on bone metabolism in adult rats. *J. Nutr.* **111**, 1721–1726.

Yang, M. T., and Yang, S. P. (1989). Effect of molybdenum supplementation on hepatic trace elements and enzymes of female rats. *J. Nutr.* **119**, 221–227.

*Selenium*

Bansal, M. P., Cook, R. G., Danielson, K. G., and Medina, D. (1989). A 14-kilodalton selenium-binding protein in mouse liver is fatty acid-binding protein. *J. Biol. Chem.* **264**, 13780–13784.

Butler, J. A., Whanger, P. D., Kaneps, A. J., and Patton, N. M. (1990). Metabolism of selenite and selenomethionine in the rhesus monkey. *J. Nutr.* **120**, 751–759.

Chambers, I., Frampton, J., Goldfarb, P., Affara, N., McBain, W., and Harrison, P. R. (1986). The structure of the mouse glutathione peroxidase gene. *EMBO J.* **5**, 1221–1227.

Chaudiere, J., Wilhelmsen, E. C., and Tappel, A. L. (1984). Mechanism of selenium-glutathione peroxidase and its inhibition by mercaptocarboxylic acids and other mercaptans. *J. Biol. Chem.* **259**, 1043–1050.

Combs, G. F., and Combs, S. B. (1984). "The Role of Selenium in Nutrition." Academic Press, San Diego.

Combs, G. F., Noguchi, T., and Scott, M. L. (1975). Mechanisms of action of selenium and vitamin E in protection of biological membranes. *Fed. Proc.* **34**, 2090–2095.

DePalo, D., Kinlaw, W. B., Zhao, C., Engelberg-Kulka, H., and St. Germain, D. L. (1994). Effect of selenium deficiency on type I 5′-deiodinase. *J. Biol. Chem.* **269**, 16223–16228.

Diplock, A. T. (1987). Trace elements in human health with special reference to selenium. *Am. J. Clin. Nutr.* **45**, 1313–1322.

Fairweather-Tait, S. J. (1997). Bioavailability of selenium. *Eur. J. Clin. Nutr., Suppl.* **51**(1), S20–S23.

Hoekstra, W. G. (1975). Biochemical function of selenium and its relationship to vitamin E. *Fed. Proc.* **34**, 2083–2089.

Leinfelder, W., Zehelein, B., Mandrand-Berthelot, M., and Bock, A. (1988). Gene for a novel tRNA species that accepts L-serine and cotranslationally inserts selenocysteine. *Nature* **331**, 723–725.

Levander, O. A. (1985). Considerations on the assessment of selenium status. *Fed. Proc.* **44**, 2579–2583.

Levander, O. A. (1987). A global view of human selenium nutrition. *Annu. Rev. Nutr.* **7**, 227–250.

Levander, O. A., Alfthan, G., Arvilommi, H., Gref, C. G., Huttunen, J. L., Kataja, M., Koivistoinen, P., and Pikkarainen, J. (1983). Bioavailability of selenium to Finnish men as assessed by platelet glutathione peroxidase activity and other blood parameters. *Am. J. Clin. Nutr.* **37**, 887–897.

Li, N., Reddy, P. S., Thyagaraju, K., Reddy, A. P., Hsu, B. L., Scholz, R. W., Tu, C., and Reddy, C. C. (1990). Elevation of rat liver mRNA for selenium-dependent glutathione peroxidase by selenium deficiency. *J. Biol. Chem.* **265**, 108–113.

McFadden, G. (1998). Even viruses can learn to cope with stress. *Science* **279**, 40–41.

Pallud, S., Lennon, A.-M., Ramauge, M., Gaveret, J.-M., Croteau, W., Pierre, M., Courtin, F., and St. Germain, D. L. (1997). Expression of the type II iodothyronine deiodinase in cultured rat astrocytes is selenium-dependent. *J. Biol. Chem.* **272**, 18104–18110.

Patterson, E. L., Milstrey, R., and Stokstad, E. L. R. (1957). Effect of selenium in preventing exudative diathesis in chicks. *Proc. Soc. Exp. Biol. Med.* **95**, 617–620.

Rayman, M. P. (1997). Dietary selenium: Time to act. *Brit. Med. J.* **314**, 387–388.

Read, R., Bellew, T., Yang, J.-G., Hill, K. E., Palmer, I. S., and Burk, R. F. (1990). Selenium and amino acid composition of selenoprotein P, the major selenoprotein in rat serum. *J. Biol. Chem.* **265**, 17899–17905.

Salbe, A. D., and Levander, O. A. (1990). Comparative toxicity and tissue retention of selenium in methionine-deficient rats fed sodium selenate or L-selenomethionine. *J. Nutr.* **120**, 207–212.

Shisler, J. L., Senkevich, T. G., Berry, M., and Moss, B. (1998). Ultraviolet-induced cell death blocked by a selenoprotein from a human dermatotropic poxvirus. *Science* **279**, 102–105.

St. Germain, D. L. (1994). Iodothyronine deiodinases. *Trends Endocrinol. Metab.* **5**, 36–42.

Sturchler, C., Westhof, E., Carbon, P., and Krol, A. (1993). Unique secondary and tertiary structural features of the eucaryotic selenocysteine tRNA. *Nucl. Acids Res.* **21**, 1073–1079.

Sunde, R. A., and Evenson, J. K. (1987). Serine incorporation into the selenocysteine moiety of glutathione peroxidase. *J. Biol. Chem.* **262**, 933–937.

Tyrala, E., Borschel, M. W., and Jacobs, J. R. (1996). Selenate forification of infant formulas improves the selenium status of preterm infants. *Am. J. Clin. Nutr.* **64**, 860–865.

Vendeland, S. C., Beilstein, M., Yeh, J., Ream, W., and Whanger, P. D. (1995). Rat skeletal muscle selenoprotein W: cDNA clone and mRNA modulation by dietary selenium. *Proc. Natl. Acad. Sci. U.S.A.* **92**, 8749–8753.

Yang, G., Wang, S., Zhou, R., and Sun, S. (1983). Endemic selenium intoxication of humans in China. *Am. J. Clin. Nutr.* **37**, 872–881.

Yeh, J.-Y., Vendeland, S., Gu, Q.-P., Butler, J.A., Ou, B.-R., and Whanger, P. D. (1997). Dietary selenium increases selenoprotein W levels in rat tissues. *J. Nutr.* **127**, 2165–2172.

## Glutathione

Bogaards, J., Verhagen, Willems, M., Poppel, G., and Bladeren, P. (1994). Consumption of Brussels sprouts results in elevated α-class glutathione *S*-transferase levels in human blood plasma. *Carcinogenesis* **15**, 1073–1075.

Boyd, S. C., Sasame, H. A., and Boyd, M. R. (1979). High concentrations of glutathione in glandular stomach: Possible implications for carcinogenesis. *Science* **205**, 1010–1012.

Burk, R. F., Lane, J. M., and Petel, K. (1984). Relationship of oxygen and glutathione in protection against carbon tetrachloride-induced hepatic microsomal lipid peroxidation and covalent binding in the rat. *J. Clin. Invest.* **74**, 1996–2001.

Carter, B. Z., Wiseman, A. L., Orkiszewski, R., Ballard, K. D., Ou, C., and Lieberman, M. W. (1997). Metabolism of leukotriene C₄ in γ-glutamyl transpeptidase-deficient mice. *J. Biol. Chem.* **272**, 12305–12310.

Dahl, N., Pigg, M., Ristoff, E., Gali, R., Carlsson, B., Mannervik, B., Larsson, A., and Board, P. (1997). Missense mutations in the human glutathione synthase gene result in severe metabolic acidosis, 5-oxoprolinuria, hemolytic anemia and neurological dysfunction. *Human Mol. Genet.* **6**, 1147–1152.

Halliwell, B., and Gutteridge, J. M. (1990). The antioxidants of human extracellular fluids. *Arch. Biochem. Biophys.* **280**, 1–8.

Harding, C. W., Williams, P., Wagner, E., Chang, D., Wild, K., Colwell, R., and Wolff, J. (1997). Mice with genetic γ-glutamyl transpeptidase deficiency exhibit glutathionuria, severe growth failure, reduced life spans, and infertility. *J. Biol. Chem.* **272**, 12560–12567.

Harrap, K. R., Jackson, R. C., Riches, P. G., Smith, C. A., and Hill, B. T. (1973). The occurrence of protein-bound mixed disulfides in rat tissues. *Biochim. Biophys. Acta* **310**, 104–110.

Ishikawa, T., and Sies, H. (1989). Glutathione as an antioxidant: Toxicological aspects. *In* "Glutathione," Part B (D. Dolphin, O. Avramovic, and R. Poulson, eds.), pp. 85–109. Wiley, New York.

Jakoby, W. B., and Ziegler, D. M. (1990). The enzymes of detoxification. *J. Biol. Chem.* **265**, 20715–20718.

Johnson, W., Ueng, Y.-F., Widersten, M., Mannervik, B., Hayes, J. D., Sherratt, P. J., Ketterer, B., and Guengerich, F. P. (1997). Conjugation of highly reactive aflatoxin B1 exo-8,9-epoxide catalyzed by rat and human glutathione transferases: Estimation of kinetic parameters. *Biochemistry* **36**, 3056–3060.

Jones, D. P., Thor, H., Andersson, B., and Orrenius, S. (1978). Detoxification reactions in isolated hepatocytes. *J. Biol. Chem.* **253**, 6031–6037.

Kaplowitz, M., Aw, T. Y., and Ookhtens, M. (1985). The regulation of hepatic glutathione. *Annu. Rev. Pharmacol. Toxicol.* **25**, 715–744.

Lee, W., Hawkins, R., Peterson, D., and Vina, J. R. (1996). Role of oxoproline in the regulation of neutral amino acid transport across the blood-brain barrier. *J. Biol. Chem.* **271**, 19129–19133.

Leeuwenburgh, C., Hollander, J., Leichtweis, S., Griffiths, M., Gore, M., and Ji, L. L. (1997). Adaptations of glutathione antioxidant system to endurance training are tissue and muscle fiber specific. *Am. J. Physiol.* **272**, R363–R369.

Lieberman, M. W., Wiseman, A., Shi. Z.,,, Carter, B., Barrios, R., Ou, C., Chavez-Barrios, P., Wang, Y., Habib, G., Goodman, J., Huang, S., Lebovitz, R., and Matzuk, M. (1996). Growth retardation and cysteine deficiency in γ-glutamyl transpeptidase-deficient mice. *Proc. Natl. Acad. Sci. U.S.A.* **93**, 7923–7926.

Lin, D., Meyer, D., Ketter, B., Lang, N., and Kadlubar, F. (1994). Effects of human and rat glutathione S-transferases on the covalent DNA binding of N-acetoxy derivatives of heterocyclic amine carcinogens *in vitro*. *Cancer Res.* **54**, 4920–4926.

Meister, A. (1988). Glutathione metabolism and its selective modification. *J. Biol. Chem.* **263**, 17205–17208.

Meister, A. (1989). Metabolism and function of glutathione. *In* "Glutathione," Part A (D. Dolphin, O. Avramovic, and R. Poulson, eds.), pp. 367–474. Wiley, New York.

Meredith, M. J., and Reed, D. J. (1982). Status of the mitochondrial pool of glutathione in the isolated hepatocyte. *J. Biol. Chem.* **257**, 3747–3753.

Nijhoff, W., Grubben, M., Nagengast, F., Jansen, J., Verhagen, H., Poppel, G., and Peters, W. (1995). Effect of consumption of Brussels sprouts on intestinal and lymphocytic glutathione S-transferases in humans. *Carcinogenesis* **16**, 2125–2128.

Sarau, H. M., Foley, J. J., Moonsammy, G., and Sachs, G. (1977). Metabolism of dog gastric mucosa. *J. Biol. Chem.* **252**, 8572–8581.

Scoggan, K. A., Jakobsson, P.-J., and Ford-Hutchinson, A. W. (1997). Production of leukotriene C4 in different human tissues is attributable to distinct membrane bound biosynthetic enzymes. *J. Biol. Chem.* **272**, 10182–10187.

Starke, P. E., and Farber, J. L. (1985). Endogenous defenses against the cytotoxicity of hydrogen peroxide in cultured rat hepatocytes. *J. Biol. Chem.* **260**, 86–92.

Thornalley, P. J. (1990). The glyoxalase system: New developments towards functional characterization of a metabolic pathway fundamental to biological life. *Biochem. J.* **269**, 1–11.

Wefers, H., and Sies, H. (1983). Oxidation of glutathione by the superoxide radical to the disulfide and the sulfonate yielding singlet oxygen. *Eur. J. Biochem.* **137**, 29–36.

Wefers, H., and Sies, H. (1988). The protection by ascorbate and glutathione against microsomal lipid peroxidation is dependent on vitamin E. *Eur. J. Biochem.* **174**, 353–357.

*Cancer, Selenium, and Glutathione*

Ames, B. N., Gold, L. S., and Willett, W. C. (1995). The causes and prevention of cancer. *Proc. Natl. Acad. Sci. U.S.A.* **92**, 5258–5265.

Ames, B. N., Profet, M., and Gold, L. S. (1990). Nature's chemicals and synthetic chemicals: Comparative toxicology. *Proc. Natl. Acad. Sci. U.S.A.* **87**, 7782–7786.

Ball, R. W., Wilson, D. W., and Coulombe, R. A. (1990). Comparative formation and removal of aflatoxin B1-DNA adducts in cultured mammalian tracheal epithelium. *Cancer Res.* **50**, 4918–4922.

Campbell, T. C., Chen, J., Liu, C., and Parpia, B. (1990). Nonassociation of aflatoxin with
primary liver cancer in cross-sectional ecology survey in the People's Republic of
China. *Cancer Res.* **50**, 6882–6893.

Clark, L., Combs, G., Turnbull, B., Slate, E., Chalker, D., Chow, J., Davis, L., Glover, R.,
Graham, G., Gross, E., Krongrad, A., Lesher, J., Park, H., Sanders, B., Smith, C., Taylor,
J.R. (1996). Effects of selenium supplementation for cancer prevention in patients with
carcinoma of the skin. *JAMA* **276**, 1957–1963.

Colditz, G. A. (1996). Selenium and cancer prevention. *JAMA* **276**, 1984–1985.

Coles, B., and Ketterer, B. (1990). The role of glutathione and glutathione transferases in
chemical carcinogenesis. *Crit. Rev. Biochem. Mol. Biol.* **25**, 47–70.

Lanfear, J., Fleming, J., Wu, L., Webster, G., and Harrison, P. R. (1994). The selenium
metabolite selenodiglutathione induces p53 and apoptosis: Relevance to the chemoprevention effects of selenium? *Carcinogenesis* **15**, 1387–1392.

Loeb, L. A. (1989). Endogenous carcinogenesis: Molecular oncology into the twenty-first
century — presidential address. *Cancer Res.* **49**, 5489–5496.

Miller, E. C., and Miller, J. A. (1986). Carcinogens and mutagens that may occur in foods.
*Cancer* **58**, 1795–1803.

Pelkonen, O., and Nebert, D. W. (1982). Metabolism of polycyclic aromatic hydrocarbons:
Etiological role in carcinogenesis. *Pharm. Rev.* **34**, 189–222.

Randerath, E., Miller, R. H., Mittal, D., Aritts, T. A., Dunsford, H. A., and Randerath, K.
(1989). Covalent DNA damage in tissues of cigarette smokers as determined by [32]P-
postlabeling assay. *J. Natl. Cancer Inst.* **81**, 341–347.

Reddy, B., Rivenson, A., El-Bayoumy, K., Upadhyaya, P., Pittman, B., and Rao, C. V. (1997).
Chemoprevention of colon cancer by organoselenium compounds and impact of high-
or low-fat diets. *J. Nat. Cancer Inst.* **89**, 506–512.

Robertson, I. G., Guthenburg, C., Mannervik, B., and Jernstrom, B. (1986). Differences in
stereoselectivity and catalytic efficiency of three human glutathione transferases in
conjunction of glutathione with 7β,8α-dihydroxy-9α,10α-oxytetrahydrobenzo(α)py-
rene. *Cancer Res.* **46**, 2220–2224.

Sell, S. (1990). Is there a liver stem cell? *Cancer Res.* **50**, 3811–3815.

Waxman, D. J. (1990). Glutathione *S*-transferases: Role in alkalating agent resistance and
possible target for modulation chemotherapy — a review. *Cancer Res.* **50**, 6449–6454.

Willett, W. C. (1997). Potential benefits of preventive nutrition strategies. *In* "Preventive
Nutrition" (A. Bendich and R. J. Deckelbaum, eds.), pp. 423–440. Humana Press, Totowa, NJ.

## Other Inorganic Nutrients

Campbell, W. W., Beard, J. L., Joseph, L. J., Davey, S. L., and Evans, W. J. (1997). Chromium
picolinate supplementation and resistive training by older men: Effects on iron-status
and hematological indexes. *Am. J. Clin. Nutr.* **66**, 944–949.

Carlisle, E. M. (1985). Silicon as an essential trace element in animal nutrition. *In* "Silicon
Biochemistry," Ciba Foundation Symposium 21, pp. 123–139. Wiley, New York.

Hallmark, M. A., Reynolds, T. H., DeSouza, C. A., Dotson, C. O., Anderson, R. A., and
Rogers, M. A. (1996). Effects of chromium and resistive training on muscle strength and
body composition. *Med. Sci. Sports Exerc.* **28**, 139–144.

Hunt, C. D., and Stoecker, B. J. (1996). Deliberations and evaluations of the approaches,
endpoints and paradigms for boron, chromium and fluoride dietary recommendations.
*J. Nutr.* **126**, 2441S–2451S.

Hunt, C. D., Herbel, J., and Nielsen, F. H. (1997). Metabolic responses of postmenopausal women to supplemental dietary boron and aluminum during usual and low magnesium intake: Boron, calcium, and magnesium absorption and retenton and blood mineral concentrations. *Am. J. Clin. Nutr.* **65**, 803–813.

Meacham, S. L., Taper, L. J., and Volpe, S. L. (1995). Effect of boron supplementation on blood and urinary calcium, magnesium, and phosphorus, and urinary boron in athletic and sedentary women. *Am. J. Clin. Nutr.* **61**, 341–345.

Mertz, W. (1993). Chromium in human nutrition: A review. *J. Nutr.* **123**, 626–633.

Patriarca, M., Lyon, T., and Fell, G. S. (1997). *Nickel metabolism in humans investigated with an oral stable isotope. Am. J. Clin. Nutr.* **66**, 616–621.

Reeves, P. G. (1997). Components of the AIN-93 diets as improvements in the AIN-76A diet. *J. Nutr.* **127**, 838S–841S.

Subiyatno, A., Mowat, D., and Yang, W. Z. (1996). Metabolite and hormonal responses to glucose or propionate infusions in periparturient dairy cows supplemented with chromium. *J. Dairy Sci.* **79**, 1436–1445.

# 11

# DIET AND CANCER

## OVERVIEW

Cancer is a disease that involves the acquisition of mutations in several specific genes in a cell, where the sum of these mutations allows the cell to divide and multiply and invade surrounding tissues and organs. Most genetic diseases, such as phenylketonuria, result from a change in a specific gene in the DNA, where the change occurs in all of the cells of the body. All of the cells of the body are affected because the change (a mutation) in the DNA was initially present in the father's germ cells or in the mother's eggs. In cancer, the genetic change arises in one cell in an existing person, resulting in only a small group of daughter cells (the cancer cells) that contains the mutation.

### Mutations in APC, RAS, p53, and Cadherin

This chapter will focus on mutations in genes that code for proteins called APC, RAS, p53, and cadherin. RAS is pronounced "rass." The name "p53" arose because it is a protein (p) with a molecular weight of 53 kDa. These proteins have vital and useful functions in the body, as do nearly all of our proteins. However, these proteins have the unfortunate property that, when altered by certain mutations, they convert their host cell to a cancer cell. The genes coding for APC, RAS, p53, and a variety of other proteins are called *proto-oncogenes* prior to mutation, and *oncogenes* after they have been mutated. In the case of any oncogene, the gene itself is totally harmless. The danger occurs because of the altered protein that is produced. Many of the proto-oncogenes were discovered and identified before their function was suspected. For this reason, some of their names are not related to their function, and may consist only of the initials of the researcher.

The functions of the normal (before mutation) and abnormal (after mutation) forms of these genes are known for several of these oncogenes. The following commentary will detail changes occurring in APC, RAS, p53, and cadherin. Mutations in the genes coding for these proteins occur in colorectal cancer, but also in other types of cancer. A mutation in any one of these genes is not sufficient to convert a normal cell to a cancer cell. The net effect of mutations occurring in several specific genes, such as these, is sufficient for conversion of a normal gut cell to a cancer cell. Mutations can involve the change of one base to another, the deletion of one or more bases, or even the permanent transfer of a gene from one chromosome to another chromosome.

## Epidemiology and Cancer

Epidemiological data have shed light on some of the mechanisms of cancers of the colon and rectum, as reviewed towards the end of this chapter. Diets high in fruits and vegetables have been associated with lower rates of colorectal cancer. Diets high in red meat have been associated with higher rates of colorectal cancer. An eventual goal is to find a consistent story, using data acquired from large populations as well as from test tubes. The common goal of epidemiologists and lab researchers is to tell each other where to look in the search for dietary risks in cancer.

## Difference between DNA Damage and Mutations

The science of DNA damage, DNA repair, and mutations constitutes a vital bridge that is needed to connect the conceptual gap between *oncogenes* and *food eating habits*. In other words, any evidence that drinking orange juice can prevent mutations in the p53 gene can best be rationalized with a commentary that involves the chemistry and enzymology of DNA damage.

## Forms of Damage Occurring in DNA

DNA consists of two molecules of single-stranded DNA (ssDNA) wrapped together in a fiber called the double-stranded helix (dsDNA). Damage can be done to DNA by a number of separate and distinct processes, for example:

1. Spontaneous hydrolysis of the amino group of cytosine. This results in cytosine (C) being converted to thymine (T) plus free ammonia.

2. Damage due to hydroxyl radicals. Small amounts of reduced iron bound to our DNA may allow the Fenton reaction to occur, thus producing HO• in the immediate vicinity of DNA. The HO• radical attacks guanine (G) residues, producing residues of 8-oxo-guanine (oxo-G).

3. Covalent damage due to naturally occurring components of the diet. Compounds in cooked meat, for example, are converted in the body to reactive chemicals, which in turn attack our DNA.

4. Covalent damage due to chemicals acquired from cigarette smoke.

5. Covalent damage due to ultraviolet rays from the sun.

6. Rare mistakes made by DNA polymerase during DNA synthesis (replication).

Damage may also be acquired from other sources. For example, workers in chemical plants, farm workers, and victims of nuclear powerplant disasters acquire damage from specific sources (synthetic chemicals and radiation) that do not affect the general population.

## How Does the Cell Respond to Damage to Its DNA?

The acquisition of DNA damage in a proto-oncogene, by itself, does not result in its conversion to an oncogene. In most cases, the damage is repaired or reversed by the cell's DNA repair system. A second type of response to DNA damage is for the cell to commit suicide and die. This orderly, controlled process of suicide is called *apoptosis*. How may apoptosis benefit the organism? It is far more desirable to lose an occasional cell by apoptosis than to retain and propagate a damaged, abnormal cell. A third response to DNA damage is for no response to occur at all, that is, for the damage to linger until the time of DNA synthesis, resulting in a permanent mutation in the daughter cells.

## Behavior of DNA Polymerase When It Encounters a Lesion

The exact nature of the mutation is a reflection of two things, namely the structure of the lesion and the response of DNA polymerase to the lesion. For convenience, any type of damage may be called a "lesion." During the process of replication, DNA polymerase may encounter a lesion and fortuitously insert the correct DNA base across from a lesion, resulting in a daughter double helix that is totally normal. However, when DNA polymerase encounters a lesion, it may also insert an incorrect DNA base. The result is that one of the two daughter double helices will contain only correct bases, while the other daughter will contain the lesion (in one strand) and an incorrect base (in the opposite strand). Damage to DNA, whether taking the form of a modified base, a hydrolyzed base, or a broken strand, is not called a mutation. Only when this damage results in a stable alteration in the DNA that can be transmitted from generation to generation is the result called a mutation.

Not all mutations are harmful. DNA bases that code for polypeptides occur as a series of triplets. Each triplet is called a codon and codes either for a specific amino acid, or for a translation stop signal. Please recall that the genetic code is degenerate and that some amino acids are coded for by more than one type of codon. Hence, a mutated base may change the codon, but with no alteration of the amino acid that is coded. Furthermore, a change in amino acid may or may not result in a change in function of the polypeptide.

In viewing dietary habits that are associated with greater or lesser rates of cancer, one might keep in mind the question, "Which of the components of food would be expected to provoke specific mutations, and which components might be expected to reduce the rate of mutations?" Various components of the diet react may directly with DNA. **Aflatoxin** is the most notorious dietary contaminant that reacts directly with DNA. Other components may react with toxic oxygen and inactivate it, thus preventing the rate of DNA damage. Still other components, which are chronically present in the diet, may interact with certain regulatory proteins and provoke sustained changes in the cell that contribute to cancer. **Phorbol ester** is the most notorious plant chemical that interacts with a regulatory protein and causes cancer. The chemical occurs in croton oil and activates protein kinase C.

## CANCER OF THE LARGE BOWEL

We will focus on cancer of the large bowel, rather than on revealing fragments of information on all of the major types of cancers. The incidence of cancer of the large bowel (colon and rectum) in the United States is 136,000 cases per year, while the incidence of death from colorectal cancer is about 55,000 per year (Lewis, 1996). Large bowel cancer develops with the appearance of a distinct number of genetic mutations, as do all other types of cancers. In the case of cancers of colorectal cancer, the steps tend to follow this progression:

1. Normal epithelial cell.
2. Polyp formation. These polyps are harmless (benign) tumors and are called *adenomas*.
3. Cancer cell. The cell continuously grows and divides.
4. Invasive cancer cell. The cell migrates and penetrates other tissues.

### Polyps Occur in the Mucosal Layer of the Large Intestines

Polyps may take two forms. They can occur with stalks or without stalks (Haggitt *et al.*, 1985). Polyps, or adenomas, with stalks are called *pedunculated*, while those without stalks are called *sessile* (Figure 11.1). When discovered in the doctor's

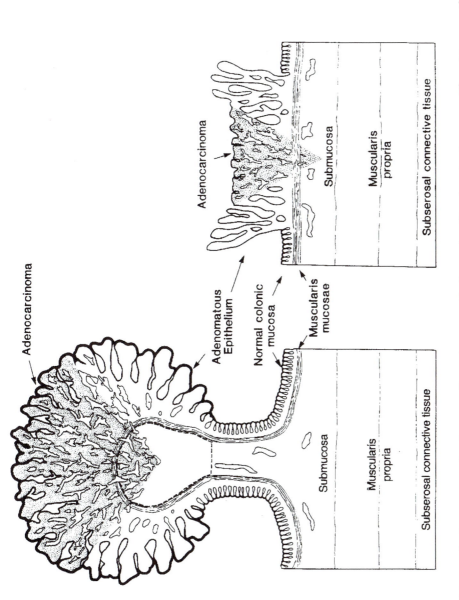

**Figure 11.1.** Pictures of adenomas — structures on the gut mucosa that occasionally give rise to cancer. The picture on the left is a pedunculated adenoma. It contains a stalk. The shaded areas depict regions of the adenoma that, at the onset of cancer, typically contain cancerous cells. As the cancer cells divide and multiply, and invade deeper regions within the gut tissue, such as the lining of muscle, the cancer becomes more difficult to remove by surgery. (Reproduced with permission from Haggitt *et al.*, 1985.)

office, these structures are often from 2 to 4 cm in diameter. Figure 11.1 reveals the structures that normally occur in the epithelial lining of the large intestines. These include the mucosa, submucosa, muscularis, and underlying connective tissues. The shaded areas indicate regions of the adenoma that may contain cancer cells during the early stages of colon cancer. Most of the cells in the shaded area are ordinary cancer cells, while only a tiny minority of the cancer cells may have acquired additional mutations that cause them to be invasive.

When cancer cells are detected in a pedunculated adenoma, the polyp can be easily removed by surgery by inserting a probe in the large intestine and snipping off the polyp by its stalk. The lack of the stalk in sessile adenomas allows any cancer cells to invade more easily the underlying structures. A cancerous adenoma that is confined to the mucosal area can be easily removed by a "polypectomy," whereas cancerous adenomas that have reached the submucosa are removed by "colectomy." Colectomy means the surgical opening of the abdomen, removing a segment of the intestines, and sewing together the shortened intestines. Colectomy can be an effective cure for colon cancer, except where the cancer cells have migrated to other organs.

Colorectal cancer is often detected via widely used screening tests. These include rectal exams, using the finger. This test detects only the colorectal adenomas and cancers that grow within reach of a finger, i.e., only 10% of all cases (Eddy, 1990). The fecal blood test is generally performed annually, after the age of 50, and detects adenomas and cancers throughout the entire large intestines. The test is useful because about one-third of adenomas bleed into the gut lumen before becoming cancerous. The bleeding occurs for about 2 years before the adenomas become cancer cells. It is interesting to note that the average time it takes a 1 cm wide adenoma to become an invasive cancer is about 7 years (Eddy, 1990).

## Metastasis

The stage at which a cancer cell invades other tissues is called **metastasis** (Gabbert, 1985). After formation of a harmless tumor, a small minority of the cells become invasive. This does not mean that they grow faster; it means that they move (migrate), they lose their adhesiveness to neighboring cells, and that they secrete proteases that break down barriers, such as collagen fibers and cell matrix proteins. Once migration starts and protease secretion occurs, the cancer cells invade nearby blood vessels and lymph ducts. Collagenase was measured in biopsies (tissue samples) taken from normal and abnormal large bowel. Samples were taken from the normal gut wall, adenomas, cancer invading the muscularis propria, and cancer invading the lymph nodes (Liabakk *et al.*, 1996) (Figure 11.2). The results show that that activity was not detectable in normal gut wall and adenoma samples, but was clearly detectable in about half of the samples taken from cancerous tissues. The data emphasize that the progression occurs in various stages and that protease secretion is a recognized and easily measured aspect of cancerous tissues.

Sometimes a cancer cell will travel alone through the bloodstream and begin growth in distant tissues. Death from cancer can result, for example, when invading cancer cells encircle a blood vessel (to acquire the nutrients in the blood) and cut off the blood supply to a vital organ.

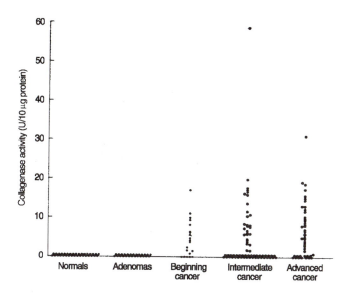

Figure 11.2.  Collagenase activity in biopsies of normal human colon, adenomas, and tissue containing various stages of cancer. Collagenase activity was expressed in terms of "specific activity." Specific activity (S.A.) is a standard term used by all biochemists to express enzyme activity. S.A. = millimoles of product/minute/mg protein, or some variation of this formula. The results show that normal tissue and adenomas contain little or no detectable activity, while a large fraction of the cancer tissues did contain activity. (Redrawn with permission from Liabakk *et al.*, 1996.)

*Possible Correlation between Gene Mutation and Cell Behavior*

The transition between each of the previous steps is thought to result from the acquisition of additional specific mutations in our DNA. In colorectal cancer, these mutations have been identified to occur in genes coding for proteins called APC, RAS, p53, and cadherin. A suggested scenario of spontaneous colorectal cancers is shown in what follows (Vogelstein and Kinzler, 1993). The scenario correlates loss in a specific gene and protein with a change in appearance of the gut. Loss of APC's function has been identified as an early step in cancer, though only loss of APC is not sufficient to result in cancer. Mutations in RAS tend to occur after mutations in APC. Mutations in p53 tend to occur at a still later step:

## GENETIC CHANGES THAT RESULT IN CANCER

Hereditary and nonhereditary forms of colon cancer exist. Nonhereditary cancers involve mutations that are said to arise spontaneously, and these mutations are called **somatic mutations**. Hereditary cancers involve mutations that are called **germline mutations**. Hereditary cancers are much less common than spontaneous cancers. About 1% of persons with colon cancer have **familial adenomatous polyposis coli**. In this inherited disease, the large bowel develops thousands of polyps at an early age. Some of these polyps become cancerous and kill the patient, unless surgically removed. The inherited defect in this disease occurs in a protein called APC. Evidence suggests that most spontaneous colon cancers also involve mutations in APC. These mutations result in conversion of normal epithelial cells to harmless polyps.

**Hereditary nonpolyposis colorectal cancer** (HNPCC) accounts for 1–5% of all colon cancers. HNPCC occurs because of a defect in DNA repair. To review the big picture, all DNA in all human cells is continuously being damaged, i.e., by spontaneous hydrolysis, by toxic oxygen, by xenobiotics, or by mistakes made by our DNA polymerase enzymes during DNA synthesis. These mistakes must be corrected by special enzymes to prevent them from causing long-lasting changes in the cell, and in daughter cells. These enzymes are called DNA repair enzymes. One such enzyme, called the **mismatch repair enzyme**, is defective in HNPCC.

## Mutations in p53

Mutations in p53 occur in spontaneous and inherited cancers. The role of p53 in DNA repair appears to be to momentarily halt the progression of the cell cycle, so as to give various DNA repair enzymes a chance to complete their job, prior to initiation of DNA synthesis. Overall, p53 is mutated in about half of all human cancers. Li–Fraumeni syndrome is a rare type of cancer that involves an inherited mutation in p53. About 50% of the affected persons acquire cancer by the age of 30 (Harvey *et al.*, 1995).

## Mutations in RAS Protein

RAS is a component of a signaling pathway, and is involved in regulating the rate of the cell's division. RAS is mutated in about 20% of all human cancers. Specific genetic diseases resulting in mutations of RAS and of cadherin seem not yet to have been identified.

## Mutations in the APC Protein

APC is a protein of 2843 amino acids. It contains short stretches of amino acids that bind catenin. These short stretches are indicated by the terms "15 amino acid repeat" and "20 amino acid repeat" in Figure 11.3. Colorectal cancer cells containing defective APC protein were studied. The APC protein was found to be mutated and inactive. Researchers used these cells as is, or introduced *normal* APC, or introduced various types of *mutated* APC into the cells (Morin *et al.*, 1997). These mutant forms were acquired from various human cancer patients, and were 331, 1309, 1941, or 2644 amino acids long. Even before looking at the data, one might expect the shorter versions to have a more impaired function and the longer versions to function at a near-normal level.

Figure 11.3 indicates the size of the normal APC protein, as well as the shortened versions of APC acquired from the various cancer patients. The results showed that the shortened (truncated) versions of the protein had lower activity in a signaling assay. The experiment provided a measure of the activity of the various versions of APC in its cell signaling pathway, and added to growing evidence that APC is used in cell signaling.

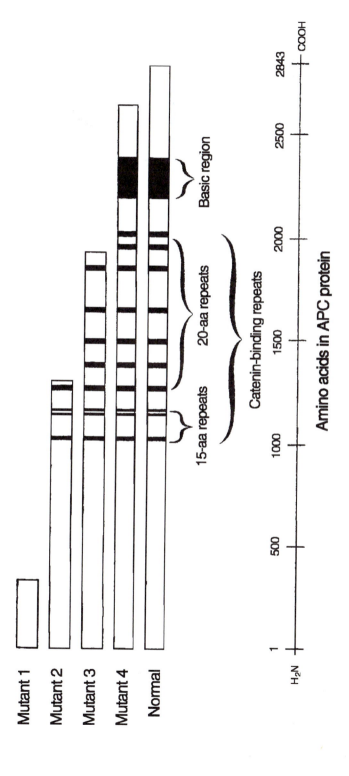

**Figure 11.3.** (A) (above) Schematic structures of normal APC protein and naturally occurring mutant APC proteins isolated from human cancer patients. The normal APC protein binds to cadherin and to catenin. The catenin-binding regions are 15 amino acids long or 20 amino acids long, as indicated. Catenin can regulate gene expression. It is thought that when APC binds catenin the event of binding prevents catenin from activating gene expression. (B) (opposite) Gene-activating properties of catenin in cultured colon cells. Cultured cancer cells contain inactive APC protein or were supplemented with normal APC protein or the indicated mutant APC proteins. Supplementation was by standard techniques of cell biology and resulted in the expression of normal or various mutant APC within the cell. After 16 hours of incubation, the influence of these types of APC on catenin's action in activating gene expression was measured. The cells containing no APC supplement showed the greatest level of gene expression, apparently because all of the catenin was free to activate the gene. The cells supplemented with normal APC showed the lowest level of gene expression, apparently because much of the catenin was tied up and bound to the APC. The results also showed that the mutant APC proteins had little, or variable, activity in controlling the activity of catenin.

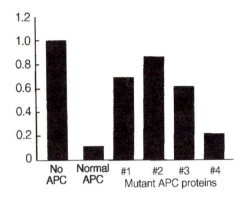

**FIGURE 11.3B**

Most of the mutations in human APC have been found to result in the conversion of CGA (cytosine–guanine–adenine), which codes for arginine, to TGA (thymine–guanine–adenine), which is a stop codon (Polakis, 1997). The stop codons in mRNA (UAA, UAG, and UGA) act as a signal during translation that halts polypeptide formation at that point. The result is synthesis of a truncated polypeptide. All types of mRNA contain stop codons; however, if a mutation results in the appearance of a stop codon in a new location, then transcription will stop at that new location.

## Heterocyclic Amines and APC

DNA damage arises via a variety of mechanisms. Some of these mechanisms are spontaneous and, for example, involve the hydrolysis of amino groups from C residues, resulting in U residues. Other mechanisms that are equally as spontaneous involve the attack of toxic oxygen at G residues, producing residues of oxo-G. DNA damage also results from the attack of xenobiotics on DNA to produce adducts. Adducts on DNA were described earlier in the section on Glutathione and Cancer. The following material illustrates how xenobiotics can attack DNA resulting in mutated APC and colon cancer.

### Source of Heterocyclic Amines in the Diet

Heterocyclic amines are generated from amino acids and other small molecules during cooking and can result in the mutation of a great number of genes in the body. Heterocyclic amines are produced when cooking meat at high temperatures. Cooked beef and pork contain the highest levels, while cooked chicken has less, and cooked fish still less (with equal cooking conditions). Cooked organ meats, eggs, cheese, and beans contain little or no heterocyclic amines. Broiling and pan-frying produces much greater levels than boiling or microwaving (Schiffman, 1990). The greatest levels of heterocyclic amines occur in charred meat. Cooking meat in aluminum foil reduces charring, and reduces the levels of these compounds. People are also exposed to heterocyclic amines when using smoky coal at home, when working in industrial coke ovens, or when working as chimney sweeps (Mumford et al., 1995).

About 20 heterocyclic amines have been identified to date. Figure 11.4 shows the structures of four of these: PhIP, IQ, MeIQ, and 4,8-DiMeIQx. These chemicals are potent mutagens and carcinogens. PhIP, for example, has been found to occur at the indicated levels in fried cod fish (70 ng/gram food), broiled beef (16 ng/g), broiled chicken (38 ng/g), and fried ground beef (0.6 ng/g) (Wakabayashi et al., 1992). Heterocyclic amines do not react with DNA as is, but must first be metabolized in the body by various enzymes. These enzymes include cytochrome P450, prostaglandin H synthase, acetyl-transferase, sulfotransferase, and other enzymes. The end-result may take the form of a nitrenium ion, which reacts with guanine residues of DNA. The adduct occurs at the C-8 position of G (Figure 11.5).

The body's goal of these enzyme-catalyzed reactions is not to cause mutations and disease, but only to modify the heterocyclic amines to forms that are more water soluble and more readily excreted in the urine. Most of the heterocyclic amines that bear acetyl groups or sulfate groups are promptly excreted. However, a minority react with small molecules and macromolecules in the cell.

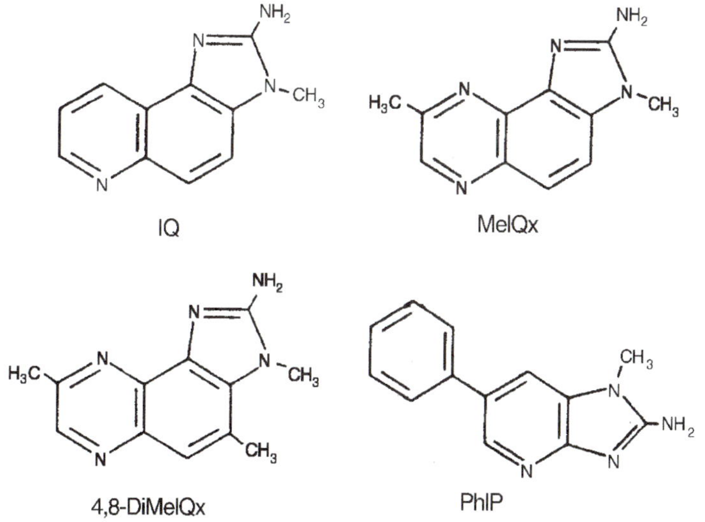

**Figure 11.4.** Structures of various heterocyclic amines. IQ is 2-amino-3-methylimidazo-quinoline. MeIQx is 2-amino-3,8-dimethylimidazo-quinoxaline. 4,8-DiMeIQx is 2-amino-3,4,8-trimethylimidazo-quinoxaline. PhIP is 2-amino-1-methyl-6-phenylimidazo-pyridine. These and other related compounds can be found in fried or broiled beef, fish, and chicken, but can also be created by cooking mixtures of creatinine, amino acids, and glucose. Microwave cooking, or cooking in aluminum foil, reduces the formation of heterocyclic amines. (From Sugimura, 1997.)

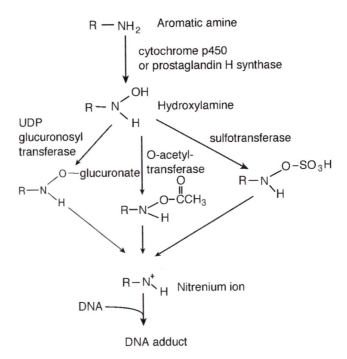

**Figure 11.5.** Some of the pathways of metabolism, and activation, of heterocyclic amines in mammals. A large number of chemicals naturally present in foods, food cooking products, and drugs are metabolized by enzymes such as cytochrome P450, glutathione *S*-transferase, and other enzymes. These chemicals tend to be harmless when consumed, but can be converted to toxic chemicals by enzymes located mainly in the liver. The goal of hydroxylation or of conjugation with glutathione, sulfate, and glucuronic acid is to make the chemicals more water soluble and more easily excreted from the body. However, this extensive metabolism may also result in the products decomposing to highly reactive nitrenium ions. Nitrenium ions can react with DNA, proteins, and other molecules in the body. Epidemiologists have taken interest in the fact that the some of these enzymes differ markedly in blacks, whites, and Asians, and may influence the forms of cancer acquired by these groups. (From Grant *et al.*, 1997.)

## Feeding Heterocyclic Amines Can Result in Mutations in APC

Let us now return to the study of the APC protein, and view mutations in this protein that arise with feeding a chemical normally present in cooked meat. Kakiuci and co-workers (1995) fed rats a diet that contained PhIP. The goal of the work was to acquire cancerous growths (tumors), to extract the DNA from the tumors, and determine if any mutations occurred in the APC gene. The researchers used very large quantities of PhIP in the diet (400 µg PhIP/g diet). This introduces the suspicion that the dietary PhIP may be metabolized by pathways that are abnormal, and in a manner that is totally irrelevant to PhIP metabolism in humans. Quite possibly, this error in assembling the diet led to misleading results. However, let us not be dissuaded from examining the data.

About half of the 20 rats acquired colon cancer within a year. (Normally, spontaneous colon cancer is rare in rats. Under 1% of rats acquire colon cancer within a year of feeding a normal diet (Ito *et al.*, 1991)). Eight of the tumors were used for analysis of the APC gene. Not all tumors contained a mutation in the APC gene. Only five mutations were found in the APC gene, and they took the form of a one-base deletion; the sequence of DNA bases was normal except that a single residue of G was missing. The sequence of a tiny part of the APC gene is shown here:

Change in gene from tumors #1 and #2:  GGTGGGATA → GGTGGATA
Change in gene from tumors #3 and #4:  AGTGGGATT → AGTGGATT
Change in gene from tumor #5:          TCCGGGAAC → TCCGGAAC

The consequence of the one-base deletions was that all bases downstream of the deletion were out of frame, thus producing a garbled series of amino acids during the process of translation. The garbling of codons downstream of the one-base deletion also resulted in generation of an early stop codon, which produced a truncated APC protein.

The results of the feeding experiment of Kakiuchi and co-workers seem consistent with the picture in humans — mutations in APC that result in a truncated protein and cancer. The results of this rat study are relevant to epidemiological data regarding high-meat diets and colorectal cancer.

## Details of p53

Let us resume our guided tour of the genes that necessarily must acquire mutations in order to provoke a normal gut cell to become a cancer cell. After the APC gene, the next gene in the sequence is the p53 gene. The p53 gene codes for a polypeptide of 393 amino acids. The p53 protein functions as a tetramer, i.e., as a complex of four of these polypeptides. Various regions of the polypeptide have different functions (Ruaro *et al.*, 1997). Amino acid residues 1–43 (N-terminal region) are used for activating transcription. p53 is a transcription factor. Residues 100–300 are used to bind to DNA. All transcription factors have regions that bind to DNA. Most of the mutations in p53, which occur in various cancers, occur in the DNA-binding region of the protein. This region binds to the *p53 response element*, a short stretch of DNA having the sequence:

Pur–Pur–Pur–C–A–T–G–Pyr–Pyr–Pyr

"Pur" means adenosine or guanine, while "Pyr" means thymine or cytosine. Continuing with the structure of p53, amino acids 320–360 are used for maintaining the adhesion of the four p53 proteins to each other (Prives, 1994). This domain is also called the "oligomerization domain."

p53 is not needed for life, since special mice prepared in the laboratory (knockout mice) that contain no p53 develop normally. On the other hand, these mice have a greatly increased number of tumors. By 6 months of age, 75% of these mice have tumors, and nearly all die by the age of 10 months (Selivanova and Wiman, 1995). What is the function of p53? p53 senses when DNA damage occurs, and responds by momentarily halting the progression of the cell cycle (Figure 11.6). When the cell decides that it is time to divide, replication occurs, as indicated by the S (S means DNA synthesis). The S phase is followed by the $G_2$ phase, and then

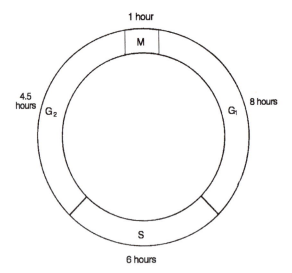

**Figure 11.6.** The cell cycle. Cells double their content of DNA during the S phase. Following the S phase is a gap period, called $G_2$, which is followed by mitosis (M), and then by another gap period, called $G_1$. Nongrowing cells are said to reside in a variation of the $G_1$ period, called $G_0$. Damage to DNA may occur at any stage of the cell cycle. Some types of damage are influenced by whether the region of DNA in question is being transcribed. Other types of damage depend on whether the region of DNA is wrapped around histones (proteins) or is exposed to the environment in the nucleus. A fair number of different enzymes catalyze the repair of DNA damage. Repair is not instantaneous, and may take many hours to accomplish. One feature of the cell cycle is that, during times of extensive DNA damage, the p53 protein functions to halt the cell cycle in the $G_1$ phase, to allow the DNA repair enzymes to finish their function prior to the S phase. Any damage that exists during the S phase might be expected to result in DNA mutations. Any mutation in DNA might be considered to be an echo or reflection of the damage that had once existed at that point.

by mitosis. Mitosis is a complex multistep process where chromosome formation, DNA recombination, and cell division takes place.

Various types of DNA damage, such as a break in one of the strands of the double helix, activates p53, and provokes the cell cycle to halt just prior to the S phase (Huang *et al.*, 1996). Once activated, p53 provokes the transcription of several genes. These genes code for proteins that interfere with cell cycle regulatory proteins and with DNA polymerase. The result is a pause in the cell cycle, which gives our DNA repair enzymes a chance to fix the damaged bases or broken strands.

## *Mutations in p53 that Are Associated with Cancer in Humans*

Mutations in p53 occur in about half of all human cancers, including colorectal cancer. Mutations in p53 constitute a necessary step in the conversion of many types of normal cells to invasive cancer cells. The defect in p53 increases the rate of accumulation of mutations in many other genes. Perhaps one of these mutations will further promote the conversion of the **fast-growing cancer cell** to a **fast-growing invasive cancer cell**. Most of the mutations in p53 occur within the DNA-binding region and tend to result in changes in amino acids 175, 248, and 273 (Hollstein *et al.*, 1991). The entire p53 protein is 393 amino acids long. A survey of the mutations occurring in human colon cancer revealed that most of the mutations involved a change in cytosine to thymine. More accurately, one might say the change is from C:G to T:A. This is because DNA consists of a double helix, where C is always paired with G, and T with A. All of the biologically possible base changes, as well the number of these base changes that were actually detected in the survey of human colon cancer, were as follows:

| Mutation in human colon cancer | Number detected |
| --- | --- |
| C:G ➡ T:A | 31 |
| C:G ➡ A:T | 0 |
| C:G ➡ G:C | 1 |
| A:T ➡ T:A | 1 |
| A:T ➡ G:C | 6 |
| A:T ➡ C:G | 0 |

The biochemical mechanisms that may have led to the C:G ➡ T:A mutation are discussed later. The mechanisms are consistent with epidemiological data suggesting that a low consumption of fruits and vegetables leads to an elevated risk for colorectal cancer. One might ask, "How can mutations involving C:G ➡ T:A, rather than other base changes, be associated with a diet low in fruits and vegetables?" Possible answers are revealed in what follows.

## Methylation and Spontaneous Deamination Can Result in a G:C ➡ A:T Mutation

DNA consists of nucleotides. Nucleotides consist of the bases A, T, G, and C, which are attached to residues of deoxyribose-phosphate. This statement is not entirely correct, since about 1.0% of the cytosine bases in human DNA is modified by a methyl group to form 5-methyl-cytosine. 5-Methyl-cytosine is formed by an enzyme that flips the cytosine residue out of the DNA helix, methylates it, and flips it back into the helix. 5-Methyl-cytosine pairs with guanine, as does cytosine.

DNA damage results when the amino group of cytosine residues spontaneously hydrolyzes in the cell, to produce a residue of uracil plus free ammonia. The amino group of 5-methyl-cytosine spontaneously hydrolyzes at a somewhat greater rate than cytosine. For any particular residue of 5-methyl-cytosine, its half-life for deamination is about 40,000 years (Shen *et al.*, 1994). The half-life for deamination of cytosine is about twice as long. These rates are very low, and researchers who study its rate have to perform their reactions for hundreds of days to be able to detect any deamination. However, when deamination does occur in our DNA, it can result in a mutation. Research has shown that the "hotspots" for mutation in p53 all actually occur at residues of 5-methyl-cytosine, and not at cytosine (Schmutte *et al.*, 1996).

The spontaneous deamination of cytosine produces a residue of uracil in the DNA helix. During replication, when DNA polymerase encounters a residue of uracil in a strand of DNA, it recognizes it as a residue of thymine. The newly synthesized strands of DNA are indicated by boldface:

Please view the following deamination example. Deaminated 5-methyl-cytosine is identical to thymine. Hence, the overall scenario of deamination of 5-methyl-cytosine followed by replication involves the conversion of 5-methyl-C:G to T:A.

Deamination of cytosine results in the formation of a U:C mispair. This mispaired region is *rapidly* processed by a special enzyme, which hydrolyzes the residue of U, and totally releasing it from the deoxyribose moiety. The enzyme is called *uracil DNA glycosylase*. The enzyme is on the small side, and has a molecular weight of only 37 kDa. The action of the enzyme in the hydrolysis of uracil is shown in Figure 11.7 (Teebor, 1995). An abasic site is formed. The subsequent action of two enzymes (AP endonuclease and 2'-deoxyribophosphodiesterase) in trimming out the remaining waste material is also shown in the figure. Please note that only one strand of the double-stranded helix is shown in Figure 11.7.

Deamination of 5-methyl-cytosine results in the formation of a T:C mispair. This mispair is *slowly* repaired by a special enzyme, which slowly hydrolyzes the residue of T. The enzyme is called **thymine DNA glycosylase** (Neddermann and Jiricny, 1993).

Let us continue to discuss the mechanisms of mutations that commonly afflict the gene for p53. Within a few paragraphs we will learn some of the most likely mechanisms that account for lower rates of cancer with diets high in fruits and vegetables.

The difference in deamination rates of C and 5-methyl-C is not enough to account for the much greater rates of mutation occurring at 5-methyl-C. Most of the difference is due to the much slower repair rate for mismatch occurring at 5-methyl-C, after deamination. Schmutte *et al.* (1995) examined the rates of enzymatic cleavage of U:G mismatches (black bars) and T:G mismatches (light bars) in normal colon biopsies taken from seven human subjects (see following diagram

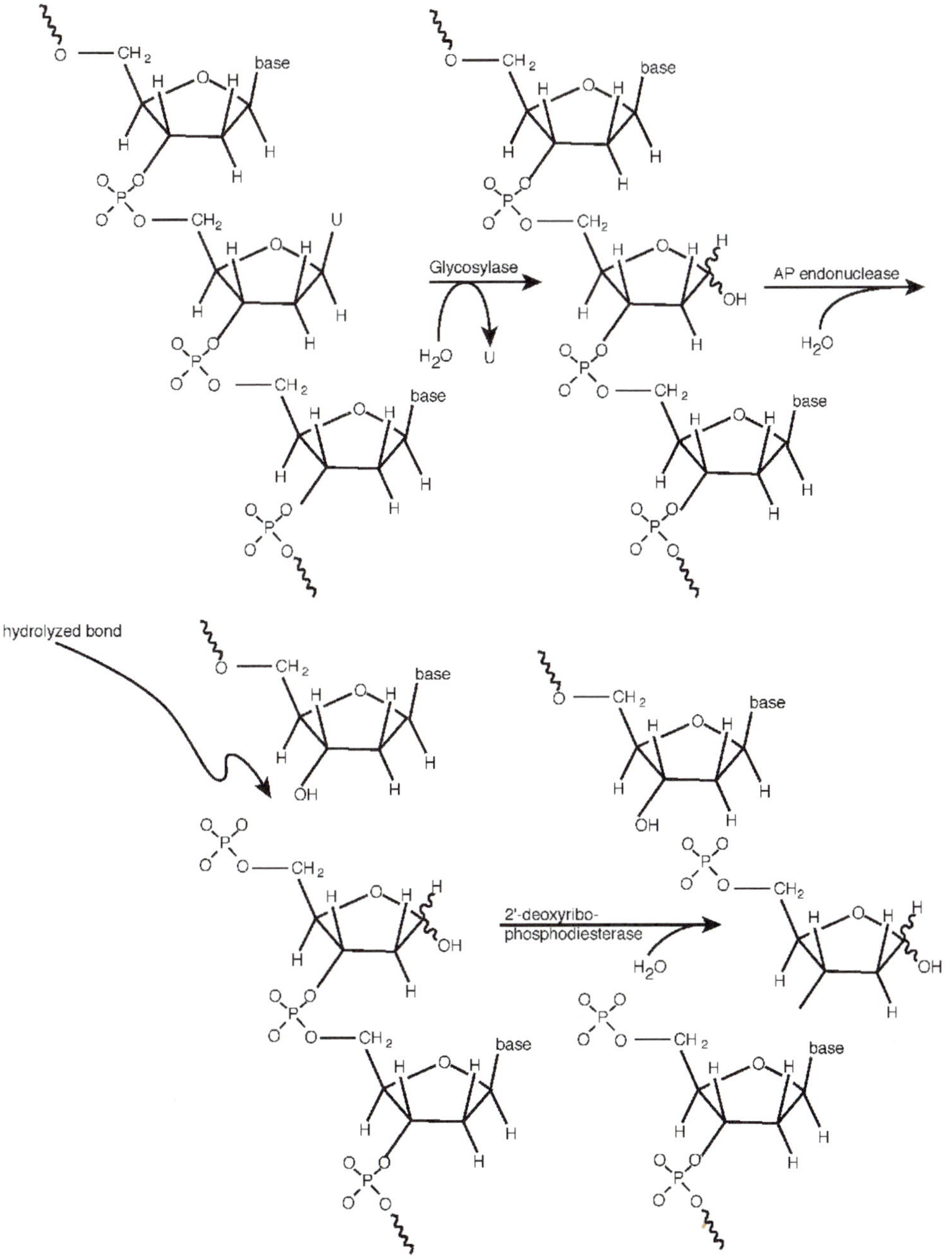

**Figure 11.7.** Enzymatic events leading to removal of DNA damage. The various types of DNA damage are removed by the action of several families of DNA repair enzymes. The result, in many cases, is a region where part of one strand of the double-stranded helix is missing. The empty region is then filled in with new nucleotides in a reaction catalyzed by DNA polymerase. The sequential activities of uracil DNA glycosylase, AP endonuclease, and 2'-deoxyribophosphodiesterase catalyze the total removal of uracil residues, as shown. Uracil DNA glycosylase hydrolyzes the uracil base from the DNA. The other two enzymes catalyze the hydrolysis of a phosphodiester bond, and then the complete removal of the sugar residue. (From Teebor, 1995.)

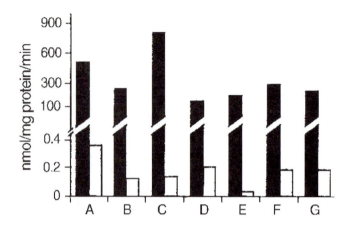

**Figure 11.8.** Activities of enzymes that remove inappropriate bases from DNA. The specific activities of uracil DNA glycosidase (black bars) and thymine DNA glycosidase (open bars) were measured in extracts of colon mucosa. Extracts were prepared from seven normal human subjects (subjects A to G). To the extracts were added synthetic strands of double-stranded DNA that contained a mismatched U:G pair or a mismatched T:G pair. The unit of specific activity used was nmol/mg protein/min. The results show, at least in tissue extracts, that the removal of U occurs at a dramatically greater rate than the removal of T.

and Figure 11.8). The researchers ground up the tissues and added a substrate. The substrate consisted of a stretch of synthetic double-stranded DNA that contained a U:G mismatch or a T:G mismatch, as shown. The X indicates the position of the U or the T. The substrate was incubated with the tissue extract, the reaction was quenched, and the product was measured:

5′-GGCTATCGTGGCXGGCCACGACGG-3′          (Double-stranded DNA
3′-CCGATAGCACCGGCCGGTGCTGCC-5′           substrate; X indicates U or T)

⇓    Cleavage of single strand by enzymes in tissue extracts

5′-GGCTATCGTGGC-3′ + 5′-XGGCCACGACGG-3′  (Products
                                          formed after
3′-CCGATAGCACCGGCCGGTGCTGCC-5′            after cleavage)

The results showed that the U:G mismatches (black bars) were processed about 1000-fold quicker than the T:G mismatches (light bars), accounting for the fact that occurrences of 5-methyl-C:G are mutational hotspots for various genes, including the p53 gene. Data from studies of biopsies of colon cancers from seven human subjects are shown in Figure 11.8. One can see the striking difference in activities in enzymes (glycosylases) that process the mismatches resulting from the deamination of cytosine and of 5-methyl-cytosine.

## Complete Repair of U:G and T:G Mismatches

The catalytic action of these glycosylases results in formation of a nucleotide that has lost its base. The result is an apyrimidinic site (AP site). The AP site consists only of a residue of deoxyribose-phosphate. What happens next in the DNA repair pathway? The deoxyribose-phosphate is subsequently snipped out of the strand of DNA by an enzyme called *endonuclease*. The full name for the enzyme is *AP endonuclease* (APE). The action of AP endonuclease results in a small gap in the double-stranded DNA helix, which is then filled in by DNA polymerase to yield repaired DNA.

A short outline of various DNA repair pathways has been presented. It is now possible to reveal several likely theories on how eating fruits and vegetables can reduce the occurrence of cancer.

## DNA Polymerase Can Make a Mistake

Folate deficiency can result in mutations at positions in DNA normally occupied by thymidine. Low levels of folate in the cell may result in accumulation of the substrate for thymidylate synthase (dUMP), and a decline in its product (dTMP) (see Folate section). This dUMP, after conversion to dUTP, is mistakenly recognized by DNA polymerase, and incorporated into DNA during replication. When DNA polymerase makes this mistake, it catalyzes the incorporation of U opposite A, as shown here:

The resulting U:A mispair is recognized by **uracil DNA glycosylase**, and the uracil base is cleaved off, as shown in Figure 11.7. Then, AP endonuclease catalyzes the cleavage of the phosphodiester bond just upstream of the abasic residue of deoxyribose. The action of AP endonuclease is followed by hydrolysis of the deoxyribose and filling in of the gap in the single strand of DNA with the correct base, as mentioned earlier. Evidence suggests that human folate deficiency may result in increased levels of dUTP, increased incorporation of uracil into DNA, and increased activity of uracil DNA glycosylase in the DNA (Blount *et al.*, 1997).

The continued accumulation of uracil in DNA, over many cell divisions, is thought to result — eventually — in the occurrence of residues of U across from each other in a double-stranded helix. Where uracil DNA glycosylase processes the U residues, the result is cleavage at both strands, and a double-stranded break. Double-stranded breaks are repaired by **recombinational repair**. Recombinational repair is not perfect, and may yield reconnected DNA that contains mutations. Very little direct evidence exists to support the previous scenario. However, it is

may account for the decreased rates of cancer found with diets that are high in fruits and vegetables.


## DNA Methylase Can Make a Mistake

5-Methyl-C residues are formed by a special enzyme called **cytosine DNA methyl-transferase**. The immediate source of the methyl group ($—CH_3$) is *S*-adenosyl-methionine (SAM). SAM is made by the 1-carbon cycle (see Folate section). Evidence suggests that, when SAM is in short supply in the cell, the methylase makes a mistake. During the normal methylation reaction, the enzyme activates the residue of C and makes the C residue responsive to the incoming methyl group. But the absence of an incoming methyl group, due to lower folate levels and lower SAM levels, can result in the deamination of C to produce U (Schmutte *et al.*, 1996) (Figure 11.9). The mechanism shown in Figure 11.9 represents the simplest explanation for the apparent finding that folate deficiency can result in mutations and cancer, but direct evidence that this scenario occurs in mammalian cells is not yet available.


EXERCISE

Please devise a genetic analysis test for determining if a human cancer arose from dietary PhIP or from the spontaneous deamination of 5-methyl-cytosine. Hint: Start with taking a biopsy (tissue sample) of the cancerous tissue. A variety of tests, not detailed in this text, are routinely used for determining the exact sequence of DNA bases occurring in any human gene, and in any mRNA produced by that gene.


EXERCISE

Please describe why mutations at 5-methyl-cytosine might be considered to be spontaneous and not related to diet. Then explain how mutations at 5-methyl-cytosine can depend on folate status.


## RAS AND THE MAP KINASE SIGNALING PATHWAY

In the process of formation of colorectal cancer, a mutation in p53 is followed at a later point in time — perhaps many years later — by a mutation in RAS. What is the normal function of RAS? Several hormones provoke growth of the their target cell via binding to membrane-bound receptors, with the consequent activation of RAS. These hormones include insulin, nerve growth factor, and epidermal growth factor. The binding of one of these growth factors to its receptor provokes the activation of RAS, of MAP kinase, and the consequent phosphorylation of a number of transcription factors. The end-result is the activation of many genes that provoke or maintain growth.

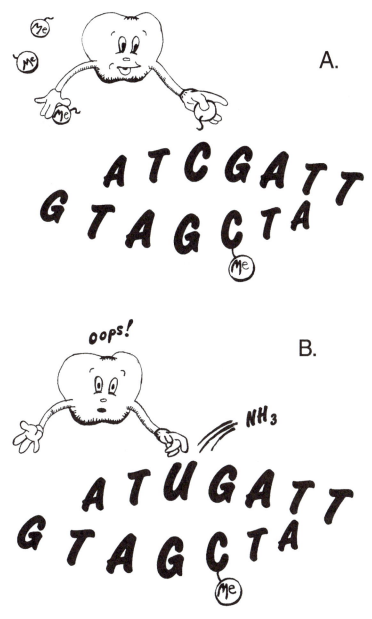

**Figure 11.9.** Cartoon representation of DNA methyltransferase. (A) Methylation under normal conditions. (B) Methylation under conditions of folate/methyl group deficiency. DNA methyltransferase utilizes SAM as a substrate and source of the methyl group, as is the case with most other methyltransferase enzymes in biology. Transfer of the methyl group to DNA produces SAH as the byproduct. The folate-dependent enzyme methionine synthase catalyzes the recycling of homocysteine back to methionine, and a separate enzyme converts methionine to SAM. Newly synthesized DNA lacks methyl groups in the newly polymerized strand of the daughter DNA, but contains methyl groups in the old strand that functioned as the template. Shortly after DNA synthesis, DNA methyltransferase scans the double-stranded helix, searching for unmodified CG sequences that reside across from G5-methyl-C sequences. DNA methyltransferase acts at the sequence CG, where the target CG sequence is base-paired across from G5-methyl-C. Evidence suggests that, when SAM is in short supply in the cell, DNA methyltransferase makes a mistake and catalyzes the conversion of the target cytosine to a uracil residue plus free ammonia. Eventually, with further DNA synthesis and cell division, the uracil base-pairs with adenine, resulting in a C:G ➡ T:A mutation.

Mutations in the genes coding for any of the signaling proteins could conceivably result in the constitutive (continual) operation of the signaling pathway, and the consequent maintenance of cell growth.

## Very Simple Version of the MAP Kinase Signaling Pathway

A very simple version of the cell-signaling pathway involving RAS and MAP kinase is shown in Figure 11.10. Step 1 depicts a hormone (in the bloodstream) binding to its receptor in the plasma membrane. Step 2 shows RAS, in the cytoplasm, binding to the inside of the plasma membrane, and the consequent activation of RAS. Step 3 shows the phosphorylation of JUN, and the production of an active transcription factor (JUN-phosphate). JUN-phosphate, in turn, binds to the *JUN response element* on DNA and provokes the activation of a nearby gene.

## Simple Version of the MAP Kinase Signaling Pathway

A more complete version of the cell-signaling pathway involving RAS and MAP kinase appears in Figure 11.11. The binding of the hormone to its receptor provokes the recruitment of the GRB2/SOS protein complex to the cell membrane. RAS binds to GRB2/SOS. When RAS is complexed in this way, RAS releases its GDP (guanosine diphosphate). A notable property of RAS is that it binds GDP or GTP. When RAS binds GDP, RAS is not active. When GDP is released, and replaced by GTP, RAS becomes active. Once activated, RAS binds MAPKKK (MAP kinase kinase kinase) and, in this way, activates MAPKKK. Then, MAPKKK phosphorylates MAPKK, which phosphorylates MAPK, which phosphorylates JUN. Hence, this signaling pathway involves a cascade of phosphorylation reactions.

It is amusing to point out that "SOS" means "son of sevenless." The name, which may be of historical interest, arose from studies of fruitflies that contained mutations in the SOS protein. The name MAP means "mitogen-activated protein." The names of these, and other signaling proteins, are not particularly relevant to our discussion. This mode of regulation involving RAS may seem excessively complicated, and may bring to mind the inventions of Rube Goldberg. However, this is the way that cell signaling actually operates.

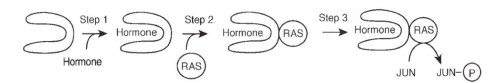

**Figure 11.10.** Very simple version of the RAS cell-signaling pathway. **Step 1.** A hormone binds to its receptor in the plasma membrane. **Step 2.** Hormone binding provokes the RAS protein to bind to the receptor, and thus adhere to the inside of the plasma membrane. This event activates RAS. **Step 3.** The activation of RAS results in the phosphorylation of JUN, a transcription factor. The activated transcription factor (JUN-phosphate) binds to special sequences of DNA upstream of certain genes, and activates the genes.

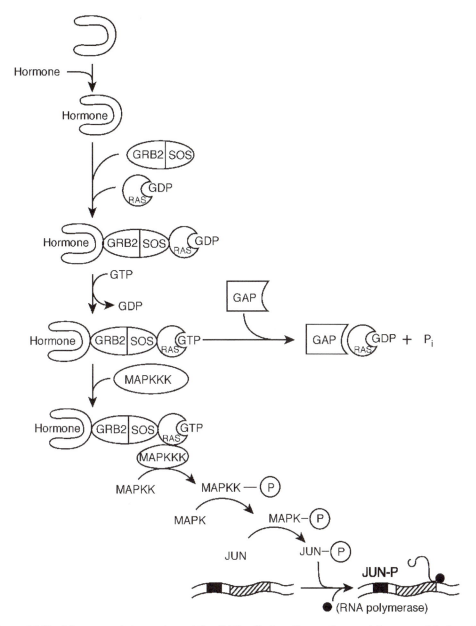

**Figure 11.11.** More complete version of the RAS cell-signaling pathway. A hormone binds to its receptor. The binding event recruits the complex of GRB2/SOS proteins to the plasma membrane. The GRB2/SOS complex acts as an adhesive and binds RAS/GDP. GDP is guanosine diphosphate. When RAS contains GDP, RAS is inactive. The binding of RAS/GDP to proteins of the plasma membrane result in the release of GDP and its replacement with GTP, and in the activation of RAS. The RAS/GTP complex (active RAS) binds MAPKKK and activates it, resulting in a cascade of phosphorylations involving various protein kinases. The final event is the phosphorylation of JUN, resulting in its conversion to an active transcription factor, and the consequent increase in the rate of transcription of specific genes. Eventually, RAS is inactivated by another protein, called GAP. The binding of GAP provokes RAS to hydrolyze the terminal phosphate of the GTP, producing GDP and free phosphate.

JUN forms a dimer with itself to form JUN–JUN, a homodimer. JUN also forms a dimer with another protein, called FOS, to generate the heterodimer JUN–FOS. In the forms JUN–JUN and JUN–FOS, these proteins function as transcription factors. To turn to the detail of phosphorylation, the activity of JUN depends on its phosphorylation at Ser 63 and Ser 73 (Kallunki *et al.*, 1996).

## Inactivation of RAS by the Hydrolysis of GTP

The cascade of phosphorylation, and consequent enzyme activation, is driven by RAS in its active form. Eventually, however, the GTP is hydrolyzed to GDP plus $P_i$, with the result that RAS reverts to its inactive form. GAP provokes the hydrolysis of GTP. In detail, GAP binds to RAS, and this event provokes RAS to catalyze the hydrolysis of GTP (Sprang, 1997). Hydrolysis of GTP is followed by the dissociation of MAPKKK, the consequent inactivation of MAPKKK, and the eventual deactivation of all downstream proteins.

## Does Anything Bad Happen When RAS Is Prevented from Hydrolyzing GTP?

Mutations in RAS are common in various types of cancer. Mutations in RAS can result in an inability to hydrolyze GTP, with the following consequence: RAS is maintained in its active form, cell signaling is constitutive (rather than momentary), and the cell is provoked to divide and grow continuously.

As mentioned earlier, GAP has the useful property of switching RAS off. Mutations in GAP cause a common disorder called **neurofibromatosis**. Here, GAP no longer can provoke RAS to catalyze GTP hydrolysis. In its most common form, neurofibromatosis results in multiple, harmless café-au-lait spots over the skin. In its most severe and rare form, the disease results in deforming tumors, as occurred in the case of the Elephant Man (Riccardi, 1981). To sum up, the tumors of the Elephant Man probably resulted because GAP failed to terminate the activity of RAS. Mutations in MAPKKK and MAPKK have also been found that result in cancer (Stang *et al.*, 1997).

## MUTATIONS IN THE RAS GENE AND CANCER

A study of colorectal adenomas and cancers in human patients revealed several types of mutations occurring in RAS. The most common mutations affected codon 12, which consists of GGT, and codes for the 12th amino acid of the polypeptide. Most of these mutations resulted in the conversion of GGT, which codes for glycine, to GAT (aspartate), to GTT (valine), and to GCT (alanine) (Zhu *et al.*, 1997). The mutations occurred at the central base of codon 12 and can be summarized as:

| Mutation in human adenomas or cancer | Number detected |
|---|---|
| G:C → A:T | 3 |
| G:C → T:A | 3 |
| G:C → C:G | 3 |

## Oxo-G Can Result in G:C ➡ T:A Mutations

The G:C ➡ T:A mutation reported in the study of RAS, as discussed earlier (Zhu *et al.*, 1997), may have been due to oxidative damage. Oxidative damage occurs at a rate of about 10,000 residues/cell per day in humans (Shigenaga *et al.*, 1989). The most common type of oxidative damage occurs at residues of guanine, and the most common product is a residue of oxo-G (Steenken, 1989). Oxo-G is short for 8-oxo-7,8-dihydroguanine. It might be noted that, at pH 7, oxo-G actually occurs as 6,8-diketoguanine (McAuley-Hecht *et al.*, 1994). However, it has become customary to refer to the product as "oxo-G." Guanine residues pair with cytosine, but oxo-G pairs with adenine (Figure 11.12).

How does oxo-G arise? Evidence suggests that atoms of iron naturally occur as contaminants on our DNA. Any ferrous iron present can catalyze the cleavage of HOOH, via the Fenton reaction, to produce hydroxyl radicals. Stuart Linn and others have provided evidence that iron can bind to DNA, support the Fenton reaction, and generate specific forms of DNA damage (Henle and Linn, 1997; Henle *et al.*, 1996; Enright *et al.*, 1992, 1996; Altman *et al.*, 1995; Yaffee *et al.*, 1996). Experiments in the laboratory showed that damage could not be prevented (the DNA could not be protected) by special molecules that are known to scavenge HO•. This resistance indicates that the HO• was produced directly on the surface of the DNA. Although ascorbate has been shown to support the Fenton reaction, Stuart Linn's group has provided evidence that NADH (the cofactor) can also reduce iron *in vivo* with the ultimate result of provoking DNA damage.

### *Repair of Oxo-Guanine*

The oxo-G in our DNA is normally repaired by the catalytic action of **oxo-guanine DNA glycosylase** (Aburatani *et al.*, 1997). The enzyme's catalytic action is similar to that of uracil DNA glycosylase. In the living cell, most of the oxo-G residues and the resulting abasic sites are enzymatically eliminated and replaced with undamaged guanosine. However, if DNA synthesis occurs before the oxo-G is repaired, this damaged base will pair with adenosine (code for adenosine instead of cytosine) and will result in a mutation in the daughter cells.

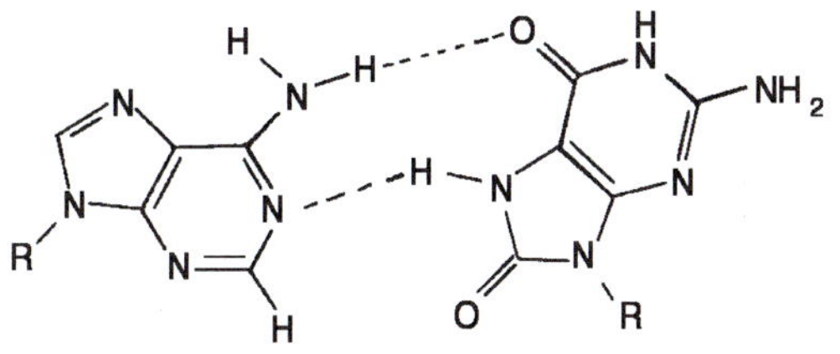

**Figure 11.12.** Orientation and binding of residues of adenine and oxo-guanine, contained in opposite strands of the DNA double helix. The R groups represent deoxyribose-phosphate, as well as opposite strands of the DNA. Compare the orientation of these bases with the normal pairing of adenine and thymine, as shown in the introductory chapter. (From McAuley-Hecht *et al.*, 1994.)

## *How Oxo-G May Miscode during DNA Synthesis*

Shibutani *et al.* (1991) detailed the coding or pairing properties of oxo-G residues. They did an experiment that involved mixing these components in a test tube: DNA polymerase, dATP, dTTP, dGTP, dCTP (substrates for DNA polymerase), double-stranded DNA, and a solution of various salts. The DNA consisted of a piece of double-stranded DNA. The mixture was incubated for a short time to allow the enzyme to settle on the DNA and begin polymerizing the dNTP residues to form a longer strand of DNA. One of the strands of the double-stranded DNA was actually short, and served as a primer to allow the enzyme to get a start in the polymerization reaction. The results using undamaged DNA were as follows, and revealed the expected normal product:

The results using oxidatively damaged DNA revealed that adenosine was placed opposite the oxo-G, as shown here:

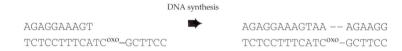

If this scenario were to occur in a living cell, subsequent cell divisions would result in the maintenance of the abnormal A, and in its coding for a residue of T. The overall scenario is summarized as:

$$G:C \rightarrow oxo\text{-}G:C \rightarrow oxo\text{-}G:A \rightarrow T:A$$

A succinct statement of the big pictures is as follows. The rate of formation of any mutation is a function of the rate of occurrence of the damage to DNA, how fast the damage is repaired, and in the behavior of DNA polymerase when it encounters the lesion during replication.

Can any component of the diet influence the rate of oxo-G formation? This is an obvious question, to be sure, but an obvious answer is not yet at hand. The control of toxic oxygen generation in the body is relevant to cancer, but also to aging. The notion that the accumulation of oxidative damage to our DNA is responsible for the aging process has stimulated a number of intriguing studies (Sohal and Weindruch, 1996). For example, restriction in the amount of food has been demonstrated to result in a marked increase in the lifespan of various animals. Evidence suggests that restriction of food also restricts the amount of oxygen consumed by the body, and thus the quantity of toxic oxygen that is produced. To resume with the topic of RAS, it has recently been found that one of the branches of the RAS signaling pathway requires the participation of toxic oxygen, as reviewed in what follows.

## One Form of RAS Protein, Called RAC, Uses Superoxide as a Signal

Superoxide and HOOH are forms of toxic oxygen. Most superoxide and HOOH arising in the cell are undesirable byproducts of the mitochondrial cytochromes. Recently, toxic oxygen has been found to be used for cell signaling (Irani *et al.*, 1997; Sundaresan *et al.*, 1996). The activation of one type of RAS protein (called RAC protein) provokes the activation of NADPH oxidase. NADPH oxidase, in turn, catalyzes the formation of superoxide. Superoxide, in turn, is converted to HOOH. Increased levels of HOOH in the cell then react with specific proteins in the cell, causing the activation of various transcription factors. Mutations in RAS that result in the constitutive activation of RAS also result in the continual elevated production of superoxide. The discovery of RAC protein, and its apparent requirement for toxic oxygen in its signaling pathway, represents a new avenue in the complex map of toxic oxygen, cell growth, DNA damage, and cancer.

## CADHERIN PROTEIN

Colorectal cancer involves the sequential accumulation of mutations in APC, p53, RAS, and possibly also in cadherin. The invasion of neighboring tissues occurs during one of the later steps in the life of the cancer cell. One of the properties of an invading cell is that it has lost its ability to adhere its sister cells (cell-to-cell adhesion).

### *Cadherin Is One of Several Proteins Used for Cell-to-Cell Adhesion*

The adhesion of cells to each other is normally due to interactions that involve a number of proteins of the extracellular matrix and the plasma membrane. Cadherin is a membrane-bound protein. The N terminus of cadherin is *extracellular*. The N termini of cadherins sticking out of adjacent cells bind to each other. This interaction requires calcium ions, hence the name **Ca**dherin. An *intracellular* interaction is also required for adhesion to occur between cells. The C terminus of cadherin contacts the cytosol and binds to a protein called **catenin** (pronounced ca-TEE-nin). Catenin, in turn, binds to the cytoskeleton. The cytoskeleton is a network of proteins that crisscross about the plasma membrane and through the cell. Defects in the cadherin gene have been found in many samples of colorectal cancer. These mutations tend to occur in the N-terminal region, i.e., in the extracellular calcium-binding domain.

Another type of mutation in cadherins is the deletion of part of chromosome #16. This mutation results in the loss of the entire cadherin gene, as well as of many other genes. Human cells contain 22 pairs of numbered chromosomes plus two sex chromosomes, giving 46 chromosomes in all. The deletion of small segments of chromosomes is a well-documented phenomenon in lung cancer cells (Minna and Mangelsdorf, 1997), in other cancers (Holliday, 1989), as well as in a number of genetic diseases, such as Angelman's Happy Puppet Syndrome (Buiting *et al.*, 1995), and most types of steroid sulfatase deficiency (Alperin and Shapiro, 1997)).

Cadherin consists of 882 amino acids. A detailed study of cadherin mutations is available from cancer cells (Risinger *et al.*, 1994). The DNA sequence for the cadherin gene from 135 tumors was analyzed. The normal codons, and mutated codons, are listed in Table 11.1. The position of the amino acid in cadherin is also listed. By convention, the codon number of the mRNA coding for all polypeptides indicates the amino acid counting from the N terminus. The results do not inform us of the function of cadherin. Nor do they provide evidence that the acquisition of a cadherin mutation occurs at an early or late step in the development of cancer. But the data do document that these mutations do occur in some cancer cells, and that such mutations are consistent with the failure of invading cancer cells to maintain a tight adhesion to each other.

EXERCISE

Please consult Table 11.1. State evidence that some mutations found in oncogenes in a cancer cell can have no relevance whatsoever to the formation of that cancer cell.

EXERCISE

The human body contains about $6 \times 10^{13}$ cells (Ames, 1989). The DNA in each human cell contains a total of about $7.8 \times 10^{9}$ base pairs. This DNA is distributed among 44 numbered chromosomes and 2 sex chromosomes. It has been estimated that oxidative damage occurs at about 10,000 bases per cell each day in the human body. The coding region for p53 is 1279 bases long. Assuming that oxidative damage is randomly distributed along our DNA, please calculate the number of oxidative lesions expected to occur in the coding region for p53 each day.

**TABLE 11.1** Mutations and Their Positions in the Cadherin Polypeptide Chain

| Number of tumors | Position of amino acid or codon | Mutation (amino acid and codon changes) |
|---|---|---|
| 1 | 711 | CTG (Leu) ➡ GTG (Val) |
| 1 | 617 | GCA (Ala) ➡ ACA (Thr) |
| 1 | 838 | AGC (Ser) ➡ GGC (Gly) |
| 1 | 145 | TCC (Ser) ➡ TCA (Ser) |
| 3 | 311 | CTC (Leu) ➡ CTG (Leu) |
| 3 | 560 | ACG (Thr) ➡ ACC (Thr) |
| 2 | 632 | CAC (His) ➡ CAT (His) |
| 64 | 692 | GCT (Ala) ➡ GCC (Ala) |
| 16 | 751 | AAC (Asn) ➡ AAT (Asn) |
| 3 | 764 | GAC (Asp) ➡ GAT (Asp) |
| 1 | 879 | GGC (Gly) ➡ GGT (Gly) |

Source: Risinger *et al.*, (1994).

## Summary

One way to view the big picture is to ask oneself the following questions:

1. For any particular cancer cell, was a researcher able to detect a mutation in any of the genes (oncogenes) associated with cancer?

2. Would the mutation be expected to result in any change in the sequence of amino acids in the protein? A change in a DNA base can result in a change, or no change, in the amino acids.

3. If a change in amino acid did occur, would the change be expected to result in a change in function of the protein? It is likely, one might note, that all existing proteins in biology can acquire slight changes in amino acid sequence with no change in the biological function of the protein.

4. Would any component of the diet be expected to attack DNA and become attached to one of the nucleoside bases? Would any constituent of the diet be expected to influence the normal process of methylation of cytosine residues on our DNA? Could any component of the diet influence the concentration of hydrogen peroxide, or other forms of toxic oxygen, in the cell? Is there any reason to believe that a particular cancer patient has a genetic defect in one of the DNA repair enzymes?

# EPIDEMIOLOGY OF DIET AND COLON CANCER

Cancer is not one disease but many. There may exist at least one kind of cancer corresponding to every type of cell in the human body. About 530,000 deaths from cancer occur each year in the United States, where cancer of the lung, colon and rectum, breast, and prostate gland account for just over half of the cases. Epidemiological studies suggest that diets rich in fruits and vegetables reduce the incidence of various cancers, whereas diets rich in red meat and animal fat increase various cancers. In focusing on the nutrients in fruits, for example, epidemiological data suggest that the folic acid and fiber in fruits may be responsible for cancer reduction. Cruciferous vegetables (broccoli, cabbage, brussels sprouts), polyphenolics (in wine and green tea), fiber, calcium, and vitamin E (Gey, 1998) have also been associated with reduced cancer rates.

An introduction to the terms and tools used to express and evaluate epidemiological data are presented here. One might consult Appendix C for definitions of various terms in epidemiology.

## Epidemiological Study of Colorectal Adenomas

A case-control study was designed to explore risk factors for colorectal adenomas. **Adenomas** are precursors to 80–90% of all colorectal cancers. The study used 236 cases (with adenomas or polyps) and 409 controls (no adenomas) (Tseng *et al.*, 1996) (Table 11.2). All subjects were patients receiving colonoscopy exams. The subjects were not recruited from the general public. A month or so after the exam, the subjects were asked to fill out a food choice questionnaire.

**TABLE 11.2** Odds Ratios for Acquiring Adenoma and Associated Estimated Nutrient Intake

|        | Folate (mg/day) | OR   | Vitamin C (mg/day) | OR   |
|--------|-----------------|------|--------------------|------|
| Women  | 44–172          | 1.0  | 10–88              | 1.0  |
|        | 172–231         | 0.59 | 89–138             | 1.01 |
|        | 231–318         | 0.54 | 139–201            | 0.63 |
|        | 319–2877        | 0.39 | 203-2322           | 0.60 |
| Men    | 5–212           | 1.0  | 2–92               | 1.0  |
|        | 212–285         | 1.86 | 93–133             | 2.4  |
|        | 285–386         | 1.17 | 134–213            | 1.35 |
|        | 386–1792        | 0.84 | 217–3163           | 0.83 |

Source: Tseng *et al.* (1996).

By viewing the indicated food choices, and using tables of nutrients in foods, the researchers were able to estimate the daily intake of various nutrients, such as folate, vitamin C, calcium, and iron. The results for folate and vitamin C are shown separately for women and men in Table 11.2. The daily intakes for the nutrients were divided into four groups, as indicated. The risk for adenomas is expressed by a special term, called the **odds ratio** (OR) (see Appendix C).

An OR of greater than 1.0 indicates that the indicated dietary practice is associated with an increased rate of adenoma. An OR of less than 1.0 means that the indicated dietary practice is associated with a decreased rate of adenoma. An OR that is greater than 2.0 or under 0.5 is considered to be impressive and convincing. An OR that is between 0.5 and 2.0 may be less convincing or, perhaps, confounding. These values are not strict cutoff points. They might be considered to be fuzzy cutoff points. The fuzzy cutoff points of 0.5 and 2.0 apply to **odds ratios**. These fuzzy cutoff points also apply to another term, the **risk ratio**.

The results show that women with the greatest estimated folate intake had a lower tendency to develop adenomas when compared to women with the lowest folate intake. In contrast, the trends in men indicated no consistent relationship between folate intake and adenomas.

Since the study was retrospective, and not prospective, a cause-and-effect relationship cannot clearly be distinguished. A source of random error is that one cannot be totally sure if the style of diet consumed by the subject, and reported in the questionnaire, was adopted years before the adenomas developed, or only in the very recent past.

A source of bias in retrospective diet studies is that sometimes people change their diet and behavior after being diagnosed with a disease. After being diagnosed with colon cancer, a person may switch to a diet that is considered to be correct and fashionable (oranges and broccoli) and quit smoking. Cancer patients may practice these new dietary practices for several years. Upon receiving a diet questionnaire from an epidemiologist, these cancer patients will state that their diet usually contained oranges at lunch and broccoli at supper. The result is a source of bias, with the consequence that the epidemiological study will reveal a trend where colon cancer is associated with oranges and broccoli.

## Epidemiological Study of Colon Cancer

A study by Giovannucci *et al.* (1995) revealed that low folate status can increase risk for colon cancer. This study was a prospective study that involved 47,931 males over the course of six years. Food intake was estimated by means of questionnaires. Low folate, alone, was found not to be associated with increased risk for colon cancer. However, when accompanied with (or combined with) low dietary methionine and regular alcohol intake, a dramatic risk for colon cancer was found. This risk ratio associated with the aforementioned three factors (low folate, low methionine, habitual alcohol) was 3.3. As noted elsewhere, risk ratios of greater than 2.0 or less than 0.5 are considered to be "convincing," and not merely "somewhat suggestive."

Is there any relationship between folate status and methionine intake? Methionine has a pro-folate effect, as revealed in the Folate section. In brief, dietary methionine spares the use of 1-carbon units that might otherwise be used for the conversion of homocysteine to methionine. Furthermore, methionine is converted in the cell to SAM, which in turn inhibits 5,10-methylene-$H_4$folate reductase. This inhibition prevents the useless accumulation or trapping of 1-carbon units in the form of 5-methyl-$H_4$folate, and allows the folate cofactors to be used for other purposes, such as the synthesis of thymidylate.

## Various Nutrients and Colon Cancer

### Alcohol and Colon Cancer

Is there any connection between alcohol intake and cancer? With alcohol drinking, a fraction of this nutrient is converted to acetaldehyde. Acetaldehyde, as most aldehyde compounds, can condense with amino groups. Evidence suggests that physiological levels of acetaldehyde can condense with the amino groups of the DNA bases, and provoke DNA damage (breakage of the strands of the double-stranded helix) (Singh and Khan, 1995). It is reasonable to believe that breakage of DNA can result in mutations, but direct evidence is not yet available.

### Red Meat, Animal Fat, and Colon Cancer

A study on diet and colon cancer was reported by W. Willett's group (Willett *et al.*, 1990) (Table 11.3). The study examined various components of the diet, such as fiber, fat, and meat. The fiber component was divided into cereal fiber and fruit fiber. The fat component was divided into meat fat, dairy fat, saturated fat, and unsaturated fat. The meat component was divided into beef, pork, and lamb, and into rare versus well-done styles of cooking. The body mass index, as defined in the Obesity chapter, was also recorded. The study was part of the Nurses' Health Study Cohort, which was initiated in 1976 and involved 121,700 female nurses. Every 2 years, the nurses filled in a questionnaire that asked about various risk factors for disease. The questionnaire asked, for example, about 61 foods and their frequency in the diet. The foods were chosen to allow epidemiologists to make broad statements regarding the component nutrients.

TABLE 11.3  Amount of Food Eaten and Risk Ratio for Colon Cancer

| | Number of main courses per month | | | | | |
|---|---|---|---|---|---|---|
| | <1 | 1–3 | 4 | 12 | 24 | >30 |
| Beef, pork, or lamb (risk ratio) | 1.0 | 1.39 | 1.39 | 1.50 | 1.84 | 2.49 |
| Processed meats (risk ratio) | 1.0 | 1.09 | 1.45 | 1.86 | 1.21 | 1.21 |
| Chicken without skin (risk ratio) | 1.0 | 1.0 | 0.96 | 0.47 | 0.47 | 0.47 |
| Fish (risk ratio) | 1.0 | 1.29 | 0.92 | 0.75 | 1.06 | 1.06 |
| | Estimate of fiber intake | | | | | |
| Fruit fiber intake (g/day) | <0.8 | 0.8–1.6 | 1.7–2.6 | 2.7–4.0 | >4.1 | |
| Fruit fiber (risk ratio) | 1.0 | 0.94 | 0.87 | 0.81 | 0.62 | |
| Cereal fiber intake (g/day) | <2.8 | 2.8–3.9 | 4.0–5.2 | 5.3–7.0 | >7.1 | |
| Cereal fiber (risk ratio) | 1.0 | 0.72 | 0.58 | 1.04 | 0.74 | |

Source: Willett et al. (1990).

Generally, 96% of the questionnaires were filled out and returned. Deaths of participants were accurately monitored by a source of information, called the National Death Index. Over a 6-year period, 150 cases of colon cancer occurred. Of the 61 foods in the questionnaire, the strongest association with colon cancer was with beef, pork, or lamb (as the main course) (Table 11.3). All of the associations are expressed in terms of the risk ratio. A risk ratio over 1.0 indicates that indicated dietary practice is associated with an increased rate of disease. A risk ratio under 1.0 means that the indicated dietary practice is associated with a decreased rate of disease.

## Components of Red Meat that May Contribute to Cancer

Several mechanisms may contribute to the association of meat with colon cancer. The heterocyclic amines and polycyclic aromatic hydrocarbons present in cooked meat are metabolized, in the body, to mutagens that may condense with DNA to form adducts. If the adduct occurs at a vital base, and if the adduct is not promptly repaired, cancer may result. Another possible mechanism is related to the enhanced excretion of bile salts into the intestines that occurs with a high-fat diet. In brief, the increased amount of bile salts (with a high-fat diet) that reaches the large intestines is metabolized by the gut microflora to an increased amount of modified bile salts. Specifically, *lithocholic acid* and *deoxycholic acid* are formed. These modified bile salts are thought to contribute to the conversion of a normal gut cell to a cancer cell. Recent studies suggest that chronic exposure of gut cells to these modified bile salts may result in chronic activation of protein kinase C and chronic stimulation of various transcription factors (Matheson *et al.*, 1996). This scenario constitutes just one of many reasons why the student of diet and cancer needs to be familiar with the collection of protein kinases that are used for relaying signals in the cell.

## Fiber and Colon Cancer

The data in the Nurses' Health Study Cohort also indicate that decreased rates of colon cancer are associated with increased fruit fiber. Fiber has enjoyed enormous publicity as an anticancer agent, though clear and direct evidence for a cause-and-effect relationship has not yet been found. Decreased rates of cancer associated with high-fiber diets may arise from the fact that fiber provokes an increase in the rate of passage of material through the intestines. This increased rate would be expected to reduce the time of exposure of bile salts and bile salt metabolites to the gut mucosa. Evidence also suggests a totally different scenario, which also involves a gut microflora/fiber relationship. In this scenario, increases in dietary fiber result in an increase in production of butyrate by the gut microflora, where this butyrate slows the growth of gut cells and impedes any tendency of normal gut cells to turn into cancer cells. What is the relationship between continued growth and the tendency of a cell to be converted to a cancer cell? One connection is that a continuously growing cell may not be capable of repairing all the damage occurring in its DNA, while a cell that divides only occasionally may have a better chance to repair most of the damage in its DNA before committing itself to cell division.

Trock *et al.* (1990) have pointed out some difficulties in pinpointing fiber as an anticancer agent. Most of the fiber in Western diets comes from vegetables, rather than from grain. Any variation in grain intake that is found in epidemiological studies may not be great enough to classify the diet as a high-fiber diet or a low-fiber diet. Another difficulty is that the protective effect of fiber may really be due to other potential anticancer agents found in vegetables, such as folate and isothiocyanates. Still another problem is that there is a tendency for the amount of dietary animal fats and red meats (associated with increased colon cancer) to vary inversely with the amount of dietary fiber. Thus, any association between lower colon cancer and higher fiber intake may really be due to lower intake of fats and red meats.

## Vegetables and Colon Cancer

A group from Atlanta, Georgia conducted a study regarding diet and death from colon cancer (Thun *et al.*, 1992) (Figure 11.12). Data on dietary habits were acquired in 1982 from 764,343 adults who did not seem to have cancer. The questionnaire asked about vegetables (i.e., carrots, tomatoes, potatoes, cabbage), fruits, grains, fatty meats, low-fat meats (fish, chicken), and dairy foods. The grain questions asked about high-fiber grains (bran muffins, brown rice, oatmeal) and low-fiber grains (white bread, spaghetti, white rice). The questionnaire also recorded other risk factors, such as body mass index (see Obesity chapter), physical activity, cancer in the family, and aspirin use.

The average age of the subjects in the group of 764,343 adults was 57 years. Between 1982 and 1988, there were some 1150 deaths from colon cancer. In 1988, a group of control subjects was chosen from the larger group. These controls were matched with the dead cancer patients. Five controls were chosen for each dead cancer patient, and they were matched according to age, race, and gender. The study was a *prospective study*, rather than a *retrospective study*, since the subjects were chosen before any of them had acquired cancer.

The investigators (Thun *et al.*, 1992) found that increased vegetable and grain intake was associated with a lower risk for death by colon cancer (Figure 11.12). Most, but not all, of the plant foods were associated with this lower risk. It is curious to point out that there was no association of colon cancer risk with the intake of potatoes and tomatoes. In contrast, most other existing epidemiological studies (50 different studies in all) that addressed all types of cancers (and asked about tomatoes) found that tomatoes have a protective effect (Table 12 in Steinmetz and Potter, 1996).

Let us digress briefly from the topic of tomatoes. The most confounding and inaccurate part of a study is the use of a single questionnaire — or perhaps even several questionnaires — to acquire data on long-term food intake. Use of a questionnaire to acquire data on the previous day's food intake is not particularly accurate. Epidemiologists attempt to use the technique of large study groups to even out the gross inaccuracies of dietary recall. Another source of variation is that various human populations actually differ from each other with respect to their spectrum of diseases. Perhaps the best course, in viewing the data of Thun *et al.* (1992), is to focus on the findings that are harmonious and consistent with other epidemiological studies, and not to dwell on apparent contradictions.

## Aspirin and Colon Cancer

In addition to finding an association between dietary vegetables and lowered rates of death from colon cancer, Thun *et al.* (1992) found a strong association with aspirin intake. Greater aspirin intake was associated with lowered rates of cancer (Figure 11.13). Data from the laboratory bench suggest that aspirin can reduce the risk of colorectal cancer by inhibiting the synthesis of certain prostaglandins, thereby reducing inflammation in the gut (Marcus, 1995). Chronic infection, for example, can result in continual inflammation of human tissues. Inflammation (by definition) involves the production of toxic oxygen by human white blood cells. The toxic oxygen may leak into nearby cells, damage the DNA, and provoke cancer (Ames *et al.*, 1996). One might note that HOOH, which is not charged, readily passes through the cell membrane, in contrast to superoxide, which is charged. A practical consequence of the previously cited work is that all persons with colorectal cancer in their family history are recommended to consume aspirin throughout their lifetime.

## Isothiocyanates: Possible Anticancer Agents from Vegetables

Broccoli, cabbage, Brussels sprouts, horseradish, and mustard contain unique chemicals called glucosinolates and isothiocyanates. These chemicals are respon-

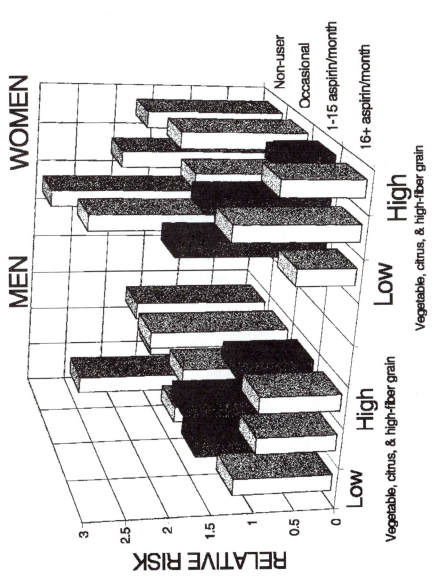

**Figure 11.13.** Results from an epidemiological study involving 764,343 adults. The subjects were followed for 6 years. During this period 1150 subjects died from colon cancer. Using techniques conventional to epidemiology, the researchers divided the subjects, and their dietary patterns, into various groups. The results indicated lower risks for colon cancer with increased aspirin consumption (in men and women), and lower risks for colon cancer with increased plant consumption (in women only). (Redrawn with permission from Thun *et al.*, 1992.)

sible for the pungent and acrid flavors of these plants. When plant cells are injured, a plant enzyme may convert the glucosinolates to the corresponding isothiocyanates plus glucose and sulfate. Glucosinolate and isothiocyanate have the following general structures:

$$R-C \overset{\displaystyle S\text{-glucose}}{\underset{\displaystyle N-O-SO^{3-}}{}}$$

Glucosinolate

$$R-N=C=S$$

Isothiocyanate

Studies with rats have revealed that dietary isothiocyanates can provoke a decrease in activity of enzymes (cytochrome P450) that activate carcinogens, and an increase in enzymes (glutathione S-transferase) used for the detoxification of carcinogens. The task of sorting out how the diet influences the metabolism of carcinogens is complicated by the existence of dozens of different types of cytochrome P450 and several types of GSH S-transferase. For example, human tissues contain the $\alpha$, $\mu$, $\pi$, and $\theta$ classes of GSH S-transferase. Another complicating factor is that the transfer of a molecule of GSH to some xenobiotics (such as heterocyclic amines) may either increase or decrease its mutagenic properties. Studies with humans are only beginning to be conducted (Nijhoff et al., 1995; Bogaards et al., 1994). The identification of which glucosinolates and isothiocyanates are actually protective will allow nutritionists to give firmer recommendations on diet. The identification of the genetics of human GSH S-transferases will allow geneticists to pinpoint which human populations are more likely to benefit from cruciferous vegetables.

An enzyme involved in cell cycle control has been identified as the target of a compound supplied by broccoli, cabbage, and Brussels sprouts. One of the glucosinolates of these foods is the source of a chemical called **indole-3-carbinol**. Indole-3-carbinol enters the cell and influences the activity of a special protein kinase that is used to control the cell cycle. The chemical prevents the synthesis of the enzyme, and thus reduces its activity. The action of this protein kinase (cyclin-dependent protein kinase 6) normally functions to make the cell embark on the $G_1$ phase of the cell cycle, and thus to stimulate progression through the cell cycle. The chronic decrease in activity of this protein kinase serves to prevent cell division, but also prevents spontaneous cancers. The effect of indole-3-carbinol has been shown to occur in cultured cells, as well as in animals (Cover et al., 1998).

## Ascorbic Acid: Possible Anticancer Agent from Fruits and Vegetables

The major antioxidants in the body include enzymes (catalase, glutathione peroxidase, superoxide dismutase), small molecules (vitamin E, vitamin C, and various sulfhydryl compounds), and the sulfhydryl groups of proteins. These antioxidants act either by destroying the toxic oxygen or by condensing with it ("soaking it up"). The terms "toxic oxygen" and "reactive oxygen species" are both used to mean the same collection of molecules, i.e., superoxide, hydroxyl radical, and hydrogen peroxide. Superoxide and HOOH are relatively unreactive, though superoxide does react with the sulfhydryl groups of proteins. On the other hand, when HOOH is mixed with reduced iron (or copper), it can result in the Fenton reaction with the production of the hydroxyl radical, which is extremely reactive.

Vitamin C has commanded some attention because it is an antioxidant present in fruits and vegetables — foods associated with lower rates of cancer. Vitamin C has clearly been shown to act as an antioxidant in studies with purified molecules in test tubes. However, the notion that vitamin C acts as a significant antioxidant inside living people remains controversial. The available studies have not supported the notion that supplements of vitamin C prevent cancer (Greenberg *et al.*, 1994; Prieme *et al.*, 1997). A fair number of studies with cultured cells or with pure components in test tubes have conclusively demonstrated that vitamin C can provoke damage to DNA (Singh, 1997; Shamberger, 1984; Anderson *et al.*, 1994). This damage is likely to be mediated via the Fenton reaction. On the other hand, it has not yet been shown that large doses of vitamin C given to living animals do result in DNA damage.

One study has revealed a protective effect of vitamin C in people. B. N. Ames and co-workers (Fraga *et al.*, 1991) maintained a group of human subjects on defined diets containing either 250 or 5 mg of vitamin C per day. The investigators measured the amount of oxidative damage to DNA (oxo-G residues) in the sperms, with the two diets. The advantage of using sperms is that they exist as a pure population of cells, and that they are freshly made and thus might be expected to be influenced by recent dietary practices.

Semen from the high-vitamin-C group contained 0.39 mM ascorbate, while that from the low-vitamin-C group contained 0.20 mM ascorbate. Sperm from the high-vitamin group contained 5.4 residues of oxo-G per 10,000 residues of guanine. The low-vitamin-C diet resulted in an increase to 10.6 residues of oxo-G/10,000 residues of guanine (Fraga *et al.*, 1991). These results will inspire further studies with human subjects.

## Ascorbic Acid and Nitrosamines

One of the more established antioxidant effects of ascorbate in the body is that of preventing nitrosamine formation. In fasting humans, saliva contains about 0.1 mM nitrate and 0.05 mM nitrite (McKnight *et al.*, 1997). This nitrate is produced by the gut microflora, absorbed into the blood, and secreted into the mouth. The nitrite in the saliva of a fasting person is created from the nitrate by rapid conversion by the bacteria living on the tongue. Nitrate ($NO_3^{3-}$) is also a naturally occurring component of the diet, as it occurs in green leafy vegetables. The bacteria in the mouth convert some of this nitrate to nitrite. About 80% of the nitrite we consume arises from conversion in the mouth. About 20% of the nitrite comes directly from the diet. Upon contact with stomach acid, the nitrite spontaneously becomes nitrous acid (HO—N=O). Nitrous acid reacts with dietary amines (R—$NH_2$) to produce nitrosamines (*N*-nitroso compounds). The nitrosamines are then absorbed into the body and converted in the liver to highly reactive compounds which react with DNA and proteins. The carcinogenic action of nitrosamines has been thoroughly established by animal studies. Ascorbate interrupts this scenario by reacting with nitrous acid, in the stomach, to yield nitrous oxide (HONO) plus dehydroascorbate. We cannot count on dietary ascorbate to protect us from all nitrosamines. This is because nitrosamines occur, pre-formed, in cigarette smoke and in certain foods (Anderson *et al.*, 1997).

## Peroxynitrite: A Form of Toxic Oxygen

Peroxynitrite (O=N—O—OH) fits well into this narrative, as this compound may be included in the list of toxic oxygens. Peroxynitrite is generated in the bloodstream of the human body by the condensation of nitric oxide (NO), a hormone, with superoxide (Pfeiffer *et al.*, 1997). Evidence suggests that the peroxynitrite damages tyrosine residues of proteins, such as low-density lipoproteins, in the bloodstream and provokes atherosclerosis (Leeuwenburgh *et al.*, 1997). Although ascorbate can react with peroxynitrite and detoxify it, the reaction rate seems much too low to have any influence on human health (Squadrito *et al.*, 1995).

## Chronic Exposure Theory

This chapter has emphasized the role of DNA mutations in the pathways leading to cancer. However, a number of chemicals are cancer-promoting purely because of their ability to activate certain enzymes, where there exists no direct or indirect influence on DNA mutations. This functional class of compounds promotes or prevents cancer only when present on a chronic basis, i.e., at elevated levels every day for many years. The notion that bile salt metabolites may result in chronic stimulation of protein kinase C was discussed earlier. The related event of fiber decreasing exposure of the colon to bile salts was also outlined. The chronic exposure theory applies to the phorbol esters that are present in certain plants. People who drink tea made from the plant *Croton flaveus* tend to acquire esophageal cancer. This cancer results from chronic exposure to phorbol esters, which occurs in the leaf extract or in an oil prepared from the plant (croton oil). The phorbol esters enter the cell, bind to protein kinase C, and activate this enzyme. Activated protein kinase C, in turn, activates the MAP kinase cascade (Ueda *et al.*, 1996). The continual activation of protein kinase C, when combined with mutations in specific proto-oncogenes, may lead to cancer and sustain the cells in the cancerous state.

Studies with animals and humans have shown that chronic treatment with estrogen can provoke the formation of certain cancers or tumors, such cancer of the breast, vagina, and endometrium (Lupulescu, 1996; Prentice *et al.*, 1990). An interesting chronic exposure relationship involving soy beans and cancer has been suggested. High consumption of soybean products has been proposed to account for the low incidence of breast cancer in Japan (Willett, 1997). Soybeans are high in isoflavonoids, compounds that have been found to counteract the action of estrogens. Hence, there is reason to suspect that soybean products may reduce the rate of estrogen-dependent cancers.

## SUMMARY

Cancer is a major public health problem throughout the world. A few dramatic epidemiological studies have revealed that the nature of various cancers differs between countries, and that these differences may be due to dietary differences. Also dramatic are epidemiologic studies that show that the types of cancers of immigrants become less similar to those of the ancestral nation, and more similar to those of the newly adopted nation. International epidemiology is, by nature, less reliable than studies performed within one nation. For this reason, only American studies were discussed in this chapter. The most basic units of measurement in epidemiology are odds ratio and risk ratio, and care was taken to define these terms in Appendix C. Epidemiological studies have found increased rates of various cancers associated with high intakes of red meat or animal fat, high intakes of overall energy, and a sedentary lifestyle. Decreased rates have been associated with high intakes of fruits, vegetables, cruciferous vegetables, cereal fiber, or fruit fiber.

One goal of epidemiology is to identify the families of chemicals responsible for the increased and decreased risks for cancer. These chemicals might be expected to exert a function on cell membranes, on regulatory proteins within the cell, on various enzymes, or directly on the DNA. Ultimately, the conversion of a normal cell to an invasive cancer cell might be expected to result largely from a collection of mutations in the DNA, and care was taken to depict the mechanisms of mutation. Many types of mutations, such as spontaneous deamination or base hydrolysis, are not expected to be influenced by lifestyle. Other mutations, such as those caused by smoking, by eating charred meat, or by chronic inflammation and infection, are a result of the environment.

As a concluding word, one should realize that the following things are different from each other: (1) a reasonable theory; (2) studies of reactions among purified components in a test tube; (3) studies of cells in culture; (4) experiments with laboratory animals, and (5) clinical or laboratory studies of human subjects. A goal in cancer research is to use theories, and data from the types of studies listed, to predict and explain each other. The researcher may find that consistency between data acquired in the aforementioned ways, and between data and theory, is not always apparent.

## REFERENCES

Aburatani, H., Hippo, Y., Ishida, T., Takashima, R., Matsuba, C., Kodama, T., Takao, M., Yasui, A., Yamamoto, K., Asano, M., Fukasawa, K., Yoshinari, T., Inoue, H., Ohtsuka, E., and Nishimura, S. (1997). Cloning and characterization of mammalian 8-hydroxy-guanine-specific DNA glycosylase/apurinic, apyrimidinic lyase. *Cancer Res.* **57**, 2151–2156.

Alperin, E. S., and Shapiro, L. J. (1997). Characterization of point mutations in patients with X-linked ichthyosis. *J. Biol. Chem.* **272**, 20756–20763.

Altman, S., Zastawny, T., Randers-Eichhorn, L., Cacciuttolo, M., Akman, S., Dizdaroglu, M., and Rao, G. (1995). Formation of DNA–protein cross-links in cultured mammalian cells upon treatment with iron atoms. *Free Rad. Biol. Med.* **19**, 897–902.

Ames, B. (1989). Endogenous oxidative DNA damage, aging, and cancer. *Free Rad. Res. Comm.* **7**, 121–128.

Ames, B. N., Gold, L. S., and Shigenaga, M. K. (1996). Cancer prevention, rodent high-dose cancer tests, and risk assessment. *Risk Anal.* **16**, 613–617.

Anderson, D., Yu, T.-W., Phillips, B. J., and Schmezer, P. (1994). The effect of various antioxidants and other modifying agents on oxygen radical generated DNA damage in human lymphocytes in the COMET assay. *Mut. Res.* **307**, 261–271.

Anderson, K. E., Hammons, G., Kadlubar, F., Potter, J. D., Kaderlik, K., Ilett, K., Minchin, R., Teitel, C., Chou, H.-C., Martin, M., Guengerich, F., Barone, G., Lang, N., and Peterson, L. (1997). Metabolic activation of aromatic amines by human pancreas. *Carcinogenesis* **18**, 1085–1092.

Blount, B. C., Mack, M. M., Wehr, C., MacGregor, J. T., Hiatt, R. A., Wang, G., Wickramas-inghe, S., Everson, R. B., and Ames, B. N. (1997). Folate deficiency causes uracil misin-corporation into human DNA and chromosome breakage. *Proc. Natl. Acad. Sci. U.S.A.* **94**, 3290–3295.

Bogaards, J., Verhagen, H., Willems, M., Poppel, G., and Bladeren, P. (1994). Consumption of Brussels sprouts results in elevated α-class glutathione S-transferase levels in human blood plasma. *Carcinogenesis* **15**, 1073–1075.

Buiting, K., Saitoh, S., Gross, S., Dittrich, B., Schwartz, S., Nicholls, R. D., and Horsthemke, B. (1995). Inherited microdeletions in the Angelman and Prader-Willi syndromes define an imprinting centre on human chromosome 15. *Nature Genet.* **9**, 395–400.

Cover, C. M., Hsieh, S., Tran, S., Hallden, G., Kim, G., Bjeldanes, L. F., and Firestone, G. L. (1998). *J. Biol. Chem.* **273**, 3838–3847.

Eddy, D. M. (1990). Screening for colorectal cancer. *Ann. Intern. Med.* **113**, 373–384.

Enright, H. U., Miller, W. J., and Hebbel, R. P. (1992). Nucleosomal histone protein protects DNA from iron-mediated damage. *Nucl. Acids Res.* **20**, 3341–3346.

Enright, H., Miller, W., Hays, R., Floyd, R., and Hebbel, R. (1996). Preferential targeting of oxidative base damage to internucleosomal DNA. *Carcinogenesis* **17**, 1175–1177.

Fraga, C. G., Motchnik, P. A., Shigenaga, M. K., Helbock, H. J., Jacob, R. A., and Ames, B. N. (1991). Ascorbic acid protects against endogenous oxidative DNA damage in human sperm. *Proc. Natl. Acad. Sci. U.S.A.* **88**, 11003–11006.

Gabbert, H. (1985). Mechanisms of tumor invasion: Evidence from *in vivo* observations. *Cancer Metastasis Rev.* **4**, 293–309.

Giovannucci, E., Rimm, E. B., Ascherio, A., Stampfer, M. J., Colditz, G., and Willett, W. C. (1995). Alcohol, low-methionine-low-folate diets and risk of colon cancer in men. *J. Natl. Cancer Inst.* **87**, 265–273.

Grant, D. M., Hughes, N., Janezic, S., Goodfellow, G. H., Chen, H., Gaedigk, A., Yu, V., and Grewal, R. (1997). Human acetyltransferase polymorphisms. *Mutat. Res.* **376**, 61–70.

Greenblatt, M. S., Bennett, W., Hollstein, M., and Harris, C. C. (1994). Mutations in the p53 tumor suppressor gene. *Cancer Res.* **54**, 4855–4878.

Haggitt, R. C., Glotzbach, R. E., Soffer, E., and Wruble, L. D. (1985). Prognostic factors in colorectal carcinomas arising in adenomas. *Gastroenterology* **89**, 328–336.

Harvey, M., Vogel, H., Morris, D., Bradley, A., Bernstein, A., and Donehower, L. A. (1995). A mutant p53 transgene accelerates tumour development in heterozygous but not nullizygous p53-deficient mice. *Nature Genet.* **9**, 305–311.

Henle, E. S., and Linn, S. (1997). Formation, prevention, and repair of DNA damage by iron/hydrogen peroxide. *J. Biol. Chem.* **272**, 19095–19098.

Henle, E. S., Luo, Y., Gassmann, W., and Linn, S. (1996). Oxidative damage to DNA con-stituents by iron-mediated Fenton reactions. *J. Biol. Chem.* **271**, 21177–21186.

Holliday, R. (1989). Chromosome error propagation and cancer. *Trends Genet.* **5**, 42–45.

Hollstein, M., Sidransky, D., Vogelstein, B., and Harris, C. (1991). p53 Mutation in human cancers. *Science* **253**, 49–53.

Irani, K., Xia, Y., Zweier, J. L., Sollott, S. J., Der, C., Fearon, E., Sundaresan, M., Finkel, T., and Goldschmidt-Clermont, P. (1997). Mitogenic signaling mediated by oxidants in ras-transformed fibroblasts. *Science* **275**, 1649–1652.

Ito, N., Hasegawa, R., Sano, M., Tamano, S., Esumi, H., Takayama, S., and Sugimura, T. (1991). A new colon and mammary carcinogen in cooked food, 2-amino-1-methyl-6-phenylimidazo[4,5-β]pyridine (PhIP). *Carcinogenesis* **12**, 1503–1506.

Kakiuchi, H., Watanabe, M., Ushijima, T., Toyota, M., Imai, K., Weisburger, J., Sugimura, T., and Nagao, M. (1995). Specific 5'-GGGA-3' to 5'-GGA-3' mutation of the Apc gene in rat colon tumors. *Proc. Natl. Acad. Sci. U.S.A.* **92**, 910–914.

Kallunki, T., Deng, T., Hibi, M., and Karin, M. (1996). c-Jun can recruit JNK to phosphorylate dimerization partners via docking interactions. *Cell* **57**, 929–939.

Leeuwenburgh, C., Hardy, M. M., Hazen, S., Wagner, P., Oh-ishi, S., Steinbrecher, U., and Heinecke, J. W. (1997). Reactive nitrogen intermediates promote low density lipoprotein oxidation in human atherosclerotic intima. *J. Biol. Chem.* **272**, 1433–1436.

Liabakk, N., Talbot, I., Smith, R., Wilkinson, K., and Balkwill, F. (1996). Matrix metalloprotease 2 (MMP-2) and matrix metalloprotease 9 (MMP-9) type IV collagenases in colorectal cancer. *Cancer Res.* **56**, 190–196.

Lewis, C. M., Neuhausen, S. L., Daley, D., Black, F., Swensen, J., Burt, R., Cannon-Albright, L., and Skolnick, M. H. (1996). Genetic heterogeneity and unmapped genes for colorectal cancer. *Cancer Res.* **56**, 1382–1385.

Lupulescu, A. P. (1996). Hormones, vitamins, and growth factors in cancer treatment and prevention. *Cancer* **78**, 2264–2280.

Marcus, A. J. (1995). Aspirin as prophylaxis against colorectal cancer. *N. Engl. J. Med.* **333**, 656–658.

Matheson, H., Branting, C., Rafter, I., Okret, S., and Rafter, J. (1996). Increased c-fos mRNA and binding to the AP-1 recognition sequence accompanies the proliferative response to deoxycholate of HT29 cells. *Carcinogenesis* **17**, 421–426.

McAuley-Hecht, K., Leonard, G. A., Gibson, N. J., Thomson, J. B., Watson, W., Hunter, W., and Brown, T. (1994). Crystal structure of a DNA duplex containing 8-hydroxydeoxyguanine–adenine base pairs. *Biochemistry* **33**, 10266–10270.

McKnight, G. M., Smith, L., Drummond, R., Duncan, C. W., Golden, M., and Benjamin, N. (1997). Chemical synthesis of nitric oxide in the stomach from dietary nitrate in humans. *Gut* **40**, 211–214.

Minna, J. D., and Mangelsdorf, D. J. (1997). Retinoic acid receptor expression abnormalities in lung cancer. *J. Natl. Cancer Inst.* **89**, 602–604.

Morin, P. J., Sparks, A. B., Korinek, V., Barker, N., Clevers, H., Vogelstein, B., and Kinzler, K. W. (1997). Activation of β-catenin-Tcf signaling in colon cancer by mutations in β-catenin or APC. *Science* **275**, 1787–1790.

Mumford, J., Li, X., Hu, F., Lu, X., and Chuang, J. (1995). Human exposure and dosimetry of polycyclic aromatic hydrocarbons in urine from Xuan Wei, China, with high lung cancer mortality associated with exposure to unvented coal smoke. *Carcinogenesis* **16**, 3031–3036.

Neddermann, P., and Jiricny, J. (1993). The purification of a mismatch-specific thymine-DNA glycosylase from HeLa cells. *J. Biol. Chem.* **268**, 21218–21224.

Nijhoff, W., Grubben, M., Nagengast, F., Jansen, Verhagen, H., Poppel, G., and Peters, W. (1995). Effects of consumption of Brussels sprouts on intestinal and lymphocytic glutathione *S*-transferases in humans. *Carcinogenesis* **16**, 2125–2128.

Pfeiffer, S., Gorren, A., Schmidt, K., Werner, E., Hansert, B., Bohle, D., and Mayer, B. (1997). Metabolic fate of peroxynitrite in aqueous solution. *J. Biol. Chem.* **272**, 3465–3470.

Polakis, P. (1997). The adenomatous polyposis coli (APC) tumor suppressor. *Biochim. Biophys. Acta* **1332**, F127–F147.

Prentice, R., Thompson, D., Clifford, C., Gorbach, S., Goldin, B., and Byar, D. (1990). Dietary fat reduction and plasma estradiol concentration in healthy postmenopausal women. *J. Natl. Cancer Inst.* **82**, 129–134.

Prieme, H., Loft, S., Nyyssonen, K., Salonen, J. T., and Poulsen, H. E. (1997). No effect of supplementation with vitamin E, ascorbic acid, or coenzyme Q10 on oxidative DNA damage estimated by 8-oxy-7,8-dihydro-2-deoxyguanosine excretion in smokers. *Am. J. Clin. Nutr.* **65**, 503–507.

Prives, C. (1994). How loops,β sheets, and α helices help us to understand p53. *Cell* **78**, 543–546.

Riccardi, V. M. (1981). Von Recklinghausen neurofibromatosis. *N. Engl. J. Med.* **305**, 1617–1626.

Risinger, J., Berchuck, A., Kohler, M., and Boyd, J. (1994). Mutations of the E-cadherin gene in human gynecologic cancers. *Nature Genet.* **7**, 98–102.

Ruaro, E. M., Collavin, L., Sal, G., Haffner, R., Oren, M., Levine, A. J., and Schneider, C. (1997). A proline-rich motif in p53 is required for transactivation-independent growth arrest as induced by Gas 1. *Proc. Natl. Acad. Sci. U.S.A.* **94**, 4675–4680.

Schiffman, M. H. (1990). Re: "Fried foods and the risk of colon cancer." *Am. J. Epidemiol.* **131**, 376–378.

Schmutte, C., Yang, A. S., Beart, R. W., and Jones, P. A. (1995). Base excision repair of U:G mismatches at a mutational hotspot in the p53 gene is more efficient than base excision repair of T:G mismatches in extracts of human colon tumors. *Cancer Res.* **55**, 3742–3746.

Schmutte, C., Yang, A. S., Nguyen, T., Beart, R., and Jones, P. A. (1996). Mechanisms for the involvement of DNA methylation in colon carcinogenesis. *Cancer Res.* **56**, 2375–2381.

Selivanova, G., and Wiman, K. (1995). p53: A cell cycle regulator activated by DNA damage. *Adv. Cancer Res.* **66**, 143–180.

Shamberger, R. J. (1984). Genetic toxicology of ascorbic acid. *Mutat. Res.* **133**, 135–159.

Shen, J.-C., Rideout, W. M., and Jones, P. A. (1994). The rate of hydrolytic deamination of 5-methylcytosine in double-stranded DNA. *Nucl. Acids Res.* **22**, 972–976.

Shibutani, S., Takeshita, M., and Grollman, A. P. (1991). Insertion of specific bases during DNA synthesis past the oxidation-damaged base 8-oxo-dG. *Nature* **349**, 431–434.

Shigenaga, M. K., Gimeno, C. J., and Ames, B. N. (1989). Urinary 8-hydroxy-2-deoxy-guanosine as a biological marker of *in vivo* oxidative DNA damage. *Proc. Natl. Acad. Sci. U.S.A.* **86**, 9697–9701.

Singh, N. P. (1997). Sodium ascorbate induces DNA single-strand breaks in human cells *in vitro*. *Mutat. Res.* **375**, 195–203.

Singh, N., and Khan, A. (1995). Acetaldehyde: Genotoxicity and cytotoxicity in human lymphocytes. *Mutat. Res.* **337**, 9–17.

Sohal, R., and Weindruch, R. (1996). Oxidative stress, caloric restriction, and aging. *Science* **273**, 59–63.

Sprang, S. R. (1997). Gap into the breach. *Science* **277**, 329–338.

Squadrito, G. L., Jin, X., and Pryor, W. A. (1995). Stopped-flow kinetic study of the reaction of ascorbic acid with peroxynitrite. *Arch. Biochem. Biophys.* **322**, 53–59.

Stang, S., Bottorff, D., and Stone, J. C. (1997). Interaction of activated ras with raf-1 alone may be sufficient for transformation of rat2 cells. *Mol. Cell. Biol.* **17**, 3047–3055.

Steenken, S. (1989). Purine bases, nucleosides, and nucleotides: Aqueous solution redox chemistry and transformation reactions of their radical cations and e⁻ and OH adducts. *Chem. Rev.* **89**, 503–520.

Steinmetz, K., and Potter, J. D. (1996). Vegetables, fruit, and cancer prevention: A review. *J. Am. Dietet. Assoc.* **96**, 1027–1039.

Sugimura, T. (1997). Overview of carcinogenic heterocyclic amines. *Mutat. Res.* **376**, 211–219.

Sundaresan, M., Yu., Z.-X., Ferrans, V., Sulciner, D. J., Gutkind, J., Irani, K., Goldschmidt-Clermont, P., and Finkel, T. (1996). Regulation of reactive oxygen species generation in fibroblasts by rac1. *Biochem. J.* **318**, 379–382.

Teebor, G. W. (1995). Excision base repair. *In* "DNA Repair Mechanisms" (J.-M. Vos, ed.), pp. 99–115. R. G. Landes, Austin, TX.

Thun, M. J., Calle, E. E., Namboodiri, M., Flanders, W., Coates, R. J., Byers, T., Boffetta, P., Garfinkel, L., and Heath, C. W. (1992). Risk factors for fatal colon cancer in a large prospective study. *J. Natl. Cancer Inst.* **84**, 1491–1500.

Trock, B., Lanza, E., and Greenwald, P. (1990). Dietary fiber, vegetables, and colon cancer: Critical review and meta-analysis of the epidemiological evidence. *J. Natl. Cancer Inst.* **82**, 650–661.

Tseng, M., Murray, S. C., Kupper, L. L., and Sandler, R. S. (1996). Micronutrients and the risk of colorectal adenomas. *Am. J. Epidemiol.* **144**, 1005–1014.

Ueda, Y., Hirai, S., Osada, S., Suzuki, A., Mizuono, K., and Ohno, S. (1996). Protein kinase C activates the MEK–ERK pathway in a manner independent of ras and dependent on raf. *J. Biol. Chem.* **271**, 23512–23519.

Vogelstein, B., and Kinzler, K. W. (1993). The multistep nature of cancer. *Trends Genet.* **9**, 138–141.

Wakabayashi, K., Nagao, M., Esumi, H., and Sugimura, T. (1992). Food-derived mutagens and carcinogens. *Cancer Res.* **52**, 2092S–2098S.

Willett, W. C. (1997). Potential benefits of preventive nutrition strategies. *In* "Preventive Nutrition" (A. Bendich and R. J. Deckelbaum, eds.), pp. 423–440. Humana Press, Totowa, NJ.

Willett, W. C., Stampfer, M. J., Colditz, G. A., Rosner, B. A., and Speizer, F. E. (1990). Relation of meat, fat, and fiber intake to the risk of colon cancer in a prospective study among women. *N. Engl. J. Med.* **323**, 1664–1672.

Yaffee, M., Walter, P., Richter, C., and Muller, M. (1996). Direct observation of iron-induced conformational changes of mitochondrial DNA by high-resolution field-emission in-lens scanning electron microscopy. *Proc. Natl. Acad. Sci. U.S.A.* **93**, 5341–5346.

Zhu, D., Keohavong, P., Finkelstein, S., Swalsky, P., Bakker, A., Weissfeld, Srivastava, S., and Whitside, T. L. (1997). K-ras gene mutations in normal colorectal tissues from K-ras mutation-positive colorectal cancer patients. *Cancer Res.* **57**, 2485–2492

# BIBLIOGRAPHY

## Medical Diagnosis and Biopsies

Eddy, D. M. (1990). Screening for colorectal cancer. *Ann. Intern. Med.* **113**, 373–384.

Gabbert, H. (1985). Mechanisms of tumor invasion: Evidence from *in vivo* observations. *Cancer Metastasis Rev.* **4**, 293–309.

Haggitt, R. C., Glotzbach, R. E., Soffer, E., and Wruble, L. D. (1985). Prognostic factors in colorectal carcinomas arising in adenomas. *Gastroenterology* **89**, 328–336.

Liabakk, N., Talbot, I., Smith, R., Wilkinson, K., and Balkwill, F. (1996). Matrix metalloprotease 2 (MMP-2) and matrix metalloprotease 9 (MMP-9) type IV collagenases in colorectal cancer. *Cancer Res.* **56**, 190–196.

Lupulescu, A. P. (1996). Hormones, vitamins, and growth factors in cancer treatment and prevention. *Cancer* **78**, 2264–2280.

Reymond, M., Sanchez, J.-C., Schneider, C., Rohwer, P., Tortola, S., Hohenberger, W., Kirchner, T., Hochstrasser, D., and Kockerling, F. (1997). Specific sample preparation in colorectal cancer. *Electrophoresis* **18**, 622–624.

Winawer, S. J., Fletcher, R., Miller, L., Godlee, F., Stolar, M., Mulrow, C., Woolf S. H., Glick, S., Baniats, T., Bond, J., Rosen, L., Zapka, J., Olsen, S., Giardiello, F., Sisk, J., Antwerp, R., Brown-Davis, C., Marciniak, D., and Mayer, R.J. (1997). Colorectal cancer screening: Clinical guidelines and rationale. *Gastroenterology* **112**, 594–642.

*Oncogenes and Their Proteins (APC, RAS, MAP kinase, p53, and Catenin)*

Aaltonen, L., Peltomaki, P., Leach, F., Sistonen, P., Pylkkanen, L., Mecklin, J., Jarvinen, H., Powell, S., Jen, J., Hamilton, S., Petersen, G., Kinzler, K., Vogelstein, B., and Chapelle, A. (1993). Clues to the pathogenesis of familial colorectal cancer. *Science* **260**, 812–819.

Bar-Sagi, D. (1994). The sos (son of sevenless) protein. *Trends Endocrinol. Metab.* **5**, 165–169.

Bollag, G., and McCormick, F. (1991). Differential regulation of ras, GAP and neurofibromatosis gene product activities. *Nature* **351**, 576–579.

Bos, J. L. (1989). Ras oncogenes in human cancer: A review. *Cancer Res.* **49**, 4682–4689.

Cover, C. M., Hsieh, S., Tran, S., Hallden, G., Kim, G., Bjeldanes, L. F., and Firestone, G. L. (1998). *J. Biol. Chem.* **273**, 3838–3847.

Dorudi, S., Hanby, A., Poulsom, R., Northover, J., and Hart, I. (1995). Level of expression of E-cadherin mRNA in colorectal cancer correlates with clinical outcome. *Brit. J. Cancer* **71**, 614–616.

Fearson, E. R., and Vogelstein, B. (1990). A genetic model for colorectal tumorigenesis. *Cell* **61**, 759–767.

Greenblatt, M. S., Bennett, W., Hollstein, M., and Harris, C. C. (1994). Mutations in the p53 tumor suppressor gene. *Cancer Res.* **54**, 4855–4878.

Hann, B. C., and Lane, D. P. (1995). The dominating effect of mutant p53. *Nature Genet.* **9**, 221–222.

Harvey, M., Vogel, H., Morris, D., Bradley, A., Bernstein, A., and Donehower, L. A. (1995). A mutant p53 transgene accelerates tumour development in heterozygous but not nullizygous p53-deficient mice. *Nature Genet.* **9**, 305–311.

Hollstein, M., Sidransky, D., Vogelstein, B., and Harris, C. (1991). p53 Mutation in human cancers. *Science* **253**, 49–53.

Janknecht, R., Cahill, M., and Nordheim, A. (1995). Signal integration and the c-fos promoter. *Carcinogenesis* **16**, 443–450.

Jankowski, J., Bedford, F., and Kim, Y. S. (1997). Changes in gene structure and regulation of E-cadherin during epithelial development, differentiation, and disease. *Progr. Nucl. Acids Res. Mol. Biol.* **57**, 187–215.

Kakiuchi, H., Watanabe, M., Ushijima, T., Toyota, M., Imai, K., Weisburger, J., Sugimura, T., and Nagao, M. (1995). Specific 5′-GGGA-3′ to 5′-GGA-3′ mutation of the Apc gene in rat colon tumors. *Proc. Natl. Acad. Sci. U.S.A.* **92**, 910–914.

Kallunki, T., Deng, T., Hibi, M., and Karin, M. (1996). c-Jun can recruit JNK to phosphorylate dimerization partners via docking interactions. *Cell* **57**, 929–939.

Kemler, R. (1993). From cadherins to catenins: Cytoplasmic protein interactions and regulation of cell adhesion. *Trends Genet.* **9**, 317–321.

Kinzler, K., and Vogelstein, B. (1996). Lessons from hereditary colorectal cancer. *Cell* **87**, 159–170.

Lemmon, M. A., Falasca, M., Ferguson, K. M., and Schlessinger, J. (1997). Regulatory recruitment of signalling molecules to the cell membrane by pleckstrin-homology domains. *Trends Cell. Biol.* **7**, 237–242.

Lewis, C. M., Neuhausen, S. L., Daley, D., Black, F., Swensen, J., Burt, R., Cannon-Albright, L., and Skolnick, M. H. (1996). Genetic heterogeneity and unmapped genes for colorectal cancer. *Cancer Res.* **56**, 1382–1385.

Magewu, A., and Jones, P. A. (1994). Ubiquitous and tenacious methylation of the CpG site in codon 248 of the p53 gene may explain its frequent appearance as a mutational hot spot in human cancer. *Mol. Cell. Biol.* **14**, 4225–4232.

Matheson, H., Branting, C., Rafter, I., Okret, S., and Rafter, J. (1996). Increased c-fos mRNA and binding to the AP-1 recognition sequence accompanies the proliferative response to deoxycholate of HT29 cells. *Carcinogenesis* **17**, 421–426.

Morin, P. J., Sparks, A. B., Korinek, V., Barker, N., Clevers, H., Vogelstein, B., and Kinzler, K. W. (1997). Activation of β-catenin-Tcf signaling in colon cancer by mutations in β-catenin or APC. *Science* **275**, 1787–1790.

Munemitsu, S., Albert, I., Souza, B., Rubinfeld, B., and Polakis, P. (1995). Regulation of intracellular β-catenin levels by the adenomatous polyposis coli (APC) tumor-suppressor protein. *Proc. Natl. Acad. Sci. U.S.A.* **92**, 3046–3050.

Polakis, P. (1997). The adenomatous polyposis coli (APC) tumor suppressor. *Biochim. Biophys. Acta* **1332**, F127–F147.

Prives, C. (1994). How loops,β sheets, and α helices help us to understand p53. *Cell* **78**, 543–546.

Riccardi, V. M. (1981). Von Recklinghausen neurofibromatosis. *N. Engl. J. Med.* **305**, 1617–1626.

Risinger, J., Berchuck, A., Kohler, M., and Boyd, J. (1994). Mutations of the E-cadherin gene in human gynecologic cancers. *Nature Genet.* **7**, 98–102.

Ruaro, E. M., Collavin, L., Sal, G., Haffner, R., Oren, M., Levine, A. J., and Schneider, C. (1997). A proline-rich motif in p53 is required for transactivation-independent growth arrest as induced by Gas 1. *Proc. Natl. Acad. Sci. U.S.A.* **94**, 4675–4680.

Rubinfeld, B., Souza, B., Albert, I., Muller, O., Chamberlain, S., Masiarz, F., Munemitsu, S., and Polakis, P. (1993). Association of the APC gene product with β-catenin. *Science* **262**, 1731–1737.

Schmutte, C., Yang, A. S., Beart, R. W., and Jones, P. A. (1995). Base excision repair of U:G mismatches at a mutational hotspot in the p53 gene is more efficient than base excision repair of T:G mismatches in extracts of human colon tumors. *Cancer Res.* **55**, 3742–3746.

Selivanova, G., and Wiman, K. (1995). p53: A cell cycle regulator activated by DNA damage. *Adv. Cancer Res.* **66**, 143–180.

Silberman, S., Janulis, M., and Schultz, R. M. (1997). Characterization of downstream ras signals that induce alternative protease-dependent invasive phenotypes. *J. Biol. Chem.* **272**, 5927–5935.

Shibata, H., Toyama, K., Shioya, H., Ito, M., Hirota, M., Hasegawa, S., Matsumoto, H., Takano, H., Akiyama, T., Toyoshima, K., Kanamuru, R., Kanegae, Y., Saito, I., Nakamura, Y., Shiba, K., and Noda, T. (1997). Rapid colorectal adenoma formation initiated by conditional targeting of the apc gene. *Science* **278**, 120–123.

Skoudy, A., Gomez, S., Fabre, M., and Herreros, G. (1996). P120-catenin expression in human colorectal cancer. *Int. J. Cancer* **68**, 14–20.

Sprang, S. R. (1997). Gap into the breach. *Science* **277**, 329–338.

Stang, S., Bottorff, D., and Stone, J. C. (1997). Interaction of activated ras with raf-1 alone may be sufficient for transformation of rat2 cells. *Mol. Cell. Biol.* **17**, 3047–3055.

Takeichi, M. (1990). Cadherins. *Annu. Rev. Biochem.* **59**, 237–252.

Ueda, Y., Hirai, S., Osada, S., Suzuki, A., Mizuono, K., and Ohno, S. (1996). Protein kinase C activates the MEK–ERK pathway in a manner independent of ras and dependent on raf. *J. Biol. Chem.* **271**, 23512–23519.

Vermeulen, S., Bruyneel, E., Bracke, M., Bruyne, G., Vennekens, K., Vleminckx, K., Berx, G., Roy, F. M., and Mareel, M. M. (1995). Transition from the noninvasive to the invasive phenotype and loss of α-catenin in human colon cancer cells. *Cancer Res.* **55**, 4722–4728.

Vleminckx, K., Vakaet, L., Mareel, M., Fiers, W., and Van Roy, F. (1991). Genetic manipulation of E-cadherin expression by epithelial tumor cells reveals an invasion suppressor role. *Cell* **66**, 107–119.

Vogelstein, B., and Kinzler, K. W. (1993). The multistep nature of cancer. *Trends Genet.* **9**, 138–141.

Wyllie, A. H., Carder, Clarke, A. R., Cripps, K. J., Gledhill, S., Greaves, M., Griffiths, S., Harrison, D. J., Hooper, M., Morris, R. G., Purdie, C. A., and Bird, C. C. (1994). Apoptosis in carcinogenesis: The role of p53. *Cold Spring Harbor Symp. Quant. Biol.* **59**, 403–409.

Zhu, D., Keohavong, P., Finkelstein, S., Swalsky, P., Bakker, A., Weissfeld, Srivastava, S., and Whitside, T. L. (1997). K-ras gene mutations in normal colorectal tissues from K-ras mutation-positive colorectal cancer patients. *Cancer Res.* **57**, 2485–2492

*Folate, DNA Methylation, and DNA Repair Enzymes*

Antequera, F., and Bird, A. (1993). Number of CpG islands and genes in human and mouse. *Proc. Natl. Acad. Sci. U.S.A.* **90**, 11995–11999.

Bandaru, B., Gopal, J., and Bhagwat, A. S. (1996). Overproduction of DNA cytosine methyltransferases causes methylation and C-T mutations at non-canonical sites. *J. Biol. Chem.* **271**, 7851–7859.

Blount, B. C., Mack, M. M., Wehr, C., MacGregor, J. T., Hiatt, R. A., Wang, G., Wickramasinghe, S., Everson, R. B., and Ames, B. N. (1997). Folate deficiency causes uracil misincorporation into human DNA and chromosome breakage. *Proc. Natl. Acad. Sci. U.S.A.* **94**, 3290–3295.

Buiting, K., Saitoh, S., Gross, S., Dittrich, B., Schwartz, S., Nicholls, R. D., and Horsthemke, B. (1995). Inherited microdeletions in the Angelman and Prader-Willi syndromes define an imprinting centre on human chromosome 15. *Nature Genet.* **9**, 395–400.

Holliday, R. (1989). Chromosome error propagation and cancer. *Trends Genet.* **5**, 42–45.

Jones, P. A. (1996). DNA methylation errors and cancer. *Cancer Res.* **56**, 2463–2467.

Jones, P. A., and Gonzalgo, M. L. (1997). Altered DNA methylation and genome instability: A new pathway to cancer? *Proc. Natl. Acad. Sci.* **94**, 2103–2105.

Neddermann, P., and Jiricny, J. (1993). The purification of a mismatch-specific thymine-DNA glycosylase from HeLa cells. *J. Biol. Chem.* **268**, 21218–21224.

Pogribny, I. P., Basnakian, A., Miller, B., Lopatina, N., Proirier, L. A., and James, S. J. (1995). Breaks in genomic DNA and within the p53 gene are associated with hypomethylation in livers of folate/methyl-deficient rats. *Cancer Res.* **55**, 1894–1901.

Shen, J.-C., Rideout, W. M., and Jones, P. A. (1994). The rate of hydrolytic deamination of 5-methylcytosine in double-stranded DNA. *Nucl. Acids Res.* **22**, 972–976.

Schmutte, C., Yang, A. S., Nguyen, T., Beart, R., and Jones, P. A. (1996). Mechanisms for the involvement of DNA methylation in colon carcinogenesis. *Cancer Res.* **56**, 2375–2381.

Slupphaug, G., Mol, C. D., Kavli, B., Arvai, A. S., Krokan, H. E., and Tainer, J. A. (1996). A nucleotide-flipping mechanism from the structure of human uracil-DNA glycosylase bound to DNA. *Nature* **384**, 87–92.

Teebor, G. W. (1995). Excision base repair. *In* "DNA Repair Mechanisms" (J.-M. Vos, ed.), pp. 99–115. R. G. Landes, Austin, TX.

*DNA Damage, Toxic Oxygen, Vitamin C, and Oxo-G*

Aburatani, H., Hippo, Y., Ishida, T., Takashima, R., Matsuba, C., Kodama, T., Takao, M., Yasui, A., Yamamoto, K., Asano, M., Fukasawa, K., Yoshinari, T., Inoue, H., Ohtsuka, E., and Nishimura, S. (1997). Cloning and characterization of mammalian 8-hydroxy-guanine-specific DNA glycosylase/apurinic, apyrimidinic lyase. *Cancer Res.* **57**, 2151–2156.

Alperin, E. S., and Shapiro, L. J. (1997). Characterization of point mutations in patients with X-linked ichthyosis. *J. Biol. Chem.* **272**, 20756–20763.

Ames, B. (1989). Endogenous oxidative DNA damage, aging, and cancer. *Free Rad. Res. Comm.* **7**, 121–128.

Anderson, D., Yu, T.-W., Phillips, B. J., and Schmezer, P. (1994). The effect of various antioxidants and other modifying agents on oxygen radical generated DNA damage in human lymphocytes in the COMET assay. *Mut. Res.* **307**, 261–271.

Beckman, K. B., and Ames, B. N. (1997). Oxidative decay of DNA. *J. Biol. Chem.* **272**, 19633–19636.

Buettner, G. R. (1993). The pecking order of free radicals and antioxidants. *Arch. Biochem. Biophys.* **300**, 535–543.

Enright, H. U., Miller, W. J., and Hebbel, R. P. (1992). Nucleosomal histone protein protects DNA from iron-mediated damage. *Nucl. Acids Res.* **20**, 3341–3346.

Fraga, C. G., Motchnik, P. A., Shigenaga, M. K., Helbock, H. J., Jacob, R. A., and Ames, B. N. (1991). Ascorbic acid protects against endogenous oxidative DNA damage in human sperm. *Proc. Natl. Acad. Sci. U.S.A.* **88**, 11003–11006.

Furge, L., and Guengerich, F. P. (1997). Analysis of nucleotide insertion and extension at 8-oxo-7,8-dihydroguanine by replicative T7 polymerase. *Biochemistry* **36**, 6475–6487.

Halliwell, B. (1996). Vitamin C: Antioxidant or pro-oxidant *in vivo*? *Free Rad. Res.* **25**, 439–454.

Halliwell, B. (1997a). Antioxidants and human disease: A general introduction. *Nutr. Rev.* **55**, S44–S52.

Halliwell, B. (1997b). Ascorbic acid: Hype, hoax, or healer? *Am. J. Clin. Nutr.* **65**, 1891–1892.

Halliwell, B., and Gutteridge, J. M. (1986). Oxygen free radicals and iron in relation to biology and medicine. *Arch. Biochem. Biophys.* **246**, 501–514.

Henle, E. S., and Linn, S. (1997). Formation, prevention, and repair of DNA damage by iron/hydrogen peroxide. *J. Biol. Chem.* **272**, 19095–19098.

Henle, E. S., Luo, Y., Gassmann, W., and Linn, S. (1996). Oxidative damage to DNA constituents by iron-mediated Fenton reactions. *J. Biol. Chem.* **271**, 21177–21186.

Irani, K., Xia, Y., Zweier, J. L., Sollott, S. J., Der, C., Fearon, E., Sundaresan, M., Finkel, T., and Goldschmidt-Clermont, P. (1997). Mitogenic signaling mediated by oxidants in ras-transformed fibroblasts. *Science* **275**, 1649–1652.

Kyrtopoloulos, S. A. (1987). Ascorbic acid and the formation of *N*-nitroso compounds: Possible role of ascorbic acid in cancer prevention. *Am. J. Clin. Nutr.* **45**, 1344–1350.

Leeuwenburgh, C., Hardy, M. M., Hazen, S., Wagner, P., Oh-ishi, S., Steinbrecher, U., and Heinecke, J. W. (1997). Reactive nitrogen intermediates promote low density lipoprotein oxidation in human atherosclerotic intima. *J. Biol. Chem.* **272**, 1433–1436.

Lindahl, T. (1993). Instability and decay of the primary structure of DNA. *Nature* **362**, 709–715.

McAuley-Hecht, K., Leonard, G. A., Gibson, N. J., Thomson, J. B., Watson, W., Hunter, W., and Brown, T. (1994). Crystal structure of a DNA duplex containing 8-hydroxydeoxyguanine–adenine base pairs. *Biochemistry* **33**, 10266–10270.

McKnight, G. M., Smith, L., Drummond, R., Duncan, C. W., Golden, M., and Benjamin, N. (1997). Chemical synthesis of nitric oxide in the stomach from dietary nitrate in humans. *Gut* **40**, 211–214.

Minna, J. D., and Mangelsdorf, D. J. (1997). Retinoic acid receptor expression abnormalities in lung cancer. *J. Natl. Cancer Inst.* **89**, 602–604.

Nakajima, M., Takeuchi, T., and Kanehisa, M. (1996). Determination of 8-hydroxydeoxyguanosine in human cells under oxygen-free conditions. *Carcinogenesis* **17**, 787–791.

Nishikimi, M., and Yagi, K. (1977). Oxidations of ascorbic acid and tocopherol by superoxide. *In* "Biochemical and Medical Aspects of Active Oxygen" (O. Hayaishi and K. Asada, ed.), pp. 79–87. University Park Press, Baltimore.

Pfeiffer, S., Gorren, A., Schmidt, K., Werner, E., Hansert, B., Bohle, D., and Mayer, B. (1997). Metabolic fate of peroxynitrite in aqueous solution. *J. Biol. Chem.* **272**, 3465–3470.

Prieme, H., Loft, S., Nyyssonen, K., Salonen, J. T., and Poulsen, H. E. (1997). No effect of supplementation with vitamin E, ascorbic acid, or coenzyme Q10 on oxidative DNA damage estimated by 8-oxy-7,8-dihydro-2-deoxyguanosine excretion in smokers. *Am. J. Clin. Nutr.* **65**, 503–507.

Shamberger, R. J. (1984). Genetic toxicology of ascorbic acid. *Mutat. Res.* **133**, 135–159.

Shibutani, S., Takeshita, M., and Grollman, A. P. (1991). Insertion of specific bases during DNA synthesis past the oxidation-damaged base 8-oxo-dG. *Nature* **349**, 431–434.

Shigenaga, M. K., Gimeno, C. J., and Ames, B. N. (1989). Urinary 8-hydroxy-2-deoxy-guanosine as a biological marker of *in vivo* oxidative DNA damage. *Proc. Natl. Acad. Sci. U.S.A.* **86**, 9697–9701.

Singh, N. P. (1997). Sodium ascorbate induces DNA single-strand breaks in human cells *in vitro*. *Mutat. Res.* **375**, 195–203.

Singh, N., and Khan, A. (1995). Acetaldehyde: Genotoxicity and cytotoxicity in human lymphocytes. *Mutat. Res.* **337**, 9–17.

Sohal, R., and Weindruch, R. (1996). Oxidative stress, caloric restriction, and aging. *Science* **273**, 59–63.

Squadrito, G. L., Jin, X., and Pryor, W. A. (1995). Stopped-flow kinetic study of the reaction of ascorbic acid with peroxynitrite. *Arch. Biochem. Biophys.* **322**, 53–59.

Steenken, S. (1989). Purine bases, nucleosides, and nucleotides: Aqueous solution redox chemistry and transformation reactions of their radical cations and e⁻ and OH adducts. *Chem. Rev.* **89**, 503–520.

Sundaresan, M., Yu., Z.-X., Ferrans, V., Sulciner, D. J., Gutkind, J., Irani, K., Goldschmidt-Clermont, P., and Finkel, T. (1996). Regulation of reactive oxygen species generation in fibroblasts by rac1. *Biochem. J.* **318**, 379–382.

Tkeshelashvili, L., McBride, T., Spence, K., and Loeb, L. A. (1991). Mutation spectrum of copper-induced DNA damage. *J. Biol. Chem.* **266**, 6401–6406.

Toyokuni, S., and Sagripanti, J.-L. (1996). Association between 8-hydroxy-2-deoxyguanos-ine formation and DNA strand breaks mediated by copper and iron. *Free Rad. Biol. Med.* **20**, 859–864.

## Chemical Carcinogens in Food

Ames, B. N., Gold, L. S., and Shigenaga, M. K. (1996). Cancer prevention, rodent high-dose cancer tests, and risk assessment. *Risk Anal.* **16**, 613–617.

Anderson, K. E., Hammons, G., Kadlubar, F., Potter, J. D., Kaderlik, K., Ilett, K., Minchin, R., Teitel, C., Chou, H.-C., Martin, M., Guengerich, F., Barone, G., Lang, N., and Peterson, L. (1997). Metabolic activation of aromatic amines by human pancreas. *Carcinogenesis* **18**, 1085–1092.

Basu, A., and Essigmann, J. M. (1995). DNA damage: Structural and functional consequences. *In* "DNA Repair Mechanisms" (J.-M. H. Vos, ed.), pp. 1–19. R. G. Landes, Austin, TX.

Bjeldanes, L. F., Kim, J.-Y., Grose, K. R., Bartholemew, J. C., and Bradfield, C. A. (1991). Aromatic hydrocarbon responsiveness-receptor agonists generated from indole-3-car-binol *in vitro* and *in vivo*: Comparisons with 2,3,7,8-tetrachlorodibenzo-p-dioxin. *Proc. Natl. Acad. Sci. U.S.A.* **88**, 9543–9547.

Felton, J. S., Knize, M. G., Roper, M., Fultz, E., Shen, N., and Turteltaub, K. W. (1992). Chemical analysis, prevention, and low-level dosimetry of heterocyclic amines from cooked food. *Cancer Res.* **52**, 2103S–2107S.

Felton, J. S., Malfatti, M. A., Knize, M., Salmon, C. P., Hopmans, E. C., and Wu, R. W. (1997). Health risks of heterocyclic amines. *Mutat. Res.* **376**, 37–41.

Gold, L. S., Slone, T. H., Manley, N. B., and Ames, B. N. (1994). Heterocyclic amines formed by cooking food. *Cancer Lett.* **83**, 21–29.

Grant, D. M., Hughes, N., Janezic, S., Goodfellow, G. H., Chen, H., Gaedigk, A., Yu, V., and Grewal, R. (1997). Human acetyltransferase polymorphisms. *Mutat. Res.* **376**, 61–70.

Ito, N., Hasegawa, R., Sano, M., Tamano, S., Esumi, H., Takayama, S., and Sugimura, T. (1991). A new colon and mammary carcinogen in cooked food, 2-amino-1-methyl-6-phenylimidazo[4,5-β]pyridine (PhIP). *Carcinogenesis* **12**, 1503–1506.

Lin, D., Meyer, D. J., Ketterer, B., Lang, N. P., and Kadlubar, F. F. (1994). Effects of human and rat glutathione *S*-transferases on the covalent DNA binding of the *N*-acetoxy derivatives of heterocyclic amine carcinogens *in vitro*. *Cancer Res.* **54**, 4920–4926.

Malfatti, M. A., Connors, M., Mauthe, R., and Felton, J. S. (1996). The capability of rat colon tissue slices to metabolize the cooked-food carcinogen 2-amino-1-methyl-6-phenylimidazo[4,5-β]pyridine. *Cancer Res.* **56**, 2550–2555.

Mumford, J., Li, X., Hu, F., Lu, X., and Chuang, J. (1995). Human exposure and dosimetry of polycyclic aromatic hydrocarbons in urine from Xuan Wei, China, with high lung cancer mortality associated with exposure to unvented coal smoke. *Carcinogenesis* **16**, 3031–3036.

Nagao, M., Ushijima, T., Toyata, M., Inoue, R., and Sugimura, T. (1997). Genetic changes induced by heterocyclic amines. *Mutat. Res.* **376**, 161–167.

Schiffman, M. H. (1990). Re: "Fried foods and the risk of colon cancer." *Am. J. Epidemiol.* **131**, 376–378.

Sugimura, T. (1997). Overview of carcinogenic heterocyclic amines. *Mutat. Res.* **376**, 211–219.

Wakabayashi, K., Nagao, M., Esumi, H., and Sugimura, T. (1992). Food-derived mutagens and carcinogens. *Cancer Res.* **52**, 2092S–2098S.

## *Anticancer Agents in Food*

Block, G., Patterson, B., and Subar, A. (1992). Fruit, vegetables, and cancer prevention: A review of the epidemiological evidence. *Nutr. Cancer* **18**, 1–29.

Bogaards, J., Verhagen, H., Willems, M., Poppel, G., and Bladeren, P. (1994). Consumption of Brussels sprouts results in elevated α-class glutathione *S*-transferase levels in human blood plasma. *Carcinogenesis* **15**, 1073–1075.

Marcus, A. J. (1995). Aspirin as prophylaxis against colorectal cancer. *N. Engl. J. Med.* **333**, 656–658.

Nijhoff, W., Grubben, M., Nagengast, F., Jansen, Verhagen, H., Poppel, G., and Peters, W. (1995). Effects of consumption of Brussels sprouts on intestinal and lymphocytic glutathione *S*-transferases in humans. *Carcinogenesis* **16**, 2125–2128.

Slavin, J., Jacobs, D., and Marquart, L. (1997). Whole-grain consumption and chronic disease: Protective mechanisms. *Nutr. Cancer* **27**, 14–21.

Steinmetz, K., and Potter, J. D. (1996). Vegetables, fruit, and cancer prevention: A review. *J. Am. Dietet. Assoc.* **96**, 1027–1039.

Verhagen, H., Vries, A., Nijhoff, W., Schouten, A., Poppel, G., Peters, W., and Berg, H. (1997). Effect of Brussels sprouts on oxidative DNA-damage in man. *Cancer Lett.* **114**, 127–130.

Zhang, Y., and Talalay, P. (1994). Anticarcinogenic activities of organic isothiocyanates: Chemistry and mechanisms. *Cancer Res.* **54**, 1976S–1981S.

Wargovich, M. J. (1997). Experimental evidence for cancer preventive elements in foods. *Cancer Lett.* **114**, 11–17.

## *Epidemiological Studies of Diet and Cancer*

Ames, B. N., Gold, L. S., and Willett, W. C. (1995). The causes and prevention of cancer. *Proc. Natl. Acad. Sci. U.S.A.* **92**, 5258–5265.

Block, G., Thompson, F. E., Hartman, A., Larkin, F. A., and Guire, K. E. (1992). Comparison of two dietary questionnaires validated against multiple dietary records collected during a 1-year period. *J. Am. Diet. Assoc.* **92**, 686–693.

Block, G., Sinha, R., and Gridley, G. (1994). Collection of dietary-supplement data and implications for analysis. *Am. J. Clin. Nutr.* **59**, 232S–239S.

Giovannucci, E., Rimm, E. B., Ascherio, A., Stampfer, M. J., Colditz, G., and Willett, W. C. (1995). Alcohol, low-methionine-low-folate diets and risk of colon cancer in men. *J. Natl. Cancer Inst.* **87**, 265–273.

Peipins, L. A., and Sandler, R. S. (1994). Epidemiology of colorectal adenomas. *Epidemiol. Rev.* **16**, 273–297.

Phillips, R. L., and Snowdon, D. A. (1985). Dietary relationships with fatal colorectal cancer among Seventh-Day Adventists. *J. Natl. Cancer Inst.* **74**, 307–317.

Potter, J. D. (1996). Nutrition and colorectal cancer. *Cancer Causes Controls* **7**, 127–146.

Potter, J. D. (1997). Cancer prevention: Epidemiology and experiment. *Cancer Lett.* **114**, 7–9.

Potter, J. D., Slattery, M. L., Bostick, R., and Gapstur, S. M. (1993). Colon cancer: A review of the epidemiology. *Epidemiol. Rev.* **15**, 499–545.

Prentice, R., Thompson, D., Clifford, C., Gorbach, S., Goldin, B., and Byar, D. (1990). Dietary fat reduction and plasma estradiol concentration in healthy postmenopausal women. *J. Natl. Cancer Inst.* **82**, 129–134.

Thun, M. J., Calle, E. E., Namboodiri, M., Flanders, W., Coates, R. J., Byers, T., Boffetta, P., Garfinkel, L., and Heath, C. W. (1992). Risk factors for fatal colon cancer in a large prospective study. *J. Natl. Cancer Inst.* **84**, 1491–1500.

Trock, B., Lanza, E., and Greenwald, P. (1990). Dietary fiber, vegetables, and colon cancer: Critical review and meta-analysis of the epidemiological evidence. *J. Natl. Cancer Inst.* **82**, 650–661.

Tseng, M., Murray, S. C., Kupper, L. L., and Sandler, R. S. (1996). Micronutrients and the risk of colorectal adenomas. *Am. J. Epidemiol.* **144**, 1005–1014.

Willett, W. C. (1997). Potential benefits of preventive nutrition strategies. *In* "Preventive Nutrition" (A. Bendich and R. J. Deckelbaum, eds.), pp. 423–440. Humana Press, Totowa, NJ.

Willett, W. C., Stampfer, M. J., Colditz, G. A., Rosner, B. A., and Speizer, F. E. (1990). Relation of meat, fat, and fiber intake to the risk of colon cancer in a prospective study among women. *N. Engl. J. Med.* **323**, 1664–1672.

Witte, J. S., Longnecker, M., Bird, C. L., Lee, E. R., Frankl, H. D., and Haile, R. W. (1996). Relation of vegetable, fruit, and grain consumption to colorectal adenomatous polyps. *Am. J. Epidemiol.* **144**, 1015–1025.

# NUTRITION
# METHODOLOGY

The researcher interested in comparing the effects of control and experimental diets, or a control diet and a diet containing a drug or toxin, needs to be aware of certain issues in raising animals. These include pair feeding, trained meal eating, the multigeneration deficiency, and the use of chemically defined diets.

Deficient diets may result in anorexia as well as impaired growth or weight loss. Do the lesser growth and associated biochemical changes result mainly from the lack of appetite or from the impairment of biochemical functions dependent on the nutrient? Pair feeding is used to distinguish between these two effects. The food consumed by the control animal (complete diet) and experimental animal (deficient diet) is weighed daily. The rat, for example, tends to eat food mainly at night, that is, in the dark. The rat is a nibbler and naturally consumes food in about 10 small meals per day. In the morning, the researcher using the pair feeding technique measures the amount of food consumed by both groups. The deficient group may have consumed less food than the controls. The researcher then attempts to equalize the amount of food to be consumed by both groups during the following evening by giving both groups an amount of food equal to that consumed by the deficient group the previous evening. In this way, the amount of food consumed can be equalized on a day-to-day basis.

## Ensuring Equal Food Intake by Control and Experimental Animals

Two examples of the use of the pair feeding technique are shown in Figures A.1 and A.2. Figure A.1 depicts the effects of a vitamin D-deficient diet, and Figure A.2 concerns a thiamin-deficient diet. Three groups of rats were used in each study: (1) rats eating the control diet *ad libitum* (—●—); (2) rats eating the deficient diet *ad libitum* (—Δ); (3) pair-fed rats consuming the control diet (—○—). The results demonstrate that the vitamin D-sufficient rats (—●—) steadily gained weight for

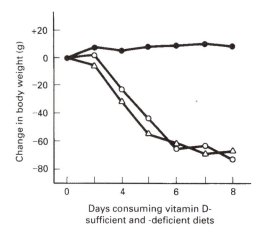

FIGURE A.1 Effects of a vitamin D-deficient diet. See text for details. [Redrawn with permission from Brommage *et al.* (1984).]

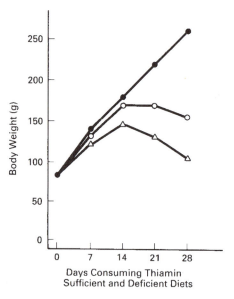

FIGURE A.2 Effects of a thiamin-deficient diet. See text for details. [Redrawn with permission from Singh (1982).]

8 days. The deficient rats (—△—) and pair-fed rats (○) consuming the sufficient diet lost weight at similar rates, indicating that the weight loss was due entirely to anorexia. The results with the thiamin-deficient diet indicate that the failure to grow was due to both anorexia and the lack of thiamin for growth and maintenance.

Changes in the diet can result in changes in feeding patterns. With pair feeding the small amount of food (complete diet) given to the control animals is usually consumed rapidly, leaving the animals hungry for the rest of the day. The experimental animals receiving the same amount of food (deficient diet) consume the food gradually over the course of the day. Differences in feeding patterns can result in changes in the body's biochemistry. For example, meal feeding (in rats) has been shown to produce an increase in body fat, when compared with nibbling (Cohn and Joseph, 1960). A number of biochemical differences between meal feeding and nibbling have been noted in humans (Wadhwa *et al.*, 1973). The requirements for certain nutrients, such as an amino acid (when supplied as a free amino acid), may be greater when consumed in a meal rather than when consumed continuously throughout the day or in many small meals (Baker, 1984).

The problem of different feeding patterns can be solved by training animals to consume their food over the same period of time. Here, control and experimental rats are trained to consume food in a short interval, for example, 2 hours (Harper, 1986). Another technique is to use a mechanical food dispenser. The dispenser monitors the amount of food consumed by the experimental animals and coincidentally delivers identical quantities of food to the control animals (Suttie, 1968; Quarterman *et al.*, 1970). A third method for ensuring equal intake of food with identical feeding patterns is force feeding. In force feeding, a lubricated tube inserted into the animal's mouth and esophagus delivers liquid diet to the stomach.

## Multigeneration Deficiency Technique Is Used When Typical Feeding Techniques Fail to Provoke Signs of Deficiency

Special techniques have been developed to induce deficiencies and assess requirements. A multigeneration deficiency is induced where a severe deficiency in a particular nutrient cannot be induced in the life span of the animal. This technique is also used to study the effect on the neonatal or young animal. The multigeneration deficiency has been used in studies of taurine (Sturman *et al.*, 1985), essential fatty acids (Tinoco *et al.*, 1978), and fluoride (Tao and Suttie, 1976).

## Nutrient Balance Is Used to Assess Requirements

The nutrient balance technique is sometimes used to determine requirements, as illustrated by the example of growing monkeys in the Protein chapter. This intake indicates only the amount required to maintain a constant level of the nutrient in the body. Under some conditions, the determination of the minimal intake that can support nutrient balance in an adult animal does not necessarily indicate the requirement (Mertz, 1987). Other techniques must then be used to determine if higher, but still constant, levels are of greater value to the animal. Here, biochemical tests may be used to assess a relevant physiological function. Kopple (1987) has pointed out a number of difficulties in using the nutrient balance technique. For example, the value for nutrient intake tends to be overestimated and that for excretion tends to be underestimated. These two factors can lead to an artifactually high value for the nutrient balance.

## Nutrient Antagonists Can Be Used to Provoke Deficiency

Nutrient antagonists are sometimes used where a deficiency cannot easily be induced by diet alone. An advantage of the use of antagonists is that they can take effect rapidly and can target specific enzymes or nutrient transport systems. A disadvantage of antagonists is that in some cases they can cause side-effects not related to the nutrient of interest. In brief, methotrexate is a folate antagonist, oxythiamin and pyrithiamin are thiamin antagonists, warfarin and chlorovitamin K are vitamin K antagonists, galactoflavin is a riboflavin antagonist, ω-methylpantothenic acid is a pantothenic acid antagonist, tungsten is a molybdenum antagonist, and ethionine is a methionine antagonist.

## Gut Microflora and Feces in Animal Nutrition

A fine point regarding fiber is that it can reduce the requirements for certain nutrients, that is, folate in animals and biotin and vitamin K in animals and humans. This is because fiber is available to the gut microflora as an energy source, thus supporting the intestinal production of vitamins, which may be released and used by the host (Harper and Elvehjem, 1957; Leoschke *et al.*, 1953; Cohen *et al.*, 1971). In addition to direct absorption by the gut, microbial vitamins can be supplied by coprophagy. Rats, along with most other mammals, practice coprophagy; that is, they eat their own feces. Feces are rich in a variety of nutrients. Coprophagy can be minimized by raising animals on screen-bottom cages, or completely prevented by raising animals in tube cages (Geyer *et al.*, 1947). Tube cages are made of chicken wire, with a feeding jar at the rat's head end. The cage must fit snuggly, but not tightly, around the rat to prevent it from reaching backward and eating its own feces. The researcher may encounter difficulty in placing the rat in the tube cage. This problem can be overcome by enshrouding the cage in a paper bag. The rat, seeing the darkened end of the tube cage, eagerly crawls inside. Another technique for eliminating the contribution of the gut microflora is use of the "germ-free" animal. Here, the neonate is delivered by cesarean birth and raised under sterile conditions. Still another technique is the use of antibiotics in the diet. Sulfa drugs have been used in a number of nutritional studies to wipe out much of the gut flora and thus minimize their contribution to the host's vitamin requirement. Sulfa drugs have been used in the study of folate nutrition, as well as in recent studies of pyrroloquinoline quinone, a compound having vitamin-like properties (Killgore *et al.*, 1989).

## Choosing the Diet

Greenfield and Briggs (1971) and Baker (1986) discussed the importance of gender, age, weight, and strain in the experimental design. They also stressed the importance of recording the duration of feeding. One important factor in raising an animal on a specific diet is that the animal may require several days to adjust to the diet. This is because changes and adjustments in the enzymes used to metabolize or transport specific nutrients of the diet may take place over several days after introduction of the new diet.

The response to a diet may change when the proportions of energy supplied as carbohydrates and fat are varied. For example, an animal may increase its consumption of food when carbohydrates are substituted in place of an equal weight of fat. This increase is due to the animal's desire to satisfy its energy requirement. The increase in food consumption results in increases in the intake of protein, vitamins, and minerals, and thus their possible oversupply. A solution to the problem of comparing physiological responses to diets of different carbohydrate/fat ratios is available. The equivalent of 1 kJ of carbohydrate can be omitted from the diet and replaced by 1 kJ of fat plus nonnutritive fiber, where the weights of the carbohydrate and fat + fiber are identical. This type of substitution maintains the nutrient density of all the other nutrients (Harper, 1986).

Commercially available animal chow diets are useful for raising animals until they are ready for the chemically defined control and experimental diets. Chow diets are generally not acceptable for use in experiments. It is the goal of the researcher to control as many aspects of the experiment as possible. Chow diets represent an unknown. Their chemical makeup is not precisely known. Chow diets can vary markedly from supplier to supplier and for one supplier from time to time. Different stock chow diets can vary in the lipid used (plant versus animal lipid). They can contain markedly different levels of sodium. Some researchers have devised studies where the control group receives chow and the experimental group receives a chow diet plus a nutrient or additive. This experimental design is generally not acceptable, as it would be expected to result in untrustworthy results. Two wrongs (chow diet for control and experimental groups) do not make a right. When using commercially available chemically defined diets, mineral mixtures, or vitamin mixtures, it is the responsibility of the researcher to ensure that the diets are complete (Greenfield and Briggs, 1971; Ad Hoc Committee, 1977).

Reeves (1997) has discussed the current recommendations of the American Institute of Nutrition (AIN) for assembling rat diets. Issues discussed include the use of sugar versus starch as the source of carbohydrate, methionine versus cystine as the source of cysteine, and the best oil for providing the essential fatty acids. In embarking on any study, the researcher needs to reevaluate the role of each of the recommended nutrients to make sure that their quantities suit the purposes of the planned experiment. For example, although the AIN recommends using casein as the source of protein, researchers interested in the interactions of calcium and phosphate need to take into account the fact that casein contains significant amounts of phosphate.

## REFERENCES/BIBLIOGRAPHY

Ad Hoc Committee (1977). Report of the American Institute of Nutrition Ad Hoc Committee on standards for nutritional studies. *J. Nutr.* **107**, 1340–1348.

Baker, D. H. (1984). Equalized versus *ad libitum* feeding. *Nutr. Rev.* **42**, 269–273.

Baker, D. H. (1986). Problems and pitfalls in animal experiments designed to establish dietary requirements for essential nutrients. *J. Nutr.* **116**, 2339–2349.

Benevenga, N. J., Gahl, M. J., Crenshaw, T. D., and Finke, M. D. (1994) Protein and amino acid requirements for maintenance and amino acid requirements for growth of laboratory rats. *J. Nutr.* **124**, 451–453.

Brommage, R., Jarnagin, K., and DeLuca, H. F. (1984). 1,25-dihydroxyvitamin $D_3$ normalizes maternal food consumption and pup growth in rats. *Am. J. Physiol.* **246**, E227–E231.

Cohen, N. L., Reyes, P. S., and Briggs, G. M. (1971). Folic acid deficiency in the golden hamster. *Lab. Anim. Sci.* **21**, 350–355.

Cohn, C., and Joseph, D. (1960). Role of rate of ingestion of diet on regulation of intermediary metabolism ("meal eating" vs. "nibbling"). *Metabolism* **9**, 492–500.

Geyer, R. P., Geyer, B. R., Derse, P. H., Zinkin, T., Elvehjem, C. A., and Hart, E. B. (1947). Growth studies with rats kept under conditions which prevent coprophagy. *J. Nutr.* **33**, 129–142.

Gill, T. J., Smith, G. J., Wissler, R. W., and Kunz, H. W. (1989). The rat as an experimental animal. *Science* **245**, 269–276.

Greenfield, H., and Briggs, G. M. (1971). Nutritional methodology in metabolic research with rats. *Annu. Rev. Biochem.* **40**, 549–572.

Harper, A. E. (1986). Nutritional control in animal experiments. *Nutr. Rev.* **44**, 175–184.

Harper, A. E., and Elvehjem, C. A. (1957). A review of the effect of different carbohydrates on vitamin and amino acid requirements. *Agric. Food Chem.* **5**, 754–758.

Killgore, J., Smidt, C., Duich, L., Romero-Chapman, N., Tinker, D., Reiser, K., Melko, M., Hyde, D., and Rucker, R. B. (1989). Nutritional importance of pyrroloquinoline quinone. *Science* **245**, 850–852.

Kopple, J. D. (1987). Uses and limitations of the balance technique. *J. Parenteral Enteral Nutr.* **11**, 79S–85S.

Leoschke, W. L., Lalor, R. J., and Elvehjem, C. A. (1953). The vitamin $B_{12}$ requirements of mink. *J. Nutr.* **49**, 541–548.

Mercer, L. P., Haijazi, H., and Hidvegi, M. (1993) Weanling rats display bioperiodicity of growth and food intake rates. *J. Nutr.* **123**, 1356–1362.

Mertz, W. (1987). Use and misuse of balance studies. *J. Nutr.* **117**, 1811–1813.

Quarterman, J., Williams, R. B., and Humphries, W. R. (1970). An apparatus for the regulation of the food supply to rats. *Br. J. Nutr.* **24**, 1049–1051.

Reeves, P. G. (1997) Components of the AIN-93 diets as improvements in the AIN-76A diet. *J. Nutr.* **127**, 838S–841S.

Singh, M. (1982). Effect of thiamin deficiency on pancreatic acinar cell function. *Am. J. Clin. Nutr.* **36**, 500–504.

Sturman, J. A., Moretz, R. C., French, J. H., and Wisniewski, H. M. (1985). Taurine deficiency in the developing cat: Persistence of the cerebellar external granule cell layer. *J. Neurochem. Res.* **13**, 405–416.

Suttie, J. W. (1968). Effect of dietary fluoride on the pattern of food intake in the rat and the development of a programmed pellet dispenser. *J. Nutr.* **96**, 529–536.

Tao, S., and Suttie, J. W. (1976). Evidence for a lack of an effect of dietary fluoride level on reproduction in mice. *J. Nutr.* **106**, 1115–1122.

Tinoco, J., Babcock, R., Hincenbergs, I., Medwadowski, B., and Miljanich, P. (1978). Linolenic acid deficiency: Changes in fatty acid patterns in female and male rats raised on a linolenic acid deficient diet for two generations. *Lipids* **13**, 6–17.

Wadhwa, P. S., Young, E. A., Schmidt, K., Elson, C. E., and Pringle, D. J. (1973). Metabolic consequences of feeding frequency in man. *Am. J. Clin. Nutr.* **26**, 823–830.

Wurtman, J. J., and Miller, S. A. (1976). Effect of litter size on weight gain in rats. *J. Nutr.* **106**, 697–701.

# Dot Blots, Molecular Cloning, and DNA Sequencing

Techniques involving the manipulation of DNA and RNA include molecular cloning, dot blots, and DNA sequencing. These and related techniques were required to provide information appearing throughout this textbook, especially in sections concerning metabolic regulation, genetic diseases, mutations, the iron response element, stem-loop structures, transcription factors, and hormone response elements. The term *molecular cloning*, indicates the cloning of DNA sequences, and is distinguished from the cloning of cells or of organisms. The term *dot blot* usually refers to the measurement of the amount of mRNA, as acquired from a biological sample, and indicates the shape of the raw data, i.e., a little blot. *DNA sequencing* refers to the determination of the order of nucleotide bases in a gene, as well as in regions of DNA that do not contain genes. The dideoxy method for DNA sequencing is one of the most powerful, and yet simple, techniques in biology.

Any regulatory event, whether involving a special element in the DNA molecule or a special structure in the mRNA molecule, should always be compared with the consequent change in the amount of polypeptide that is produced. Enzymes are one type of polypeptide, and since enzyme activities are generally easy to measure, the discussion will begin with an outline of the task of measuring enzyme activities.

## ASSESSING METABOLIC REGULATION BY MEASURING CHANGES IN ENZYME ACTIVITY

A common question in metabolic control is whether a change in enzyme activity is provoked by regulation at the enzymatic level or at the genetic level. Regulation at the enzymatic level, for example, may involve enzyme phosphorylation, binding of cyclic AMP, or binding of the calmodulin–$Ca^{2+}$ complex to the enzyme. Regulation at the genetic level usually entails a change in the rate of transcription of the gene coding for the enzyme of interest. Transcription is the production of messenger RNA. A regulatory action that provokes an increase in the rate of transcription, results in an increase in the amount of protein encoded by that gene.

Determination of the activity of an enzyme in an organ or tissue involves the following steps:

1. *Homogenization*: The tissue is homogenized. This means that the cells are broken by disruption while suspended in a solution of buffer and salts. A blender is used. In the case of a liver homogenate the resulting material is pink and somewhat viscous.

2. *Assay mixture*: The enzyme assay mixture is assembled. To a test tube are added a known amount of tissue homogenate, a known concentration of substrate, and specific salts, buffers, and other agents known to support the maximal activity of the enzyme. The homogenate is usually treated to remove debris and particulate matter prior to its addition to the assay mixture.

3. *Incubation of assay mixture*: The enzyme assay mixture is incubated for a specific period. Incubation gives the enzyme an opportunity to catalyze the conversion of a small proportion of the substrate to the product. Enzyme incubations are usually carried out at body temperature ($37°C$). Catalysis is terminated by adding strong acid or strong alkali, by immersing the tube in boiling water, or by adding an appropriate inhibitor of the enzyme of interest.

4. *Measurement of the product*: The terminated enzyme assay mixture contains a certain amount of product. The product is measured by an appropriate technique. The detection of enzyme control at the genetic level may be assessed by measuring the amount of mRNA coding for the enzyme or protein of interest. Measurements may be carried out using tissue samples taken from animals exposed to different dietary or hormonal conditions.

## ASSESSING METABOLIC REGULATION BY MEASURING LEVELS OF MESSENGER RNA (DOT BLOTS)

Determination of the amount of mRNA molecules coding for a protein of interest involves the following steps:

1. *Homogenization*: The tissue is homogenized to break up the cells and release all of the types of mRNA.

2. *Blotting*: A known quantity of the homogenate is placed on a special type of paper called nitrocellulose. The mRNA molecules bind tightly to the paper. All types of mRNA bind to the nitrocellulose, not just that coding for the protein of interest.

3. *Addition of DNA probe to the blot*: The amount of mRNA coding specifically for the protein of interest can be measured by use of a DNA probe. The DNA

probe is mixed with the nitrocellulose paper in a bath of water. The probe is allowed to contact all regions of the paper until it binds to the mRNA of interest. This process of binding is called hybridization. The DNA probe that does not hybridize to the mRNA of interest is washed away. The DNA probe binds to the mRNA of interest because these molecules contain nucleotide bases that are complementary to each other, as explained later.

4. *Measurement of bound DNA probe*: The total amount of DNA probe that is bound to the paper reflects the amount of mRNA coding for the protein of interest. The quantity of DNA probe that is bound to the paper can be measured because the DNA probe is radioactive, and no component of the tissue extract is radioactive. This is the dot blot technique. In cases where the various species of mRNA extracted from the tissue are separated from each other prior to transfer to the nitrocellulose, the technique is called the northern blot technique. Here, separation is by gel electrophoresis.

## Hybridization

DNA contains the bases adenine (A), thymine (T), guanine (G), and cytosine (C). These bases are covalently attached to the deoxyribose phosphate to form nucleotides. The respective nucleotides are called deoxyadenylic acid (dA), deoxythymidylic acid (dT), deoxyguanylic acid (dG), and deoxycytidylic acid (dC). The prefix "deoxy" indicates that the sugar is deoxyribose, rather than ribose. The word "acid" indicates that the ribose contains a covalently bound phosphoric acid group.

RNA contains the nucleotide bases adenine (A), uracil (U), guanine (G), and cytosine (C). These bases are covalently attached to the ribose phosphate to form nucleotides. The respective nucleotides are called adenylic acid (A), uridylic acid (U), guanylic acid (G), and cytidylic acid (C).

A molecule of single-stranded DNA (ssDNA) is comprised by a polymer of nucleotides:

dA–dA–dA–dG–dG–dC–dC–dA–dA–dA–dA–dG–dT–dG–dT–dC

A molecule of mRNA is made of a polymer of nucleotides:

C–C–G–G–G–U–G–G–G–G–U–U–A–C–A–C–G–G–G–G–A–A–C

A molecule of double-stranded DNA (dsDNA) contains two polymers of nucleotides. Each of these polymers may be called a DNA strand. The nucleotides in one strand bind to the nucleotides in the associated opposite strand. This binding is not covalent. The binding occurs in an orderly manner. The bases of A in one strand bind only to the bases of T in the opposite strand. The bases of G in one strand bind only to the bases of C in the other. All of the nucleotide bases in one strand occur in an order that is complementary to the bases in the opposite strand. To summarize, A binds to T and G binds to C.

The dot blot technique takes advantage of the fact that the bases in DNA can also bind to bases in mRNA. Here, as before, A binds to T and G binds to C. The

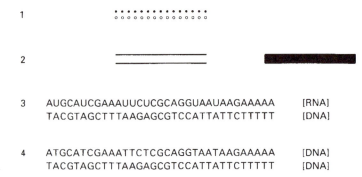

FIGURE B.1  Abbreviations for double-stranded nucleic acids.

base of uracil in mRNA binds to the base of A in DNA. The stable complex of a strand of DNA with a complementary strand of RNA is a hybrid. The strands have hybridized with each other. Complementary strands of nucleic acid that are bound to each other may be abbreviated in a number of ways (Figure B.1).

## Nitrocellulose

Figure B.2 shows a disk of nitrocellulose with mRNAs tightly adhering to it. Part of the mRNA coding for the protein of interest, acetyl-CoA carboxylase, and part of an unrelated mRNA are shown bound to the disk. In practice, hundreds of different types of mRNAs are bound to the disk. Also shown are molecules of the radioactively labeled DNA probe. The probe contains an atom of phosphorus-32,

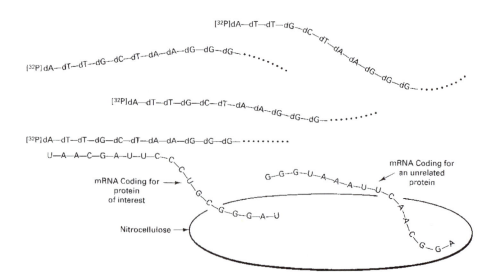

FIGURE B.2  Hybridization of the DNA probe to messenger RNA adhering to a nitrocellulose disk.

abbreviated $^{32}$P. One of the molecules of probe is shown bound to the mRNA of interest; the others remain unhybridized and in solution.

The mRNA coding for the protein of interest is chemically quite similar to those coding for unrelated proteins; however, the sequence of bases in the mRNA of interest is highly unique. The DNA probe contains a sequence of bases that is complementary to part of the mRNA of interest and hence binds to, or hybridizes with, this mRNA. A short DNA probe, perhaps consisting of less than a dozen nucleotides, binds only weakly and nonspecifically and is easily washed away. A long probe, that is, one containing 50 or more nucleotides, binds avidly and in a highly specific manner. In an analysis for mRNA coding for acetyl-CoA carboxylase, for example, from an extract of the liver of a rat, one must prepare a liver extract, purify the mixture of all mRNAs, treat the collection of mRNAs with formaldehyde, and finally associate with the disk of nitrocellulose. Formaldehyde promotes near-irreversible adhesion of the mRNA to the disk. The [$^{32}$P]DNA probe is then mixed with the disk and incubated to allow a maximal level of hybridization. Nonbinding probe is carefully washed away, and the amount of bound probe is then measured. In the event that samples from 10 different livers are used, 10 different disks must be prepared.

## Obtaining the [$^{32}$P]DNA Probe

A long radioactively labeled DNA probe can be synthesized using cloning techniques, as in the 10-step procedure illustrated in Figure B.3. The material needed to synthesize the probe includes a number of DNA-metabolizing enzymes, a plasmid, a DNA primer molecule, and *Escherichia coli*. Also needed is a source of mRNA coding for the protein of interest, that is, rat liver extract. The liver extract contains many different types of mRNA, and these are present in only small quantities. The power behind the cloning procedure is that it enables one to obtain the information in any particular mRNA molecule in large quantities (several milligrams) in pure form:

1. *Homogenize the tissue*: The tissue is homogenized to extract all of the mRNAs.

2. *Enzymatically produce mRNA/DNA hybrids*: The mRNAs in the extract are used as a template, or guide, for the production of complementary DNA. Reverse transcriptase uses mRNA as a template and catalyzes the synthesis of a strand of DNA bearing a sequence of nucleotide bases complementary to that of the mRNA. The reaction is carried out in a test tube. At this point, one has a hybrid structure containing a strand of mRNA and a strand of DNA. The cloning procedure, on the other hand, requires the use of dsDNA, not the mRNA/DNA hybrid. The reaction mixture is then made mildly alkaline and incubated. This procedure destroys (hydrolyzes) the mRNA but not the DNA.

3. *Produce dsDNA*: The strands of DNA in the test tube, after treatment of alkali, occur as ssDNA. Each molecule of ssDNA may be converted to dsDNA by adding DNA polymerase I. DNA polymerase I catalyzes the synthesis of a strand of DNA that is complementary to the original strand of ssDNA. The dsDNA prepared in this way cannot easily be introduced into the bacterium,

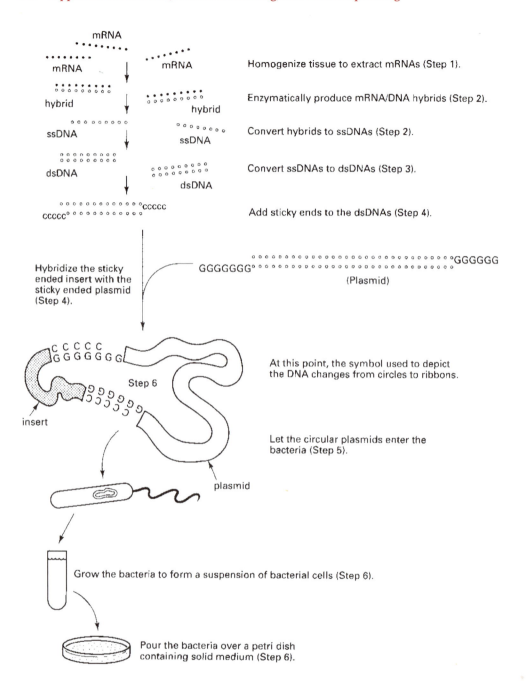

mRNA

mRNA                          mRNA                    Homogenize tissue to extract mRNAs (Step 1).

hybrid                        hybrid                  Enzymatically produce mRNA/DNA hybrids (Step 2).

ssDNA                         ssDNA                   Convert hybrids to ssDNAs (Step 2).

dsDNA                         dsDNA                   Convert ssDNAs to dsDNAs (Step 3).

ccccc                         cccc                    Add sticky ends to the dsDNAs (Step 4).

Hybridize the sticky                    GGGGGGG°           °GGGGGG
ended insert with the                              (Plasmid)
sticky ended plasmid
(Step 4).

C C C C C
G G G G G G G                                       At this point, the symbol used to depict
                          Step 6                     the DNA changes from circles to ribbons.

insert

                                                    Let the circular plasmids enter the
                                                    bacteria (Step 5).

                                        plasmid

Grow the bacteria to form a suspension of bacterial cells (Step 6).

Pour the bacteria over a petri dish
containing solid medium (Step 6).

**FIGURE B.3** Use of cloning techniques for the synthesis of a long DNA probe. The procedure, shown in 10 steps, begins with a mixture of mRNA extracted from tissue. The information in the mRNA is converted to a form that can be propagated in *E. coli*, i.e., a DNA insert connected to a plasmid. The *E. coli* are grown and then spread in a dilute smear over petri dishes. This step constitutes the event that separates the information present in heterogeneous collections of cellular mRNAs from each other. The information contained in each bacterial colony represents that from one single type of mRNA.

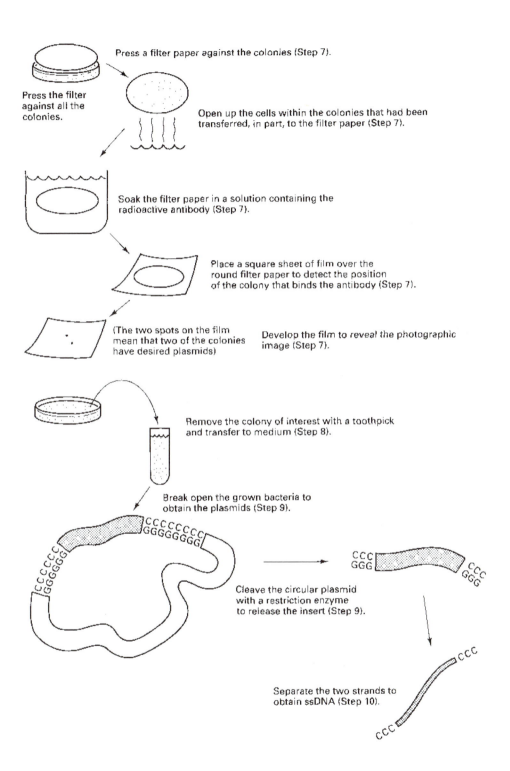

Press a filter paper against the colonies (Step 7).

Press the filter against all the colonies.

Open up the cells within the colonies that had been transferred, in part, to the filter paper (Step 7).

Soak the filter paper in a solution containing the radioactive antibody (Step 7).

Place a square sheet of film over the round filter paper to detect the position of the colony that binds the antibody (Step 7).

(The two spots on the film mean that two of the colonies have desired plasmids)

Develop the film to reveal the photographic image (Step 7).

Remove the colony of interest with a toothpick and transfer to medium (Step 8).

Break open the grown bacteria to obtain the plasmids (Step 9).

Cleave the circular plasmid with a restriction enzyme to release the insert (Step 9).

Separate the two strands to obtain ssDNA (Step 10).

**FIGURE B.3** continued

in preparation for cloning, because it is a linear molecule. Linear dsDNA is degraded by nucleases if it enters *E. coli* (Smith, 1988). The DNA should be in the form of a loop for use in the cloning procedure. A second problem is that, even if the dsDNA were to enter the bacterium, it would not be able to reproduce itself during growth and reproduction of the bacterial cell. The problems of entry and subsequent reproduction are solved by using a plasmid. A plasmid is a type of double-stranded DNA that is able to reproduce itself within a living cell. A plasmid shares some of the qualities of virus DNA.

4. *Attach the dsDNA to the plasmid*: The dsDNA prepared by use of DNA polymerase I is linked with the plasmid, which also takes the form of linear dsDNA. When linked together, a circle of dsDNA forms. The dsDNA bearing the information of interest is called an insert, or passenger DNA. The plasmid contains information that ensures its survival in the bacterial cell, allows its reproduction along with cell division, and guides the transcription of the DNA of the insert.

Note that the insert and plasmid need to be specially modified to link together. "Sticky ends" must first be attached to the insert and the plasmid. The sticky ends attached to the linear insert consist of stretches of deoxycytidylic acid (CCCCCCC). The sticky ends attached to the linear plasmid consist of stretches of deoxyguanidylic acid (GGGGGGGG). The sticky ends consisting of "CCCCCCC" can hybridize with those consisting of "GGGGGGGG." The result is a closed, circular dsDNA. The sticky ends are attached to the insert and plasmid by terminal transferase. Terminal transferase catalyzes the attachment of nucleotides to one end of a linear DNA molecule.

5. *Introduce the plasmid into the bacterium*: At this point, the test tube contains thousands of different plasmids, each of which contains an insert derived from a different molecule of mRNA. Perhaps only a dozen plasmids in the test tube contain an insert coding for the enzyme of interest. The plasmids are mixed with a suspension of bacteria. Each bacterium tends to accept only a single plasmid molecule. Normally, bacteria do not readily take up plasmids; however, the simple procedure of treating the bacterial suspension with calcium ions enhances the cell's ability to take up the dsDNAs, without harming the bacteria (Hanahan, 1983). This technique is called the calcium shock technique. An alternate and more powerful technique is electroporation. Electroporation involves exposing the suspension of bacteria and plasmid to a brief, high-voltage electric field. The field induces the temporary formation of pores in the cell, allowing uptake of DNA that may be adhering to the cell surface. The fact that linear strands of DNA cannot survive in the bacterium simplifies the cloning procedure. The survival of plasmids not containing an insert would result in generation of unwanted varieties of *E. coli*, thus complicating the procedure. Only the plasmids containing the insert are able to circularize and thus be propagated in the cells.

6. *Grow a colony from each bacterial cell*: The growth of the bacteria is initiated. Each plasmid within the cell is given a chance to reproduce itself, generating perhaps several dozen identical copies of itself within the cell. The suspen-

sion of growing cells is then diluted and poured on a bed of solid nutrient medium. The solid medium is contained in a petri dish. The solid medium contains a drug that kills bacteria that do not contain a plasmid. Only cells that contain the plasmid survive in the petri dish. The amount of cells poured over the solid medium is adjusted so that only about 100 bacterial cells containing plasmids are in the petri dish. The cells, which are distributed evenly over the surface of the solid medium, are permitted to continue growing. This growth results in the formation of a tiny clump of cells originating from a single cell. The tiny clump is a bacterial colony.

7. *Detect the colony of interest by using an antibody*: A bacterium containing a plasmid that harbors an insert also may produce the enzyme coded for by the insert. Hence, any procedure that can detect the enzyme of interest can also be used to determine which colony harbors the insert of interest. One of the most sensitive techniques for detecting a protein uses a $^{125}$I-labeled antibody. Iodine-125 is a radioactive isotope of iodine. The colonies are exposed to chloroform vapors. The vapors induce the opening up of the cell membranes of the bacteria in each colony. This results in release of all of the enzymes of the cells in the vicinity of the colony. The enzyme or protein of interest can be detected by use of an antibody that recognizes and binds to the protein. The antibody is specially tagged with an atom of iodine-125. The opened colonies are mixed with a solution containing $^{125}$I-labeled antibody. The antibody binds tightly only to the colonies that had manufactured the protein of interest. The nonbinding antibodies are then washed away. The location of the colony containing the bound $^{125}$I-labeled antibody is a simple task. A piece of photographic film is placed over all of the colonies. Only the colonies containing the radioactive iodine will produce an image on the film. Note that the chloroform treatment and the $^{125}$I-labeled antibody treatment are not performed using the original colonies. These treatments are performed on a piece of filter paper that had been carefully dabbed on the bed of colonies. The paper picks up a mirror image of the original colonies. Hence, the original colonies are not damaged by the chloroform vapors. The photographic results are used to locate the correct colony corresponding to the image on the film.

8. *Grow large quantities of the colony of interest*: The colony containing the desired insert is removed from the petri disk with a toothpick. It is transferred to a flask containing bacterial growth medium. Large quantities of the bacterium are grown. Each and every one of the bacterial cells contains the same plasmid which bears the identical insert.

9. *Harvest the plasmid from the cells and remove the insert from the plasmid*: The cells are broken up with a homogenizer, releasing the plasmids. The insert, at this point, is still a continuous part of the plasmid. The boundary between the insert and the original plasmid still consists of the sticky ends. The sticky end region is then cleaved by an enzyme that catalyzes the hydrolysis of dsDNA in the region of the sticky ends. This enzyme is called a restriction enzyme. Restriction enzymes recognize specific sequences of bases in the dsDNA and cleanly cut through the dsDNA in this sequence without damaging other regions of the DNA. The restriction enzyme used at this point

must be chosen carefully, as it must recognize only a region of dsDNA in or very close to the sticky ends. Hydrolysis results in liberation of the insert from the plasmid. The insert is much smaller than the plasmid and can easily be separated from it and obtained in pure form.

10. *Convert the insert to the DNA probe*: At this point, the test tube contains a large quantity of the insert. The goal is to convert the insert to a form usable as a probe. T4 polynucleotide kinase is used to catalyze the attachment of radioactive phosphate to the ends of the DNA:

The product of the reaction takes the form of dsDNA. A separate procedure is then used to separate the strands. Only one of the strands is appropriate for use in the dot blot procedure.

## Reverse Transcriptase and Terminal Transferase

Steps 2 through 4 of the cloning procedure are reviewed in detail in Figure B.4. The overall goal of these steps is to produce a collection of dsDNAs corresponding to the collection of mRNAs in the tissue extract. The messenger RNAs cannot be used directly for cloning; however, dsDNAs can be inserted into plasmids, and then plasmids inserted into bacteria and cloned.

The mRNAs coding for nearly all mammalian proteins contain a tail of adenylate molecules. This tail is called **poly(A)**. The tail occurs only at one end of the mRNA molecule. Conversion of the mRNA to the mRNA/DNA hybrid takes advantage of poly(A). This conversion is catalyzed by reverse transcriptase. Reverse transcriptase cannot bind to mRNA as is. It requires a primer for activation. The primer can take the form of a short strand of poly(dT). The poly(dT) primer binds to the poly(A) tail of the mRNA, forming a small hybridized region. Formation of the hybrid is possible because A binds to T.

Reverse transcriptase catalyzes the formation of ssDNA, in the form of the RNA/DNA hybrid. The RNA is then hydrolyzed nonenzymatically, leaving the ssDNA intact. The ssDNA is converted to dsDNA using DNA polymerase I. Here, a DNA/DNA hybrid is formed. The plasmid also occurs as dsDNA. Terminal transferase is used to catalyze the addition of poly(dC) sticky ends to the insert (Figure B.4) and poly(dG) sticky ends to the plasmid (Figure B.5). The modified insert and plasmid are then mixed together, allowing hybridization of the sticky ends. Annealing of the complementary sticky ends results in the formation of circular dsDNA containing one copy each of the insert and plasmid.

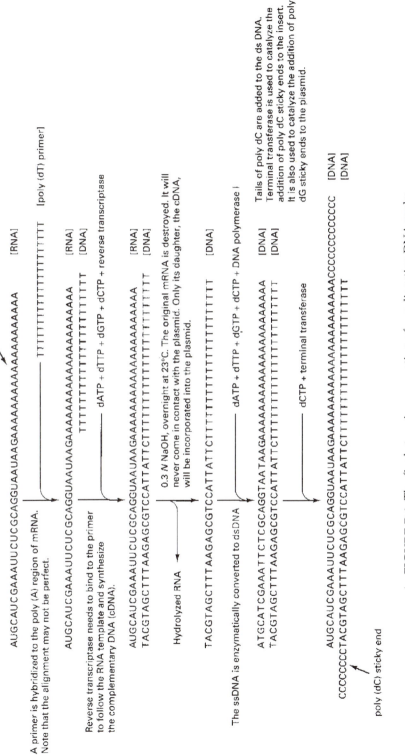

poly A tail of the message

AUGCAUCGAAAUUCUCGCAGGUAAUAAGAAAAAAAAAAAAA    [RNA]

A primer is hybridized to the poly (A) region of mRNA.
Note that the alignment may not be perfect.

TTTTTTTTTTTTTTTTTTT    [poly (dT) primer]

AUGCAUCGAAAUUCUCGCAGGUAAUAAGAAAAAAAAAAAAA    [RNA]
                              TTTTTTTTTTTTTTTTTTT    [DNA]

Reverse transcriptase needs to bind to the primer
to follow the RNA template and synthesize
the complementary DNA (cDNA).

dATP + dTTP + dGTP + dCTP + reverse transcriptase

AUGCAUCGAAAUUCUCGCAGGUAAUAAGAAAAAAAAAAAAA    [RNA]
TACGTAGCTTTAAGAGCGTCCATTATTCTTTTTTTTTTTTTT    [DNA]

Hydrolyzed RNA

0.3 N NaOH, overnight at 23°C. The original mRNA is destroyed. It will
never come in contact with the plasmid. Only its daughter, the cDNA,
will be incorporated into the plasmid.

TACGTAGCTTTAAGAGCGTCCATTATTCTTTTTTTTTTTTTT    [DNA]

The ssDNA is enzymatically converted to dsDNA

dATP + dTTP + dGTP + dCTP + DNA polymerase I

ATGCATCGAAATTCTCGCAGGTAATAAGAAAAAAAAAAAAA    [DNA]
TACGTAGCTTTAAGAGCGTCCATTATTCTTTTTTTTTTTTTT    [DNA]

Tails of poly dC are added to the ds DNA.
Terminal transferase is used to catalyze the
addition of poly dC sticky ends to the insert.
It is also used to catalyze the addition of poly
dG sticky ends to the plasmid.

dCTP + terminal transferase

AUGCAUCGAAAUUCUCGCAGGUAAUAAGAAAAAAAAAAAAACCCCCCCCCCC    [DNA]
CCCCCCCTACGTAGCTTTAAGAGCGTCCATTATTCTTTTTTTTTTTTTT    [DNA]

poly (dC) sticky end

**FIGURE B.4**  The final steps in preparation of a radioactive DNA probe.

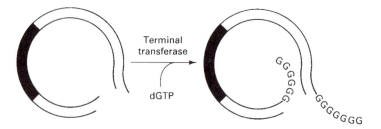

**FIGURE B.5** Details of the cloning procedure. The steps involving reverse transcriptase, DNA polymerase I, and terminal transferase are shown. The overall goal is to produce dsDNA suitable for insertion into the cloning plasmid, using mRNA as the starting material.

## USING DRUG RESISTANCE AS A SELECTION TECHNIQUE

The goal of step 6 (see Figure B.3) of the cloning procedure is to grow and clone colonies of the bacteria that had accepted a plasmid, while killing cells that did not contain a plasmid. The continued viability and growth of bacteria that do not contain a plasmid greatly complicates the search for colonies containing the insert of interest. However, this complicating factor is easily overcome in daily practice by use of the drug, ampicillin.

The solid nutrient medium in the petri dish contains the antibacterial drug ampicillin. This drug kills *E. coli*; however, the cells that harbor the plasmid readily grow in the presence of the drug because part of the plasmid contains a gene that codes for a drug resistance factor. A bacterium that has accepted a plasmid is said to have acquired ampicillin resistance.

The plasmid, in linear form, is depicted in the process of receiving sticky ends as part of the cloning procedure in Figure B.5. The region containing the gene coding for drug resistance appears shaded. This region remains part of the plasmid during the entire cloning procedure. One plasmid widely used in cloning is pBR322, the letter p standing for "plasmid." pBR322 consists of 4361 base pairs, that is, two strands of DNA, each containing 4361 nucleotides. The region coding for antibiotic resistance consists of 1936 base pairs and codes for β-lactamase, a protein of 286 amino acids. The amino terminus of this protein, as well as of other secreted proteins, consists of a stretch of some 20 amino acids that guides the movement of the protein through the cell membrane and into the periplasmic space. This region of the protein is called a signal sequence. β-lactamase, in this space, catalyzes the hydrolysis and inactivation of ampicillin, as well as of related antibiotics such as penicillin. In cloning experiments, ampicillin is included in growth medium at a level of 0.05 mg/ml. The antibiotic totally prevents growth of *E. coli* unless the bacterium synthesizes β-lactamase.

Note that plasmids exist as circular dsDNA when they occur in bacteria, that is, when they are replicated and transcribed *in vivo*. When plasmids are extracted from bacteria they also occur as circular dsDNA, but they can easily be converted to linear dsDNA in a test tube for the purpose of receiving inserts. Special enzymes called restriction enzymes recognize specific and rarely occurring sequences of base pairs and catalyze the hydrolysis of the dsDNA at a specific point within or near the sequence. In this way, a plasmid consisting of several thousand base pairs can be cut at one site only and converted to a single piece of linear dsDNA.

## RESTRICTION SITES AND RESTRICTION ENZYMES

The commentary on dot blots is self-sufficient, but one shortcut was taken, namely omission of details of restriction enzymes. Restriction enzymes are used for most, or all, DNA cloning procedures. DNA cloning by use of plasmids serves several purposes. These include: (1) generating large levels of a specific sequence of DNA to allow the sequencing of that DNA; (2) generating large levels of a specific sequence of DNA to allow generation of an mRNA probe; (3) generating a population of *E. coli* cells that contain the specific sequence of DNA, thus allowing synthesis of large amounts of protein from that DNA. For this purpose, a special plasmid must be used, namely, an "expression plasmid." The expression plasmid allows for propagation and cloning of the gene of interest, but also allows transcription and translation of the gene within the growing *E. coli* cell. All of the above activities are facilitated by use of restriction enzymes and restriction sites.

**Restriction enzymes** are nucleases that are made by bacteria to protect themselves from viruses (Bickle and Kruger, 1993). Bacteria that live in the ocean, inside the intestines of ladybugs, or in human milk, all make restriction enzymes. These enzymes are part of the bacteria's immune system. These enzymes can digest the DNA of a virus that attempts to inject its DNA into the bacterial cell, and attempts to reproduce within the cell. The enzyme action results in the death of the invading virus. Molecular biologists have purified these enzymes for use in cloning. Each enzyme digests DNA at specific sequences that occur at rare and infrequent positions in the genomes of all living organisms. The use of restriction enzymes, as well as of short stretches of DNA bearing the target cutting sites, has replaced the use of the tailing reactions and sticky ends, mentioned above.

All restriction enzymes are named after the bacterium where they originated. The cutting sites (restriction sites) are also named, and they bear the same name as the relevant enzyme. Table B.1 shows several commonly used restriction enzymes, the structures of the target site that is hydrolyzed by the enzyme, and the organisms that originally produced the enzymes in nature (Bhagwat, 1992).

How can a small piece of DNA replace the sticky poly C and poly G tails, for use in attaching two pieces of DNA together? The small piece of DNA containing the restriction site is called a **linker**. Linkers can be attached to the ends of pieces of double-stranded DNA, and to the ends of a linear plasmid, as shown here:

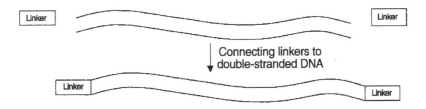

Then, the ends are digested with a restriction enzyme. The substrate specificity of the enzyme must be one that matches the structure of the linker. In addition, the enzyme must be chosen so that it does not hydrolyze a restriction site that may reside in the sequence of the gene, or in the plasmid. Digestion creates complementary ends at each end of each double-stranded DNA, i.e., at each end of the passenger DNA, and at each end of the plasmid:

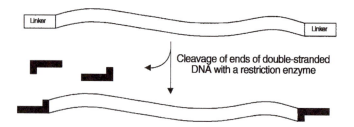

The enzyme-treated gene and plasmid are then hybridized together (annealed), and allowed to form a large circle. The process is exactly as shown at step 4 of Figure B.3, except that the cleaved linkers replace poly G and poly C. After amplifying the copies of plasmid via culturing in bacteria (step 6, Figure B.3), and using antibody probing to locate colonies that actually contain the passenger DNA (step 7, Figure B.3), the gene can be removed by treatment with the original restriction enzyme (step 9, Figure B.3).

Linkers have several advantages over connections made of poly C and poly G. One advantage is that restriction enzyme treatment can be used as a screening technique for locating the bacterial colony that contains the gene of interest. Another advantage is that restriction enzymes can be used to attach two different genes together at precisely known positions, to create a new gene. Still another advantage of using restriction enzymes is that they can be used for transferring the gene of interest from a plasmid that is good for DNA cloning in bacterial cells, to a different type of plasmid that is good for DNA cloning in mammalian cells.

What is the purpose of the variety and number of available restriction enzymes (see Table B.1)? Not I is an especially useful site, when performing the initial step of annealing the insert DNA, coding for the gene of interest, with the plasmid. This is because Not I contains many guanines and cytosines. The ionic interaction between G and C is about twice as strong as between A and T. Thus, two pieces of DNA that have been pre-treated with Not I, tend to form a strong bond, in comparison to DNA that had been cut with other restriction enzymes. A plasmid and insert DNA that are both treated with Not I can anneal together more readily and predictably than when treatment is with any other restriction enzyme.

Nde I is especially useful where the goal is to attach a promoter (promoters are binding sites for RNA polymerase) to a coding region (coding regions always begin with the ATG sequence). Please note that the Nde I site contains an ATG sequence.

All of the restriction enzymes have proven useful for screening plasmids contained within bacterial colonies. Take the example where initial screening (with a radioactive antibody or radioactive DNA probe) has given 50 candidate plasmids. Screening with a probe does not automatically ensure that every plasmid contains the desired gene. In fact, most of the plasmids acquired with initial screening contain irrelevant genes, irrelevant pieces of passenger DNA. The use of restriction enzymes (restriction screening) is a quick and inexpensive way of creating a "fingerprint" of 50 candidate plasmids.

A knowledge of the restriction site used as the cloning site, and of possible restriction sites within the passenger DNA, is used to interpret the fingerprints.

**TABLE B.1**  Restriction Enzymes and Target Cleavage Sites

| Restriction enzyme | Cleavage reaction in DNA | | | Organism used as source of enzyme |
|---|---|---|---|---|
| Aat II | GACGTC CTGCAG ⇓ | | | *Acetobacter aceti* |
| | GACGT C | + | C TGCAG | |
| Bam HI | GGATCC CCTAGG ⇓ | | | *Bacillus amyloliquefaciens* |
| | G CCTAG | + | GATCC G | |
| Eco RI | GAATTC CTTAAG ⇓ | | | *Escherichia coli* |
| | G CTTAA | + | AATTC G | |
| Hind III | AAGCTT TTCGAA ⇓ | | | *Haemophilus influenzae* |
| | A TTCGA | + | AGCTT A | |
| Nde I | CATATG GTATAC ⇓ | | | *Neisseria denitrificans* |
| | CA GTAT | + | TATG AC | |
| Not I | GCGGCCGC CGCCGGCG ⇓ | | | *Nocardia otitidis-caviarum* |
| | GC CGCCGG | + | GGCCGC CG | |
| Xma I | CCCGGG GGGCCC ⇓ | | | *Xanhomonas malvacaerum* |
| | C GGGCC | + | CCGGG C | |

# PHAGES, PHAGEMIDS, AND PLASMIDS ARE USED TO ACQUIRE DNA FOR SEQUENCING

Nucleic acid sequences were presented in almost every chapter in this textbook. Determination of these sequences requires a DNA sequencing technique. The technique most often used is the **dideoxy technique**. Acquisition of the DNA utilized in this method requires several steps, namely the construction of a DNA library, screening the DNA library to acquire the desired passenger DNA, and amplification of the desired DNA by many billion-fold via growth of a plasmid in bacteria. A number of different pathways can be used to acquire a DNA sample suitable for sequencing. Sometimes several approaches are taken over the course of many years in order to get the DNA sample. With luck, the procedure may take only a few weeks.

The following material details a frequently used approach for acquiring a homogeneous collection of DNA molecules for use in DNA sequencing. DNA sequencing will then be described.

## PHAGES, PLASMIDS, AND PHAGEMIDS

### DNA Libraries Contained in a Phage Can be Screened with Radioactive DNA, Where the Screening Process Does Not Depend on Transcription and Translation of the Gene of Interest

Figure B.3 depicts the creation of a "DNA library" whose composition is intended to reflect the sequences of all the mRNA molecules made in a particular living cell or tissue. **DNA library** is the term for a plasmid (or other cloning device) that contains a collection of DNA, where the collection reflects all the types of mRNA in a particular tissue. The tissue may be liver, breast cancer cells, fetal brain, or silkworms.

DNA libraries can be created within (kept and stored within) a plasmid, but more often they are initially stored inside a bacterial virus (phage, bacteriophage). The phage allows propagation of all the genes of the original tissue of interest, and greatly facilitates screening the DNA library for the gene of interest. Plasmids can also be used for propagating DNA libraries, and screening DNA libraries, as was shown above using the example of the radioactive antibody (step 7, Figure B.3). The antibody method of screening is difficult and time-consuming, and depends on an assumption that the DNA will be utilized within the bacterium for transcription and translation. There is no guarantee that any type of foreign DNA, when placed inside a bacterium, will be used to create a protein. Often, foreign DNA fails to be used for protein synthesis, unless the researcher is able to make fine adjustments in the structure of the foreign DNA. The process of DNA library screening is much easier where a virus is used to contain the DNA library.

The tool used for screening a DNA library contained in a virus is a radioactive DNA probe. This DNA probe is designed, by the researcher, so that its sequence closely or exactly matches the sequence expected in the gene of interest. The DNA probe consists of single-stranded DNA. It may be only 25 deoxynucleotides long, or it can be several hundred nucleotides long.

Use of a phage greatly facilitates probing for the region of the agar on the petri dish containing the phage colony (phage plaque) of interest, because of the huge concentration of target DNA contained in the phage plaque.

## Advantages of Placing the DNA Library Inside a Phage

One phage, called λZAP (lambda ZAP), has proved to be a useful cloning device. Screening a DNA library contained in λZAP is easier than screening a DNA library contained in a plasmid. In addition, λZAP has special properties that make it easier to work with than other lambda phages. λZAP facilitates acquisition of a pure population of plasmids bearing the gene of interest. What is this property? The answer is that λZAP has the properties of a phage and of a plasmid. In fact, λZAP is called a **phagemid**.

## Advantage of Placing the DNA Library Inside a Phagemid

In contrast to a typical phage, bearing a genome of about 50 kilobases, λZAP contains, within its genome, a plasmid. This plasmid is named pBluescript. With construction of the DNA library, the genes are inserted into pBluescript, where pBluescript is nested inside the genome of λZAP. To view some of the numbers, λZAP contains a region (40.8 kb) that consist of phage DNA, and a region (2960 bases) that consists of plasmid DNA. The plasmid can hold up to 10.2 kb of insert DNA (passenger DNA). The insert may consist of a gene, or of non-coding DNA.

## Screening the DNA Library Contained in λZAP

The phage DNA library is analogous to a thousand little ferry boats, each transporting an automobile, where each automobile holds a single man. The collection of ferries carries thousands of automobiles, and thus thousands of different men. The goal is to detect and acquire a single ferry (containing the man of interest), and to discard the unwanted ferries. Once the correct automobile has been found, it is child's play to open the door, to let the man out, and to study the man.

In this analogy, the phage library (consisting of thousands of phage particles suspended in a test tube) is analogous to the collection of ferry boats, and the automobile is analogous to the plasmid. The man is analogous to the passenger DNA. Once the screening technique has located the phage that contains the desired passenger DNA, a mechanism (gene II protein) is used to initiate formation of multiple independent copies of the plasmid. Gene II protein might be analogous to the gangplank of the ferry boat — it allows the automobile to come out. The gene II protein is not coded by the genome of the host cell, the phage DNA, or by pBluescript's DNA. The gene II protein is supplied by infecting the bacterial cells with a **helper phage**.

Only after screening the phage library, and acquiring a single phage plaque, and infecting bacteria with this single phage plaque is the helper phage used. The purified phage plaque and the helper phage are both added at the same time to infect the bacterial cell.

The process of inserting a DNA library into λZAP, infecting bacteria with this phage, viewing the agar plate containing phage plaques, screening the agar plate of phage plaques, acquiring the positive plaque or plaques, infecting host cells with the positive plaque, and acquiring pBluescript, are outlined in Figure B.6.

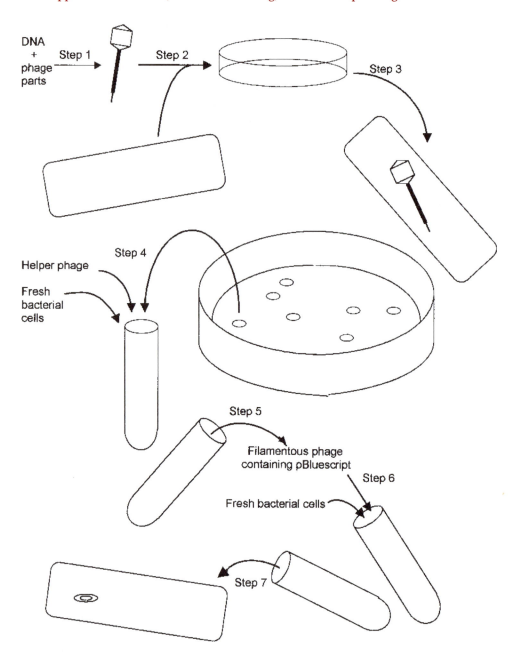

**FIGURE B.6**   Acquisition of pBluescript-containing homogeneous DNA insert. A DNA library, within lambda phage, was created just prior to step 1. The starting material in the sequence of steps 1–7 is a collection of double stranded DNA, made from all of the mRNAs present in the cell or tissue of interest. The starting material also includes the separate parts of a lambda phage, which can be assembled in a test tube. The assembled lambda phage contains, within its genome, the **heterogeneous** collection of dsDNA molecules acquired from the heterogeneous collection of mRNA molecules. The final material in the sequence is a plasmid, containing a **homogeneous** piece of DNA (coding for only one gene), which originated from a single type of mRNA from the extracted cell or tissue. The plasmid can be easily extracted from a colony of the bacterium, in a step that takes 30 min, and utilized in a DNA sequencing procedure. The DNA sequencing procedure requires only a few hours of actual work, and the actual DNA sequence can usually be acquired 1 day after performing the laboratory work. Steps 1–7 refer to a procedure that can be followed with λZAP (and pBluescript), or with similar biotechnology products. Step 1 shows the assembly of the phage in the test tube. The phage has a head, body, and legs, as do all normal lambda phages. The DNA resides in

With detection of the phage of interest (residing at a specific point in the agar on a petri dish), the next task is to remove the plasmid (pBluescript) from the phage, and to propagate this plasmid (in total absence of phage) in bacteria. The removal of the plasmid, from its home in the lambda phage, is initiated by **gene II protein**.

After removal of the plasmid from the phage, the plasmid can be used in reaction mixtures used for DNA sequencing. The dideoxy technique for sequencing DNA is shown below.

How does λZAP differ from lambda phages occurring naturally in the wild? Researchers removed much of the genome of the lambda phage. This removal prevents the lambda phage from engaging in the lysogenic life cycle. Researchers also removed many naturally occurring restriction sites from the phage. Finally, pBluescript was inserted into the genome of λZAP.

How does pBluescript differ from a plasmid in the wild? pBluescript is a modified version of pBR322. Researchers removed many naturally occurring restriction sites from the plasmid. They added special DNA sequences, acquired from a totally different phage (filamentous phage), called the **initiation signal** and **termination signal**. These special signals bracket the entire molecule of pBluescript (when the plasmid occurs in a linear manner within the genome of the lambda phage). These signals are the targets of the gene II protein (supplied by the helper phage), and guide the synthesis of many copies of the plasmid which are free and independent of the lambda phage.

pBluescript also contains a gene for ampicillin resistance. The transcription and translation of this gene, within the bacterial cell, allows the bacterial cell containing the plasmid to live, but kills all other bacterial cells with amazing efficiency. Hence, ampicillin is used as a selection technique.

pBluescript also contains a set of about 20 restriction sites, occurring in a row. This region could be considered to be the "grasping hands" of pBluescript, as it is

---

the head, and contains the phage's genome, the plasmid, and the DNA insert, all in one continuous DNA molecule. Step 2 shows the phage being mixed with *E. coli* cells, and then poured on a plate containing nutrient agar. During this period, the phage binds to the surface of the *E. coli*, and injects its DNA into the bacterial cell. The number of phages is adjusted, by the researcher, to about a thousand phages before pouring over the nutrient agar. The purpose of this adjustment is to allow about a thousand evenly spaced regions (on the agar plate), each containing a single phage. The number of *E. coli* cells is adjusted so that they form a continuous, even layer of growing bacteria. This continuous layer of cells is called a **lawn**, as in a grass lawn. Step 3 shows the production of an *E. coli* cell containing a phage particle (actually, each *E. coli* cell contains many phage particles), and the generation of a bacterial lawn bearing seven plaques. Each plaque arose from a single phage particle, and consists of an area, about 1 mm wide, containing dead bacteria and large concentrations of phage, where each phage particle contains the exact same DNA insert. The plaque looks clear compared to regions of the plate containing healthy uninfected bacteria. Note that if the researcher had mixed 10 million phages with the bacteria and then poured them on the plate, it would not have been possible to distinguish separate discrete plaques. Step 4 shows the occurrence of seven plaques. Screening of all seven plaques (in actual practice, over 1000 plaques occur on the agar) is performed in a manner similar to that shown in Figure B.3 (step 7), except that instead of probing with a radioactive antibody, probing is performed with a piece of radioactive DNA. The plaques that give a positive signal are then picked. Step 4 shows one of the plaques being picked (with a toothpick) and placed into a test tube containing a suspension of fresh bacterial cells and helper phage. The test tube is incubated to allow each *E. coli* cell to be infected with the harvested phage and with the helper phage (co-infection). Between steps 4 and 5 (during the incubation period), large quantities of pBluescript (bearing the insert) are produced, packaged in the filamentous phage, and excreted from the *E. coli* into the medium. Step 5 shows the filamentous phage taken out of the tube, and step 6 shows it being placed into another test tube and being mixed with fresh *E. coli*. Step 7 shows the cells, containing pBluescript, being poured out of the test tube. Typically, each bacterial cell might contain 10–30 copies of the plasmid.

the only part of pBluescript that contacts the passenger DNA (the insert DNA). The researcher has the option of inserting the foreign DNA into any of the restriction sites occurring in the linker region; however, use of the Not I site allows ease of connecting the foreign DNA into the plasmid because of the large number of G and C bases.

## PROPERTIES OF PHAGE LAMBDA AND OF FILAMENTOUS PHAGE

An outline of the characteristics of naturally occurring lambda and filamentous phages may bring into sharper focus the altered types of lambda and filamentous phage described in this appendix. To recapitulate, the altered lambda phage is λZAP. Bits of filamentous phage occur within pBluescript as special sequences of DNA called the **initiation signal** and **termination signal**. In addition, the helper phage is an altered filamentous phage.

### Filamentous Phage

The filamentous phages include strains called M13, f1, and fd. These phages consist of a genome of about 6400 bases, which codes for 10 genes. The genome, as it occurs within the phage, consists of one piece of single-stranded DNA. This DNA is coated with 2800 copies of a protein called gene VIII protein. The coat takes the form of a tube that is about 930 nm long and 6.5 nm wide. One end of the tube contains a few copies of the gene III protein. Gene III protein mediates absorption of the phage on target E. coli cells. Specifically, gene III protein binds to a structure of the bacterium called a pilus. In normal bacterial biology, the pilus serves the function of a penis, and allows the donor bacterial (male) to transfer some of its DNA to a recipient bacterium. In phage biology, the filamentous phage utilizes the pilus as a point of entry. The phage injects its DNA into the E. coli, and the phage's coat proteins remain behind. Once the phage's DNA is inside the cell, the phage's gene II protein is used to begin amplification of the number of copies of the phage's DNA within the host cell. DNA polymerase, and other enzymes of the E. coli, are necessary for helping make copies of the phage's genome. At one end of the phage's genome is a short DNA sequence called the "initiation signal," while at the other end is a short DNA sequence called the "termination signal."

The researcher who takes advantage of filamentous phage biology for cloning purposes might take note of the following facts. (1) The filamentous phage is capable of holding DNA of a much greater size than normal, i.e., many times the size of the normal genome. This fact is beneficial to the researcher, as it allows one to include a human gene as part of the phage's genome, for the purpose of cloning. (2) The researcher must be sure that recipient bacterial cells are male, when attempting to infect E. coli cells with the filamentous phage. On the other hand, once the filamentous phage's genome is inside the E. coli, it is not necessary for the E. coli to keep the pilus. (3) The researcher who uses pBluescript might like to know that most of this plasmid was derived from pBR322; however, the ends of pBluescript consist of the initiation and termination signals derived from the filamentous phage.

## Lambda Phage

Lambda phage consists of a head, tail, and tail fiber. The head contains a genome consisting of 48,502 base pairs (48.5 kb) of DNA. The DNA is linear and double stranded. At each end of the DNA is a short piece of single-stranded DNA, extending by 12 nucleotides. Phage development starts when the phage attaches to the bacterium with the tip of its tail fiber. The DNA is injected into the bacterium, leaving the phage's body behind. Immediately after injection, the DNA circularizes by means of the 12-base overhangs. The two 12-base overhangs are complementary, and they readily anneal together (once outside the phage particle). The genome codes for about 50 different proteins. These proteins, as well as some of the bacteria's own proteins, are used for reproduction of the phage's genome and for creation of new phage particles. For example, three of the phage-coded proteins catalyze destruction of the bacterial membrane. Two of the phage-coated proteins are used to form **terminase**, an enzyme that catalyzes cleavage of the circular lambda phage genome at one spot, thus creating a linear genome suitable for packaging into newly created produced phages. Terminase also serves to insert the genome into the empty heads. After 40 minutes of injection of the phage genome, about 100 new phage particles are produced, and the host cell is burst. The above events are called the **lytic cycle**. An alternate lifestyle is often practiced by lambda phage. In this lifestyle, which is called the **lysogenic cycle**, the phage incorporates its genome into the bacterial genome, and remains indefinitely dormant. Certain types of environmental stress can provoke the phage's genome to pop out of the bacteria's genome, to create daughter phage particles, and to burst the bacterial cell.

Lambda phage can exist with a genome that is only in a specific size range, that is, 38–53 kb long. This means that the maximal size compatible with packaging or enclosure of the DNA within the phage coat is 53 kb. One-third of the phage's genome can be totally eliminated by the researcher, and replaced with DNA that is irrelevant to phage biology. The resulting phage is capable of engaging in the lytic life cycle, but can no longer engage in the lysogenic life cycle. λZAP is a lambda phage that has the central section of DNA removed and partially replaced with plasmid DNA (pBluescript), and with the capacity of accepting a further 10 kb of insert DNA. λZAP phage can engage in the lytic life cycle, but not the lysogenic life cycle.

## Comparison of Filamentous Phage and Lambda Phage

Lambda phage is closely related to other bacteriophages, and these are called T3, T7, T5, Mu, T2, and T4. All of these phages consist of a double-stranded DNA genome, an icosahedral head, a long cylindrical tail, and a variable number of tail filaments. The related filamentous phages are M13, f1, and fd.

At the start of the infectious process, lambda phage contacts a nutrient transport protein (the maltose transporter) on the membrane of the *E. coli*. Without this membrane-bound protein, the phage cannot infect the *E. coli*. At the start of the infectious process, filamentous phage contacts the pilus of the *E. coli*. Without this extracellular structure, the phage cannot infect the *E. coli*. When lambda phage reproduces, it causes the bacterial cell to burst, and the *E. coli* is killed. When filamentous phage reproduces, it leaves the *E. coli* in a harmless budding process, and the *E. coli* survives.

## SEQUENCING OF DNA

Figure B.7 depicts the dideoxy method of DNA sequencing. This method was devised by Fred Sanger. The procedure is bipartite: (1) utilization of plasmid DNA and DNA polymerase to create a few hundred different pieces of DNA, where each piece differs in length (from the other pieces) by one nucleotide; and (2) use of gel

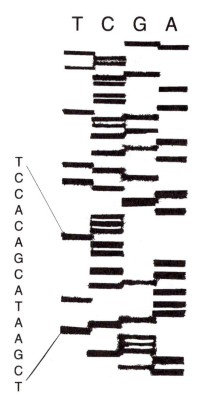

**FIGURE B.7** Polyacrylamide gel containing bands of radioactive DNA. The figure shows a radioautographic image of a portion of a polyacrylamide gel that was used for determining a DNA sequence. Samples taken from four separate reactions, described in the text, were applied to four wells, and these wells were named T, C, G, and A. The corresponding lanes containing the ladder pattern were also named T, C, G, and A. The order of application of the four samples is up to the discretion of the researcher, as this order may also be A–T–G–C, for example. Each band contains a homogeneous population of radioactive single-stranded DNA. All of the DNA molecules in the first lane begin with the primer sequence, and end in T. All of the DNA molecules in the second lane begin with the primer sequence, and end in C, and so on. The figure depicts the image acquired after exposing the gel to the X-ray film for 1 day. This technique is called **autoradiography**, because the sample of interest (the DNA molecules embedded in the gel) are radioactive and provoke a graphic image in the film. The electric field, applied to the gel, forces the negatively charged molecules of DNA to migrate downward through the polyacrylamide gel. The separation of the DNA molecules from each other is based on their size, not charge. The polyacrylamide gel acts as a sieve, allowing smaller DNA molecules to migrate faster through the gel matrix, and larger DNA molecules to migrate more slowly. The DNA molecules tend to stay in their own lanes, and tend not to cross-contaminate adjacent lanes. The sequence of DNA is read starting from the bottom. The indicated sequence is TCGAATACGACACCT. The image in the figure represents an "in-flight" picture of the DNA sequences, which captures their relative positions when they have migrated only partly through the length of the foot-long polyacrylamide gel. In determining the sequence of an insert contained in pBluescript, or within any other plasmid, the researcher can design a primer that binds to either strand of the double-stranded DNA. By convention, one strand is called the **coding strand**, while the other strand is called the **template strand**. The coding strand bears the same sequence of bases as found in the messenger RNA. The template strand is the strand that is physically contacted by the various transfer RNAs, during translation. The careful

electrophoresis to separate the different pieces of DNA, to provide an image consisting of a few hundred signals (shaped like tiny lines) that are stacked on top of each other.

## Chemicals

The chemical components of the procedure include:

1. *DNA polymerase*. All organisms contain DNA polymerases that have different properties. Even some bacteriophages code for their own DNA polymerases. The type of DNA polymerase used in the dideoxy technique was chosen because it works the best.

2. *Deoxyribonucleotides* (dATP, dTTP, dGTP, dCTP). Deoxyribonucleotides are the substrates of all DNA polymerases. The dideoxy method utilizes reactions, where the enzyme catalyzes polymerization of these nucleotides into new strands of DNA. The dATP that is used actually is not dATP, but a similar compound that is radioactive. This compound is sulfur-35-labeled dATP. Only the dATP is spiked with $^{35}$S. The other deoxyribonucleotides are not radioactive.

3. *DNA primer*. The DNA primer consists of a short piece of single-stranded DNA that is used to give DNA polymerase its start. The DNA primer also serves the vital function of assuring that all DNA sequences that are made begin at exactly the same point in the template DNA.

The researcher chooses the sequence of the primer to be used, and orders their synthesis from a vendor. Sometimes the sequence of the primer corresponds to part of the insert; sometimes the sequence of the primer corresponds to part of pBluescript that occurs immediately adjacent to the cloning site.

4. *Template DNA*. The template DNA consists of a preparation of pBluescript containing the DNA insert, i.e., the stretch of DNA that needs to be sequenced.

The researcher has no need to determine the sequence in pBluescript, as its sequence is known with certainty. The template DNA is briefly exposed to mild alkali (sodium hydroxide) in order to separate the two complementary strands of the double-stranded plasmid. Without alkali-treatment, the primer would not be able to bind to the correct starting place. Alkali treatment converts the double-

---

researcher always performs sequencing reactions on both strands, and thus needs to use two types of primers. Typically, a researcher may spend a week or so in sequencing one strand, and then devote a week or so to sequencing the opposite strand. The determination of the sequence for the entire coding region of a gene may involve use of several different plasmids, bearing overlapping bits of the gene, and a dozen different primers. Although sequencing a DNA insert contained in a plasmid may take a few weeks, sequencing an entire gene can take several months. Almost every known gene, whether occurring in viruses, bacteria, maize, worms, cancer cells, or humans, has been (or will be) sequenced by a protocol similar to that depicted here. Data from Brody *et al.* (1995).

stranded target DNA into two separated pieces of single-stranded DNA. Only one of these two strands contains the binding site for the primer.

5. *Dideoxyribonucleotides* (ddATP, ddTTP, ddGTP, ddCTP). Each of the above deoxyribonucleotides (dATP, dTTP, dGTP, dCTP) are spiked with low levels of the corresponding dideoxyribonucleotides (ddATP, ddTTP, ddGTP, ddCTP). Deoxyribonucleotides are the normal substrates of DNA polymerase. The deoxyribose moiety of these substrates contains one free hydroxyl group. This free hydroxyl group is used for attachment of further deoxyribonucleotides, allowing elongation of the growing DNA chain.

Dideoxyribonucleotides lack the free hydroxyl group, and contain a hydrogen atom in its place. DNA polymerase is able to recognize a dideoxyribonucleotide as a substrate, and is able to catalyze its addition to the end of a daughter DNA chain. However, where the terminus that is encountered consists of a dideoxyribonucleotide, the DNA polymerase is prevented from adding more deoxyribonucleotides to this end. The result is a dead-end product.

All of the dead-end products consist of exactly the same starting point (the 5'-end of the DNA). However, each of the dead-end products has a different length and a different termination point (the 3'-end of the DNA). The dead-end products consist of a mixture DNA molecules that are from about 20, 21, 22, 23, 24, 25, 26, 27, 28, 29, 30, 31, 32, and so on up to about 400 deoxyribonucleotides long. Thus, each of the dead-end products has a different length.

## Protocol for Using the Template, Primer, Enzyme, and Substrates

Four separate test tubes are used, and four separate reactions are performed. Each of the four test tubes contains all of the ingredients listed above, with one exception. The first test tube contains ddATP, the second ddTTP, the third ddGTP, and the fourth ddCTP. Each test tube contains equal quantities of enzyme, template, and primer.

With assembly of the four reaction mixtures, the primer rapidly diffuses throughout the solution, and within seconds finds its resting place on the DNA template. The primer that is 15 deoxyribonucleotides long is shown in the following diagram. The 5'- and 3'-ends of the primer are indicated. The process of binding of the primer to the template is called **annealing** or **hybridization**, as indicated in the diagram. Please convince yourself that every A anneals to a T, and that every G anneals to a C:

As soon as the primer binds, DNA polymerase locates the 3'-end of the primer and catalyzes the covalent attachment of deoxynucleotides to this end, resulting in a growing chain of single-stranded DNA, as shown in the following diagram:

5′ –**GCTGGCTGGCCCACA**GATCCTGCCACCATGCCGCAGC
••••CGACCGACCGGGTGTCTAGGACGGTGGTACGGCGTCGTAGCAGTCCTGGGAGGTGGTCGTATT••••••••

A few of the growing chains will grow to about 400 deoxyribonucleotides in length. A few will grow to 399 deoxyribonucleotides. A few will grow to only 30 or 40 deoxyribonucleotides. The eventual length of the daughter strand of DNA depends on how soon DNA polymerase encounters a dideoxy substrate (ddNTP), and catalyzes its incorporation into the growing strand of DNA. The conditions of the reaction mixtures assures that chain termination occurs evenly throughout the length of the entire DNA template. The diagram depicts the termination of a daughter strand of DNA caused by the incorporation of a molecule of dideoxyG. The dideoxyG (G–H) was incorporated opposite the complementary deoxyribonucleotide (C). The H in the symbol "G–H" indicates the hydrogen atom in dideoxyguanosine that replaces the hydroxyl group of deoxyguanosine:

5′ –**GCTGGCTGGCCCACA**GATCCTGCCACCATGCCGCAGCATCG–H
••••CGACCGACCGGGTGTCTAGGACGGTGGTACGGCGTCGTAGCAGTCCTGGGAGGTGGTCGTATT••••••••

The first test tube contains a collection of daughter DNA molecules, of various lengths, where each molecule *begins with the primer and ends with ddAMP*. The second test tube contains DNA products that *begin with the primer and end with ddTMP*. The third test tube contains DNA that *begins with the primer and ends with ddGMP*. The fourth tube contains DNA product that *begins with primer and ends with ddCMP*. All of the DNA products contain radioactivity ($^{35}$S) due to the presence of radioactively labeled dAMP groups occurring periodically and at various intervals in the daughter DNA.

The contents of the four respective test tubes are loaded into four separate wells located at the top of a slab of polyacrylamide gel. The polyacrylamide gel may be a few inches wide and a foot long. The wells consist of small indentations (wells) at the top of the gel. These indentations function as chambers that hold the reaction mixture until an electric field draws the DNA molecules into the gel. Electrodes are fastened to the top of the gel and to the bottom of the gel, and the power is turned on.

The electric field draws the negatively charged DNA molecules into the gel. The shorter DNA molecules migrate at a faster rate, as they can move quickly through the pores of the gel. The longer DNA molecules migrate at slower rates. This method of separation is extremely sensitive, as it can easily separate DNA strands that differ in length by only one deoxyribonucleotide. For example, the method can separate a molecule that is 387 deoxyribonucleotides long from one that is 388 deoxyribonucleotides long. The power is turned off just before the shortest DNA products reach the bottom edge of the polyacrylamide gel.

### Interpretation of the Pattern on the Polyacrylamide Gel

The polyacrylamide gel is placed on top of a sheet of X-ray film for a day or so. By allowing close contact between the gel and the film, the radioactive DNA molecules that reside in the gel produce a corresponding pattern of bands on the X-ray film. After the exposure period, the X-ray film is developed in a manner similar to that used for developing ordinary black-and-white film. The result is four ladder-like stacks of spots, or bands. The four ladders reside side by side and correspond to the four wells. All of the bands occur at unique distances from the top edge of the gel.

In interpreting the pattern of the gel, one's eyes need to play a game of hopscotch. This game of hopscotch is played as follows. First, one views the lowest band. The lowest band may be in the first, second, third, or fourth ladder (lane). The four lanes are labeled according to the particular dideoxyribonucleotide that was used for spiking the respective reaction mixture (Figure B.7). In Figure B.7 the lowest band is in the fourth lane. Thus, one should write A on a piece of paper. Then one searches for the next band up. The next band up resides in the third lane, which is the G lane. One can then write G on the paper, and so on. Part of the DNA sequence revealed by the gel is indicated in Figure B.7.

## SUMMARY

The dideoxy technique of DNA sequencing depends on the ability of DNA polymerase to make a complementary copy of the DNA template, on the ability of the enzyme to catalyze the incorporation of an occasional residue of dideoxyribonucleotide, and on the fact that this nucleotide brings a halt to further polymerization. The conditions of the enzymatic reaction are such that a separate daughter chain is created that terminates with a base corresponding to every one of the several hundred bases in the target DNA. In other words, chain termination occurs to about the same extent at every base of the template. Finally, the dideoxy method depends on the ability of gel electrophoresis to separate, by a fraction of a millimeter, daughter DNA products that differ in length by only a single deoxyribonucleotide. Daughter DNAs that are 20–100 deoxyribonucleotides long and one base apart in length are clearly resolved from each other on the gel. However, daughter DNAs that are 300–400 deoxyribonucleotides long and one base apart in length may be poorly resolved from each other, i.e., the bands are squeezed together.

EXERCISE

With use of the dideoxy technique, and with viewing the autoradiographic gel (the "autorad"), would you expect to be able to read and determine the DNA sequence that occurs in the primer?

What activities follow the determination of a DNA sequence? The first task is to determine the region of DNA that codes for amino acids, and which regions are likely to be nontranslated regions. The coding region can be found by determining the sequence of part of the polypeptide in question. Usually the sequence of some of the amino acids in the polypeptide is known prior to embarking on the DNA cloning project. Once 10 consecutive amino acids have been identified by direct sequencing, the corresponding 30 consecutive DNA bases can be inferred, by consulting the genetic code (see Chapter 1). The "best guess" for these 30 DNA bases is then entered into the computer and compared with the DNA sequence acquired from the polyacrylamide gel. The computer will then highlight the region on the gel that codes for the known amino acids.

The above task, once accomplished, allows the researcher to view the sequence of DNA bases (read from the polyacrylamide gel) in terms of a series of consecutive codons (triplets of bases that code for amino acids). The next task is to search for the translation start site and the translation stop site. Once the entire sequence of the polypeptide has been acquired from the DNA sequence, the researcher may compare the amino acid sequence with those of all known proteins. This task, which is performed on the computer, can reveal whether or not the polypeptide being studied is closely related to another, better characterized polypeptide. If a close match is found, the researcher may gain insight into the functions of the gene being sequenced.

## Acquisition of DNA and Protein Sequences Using the Internet

The sequences of a great number of genes have been determined and stored electronically in several databanks. Where it was possible to derive protein sequences from the coding regions of the genes, the derived protein sequences were also stored in various databanks. By utilizing one's computer and a search engine, one can access the information within a databank. By entering the name of a protein of interest, one can acquire a display of its amino acid sequence, or of the DNA sequence of its gene.

The National Institutes of Health maintains a databank called Genbank. Its address on the internet is http://www.ncbi.nlm.nih.gov. Once the homepage has been displayed on the screen, one can highlight **Nucleotide Sequence Search**, type the name of the protein coded by the nucleotides, and access the nucleotide sequence. Alternatively, one can highlight **Protein Sequence Search**, type the name of the protein, and acquire the amino acid sequence. If one knows the partial DNA sequence, i.e., about 50 consecutive nucleotides, one can enter the sequence and, in this way, acquire a display of any gene that contains the entered sequence. If one knows the partial amino acid sequence, i.e., about 15 consecutive amino acids, one can also enter this sequence and get a display of any polypeptide containing the entered sequence.

Another DNA sequence databank aside from Genbank is that of the European Molecular Biology Laboratory (EMBL) (the corresponding protein sequence databank is called SWISS-PROT). There is also the DNA Databank of Japan (DDBJ). **Bioinformatics** means the use of computers for the creation and storage of great quantities of DNA sequences, and the use of special applications for the analysis and comparison of these sequences. Benton (1996) and Fishman (1996) have outlined the field of bioinformatics. A related term, **genomics**, means the study of the genomes of various organisms, an activity that usually involves the use of computers.

## REFERENCES/BIBLIOGRAPHY

Alting-Mees, M., Sorge, J. A., and Short, J. M. (1992). pBluescript II: multifunctional cloning and mapping vectors. *Methods Enzymol.* **216**, 483–495.

Ansorge, W., Voss, H., and Zimmermann, J. (1997). "DNA Sequencing Strategies." Wiley, New York.

Benton, D. (1996). Bioinformatics: Principles and potential of a new multidisciplinary tool. *Trends Biotechnol.* **14**, 261–272.

Bhagwat, A. S. (1992). Restriction enzymes: Properties and use. *Methods Enzymol.* **216**, 199–224.

Brody, T., Keeney, S., and Linn, S. (1995). Human damage-specific DNA binding protein p48 subunit mRNA. Genbank, Accession #U18299.

Chauthaiwale, V., Therwath, A., and Deshpande, V. (1992). Bacteriophage lambda as a cloning vector. *Microbiol. Rev.* **56**, 577–591.

Cohen, S. N., Chang, A. C., and Hsu, L. (1972). Nonchromosomal antibiotic resistance in bacteria: Genetic transformation of *Escherichia coli* by R-factor DNA. *Proc. Natl. Acad. Sci. U.S.A.* **69**, 2110–2114.

Deng, G.-R., and Wu, R. (1983). Terminal transferase: Use in the tailing of DNA and for *in vitro* mutagenesis. *In* "Methods in Enzymology" (R. Wu, L. Grossman, and K. Moldave, eds.), Vol. 100, pp. 96–116. Academic Press, San Diego.

Fishman, J. (1996). Working the web with a virtual lab and some java. *Science* **273**, 591–593.

Hanahan, D. (1983). Studies on transformation of *Escherichia coli* with plasmids. *J. Mol. Biol.* **166**, 557–580.

Harth, G., Baumel, I., Meyer, T., and Geider, K. (1981). Bacteriophage fd gene-2 protein. *Eur. J. Biochem.* **119**, 663–668.

Helfman, D. M., Feramisco, J. R., Fiddes, J. C., Thomas, G. P., and Hughes, S. H. (1983). Identification of clones that encode chicken tropomyosin by direct immunological screening of a cDNA expression library. *Proc. Natl. Acad. Sci. U.S.A.* **80**, 31–35.

Krieg, P. A. (1996). "A Laboratory Guide to RNA." Wiley-Liss, New York.

Meyer, T. F., and Geider, K. (1979). Bacteriophage fd gene II-protein. *J. Biol. Chem.* **254**, 12636–12641.

Meyer, T. F., Baumel, I., Geider, K., and Bedinger, P. (1981). Replication of phage fd RF with gene 2 protein and phage T4 enzymes. *J. Biol. Chem.* **256**, 5810–5813.

Marvin, D. A., and Hohn, B. (1969). Filamentous bacterial viruses. *Bacteriol. Rev.* **33**, 172–209.

Minth, C. D., Taylor, W. L. Magazin, M., Tavianini, M. A., Collier, K., Weith, H. L., and Dixon, J. E. (1982). The structure of cloned DNA complementary to catfish pancreatic somatostatin-14 messenger RNA. *J. Biol. Chem.* **257**, 10372–10377.

Murialdo, H. (1991). Bacteriophage lambda DNA maturation and packaging. *Annu. Rev. Biochem.* **60**, 125–153.

Okayama, H., and Berg, P. (1982). High-efficiency cloning of full-length cDNA. *Mol. Cell. Biol.* **2**, 161–170.

Potter, H. (1992). Application of electroporation in recombinant DNA technology. *In* "Methods in Enzymology" (R. Wu, ed.), Vol. 217, pp. 461–478. Academic Press, San Diego.

Sambrook, J., Fritsch, E. F., and Maniatis, T. (1989). "Molecular Cloning," 2nd ed. Cold Spring Harbor Laboratory Press, Cold Spring Harbor, NY.

Short, J. M., and Sorge, J. A. (1992). *In vivo* excision properties of bacteriophage expression vectors. *Methods Enzymol.* **216**, 495–508.

Short, J. M., Fernandez, J. M., Sorge, J. A., and Huse, W. D. (1988). λZAP: A bacteriophage λ expression vector with *in vivo* excision properties. *Nucl. Acids Res.* **16**, 783–7600.

Smith, G. R. (1988). Homologous recombination in procaryotes. *Microbiol. Rev.* **52**, 1–28.

Szybalski, E., and Szybalski, W. (1979). A comprehensive map of bacteriophage lambda. *Gene* **7**, 217–270.

Watson, N. (1988). A new revision of the sequence of plasmid pBR322. *Gene* **70**, 399–403.

Young, R. (1992). Bacteriophage lysis: Mechanism and regulation. *Microbiol. Rev.* **56**, 430–481.

# APPENDIX C

# METHODS AND TERMS
# IN EPIDEMIOLOGY

Epidemiology is a vital tool for guiding nutritionists who need to advise the public on dietary choices. Epidemiology has also proven helpful to molecular biologists who find inspiration in data suggesting that specific nutrients can influence the activity or behavior of specific genes. All epidemiological studies involve large groups of people, i.e., from a thousand to a million. Many problems studied by epidemiologists occur at low levels in the population, i.e., 0.01 to 1.0%, and hence require the study of large groups. Two terms, *incidence* and *prevalence*, are used to refer to the levels of a disease in any population. Incidence means the number of new cases to arise during a particular time interval. Prevalence means the number of cases existing in a specific instant of time, i.e., on September 2, 1996.

The following definitions are needed for a clear understanding of the presentation of colorectal cancer that appeared earlier in the text. Several types of studies are used by epidemiologists. Two general studies are *cross-sectional studies* and *longitudinal studies*. Cross-sectional studies are relatively inexpensive, and usually take note of the number of occurrences of a specific disease, and risk factors, existing during a specific small interval of time (prevalence). Cross-sectional studies can also take note of the number of new occurrences of a disease, and risk factors, occurring in a specific interval of time (incidence). Cross-sectional studies cannot be used to determine cause and effect. For example, a cross-sectional study may produce data that associates obesity with inactivity, but it will not tell you which came first, i.e., obesity or the sedentary lifestyle. The terms **cross-sectional study** and **prevalence study** generally mean the same thing. The results from a prevalence study have limited value, but they may be used to justify setting up a specific screening test in a clinic, or for acquiring funding for a longitudinal study.

A problem with *prevalence* studies is that they may underestimate diseases having a rapid, early fatality rate. Bacterial diarrhea in infants may be detected and then kill an infant within a week or so. Please imagine a tropical village where 10% of the persons die from this disease in 1 year. However, prevalence data may reveal that none of the persons had this disease during the first week of July. The use of prevalence data to assess a rapidly fatal disease in a given population may therefore grossly underestimate the seriousness of this disease. Hence, one might prefer to use incidence data for acquiring information on infectious diarrhea.

**Longitudinal studies** follow a certain property *over the course of time*, taking data points at various times. Longitudinal studies can be prospective (cohort studies) or retrospective (case control studies). In a **prospective study**, the investigator contacts two or more groups, one group exposed to a certain risk factor and another group not exposed to the factor, follows both groups for several years, and eventually determines which group was associated with the risk factor. **Prospective studies** are also called **cohort studies**. The term *cohort* means the entire group of people that is contacted, i.e., those exposed to various suspected risk factors as well as those not exposed to risk factors. The subjects are classed according to exposure to such things as diet, alcohol intake, radiation exposure, age at first pregnancy, use of hair dye, smoking, and frequency of exercise. Only after the subjects are classed according to relevant risk factors does the epidemiologist maintain communication with the subjects and determine if they acquire a specific disease. Cohort studies are expensive because one might need to contact at least 10,000 persons, and wait 5 years, before 10 persons acquire the disease of interest.

The most conclusive, and most expensive, type of study is a prospective study where the epidemiologist intervenes with the subjects. This type of study is called an **intervention study**. For example, 5000 persons may be given pills containing folic acid, while 5000 persons may be given pills containing only filler, where the pills are consumed for 10 years and the subjects are evaluated for cancer. In this manner, the investigator exerts some control over the risk factors of interest.

One problem with cohort studies is that often a third of the subjects will drop out of the study. However, this should lead to error only if the reason for dropping out is somehow connected to the disease. Another problem with cohort studies is nonparticipation. For example, evidence suggests that smokers are more likely to refuse to participate in epidemiological studies than nonsmokers. This trend to refusal could result in a failure to correlate smoking in a cohort study of cancer. In short, a significant number of data points taken from smokers would not be available. The problem of nonparticipation and dropouts can be reduced by selecting subjects from one company, such as an oil refinery. Persons exposed to various risk factors will include those directly involved in manufacturing, while those not exposed to risk factors will include those involved in clerical and administrative tasks. The boon to the epidemiologist is that employees and retirees from a specific company (e.g., General Electric), or with a specific profession (nurses), are less likely to drop out than subjects taken from the general public.

Still another problem with **cohort studies** is that a person with a specified risk factor (smoking, oral contraceptives, wealthy social classes) may be more likely to seek medical attention than others. An increased rate of medical attention given to smokers, for example, could result in an increased rate of detection of certain relevant (bronchitis) or irrelevant (baldness) diseases in this group. An increased rate of medical attention given to contraceptive users or wealthy persons could also lead to increased rates of detection of bronchitis and baldness in these groups.

In a **retrospective study**, the epidemiologist first contacts persons with a certain trait (and controls without it) and then seeks to discover the risk factor in their past history. Retrospective studies are also called **case-control studies**. For example, one might choose 500 subjects with colon cancer and 500 subjects selected at random from the general population. Where the epidemiologist is studying colon cancer, for example, subjects with colon cancer are acquired from records of patients admitted for treatment of this disease. Controls are also often recruited from patients admitted for medical care, i.e., for routine tests such as pelvic and rectal exams. Only after acquiring the two groups does the epidemiologist acquire information from medical records and interviews on likely risk factors, such as diet, smoking, and coffee.

Case-control studies are relatively inexpensive because the epidemiologist is able to select a large population of persons who already have the disease of interest. Even diseases that are rare, and strike only one person in 100,000, can be studied by case-control studies, simply by scanning medical records. In contrast, the epidemiologist embarking on a cohort study does not have the privilege of knowing who will contract the disease of interest.

A source of error in case-control studies is where the disease of interest did *not have a clear beginning point*. For example, hypertension is a disease that may be undiagnosed for many years. As the goal of the study is to determine possible risk factors for hypertension (such as smoking, obesity, diet), any data regarding hypertension is confused by the lack of knowledge of which came first, the disease or the possible risk factors.

## ALGEBRA USED FOR CALCULATING RISK RATIO AND ODDS RATIO

**Risk** and **odds** are two terms used to express, in numbers, the frequency of association of a risk factor with a disease. **Risk ratio** (relative risk) and **odds ratio** are two terms that compare the rate of occurrence of a disease in persons exposed and not exposed to a specific risk factor. **Odds ratio** is commonly used in case-control studies. **Risk ratio** tends to be used in cohort studies. "Risk" and "risk ratio" are concepts that are easy to grasp, while "odds" and "odds ratio" are somewhat more difficult.

An imaginary study involved smokers and lung cancer. The cases (persons with disease) consisted of 75 smokers and 20 nonsmokers. The controls (no disease) consisted of 175 smokers and 180 nonsmokers. The **risk** of cancer in *smokers* was

$$(75) / (75 + 175) = 0.3$$

When somebody asks, "What is my risk of cancer if I smoke?" this is a request for a comparison of the number of smokers with the disease to the number of all smokers (healthy and nonhealthy). The question can be answered by "Your risk is 75 out of 250, which is 75/250, or 0.3." However, the value for risk ignores the possibility that nonsmokers may have the identical risk for cancer. The value for risk does not tell us if smoking is risky. The value does not involve a comparison of risk for smokers and nonsmokers. The **risk** of cancer in *nonsmokers* was

$$(20) / (180 + 20) = 0.1$$

To compare the risk of cancer in smokers to the risk in nonsmokers, one can calculate the **risk ratio**. The risk ratio, from the above study, is 0.3/0.1 = 3.0. The **risk ratio** is a number that contains two different rates of risk. **Risk ratio** is also called **relative risk**.

Odds is another style of expressing the association of pre-existing facts with an outcome. The *odds* of cancer with smoking is calculated from Formula A. Formula A gives the odds of finding cancer in people who smoke. The request "What are my odds for getting cancer if I smoke?" is a request for an answer that takes these forms: "So you want to know your odds for getting cancer if you smoke? Your odds are 75 to 20. Your odds are 75/20. Your odds are 3.75."

FORMULA A

$$\frac{\text{(Probability of cancer with smoking)}}{\text{(Probability of health with smoking)}} = \frac{0.3}{0.7} = 0.429$$

The value for odds may tell a smoker if he should expect to get cancer, or if he should not worry about cancer. However, the value for odds, whether high or low, can never supply motivation for quitting smoking. This motivation can be acquired by also calculating the **odds** of cancer with no smoking. Formula B gives the **odds** of finding cancer in people who do not smoke:

FORMULA B

$$\frac{\text{(Probability of cancer with no smoking)}}{\text{(Probability of health with no smoking)}} = \frac{0.1}{0.9} = 0.111$$

The odds of finding cancer in smokers (0.429) is greater than the odds of finding cancer in nonsmokers (0.111). These two numbers are used to create a ratio. The odds ratio (OR) is a number that is found by dividing the two sets of **odds**, as follows:

$$\text{Odds ratio} = \text{(Formula A) / (Formula B)} = 0.429 / 0.111 = 3.86$$

The equation OR = (Formula A)/(Formula B) can also be written in another way, as one can see by using simple algebra. Odds ratio can also be calculated by

$$\text{OR} = \frac{\text{(\# cases with specific risk factor) (\# controls without risk factor)}}{\text{(\# cases without risk factor) (\# controls with risk factor)}} = 3.86$$

Stepping back to view the big picture, one can see that the values for the risk ratio (3.0) and odds ratio (3.86), as calculated above, are not exactly the same. For both risk ratio and odds ratio, a value of 1.0 corresponds to a situation where no association exists between the risk factor and the outcome. The use of the risk ratio or odds ratio for expressing epidemiological data can be a matter of personal taste and custom. However, it should be noted that odds ratio has a valuable property not shared by risk ratio, as illustrated with imaginary data involving cholesterol and diabetes (Feinstein, 1985).

In setting up an imaginary epidemiological study, 80 diabetics were selected from 497 diabetics, and 80 controls were selected from 9503 available controls. The occurrence of high or normal cholesterol in the small group (Table C.1) and big group (Table C.2) are shown.

For the small group, the risk for diabetes in high-cholesterol subjects is

**TABLE C.1**

|  | Diabetes | No diabetes (controls) |
|---|---|---|
| High-cholesterol subjects | 21 | 6 |
| Normal-cholesterol subjects | 59 | 74 |

$$(21) / (21 + 6) = 0.777$$

The risk for diabetes in normal-cholesterol subjects is 0.443. The *risk ratio* is (Feinstein, 1985)

$$(0.777) / (0.443) = 1.753$$

The *odds ratio* is

$$(21) (74) / (6) (59) = 1554 / 354 = 4.39$$

For the big group, the risk for diabetes in high-cholesterol subjects is 0.154, while

**TABLE C.2**

|  | Diabetes | No diabetes (controls) |
|---|---|---|
| High-cholesterol subjects | 130 | 713 |
| Normal-cholesterol subjects | 367 | 8790 |

the risk for diabetes in normal-cholesterol subjects is 0.040. The **risk ratio** is

$$(0.154) / (0.040) = 3.847$$

The *odds ratio* is:

$$(130) (8790) / (713) (367) = 4.37$$

For the small and big groups, the risk ratios and odds ratios are well over 1.0, and reveal that high cholesterol is a risk factor for diabetes. However, the purpose of this exercise is to demonstrate that **risk ratios** change, according to the size of the group chosen for making the calculation, but the **odds ratios** do not change. Again, one might note that, in the small group and in the big group, 10 out of 28 persons with *high cholesterol* had diabetes, and that 10 out of 123 persons with *normal cholesterol* had diabetes. It might be useful, but not essential, for the measure of risk to have the property of being not influenced when perusing population groups of different sizes. The odds ratio has this property. The risk ratio does not.

This concerns whether one should use risk ratio or odds ratio for expressing results when evaluating a retrospective or prospective study. It is possible to calculate risk and risk ratio using data from a case-control study, and to acquire mathematical results, but it is not correct to do so. This is because the denominator (the number in the denominator) in the risk formula does not represent a broad random sampling of the population. The number in the denominator represents two groups of subjects who were screened and recruited into the study. Risk and risk ratio are used for prospective studies. Odds ratio can be accurately used for both retrospective and prospective studies but is customarily only used for retrospective studies.

## ACQUIRING DATA ON FOOD AND NUTRIENT INTAKE

Food consumption can be measured most accurately by confining human subjects to an apartment for several days or months, with feeding accurately measured, chemically defined diets. This type of study is very expensive, and is necessarily used when conducting nutrient balance studies. Somewhat less accurate is the food diary. Here, subjects record the quantity of all foods eaten (at home) over the course of a week. Subjects use measuring cups to record the intake of liquid foods and flour, and a balance to record meat and plant consumption. Even less accurate is the use of questionnaires, which are either filled out along with an interview, or by mail. The type of questionnaire chosen depends on the goal of the epidemiologist or clinician. Sometimes questionnaires are given once, and they may ask about general frequencies of food intake during the previous year. Sometimes questionnaires are given four times during a year, and they ask about food intake only during the previous 24 hours.

Questionnaires generally list some 50–100 foods and ask the subject to indicate portion size and frequency in the diet. When interviews are given, the interviewer may estimate portion sizes by displaying photographs or plastic models of different portions of food. In this way, one reduces variation due to different perceptions of what is meant by a "large serving of potatoes" or a "small fish." Some questionnaires are designed to measure a broad variety of nutrients, while others are designed to pinpoint intake of a specific nutrient, i.e., how much fat is eaten.

Gladys Block and co-workers have devised a questionnaire for collecting data on the general diet, but also on suspected risk factors for cancer. Part of this questionnaire is shown in Figure C.1. Block's questionnaire takes care to include artificial sources of vitamins and minerals, which can be major sources of these nutrients in the diet. These include vitamin supplements (e.g., One-A-Day vitamins), mineral supplements (e.g., Tums, calcium), and fortified foods, such as "breakfast bars."

EXERCISE

Look at the questionnaire shown in Figure C.1. Why does it ask about "oranges (not including juice)"? What do oranges contain that juice does not contain?

EXERCISE

Which nutrients would be expected to vary most (and least) from day to day in the diet: vitamin $B_{12}$, energy, protein, and sodium? See Block and Hartman (1989).

Having conducted a dietary survey, an epidemiologist may be able to correlate the risk for certain diseases with certain dietary habits. To inspire and guide the activities of research scientists, epidemiologists often try to correlate the risk for disease with the actual chemical constituents of the diet. The connection between food eating habits and these chemicals is made possible by using food tables. The most comprehensive food table available is *Handbook No. 8: Composition of Foods, Raw, Processed and Prepared*, written by the U.S. Department of Agriculture (USDA).

The diverse collection of eating habits of 10,000 persons may be confounding to an investigator when expressed in terms of fish, apples, wine, nuts, milk, pizza, and hundreds of other foods. However, correlations that are striking and useful may materialize when these foods, and their frequencies and portions, are entered into a computer software application that converts the intake of various foods into the intake of the chemical constituents of foods. To summarize, a simple survey may correlate a lower incidence of a specific type of cancer with diets of frozen orange juice, broccoli, and omelets (a bewildering array of foods), but no correlation with diets of canned peas, fried potatoes, and soda crackers (another bewildering collection of foods). By inputting these foods into a software application that contains a comprehensive food table, one may discover, for example, that the lower incidence of cancer can be correlated with an increased intake of folate.

One might note that food tables are not extremely accurate. The levels of specific nutrients can vary by several-fold, depending on methods for storage and cooking of food, plant variety, breed of farm animal, minerals in the soil used to grow the plant food, and method of nutrient analysis.

## REFERENCES/BIBLIOGRAPHY

Austin, H., Hill, H., Flanders, D., and Greenberg, R. S. (1994). Limitations in the application of case-control methodology. *Epidemiol. Rev.* **16**, 65–76.

Block, G., and Hartman, A. (1989). Issues in reproducibilty and reliability of dietary studies. *Am. J. Clin. Nutr.* **50**, 1133–1138.

Block, G., Patterson, B., and Subar, A. (1992). Fruit, vegetables, and cancer prevention: A review of the epidemiological evidence. *Nutr. Cancer* **18**, 1–29.

Elwood, J. M. (1988). "Causal Relationships in Medicine." Oxford Univ. Press, New York.

Feinstein, A. R. (1985). "Clinical Epidemiology." Saunders, Philadelphia.

Munro, B. H. (1997). "Statistical Methods for Health Care Research," 3rd ed. Lippincott, Philadelphia.

Kahn, H. A. (1983). "An Introduction to Epidemiological Methods." Oxford Univ. Press, New York.

Pandey, D. K., Shekelle, R., Selwyn, B. J., Tangney, C., and Stamler, J. (1995). Dietary vitamin C and β-carotene and risk of death in middle-aged men. *Am. J. Epidemiol.* **142**, 1269–1278.

**FIGURE C.1** Food questionnaire used to assess long-term intake of all foods, with the intention of assessing intake of nutrients thought to be associated with increased or decreased risk for cancer (pp. 971–973). These nutrients include vitamin C, β-carotene, calcium, fat, and folate. Only three of the eight total pages are shown. The questionnaire asks about intake of vitamin pills, bananas, oranges, pancakes, eggs, beans, tomatoes, soy meat substitutes, hamburgers, oysters, pasta, various types of bread, children's drinks containing vitamin C, high-fat foods such as bacon, doughnuts, pastry, ice cream, and cream, and use of cigarettes. (Reproduced with permission from Dr. Gladys Block, the University of California at Berkeley.)

| TYPE OF FOOD | HOW OFTEN | | | | | | | | HOW MUCH | | | |
| | Never or less than once per per month | 1 per mon. | 2–3 per mon. | 1 per week | 2 per week | 3–4 per week | 5–6 per week | Every day | MEDIUM SERVING | SERVING SIZE S | M | L |
| --- | --- | --- | --- | --- | --- | --- | --- | --- | --- | --- | --- | --- |
| EXAMPLE: Bananas | | | | ✓ | | | | | 1 medium | | ✓ | |
| Bananas | | | | | | | | | 1 medium | | | |
| Apples, applesauce | | | | | | | | | 1 medium | | | |
| Oranges (not including juice) | | | | | | | | | 1 medium | | | |
| Grapefruit (not including juice) | | | | | | | | | 1/2 medium | | | |
| Cantaloupe | | | | | | | | | 1/4 medium | | | |
| Peaches, apricots (fresh, in season) | | | | | | | | | 1 medium | | | |
| Peaches, apricots (canned or dried) | | | | | | | | | 1 medium or 1/2 cup | | | |
| Prunes, or prune juice | | | | | | | | | 1/2 cup | | | |
| Watermelon (in season) | | | | | | | | | 1 slice | | | |
| Strawberries, other berries (in season) | | | | | | | | | 1/2 cup | | | |
| Any other fruit, including kiwi, fruit cocktail, grapes, raisins, mangoes | | | | | | | | | 1/2 cup | | | |
| Fiber cereals like raisin bran, granola or shredded wheat | | | | | | | | | 1 medium bowl | | | |
| Sweetened cereals like frosted flakes | | | | | | | | | 1 medium bowl | | | |
| Other cold cereals like corn flakes or cheerios | | | | | | | | | 1 medium bowl | | | |
| Cooked cereal like oatmeal, oat bran or grits | | | | | | | | | 1 medium bowl | | | |
| Milk on cereal | | | | | | | | | 1/2 cup | | | |
| Breakfast bars, granola bars, power bars | | | | | | | | | 1 serving | | | |
| Breakfast shakes, diet shakes | | | | | | | | | 1 serving | | | |
| Pancakes; or waffles | | | | | | | | | 2 med. | | | |
| Eggs | | | | | | | | | 1 egg | | | |
| Egg substitutes, Egg Beaters, egg whites | | | | | | | | | 2 eggs | | | |
| Sausage or bacon | | | | | | | | | 2 patties | | | |
| Cottage cheese | | | | | | | | | 1/2 cup | | | |
| Other cheeses and cheese spreads (regular or low-fat) | | | | | | | | | 2 slices or 2 ounces | | | |
| Yogurt, frozen yogurt (regular or low-fat) | | | | | | | | | 8 oz. container | | | |

| TYPE OF FOOD | HOW OFTEN: Never or less than once per per month | 1 per mon. | 2–3 per mon. | 1 per week | 2 per week | 3–4 per week | 5–6 per week | Every day | HOW MUCH: MEDIUM SERVING | SERVING SIZE: S | M | L |
|---|---|---|---|---|---|---|---|---|---|---|---|---|
| VEGETABLES | | | | | | | | | | | | |
| String beans, green beans | | | | | | | | | 1/2 cup | | | |
| Peas | | | | | | | | | 1/2 cup | | | |
| Chili with beans (with or without meat) | | | | | | | | | 1 cup | | | |
| Other beans such as baked beans, pintos, kidney (not including soup) | | | | | | | | | 3/4 cup | | | |
| Corn | | | | | | | | | 1/2 cup | | | |
| Alfalfa sprouts, including on sandwiches | | | | | | | | | 1/2 cup | | | |
| Tomatoes, tomato juice | | | | | | | | | 1 med. or 6 oz. glass | | | |
| Salsa, ketchup, taco sauce | | | | | | | | | 2 tbsp. | | | |
| Broccoli | | | | | | | | | 1/2 cup | | | |
| Cauliflower or brussels sprouts | | | | | | | | | 1/2 cup | | | |
| Spinach (cooked or raw) | | | | | | | | | 1/2 cup | | | |
| Mustard greens, turnip greens, collards | | | | | | | | | 1/2 cup | | | |
| Cole slaw, cabbage | | | | | | | | | 1/2 cup | | | |
| Carrots, or mixed vegetables containing carrots | | | | | | | | | 1/2 cup | | | |
| Green salad | | | | | | | | | 1 med. bowl | | | |
| Salad dressing & mayonnaise (regular or low-fat) | | | | | | | | | 2 tbsp. | | | |
| French fries and fried potatoes | | | | | | | | | 3/4 cup | | | |
| White potatoes not fried, including boiled, baked, mashed and in potato salad | | | | | | | | | 1 med. or 1/2 cup | | | |
| Sweet potatoes, yams | | | | | | | | | 1/2 cup | | | |
| Any other vegetable, such as cooked onions, summer squash | | | | | | | | | 1/2 cup | | | |
| Butter, margarine or other fat added to veg., potatoes, etc. | | | | | | | | | 2 pats | | | |
| Tofu, bean curd | | | | | | | | | 1/2 cup | | | |
| Meat substitutes made from soy | | | | | | | | | 1 cup or patty | | | |

| SUMMARY QUESTIONS | AVERAGE USE LAST YEAR | | | | | | | | |
|---|---|---|---|---|---|---|---|---|---|
| | Less than once per week | 1–2 per week | 3–4 per week | 5–6 per week | 1 per day | 1.5 per day | 2 per day | 3 per day | 4+ per day |
| a. How often do you use fat or oil in cooking? | | | | | | | | | |
| b. About how many servings of vegetables do you eat, not counting salad or potatoes? | | | | | | | | | |
| c. About how many servings of fruit do you eat, not counting juices? | | | | | | | | | |
| d. About how many servings of cold cereal do you eat? | | | | | | | | | |
| e. About how many glasses of milk (or chocolate milk) do you drink? | | | | | | | | | |

**What kinds of fat do you usually use in cooking (to fry or stir-fry)?**
**Mark the one or two you use most often:**

| | Don't know | | Pam or no oil | | Lard, fatback, baconfat | | Crisco shortening |
|---|---|---|---|---|---|---|---|
| | Stick margarine | | Soft tub margarine | | Low calorie margarine | | |
| | Butter | | Olive oil or canola oil | | Corn oil, vegetable oil | | |

**What kinds of fat do you usually add to vegetables, potatoes, etc.?**
**Mark the one or two you use most often:**

| | Don't add fat | | Butter | | Soft tub margarine | | Crisco shortening | | Olive oil |
|---|---|---|---|---|---|---|---|---|---|
| | Stick margarine | | Lard, fatback, baconfat | | Whipped butter | | Low calorie margarine | | |

**When you eat the following foods, how often do you eat a low-fat or non-fat version of that food?**

| Cheese | | Always low-fat | | Sometimes | | Rarely low-fat |
|---|---|---|---|---|---|---|
| Ice cream or yogurt | | Always low-fat | | Sometimes | | Rarely low-fat |
| Salad dressing | | Always low-fat | | Sometimes | | Rarely low-fat |
| Cake or cookies | | Always low-fat | | Sometimes | | Rarely low-fat |

| How often do you add salt to your food? | | Seldom | | Sometimes | | Often |
|---|---|---|---|---|---|---|
| How often do you eat the skin on chicken? | | Seldom | | Sometimes | | Often |
| How often do you eat the fat on meat? | | Seldom | | Sometimes | | Often |
| How do you like your meat cooked? | | Rare | | Medium | | Well done |

# INDEX

## FOOD AND NUTRITION BOARD, NATIONAL ACADEMY OF SCIENCES—NATIONAL RE

| | Age | Weight (kg) | Weight (lb) | Height (cm) | Height (in) | Protein (g) | Vitamin A (µg RE)[a] | Vitamin D (µg)[b] | Vitamin E (mg α-TE)[c] | Vitamin K (µg) | Vitamin C (mg) |
|---|---|---|---|---|---|---|---|---|---|---|---|
| INFANTS | 0–0.5 | 6 | 13 | 60 | 24 | 13 | 375 | 7.5 | 3 | 5 | 30 |
| | 0.5–1.0 | 9 | 20 | 71 | 28 | 14 | 375 | 10 | 4 | 10 | 35 |
| CHILDREN | 1–3 | 13 | 29 | 90 | 35 | 16 | 400 | 10 | 6 | 15 | 40 |
| | 4–6 | 20 | 44 | 112 | 44 | 24 | 500 | 10 | 7 | 20 | 45 |
| | 7–10 | 28 | 62 | 132 | 52 | 28 | 700 | 10 | 7 | 30 | 45 |
| MALES | 11–14 | 45 | 99 | 157 | 62 | 45 | 1000 | 10 | 10 | 45 | 50 |
| | 15–18 | 66 | 145 | 176 | 69 | 59 | 1000 | 10 | 10 | 65 | 60 |
| | 19–24 | 72 | 160 | 177 | 70 | 58 | 1000 | 10 | 10 | 70 | 60 |
| | 25–50 | 79 | 174 | 176 | 70 | 63 | 1000 | 5 | 10 | 80 | 60 |
| | 51+ | 77 | 170 | 173 | 68 | 63 | 1000 | 5 | 10 | 80 | 60 |
| FEMALES | 11–14 | 46 | 101 | 157 | 62 | 46 | 800 | 10 | 8 | 45 | 50 |
| | 15–18 | 55 | 120 | 163 | 64 | 44 | 800 | 10 | 8 | 55 | 60 |
| | 19–24 | 58 | 128 | 164 | 65 | 46 | 800 | 10 | 8 | 60 | 60 |
| | 25–50 | 63 | 138 | 163 | 64 | 50 | 800 | 5 | 8 | 65 | 60 |
| | 51+ | 65 | 143 | 160 | 63 | 50 | 800 | 5 | 8 | 65 | 60 |
| Pregnant | | | | | | 60 | 800 | 10 | 10 | 65 | 70 |
| Lactating | 1st 6 mo | | | | | 65 | 1300 | 10 | 12 | 65 | 95 |
| | 2nd 6 mo | | | | | 62 | 1200 | 10 | 11 | 65 | 90 |

Source: Food and Nutrition Board (1989). "Recommended Dietary Allowances," 10th ed. National Academy Press, Washing

[a]Retinol equivalents. 1 retinol equivalent=1 µg retinol or 6 µg β-carotene.
[b]As cholecalciferol. 10 µg cholecalciferol=400 IU vitamin D.
[c]α-Tocopherol equivalents. 1 mg d-α–tocopherol=1 α-TE.
[d]Niacin equivalents. 1 NE=1 mg niacin or 60 mg dietary tryptophan.